M. C. Escher utilized both Euclidean and non-Euclidean geometry in these works of art, as discussed in Section 8.7 on pages 607–609. (M. C. Escher's *Symmetry Drawing E45,* © 2006 The M. C. Escher Company—Holland. All rights reserved.)

Preface

Goal

The goal of *Mathematics: A Practical Odyssey;* Sixth Edition, is to expose students to topics in mathematics that are usable and relevant to any educated person. It is also our hope that students will encounter topics that will be useful at some time during their lives. In addition, students are encouraged to recognize the relevance of mathematics to a well-rounded education and to appreciate the creative, human aspect of mathematics.

Liberal Arts Mathematics

This text is designed for students in liberal arts programs and other fields that do not require a core of mathematics. The term "liberal arts" is a translation of a Latin phrase that means "studies befitting a free person." It was applied during the Middle Ages to seven branches of learning: arithmetic, geometry, logic, grammar, rhetoric, astronomy, and music. You may be surprised to learn that almost half of the original liberal arts are mathematical subjects.

In accordance with the tradition handed down from the Middle Ages that a broad-based education includes some mathematics, many institutions of higher education require their students to complete a college-level mathematics course. These schools award a bachelor's degree to a person who has not only acquired a detailed knowledge of his or her field but also has a broad background in the liberal arts.

Topics

Algebra is the language of mathematics. A background in algebra allows you to learn mathematics that is usable and relevant to any educated person. In particular, it allows you to do the following:

- Learn enough about LOGIC to analyze the validity of an argument.
- Learn enough about SETS, COUNTING and PROBABILITY to understand the risks of inherited diseases and to realize what an incredibly bad bet a lottery ticket is.
- Learn enough about STATISTICS to understand the accuracy and validity of a public opinion poll.
- Learn enough about FINANCE to calculate the monthly payment required by a car loan or a home loan and to understand loans well enough to make an educated decision when selecting one.
- Learn enough about VOTING AND APPORTIONMENT to understand that there is no perfect voting system or method of apportionment; all methods have inherent flaws.

- Learn enough about NUMBER SYSTEMS AND NUMBER THEORY to understand why our commonly used base ten and base two number systems work, and to appreciate the prevalence of Fibonacci numbers in nature and the use of the golden ratio in art.
- Learn enough about GEOMETRY to understand its place in the history of Western civilization.
- Learn enough about GRAPH THEORY to be able to use it in scheduling and to create networks.
- Learn enough about EXPONENTIAL AND LOGARITHMIC FUNCTIONS to understand how populations grow, how radiocarbon dating works, how the Richter scale measures earthquakes, and how sound is measured in decibels.
- Learn enough about MATRICES AND MARKOV CHAINS to understand how manufacturers can predict their products' success or failure in the marketplace.
- Learn enough about LINEAR PROGRAMMING to understand how a small business can determine how to utilize its limited resources in order to maximize its profit.
- Learn enough about CALCULUS to understand just what the subject is and why it is so important.

Course Prerequisite

Mathematics: A Practical Odyssey is written for the student who has successfully completed a course in intermediate algebra, not the student who excelled in it. It would be difficult for a student without background in intermediate algebra to succeed in a course using this book. However, some chapters are not algebra-based. These chapters require a level of critical thinking and mathematical maturity more commonly found in students who have taken intermediate algebra.

New Chapters in the Sixth Edition

In response to many users' requests, we have added three new chapters.

Chapter 6: Voting and Apportionment
- Voting Systems
- Methods of Apportionment
- Flaws of Apportionment

Chapter 7: Number Theory
- Place Systems
- Arithmetic in Different Bases
- Prime Numbers and Perfect Numbers
- Fibonacci Numbers and the Golden Ratio

Chapter 9: Graph Theory
- A Walk Through Konigsberg
- Graphs and Euler Trails
- Hamilton Circuits
- Networks
- Scheduling

New Features of the Sixth Edition

"Topic X" and the Real World

Discussing current, powerful, real-world uses of the text's topics provides an opportunity to reinforce the practical emphasis of the text. To create the extended applications for *Topic X and the Real World,* we researched articles in professional journals, magazines, newspapers and the web, and distilled the information so that it is accessible to the liberal arts mathematics student. These highly practical applications are used as jumping-off points for both reality-based conceptual and computational exercises.

- Logic and the Real World: Su Doku (Section 1.1, page 10)
- Set Theory and the Real World: Blood Types (Section 2.2, page 82)
- Probabilities and the Real World: The Business of Gambling (Section 3.4, page 167)
- Probabilities and the Real World: HIV/AIDS (Section 3.6, page 194)
- Statistics and the Real World: Random Sampling in Opinion Polls (Section 4.5, page 303)
- Compound Interest and the Real World: Benjamin Franklin's Gift (Section 5.2, page 347)
- Voting Systems and the Real World: Apportionment of the U.S. House of Representatives (Section 6.2, page 452)
- The Binary System in the Real World: Computer Imaging (Section 7.1, page 486)
- Geometry in the Real World (and Universe): Astronomical Measurement (Section 8.1, page 533)
- Fractals in the Real World: Computer Graphics in Movies (Section 8.8, page 621)
- PERT Charts in the Real World: NASA (Section 9.5, page 709)
- Radioactivity the Real World: Nuclear Medicine (Section 10.2, page 777)
- Logarithms and the Real World: The Negative Impact of Sounds (Section 10.3, page 794)
- Markov Chains and the Real World: The Suburbanization of America (Section 11.5, page 849)
- Linear Programming and the Real World: Pilkington Libbey-Owens-Ford (Section 12.1, page 880)

Web Projects

These new projects include links to web pages that the students can use as starting points in their research.

Chapter 1: Logic
- Investigate several su doku websites. (Section 1.1, Exercise 62, page 19)
- Write a research paper on an historic topic. (Section 1.5, Exercise 46, page 58)

Chapter 2: Sets and Counting
- Determine the compatibility of the various blood types, including Rh factors. (Section 2.2, Exercise 51, page 88)
- Write a research paper on an historic topic. (Section 2.4, Exercise 62, page 111; and Section 2.5, Exercise 28, page 122)

Chapter 3: Probability
- Write a research paper on an historic topic. (Section 3.1, Exercise 46, page 133)
- Write a research paper on the successful screening of Jews for Tay-Sachs disease and the unsuccessful screening of blacks for sickle-cell anemia. (Section 3.2, Exercise 93, page 152)
- Determine the prevalence of the various blood types in the United States, the percentage of U.S. residents that can donate blood to people of each of the various blood types, and the percentage of U.S. residents that can receive blood from people of each of the various blood types. (Section 3.6, Exercise 71, page 200)
- Investigate automobile rollovers. Determine which type of vehicle—a sedan, an SUV or a van—is more prone to rollovers. (Section 3.6, Exercise 72, page 200)

Chapter 4: Statistics
- Analyze daily high and low temperatures in a favorite city. (Section 4.2, Exercise 37, page 258; and Section 4.3, Exercise 27, page 271)
- Obtain current national opinions concerning major issues in the headlines. (Section 4.5, Exercise 35, page 306)
- Write an essay on the use of random digit dialing in national surveys. (Section 4.5, Exercise 36, page 306)

Chapter 5: Finance
- Compare different banks' CD interest rates, compounding frequencies, and annual yields. (Section 5.2, Exercise 71, page 351)
- Use different banks' on-line information to determine the cost of financing a home, car, boat or motorcycle. (Section 5.3, Exercise 49, page 366; and Section 5.4, Exercises 47 and 48, page 380)
- Predict the cost of your baby's college education, and use different banks' on-line information to determine the necessary annuity payment that will allow you to pay for this education. (Section 5.3, Exercise 50, page 366)
- Use web sources to find the annual rate of inflation for the last ten years. Use that to predict your needed income upon retiring. Then determine the monthly annuity payments that would allow you to save the necessary amount. (Section 5.6, Exercise 21, page 404)

Chapter 6: Voting and Apportionment
- Analyzing the change in the "number of people represented per seat" for the members of the House of Representatives from 1790 to 2000. (Section 6.2, Exercise 60, page 460)

Chapter 7: Number Systems and Number Theory
- Write a research paper on the history of the abacus, calculations using the abacus, the abacus versus the calculator, the Roman abacus, the Chinese suan pan, the Mesoamerican abacus, or the Mayan base 20. (Section 7.1, Exercise 106, page 490)
- Write a research paper on Mersenne primes, the great Internet Mersenne prime search, large prime numbers, perfect numbers, or mysticism and numbers. (Section 7.3, Exercise 50, page 510)
- Write a research paper on Fibonacci numbers in nature, Fibonacci puzzles, Fibonacci numbers and the golden rectangle, the mathematics of Fibonacci numbers, Fibonacci numbers and continued fractions, Fibonacci's biography, the golden rectangle and art, da Vinci's collaborative efforts with Fibonacci, Fibonacci numbers and the stock market, Fibonacci numbers and music, the

golden ratio and Debussy, Mozart and Bartok. (Section 7.4, Exercise 26, page 519)

Chapter 8: Geometry

- Write a research paper on an historic topic. (Section 8.4, Exercises 31 and 32, page 573)
- Investigate the artwork of M.C. Escher. (Section 8.7, Exercise 27, page 610)
- Write an essay on the use of fractals by the film industry in creating computer graphics. (Section 8.8, Exercise 35, page 632)
- Investigate several fractal websites. (Section 8.8, Exercise 36, page 632)

Chapter 9: Graph Theory

- Investigate some of the unsolved problems in graph theory. (Section 9.2, Exercise 33, page 668)
- Write an essay on the four-color map problem. (Section 9.2, Exercise 34, page 668)
- Investigate the traveling salesman problem on the web. Summarize their history, and describe specific problems that led to progress. (Section 9.3, Exercise 32, page 682)

Chapter 10: Exponential and Logarithmic Functions

- Use U.S. Census population clock to create a model, and use the model to make a prediction. Then determine the prediction's accuracy. (Section 10.1, Exercise 41, page 763)
- Investigating methods other than radiocarbon dating used in determining the age of artifacts. (Section 10.2, Exercise 51, page 781)

Chapter 11: Matrices and Markov Chains

- Analyze the game of Monopoly using Markov chains. (Section 11.3, Exercise 10, page 840)

Chapter 12: Linear Programming

- Write an essay on why George Dantzig did not receive a Nobel Prize for his invention of linear programming. (Section 12.1, Exercise 39, page 885)
- Research linear programming software packages. (Section 12.1, Exercise 40, page 885)

Chapter 13: Calculus

- Write a research paper on an historical topic, or on Fermat's last theorem. (Section 13.1, Exercise 44, page 13-13, and Section 13.2, Exercises 43 and 44, page 13-42)

Features Continued from the Fifth Edition

Graphing and Scientific Calculators

Calculators are useful and powerful tools that have become an integral part of many classrooms. However, many students are unable to use their calculators effectively. Therefore, we have included detailed instructions on scientific and graphing calculator use throughout the text, and we have included several appendices that help students master particular aspects.

Furthermore, a number of optional subsections are included that address some of the more advanced capabilities of graphing calculators. These subsections allow instructors to incorporate technology into their class if they so desire, but they are entirely optional, and the book is in no way calculator-dependent. The graphing calculator subsections were specifically written for the Texas Instruments TI-83, TI-84, and TI-86 calculators, and Casio graphing calculators. The following graphing calculator topics are addressed:

- Section 3.3: Fractions on a Graphing Calculator (page 163)
- Section 4.1: Histograms on a Graphing Calculator (page 236)
- Section 4.3: Measures of Central Tendency and Dispersion on a Graphing Calculator (page 271)
- Section 4.6: Linear Regression on a Graphing Calculator (page 317)
- Section 5.2: Doubling Time on a Graphing Calculator (page 352)
- Section 5.3: Annuities on a Graphing Calculator (page 367)
- Section 5.5: Annual Percentage Rate on a Graphing Calculator (page 387)
- Section 11.0: Matrix Multiplication on a Graphing Calculator (page 813)
- Section 11.2: Solving Systems of Two Linear Equations in Two Unknowns on a Graphing Calculator (page 834)
- Section 11.4: Solving Larger Systems of Linear Equations on a Graphing Calculator (page 844)
- Section 12.0: Graphing Linear Inequalities on a Graphing Calculator (page 864)

- Section 12.3: The Row Operations on a Graphing Calculator (page 12-24)
- Appendix A. Using a Scientific Calculator (page A-1)
- Appendix B. Using a Graphing Calculator (page A-10)
- Appendix C. Graphing with a Graphing Calculator (page A-21)
- Appendix D. Finding Points of Intersection with a Graphing Calculator (page A-26)

Excel and Computers

Computers are ubiquitous at the workplace, and are becoming increasingly common in the classroom. However, many students have no mathematical experience using computer software such as Microsoft Excel. For this reason, we have included a number of subsections that give instructions on the use of Excel. We have also included a number of optional subsections that give instructions on the use of *Amortrix,* custom text-specific software that is available on the text website (go to **www.thomsonedu.mathematics/johnson**). The following topics are addressed:

- Section 4.1: Histograms and Pie Charts on a Computerized Spreadsheet (page 238)
- Section 4.3: Measures of Central Tendency and Dispersion on Excel (page 273)
- Section 5.4: Amortization Schedules on Amortrix (page 382)
- Section 5.4: Amortization Schedules on Excel (page 383)

- Section 12.3: The Row Operations on Amortrix (page 12-30)

These subsections allow instructors to incorporate the computer into their class if they so desire, but they are entirely optional, and the book is in no way computer dependent. The subsections do not assume any previous experience with Excel.

Projects

Chapter 3: Probability
- Relative frequency of an odd phone number. (Section 3.2, Exercise 88, page 151)
- Compare relative frequency and theoretical probability with a coin toss. (Section 3.2, Exercise 89, page 151)
- Compare relative frequency and theoretical probability with a roll of a die. (Section 3.2, Exercise 90, page 152)

- Determine if the outcomes are equally likely with coin experiments. (Section 3.2, Exercises 91 and 92, page 152)
- Design a game of chance. Use probabilities and expected values to set house odds. (Section 3.5, Exercise 50, page 186)
- Investigate seemingly contradictory information regarding sex bias in graduate school and/or tuberculosis incidence in New York City and Richmond. (Section 3.6, Exercises 73 and 74, page 201)

Chapter 4: Statistics
- Variation in the pricing of a common commodity. (Section 4.3, Exercises 25 and 26, pages 270 and 271)
- Conducting an opinion poll. (Section 4.5, Exercise 35, page 306)
- Examining the correlation between height and weight. (Section 4.6, Exercise 17, page 317)

Chapter 5: Finance
- Use graphs to compare simple interest, interest compounded annually, and interest compounded daily. (Section 5.2, Exercise 72, page 351)
- Compute the monthly payment for an adjustable rate mortgage at different points in time with varying assumptions about the interest rate. (Section 5.4, Exercises 49–51, page 381)

Chapter 7: Number Systems and Number Theory
- Investigate the occurrence of Fibonacci numbers in nature—pineapples, cauliflowers, sunflowers, and other flowers. (Section 7.4, Exercise 24, page 519)
- Investigate the use of the golden rectangle in art—da Vinci, Boticelli, and others. (Section 7.4, Exercise 25, page 519)

Chapter 8: Geometry
- Verifying the Pythagorean Theorem. (Section 8.1, Exercise 43, page 538)
- Calculating the cost of materials to resurface a floor. (Section 8.1, Exercise 44, page 538)
- Converting fluid ounces to cubic inches. (Section 8.2, Exercise 35, page 549)
- Calculating the height of a tall building by using similar triangles. (Section 8.4, Exercise 30, page 573)
- Calculating the height of a tall building by measuring an angle of elevation. (Section 8.5, Exercise 49, page 588)
- Design your own fractal. (Section 8.8, Exercise 33, page 632; and Section 8.9, Exercise 10, page 641)
- Create your own coastline. (Section 8.8, Exercise 34, page 632)
- Approximate the area of a Koch snowflake. (Section 8.9, Exercise 11, page 641)

Chapter 9: Graph Theory
- Find some appropriate information in the media and present it as a graph. (Section 9.1, Exercise 20, page 654)
- Create a PERT chart and a Gantt chart for a project that you are familiar with. (Section 9.5, Exercise 35, page 714)

Chapter 10: Exponential and Logarithmic Functions
- Calculate the growth rate of your community. Predict its future size and calculate its doubling time. (Section 10.1, Exercise 40, page 762)

Chapter 11: Matrices and Markov Chains
- Investigate the U.S. population shifts between cities, suburbs and rural areas. (Section 11.1, Exercise 21, page 826 and Section 11.5, Exercise 9, page 851)

- Investigate the U.S. population shifts between the Northeast, Midwest, South and West. (Section 11.1, Exercise 22, page 827; and Section 11.5, Exercise 12, page 852)
- Investigate the changes in U.S. land use. (Section 11.1, Exercise 23, page 828)
- Use Punnett squares and Markov chains to investigate genetics. (Section 11.5, Exercise 10, page 851)
- Investigate the intergenerational transition between income groups in the U.S. (Section 11.5, Exercise 11, page 851)

Chapter 12: Linear Programming

- Two different long, detailed linear programming problems. (Section 12.1, Exercises 37 and 38, page 884 and Section 12.3, Exercise 36, page 12-33)

Chapter 13: Calculus

- Investigate and discuss Descartes' use of language and symbols in algebra. (Section 13.1, Exercise 40, page 13-26)
- Trace the evolution of algebra and/or analytic geometry. (Section 13.1, Exercise 41, page 13-27)
- Write a research paper on the contributions of Apollonius, Oresme, Descartes and Fermat to analytic geometry. (Section 13.1, Exercise 42, page 13-27)
- Compare and contrast the algebras of Apollonius, al-Khowarizmi, and Descartes. (Section 13.1, Exercise 42, page 13-27)
- Use dimensional analysis to convert the distance and speed falling object formulas from the English system to the metric system. (Section 13.7, Exercise 1, page 13-87)
- Use calculus to sketch the graph of a given polynomial function. (Section 13.7, Exercise 2, page 13-88)
- Use calculus to find the areas of some complicated shapes. (Section 13.7, Exercises 3–6, pages 13-88–13-89)

Algebra Review

When appropriate, algebraic topics are reviewed, but in a very selective and focused manner. Only those topics that are used in the book are reviewed. There is no "Review of Algebra" chapter. Instead, the reviews are placed as close as possible to the topics that utilize them, usually in a Section 0 at the beginning of the chapter. These review sections are direct and to the point. They do not include any applications of the algebra; those are covered in the subsequent sections. They do not attempt to provide a thorough treatment of the algebra in question but rather focus on the algebra that will be used in the sections that follow.

Intermediate algebra courses vary significantly from school to school. Among the Section 0 topics are some that some students have never seen before, such as matrices and logarithms. In these cases, the reviews are more detailed and assume lass prior knowledge.

History

The history of the subject matter is interwoven throughout most chapters. In addition, Historical Notes give in-depth biographies of the prominent people involved. It is our hope that students will see the human side of mathematics. After all, mathematics was invented by real people for real purposes and is a part of our culture. Interesting research topics are given, and writing assignments are suggested.

Bridges Between Chapters

As much as possible, this book has been written so that its chapters are independent of each other. The instructor therefore has wide latitude in selecting topics to cover and can teach a course that is responsive to the needs of his or her students and institution. Sometimes, however, this independence is not desirable, because it does not allow connections to be made between seemingly unrelated topics. For this reason, the text features a number of bridges between chapters. These bridges feature both discussion and exercises. They are clearly labeled; that is, a bridge between an earlier Chapter X and a later Chapter Y is labeled "for those students who have completed Chapter X."

- There is a bridge between Chapter 1 (Logic) and Chapter 2 (Sets and Counting). At the end of Section 2.1, the similarities between the respective concepts and notation of logic and sets are discussed.
- There is a bridge between Chapter 5 (Finance) and Chapter 10 (Exponential and Logarithmic Functions). That bridge discusses the relationship between the exponential growth model ($y = ae^{bt}$) and the compound interest model [$FV = P(I + i)^n$], and the circumstances under which one model can be used in place of the other.
- There is a bridge between Chapter 8 (Geometry) and Chapter 13 (Calculus). That bridge discusses the use of the trigonometric functions to determine the equation of a trajectory when the initial angle of elevation is other than $30°$, $45°$, or $60°$.

Topic Selection and Prerequisite Maps

Obviously, the book contains more material than you could ever cover in a one-semester or one-quarter course. It is written in such a way that most chapters are independent of each other. The Instructor's Manual includes a "prerequisite map" so that you can easily tell which earlier topics must be covered. It also includes sample course outlines, suggesting specific chapters to be used in forming a typical course.

Flexibility and Course Level

The book contains a wide range of topics, varying in level of sophistication and difficulty. Chapters 1 through 9 cover topics that are not uncommon to beginning algebra-based liberal arts mathematics texts. However, because of the intermediate algebra prerequisite, the topics are covered more thoroughly, and the acquisition of problem-solving skills is emphasized. Chapters 10 through 13 are more sophisticated, covering topics not commonly found in liberal arts texts. However, the treatment is such that the material is accessible to the students who enroll in this course.

A chapter need not be covered in its entirety; topics can easily be left out. In most cases, a chapter has a suggested core of key sections as well as a selection of optional sections, which tend to be more sophisticated. They are not labeled "optional" in the text; rather, this distinction is made in the prerequisite map in the Instructor's Resource Manual.

This text is designed so that the instructor determines the difficulty of the course by selecting the chapters and sections to be covered. The instructor can thus create his or her own course-one that will fit students' needs.

Usability

This book is user-friendly:

- The examples don't skip steps.
- Key points are boxed for emphasis.
- Step-by-step procedures are given.
- There is an abundance of exposition.

Exercises

The exercises in this text are designed to solidify the students' understanding of the material and make them proficient in the calculations involved. It is assumed that most students who complete this course will not continue in their formal study of mathematics. Consequently, neither the exposition nor the exercises are designed to expose the students to all aspects of the topic.

The exercises vary in difficulty. Some are exactly like the examples, and others demand more of the students. The exercises are not explicitly graded into A, B and C categories, nor are any marked "optional;" students in this audience tend to react negatively if asked to do anything labeled in this manner. The more difficult exercises are indicated in the Instructor's Resource Manual.

The short-answer historical questions are meant to focus and reinforce the students' understanding of the historical material. They also serve to warn them that history questions may appear on exams. The essay questions can be used as an integral part of the students' grades, as background for classroom discussion, or for extra credit work. Most are research topics and are kept as open-ended as possible.

Answers to the odd-numbered exercises are given in the back of the book, with two exceptions:

- Answers to historical questions and essay questions are not given
- Answers are not given when the exercises instruct the student to check the answers themselves.

Complete solutions to every other odd-numbered exercise are given in the student's solutions manual, with the above two exceptions.

Answer Checking

Throughout the text, we emphasize the importance of checking ones answers. It's important that students learn to evaluate the reasonableness of their answers, rather than accepting them on face value. To this end, some exercises do not have answers in the back of the book when students are instructed to check answers for themselves.

Standards

This text is well-suited in addressing the AMATYC and NCTM standards.

Regarding *Standards for Intellectual Development,* the text contains a wide range of exercises designed to engage students in mathematical problem solving and modeling real-world situations. In addition, the exercises provide students an opportunity to expand their mathematical reasoning skills and to communicate their results effectively. Also, the text emphasizes the interrelationships between mathematics, human culture, and other disciplines. The use of appropriate technology is woven throughout the text, along with material that encourages independent exploration and confidence building in mathematics.

Regarding *Standards for Content,* the text contains material covering all requisite topics including number sense, symbolism and algebra, geometry, function, combinatorics, probability and statistics, and deductive proof.

Regarding *Standards for Pedagogy,* the text provides ample opportunity for faculty to use technology in the teaching of mathematics, to foster interactive learning through collaborative activities and effective communication, to make connections between various branches of mathematics and between mathematics and the students' lives, and to use numerical, graphical, symbolic, and verbal approaches in the teaching of mathematics.

For the Instructor

- **Instructor's Edition:** This special version of the complete student text contains a complete set of answers printed in the back of the text. ISBN 0495017833
- **Instructor's Resource Manual:** The Instructor's Resource Manual contains additional resources for the instructor and provides worked-out solutions to all problems in the text. ISBN 0495018058
- **Test Bank:** The Test Bank includes six tests per chapter as well as three final exams. The tests are made up of a combination of multiple-choice, free-response, and fill-in-the-blank questions. ISBN 0495018066
- **ThomsonNOW:** An integrated testing, tutorial, and class management system that is efficient and versatile, **ThomsonNOW**™ transforms the teaching and learning experience. Powerful and easy to use, **ThomsonNOW** saves time and improves performance! Providing instructors and students with unsurpassed control, variety, and all-in-one utility, **ThomsonNOW** is a powerful and fully integrated teaching and learning system.

  Now featuring a more intuitive design and faster speed in release 5.0, **ThomsonNOW** ties together fundamental learning activities such as tutorials, homework, quizzing, testing, and diagnostics. Easy to use, **ThomsonNOW** offers instructors complete control when creating assessments, whether drawing from provided tests or creating their own. **ThomsonNOW** features the greatest variety of problem types—allowing instructors to assess the way they teach! A real timesaver for instructors, **ThomsonNOW** offers automatic grading of homework, quizzes, and tests with results flowing directly into the gradebook. Designed for maximum flexibility that enhances all teaching and learning styles, **ThomsonNOW** helps instructors and students succeed!

  Available on request: **ThomsonNOW** can be integrated with **Blackboard**® and **WebCT**™. Ask your Thomson representative for more details. ISBN–0495126683

- **Personalized Study Plan:** A diagnostic tool that helps your students make the most of their study time. **ThomsonNOW** offers a powerful diagnostic tool and online learning companion that helps students identify their unique study needs and improve performance in your course. Students can use this tool—which is seamlessly integrated with content from **ThomsonNOW** tutorials—on their own, with no instructor setup or involvement (which saves YOU time). Or, you can assign the self-grading quizzes and tutorials as homework, and the results can flow automatically into your **ThomsonNOW** gradebook.

For the Students

- **Student Solutions Manual:** The Student Solutions Manual provides worked out solutions to every other odd-numbered problem in the text. ISBN 0495016683
- **Interactive Video Skillbuilder CD:** The Interactive Video Skillbuilder CD–ROM contains hours of video instruction. To help students evaluate their progress, each section contains a 10–question web quiz and each chapter contains a chapter test with answers to each problem on each test. Also includes MathCue Tutorial, dual platform software that presents and scores problems and tutors students by displaying annotated, step-by-step solutions. Problem sets may be customized as desired. ISBN 049501785X
- **ThomsonNOW: Make the most of your study time with ThomsonNOW! ThomsonNOW** provides text-specific, interactive, web-based tutorials for students. The browser-based interface of **ThomsonNOW** makes it an intuitive mathematical guide even for students with little technological proficiency. Simple to use, **ThomsonNOW** allows students to work with real math notation in real time, providing instant analysis and feedback. The entire textbook is available in PDF format through **ThomsonNOW,** as practice problems and additional student resources.

And, when you get stuck on a particular problem or concept, you need only log on to **vMentor™,** accessed through **ThomsonNOW,** where you can talk (using your own computer microphone) to **vMentor** tutors who will skillfully guide you through the problem using an interactive whiteboard for illustration.

Study Smarter—and Make Every Minute Count!

If an access card came with this book, you can start using **ThomsonNOW** right away by following the directions on the card. If a card was not included, you may visit http://www.thomsonedu.com to purchase electronic access. ISBN 0495126691

Web Site

Located at www.thomsonedu.com/mathematics/johnson, the book companion web site offers book-specific student and instructor resources, as well as discipline-specific links. Student resources at the web site include a web-based version of Amortrix (the software used in Chapters 5, 7, and 8), downloadable partially completed Excel spreadsheets that are keyed to examples in the text, graphing and calculation tools, and Internet links for further research. The web site also has quizzes by chapter, whose results can be e-mailed to the instructor.

Acknowledgments

The authors would like to thank John-Paul Ramin, Carolyn Crockett, Katie Cook and all the fine people at Thomson Publishing; Laura Horowitz at Hearthside Publishing Services; Shona Burke; Leslie Month; and Sudhir Goel. Special thanks go to the users of the text and reviewers who evaluated the manuscript for this edition, as well as those who offered comments on previous editions.

Reviewers

Dennis Airey, *Rancho Santiago College*

Francisco E. Alarcon, *Indiana University of Pennsylvania*

Judith Arms, *University of Washington*

Bruce Atkinson, *Palm Beach Atlantic College*

Wayne C. Bell, *Murray State University*

Wayne Bishop, *California State University—Los Angeles*

David Boliver, *Trenton State College*

Stephen Brick, *University of South Alabama*

Barry Brunson, *Western Kentucky University*

Frank Burk, *California State University—Chico*

Laura Cameron, *University of New Mexico*

Jack Carter, *California State University—Hayward*

Timothy D. Cavanaugh, *University of Northern Colorado*

Joseph Chavez, *California State University—San Bernardino*

Eric Clarkson, *Murray State University*

Rebecca Conti, *State University of New York at Fredonia*

S. G. Crossley, *University of Southern Alabama*

Ben Divers, Jr., *Ferrum College*

Al Dixon, *College of the Ozarks*

Joe. S. Evans, *Middle Tennessee State University*

Hajrudin Fejzie, *California State University—San Bernardino*

Lloyd Gavin, *California State University—Sacramento*

William Greiner, *McLennan Community College*

Martin Haines, *Olympic College*

Ray Hamlett, *East Central University*

Virginia Hanks, *Western Kentucky University*

Anne Herbst, *Santa Rosa Junior College*

Linda Hinzman, *Pasadena City College*

Thomas Hull, *University of Rhode Island*

Robert W. Hunt, *Humboldt State University*

Robert Jajcay, *Indiana State University*

Irja Kalantari, *Western Illinois University*

Daniel Katz, *University of Kansas*

Katalin Kolossa, *Arizona State University*

Donnald H. Lander, *Brevard College*

Lee LaRue, *Paris Junior College*

Thomas McCready, *California State University—Chico*

Vicki McMillian, *Stockton State College*

Narendra L. Maria, *California State University—Stanislaus*

John Martin, *Santa Rosa Junior College*

Gael Mericle, *Mankato State University*

Robert Morgan, *Pima Community College*

Pamela G. Nelson, *Panhandle State University*

Carol Oelkers, *Fullerton College*

Michael Olinick, *Middlebury College*

Matthew Pickard, *University of Puget Sound*

Joan D. Putnam, *University of Northern Colorado*

J. Doug Richey, *Northeast Texas Community College*

Stewart Robinson, *Cleveland State University*

Eugene P. Schlereth, *University of Tennessee at Chattanooga*

Lawrence Somer, *Catholic University of America*

Michael Trapuzzano, *Arizona State University*

Pat Velicky, *Mid-Plains Community College*

Karen M. Walters, *University of Northern California*

Dennis W. Watson, *Clark College*

Charles Ziegenfus, *James Madison University*

Contents

*Sections in color can be found at www.thomson.edu/mathematics/johnson

1

Logic

© The Granger Collection

When writer Lewis Carroll took Alice on her journeys "through the looking glass" to Wonderland, she had many fantastic encounters with the hookah-smoking Caterpillar, the White Rabbit, the Red and White Queens, and the Cheshire Cat. On the surface, Carroll's writings seem to be delightful nonsense and mere children's entertainment. However, they contain hints of deeper roots and meanings. During one of her many surreal adventures, Alice came upon Tweedledum and Tweedledee. The two sarcastic twins taunted her with the following: "'I know what you're thinking about,' said Tweedledum; 'but it isn't so, nohow.' 'Contrariwise,' continued Tweedledee, 'if it was so, it might be; and if it were so, it would be; but as it isn't, it ain't. That's logic.'"

Many people are quite surprised to learn that *Alice's Adventures in Wonderland* is as much an exercise in logic as it is a fantasy and that Lewis Carroll was actually Charles Dodgson, an Oxford mathematician. Dodgson's many writings include the whimsical *The Game of Logic* and the brilliant *Symbolic Logic,* in addition to *Alice's Adventures in Wonderland* and *Through the Looking Glass.*

Webster's dictionary defines **logic** as "the science of correct reasoning; the science which describes relationships among propositions in terms of implication, contradiction, contrariety, and conversion." In addition to being flaunted in Mr. Spock's claim that "your human emotions have drawn you to an illogical conclusion" and in Sherlock Holmes's immortal phrase "elementary, my dear Watson," logic is fundamental both to critical thinking and to problem solving. In this world of misleading commercial claims, innuendo, and political rhetoric, the ability to distinguish between valid and invalid arguments is important. Cold, emotionless Vulcans and eccentric, violin-playing detectives are not the only ones who can benefit from logic. Armed with the fundamentals of logic, we can surely join Spock and "live long and prosper!"

ThomsonNOW™ **for Liberal Arts Mathematics**
Throughout this chapter, the ThomsonNOW logo indicates an opportunity for online self-study, which:
- Assesses your knowledge of important concepts and provides a personalized study plan.
- Links you to interactive tutorials and videos to help you study.
- Helps to test your knowledge of material and prepare for exams.
Visit ThomsonNOW at **www.thomsonedu.com/1pass** to access these resources.

1

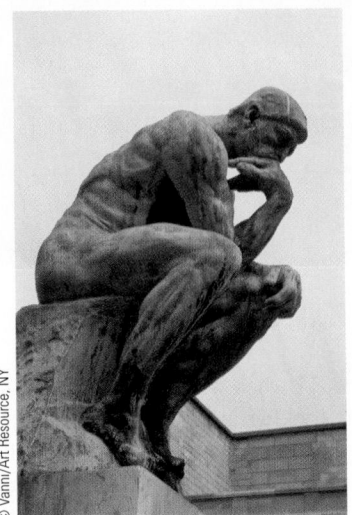

© Vanni/Art Resource, NY

Logic is the science of correct reasoning. Auguste Rodin captured this ideal in his bronze sculpture *The Thinker.*

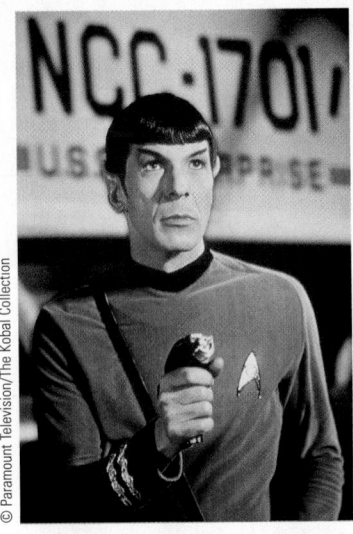

© Paramount Television/The Kobal Collection

In their quest for logical perfection, the Vulcans of *Star Trek* abandoned all emotion. Mr. Spock's frequent proclamation that "emotions are illogical" typified this attitude.

1.1 Deductive versus Inductive Reasoning

Logic is the science of correct reasoning. Webster's dictionary defines **reasoning** as "the drawing of inferences or conclusions from known or assumed facts." Reasoning is an integral part of our daily lives; we take appropriate actions based on our perceptions and experiences. For instance, if the sky is heavily overcast this morning, you might assume that it will rain today and take your umbrella to work.

Problem Solving

Logic and reasoning are associated with the phrases *problem solving* and *critical thinking.* If we are faced with a problem, puzzle, or dilemma, we attempt to reason through it in hopes of arriving at a solution.

The first step in solving any problem is to define the problem in a thorough and accurate manner. Although this might sound like an obvious step, it is often overlooked. Always ask yourself, "What am I being asked to do?" Before you can solve a problem, you must understand the question. Once the problem has been defined, all known information that is relevant to it must be gathered, organized, and analyzed. This analysis should include a comparison of the present problem to previous ones. How is it similar? How is it different? Does a previous method of solution apply? If it seems appropriate, draw a picture of the problem; visual representations often provide insight into the interpretation of clues.

Before using any specific formula or method of solution, determine whether its use is valid for the situation at hand. A common error is to use a formula or method of solution when it does not apply. If a past formula or method of solution is appropriate, use it; if not, explore standard options and develop creative alternatives. Do not be afraid to try something different or out of the ordinary. "What if I try this . . . ?" may lead to a unique solution.

Using his extraordinary powers of logical deduction, Sherlock
Holmes solves another mystery. "Finding the villain was elementary,
my dear Watson."

Deductive Reasoning

Once a problem has been defined and analyzed, it might fall into a known category
of problems, so a common method of solution may be applied. For instance, when
one is asked to solve the equation $x^2 = 2x + 1$, realizing that it is a second-degree
equation (that is, a quadratic equation) leads one to put it into the standard form
$(x^2 - 2x - 1 = 0)$ and apply the Quadratic Formula.

EXAMPLE 1   Solve the equation $x^2 = 2x + 1$.

Solution   The given equation is a second-degree equation in one variable. We know that all
second-degree equations in one variable (in the form $ax^2 + bx + c = 0$) can be
solved by applying the Quadratic Formula:

$$x = \frac{-b \pm \sqrt{b^2 - 4ac}}{2a}$$

Therefore, $x^2 = 2x + 1$ can be solved by applying the Quadratic Formula:

$$x^2 = 2x + 1$$

$$x^2 - 2x - 1 = 0$$

$$x = \frac{-(-2) \pm \sqrt{(-2)^2 - (4)1(-1)}}{2(1)}$$

$$x = \frac{2 \pm \sqrt{4 + 4}}{2}$$

$$x = \frac{2 \pm \sqrt{8}}{2}$$

$$x = \frac{2 \pm 2\sqrt{2}}{2}$$

$$x = \frac{2(1 \pm \sqrt{2})}{2}$$

$$x = 1 \pm \sqrt{2}$$

The solutions are $x = 1 + \sqrt{2}$ or $x = 1 - \sqrt{2}$.   ■

In Example 1, we applied a general rule to a specific case; we reasoned that it was valid to apply the (general) Quadratic Formula to the (specific) equation $x^2 = 2x + 1$. This type of logic is known as **deductive reasoning**—that is, the application of a general statement to a specific instance.

Deductive reasoning and the formal structure of logic have been studied for thousands of years. One of the earliest logicians, and one of the most renowned, was Aristotle (384–322 B.C.). He was the student of the great philosopher Plato and the tutor of Alexander the Great, the conqueror of all the land from Greece to India. Aristotle's philosophy is pervasive; it influenced Roman Catholic theology through St. Thomas Aquinas and continues to influence modern philosophy. For centuries, Aristotelian logic was part of the education of lawyers and politicians and was used to distinguish valid arguments from invalid ones.

For Aristotle, logic was the necessary tool for any inquiry, and the syllogism was the sequence followed by all logical thought. A **syllogism** is an argument composed of two statements, or **premises** (the major and minor premises), followed by a **conclusion.** For any given set of premises, if the conclusion of an argument is guaranteed (that is, if it is inescapable in all instances), the argument is **valid.** If the conclusion is not guaranteed (that is, if there is at least one instance in which it does not follow), the argument is **invalid.**

Perhaps the best known of Aristotle's syllogisms is the following:

1. All men are mortal.   major premise

2. Socrates is a man.   minor premise

Therefore, Socrates is mortal.   conclusion

Applying the major premise to the minor premise, the conclusion is inescapable; the argument is valid.

Notice that the deductive reasoning used in the analysis of Example 1 has exactly the same structure as Aristotle's syllogism concerning Socrates:

1. All second-degree equations in one variable can be solved by applying the Quadratic Formula.   major premise

2. $x^2 = 2x + 1$ is a second-degree equation in one variable.   minor premise

Therefore, $x^2 = 2x + 1$ can be solved by applying the Quadratic Formula.   conclusion

Each of these syllogisms is of the following general form:

1. If A, then B.   All A are B. (major premise)

2. x is A.   We have A. (minor premise)

Therefore, x is B.   Therefore, we have B. (conclusion)

Historically, this valid pattern of deductive reasoning is known as *modus ponens.*

Deductive Reasoning and Venn Diagrams

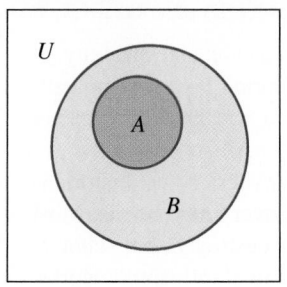

Figure 1.1
All *A* are *B*. (If *A*, then *B*.)

The validity of a deductive argument can be shown by use of a Venn diagram. A **Venn diagram** is a diagram consisting of various overlapping figures contained within a rectangle (called the "universe"). To depict a statement of the form "All *A* are *B*" (or equivalently, "If *A*, then *B*"), we draw two circles, one inside the other; the inner circle represents *A*, the outer circle represents *B*. This relationship is shown in Figure 1.1.

Venn diagrams depicting "No *A* are *B*" and "Some *A* are *B*" are shown in Figures 1.2 and 1.3, respectively.

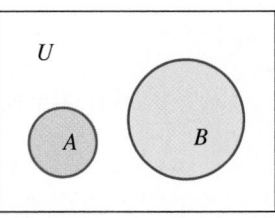

Figure 1.2
No *A* are *B*.

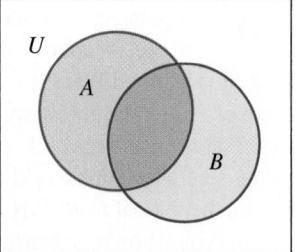

Figure 1.3
Some *A* are *B*. (At least one *A* is *B*.)

ThomsonNOW™
for Liberal Arts Mathematics
www.thomsonedu.com/1pass.

EXAMPLE 2   Construct a Venn diagram to verify the validity of the following argument:

    1. All men are mortal.

    2. Socrates is a man.

    Therefore, Socrates is mortal.

Solution   Premise 1 is of the form "All *A* are *B*" and can be represented by a diagram like that shown in Figure 1.4.

Premise 2 refers to a specific man, namely, Socrates. If we let *x* = Socrates, the statement "Socrates is a man" can then be represented by placing *x* within the circle labeled "men," as shown in Figure 1.5. Because we placed *x* within the "men" circle, and all of the "men" circle is inside the "mortal" circle, the conclusion "Socrates is mortal" is inescapable; the argument is valid.

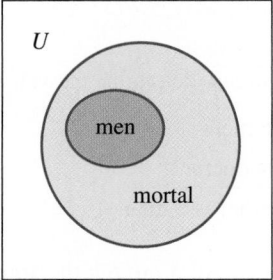

Figure 1.4
All men are mortal.

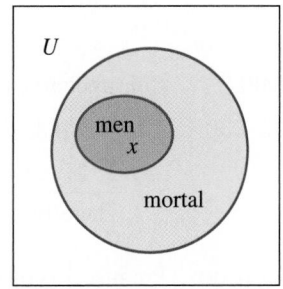

x = Socrates
Figure 1.5

■

Historical Note   Aristotle 384–322 B.C.

Museo Archaeologico Nazionale, Naples, © Scala/Art Resource, NY

Aristotle was born in 384 B.C. in the small Macedonian town of Stagira, 200 miles north of Athens, on the shore of the Aegean Sea. Aristotle's father was the personal physician of King Amyntas II, ruler of Macedonia. When he was seventeen, Aristotle enrolled at the Academy in Athens and became a student of the famed Plato.

Aristotle was one of Plato's brightest students; he frequently questioned Plato's teachings and openly disagreed with him. Whereas Plato emphasized the study of abstract ideas and mathematical truth, Aristotle was more interested in observing the "real world" around him. Plato often referred to Aristotle as "the brain" or "the mind of the school." Plato commented, "Where others need the spur, Aristotle needs the rein."

Aristotle stayed at the Academy for twenty years, until the death of Plato. Then the king of Macedonia invited Aristotle to supervise the education of his son Alexander. Aristotle accepted the invitation and taught Alexander until he succeeded his father as ruler. At that time, Aristotle founded a school known as the Lyceum, or Peripatetic School. The school had a large library with many maps, as well as botanical gardens containing an extensive collection of plants and animals. Aristotle and his students would walk about the grounds of the Lyceum while discussing various subjects (*peripatetic* is from the Greek word meaning "to walk").

Many consider Aristotle to be a founding father of the study of biology and of science in general; he observed and classified the behavior and anatomy of hundreds of living creatures. During his many military campaigns, Alexander the Great had his troops gather specimens from distant places for Aristotle to study.

Aristotle was a prolific writer; some historians credit him with the writing of over 1,000 books. Most of his works have been lost or destroyed, but scholars have recreated some of his more influential works, including *Organon*.

Collections of the New York Public Library

THE
ORGANON;
OR,
LOGICAL TREATISES
OF
ARISTOTLE.
TRANSLATED FROM THE GREEK.
WITH
COPIOUS ELUCIDATIONS,
FROM
THE COMMENTARIES OF AMMONIUS AND SIMPLICIUS.
BY *THOMAS TAYLOR*.
JOVE HONOURS ME, AND FAVOURS MY DESIGNS.
Pope's Homer's Iliad, Book 9b, v. 737.

LONDON:
PRINTED FOR THE TRANSLATOR,
MANOR-PLACE, WALWORTH, SURREY;
BY ROBERT WILKS, 89, CHANCERY-LANE, FLEET-STREET.
1807.

Aristotle's collective works on syllogisms and deductive logic are known as *Organon,* meaning "instrument," for logic is the instrument used in the acquisition of knowledge.

EXAMPLE 3   Construct a Venn diagram to determine the validity of the following argument:

    1. All doctors are men.

    2. My mother is a doctor.

    Therefore, my mother is a man.

Solution   Premise 1 is of the form "All A are B"; the argument is depicted in Figure 1.6.

No matter where x is placed within the "doctors" circle, the conclusion "My mother is a man" is inescapable; the argument is valid. ∎

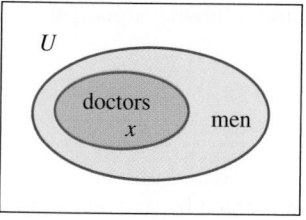

$x =$ My mother
Figure 1.6

Saying that an argument is valid does not mean that the conclusion is true. The argument given in Example 3 *is* valid, but the conclusion is *false.* One's mother cannot be a man! Validity and truth do not mean the same thing. An argument is valid if the conclusion is inescapable, *given the premises.* Nothing is said about the truth of the premises. Thus, when examining the validity of an argument, we are not determining whether the conclusion is true or false. Saying that an argument is valid merely means that, *given the premises,* the reasoning used to obtain the conclusion is logical. However, if the premises of a valid argument are true, then the conclusion will also be true.

EXAMPLE 4   Construct a Venn diagram to determine the validity of the following argument:

1. All professional wrestlers are actors.
2. The Rock is an actor.

Therefore, The Rock is a professional wrestler.

Solution   Premise 1 is of the form "All *A* are *B*"; the "circle of professional wrestlers" is contained within the "circle of actors." If we let *x* represent The Rock, premise 2 simply requires that we place *x* somewhere within the actor circle; *x* could be placed in either of the two locations shown in Figures 1.7 and 1.8.

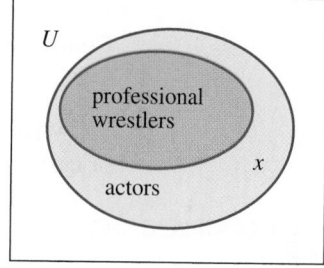

$x =$ The Rock
Figure 1.7

$x =$ The Rock
Figure 1.8

If *x* is placed as in Figure 1.7, the argument would appear to be valid; the figure supports the conclusion "The Rock is a professional wrestler." However, the placement of *x* in Figure 1.8 does not support the conclusion; given the premises, we cannot *logically* deduce that "The Rock is a professional wrestler." Since the conclusion is *not* inescapable, the argument is invalid.   ■

Saying that an argument is invalid does not mean that the conclusion is false. Example 4 demonstrates that an invalid argument can have a true conclusion; even though The Rock is a professional wrestler, the argument used to obtain the conclusion is invalid. In logic, validity and truth do not have the same meaning. *Validity* refers to the process of reasoning used to obtain a conclusion; *truth* refers to conformity with fact or experience.

Venn Diagrams and Invalid Arguments

To show that an argument is invalid, you must construct a Venn diagram in which the premises are met yet the conclusion does not necessarily follow.

Even though The Rock *is* a professional wrestler, the argument used to obtain the conclusion is invalid.

EXAMPLE 5 Construct a Venn diagram to determine the validity of the following argument:

    1. Some plants are poisonous.

    2. Broccoli is a plant.

    Therefore, broccoli is poisonous.

Solution   Premise 1 is of the form "Some *A* are *B*"; it can be represented by two overlapping circles (as in Figure 1.3). If we let *x* represent broccoli, premise 2 requires that we place *x* somewhere within the plant circle. If *x* is placed as in Figure 1.9, the argument would appear to be valid. However, if *x* is placed as in Figure 1.10, the conclusion does not follow. Because we can construct a Venn diagram in which the premises are met yet the conclusion does not follow (Figure 1.10), the argument is invalid.

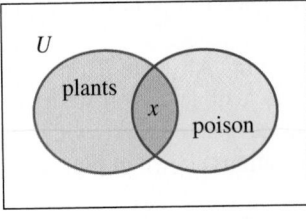

x = broccoli

Figure 1.9

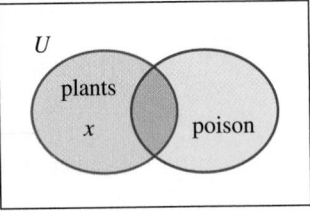

x = broccoli

Figure 1.10 ■

    When analyzing an argument via a Venn diagram, you might have to draw three or more circles, as in the next example.

EXAMPLE 6 Construct a Venn diagram to determine the validity of the following argument:

    1. No snake is warm-blooded.

    2. All mammals are warm-blooded.

    Therefore, snakes are not mammals.

Solution   Premise 1 is of the form "No *A* are *B*"; it is depicted in Figure 1.11. Premise 2 is of the form "All *A* are *B*"; the "mammal circle" must be drawn within the "warm-blooded circle." Both premises are depicted in Figure 1.12.

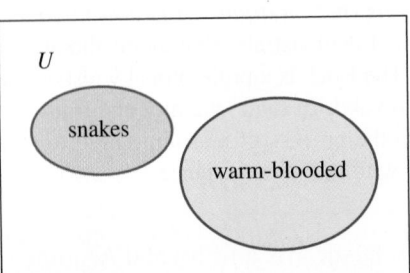

Figure 1.11
No snake is warm-blooded.

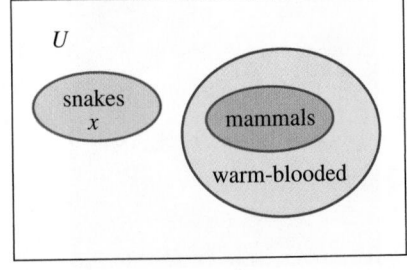

x = snake

Figure 1.12
All mammals are warm-blooded.

Because we placed x (= snake) within the "snake" circle, and the "snake" circle is outside the "warm-blooded" circle, x cannot be within the "mammal" circle (which is inside the "warm-blooded" circle). Given the premises, the conclusion "Snakes are not mammals" is inescapable; the argument is valid. ∎

You might have encountered Venn diagrams when you studied sets in your algebra class. The academic fields of set theory and logic are historically intertwined; set theory was developed in the late nineteenth century as an aid in the study of logical arguments. Today, set theory and Venn diagrams are applied to areas other than the study of logical arguments; we will utilize Venn diagrams in our general study of set theory in Chapter 2.

Inductive Reasoning

The conclusion of a valid deductive argument (one that goes from general to specific) is guaranteed: Given true premises, a true conclusion must follow. However, there are arguments in which the conclusion is not guaranteed even though the premises are true. Consider the following:

1. Joe sneezed after petting Frako's cat.
2. Joe sneezed after petting Paulette's cat.

Therefore, Joe is allergic to cats.

Is the conclusion guaranteed? If the premises are true, they certainly *support* the conclusion, but we cannot say with 100% certainty that Joe is allergic to cats. The conclusion is *not* guaranteed. Maybe Joe is allergic to the flea powder that the cat owners used; maybe he is allergic to the dust that is trapped in the cat's fur; or maybe he has a cold!

Reasoning of this type is called inductive reasoning. **Inductive reasoning** involves going from a series of specific cases to a general statement (see Figure 1.13). Although it may seem to follow and may in fact be true, *the conclusion in an inductive argument is never guaranteed.*

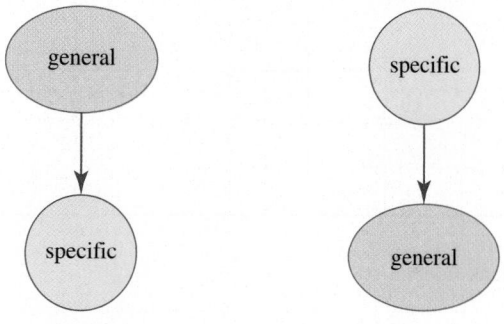

Deductive Reasoning
(Conclusion is guaranteed.)

Inductive Reasoning
(Conclusion may be probable but is not guaranteed.)

Figure 1.13

EXAMPLE 7   What is the next number in the sequence 1, 8, 15, 22, 29, . . . ?

Solution   Noticing that the difference between consecutive numbers in the sequence is 7, we may be tempted to say that the next term is $29 + 7 = 36$. Is this conclusion guaranteed? No! Another sequence in which numbers differ by 7 are dates of a given day of the week. For instance, the dates of the Saturdays in the year 2005 are (January) 1, 8, 15, 22, 29, (February) 5, 12, 19, 26, ... Therefore, the next number in the sequence 1, 8, 15, 22, 29, . . . might be 5. Without further information, we cannot determine the next number in the given sequence. We can only use inductive reasoning and give one or more *possible* answers.   ■

Logic in the Real World: Su doku

Throughout history, people have always been attracted to puzzles, mazes, and brainteasers. Who can deny the inherent satisfaction of solving a seemingly unsolvable or perplexing riddle? A popular new addition to the world of puzzle solving is *su doku,* a numbers puzzle. Loosely translated from Japanese, *su doku* means "single number"; a su doku puzzle simply involves placing the digits 1 through 9 in a grid containing 9 rows and 9 columns. In addition, the 9 by 9 grid of squares is subdivided into nine 3 by 3 grids, or "boxes," as shown in Figure 1.14.

The rules of su doku are quite simple: Each row, each column, and each box must contain the digits 1 through 9; and no row, column, or box can contain 2 squares with the same number. Consequently, su doku does not require any arithmetic or mathematical skill; su doku requires logic only. In solving a puzzle, a common thought is "What happens if I put this number here?"

Like crossword puzzles, su doku puzzles are printed daily in many newspapers across the country and around the world. Web sites containing su doku puzzles and strategies provide an endless source of new puzzles and help. See Exercise 62 to find links to popular sites.

Figure 1.14
A blank su doku grid.

EXAMPLE 8   Solve the su doku puzzle given in Figure 1.15.

		2		6	4	8	5	
3	8						6	
			8	5			3	
			6		5	9		
6						1		5
7		8			1			
	6			7	9			
	4		1				9	8
	9	3	4	2		7		6

Figure 1.15
A su doku puzzle

Solution   Recall that each 3 by 3 grid is referred to as a box. For convenience, the boxes are numbered 1 through 9, starting in the upper left-hand corner and moving from left to right, and each square can be assigned coordinates (x, y) based on its row number x and column number y as shown in Figure 1.16.

Figure 1.16
Box numbers and coordinate system in su doku

For example, the digit 2 in Figure 1.16 is in box 1 and has coordinates $(1, 3)$, the digit 8 is in box 3 and has coordinates $(1, 7)$, the digit 6 is in box 4 and has coordinates $(5, 1)$ and the digit 7 is in box 9 and has coordinates $(9, 7)$.

When you are first solving a su doku puzzle, concentrate on only a few boxes rather than the puzzle as a whole. For instance, looking at boxes 1, 4, and 7, we see that boxes 4 and 7 each contain the digit 6, whereas box 1 does not. Consequently, the 6 in box 1 must be placed in column 3 because (shaded) columns 1 and 2 already have a 6. However, (shaded) row 2 already has a 6, so we can deduce that 6 must be placed in row 3, column 3, that is, in square $(3, 3)$ as shown in Figure 1.17.

		2		6	4	8	5	
3	8						**6**	
		6	8	5			3	
			6		5	9		
6						1		5
7		8			1			
	6			7	9			
	4		1				9	8
	9	3	4	2		7		6

Figure 1.17
The 6 in box 1 must be placed in square (3, 3)

Examining, boxes 1, 2, and 3, we see that boxes 2 and 3 each contain the digit 5, whereas box 1 does not. We deduce that 5 must be placed in square (2, 3) because rows 1 and 3 already have a 5. In a similar fashion, square (1, 4) must contain 3. See Figure 1.18.

		2	**3**	6	4	8	**5**	
3	8	**5**					6	
		6	8	**5**			**3**	
			6		5	9		
6						1		5
7		8			1			
	6			7	9			
	4		1				9	8
	9	3	4	2		7		6

Figure 1.18
Analyzing boxes 1, 2, and 3; placing the digits 3 and 5

Because we have placed two new digits in box 1, we might wish to focus on the remainder of (shaded) box 1. Notice that the digit 4 can be placed only in square (3, 1), as row 1 and column 2 already have a 4 in each of them; likewise, the digit 9 can be placed only in square (1, 1) because column 2 already has a 9. Finally, either of the digits 1 or 7 can be placed in square (1, 2) or (3, 2) as shown in Figure 1.19. At some point later in the solution, we will be able to determine the exact values of squares (1, 2) and (3, 2), that is, which square receives a 1 and which receives a 7.

Using this strategy of analyzing the contents of three consecutive boxes, we deduce the following placement of digits: 1 must go in (2, 5), 6 must go in (6, 7), 6 must go in (8, 6), 7 must go in (8, 3), 3 must go in (8, 5), 8 must go in (9, 6), and 5 must go in (7, 4). At this point, box 8 is complete as shown in Figure 1.20. (Remember, each box must contain each of the digits 1 through 9.)

9	1,7	2	3	6	4	8	5	
3	8	5					6	
4	1,7	6	8	5			3	
			6		5	9		
6						1		5
7		8			1			
	6			7	9			
	4		1				9	8
	9	3	4	2		7		6

Figure 1.19
Focusing on box 1

9	1,7	2	3	6	4	8	5	
3	8	5		1			6	
4	1,7	6	8	5			3	
			6		5	9		
6						1		5
7		8			1	6		
	6		5	7	9			
	4	7	1	3	6		9	8
	9	3	4	2	8	7		6

Figure 1.20
Box 8 is complete.

Once again, we use the three consecutive box strategy and deduce the following placement of digits: 5 must go in (8, 7), 5 must go in (9, 1), 8 must go in (7, 1), 2 must go in (8, 1), and 1 must go in (7, 3). At this point, box 7 is complete as shown in Figure 1.21.

9	1,7	2	3	6	4	8	5	
3	8	5		1			6	
4	1,7	6	8	5			3	
			6		5	9		
6						1		5
7		8			1	6		
8	6	1	5	7	9			
2	4	7	1	3	6	5	9	8
5	9	3	4	2	8	7		6

Figure 1.21
Box 7 is complete.

We now focus on box 4 and deduce the following placement of digits: 1 must go in (4, 1), 9 must go in (5, 3), 4 must go in (4, 3), 5 must go in (6, 2), 3 must go in (4, 2), and 2 must go in (5, 2). At this point, box 4 is complete. In addition, we deduce that 1 must go in (9, 8), and 3 must go in (5, 6) as shown in Figure 1.22.

9	1,7	2	3	6	4	8	5	
3	8	5		1			6	
4	1,7	6	8	5			3	
1	3	4	6		5	9		
6	2	9			3	1		5
7	5	8			1	6		
8	6	1	5	7	9			
2	4	7	1	3	6	5	9	8
5	9	3	4	2	8	7	1	6

Figure 1.22
Box 4 is complete.

Once again, we use the three consecutive box strategy and deduce the following placement of digits: 3 must go in (6, 9), 3 must go in (7, 7), 9 must go in (2, 4), and 9 must go in (6, 5). Now, to finish row 6, we place 4 in (6, 8) and 2 in (6, 4) as shown in Figure 1.23. (Remember, each row must contain each of the digits 1 through 9.)

9	1,7	2	3	6	4	8	5	
3	8	5	9	1			6	
4	1,7	6	8	5			3	
1	3	4	6		5	9		
6	2	9			3	1		5
7	5	8	2	9	1	6	4	3
8	6	1	5	7	9	3		
2	4	7	1	3	6	5	9	8
5	9	3	4	2	8	7	1	6

Figure 1.23
Row 6 is complete.

After we place 7 in (5, 4), column 4 is complete. (Remember, each column must contain each of the digits 1 through 9.) This leads to placing 4 in (5, 5) and 8 in (4, 5), thus completing box 5; row 5 is finalized by placing 8 in (5, 8) as shown in Figure 1.24.

Now column 7 is completed by placing 4 in (2, 7) and 2 in (3, 7); placing 2 in (2, 6) and 7 in (3, 6) completes column 6 as shown in Figure 1.25.

At this point, we deduce that the digit in (3, 2) must be 1 because row 3 cannot have two 7's. This in turn reveals that 7 must go in (1, 2), and box 1 is now complete. To complete row 7, we place 4 in (7, 9) and 2 in (7, 8); row 4 is finished with 2 in (4, 9) and 7 in (4, 8). See Figure 1.26.

9	1,7	2	3	6	4	8	5	
3	8	5	9	1			6	
4	1,7	6	8	5			3	
1	3	4	6	**8**	5	9		
6	2	9	**7**	**4**	3	1	**8**	5
7	5	8	2	9	1	6	4	3
8	6	1	5	7	9	3		
2	4	7	1	3	6	5	9	8
5	9	3	4	2	8	7	1	6

Figure 1.24
Column 4, box 5, and row 5 are complete.

9	1,7	2	3	6	4	8	5	
3	8	5	9	1	**2**	**4**	6	
4	1,7	6	8	5	**7**	**2**	3	
1	3	4	6	8	5	9		
6	2	9	7	4	3	1	8	5
7	5	8	2	9	1	6	4	3
8	6	1	5	7	9	3		
2	4	7	1	3	6	5	9	8
5	9	3	4	2	8	7	1	6

Figure 1.25
Columns 7 and 6 are complete.

9	**7**	2	3	6	4	8	5	
3	8	5	9	1	2	4	6	
4	**1**	6	8	5	7	2	3	
1	3	4	6	8	5	9	**7**	**2**
6	2	9	7	4	3	1	8	5
7	5	8	2	9	1	6	4	3
8	6	1	5	7	9	3	**2**	**4**
2	4	7	1	3	6	5	9	8
5	9	3	4	2	8	7	1	6

Figure 1.26
Box 1, row 7, and row 4 are complete.

To finish rows 1, 2, and 3, 1 must go in (1, 9), 7 must go in (2, 9), and 9 must go in (3, 9). The puzzle is now complete as shown in Figure 1.27.

9	7	2	3	6	4	8	5	1
3	8	5	9	1	2	4	6	7
4	1	6	8	5	7	2	3	9
1	3	4	6	8	5	9	7	2
6	2	9	7	4	3	1	8	5
7	5	8	2	9	1	6	4	3
8	6	1	5	7	9	3	2	4
2	4	7	1	3	6	5	9	8
5	9	3	4	2	8	7	1	6

Figure 1.27
A completed su doku puzzle

As a final check, we scrutinize each box, row, and column to verify that no box, row, or column contains the same digit twice. Congratulations, the puzzle has been solved! ∎

1.1 Exercises

In Exercises 1–20, construct a Venn diagram to determine the validity of the given argument.

1. **a.** 1. All master photographers are artists.
  2. Ansel Adams is a master photographer.

  Therefore, Ansel Adams is an artist.

  b. 1. All master photographers are artists.
  2. Ansel Adams is an artist.

  Therefore, Ansel Adams is a master photographer.

2. **a.** 1. All Olympic gold medal winners are role models.
  2. Apolo Ohno is an Olympic gold medal winner.

  Therefore, Apolo Ohno is a role model.

  b. 1. All Olympic gold medal winners are role models.
  2. Apolo Ohno is a role model.

  Therefore, Apolo Ohno is an Olympic gold medal winner.

3. **a.** 1. All homeless people are unemployed.
  2. Bill Gates is not a homeless person.

  Therefore, Bill Gates is not unemployed.

  b. 1. All homeless people are unemployed.
  2. Bill Gates is not unemployed.

  Therefore, Bill Gates is not a homeless person.

4. **a.** 1. All professional wrestlers are actors.
  2. Ralph Nader is not an actor.

  Therefore, Ralph Nader is not a professional wrestler.

  b. 1. All professional wrestlers are actors.
  2. Ralph Nader is not a professional wrestler.

  Therefore, Ralph Nader is not an actor.

5. 1. All pesticides are harmful to the environment.
  2. No fertilizer is a pesticide.

  Therefore, no fertilizer is harmful to the environment.

6. 1. No one who can afford health insurance is unemployed.
  2. All politicians can afford health insurance.

  Therefore, no politician is unemployed.

7. 1. No vegetarian owns a gun.
  2. All policemen own guns.

  Therefore, no policeman is a vegetarian.

8. 1. No professor is a millionaire.
    2. No millionaire is illiterate.

Therefore, no professor is illiterate.

9. 1. All poets are loners.
    2. All loners are taxi drivers.

Therefore, all poets are taxi drivers.

10. 1. All forest rangers are environmentalists.
    2. All forest rangers are storytellers.

Therefore, all environmentalists are storytellers.

11. 1. Real men don't eat quiche.
    2. Clint Eastwood is a real man.

Therefore, Clint Eastwood doesn't eat quiche.

12. 1. Real men don't eat quiche.
    2. Oscar Meyer eats quiche.

Therefore, Oscar Meyer isn't a real man.

13. 1. All roads lead to Rome.
    2. Route 66 is a road.

Therefore, Route 66 leads to Rome.

14. 1. All smiling cats talk.
    2. The Cheshire Cat smiles.

Therefore, the Cheshire Cat talks.

15. 1. Some animals are dangerous.
    2. A tiger is an animal.

Therefore, a tiger is dangerous.

16. 1. Some professors wear glasses.
    2. Mr. Einstein wears glasses.

Therefore, Mr. Einstein is a professor.

17. 1. Some women are police officers.
    2. Some police officers ride motorcycles.

Therefore, some women ride motorcycles.

18. 1. All poets are eloquent.
    2. Some poets are wine connoisseurs.

Therefore, some wine connoisseurs are eloquent.

19. 1. All squares are rectangles.
    2. Some quadrilaterals are squares.

Therefore, some quadrilaterals are rectangles.

20. 1. All squares are rectangles.
    2. Some quadrilaterals are rectangles.

Therefore, some quadrilaterals are squares.

21. Classify each argument as deductive or inductive.
  a. 1. My television set did not work two nights ago.
      2. My television set did not work last night.

Therefore, my television set is broken.

b. 1. All electronic devices give their owners grief.
    2. My television set is an electronic device.

Therefore, my television set gives me grief.

22. Classify each argument as deductive or inductive.
  a. 1. I ate a chili dog at Joe's and got indigestion.
      2. I ate a chili dog at Ruby's and got indigestion.

Therefore, chili dogs give me indigestion.

b. 1. All spicy foods give me indigestion.
    2. Chili dogs are spicy food.

Therefore, chili dogs give me indigestion.

In Exercises 23–32, fill in the blank with what is most likely to be the next number. Explain (using complete sentences) the pattern generated by your answer.

23. 3, 8, 13, 18, _____
24. 10, 11, 13, 16, _____
25. 0, 2, 6, 12, _____
26. 1, 2, 5, 10, _____
27. 1, 4, 9, 16, _____
28. 1, 8, 27, 64, _____
29. 2, 3, 5, 7, 11, _____
30. 1, 1, 2, 3, 5, _____
31. 5, 8, 11, 2, _____
32. 12, 5, 10, 3, _____

In Exercises 33–36, fill in the blanks with what are most likely to be the next letters. Explain (using complete sentences) the pattern generated by your answers.

33. O, T, T, F, _____, _____
34. T, F, S, E, _____, _____
35. F, S, S, M, _____, _____
36. J, F, M, A, _____, _____

In Exercises 37–42, explain the general rule or pattern used to assign the given letter to the given word. Fill in the blank with the letter that fits the pattern.

37.

circle	square	trapezoid	octagon	rectangle
c	s	t	o	_____

38.

circle	square	trapezoid	octagon	rectangle
i	u	a	o	_____

39.

circle	square	trapezoid	octagon	rectangle
j	v	b	p	_____

40.

circle	square	trapezoid	octagon	rectangle
c	r	p	g	_____

41.

banana	strawberry	asparagus	eggplant	orange
b	z	t	u	_____

42.

banana	strawberry	asparagus	eggplant	orange
y	r	g	p	_____

43. Find two different numbers that could be used to fill in the blank.

1, 4, 7, 10, _____

Explain the pattern generated by each of your answers.

44. Find five different numbers that could be used to fill in the blank.

7, 14, 21, 28, _____

Explain the pattern generated by each of your answers.

45. Example 1 utilized the Quadratic Formula. Verify that

$$x = \frac{-b + \sqrt{b^2 - 4ac}}{2a}$$

is a solution of the equation $ax^2 + bx + c = 0$.

HINT: Substitute the fraction for x in $ax^2 + bx + c$ and simplify.

46. Example 1 utilized the Quadratic Formula. Verify that

$$x = \frac{-b - \sqrt{b^2 - 4ac}}{2a}$$

is a solution of the equation $ax^2 + bx + c = 0$.

HINT: Substitute the fraction for x in $ax^2 + bx + c$ and simplify.

47. As a review of algebra, use the Quadratic Formula to solve

$$x^2 - 6x + 7 = 0$$

48. As a review of algebra, use the Quadratic Formula to solve

$$x^2 - 2x - 4 = 0$$

Solve the su doku puzzles in Exercises 49–54.

49.

	5							9
		7				3	4	
					2	1	5	6
6		1	8	9				
	3						2	
				3		6		5
5	9	2	1		4			
	6	3				8		
4						1		

50.

4		7					3	
	2		5		1		7	
					4	5		
	7		9					2
2		9		4		6		3
3					6		9	
		2	4					
	8		6		5		4	
	6					2		9

51.

	5		3			2	8	
		3			7			9
		7		9	2			
	8	7						
5	9			6			1	3
					4	6		
			8		9		2	
2			5			6		
	3	9			6		7	

52.

		2				1		
6	8			2			5	4
		3	8		6	2		
7			6		4			5
2				3		9		7
		6	7		5	4		
3	5			1			8	6
		7			5			

53.

7	6		5					
		2	7					8
		4	6				5	
	7	1			3		2	
5								9
	9		8			7	6	
	8				1	3		
3					6	1		
					7		9	6

54.

7	1	2					8	6
				7				
		4				1		
4		3	5	8				
1			2		4			8
				9	1	2		5
	8					4		
					3			
6	3					8	2	9

Answer the following questions using complete sentences and your own words.

CONCEPT QUESTIONS
55. Explain the difference between deductive and inductive reasoning.
56. Explain the difference between truth and validity.
57. What is a syllogism? Give an example of a syllogism that relates to your life.

HISTORY QUESTIONS
58. From the days of the ancient Greeks, the study of logic has been mandatory in what two professions? Why?
59. Who developed a formal system of deductive logic based on arguments?
60. What was the name of the school Aristotle founded? What does it mean?
61. How did Aristotle's school of thought differ from Plato's?

 WEB PROJECT
62. Obtain a su doku puzzle and its solution from a popular web site. Some useful links for this web project are listed on the text web site:
www.thomsonedu.com/mathematics/johnson

1.2 Symbolic Logic

The syllogism ruled the study of logic for nearly 2,000 years and was not supplanted until the development of symbolic logic in the late seventeenth century. As its name implies, symbolic logic involves the use of symbols and algebraic manipulations in logic.

Statements

All logical reasoning is based on statements. A **statement** is a sentence that is either true or false.

EXAMPLE 1   Which of the following are statements? Why or why not?

a. Apple manufactures computers.
b. Apple manufactures the world's best computers.
c. Did you buy an IBM?
d. A $2,000 computer that is discounted 25% will cost $1,000.
e. I am telling a lie.

Solution   **a.** The sentence "Apple manufactures computers" is true; therefore, it is a statement.
b. The sentence "Apple manufactures the world's best computers" is an opinion, and as such, it is neither true nor false. It is true for some people and false for others. Therefore, it is not a statement.

c. The sentence "Did you buy an IBM?" is a question. As such, it is neither true nor false; it is not a statement.

d. The sentence "A $2,000 computer that is discounted 25% will cost $1,000" is false; therefore, it is a statement. (A $2,000 computer that is discounted 25% would cost $1,500.)

e. The sentence "I am telling a lie" is a self-contradiction, or paradox. If it were true, the speaker would be telling a lie, but in telling the truth, the speaker would be contradicting the statement that he or she was lying; if it were false, the speaker would not be telling a lie, but in not telling a lie, the speaker would be contradicting the statement that he or she was lying. The sentence is not a statement.   ■

By tradition, symbolic logic uses lowercase letters as labels for statements. The most frequently used letters are *p, q, r, s,* and *t.* We can label the statement "It is snowing" as statement *p* in the following manner:

p: It is snowing.

If it *is* snowing, *p* is labeled true, whereas if it is *not* snowing, *p* is labeled false.

Compound Statements and Logical Connectives

It is easy to determine whether a statement such as "Charles donated blood" is true or false; either he did or he didn't. However, not all statements are so simple; some are more involved. For example, the truth of "Charles donated blood and did not wash his car, or he went to the library," depends on the truth of the individual pieces that make up the larger, compound statement. A **compound statement** is a statement that contains one or more simpler statements. A compound statement can be formed by inserting the word *not* into a simpler statement or by joining two or more statements with connective words such as *and, or, if . . . then . . . , only if,* and *if and only if.* The compound statement "Charles did *not* wash his car" is formed from the simpler statement "Charles did wash his car." The compound statement "Charles donated blood *and* did *not* wash his car, *or* he went to the library" consists of three statements, each of which may be true or false.

Figure 1.28 diagrams two equivalent compound statements.

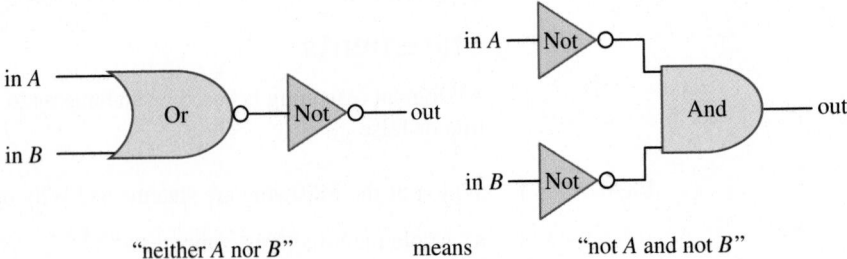

Figure 1.28
Technicians and engineers use compound statements and logical connectives to study the flow of electricity through switching circuits.

When is a compound statement true? Before we can answer this question, we must first examine the various ways in which statements can be connected. Depending on how the statements are connected, the resulting compound statement can be a *negation,* a *conjunction,* a *disjunction,* a *conditional,* or any combination thereof.

The Negation ~p

The **negation** of a statement is the denial of the statement and is represented by the symbol ~. The negation is frequently formed by inserting the word *not.* For example, given the statement "*p:* It is snowing," the negation would be "~*p:* It is not snowing." If it *is* snowing, *p* is true and ~*p* is false. Similarly, if it is *not* snowing, *p* is false and ~*p* is true. A statement and its negation always have opposite truth values; when one is true, the other is false. Because the truth of the negation depends on the truth of the original statement, a negation is classified as a compound statement.

ThomsonNOW™
for Liberal Arts Mathematics
www.thomsonedu.com/1pass.

EXAMPLE 2 Write a sentence that represents the negation of each statement:

a. The senator is a Democrat.   **b.** The senator is not a Democrat.
c. Some senators are Republicans.   **d.** All senators are Republicans.
e. No senator is a Republican.

Solution   **a.** The negation of "The senator is a Democrat" is "The senator is not a Democrat."
b. The negation of "The senator is not a Democrat" is "The senator is a Democrat."
c. A common error would be to say that the negation of "Some senators are Republicans" is "Some senators are not Republicans." However, "Some senators are Republicans" is not denied by "Some senators are not Republicans." The statement "Some senators *are* Republicans" implies that at least one senator is a Republican. The negation of this statement is "It is not the case that at least one senator is a Republican," or (more commonly phrased) the negation is "No senator is a Republican."
d. The negation of "All senators are Republicans" is "It is not the case that all senators are Republicans," or "There exists a senator who is not a Republican," or (more commonly phrased) "Some senators are not Republicans."
e. The negation of "No senator is a Republican" is "It is not the case that no senator is a Republican" or, in other words, "There exists at least one senator who *is* a Republican." If "some" is interpreted as meaning "at least one," the negation can be expressed as "Some senators are Republicans." ∎

The words *some, all,* and *no* (or *none*) are referred to as **quantifiers.** Parts (c) through (e) of Example 2 contain quantifiers. The linked pairs of quantified statements shown in Figure 1.29 are negations of each other.

Figure 1.29
Negations of statements containing quantifiers

The Conjunction p ∧ q

Consider the statement "Norma Rae is a union member and she is a Democrat." This is a compound statement, because it consists of two statements—"Norma Rae is a union member" and "she (Norma Rae) is a Democrat"—and the connective word *and.* Such a compound statement is referred to as a conjunction. A **conjunction** consists of two or more statements connected by the word *and.* We use the symbol ∧ to represent the word *and;* thus, the conjunction "$p \wedge q$" represents the compound statement "p and q."

EXAMPLE 3   Using the symbolic representations

  p: Norma Rae is a union member.

  q: Norma Rae is a Democrat.

express the following compound statements in symbolic form:

a. Norma Rae is a union member and she is a Democrat.
b. Norma Rae is a union member and she is not a Democrat.

Solution   **a.** The compound statement "Norma Rae is a union member and she is a Democrat" can be represented as $p \wedge q$.
b. The compound statement "Norma Rae is a union member and she is not a Democrat" can be represented as $p \wedge \sim q$. ∎

The Disjunction $p \vee q$

When statements are connected by the word *or,* a **disjunction** is formed. We use the symbol $\vee$ to represent the word *or.* Thus, the disjunction "$p \vee q$" represents the compound statement "p or q." We can interpret the word *or* in two ways. Consider the statements

  p: Kaitlin is a registered Republican.

  q: Paki is a registered Republican.

The statement "Kaitlin is a registered Republican or Paki is a registered Republican" can be symbolized as $p \vee q$. Notice that it is possible that *both* Kaitlin and Paki are registered Republicans. In this example, *or* includes the possibility that both things may happen. In this case, we are working with the **inclusive *or.***
 Now consider the statements

  p: Kaitlin is a registered Republican.

  q: Kaitlin is a registered Democrat.

The statement "Kaitlin is a registered Republican or Kaitlin is a registered Democrat" does *not* include the possibility that both may happen; one statement *excludes* the other. When this happens, we are working with the **exclusive *or.*** In our study of symbolic logic (as in most mathematics), we will always use the *inclusive or.* Therefore, "p or q" means "p or q or both."

EXAMPLE 4   Using the symbolic representations

  p: Juanita is a college graduate.

  q: Juanita is employed.

express the following compound statements in words:

a. $p \vee q$ **b.** $p \wedge q$
c. $p \vee \sim q$ **d.** $\sim p \wedge q$

Historical Note   Gottfried Wilhelm Leibniz 1646–1716

In addition to cofounding calculus (see Chapter 13), the German-born Gottfried Wilhelm Leibniz contributed much to the development of symbolic logic. A precocious child, Leibniz was self-taught in many areas. He taught himself Latin at the age of eight and began the study of Greek when he was twelve. In the process, he was exposed to the writings of Aristotle and became intrigued by formalized logic.

At the age of fifteen, Leibniz entered the University of Leipzig to study law. He received his bachelor's degree two years later, earned his master's degree the following year, and then transferred to the University of Nuremberg.

Leibniz received his doctorate in law within a year and was immediately offered a professorship but refused it, saying that he had "other things in mind." Besides law, these "other things" included politics, religion, history, literature, metaphysics, philosophy, logic, and mathematics. Thereafter, Leibniz worked under the sponsorship of the courts of various nobles, serving as lawyer, historian, and librarian to the elite. At one point, Leibniz was offered the position of librarian at the Vatican but declined the offer.

Leibniz's affinity for logic was characterized by his search for a *characteristica universalis,* or "universal character." Leibniz believed that by combining logic and mathematics, a general symbolic language could be created in which all scientific problems could be solved with a minimum of effort. In this universal language, statements and the logical relationships between them would be represented by letters and symbols. In Leibniz's words, "All truths of reason would be reduced to a kind of calculus, and the errors would only be errors of computation." In essence, Leibniz believed that once a problem had been translated into this universal language of symbolic logic, it would be solved automatically by simply applying the mathematical rules that governed the manipulation of the symbols.

Leibniz's work in the field of symbolic logic did not arouse much academic curiosity; many say that it was too far ahead of its time. The study of symbolic logic was not systematically investigated again until the nineteenth century.

In the early 1670s, Leibniz invented one of the world's first mechanical calculating machines. Leibniz's machine could multiply and divide, whereas an earlier machine invented by Blaise Pascal (see Chapter 3) could only add and subtract.

Solution

a. $p \vee q$ represents the statement "Juanita is a college graduate or Juanita is employed (or both)."

b. $p \wedge q$ represents the statement "Juanita is a college graduate and Juanita is employed."

c. $p \vee \sim q$ represents the statement "Juanita is a college graduate or Juanita is not employed."

d. $\sim p \wedge q$ represents the statement "Juanita is not a college graduate and Juanita is employed." ∎

The Conditional $p \rightarrow q$

Consider the statement "If it is raining, then the streets are wet." This is a compound statement because it connects two statements, namely, "it is raining" and "the streets are wet." Notice that the statements are connected with "if . . . then . . ." phrasing. Any statement of the form "if p then q" is called a **conditional** (or an **implication**); p is called the **hypothesis** (or **premise**) of the conditional, and q is called the **conclusion** of the conditional. The conditional "if p then q" is represented by the symbols "$p \rightarrow q$" (p implies q). When people use conditionals in everyday speech, they often omit the word *then,* as in "If it is raining, the streets are wet." Alternatively, the conditional "if p then q" may be phrased as "q if p" ("The streets are wet if it is raining").

ThomsonNOW™
for Liberal Arts Mathematics
www.thomsonedu.com/1pass.

EXAMPLE 5   Using the symbolic representations

> $p:$ I am healthy.
>
> $q:$ I eat junk food.
>
> $r:$ I exercise regularly.

express the following compound statements in symbolic form:

a. I am healthy if I exercise regularly.
b. If I eat junk food and do not exercise, then I am not healthy.

Solution   **a.** "I am healthy if I exercise regularly" is a conditional (*if . . . then . . .*) and can be rephrased as follows:

"[If] I exercise regularly, [then] I am healthy."

Statement r Statement p
is the premise.   is the conclusion.

The given compound statement can be expressed as $r \rightarrow p$.

b. "If I eat junk food and do not exercise, then I am not healthy" is a conditional (*if . . . then . . .*) that contains a conjunction (*and*) and two negations (*not*):

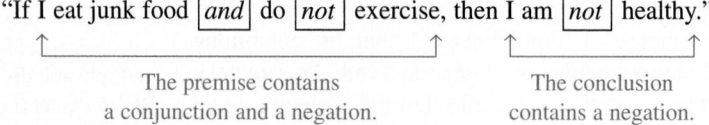

"If I eat junk food [and] do [not] exercise, then I am [not] healthy."

The premise contains   The conclusion
a conjunction and a negation.   contains a negation.

The premise of the conditional can be represented by $q \wedge \sim r$, while the conclusion can be represented by $\sim p$. Thus, the given compound statement has the symbolic form $(q \wedge \sim r) \rightarrow \sim p$. ∎

EXAMPLE 6   Express the following statements in symbolic form:

a. All mammals are warm-blooded.
b. No snake is warm-blooded.

Solution   **a.** The statement "All mammals are warm-blooded" can be rephrased as "If it is a mammal, then it is warm-blooded." Therefore, if we define p and q as

> $p:$ It is a mammal.
>
> $q:$ It is warm-blooded.

the statement can be expressed as $p \rightarrow q$. In general, the statement "All p are q" can be symbolized as $p \rightarrow q$.

b. The statement "No snake is warm-blooded" can be rephrased as "If it is a snake, then it is not warm-blooded." Therefore, if we define p and q as

> p: It is a snake.
>
> q: It is warm-blooded.

the statement can be expressed as $p \rightarrow \sim q$. In general, the statement "No p is q" can be symbolized as $p \rightarrow \sim q$. ∎

We have seen that a statement is a sentence that is either true or false and that connecting two or more statements forms a compound statement. Figure 1.30 summarizes the logical connectives and symbols introduced in this section. The various connectives have been defined; we can now proceed in our analysis of the conditions under which a compound statement is true. This analysis is carried out in the next section.

Statement	Symbol	Read as ...
negation	~	not
conjunction	∧	and
disjunction	∨	(inclusive) or
conditional (implication)	→	if ... then ...

Figure 1.30
Logical connectives

1.2 Exercises

1. Which of the following are statements? Why or why not?
   a. George Washington was the first president of the United States.
   b. Abraham Lincoln was the second president of the United States.
   c. Who was the first vice president of the United States?
   d. Abraham Lincoln was the best president.

2. Which of the following are statements? Why or why not?
   a. $3 + 5 = 6$
   b. Solve the equation $2x + 5 = 3$.
   c. $x^2 + 1 = 0$ has no solution.
   d. $x^2 - 1 = (x + 1)(x - 1)$
   e. Is $\sqrt{2}$ a rational number?

3. Determine which pairs of statements are negations of each other.
   a. All of the fruits are red.
   b. None of the fruits is red.
   c. Some of the fruits are red.
   d. Some of the fruits are not red.

4. Determine which pairs of statements are negations of each other.
   a. Some of the beverages contain caffeine.
   b. Some of the beverages do not contain caffeine.
   c. None of the beverages contain caffeine.
   d. All of the beverages contain caffeine.

5. Write a sentence that represents the negation of each statement.
   a. Her dress is not red.
   b. Some computers are priced under $100.

    c. All dogs are four-legged animals.

    d. No sleeping bag is waterproof.

6. Write a sentence that represents the negation of each statement.

    a. She is not a vegetarian.

    b. Some elephants are pink.

    c. All candy promotes tooth decay.

    d. No lunch is free.

7. Using the symbolic representations

    p: The lyrics are controversial.

    q: The performance is banned.

    express the following compound statements in symbolic form.

    a. The lyrics are controversial and the performance is banned.

    b. If the lyrics are not controversial, the performance is not banned.

    c. It is not the case that the lyrics are controversial or the performance is banned.

    d. The lyrics are controversial and the performance is not banned.

8. Using the symbolic representations

    p: The food is spicy.

    q: The food is aromatic.

    express the following compound statements in symbolic form.

    a. The food is aromatic and spicy.

    b. If the food isn't spicy, it isn't aromatic.

    c. The food is spicy and it isn't aromatic.

    d. The food isn't spicy or aromatic.

9. Using the symbolic representations

    p: A person plays the guitar.

    q: A person rides a motorcycle.

    r: A person wears a leather jacket.

    express the following compound statements in symbolic form.

    a. If a person plays the guitar or rides a motorcycle, then the person wears a leather jacket.

    b. A person plays the guitar, rides a motorcycle, and wears a leather jacket.

    c. A person wears a leather jacket and doesn't play the guitar or ride a motorcycle.

    d. All motorcycle riders wear leather jackets.

10. Using the symbolic representations

    p: The car costs $40,000.

    q: The car goes 140 mph.

    r: The car is red.

    express the following compound statements in symbolic form.

    a. All red cars go 140 mph.

    b. The car is red, goes 140 mph, and does not cost $40,000.

    c. If the car does not cost $40,000, it does not go 140 mph.

    d. The car is red and it does not go 140 mph or cost $40,000.

In Exercises 11–26, translate the sentence into symbolic form. Be sure to define each letter you use. (More than one answer is possible.)

11. All squares are rectangles.

12. No square is a triangle.

13. All whole numbers are even or odd.

14. No whole number is greater than 3 and less than 4.

15. All people born in the United States are American citizens.

16. No convicted felon is eligible to vote.

17. All muscle cars from the Sixties are polluters.

18. No electric-powered car is a polluter.

19. I do not sleep soundly if I drink coffee or eat chocolate.

20. I sleep soundly if I do not drink coffee or eat chocolate.

21. Your check is not accepted if you do not have a driver's license or a credit card.

22. Your check is accepted if you have a driver's license or a credit card.

23. If you drink and drive, you are fined or you go to jail.

24. If you are rich and famous, you have many friends and enemies.

25. You get a refund or a store credit if the product is defective.

26. The streets are slippery if it is raining or snowing.

27. Using the symbolic representations

    p: I am an environmentalist.

    q: I recycle my aluminum cans.

    express the following in words.

    a. $p \wedge q$ **b.** $p \rightarrow q$

    c. $\sim q \rightarrow \sim p$ **d.** $q \vee \sim p$

28. Using the symbolic representations

    p: I am innocent.

    q: I have an alibi.

    express the following in words.

    a. $p \wedge q$ **b.** $p \rightarrow q$

    c. $\sim q \rightarrow \sim p$ **d.** $q \vee \sim p$

29. Using the symbolic representations

    p: I am an environmentalist.

    q: I recycle my aluminum cans.

    r: I recycle my newspapers.

    express the following in words.

a. $(q \vee r) \rightarrow p$ **b.** $\sim p \rightarrow \sim(q \vee r)$
c. $(q \wedge r) \vee \sim p$ **d.** $(r \wedge \sim q) \rightarrow \sim p$

30. Using the symbolic representations

p: I am innocent.
q: I have an alibi.
r: I go to jail.

express the following in words.

a. $(p \vee q) \rightarrow \sim r$ **b.** $(p \wedge \sim q) \rightarrow r$
c. $(\sim p \wedge q) \vee r$ **d.** $(p \wedge r) \rightarrow \sim q$

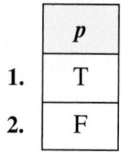 *Answer the following questions using complete sentences and your own words.*

CONCEPT QUESTIONS

31. What is a negation?
32. What is a conjunction?

33. What is a disjunction?
34. What is a conditional?
35. What is the difference between the inclusive *or* and the exclusive *or*?
36. Create a sentence that is a self-contradiction, or paradox, as in part (e) of Example 1.

HISTORY QUESTIONS

37. In what academic field did Gottfried Leibniz receive his degrees? Why is the study of logic important in this field?
38. Who developed a formal system of logic based on syllogistic arguments?
39. What is meant by *characteristica universalis?* Who proposed this theory?

1.3 Truth Tables

Suppose your friend Maria is a doctor, and you know that she is a Democrat. If someone told you, "Maria is a doctor and a Republican," you would say that the statement was false. On the other hand, if you were told, "Maria is a doctor or a Republican," you would say that the statement was true. Each of these statements is a compound statement—the result of joining individual statements with connective words. When is a compound statement true, and when is it false? To answer these questions, we must examine whether the individual statements are true or false and the manner in which the statements are connected.

The **truth value** of a statement is the classification of the statement as true or false and is denoted by T or F. For example, the truth value of the statement "Santa Fe is the capital of New Mexico" is T. (The statement is true.) In contrast, the truth value of "Memphis is the capital of Tennessee" is F. (The statement is false.)

A convenient way of determining whether a compound statement is true or false is to construct a truth table. A **truth table** is a listing of all possible combinations of the individual statements as true or false, along with the resulting truth value of the compound statement. As we will see, truth tables also allow us to distinguish valid arguments from invalid arguments.

The Negation $\sim p$

The **negation** of a statement is the denial, or opposite, of the statement. (As was stated in the previous section, because the truth value of the negation depends on the truth value of the original statement, a negation can be classified as a compound statement.) To construct the truth table for the negation of a statement, we must first examine the original statement. A statement p may be true or false, as shown in Figure 1.31. If the statement p is true, the negation $\sim p$ is false; if p is false, $\sim p$ is true. The truth table for the compound statement $\sim p$ is given in Figure 1.32. Row 1 of the table is read "$\sim p$ is false when p is true." Row 2 is read "$\sim p$ is true when p is false."

	p
1.	T
2.	F

Figure 1.31
Truth values for a statement p

	p	$\sim p$
1.	T	F
2.	F	T

Figure 1.32
Truth table for a negation $\sim p$

	p	*q*
1.	T	T
2.	T	F
3.	F	T
4.	F	F

Figure 1.33
Truth values for two statements

	p	*q*	*p* ∧ *q*
1.	T	T	T
2.	T	F	F
3.	F	T	F
4.	F	F	F

Figure 1.34
Truth table for a conjunction
p ∧ *q*

	p	*q*	*p* ∨ *q*
1.	T	T	T
2.	T	F	T
3.	F	T	T
4.	F	F	F

Figure 1.35
Truth table for a disjunction
p ∨ *q*

The Conjunction *p* ∧ *q*

A **conjunction** is the joining of two statements with the word *and*. The compound statement "Maria is a doctor and a Republican" is a conjunction with the following symbolic representation:

p: Maria is a doctor.

q: Maria is a Republican.

p ∧ *q:* Maria is a doctor and a Republican.

The truth value of a compound statement depends on the truth values of the individual statements that make it up. How many rows will the truth table for the conjunction *p* ∧ *q* contain? Because *p* has two possible truth values (T or F) and *q* has two possible truth values (T or F), we need four (2 · 2) rows in order to list all possible combinations of Ts and Fs, as shown in Figure 1.33.

For the conjunction *p* ∧ *q* to be true, the components *p* and *q* must *both* be true; the conjunction is false otherwise. The completed truth table for the conjunction *p* ∧ *q* is given in Figure 1.34. The symbols *p* and *q* can be replaced by any statements. The table gives the truth value of the statement "*p* and *q*," dependent upon the truth values of the individual statements "*p*" and "*q*." For instance, row 3 is read "The conjunction *p* ∧ *q* is false when *p* is false and *q* is true." The other rows are read in a similar manner.

The Disjunction *p* ∨ *q*

A **disjunction** is the joining of two statements with the word *or*. The compound statement "Maria is a doctor or a Republican" is a disjunction (the *inclusive or*) with the following symbolic representation:

p: Maria is a doctor.

q: Maria is a Republican.

p ∨ *q:* Maria is a doctor or a Republican.

Even though your friend Maria the doctor is not a Republican, the disjunction "Maria is a doctor or a Republican" is true. For a disjunction to be true, *at least one* of the components must be true. A disjunction is false only when *both* components are false. The truth table for the disjunction *p* ∨ *q* is given in Figure 1.35.

 EXAMPLE 1
for Liberal Arts Mathematics
www.thomsonedu.com/1pass.

Under what specific conditions is the following compound statement true? "I have a high school diploma, or I have a full-time job and no high school diploma."

Solution   First, we translate the statement into symbolic form, and then we construct the truth table for the symbolic expression. Define *p* and *q* as

p: I have a high school diploma.

q: I have a full-time job.

The given statement has the symbolic representation *p* ∨ (*q* ∧ ~*p*).

Because there are two letters, we need 2 · 2 = 4 rows. We need to insert a column for each connective in the symbolic expression *p* ∨ (*q* ∧ ~*p*). As in algebra, we start inside any grouping symbols and work our way out. Therefore, we need a column for $\sim p$, a column for $q \wedge \sim p$, and a column for the entire expression $p \vee (q \wedge \sim p)$, as shown in Figure 1.36.

	p	q	$\sim p$	$q \wedge \sim p$	$p \vee (q \wedge \sim p)$
1.	T	T			
2.	T	F			
3.	F	T			
4.	F	F			

Figure 1.36
Required columns in the truth table

In the $\sim p$ column, fill in truth values that are opposite those for p. Next, the conjunction $q \wedge \sim p$ is true only when both components are true; enter a T in row 3 and Fs elsewhere. Finally, the disjunction $p \vee (q \wedge \sim p)$ is false only when both components p and $(q \wedge \sim p)$ are false; enter an F in row 4 and Ts elsewhere. The completed truth table is shown in Figure 1.37.

	p	q	$\sim p$	$q \wedge \sim p$	$p \vee (q \wedge \sim p)$
1.	T	T	F	F	T
2.	T	F	F	F	T
3.	F	T	T	T	T
4.	F	F	T	F	F

Figure 1.37
Truth table for $p \vee (q \wedge \sim p)$

As is indicated in the truth table, the symbolic expression $p \vee (q \wedge \sim p)$ is true under all conditions except one: row 4; the expression is false when both p and q are false. Therefore, the statement "I have a high school diploma, or I have a full-time job and no high school diploma" is true in every case except when the speaker has no high school diploma and no full-time job. ∎

If the symbolic representation of a compound statement consists of two different letters, its truth table will have $2 \cdot 2 = 4$ rows. How many rows are required if a compound statement consists of three letters—say, p, q, and r? Because each statement has two possible truth values (T and F), the truth table must contain $2 \cdot 2 \cdot 2 = 8$ rows. In general, each time a new statement is added, the number of rows doubles.

Number of Rows

If a compound statement consists of n individual statements, each represented by a different letter, the number of rows required in its truth table is 2^n.

EXAMPLE 2   Under what specific conditions is the following compound statement true? "I own a handgun, and it is not the case that I am a criminal or police officer."

Solution   First, we translate the statement into symbolic form, and then we construct the truth table for the symbolic expression. Define the three simple statements as follows:

> *p:* I own a handgun.
>
> *q:* I am a criminal.
>
> *r:* I am a police officer.

The given statement has the symbolic representation $p \wedge \sim(q \vee r)$. Since there are three letters, we need $2^3 = 8$ rows. We start with three columns, one for each letter. To account for all possible combinations of *p, q,* and *r* as true or false, proceed as follows:

1. Fill the first half (four rows) of column 1 with Ts and the rest with Fs, as shown in Figure 1.38(a).

2. In the next column, split each half into halves, the first half receiving Ts and the second Fs. In other words, alternate two Ts and two Fs in column 2, as shown in Figure 1.38(b).

3. Again, split each half into halves; the first half receives Ts, and the second receives Fs. Because we are dealing with the third (last) column, the Ts and Fs will alternate, as shown in Figure 1.38(c).

	p	*q*	*r*
1.	T		
2.	T		
3.	T		
4.	T		
5.	F		
6.	F		
7.	F		
8.	F		

(a)

	p	*q*	*r*
1.	T	T	
2.	T	T	
3.	T	F	
4.	T	F	
5.	F	T	
6.	F	T	
7.	F	F	
8.	F	F	

(b)

	p	*q*	*r*
1.	T	T	T
2.	T	T	F
3.	T	F	T
4.	T	F	F
5.	F	T	T
6.	F	T	F
7.	F	F	T
8.	F	F	F

(c)

Figure 1.38
Truth values for three statements

(This process of filling the first half of the first column with Ts and the second half with Fs and then splitting each half into halves with blocks of Ts and Fs applies to all truth tables.)

We need to insert a column for each connective in the symbolic expression $p \wedge \sim(q \vee r)$, as shown in Figure 1.39.

Now fill in the appropriate symbol in the column under $q \vee r$. Enter F if *both q* and *r* are false; enter T otherwise (that is, if at least one is true). In the $\sim(q \vee r)$ column, fill in truth values that are opposite those for $q \vee r$, as in Figure 1.40.

The conjunction $p \wedge \sim(q \vee r)$ is true only when *both p* and $\sim(q \vee r)$ are true; enter a T in row 4 and Fs elsewhere. The truth table is shown in Figure 1.41.

	p	q	r	$q \vee r$	$\sim(q \vee r)$	$p \wedge \sim(q \vee r)$
1.	T	T	T			
2.	T	T	F			
3.	T	F	T			
4.	T	F	F			
5.	F	T	T			
6.	F	T	F			
7.	F	F	T			
8.	F	F	F			

Figure 1.39
Required columns in the truth table

	p	q	r	$q \vee r$	$\sim(q \vee r)$	$p \wedge \sim(q \vee r)$
1.	T	T	T	T	F	
2.	T	T	F	T	F	
3.	T	F	T	T	F	
4.	T	F	F	F	T	
5.	F	T	T	T	F	
6.	F	T	F	T	F	
7.	F	F	T	T	F	
8.	F	F	F	F	T	

Figure 1.40
Truth values of the expressions $q \vee r$ and $\sim(q \vee r)$

	p	q	r	$q \vee r$	$\sim(q \vee r)$	$p \wedge \sim(q \vee r)$
1.	T	T	T	T	F	F
2.	T	T	F	T	F	F
3.	T	F	T	T	F	F
4.	T	F	F	F	T	T
5.	F	T	T	T	F	F
6.	F	T	F	T	F	F
7.	F	F	T	T	F	F
8.	F	F	F	F	T	F

Figure 1.41
Truth table for $p \wedge \sim(q \vee r)$

As indicated in the truth table, the expression $p \wedge \sim(q \vee r)$ is true only when p is true and both q and r are false. Therefore, the statement "I own a handgun, and it is not the case that I am a criminal or police officer" is true only when the speaker owns a handgun, is not a criminal, and is not a police officer—in other words, the speaker is a law-abiding citizen who owns a handgun. ∎

The Conditional $p \rightarrow q$

A **conditional** is a compound statement of the form "If p, then q" and is symbolized $p \rightarrow q$. Under what circumstances is a conditional true, and when is it false? Consider the following (compound) statement: "If you give me \$50, then I will give you a ticket to the ballet." This statement is a conditional and has the following representation:

p: You give me \$50.

q: I give you a ticket to the ballet.

$p \rightarrow q$: If you give me \$50, then I will give you a ticket to the ballet.

The conditional can be viewed as a promise: *If* you give me \$50, *then* I will give you a ticket to the ballet. Suppose you give me \$50; that is, suppose p is true. I have two options: Either I give you a ticket to the ballet (q is true), or I do not (q is false). If I do give you the ticket, the conditional $p \rightarrow q$ is true (I have kept my promise); if I do not give you the ticket, the conditional $p \rightarrow q$ is false (I have not kept my promise). These situations are shown in rows 1 and 2 of the truth table in Figure 1.42. Rows 3 and 4 require further analysis.

Suppose you do not give me \$50; that is, suppose p is false. Whether or not I give you a ticket, you cannot say that I broke my promise; that is, you cannot say that the conditional $p \rightarrow q$ is false. Consequently, since a statement is either true or false, the conditional is labeled true (by default). In other words, when the premise p of a conditional is false, it does not matter whether the conclusion q is true or false. In both cases, the conditional $p \rightarrow q$ is automatically labeled true, because it is not false.

The completed truth table for a conditional is given in Figure 1.43. Notice that the only circumstance under which a conditional is false is when the premise p is true and the conclusion q is false, as shown in row 2.

	p	q	$p \rightarrow q$
1.	T	T	T
2.	T	F	F
3.	F	T	?
4.	F	F	?

Figure 1.42
What if p is false?

	p	q	$p \rightarrow q$
1.	T	T	T
2.	T	F	F
3.	F	T	T
4.	F	F	T

Figure 1.43
Truth table for $p \rightarrow q$

EXAMPLE 3 Under what conditions is the symbolic expression $q \rightarrow \sim p$ true?

Solution Our truth table has $2^2 = 4$ rows and contains a column for p, q, $\sim p$, and $q \rightarrow \sim p$, as shown in Figure 1.44.

	p	q	$\sim p$	$q \rightarrow \sim p$
1.	T	T		
2.	T	F		
3.	F	T		
4.	F	F		

Figure 1.44
Required columns in the truth table

	p	q	$\sim p$	$q \rightarrow \sim p$
1.	T	T	F	F
2.	T	F	F	T
3.	F	T	T	T
4.	F	F	T	T

Figure 1.45
Truth table for $q \rightarrow \sim p$

In the $\sim p$ column, fill in truth values that are opposite those for p. Now, a conditional is false only when its premise (in this case, q) is true and its conclusion (in this case, $\sim p$) is false. Therefore, $q \rightarrow \sim p$ is false only in row 1; the conditional $q \rightarrow \sim p$ is true under all conditions except the condition that both p and q are true. The completed truth table is shown in Figure 1.45. ■

EXAMPLE 4 Construct a truth table for the following compound statement: "I walk up the stairs if I want to exercise or if the elevator isn't working."

Solution Rewriting the statement so the word *if* is first, we have "If I want to exercise or (if) the elevator isn't working, then I walk up the stairs."

Now we must translate the statement into symbols and construct a truth table. Define the following:

p: I want to exercise.

q: The elevator is working.

r: I walk up the stairs.

The statement now has the symbolic representation $(p \vee \sim q) \rightarrow r$. Because we have three letters, our table must have $2^3 = 8$ rows. Inserting a column for each letter and a column for each connective, we have the initial setup shown in Figure 1.46.

	p	*q*	*r*	*~q*	*p ∨ ~q*	*(p ∨ ~q) → r*
1.	T	T	T			
2.	T	T	F			
3.	T	F	T			
4.	T	F	F			
5.	F	T	T			
6.	F	T	F			
7.	F	F	T			
8.	F	F	F			

Figure 1.46
Required columns in the truth table

In the column labeled $\sim q$, enter truth values that are the opposite of those of q. Next, enter the truth values of the disjunction $p \vee \sim q$ in column 5. Recall that a disjunction is false only when both components are false and is true otherwise. Consequently, enter Fs in rows 5 and 6 (since both p and $\sim q$ are false) and Ts in the remaining rows, as shown in Figure 1.47.

The last column involves a conditional; it is false only when its premise is true and its conclusion is false. Therefore, enter Fs in rows 2, 4, and 8 (since $p \vee \sim q$ is true and r is false) and Ts in the remaining rows. The truth table is shown in Figure 1.48.

	p	**q**	**r**	**~q**	**p ∨ ~q**	**(p ∨ ~q) → r**
1.	T	T	T	F	T	
2.	T	T	F	F	T	
3.	T	F	T	T	T	
4.	T	F	F	T	T	
5.	F	T	T	F	F	
6.	F	T	F	F	F	
7.	F	F	T	T	T	
8.	F	F	F	T	T	

Figure 1.47
Truth values of the expressions

	p	**q**	**r**	**~q**	**p ∨ ~q**	**(p ∨ ~q) → r**
1.	T	T	T	F	T	T
2.	T	T	F	F	T	F
3.	T	F	T	T	T	T
4.	T	F	F	T	T	F
5.	F	T	T	F	F	T
6.	F	T	F	F	F	T
7.	F	F	T	T	T	T
8.	F	F	F	T	T	F

Figure 1.48
Truth table for $(p \vee \sim q) \rightarrow r$

As Figure 1.48 shows, the statement "I walk up the stairs if I want to exercise or if the elevator isn't working" is true in all situations except those listed in rows 2, 4, and 8. For instance, the statement is false (row 8) when the speaker does not want to exercise, the elevator is not working, and the speaker does not walk up the stairs—in other words, the speaker stays on the ground floor of the building when the elevator is broken.   ■

Equivalent Expressions

When you purchase a car, the car is either new or used. If a salesperson told you, "It is not the case that the car is not new," what condition would the car be in? This compound statement consists of one individual statement ("*p:* The car is new") and two negations:

"It is not the case that the car is not new."

$\sim$ $\sim p$

Figure 1.49
Truth values of p

	p	$\sim p$	$\sim(\sim p)$
1.	T	F	T
2.	F	T	F

Figure 1.50
Truth table for $\sim(\sim p)$

Does this mean that the car is new? To answer this question, we will construct a truth table for the symbolic expression $\sim(\sim p)$ and compare its truth values with those of the original p. Because there is only one letter, we need $2^1 = 2$ rows, as shown in Figure 1.49.

We must insert a column for $\sim p$ and a column for $\sim(\sim p)$. Now, $\sim p$ has truth values that are opposite those of p, and $\sim(\sim p)$ has truth values that are opposite those of $\sim p$, as shown in Figure 1.50.

Notice that the values in the column labeled $\sim(\sim p)$ are identical to those in the column labeled p. Whenever this happens, the expressions are said to be equivalent and may be used interchangeably. Therefore, the statement "It is not the case that the car is not new" is equivalent in meaning to the statement "The car is new."

Equivalent expressions are symbolic expressions that have identical truth values in each corresponding entry. The expression $p \equiv q$ is read "p is equivalent to q" or "p and q are equivalent." As we can see in Figure 1.50, an expression and its double negation are logically equivalent. This relationship can be expressed as $p \equiv \sim(\sim p)$.

EXAMPLE 5   Are the statements "If I am a homeowner, then I pay property taxes" and "I am a homeowner and I do not pay property taxes" equivalent?

Solution   We begin by defining the statements:

p: I am a homeowner.

q: I pay property taxes.

$p \rightarrow q$: If I am a homeowner, then I pay property taxes.

$p \wedge \sim q$: I am a homeowner and I do not pay property taxes.

The truth table contains $2^2 = 4$ rows, and the initial setup is shown in Figure 1.51.

Now enter the appropriate truth values under $\sim q$ (the opposite of q). Because the conjunction $p \wedge \sim q$ is true only when both p and $\sim q$ are true, enter a T in row 2 and Fs elsewhere. The conditional $p \rightarrow q$ is false only when p is true and q is false; therefore, enter an F in row 2 and Ts elsewhere. The completed truth table is shown in Figure 1.52.

Because the entries in the columns labeled $p \wedge \sim q$ and $p \rightarrow q$ are not the same, the statements are not equivalent. "If I am a homeowner, then I pay property taxes" is *not* equivalent to "I am a homeowner and I do not pay property taxes."

	p	q	$\sim q$	$p \wedge \sim q$	$p \rightarrow q$
1.	T	T			
2.	T	F			
3.	F	T			
4.	F	F			

Figure 1.51
Required columns in the truth table

	p	q	$\sim q$	$p \wedge \sim q$	$p \rightarrow q$
1.	T	T	F	F	T
2.	T	F	T	T	F
3.	F	T	F	F	T
4.	F	F	T	F	T

Figure 1.52
Truth table for $p \rightarrow q$

Historical Note   George Boole, 1815–1864

© Corbis

George Boole is called "the father of symbolic logic." Computer science owes much to this self-educated mathematician. Born the son of a poor shopkeeper in Lincoln, England, Boole had very little formal education, and his prospects for rising above his family's lower-class status were dim. Like Leibniz, he taught himself Latin; at the age of twelve, he translated an ode of Horace into English, winning the attention of the local schoolmasters. (In his day, the knowledge of Latin was a prerequisite to scholarly endeavors and to becoming a socially accepted gentleman.) After that, his academic desires were encouraged, and at the age of fifteen, he began his long teaching career. While teaching arithmetic, he studied advanced mathematics and physics.

In 1849, after nineteen years of teaching at elementary schools, Boole received his big break: He was appointed professor of mathematics at Queen's College in the city of Cork, Ireland. At last, he was able to research advanced mathematics, and he became recognized as a first-class mathematician. This was a remarkable feat, considering Boole's lack of formal training and degrees.

Boole's most influential work, *An Investigation of the Laws of Thought, on Which Are Founded the Mathematical Theories of Logic and Probabilities,* was published in 1854. In it he wrote, "There exist certain general principles founded in the very nature of language and logic that exhibit laws as identical in form as with the laws of the general symbols of algebra." With this insight, Boole had taken a big step into the world of logical reasoning and abstract mathematical analysis.

Perhaps because of his lack of formal training, Boole challenged the status quo, including the Aristotelian assumption that *all* logical arguments could be reduced to syllogistic arguments. In doing so, he employed symbols to represent concepts, as did Leibniz, but he also developed systems of algebraic manipulation to accompany these symbols. Thus, Boole's creation is a marriage of logic and mathematics. However, as is the case with almost all new theories, Boole's symbolic logic was not met with total approbation. In particular, one staunch opponent of his work was Georg Cantor, whose work on the origins of set theory and the magnitude of infinity will be investigated in Chapter 2.

In the many years since Boole's original work was unveiled, various scholars have modified, improved, generalized, and extended its central concepts. Today, Boolean algebras are the essence of computer software and circuit design. After all, a computer merely manipulates predefined symbols and conforms to a set of preassigned algebraic commands.

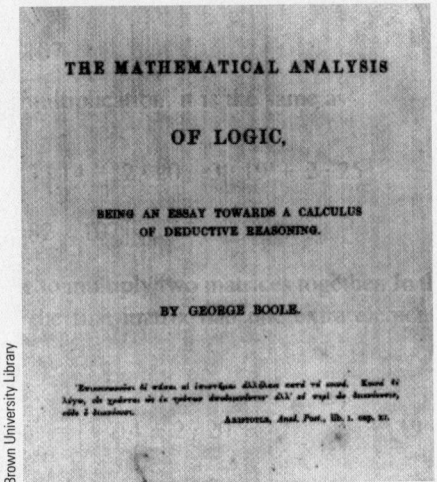

Brown University Library

THE MATHEMATICAL ANALYSIS

OF LOGIC,

BEING AN ESSAY TOWARDS A CALCULUS
OF DEDUCTIVE REASONING.

BY GEORGE BOOLE.

Through an algebraic manipulation of logical symbols, Boole revolutionized the age-old study of logic. His essay *The Mathematical Analysis of Logic* laid the foundation for his later book *An Investigation of the Laws of Thought.*

Notice that the truth values in the columns under $p \wedge \sim q$ and $p \rightarrow q$ in Figure 1.52 are exact opposites; when one is T, the other is F. Whenever this happens, one statement is the negation of the other. Consequently, $p \wedge \sim q$ is the negation of $p \rightarrow q$ (and vice versa). This can be expressed as $p \wedge \sim q \equiv \sim(p \rightarrow q)$. The negation of a conditional is logically equivalent to the conjunction of the premise and the negation of the conclusion.

Statements that look or sound different may in fact have the same meaning. For example, "It is not the case that the car is not new" really means the same as "The car is new," and "It is not the case that if I am a homeowner, then I pay property taxes" actually means the same as "I am a homeowner and I do not pay property taxes." When we are working with equivalent statements, we can substitute either statement for the other without changing the truth value.

De Morgan's Laws

Earlier in this section, we saw that the negation of a negation is equivalent to the original statement; that is, $\sim(\sim p) \equiv p$. Another negation "formula" that we discovered was $\sim(p \rightarrow q) \equiv p \wedge \sim q$, that is, the negation of a conditional. Can we find similar "formulas" for the negations of the other basic connectives, namely, the conjunction and the disjunction? The answer is yes, and the results are credited to the English mathematician and logician Augustus De Morgan.

De Morgan's Laws

The negation of the conjunction $p \wedge q$ is given by $\sim(p \wedge q) \equiv \sim p \vee \sim q$.

"Not p and q" is equivalent to "not p or not q."

The negation of the disjunction $p \vee q$ is given by $\sim(p \vee q) \equiv \sim p \wedge \sim q$.

"Not p or q" is equivalent to "not p and not q."

De Morgan's Laws are easily verified through the use of truth tables and will be addressed in the exercises (see Exercises 47 and 48).

ThomsonNOW™
for Liberal Arts Mathematics
www.thomsonedu.com/1pass.

EXAMPLE 6   Using De Morgan's Laws, find the negation of each of the following:

a. It is Friday and I receive a paycheck.

b. You are correct or I am crazy.

Solution   **a.** The symbolic representation of "It is Friday and I receive a paycheck" is

p: It is Friday.

q: I receive a paycheck.

$p \wedge q$: It is Friday and I receive a paycheck.

Therefore, the negation is $\sim(p \wedge q) \equiv \sim p \vee \sim q$, that is, "It is not Friday or I do not receive a paycheck."

b. The symbolic representation of "You are correct or I am crazy" is

p: You are correct.

q: I am crazy.

$p \vee q$: You are correct or I am crazy.

Therefore, the negation is $\sim(p \vee q) \equiv \sim p \wedge \sim q$, that is, "You are not correct and I am not crazy."   ∎

As we have seen, the truth value of a compound statement depends on the truth values of the individual statements that make it up. The truth tables of the basic connectives are summarized in Figure 1.53.

Equivalent statements are statements that have the same meaning. Equivalent statements for the negations of the basic connectives are given in Figure 1.54.

	p	$\sim p$
1.	T	F
2.	F	T

Negation

	p	q	$p \wedge q$
1.	T	T	T
2.	T	F	F
3.	F	T	F
4.	F	F	F

Conjunction

	p	q	$p \vee q$
1.	T	T	T
2.	T	F	T
3.	F	T	T
4.	F	F	F

Disjunction

	p	q	$p \rightarrow q$
1.	T	T	T
2.	T	F	F
3.	F	T	T
4.	F	F	T

Conditional

Figure 1.53
Truth tables for the basic connectives

1. $\sim(\sim p) \equiv p$	the negation of a negation
2. $\sim(p \wedge q) \equiv \sim p \vee \sim q$	the negation of a conjunction
3. $\sim(p \vee q) \equiv \sim p \wedge \sim q$	the negation of a disjunction
4. $\sim(p \rightarrow q) \equiv p \wedge \sim q$	the negation of a conditional

Figure 1.54
Negations of the basic connectives

1.3   Exercises

In Exercises 1–20, construct a truth table for the symbolic expressions.

1. $p \vee \sim q$
2. $p \wedge \sim q$
3. $p \vee \sim p$
4. $p \wedge \sim p$
5. $p \rightarrow \sim q$
6. $\sim p \rightarrow q$
7. $\sim q \rightarrow \sim p$
8. $\sim p \rightarrow \sim q$
9. $(p \vee q) \rightarrow \sim p$
10. $(p \wedge q) \rightarrow \sim q$
11. $(p \vee q) \rightarrow (p \wedge q)$
12. $(p \wedge q) \rightarrow (p \vee q)$
13. $p \wedge \sim(q \vee r)$
14. $p \vee \sim(q \vee r)$
15. $p \vee (\sim q \wedge r)$
16. $\sim p \vee \sim(q \wedge r)$
17. $(\sim r \vee p) \rightarrow (q \wedge p)$
18. $(q \wedge p) \rightarrow (\sim r \vee p)$
19. $(p \vee r) \rightarrow (q \wedge \sim r)$
20. $(p \wedge r) \rightarrow (q \vee \sim r)$

In Exercises 21–36, translate the compound statement into symbolic form and then construct the truth table for the expression.

21. If it is raining, then the streets are wet.
22. If the lyrics are not controversial, the performance is not banned.
23. The water supply is rationed if it does not rain.
24. The country is in trouble if he is elected.
25. All squares are rectangles.
26. All muscle cars from the Sixties are polluters.
27. No square is a triangle.
28. No electric-powered car is a polluter.

29. Your check is accepted if you have a driver's license or a credit card.

30. You get a refund or a store credit if the product is defective.

31. If leaded gasoline is used, the catalytic converter is damaged and the air is polluted.

32. If he does not go to jail, he is innocent or has an alibi.

33. I have a college degree and I do not have a job or own a house.

34. I surf the Internet and I make purchases and do not pay sales tax.

35. If Proposition A passes and Proposition B does not, jobs are lost or new taxes are imposed.

36. If Proposition A does not pass and the legislature raises taxes, the quality of education is lowered and unemployment rises.

In Exercises 37–46, construct a truth table to determine whether the statements in each pair are equivalent.

37. The streets are wet or it is not raining.
   If it is raining, then the streets are wet.

38. The streets are wet or it is not raining.
   If the streets are not wet, then it is not raining.

39. He has a high school diploma or he is unemployed.
   If he does not have a high school diploma, then he is unemployed.

40. She is unemployed or she does not have a high school diploma.
   If she is employed, then she does not have a high school diploma.

41. If handguns are outlawed, then outlaws have handguns.
   If outlaws have handguns, then handguns are outlawed.

42. If interest rates continue to fall, then I can afford to buy a house.
   If interest rates do not continue to fall, then I cannot afford to buy a house.

43. If the spotted owl is on the endangered species list, then lumber jobs are lost.
   If lumber jobs are not lost, then the spotted owl is not on the endangered species list.

44. If I drink decaffeinated coffee, then I do not stay awake.
   If I do stay awake, then I do not drink decaffeinated coffee.

45. The plaintiff is innocent or the insurance company does not settle out of court.
   The insurance company settles out of court and the plaintiff is not innocent.

46. The plaintiff is not innocent and the insurance company settles out of court.
   It is not the case that the plaintiff is innocent or the insurance company does not settle out of court.

47. Using truth tables, verify De Morgan's Law
$$\sim(p \wedge q) \equiv \sim p \vee \sim q.$$

48. Using truth tables, verify De Morgan's Law
$$\sim(p \vee q) \equiv \sim p \wedge \sim q.$$

In Exercises 49–56, write the statement in symbolic form, construct the negation of the expression (in simplified symbolic form), and express the negation in words.

49. I have a college degree and I am not employed.

50. It is snowing and classes are canceled.

51. The television set is broken or there is a power outage.

52. The freeway is under construction or I do not ride the bus.

53. If the building contains asbestos, the original contractor is responsible.

54. If the legislation is approved, the public is uninformed.

55. The First Amendment has been violated if the lyrics are censored.

56. Your driver's license is taken away if you do not obey the laws.

Answer the following questions using complete sentences and your own words.

CONCEPT QUESTIONS

57. a. Under what conditions is a disjunction true?
   b. Under what conditions is a disjunction false?

58. a. Under what conditions is a conjunction true?
   b. Under what conditions is a conjunction false?

59. a. Under what conditions is a conditional true?
   b. Under what conditions is a conditional false?

60. a. Under what conditions is a negation true?
   b. Under what conditions is a negation false?

61. What are equivalent expressions?

62. What is a truth table?

63. When constructing a truth table, how do you determine how many rows to create?

HISTORY QUESTIONS

64. Who is considered "the father of symbolic logic"?

65. Boolean algebra is a combination of logic and mathematics. What is it used for?

1.4 More on Conditionals

Conditionals differ from conjunctions and disjunctions with regard to the possibility of changing the order of the statements. In algebra, the sum $x + y$ is equal to the sum $y + x$; that is, addition is commutative. In everyday language, one realtor might say, "The house is perfect and the lot is priceless," while another says, "The lot is priceless and the house is perfect." Logically, their meanings are the same, since $(p \wedge q) \equiv (q \wedge p)$. The order of the components in a conjunction or disjunction makes no difference in regard to the truth value of the statement. This is not so with conditionals.

Variations of a Conditional

Given two statements p and q, various "if . . . then . . ." statements can be formed.

ThomsonNOW™
for Liberal Arts Mathematics
www.thomsonedu.com/1pass.

EXAMPLE 1 Using the statements

> p: You are compassionate.
>
> q: You contribute to charities.

write the sentence represented by each of the following:

a. $p \to q$ **b.** $q \to p$ **c.** $\sim p \to \sim q$ **d.** $\sim q \to \sim p$

Solution **a.** $p \to q$: If you are compassionate, then you contribute to charities.
b. $q \to p$: If you contribute to charities, then you are compassionate.
c. $\sim p \to \sim q$: If you are not compassionate, then you do not contribute to charities.
d. $\sim q \to \sim p$: If you do not contribute to charities, then you are not compassionate. ∎

Each part of Example 1 contains an "if . . . then . . ." statement and is called a conditional. Any given conditional has three variations: a converse, an inverse, and a contrapositive. The **converse** of the conditional "if p, then q" is the compound statement "if q, then p" That is, we form the converse of the conditional by interchanging the premise and the conclusion; $q \to p$ is the converse of $p \to q$. The statement in part (b) of Example 1 is the converse of the statement in part (a).

The **inverse** of the conditional "if p, then q" is the compound statement "if not p, then not q." We form the inverse of the conditional by negating both the premise and the conclusion; $\sim p \to \sim q$ is the inverse of $p \to q$. The statement in part (c) of Example 1 is the inverse of the statement in part (a).

The **contrapositive** of the conditional "if p, then q" is the compound statement "if not q, then not p." We form the contrapositive of the conditional by negating *and* interchanging both the premise and the conclusion; $\sim q \to \sim p$ is the contrapositive of $p \to q$. The statement in part (d) of Example 1 is the contrapositive of the statement in part (a). The variations of a given conditional are summarized in Figure 1.55. As we will see, some of these variations are equivalent, and some are not. Unfortunately, many people incorrectly treat them all as equivalent.

ThomsonNOW™
for Liberal Arts Mathematics
www.thomsonedu.com/1pass.

EXAMPLE 2 Given the conditional "You did not receive the proper refund if you prepared your own income tax form," write the sentence that represents each of the following.

a. the converse of the conditional
b. the inverse of the conditional
c. the contrapositive of the conditional

Name	Symbolic Form	Read As . . .
a (given) conditional	$p \rightarrow q$	If p, then q.
the converse (of $p \rightarrow q$)	$q \rightarrow p$	If q, then p.
the inverse (of $p \rightarrow q$)	$\sim p \rightarrow \sim q$	If not p, then not q.
the contrapositive (of $p \rightarrow q$)	$\sim q \rightarrow \sim p$	If not q, then not p.

Figure 1.55
Variations of a conditional

Solution

a. Rewriting the statement in the standard "if . . . then . . ." form, we have the conditional "If you prepared your own income tax form, then you did not receive the proper refund." The converse is formed by interchanging the premise and the conclusion. Thus, the converse is written as "If you did not receive the proper refund, then you prepared your own income tax form."

b. The inverse is formed by negating both the premise and the conclusion. Thus, the inverse is written as "If you did not prepare your own income tax form, then you received the proper refund."

c. The contrapositive is formed by negating *and* interchanging the premise and the conclusion. Thus, the contrapositive is written as "If you received the proper refund, then you did not prepare your own income tax form." ∎

Equivalent Conditionals

We have seen that the conditional $p \rightarrow q$ has three variations: the converse $q \rightarrow p$, the inverse $\sim p \rightarrow \sim q$, and the contrapositive $\sim q \rightarrow \sim p$. Do any of these "if . . . then . . ." statements convey the same meaning? In other words, are any of these compound statements equivalent?

EXAMPLE 3

Determine which (if any) of the following are equivalent: a conditional $p \rightarrow q$, the converse $q \rightarrow p$, the inverse $\sim p \rightarrow \sim q$, and the contrapositive $\sim q \rightarrow \sim p$.

Solution

To investigate the possible equivalencies, we must construct a truth table that contains all the statements. Because there are two letters, we need $2^2 = 4$ rows. The table must have a column for $\sim p$, one for $\sim q$, one for the conditional $p \rightarrow q$, and one for each variation of the conditional. The truth values of the negations $\sim p$ and $\sim q$ are readily entered, as shown in Figure 1.56.

	p	q	$\sim p$	$\sim q$	$p \rightarrow q$	$q \rightarrow p$	$\sim p \rightarrow \sim q$	$\sim q \rightarrow \sim p$
1.	T	T	F	F				
2.	T	F	F	T				
3.	F	T	T	F				
4.	F	F	T	T				

Figure 1.56
Required columns in the truth table

An "if . . . then . . ." statement is false only when the premise is true and the conclusion is false. Consequently, $p \rightarrow q$ is false only when p is T and q is F; enter an F in row 2 and Ts elsewhere in the column under $p \rightarrow q$.

Likewise, the converse $q \rightarrow p$ is false only when q is T and p is F; enter an F in row 3 and Ts elsewhere.

In a similar manner, the inverse $\sim p \rightarrow \sim q$ is false only when $\sim p$ is T and $\sim q$ is F; enter an F in row 3 and Ts elsewhere.

Finally, the contrapositive $\sim q \rightarrow \sim p$ is false only when $\sim q$ is T and $\sim p$ is F; enter an F in row 2 and Ts elsewhere.

The completed truth table is shown in Figure 1.57. Examining the entries in Figure 1.57, we can see that the columns under $p \rightarrow q$ and $\sim q \rightarrow \sim p$ are identical; each has an F in row 2 and Ts elsewhere. Consequently, a conditional and its contrapositive are equivalent: $p \rightarrow q \equiv \sim q \rightarrow \sim p$.

Likewise, we notice that $q \rightarrow p$ and $\sim p \rightarrow \sim q$ have identical truth values; each has an F in row 3 and Ts elsewhere. Thus, the converse and the inverse of a conditional are equivalent: $q \rightarrow p \equiv \sim p \rightarrow \sim q$.

	p	q	$\sim p$	$\sim q$	$p \rightarrow q$	$q \rightarrow p$	$\sim p \rightarrow \sim q$	$\sim q \rightarrow \sim p$
1.	T	T	F	F	T	T	T	T
2.	T	F	F	T	F	T	T	F
3.	F	T	T	F	T	F	F	T
4.	F	F	T	T	T	T	T	T

Figure 1.57
Truth table for a conditional and its variations

We have seen that different "if . . . then . . ." statements can convey the same meaning—that is, that certain variations of a conditional are equivalent (see Figure 1.58). For example, the compound statements "If you are compassionate, then you contribute to charities" and "If you do not contribute to charities, then you are not compassionate" convey the same meaning. (The second conditional is the contrapositive of the first.) Regardless of its specific contents (p, q, $\sim p$, or $\sim q$), every "if . . . then . . ." statement has an equivalent variation formed by negating *and* interchanging the premise and the conclusion of the given conditional statement.

Equivalent Statements	Symbolic Representations
a conditional and its contrapositive	$(p \rightarrow q) \equiv (\sim q \rightarrow \sim p)$
the converse and the inverse (of the conditional $p \rightarrow q$)	$(q \rightarrow p) \equiv (\sim p \rightarrow \sim q)$

Figure 1.58
Equivalent "if . . . then . . ." statements

The "Only If" Connective

Consider the statement "A prisoner is paroled only if the prisoner obeys the rules." What is the premise, and what is the conclusion? Rather than using p and q (which might bias our investigation), we define

> *r:* A prisoner is paroled.

> *s:* A prisoner obeys the rules.

The given statement is represented by "*r* only if *s*." Now, "*r* only if *s*" means that *r* can happen *only* if *s* happens. In other words, if *s* does not happen, then *r* does not happen, or $\sim s \to \sim r$. We have seen that $\sim s \to \sim r$ is equivalent to $r \to s$. Consequently, "*r* only if *s*" is equivalent to the conditional $r \to s$. The premise of the statement "A prisoner is paroled only if the prisoner obeys the rules" is "A prisoner is paroled," and the conclusion is "The prisoner obeys the rules."

The conditional $p \to q$ can be phrased "*p* only if *q*." Even though the word *if* precedes *q*, *q* is not the premise. *Whatever follows the connective "only if" is the conclusion of the conditional.*

EXAMPLE 4   For the compound statement "You receive a federal grant only if your artwork is not obscene," do the following:

a. Determine the premise and the conclusion.
b. Rewrite the compound statement in the standard "if . . . then . . ." form.
c. Interpret the conditions that make the statement false.

Solution   **a.** Because the compound statement contains an "only if" connective, the statement that follows "only if" is the conclusion of the conditional. The premise is "You receive a federal grant." The conclusion is "Your artwork is not obscene."
b. The given compound statement can be rewritten as "If you receive a grant, then your artwork is not obscene."
c. First we define the symbols.

> *p:* You receive a federal grant.

> *q:* Your artwork is obscene.

	p	*q*	~*q*	*p* → ~*q*
1.	T	T	F	F
2.	T	F	T	T
3.	F	T	F	T
4.	F	F	T	T

Figure 1.59
Truth table for the conditional
$p \to \sim q$

Then the statement has the symbolic representation $p \to \sim q$. The truth table for $p \to \sim q$ is given in Figure 1.59.

The expression $p \to q$ is false under the conditions listed in row 1 (when *p* and *q* are both true). Therefore, the statement "You receive a federal grant only if your artwork is not obscene" is false when an artist *does* receive a federal grant *and* the artist's artwork *is* obscene.   ∎

The Biconditional $p \leftrightarrow q$

What do the words *bicycle, binomial,* and *bilingual* have in common? Each word begins with the prefix *bi,* meaning "two." Just as the word *bilingual* means "two languages," the word *biconditional* means "two conditionals."

In everyday speech, conditionals often get "hooked together" in a circular fashion. For instance, someone might say, "If I am rich, then I am happy, and if I am happy, then I am rich." Notice that this compound statement is actually the conjunction (*and*) of a conditional (if rich, then happy) and its converse (if happy, then rich). Such a statement is referred to as a biconditional. A **biconditional** is a statement of the form $(p \to q) \land (q \to p)$ and is symbolized as $p \leftrightarrow q$. The symbol $p \leftrightarrow q$ is read "*p* if and only if *q*" and is frequently abbreviated "*p* iff *q*." A biconditional is equivalent to the conjunction of two conversely related conditionals: $p \leftrightarrow q \equiv [(p \to q) \land (q \to p)]$.

EXAMPLE 5   Express the biconditional "A citizen is eligible to vote if and only if the citizen is at least eighteen years old" as the conjunction of two conditionals.

Solution   The given biconditional is equivalent to "If a citizen is eligible to vote, then the citizen is at least eighteen years old, *and* if a citizen is at least eighteen years old, then the citizen is eligible to vote." ∎

Under what circumstances is the biconditional $p \leftrightarrow q$ true, and when is it false? To find the answer, we must construct a truth table. Utilizing the equivalence $p \leftrightarrow q \equiv [(p \rightarrow q) \wedge (q \rightarrow p)]$, we get the completed table shown in Figure 1.60. (Recall that a conditional is false only when its premise is true and its conclusion is false and that a conjunction is true only when both components are true.) We can see that a biconditional is true only when the two components p and q have the same truth value—that is, when p and q are both true or when p and q are both false.

	p	q	$p \rightarrow q$	$q \rightarrow p$	$(p \rightarrow q) \wedge (q \rightarrow p)$
1.	T	T	T	T	T
2.	T	F	F	T	F
3.	F	T	T	F	F
4.	F	F	T	T	T

Figure 1.60
Truth table for a biconditional $p \leftrightarrow q$

Many theorems in mathematics can be expressed as biconditionals. For example, when solving a quadratic equation, we have the following: "The equation $ax^2 + bx + c = 0$ has exactly one solution if and only if the discriminant $b^2 - 4ac = 0$." Recall that the solutions of a quadratic equation are

$$x = \frac{-b \pm \sqrt{b^2 - 4ac}}{2a}$$

This biconditional is equivalent to "If the equation $ax^2 + bx + c = 0$ has exactly one solution, then the discriminant $b^2 - 4ac = 0$, and if the discriminant $b^2 - 4ac = 0$, then the equation $ax^2 + bx + c = 0$ has exactly one solution"—that is, one condition implies the other.

1.4 Exercises

In Exercises 1–2, using the given statements, write the sentence represented by each of the following.

a. $p \rightarrow q$ **b.** $q \rightarrow p$
c. $\sim p \rightarrow \sim q$ **d.** $\sim q \rightarrow \sim p$
e. Which of parts (a)–(d) are equivalent? Why?
1. *p:* She is a police officer.
     q: She carries a gun.
2. *p:* I am a multimillion-dollar lottery winner.
     q: I am a world traveler.

In Exercises 3–4, using the given statements, write the sentence represented by each of the following.

a. $p \rightarrow \sim q$ **b.** $\sim q \rightarrow p$
c. $\sim p \rightarrow q$ **d.** $q \rightarrow \sim p$
e. Which of parts (a)–(d) are equivalent? Why?
3. *p:* I watch television.
     q: I do my homework.
4. *p:* He is an artist.
     q: He is a conformist.

In Exercises 5–10, form (a) the inverse, (b) the converse, and (c) the contrapositive of the given conditional.

5. If you pass this mathematics course, then you fulfill a graduation requirement.
6. If you have the necessary tools, assembly time is less than thirty minutes.
7. The television set does not work if the electricity is turned off.
8. You do not win if you do not buy a lottery ticket.
9. You are a vegetarian if you do not eat meat.
10. If chemicals are properly disposed of, the environment is not damaged.

In Exercises 11–16, (a) determine the premise and conclusion, (b) rewrite the compound statement in the standard "if . . . then . . ." form, and (c) interpret the conditions that make the statement false.

11. I take public transportation only if it is convenient.
12. I eat raw fish only if I am in a Japanese restaurant.
13. I buy foreign products only if domestic products are not available.
14. I ride my bicycle only if it is not raining.
15. You may become a U.S. senator only if you are at least thirty years old and have been a citizen for nine years.
16. You may become the president of the United States only if you are at least thirty-five years old and were born a citizen of the United States.

In Exercises 17–22, express the given biconditional as the conjunction of two conditionals.

17. You obtain a refund if and only if you have a receipt.
18. We eat at Burger World if and only if Ju Ju's Kitsch-Inn is closed.
19. The quadratic equation $ax^2 + bx + c = 0$ has two distinct real solutions if and only if $b^2 - 4ac > 0$.
20. The quadratic equation $ax^2 + bx + c = 0$ has complex solutions if and only iff $b^2 - 4ac < 0$.
21. A polygon is a triangle iff the polygon has three sides.
22. A triangle is isosceles iff the triangle has two equal sides.

In Exercises 23–28, translate the two statements into symbolic form and use truth tables to determine whether the statements are equivalent.

23. I cannot have surgery if I do not have health insurance.
If I can have surgery, then I do have health insurance.
24. If I am illiterate, I cannot fill out an application form.
I can fill out an application form if I am not illiterate.

25. If you earn less than $12,000 per year, you are eligible for assistance.
If you are not eligible for assistance, then you earn at least $12,000 per year.
26. If you earn less than $12,000 per year, you are eligible for assistance.
If you earn at least $12,000 per year, you are not eligible for assistance.
27. I watch television only if the program is educational.
I do not watch television if the program is not educational.
28. I buy seafood only if the seafood is fresh.
If I do not buy seafood, the seafood is not fresh.

In Exercises 29–34, write an equivalent variation of the given conditional.

29. If it is not raining, I walk to work.
30. If it makes a buzzing noise, it is not working properly.
31. It is snowing only if it is cold.
32. You are a criminal only if you do not obey the law.
33. You are not a vegetarian if you eat meat.
34. You are not an artist if you are not creative.

In Exercises 35–40, determine which pairs of statements are equivalent.

35. **i.** If Proposition 111 passes, freeways are improved.
ii. If Proposition 111 is defeated, freeways are not improved.
iii. If the freeways are improved, Proposition 111 passes.
iv. If the freeways are not improved, Proposition 111 does not pass.
36. **i.** If the Giants win, then I am happy.
ii. If I am happy, then the Giants win.
iii. If the Giants lose, then I am unhappy.
iv. If I am unhappy, then the Giants lose.
37. **i.** I go to church if it is Sunday.
ii. I go to church only if it is Sunday.
iii. If I do not go to church, it is not Sunday.
iv. If it is not Sunday, I do not go to church.
38. **i.** I am a rebel if I do not have a cause.
ii. I am a rebel only if I do not have a cause.
iii. I am not a rebel if I have a cause.
iv. If I am not a rebel, I have a cause.
39. **i.** If line 34 is greater than line 29, I use Schedule X.
ii. If I use Schedule X, then line 34 is greater than line 29.
iii. If I do not use Schedule X, then line 34 is not greater than line 29.
iv. If line 34 is not greater than line 29, then I do not use Schedule X.

40.

   i. If you answer yes to all of the above, then you complete Part II.

   ii. If you answer no to any of the above, then you do not complete Part II.

   iii. If you completed Part II, then you answered yes to all of the above.

   iv. If you did not complete Part II, then you answered no to at least one of the above.

Answer the following questions using complete sentences and your own words.

CONCEPT QUESTIONS

41. What is a contrapositive?

42. What is a converse?

43. What is an inverse?

44. What is a biconditional?

45. How is an "if . . . then . . ." statement related to an "only if" statement?

1.5   Analyzing Arguments

Lewis Carroll's Cheshire Cat told Alice that he was mad (crazy). Alice then asked, "'And how do you know that you're mad?' 'To begin with,' said the cat, 'a dog's not mad. You grant that?' 'I suppose so,' said Alice. 'Well, then,' the cat went on, 'you see a dog growls when it's angry, and wags its tail when it's pleased. Now *I* growl when I'm pleased, and wag my tail when I'm angry. Therefore I'm mad!'"

Does the Cheshire Cat have a valid deductive argument? Does the conclusion follow logically from the hypotheses? To answer this question, and others like it, we will utilize symbolic logic and truth tables to account for all possible combinations of the individual statements as true or false.

Valid Arguments

When someone makes a sequence of statements and draws some conclusion from them, he or she is presenting an argument. An **argument** consists of two components: the initial statements, or hypotheses, and the final statement, or conclusion. When presented with an argument, a listener or reader may ask, "Does this person have a logical argument? Does his or her conclusion necessarily follow from the given statements?"

An argument is **valid** if the conclusion of the argument is guaranteed under its given set of hypotheses. (That is, the conclusion is inescapable in all instances.) For example, the argument

"All men are mortal.   ⎫
                                    ⎬ the hypotheses
Socrates is a man.   ⎭

Therefore, Socrates is mortal."   } the conclusion

is a valid argument. Given the hypotheses, the conclusion is guaranteed. The term *valid* does not mean that all the statements are true but merely that the conclusion was reached via a proper deductive process. The argument

"All doctors are men.   ⎫
                                    ⎬ the hypotheses
My mother is a doctor. ⎭

Therefore, my mother is a man."   } the conclusion

is valid. Even though the conclusion is obviously false, the conclusion is guaranteed, *given the hypotheses.*

Using a logical argument, Lewis Carroll's Cheshire Cat tried to convince Alice that he was crazy. Was his argument valid?

The hypotheses in a given logical argument may consist of several interrelated statements, each containing negations, conjunctions, disjunctions, and conditionals. By joining all the hypotheses in the form of a conjunction, we can form a single conditional that represents the entire argument. That is, if an argument has n hypotheses $(h_1, h_2, \ldots, h_n)$ and conclusion c, the argument will have the form "if $(h_1$ and $h_2 \ldots$ and $h_n)$, then c."

Conditional Representation of an Argument

An argument having n hypotheses $h_1, h_2, \ldots, h_n$ and conclusion c can be represented by the conditional $[h_1 \wedge h_2 \wedge \ldots \wedge h_n] \rightarrow c$.

If the conditional representation of an argument is always true (regardless of the actual truthfulness of the individual statements), the argument is valid. If there is at least one instance in which the conditional is false, the argument is invalid.

EXAMPLE 1   Determine whether the following argument is valid:

"If he is illiterate, he cannot fill out the application.

He can fill out the application.

Therefore, he is not illiterate."

Solution   First, number the hypotheses and separate them from the conclusion with a line:

1. If he is illiterate, he cannot fill out the application.
2. He can fill out the application.

Therefore, he is not illiterate.

Church Carving May Be Original 'Cheshire Cat'

Reprinted with permission from Reuters.

LONDON—Devotees of writer Lewis Carroll believe they have found what inspired his grinning Cheshire Cat, made famous in his book "Alice's Adventures in Wonderland."

Members of the Lewis Carroll Society made the discovery over the weekend in a church at which the author's father was once rector in the Yorkshire village of Croft in northern England.

It is a rough-hewn carving of a cat's head smiling near an altar, probably dating to the 10th century. Seen from below and from the perspective of a small boy, all that can be seen is the grinning mouth.

Carroll's Alice watched the Cheshire Cat disappear "ending with the grin, which remained for some time after the rest of the head had gone."

Alice mused: "I have often seen a cat without a grin, but not a grin without a cat. It is the most curious thing I have seen in all my life."

Now use symbols to represent each different component in the statements:

> p: He is illiterate.
>
> q: He can fill out the application.

We could have defined q as "He *cannot* fill out the application" (as stated in premise 1), but it is customary to define the symbols with a positive sense. Symbolically, the argument has the form

$$
\begin{array}{l}
1.\ p \rightarrow \sim q \\
\underline{2.\ q} \\
\therefore \sim p
\end{array}
\quad
\begin{array}{l}
\} \text{ the hypotheses} \\
\\
\} \text{ conclusion}
\end{array}
$$

and is represented by the conditional $[(p \rightarrow \sim q) \wedge q] \rightarrow \sim p$. The symbol $\therefore$ is read "therefore."

To construct a truth table for this conditional, we need $2^2 = 4$ rows. A column is required for the following: each negation, each hypothesis, the conjunction of the hypotheses, the conclusion, and the conditional representation of the argument. The initial setup is shown in Figure 1.61.

	p	q	$\sim q$	Hypothesis 1	Hypothesis 2	Column Representing All the Hypotheses	Conclusion c	Conditional Representation of the Argument
	p	q	$\sim q$	$p \to \sim q$	q	$1 \wedge 2$	$\sim p$	$(1 \wedge 2) \to c$
1.	T	T						
2.	T	F						
3.	F	T						
4.	F	F						

Figure 1.61
Required columns in the truth table

Fill in the truth table as follows:

$\sim q$: A negation has the opposite truth values; enter a T in rows 2 and 4 and an F in rows 1 and 3.

Hypothesis 1: A conditional is false only when its premise is true and its conclusion is false; enter an F in row 1 and Ts elsewhere.

Hypothesis 2: Recopy the q column.

$\mathbf{1 \wedge 2}$: A conjunction is true only when both components are true; enter a T in row 3 and Fs elsewhere.

Conclusion c: A negation has the opposite truth values; enter an F in rows 1 and 2 and a T in rows 3 and 4.

At this point, all that remains is the final column (see Figure 1.62).

				1	**2**		c	
	p	q	$\sim q$	$p \to \sim q$	q	$1 \wedge 2$	$\sim p$	$(1 \wedge 2) \to c$
1.	T	T	F	F	T	F	F	
2.	T	F	T	T	F	F	F	
3.	F	T	F	T	T	T	T	
4.	F	F	T	T	F	F	T	

Figure 1.62
Truth values of the expressions

The last column in the truth table is the conditional that represents the entire argument. A conditional is false only when its premise is true and its conclusion is false. The only instance in which the premise $(1 \wedge 2)$ is true is row 3. Corresponding to this entry, the conclusion $\sim p$ is also true. Consequently, the conditional $(1 \wedge 2) \to c$ is true in row 3. Because the premise $(1 \wedge 2)$ is false in rows 1, 2, and 4, the conditional $(1 \wedge 2) \to c$ is automatically true in those rows as well. The completed truth table is shown in Figure 1.63.

The completed truth table shows that the conditional $[(p \to \sim q) \wedge q] \to \sim p$ is always true. The conditional represents the argument "If he is illiterate, he cannot fill out the application. He can fill out the application. Therefore, he is not illiterate." Thus, the argument is valid.

	p	q	$\sim q$	**1** $p \to \sim q$	**2** q	$1 \wedge 2$	c $\sim p$	$(1 \wedge 2) \to c$
1.	T	T	F	F	T	F	F	T
2.	T	F	T	T	F	F	F	T
3.	F	T	F	T	T	T	T	T
4.	F	F	T	T	F	F	T	T

Figure 1.63
Truth table for the argument $[(p \to \sim q) \wedge q] \to \sim p$ ∎

Tautologies

A **tautology** is a statement that is always true. For example, the statement

"$(a + b)^2 = a^2 + 2ab + b^2$"

is a tautology.

EXAMPLE 2 Determine whether the statement $(p \wedge q) \to (p \vee q)$ is a tautology.

Solution We need to construct a truth table for the statement. Because there are two letters, the table must have $2^2 = 4$ rows. We need a column for $(p \wedge q)$, one for $(p \vee q)$, and one for $(p \wedge q) \to (p \vee q)$. The completed truth table is shown in Figure 1.64.

	p	q	$p \wedge q$	$p \vee q$	$(p \wedge q) \to (p \vee q)$
1.	T	T	T	T	T
2.	T	F	F	T	T
3.	F	T	F	T	T
4.	F	F	F	F	T

Figure 1.64
Truth table for the statement $(p \wedge q) \to (p \vee q)$

Because $(p \wedge q) \to (p \vee q)$ is always true, it is a tautology. ∎

As we have seen, an argument can be represented by a single conditional. If this conditional is always true, the argument is valid (and vice versa).

> ### Validity of an Argument
>
> An argument having n hypotheses $h_1, h_2, \ldots, h_n$ and conclusion c is valid if and only if the conditional $[h_1 \wedge h_2 \wedge \ldots \wedge h_n] \to c$ is a tautology.

ThomsonNOW™
for Liberal Arts Mathematics
www.thomsonedu.com/1pass.

EXAMPLE 3 Determine whether the following argument is valid:
"If the defendant is innocent, the defendant does not go to jail. The defendant does not go to jail. Therefore, the defendant is innocent."

Solution Separating the hypotheses from the conclusion, we have

1. If the defendant is innocent, the defendant does not go to jail.
2. The defendant does not go to jail.

Therefore, the defendant is innocent.

Now we define symbols to represent the various components of the statements:

p: The defendant is innocent.

q: The defendant goes to jail.

Symbolically, the argument has the form

1. $p \rightarrow \sim q$
2. $\sim q$

$\therefore p$

and is represented by the conditional $[(p \rightarrow \sim q) \wedge \sim q] \rightarrow p$.

Now we construct a truth table with four rows, along with the necessary columns. The completed table is shown in Figure 1.65.

			2	**1**		**c**	
	p	*q*	$\sim q$	$p \rightarrow \sim q$	$1 \wedge 2$	*p*	$(1 \wedge 2) \rightarrow c$
1.	T	T	F	F	F	T	T
2.	T	F	T	T	T	T	T
3.	F	T	F	T	F	F	T
4.	F	F	T	T	T	F	F

Figure 1.65
Truth table for the argument $[(p \rightarrow \sim q) \wedge \sim q] \rightarrow p$

The column representing the argument has an F in row 4; therefore, the conditional representation of the argument is *not* a tautology. In particular, the conclusion does not logically follow the hypotheses when both *p* and *q* are false (row 4). The argument is not valid. Let us interpret the circumstances expressed in row 4, the row in which the argument breaks down. Both *p* and *q* are false—that is, the defendant is guilty and the defendant does *not* go to jail. Unfortunately, this situation can occur in the real world; guilty people do not *always* go to jail! As long as it is possible for a guilty person to avoid jail, the argument is invalid. ∎

The following argument was presented as Example 6 in Section 1.1. In that section, we constructed a Venn diagram to show that the argument was in fact valid. We now show an alternative method; that is, we construct a truth table to determine whether the argument is valid.

Historical Note   Charles Lutwidge Dodgson, 1832–1898

© Bettmann/Corbis

To those who assume that it is impossible for a person to excel both in the creative worlds of art and literature and in the disciplined worlds of mathematics and logic, the life of Charles Lutwidge Dodgson is a wondrous counterexample. Known the world over as Lewis Carroll, Dodgson penned the nonsensical classics *Alice's Adventures in Wonderland* and *Through the Looking Glass*. However, many people are surprised to learn that Dodgson (from age eighteen to his death) was a permanent resident at the University at Oxford, teaching mathematics and logic. And as if that were not enough, Dodgson is now recognized as one of the leading portrait photographers of the Victorian era.

The eldest son in a family of eleven children, Charles amused his younger siblings with elaborate games, poems, stories, and humorous drawings. This attraction to entertaining children with fantastic stories manifested itself in much of his later work as Lewis Carroll. Besides his obvious interest in telling stories, the young Dodgson was also intrigued by mathematics. At the age of eight, Charles asked his father to explain a book on logarithms. When told that he was too young to understand, Charles persisted, "But please, explain!"

The Dodgson family had a strong ecclesiastical tradition; Charles's father, great-grandfather, and great-great-grandfather were all clergymen. Following in his father's footsteps, Charles attended Christ Church, the largest and most celebrated of all the Oxford colleges. After graduating in 1854, Charles remained at Oxford, accepting the position of mathematical lecturer in 1855. However, appointment to this position was conditional upon his taking Holy Orders in the Anglican church and upon his remaining celibate. Dodgson complied and was named a deacon in 1861.

The year 1856 was filled with events that had lasting effects on Dodgson. Charles Lutwidge created his pseudonym by translating his first and middle names into Latin (Carolus Ludovic), reversing their order (Ludovic Carolus), and translating them back into English (Lewis Carroll). In this same year, Dodgson began his "hobby" of photography. He is considered by many to have been an artistic pioneer in this new field (photography was invented in 1839). Most of Dodgson's work consists of portraits that chronicle the Victorian era, and over 700 photographs taken by Dodgson have been preserved. His favorite subjects were children, especially young girls.

© Bettmann/Corbis

Young Alice Liddell inspired Lewis Carroll to write *Alice's Adventures in Wonderland*. This photo is one of the many Carroll took of Alice.

EXAMPLE 4   Determine whether the following argument is valid: "No snake is warm-blooded. All mammals are warm-blooded. Therefore, snakes are not mammals."

Solution   Separating the hypotheses from the conclusion, we have

1. No snake is warm-blooded.

2. All mammals are warm-blooded.

Therefore, snakes are not mammals.

Dodgson's affinity for children brought about a meeting in 1856 that would eventually establish his place in the history of literature. Early in the year, Dodgson met the four children of the dean of Christ Church: Harry, Lorina, Edith, and Alice Liddell. He began seeing the children on a regular basis, amusing them with stories and photographing them. Although he had a wonderful relationship with all four, Alice received his special attention.

On July 4, 1862, while rowing and picnicking with Alice and her sisters, Dodgson entertained the Liddell girls with a fantastic story of a little girl named Alice who fell into a rabbit hole. Captivated by the story, Alice Liddell insisted that Dodgson write it down for her. He complied, initially titling it *Alice's Adventure Underground.*

Dodgson's friends subsequently encouraged him to publish the manuscript, and in 1865, after editing and inserting new episodes, Lewis Carroll gave the world *Alice's Adventures in Wonderland.* Although the book appeared to be a whimsical excursion into chaotic nonsense, Dodgson's masterpiece contained many exercises in logic and metaphor. The book was a success, and in 1871 a sequel, *Through the Looking Glass,* was printed. When asked to comment on the meaning of his writings, Dodgson replied, "I'm very much afraid I didn't mean anything but nonsense! Still, you know, words mean more than we mean to express when we use them; so a whole book ought to mean a great deal more than the writer means. So, whatever good meanings are in the book, I'm glad to accept as the meaning of the book."

In addition to writing "children's stories," Dodgson wrote numerous mathematics essays and texts, including *The Fifth Book of Euclid Proved Algebraically, Formulae of Plane Trigonometry, A Guide to the Mathematical Student,* and *Euclid and His Modern Rivals.* In the field of formal logic, Dodgson's books *The Game of Logic* (1887) and *Symbolic Logic* (1896) are still used as sources of inspiration in numerous schools worldwide.

Brown University Library

Carroll's book *The Game of Logic* presents the study of formalized logic in a gamelike fashion. After listing the "rules of the game" (complete with gameboard and markers), Carroll captures the reader's interest with nonsensical syllogisms.

These statements can be rephrased as follows:

1. If it is a snake, then it is not warm-blooded.
2. If it is a mammal, then it is warm-blooded.

Therefore, if it is a snake, then it is not a mammal.

Now we define symbols to represent the various components of the statements:

p: It is a snake.

q: It is warm-blooded.

r: It is a mammal.

Symbolically, the argument has the form

1. $p \to \sim q$
2. $r \to q$
$\therefore p \to \sim r$

and is represented by the conditional $[(p \to \sim q) \wedge (r \to q)] \to (p \to \sim r)$.

Now we construct a truth table with eight rows ($2^3 = 8$), along with the necessary columns. The completed table is shown in Figure 1.66.

	p	q	r	$\sim q$	$\sim r$	1 $p \to \sim q$	2 $r \to q$	$1 \wedge 2$	c $p \to \sim r$	$(1 \wedge 2) \to c$
1.	T	T	T	F	F	F	T	F	F	T
2.	T	T	F	F	T	F	T	F	T	T
3.	T	F	T	T	F	T	F	F	F	T
4.	T	F	F	T	T	T	T	T	T	T
5.	F	T	T	F	F	T	T	T	T	T
6.	F	T	F	F	T	T	T	T	T	T
7.	F	F	T	T	F	T	F	F	T	T
8.	F	F	F	T	T	T	T	T	T	T

Figure 1.66
Truth table for the argument $[(p \to \sim q) \wedge (r \to q)] \to (p \to \sim r)$

The last column of the truth table represents the argument and contains all T's. Consequently, the conditional $[(p \to \sim q) \wedge (r \to q)] \to (p \to \sim r)$ is a tautology; the argument is valid. ■

The preceding examples contained relatively simple arguments, each consisting of only two hypotheses and two simple statements (letters). In such cases, many people try to employ "common sense" to confirm the validity of the argument. For instance, the argument "If it is raining, the streets are wet. It is raining. Therefore, the streets are wet" is obviously valid. However, it might not be so simple to determine the validity of an argument that contains several hypotheses and many simple statements. Indeed, in such cases, the argument's truth table might become quite lengthy, as in the next example.

EXAMPLE 5   The following whimsical argument was written by Lewis Carroll and appeared in his 1896 book *Symbolic Logic:*

"No ducks waltz. No officers ever decline to waltz. All my poultry are ducks. Therefore, my poultry are not officers."

Construct a truth table to determine whether the argument is valid.

Solution   Separating the hypotheses from the conclusion, we have

1. No ducks waltz.
2. No officers ever decline to waltz.
3. All my poultry are ducks.

Therefore, my poultry are not officers.

These statements can be rephrased as

1. If it is a duck, then it does not waltz.
2. If it is an officer, then it does not decline to waltz.
   (Equivalently, "If it is an officer, then it will waltz.")
3. If it is my poultry, then it is a duck.

Therefore, if it is my poultry, then it is not an officer.

Now we define symbols to represent the various components of the statements:

p: It is a duck.

q: It will waltz.

r: It is an officer.

s: It is my poultry.

Symbolically, the argument has the form

1. $p \rightarrow \sim q$
2. $r \rightarrow q$
3. $s \rightarrow p$

$\therefore s \rightarrow \sim r$

Now we construct a truth table with sixteen rows ($2^4 = 16$), along with the necessary columns. The completed table is shown in Figure 1.67.

	p	*q*	*r*	*s*	*~q*	*~r*	**1** $p \rightarrow \sim q$	**2** $r \rightarrow q$	**3** $s \rightarrow p$	$1 \wedge 2 \wedge 3$	***c*** $s \rightarrow \sim r$	$(1 \wedge 2 \wedge 3) \rightarrow c$
1.	T	T	T	T	F	F	F	T	T	F	F	T
2.	T	T	T	F	F	F	F	T	T	F	T	T
3.	T	T	F	T	F	T	F	T	T	F	T	T
4.	T	T	F	F	F	T	F	T	T	F	T	T
5.	T	F	T	T	T	F	T	F	T	F	F	T
6.	T	F	T	F	T	F	T	F	T	F	T	T
7.	T	F	F	T	T	T	T	T	T	T	T	T
8.	T	F	F	F	T	T	T	T	T	T	T	T
9.	F	T	T	T	F	F	T	T	F	F	F	T
10.	F	T	T	F	F	F	T	T	T	T	T	T
11.	F	T	F	T	F	T	T	T	F	F	T	T
12.	F	T	F	F	F	T	T	T	T	T	T	T
13.	F	F	T	T	T	F	T	F	F	F	F	T
14.	F	F	T	F	T	F	T	F	T	F	T	T
15.	F	F	F	T	T	T	T	T	F	F	T	T
16.	F	F	F	F	T	T	T	T	T	T	T	T

Figure 1.67
Truth table for the argument $[(p \rightarrow \sim q) \wedge (r \rightarrow q) \wedge (s \rightarrow p)] \rightarrow (s \rightarrow \sim r)$

The last column of the truth table represents the argument and contains all T's. Consequently, the conditional $[(p \rightarrow \sim q) \land (r \rightarrow q) \land (s \rightarrow p)] \rightarrow (s \rightarrow \sim r)$ is a tautology; the argument is valid. ∎

1.5 Exercises

In Exercises 1–6, use the given symbols to rewrite the argument in symbolic form.

1. p: It is raining.
$\quad q$: The streets are wet. } Use these symbols.

    1. If it is raining, then the streets are wet.
    2. It is raining.

    Therefore, the streets are wet.

2. p: I have a college degree.
$\quad q$: I am lazy. } Use these symbols.

    1. If I have a college degree, I am not lazy.
    2. I do not have a college degree.

    Therefore, I am lazy.

3. p: It is Tuesday.
$\quad q$: The tour group is in Belgium. } Use these symbols.

    1. If it is Tuesday, then the tour group is in Belgium.
    2. The tour group is not in Belgium.

    Therefore, it is not Tuesday.

4. p: You are a gambler.
$\quad q$: You have financial security. } Use these symbols.

    1. You do not have financial security if you are a gambler.
    2. You do not have financial security.

    Therefore, you are a gambler.

5. p: You exercise regularly.
$\quad q$: You are healthy. } Use these symbols.

    1. You exercise regularly only if you are healthy.
    2. You do not exercise regularly.

    Therefore, you are not healthy.

6. p: The senator supports new taxes.
$\quad q$: The senator is reelected. } Use these symbols.

    1. The senator is not reelected if she supports new taxes.
    2. The senator does not support new taxes.

    Therefore, the senator is reelected.

In Exercises 7–12, use a truth table to determine the validity of the argument specified. If the argument is invalid, interpret the specific circumstances that cause it to be invalid.

  7. the argument in Exercise 1
  8. the argument in Exercise 2
  9. the argument in Exercise 3
10. the argument in Exercise 4
11. the argument in Exercise 5
12. the argument in Exercise 6

In Exercises 13–30, define the necessary symbols, rewrite the argument in symbolic form, and use a truth table to determine whether the argument is valid. If the argument is invalid, interpret the specific circumstances that cause the argument to be invalid.

13. 1. If the Democrats have a majority, Smith is appointed and student loans are funded.
    2. Smith is appointed or student loans are not funded.

    Therefore, the Democrats do not have a majority.

14. 1. If you watch television, you do not read books.
    2. If you read books, you are wise.

    Therefore, you are not wise if you watch television.

15. 1. If you argue with a police officer, you get a ticket.
    2. If you do not break the speed limit, you do not get a ticket.

    Therefore, if you break the speed limit, you argue with a police officer.

16. 1. If you do not recycle newspapers, you are not an environmentalist.
    2. If you recycle newspapers, you save trees.

    Therefore, you are an environmentalist only if you save trees.

17. 1. All pesticides are harmful to the environment.
    2. No fertilizer is a pesticide.

    Therefore, no fertilizer is harmful to the environment.

18. 1. No one who can afford health insurance is unemployed.
2. All politicians can afford health insurance.

Therefore, no politician is unemployed.

19. 1. All poets are loners.
2. All loners are taxi drivers.

Therefore, all poets are taxi drivers.

20. 1. All forest rangers are environmentalists.
2. All forest rangers are storytellers.

Therefore, all environmentalists are storytellers.

21. 1. No professor is a millionaire.
2. No millionaire is illiterate.

Therefore, no professor is illiterate.

22. 1. No artist is a lawyer.
2. No lawyer is a musician.

Therefore, no artist is a musician.

23. 1. All lawyers study logic.
2. You study logic only if you are a scholar.
3. You are not a scholar.

Therefore, you are not a lawyer.

24. 1. All licensed drivers have insurance.
2. You obey the law if you have insurance.
3. You obey the law.

Therefore, you are a licensed driver.

25. If the defendant is innocent, he does not go to jail. The defendant goes to jail. Therefore, the defendant is guilty.

26. If the defendant is innocent, he does not go to jail. The defendant is guilty. Therefore, the defendant goes to jail.

27. If you are not in a hurry, you eat at Lulu's Diner. If you are in a hurry, you do not eat good food. You eat at Lulu's. Therefore, you eat good food.

28. If you give me a hamburger today, I pay you tomorrow. If you are a sensitive person, you give me a hamburger today. You are not a sensitive person. Therefore, I do not pay you tomorrow.

29. If you listen to rock and roll, you do not go to heaven. If you are a moral person, you go to heaven. Therefore, you are not a moral person if you listen to rock and roll.

30. If you follow the rules, you have no trouble. If you are not clever, you have trouble. You are clever. Therefore, you do not follow the rules.

The arguments given in Exercises 31–38 were written by Lewis Carroll and appeared in his 1896 book Symbolic Logic. *For each argument, define the necessary symbols, rewrite the argument in symbolic form, and use a truth table to determine whether the argument is valid.*

31. 1. All medicine is nasty.
2. Senna is a medicine.

Therefore, senna is nasty.

NOTE: Senna is a laxative extracted from the dried leaves of cassia plants.

32. 1. All pigs are fat.
2. Nothing that is fed on barley-water is fat.

Therefore, pigs are not fed on barley-water.

33. 1. Nothing intelligible ever puzzles me.
2. Logic puzzles me.

Therefore, logic is unintelligible.

34. 1. No misers are unselfish.
2. None but misers save eggshells.

Therefore, no unselfish people save eggshells.

35. 1. No Frenchmen like plum pudding.
2. All Englishmen like plum pudding.

Therefore, Englishmen are not Frenchmen.

36. 1. A prudent man shuns hyenas.
2. No banker is imprudent.

Therefore, no banker fails to shun hyenas.

37. 1. All wasps are unfriendly.
2. No puppies are unfriendly.

Therefore, puppies are not wasps.

38. 1. Improbable stories are not easily believed.
2. None of his stories are probable.

Therefore, none of his stories are easily believed.

Answer the following questions using complete sentences and your own words.

CONCEPT QUESTIONS

39. What is a tautology?
40. What is the conditional representation of an argument?
41. Find a "logical" argument in a newspaper article, an advertisement, or elsewhere in the media. Analyze that argument and discuss the implications.

HISTORY QUESTIONS

42. What was Charles Dodgson's pseudonym? How did he get it? What classic "children's stories" did he write?
43. What did Charles Dodgson contribute to the study of formal logic?

44. Charles Dodgson was a pioneer in what artistic field?

45. Who was Alice Liddell?

 WEB PROJECT

46. Write a research paper on any historical topic referred to in this chapter or a related topic. Below is a partial list of topics.
- Aristotle
- George Boole

- Augustus De Morgan
- Charles Dodgson/Lewis Carroll
- Gottfried Wilhelm Leibniz

Some useful links for this web project are listed on the text web site: **www.thomsonedu.com/ mathematics/johnson**

Chapter 1   Review

ThomsonNOW™ for Liberal Arts Mathematics
Preparing for an exam? Test yourself on key material by visiting ThomsonNOW at **www.thomsonedu.com/1pass.**

TERMS

argument	deductive reasoning	invalid argument	su doku
biconditional	disjunction	inverse	syllogism
compound statement	equivalent expressions	logic	tautology
conclusion	exclusive *or*	negation	truth table
conditional	hypothesis	premise	truth value
conjunction	implication	quantifier	valid argument
contrapositive	inclusive *or*	reasoning	Venn diagram
converse	inductive reasoning	statement	

REVIEW EXERCISES

1. Classify each argument as deductive or inductive.

a. 1. Hitchcock's "Psycho" is a suspenseful movie.
2. Hitchcock's "The Birds" is a suspenseful movie.

Therefore, all Hitchcock movies are suspenseful.

b. 1. All Hitchcock movies are suspenseful.
2. "Psycho" is a Hitchcock movie.

Therefore, "Psycho" is suspenseful.

2. Explain the general rule or pattern used to assign the given letter to the given word. Fill in the blank with the letter that fits the pattern.

day	morning	afternoon	dusk	night
y	r	t	s	_____

3. Fill in the blank with what is most likely to be the next number. Explain the pattern generated by your answer.

1, 6, 11, 4, _____

In Exercises 4–9, construct a Venn diagram to determine the validity of the given argument.

4. 1. All truck drivers are union members.
2. Rocky is a truck driver.

Therefore, Rocky is a union member.

5. 1. All truck drivers are union members.
2. Rocky is not a truck driver.

Therefore, Rocky is not a union member.

6. 1. All mechanics are engineers.
2. Casey Jones is an engineer.

Therefore, Casey Jones is a mechanic.

7. 1. All mechanics are engineers.
2. Casey Jones is not an engineer.

Therefore, Casey Jones is not a mechanic.

8. 1. Some animals are dangerous.
2. A gun is not an animal.

Therefore, a gun is not dangerous.

9. 1. Some contractors are electricians.
2. All contractors are carpenters.

Therefore, some electricians are carpenters.

10. Solve the following su doku puzzle.

	2					3	4	
		3		6				7
		7			9			8
		4			1	7		
5			4	7	8			3
		1	6			2		
2		6			4			
3			5		7			
	7	8					3	

11. Explain why each of the following is or is not a statement.
   a. The Golden Gate Bridge spans Chesapeake Bay.
   b. The capital of Delaware is Dover.
   c. Where are you spending your vacation?
   d. Hawaii is the best place to spend a vacation.

12. Determine which pairs of statements are negations of each other.
   a. All of the lawyers are ethical.
   b. Some of the lawyers are ethical.
   c. None of the lawyers is ethical.
   d. Some of the lawyers are not ethical.

13. Write a sentence that represents the negation of each statement.
   a. His car is not new.
   b. Some buildings are earthquakeproof.
   c. All children eat candy.
   d. I never cry in a movie theater.

14. Using the symbolic representations

p: The television program is educational.
q: The television program is controversial.

express the following compound statements in symbolic form.
   a. The television program is educational and controversial.
   b. If the television program isn't controversial, it isn't educational.
   c. The television program is educational and it isn't controversial.
   d. The television program isn't educational or controversial.

15. Using the symbolic representations

p: The advertisement is effective.
q: The advertisement is misleading.
r: The advertisement is outdated.

express the following compound statements in symbolic form.

   a. All misleading advertisements are effective.
   b. It is a current, honest, effective advertisement.
   c. If an advertisement is outdated, it isn't effective.
   d. The advertisement is effective and it isn't misleading or outdated.

16. Using the symbolic representations

p: It is expensive.
q: It is undesirable.

express the following in words.
   a. $p \to \sim q$ **b.** $q \leftrightarrow \sim p$
   c. $\sim(p \vee q)$ **d.** $(p \wedge \sim q)/(\sim p \wedge q)$

17. Using the symbolic representations

p: The movie is critically acclaimed.
q: The movie is a box office hit.
r: The movie is available on videotape.

express the following in words.
   a. $(p \vee q) \to r$ **b.** $(p \wedge \sim q) \to \sim r$
   c. $\sim(p \vee q) \wedge r$ **d.** $\sim r \to (\sim p \wedge \sim q)$

In Exercises 18–25, construct a truth table for the compound statement.

18. $p \vee \sim q$ **19.** $p \wedge \sim q$
20. $\sim p \to q$ **21.** $(p \wedge q) \to \sim q$
22. $q \vee \sim(p \vee r)$ **23.** $\sim p \to (q \vee r)$
24. $(q \wedge p) \to (\sim r \vee p)$ **25.** $(p \vee r) \to (q \wedge \sim r)$

In Exercises 26–29, construct a truth table to determine whether the statements in each pair are equivalent.

26. The car is unreliable or expensive.
   If the car is reliable, then it is expensive.
27. If I get a raise, I will buy a new car.
   If I do not get a raise, I will not buy a new car.
28. She is a Democrat or she did not vote.
   She is not a Democrat and she did vote.
29. The raise is not unjustified and the management opposes it.
   It is not the case that the raise is unjustified or the management does not oppose it.

In Exercises 30–35, write a sentence that represents the negation of each statement.

30. Jesse had a party and nobody came.
31. You do not go to jail if you pay the fine.
32. I am the winner or you are blind.
33. He is unemployed and he did not apply for financial assistance.
34. The selection procedure has been violated if his application is ignored.

35. The jackpot is at least $1 million.

36. Given the statements

p: You are an avid jogger.

q: You are healthy.

write the sentence represented by each of the following.

a. $p \rightarrow q$ **b.** $q \rightarrow p$

c. $\sim p \rightarrow \sim q$ **d.** $\sim q \rightarrow \sim p$

e. $p \leftrightarrow q$

37. Form (a) the inverse, (b) the converse, and (c) the contrapositive of the conditional "If he is elected, the country is in big trouble."

In Exercises 38–41, (a) determine the premise and conclusion and (b) rewrite the compound statement in the standard "if . . . then . . ." form.

38. The economy improves only if unemployment goes down.

39. The economy improves if unemployment goes down.

40. No computer is unrepairable.

41. All gemstones are valuable.

In Exercises 42 and 43, translate the two statements into symbolic form and use truth tables to determine whether the statements are equivalent.

42. If you are allergic to dairy products, you cannot eat cheese.

If you cannot eat cheese, then you are allergic to dairy products.

43. You are a fool if you listen to me.

You are not a fool only if you do not listen to me.

44. Which pairs of statements are equivalent?

  i. If it is not raining, I ride my bicycle to work.

 ii. If I ride my bicycle to work, it is not raining.

iii. If I do not ride my bicycle to work, it is raining.

 iv. If it is raining, I do not ride my bicycle to work.

In Exercises 45–48, define the necessary symbols, rewrite the argument in symbolic form, and use a truth table to determine whether the argument is valid.

45. 1. If you do not make your loan payment, your car is repossessed.

   2. Your car is repossessed.

Therefore, you did not make your loan payment.

46. 1. If you do not pay attention, you do not learn the new method.

   2. You do learn the new method.

Therefore, you do pay attention.

47. 1. If you rent videocassettes, you will not go to the movie theater.

   2. If you go to the movie theater, you pay attention to the movie.

Therefore, you do not pay attention to the movie if you rent videocassettes.

48. 1. If the Republicans have a majority, Farnsworth is appointed and no new taxes are imposed.

   2. New taxes are imposed.

Therefore, the Republicans do not have a majority or Farnsworth is not appointed.

In Exercises 49–55, define the necessary symbols, rewrite the argument in symbolic form, and use a truth table to determine whether the argument is valid.

49. If the defendant is guilty, he goes to jail. The defendant does not go to jail. Therefore, the defendant is not guilty.

50. I will go to the concert only if you buy me a ticket. You bought me a ticket. Therefore, I will go to the concert.

51. If tuition is raised, students take out loans or drop out. If students do not take out loans, they drop out. Students do drop out. Therefore, tuition is raised.

52. If our oil supply is cut off, our economy collapses. If we go to war, our economy doesn't collapse. Therefore, if our oil supply isn't cut off, we do not go to war.

53. No professor is uneducated. No monkey is educated. Therefore, no professor is a monkey.

54. No professor is uneducated. No monkey is a professor. Therefore, no monkey is educated.

55. Vehicles stop if the traffic light is red. There is no accident if vehicles stop. There is an accident. Therefore, the traffic light is not red.

Determine the validity of the arguments in Exercises 56 and 57 by constructing a

  a. *Venn diagram and a*

  b. *truth table.*

  c. *How do the answers to parts (a) and (b) compare? Why?*

56. 1. If you own a hybrid vehicle, then you are an environmentalist.

   2. You are not an environmentalist.

Therefore, you do not own a hybrid vehicle.

57. 1. If you own a hybrid vehicle, then you are an environmentalist.

   2. You are an environmentalist.

Therefore, you own a hybrid vehicle.

Answer the following questions using complete sentences and your own words.

CONCEPT QUESTIONS

58. What is a statement?

59. **a.** What is a disjunction? Under what conditions is a disjunction true?

b. What is a conjunction? Under what conditions is a conjunction true?

c. What is a conditional? Under what conditions is a conditional true?

d. What is a negation? Under what conditions is a negation true?

60. What is a tautology?

61. When constructing a truth table, how do you determine how many rows to create?

HISTORY QUESTIONS

62. What role did the following people play in the development of formalized logic?

- Aristotle
- George Boole
- Augustus De Morgan
- Charles Dodgson
- Gottfried Wilhelm Leibniz

2

Sets and Counting

© Terry Williams/The Image Bank/Getty Images

R ecently, 1,000 college seniors were asked whether they favored increasing the state's gasoline tax to generate funds to improve highways and whether they favored increasing the state's alcohol tax to generate funds to improve the public education system. The responses were tallied, and the following results were printed in the campus newspaper: 750 favored an increase in the gasoline tax, 600 favored an increase in the alcohol tax, and 450 favored increases in both taxes. How many of these 1,000 students favored an increase in at least one of the taxes? How many favored increasing only the gasoline tax, increasing only the alcohol tax, or increasing neither tax? The mathematical tool designed to answer questions like these is the *set*. Although you might be able to answer the given questions without any formal knowledge of sets, the mental reasoning involved in obtaining your answers uses some of the basic principles of sets. (Incidentally, the answers are 900, 300, 150, and 100, respectively.)

The branch of mathematics that deals with sets is called **set theory.** Set theory can be helpful in solving both mathematical and nonmathematical problems. It is an important tool in analyzing the results of consumer surveys, marketing analyses, and political polls. Standardized admissions tests such as the Graduate Record Examination (G.R.E.) ask questions that can be answered with set theory. In this text, we use sets extensively in Chapter 3 on probability.

ThomsonNOW™ **for Liberal Arts Mathematics**
Throughout this chapter, the ThomsonNOW logo indicates an opportunity for online self-study, linking you to interactive tutorials and videos based on your level of understanding.

62

Sets and Set Operations

A **set** is a collection of objects or things. The objects or things in the set are called **elements** (or *members*) of the set. In our example above, we could talk about the *set* of students who favor increasing only the gasoline tax or the *set* of students who do not favor increasing either tax. In geography, we can talk about the *set* of all state capitals or the *set* of all states west of the Mississippi. It is easy to determine whether something is in these sets; for example, Des Moines is an element of the set of state capitals, whereas Dallas is not. Such sets are called **well-defined** because there is a way of determining for sure whether a particular item is an element of the set.

EXAMPLE 1 Which of the following sets are well-defined?

a. the set of all movies directed by Alfred Hitchcock
b. the set of all great rock-and-roll bands
c. the set of all possible two-person committees selected from a group of five people

Solution **a.** This set is well-defined; either a movie was directed by Hitchcock, or it was not.
b. This set is *not* well-defined; membership is a matter of opinion. Some people would say that the Ramones (one of the pioneer punk bands of the late 1970s) are a member, while others might say they are not. (Note: The Ramones were inducted into the Rock and Roll Hall of Fame in 2002.)
c. This set is well-defined; either the two people are from the group of five, or they are not. ∎

Notation

By tradition, a set is denoted by a capital letter, frequently one that will serve as a reminder of the contents of the set. **Roster notation** (also called *listing notation*) is a method of describing a set by listing each element of the set inside the symbols { and }, which are called *set braces*. In a listing of the elements of a set, each distinct element is listed only once, and the order of the elements doesn't matter.

The symbol $\in$ stands for the phrase *is an element of,* and $\notin$ stands for *is not an element of.* The **cardinal number** of a set A is the number of elements in the set and is denoted by $n(A)$. Thus, if R is the set of all letters in the name "Ramones," then $R = \{r, a, m, o, n, e, s\}$. Notice that m is an element of the set R, x is not an element of R, and R has 7 elements. In symbols, $m \in R$, $x \notin R$, and $n(R) = 7$.

Two sets are **equal** if they contain exactly the same elements. *The order in which the elements are listed does not matter.* If M is the set of all letters in the name "Moaners," then $M = \{m, o, a, n, e, r, s\}$. This set contains exactly the same elements as the set R of letters in the name "Ramones." Therefore, $M = R = \{a, e, m, n, o, r, s\}$.

Often, it is not appropriate or not possible to describe a set in roster notation. For extremely large sets, such as the set V of all registered voters in Detroit, or for sets that contain an infinite number of elements, such as the set G of all negative real numbers, the roster method would be either too cumbersome or impossible to use. Although V could be expressed via the roster method (since each county compiles a list of all registered voters in its jurisdiction), it would take hundreds or even thousands of pages to list everyone who is registered to vote in Detroit! In the case of the set G of all negative real numbers, no list, no matter how long, is capable of listing all members of the set; there is an infinite number of negative numbers.

The "Ramones" or The "Moaners"? The set R of all letters in the name "Ramones" is the same as the set M of all letters in the name "Moaners." Consequently, the sets are equal; $M = R = \{a, e, m, n, o, r, s\}$. (R.I.P. Joey Ramone 1951–2001, Dee Dee Ramone 1952–2002, Johnny Ramone 1948–2004)

In such cases, it is often necessary, or at least more convenient, to use **set-builder notation,** which lists the rules that determine whether an object is an element of the set rather than the actual elements. A set-builder description of set G on page 63 is

$$G = \{x \mid x < 0 \quad \text{and} \quad x \in \Re\}$$

which is read as "the set of all x such that x is less than zero and x is a real number." A set-builder description of set V above is

$$V = \{\text{persons} \mid \text{the person is a registered voter in Detroit}\}$$

which is read as "the set of all persons such that the person is a registered voter in Detroit." In set-builder notation, the vertical line stands for the phrase "such that." Whatever is on the left side of the line is the general type of thing in the set, while the rules about set membership are listed on the right.

EXAMPLE 2   Describe each of the following in words.

a. $\{x \mid x > 0 \text{ and } x \in \Re\}$
b. $\{\text{persons} \mid \text{the person is a living former U.S. president}\}$
c. $\{\text{women} \mid \text{the woman is a former U.S. president}\}$

Solution   **a.** the set of all x such that x is a positive real number
b. the set of all people such that the person is a living former U.S. president
c. the set of all women such that the woman is a former U.S. president   ∎

The set listed in part (c) of Example 2 has no elements; there are no women who are former U.S. presidents. If we let W equal "the set of all women such that the woman is a former U.S. president," then $n(W) = 0$. A set that has no elements is called an **empty set** and is denoted by $\varnothing$ or by $\{\}$. Notice that since the empty set has no elements, $n(\varnothing) = 0$. In contrast, the set $\{0\}$ is not empty; it has one element, the number zero, so $n(\{0\}) = 1$.

Universal Set and Subsets

When we work with sets, we must define a universal set. For any given problem, the **universal set,** denoted by U, is the set of all possible elements of any set used in the problem. For example, when we spell words, U is the set of all letters in the alphabet. When every element of one set is also a member of another set, we say that the first set is a *subset* of the second; for instance, {p, i, n} is a subset of {p, i, n, e}. In general, we say that A is a **subset** of B, denoted by $A \subseteq B$, if for every $x \in A$ it follows that $x \in B$. Alternatively, $A \subseteq B$ if A contains no elements that are not in B. If A contains an element that is not in B, then A is not a subset of B (symbolized as $A \nsubseteq B$).

EXAMPLE 3   Let $B = \{\text{countries} \mid \text{the country has a permanent seat on the U.N. Security Council}\}$. Determine if A is a subset of B.

a. $A = \{\text{Russia, United States}\}$
b. $A = \{\text{China, Japan}\}$
c. $A = \{\text{United States, France, China, United Kingdom, Russia}\}$
d. $A = \{\ \}$

Solution   We use the roster method to list the elements of set B.

$$B = \{\text{China, France, Russia, United Kingdom, United States}\}$$

a. Since every element of A is also an element of B, A is a subset of B; $A \subseteq B$.
b. Since A contains an element (Japan) that is not in B, A is not a subset of B; $A \nsubseteq B$.
c. Since every element of A is also an element of B (note that $A = B$), A is a subset of B (and B is a subset of A); $A \subseteq B$ (and $B \subseteq A$). In general, every set is a subset of itself; $A \subseteq A$ for any set A.
d. Does A contain an element that is not in B? No! Therefore, A (an empty set) is a subset of B; $A \subseteq B$. In general, the empty set is a subset of all sets; $\varnothing \subseteq A$ for any set A.   ∎

We can express the relationship $A \subseteq B$ visually by drawing a Venn diagram, as shown in Figure 2.1. A **Venn diagram** consists of a rectangle, representing the universal set, and various closed figures within the rectangle, each representing a set. Recall that Venn diagrams were used in Section 1.1 to determine whether an argument was valid.

If two sets are equal, they contain exactly the same elements. It then follows that each is a subset of the other. For example, if $A = B$, then every element of A is an element of B (and vice versa). In this case, A is called an **improper subset** of B. (Likewise, B is an improper subset of A.) Every set is an improper subset of itself; for example, $A \subseteq A$. On the other hand, if A is a subset of B and B contains an element not in A (that is, $A \neq B$), then A is called a **proper subset** of B. To indicate a

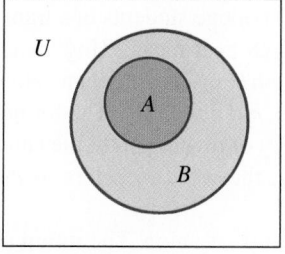

Figure 2.1
A is a subset of B. $A \subseteq B$

proper subset, the symbol $\subset$ is used. While it is acceptable to write $\{1, 2\} \subseteq \{1, 2, 3\}$, the relationship of a proper subset is stressed when it is written $\{1, 2\} \subset \{1, 2, 3\}$. Notice the similarities between the subset symbols, $\subset$ and $\subseteq$, and the inequality symbols, $<$ and $\leq$, used in algebra; it is acceptable to write $1 \leq 3$, but writing $1 < 3$ is more informative.

Intersection of Sets

Sometimes an element of one set is also an element of another set; that is, the sets may overlap. This overlap is called the **intersection** of the sets. If an element is in two sets *at the same time,* it is in the intersection of the sets.

Intersection of Sets

The **intersection** of set A and set B, denoted by $A \cap B$, is

$$A \cap B = \{x \mid x \in A \quad \text{and} \quad x \in B\}$$

The intersection of two sets consists of those elements that are common to both sets.

For example, given the sets $A = \{$Buffy, Spike, Willow, Xander$\}$ and $B = \{$Angel, Anya, Buffy, Giles, Spike$\}$, their intersection is $A \cap B = \{$Buffy, Spike$\}$.

Venn diagrams are useful in depicting the relationship between sets. The Venn diagram in Figure 2.2 illustrates the intersection of two sets; the shaded region represents $A \cap B$.

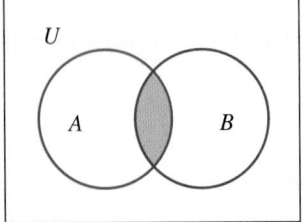

Figure 2.2
The intersection $A \cap B$

Mutually Exclusive Sets

Sometimes a pair of sets has no overlap. Consider an ordinary deck of playing cards. Let $D = \{$cards $\mid$ the card is a diamond$\}$ and $S = \{$cards $\mid$ the card is a spade$\}$. Certainly, *no* cards are both diamonds and spades *at the same time;* that is, $S \cap D = \varnothing$.

Two sets A and B are **mutually exclusive** (or *disjoint*) if they have no elements in common, that is, if $A \cap B = \varnothing$. The Venn diagram in Figure 2.3 illustrates mutually exclusive sets.

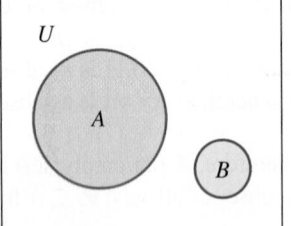

Figure 2.3
Mutually exclusive sets
$(A \cap B = \varnothing)$

Union of Sets

What does it mean when we ask, "How many of the 500 college students in a transportation survey own an automobile or a motorcycle?" Does it mean "How many students own either an automobile or a motorcycle *or both*?" or does it mean "How many students own either an automobile or a motorcycle, *but not both*?" The former is called the *inclusive or,* because it includes the possibility of owning both; the latter is called the *exclusive or.* In logic and in mathematics, the word *or* refers to the *inclusive or,* unless you are told otherwise.

The meaning of the word *or* is important to the concept of union. The **union** of two sets is a new set formed by joining those two sets together, just as the union of the states is the joining together of 50 states to form one nation.

> **Union of Sets**
>
> The **union** of set A and set B, denoted by $A \cup B$, is
>
> $$A \cup B = \{x \mid x \in A \quad \text{or} \quad x \in B\}$$
>
> The union of A and B consists of all elements that are in either A or B or both, that is, all elements that are in at least one of the sets.

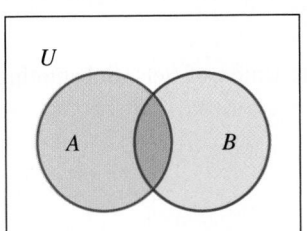

Figure 2.4
The union $A \cup B$

For example, given the sets $A = \{\text{David, Jay}\}$ and $B = \{\text{Conan, Geraldo, Oprah, Jerry}\}$, their union is $A \cup B = \{\text{Conan, David, Geraldo, Jay, Oprah, Jerry}\}$, and their intersection is $A \cap B = \varnothing$. The Venn diagram in Figure 2.4 illustrates the union of two sets; the shaded region represents $A \cup B$.

ThomsonNOW
for Liberal Arts Mathematics
www.thomsonedu.com/1pass.

EXAMPLE 4   Given the sets $A = \{1, 2, 3\}$ and $B = \{2, 4, 6\}$, find the following.

a. $A \cap B$ (the intersection of A and B)
b. $A \cup B$ (the union of A and B)

Solution   **a.** The intersection of two sets consists of those elements that are common to both sets; therefore, we have

$$A \cap B = \{1, 2, 3\} \cap \{2, 4, 6\}$$
$$= \{2\}$$

b. The union of two sets consists of all elements that are in at least one of the sets; therefore, we have

$$A \cup B = \{1, 2, 3\} \cup \{2, 4, 6\}$$
$$= \{1, 2, 3, 4, 6\} \qquad \blacksquare$$

Because $A \cup B$ consists of all elements that are in A or B (or both), to find $n(A \cup B)$, we add $n(A)$ plus $n(B)$. However, doing so results in an answer that might be too big; that is, if A and B have elements in common, these elements will be counted twice (once as a part of A and once as a part of B). Therefore, to find the cardinal number of $A \cup B$, we add the cardinal number of A to the cardinal number of B and then *subtract* the cardinal number of $A \cap B$ (so that the overlap is not counted twice).

> **Cardinal Number Formula for the Union of Sets**
>
> For any two sets A and B, the number of elements in their union is $n(A \cup B)$, where
>
> $$n(A \cup B) = n(A) + n(B) - n(A \cap B)$$

As long as any three of the four quantities in the general formula are known, the missing quantity can be found by algebraic manipulation.

Thomson**NOW**
for Liberal Arts Mathematics
www.thomsonedu.com/1pass.

EXAMPLE 5   Given $n(U) = 169$, $n(A) = 81$, and $n(B) = 66$, find the following.

a. If $n(A \cap B) = 47$, find $n(A \cup B)$ and draw a Venn diagram depicting the composition of the universal set.

b. If $n(A \cup B) = 147$, find $n(A \cap B)$ and draw a Venn diagram depicting the composition of the universal set.

Solution   **a.** We must use the Cardinal Number Formula for the Union of Sets. Substituting the three given quantities, we have

$$n(A \cup B) = n(A) + n(B) - n(A \cap B)$$
$$= 81 + 66 - 47$$
$$= 100$$

The Venn diagram in Figure 2.5 illustrates the composition of U.

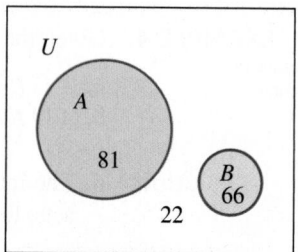

Figure 2.5
$n(A \cap B) = 47$.

Figure 2.6
$n(A \cup B) = 147$.

b. We must use the Cardinal Number Formula for the Union of Sets. Substituting the three given quantities, we have

$$n(A \cup B) = n(A) + n(B) - n(A \cap B)$$
$$147 = 81 + 66 - n(A \cap B)$$
$$147 = 147 - n(A \cap B)$$
$$n(A \cap B) = 147 - 147$$
$$n(A \cap B) = 0$$

Therefore, A and B have no elements in common; they are mutually exclusive. The Venn diagram in Figure 2.6 illustrates the composition of U.   ■

EXAMPLE 6   A recent transportation survey of 500 college students (the universal set U) yielded the following information: 291 own an automobile (A), 179 own a motorcycle (M), and 85 own both an automobile and a motorcycle ($A \cap M$). What percent of these students own an automobile or a motorcycle?

Solution   Recall that "automobile or motorcycle" means "automobile or motorcycle or both" (the inclusive *or*) and that *or* implies union. Hence, we must find $n(A \cup M)$, the cardinal number of the union of sets A and M. We are given that $n(A) = 291$, $n(M) = 179$, and $n(A \cap M) = 85$. Substituting the given values into the Cardinal Number Formula for the Union of Sets, we have

$$n(A \cup M) = n(A) + n(M) - n(A \cap M)$$
$$= 291 + 179 - 85$$
$$= 385$$

Therefore, 385 of the 500 students surveyed own an automobile or a motorcycle. Expressed as a percent, $385/500 = 0.77$; therefore, 77% of the students own an automobile or a motorcycle (or both). ■

Complement of a Set

In certain situations, it might be important to know how many things are *not* in a given set. For instance, when playing cards, you might want to know how many cards are not ranked lower than a five; or when taking a survey, you might want to know how many people did not vote for a specific proposition. The set of all elements in the universal set that are *not* in a specific set is called the *complement* of the set.

Complement of a Set

The **complement** of set A, denoted by A' (read "A prime" or "the complement of A"), is

$$A' = \{x \mid x \in U \quad \text{and} \quad x \notin A\}$$

The complement of a set consists of all elements that are in the universal set but not in the given set.

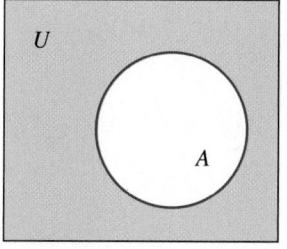

Figure 2.7
The complement A'

For example, given that $U = \{1, 2, 3, 4, 5, 6, 7, 8, 9\}$ and $A = \{1, 3, 5, 7, 9\}$, the complement of A is $A' = \{2, 4, 6, 8\}$. What is the complement of A'? Just as $-(-x) = x$ in algebra, $(A')' = A$ in set theory. The Venn diagram in Figure 2.7 illustrates the complement of set A; the shaded region represents A'.

Suppose A is a set of elements, drawn from a universal set U. If x is an element of the universal set ($x \in U$), then exactly one of the following must be true: (1) x is an element of A ($x \in A$), or (2) x is not an element of A ($x \notin A$). Since no element of the universal set can be in both A and A' at the same time, it follows that A and A' are mutually exclusive sets whose union equals the entire universal set. Therefore, the sum of the cardinal numbers of A and A' equals the cardinal number of U.

It is often quicker to count the elements that are *not* in a set than to count those that are. Consequently, to find the cardinal number of a set, we can subtract the cardinal number of its complement from the cardinal number of the universal set; that is, $n(A) = n(U) - n(A')$.

Cardinal Number Formula for the Complement of a Set

For any set A and its complement A',

$$n(A) + n(A') = n(U)$$

where U is the universal set.
Alternatively,

$$n(A) = n(U) - n(A') \quad \text{and} \quad n(A') = n(U) - n(A)$$

Historical Note   John Venn, 1834–1923

Courtesy The Masters and Fellows of Gonville and Caius College, Cambridge

John Venn is considered by many to be one of the originators of modern symbolic logic. Venn received his degree in mathematics from the University at Cambridge at the age of twenty-three. He was then elected a fellow of the college and held this fellowship until his death, some 66 years later. Two years after receiving his degree, Venn accepted a teaching position at Cambridge: college lecturer in moral sciences.

During the latter half of the nineteenth century, the study of logic experienced a rebirth in England. Mathematicians were attempting to symbolize and quantify the central concepts of logical thought. Consequently, Venn chose to focus on the study of logic during his tenure at Cambridge. In addition, he investigated the field of probability and published *The Logic of Chance,* his first major work, in 1866.

Venn was well read in the works of his predecessors, including the noted logicians Augustus De Morgan, George Boole, and Charles Dodgson (a.k.a. Lewis Carroll). Boole's pioneering work on the marriage of logic and algebra proved to be a strong influence on Venn; in fact, Venn used the type of diagram that now bears his name in an 1876 paper in which he examined Boole's system of symbolic logic.

Venn was not the first scholar to use the diagrams that now bear his name. Gottfried Leibniz, Leonhard Euler, and others utilized similar diagrams years before Venn did. Examining each author's diagrams, Venn was critical of their lack of uniformity. He developed a consistent, systematic explanation of the general use of geometrical figures in the analysis of logical arguments. Today, these geometrical figures are known by his name and are used extensively in elementary set theory and logic.

Venn's writings were held in high esteem. His textbooks, *Symbolic Logic* (1881) and *The Principles of Empirical Logic* (1889), were used during the late nineteenth and early twentieth centuries. In addition to his works on logic and probability, Venn also conducted much research into historical records, especially those of his college and those of his family.

THE

LOGIC OF CHANCE

AN ESSAY

ON THE FOUNDATIONS AND PROVINCE OF
THE THEORY OF PROBABILITY,

WITH ESPECIAL REFERENCE TO ITS APPLICATION TO

MORAL AND SOCIAL SCIENCE.

BY

JOHN VENN, M.A.

FELLOW OF GONVILLE AND CAIUS COLLEGE, CAMBRIDGE.

"So careful of the type she seems
So careless of the single life."

London and Cambridge:
MACMILLAN AND CO.
1866

Brown University Library

Set theory and the cardinal numbers of sets are used extensively in the study of probability. Although he was a professor of logic, Venn investigated the foundations and applications of theoretical probability. Venn's first major work, *The Logic of Chance,* exhibited the diversity of his academic interests.

EXAMPLE 7   How many letters in the alphabet precede the letter w?

Solution   Rather than counting all the letters that precede w, we will take a shortcut by counting all the letters that do *not* precede w. Let L = {letters | the letter precedes w}. Therefore, L' = {letter | the letter does not precede w}. Now L' = {w, x, y, z} and

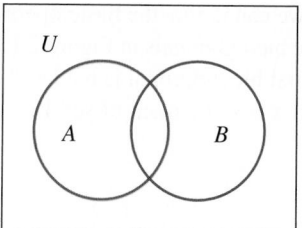

Figure 2.8
Two overlapping circles.

$n(L') = 4$; therefore, we have

$$n(L) = n(U) - n(L') \quad \text{Cardinal Number Formula for the Complement of a Set}$$
$$= 26 - 4$$
$$= 22$$

There are 22 letters preceding the letter w. ∎

Shading Venn Diagrams

In an effort to visualize the results of operations on sets, it may be necessary to shade specific regions of a Venn diagram. The following example shows a systematic method for shading the intersection or union of any two sets.

ThomsonNOW™
for Liberal Arts Mathematics
www.thomsonedu.com/1pass.

EXAMPLE 8   On a Venn diagram, shade in the region corresponding to the indicated set.

a. $A \cap B'$ **b.** $A \cup B'$

Solution

a. First, draw and label two overlapping circles as shown in Figure 2.8. The two "components" of the operation $A \cap B'$ are "A" and "B'." Shade each of these components in contrasting ways; shade one of them, say A, with horizontal lines, and the other with vertical lines as in Figure 2.9. Be sure to include a legend, or key, identifying each type of shading.

To be in the intersection of two sets, an element must be in *both* sets at the same time. Therefore, the intersection of A and B' is the region that is shaded in *both* directions (horizontal and vertical) at the same time. A final diagram depicting $A \cap B'$ is shown in Figure 2.10.

b. Refer to Figure 2.9. To be in the union of two sets, an element must be in *at least one* of the sets. Therefore, the union of A and B' consists of all regions that are shaded in *any* direction whatsoever (horizontal or vertical or both). A final diagram depicting $A \cup B'$ is shown in Figure 2.11.

Figure 2.9

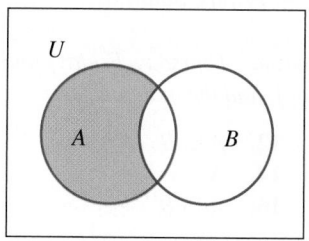

$A \cap B'$

Figure 2.10

Set Theory and Logic

If you have read Chapter 1, you have probably noticed that set theory and logic have many similarities. For instance, the union symbol $\cup$ and the disjunction symbol $\vee$ have the same meaning, but they are used in different circumstances; $\cup$ goes between sets, while $\vee$ goes between logical expressions. The $\cup$ and $\vee$ symbols are similar in appearance because their usages are similar. A comparison of the terms and symbols used in set theory and logic is given in Figure 2.12.

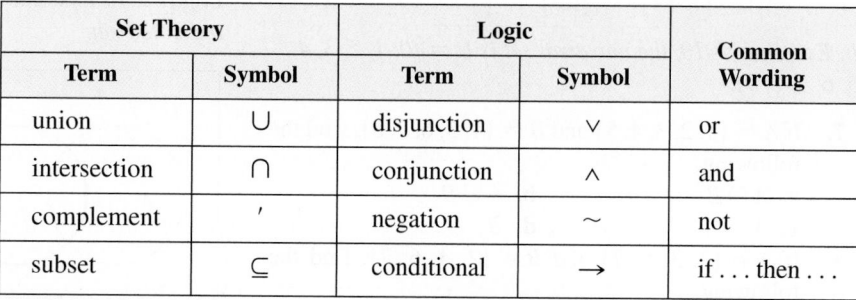

Set Theory		Logic		Common Wording
Term	**Symbol**	**Term**	**Symbol**	
union	$\cup$	disjunction	$\vee$	or
intersection	$\cap$	conjunction	$\wedge$	and
complement	$'$	negation	$\sim$	not
subset	$\subseteq$	conditional	$\rightarrow$	if . . . then . . .

Figure 2.12
Comparison of terms and symbols used in set theory and logic

$A \cup B'$

Figure 2.11

Applying the concepts and symbols of Chapter 1, we can define the basic operations of set theory in terms of logical biconditionals. The biconditionals in Figure 2.13 are tautologies (expressions that are always true); the first biconditional is read as "x is an element of the union of sets A and B if and only if x is an element of set A or x is an element of set B."

Basic Operations in Set Theory	Logical Biconditional
union	$[x \in (A \cup B)] \leftrightarrow [x \in A \vee x \in B]$
intersection	$[x \in (A \cap B)] \leftrightarrow [x \in A \wedge x \in B]$
complement	$(x \in A') \leftrightarrow \sim (x \in A)$
subset	$(A \subseteq B) \leftrightarrow (x \in A \rightarrow x \in B)$

Figure 2.13
Set theory operations as logical biconditionals

2.1   Exercises

1. State whether the given set is well defined.
   a. the set of all black automobiles
   b. the set of all inexpensive automobiles
   c. the set of all prime numbers
   d. the set of all large numbers
2. Suppose $A = \{2, 5, 7, 9, 13, 25, 26\}$.
   a. Find $n(A)$
   b. True or false: $7 \in A$
   c. True or false: $9 \notin A$
   d. True or false: $20 \notin A$

In Exercises 3–6, list all subsets of the given set. Identify which subsets are proper and which are improper.

3. $B = \{Lennon, McCartney\}$
4. $N = \{0\}$
5. $S = \{yes, no, undecided\}$
6. $M = \{classical, country, jazz, rock\}$

In Exercises 7–10, the universal set is $U = \{0, 1, 2, 3, 4, 5, 6, 7, 8, 9\}$.

7. If $A = \{1, 2, 3, 4, 5\}$ and $B = \{4, 5, 6, 7, 8\}$, find the following.
   a. $A \cap B$ b. $A \cup B$
   c. A' d. B'
8. If $A = \{2, 3, 5, 7\}$ and $B = \{2, 4, 6, 7\}$, find the following.
   a. $A \cap B$ b. $A \cup B$
   c. A' d. B'

9. If $A = \{1, 3, 5, 7, 9\}$ and $B = \{0, 2, 4, 6, 8\}$, find the following.
   a. $A \cap B$ b. $A \cup B$
   c. A' d. B'
10. If $A = \{3, 6, 9\}$ and $B = \{4, 8\}$, find the following.
    a. $A \cap B$ b. $A \cup B$
    c. A' d. B'

In Exercises 11–16, the universal set is $U = \{Monday, Tuesday, Wednesday, Thursday, Friday, Saturday, Sunday\}$. If $A = \{Monday, Tuesday, Wednesday, Thursday, Friday\}$ and $B = \{Friday, Saturday, Sunday\}$, find the indicated set.

11. $A \cap B$ 12. $A \cup B$
13. B' 14. A'
15. $A' \cup B$ 16. $A \cap B'$

In Exercises 17–26, use a Venn diagram like the one in Figure 2.14 to shade in the region corresponding to the indicated set.

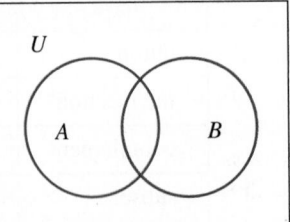

Figure 2.14
Two overlapping circles.

17. $A \cap B$ **18.** $A \cup B$
19. A' **20.** B'
21. $A \cup B'$ **22.** $A' \cup B$
23. $A' \cap B$ **24.** $A \cap B'$
25. $A' \cup B'$ **26.** $A' \cap B'$

27. Suppose $n(U) = 150$, $n(A) = 37$, and $n(B) = 84$.
  a. If $n(A \cup B) = 100$, find $n(A \cap B)$ and draw a Venn diagram illustrating the composition of U.
  b. If $n(A \cup B) = 121$, find $n(A \cap B)$ and draw a Venn diagram illustrating the composition of U.

28. Suppose $n(U) = w$, $n(A) = x$, $n(B) = y$, and $n(A \cup B) = z$.
  a. Why must x be less than or equal to z?
  b. If $A \neq U$ and $B \neq U$, fill in the blank with the most appropriate symbol: $<$, $>$, $\leq$, or $\geq$.
    w____z, w____y, y____z, x____w
  c. Find $n(A \cap B)$ and draw a Venn diagram illustrating the composition of U.

29. In a recent transportation survey, 500 high school seniors were asked to check the appropriate box or boxes on the following form:

☐ I own an automobile.
☐ I own a motorcycle.

The results were tabulated as follows: 91 checked the automobile box, 123 checked the motorcycle box, and 29 checked both boxes.
  a. Draw a Venn diagram illustrating the results of the survey.
  b. What percent of these students own an automobile or a motorcycle?

30. In a recent market research survey, 500 married couples were asked to check the appropriate box or boxes on the following form:

☐ We own a VCR.
☐ We own a microwave oven.

The results were tabulated as follows: 248 checked the VCR box, 314 checked the microwave oven box, and 166 checked both boxes.
  a. Draw a Venn diagram illustrating the results of the survey.
  b. What percent of these couples own a VCR or a microwave oven?

31. In a recent socioeconomic survey, 750 married women were asked to check the appropriate box or boxes on the following form:

☐ I have a career.
☐ I have a child.

The results were tabulated as follows: 529 checked the child box, 213 checked the career box, and 143 were blank (no boxes were checked).
  a. Draw a Venn diagram illustrating the results of the survey.
  b. What percent of these women had both a child and a career?

32. In a recent health survey, 750 single men in their twenties were asked to check the appropriate box or boxes on the following form:

☐ I am a member of a private gym.
☐ I am a vegetarian.

The results were tabulated as follows: 374 checked the gym box, 92 checked the vegetarian box, and 332 were blank (no boxes were checked).
  a. Draw a Venn diagram illustrating the results of the survey.
  b. What percent of these men were both members of a private gym and vegetarians?

For Exercises 33–36, let

$U = \{x \mid x$ is the name of one of the states in the United States$\}$
$A = \{x \mid x \in U$ and x begins with the letter A$\}$
$I = \{x \mid x \in U$ and x begins with the letter I$\}$
$M = \{x \mid x \in U$ and x begins with the letter M$\}$
$N = \{x \mid x \in U$ and x begins with the letter N$\}$
$O = \{x \mid x \in U$ and x begins with the letter O$\}$

33. Find $n(M')$. **34.** Find $n(A \cup N)$.
35. Find $n(I' \cap O')$. **36.** Find $n(M \cap I)$.

For Exercises 37–40, let

$U = \{x \mid x$ is the name of one of the months in a year$\}$
$J = \{x \mid x \in U$ and x begins with the letter J$\}$
$Y = \{x \mid x \in U$ and x ends with the letter Y$\}$
$V = \{x \mid x \in U$ and x begins with a vowel$\}$
$R = \{x \mid x \in U$ and x ends with the letter R$\}$

37. Find $n(R')$.   **38.** Find $n(J \cap V)$.
39. Find $n(J \cup Y)$.   **40.** Find $n(V \cap R)$.

In Exercises 41–50, determine how many cards, in an ordinary deck of 52, fit the description. (If you are unfamiliar with playing cards, see the end of Section 3.1 for a description of a standard deck.)

41. spades or aces   **42.** clubs or twos
43. face cards or black   **44.** face cards or diamonds
45. face cards and black   **46.** face cards and diamonds
47. aces or eights   **48.** threes or sixes
49. aces and eights   **50.** threes and sixes
51. Suppose $A = \{1, 2, 3\}$ and $B = \{1, 2, 3, 4, 5, 6\}$.
  a. Find $A \cap B$.
  b. Find $A \cup B$.
  c. In general, if $E \cap F = E$, what must be true concerning sets E and F?
  d. In general, if $E \cup F = F$, what must be true concerning sets E and F?
52. Fill in the blank, and give an example to support your answer.
  a. If $A \subset B$, then $A \cap B =$ _____.
  b. If $A \subset B$, then $A \cup B =$ _____.
53. **a.** List all subsets of $A = \{a\}$. How many subsets does A have?
  b. List all subsets of $A = \{a, b\}$. How many subsets does A have?
  c. List all subsets of $A = \{a, b, c\}$. How many subsets does A have?
  d. List all subsets of $A = \{a, b, c, d\}$. How many subsets does A have?
  e. Is there a relationship between the cardinal number of set A and the number of subsets of set A?
  f. How many subsets does $A = \{a, b, c, d, e, f\}$ have?

  HINT: Use your answer to part (e).

54. Prove the Cardinal Number Formula for the Complement of a Set.

  HINT: Apply the Cardinal Number Formula for the Union of Sets to A and A'.

Answer the following questions using complete sentences and your own words.

CONCEPT QUESTIONS
55. If $A \cap B = \emptyset$, what is the relationship between sets A and B?
56. If $A \cup B = \emptyset$, what is the relationship between sets A and B?

57. Explain the difference between $\{0\}$ and $\emptyset$.
58. Explain the difference between 0 and $\{0\}$.
59. Is it possible to have $A \cap A = \emptyset$?
60. What is the difference between proper and improper subsets?
61. A set can be described by two methods: the roster method and set-builder notation. When is it advantageous to use the roster method? When is it advantageous to use set-builder notation?
62. Translate the following symbolic expressions into English sentences.
  a. $x \in (A \cap B) \leftrightarrow (x \in A \wedge x \in B)$
  b. $(x \in A') \leftrightarrow \sim (x \in A)$
  c. $(A \subseteq B) \leftrightarrow (x \in A \rightarrow x \in B)$

HISTORY QUESTIONS
63. In what academic field was John Venn a professor? Where did he teach?
64. What was one of John Venn's main contributions to the field of logic? What new benefits did it offer?

Many graduate schools require applicants to take the Graduate Record Examination (G.R.E.). This exam is intended to measure verbal, quantitative, and analytical skills that have developed throughout a person's life. Many classes and study guides are available to help people prepare for the exam. The remaining questions are typical of those found in the study guides and on the exam itself.

Exercises 65–69 refer to the following: Two collectors, John and Juneko, are each selecting a group of three posters from a group of seven movie posters: J, K, L, M, N, O, and P. No poster can be in both groups. The selections made by John and Juneko are subject to the following restrictions:
- If K is in John's group, M must be in Juneko's group.
- If N is in John's group, P must be in Juneko's group.
- J and P cannot be in the same group.
- M and O cannot be in the same group.

65. Which of the following pairs of groups selected by John and Juneko conform to the restrictions?

	John	Juneko
a.	J, K, L	M, N, O
b.	J, K, P	L, M, N
c.	K, N, P	J, M, O
d.	L, M, N	K, O, P
e.	M, O, P	J, K, N

66. If N is in John's group, which of the following could not be in Juneko's group?
  a. J   **b.** K   **c.** L   **d.** M   **e.** P

67. If K and N are in John's group, Juneko's group must consist of which of the following?

    a. J, M, and O   **b.** J, O, and P

    c. L, M, and P   **d.** L, O, and P

    e. M, O, and P

68. If J is in Juneko's group, which of the following is true?

    a. K cannot be in John's group.

    b. N cannot be in John's group.

    c. O cannot be in Juneko's group.

    d. P must be in John's group.

    e. P must be in Juneko's group.

69. If K is in John's group, which of the following is true?

    a. J must be in John's group.

    b. O must be in John's group.

    c. L must be in Juneko's group.

    d. N cannot be in John's group.

    e. O cannot be in Juneko's group.

2.2   Applications of Venn Diagrams

As we have seen, Venn diagrams are very useful tools for visualizing the relationships between sets. They can be used to establish general formulas involving set operations and to determine the cardinal numbers of sets. Venn diagrams are particularly useful in survey analysis.

Surveys

Surveys are often used to divide people or objects into categories. Because the categories sometimes overlap, people can fall into more than one category. Venn diagrams and the formulas for cardinal numbers can help researchers organize the data.

EXAMPLE 1   Has the advent of the DVD affected attendance at movie theaters? To study this question, Professor Redrum's film class conducted a survey of people's movie-watching habits. He had his students ask hundreds of people between the ages of sixteen and forty-five to check the appropriate box or boxes on the following form:

> ☐   I watched a movie in a theater during the past month.
> ☐   I watched a movie on a DVD during the past month.

After the professor had collected the forms and tabulated the results, he told the class that 388 people checked the theater box, 495 checked the DVD box, 281 checked both boxes, and 98 of the forms were blank. Giving the class only this information, Professor Redrum posed the following three questions.

    a. What percent of the people surveyed watched a movie in a theater or on a DVD during the past month?

    b. What percent of the people surveyed watched a movie in a theater only?

    c. What percent of the people surveyed watched a movie on a DVD only?

Solution   **a.** To calculate the desired percentages, we must determine $n(U)$, the total number of people surveyed. This can be accomplished by drawing a Venn diagram. Because the survey divides people into two categories (those who watched a movie in a theater and those who watched a movie on a DVD), we need to define two sets. Let

T = {people | the person watched a movie in a theater}

D = {people | the person watched a movie on a DVD}

Now translate the given survey information into the symbols for the sets and attach their given cardinal numbers: $n(T) = 388$, $n(D) = 495$, and $n(T \cap D) = 281$.

Our first goal is to find $n(U)$. To do so, we will fill in the cardinal numbers of all regions of a Venn diagram consisting of two overlapping circles (because we are dealing with two sets). The intersection of T and D consists of 281 people, so we draw two overlapping circles and fill in 281 as the number of elements in common (see Figure 2.15).

Because we were given $n(T) = 388$ and know that $n(T \cap D) = 281$, the difference $388 - 281 = 107$ tells us that 107 people watched a movie in a theater but did not watch a movie on a DVD. We fill in 107 as the number of people who watched a movie only in a theater (see Figure 2.16).

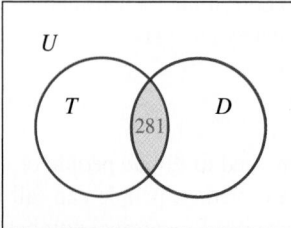

Figure 2.15
$n(T \cap D) = 281$.

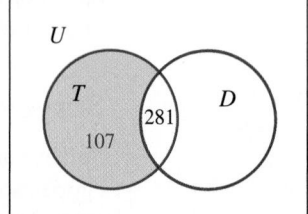

Figure 2.16
$n(T \cap D') = 107$.

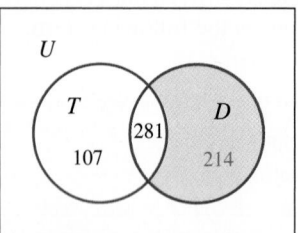

Figure 2.17
$n(D \cap T') = 214$

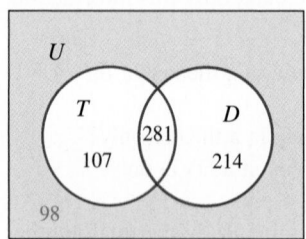

Figure 2.18
Completed Venn diagram.

Because $n(D) = 495$, the difference $495 - 281 = 214$ tells us that 214 people watched a movie on a DVD but not in a theater. We fill in 214 as the number of people who watched a movie only on a DVD (see Figure 2.17).

The only region remaining to be filled in is the region outside both circles. This region represents people who didn't watch a movie in a theater or on a DVD and is symbolized by $(T \cup D)'$. Because 98 people didn't check either box on the form, $n[(T \cup D)'] = 98$ (see Figure 2.18).

After we have filled in the Venn diagram with all the cardinal numbers, we readily see that $n(U) = 98 + 107 + 281 + 214 = 700$. Therefore, 700 people were in the survey.

To determine what *percent* of the people surveyed watched a movie in a theater *or* on a DVD during the past month, simply divide $n(T \cup D)$ by $n(U)$:

$$\frac{n(T \cup D)}{n(U)} = \frac{107 + 281 + 214}{700}$$

$$= \frac{602}{700}$$

$$= 0.86$$

Therefore, exactly 86% of the people surveyed watched a movie in a theater or on a DVD during the past month.

b. To find what *percent* of the people surveyed watched a movie in a theater only, divide 107 (the number of people who watched a movie in a theater only) by $n(U)$:

$$\frac{107}{700} = 0.152857142\ldots$$

Approximately 15.3% of the people surveyed watched a movie in a theater only.

c. Because 214 people watched a movie on DVD only, $214/700 = 0.305714285\ldots$, or approximately 30.6%, of the people surveyed watched a movie on DVD only. ∎

When you solve a cardinal number problem (a problem that asks, "How many?" or "What percent?") involving a universal set that is divided into various categories (for instance, a survey), use the following general steps.

Solving a Cardinal Number Problem

A cardinal number problem is a problem in which you are asked, "How many?" or "What percent?"

1. Define a set for each category in the universal set. If a category and its negation are both mentioned, define one set A and utilize its complement A'.
2. Draw a Venn diagram with as many overlapping circles as the number of sets you have defined.
3. Write down all the given cardinal numbers corresponding to the various given sets.
4. Starting with the innermost overlap, fill in each region of the Venn diagram with its cardinal number.

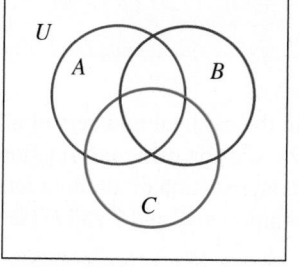

Figure 2.19
Three overlapping circles.

When we are working with three sets, we must account for all possible intersections of the sets. Hence, in such cases, we will use the Venn diagram shown in Figure 2.19.

EXAMPLE 2   A consumer survey was conducted to examine patterns in ownership of personal computers, cellular telephones, and DVD players. The following data were obtained: 213 people had personal computers, 294 had cellular telephones, 337 had DVD players, 109 had all three, 64 had none, 198 had cell phones and DVD players, 382 had cell phones or computers, and 61 had computers and DVD players but no cell phones.

a. What percent of the people surveyed owned a computer but no DVD player or cell phone?

b. What percent of the people surveyed owned a DVD player but no computer or cell phone?

Solution   **a.** To calculate the desired percentages, we must determine $n(U)$, the total number of people surveyed. This can be accomplished by drawing a Venn diagram. Because the survey divides people into three categories (those who own a computer, those who own a cell phone, and those who own a DVD player), we need to define three sets. Let

C = {people | the person owns a personal computer}

T = {people | the person owns a cellular telephone}

D = {people | the person owns a DVD player}

Now translate the given survey information into the symbols for the sets and attach their given cardinal numbers:

213 people had computers —————————→	$n(C) = 213$
294 had cellular telephones —————————→	$n(T) = 294$
337 had DVD players —————————→	$n(D) = 337$
109 had all three —————————→ (C and T and D)	$n(C \cap T \cap D) = 109$
64 had none —————————→ (not C and not T and not D)	$n(C' \cap T' \cap D') = 64$
198 had cell phones and DVD players —————→ (T and D)	$n(T \cap D) = 198$
382 had cell phones or computers —————→ (T or C)	$n(T \cup C) = 382$
61 had computers and DVD players but no cell phones ——→ (C and D and not T)	$n(C \cap D \cap T') = 61$

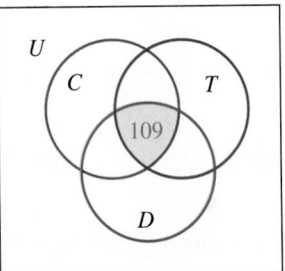

Figure 2.20
$n(C \cap T \cap D) = 109$.

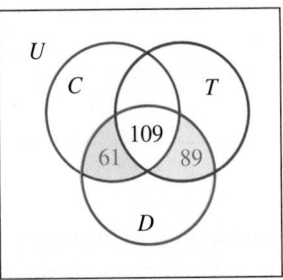

Figure 2.21
Determining cardinal numbers in a Venn diagram.

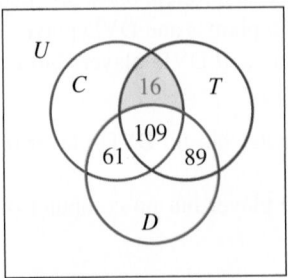

Figure 2.22
Determining cardinal numbers in a Venn diagram.

Our first goal is to find $n(U)$. To do so, we will fill in the cardinal numbers of all regions of a Venn diagram like that in Figure 2.19. We start by using information concerning membership in all three sets. Because the intersection of all three sets consists of 109 people, we fill in 109 in the region common to C and T and D (see Figure 2.20).

Next, we utilize any information concerning membership in two of the three sets. Because $n(T \cap D) = 198$, a total of 198 people are common to both T and D; some are in C, and some are not in C. Of these 198 people, 109 are in C (see Figure 2.20). Therefore, the difference $198 - 109 = 89$ gives the number not in C. Eighty-nine people are in T and D and *not* in C; that is, $n(T \cap D \cap C') = 89$. Concerning membership in the two sets C and D, we are given $n(C \cap D \cap T') = 61$. Therefore, we know that 61 people are in C and D and not in T (see Figure 2.21).

We are given $n(T \cup C) = 382$. From this number, we can calculate $n(T \cap C)$ by using the Cardinal Number Formula for the Union of Sets:

$$n(T \cup C) = n(T) + n(C) - n(T \cap C)$$
$$382 = 294 + 213 - n(T \cap C)$$
$$n(T \cap C) = 125$$

Therefore, a total of 125 people are in T and C; some are in D, and some are not in D. Of these 125 people, 109 are in D (see Figure 2.20). Therefore, the difference $125 - 109 = 16$ gives the number not in D. Sixteen people are in T and C and *not* in D; that is, $n(C \cap T \cap D') = 16$ (see Figure 2.22).

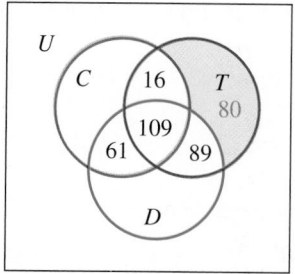

Figure 2.23
Determining cardinal numbers in a Venn diagram.

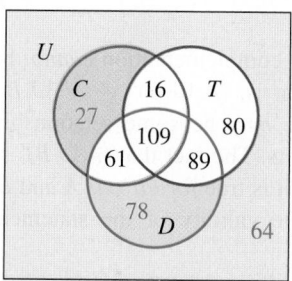

Figure 2.24
A completed Venn diagram.

Knowing that a total of 294 people are in T (given $n(T) = 294$), we are now able to fill in the last region of T. The missing region (people in T only) has $294 - 109 - 89 - 16 = 80$ members; $n(T \cap C' \cap D') = 80$ (see Figure 2.23).

In a similar manner, we subtract the known pieces of C from $n(C) = 213$, which is given, and obtain $213 - 61 - 109 - 16 = 27$; therefore, 27 people are in C only. Likewise, to find the last region of D, we use $n(D) = 337$ (given) and obtain $337 - 89 - 109 - 61 = 78$; therefore, 78 people are in D only. Finally, the 64 people who own none of the items are placed "outside" the three circles (see Figure 2.24).

By adding up the cardinal numbers of all the regions in Figure 2.24, we find that the total number of people in the survey is 524; that is, $n(U) = 524$.

Now, to determine what *percent* of the people surveyed owned only a computer (no DVD player and no cell phone), we simply divide $n(C \cap D' \cap T')$ by $n(U)$:

$$\frac{n(C \cap D' \cap T')}{n(U)} = \frac{27}{524}$$

$$= 0.051526717\ldots$$

Approximately 5.2% of the people surveyed owned a computer and did not own a DVD player or a cellular telephone.

b. To determine what *percent* of the people surveyed owned only a DVD player (no computer and no cell phone), we divide $n(D \cap C' \cap T')$ by $n(U)$:

$$\frac{n(D \cap C' \cap T')}{n(U)} = \frac{78}{524}$$

$$= 0.148854961\ldots$$

Approximately 14.9% of the people surveyed owned a DVD player and did not own a computer or a cellular telephone. ∎

De Morgan's Laws

One of the basic properties of algebra is the distributive property:

$$a(b + c) = ab + ac$$

Given $a(b + c)$, the operation outside the parentheses can be distributed over the operation inside the parentheses. It makes no difference whether you add b and c first and then multiply the sum by a or first multiply each pair, a and b, a and c, and then add their products; the same result is obtained. Is there a similar property for the complement, union, and intersection of sets?

ThomsonNOW™
for Liberal Arts Mathematics
www.thomsonedu.com/1pass.

EXAMPLE 3   Suppose $U = \{1, 2, 3, 4, 5\}$, $A = \{1, 2, 3\}$, and $B = \{2, 3, 4\}$.

a. For the given sets, does $(A \cup B)' = A' \cup B'$?
b. For the given sets, does $(A \cup B)' = A' \cap B'$?

Solution   **a.** To find $(A \cup B)'$, we must first find $A \cup B$:

$$A \cup B = \{1, 2, 3\} \cup \{2, 3, 4\}$$
$$= \{1, 2, 3, 4\}$$

The complement of $A \cup B$ (relative to the given universal set U) is

$$(A \cup B)' = \{5\}$$

To find $A' \cup B'$, we must first find A' and B':

$$A' = \{4, 5\} \quad \text{and} \quad B' = \{1, 5\}$$

The union of A' and B' is

$$A' \cup B' = \{4, 5\} \cup \{1, 5\}$$
$$= \{1, 4, 5\}$$

Now, $\{5\} \neq \{1, 4, 5\}$; therefore, $(A \cup B)' \neq A' \cup B'$.

b. We find $(A \cup B)'$ as in part (a): $(A \cup B)' = \{5\}$. Now,

$$A' \cap B' = \{4, 5\} \cap \{1, 5\}$$
$$= \{5\}$$

For the given sets, $(A \cup B)' = A' \cap B'$. ∎

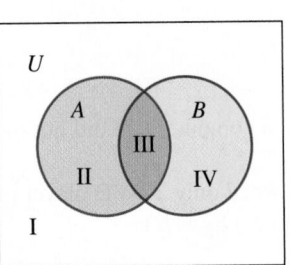

Figure 2.25
Four regions in a universal set U.

Part (a) of Example 3 shows that the operation of complementation *cannot* be explicitly distributed over the operation of union; that is, $(A \cup B)' \neq A' \cup B'$. However, part (b) of the example implies that there *may* be some relationship between the complement, union, and intersection of sets. The fact that $(A \cup B)' = A' \cap B'$ *for the given sets* A and B does not mean that it is true *for all sets* A and B. We will use a general Venn diagram to examine the validity of the statement $(A \cup B)' = A' \cap B'$.

When we draw two overlapping circles within a universal set, four regions are formed. Every element of the universal set U is in exactly one of the following regions, as shown in Figure 2.25:

I   in neither A nor B

II   in A and not in B

III   in both A and B

IV   in B and not in A

The set $A \cup B$ consists of all elements in regions II, III, and IV. Therefore, the complement $(A \cup B)'$ consists of all elements in region I. A' consists of all elements in regions I and IV, and B' consists of the elements in regions I and II. Therefore, the elements common to both A' and B' are those in region I; that is, the set $A' \cap B'$ consists of all elements in region I. Since $(A \cup B)'$ and $A' \cap B'$ contain exactly the same elements (those in region I), the sets are equal; that is, $(A \cup B)' = A' \cap B'$ is true for all sets A and B.

The relationship $(A \cup B)' = A' \cap B'$ is known as one of **De Morgan's Laws.** Simply stated, "the complement of a union is the intersection of the complements." In a similar manner, it can be shown that $(A \cap B)' = A' \cup B'$ (see Exercise 31).

De Morgan's Laws

For any sets A and B,

$$(A \cup B)' = A' \cap B'$$

That is, the complement of a union is the intersection of the complements. Also,

$$(A \cap B)' = A' \cup B'$$

That is, the complement of an intersection is the union of the complements.

Historical Note   Augustus De Morgan, 1806–1871

© Bettmann/Corbis

Being born blind in one eye did not stop Augustus De Morgan from becoming a well-read philosopher, historian, logician, and mathematician. De Morgan was born in Madras, India, where his father was working for the East India Company. On moving to England, De Morgan was educated at Cambridge, and at the age of twenty-two, he became the first professor of mathematics at the newly opened University of London (later renamed University College).

De Morgan viewed all of mathematics as an abstract study of symbols and of systems of operations applied to these symbols. While studying the ramifications of symbolic logic, De Morgan formulated the general properties of complementation that now bear his name. Not limited to symbolic logic, De Morgan's many works include books and papers on the foundations of algebra, differential calculus, and probability. He was known to be a jovial person who was fond of puzzles, and his witty and amusing book *A Budget of Paradoxes* still entertains readers today. Besides his accomplishments in the academic arena, De Morgan was an expert flutist, spoke five languages, and thoroughly enjoyed big-city life.

Knowing of his interest in probability, an actuary (someone who studies life expectancies and determines payments of premiums for insurance companies) once asked De Morgan a question concerning the probability that a certain group of people would be alive at a certain time. In his response, De Morgan employed a formula containing the number π. In amazement, the actuary responded, "That must surely be a delusion! What can a circle have to do with the number of people alive at a certain time?" De Morgan replied that π has numerous applications and occurrences in many diverse areas of mathematics. Because it was first defined and used in geometry, people are conditioned to accept the mysterious number only in reference to a circle. However, in the history of mathematics, if probability had been systematically studied before geometry and circles, our present-day interpretation of the number π would be entirely different. In addition to his accomplishments in logic and higher-level mathematics, De Morgan introduced a convention with which we are all familiar: In a paper written in 1845, he suggested the use of a slanted line to represent a fraction, such as 1/2 or 3/4.

De Morgan was a staunch defender of academic freedom and religious tolerance. While a student at Cambridge, his application for a fellowship was refused because he would not take and sign a theological oath. Later in life, he resigned his professorship as a protest against religious bias. (University College gave preferential treatment to members of the Church of England when textbooks were selected and did not have an open policy on religious philosophy.) Augustus De Morgan was a man who was unafraid to take a stand and make personal sacrifices when it came to principles he believed in.

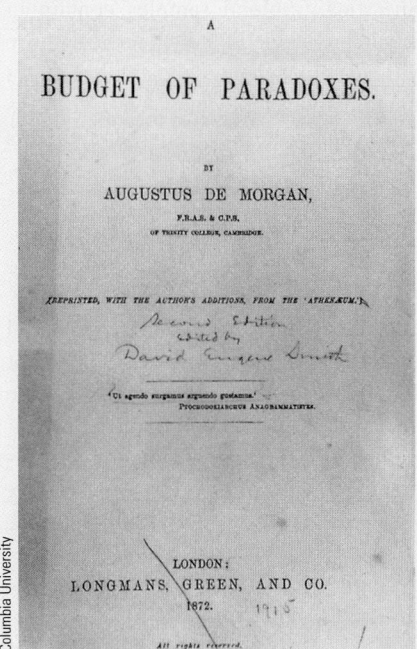

Columbia University

Gematria is a mystic pseudoscience in which numbers are substituted for the letters in a name. De Morgan's book *A Budget of Paradoxes* contains several gematria puzzles, such as, "Mr. Davis Thom found a young gentleman of the name of St. Claire busy at the Beast number: he forthwith added the letters in στκλαιρε (the Greek spelling of St. Claire) and found 666." (Verify this by using the Greek numeral system.)

Set Theory in the Real World: Blood Types

Human blood types are a classic example of set theory. As you may know, there are four categories (or sets) of blood types: A, B, AB, and O. Knowing someone's blood type is extremely important in case a blood transfusion is required; if blood of two different types is combined, the blood cells may begin to clump together, with potentially fatal consequences! (Do you know your blood type?)

What exactly are "blood types"? In the early 1900s, the Austrian scientist Karl Landsteiner observed the presence (or absence) of two distinct chemical molecules on the surface of all red blood cells in numerous samples of human blood. Consequently, he labeled one molecule "A" and the other "B." The presence or absence of these specific molecules is the basis of the universal classification of blood types. Specifically, blood samples containing only the A molecule are labeled type A, whereas those containing only the B molecule are labeled type B. If a blood sample contains both molecules (A and B) it is labeled type AB; and if neither is present, the blood is typed as O. The presence (or absence) of these molecules can be depicted in a standard Venn diagram as shown in Figure 2.26. In the notation of set operations, type A blood is denoted $A \cap B'$, type B is $B \cap A'$, type AB is $A \cap B$, and type O is $A' \cap B'$.

If a specific blood sample is mixed with blood containing a blood molecule (A or B) that it does not already have, the presence of the foreign molecule may cause the mixture of blood to clump. For example, type A blood cannot be mixed with any blood containing the B molecule (type B or type AB). Therefore, a person with type A blood can only receive a transfusion of type A or type O blood. Consequently, a person with type AB blood may receive a transfusion of any blood type; type AB is referred to as the "universal receiver." Because type O blood contains neither the A nor the B molecule, all blood types are compatible with type O blood; type O is referred to as the "universal donor."

It is not uncommon for scientists to study rhesus monkeys in an effort to learn more about human physiology. In so doing, a certain blood protein was discovered in rhesus monkeys. Subsequently, scientists found that the blood of some people contained this protein, whereas the blood of others did not. The presence, or absence, of this protein in human blood is referred to as the *Rh factor*; blood containing the protein is labeled "Rh+", whereas "Rh−" indicates the absence of the protein. The Rh factor of human blood is especially important for expectant mothers; a fetus can develop problems if its parents have opposite Rh factors.

When a person's blood is typed, the designation includes both the regular blood type and the Rh factor. For instance, type AB− indicates the presence of both the A and B molecules (type AB), along with the absence of the rhesus protein; type O+ indicates the absence of both the A and B molecules (type O), along with the presence of the rhesus protein. Utilizing the Rh factor, there are eight possible blood types as shown in Figure 2.27.

We will investigate the occurrence and compatibility of the various blood types in Example 5 and in Exercises 33–42.

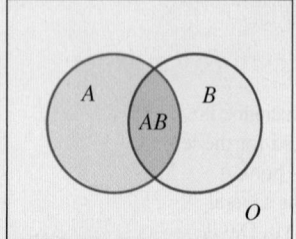

Figure 2.26
Blood types and the presence of the A and B molecules

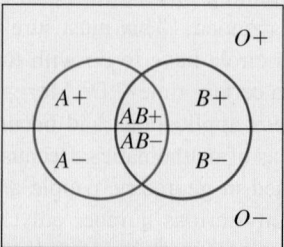

Figure 2.27
Blood types combined with the Rh factor

ThomsonNOW
for Liberal Arts Mathematics
www.thomsonedu.com/1pass.

EXAMPLE 4 Suppose $U = \{0, 1, 2, 3, 4, 5, 6, 7, 8, 9\}$, $A = \{2, 3, 7, 8\}$, and $B = \{0, 4, 5, 7, 8, 9\}$. Use De Morgan's Law to find $(A' \cup B)'$.

Solution The complement of a union is equal to the intersection of the complements; therefore, we have

$$(A' \cup B)' = (A')' \cap B' \qquad \text{De Morgan's Law}$$
$$= A \cap B' \qquad (A')' = A$$
$$= \{2, 3, 7, 8\} \cap \{1, 2, 3, 6\}$$
$$= \{2, 3\}$$

Notice that this problem could be done without using De Morgan's Law, but solving it would then involve finding first A', then $A' \cup B$, and finally $(A' \cup B)'$. This method would involve more work. (Try it!) ∎

EXAMPLE 5 The American Red Cross has compiled a massive database of the occurrence of blood types in the United States. Their data indicate that on average, out of every 100 people in the United States, 44 have the A molecule, 15 have the B molecule, and 45 have neither the A nor the B molecule. What percent of the U.S. population have the following blood types?

a. Type O?   **b.** Type AB?   **c.** Type A?   **d.** Type B?

Solution **a.** First, we define the appropriate sets. Let

$$A = \{\text{Americans} \mid \text{the person has the A molecule}\}$$
$$B = \{\text{Americans} \mid \text{the person has the B molecule}\}$$

We are given the following cardinal numbers: $n(U) = 100$, $n(A) = 44$, $n(B) = 15$, and $n(A' \cap B') = 45$. Referring to Figure 2.26, and given that 45 people (out of 100) have neither the A molecule nor the B molecule, we conclude that 45 of 100 people, or 45%, have type O blood as shown in Figure 2.28.

b. Applying De Morgan's Law to the Cardinal Number Formula for the Complement of a Set, we have the following.

$$n(A \cup B) + n[(A \cup B)'] = n(U) \quad \text{\small Cardinal Number Formula for the Complement of a Set}$$
$$n(A \cup B) + n(A' \cap B') = n(U) \quad \text{\small applying De Morgan's Law}$$
$$n(A \cup B) + 45 = 100 \quad \text{\small substituting known values}$$

Therefore, $n(A \cup B) = 55$.

Now, use the Cardinal Number Formula for the Union of Sets.

$$n(A \cup B) = n(A) + n(B) - n(A \cap B) \quad \text{\small Cardinal Number Formula for the Union of Sets}$$
$$55 = 44 + 15 - n(A \cap B) \quad \text{\small substituting known values}$$
$$n(A \cap B) = 44 + 15 - 55 \quad \text{\small adding } n(A \cap B) \text{ and subtracting 55}$$

Therefore, $n(A \cap B) = 4$. This means that 4 people (of 100) have both the A and the B molecules; that is, 4 of 100 people, or 4%, have type AB blood. See Figure 2.29.

c. Knowing that a total of 44 people have the A molecule, that is, $n(A) = 44$, we subtract $n(A \cap B) = 4$ and conclude that 40 have *only* the A molecule.

Therefore, $n(A \cap B') = 40$. This means that 40 people (of 100) have only the A molecule; that is, 40 of 100 people, or 40%, have type A blood. See Figure 2.30.

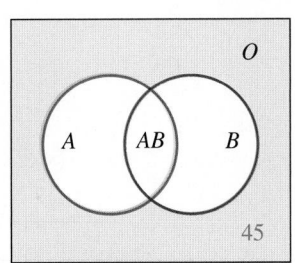

Figure 2.28
Forty-five of 100 (45%) have type O blood.

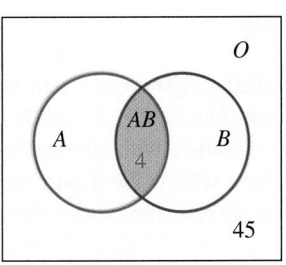

Figure 2.29
Four in 100 (4%) have type AB blood.

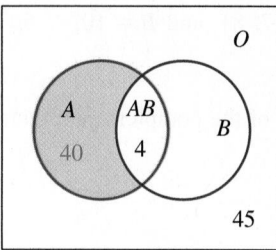

Figure 2.30
Forty in 100 (40%) have type
A blood.

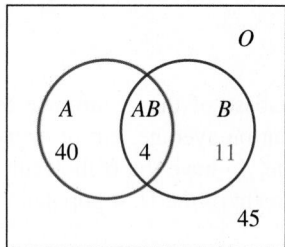

Figure 2.31
Eleven in 100 (11%) have
type B blood.

d. Knowing that a total of 15 people have the B molecule, that is, $n(B) = 15$, we subtract $n(A \cap B) = 4$ and conclude that 11 have *only* the B molecule.

Therefore, $n(B \cap A') = 11$. This means that 11 people (of 100) have only the B molecule; that is, 11 of 100 people, or 11%, have type B blood. See Figure 2.31.

The occurrence of blood types in the United States is summarized in Figure 2.32.

Blood Type	O	A	B	AB
Occurrence	45%	40%	11%	4%

Figure 2.32
Occurrence of blood types in the United States ∎

The occurrence of blood types given in Figure 2.32 can be further categorized by including the Rh factor. According to the American Red Cross, out of every 100 people in the United States, blood types and Rh factors occur at the rates shown in Figure 2.33.

Blood Type	O+	O−	A+	A−	B+	B−	AB+	AB−
Occurrence	38	7	34	6	9	2	3	1

Figure 2.33
Blood types per 100 people in the United States
(*Source:* American Red Cross)

2.2 Exercises

1. A survey of 200 people yielded the following information: 94 owned a DVD, 127 owned a microwave oven, and 78 owned both. How many people owned the following?
    a. a DVD or a microwave oven
    b. a DVD but not a microwave oven
    c. a microwave oven but not a DVD
    d. neither a DVD nor a microwave oven

2. A survey of 300 workers yielded the following information: 231 belonged to a union, and 195 were Democrats. If 172 of the union members were Democrats, how many workers were in the following situations?
    a. belonged to a union or were Democrats
    b. belonged to a union but were not Democrats
    c. were Democrats but did not belong to a union
    d. neither belonged to a union nor were Democrats

3. The records of 1,492 high school graduates were examined, and the following information was obtained: 1,072 took biology, and 679 took geometry. If 271 of those who took geometry did not take biology, how many graduates took the following?
    a. both classes
    b. at least one of the classes
    c. biology but not geometry
    d. neither class

4. A department store surveyed 428 shoppers, and the following information was obtained: 214 made a purchase, and 299 were satisfied with the service they received. If 52 of those who made a purchase were not satisfied with the service, how many shoppers did the following?
    a. made a purchase and were satisfied with the service
    b. made a purchase or were satisfied with the service

c. were satisfied with the service but did not make a purchase

d. were not satisfied and did not make a purchase

5. In a survey, 674 adults were asked what television programs they had recently watched. The following information was obtained: 226 watched neither the Big Game nor the New Movie, and 289 watched the New Movie. If 183 of those who watched the New Movie did not watch the Big Game, how many of the surveyed adults watched the following?

a. both programs

b. at least one program

c. the Big Game

d. the Big Game but not the New Movie

6. A survey asked 816 college freshmen whether they had been to a movie or eaten in a restaurant during the past week. The following information was obtained: 387 had been to neither a movie nor a restaurant, and 266 had been to a movie. If 92 of those who had been to a movie had not been to a restaurant, how many of the surveyed freshmen had been to the following?

a. both a movie and a restaurant

b. a movie or a restaurant

c. a restaurant

d. a restaurant but not a movie

7. A recent survey of w shoppers (that is, $n(U) = w$) yielded the following information: x shopped at Sears, y shopped at JCPenney's, and z shopped at both. How many people shopped at the following?

a. Sears or JCPenney's

b. only Sears

c. only JCPenney's

d. neither Sears nor JCPenney's

8. A recent transportation survey of w urban commuters (that is, $n(U) = w$) yielded the following information: x rode neither trains nor buses, y rode trains, and z rode only trains. How many people rode the following?

a. trains and buses   b. only buses

c. buses   d. trains or buses

9. A consumer survey was conducted to examine patterns in ownership of personal computers, cellular telephones, and DVDs. The following data were obtained: 313 people had personal computers, 232 had cellular telephones, 269 had DVDs, 69 had all three, 64 had none, 98 had cell phones and DVDs, 57 had cell phones but no computers or DVDs, and 104 had computers and DVDs but no cell phones.

a. What percent of the people surveyed owned a cell phone?

b. What percent of the people surveyed owned only a cell phone?

10. In a recent survey of monetary donations made by college graduates, the following information was obtained: 95 had donated to a political campaign, 76 had donated to assist medical research, 133 had donated to help preserve the environment, 25 had donated to all three, 22 had donated to none of the three, 38 had donated to a political campaign and to medical research, 46 had donated to medical research and to preserve the environment, and 54 had donated to a political campaign and to preserve the environment.

a. What percent of the college graduates donated to none of the three listed causes?

b. What percent of the college graduates donated to exactly one of the three listed causes?

11. Recently, Audioslave, Coldplay, and White Stripes had concert tours in the United States. A large group of college students was surveyed, and the following information was obtained: 825 saw Coldplay, 1,033 saw Audioslave, 1,247 saw White Stripes, 211 saw all three, 514 saw none, 240 saw only White Stripes, 677 saw White Stripes and Audioslave, and 201 saw Audioslave and Coldplay but not White Stripes.

a. What percent of the college students saw at least one of the bands?

b. What percent of the college students saw exactly one of the bands?

12. Dr. Hawk works in an allergy clinic, and his patients have the following allergies: 68 are allergic to dairy products, 93 are allergic to pollen, 91 are allergic to animal fur, 31 are allergic to all three, 29 are allergic only to pollen, 12 are allergic only to dairy products, 40 are allergic to dairy products and pollen.

a. What percent of Dr. Hawk's patients are allergic to animal fur?

b. What percent of Dr. Hawk's patients are allergic only to animal fur?

13. When the members of the Eye and I Photo Club discussed what type of film they had used during the past month, the following information was obtained: 77 used black and white, 24 used only black and white, 65 used color, 18 used only color, 101 used black and white or color, 27 used infrared, 9 used all three types, and 8 didn't use any film during the past month.

a. What percent of the members used only infrared film?

b. What percent of the members used at least two of the types of film?

14. After leaving the polls, many people are asked how they voted. (This is called an *exit poll*.) Concerning Propositions A, B, and C, the following information was obtained: 294 voted yes on A, 90 voted yes only on A, 346 voted yes on B, 166 voted yes only on B, 517 voted yes on A or B, 339 voted yes on C, no one voted yes on all three, and 72 voted no on all three.
  a. What percent of the voters in the exit poll voted no on A?
  b. What percent of the voters voted yes on more than one proposition?

15. In a recent survey, consumers were asked where they did their gift shopping. The following results were obtained: 621 shopped at Macy's, 513 shopped at Emporium, 367 shopped at Nordstrom, 723 shopped at Emporium or Nordstrom, 749 shopped at Macy's or Nordstrom, 776 shopped at Macy's or Emporium, 157 shopped at all three, 96 shopped at neither Macy's nor Emporium nor Nordstrom.
  a. What percent of the consumers shopped at more than one store?
  b. What percent of the consumers shopped exclusively at Nordstrom?

16. A company that specializes in language tutoring lists the following information concerning its English-speaking employees: 23 speak German, 25 speak French, 31 speak Spanish, 43 speak Spanish or French, 38 speak French or German, 46 speak German or Spanish, 8 speak Spanish, French, and German, and 7 office workers and secretaries speak English only.
  a. What percent of the employees speak at least one language other than English?
  b. What percent of the employees speak at least two languages other than English?

17. In a recent survey, people were asked which radio station they listened to on a regular basis. The following results were obtained: 140 listened to WOLD (oldies), 95 listened to WJZZ (jazz), 134 listened to WTLK (talk show news), 235 listened to WOLD or WJZZ, 48 listened to WOLD and WTLK, 208 listened to WTLK or WJZZ, and 25 listened to none.
  a. What percent of people in the survey listened only to WTLK on a regular basis?
  b. What percent of people in the survey did not listen to WTLK on a regular basis?

18. In a recent health insurance survey, employees at a large corporation were asked, "Have you been a patient in a hospital during the past year, and if so, for what reason?" The following results were obtained: 494 had an injury, 774 had an illness, 1,254 had tests, 238 had an injury and an illness and tests, 700 had an illness and tests, 501 had tests and no injury or illness, 956 had an injury or illness, and 1,543 had not been a patient.
  a. What percent of the employees had been patients in a hospital?
  b. What percent of the employees had tests in a hospital?

In Exercises 19 and 20, use a Venn diagram like the one in Figure 2.34.

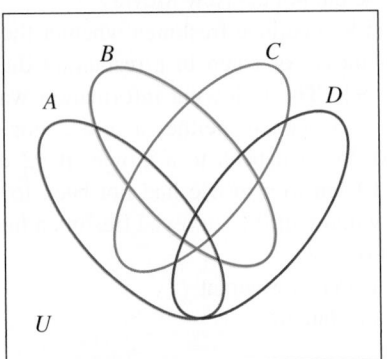

Figure 2.34
Four overlapping reasons.

19. A survey of 136 pet owners yielded the following information: 49 own fish; 55 own a bird; 50 own a cat; 68 own a dog; 2 own all four; 11 own only fish; 14 own only a bird; 10 own fish and a bird; 21 own fish and a cat; 26 own a bird and a dog; 27 own a cat and a dog; 3 own fish, a bird, a cat, and no dog; 1 owns fish, a bird, a dog, and no cat; 9 own fish, a cat, a dog, and no bird; and 10 own a bird, a cat, a dog, and no fish. How many of the surveyed pet owners have no fish, no birds, no cats, and no dogs? (They own other types of pets.)

20. An exit poll of 300 voters yielded the following information regarding voting patterns on Propositions A, B, C, and D: 119 voted yes on A; 163 voted yes on B; 129 voted yes on C; 142 voted yes on D; 37 voted yes on all four; 15 voted yes on A only; 50 voted yes on B only; 59 voted yes on A and B; 70 voted yes on A and C; 82 voted yes on B and D; 93 voted yes on C and D; 10 voted yes on A, B, and C and no on D; 2 voted yes on A, B, and D and no on C; 16 voted yes on A, C, and D and no on B; and 30 voted yes on B, C, and D and no on A. How many of the surveyed voters voted no on all four propositions?

In Exercises 21–24, given the sets U = {0, 1, 2, 3, 4, 5, 6, 7, 8, 9}, A = {0, 2, 4, 5, 9}, and B = {1, 2, 7, 8, 9}, use De Morgan's laws to find the indicated sets.

21. $(A' \cup B)'$ **22.** $(A' \cap B)'$
23. $(A \cap B')'$ **24.** $(A \cup B')'$

In Exercises 25–30, use a Venn diagram like the one in Figure 2.35 to shade in the region corresponding to the indicated set.

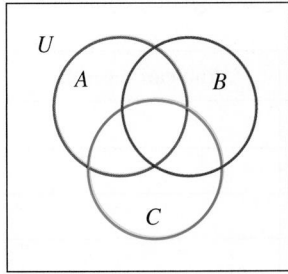

Figure 2.35
Three overlapping circles.

25. $A \cap B \cap C$ **26.** $A \cup B \cup C$
27. $(A \cup B)' \cap C$ **28.** $A \cap (B \cup C)'$
29. $B \cap (A \cup C')$ **30.** $(A' \cup B) \cap C'$
31. Using Venn diagrams, prove De Morgan's Law $(A \cap B)' = A' \cup B'$.
32. Using Venn diagrams, prove $A \cup (B \cap C) = (A \cup B) \cap (A \cup C)$.

Use the data in Figure 2.33 to complete Exercises 33–37. Round off your answers to a tenth of a percent.

33. What percent of all people in the United States have blood that is
    a. Rh positive?   **b.** Rh negative?
34. Of all people in the United States who have type O blood, what percent are
    a. Rh positive?   **b.** Rh negative?
35. Of all people in the United States who have type A blood, what percent are
    a. Rh positive?   **b.** Rh negative?
36. Of all people in the United States who have type B blood, what percent are
    a. Rh positive?   **b.** Rh negative?
37. Of all people in the United States who have type AB blood, what percent are
    a. Rh positive?   **b.** Rh negative?
38. If a person has type A blood, what blood types may they receive in a transfusion?
39. If a person has type B blood, what blood types may they receive in a transfusion?

40. If a person has type AB blood, what blood types may they receive in a transfusion?
41. If a person has type O blood, what blood types may they receive in a transfusion?

Answer the following questions using complete sentences and your own words.

HISTORY QUESTIONS
42. What notation did De Morgan introduce in regard to fractions?
43. Why did De Morgan resign his professorship at University College?

Many graduate schools require applicants to take the Graduate Record Examination (G.R.E.). This exam is intended to measure verbal, quantitative, and analytical skills that have developed throughout a person's life. Many classes and study guides are available to help people prepare for the exam. The remaining questions are typical of those found in the study guides and on the exam itself.

Exercises 44–50 refer to the following: A nonprofit organization's board of directors, composed of four women (Angela, Betty, Carmine, and Delores) and three men (Ed, Frank, and Grant), holds frequent meetings. A meeting can be held at Betty's house, at Delores's house, or at Frank's house.

- Delores cannot attend any meetings at Betty's house.
- Carmine cannot attend any meetings on Tuesday or on Friday.
- Angela cannot attend any meetings at Delores's house.
- Ed can attend only those meetings that Grant also attends.
- Frank can attend only those meetings that both Angela and Carmine attend.

44. If all members of the board are to attend a particular meeting, under which of the following circumstances can it be held?
    a. Monday at Betty's
    b. Tuesday at Frank's
    c. Wednesday at Delores's
    d. Thursday at Frank's
    e. Friday at Betty's
45. Which of the following can be the group that attends a meeting on Wednesday at Betty's?
    a. Angela, Betty, Carmine, Ed, and Frank
    b. Angela, Betty, Ed, Frank, and Grant
    c. Angela, Betty, Carmine, Delores, and Ed

d. Angela, Betty, Delores, Frank, and Grant

e. Angela, Betty, Carmine, Frank, and Grant

46. If Carmine and Angela attend a meeting but Grant is unable to attend, which of the following could be true?

   a. The meeting is held on Tuesday.

   b. The meeting is held on Friday.

   c. The meeting is held at Delores's.

   d. The meeting is held at Frank's.

   e. The meeting is attended by six of the board members.

47. If the meeting is held on Tuesday at Betty's, which of the following pairs can be among the board members who attend?

   a. Angela and Frank   **b.** Ed and Betty

   c. Carmine and Ed   **d.** Frank and Delores

   e. Carmine and Angela

48. If Frank attends a meeting on Thursday that is not held at his house, which of the following must be true?

   a. The group can include, at most, two women.

   b. The meeting is at Betty's house.

   c. Ed is not at the meeting.

   d. Grant is not at the meeting.

   e. Delores is at the meeting.

49. If Grant is unable to attend a meeting on Tuesday at Delores's, what is the largest possible number of board members who can attend?

   a. 1   **b.** 2   **c.** 3

   d. 4   **e.** 5

50. If a meeting is held on Friday, which of the following board members *cannot* attend?

   a. Grant   **b.** Delores   **c.** Ed

   d. Betty   **e.** Frank

WEB PROJECT

51. A person's Rh factor will limit their options regarding the blood types they may receive during a transfusion. Fill in the following chart. How does a person's Rh factor limit their options regarding compatible blood?

If your blood type is:	You can receive:
O+	
O−	
A+	
A−	
B+	
B−	
AB+	
AB−	

Some useful links for this web project are listed on the text web site:

www.thomsonedu.com/mathematics/johnson

2.3   Introduction to Combinatorics

If you went on a shopping spree and bought two pairs of jeans, three shirts, and two pairs of shoes, how many new outfits (consisting of a new pair of jeans, a new shirt, and a new pair of shoes) would you have? A compact disc buyers' club sends you a brochure saying that you can pick any five CDs from a group of 50 of today's hottest sounds for only $1.99. How many different combinations can you choose? Six local bands have volunteered to perform at a benefit concert, and there is some concern over the order in which the bands will perform. How many different lineups are possible? The answers to questions like these can be obtained by listing all the possibilities or by using three shortcut counting methods: the **Fundamental Principle of Counting, combinations,** and **permutations.** Collectively, these methods are known as **combinatorics.** (Incidentally, the answers to the questions above are 12 outfits, 2,118,760 CD combinations, and 720 lineups.) In this section, we consider the first shortcut method.

The Fundamental Principle of Counting

Daily life requires that we make many decisions. For example, we must decide what food items to order from a menu, what items of clothing to put on in the morning, and what options to order when purchasing a new car. Often, we are asked to make a series of decisions: "Do you want soup or salad? What type of dressing? What type of vegetable? What entrée? What beverage? What dessert?" These individual components of a complete meal lead to the question, "Given all the choices of soups, salads, dressings, vegetables, entrées, beverages, and desserts, what is the total number of possible dinner combinations?"

When making a series of decisions, how can you determine the total number of possible selections? One way is to list all the choices for each category and then match them up in all possible ways. To ensure that the choices are matched up in all possible ways, you can construct a **tree diagram.** A tree diagram consists of clusters of line segments, or *branches,* constructed as follows: A cluster of branches is drawn for each decision to be made such that the number of branches in each cluster equals the number of choices for the decision. For instance, if you must make two decisions and there are two choices for decision 1 and three choices for decision 2, the tree diagram would be similar to the one shown in Figure 2.36.

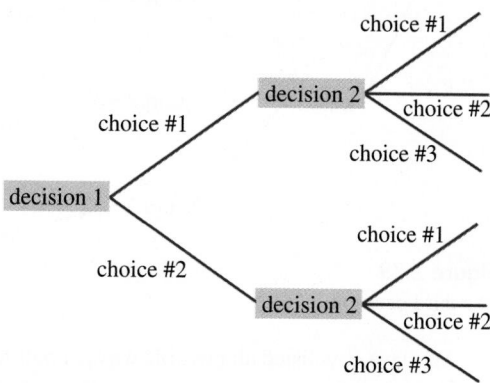

Figure 2.36
A tree diagram.

Although this method can be applied to all problems, it is very time consuming and impractical when you are dealing with a series of many decisions, each of which contains numerous choices. Instead of actually listing all possibilities via a tree diagram, using a shortcut method might be desirable. The following example gives a clue to finding such a shortcut.

EXAMPLE 1   If you buy two pairs of jeans, three shirts, and two pairs of shoes, how many new outfits (consisting of a new pair of jeans, a new shirt, and a new pair of shoes) would you have?

Solution   Because there are three categories, selecting an outfit requires a series of three decisions: You must select one pair of jeans, one shirt, and one pair of shoes. We will make our three decisions in the following order: jeans, shirt, and shoes. (The order in which the decisions are made does not affect the overall outfit.)

Our first decision (jeans) has two choices (jeans 1 or jeans 2); our tree starts with two branches, as in Figure 2.37.

Our second decision is to select a shirt, for which there are three choices. At each pair of jeans on the tree, we draw a cluster of three branches, one for each shirt, as in Figure 2.38.

Our third decision is to select a pair of shoes, for which there are two choices. At each shirt on the tree, we draw a cluster of two branches, one for each pair of shoes, as in Figure 2.39.

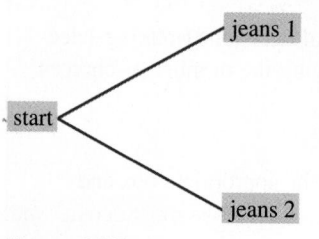

Figure 2.37
The first decision.

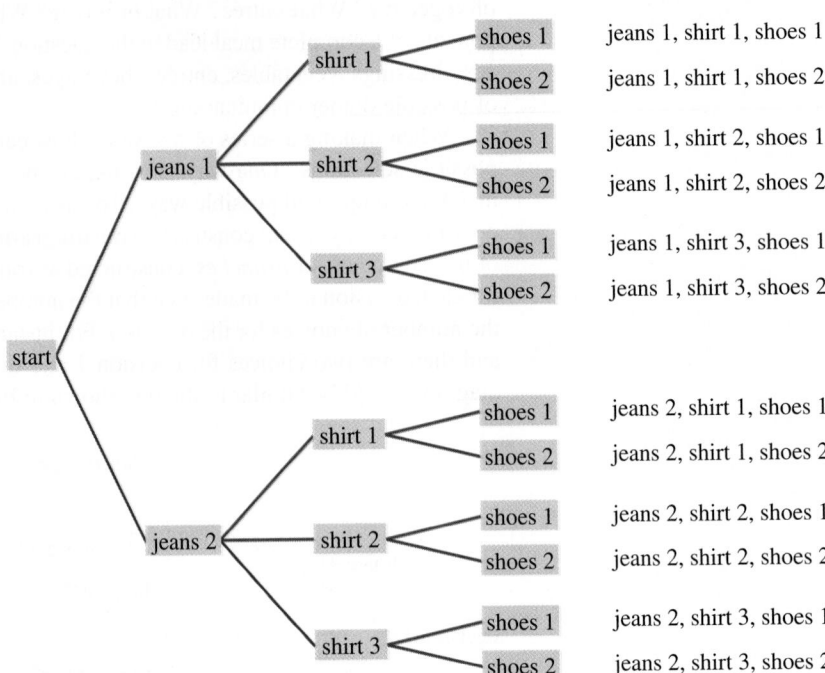

Figure 2.39
The third decision.

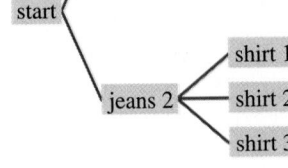

Figure 2.38
The second decision.

We have now listed all possible ways of putting together a new outfit; twelve outfits can be formed from two pairs of jeans, three shirts, and two pairs of shoes. ■

Referring to Example 1, note that each time a decision had to be made, the number of branches on the tree diagram was *multiplied* by a factor equal to the number of choices for the decision. Therefore, the total number of outfits could have been obtained by *multiplying* the number of choices for each decision:

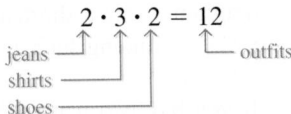

The generalization of this process of multiplication is called the Fundamental Principle of Counting.

The Fundamental Principle of Counting

The total number of possible outcomes of a series of decisions (making selections from various categories) is found by multiplying the number of choices for each decision (or category) as follows:

1. Draw a box for each decision.
2. Enter the number of choices for each decision in the appropriate box and multiply.

Thomson NOW™
for Liberal Arts Mathematics
www.thomsonedu.com/1pass.

EXAMPLE 2 A serial number consists of two consonants followed by three nonzero digits followed by a vowel (A, E, I, O, U): for example, "ST423E" and "DD666E." Determine how many serial numbers are possible given the following conditions.

a. Letters and digits cannot be repeated in the same serial number.
b. Letters and digits can be repeated in the same serial number.

Solution **a.** Because the serial number has six symbols, we must make six decisions. Consequently, we must draw six boxes:

□ □ □ □ □ □

There are 21 different choices for the first consonant. Because the letters cannot be repeated, there are only 20 choices for the second consonant. Similarly, there are nine different choices for the first nonzero digit, eight choices for the second, and seven choices for the third. There are five different vowels, so the total number of possible serial numbers is

$$\boxed{21} \times \boxed{20} \times \boxed{9} \times \boxed{8} \times \boxed{7} \times \boxed{5} = 1{,}058{,}400$$

consonants   nonzero digits   vowel

There are 1,058,400 possible serial numbers when the letters and digits cannot be repeated within a serial number.

b. Because letters and digits can be repeated, the number of choices does not decrease by one each time as in part (a). Therefore, the total number of possibilities is

$$\boxed{21} \times \boxed{21} \times \boxed{9} \times \boxed{9} \times \boxed{9} \times \boxed{5} = 1{,}607{,}445$$

consonants   nonzero digits   vowel

There are 1,607,445 possible serial numbers when the letters and digits can be repeated within a serial number. ∎

Factorials

EXAMPLE 3 Three students rent a three-bedroom house near campus. One of the bedrooms is very desirable (it has its own bath), one has a balcony, and one is undesirable (it is very small). In how many ways can the housemates choose the bedrooms?

Solution Three decisions must be made: who gets the room with the bath, who gets the room with the balcony, and who gets the small room. Using the Fundamental Principle of Counting, we draw three boxes and enter the number of choices for each decision. There are three choices for who gets the room with the bath. Once that decision has been made, there are two choices for who gets the room with the balcony, and finally, there is only one choice for the small room.

$$\boxed{3} \times \boxed{2} \times \boxed{1} = 6$$

There are six different ways in which the three housemates can choose the three bedrooms. ∎

Combinatorics often involve products of the type $3 \cdot 2 \cdot 1 = 6$, as seen in Example 3. This type of product is called a **factorial,** and the product $3 \cdot 2 \cdot 1$ is written as 3!. In this manner, $4! = 4 \cdot 3 \cdot 2 \cdot 1\ (= 24)$, and $5! = 5 \cdot 4 \cdot 3 \cdot 2 \cdot 1\ (= 120)$.

Factorials

If n is a positive integer, then n *factorial,* denoted by $n!$, is the product of all positive integers less than or equal to n.

$$n! = n \cdot (n-1) \cdot (n-2) \cdot \cdots \cdot 2 \cdot 1$$

As a special case, we define $0! = 1$.

Many scientific calculators have a button that will calculate a factorial. Depending on your calculator, the button will look like $\boxed{x!}$ or $\boxed{n!}$, and you might have to press a $\boxed{\text{shift}}$ or $\boxed{\text{2nd}}$ button first. For example, to calculate 6!, type the number 6, press the factorial button, and obtain 720. To calculate a factorial on most graphing calculators, do the following:

- Type the value of n. (For example, type the number 6.)
- Press the $\boxed{\text{MATH}}$ button.
- Press the right arrow button $\boxed{\rightarrow}$ as many times as necessary to highlight $\boxed{\text{PRB}}$.
- Press the down arrow $\boxed{\downarrow}$ as many times as necessary to highlight the "!" symbol, and press $\boxed{\text{ENTER}}$.
- Press $\boxed{\text{ENTER}}$ to execute the calculation.

To calculate a factorial on a Casio graphing calculator, do the following:

- Press the $\boxed{\text{MENU}}$ button; this gives you access to the main menu.
- Press 1 to select the RUN mode; this mode is used to perform arithmetic operations.
- Type the value of n. (For example, type the number 6.)
- Press the $\boxed{\text{OPTN}}$ button; this gives you access to various options displayed at the bottom of the screen.
- Press the $\boxed{\text{F6}}$ button to see more options (i.e., $\boxed{\rightarrow}$).
- Press the $\boxed{\text{F3}}$ button to select probability options (i.e., $\boxed{\text{PROB}}$).
- Press the $\boxed{\text{F1}}$ button to select factorial (i.e., $\boxed{x!}$).
- Press the $\boxed{\text{EXE}}$ button to execute the calculation.

The factorial symbol "$n!$" was first introduced by Christian Kramp (1760–1826) of Strasbourg in his *Élements d'Arithmétique Universelle* (1808). Before the introduction of this "modern" symbol, factorials were commonly denoted by $\underline{\mathrm{n}}$. However, printing presses of the day had difficulty printing this symbol; consequently, the symbol $n!$ came into prominence because it was relatively easy for a typesetter to use.

EXAMPLE 4   Find the following values.

    a. 6!   **b.** $\dfrac{8!}{5!}$ **c.** $\dfrac{8!}{3! \cdot 5!}$

Solution   **a.** $6! = 6 \cdot 5 \cdot 4 \cdot 3 \cdot 2 \cdot 1$
                  $= 720$

        Therefore, $6! = 720$.

b. $\dfrac{8!}{5!} = \dfrac{8 \cdot 7 \cdot 6 \cdot 5 \cdot 4 \cdot 3 \cdot 2 \cdot 1}{5 \cdot 4 \cdot 3 \cdot 2 \cdot 1}$

$\qquad = \dfrac{8 \cdot 7 \cdot 6 \cdot \cancel{5} \cdot \cancel{4} \cdot \cancel{3} \cdot \cancel{2} \cdot \cancel{1}}{\cancel{5} \cdot \cancel{4} \cdot \cancel{3} \cdot \cancel{2} \cdot \cancel{1}}$

$\qquad = 8 \cdot 7 \cdot 6$

$\qquad = 336$

Therefore, $\frac{8!}{5!} = 336$.

Using a calculator, we obtain the same result.

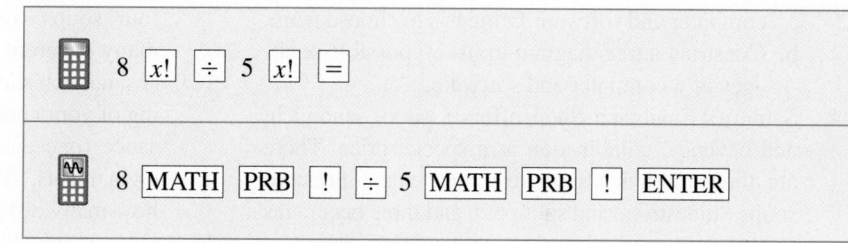

c. $\dfrac{8!}{3! \cdot 5!} = \dfrac{8 \cdot 7 \cdot 6 \cdot 5 \cdot 4 \cdot 3 \cdot 2 \cdot 1}{(3 \cdot 2 \cdot 1)(5 \cdot 4 \cdot 3 \cdot 2 \cdot 1)}$

$\qquad = \dfrac{8 \cdot 7 \cdot 6 \cdot \cancel{5} \cdot \cancel{4} \cdot \cancel{3} \cdot \cancel{2} \cdot \cancel{1}}{(3 \cdot 2 \cdot 1)\,(\cancel{5} \cdot \cancel{4} \cdot \cancel{3} \cdot \cancel{2} \cdot \cancel{1})}$

$\qquad = \dfrac{8 \cdot 7 \cdot 6}{3 \cdot 2 \cdot 1}$

$\qquad = 56$

Therefore, $\frac{8!}{3! \cdot 5!} = 56$.

Using a calculator, we obtain the same result.

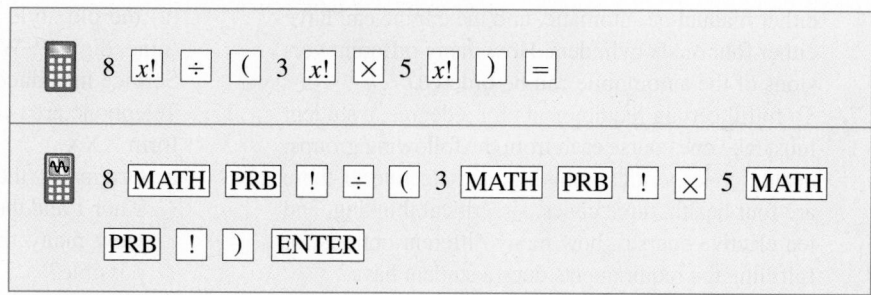

2.3 Exercises

1. A nickel, a dime, and a quarter are tossed.
   a. Use the Fundamental Principle of Counting to determine how many different outcomes are possible.
   b. Construct a tree diagram to list all possible outcomes.

2. A die is rolled, and a coin is tossed.
   a. Use the Fundamental Principle of Counting to determine how many different outcomes are possible.
   b. Construct a tree diagram to list all possible outcomes.

3. Jamie has decided to buy either a Mega or a Better Byte personal computer. She also wants to purchase either Big Word, Word World, or Great Word word-processing software and either Big Number or Number World spreadsheet software.
   a. Use the Fundamental Principle of Counting to determine how many different packages of a computer and software Jamie has to choose from.
   b. Construct a tree diagram to list all possible packages of a computer and software.

4. Sammy's Sandwich Shop offers a soup, sandwich, and beverage combination at a special price. There are three sandwiches (turkey, tuna, and tofu), two soups (minestrone and split pea), and three beverages (coffee, milk, and mineral water) to choose from.
   a. Use the Fundamental Principle of Counting to determine how many different meal combinations are possible.
   b. Construct a tree diagram to list all possible soup, sandwich, and beverage combinations.

5. If you buy three pairs of jeans, four sweaters, and two pairs of boots, how many new outfits (consisting of a new pair of jeans, a new sweater, and a new pair of boots) will you have?

6. A certain model of automobile is available in six exterior colors, three interior colors, and three interior styles. In addition, the transmission can be either manual or automatic, and the engine can have either four or six cylinders. How many different versions of the automobile can be ordered?

7. To fulfill certain requirements for a degree, a student must take one course each from the following groups: health, civics, critical thinking, and elective. If there are four health, three civics, six critical thinking, and ten elective courses, how many different options for fulfilling the requirements does a student have?

8. To fulfill a requirement for a literature class, a student must read one short story by each of the following authors: Stephen King, Clive Barker, Edgar Allan Poe, and H. P. Lovecraft. If there are twelve King, six Barker, eight Poe, and eight Lovecraft stories to choose from, how many different combinations of reading assignments can a student choose from to fulfill the reading requirement?

9. A sporting goods store has fourteen lines of snow skis, seven types of bindings, nine types of boots, and three types of poles. Assuming that all items are compatible with each other, how many different complete ski equipment packages are available?

10. An audio equipment store has ten different amplifiers, four tuners, six turntables, eight tape decks, six compact disc players, and thirteen speakers. Assuming that all components are compatible with each other, how many different complete stereo systems are available?

11. A cafeteria offers a complete dinner that includes one serving each of appetizer, soup, entrée, and dessert for $6.99. If the menu has three appetizers, four soups, six entrées, and three desserts, how many different meals are possible?

12. A sandwich shop offers a "U-Chooz" special consisting of your choice of bread, meat, cheese, and special sauce (one each). If there are six different breads, eight meats, five cheeses, and four special sauces, how many different sandwiches are possible?

13. How many different Social Security numbers are possible? (A Social Security number consists of nine digits that can be repeated.)

14. To use an automated teller machine (ATM), a customer must enter his or her four-digit Personal Identification Number (PIN). How many different PINs are possible?

15. Every book published has an International Standard Book Number (ISBN). The number is a code used to identify the specific book and is of the form X-XXX-XXXXX-X, where X is one of digits 0, 1, 2, . . . , 9. How many different ISBNs are possible?

16. How many different Zip Codes are possible using (a) the old style (five digits) and (b) the new style (nine digits)? Why do you think the U.S. Postal Service introduced the new system?

17. Telephone area codes are three-digit numbers of the form XXX.
   a. Originally, the first and third digits were neither 0 nor 1 and the second digit was always a 0 or a 1. How many three-digit numbers of this type are possible?
   b. Over time, the restrictions listed in part (a) have been altered; currently, the only requirement is

that the first digit is neither 0 nor 1. How many three-digit numbers of this type are possible?

  c. Why were the original restrictions listed in part (a) altered?

18. Major credit cards such as VISA and MasterCard have a sixteen-digit account number of the form XXXX-XXXX-XXXX-XXXX. How many different numbers of this type are possible?

19. The serial number on a dollar bill consists of a letter followed by eight digits and then a letter. How many different serial numbers are possible, given the following conditions?

  a. Letters and digits cannot be repeated.
  b. Letters and digits can be repeated.
  c. The letters are nonrepeated consonants and the digits can be repeated.

20. The serial number on a new twenty-dollar bill consists of two letters followed by eight digits and then a letter. How many different serial numbers are possible, given the following conditions?

  a. Letters and digits cannot be repeated.
  b. Letters and digits can be repeated.
  c. The first and last letters are repeatable vowels, the second letter is a consonant, and the digits can be repeated.

21. Each student at State University has a student I.D. number consisting of four digits (the first digit is nonzero, and digits may be repeated) followed by three of the letters A, B, C, D, and E (letters may not be repeated). How many different student numbers are possible?

22. Each student at State College has a student I.D. number consisting of five digits (the first digit is nonzero, and digits may be repeated) followed by two of the letters A, B, C, D, and E (letters may not be repeated). How many different student numbers are possible?

In Exercises 23–38, find the indicated value.

23. $4!$

24. $5!$

25. $10!$

26. $8!$

27. $20!$

28. $25!$

29. $6! \cdot 4!$

30. $8! \cdot 6!$

31. **a.** $\dfrac{6!}{4!}$ **b.** $\dfrac{6!}{2!}$

32. **a.** $\dfrac{8!}{6!}$ **b.** $\dfrac{8!}{2!}$

33. $\dfrac{8!}{5! \cdot 3!}$

34. $\dfrac{9!}{5! \cdot 4!}$

35. $\dfrac{8!}{4! \cdot 4!}$

36. $\dfrac{6!}{3! \cdot 3!}$

37. $\dfrac{82!}{80! \cdot 2!}$

38. $\dfrac{77!}{74! \cdot 3!}$

39. Find the value of $\dfrac{n!}{(n-r)!}$ when $n=16$ and $r=14$.

40. Find the value of $\dfrac{n!}{(n-r)!}$ when $n=19$ and $r=16$.

41. Find the value of $\dfrac{n!}{(n-r)!}$ when $n=5$ and $r=5$.

42. Find the value of $\dfrac{n!}{(n-r)!}$ when $n=r$.

43. Find the value of $\dfrac{n!}{(n-r)!r!}$ when $n=7$ and $r=3$.

44. Find the value of $\dfrac{n!}{(n-r)!r!}$ when $n=7$ and $r=4$.

45. Find the value of $\dfrac{n!}{(n-r)!r!}$ when $n=5$ and $r=5$.

46. Find the value of $\dfrac{n!}{(n-r)!r!}$ when $n=r$.

Answer the following questions using complete sentences and your own words.

CONCEPT QUESTIONS

47. What is the Fundamental Principle of Counting? When is it used?

48. What is a factorial?

HISTORY QUESTIONS

49. Who invented the modern symbol denoting a factorial? What symbol did it replace? Why?

Many graduate schools require applicants to take the Graduate Record Examination (G.R.E.). This exam is intended to measure verbal, quantitative, and analytical skills that have developed throughout a person's life. Many classes and study guides are available to help people prepare for the exam. The remaining questions are typical of those found in the study guides and on the exam itself.

Exercises 50–54 refer to the following: In an executive parking lot, there are six parking spaces in a row, labeled 1 through 6. Exactly five cars of five different colors—black, gray, pink, white, and yellow—are to be parked in the spaces. The cars can park in any of the spaces as long as the following conditions are met:

- The pink car must be parked in space 3.
- The black car must be parked in a space next to the space in which the yellow car is parked.
- The gray car cannot be parked in a space next to the space in which the white car is parked.

50. If the yellow car is parked in space 1, how many acceptable parking arrangements are there for the five cars?

  a. 1   **b.** 2   **c.** 3   **d.** 4   **e.** 5

51. Which of the following must be true of any acceptable parking arrangement?
  a. One of the cars is parked in space 2.
  b. One of the cars is parked in space 6.
  c. There is an empty space next to the space in which the gray car is parked.
  d. There is an empty space next to the space in which the yellow car is parked.
  e. Either the black car or the yellow car is parked in a space next to space 3.

52. If the gray car is parked in space 2, none of the cars can be parked in which space?
  a. 1   **b.** 3   **c.** 4   **d.** 5   **e.** 6

53. The white car could be parked in any of the spaces except which of the following?
  a. 1   **b.** 2   **c.** 4   **d.** 5   **e.** 6

54. If the yellow car is parked in space 2, which of the following must be true?
  a. None of the cars is parked in space 5.
  b. The gray car is parked in space 6.
  c. The black car is parked in a space next to the space in which the white car is parked.
  d. The white car is parked in a space next to the space in which the pink car is parked.
  e. The gray car is parked in a space next to the space in which the black car is parked.

2.4 Permutations and Combinations

The Fundamental Principle of Counting allows us to determine the total number of possible outcomes when a series of decisions (making selections from various categories) must be made. In Section 2.3, the examples and exercises involved selecting *one item each* from various categories; if you buy two pairs of jeans, three shirts, and two pairs of shoes, you will have twelve ($2 \cdot 3 \cdot 2 = 12$) new outfits (consisting of a new pair of jeans, a new shirt, and a new pair of shoes). In this section, we examine the situation when *more than one* item is selected from a category. If more than one item is selected, the selections can be made either *with* or *without* replacement.

With versus Without Replacement

Selecting items *with replacement* means that the same item *can* be selected more than once; after a specific item has been chosen, it is put back into the pool of future choices. Selecting items *without replacement* means that the same item *cannot* be selected more than once; after a specific item has been chosen, it is not replaced.

Suppose you must select a four-digit Personal Identification Number (PIN) for a bank account. In this case, the digits are selected with replacement; each time a specific digit is selected, the digit is put back into the pool of choices for the next selection. (Your PIN can be 3666; the same digit can be selected more than once.) When items are selected with replacement, we use the Fundamental Principle of Counting to determine the total number of possible outcomes; there are $10 \cdot 10 \cdot 10 \cdot 10 = 10{,}000$ possible four-digit PINs.

In many situations, items cannot be selected more than once. For instance, when selecting a committee of three people from a group of twenty, you cannot select the same person more than once. Once you have selected a specific person (say, Lauren), you do not put her back into the pool of choices. When selecting items without replacement, depending on whether the order of selection is important, *permutations* or *combinations* are used to determine the total number of possible outcomes.

Permutations

When more than one item is selected (without replacement) from a single category, and the order of selection *is* important, the various possible outcomes are called **permutations.** For example, when the rankings (first, second, and third place) in a talent contest are announced, the order of selection is important; Monte in first, Lynn in second, and Ginny in third place is different from Ginny in first, Monte in second, and Lynn in third. "Monte, Lynn, Ginny" and "Ginny, Monte, Lynn" are different permutations of the contestants. Naturally, these selections are made without replacement; we cannot select Monte for first place and reselect him for second place.

EXAMPLE 1   Six local bands have volunteered to perform at a benefit concert, but there is enough time for only four bands to play. There is also some concern over the order in which the chosen bands will perform. How many different lineups are possible?

Solution   We must select four of the six bands and put them in a specific order. The bands are selected without replacement; a band cannot be selected to play and then be reselected to play again. Because we must make four decisions, we draw four boxes and put the number of choices for each decision in each appropriate box. There are six choices for the opening band. Naturally, the opening band could not be the followup act, so there are only five choices for the next group. Similarly, there are four candidates for the third group and three choices for the closing band. The total number of different lineups possible is found by multiplying the number of choices for each decision:

$$\boxed{6} \times \boxed{5} \times \boxed{4} \times \boxed{3} = 360$$

$$\underset{\substack{\uparrow \\ \text{opening} \\ \text{band}}}{\quad} \qquad \underset{\substack{\uparrow \\ \text{closing} \\ \text{band}}}{\quad}$$

With four out of six bands playing in the performance, 360 lineups are possible. *Because the order of selecting the bands is important, the various possible outcomes, or lineups, are called permutations;* there are 360 permutations of six items when the items are selected four at a time.   ∎

The computation in Example 1 is similar to a factorial, but the factors do not go all the way down to 1; the product $6 \cdot 5 \cdot 4 \cdot 3$ is a "truncated" (cut-off) factorial. We can change this truncated factorial into a complete factorial in the following manner:

$$6 \cdot 5 \cdot 4 \cdot 3 = \frac{6 \cdot 5 \cdot 4 \cdot 3 \cdot (2 \cdot 1)}{(2 \cdot 1)} \qquad \text{multiplying by } \frac{2}{2} \text{ and } \frac{1}{1}$$

$$= \frac{6!}{2!}$$

Notice that this last expression can be written as $\frac{6!}{2!} = \frac{6!}{(6-4)!}$. (Recall that we were selecting four out of six bands.) This result is generalized as follows.

Permutation Formula

The number of **permutations,** or arrangements, of r items selected without replacement from a pool of n items ($r \leq n$), denoted by $_nP_r$, is

$$_nP_r = \frac{n!}{(n-r)!}$$

Permutations are used whenever more than one item is selected (without replacement) from a category and the order of selection is important.

Using the notation above and referring to Example 1, note that 360 possible lineups of four bands selected from a pool of six can be denoted by $_6P_4 = \frac{6!}{(6-4)!} = 360$. Other notations can be used to represent the number of permutations of a group of items. In particular, the notations $_nP_r$, $P(n, r)$, P_r^n, and $P_{n,\,r}$ all represent the number of possible permutations (or arrangements) of r items selected (without replacement) from a pool of n items.

ThomsonNOW
for Liberal Arts Mathematics
www.thomsonedu.com/1pass.

EXAMPLE 2   Three door prizes (first, second, and third) are to be awarded at a ten-year high school reunion. Each of the 112 attendees puts his or her name in a hat. The first name drawn wins a two-night stay at the Chat 'n' Rest Motel, the second name wins dinner for two at Juju's Kitsch-Inn, and the third wins a pair of engraved mugs. In how many different ways can the prizes be awarded?

Solution   We must select 3 out of 112 people (without replacement), and the order in which they are selected *is* important. (Winning dinner is different from winning the mugs.) Hence, we must find the number of permutations of 3 items selected from a pool of 112:

$$_{112}P_3 = \frac{112!}{(112-3)!}$$
$$= \frac{112!}{109!}$$
$$= \frac{112 \cdot 111 \cdot 110 \cdot 109 \cdot 108 \cdots \cdots 2 \cdot 1}{109 \cdot 108 \cdots \cdots 2 \cdot 1}$$
$$= 112 \cdot 111 \cdot 110$$
$$= 1,367,520$$

There are 1,367,520 different ways in which the three prizes can be awarded to the 112 people.   ∎

In Example 2, if you try to use a calculator to find $\frac{112!}{109!}$ directly, you will not obtain an answer. Entering 112 and pressing $\boxed{x!}$ results in a calculator error. (Try it.) Because factorials get very large very quickly, most calculators are not able to find any factorial over 69!. ($69! = 1.711224524 \times 10^{98}$.)

ThomsonNOW
for Liberal Arts Mathematics
www.thomsonedu.com/1pass.

EXAMPLE 3   A bowling league has ten teams. How many different ways can the teams rank in the standings at the end of a tournament? (Ties are not allowed.)

Solution   Because order is important, we find the number of permutations of ten items selected from a pool of ten items:

$$_{10}P_{10} = \frac{10!}{(10-10)!}$$
$$= \frac{10!}{0!} \qquad \text{Recall that } 0! = 1.$$
$$= \frac{10!}{1}$$
$$= 3,628,800$$

In a league containing ten teams, there are 3,628,800 different standings possible at the end of a tournament.   ∎

Combinations

When items are selected from a group, the order of selection may or may not be important. If the order is important (as in Examples 1, 2, and 3), permutations are used to determine the total number of selections possible. What if the order of selection is *not* important? When more than one item is selected (without replacement) from a single category and the order of selection is not important, the various possible outcomes are called **combinations.**

EXAMPLE 4   Two adults are needed to chaperone a daycare center's field trip. Marcus, Vivian, Frank, and Keiko are the four managers of the center. How many different groups of chaperones are possible?

Solution   In selecting the chaperones, the order of selection is *not* important; "Marcus and Vivian" is the same as "Vivian and Marcus." Hence, the permutation formula cannot be used. Because we do not yet have a shortcut for finding the total number of possibilities when the order of selection is not important, we must list all the possibilities:

Marcus and Vivian	Marcus and Frank	Marcus and Keiko
Vivian and Frank	Vivian and Keiko	Frank and Keiko

Therefore, six different groups of two chaperones are possible from the group of four managers. Because the order in which the people are selected is not important, the various possible outcomes, or groups of chaperones, are called *combinations;* there are six combinations when two items are selected from a pool of four.   ∎

Just as $_nP_r$ denotes the number of *permutations* of r elements selected from a pool of n elements, $_nC_r$ denotes the number of *combinations* of r elements selected from a pool of n elements. In Example 4, we found that there are six combinations of two people selected from a pool of four by listing all six of the combinations; that is, $_4C_2 = 6$. If we had a larger pool, listing each combination to find out how many there are would be extremely time consuming and tedious! Instead of listing, we take a different approach. We first find the number of permutations (with the permutation formula) and then alter that number to account for the distinction between permutations and combinations.

To find the number of combinations of two people selected from a pool of four, we first find the number of permutations:

$$_4P_2 = \frac{4!}{(4-2)!} = \frac{4!}{2!} = 12$$

This figure of 12 must be altered to account for the distinction between permutations and combinations.

In Example 4, we listed combinations; one such combination was "Marcus and Vivian." If we had listed permutations, we would have had to list both "Marcus and

Vivian" and "Vivian and Marcus," because the *order* of selection matters with permutations. In fact, each combination of two chaperones listed in Example 4 generates two permutations; each pair of chaperones can be given in two different orders. Thus, there are twice as many permutations of two people selected from a pool of four as there are combinations. Alternatively, there are half as many combinations of two people selected from a pool of four as there are permutations. We used the permutation formula to find that $_4P_2 = 12$; thus,

$$_4C_2 = \frac{1}{2} \cdot {}_4P_2 = \frac{1}{2}(12) = 6$$

This answer certainly fits with Example 4; we listed exactly six combinations.

What if three of the four managers were needed to chaperone the daycare center's field trip? Rather than finding the number of combinations by listing each possibility, we first find the number of permutations and then alter that number to account for the distinction between permutations and combinations.

The number of permutations of three people selected from a pool of four is

$$_4P_3 = \frac{4!}{(4-3)!} = \frac{4!}{1!} = 24$$

We know that some of these permutations represent the same combination. For example, the combination "Marcus and Vivian and Keiko" generates $3! = 6$ different permutations (using initials, they are: MVK, MKV, KMV, KVM, VMK, VKM). Because each combination of three people generates six different permutations, there are one-sixth as many combinations as permutations. Thus,

$$_4C_3 = \frac{1}{6} \cdot {}_4P_3 = \frac{1}{6}(24) = 4$$

This means that if three of the four managers were needed to chaperone the daycare center's field trip, there would be $_4C_3 = 4$ possible combinations.

We just saw that when two items are selected from a pool of n items, each combination of two generates $2! = 2$ permutations, so

$$_nC_2 = \frac{1}{2!} \cdot {}_nP_2$$

We also saw that when three items are selected from a pool of n items, each combination of three generates $3! = 6$ permutations, so

$$_nC_3 = \frac{1}{3!} \cdot {}_nP_3$$

More generally, when r items are selected from a pool of n items, each combination of r items generates $r!$ permutations, so

$$_nC_r = \frac{1}{r!} \cdot {}_nP_r$$

$$= \frac{1}{r!} \cdot \frac{n!}{(n-r)!} \qquad \text{using the Permutation Formula}$$

$$= \frac{n!}{r! \cdot (n-r)!} \qquad \text{multiplying the fractions together}$$

Combination Formula

The number of distinct **combinations** of r items selected without replacement from a pool of n items ($r \leq n$), denoted by $_nC_r$, is

$$_nC_r = \frac{n!}{(n-r)!\, r!}$$

Combinations are used whenever one or more items are selected (without replacement) from a category and the order of selection is not important.

EXAMPLE 5   A compact disc club sends you a brochure that offers any five CDs from a group of 50 of today's hottest releases. How many different selections can you make?

Solution   Because the order of selection is *not* important, we find the number of combinations when five items are selected from a pool of 50:

$$_{50}C_5 = \frac{50!}{(50-5)!\, 5!}$$

$$= \frac{50!}{45!\, 5!}$$

$$= \frac{50 \cdot 49 \cdot 48 \cdot 47 \cdot 46}{5 \cdot 4 \cdot 3 \cdot 2 \cdot 1}$$

$$= 2{,}118{,}760$$

Graphing calculators have buttons that will calculate $_nP_r$ and $_nC_r$. To use them, do the following:

- Type the value of n. (For example, type the number 50.)
- Press the $\boxed{\text{MATH}}$ button.
- Press the right arrow button $\boxed{\rightarrow}$ as many times as necessary to highlight $\boxed{\text{PRB}}$.
- Press the down arrow button $\boxed{\downarrow}$ as many times as necessary to highlight the appropriate symbol—$\boxed{_nP_r}$ for permutations, $\boxed{_nC_r}$ for combinations—and press $\boxed{\text{ENTER}}$.
- Type the value of r. (For example, type the number 5.)
- Press $\boxed{\text{ENTER}}$ to execute the calculation.

On a Casio graphing calculator, do the following:

- Press the $\boxed{\text{MENU}}$ button; this gives you access to the main menu.
- Press 1 to select the RUN mode; this mode is used to perform arithmetic operations.
- Type the value of n. (For example, type the number 50.)
- Press the $\boxed{\text{OPTN}}$ button; this gives you access to various options displayed at the bottom of the screen.
- Press the $\boxed{\text{F6}}$ button to see more options (i.e., $\boxed{\rightarrow}$).
- Press the $\boxed{\text{F3}}$ button to select probability options (i.e., $\boxed{\text{PROB}}$).

- Press the $\boxed{F3}$ button to select combinations (i.e., $_nC_r$) or the $\boxed{F2}$ button to select permutations (i.e., $_nP_r$).
- Type the value of r. (For example, type the number 5.)
- Press the $\boxed{EXE}$ button to execute the calculation.

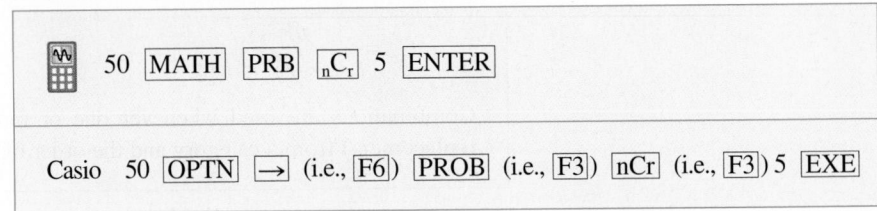

In choosing five out of 50 CDs, 2,118,760 combinations are possible. ■

EXAMPLE 6   A group consisting of twelve women and nine men must select a five-person committee. How many different committees are possible if it must consist of the following?

a. three women and two men   **b.** any mixture of men and women

Solution   **a.** Our problem involves two categories: women and men. The Fundamental Principle of Counting tells us to draw two boxes (one for each category), enter the number of choices for each, and multiply:

$$\boxed{\begin{array}{c}\text{the number of ways}\\ \text{we can select}\\ \text{three out}\\ \text{of twelve}\\ \text{women}\end{array}} \times \boxed{\begin{array}{c}\text{the number of ways}\\ \text{we can select}\\ \text{two out}\\ \text{of nine}\\ \text{men}\end{array}} = ?$$

Because the order of selecting the members of a committee is not important, we will use combinations:

$$(_{12}C_3) \cdot (_9C_2) = \frac{12!}{(12-3)! \cdot 3!} \cdot \frac{9!}{(9-2)! \cdot 2!}$$

$$= \frac{12!}{9! \cdot 3!} \cdot \frac{9!}{7! \cdot 2!}$$

$$= \frac{12 \cdot 11 \cdot 10}{3 \cdot 2 \cdot 1} \cdot \frac{9 \cdot 8}{2 \cdot 1}$$

$$= 220 \cdot 36$$

$$= 7{,}920$$

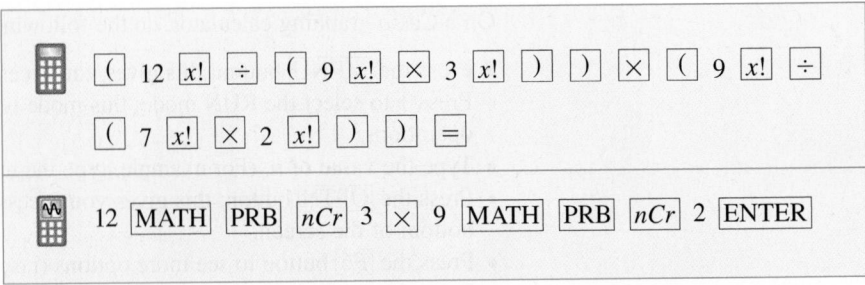

There are 7,920 different committees consisting of three women and two men.

b. Because the gender of the committee members doesn't matter, our problem involves only one category: people. We must choose five out of the 21 people, and the order of selection is not important:

$$_{21}C_5 = \frac{21!}{(21-5)! \cdot 5!}$$

$$= \frac{21!}{16! \cdot 5!}$$

$$= \frac{21 \cdot 20 \cdot 19 \cdot 18 \cdot 17}{5 \cdot 4 \cdot 3 \cdot 2 \cdot 1}$$

$$= 20,349$$

There are 20,349 different committees consisting of five people. ∎

EXAMPLE 7   Find the value of $_5C_r$ for the following values of r:

a. $r = 0$ **b.** $r = 1$ **c.** $r = 2$ **d.** $r = 3$ **e.** $r = 4$ **f.** $r = 5$

Solution   **a.** $_5C_0 = \dfrac{5!}{(5-0)! \cdot 0!} = \dfrac{5!}{5! \cdot 0!} = 1$

b. $_5C_1 = \dfrac{5!}{(5-1)! \cdot 1!} = \dfrac{5!}{4! \cdot 1!} = 5$

c. $_5C_2 = \dfrac{5!}{(5-2)! \cdot 2!} = \dfrac{5!}{3! \cdot 2!} = 10$

d. $_5C_3 = \dfrac{5!}{(5-3)! \cdot 3!} = \dfrac{5!}{2! \cdot 3!} = 10$

e. $_5C_4 = \dfrac{5!}{(5-4)! \cdot 4!} = \dfrac{5!}{1! \cdot 4!} = 5$

f. $_5C_5 = \dfrac{5!}{(5-5)! \cdot 5!} = \dfrac{5!}{0! \cdot 5!} = 1$ ∎

The combinations generated in Example 7 exhibit a curious pattern. Notice that the values of $_5C_r$ are symmetric: $_5C_0 = {}_5C_5, {}_5C_1 = {}_5C_4$, and $_5C_2 = {}_5C_3$. Now examine the diagram in Figure 2.40. Each number in this "triangle" of numbers is the sum of two numbers in the row immediately above it. For example, $2 = 1 + 1$ and $10 = 4 + 6$, as shown by the inserted arrows. It is no coincidence that the values of $_5C_r$ found in Example 7 also appear as a row of numbers in this "magic" triangle. In fact, the 6th row contains all the values of $_5C_r$ for $r = 0, 1, 2, 3, 4,$ and 5. In general, the $(n + 1)^{th}$ row of the triangle contains all the values of $_nC_r$ for $r = 0, 1, 2, \ldots, n$; alternatively, the nth row of the triangle contains all the values of $_{n-1}C_r$ for $r = 0, 1, 2, \ldots, n-1$. For example, the values of $_9C_r$, for $r = 0, 1, 2, \ldots, 9$, are in the 10th row, and vice versa, the entries in the 10th row are the values of $_9C_r$, for $r = 0, 1, 2, \ldots, 9$.

Historically, this triangular pattern of numbers is referred to as *Pascal's Triangle,* in honor of the French mathematician, scientist, and philosopher Blaise

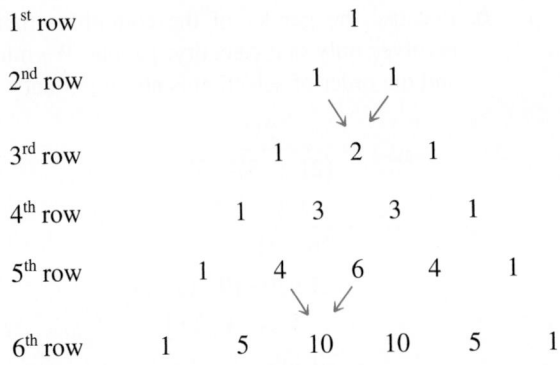

and so on
Figure 2.40
Pascal's triangle.

Pascal (1623–1662). Pascal is a cofounder of probability theory (see the Historical Note in Section 3.1). Although the triangle has Pascal's name attached to it, this "magic" arrangement of numbers was known to other cultures hundreds of years before Pascal's time.

The most important part of any problem involving combinatorics is deciding which counting technique (or techniques) to use. The following list of general steps and the flowchart in Figure 2.41 can help you to decide which method or methods to use in a specific problem.

Which Counting Technique?

1. What is being selected?
2. If the selected items can be repeated, use the **Fundamental Principle of Counting** and multiply the number of choices for each category.
3. If there is only one category, use:
   combinations if the order of selection does not matter—that is, r items can be selected from a pool of n items in $_nC_r = \frac{n!}{(n-r)! \cdot r!}$ ways.
   permutations if the order of selection does matter—that is, r items can be selected from a pool of n items in $_nP_r = \frac{n!}{(n-r)!}$ ways.
4. If there is more than one category, use the **Fundamental Principle of Counting** with one box per category.
   a. If you are selecting one item per category, the number in the box for that category is the number of choices for that category.
   b. If you are selecting more than one item per category, the number in the box for that category is found by using step 3.

EXAMPLE 8   A standard deck of playing cards contains 52 cards.

   a. How many different five-card hands containing four kings are possible?
   b. How many different five-card hands containing four queens are possible?
   c. How many different five-card hands containing four kings or four queens are possible?
   d. How many different five-card hands containing four of a kind are possible?

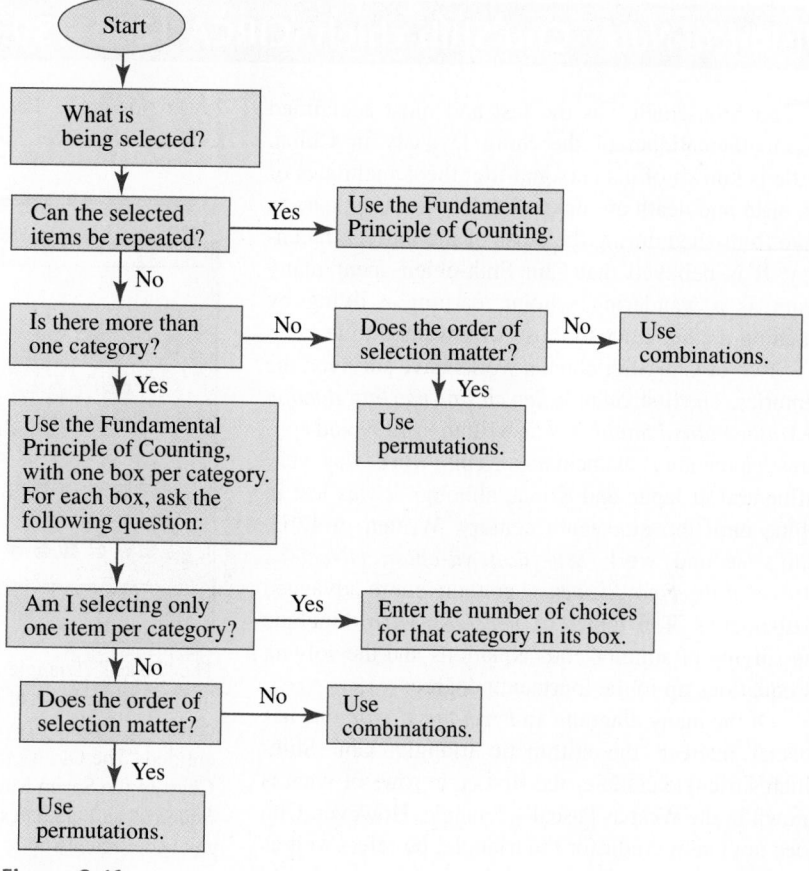

Figure 2.41
Which counting technique?

Solution **a.** We use the flowchart in Figure 2.41 and answer the following questions.

Q. What is being selected?

A. Playing cards.

Q. Can the selected items be repeated?

A. No.

Q. Is there more than one category?

A. Yes: Because we must have five cards, we need four kings and one non-king. Therefore, we need two boxes:

$$\boxed{\text{kings}} \times \boxed{\text{non-kings}}$$

Q. Am I selecting only one item per category?

A. *Kings*: no. Does the order of selection matter? No: Use combinations. Because there are $n = 4$ kings in the deck and we want to select $r = 4$, we must compute $_4C_4$. *Non-kings*: yes. Enter the number of choices for that category: There are 48 non-kings.

Historical Note Chu Shih-chieh, CIRCA 1280–1303

Chu Shih-chieh was the last and most acclaimed mathematician of the Sung Dynasty in China. Little is known of his personal life; the actual dates of his birth and death are unknown. His work appears to have flourished during the close of the thirteenth century. It is believed that Chu Shih-chieh spent many years as a wandering scholar, earning a living by teaching mathematics to those who wanted to learn.

Two of Chu Shih-chieh's works have survived the centuries. The first, *Suan-hsüeh ch'i-meng (Introduction to Mathematical Studies),* was written in 1299 and contains elementary mathematics. This work was very influential in Japan and Korea, although it was lost in China until the nineteenth century. Written in 1303, Chu's second work *Ssu-yüan yü-chien (Precious Mirror of the Four Elements)* contains more advanced mathematics. The topics of *Precious Mirror* include the solving of simultaneous equations and the solving of equations up to the fourteenth degree.

Of the many diagrams in *Precious Mirror,* one has special interest: the arithmetic triangle. Chu Shih-chieh's triangle contains the first eight rows of what is known in the West as Pascal's Triangle. However, Chu does not claim credit for the triangle; he refers to it as "a diagram of the *old* method for finding eighth and

lower powers." "Pascal's" Triangle was known to the Chinese well over 300 years before Pascal was born!

Brown University Library

The "Pascal" Triangle as depicted in 1303 at the front of Chu Shih-chieh's *Ssu-yüan yü-chien.* It is entitled "The Old Method Chart of the Seven Multiplying Squares" and tabulates the binomial coefficients up to the eighth power.

$$\boxed{\text{kings}} \times \boxed{\text{non-kings}} = \boxed{{}_4C_4} \times \boxed{48}$$
$$= \frac{4!}{(4-4)! \cdot 4!} \cdot 48$$
$$= \frac{4!}{0! \cdot 4!} \cdot 48$$
$$= 1 \cdot 48$$
$$= 48$$

There are 48 different five-card hands containing four kings.

b. Using the same method as in part (a), we would find that there are 48 different five-card hands containing four queens; the number of five-card hands containing four queens is the same as the number of five-card hands containing four kings.

c. To find the number of five-card hands containing four kings or four queens, we define the following sets:

$A = \{$five card hands $|$ the hand contains four kings$\}$

$B = \{$five card hands $|$ the hand contains four queens$\}$

Consequently,

$A \cup B = \{$five card hands $\mid$ the hand contains 4 kings *or* 4 queens$\}$

$A \cap B = \{$five card hands $\mid$ the hand contains 4 kings *and* 4 queens$\}$

(Recall that the union symbol, $\cup$, may be interpreted as the word "or," while the intersection symbol, $\cap$, may be interpreted as the word "and." See Figure 2.12 for a comparison of set theory and logic.)

Because there are no five-card hands that contain four kings *and* four queens, we note that $n(A \cap B) = 0$.

Using the Cardinal Number Formula for the Union of Sets, we obtain

$$n(A \cup B) = n(A) + n(B) - n(A \cap B)$$
$$= 48 + 48 - 0$$
$$= 96$$

There are 96 different five-card hands containing four kings or four queens.

d. *Four of a kind* means four cards of the same "denomination," that is, four twos, or four threes, or four fours, or . . . , or four kings, or four aces. Now, regardless of the denomination of the card, there are 48 different five-card hands that contain four of any specific denomination; there are 48 different five-card hands that contain four twos, there are 48 different five-card hands that contain four threes, there are 48 different five-card hands that contain four fours, and so on. As is shown in part (c), the word "or" implies that we *add* cardinal numbers. Consequently,

$$n(\text{four of a kind}) = n(4 \text{ twos or } 4 \text{ threes or } \ldots \text{ or } 4 \text{ kings or } 4 \text{ aces})$$
$$= n(4 \text{ twos}) + n(4 \text{ threes} + \cdots + n(4 \text{ kings}) + n(4 \text{ aces})$$
$$= 48 + 48 + \cdots + 48 + 48 \quad (13 \text{ times})$$
$$= 13 \times 48$$
$$= 624$$

There are 624 different five-card hands containing four of a kind. ∎

As is shown in Example 8, there are 624 possible five-card hands that contain four of a kind. When you are dealt five cards, what is the likelihood (or probability) that you will receive one of these hands? This question, and its answer, will be explored in Section 3.4, "Combinatorics and Probability."

Permutations of Identical Items

In how many different ways can the three letters in the word "SAW" be arranged? As we know, arrangements are referred to as *permutations,* so we can apply the Permutation Formula, $_nP_r = \frac{n!}{(n - r)!}$.

Therefore,

$$_3P_3 = \frac{3!}{(3 - 3)!} = \frac{3!}{0!} = \frac{3 \cdot 2 \cdot 1}{1} = 6$$

The six permutations of the letters in SAW are:

SAW   SWA   AWS   ASW   WAS   WSA.

In general, if we have three *different* items (the letters in SAW), we can arrange them in $3! = 6$ ways. However, this method applies only if the items are all different (distinct).

What happens if some of the items are the same (identical)? For example, in how many different ways can the three letters in the word "SEE" be arranged? Because two of the letters are identical (E), we cannot use the Permutation Formula directly; we take a slightly different approach. Temporarily, let us assume that the E's are written in different colored inks, say, red and blue. Therefore, SEE could be expressed as SEE. These three symbols could be arranged in 3! = 6 ways as follows:

> SEE   SEE   ESE   ESE   EES   EES

If we now remove the color, the arrangements are

> SEE   SEE   ESE   ESE   EES   EES

Some of these arrangements are duplicates of others; as we can see, there are only three different or **distinguishable permutations,** namely, SEE, ESE, and EES. Notice that when $n = 3$ (the total number of letters in SEE) and $x = 2$ (the number of identical letters), we can divide $n!$ by $x!$ to obtain the number of distinguishable permutations; that is,

$$\frac{n!}{x!} = \frac{3!}{2!} = \frac{3\cdot2\cdot1}{2\cdot1} = \frac{3\cdot2\cdot\cancel{1}}{2\cdot\cancel{1}} = 3$$

This method is applicable because dividing by the factorial of the repeated letter eliminates the duplicate arrangements; the method may by generalized as follows.

> ### Distinguishable Permutations of Identical Items
>
> The number of **distinguishable permutations** (or arrangements) of n items in which x items are identical, y items are identical, z items are identical, and so on, is $\frac{n!}{x!y!z!\cdots}$. That is, to find the number of distinguishable permutations, divide the total factorial by the factorial of each repeated item.

ThomsonNOW
for Liberal Arts Mathematics
www.thomsonedu.com/1pass.

EXAMPLE 9   Find the number of distinguishable permutations of the letters in the word "MISSISSIPPI."

Solution   The word "MISSISSIPPI" has $n = 11$ letters; I is repeated $x = 4$ times, S is repeated $y = 4$ times, and P is repeated $z = 2$ times. Therefore, we divide the total factorial by the factorial of each repeated letter and obtain

$$\frac{n!}{x!y!z!} = \frac{11!}{4!4!2!} = 34{,}650$$

the letters in the word MISSISSIPPI can be arranged in 34,650 ways. (Note that if the 11 letters were all different, there would be $11! = 39{,}96{,}800$ permutations.) ■

2.4 Exercises

In Exercises 1–12, find the indicated value:

1. a. $_7P_3$ **b.** $_7C_3$
3. a. $_5P_5$ **b.** $_5C_5$
5. a. $_{14}P_1$ **b.** $_{14}C_1$
7. a. $_{100}P_3$ **b.** $_{100}C_3$

2. a. $_8P_4$ **b.** $_8C_4$
4. a. $_9P_0$ **b.** $_9C_0$
6. a. $_{13}C_3$ **b.** $_{13}C_{10}$
8. a. $_{80}P_4$ **b.** $_{80}C_4$

9. a. $_xP_{x-1}$ **b.** $_xC_{x-1}$
10. a. $_xP_1$ **b.** $_xC_1$
11. a. $_xP_2$ **b.** $_xC_2$
12. a. $_xP_{x-2}$ **b.** $_xC_{x-2}$
13. a. Find $_3P_2$.
b. List all of the permutations of {a, b, c} when the elements are taken two at a time.

14. **a.** Find $_3C_2$.
     b. List all of the combinations of {a, b, c} when the elements are taken two at a time.

15. **a.** Find $_4C_2$.
     b. List all of the combinations of {a, b, c, d} when the elements are taken two at a time.

16. **a.** Find $_4P_2$.
     b. List all of the permutations of {a, b, c, d} when the elements are taken two at a time.

17. An art class consists of 12 students. All of them must present their portfolios and explain their work to the instructor and their classmates at the end of the semester.
     a. If their names are drawn from a hat to determine who goes first, second, and so on, how many presentation orders are possible?
     b. If their names are put in alphabetical order to determine who goes first, second, and so on, how many presentation orders are possible?

18. An English class consists of 24 students, and three are to be chosen to give speeches in a school competition. In how many different ways can the teacher choose the team, given the following conditions?
     a. The order of the speakers is important.
     b. The order of the speakers is not important.

19. In how many ways can the letters in the word "STOP" be arranged? (See the photograph below.)

20. A committee of four is to be selected from a group of 15 people. How many different committees are possible, given the following conditions?
     a. There is no distinction between the responsibilities of the members.
     b. One person is the chair, and the rest are general members.
     c. One person is the chair, one person is the secretary, one person is responsible for refreshments, and one person cleans up after meetings.

21. A softball league has 14 teams. If every team must play every other team once in the first round of league play, how many games must be scheduled?

22. In a group of 19 people, each person shakes hands once with each other person in the group. How many handshakes will occur?

23. A softball league has 14 teams. How many different end-of-the-season rankings of first, second, and third place are possible (disregarding ties)?

24. Two hundred people buy raffle tickets. Three winning tickets will be drawn at random.
     a. If first prize is $100, second prize is $50, and third prize is $20, in how many different ways can the prizes be awarded?
     b. If each prize is $50, in how many different ways can the prizes be awarded?

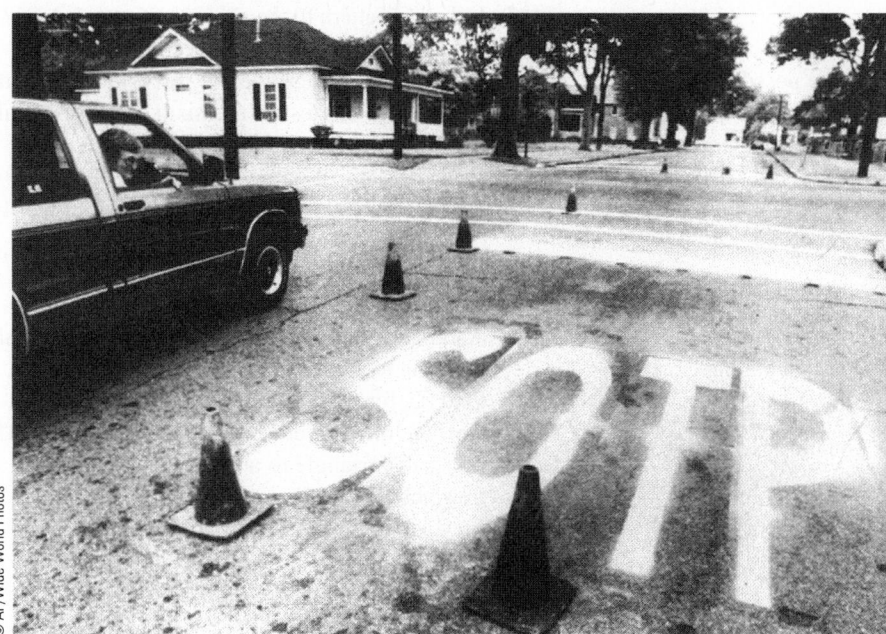

Exercise 19: Right Letters, Wrong Order. A woman in a passing vehicle got a chuckle out of the freshly painted 'SOTP' sign at a street corner in Anniston, Ala. A city paint crew quickly corrected the misspelling.

25. A group of eight women and six men must select a four-person committee. How many committees are possible if it must consist of the following?
  a. two women and two men
  b. any mixture of men and women
  c. a majority of women

26. A group of ten seniors, eight juniors, five sophomores, and five freshmen must select a committee of four. How many committees are possible if the committee must contain the following:
  a. one person from each class
  b. any mixture of the classes
  c. exactly two seniors

Exercises 27–32 refer to a deck of 52 playing cards (jokers not allowed). If you are unfamiliar with playing cards, see the end of Section 3.1 for a description of a standard deck.

27. How many five-card poker hands are possible?

28. **a.** How many five-card poker hands consisting of all hearts are possible?
  b. How many five-card poker hands consisting of all cards of the same suit are possible?

29. **a.** How many five-card poker hands containing exactly three aces are possible?
  b. How many five-card poker hands containing three of a kind are possible?

30. **a.** How many five-card poker hands consisting of three kings and two queens are possible?
  b. How many five-card poker hands consisting of three of a kind and a pair (a *full house*) are possible?

31. How many five-card poker hands containing two pair are possible?

  HINT: You must select two of the 13 ranks, then select a pair of each, and then one of the remaining cards.

32. How many five-card poker hands containing exactly one pair are possible?

  HINT: After selecting a pair, you must select three of the remaining 12 ranks and then select one card of each.

33. A 6/53 lottery requires choosing six of the numbers 1 through 53. How many different lottery tickets can you choose? (Order is not important, and the numbers do not repeat.)

34. A 7/39 lottery requires choosing seven of the numbers 1 through 39. How many different lottery tickets can you choose? (Order is not important, and the numbers do not repeat.)

35. A 5/36 lottery requires choosing five of the numbers 1 through 36. How many different lottery tickets can you choose? (Order is not important, and the numbers do not repeat.)

36. A 6/49 lottery requires choosing six of the numbers 1 through 49. How many different lottery tickets can you choose? (Order is not important, and the numbers do not repeat.)

37. Which lottery would be easier to win, a 6/53 or a 5/36? Why?

  HINT: See Exercises 33 and 35.

38. Which lottery would be easier to win, a 6/49 or a 7/39? Why?

  HINT: See Exercises 34 and 36.

39. **a.** Find the sum of the entries in the first row of Pascal's Triangle.
  b. Find the sum of the entries in the second row of Pascal's Triangle.
  c. Find the sum of the entries in the third row of Pascal's Triangle.
  d. Find the sum of the entries in the fourth row of Pascal's Triangle.
  e. Find the sum of the entries in the fifth row of Pascal's Triangle.
  f. Is there a pattern to the answers to parts (a)–(e)? If so, describe the pattern you see.
  g. Use the pattern described in part (f) to predict the sum of the entries in the sixth row of Pascal's triangle.
  h. Find the sum of the entries in the sixth row of Pascal's Triangle. Was your prediction in part (g) correct?
  i. Find the sum of the entries in the nth row of Pascal's Triangle.

40. **a.** Add adjacent entries of the sixth row of Pascal's Triangle to obtain the seventh row.
  b. Find $_6C_r$ for $r = 0, 1, 2, 3, 4, 5,$ and 6.
  c. How are the answers to parts (a) and (b) related?

41. Use Pascal's Triangle to answer the following.
  a. In which row would you find the value of $_4C_2$?
  b. In which row would you find the value of $_nC_r$?
  c. Is $_4C_2$ the second number in the fourth row?
  d. Is $_4C_2$ the third number in the fifth row?
  e. What is the location of $_nC_r$? Why?

42. Given the set $S = \{a, b, c, d\}$, answer the following.
  a. How many one-element subsets does S have?
  b. How many two-element subsets does S have?

c. How many three-element subsets does S have?
d. How many four-element subsets does S have?
e. How many zero-element subsets does S have?
f. How many subsets does S have?
g. If $n(S) = k$, how many subsets will S have?

In Exercises 43–50, find the number of permutations of the letters in each word.

43. ALASKA
44. ALABAMA
45. ILLINOIS
46. HAWAII
47. INDIANA
48. TENNESSEE
49. TALLAHASSEE
50. PHILADELPHIA

The words in each of Exercises 51–54 are homonyms *(words that are pronounced the same but have different meanings). Find the number of permutations of the letters in each word.*

51. **a.** PIER   **b.** PEER
52. **a.** HEAR   **b.** HERE
53. **a.** STEAL   **b.** STEEL
54. **a.** SHEAR   **b.** SHEER

━━━━━➤ *Answer the following questions using complete sentences and your own words.*

CONCEPT QUESTIONS

55. Suppose you want to know how many ways r items can be selected from a group of n items. What determines whether you should calculate $_nP_r$ or $_nC_r$?

56. For any given values of n and r, which is larger, $_nP_r$ or $_nC_r$? Why?

Many graduate schools require applicants to take the Graduate Record Examination (G.R.E.). This exam is intended to measure verbal, quantitative, and analytical skills that have developed throughout a person's life. Many classes and study guides are available to help people prepare for the exam. The remaining questions are typical of those found in the study guides and on the exam itself.

Exercises 57–61 refer to the following: A baseball league has six teams: A, B, C, D, E, and F. All games are played at 7:30 P.M. on Fridays, and there are sufficient

fields for each team to play a game every Friday night. Each team must play each other team exactly once, and the following conditions must be met:

- Team A plays team D first and team F second.
- Team B plays team E first and team C third.
- Team C plays team F first.

57. What is the total number of games that each team must play during the season?
a. 3   **b.** 4   **c.** 5   **d.** 6   **e.** 7

58. On the first Friday, which of the following pairs of teams play each other?
a. A and B; C and F; D and E
b. A and B; C and E; D and F
c. A and C; B and E; D and F
d. A and D; B and C; E and F
e. A and D; B and E; C and F

59. Which of the following teams must team B play second?
a. A   **b.** C   **c.** D   **d.** E   **e.** F

60. The last set of games could be between which teams?
a. A and B; C and F; D and E
b. A and C; B and F; D and E
c. A and D; B and C; E and F
d. A and E; B and C; D and F
e. A and F; B and E; C and D

61. If team D wins five games, which of the following must be true?
a. Team A loses five games.
b. Team A wins four games.
c. Team A wins its first game.
d. Team B wins five games.
e. Team B loses at least one game.

WEB PROJECT

62. Write a research paper on any historical topic referred to in this section or in a previous section. Following is a partial list of topics:
- John Venn
- Augustus De Morgan
- Chu Shih-chieh

Some useful links for this web project are listed on the text web site: **www.thomsonedu.com/ mathematics/johnson**

2.5 Infinite Sets

> **WARNING:** Many leading nineteenth-century mathematicians and philosophers claim that the study of infinite sets may be dangerous to your mental health.

Consider the sets E and N, where $E = \{2, 4, 6, \ldots\}$ and $N = \{1, 2, 3, \ldots\}$. Both are examples of infinite sets (they "go on forever"). E is the set of all even counting numbers, and N is the set of all counting (or natural) numbers. Because every element of E is an element of N, E is a subset of N. In addition, N contains elements not in E; therefore, E is a *proper* subset of N. Which set is "bigger," E or N? Intuition may lead many people to think that N is twice as big as E because N contains all the even counting numbers *and* all the odd counting numbers. Not so! According to the work of Georg Cantor (considered by many to be the father of set theory), N and E have exactly the same number of elements! This seeming paradox, a *proper* subset that has the *same number* of elements as the set from which it came, caused a philosophic uproar in the late nineteenth century. (Hence, the warning at the beginning of this section.) To study Cantor's work (which is now accepted and considered a cornerstone in modern mathematics), we must first investigate the meaning of a one-to-one correspondence and equivalent sets.

One-to-One Correspondence

Is there any relationship between the sets $A = \{$one, two, three$\}$ and $B = \{$Pontiac, Chevrolet, Ford$\}$? Although the sets contain different types of things (numbers versus automobiles), each contains the same number of things; they are the same size. This relationship (being the same size) forms the basis of a one-to-one correspondence. A **one-to-one correspondence** between the sets A and B is a pairing up of the elements of A and B such that each element of A is paired up with exactly one element of B, and vice versa, with no element left out. For instance, the elements of A and B might be paired up as follows:

(Other correspondences, or matchups, are possible.) If two sets have the same cardinal number, their elements can be put into a one-to-one correspondence. Whenever a one-to-one correspondence exists between the elements of two sets A and B, the sets are **equivalent** (denoted by $A \sim B$). Hence, equivalent sets have the same number of elements.

If two sets have different cardinal numbers, it is not possible to construct a one-to-one correspondence between their elements. The sets $C = \{$one, two$\}$ and $B = \{$Pontiac, Chevrolet, Ford$\}$ do *not* have a one-to-one correspondence; no matter how their elements are paired up, one element of B will always be left over (B has more elements; it is "bigger"):

one   two
$\updownarrow$ $\updownarrow$
Pontiac   Chevrolet   Ford

The sets C and B are *not* equivalent.

Historical Note   Georg Cantor, 1845–1918

© Corbis

Georg Ferdinand Ludwig Philip Cantor was born in St. Petersburg, Russia. His father was a stockbroker and wanted his son to become an engineer; his mother was an artist and musician. Several of Cantor's maternal relatives were accomplished musicians; in his later years, Cantor often wondered how his life would have turned out if he had become a violinist instead of pursuing a controversial career in mathematics.

Following his father's wishes, Cantor began his engineering studies at the University of Zurich in 1862. However, after one semester, he decided to study philosophy and pure mathematics. He transferred to the prestigious University of Berlin, studied under the famed mathematicians Karl Weierstrass, Ernst Kummer, and Leopold Kronecker, and received his doctorate in 1867. Two years later, Cantor accepted a teaching position at the University of Halle and remained there until he retired in 1913.

Cantor's treatises on set theory and the nature of infinite sets were first published in 1874 in *Crelle's Journal,* which was influential in mathematical circles. On their publication, Cantor's theories generated much controversy among mathematicians and philosophers. Paradoxes concerning the cardinal numbers of infinite sets, the nature of infinity, and Cantor's form of logic were unsettling to many, including Cantor's former teacher Leopold Kronecker. In fact, some felt that Cantor's work was not just revolutionary but actually dangerous.

Kronecker led the attack on Cantor's theories. He was an editor of *Crelle's Journal* and held up the publication of one of Cantor's subsequent articles for so long that Cantor refused to publish ever again in the *Journal.* In addition, Kronecker blocked Cantor's efforts to obtain a teaching position at the University of Berlin. Even though Cantor was attacked by Kronecker and his followers, others respected him. Realizing the importance of communication among scholars, Cantor founded the Association of German Mathematicians in 1890 and served as its president for many years. In addition, Cantor was instrumental in organizing the first International Congress of Mathematicians, held in Zurich in 1897.

As a result of the repeated attacks on him and his work, Cantor suffered many nervous breakdowns, the first when he was thirty-nine. He died in a mental hospital in Halle at the age of seventy-three, never having received proper recognition for the true value of his discoveries. Modern mathematicians believe that Cantor's form of logic and his concepts of infinity revolutionized all of mathematics, and his work is now considered a cornerstone in its development.

Brown University Library

Written in 1874, Cantor's first major paper on the theory of sets, *Über eine Eigenshaft des Inbegriffes aller reellen algebraischen Zahlen* (On a Property of the System of All the Real Algebraic Numbers), sparked a major controversy concerning the nature of infinite sets. To gather international support for his theory, Cantor had his papers translated into French. This 1883 French version of Cantor's work was published in the newly formed journal *Acta Mathematica.* Cantor's works were translated into English during the early twentieth century.

Given two sets A and B, if any one of the following statements is true, then the other statements are also true:

1. There exists a one-to-one correspondence between the elements of A and B.

2. A and B are equivalent sets.

3. A and B have the same cardinal number; that is, $n(A) = n(B)$.

ThomsonNOW
for Liberal Arts Mathematics
www.thomsonedu.com/1pass.

EXAMPLE 1   Determine whether the sets in each of the following pairs are equivalent. If they are equivalent, list a one-to-one correspondence between their elements.

a. $A = \{$John, Paul, George, Ringo$\}$;
$B = \{$Lennon, McCartney, Harrison, Starr$\}$
b. $C = \{\alpha, \beta, \chi, \delta\}; D = \{$I, O, $\Delta\}$
c. $A = \{1, 2, 3, \ldots, 48, 49, 50\}; B = \{1, 3, 5, \ldots, 95, 97, 99\}$

Solution   **a.** If sets have the same cardinal number, they are equivalent. Now, $n(A) = 4$ and $n(B) = 4$; therefore, $A \sim B$.

Because A and B are equivalent, their elements can be put into a one-to-one correspondence. One such correspondence follows:

John	Paul	George	Ringo
$\updownarrow$	$\updownarrow$	$\updownarrow$	$\updownarrow$
Lennon	McCartney	Harrison	Starr

b. Because $n(C) = 4$ and $n(D) = 3$, C and D are not equivalent.

c. A consists of all natural numbers from 1 to 50, inclusive. Hence, $n(A) = 50$. B consists of all odd natural numbers from 1 to 99, inclusive. Since half of the natural numbers from 1 to 100 are odd (and half are even), there are fifty ($100 \div 2 = 50$) odd natural numbers less than 100; that is, $n(B) = 50$. Because A and B have the same cardinal number, $A \sim B$.

Many different one-to-one correspondences may be established between the elements of A and B. One such correspondence follows:

$$A = \{1, 2, 3, \ldots, \quad n, \quad \ldots, 48, 49, 50\}$$
$$\updownarrow \updownarrow \updownarrow \ldots \qquad \updownarrow \qquad \ldots \updownarrow \updownarrow \updownarrow$$
$$B = \{1, 3, 5, \ldots, (2n - 1), \ldots, 95, 97, 99\}$$

That is, each natural number $n \in A$ is paired up with the odd number $(2n - 1) \in B$. The $n \leftrightarrow (2n - 1)$ part is crucial because it shows *each* individual correspondence. For example, it shows that $13 \in A$ corresponds to $25 \in B$ ($n = 13$, so $2n = 26$ and $2n - 1 = 25$). Likewise, $69 \in B$ corresponds to $35 \in A$ ($2n - 1 = 69$, so $2n = 70$ and $n = 35$). ∎

As we have seen, if two sets have the same cardinal number, they are equivalent, and their elements can be put into a one-to-one correspondence. Conversely, if the elements of two sets can be put into a one-to-one correspondence, the sets have the same cardinal number and are equivalent. Intuitively, this result appears to be quite obvious. However, when Georg Cantor applied this relationship to infinite sets, he sparked one of the greatest philosophical debates of the nineteenth century.

Countable Sets

Consider the set of all counting numbers $N = \{1, 2, 3, \ldots\}$, which consists of an infinite number of elements. Each of these numbers is either odd or even. Defining

O and E as $O = \{1, 3, 5, \ldots\}$ and $E = \{2, 4, 6, \ldots\}$, we have $O \cap E = \varnothing$ and $O \cup E = N$; the sets O and E are mutually exclusive, and their union forms the entire set of all counting numbers. Obviously, N contains elements that E does not. As we mentioned earlier, the fact that E is a *proper* subset of N might lead people to think that N is "bigger" than E. In fact, N and E are the "same size"; N and E each contain the same number of elements.

Recall that two sets are equivalent and have the same cardinal number if the elements of the sets can be matched up via a one-to-one correspondence. To show the existence of a one-to-one correspondence between the elements of two sets of numbers, we must find an explicit correspondence between the general elements of the two sets. In Example 1(c), we expressed the general correspondence as $n \leftrightarrow (2n - 1)$.

ThomsonNOW™
for Liberal Arts Mathematics
www.thomsonedu.com/1pass.

EXAMPLE 2

a. Show that $E = \{2, 4, 6, 8, \ldots\}$ and $N = \{1, 2, 3, 4, \ldots\}$ are equivalent sets.
b. Find the element of N that corresponds to $1430 \in E$.
c. Find the element of N that corresponds to $x \in E$.

Solution

a. To show that $E \sim N$, we must show that there exists a one-to-one correspondence between the elements of E and N. The elements of E and N can be paired up as follows:

$$N = \{1, 2, 3, 4, \ldots, n, \ldots\}$$
$$\updownarrow \updownarrow \updownarrow \updownarrow \ldots \quad \updownarrow$$
$$E = \{2, 4, 6, 8, \ldots, 2n, \ldots\}$$

Any natural number $n \in N$ corresponds with the even natural number $2n \in E$. Because there exists a one-to-one correspondence between the elements of E and N, the sets E and N are equivalent; that is, $E \sim N$.

b. $1430 = 2n \in E$, so $n = \frac{1430}{2} = 715 \in N$. Therefore, $715 \in N$ corresponds to $1430 \in E$.

c. $x = 2n \in E$, so $n = \frac{x}{2} \in N$. Therefore, $n = \frac{x}{2} \in N$ corresponds to $x = 2n \in E$. ∎

We have just seen that the set of *even* natural numbers is equivalent to the set of *all* natural numbers. This equivalence implies that the two sets have the same number of elements! Although E is a proper subset of N, both sets have the same cardinal number; that is, $n(E) = n(N)$. Settling the controversy sparked by this seeming paradox, mathematicians today define a set to be an **infinite set** if it can be placed in a one-to-one correspondence with a proper subset of itself.

How many counting numbers are there? How many even counting numbers are there? We know that each set contains an infinite number of elements and that $n(N) = n(E)$, but how many is that? In the late nineteenth century, Georg Cantor defined the cardinal number of the set of counting numbers to be $\aleph_0$ (read "**aleph-null**"). Cantor utilized Hebrew letters, of which aleph, $\aleph$, is the first. Consequently, the proper response to "How many counting numbers are there?" is "There are aleph-null of them"; $n(N) = \aleph_0$. Any set that is equivalent to the set of counting numbers has cardinal number $\aleph_0$. A set is **countable** if it is finite or if it has cardinality $\aleph_0$.

Cantor was not the first to ponder the paradoxes of infinite sets. Hundreds of years before, Galileo observed that part of an infinite set contained as many elements as the whole set. In his monumental *Dialogue Concerning the Two Chief World Systems* (1632), Galileo made a prophetic observation: "There are as many (perfect) squares as there are (natural) numbers because they are just as numerous as their roots." In other words, the elements of the sets $N = \{1, 2, 3, \ldots, n, \ldots\}$ and $S = \{1^2, 2^2, 3^2, \ldots, n^2, \ldots\}$ can be put into a one-to-one correspondence $(n \leftrightarrow n^2)$. Galileo pondered which of the sets (perfect squares or natural numbers) was "larger" but abandoned the subject because he could find no practical application of this puzzle.

EXAMPLE 3   Consider the following one-to-one correspondence between the set I of all integers and the set N of all natural numbers:

$$N = \{1, 2, \quad 3, \quad 4, \quad 5, \ldots\}$$
$$\updownarrow \updownarrow \quad \updownarrow \quad \updownarrow \quad \updownarrow$$
$$I = \{0, 1, -1, 2, -2, \ldots\}$$

where an odd natural number n corresponds to a nonpositive integer $\frac{1-n}{2}$ and an even natural number n corresponds to a positive integer $\frac{n}{2}$.

a. Find the 613th integer; that is, find the element of I that corresponds to $613 \in N$.
b. Find the element of N that corresponds to $853 \in I$.
c. Find the element of N that corresponds to $-397 \in I$.
d. Find $n(I)$, the cardinal number of the set I of all integers.
e. Is the set of integers countable?

Solution   a. $613 \in N$ is odd, so it corresponds to $\frac{1 - 613}{2} = \frac{-612}{2} = -306$. If you continued counting the integers as shown in the above correspondence, -306 would be the 613th integer in your count.

b. $853 \in I$ is positive, so

$$853 = \frac{n}{2} \qquad \text{multiplying by 2}$$
$$n = 1{,}706$$

$1{,}706 \in N$ corresponds to $853 \in I$.
   This means that 853 is the 1,706th integer.

c. $-397 \in I$ is negative, so

$$-397 = \frac{1 - n}{2}$$
$$-794 = 1 - n \qquad \text{multiplying by 2}$$
$$-795 = -n \qquad \text{subtracting 1}$$
$$n = 795 \qquad \text{multiplying by } -1$$

$795 \in N$ corresponds to $-397 \in I$.
   This means that -397 is the 795th integer.

d. The given one-to-one correspondence shows that I and N have the same (infinite) number of elements; $n(I) = n(N)$. Because $n(N) = \aleph_0$, the cardinal number of the set of all integers is $n(I) = \aleph_0$.

e. By definition, a set is called countable if it is finite or if it has cardinality $\aleph_0$. The set of integers has cardinality $\aleph_0$, so it is countable. This means that we can "count off" all of the integers, as we did in parts (a), (b), and (c). ∎

We have seen that the sets N (all counting numbers), E (all even counting numbers), and I (all integers) contain the same number of elements, $\aleph_0$. What about a set containing fractions?

EXAMPLE 4   Determine whether the set P of all positive rational numbers is countable.

Solution   The elements of P can be systematically listed in a table of rows and columns as follows: All positive rational numbers whose denominator is 1 are listed in the first row, all positive rational numbers whose denominator is 2 are listed in the second row, and so on, as shown in Figure 2.42.

Each positive rational number will appear somewhere in the table. For instance, $\frac{125}{66}$ will be in row 66 and column 125. Note that not all the entries in Figure 2.42 are in lowest terms; for instance, $\frac{2}{4}$, $\frac{3}{6}$, $\frac{4}{8}$, and so on are all equal to $\frac{1}{2}$.

Figure 2.42
A list of all positive rational numbers.

Figure 2.43
The circled rational numbers are not in lowest terms; they are omitted from the list.

Consequently, to avoid listing the same number more than once, an entry that is not in lowest terms must be eliminated from our list. To establish a one-to-one correspondence between P and N, we can create a zigzag diagonal pattern as shown by the arrows in Figure 2.43. Starting with $\frac{1}{1}$, we follow the arrows and omit any number that is not in lowest terms (the circled numbers in Figure 2.43). In this manner, a list of all positive rational numbers with no repetitions is created. Listing the elements of P in this order, we can put them in a one-to-one correspondence with N:

$$N = \{1, 2, 3, 4, 5, 6, 7, 8, 9, 10, 11, \ldots\}$$
$$\updownarrow \ \updownarrow \ \updownarrow \ \updownarrow \ \updownarrow \ \updownarrow \ \updownarrow \ \updownarrow \ \ \updownarrow \ \ \updownarrow$$
$$P = \left\{1, 2, \frac{1}{2}, \frac{1}{3}, 3, 4, \frac{3}{2}, \frac{2}{3}, \frac{1}{4}, \frac{1}{5}, 5, \ldots\right\}$$

Any natural number n is paired up with the positive rational number found by counting through the "list" given in Figure 2.43. Conversely, any positive rational number is located somewhere in the list and is paired up with the counting number corresponding to its place in the list.

Therefore, $P \sim N$, so the set of all positive rational numbers is countable.   ∎

Uncountable Sets

Every infinite set that we have examined so far is countable; each can be put into a one-to-one correspondence with the set of all counting numbers and consequently has cardinality $\aleph_0$. Do not be misled into thinking that all infinite sets are countable! By utilizing a "proof by contradiction," Georg Cantor showed that the infinite set $A = \{x \mid 0 \le x < 1\}$ is *not* countable. This proof involves logic that is different from what you are used to. Do not let that intimidate you.

Assume that the set $A = \{x \mid 0 \le x < 1\}$ is countable; that is, assume that $n(A) = \aleph_0$. This assumption implies that the elements of A and N can be put into a one-to-one correspondence; each $a \in A$ can be listed and counted. Because the elements of A are nonnegative real numbers less than 1, each $a_n = 0.\square\square\square\square\square\ldots$. Say, for instance, the numbers in our list are

$a_1 = 0.3750000 \ldots$ the first element of A

$a_2 = 0.7071067 \ldots$ the second element of A

$a_3 = 0.5000000 \ldots$ the third element of A

$a_4 = 0.6666666 \ldots$ and so on.

The *assumption* that A is countable implies that every element of A appears somewhere in the above list. However, we can create an element of A (call it b) that is *not* in the list. We build b according to the "diagonal digits" of the numbers in our list and the following rule: If the digit "on the diagonal" is not zero, put a 0 in the corresponding place in b; if the digit "on the diagonal" is zero, put a 1 in the corresponding place in b.

The "diagonal digits" of the numbers in our list are as follows:

$a_1 = 0.\boxed{3}750000 \ldots$

$a_2 = 0.7\boxed{0}71067 \ldots$

$a_3 = 0.50\boxed{0}0000 \ldots$

$a_4 = 0.666\boxed{6}666 \ldots$

Because the first digit on the diagonal is 3, the first digit of b is 0. Because the second digit on the diagonal is 0, the second digit of b is 1. Using all the "diagonal digits" of the numbers in our list, we obtain $b = 0.0110\ldots$. Because $0 \le b < 1$, it follows that $b \in A$. However, the number b is not on our list of all elements of A. This is because

$b \ne a_1$ (b and a_1 differ in the first decimal place)

$b \ne a_2$ (b and a_2 differ in the second decimal place)

$b \ne a_3$ (b and a_3 differ in the third decimal place), and so on

This contradicts the assumption that the elements of A and N can be put into a one-to-one correspondence. Since the assumption leads to a contradiction, the assumption must be false; $A = \{x \mid 0 \le x < 1\}$ is not countable. Therefore, $n(A) \ne n(N)$. That is, A is an infinite set and $n(A) \ne \aleph_0$.

An infinite set that cannot be put into a one-to-one correspondence with N is said to be **uncountable.** Consequently, an uncountable set has *more* elements than the set of all counting numbers. This implies that there are different magnitudes of infinity! To distinguish the magnitude of A from that of N, Cantor denoted the cardinality of $A = \{x \mid 0 \le x < 1\}$ as $n(A) = c$ (c for **continuum**). Thus, Cantor showed that $\aleph_0 < c$. Cantor went on to show that A was equivalent to the entire set of all real numbers, that is, $A \sim \Re$. Therefore, $n(\Re) = c$.

Although he could not prove it, Cantor hypothesized that no set could have a cardinality between $\aleph_0$ and c. This famous unsolved problem, labeled the *Continuum Hypothesis,* baffled mathematicians throughout the first half of the twentieth century. It is said that Cantor suffered a devastating nervous breakdown in 1884 when he announced that he had a proof of the Continuum Hypothesis only to declare the next day that he could show the Continuum Hypothesis to be false!

The problem was finally "solved" in 1963. Paul J. Cohen demonstrated that the Continuum Hypothesis is independent of the entire framework of set theory; that is, it can be neither proved nor disproved by using the theorems of set theory. Thus, the Continuum Hypothesis is not provable.

Although no one has produced a set with cardinality between $\aleph_0$ and c, many sets with cardinality greater than c have been constructed. In fact, modern mathematicians have shown that there are *infinitely* many magnitudes of infinity! Using subscripts, these magnitudes, or cardinalities, are represented by $\aleph_0, \aleph_1, \aleph_2, \ldots$ and have the property that $\aleph_0 < \aleph_1 < \aleph_2 < \ldots$. In this sense, the set N of all natural numbers forms the "smallest" infinite set. Using this subscripted notation, the Continuum Hypothesis implies that $c = \aleph_1$; that is, given that N forms the smallest infinite set, the set $\Re$ of all real numbers forms the next "larger" infinite set.

Points on a Line

Figure 2.44
The real number line.

When students are first exposed to the concept of the real number system, a number line like the one in Figure 2.44 is inevitably introduced. The real number system, denoted by $\Re$, can be put into a one-to-one correspondence with all points on a line, such that every real number corresponds to exactly one point on a line and every point on a line corresponds to exactly one real number. Consequently, any (infinite) line contains c points. What about a line segment? For example, how many points does the segment $[0, 1]$ contain? Does the segment $[0, 2]$ contain twice as many points as the segment $[0, 1]$? Once again, intuition can lead to erroneous conclusions when people are dealing with infinite sets.

EXAMPLE 5   Show that the line segments $[0, 1]$ and $[0, 2]$ are equivalent sets of points.

Solution   Because the segment $[0, 2]$ is twice as long as the segment $[0, 1]$, intuition might tell us that it contains twice as many points. Not so! Recall that two sets are equivalent (and have the same cardinal number) if their elements can be put into a one-to-one correspondence.

On a number line, let A represent the point 0, let B represent 1, and let C represent 2, as shown in Figure 2.45. Our goal is to develop a one-to-one correspondence between the elements of the segments AB and AC. Now draw the segments separately, with AB above AC, as shown in Figure 2.46. (To distinguish the segments from each other, point A of segment AB has been relabeled as point A'.)

Extend segments AA' and CB so that they meet at point D, as shown in Figure 2.47. Any point E on $A'B$ can be paired up with the unique point F on AC formed by the intersection of lines DE and AC, as shown in Figure 2.48. Conversely, any point F on segment AC can be paired up with the unique point E on $A'B$ formed by the intersection of lines DF and $A'B$. Therefore, a one-to-one correspondence exists between the two segments, so $[0, 1] \sim [0, 2]$. Consequently, the interval $[0, 1]$ contains exactly the same number of points as the interval $[0, 2]$!

Figure 2.45
The interval $[0, 2]$.

Figure 2.46
The intervals $[A, B]$ and $[A, C]$.

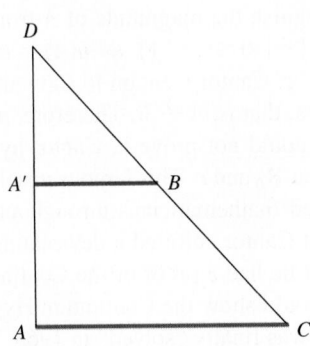

Figure 2.47
Extending AA' and CB to form point D.

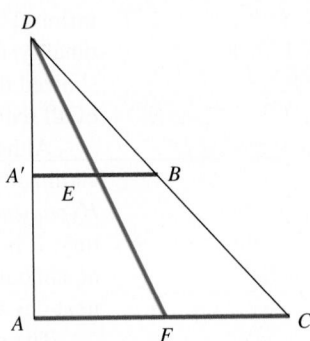

Figure 2.48
A one-to-one correspondence between line segments. ∎

Even though the segment [0, 2] is twice as long as the segment [0, 1], each contains exactly the same number of points. The method used in Example 5 can be applied to any two line segments. Consequently, all line segments, regardless of their length, contain exactly the same number of points; a line segment 1 inch long has exactly the same number of points as a segment 1 mile long! Once again, it is easy to see why Cantor's work on the magnitude of infinity was so unsettling to many scholars.

Having concluded that all line segments contain the same number of points, we might ask how many points that is. What is the cardinal number? Given any line segment AB, it can be shown that $n(AB) = c$; the points of a line segment can be put into a one-to-one correspondence with the points of a line. Consequently, the interval [0, 1] contains the same number of elements as the entire real number system.

If things seem rather strange at this point, keep in mind that Cantor's pioneering work produced results that puzzled even Cantor himself. In a paper written in 1877, Cantor constructed a one-to-one correspondence between the points in a square (a two-dimensional figure) and the points on a line segment (a one-dimensional figure). Extending this concept, he concluded that a line segment and the entire two-dimensional plane contain exactly the same number of points, c. Communicating with his colleague Richard Dedekind, Cantor wrote, "I see it, but I do not believe it." Subsequent investigation has shown that the number of points contained in the interval [0, 1] is the same as the number of points contained in all of three-dimensional space! Needless to say, Cantor's work on the cardinality of infinity revolutionized the world of modern mathematics.

2.5 Exercises

In Exercises 1–10, find the cardinal numbers of the sets in each given pair to determine whether the sets are equivalent. If they are equivalent, list a one-to-one correspondence between their elements.

1. S = {Sacramento, Lansing, Richmond, Topeka}
   C = {California, Michigan, Virginia, Kansas}
2. T = {Wyoming, Ohio, Texas, Illinois, Colorado}
   P = {Cheyenne, Columbus, Austin, Springfield, Denver}

3. R = {a, b, c}; G = {$\alpha, \beta, \chi, \delta$}
4. W = {I, II, III}; H = {one, two}
5. C = {3, 6, 9, 12, . . . , 63, 66}
   D = {4, 8, 12, 16, . . . , 84, 88}
6. A = {2, 4, 6, 8, . . . , 108, 110}
   B = {5, 10, 15, 20, . . . , 270, 275}
7. G = {2, 4, 6, 8, . . . , 498, 500}
   H = {1, 3, 5, 7, . . . , 499, 501}
8. E = {2, 4, 6, 8, . . . , 498, 500}
   F = {3, 6, 9, 12, . . . , 750, 753}

9. $A = \{1, 3, 5, \ldots, 121, 123\}$
$B = \{125, 127, 129, \ldots, 245, 247\}$

10. $S = \{4, 6, 8, \ldots, 664, 666\}$
$T = \{5, 6, 7, \ldots, 335, 336\}$

11. a. Show that the set O of all odd counting numbers, $O = \{1, 3, 5, 7, \ldots\}$, and $N = \{1, 2, 3, 4, \ldots\}$ are equivalent sets.
   b. Find the element of N that corresponds to $1{,}835 \in O$.
   c. Find the element of N that corresponds to $x \in O$.
   d. Find the element of O that corresponds to $782 \in N$.
   e. Find the element of O that corresponds to $n \in N$.

12. a. Show that the set W of all whole numbers, $W = \{0, 1, 2, 3, \ldots\}$, and $N = \{1, 2, 3, 4, \ldots\}$ are equivalent sets.
   b. Find the element of N that corresponds to $932 \in W$.
   c. Find the element of N that corresponds to $x \in W$.
   d. Find the element of W that corresponds to $932 \in N$.
   e. Find the element of W that corresponds to $n \in N$.

13. a. Show that the set T of all multiples of 3, $T = \{3, 6, 9, 12, \ldots\}$, and $N = \{1, 2, 3, 4, \ldots\}$ are equivalent sets.
   b. Find the element of N that corresponds to $936 \in T$.
   c. Find the element of N that corresponds to $x \in T$.
   d. Find the element of T that corresponds to $936 \in N$.
   e. Find the element of T that corresponds to $n \in N$.

14. a. Show that the set F of all multiples of 5, $F = \{5, 10, 15, 20, \ldots\}$, and $N = \{1, 2, 3, 4, \ldots\}$ are equivalent sets.
   b. Find the element of N that corresponds to $605 \in F$.
   c. Find the element of N that corresponds to $x \in F$.
   d. Find the element of F that corresponds to $605 \in N$.
   e. Find the element of F that corresponds to $n \in N$.

15. Consider the following one-to-one correspondence between the set A of all even integers and the set N of all natural numbers:

$$N = \{1, 2, \quad 3, \quad 4, \quad 5, \ldots\}$$
$$\updownarrow\updownarrow \quad \updownarrow \quad \updownarrow \quad \updownarrow$$
$$A = \{0, 2, -2, 4, -4, \ldots\}$$

where an odd natural number n corresponds to the nonpositive integer $1 - n$ and an even natural number n corresponds to the positive even integer n.
   a. Find the 345th even integer; that is, find the element of A that corresponds to $345 \in N$.
   b. Find the element of N that corresponds to $248 \in A$.
   c. Find the element of N that corresponds to $-754 \in A$.
   d. Find $n(A)$.

16. Consider the following one-to-one correspondence between the set B of all odd integers and the set N of all natural numbers:

$$N = \{1, \quad 2, \quad 3, \quad 4, \quad 5, \ldots\}$$
$$\updownarrow \quad \updownarrow\updownarrow \quad \updownarrow \quad \updownarrow$$
$$B = \{1, -1, 3, -3, 5, \ldots\}$$

where an even natural number n corresponds to the negative odd integer $1 - n$ and an odd natural number n corresponds to the odd integer n.
   a. Find the 345th odd integer; that is, find the element of B that corresponds to $345 \in N$.
   b. Find the element of N that corresponds to $241 \in B$.
   c. Find the element of N that corresponds to $-759 \in B$.
   d. Find $n(B)$.

In Exercises 17–22, show that the given sets of points are equivalent by establishing a one-to-one correspondence.

17. the line segments [0, 1] and [0, 3]

18. the line segments [1, 2] and [0, 3]

19. the circle and square shown in Figure 2.49

   HINT: Draw one figure inside the other.

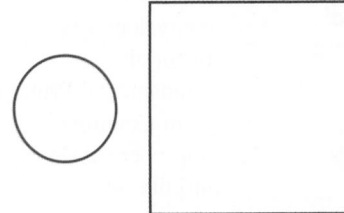

Figure 2.49

20. the rectangle and triangle shown in Figure 2.50

   HINT: Draw one figure inside the other.

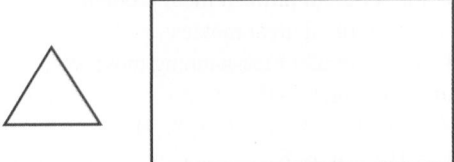

Figure 2.50

21. a circle of radius 1 cm and a circle of 5 cm

   HINT: Draw one figure inside the other.

22. a square of side 1 cm and a square of side 5 cm

HINT: Draw one figure inside the other.

23. Show that the set of all real numbers between 0 and 1 has the same cardinality as the set of all real numbers.

HINT: Draw a semicircle to represent the set of real numbers between 0 and 1 and a line to represent the set of all real numbers, and use the method of Example 5.

▬▬▬▶ *Answer the following questions using complete sentences and your own words.*

CONCEPT QUESTIONS

24. What is the cardinal number of the "smallest" infinite set? What set or sets have this cardinal number?

HISTORY QUESTIONS

25. What aspect of Georg Cantor's set theory caused controversy among mathematicians and philosophers?

26. What contributed to Cantor's breakdown in 1884?

27. Who demonstrated that the Continuum Hypothesis cannot be proven? When?

 ## WEB PROJECT

28. Write a research paper on any historical topic referred to in this section. Following is a partial list of topics:
- Georg Cantor
- Richard Dedekind
- Paul J. Cohen
- Bernhard Bolzano
- Leopold Kronecker
- the Continuum Hypothesis

Some useful links for this web project are listed on the text web site: **www.thomsonedu.com/mathematics/johnson**

Chapter 2   Review

ThomsonNOW™ **for Liberal Arts Mathematics**
Preparing for an exam? Test yourself on key material by visiting ThomsonNOW at **www.thomsonedu.com/1pass.**

TERMS

aleph-null	element	intersection	subset
cardinal number	empty set	mutually exclusive	tree diagram
combination	equal sets	one-to-one correspondence	uncountable set
combinatorics	equivalent sets	permutation	union
complement	factorial	proper subset	universal set
continuum	Fundamental Principle	roster notation	Venn diagram
countable set	of Counting	set	well-defined set
De Morgan's Laws	improper subset	set-builder notation	
distinguishable	infinite set	set theory	
permutations			

REVIEW EXERCISES

1. State whether the given set is well-defined.
  a. the set of all multiples of 5.
  b. the set of all difficult math problems
  c. the set of all great movies
  d. the set of all Oscar-winning movies

2. Given the sets
  $U = \{0, 1, 2, 3, 4, 5, 6, 7, 8, 9\}$
  $A = \{0, 2, 4, 6, 8\}$
  $B = \{1, 3, 5, 7, 9\}$
  find the following using the roster method.
  a. A' **b.** B'
  c. $A \cup B$ **d.** $A \cap B$

3. Given the sets $A = \{$Maria, Nobuko, Leroy, Mickey, Kelly$\}$ and $B = \{$Rachel, Leroy, Deanna, Mickey$\}$, find the following.
  a. $A \cup B$ **b.** $A \cap B$

4. List all subsets of $C = \{$Dallas, Chicago, Tampa$\}$. Identify which subsets are proper and which are improper.

5. Given $n(U) = 61$, $n(A) = 32$, $n(B) = 26$, and $n(A \cup B) = 40$, do the following.
  a. Find $n(A \cap B)$.
  b. Draw a Venn diagram illustrating the composition of U.

6. A survey of 2,000 college seniors yielded the following information: 1,324 favored capital punishment, 937 favored stricter gun control, and 591 favored both.
   a. How many favored capital punishment or stricter gun control?
   b. How many favored capital punishment but not stricter gun control?
   c. How many favored stricter gun control but not capital punishment?
   d. How many favored neither capital punishment nor stricter gun control?

7. An exit poll yielded the following information concerning people's voting patterns on Propositions A, B, and C: 305 voted yes on A, 95 voted yes only on A, 393 voted yes on B, 192 voted yes only on B, 510 voted yes on A or B, 163 voted yes on C, 87 voted yes on all three, and 213 voted no on all three. What percent of the voters voted yes on more than one proposition?

8. Given the sets $U = \{a, b, c, d, e, f, g, h, i\}$, $A = \{b, d, f, g\}$, and $B = \{a, c, d, g, i\}$, use De Morgan's Laws to find the following.
   a. $(A' \cup B)'$ b. $(A \cap B')'$

9. Refer to the Venn diagram depicted in Figure 2.31.
   a. In a group of 100 Americans, how many have type O or type A blood?
   b. In a group of 100 Americans, how many have type O and type A blood?
   c. In a group of 100 Americans, how many have neither type O nor type A blood?

10. Refer to the Venn diagram depicted in Figure 2.27.
   a. For a typical group of 100 Americans, fill in the cardinal number of each region in the diagram.
   b. In a group of 100 Americans, how many have type O blood or are Rh+?
   c. In a group of 100 Americans, how many have type O blood and are Rh+?
   d. In a group of 100 Americans, how many have neither type O blood nor are Rh+?

11. Sid and Nancy are planning their anniversary celebration, which will include viewing an art exhibit, having dinner, and going dancing. They will go to either the Museum of Modern Art or the New Photo Gallery; dine either at Stars, at Johnny's, or at the Chelsea; and go dancing either at Le Club or at Lizards.
   a. How many different ways can Sid and Nancy celebrate their anniversary?
   b. Construct a tree diagram to list all possible ways in which Sid and Nancy can celebrate their anniversary.

12. A certain model of pickup truck is available in five exterior colors, three interior colors, and three interior styles. In addition, the transmission can be either manual or automatic, and the truck can have either two-wheel or four-wheel drive. How many different versions of the pickup truck can be ordered?

13. Each student at State University has a student I.D. number consisting of five digits (the first digit is nonzero, and digits can be repeated) followed by two of the letters A, B, C, and D (letters cannot be repeated). How many different student numbers are possible?

14. Find the value of each of the following.
   a. $(17 - 7)!$ b. $(17 - 17)!$
   c. $\dfrac{82!}{79!}$ d. $\dfrac{27!}{20!7!}$

15. In how many ways can you select three out of eleven items under the following conditions?
   a. Order of selection is not important.
   b. Order of selection is important.

16. Find the value of each of the following.
   a. $_{15}P_4$ b. $_{15}C_4$ c. $_{15}P_{11}$

17. A group of ten women and twelve men must select a three-person committee. How many committees are possible if it must consist of the following?
   a. one woman and two men
   b. any mixture of men and women
   c. a majority of men

18. A volleyball league has ten teams. If every team must play every other team once in the first round of league play, how many games must be scheduled?

19. A volleyball league has ten teams. How many different end-of-the-season rankings of first, second, and third place are possible (disregarding ties)?

20. Using a standard deck of 52 cards (no jokers), how many seven-card poker hands are possible?

21. Using a standard deck of 52 cards and two jokers, how many seven-card poker hands are possible?

22. A 6/42 lottery requires choosing six of the numbers 1 through 42. How many different lottery tickets can you choose?

In Exercises 23–24, find the number of permutations of the letters in each word.

23. a. FLORIDA   b. ARIZONA   c. MONTANA
24. a. REAL   b. REEL
25. What is the major difference between permutations and combinations?

26. Use Pascal's Triangle to answer the following.
    a. Which entry in which row would you find the value of $_7C_3$?
    b. Which entry in which row would you find the value of $_7C_4$?
    c. How is the value of $_7C_3$ related to the value of $_7C_4$? Why?
    d. What is the location of $_nC_r$? Why?

27. Given the set $S = \{a, b, c\}$, answer the following.
    a. How many one-element subsets does S have?
    b. How many two-element subsets does S have?
    c. How many three-element subsets does S have?
    d. How many zero-element subsets does S have?
    e. How many subsets does S have?
    f. How is the answer to part (e) related to $n(S)$?

In Exercises 28–30, find the cardinal numbers of the sets in each given pair to determine whether the sets are equivalent. If they are equivalent, list a one-to-one correspondence between their elements.

28. $A = \{I, II, III, IV, V\}$ and $B = \{one, two, three, four, five\}$

29. $C = \{3, 5, 7, \ldots, 899, 901\}$ and $D = \{2, 4, 6, \ldots, 898, 900\}$

30. $E = \{Ronald\}$ and $F = \{Reagan, McDonald\}$

31. **a.** Show that the set S of perfect squares, $S = \{1, 4, 9, 16, \ldots\}$, and $N = \{1, 2, 3, 4, \ldots\}$ are equivalent sets.
    b. Find the element of N that corresponds to $841 \in S$.
    c. Find the element of N that corresponds to $x \in S$.
    d. Find the element of S that corresponds to $144 \in N$.
    e. Find the element of S that corresponds to $n \in N$.

32. Consider the following one-to-one correspondence between the set A of all integer multiples of 3 and the set N of all natural numbers:

$$N = \{1, 2, \ 3, \ 4, \ 5, \ldots\}$$
$$\updownarrow \ \updownarrow \ \ \updownarrow \ \ \updownarrow \ \ \updownarrow$$
$$A = \{0, 3, -3, 6, -6, \ldots\}$$

where an odd natural number n corresponds to the nonpositive integer $\frac{3}{2}(1 - n)$, and an even natural number n corresponds to the positive even integer $\frac{3}{2}n$.
    a. Find the element of A that corresponds to $396 \in N$.
    b. Find the element of N that corresponds to $396 \in A$.
    c. Find the element of N that corresponds to $-153 \in A$.
    d. Find $n(A)$.

33. Show that the line segments $[0, 1]$ and $[0, \pi]$ are equivalent sets of points by establishing a one-to-one correspondence.

▬▬▶ *Answer the following questions using complete sentences and your own words.*

CONCEPT QUESTIONS

34. What is the difference between proper and improper subsets?

35. Explain the difference between $\{0\}$ and $\emptyset$.

36. What is a factorial?

37. What is the difference between permutations and combinations?

HISTORY QUESTIONS

38. What roles did the following people play in the development of set theory and combinatorics?
- Georg Cantor
- Augustus De Morgan
- Christian Kramp
- Chu Shih-chieh
- John Venn

3

Probability

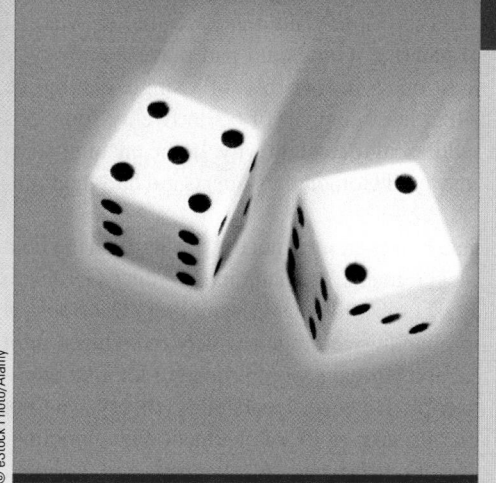

© eStock Photo/Alamy

Uncertainty is a part of our lives. When we wake up, we check the weather report to find out whether it might rain. We check the traffic report to find out whether we might get stuck in a traffic jam. We check to find out whether the interest rate changed for that car, boat, or home loan. Probability theory is the branch of mathematics that analyzes uncertainty.

Businesses have long used probability theory to help determine the best way to deal with uncertainty. This began in the 1600s, when the first life insurance companies used probability theory to analyze death records and help to determine appropriate fees. Today, all insurance companies use probability theory to determine fees for all types of insurance. Investment firms use probability theory to assess financial risk. Medical firms use probability theory in diagnosing ailments and determining appropriate treatments.

Probability theory began with a roll of the dice. The Chevalier de Méré, a seventeenth-century French nobleman and generally successful gambler, had lost too much money with a certain dice bet, and he asked his friend, the French mathematician Blaise Pascal, to explain why. In answering his friend's question, Pascal laid the foundation of probability theory.

Today, casinos use probability theory to set their house odds so that the casinos are assured of making a profit. In this chapter, we will use probability theory to determine whether there are any good bets in casino games of chance.

Probability theory has an interesting history filled with interesting people. Gerolamo Cardano was one of these creators of probability theory; he was also a compulsive gambler, and he might well have plagiarized much of his mathematics from one of his servants. He went on to become one of Europe's most successful physicians. Blaise Pascal was a very religious man and not a gambler. However, he was willing to use mathematics to help his friend win. Later, Pascal gave up mathematics and the sciences and devoted his intellect to defending the Christian sect of Jansenism.

For a long time, probability theory was viewed as a rather disreputable branch of mathematics because of its association with gambling. Gregor Mendel helped to change this view. He used probability theory to analyze the effect of randomness on heredity, thus founding the science of genetics. We will use probability theory to analyze the transmission of hair color: If you know the color of the parents' hair, you can predict the color of their child's hair. We will also analyze the transmission of inherited diseases such as sickle-cell anemia, cystic fibrosis, Huntington's disease, and Tay-Sachs disease. You should understand what probability theory predicts about such diseases because your child could acquire one of them even if you and your spouse are disease-free.

ThomsonNOW™ for Liberal Arts Mathematics

3.1   History of Probability

Probability theory is the field of mathematics that measures the likelihood of an event. It is generally considered to have originated in 1654 with the written correspondence between French mathematicians Blaise Pascal and Pierre de Fermat about several problems involving dice games. The correspondence was in response to a problem posed by Pascal's friend Antoine Gombauld, the Chevalier de Méré, a French nobleman and rather successful gambler. Gombauld had made money over the years betting that he could roll at least one 6 in four rolls of a single die, but he had lost money betting that he could roll at least one pair of 6's in twenty-four rolls of a pair of dice. He challenged his friend Pascal to explain why. In the ensuing communications, the two mathematicians answered Gombauld's query and laid the foundation of probability theory.

The letters between Pascal and Fermat, however, were not the first entry in the field of probability. In 1545, the Italian physician, mathematician, and gambler Gerolamo Cardano wrote the first theoretical study of probabilities and gambling, *The Book on Games of Chance.* Cardano's ideas on probabilities were not widely recognized until after his death. Interest aroused by Pascal and Fermat's work prompted the first publication of Cardano's century-old manuscript. However, probability theory was not viewed as a serious branch of mathematics because of its involvement with gambling.

In 1662, the Englishman John Graunt published his *Natural and Political Observations on the Bills of Mortality,* which used probabilities to analyze death records and was the first important incidence of the use of probabilities for a purpose other than gambling. English firms were selling the first life insurance policies, and they used mortality tables to make their fees appropriate to the risks involved.

Probability theory was shown to be a serious area of interest when Jacob Bernoulli's *Ars Conjectandi (Art of Guessing)* was published in 1713. Although its focus was on gambling, it also suggested applications of probability to government, economics, law, and genetics.

In 1812, Pierre-Simon, the Marquis de Laplace, published his *Théorie Analytique des Probabilités.* This work, written in conjunction with his astronomical studies, showed probability to be a respectable and significant field of mathematics by demonstrating its usefulness in interpreting scientific data. Laplace showed how probabilities can be used to determine the most probable true size of a quantity when repeated measurements of that quantity vary somewhat.

The work of Gregor Mendel, an Austrian monk, was especially important in the history of probability theory. In 1865, he published the results of his experiments on pea plants in the abbey garden. These results allowed for randomness in genetics and used probabilities to analyze the effect of that randomness. Ignored during his lifetime, Mendel's work is considered the foundation of genetics.

Probability theory was originally created to aid gamblers, and games of chance are the most easily and universally understood topic to which probability theory can be applied. Learning about probability and games of chance may convince you not to be a gambler. As we will see in Section 3.5, there is not a single good bet in a casino game of chance.

This **Bills of Mortality** (records of death) was published shortly after John Graunt's analysis.

Roulette

Roulette is the oldest casino game still being played. Its invention has variously been credited to Pascal, the ancient Chinese, a French monk, and the Italian mathematician

Historical Note Blaise Pascal, 1623–1662

© Sheila Terry/Photo Researchers, Inc.

As a child, Blaise Pascal showed an early aptitude for science and mathematics, even though he was discouraged from studying in order to protect his poor health. Acute digestive problems and chronic insomnia made his life miserable. Few of Pascal's days were without pain.

At age sixteen, Pascal wrote a paper on geometry that won the respect of the French mathematical community and the jealousy of the prominent French mathematician René Descartes. It has been suggested that the animosity between Descartes and Pascal was in part due to their religious differences. Descartes was a Jesuit, and Pascal was a Jansenist. Although Jesuits and Jansenists were both Roman Catholics, Jesuits believed in free will and supported the sciences, while Jansenists believed in predestination and mysticism and opposed the sciences and the Jesuits.

At age nineteen, to assist his father, a tax administrator, Pascal invented the first calculating machine.

Besides cofounding probability theory with Pierre de Fermat, Pascal contributed to the advance of calculus, and his studies in physics culminated with a law on the effects of pressure on fluids that bears his name.

At age thirty-one, after a close escape from death in a carriage accident, Pascal turned his back on mathematics and the sciences and focused on defending Jansenism against the Jesuits. Pascal came to be the greatest Jansenist, and he aroused a storm with his anti-Jesuit *Provincial Letters*. This work is still famous for its polite irony.

Pascal turned so far from the sciences that he came to believe that reason is inadequate to solve humanity's difficulties or to satisfy our hopes, and he regarded the pursuit of science as a vanity. Even so, he still occasionally succumbed to its lure. Once, to distract himself from pain, he concentrated on a geometry problem. When the pain stopped, he decided that it was a signal from God that he had not sinned by thinking of mathematics instead of his soul. This incident resulted in the only scientific work of his last few years—and his last work. He died later that year, at age thirty-nine.

© SSPL/The Image Works

Pascal's calculator, the Pascaline

Historical Note   Gerolamo Cardano, 1501–1576

North Wind Picture Archives

Gerolamo Cardano is the subject of much disagreement among historians. Some see him as a man of tremendous accomplishments, while others see him as a plagiarist and a liar. All agree that he was a compulsive gambler.

Cardano was trained as a medical doctor, but he was initially denied admission to the College of Physicians of Milan. That denial was ostensibly due to his illegitimate birth, but some suggest that the denial was in fact due to his unsavory reputation as a gambler, since illegitimacy was neither a professional nor a social obstacle in sixteenth-century Italy. His lack of professional success left him with much free time, which he spent gambling and reading. It also resulted in a stay in the poorhouse.

Cardano's luck changed when he obtained a lectureship in mathematics, astronomy, and astrology at the University of Milan. He wrote a number of books on mathematics and became famous for publishing a method of solving third-degree equations. Some claim that Cardano's mathematical success was due not to his own abilities but rather to those of Ludovico Ferrari, a servant of his who went on to become a mathematics professor.

While continuing to teach mathematics, Cardano returned to the practice of medicine. (He was finally allowed to join the College of Physicians, perhaps owing to his success as a mathematician.) Cardano wrote books on medicine and the natural sciences that were well thought of. He became one of the most highly regarded physicians in Europe and counted many prominent people among his patients. He designed a tactile system, somewhat like braille, for the blind. He also designed an undercarriage suspension device that was later adapted as a universal joint for automobiles and that is still called a *cardan* in Europe.

Cardano's investment in gambling was enormous. He not only wagered (and lost) a great deal of money but also spent considerable time and effort calculating probabilities and devising strategies. His *Book on Games of Chance* contains the first correctly calculated theoretical probabilities.

Cardano's autobiography, *The Book of My Life,* reveals a unique personality. He admitted that he loved talking about himself, his accomplishments, and his illnesses and diseases. He frequently wrote of injuries done him by others and followed these complaints with gleeful accounts of his detractors' deaths. Chapter titles include "Concerning my friends and patrons," "Calumny, defamations, and treachery of my unjust accusers," "Gambling and dicing," "Religion and piety," "The disasters of my sons," "Successes in my practice," "Things absolutely supernatural," and "Things of worth which I have achieved in various studies."

Don Pasquale. It became popular when a French policeman introduced it to Paris in 1765 in an attempt to take the advantage away from dishonest gamblers. When the Monte Carlo casino opened in 1863, roulette was the most popular game, especially among the aristocracy. The American roulette wheel has 38 numbered compartments around its circumference. Thirty-six of these compartments are numbered from 1 to 36 and are colored red or black. The remaining two are numbered 0 and 00 and are colored green. Players place their bets by putting their chips on an appropriate spot on the roulette table (see Figure 3.1). The dealer spins the wheel and then drops a ball onto the spinning wheel. The ball eventually comes to rest in one of the compartments, and that compartment's number is the winning number. (See page 177 for a photo of the roulette wheel and table.)

For example, if a player wanted to bet $10 that the ball lands in compartment number 7, she would place $10 worth of chips on the number 7 on the roulette table. This is a single-number bet, so house odds are 35 to 1 (see Figure 3.1 for house odds). This means that if the player wins, she wins $10 · 35 = $350, and if she loses, she loses her $10.

Roulette quickly became a favorite game of the French upper class, as shown in this 1910 photo of Monte Carlo.

Similarly, if a player wanted to bet \$5 that the ball lands either on 13 or 14, he would place \$5 worth of chips on the line separating numbers 13 and 14 on the roulette table. This is a two-numbers bet, so house odds are 17 to 1. This means that if the player wins, he wins $\$5 \cdot 17 = \$85$, and if he loses, he loses his \$5.

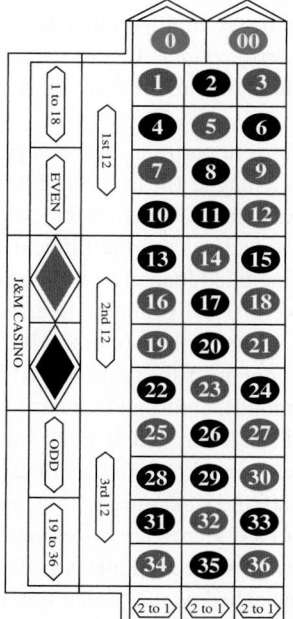

Bet	House Odds
single number	35 to 1
two numbers ("split")	17 to 1
three numbers ("street")	11 to 1
four numbers ("square")	8 to 1
five numbers ("line")	6 to 1
six numbers ("line")	5 to 1
twelve numbers (column or section)	2 to 1
low or high (1 to 18 or 19 to 36, respectively)	1 to 1
even or odd (0 and 00 are neither even nor odd)	1 to 1
red or black	1 to 1

Figure 3.1
The roulette table and house odds for the various roulette bets

A die. The other three faces have four, five, and six spots.

A sixteenth-century two of swords

A sixteenth-century king of clubs

Dice and Craps

Dice have been cast since the beginning of time, for both divination and gambling purposes. The earliest **die** (singular of *dice*) was an animal bone, usually a knucklebone or foot bone. The Romans were avid dice players. The Roman emperor Claudius I wrote a book titled *How to Win at Dice*. During the Middle Ages, dicing schools and guilds of dicers were quite popular among the knights and ladies.

Hazard, an ancestor of the dice game craps, is an English game that was supposedly invented by the Crusaders in an attempt to ward off boredom during long, drawn-out sieges. It became quite popular in England and France in the nineteenth century. The English called a throw of 2, 3, or 12 *crabs,* and it is believed that *craps* is a French mispronunciation of that term. The game came to America with the French colonization of New Orleans and spread up the Mississippi.

Cards

The invention of playing cards has been credited to the Indians, the Arabs, the Egyptians, and the Chinese. During the Crusades, Arabs endured lengthy sieges by playing card games. Their European foes acquired the cards and introduced them to their homelands. Cards, like dice, were used for divination as well as gambling. In fact, the modern deck is derived from the Tarot deck, which is composed of four suits plus twenty-two *atouts* that are not part of any suit. Each suit represents a class of medieval society: swords represent the nobility; coins, the merchants; batons or clubs, the peasants; and cups or chalices, the church. These suits are still used in regular playing cards in southern Europe. The Tarot deck also includes a joker and, in each suit, a king, a queen, a knight, a knave, and ten numbered cards.

Around 1500, the French dropped the knights and *atouts* from the deck and changed the suits from swords, coins, clubs, and chalices to *piques* (soldiers' pikes), *carreaux* (diamond-shaped building tiles), *trèfles* (clover leaf-shaped trefoils), and *coeurs* (hearts). In sixteenth-century Spain, *piques* were called *espados,* from which we get our term *spades.* Our diamonds are so named because of the shape of the carreaux. Clubs was an original Tarot suit, and hearts is a translation of *coeurs.*

The pictures on the cards were portraits of actual people. In fourteenth-century Europe, the kings were Charlemagne (hearts), the biblical David (spades), Julius Caesar (diamonds), and Alexander the Great (clubs); the queens included Helen of Troy (hearts), Pallas Athena (spades), and the biblical Rachel (diamonds). Others honored as "queen for a day" included Joan of Arc, Elizabeth I, and Elizabeth of York, wife of Henry VII. Jacks were usually famous warriors, including Sir Lancelot (clubs) and Roland, Charlemagne's nephew (diamonds).

A modern deck of cards contains fifty-two cards (thirteen in each of four suits). The four suits are hearts, diamonds, clubs, and spades (♥, ♦, ♣, ♠). Hearts and diamonds are red, and clubs and spades are black. Each suit consists of cards labeled 2 through 10, followed by jack, queen, king, and ace. **Face cards** are the jack, queen, and king; and **picture cards** are the jack, queen, king, and ace.

Two of the most popular card games are poker and blackjack. Poker's ancestor was a Persian game called *dsands,* which became popular in eighteenth-century Paris. It was transformed into a game called *poque,* which spread to America via the French colony in New Orleans. *Poker* is an American mispronunciation of the word *poque.*

The origins of blackjack (also known as *vingt-et-un* or twenty-one) are unknown. The game is called twenty-one because high odds are paid if a player's first two cards total 21 points (an ace counts either 1 or 11, and the ten, jack, queen, and king each count 10). A special bonus used to be paid if those two cards were a black jack and a black ace, hence the name *blackjack*.

A modern deck of cards

3.1 Exercises

1. Roll a single die four times, and record the number of times a six comes up. Repeat this ten times. If you had made the Chevalier de Méré's favorite bet (at $10 per game), would you have won or lost money? How much?

2. Roll a pair of dice twenty-four times, and record the number of times a pair of sixes comes up. Repeat this five times. If you had made the Chevalier de Méré's bet (at $10 per game), would you have won or lost money? How much?

3. **a.** If you were to flip a pair of coins thirty times, approximately how many times do you think a pair of heads would come up? A pair of tails? One head and one tail?
   b. Flip a pair of coins thirty times, and record the number of times a pair of heads comes up, the number of times a pair of tails comes up, and the number of times one head and one tail come up. How closely do the results agree with your guess?

4. **a.** If you were to flip a single coin twenty times, approximately how many times do you think heads would come up? Tails?
   b. Flip a single coin twenty times, and record the number of times heads comes up and the number of times tails comes up. How closely do the results agree with your guesses?

5. **a.** If you were to roll a single die twenty times, approximately how many times do you think an even number would come up? An odd number?
   b. Roll a single die twenty times, and record the number of times an even number comes up and the number of times an odd number comes up. How closely do the results agree with your guess?

6. **a.** If you were to roll a pair of dice thirty times, approximately how many times do you think the total would be 7? Approximately how many times do you think the total would be 12?
   b. Roll a single die thirty times, and record the number of times the total is 7 and the number of

times the total is 12. How closely do the results agree with your guess?

7. **a.** If you were to deal twenty-six cards from a complete deck (without jokers), approximately how many cards do you think would be red? Approximately how many do you think would be aces?

   b. Deal twenty-six cards from a complete deck (without jokers), and record the number of times a red card is dealt and the number of times an ace is dealt. How closely do the results agree with your guess?

8. **a.** If you were to deal twenty-six cards from a complete deck (without jokers), approximately how many cards do you think would be black? Approximately how many do you think would be jacks, queens, or kings?

   b. Deal twenty-six cards from a complete deck (without jokers), and record the number of times a black card is dealt and the number of times a jack, queen, or king is dealt. How closely do the results agree with your guess?

In Exercises 9–24, use Figure 3.1 to find the outcome of the bets in roulette, given the results listed.

9. You bet $10 on the 25.
   a. The ball lands on number 25.
   b. The ball lands on number 14.

10. You bet $15 on the 17.
    a. The ball lands on 18.
    b. The ball lands on 17.

11. You bet $5 on 17-20 split.
    a. The ball lands on number 17.
    b. The ball lands on number 20.
    c. The ball lands on number 32.

12. You bet $30 on the 22-23-24 street.
    a. The ball lands on number 19.
    b. The ball lands on number 22.
    c. The ball lands on number 0.

13. You bet $20 on the 8-9-11-12 square.
    a. The ball lands on number 15.
    b. The ball lands on number 9.
    c. The ball lands on number 00.

14. You bet $100 on the 0-00-1-2-3 line (the only five-number line on the table).
    a. The ball lands on number 29.
    b. The ball lands on number 2.

15. You bet $10 on the 31-32-33-34-35-36 line.
    a. The ball lands on number 5.
    b. The ball lands on number 33.

16. You bet $20 on the 13 through 24 section.
    a. The ball lands on number 00.
    b. The ball lands on number 15.

17. You bet $25 on the first column.
    a. The ball lands on number 13.
    b. The ball lands on number 14.

18. You bet $30 on the low numbers.
    a. The ball lands on number 8.
    b. The ball lands on number 30.

19. You bet $50 on the odd numbers.
    a. The ball lands on number 00.
    b. The ball lands on number 5.

20. You bet $20 on the black numbers.
    a. The ball lands on number 11.
    b. The ball lands on number 12.

21. You make a $20 single-number bet on number 14 and also a $25 single-number bet on number 15.
    a. The ball lands on number 16.
    b. The ball lands on number 15.
    c. The ball lands on number 14.

22. You bet $10 on the low numbers and also bet $20 on the 16-17-19-20 square.
    a. The ball lands on number 16.
    b. The ball lands on number 19.
    c. The ball lands on number 14.

23. You bet $30 on the 1-2 split and also bet $15 on the even numbers.
    a. The ball lands on number 1.
    b. The ball lands on number 2.
    c. The ball lands on number 3.
    d. The ball lands on number 4.

24. You bet $40 on the 1-12 section and also bet $10 on number 10.
    a. The ball lands on number 7.
    b. The ball lands on number 10.
    c. The ball lands on number 21.

25. How much must you bet on a single number to be able to win at least $100? (Bets must be in $1 increments.)

26. How much must you bet on a two-number split to be able to win at least $200? (Bets must be in $1 increments.)

27. How much must you bet on a twelve-number column to be able to win at least $1000? (Bets must be in $1 increments.)

28. How much must you bet on a four-number square to win at least $600? (Bets must be in $1 increments.)

29. **a.** How many hearts are there in a deck of cards?
    b. What fraction of a deck is hearts?

30. **a.** How many red cards are there in a deck of cards?
    b. What fraction of a deck is red?
31. **a.** How many face cards are there in a deck of cards?
    b. What fraction of a deck is face cards?
32. **a.** How many black cards are there in a deck of cards?
    b. What fraction of a deck is black?
33. **a.** How many kings are there in a deck of cards?
    b. What fraction of a deck is kings?

———— *Answer the following questions using complete sentences and your own words.*

HISTORY QUESTIONS
34. Who started probability theory? How?
35. Why was probability theory not considered a serious branch of mathematics?
36. Which authors established probability theory as a serious area of interest? What are some of the areas to which these authors applied probability theory?
37. What did Gregor Mendel do with probabilities?
38. Who was Antoine Gombauld, and what was his role in probability theory?
39. Who was Gerolamo Cardano, and what was his role in probability theory?

40. Which games of chance came to America from France via New Orleans?
41. Which implements of gambling were also used for divination?
42. What is the oldest casino game still being played?
43. How were cards introduced to Europe?
44. What is the modern deck of cards derived from?
45. What game was supposed to take the advantage away from dishonest gamblers?

WEB PROJECT
46. Write a research paper on any historical topic referred to in this section or a related topic. Following is a partial list of topics:
    - Jacob Bernoulli
    - Gerolamo Cardano
    - Pierre de Fermat
    - Blaise Pascal
    - The Marquis de Laplace
    - John Graunt
    - Gregor Mendel

    Some useful links for this web project are listed on the text web site: **www.thomsonedu.com/ mathematics/johnson**

3.2   Basic Terms of Probability

Much of the terminology and many of the computations of probability theory have their basis in set theory, because set theory contains the mathematical way of describing collections of objects and the size of those collections.

> **Basic Probability Terms**
>
> **experiment:** a process by which an observation, or **outcome,** is obtained
>
> **sample space:** the set S of all possible outcomes of an experiment
>
> **event:** any subset E of the sample space S

If a single die is rolled, the *experiment* is the rolling of the die. The possible *outcomes* are 1, 2, 3, 4, 5, and 6. The *sample space* (set of all possible outcomes) is $S = \{1, 2, 3, 4, 5, 6\}$. (The term *sample space* really means the same thing as *universal set;* the only distinction between the two ideas is that *sample space* is used only in probability theory, while *universal set* is used in any situation in which sets are used.) There are several possible *events* (subsets of the sample space), including the following:

$E_1 = \{3\}$ "a three comes up"

$E_2 = \{2, 4, 6\}$ "an even number comes up"

$E_3 = \{1, 2, 3, 4, 5, 6\}$ "a number between 1 and 6 inclusive comes up"

Notice that an event is not the same as an outcome. An event is a subset of the sample space; an outcome is an element of the sample space. "Rolling an odd number" is an event, not an outcome. It is the set $\{1, 3, 5\}$ that is composed of three separate outcomes. Some events are distinguished from outcomes only in that set brackets are used with events and not with outcomes. For example, $\{5\}$ is an event, and 5 is an outcome; either refers to "rolling a five."

The event E_3 ("a number between 1 and 6 inclusive comes up") is called a *certain event*, since $E_3 = S$. That is, E_3 is a sure thing. "Getting 17" is an *impossible event*. No outcome in the sample space $S = \{1, 2, 3, 4, 5, 6\}$ would result in a 17, so this event is actually the empty set.

THE FAR SIDE® BY GARY LARSON

Early shell games

An early certain event

> **More Probability Terms**
>
> A **certain event** is an event that is equal to the sample space.
> An **impossible event** is an event that is equal to the empty set.

Finding Probabilities and Odds

The **probability** of an event is a measure of the likelihood that the event will occur. If a single die is rolled, the outcomes are equally likely; a three is just as likely to come up as any other number. There are six possible outcomes, so a three should come up about one out of every six rolls. That is, the probability of event E_1 ("a three comes up") is $\frac{1}{6}$. The 1 in the numerator is the number of elements in $E_1 = \{3\}$. The 6 in the denominator is the number of elements in $S = \{1, 2, 3, 4, 5, 6\}$.

If an experiment's outcomes are equally likely, then the probability of an event E is the number of outcomes in the event divided by the number of outcomes in the sample space, or $n(E)/n(S)$. (In this chapter, we discuss only experiments with equally likely outcomes.) Probability can be thought of as "success over a total."

> **Probability of an Event**
>
> The **probability** of an event E, denoted by $p(E)$, is
>
> $$p(E) = \frac{n(E)}{n(S)}$$
>
> if the experiment's outcomes are equally likely.
> (*Think: success over total.*)

Many people use the words *probability* and *odds* interchangeably. However, the words have different meanings. The **odds** in favor of an event are the number of ways the event can occur compared to the number of ways the event *can fail to occur,* or "success compared to *failure*" (if the experiment's outcomes are equally likely). The odds of event E_1 ("a three comes up") are 1 to 5 (or 1:5), since a three can come up in one way and can fail to come up in five ways. Similarly, the odds of event E_3 ("a number between 1 and 6 inclusive comes up") are 6 to 0 (or 6:0), since a number between 1 and 6 inclusive can come up in six ways and can fail to come up in zero ways.

> **Odds of an Event**
>
> The **odds** of an event E with equally likely outcomes, denoted by $o(E)$, are given by
>
> $$o(E) = n(E):n(E')$$
>
> (*Think: success compared with failure.*)

In addition to the above meaning, the word *odds* can also refer to "house odds," which has to do with how much you will be paid if you win a bet at a casino. The odds of an event are sometimes called the **true odds** to distinguish them from the house odds.

EXAMPLE 1   A die is rolled. Find the following.

    a. the probability of rolling a five
    b. the odds of rolling a five
    c. the probability of rolling a number below 5
    d. the odds of rolling a number below 5

Solution   **a.** The sample space is $S = \{1, 2, 3, 4, 5, 6\}$. $E_1 = \{5\}$ ("rolling a five"). The probability of E_1 is

$$p(E_1) = \frac{n(E_1)}{n(S)} = \frac{1}{6}$$

This means that one out of every six possible outcomes is a success (that is, a five).
    b. $E_1' = \{1, 2, 3, 4, 6\}$. The odds of E_1 are

$$o(E_1) = n(E_1):n(E_1') = 1:5$$

This means that there is one possible success for every five possible failures.
    c. $E_2 = \{1, 2, 3, 4\}$ ("rolling a number below 5"). The probability of E_2 is

$$p(E_2) = \frac{n(E_2)}{n(S)} = \frac{4}{6} = \frac{2}{3}$$

This means that two out of every three possible outcomes are a success.
    d. $E_2' = \{5, 6\}$. The odds of E_2 are

$$o(E_2) = n(E_2):n(E_2') = 4:2 = 2:1$$

This means that there are two possible successes for every one possible failure. Notice that odds are reduced in the same manner that a fraction is reduced.   ■

EXAMPLE 2   A coin is flipped. Find the following.

    a. the sample space
    b. the probability of event E_1, "getting heads"
    c. the odds of event E_1, "getting heads"
    d. the probability of event E_2, "getting heads or tails"
    e. the odds of event E_2, "getting heads or tails"

Solution   **a.** *Finding the sample space S:* The experiment is flipping a coin. The only possible outcomes are heads and tails. The sample space S is the set of all possible outcomes, so $S = \{h, t\}$.
    b. *Finding the probability of heads:*

$$E_1 = \{h\} \text{ ("getting heads")}$$
$$p(E_1) = \frac{n(E_1)}{n(S)} = \frac{1}{2}$$

This means that one out of every two possible outcomes is a success.

c. *Finding the odds of heads:*

$$E_1' = \{t\}$$
$$o(E_1) = n(E_1) : n(E_1') = 1 : 1$$

This means that for every one possible success, there is one possible failure.

d. *Finding the probability of heads or tails:*

$$E_2 = \{h, t\}$$
$$p(E_2) = \frac{n(E_2)}{n(S)} = \frac{2}{2} = \frac{1}{1}$$

This means that every outcome is a success. Notice that E_2 is a certain event.

e. *Finding the odds of heads or tails:*

$$E_2' = \varnothing$$
$$o(E_2) = n(E_2) : n(E_2') = 2 : 0 = 1 : 0$$

This means that there are no possible failures.   ∎

Relative Frequency versus Probability

So far, we have discussed probabilities only in a theoretical way. When we found that the probability of heads was $\frac{1}{2}$, we never actually tossed a coin. It does not always make sense to calculate probabilities theoretically; sometimes they must be found empirically, the way a batting average is calculated. For example, in 8,389 times at bat, Babe Ruth had 2,875 hits. His batting average was $\frac{2,875}{8,389} \approx 0.343$. In other words, his probability of getting a hit was 0.343.

Sometimes a probability can be found either theoretically or empirically. We have already found that the theoretical probability of heads is $\frac{1}{2}$. We could also flip a coin a number of times and calculate (number of heads)/(total number of flips); this can be called the **relative frequency** of heads, to distinguish it from the theoretical probability of heads.

Usually, the relative frequency of an outcome is not equal to its probability, but if the number of trials is large, the two tend to be close. If you tossed a coin a couple of times, anything could happen, and the fact that the probability of heads is $\frac{1}{2}$ would have no impact on the results. However, if you tossed a coin 100 times, you would probably find that the relative frequency of heads was close to $\frac{1}{2}$. If your friend tossed a coin 1,000 times, she would probably find the relative frequency of heads to be even closer to $\frac{1}{2}$ than in your experiment. This relationship between probabilities and relative frequencies is called the **Law of Large Numbers.**

Law of Large Numbers

If an experiment is repeated a large number of times, the relative frequency of an outcome will tend to be close to the probability of that outcome.

The graph in Figure 3.2 shows the result of a simulated coin toss, using a computer and a random number generator rather than an actual coin. Notice that when the number of tosses is small, the relative frequency achieves values such as 0, 0.67, and 0.71. These values are not that close to the theoretical probability of 0.5. However, when the number of tosses is large, the relative frequency achieves values such as 0.48. These values are very close to the theoretical probability.

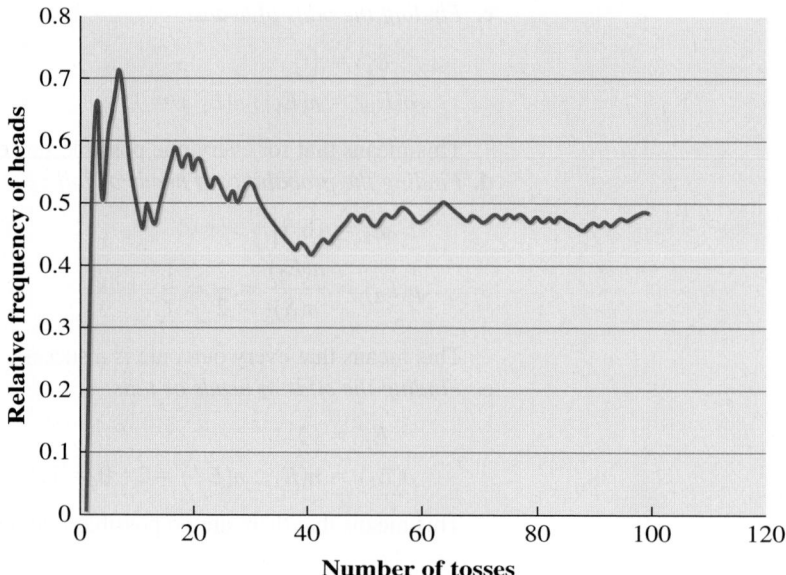

Figure 3.2
The relative frequency of heads after 100 simulated coin tosses

What if we used a real coin, rather than a computer, and we tossed the coin a lot more? Three different mathematicians have performed such an experiment:

- In the eighteenth century, Count Buffon tossed a coin 4,040 times. He obtained 2,048 heads, for a relative frequency of $2{,}048/4{,}040 \approx 0.5069$.
- During World War II, the South African mathematician John Kerrich tossed a coin 10,000 times while he was imprisoned in a German concentration camp. He obtained 5,067 heads, for a relative frequency of $5{,}067/10{,}000 = 0.5067$.
- In the early twentieth century, the English mathematician Karl Pearson tossed a coin 24,000 times! He obtained 12,012 heads, for a relative frequency of $12{,}012/24{,}000 = 0.5005$.

At a casino, probabilities are much more useful to the casino (the "house") than to an individual gambler, because the house performs the experiment for a much larger number of trials (in other words, plays the game more often). In fact, the house plays the game so many times that the relative frequencies will be almost exactly the same as the probabilities, with the result that the house is not gambling at all—it knows what is going to happen. Similarly, a gambler with a "system" has to play the game for a long time for the system to be of use.

EXAMPLE 3   A pair of coins is flipped.

a. Find the sample space.
b. Find the event E "getting exactly one heads."
c. Find the probability of event E.
d. Use the law of large numbers to interpret the probability of event E.

Solution   **a.** The experiment is the flipping of a pair of coins. One possible outcome is that one coin is heads and the other is tails. A second and *different* outcome is that one coin is tails and the other is heads. These two outcomes seem the same. However, if

The outcomes (h, h), (h, t), (t, h), and (t, t).

one coin were painted, it would be easy to tell them apart. Outcomes of the experiment can be described by using ordered pairs in which the first component refers to the first coin and the second component refers to the second coin. The two different ways of getting one heads and one tails are (h, t) and (t, h).

The sample space, or set of all possible outcomes, is the set $S = \{(h, h), (t, t), (h, t), (t, h)\}$. These outcomes are equally likely.

b. The event E "getting exactly one heads" is

$$E = \{(h, t), (t, h)\}$$

c. The probability of event E is

$$P(E) = \frac{n(E)}{n(S)} = \frac{2}{4} = \frac{1}{2}$$

d. According to the law of large numbers, if an experiment is repeated a large number of times, the relative frequency of that outcome will tend to be close to the probability of the outcome. Here, that means that if we were to toss a pair of coins many times, we should expect to get exactly one heads about half the time. Realize that this is only a prediction, and we might never get exactly one heads.   ■

Mendel's Use of Probabilities

In his experiments with plants, Gregor Mendel pollinated peas until he produced pure-red plants (that is, plants that would produce only red-flowered offspring) and pure-white plants. He then cross-fertilized these pure reds and pure whites and obtained offspring that had only red flowers. This amazed him, because the accepted theory of the day incorrectly predicted that these offspring would all have pink flowers.

He explained this result by postulating that there is a "determiner" that is responsible for flower color. These determiners are now called **genes.** Each plant has two flower color genes, one from each parent. Mendel reasoned that these offspring had to have inherited a red gene from one parent and a white gene from the other. These plants had one red flower gene and one white flower gene, but they had red flowers. That is, the red flower gene is **dominant,** and the white flower gene is **recessive.**

If we use R to stand for the red gene and w to stand for the white gene (with the capital letter indicating dominance), then the results can be described with a **Punnett square** in Figure 3.3.

	R	**R**	
			← first parent's genes
w	(R, w)	(R, w)	← possible offspring
w	(R, w)	(R, w)	← possible offspring

↑
second parent's genes

Figure 3.3
A Punnett square for the first generation

When the offspring of this experiment were cross-fertilized, Mendel found that approximately three-fourths of the offspring had red flowers and one-fourth had white flowers (see Figure 3.4). Mendel successfully used probability theory to analyze this result.

Historical Note   Gregor Johann Mendel, 1822–1884

Johann Mendel was born to an Austrian peasant family. His interest in botany began on the family farm, where he helped his father graft fruit trees. He studied philosophy, physics, and mathematics at the University Philosophical Institute in Olmütz. He was unsuccessful in finding a job, so he quit school and returned to the farm. Depressed by the prospects of a bleak future, he became ill and stayed at home for a year.

Mendel later returned to Olmütz. After two years of study, he found the pressures of school and work to be too much, and his health again broke down. On the advice of his father and a professor, he entered the priesthood, even though he did not feel called to serve the church. His name was changed from Johann to Gregor.

Relieved of his financial difficulties, he was able to continue his studies. However, his nervous disposition interfered with his pastoral duties, and he was assigned to substitute teaching. He enjoyed this work and was popular with the staff and students, but he failed the examination for certification as a teacher. Ironically, his lowest grades were in biology. The

Augustinians then sent him to the University of Vienna, where he became particularly interested in his plant physiology professor's unorthodox belief that new plant varieties can be caused by naturally arising variations. He was also fascinated by his classes in physics, where he was exposed to the physicists' experimental and mathematical approach to their subject.

After further breakdowns and failures, Mendel returned to the monastery and was assigned the low-stress job of keeping the abbey garden. There he combined the experimental and mathematical approach of a physicist with his background in biology and performed a series of experiments designed to determine whether his professor was correct in his beliefs regarding the role of naturally arising variants in plants.

Mendel studied the transmission of specific traits of the pea plant—such as flower color and stem length—from parent plant to offspring. He pollinated the plants by hand and separated them until he had isolated each trait. For example, in his studies of flower color, he pollinated the plants until he produced pure-red plants (plants that would produce only red-flowered offspring) and pure-white plants.

At the time, the accepted theory of heredity was that of blending. In this view, the characteristics of

	R	**w**	
R	(R, R)	(w, R)	← first parent's genes
w	(R, w)	(w, w)	← possible offspring
			← possible offspring

↑
second parent's genes

Figure 3.4
A Punnett square for the second generation

Only one of the four possible outcomes, (w, w), results in a white-flowered plant, so event E_1 that the plant has white flowers is $E_1 = \{(w, w)\}$; therefore,

$$p(E_1) = \frac{n(E_1)}{n(S)} = \frac{1}{4}$$

This means that we should expect the actual relative frequency of white-flowered plants to be close to $\frac{1}{4}$.

A nineteenth-century drawing illustrating Mendel's pea plants, showing the original cross, the first generation, and the second generation.

both parents blend together to form an individual. Mendel reasoned that if the blending theory was correct, the union of a pure-red pea plant and a pure-white pea plant would result in a pink-flowered offspring. However, his experiments showed that such a union consistently resulted in red-flowered offspring.

Mendel crossbred a large number of peas that had different characteristics. In many cases, an offspring would have a characteristic of one of its parents, undiluted by that of the other parent. Mendel concluded that the question of which parent's characteristics would be passed on was a matter of chance, and he successfully used probability theory to estimate the frequency with which characteristics would be passed on. In so doing, Mendel founded modern genetics. Mendel attempted similar experiments with bees, but these experiments were unsuccessful because he was unable to control the mating behavior of the queen bee.

Mendel was ignored when he published his paper "Experimentation in Plant Hybridization." Sixteen years after his death, his work was rediscovered by three European botanists who had reached similar conclusions in plant breeding, and the importance of Mendel's work was finally recognized.

Each of the other three outcomes, (R, R), (R, w), and (w, R), results in a red-flowered plant, because red dominates white. The event E_2 that the plant has red flowers is $E_2 = \{(R, R), (R, w), (w, R)\}$; therefore,

$$p(E_2) = \frac{n(E_2)}{n(S)} = \frac{3}{4}$$

Thus, we should expect the actual relative frequency of red-flowered plants to be close to $\frac{3}{4}$.

Outcomes (R, w) and (w, R) are genetically identical; it does not matter which gene is inherited from which parent. For this reason, geneticists do not use the ordered-pair notation and instead refer to each of these two outcomes as "Rw." The only difficulty with this convention is that it makes the sample space appear to be S = {RR, Rw, ww}, which consists of only three elements, when in fact it consists of four elements. This distinction is important; if the sample space consisted of three equally likely elements, then the probability of a red-flowered offspring would be $\frac{2}{3}$ rather than $\frac{3}{4}$. Mendel knew that the sample space had to have four elements, because his cross-fertilization experiments resulted in a relative frequency very close to $\frac{3}{4}$, not $\frac{2}{3}$.

Ronald Fisher, a noted British statistician, used statistics to deduce that Mendel fudged his data. Mendel's relative frequencies were unusually close to the theoretical probabilities, and Fisher found that there was only about a 0.00007 chance of such close agreement. Others have suggested that perhaps Mendel did not willfully change his results but rather continued collecting data until the numbers were in close agreement with his expectations.*

Probabilities in Genetics

Cystic fibrosis is an inherited disease characterized by abnormally functioning exocrine glands that secrete a thick mucus, clogging the pancreatic ducts and lung passages. Most patients with cystic fibrosis die of chronic lung disease; until recently, most died in early childhood. This early death makes it extremely unlikely that an afflicted person would ever parent a child. Only after the advent of Mendelian genetics did it become clear how a child could inherit the disease from two healthy parents.

In 1989, a team of Canadian and American doctors announced the discovery of the gene that is responsible for most cases of cystic fibrosis. As a result of that discovery, a new therapy for cystic fibrosis is being developed. Researchers splice a therapeutic gene into a cold virus and administer it through an affected person's nose. When the virus infects the lungs, the gene becomes active. It is hoped that this will result in normally functioning cells, without the damaging mucus.

In April 1993, a twenty-three-year-old man with advanced cystic fibrosis became the first patient to receive this therapy. In September 1996, a British team announced that eight volunteers with cystic fibrosis had received this therapy; six were temporarily cured of the disease's debilitating symptoms. In March 1999, another British team announced a new therapy that involves administering the therapeutic gene through an aerosol spray.

Cystic fibrosis occurs in about 1 out of every 2,000 births in the Caucasian population and only in about 1 in 250,000 births in the non-Caucasian population. It is one of the most common inherited diseases in North America. One in 25 Americans carries a single gene for cystic fibrosis. Children who inherit two such genes develop the disease; that is, cystic fibrosis is recessive.

There are tests that can be used to determine whether a person carries the gene. However, they are not accurate enough to use for the general population. Instead, they are typically used in families with a previously affected child.

ThomsonNOW™
for Liberal Arts Mathematics
www.thomsonedu.com/1pass.

EXAMPLE 4 Each of two prospective parents carries one cystic fibrosis gene.

a. Find the probability that their child would have cystic fibrosis.
b. Find the probability that their child would be free of symptoms.
c. Find the probability that their child would be free of symptoms but could pass the cystic fibrosis gene on to his or her own child.

Solution We will denote the recessive cystic fibrosis gene with the letter c and the normal disease-free gene with an N. Each parent is Nc and thus does not have the disease. Figure 3.5 shows the Punnett square for the child.

---

*R. A. Fisher, "Has Mendel's Work Been Rediscovered?" *Annals of Science* 1, 1936, pp. 115–137.

	N	c
N	(N, N)	(c, N)
c	(N, c)	(c, c)

Figure 3.5
A Punnett square for Example 4

a. Cystic fibrosis is recessive, so only the (c, c) child will have the disease. The probability of such an event is 1/4.

b. The (N, N), (c, N), and (N, c) children will be free of symptoms. The probability of this event is

$$p(\text{healthy}) = p((N, N)) + p((c, N)) + p((N, c))$$
$$= \frac{1}{4} + \frac{1}{4} + \frac{1}{4} = \frac{3}{4}$$

c. The (c, N) and (N, c) children would never suffer from any symptoms but could pass the cystic fibrosis gene on to his or her own child. The probability of this event is

$$p((c, N)) + p((N, c)) = \frac{1}{4} + \frac{1}{4} = \frac{1}{2}$$ ∎

In Example 4, the Nc child is called a **carrier** because that child would never suffer from any symptoms but could pass the cystic fibrosis gene on to his or her own child. Both of the parents were carriers.

Sickle-cell anemia is an inherited disease characterized by a tendency of the red blood cells to become distorted and deprived of oxygen. Although it varies in severity, the disease can be fatal in early childhood. More often, patients have a shortened life span and chronic organ damage. Newborns are now routinely screened for sickle-cell disease. The only true cure is a bone marrow transplant from a sibling without sickle-cell anemia; however, this can cause the patient's death, so it is done only under certain circumstances. There are also medications that can decrease the episodes of pain. Approximately 1 in every 500 black babies is born with sickle-cell anemia, but only 1 in 160,000 nonblack babies has the disease. This disease is **codominant:** A person with two sickle-cell genes will have the disease, while a person with one sickle-cell gene will have a mild, nonfatal anemia called **sickle-cell trait.** Approximately 8–10% of the black population has sickle-cell trait.

Huntington's disease, caused by a dominant gene, is characterized by nerve degeneration causing spasmodic movements and progressive mental deterioration. The symptoms do not usually appear until well after reproductive age has been reached; the disease usually hits people in their 40s. Death typically follows 12 to 15 years after the onset of the symptoms. No effective treatment is available, but physicians can now assess with certainty whether someone will develop the disease, and they can estimate when the disease will strike. Many of those who are at risk choose not to undergo the test, especially if they have already had children. Folk singer Arlo Guthrie is in this situation; his father, Woody Guthrie, died of Huntington's disease.

Historical Note   Nancy Wexler

Courtesy of Dr. Nancy Wexler

In 1993, scientists working together at six major research centers located most genes that cause Huntington's disease. This discovery will enable people to learn whether they carry a Huntington's gene, and it will allow pregnant women to determine whether their child carries the gene. The discovery could eventually lead to a treatment.

The collaboration of research centers was organized largely by Nancy Wexler, a Columbia University professor of neuropsychology, who is herself at risk for Huntington's disease—her mother died of it in 1978. Dr. Wexler, President of the Hereditary Disease Foundation, has made numerous trips to study and aid the people of the Venezuelan village of Lake Maracaibo, many of whom suffer from the disease or are at risk for it. All are related to one woman who died of the disease in the early 1800s. Wexler took blood and tissue samples and gave neurological and psychoneurological tests to the inhabitants of the village. The samples and test results enabled the researchers to find the single gene that causes Huntington's disease.

In October 1993, Wexler received an Albert Lasker Medical Research Award, a prestigious honor that is often a precursor to a Nobel Prize. The award was given in recognition for her contribution to the international effort that culminated in the discovery of the Huntington's disease gene. At the awards ceremony, she explained to then first lady Hillary Clinton that her genetic heritage has made her uninsurable—she would lose her health coverage if she switched jobs. She told Mrs. Clinton that more Americans will be in the same situation as more genetic discoveries are made, unless the health care system is reformed. The then first lady incorporated this information into her speech at the awards ceremony: "It is likely that in the next years, every one of us will have a pre-existing condition and will be uninsurable. . . . What will happen as we discover those genes for breast cancer, or prostate cancer, or osteoporosis, or any of the thousands of other conditions that affect us as human beings?"

© AP/Wide World Photos

Woody Guthrie's most famous song is "This Land is Your Land." This folksinger, guitarist, and composer was a friend of Leadbelly, Pete Seeger, and Ramblin' Jack Elliott and exerted a strong influence on Bob Dylan. Guthrie died at the age of 55 of Huntington's disease.

In August 1999, researchers in Britain, Germany, and the United States discovered what causes brain cells to die in people with Huntington's disease. This discovery may eventually lead to a treatment.

Genetic Screening

At this time, there are no conclusive tests that will tell a parent whether he or she is a cystic fibrosis carrier, nor are there conclusive tests that will tell whether a fetus has the disease. A new test resulted from the 1989 discovery of the location of most cystic fibrosis genes, but that test will detect only 85% to 95% of the cystic fibrosis genes, depending on the individual's ethnic background. The extent to which this test will be used has created quite a controversy.

Individuals who have relatives with cystic fibrosis are routinely informed about the availability of the new test. The controversial question is whether a massive genetic screening program should be instituted to identify cystic fibrosis carriers in the general population, regardless of family history. This is an important question, considering that four in five babies with cystic fibrosis are born to couples with no previous family history of the condition.

Opponents of routine screening cite a number of important concerns. The existing test is rather inaccurate; 5% to 15% of the cystic fibrosis carriers would be missed. It is not known how health insurers would use this genetic information—insurance firms could raise rates or refuse to carry people if a screening test indicated a presence of cystic fibrosis. Also, some experts question the adequacy of quality assurance for the diagnostic facilities and for the tests themselves.

The Row over Sickle-Cell

From: Newsweek, February 12, 1973.
© 1973 Newsweek, Inc. All rights reserved.
Reprinted by permission.

. . . Two years ago, President Nixon listed sickle-cell anemia along with cancer as diseases requiring special Federal attention. . . . Federal spending for sickle-cell anemia programs has risen from a scanty $1 million a year to $15 million for 1973. At the same time, in what can only be described as a head-long rush, at least a dozen states have passed laws requiring sickle-cell screening for blacks.

While all these efforts have been undertaken with the best intentions of both whites and blacks, in recent months the campaign has begun to stir widespread and bitter controversy. Some of the educational programs have been riddled with misinformation and have unduly frightened the black community. To quite a few Negroes, the state laws are discriminatory—and to the extent that they might inhibit childbearing, even genocidal. . . .

Parents whose children have the trait often misunderstand and assume they have the disease. In some cases, airlines have allegedly refused to hire black stewardesses who have the trait, and some carriers have been turned down by life-insurance companies—or issued policies at high-risk rates.

Because of racial overtones and the stigma that attaches to persons found to have the sickle-cell trait, many experts seriously object to mandatory screening programs. They note, for example, that there are no laws requiring testing for Cooley's anemia [or other disorders that have a hereditary basis]. . . . Moreover, there is little that a person who knows he has the disease or the trait can do about it. "I don't feel," says Dr. Robert L. Murray, a black geneticist at Washington's Howard University, "that people should be required by law to be tested for something that will provide information that is more negative than positive." . . . Fortunately, some of the mandatory laws are being repealed.

Supporters of routine testing say that the individual should be allowed to decide whether to be screened. Failing to inform people denies them the opportunity to make a personal choice about their reproductive future. An individual who is found to be a carrier could choose to avoid conception, to adopt, to use artificial insemination by a donor, or to use prenatal testing to determine whether a fetus is affected—at which point the additional controversy regarding abortion could enter the picture.

The history of genetic screening programs is not an impressive one. In the 1970s, mass screening of blacks for sickle-cell anemia was instituted. This program caused unwarranted panic; those who were told they had sickle-cell trait feared that they would develop symptoms of the disease and often did not understand the probability that their children would inherit the disease (see Exercises 73 and 74). Some people with sickle-cell trait were denied health insurance and life insurance.

3.2 Exercises

In Exercises 1–14, use this information: A jar on your desk contains twelve black, eight red, ten yellow, and five green jellybeans. You pick a jellybean without looking.

1. What is the experiment?
2. What is the sample space?

In Exercises 3–14, find the following:

3. the probability that it is black
4. the probability that it is green
5. the probability that it is red or yellow
6. the probability that it is red or black
7. the probability that it is not yellow
8. the probability that it is not red
9. the probability that it is white
10. the probability that it is not white
11. the odds of picking a black jellybean
12. the odds of picking a green jellybean
13. the odds of picking a red or yellow jellybean
14. the odds of picking a red or black jellybean

In Exercises 15–28, one card is drawn from a well-shuffled deck of 52 cards (no jokers).

15. What is the experiment?
16. What is the sample space?

In Exercises 17–28, (a) find the probability and (b) the odds of drawing the given cards. Also, (c) use the Law of Large Numbers to interpret both the probability and the odds. (You might want to review the makeup of a deck of cards in Section 3.1.)

17. a black card
18. a heart
19. a queen
20. a two of clubs
21. a queen of spades
22. a club
23. a card below a 5 (count an ace as high)
24. a card below a 9 (count an ace as high)
25. a card above a 4 (count an ace as high)
26. a card above an 8 (count an ace as high)
27. a face card
28. a picture card

In Exercises 29–38, (a) find the probability and (b) the odds of winning the given bet in roulette. Also, (c) use the Law of Large Numbers to interpret both the probability and the odds. (You might want to review the description of the game in Section 3.1.)

29. the single-number bet
30. the two-number bet
31. the three-number bet
32. the four-number bet
33. the five-number bet
34. the six-number bet
35. the twelve-number bet
36. the low-number bet
37. the even-number bet
38. the red-number bet

39. Use the information in Figure 3.6 from the U.S. Census Bureau to answer the following questions.

Age	0–4	5–19	20–44	45–64	65–84	85+	total
Male	9,831	31,454	52,294	30,381	13,212	1,240	138,411
Female	9,387	29,877	51,781	32,059	17,582	3,028	143,713

Figure 3.6
2000 U.S. population, in thousands, by age and gender
Source: 2000 Census, U.S. Bureau of the Census

a. Find the probability that in the year 2000, a U.S. resident was female.

b. Find the probability that in the year 2000, a U.S. resident was male and between 20 and 44 years of age, inclusive.

40. Use the information in Figure 3.7 from the U.S. Census Bureau to answer the following questions.

a. Find the probability that in the year 2005, a U.S. resident was American Indian, Eskimo, or Aleut.

b. Find the probability that in the year 2005, a U.S. resident was Asian or Pacific Islander and between 20 and 44 years of age.

41. The dartboard in Figure 3.8 is composed of circles with radii 1 inch, 3 inches, and 5 inches.

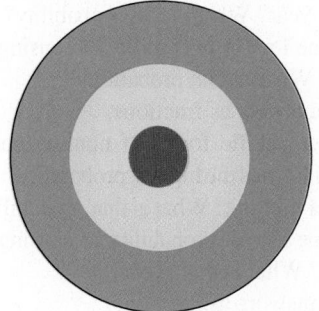

Figure 3.8
A dartboard for Exercise 41

a. What is the probability that a dart hits the red region if the dart hits the target randomly?

b. What is the probability that a dart hits the yellow region if the dart hits the target randomly?

c. What is the probability that a dart hits the green region if the dart hits the target randomly?

d. Use the Law of Large Numbers to interpret the probabilities in parts (a), (b), and (c).

42. a. What is the probability of getting red on the spinner shown in Figure 3.9?

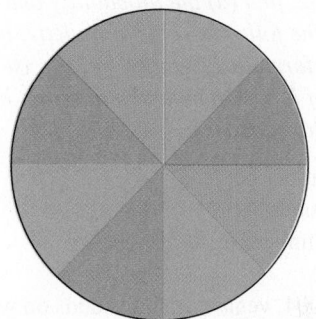

Figure 3.9
All sectors are equal in size and shape.

b. What is the probability of getting blue?

c. Use the Law of Large Numbers to interpret the probabilities in parts (a) and (b).

Age	White	Black	American Indian, Eskimo, Aleut	Asian, Pacific Islander
0–4	15,041	2,907	222	1,042
5–19	47,556	9,445	709	3,036
20–44	79,593	14,047	1,005	5,165
45–64	60,139	8,034	497	2,911
65–84	27,426	2,826	166	987
85+	4,467	362	29	110

Figure 3.7
2005 U.S. population (projected), in thousands, by age and race
Source: Annual Population Estimates by Age Group and Sex, U.S. Bureau of the Census

43. Amtrack's *Empire Service* train starts in Albany, New York, at 6:00 A.M., Mondays through Fridays. It arrives at New York City at 8:25 A.M. If the train breaks down, find the probability that it breaks down in the first half hour of its run. Assume that the train breaks down at random times.

44. Amtrack's *Downeaster* train leaves Biddeford, Maine, at 6:42 A.M., Mondays through Fridays. It arrives at Boston at 8:50 A.M. If the train breaks down, find the probability that it breaks down in the last 45 minutes of its run. Assume that the train breaks down at random times.

45. If $p(E) = \frac{1}{5}$, find $o(E)$.
46. If $p(E) = \frac{8}{9}$, find $o(E)$.
47. If $o(E) = 3:2$, find $p(E)$.
48. If $o(E) = 4:7$, find $p(E)$.
49. If $p(E) = \frac{a}{b}$, find $o(E)$.

In Exercises 50–52, find (a) the probability and (b) the odds of winning the following bets in roulette. In finding the odds, use the formula developed in Exercise 49. Also, (c) use the Law of Large Numbers to interpret both the probability and the odds.

50. the high-number bet
51. the odd-number bet
52. the black-number bet

53. In March 2004, vegas.com gave odds on who would win the 2004 World Series. They gave the New York Yankees 2:1 odds and gave the Boston Red Sox 5:2 odds.
    a. Are these house odds or true odds? Why?
    b. Convert the odds to probabilities.
    c. Who did they think was more likely to win the World Series: the Yankees or the Red Sox? Why?

54. In March 2004, linesmaker.com gave odds on who would win the 2004 World Series. They gave the New York Yankees 13:5 odds and gave the Boston Red Sox 9:2 odds.
    a. Are these house odds or true odds? Why?
    b. Convert the odds to probabilities.
    c. Who did they think was more likely to win the World Series: the Yankees or the Red Sox? Why?
    d. Use the information in Exercise 53 to determine which source thought it was more probable that the Yankees would win the World Series: vegas.com or linesmaker.com.

In Exercises 55–60, use Figure 3.10, on page 149.

55. a. Of the specific causes listed, which is the most likely cause of death in one year?
    b. Which is the most likely cause of death in a lifetime?
       Justify your answers.

56. a. Of the causes listed, which is the least likely cause of death in one year?
    b. Which is the least likely cause of death in a lifetime?
       Justify your answers.

57. a. What is the probability that a person will die of a car transportation accident in one year?
    b. What is the probability that a person will die of an airplane transportation accident in one year?
       Write your answers as fractions.

58. a. What is the probability that a person will die from lightning in a lifetime?
    b. What is the probability that a person will die from an earthquake in a lifetime?
       Write your answers as fractions.

59. a. Of all of the specific forms of transportation accidents, which has the highest probability of causing death in one year? What is that probability?
    b. Which has the lowest probability of causing death in one year? What is that probability?
       Write your answers as fractions.

60. a. Of all of the specific forms of nontransportation accidents, which has the highest probability of causing death in a lifetime? What is that probability?
    b. Which has the lowest probability of causing death in a lifetime? What is that probability?
       Write your answers as fractions.

61. A family has two children. Using b to stand for boy and g for girl in ordered pairs, give each of the following.
    a. the sample space
    b. the event E that the family has exactly one daughter
    c. the event F that the family has at least one daughter
    d. the event G that the family has two daughters
    e. $p(E)$ f. $p(F)$ g. $p(G)$
    h. $o(E)$ i. $o(F)$ j. $o(G)$
    (Assume that boys and girls are equally likely.)

62. Two coins are tossed. Using ordered pairs, give the following.
    a. the sample space
    b. the event E that exactly one is heads
    c. the event F that at least one is heads
    d. the event G that two are heads

Type of accident or injury	Odds of dying in 1 year	Lifetime odds of dying
All transportation accidents	1 : 5953	1 : 77
Pedestrian transportation accident	1 : 47,273	1 : 612
Bicyclist transportation accident	1 : 375,412	1 : 4,857
Motorcyclist transportation accident	1 : 89,562	1 : 1,159
Car transportation accident	1 : 17,625	1 : 228
Airplane transportation accident	1 : 440,951	1 : 5,704
All non-transportation accidents	1 : 4,933	1 : 64
Falling	1 : 17,712	1 : 229
Drowning	1 : 83,534	1 : 1,081
Fire	1 : 91,149	1 : 1,179
Venomous animals and poisonous plants	1 : 3,788,700	1 : 49,013
Lightning	1 : 4,362,746	1 : 56,439
Earthquake	1 : 9,288,426	1 : 120,161
Storm	1 : 4,570,496	1 : 59,127
Intentional self-harm	1 : 9,096	1 : 118
Assault	1 : 16,325	1 : 211
Legal execution	1 : 4,297,630	1 : 55,597
War	1 : 14,397,061	1 : 186,299
Complications of medical care	1 : 101,281	1 : 1,310

Figure 3.10

The odds of dying due to an accident or injury in the United States in 2002

Source: What are the odds of dying?, 2004 edition, National Safety Council, www.nsc.org

e. $p(E)$ **f.** $p(F)$ **g.** $p(G)$
h. $o(E)$ **i.** $o(F)$ **j.** $o(G)$

63. A family has three children. Using b to stand for boy and g for girl, and using ordered triples such as (b, b, g), give the following.
  a. the sample space
  b. the event E that the family has exactly two daughters
  c. the event F that the family has at least two daughters
  d. the event G that the family has three daughters
  e. $p(E)$ **f.** $p(F)$ **g.** $p(G)$
  h. $o(E)$ **i.** $o(F)$ **j.** $o(G)$
  (Assume that boys and girls are equally likely.)

64. Three coins are tossed. Using ordered triples, give the following.
  a. the sample space
  b. the event E that exactly two are heads
  c. the event F that at least two are heads
  d. the event G that all three are heads
  e. $p(E)$ **f.** $p(F)$ **g.** $p(G)$
  h. $o(E)$ **i.** $o(F)$ **j.** $o(G)$

65. A couple plans on having two children.
  a. Find the probability of having two girls.
  b. Find the probability of having one girl and one boy.
  c. Find the probability of having two boys.
  d. Which is more likely: having two children of the same sex or two of different sexes? Why?
  (Assume that boys and girls are equally likely.)

66. Two coins are tossed.
  a. Find the probability that both are heads.
  b. Find the probability that one is heads and one is tails.
  c. Find the probability that both are tails.
  d. Which is more likely: that the two coins match or that they don't match? Why?

67. A couple plans on having three children. Which is more likely: having three children of the same sex or of different sexes? Why? (Assume that boys and girls are equally likely.)

68. Three coins are tossed. Which is more likely: that the three coins match or that they don't match? Why?

69. A pair of dice is rolled. Using ordered pairs, give the following.
  a. the sample space

  Hint: S has 36 elements, one of which is (1, 1).

  b. the event E that the sum is 7
  c. the event F that the sum is 11

d. the event G that the roll produces doubles
  e. $p(E)$ **f.** $p(F)$ **g.** $p(G)$
  h. $o(E)$ **i.** $o(F)$ **j.** $o(G)$

70. Mendel found that snapdragons have no color dominance; a snapdragon with one red gene and one white gene will have pink flowers. If a pure-red snapdragon is crossed with a pure-white one, find the probability of the following.
  a. a red offspring
  b. a white offspring
  c. a pink offspring

71. If two pink snapdragons are crossed (see Exercise 70), find the probability of the following.
  a. a red offspring
  b. a white offspring
  c. a pink offspring

72. One parent is a cystic fibrosis carrier, and the other has no cystic fibrosis gene. Find the probability of each of the following.
  a. The child would have cystic fibrosis.
  b. The child would be a carrier.
  c. The child would not have cystic fibrosis and not be a carrier.
  d. The child would be healthy (i.e., free of symptoms).

73. If carrier-detection tests show that two prospective parents have sickle-cell trait (and are therefore carriers), find the probability of each of the following.
  a. Their child would have sickle-cell anemia.
  b. Their child would have sickle-cell trait.
  c. Their child would be healthy (i.e., free of symptoms).

74. If carrier-detection tests show that one prospective parent is a carrier of sickle-cell anemia and the other has no sickle-cell gene, find the probability of each of the following.
  a. The child would have sickle-cell anemia.
  b. The child would have sickle-cell trait.
  c. The child would be healthy (i.e., free of symptoms).

Tay-Sachs disease is a recessive disease characterized by an abnormal accumulation of certain fat compounds in the spinal cord and brain, resulting in paralysis, severe mental impairment, and blindness. There is no effective treatment, and death usually occurs before the age of five. The disease occurs once in 3,600 births among Ashkenazi Jews (Jews from central and eastern Europe) but only once in 600,000 births in other populations. Carrier-detection tests and fetal-monitoring tests *are available. The successful use of these tests, coupled with an aggressive counseling program, has resulted in a decrease of 90% in the incidence of this disease.*

75. If carrier-detection tests show that one prospective parent is a carrier of Tay-Sachs and the other has no Tay-Sachs gene, find the probability of each of the following.
   a. The child would have the disease.
   b. The child would be a carrier.
   c. The child would be healthy (i.e., free of symptoms).

76. If carrier-detection tests show that both prospective parents are carriers of Tay-Sachs, find the probability of each of the following.
   a. Their child would have the disease.
   b. Their child would be a carrier.
   c. Their child would be healthy (i.e., free of symptoms).

77. If a parent started to exhibit the symptoms of Huntington's disease after the birth of his or her child, find the probability of each of the following. (Assume that one parent carries a single gene for Huntington's disease and the other carries no such gene.)
   a. The child would have the disease.
   b. The child would be a carrier.
   c. The child would be healthy (i.e., free of symptoms).

▬▬▬▶ *Answer the following questions using complete sentences and your own words.*

CONCEPT QUESTIONS

78. Explain how you would find the theoretical probability of rolling an even number on a single die. Explain how you would find the relative frequency of rolling an even number on a single die.

79. Give five examples of events whose probabilities must be found empirically rather than theoretically.

80. Does the theoretical probability of an event remain unchanged from experiment to experiment? Why? Does the relative frequency of an event remain unchanged from experiment to experiment? Why?

81. Consider a "weighted die"—one that has a small weight in its interior. Such a weight would cause the face closest to the weight to come up less frequently and the face farthest from the weight to come up more frequently. Would the probabilities computed in Example 1 still be correct? Why or why not? Would the definition $p(E) = \frac{n(E)}{n(S)}$ still be appropriate? Why or why not?

82. Some dice have spots that are small indentations; other dice have spots that are filled with the same material but of a different color. Which of these two types of dice is not fair? Why? What would be the most likely outcome of rolling this type of die? Why?
   Hint: 1 and 6 are on opposite faces, as are 2 and 5, and 3 and 4.

83. In the United States, 52% of the babies are boys and 48% are girls. Do these percentages contradict an assumption that boys and girls are equally likely? Why?

84. Compare and contrast theoretical probability and relative frequency. Be sure you discuss both the similarities and differences between the two.

85. Compare and contrast probability and odds. Be sure you discuss both the similarities and differences between the two.

HISTORY QUESTIONS

86. What prompted Dr. Nancy Wexler's interest in Huntington's disease?

87. What resulted from Dr. Nancy Wexler's interest in Huntington's disease?

PROJECTS

88.
   a. In your opinion, what is the probability that the last digit of a phone number is odd? Justify your answer.
   b. Randomly choose one page from the residential section of your local phone book. Count how many phone numbers on that page have an odd last digit and how many have an even last digit. Then compute (number of phone numbers with odd last digits)/(total number of phone numbers).
   c. Is your answer to part (b) a theoretical probability or a relative frequency? Justify your answer.
   d. How do your answers to parts (a) and (b) compare? Are they exactly the same, approximately the same, or dissimilar? Discuss this comparison, taking into account the ideas of probability theory.

89.
   a. Flip a coin ten times, and compute the relative frequency of heads.
   b. Repeat the experiment described in part (a) nine more times. Each time, compute the relative frequency of heads. After finishing parts (a) and (b), you will have flipped a coin 100 times, and you will have computed ten different relative frequencies.

c. Discuss how your ten relative frequencies compare with each other and with the theoretical probability heads. In your discussion, use the ideas discussed under the heading "Relative Frequency versus Probability" in this section. Do not plagiarize; use your own words.

d. Combine the results of parts (a) and (b), and find the relative frequency of heads for all 100 coin tosses. Discuss how this relative frequency compares with those found in parts (a) and (b) and with the theoretic probability of heads. Be certain to incorporate the Law of Large Numbers into your discussion.

90. a. Roll a single die twelve times, and compute the relative frequency with which you rolled a number below a 3.

b. Repeat the experiment described in part (a) seven more times. Each time, compute the relative frequency with which you rolled a number below a 3. After finishing parts (a) and (b), you will have rolled a die ninety-six times, and you will have computed eight different relative frequencies.

c. Discuss how your eight relative frequencies compare with each other and with the theoretical probability rolling a number below a 3. In your discussion, use the ideas discussed under the heading "Relative Frequency versus Probability" in this section. Do not plagiarize; use your own words.

d. Combine the results of parts (a) and (b), and find the relative frequency of rolling a number below a 3 for all ninety-six die rolls. Discuss how this relative frequency compares with those found in parts (a) and (b) and with the theoretic probability of rolling a number less than a 3. Be certain to incorporate the Law of Large Numbers into your discussion.

91. Stand a penny upright on its edge on a smooth, hard, level surface. Then spin the penny on its edge. To do this, gently place a finger on the top of the penny. Then snap the penny with another finger (and immediately remove the holding finger) so that the penny spins rapidly before falling. Repeat this experiment fifty times, and compute the relative frequency of heads. Discuss whether or not the outcomes of this experiment are equally likely.

92. Stand a penny upright on its edge on a smooth, hard, level surface. Pound the surface with your hand so that the penny falls over. Repeat this experiment fifty times, and compute the relative frequency of heads. Discuss whether or not the outcomes of this experiment are equally likely.

 WEB PROJECT

93. In the 1970s, there was a mass screening of blacks for sickle-cell anemia and a mass screening of Jews for Tay-Sachs disease. One of these was a successful program. One was not. Write a research paper on these two programs.

Some useful links for this web project are listed on the text web site:
www.thomsonedu.com/mathematics/johnson

3.3 Basic Rules of Probability

How Big or Small Can a Probability Be?

No event occurs less than 0% of the time. How could an event occur negative 15% of the time? Also, no event occurs more than 100% of the time. How could an event occur 125% of the time? Every event must occur between 0% and 100% of the time. For every event E,

$$0\% \leq p(E) \leq 100\%$$
$$0 \leq p(E) \leq 1 \qquad \text{converting from percents to decimals}$$

This means that *if you ever get a negative answer or an answer greater than 1 when you calculate a probability, go back and find your error.*

What type of event occurs 100% of the time? That is, what type of event has a probability of 1? Such an event must include *all possible* outcomes; if any possible outcome is left out, the event could not occur 100% of the time. An event that has a probability of 1 must be the sample space, because the sample space is the set of all possible outcomes. As we discussed earlier, such an event is called a *certain event*.

What type of event occurs 0% of the time? That is, what type of event has a probability of 0? Such an event must not include any possible outcome; if a possible outcome is included, the event would not have a probability of 0. An event with a probability of 0 must be the null set. As we discussed earlier, such an event is called an *impossible event*.

Probability Rules

Rule 1	$p(\varnothing) = 0$	The probability of the null set is 0.
Rule 2	$p(S) = 1$	The probability of the sample space is 1.
Rule 3	$0 \le p(E) \le 1$	Probabilities are between 0 and 1 (inclusive).

Probability Rules 1, 2, and 3 can be formally verified as follows:

Rule 1: $\quad p(\varnothing) = \dfrac{n(\varnothing)}{n(S)} = \dfrac{0}{n(S)} = 0$

Rule 2: $\quad p(S) = \dfrac{n(S)}{n(S)} = 1$

Rule 3: $\quad E$ is a subset of S; therefore,

$$0 \le n(E) \le n(S)$$

$$\frac{0}{n(S)} \le \frac{n(E)}{n(S)} \le \frac{n(S)}{n(S)} \qquad \text{dividing by } n(S)$$

$$0 \le p(E) \le 1$$

ThomsonNOW™
for Liberal Arts Mathematics
www.thomsonedu.com/1pass.

EXAMPLE 1   A single die is rolled once. Find the probability of:

a. event E, "a 15 is rolled"
b. event F, "a number between 1 and 6 (inclusive) is rolled"
c. event G, "a 3 is rolled"

Solution   The sample space is $S = \{1, 2, 3, 4, 5, 6\}$, and $n(S) = 6$.

a. It is *wrong* to say that $E = \{15\}$. Remember, we are talking about rolling a single die. Rolling a 15 is not one of the possible outcomes. That is, $15 \notin S$. There are no possible outcomes that result in rolling a 15, so the number of outcomes in event E is $n(E) = 0$. Thus,

$$p(E) = n(E)/n(S) = \frac{0}{6} = 0$$

The number 15 will be rolled 0% of the time. Event E is an impossible event.

Mathematically, we say that event E consists of no possible outcomes, so $E = \varnothing$. This agrees with rule 1, since

$$p(E) = p(\varnothing) = 0$$

b. $F = \{1, 2, 3, 4, 5, 6\}$, so $n(F) = 6$. Thus,

$$p(F) = n(F)/n(S) = \frac{6}{6} = 1$$

A number between 1 and 6 (inclusive) will be rolled 100% of the time. Event F is a certain event.

Mathematically, we say that event F consists of every possible outcome, so $F = S$. This agrees with rule 2, since

$$p(F) = p(S) = 1$$

c. $G = \{3\}$, so $n(G) = 1$. Thus

$$p(G) = n(G)/n(S) = 1/6$$

This agrees with rule 3, since

$$0 \leq p(G) \leq 1$$

■

Mutually Exclusive Events

Two events that cannot both occur at the same time are called **mutually exclusive.** In other words, E and F are mutually exclusive if and only if $E \cap F = \varnothing$.

EXAMPLE 2 A die is rolled. Let E be the event "an even number comes up," F the event "a number greater than 3 comes up," and G the event "an odd number comes up."

a. Are E and F mutually exclusive?
b. Are E and G mutually exclusive?

Solution **a.** $E = \{2, 4, 6\}$, $F = \{4, 5, 6\}$, and $E \cap F = \{4, 6\} \neq \varnothing$ (see Figure 3.11). Therefore, E and F are *not* mutually exclusive; the number that comes up could be *both* even *and* greater than 3. In particular, it could be 4 or 6.

b. $E = \{2, 4, 6\}$, $G = \{1, 3, 5\}$, and $E \cap G = \varnothing$. Therefore, E and G *are* mutually exclusive; the number that comes up could *not* be both even and odd.

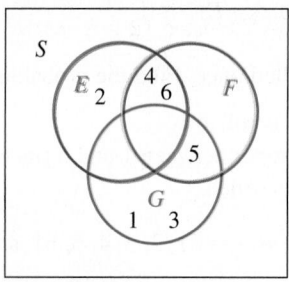

Figure 3.11
A Venn diagram for Example 2

■

EXAMPLE 3 Let M be the event "being a mother," F the event "being a father," and D the event "being a daughter."

a. Are events M and D mutually exclusive?
b. Are events M and F mutually exclusive?

Solution **a.** M and D are mutually exclusive if $M \cap D = \varnothing$. $M \cap D$ is the set of all people who are both mothers and daughters, and that set is not empty. A person can be a mother and a daughter at the same time. M and D are not mutually exclusive because being a mother does not exclude being a daughter.

b. M and F are mutually exclusive if $M \cap F = \varnothing$. $M \cap F$ is the set of all people who are both mothers and fathers, and that set is empty. A person cannot be a mother and a father at the same time. *M* and *F* are mutually exclusive because being a mother does exclude the possibility of being a father.   ■

Pair-of-Dice Probabilities

To find probabilities involving the rolling of a pair of dice, we must first determine the sample space. The *sum* can be anything from 2 to 12, but we will not use {2, 3, 4, 5, 6, 7, 8, 9, 10, 11, 12} as the sample space because those outcomes are not equally likely. Compare a sum of 2 with a sum of 7. There is only one way in which the sum can be 2, and that is if each die shows a 1. There are many ways in which the sum can be a 7, including:

- a 1 and a 6
- a 2 and a 5
- a 3 and a 4

What about a 4 and a 3? Is that the same as a 3 and a 4? They certainly *appear* to be the same. But if one die were blue and the other were white, it would be quite easy to tell them apart. So there are other ways in which the sum can be a 7:

- a 4 and a 3
- a 5 and a 2
- a 6 and a 1

The event (3, 4).

The event (4, 3).

Altogether, there are six ways in which the sum can be a 7. There is only one way in which the sum can be a 2. The outcomes "the sum is 2" and "the sum is 7" are not equally likely.

   To have equally likely outcomes, we must use outcomes such as "rolling a 4 and a 3" and "rolling a 3 and a 4." We will use ordered pairs as a way of abbreviating these longer descriptions. So we will denote the outcome "rolling a 4 and a 3" with the ordered pair (4, 3), and we will denote the outcome "rolling a 3 and a 4" with the ordered pair (3, 4). Figure 3.12 lists all possible outcomes and the resulting sums. Notice that $n(S) = 6 \cdot 6 = 36$.

Figure 3.12
Outcomes of rolling two dice

ThomsonNOW™
for Liberal Arts Mathematics
www.thomsonedu.com/1pass.   **EXAMPLE 4**   A pair of dice is rolled. Find the probability of each of the following events.

a. The sum is 7.   **b.** The sum is greater than 9.
c. The sum is even.   **d.** The sum is not greater than 9.
e. The sum is greater than 9 and even.   **f.** The sum is greater than 9 or even.

Solution   **a.** To find the probability that the sum is 7, let D be the event "the sum is 7." From Figure 3.12, $D = \{(1, 6), (2, 5), (3, 4), (4, 3), (5, 2), (6, 1)\}$, so $n(D) = 6$; therefore,

$$p(D) = \frac{n(D)}{n(S)} = \frac{6}{36} = \frac{1}{6}$$

This means that if we were to roll a pair of dice a large number of times, we should expect to get a sum of 7 approximately one-sixth of the time.

 Notice that $p(D) = \frac{1}{6}$ is between 0 and 1, as are all probabilities.

b. To find the probability that the sum is greater than 9, let E be the event "the sum is greater than 9."

$E = \{(4, 6), (5, 5), (6, 4), (5, 6), (6, 5), (6, 6)\}$, so $n(E) = 6$; therefore,

$$p(E) = \frac{n(E)}{n(S)} = \frac{6}{36} = \frac{1}{6}$$

This means that if we were to roll a pair of dice a large number of times, we should expect to get a sum greater than 9 approximately one-sixth of the time.

c. Let F be the event "the sum is even."

$F = \{(1, 1), (1, 3), (2, 2), (3, 1), \ldots, (6, 6)\}$, so $n(F) = 18$ (refer to Figure 3.12); therefore,

$$p(F) = \frac{n(F)}{n(S)} = \frac{18}{36} = \frac{1}{2}$$

This means that if we were to roll a pair of dice a large number of times, we should expect to get an even sum approximately half of the time.

© Scala/Art Resource, NY

A Roman painting on marble of the daughters of Niobe using knucklebones as dice. This painting was found in the ruins of Herculaneum, a city that was destroyed along with Pompeii by the eruption of Vesuvius.

d. We could find the probability that the sum is not greater than 9 by counting, as in (a), (b), and (c), but the counting would be rather excessive. It is easier to use one of the Cardinal Number Formulas from Chapter 2 on sets. The event "the sum is not greater than 9" is the complement of event E ("the sum is greater than 9") and can be expressed as E'.

$$n(E') = n(U) - n(E) \quad \text{Cardinal Number Formula}$$
$$= n(S) - n(E) \quad \text{"universal set" and "sample space"}$$
$$= 36 - 6 = 30 \quad \text{represent the same idea.}$$
$$p(E') = \frac{n(E')}{n(S)} = \frac{30}{36} = \frac{5}{6}$$

This means that if we were to roll a pair of dice a large number of times, we should expect to get a sum that's not greater than 9 approximately five-sixths of the time.

e. The event "the sum is greater than 9 and even" can be expressed as the event $E \cap F$. $E \cap F = \{(4, 6), (5, 5), (6, 4), (6, 6)\}$, so $n(E \cap F) = 4$; therefore,

$$p(E \cap F) = \frac{n(E \cap F)}{n(S)} = \frac{4}{36} = \frac{1}{9}$$

This means that if we were to roll a pair of dice a large number of times, we should expect to get a sum that's both greater than 9 and even approximately one-ninth of the time.

f. Finding the probability that the sum is greater than 9 or even by counting would require an excessive amount of counting. It is easier to use one of the Cardinal Number Formulas from Chapter 2. The event "the sum is greater than 9 or even" can be expressed as the event $E \cup F$.

$$n(E \cup F) = n(E) + n(F) - n(E \cap F) \quad \text{Cardinal Number Formula}$$
$$= 6 + 18 - 4 \quad \text{from parts (b), (c), and (e)}$$
$$= 20$$
$$p(E \cup F) = \frac{n(E \cup F)}{n(S)} = \frac{20}{36} = \frac{5}{9}$$

This means that if we were to roll a pair of dice a large number of times, we should expect to get a sum that's either greater than 9 or even approximately five-ninths of the time. ∎

More Probability Rules

In part (b) of Example 4, we found that $p(E) = \frac{1}{6}$, and in part (d), we found that $p(E') = \frac{5}{6}$. Notice that $p(E) + p(E') = \frac{1}{6} + \frac{5}{6} = 1$. This should make sense to you. It just means that if E happens one-sixth of the time, then E' has to happen the other five-sixths of the time. This always happens—for any event E, $p(E) + p(E') = 1$.

As we will see in Exercise 79, the fact that $p(E) + p(E') = 1$ is closely related to the Cardinal Number Formula $n(E) + n(E') = n(S)$. The main difference is that one is expressed in the language of probability theory and the other is expressed in the language of set theory. In fact, the following three rules are all set theory rules (from Chapter 2) rephrased so that they use the language of probability theory rather than the language of set theory.

More Probability Rules

Rule 4 $p(E \cup F) = p(E) + p(F) - p(E \cap F)$

Rule 5 $p(E \cup F) = p(E) + p(F),$ if E and F are mutually exclusive

Rule 6 $p(E) + p(E') = 1$ or, equivalently, $p(E') = 1 - p(E)$

In Example 4, we used some Cardinal Number Formulas from Chapter 2 to avoid excessive counting. Some find it easier to use Probability Rules to calculate probabilities, rather than Cardinal Number Formulas.

EXAMPLE 5   Use a Probability Rule rather than a Cardinal Number Formula to find

a. the probability that the sum is not greater than 9
b. the probability that the sum is greater than 9 or even

Solution   We will use the results of Example 4.

a. $p(E') = 1 - p(E)$ Probability Rule 6

$$= 1 - \frac{1}{6}$$ from Example 4

$$= \frac{5}{6}$$

b. $p(E \cup F) = p(E) + p(F) - p(E \cap F)$ Probability Rule 4

$$= \frac{1}{6} + \frac{1}{2} - \frac{1}{9}$$ from Example 4 parts (b), (c), and (e)

$$= \frac{3}{18} + \frac{9}{18} - \frac{2}{18} = \frac{5}{9}$$ ∎

Probabilities and Venn Diagrams

Venn diagrams can be used to illustrate probabilities in the same way in which they are used in set theory. In this case, we label each region with its probability rather than its cardinal number.

EXAMPLE 6   Zaptronics manufactures compact discs and their cases for several major labels. A recent sampling of Zaptronics' products has indicated that 5% have a defective case, 3% have a defective disc, and 7% have at least one of the two defects.

a. Find the probability that a Zaptronics product has both defects.
b. Draw a Venn diagram that shows the probabilities of each of the basic regions.
c. Use the Venn diagram to find the probability that a Zaptronics product has neither defect.
d. Find the probability that a Zaptronics product has neither defect, using probability rules rather than a Venn diagram.

Solution   **a.** Let C be the event that the case is defective, and let D be the event that the disc is defective. We are given that $p(C) = 5\% = 0.05$, $p(D) = 3\% = 0.03$, and $p(C \cup D) = 7\% = 0.07$, and we are asked to find $p(C \cap D)$. To do this, substitute into Probability Rule 4.

$$p(C \cup D) = p(C) + p(D) - p(C \cap D) \quad \text{Probability Rule 4}$$
$$0.07 = 0.05 + 0.03 - p(C \cap D) \quad \text{substituting}$$
$$0.07 = 0.08 - p(C \cap D)$$
$$p(C \cap D) = 0.01 = 1\%$$

This means that 1% of Zaptronics' products have a defective case *and* a defective disc.

b. The Venn diagram for this is shown in Figure 3.13. The 0.93 probability in the lower right corner is obtained by first finding the sum of the probabilities of the other three basic regions:

$$0.04 + 0.01 + 0.02 = 0.07$$

Figure 3.13
A Venn diagram for Example 6

The rectangle itself corresponds to the sample space, and its probability is 1. Thus, the probability of the outer region, the region outside of the C and D circles, is $1 - 0.07 = 0.93$.

c. The only region that involves no defects is the outer region, the one whose probability is 0.93. This means that 93% of Zaptronics' products have neither defect.

d. If we are to not use the Venn diagram, we must first rephrase the event "a Zaptronics product has neither defect" as "a product does not have either defect." The "does not" part of this means *complement,* and the "either defect" means *either C or D.* This means that

- "has either defect" translates to $C \cup D$
- "does not have either defect" translates to $(C \cup D)'$

So we are asked to find $P((C \cup D)')$.

$$p((C \cup D)') = 1 - p(C \cup D) \quad \text{Probability Rule 6}$$
$$= 1 - 0.07 \quad \text{substituting}$$
$$= 0.93 = 93\%$$

This means that 93% of Zaptronics' products are defect-free. ■

3.3 Exercises

In Exercises 1–10, determine whether E and F are mutually exclusive. Write a sentence justifying your answer.

1. E is the event "being a doctor," and F is the event "being a woman."

2. E is the event "it is raining," and F is the event "it is sunny."

3. E is the event "being single," and F is the event "being married."

4. E is the event "having naturally blond hair," and F is the event "having naturally black hair."

5. E is the event "having brown hair," and F is the event "having gray hair."

6. E is the event "being a plumber," and F is the event "being a stamp collector."

7. E is the event "wearing boots," and F is the event "wearing sandals."

8. E is the event "wearing shoes," and F is the event "wearing socks."

9. If a die is rolled once, E is the event "getting a four," and F is the event "getting an odd number."

10. If a die is rolled once, E is the event "getting a four," and F is the event "getting an even number."

In Exercises 11–18, a card is dealt from a complete deck of 52 playing cards (no jokers). Use probability rules (when appropriate) to find the probability that the card is as stated. (Count an ace as high.)

11. **a.** a jack and red   **b.** a jack or red
       c. not a red jack

12. **a.** a jack and a heart   **b.** a jack or a heart
       c. not a jack of hearts

13. **a.** a ten and a spade **b.** a ten or a spade
    c. not a ten of spades
14. **a.** a five and black **b.** a five or black
    c. not a black five
15. **a.** under a four
    b. above a nine
    c. both under a four and above a nine
    d. either under a four or above a nine
16. **a.** above a jack
    b. below a three
    c. both above a jack and below a three
    d. either above a jack or below a three
17. **a.** above a five
    b. below a ten
    c. both above a five and below a ten
    d. either above a five or below a ten
18. **a.** above a seven
    b. below a queen
    c. both above a seven and below a queen
    d. either above a seven or below a queen

In Exercises 19–26, use complements to find the probability that a card dealt from a full deck (no jokers) is as stated. (Count an ace as high.)

19. not a queen   20. not a seven
21. not a face card   22. not a heart
23. above a three   24. below a queen
25. below a jack   26. above a five

27. If $o(E) = 5:9$, find $o(E')$.
28. If $o(E) = 1:6$, find $o(E')$.
29. If $p(E) = \frac{2}{7}$, find $o(E)$ and $o(E')$.
30. If $p(E) = \frac{3}{8}$, find $o(E)$ and $o(E')$.
31. If $o(E) = a:b$, find $o(E')$.
32. If $p(E) = \frac{a}{b}$, find $o(E')$.

    Hint: Use Exercise 49 from Section 3.2 and Exercise 31 above.

In Exercises 33–38, use Exercise 32 above to find the odds that a card dealt from a full deck (no jokers) is as stated.

33. not a king   34. not an eight
35. not a face card   36. not a club
37. above a four   38. below a king

In Exercises 39–42, use the following information: To determine the effect their salespersons have on purchases, a department store polled 700 shoppers regarding whether or not they made a purchase and whether or not they were pleased with the service they

received. Of those who made a purchase, 151 were happy with the service and 133 were not. Of those who made no purchase, 201 were happy with the service and 215 were not. Use probability rules (when appropriate) to find the probability of the event stated.

39. **a.** A shopper made a purchase.
    b. A shopper did not make a purchase.
40. **a.** A shopper was happy with the service received.
    b. A shopper was unhappy with the service received.
41. **a.** A shopper made a purchase and was happy with the service.
    b. A shopper made a purchase or was happy with the service.
42. **a.** A shopper made no purchase and was unhappy with the service.
    b. A shopper made no purchase or was unhappy with the service.

In Exercises 43–46, use the following information: A supermarket polled 1,000 customers regarding the size of their bill. The results are given in Figure 3.14.

Size of Bill	Number of Customers
below $20.00	208
$20.00–$39.99	112
$40.00–$59.99	183
$60.00–$79.99	177
$80.00–$99.99	198
$100.00 or above	122

Figure 3.14
Supermarket bills

Use probability rules (when appropriate) to find the relative frequency with which a customer's bill is as stated.

43. **a.** less than $40.00   **b.** $40.00 or more
44. **a.** less than $80.00   **b.** $80.00 or more
45. **a.** between $40.00 and $79.99
    b. not between $40.00 and $79.99
46. **a.** between $20.00 and $79.99
    b. not between $20.00 and $79.99

In Exercises 47–54, find the probability that the sum is as stated when a pair of dice is rolled.

47. **a.** 7   **b.** 9   **c.** 11
48. **a.** 2   **b.** 4   **c.** 6
49. **a.** 7 or 11 **b.** 7 or 11 or doubles
50. **a.** 8 or 10 **b.** 8 or 10 or doubles

51. a. odd and greater than 7
   b. odd or greater than 7
52. a. even and less than 5
   b. even or less than 5
53. a. even and doubles   **b.** even or doubles
54. a. odd and doubles   **b.** odd or doubles

In Exercises 55–58, use the following information: After examining their clients' records, Crashorama Auto Insurance calculated the probabilities in Figure 3.15.

	Miles/year < 10,000	10,000 ≤ miles/yr < 20,000	20,000 ≤ miles/yr
Accident	0.05	0.10	0.3
No accident	0.20	0.15	0.2

Figure 3.15
Crashorama's Accident Incidence

55. a. Find the probability that a client drives 20,000 miles/year or more or has an accident.
   b. Find the probability that a client drives 20,000 miles/year or more and has an accident.
   c. Find the probability that the client does not drive 20,000 miles/year or more and does not have an accident.
56. a. Find the probability that the client drives less than 10,000 miles/year and has an accident.
   b. Find the probability that a client drives less than 10,000 miles/year or has an accident.
   c. Find the probability that a client does not drive less than 10,000 miles per year and does not have an accident.
57. Are the probabilities in Figure 3.15 theoretical probabilities or relative frequencies? Why?
58. a. Find the probability that a client drives either less than 10,000 miles/year or 20,000 miles/year or more.
   b. Find the probability that a client both has an accident and drives either less than 10,000 miles/year or 20,000 miles/year or more.

In Exercises 59–62, use the following information: After interviewing randomly selected voters by calling them at home during dinner, Polls-R-Us calculated the probabilities in Figure 3.16.

59. a. Find the probability that a voter supported Bush or is a Republican.
   b. Find the probability that a voter supported Bush and is a Republican.

	Democrats	Republicans	Green Party
Bush	0.08	0.33	0.02
Kerry	0.38	0.13	0.06

Figure 3.16
Telephone poll results

   c. Find the probability that a voter didn't support Bush and is not a Republican.
60. a. Find the probability that a voter supported Kerry and is a Democrat
   b. Find the probability that a voter supported Kerry or is a Democrat.
   c. Find the probability that a voter did not support Kerry or is not a Democrat.
61. Are the probabilities in Figure 3.16 theoretical probabilities or relative frequencies? Why?
62. a. Find the probability that a voter is a Democrat or a Green Party member.
   b. Find the probability that a voter both supports Kerry and is either a Democrat or a Green Party member.
63. Fried Foods hosted a group of twenty-five consumers at a tasting session. The consumers were asked to taste two new products and state whether or not they would be interested in buying them. Fifteen consumers said that they would be interested in buying Snackolas, twelve said that they would be interested in buying Chippers, and ten said that they would be interested in buying both.
   a. What is the probability that one of the tasters is interested in buying at least one of the two new products?
   b. What is the probability that one of the tasters is not interested in buying Snackolas but is interested in buying Chippers?
64. Ink Inc., a publishing firm, offers its 899 employees a cafeteria approach to benefits, in which employees can enroll in the benefit plan of their choice. Seven hundred thirteen employees have health insurance, 523 have dental insurance, and 489 have both health and dental insurance.
   a. What is the probability that one of the employees has either health or dental insurance?
   b. What is the probability that one of the employees has health insurance but not dental insurance?
65. The Central State Bell Corporation is considering offering DSL connections to the World Wide Web. They asked 3,000 customers about their use of the Web at home, and they received 1,451 responses. They

found that 1,230 customers are currently connecting to the Web with a telephone modem, 726 are interested in obtaining a DSL connection, and 514 are interested in switching from a telephone modem to DSL.

a. What is the probability that one of the respondents is interested in obtaining a DSL connection and is not currently using a telephone modem?

b. What is the probability that one of the respondents is interested in obtaining a DSL connection and is currently using a telephone modem?

66. The Video Emporium rents DVDs and videotapes only. They surveyed their 1,167 rental receipts for the last two weeks. Eight hundred thirty-two customers rented DVDs, and 692 rented videotapes.

a. What is the probability that a customer rents DVDs only?

b. What is the probability that a customer rents DVDs and videotapes?

67. *Termiyak Magazine* conducted a poll of its readers, asking them about their telephones. Six hundred ninety-six readers responded. Five hundred seventy-two said they have a cell phone. Six hundred twelve said that they have a land line (a traditional, non-wireless telephone) at their home. Everyone has either a cell phone or a land line.

a. What is the probability that one of the readers has only a cell phone?

b. What is the probability that one of the readers has only a land line?

68. Which of the probabilities in Exercises 63–67 are theoretical probabilities and which are relative frequencies? Why?

69. Use probability rules to find the probability that a child will either have Tay-Sachs disease or be a carrier if (a) each parent is a Tay-Sachs carrier, (b) one parent is a Tay-Sachs carrier and the other parent has no Tay-Sachs gene, (c) one parent has Tay-Sachs and the other parent has no Tay-Sachs gene.

70. Use probability rules to find the probability that a child will have either sickle-cell anemia or sickle-cell trait if (a) each parent has sickle-cell trait, (b) one parent has sickle-cell trait and the other parent has no sickle-cell gene, (c) one parent has sickle-cell anemia and the other parent has no sickle-cell gene.

71. Use probability rules to find the probability that a child will neither have Tay-Sachs disease nor be a carrier if (a) each parent is a Tay-Sachs carrier, (b) one parent is a Tay-Sachs carrier and the other parent has no Tay-Sachs gene, (c) one parent has Tay-Sachs and the other parent has no Tay-Sachs gene.

72. Use probability rules to find the probability that a child will have neither sickle-cell anemia nor sickle-cell trait if (a) each parent has sickle-cell trait, (b) one parent has sickle-cell trait and the other parent has no sickle-cell gene, (c) one parent has sickle-cell anemia and the other parent has no sickle-cell gene.

73. Mary is taking two courses, photography and economics. Student records indicate that the probability of passing photography is 0.75, that of failing economics is 0.65, and that of passing at least one of the two courses is 0.85. Find the probability of the following.

a. Mary will pass economics.

b. Mary will pass both courses.

c. Mary will fail both courses.

d. Mary will pass exactly one course.

74. Alex is taking two courses, algebra and U.S. history. Student records indicate that the probability of passing algebra is 0.35, that of failing U.S. history is 0.35, and that of passing at least one of the two courses is 0.80. Find the probability of the following.

a. Alex will pass history.

b. Alex will pass both courses.

c. Alex will fail both courses.

d. Alex will pass exactly one course.

75. Of all the flashlights in a large shipment, 15% have a defective bulb, 10% have a defective battery, and 5% have both defects. If you purchase one of the flashlights in this shipment, find the probability that it has the following.

a. a defective bulb or a defective battery

b. a good bulb or a good battery

c. a good bulb and a good battery

76. Of all the videotapes in a large shipment, 20% have a defective tape, 15% have a defective case, and 10% have both defects. If you purchase one of the videotapes in this shipment, find the probability that it has the following.

a. a defective tape or a defective case

b. a good tape or a good case

c. a good tape and a good case

77. Verify Probability Rule 4.

Hint: Divide the Cardinal Number Formula for the Union of Sets from Section 2.1 by $n(S)$.

78. Verify Probability Rule 5.

Hint: Start with Probability Rule 4. Then use the fact that E and F are mutually exclusive.

79. Verify Probability Rule 6.

Hint: Are E and E' mutually exclusive?

Answer the following questions using complete sentences and your own words.

CONCEPT QUESTIONS

80. What is the complement of a certain event? Justify your answer.

81. Compare and contrast mutually exclusive events and impossible events. Be sure to discuss both the similarities and the differences between these two concepts.

82. Explain why it is necessary to subtract $p(E \cap F)$ in Probability Rule 4. In other words, explain why Probability Rule 5 is not true for all events.

 ## Fractions on a Graphing Calculator

Some graphing calculators—including the TI-83, TI-84, TI-86, and Casio—will add, subtract, multiply, and divide fractions and will give answers in reduced fractional form.

Reducing Fractions

The fraction 42/70 reduces to 3/5. To do this on your calculator, you must make your screen read "42/70 ▶ Frac" ("42 ⌐70" on a Casio). The way that you do this varies.

TI-83/84	• Type 42 ÷ 70, but do not press ENTER. This causes "42/70" to appear on the screen. • Press the MATH button. • Highlight option 1, "▶Frac." (Option 1 is automatically highlighted. If we were selecting a different option, we would use the ▲ and ▼ buttons to highlight it.) • Press ENTER. This causes "42/70▶Frac" to appear on the screen. • Press ENTER. This causes 3/5 to appear on the screen.
TI-86	• Type 42 ÷ 70, but do not press ENTER. This causes "42/70" to appear on the screen. • Press 2nd MATH. • Press MISC (i.e., F5). • Press MORE until "▶Frac" appears. • Press ▶Frac (i.e., F1). This causes "42/70▶Frac" to appear on the screen. • Press ENTER. This causes 3/5 to appear on the screen.
CASIO	• Type 42 ab/c 70. • Type EXE, and your display will read "3⌐5," which should be interpreted as "3/5."

EXAMPLE 5   Use your calculator to compute

$$\frac{1}{6} + \frac{1}{2} - \frac{1}{9}$$

and give your answer in reduced fractional form.

Solution   On a TI, make your screen read "1/6 + 1/2 − 1/9 ▶Frac" by typing

1 ÷ 6 + 1 ÷ 2 − 1 ÷ 9

and then inserting the "▶Frac" command, as described on page 163. Once you press ENTER, the screen will read "5/9."

On a Casio, make your screen read "1⌐6 + 1⌐2 − 1⌐9" by typing

1 a^b_c 6 + 1 a^b_c 2 − 1 a^b_c 9

Once you press EXE, the screen will read "5⌐9," which should be interpreted as "5/9."   ∎

Exercises

In Exercises 83–84, reduce the given fractions to lowest terms, both (a) by hand and (b) with a calculator. Check your work by comparing the two answers.

83. $\dfrac{18}{33}$ **84.** $-\dfrac{42}{72}$

In Exercises 85–90, perform the indicated operations, and reduce the answers to lowest terms, both (a) by hand and (b) with a calculator. Check your work by comparing the two answers.

85. $\dfrac{6}{15} \cdot \dfrac{10}{21}$

86. $\dfrac{6}{15} \div \dfrac{10}{21}$

87. $\dfrac{6}{15} + \dfrac{10}{21}$

88. $\dfrac{6}{15} - \dfrac{10}{21}$

89. $\dfrac{7}{6} - \dfrac{5}{7} + \dfrac{9}{14}$

90. $\dfrac{-8}{5} - \left(\dfrac{-3}{28} + \dfrac{5}{21} \right)$

91. How could you use your calculator to get decimal answers to the above exercises rather than fractional answers?

3.4 Combinatorics and Probability

Finding a probability involves finding the number of outcomes in an event and the number of outcomes in the sample space. So far, we have used probability rules as an alternative to excessive counting. Another alternative is combinatorics—that is, permutations, combinations, and the fundamental principle of counting, as covered in Sections 2.3 and 2.4. The flowchart used in Chapter 2 is summarized on page 165.

EXAMPLE 1   A group of three people is selected at random. What is the probability that at least two of them will have the same birthday?

Solution   We will assume that all birthdays are equally likely, and for the sake of simplicity, we will ignore leap-year day (February 29). The experiment is to ask three people their birthdays. One possible outcome is (May 1, May 3, August 23). The sample

> **Which Combinatorics Method?**
>
> **1.** If the selection is done with replacement, use the Fundamental Principle of Counting.
> **2.** If the selection is done without replacement and there is only one category, use:
>   **a.** permutations if the order of selection does matter:
> $$_nP_r = \frac{n!}{(n-r)!}$$
>   **b.** combinations if the order of selection does not matter:
> $$_nC_r = \frac{n!}{(n-r)!\,r!}$$
> **3.** If there is more than one category, use the Fundamental Principle of Counting with one box per category. Use (2) above to determine the numbers that go in the boxes.

space is the set of all possible lists of three birthdays. Finding $n(S)$ by counting the elements in S is impractical, so we will use combinatorics and the above "Which Combinatorics Method?" chart.

The selected items are birthdays. They are selected with replacement because people can share the same birthday. This means that we must use the Fundamental Principle of Counting. We make three boxes, one for each birthday. The first birthday may be selected in any of 365 different ways, since there are 365 different days in a year. The second birthday may be selected in any of 365 ways also, as can the third birthday, because people can share the same birthday. Thus, the number of elements in the sample space is

$$n(S) = \boxed{365} \cdot \boxed{365} \cdot \boxed{365}$$

The event E is the set of all possible lists of three birthdays in which at least two of those birthdays are the same. It is rather difficult to compute $n(E)$ directly. Instead, we will compute $n(E')$ and then use Probability Rule 6, the complement rule.

E' is the set of all possible lists of three birthdays in which no two of those birthdays are the same. To find the number of elements in E', we again follow the flowchart. The birthdays *cannot* be repeated, there is only one category (birthdays), and the order of selection does matter [(May 1, May 3, August 23) is a different list from (August 23, May 3, May 1)]. Thus, we use permutations, and the number of elements in E' is

$$n(E') = {}_{365}P_3$$
$$p(E') = \frac{n(E')}{n(S)}$$
$$= \frac{{}_{365}P_3}{365^3}$$
$$p(E) = 1 - p(E') \qquad \text{Probability Rule 6}$$
$$= 1 - \frac{{}_{365}P_3}{365^3}$$
$$= 0.008204\ldots \approx 0.008$$

$$\boxed{\text{(calc)} \quad 1 \;\boxed{-}\; 365 \;\boxed{\text{nPr}}\; 3 \;\boxed{\div}\; 365 \;\boxed{\wedge}\; 3 \;\boxed{\text{ENTER}}}$$

where nPr refers to the nPr command on the "MATH" menu, discussed in Section 2.4.

This result is not at all surprising. It means that it is extremely unlikely that two or more people in a group of three share a birthday. However, it is quite *likely* that two or more people in a group of thirty share a birthday. See Exercise 1. ∎

Lotteries

ThomsonNOW™
for Liberal Arts Mathematics
www.thomsonedu.com/1pass.

EXAMPLE 2   Connecticut, Louisiana, and Virginia all operate 6/44 lotteries. In them, the player selects any six of the numbers from 1 to 44. The state determines six winning numbers, usually by having a mechanical device choose six balls from a container filled with balls numbered from 1 to 44. If a player's six selections match the six winning numbers, the player wins first prize. Smaller prizes are awarded to players who select five or four of the winning numbers.

a. Find the probability of winning first prize.
b. Find the probability of winning second prize.

Solution   **a.** Let E be the event "winning first prize." To find the probability, we must find the number of outcomes in event E and the number of outcomes in the sample space S.

The sample space is the set of all possible lottery tickets. That is, it is the set of all possible choices of six numbers chosen from the numbers 1 through 44.

Typical lottery tickets

Probabilities and the Real World: The Business of Gambling

It used to be that legal commercial gambling was not common. Nevada made casino gambling legal in 1931, and for more than thirty years, it was the only place in the United States that had legal commercial gambling. Then in 1964, New Hampshire instituted the first lottery in the United States since 1894. In 1978, New Jersey became the second state to legalize casino gambling. Now, state-sponsored lotteries are common, Native American tribes have casinos in more than half the states, and many states have casinos.

It is very likely that you will be exposed to gambling if you have not been already. You should approach gambling with an educated perspective. If you are considering gambling, know what you are up against. The casinos all use probabilities and combinatorics in designing their games *to ensure that they make a consistent profit*. Learn this mathematics so that they do not take advantage of you.

Public lotteries have a long history in the United States. The settlement of Jamestown was financed in part by an English lottery. George Washington managed a lottery that paid for a road through the Cumberland Mountains. Benjamin Franklin used lotteries to finance cannons for the Revolutionary War, and John Hancock used lotteries to rebuild Faneuil Hall in Boston. Several universities, including Harvard, Dartmouth, Yale, and Columbia, were partly financed by lotteries. The U.S. Congress operated a lottery to help finance the Revolutionary War.

Now, public lotteries are a very big business. They are quite lucrative for the states that offer them and for the corporations the states hire to administer them. Since 1964, when New Hampshire became the first state to offer a state-sponsored lottery, Americans have paid almost $428 billion for lottery tickets. In 2001 alone, Americans paid more than $39 billion for lottery tickets. Overall, 52% of this money was paid out in prizes, 35% went to the government, and 13% was spent in administering the lotteries. Of course, these numbers vary from state to state. In Delaware, only 26% was paid out in prizes, while in South Dakota, 64% was paid out in prizes.

Lottery players go berserk when the jackpots accumulate, partially because of the amazingly large winnings but also because of a lack of understanding of how unlikely it is that they will actually win. The largest cumulative jackpot was $363 million, which was split by winners in Michigan and Illinois in 2002. The largest single-winner jackpot was $315 million in West Virginia in 2000.

In this section, we will discuss probabilities and gambling. Specifically, we will explore lotteries, keno, and card games. See Examples 2–6, and Exercises 5–18 and 29.

In Section 3.5, we will discuss how much money you can expect to win or lose when gambling. We will continue to discuss lotteries, keno, and card games, but we will also discuss roulette and raffles. See Exercises 1–12, 28–32, and 37–47.

United States Lottery. N.

CLASS the FIRST

THIS TICKET entitles the Bearer to receive such Prize as may be drawn against its Number, according to a *Resolution* of CONGRESS, passed at *Philadelphia*, November 18, 1776.

N.

FAC-SIMILE OF A CONTINENTAL LOTTERY TICKET.

One possible selection is

> 1   32   43   4   15   26

Another is

> 41   12   4   24   25   37

Finding $n(S)$ by counting the elements in S is impractical, so we will use combinatorics and the above "Which Combinatorics Method?" chart. Selection is done without replacement, because you can't select the same lottery number twice. Order does not matter, because the player can choose the six numbers in any order. Therefore, we use *combinations*.

$$n(S) = {}_{44}C_6 = \frac{44!}{6! \cdot (44 - 6)!} = 7{,}059{,}052$$

As you can see, the sample space is amazingly huge. There are more than seven million possible lottery tickets.

We do not need to use combinatorics to find $n(E)$, because there is only one winning lottery ticket—that is, there is only one winning combination of numbers. So $n(E) = 1$.

The probability of selecting all six winning numbers is

$$p(E) = \frac{n(E)}{n(S)} = \frac{1}{7{,}059{,}052} \approx 0.00000014$$

This means that only one out of approximately *seven million* combinations is the first-prize-winning combination. This is an incredibly unlikely event. It is about as probable as dying by being hit by lightning in one year.

b. Let F be the event "winning second prize." To find the probability of F, we must find the number of outcomes in event F and divide this by the number of outcomes in the sample space S.

The sample space is the same as in part (a), so $n(S) = {}_{44}C_6 = 7{,}059{,}052$.

To win second prize, five of the numbers you select must be winning numbers, and one must be a losing number. This involves two categories (winning numbers are losing numbers), so we will use the Fundamental Principle of Counting. In each box, we will use combinations, for the same reasons that we used combinations to find $n(S)$.

win	lose

The state selects six winning numbers, and the second-prize winning player must select five of them, so there are ${}_6C_5$ ways of selecting the five winning numbers. Also, there are $44 - 6 = 38$ losing numbers, of which the player selects 1, so there are ${}_{38}C_1$ ways of selecting one losing number.

win	lose
${}_6C_5$	${}_{38}C_1$

We have found that

$$p(F) = \frac{n(F)}{n(S)} = \frac{{}_6C_5 \cdot {}_{38}C_1}{{}_{44}C_6} = \frac{6 \cdot 38}{7,059,059} \approx 0.000032298$$

How do you make sense of a decimal with so many zeros in front of it? One way is to round it off at the first nonzero digit and convert the result to a fraction:

$$0.000032298 \approx 0.00003 = 3/100,000$$

This means that if you buy a lot of 6/44 lottery tickets, you will win second prize approximately 3 times out of every 100,000 times you play.

> ✓ To check your work, notice that in the event, there is a distinction between two categories (winning numbers and losing numbers); in the sample space, there is no such distinction. Thus, the numerator of
>
> $$p(E) = \frac{{}_6C_5 \cdot {}_{38}C_1}{{}_{44}C_6}$$
>
> has two parts (one for each category), and the denominator has one part. Also, the numbers in front of the Cs add correctly (6 winning numbers + 38 losing numbers = 44 total numbers to choose from), and the numbers after the Cs add correctly (5 winning numbers + 1 losing number = 6 total numbers to select).

■

Keno

The game of keno is a casino version of the lottery. In this game, the casino has a container filled with balls numbered from 1 to 80. The player buys a keno ticket, with which he or she selects anywhere from 1 to 15 (usually 6, 8, 9, or 10) of those 80 numbers; the player's selections are called "spots." The casino chooses 20 winning numbers, using a mechanical device to ensure a fair game. If a sufficient number of the player's spots are winning numbers, the player receives an appropriate payoff.

EXAMPLE 3   In the game of keno, if eight spots are marked, the player wins if five or more of his or her spots are selected. Find the probability of having five winning spots.

Solution   The sample space S is the set of all ways in which the player can select eight numbers from the 80 numbers in the game. Selection is made without replacement, because you cannot select the same keno number twice. Order does not matter, because the player can choose the eight numbers in any order. Therefore, using combinations,

$$n(S) = {}_{80}C_8 = \frac{80!}{8! \cdot (80 - 8)!} = 28,987,537,150$$

Let E represent the event "having five winning spots (and three losing spots)." This involves two categories (winning spots and losing spots), so we will use the Fundamental Principle of Counting.

win   lose
□   □

A typical keno ticket

The casino selects 20 winning numbers, from which the player is to select five winning numbers, so there are $_{20}C_5$ ways of getting five winning spots. Also, there are $80 - 20 = 60$ losing numbers, from which the player selects three, so there are $_{60}C_3$ ways of getting three losing spots.

<table>
<tr><td>win</td><td>lose</td></tr>
<tr><td>$_{20}C_5$</td><td>$_{60}C_3$</td></tr>
</table>

We have found that

$$p(E) = \frac{_{20}C_5 \cdot {_{60}C_3}}{_{80}C_8} = \frac{15{,}504 \cdot 34{,}220}{28{,}987{,}537{,}150} \approx 0.018303$$

This means that if you play eight-spot keno a lot, you will have five winning spots about 1.8% of the time.

> ✔ To check your work, notice that in the event, there is a distinction between two categories (winning spots and losing spots); in the sample space, there is no such distinction. Thus, the numerator of
>
> $$p(E) = \frac{_{20}C_5 \cdot {_{60}C_3}}{_{80}C_8}$$
>
> has two parts (one for each category), and the denominator has one part. Also, the numbers in front of the Cs add correctly (20 winning spots + 60 losing spots = 80 total spots to choose from), and the numbers after the Cs add correctly (5 winning spots + 3 losing spots = 8 total spots to select).

Cards

One common form of poker is five-card draw, in which each player is dealt five cards. The order in which the cards are dealt is unimportant, so we compute probabilities with combinations rather than permutations.

EXAMPLE 4   Find the probability of being dealt four aces.

Solution   The sample space consists of all possible five-card hands than can be dealt from a deck of fifty-two cards. There are

$$_{52}C_5 = \frac{52!}{5! \cdot 47!} = 2{,}598{,}960$$

possible hands.

The event consists of all possible five-card hands that include four aces and one non-ace. This involves two categories (aces and non-aces), so we will use the Fundamental Counting Principle and multiply the number of ways of getting four aces and the number of ways of getting one non-ace. There is

$$_4C_4 = \frac{4!}{4! \cdot 0!} = 1$$

way of getting four aces, and there are

$$_{48}C_1 = \frac{48!}{1! \cdot 47!} = 48$$

ways of getting one non-ace. (These numbers could certainly be obtained with common sense rather than combinations.) Thus, the event consists of

$$_4C_4 \cdot {}_{48}C_1 = 1 \cdot 48 = 48$$

elements, and the probability of being dealt four aces is

$$p(E) = \frac{_4C_4 \cdot {}_{48}C_1}{_{52}C_5} = \frac{48}{2{,}598{,}960} \approx 0.00001847$$

In the event, there is a distinction between two categories (aces and non-aces); in the sample space, there is no such distinction. Thus, the numerator of

$$p(E) = \frac{_4C_4 \cdot {}_{48}C_1}{_{52}C_5}$$

has two parts (one for each category), and the denominator has one part. Also, the numbers in front of the Cs add correctly (4 aces + 48 non-aces = 52 cards to choose from), and the numbers after the Cs add correctly (4 aces + 1 non-ace = 5 cards to select).

ThomsonNOW™
for Liberal Arts Mathematics
at www.thomsonedu.com/1pass.

EXAMPLE 5   Find the probability of being dealt four of a kind.

Solution   The sample space is the same as that in Example 4. The event "being dealt four of a kind" means "being dealt four twos or being dealt four threes or being dealt four fours . . . or being dealt four kings or being dealt four aces." These latter events ("four twos," "four threes," etc.) are all mutually exclusive, so we can use Probability Rule 5:

p (four of a kind)

$\quad = p(\text{four twos} \cup \text{four threes} \cup \cdots \cup \text{four kings} \cup \text{four aces})$

$\quad = p(\text{four twos}) + p(\text{four threes}) + \cdots + p(\text{four kings}) + p(\text{four aces})$

using Probability Rule 5

Furthermore, these probabilities are all the same

$$p(\text{four twos}) = p(\text{four threes}) = \cdots = p(\text{four aces}) = \frac{_4C_4 \cdot _{48}C_1}{_{52}C_5}$$

This means that the probability of being dealt four of a kind is

$$p(\text{four of a kind}) = p(\text{four twos}) + p(\text{four threes}) + \cdots + p(\text{four kings}) + p(\text{four aces})$$

$$= \frac{_4C_4 \cdot _{48}C_1}{_{52}C_5} + \frac{_4C_4 \cdot _{48}C_1}{_{52}C_5} + \cdots + \frac{_4C_4 \cdot _{48}C_1}{_{52}C_5} + \frac{_4C_4 \cdot _{48}C_1}{_{52}C_5}$$

$$= 13 \cdot \frac{_4C_4 \cdot _{48}C_1}{_{52}C_5} \quad \text{there are 13 denominations (two through ace)}$$

$$= 13 \cdot \frac{1 \cdot 48}{2{,}598{,}960} \quad \text{from Example 4}$$

$$= \frac{624}{2{,}598{,}960} \approx 0.0002401$$

See Example 8 in Section 2.4.   ∎

EXAMPLE 6   Find the probability of being dealt five hearts.

Solution   The sample space is the same as in Example 4. The event consists of all possible five-card hands that include five hearts and no non-hearts. This involves two categories (hearts and non-hearts), so we will use the Fundamental Counting Principle and multiply the number of ways of getting five hearts and the number of ways of getting no non-hearts. There are

$$_{13}C_5 = \frac{13!}{5! \cdot 8!} = 1{,}287$$

ways of getting five hearts, and there is

$$_{39}C_0 = \frac{39!}{0! \cdot 39!} = 1$$

way of getting no non-hearts. Thus, the probability of being dealt five hearts is

$$p(E) = \frac{_{13}C_5 \cdot _{39}C_0}{_{52}C_5} = \frac{1287 \cdot 1}{2{,}598{,}960} = 0.000495198$$

> ✓   In the event, there is a distinction between two categories (hearts and non-hearts); in the sample space, there is no such distinction. Thus, the numerator of
>
> $$p(E) = \frac{{}_{13}C_5 \cdot {}_{39}C_0}{{}_{52}C_5}$$
>
> has two parts (one for each category) and the denominator has one part. Also, the numbers in front of the Cs add correctly (13 hearts + 39 non-hearts = 52 cards to choose from), and the numbers after the Cs add correctly (5 hearts + 0 non-hearts = 5 cards to select).

Notice that in Example 6, we could argue that since we're selecting only hearts, we can disregard the non-hearts. This would lead to the answer obtained in Example 6:

$$p(E) = \frac{{}_{13}C_5}{{}_{52}C_5} = \frac{1287}{2,598,960} = 0.000495198$$

However, this approach would not allow us to check our work in the manner described above; the numbers in front of the Cs don't add correctly, nor do the numbers after the Cs.

3.4   Exercises

1. A group of thirty people is selected at random. What is the probability that at least two of them will have the same birthday?
2. A group of sixty people is selected at random. What is the probability that at least two of them will have the same birthday?
3. How many people would you have to have in a group so that there is a probability of at least 0.5 that at least two of them will have the same birthday?
4. How many people would you have to have in a group so that there is a probability of at least 0.9 that at least two of them will have the same birthday?
5. In 1990, California switched from a 6/49 lottery to a 6/53 lottery. Later, the state switched again to a 6/51 lottery.
   a. Find the probability of winning first prize in a 6/49 lottery.
   b. Find the probability of winning first prize in a 6/53 lottery.
   c. Find the probability of winning first prize in a 6/51 lottery.

   d. How much more probable is it that one will win the 6/49 lottery than the 6/53 lottery?
   e. Why do you think California switched from a 6/49 lottery to a 6/53 lottery? And why do you think the state then switched to a 6/51 lottery? (Answer using complete sentences.)
6. Find the probability of winning second prize—that is, picking five of the six winning numbers—with a 6/53 lottery.
7. Find the probability of winning second prize (that is, picking five of the six winning numbers) with a 6/44 lottery, as played in Connecticut, Missouri, Oregon, and Virginia.
8. Find the probability of winning third prize—that is, picking four of the six winning numbers—with a 6/44 lottery.
9. Currently, the second most popular type of lottery is the 5/35 lottery. It is played in Arizona, Massachusetts, Connecticut, Iowa, South Dakota, and Nebraska.
   a. Find the probability of winning first prize.
   b. Find the probability of winning second prize.

10. The most popular type of lottery is the 6/49 lottery. It is currently played in Massachusetts, Maryland, New Jersey, Ohio, Pennsylvania, Washington, and Wisconsin.
  a. Find the probability of winning first prize.
  b. Find the probability of winning second prize.
11. The 5/39 lottery is currently played in California, Georgia, New York, Pennsylvania, and Texas.
  a. Find the probability of winning first prize.
  b. Find the probability of winning second prize.
12. The 6/42 lottery is currently played in Colorado, Massachusetts, Kentucky, Maine, New Hampshire, and Vermont.
  a. Find the probability of winning first prize.
  b. Find the probability of winning second prize.
13. There is an amazing variety of lotteries played in the United States. Currently, the following lotteries are played: 6/25, 5/26, 5/30, 6/30, 5/31, 7/31, 5/32, 5/33, 5/34, 5/35, 6/35, 5/36, 6/36, 5/37, 5/38, 4/39, 5/39, 6/39, 6/40, 6/41, 6/42, 5/44, 6/44, 6/46, 6/48, 6/49, 4/50, 6/50, 6/51, 5/52, 6/52, 6/53, and 4/100. Which is the easiest to win? Which is the hardest to win? Explain your reasoning.
  Hint: It isn't necessary to compute every single probability.
14. In the game of keno, if six spots are marked, the player wins if four or more of his or her spots are selected. Complete the chart in Figure 3.17.

Outcome	Probability
6 winning spots	
5 winning spots	
4 winning spots	
3 winning spots	
fewer than 3 winning spots	

Figure 3.17
Chart for Exercise 14

15. In the game of keno, if eight spots are marked, the player wins if five or more of his or her spots are selected. Complete the chart in Figure 3.18.
16. In the game of keno, if nine spots are marked, the player wins if six or more of his or her spots are selected. Complete the chart in Figure 3.19.
17. **a.** Find the probability of being dealt five spades when playing five-card draw poker.

Outcome	Probability
8 winning spots	
7 winning spots	
6 winning spots	
5 winning spots	
4 winning spots	
fewer than 4 winning spots	

Figure 3.18
Chart for Exercise 15

Outcome	Probability
9 winning spots	
8 winning spots	
7 winning spots	
6 winning spots	
5 winning spots	
fewer than 5 winning spots	

Figure 3.19
Chart for Exercise 16

  b. Find the probability of being dealt five cards of the same suit when playing five-card draw poker.
  c. When you are dealt five cards of the same suit, you have either a *flush* (if the cards are not in sequence) or a *straight flush* (if the cards are in sequence). For each suit, there are ten possible straight flushes ("ace, two, three, four, five," through "ten, jack, queen, king, ace"). Find the probability of being dealt a straight flush.
  d. Find the probability of being dealt a flush.
18. **a.** Find the probability of being dealt an "aces over kings" full house (three aces and two kings).
  b. Why are there 13 · 12 different types of full houses?
  c. Find the probability of being dealt a full house. (Round each answer off to six decimal places.)

You order twelve burritos to go from a Mexican restaurant, five with hot peppers and seven without. However, the restaurant forgot to label them. If you pick three burritos at random, find the probability of each event in Exercises 19–26.

19. All have hot peppers.
20. None has hot peppers.

21. Exactly one has hot peppers.
22. Exactly two have hot peppers.
23. At most one has hot peppers.
24. At least one has hot peppers.
25. At least two have hot peppers.
26. At most two have hot peppers.

27. Two hundred people apply for two jobs. Sixty of the applicants are women.
   a. If two people are selected at random, what is the probability that both are women?
   b. If two people are selected at random, what is the probability that only one is a woman?
   c. If two people are selected at random, what is the probability that both are men?
   d. If you were an applicant and the two selected people were not of your gender, do you think that the above probabilities would indicate the presence or absence of gender discrimination in the hiring process? Why?

28. Two hundred people apply for three jobs. Sixty of the applicants are women.
   a. If three persons are selected at random, what is the probability that all are women?
   b. If three persons are selected at random, what is the probability that two are women?
   c. If three persons are selected at random, what is the probability that one is a woman?
   d. If three persons are selected at random, what is the probability that none is a woman?
   e. If you were an applicant, and the three selected people were not of your gender, should the above probabilities have an impact on your situation? Why?

29. Powerball is a complicated lottery game played in Arizona, Connecticut, the District of Columbia, Delaware, Idaho, Indiana, Iowa, Kansas, Kentucky, Louisiana, Minnesota, Missouri, Nebraska, New Hampshire, New Mexico, Oregon, Rhode Island, South Dakota, West Virginia, and Wisconsin. This popular game involves selecting any five of the numbers from 1 to 49 plus any one of the numbers from 1 to 42.
   a. Find the probability of winning first prize in Powerball.

   b. The most similar simple lottery game involves selecting any six of the numbers from 1 to 49. Find the probability of winning first prize in a 6/49 game.
   c. Which is more likely: winning first prize in Powerball or winning first prize in a 6/49 game? How many times more likely is it?
   d. Do you think that a typical lottery player would think that Powerball or a 6/49 game is easier to win? Why?
   e. Why do you think that Powerball is structured the way it is?

30. In Example 2, $n(E) = 1$ because only one of the 7,059,052 possible lottery tickets is the first prize winner. Use combinations to show that $n(E) = 1$.

Answer the following questions using complete sentences and your own words.

CONCEPT QUESTIONS

31. Explain why, in Example 5, p(four twos) = p(four threes) = $\cdots$ = p(four kings) = p(four aces).
32. Do you think a state lottery is a good thing for the state's citizens? Why? Be certain to include a discussion of both the advantages and disadvantages of a state lottery to its citizens.
33. Why are probabilities for most games of chance calculated with combinations, rather than permutations?
34. Suppose a friend or relative of yours regularly spends (and loses) a good deal of money on lotteries. How would you explain to this person why he or she loses so frequently?
35. In Example 2, $n(E) = 1$ because only one of the 7,059,052 possible lottery tickets is the first prize winner. Does this mean that it is impossible for two people to each buy a first prize winning lottery ticket? Explain.

HISTORY QUESTION

36. Are public lotteries relative newcomers to the American scene?

3.5 Expected Value

Suppose you are playing roulette, concentrating on the $1 single-number bet. At one point, you were $10 ahead, but now you are $14 behind. How much should you expect to win or lose, on average, if you place the bet many times?

The probability of winning a single-number bet is $\frac{1}{38}$, because there are 38 numbers on the roulette wheel and only one of them is the subject of the bet. This means that if you place the bet a large number of times, it is most likely that you will win once for every 38 times you place the bet (and lose the other 37 times). When you win, you win \$35, because the house odds are 35 to 1. When you lose, you lose \$1. Your average profit would be

$$\frac{\$35 + 37 \cdot (-\$1)}{38} = \frac{-\$2}{38} \approx -\$0.053$$

per game. This is called the *expected value* of a \$1 single-number bet, because you should expect to lose about a nickel for every dollar you bet if you play the game a long time. If you play a few times, anything could happen—you could win every single bet (though it is not likely). The house makes the bet so many times that it can be certain that its profit will be \$0.053 per dollar bet.

The standard way to find the **expected value** of an experiment is to multiply the value of each outcome of the experiment by the probability of that outcome and add the results. Here the experiment is placing a \$1 single-number bet in roulette. The outcomes are winning the bet and losing the bet. The values of the outcomes are +\$35 (if you win) and −\$1 (if you lose); the probabilities of the outcomes are $\frac{1}{38}$ and $\frac{37}{38}$, respectively (see Figure 3.20). The expected value would then be

$$35 \cdot \frac{1}{38} + (-1) \cdot \frac{37}{38} \approx -\$0.053$$

It is easy to see that this calculation is algebraically equivalent to the calculation done above.

Outcome	Value	Probability
winning	35	$\frac{1}{38}$
losing	−1	$\frac{37}{38}$

Figure 3.20
Finding the expected value

Finding an expected value of a bet is very similar to finding your average test score in a class. Suppose that you are a student in a class in which you have taken four tests. If your scores were 80%, 76%, 90%, and 90%, your average test score would be

$$\frac{80 + 76 + 2 \cdot 90}{4} = 84\%$$

or, equivalently,

$$80 \cdot \frac{1}{4} + 76 \cdot \frac{1}{4} + 90 \cdot \frac{2}{4}$$

The difference between finding an average test score and finding the expected value of a bet is that with the average test score you are summarizing what *has* happened, whereas with a bet you are using probabilities to project what *will* happen.

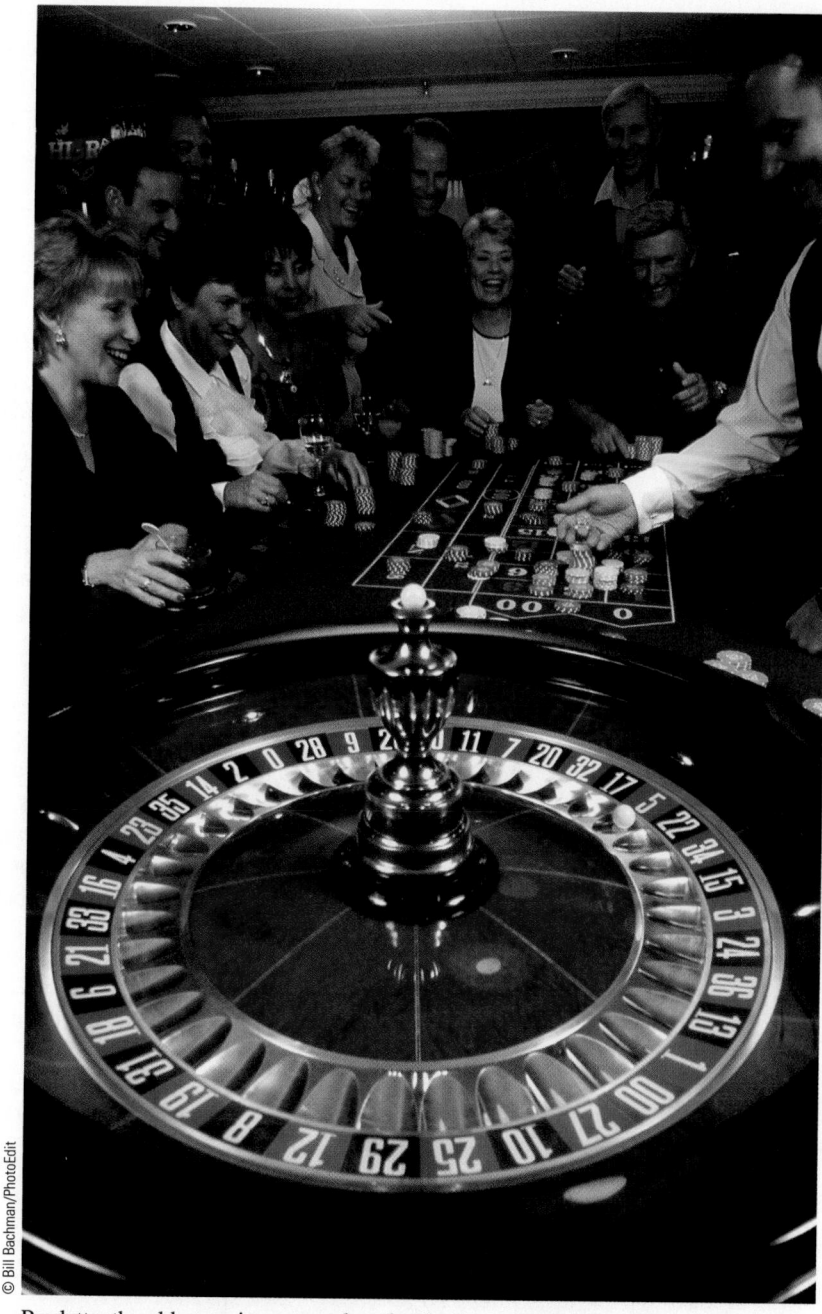

Roulette, the oldest casino game played today, has been popular since it was introduced to Paris in 1765. Does this game have any good bets?

Expected Value

To find the **expected value** (or "long-term average") of an experiment, multiply the value of each outcome of the experiment by its probability and add the results.

EXAMPLE 1 By analyzing her sales records, a saleswoman has found that her weekly commissions have the probabilities in Figure 3.21. Find the saleswoman's expected commission.

Commission	0	$100	$200	$300	$400
Probability	0.05	0.15	0.25	0.45	0.1

Figure 3.21
Commission data for Example 1

Solution To find the expected commission, we multiply each possible commission by its probability and add the results. Therefore,

$$\text{expected commission} = (0)(0.05) + (100)(0.15) + (200)(0.25)$$
$$+ (300)(0.45) + (400)(0.1)$$
$$= 240$$

On the basis of her history, the saleswoman should expect to average $240 per week in future commissions. Certainly, anything can happen in the future—she could receive a $700 commission (it's not likely, though, because it has never happened before). ∎

Why the House Wins

Four of the "best" bets that can be made in a casino game of chance are the pass, don't pass, come, and don't come bets in craps. They all have the same expected value, −$0.014. In the long run, *there isn't a single bet in any game of chance with which you can expect to break even, let alone make a profit.* After all, the casinos are out to make money. The expected values for $1 bets in the more common games are shown in Figure 3.22.

Game	Expected Value of $1 Bet
baccarat	−$0.014
blackjack	−$0.06 to +$0.10 (varies with strategies)
craps	−$0.014 for pass, don't pass, come, don't come bets *only*
slot machines	−$0.13 to ? (varies)
keno (eight-spot ticket)	−$0.29
average state lottery	−$0.48

Figure 3.22
Expected values of common games of chance

It is possible to achieve a positive expected value in blackjack and other card games in which a number of hands are played from the same deck without reshuffling before each hand. To do this, the player must memorize which cards have been played in previous hands. Many casinos use four decks at once to discourage memorization. Some people try to sneak small homemade computers into the casino to aid them. Naturally, the casinos forbid this.

Decision Theory

Which is the better bet: a $1 single-number bet in roulette or a lottery ticket? Each costs $1. The roulette bet pays $35, but the lottery ticket might pay several million dolllars. Lotteries are successful in part because the possibility of winning a large amount of money distracts people from the fact that winning is extremely unlikely. In Example 2 of Section 3.4, we found that the probability of winning first prize in many state lotteries is $\frac{1}{7,059,052} \approx 0.00000014$. At the beginning of this section, we found that the probability of winning the roulette bet is $\frac{1}{38} \approx 0.03$.

A more informed decision would take into account not only the potential winnings and losses but also their probabilities. The expected value of a bet does just that, since its calculation involves both the value and the probability of each outcome. We found that the expected value of a $1 single-number bet in roulette is about $-\$0.053$. The expected value of the average state lottery is $-\$0.48$ (see Figure 3.22). The roulette bet is a much better bet than is the lottery. (Of course, there is a third option, which has an even better expected value of 0.00: not gambling!)

A decision always involves choosing between various alternatives. If you compare the expected values of the alternatives, then you are taking into account the alternatives' potential advantages and disadvantages as well as their probabilities. This form of decision making is called **decision theory.**

EXAMPLE 2 The saleswoman in Example 1 has been offered a new job that has a fixed weekly salary of $290. Financially, which is the better job?

Solution In Example 1, we found that her expected weekly commission was $240. The new job has a guaranteed weekly salary of $290. Financial considerations indicate that she should take the new job. ∎

Betting Strategies

One very old betting strategy is to "cover all the numbers." In 1729, the French philosopher and writer François Voltaire organized a group that successfully implemented this strategy to win the Parisian city lottery by buying most if not all of the tickets. Their strategy was successful because, owing to a series of poor financial decisions by the city of Paris, the total value of the prizes was greater than the combined price of all of the tickets! Furthermore, there were not a great number of tickets to buy. This strategy is still being used. (See the newspaper article on page 180 on its use in 1992 in Virginia.)

A **martingale** is a gambling strategy in which the gambler doubles his or her bet after each loss. A person using this strategy in roulette, concentrating on the black numbers bet (which has 1-to-1 house odds), might lose three times before winning. This strategy would result in a net gain, as illustrated in Figure 3.23 on page 180.

This seems to be a great strategy. Sooner or later, the player will win a bet, and because each bet is larger than the player's total losses, he or she has to come out ahead! We will examine this strategy further in Exercises 45 and 46.

Virginia lottery hedges on syndicate's big win

© 1992, The Washington Post,
reprinted with permission.

RICHMOND, VA.—Virginia lottery officials confirmed yesterday that an Australian gambling syndicate won last month's record $27 million jackpot after executing a massive block-buying operation that tried to cover all 7 million possible ticket combinations.

But lottery director Kenneth Thorson said the jackpot may not be awarded because the winning ticket may have been bought in violation of lottery rules.

The rules say tickets must be paid for at the same location where they are issued. The Australian syndicate, International Lotto Fund, paid for many of its tickets at the corporate offices of Farm Fresh Inc. grocery stores, rather than at the Farm Fresh store in Chesapeake where the winning ticket was issued, Thorson said.

"We have to validate who bought the ticket, where the purchase was made and how the purchase was made," Thorson said. "It's just as likely that we will honor the ticket as we won't honor the ticket." He said he may not decide until the end of next week.

Two Australians representing the fund, Joseph Franck and Robert Hans Roos, appeared at lottery headquarters yesterday to claim the prize.

The group succeeded in buying about 5 million of the more than 7 million possible numerical combinations before the February 15 drawing. The tactic is not illegal, although lottery officials announced new rules earlier this week aimed at making such block purchases more difficult.

The Australian fund was started last year and raised about $13 million from an estimated 2,500 shareholders who each paid a minimum of $4,000, according to Tim Phillipps of the Australian Securities Commission.

Half the money went for management expenses, much of that to Pacific Financial Resources, a firm controlled by Stefan Mandel, who won fame when he covered all the numbers in a 1986 Sydney lottery. Roos owns 10 percent of Pacific Financial Resources.

Australian Securities Commission officials said last week that the fund is under investigation for possible violations of Australian financial laws.

Bet Number	Bet	Result	Total Winnings/Losses
1	$1	lose	−$1
2	$2	lose	−$3
3	$4	lose	−$7
4	$8	win	+$1

Figure 3.23
Analyzing the martingale strategy

3.5   Exercises

In Exercises 1–10, (a) find and (b) use the Law of Large Numbers to interpret the expected value of each $1 bet in roulette.

1. the two-number bet
2. the three-number bet
3. the four-number bet
4. the five-number bet
5. the six-number bet
6. the twelve-number bet
7. the low-number bet
8. the even-number bet
9. the red-number bet
10. the black-number bet
11. Using the expected values obtained in the text and in the preceding odd-numbered exercises, determine a

casino's expected net income from a 24-hour period at a single roulette table if the casino's total overhead for the table is $50 per hour and if customers place a total of $7,000 on single-number bets, $4,000 on two-number bets, $4,000 on four-number bets, $3,000 on six-number bets, $7,000 on low-number bets, and $8,000 on red-number bets.

12. Using the expected values obtained in the text and in the preceding even-numbered exercises, determine a casino's expected net income from a 24-hour period at a single roulette table if the casino's total overhead for the table is $50 per hour and if customers place a total of $8,000 on single-number bets, $3,000 on three-number bets, $4,000 on five-number bets, $4,000 on twelve-number bets, $8,000 on even-number bets, and $9,000 on black-number bets.

13. On the basis of his previous experience, the public librarian at Smallville knows that the number of books checked out by a person visiting the library has the following probabilities:

Number of Books	0	1	2	3	4	5
Probability	0.15	0.35	0.25	0.15	0.05	0.05

Figure 3.24
Probabilities for Exercise 13

Find the expected number of books checked out by a person visiting this library.

14. On the basis of his sale records, a salesman knows that his weekly commissions have the following probabilities:

Commission	0	$1,000	$2,000	$3,000	$4,000
Probability	0.15	0.2	0.45	0.1	0.1

Figure 3.25
Probabilities for Exercise 14

Find the salesman's expected commission.

15. Of all workers at a certain factory, the proportions earning certain hourly wages are as shown in Figure 3.26

Hourly Wage	$8.50	$9.00	$9.50	$10.00	$12.50	$15.00
Proportion	20%	15%	25%	20%	15%	5%

Figure 3.26
Data for Exercise 15

Find the expected hourly wage that a worker at this factory makes.

16. Of all students at the University of Metropolis, the proportions taking certain numbers of units are as shown in Figure 3.27. Find the expected number of units that a student at U.M. takes.

Units	3	4	5	6	7	8
Proportion	3%	4%	5%	6%	5%	4%

Units	9	10	11	12	13	14
Proportion	8%	12%	13%	13%	15%	12%

Figure 3.27
Data for Exercise 16

17. You have been asked to play a dice game. It works like this:
    - If you roll a 1 or 2, you win $50.
    - If you roll a 3, you lose $20.
    - If you roll a 4, 5, or 6, you lose $30.
    Should you play the game? Use expected values to justify your answer.

18. You have been asked to play a dice game. It works like this:
    - If you roll a 1, 2, 3, or 4, you win $50.
    - If you roll a 5 or 6, you lose $80.
    Should you play the game? Use expected values to justify your answer.

19. You are on a TV show. You have been asked to either play a dice game ten times or accept a $100 bill. The dice game works like this:
    - If you roll a 1 or 2, you win $50.
    - If you roll a 3, you win $20.
    - If you roll a 4, 5, or 6, you lose $30.
    Should you play the game? Use expected values and decision theory to justify your answer.

20. You are on a TV show. You have been asked to either play a dice game five times or accept a $50 bill. The dice game works like this:
    - If you roll a 1, 2, or 3, you win $50.
    - If you roll a 4 or 5, you lose $25.
    - If you roll a 6, you lose $90.
    Should you play the game? Use expected values and decision theory to justify your answer.

21. Show why the calculation at the top of page 176 is algebraically equivalent to the calculation in the middle of page 176.

22. In Example 1, the saleswoman's most likely weekly commission was $300. With her new job (in Example 2), she will always make $290 per week. This implies that she would be better off with the old job. Is this reasoning more or less valid than that used in Example 2? Why?

23. Maria just inherited $10,000. Her bank has a savings account that pays 4.1% interest per year. Some of her friends recommended a new mutual fund, which has been in business for three years. During its first year, the fund went up in value by 10%; during the second year, it went down by 19%; and during its third year, it went up by 14%. Maria is attracted by the mutual fund's potential for relatively high earnings but concerned by the possibility of actually losing some of her inheritance. The bank's rate is low, but it is insured by the federal government. Use decision theory to find the best investment. (Assume that the fund's past behavior predicts its future behavior.)

24. Trang has saved $8,000. It is currently in a bank savings account that pays 3.9% interest per year. He is considering putting the money into a speculative investment that would either earn 20% in one year if the investment succeeds or lose 18% in one year if it fails. At what probability of success would the speculative investment be the better choice?

25. Erica has her savings in a bank account that pays 4.5% interest per year. She is considering buying stock in a pharmaceuticals company that is developing a cure for cellulite. Her research indicates that she could earn 50% in one year if the cure is successful or lose 60% in one year if it is not. At what probability of success would the pharmaceuticals stock be the better choice?

26. Debra is buying prizes for a game at her school's fundraiser. The game has three levels of prizes, and she has already bought the second and third prizes. She wants the first prize to be nice enough

Prize	Cost of Prize	Probability
1st	?	.15
2nd	$1.25	.30
3rd	$0.75	.45

Figure 3.28
Data for Exercise 26

to attract people to the game. The game's manufacturer has supplied her with the probabilities of winning first, second, and third prizes. Tickets cost $3 each, and she wants the school to profit an average of $1 per ticket. How much should she spend on each first prize?

27. Few students manage to complete their schooling without taking a standardized admissions test such as the Scholastic Achievement Test, or S.A.T. (used for admission to college); the Law School Admissions Test, or L.S.A.T.; and the Graduate Record Exam, or G.R.E. (used for admission to graduate school). Sometimes, these multiple-choice tests discourage guessing by subtracting points for wrong answers. In particular, a correct answer will be worth $+1$ point, and an incorrect answer on a question with 5 listed answers (a through e) will be worth $-\frac{1}{4}$ point.

    a. Find the expected value of a random guess.
    b. Find the expected value of eliminating one answer and guessing among the remaining four possible answers.
    c. Find the expected value of eliminating three answers and guessing between the remaining two possible answers.
    d. Use decision theory and your answers to parts (a), (b), and (c) to create a guessing strategy for standardized tests such as the S.A.T.

28. Find the expected value of a $1 bet in six-spot keno if three winning spots pays $1 (but you pay $1 to play, so you actually break even), four winning spots pays $3 (but you pay $1 to play, so you profit $2), five pays $100, and six pays $2,600. (You might want to use the probabilities computed in Exercise 14 of Section 3.4.)

29. Find the expected value of a $1 bet in eight-spot keno if four winning spots pays $1 (but you pay $1 to play, so you actually break even), five winning spots pays $5 (but you pay $1 to play, so you profit $4), six winning spots pays $100, seven winning spots pays $1,480, and eight winning spots pays $19,000. (You might want to use the probabilities computed in Exercise 15 of Section 3.4.)

30. Find the expected value of a $1 bet in nine-spot keno if five winning spots pays $1 (but you pay $1 to play, so you actually break even), six winning spots pays $50 (but you pay $1 to play, so you profit $49), seven pays $390, eight pays $6,000, and nine pays $25,000. (You might want to use the probabilities computed in Exercise 16 of Section 3.4.)

31. Arizona's "Fantasy Five" is a 5/35 lottery. It differs from many other state lotteries in that its payouts are set; they do not vary with sales. To win first prize, you must select all five of the winning numbers. To win second prize, you must select any four of the five winning numbers; to win third prize, you must select any three of the five winning numbers. The first prize jackpot is $50,000 (but you pay $1 to play, so you profit $49,999). Second prize pays $500, and third prize pays $5. If you select two or fewer winning numbers, you lose your $1. Find the expected value of the Fantasy Five. (You might want to use the probabilities computed in Exercise 9 of Section 3.4.)

32. Write a paragraph in which you compare the states' fiscal policies concerning their lotteries with the casinos' fiscal policies concerning their keno games. Assume that the expected value of Arizona's Fantasy Five as described in Exercise 31 is representative of that of the other states' lotteries, and assume that the expected value of a $1 keno bet as described in Exercises 28, 29 and/or 30 is representative of other keno bets.

33. Trustworthy Insurance Co. estimates that a certain home has a 1% chance of burning down in any one year. They calculate that it would cost $120,000 to rebuild that home. Use expected values to determine the annual insurance premium.

34. Mr. and Mrs. Trump have applied to the Trustworthy Insurance Co. for insurance on Mrs. Trump's diamond tiara. The tiara is valued at $97,500. Trustworthy estimates that the jewelry has a 2.3% chance of being stolen in any one year. Use expected values to determine the annual insurance premium.

35. The Black Gold Oil Co. is considering drilling either in Jed Clampett's back yard or his front yard. After thorough testing and analysis, they estimate that there is a 30% chance of striking oil in the back yard and a 40% chance in the front yard. They also estimate that the back yard site would either net $60 million (if oil is found) or lose $6 million (if oil is not found), and the front yard site would either net $40 million or lose $6 million. Use decision theory to determine where they should drill.

36. If in Exercise 35, Jed Clampett rejected the use of decision theory, where would he drill if he were an optimist? What would he do if he were a pessimist?

37. A community youth group is having a raffle to raise funds. Several community businesses have donated prizes. The prizes and their retail value are listed in Figure 3.29. Each prize will be given away, regardless of the number of raffle tickets sold. Tickets are sold for $15 each. Determine the expected value of a ticket, and discuss whether it would be to your financial advantage to buy a ticket under the given circumstances.
  a. 1000 tickets are sold.
  b. 2000 tickets are sold.
  c. 3000 tickets are sold.

38. The Centerville High School PTA is having a raffle to raise funds. Several community businesses have donated prizes. The prizes and their retail value are listed in Figure 3.30 on page 184. Each prize will be given away, regardless of the number of raffle tickets sold. Tickets are sold for $30 each. Determine the expected value of a ticket, and discuss whether it

Prize	Retail Value	Number of These Prizes to Be Given Away
new car	$21,580	1
a cell phone and a one-year subscription	$940	1
a one-year subscription to an Internet service provider	$500	2
dinner for two at Spiedini's restaurant	$100	2
a one-year subscription to the local newspaper	$180	20

Figure 3.29
Data for Exercise 37

Prize	Retail Value	Number of These Prizes to Be Given Away
one week in a condo in Hawaii, and airfare for two	$2,575	1
tennis lessons for two	$500	1
a one-year subscription to an Internet service provider	$500	2
dinner for two at Haute Stuff restaurant	$120	2
a one-year subscription at the Centerville Skate Park	$240	5
a copy of *Centerville Cooks,* a cookbook containing PTA members' favorite recipes	$10	20

Figure 3.30
Data for Exercise 38

would be to your financial advantage to buy a ticket under the given circumstances.

a. 100 tickets are sold.

b. 200 tickets are sold.

c. 300 tickets are sold.

39. The Central State University Young Republicans Club is having a raffle to raise funds. Several community businesses have donated prizes. The prizes and their retail value are listed in Figure 3.31. Each prize will be given away, regardless of the number of raffle tickets sold. Tickets are sold for $5 each. Determine the expected value of a ticket, and discuss whether it would be to your financial advantage to buy a ticket under the given circumstances.

a. 1000 tickets are sold.

b. 2000 tickets are sold.

c. 3000 tickets are sold.

Prize	Retail Value	Number of These Prizes to Be Given Away
laptop computer	$2,325	1
MP3 player	$425	2
20 CDs at Einstein Entertainment	$320	3
a giant pizza and your choice of beverage at Freddie's Pizza	$23	4
a one-year subscription to the *Young Republican Journal*	$20	15

Figure 3.31
Data for Exercise 39

Prize	Retail Value	Number of These Prizes to Be Given Away
one-week ecovacation to Costa Rica, including airfare for two	$2,990	1
Sierra Designs tent	$750	1
REI backpack	$355	2
Jansport daypack	$75	5
fleece jacket with Ecology Club logo	$50	10
Ecology Club T-shirt	$18	20

Figure 3.32
Data for Exercise 40

40. The Southern State University Ecology Club is having a raffle to raise funds. Several community businesses have donated prizes. The prizes and their retail value are listed in Figure 3.32. Each prize will be given away, regardless of the number of raffle tickets sold. Tickets are sold for $20 each. Determine the expected value of a ticket, and discuss whether it would be to your financial advantage to buy a ticket under the given circumstances.
  a. 100 tickets are sold.
  b. 200 tickets are sold.
  c. 300 tickets are sold.

41. Find the expected value of the International Lotto Fund's application of the "cover all of the numbers" strategy from the newspaper article on page 180. Assume that $5 million was spent on lottery tickets, that half of the $13 million raised went for management expenses, and that the balance was never spent. Also assume that Virginia honors the winning ticket.

42. One application of the "cover all the numbers" strategy would be to bet $1 on every single number in roulette.
  a. Find the results of this strategy.
  b. How could you use the expected value of the $1 single-number bet $\left(\frac{-\$2}{38}\right)$ to answer part (a)?

43. The application of the "cover all the numbers" strategy to a modern state lottery would involve the purchase of a large number of tickets.
  a. How many tickets would have to be purchased if you were in a state that has a 6/49 lottery (the player selects 6 out of 49 numbers)?
  b. How much would it cost to purchase these tickets if each costs $1?

c. If you organized a group of 100 people to purchase these tickets and it takes 1 minute to purchase each ticket, how many days would it take to purchase the required number of tickets?

44. The application of the "cover all the numbers" strategy to Keno would involve the purchase of a large number of tickets.
  a. How many tickets would have to be purchased if you were playing eight-spot keno?
  b. How much would it cost to purchase these tickets if each costs $5?
  c. If it takes 5 seconds to purchase one ticket and the average keno game lasts 20 minutes, how many people would it take to purchase the required number of tickets?

45. If you had $100 and were applying the martingale strategy to the black-number bet in roulette and you started with a $1 bet, how many successive losses could you afford? How large would your net profit be if you lost each bet except for the last one?

46. If you had $10,000 and were applying the martingale strategy to the black-number bet in roulette and you started with a $1 bet, how many successive losses could you afford? How large would your net profit be if you lost each bet except for the last one?

Answer the following questions using complete sentences and your own words.

CONCEPT QUESTIONS

47. Discuss the meaningfulness of the concept of expected value to three different people: an occasional gambler, a regular gambler, and a casino owner. To whom is the concept most meaningful? To whom is it least meaningful? Why?

48. Discuss the advantages and disadvantages of decision theory. Consider the application of decision theory to a nonrecurring situation and to a recurring situation. Also consider Jed Clampett and the Black Gold Oil Co. in Exercises 35 and 36.

49. The "Probabilities in the Real World" article in Section 3.4 states that "In 2001 alone, Americans paid more than $39 billion for lottery tickets. Overall, 52% of this money was paid out in prizes, 35% went to the government, and 13% was spent in administering the lotteries."

   a. Use this information to find the expected value of an American lottery ticket in 2001.

   b. Is this method of computing an expected value more or less valid than that used in Exercises 28–31? Why?

PROJECTS

50. Design a game of chance. Use probabilities and expected values to set the house odds so that the house will make a profit. However, be certain that your game is not so obviously pro-house that no one would be willing to play. Your project should include the following:
   • a complete description of how the game is played
   • a detailed mathematical analysis of the expected value of the game (or of each separate bet in the game, whichever is appropriate)
   • a complete description of the bet(s) and the house odds

3.6 Conditional Probability

Public opinion polls, such as those found in newspapers and magazines and on television, frequently categorize the respondents by such groups as sex, age, race, or level of education. This is done so that the reader or listener can make comparisons and observe trends, such as "people over 40 are more likely to support the Social Security system than are people under 40." The tool that enables us to observe such trends is conditional probability.

Probabilities and Polls

In a newspaper poll concerning violence on television, 600 people were asked, "What is your opinion of the amount of violence on prime-time television—is there too much violence on television?" Their responses are indicated in Figure 3.33.

	Yes	No	Don't Know	Total
Men	162	95	23	280
Women	256	45	19	320
Total	418	140	42	600

Figure 3.33
Results of "Violence on Television" poll

Six hundred people were surveyed in this poll; that is, the sample space consists of 600 responses. Of these, 418 said they thought there was too much violence on television, so the probability of a "yes" response is $\frac{418}{600}$, or about 0.70 = 70%. The probability of a "no" response is $\frac{140}{600}$, or about 0.23 = 23%.

If we are asked to find the probability that a *woman* responded yes, we do not consider all 600 responses but instead limit the sample space to only the responses from women. See Figure 3.34.

	Yes	No	Don't Know	Total
Women	256	45	19	320

Figure 3.34
Women and violence on television

The probability that a woman responded yes is $\frac{256}{320} = 0.80 = 80\%$.

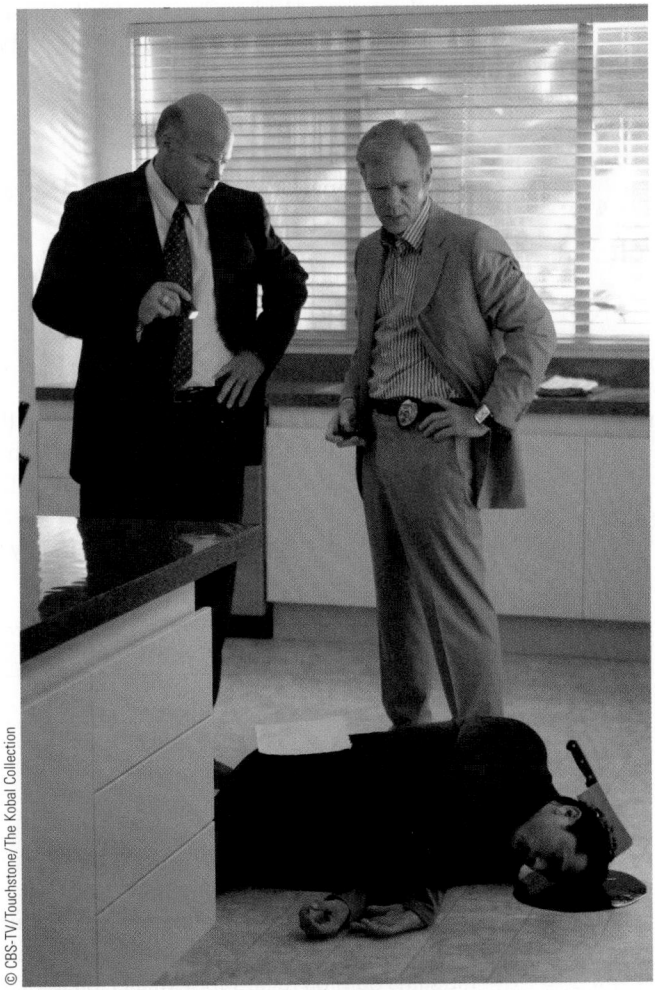

Is there too much violence on TV?

Suppose we label the events in the following manner: W is the event that a response is from a woman, M is the event that a response is from a man, Y is the event that a response is yes, and N is the event that a response is no. Then the event that a woman responded yes would be written as

$Y \mid W$

The vertical bar stands for the phrase "given that"; the event $Y \mid W$ is read "a response is yes, given that the response is from a woman." The probability of this event is called a *conditional probability:*

$$p(Y \mid W) = \frac{256}{320} = \frac{4}{5} = 0.80 = 80\%$$

The numerator of this probability, 256, is the number of responses that are yes and are from women; that is, $n(Y \cap W) = 256$. The denominator, 320, is the number of responses that are from women; that is, $n(W) = 320$. A **conditional probability** is a probability whose sample space has been limited to only those outcomes that fulfill a certain condition. Because an event is a subset of the sample space, the event must also fulfill that condition. The numerator of $p(Y \mid W)$ is 256 rather than 418 even though there were 418 "yes" responses, because many of those 418 responses were made by men; we are interested only in the probability that a woman responded yes.

Conditional Probability Definition

The **conditional probability** of event A, given event B, is

$$p(A \mid B) = \frac{n(A \cap B)}{n(B)}$$

EXAMPLE 1   Using the data in Figure 3.33, find the following.

a. the probability that a response is yes, given that the response is from a man.
b. the probability that a response is from a man, given that the response is yes.
c. the probability that a response is yes and is from a man.

Solution   **a.** *Finding $p(Y \mid M)$:* We are told to consider only the male responses—that is, to limit our sample space to men. See Figure 3.35.

	Yes	No	Don't Know	Total
Men	162	95	23	280

Figure 3.35
Men and violence on television

$$p(Y \mid M) = \frac{n(Y \cap M)}{n(M)} = \frac{162}{280} \approx 0.58 = 58\%$$

In other words, approximately 58% of the men responded yes. (Recall that 80% of the women responded yes. This poll indicates that men and women do not have the same opinion regarding violence on television and, in particular, that a woman is more likely to oppose the violence.)

b. *Finding $p(M \mid Y)$:* We are told to consider only the "yes" responses shown in Figure 3.36.

$$p(M \mid Y) = \frac{n(M \cap Y)}{n(Y)} = \frac{162}{418} \approx 0.39 = 39\%$$

	Yes
Men	162
Women	256
Total	418

Figure 3.36
Limiting our sample space to "yes" responses

Therefore, of those who responded yes, approximately 39% were male.

c. *Finding* $p(Y \cap M)$: This is *not* a conditional probability (there is no vertical bar), so we do *not* limit our sample space.

$$p(Y \cap M) = \frac{n(Y \cap M)}{n(S)} = \frac{162}{600} = 0.27 = 27\%$$

Therefore, of all those polled, 27% were men who responded yes.   ■

Each of the above three probabilities has the *same numerator,* $n(Y \cap M) = 162$. This is the number of responses that are yes and are from men.

But the three probabilities have *different denominators*. This means that we are comparing the group of men who said yes with three different larger groups. See Figure 3.37.

In the probability:	the denominator is:	so we're comparing the group of men who said yes with:
$p(Y \mid M)$	$n(M)$, the number of male responses	all of the men
$p(M \mid Y)$	$n(Y)$, the number of yes responses	all of the yes responses
$p(Y \cap M)$	$n(S)$, the number of responses	all of the people polled

Figure 3.37
The impact of the different denominators

The Product Rule

If two cards are dealt from a full deck (no jokers), how would you find the probability that both are hearts? The probability that the first card is a heart is easy to find—it is $\frac{13}{52}$, because there are 52 cards in the deck and 13 of them are hearts. The probability that the second card is a heart is more difficult to find. There are only 51 cards left in the deck (one was already dealt), but how many of these are hearts? The number of hearts left in the deck depends on the first card that was dealt. If it was a heart, then there are 12 hearts left in the deck; if it was not a heart, then there are 13 hearts left. We could certainly say that the probability that the second card is a heart, *given that the first card was a heart,* is $\frac{12}{51}$.

Therefore, the probability that the first card is a heart is $\frac{13}{52}$ and the probability that the second card is a heart, given that the first was a heart, is $\frac{12}{51}$. How do we put these two probabilities together to find the probability that *both* the first and the second cards are hearts? Should we add them? Subtract them? Multiply them? Divide them?

The answer is obtained by algebraically rewriting the Conditional Probability Definition to obtain what is called the *Product Rule*:

$$p(A \mid B) = \frac{n(A \cap B)}{n(B)} \quad \text{Conditional Probability Definition}$$

$$p(A \mid B) \cdot n(B) = n(A \cap B) \quad \text{multiplying by } n(B)$$

$$\frac{p(A \mid B) \cdot n(B)}{n(S)} = \frac{n(A \cap B)}{n(S)} \quad \text{dividing by } n(S)$$

$$\frac{p(A \mid B)}{1} \cdot \frac{n(B)}{n(S)} = \frac{n(A \cap B)}{n(S)} \quad \text{since } 1 \cdot n(S) = n(S)$$

$$p(A \mid B) \cdot p(B) = p(A \cap B) \quad \text{definition of probability}$$

Product Rule

For any events A and B, the probability of A and B is

$$p(A \cap B) = p(A \mid B) \cdot p(B)$$

EXAMPLE 2   If two cards are dealt from a full deck, find the probability that both are hearts.

Solution

$$p(A \cap B) = p(A \mid B) \cdot p(B)$$

$$p(\text{2nd heart and 1st heart}) = p(\text{2nd heart} \mid \text{1st heart}) \cdot p(\text{1st heart})$$

$$= \frac{12}{51} \cdot \frac{13}{52}$$

$$= \frac{4}{17} \cdot \frac{1}{4}$$

$$= \frac{1}{17} \approx 0.06 = 6\%$$

Therefore, there is a 6% probability that both cards are hearts.   ■

Tree Diagrams

Many people find that a *tree diagram* helps them understand problems like the one in Example 2, in which an experiment is performed in stages over time. Figure 3.38 shows the tree diagram for Example 2. The first column gives a list of the possible outcomes of the first stage of the experiment; in Example 2, the first stage is dealing

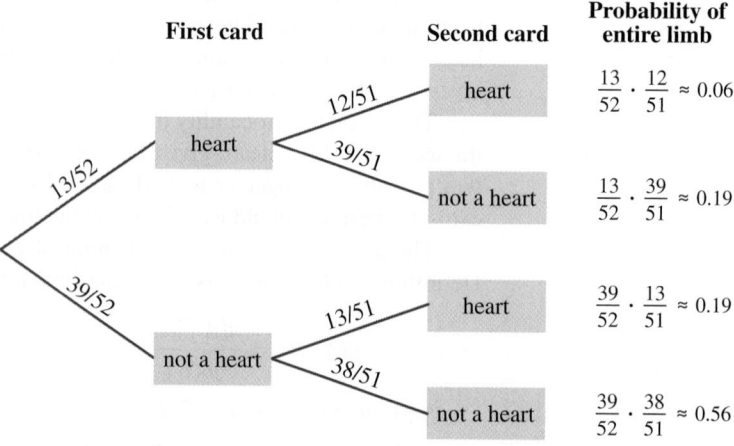

Figure 3.38
A tree diagram of dealing two hearts

the first card, and its outcomes are "heart," and "not a heart." The branches leading to those outcomes represent their probabilities. The second column gives a list of the possible outcomes of the second stage of the experiment; in Example 2, the second stage is dealing the second card. A branch leading from a first-stage outcome to a second-stage outcome is the conditional probability p(2nd stage outcome | 1st stage outcome).

Looking at the top pair of branches, we see that the first branch stops at "first card is a heart" and the probability is $p(\text{1st heart}) = \frac{13}{52}$. The second branch starts at "first card is a heart" and stops at "second card is a heart" and gives the conditional probability $p(\text{2nd heart} \mid \text{1st heart}) = \frac{12}{51}$. The probability that we were asked to calculate in Example 2, p(1st heart and 2nd heart), is that of the top limb:

$$p(\text{1st heart and 2nd heart}) = p(\text{2nd heart} \mid \text{1st heart}) \cdot p(\text{1st heart})$$

$$= \frac{12}{51} \cdot \frac{13}{52}$$

(We use the word *limb* to refer to a sequence of branches that starts at the beginning of the tree.) Notice that the sum of the probabilities of the four limbs is 1.00. Because the four limbs are the only four possible outcomes of the experiment, they must add up to 1.

Conditional probabilities always start at their condition, never at the beginning of the tree. For example, $p(\text{2nd heart} \mid \text{1st heart})$ is a conditional probability; its condition is that the first card is a heart. Thus, its branch starts at the box "first card is a heart." However, $p(\text{1st heart})$ is not a conditional probability, so it starts at the beginning of the tree. Similarly, p(1st heart and 2nd heart) is not a conditional probability, so it too starts at the beginning of the tree. The product rule tells us that

$$p(\text{1st heart and 2nd heart}) = p(\text{2nd heart} \mid \text{1st heart}) \cdot p(\text{1st heart})$$

That is, the product rule tells us to multiply the branches that make up the top horizontal limb. In fact, "Multiply when moving horizontally across a limb" is a restatement of the product rule.

EXAMPLE 3   Two cards are drawn from a full deck. Use the tree diagram in Figure 3.38 to find the probability that the second card is a heart.

Solution   The second card can be a heart if the first card is a heart *or* if it is not. The event "the second card is a heart" is the union of the following two mutually exclusive events:

$E = $ 1st heart and 2nd heart

$F = $ 1st not heart and 2nd heart

We previously used the tree diagram to find that

$$p(E) = \frac{13}{52} \cdot \frac{12}{51}$$

Similarly,

$$p(F) = \frac{39}{52} \cdot \frac{13}{51}$$

Thus, we add the probabilities of limbs that result in the second card being a heart:

$$p(\text{2nd heart}) = p(E \cup F)$$
$$= p(E) + p(F) \qquad \text{Probability Rule 5}$$
$$= \frac{13}{52} \cdot \frac{12}{51} + \frac{39}{52} \cdot \frac{13}{51} = 0.25 \qquad \blacksquare$$

In Example 3, the first and third limbs represent the only two ways that the second card can be a heart. These two limbs represent mutually exclusive events, so we used Probability Rule 5 $[p(E \cup F) = p(E) + p(F)]$ to add their probabilities. In fact, "add when moving vertically from limb to limb" is a good restatement of Probability Rule 5.

Tree Diagram Summary

- Conditional probabilities start at their condition.
- Nonconditional probabilities start at the beginning of the tree.
- Multiply when moving horizontally across a limb.
- Add when moving vertically from limb to limb.

ThomsonNOW for Liberal Arts Mathematics www.thomsonedu.com/1pass.

EXAMPLE 4 Big Fun Bicycles manufactures its product at two plants, one in Korea and one in Peoria. The Korea plant manufactures 60% of the bicycles; 4% of the Korean bikes are defective; and 5% of the Peorian bikes are defective.

a. Draw a tree diagram that shows this information.
b. Use the tree diagram to find the probability that a bike is defective and came from Korea.
c. Use the tree diagram to find the probability that a bike is defective.
d. Use the tree diagram to find the probability that a bike is defect-free.

Solution a. First, we need to determine which probabilities have been given and find their complements, as shown in Figure 3.39.

Probabilities Given	Complements of These Probabilities
$p(\text{Korea}) = 60\% = 0.60$	$p(\text{Peoria}) = p(\text{not Korea}) = 1 - 0.60 = 0.40$
$p(\text{defective} \mid \text{Korea}) = 4\% = 0.04$	$p(\text{not defective} \mid \text{Korea}) = 1 - 0.04 = 0.96$
$p(\text{defective} \mid \text{Peoria}) = 5\% = 0.05$	$p(\text{not defective} \mid \text{Peoria}) = 1 - 0.05 = 0.95$

Figure 3.39
Probabilities for Example 4

The first two of these probabilities [$p(\text{Korea})$ and $p(\text{Peoria})$] are not conditional, so they start at the beginning of the tree. The next two probabilities [$p(\text{defective} \mid \text{Korea})$ and $p(\text{not defective} \mid \text{Korea})$] are conditional, so they start at their condition (Korea). Similarly, the last two probabilities are conditional, so they start at their condition (Peoria). This placement of the probabilities yields the tree diagram in Figure 3.40.

Plant		Defect	Probability

Korea

defective 0.60 · 0.04 = 0.024

defect-free 0.60 · 0.96 = 0.576

Peoria

defective 0.40 · 0.05 = 0.02

defect-free 0.40 · 0.95 = 0.38

Figure 3.40
A tree diagram for Example 4

b. The probability that a bike is defective and came from Korea is a nonconditional probability, so it starts at the beginning of the tree. Do not confuse it with the conditional probability that a bike is defective, *given that* it came from Korea, which starts at its condition (Korea). The former is the limb that goes through "Korea" and stops at "defective"; the latter is one branch of that limb. We use the product rule to multiply when moving horizontally across a limb:

$$p(\text{defective and Korea}) = p(\text{defective} \mid \text{Korea}) \cdot p(\text{Korea})$$
$$= 0.04 \cdot 0.60 = 0.024 \qquad \text{Product Rule}$$

This means that 2.4% of all of Big Fun's bikes are defective bikes manufactured in Korea.

c. The event that a bike is defective is the union of two mutually exclusive events:

The bike is defective and came from Korea.

The bike is defective and came from Peoria.

These two events are represented by the first and third limbs of the tree. We use Probability Rule 5 to add when moving vertically from limb to limb:

$$p(\text{defective}) = p(\text{defective and Korea} \cup \text{defective and Peoria})$$
$$= p(\text{defective and Korea}) + p(\text{defective and Peoria})$$
$$= 0.024 + 0.02 = 0.044$$

This means that 4.4% of Big Fun's bicycles are defective.

d. The probability that a bike is defect-free is the complement of part (c).

$$p(\text{defect-free}) = p(\text{not defective}) = 1 - 0.044 = 0.956$$

Alternatively, we can find the sum of all the limbs that stop at "defect-free":

$$p(\text{defect-free}) = 0.576 + 0.38 = 0.956$$

This means that 95.6% of Big Fun's bicycles are defect-free. ∎

Probabilities in the Real World: HIV/AIDS

The human immunodeficiency virus (or HIV) is the virus that causes AIDS. The Centers for Disease Control estimates that in 2005, 950,000 Americans had HIV/AIDS. They are of both genders, all ages and sexual orientations. Worldwide, nearly half of the 38 million people living with HIV/AIDS are women between 15 and 24 years old.

Most experts agree that the HIV/AIDS epidemic is in its early stages, and that a vaccine is not on the horizon. The only hope of stemming the infection lies in education and prevention. Education is particularly important, because an infected person can be symptom-free for eight years or more. You may well know many people who are unaware that they are infected, because they have no symptoms.

Health organizations such as the Centers for Disease Control routinely use probabilities to determine whether men or women are more likely to get HIV, which age groups are more likely to develop AIDS, and the different sources of exposure to HIV. See Exercises 49 and 50. In Section 3.7, we will investigate the accuracy of some HIV/AIDS tests. See Exercises 33 and 34 in Section 3.7.

3.6   Exercises

In Example 1, we wrote "the probability that a response is yes, given that it is from a man" as $p(Y\,|\,M)$, and we wrote "the probability that a response is yes and is from a man" as $p(Y \cap M)$. In Exercises 1–4, write the given probabilities in a similar manner. Also, identify which are conditional and which are not conditional.

1. Let H be the event that a job candidate is hired, and let Q be the event that a job candidate is well qualified. Use the symbols H, Q, $|$, and $\cap$ to write the following probabilities.
   a. The probability that a job candidate is hired given that the candidate is well qualified.
   b. The probability that a job candidate is hired and the candidate is well qualified.

2. Let W be the event that a gambler wins a bet, and let L be the event that a gambler is feeling lucky. Use the symbols W, L, $|$, and $\cap$ to write the following probabilities.
   a. The probability that a gambler wins a bet and is feeling lucky.
   b. The probability that a gambler wins a bet given that the gambler is feeling lucky.

3. Let S be the event that a cell phone user switches carriers, and let D be the event that a cell phone user gets dropped a lot. Use the symbols S, D, $|$, and $\cap$ to write the following probabilities.
   a. The probability that a cell phone user switches carriers given that she gets dropped a lot.
   b. The probability that a cell phone user switches carriers and gets dropped a lot.

c. The probability that a cell phone user gets dropped a lot given that she switches carriers.

4. Let P be the event that a student passes the course, and S be the event that a student studies hard. Use the symbols P, S, $|$, and $\cap$ to write the following probabilities.
   a. The probability that a student passes the course given that the student studies hard.
   b. The probability that a student studies hard given that the student passes the course.
   c. The probability that a student passes the course and studies hard.

In Exercises 5–8, use Figure 3.41.

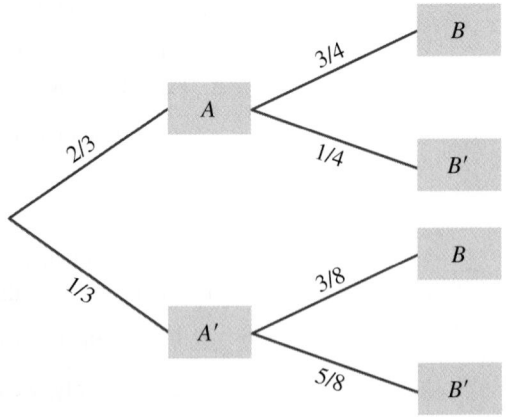

Figure 3.41
A tree diagram for Exercises 5–8

5. **a.** Find $p(B \mid A)$
    b. Find $p(B \cap A)$
6. **a.** Find $p(B' \mid A)$
    b. Find $p(B' \cap A)$
7. **a.** Find $p(B \mid A')$
    b. Find $p(B \cap A')$
8. **a.** Find $p(B' \mid A')$
    b. Find $p(B' \cap A')$

9. Use the data in Figure 3.33 on page 186 to find the given probabilities. Also, write a sentence explaining what each means.
    a. $p(N)$ **b.** $p(W)$ **c.** $p(N \mid W)$
    d. $p(W \mid N)$ **e.** $p(N \cap W)$ **f.** $p(W \cap N)$
10. Use the data in Figure 3.33 on page 186 to find the given probabilities. Also, write a sentence explaining what each means.
    a. $p(N)$ **b.** $p(M)$ **c.** $p(N \mid M)$
    d. $p(M \mid N)$ **e.** $p(N \cap M)$ **f.** $p(M \cap N)$

Use the information in Figure 3.10 on page 149 to answer Exercises 11–14. Round your answers off to the nearest hundredth. Also interpret each of your answers using percentages and everyday English.

11. **a.** Find and interpret the probability that a U.S. resident dies of a pedestrian transportation accident, given that the person dies of a transportation accident in one year.
    b. Find and interpret the probability that a U.S. resident dies of a pedestrian transportation accident, given that the person dies of a transportation accident, in a lifetime.
    c. Find and interpret the probability that a U.S. resident dies of a pedestrian transportation accident, given that the person dies of a non-transportation accident, in a lifetime.
12. **a.** Find and interpret the probability that a U.S. resident dies from lightning, given that the person dies of a non-transportation accident, in one year.
    b. Find and interpret the probability that a U.S. resident dies from lightning, given that the person dies of a non-transportation accident, in a lifetime.
    c. Find and interpret the probability that a U.S. resident dies from lightning, given that the person dies of a transportation accident, in a lifetime.
13. **a.** Find and interpret the probability that a U.S. resident dies from an earthquake, given that the person dies from a non-transportation accident, in a lifetime.

    b. Find and interpret the probability that a U.S. resident dies from an earthquake, given that the person dies from a non-transportation accident, in one year.
    c. Find and interpret the probability that a U.S. resident dies from an earthquake, given that the person dies from an external cause, in one year.
14. **a.** Find and interpret the probability that a U.S. resident dies from a motorcyclist transportation accident, given that the person dies from a transportation accident, in a lifetime.
    b. Find and interpret the probability that a U.S. resident dies from a motorcyclist transportation accident, given that the person dies from a transportation accident, in one year.
    c. Find and interpret the probability that a U.S. resident dies from a motorcyclist transportation accident, given that the person dies from an external cause, in one year.

In Exercises 15–18, cards are dealt from a full deck of 52. Find the probabilities of the given events.

15. **a.** The first card is a club.
    b. The second card is a club, given that the first was a club.
    c. The first and second cards are both clubs.
    d. Draw a tree diagram illustrating this.
16. **a.** The first card is a king.
    b. The second card is a king, given that the first was a king.
    c. The first and second cards are both kings.
    d. Draw a tree diagram illustrating this. (Your diagram need not be a complete tree. It should have all the branches referred to in parts (a), (b), and (c), but it does not need other branches.)
17. **a.** The first card is a diamond.
    b. The second card is a spade, given that the first was a diamond.
    c. The first card is a diamond and the second is a spade.
    d. Draw a tree diagram illustrating this. (Your diagram need not be a complete tree. It should have all the branches referred to in parts (a), (b), and (c), but it does not need other branches.)
18. **a.** The first card is a jack.
    b. The second card is an ace, given that the first card was a jack.
    c. The first card is a jack and the second is an ace.
    d. Draw a tree diagram illustrating this.

*In Exercises 19 and 20, determine which probability the indicated branch in Figure 3.42 refers to. For example, the branch labeled * refers to the probability p(A).*

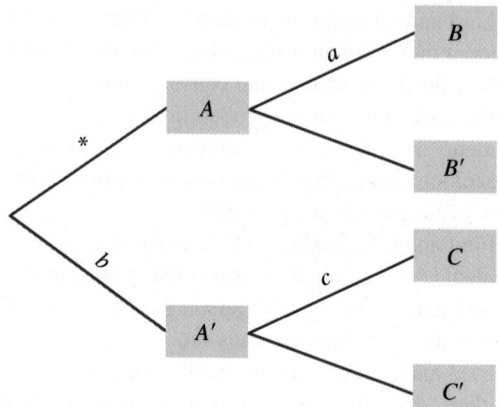

Figure 3.42
A tree diagram for Exercises 19 and 20

19. **a.** the branch labeled a
     b. the branch labeled b
     c. the branch labeled c
20. **a.** Should probabilities (*) and (*a*) be added or multiplied? What rule tells us that? What is the result of combining them?
     b. Should probabilities (*b*) and (*c*) be added or multiplied? What rule tells us that? What is the result of combining them?
     c. Should the probabilities that result from parts (*a*) and (*b*) of this exercise be added or multiplied? What rule tells us that? What is the result of combining them?

In Exercises 21 and 22, a single die is rolled. Find the probabilities of the given events.

21. **a.** rolling a 6
     b. rolling a 6, given that the number rolled is even
     c. rolling a 6, given that the number rolled is odd
     d. rolling an even number, given that a 6 was rolled
22. **a.** rolling a 5
     b. rolling a 5, given that the number rolled is even
     c. rolling a 5, given that the number rolled is odd
     d. rolling an odd number, given that a 5 was rolled

In Exercises 23–26, a pair of dice is rolled. Find the probabilities of the given events.

23. **a.** The sum is 6.
     b. The sum is 6, given that the sum is even.
     c. The sum is 6, given that the sum is odd.
     d. The sum is even, given that the sum is 6.

24. **a.** The sum is 12.
     b. The sum is 12, given that the sum is even.
     c. The sum is 12, given that the sum is odd.
     d. The sum is even, given that the sum was 12.
25. **a.** The sum is 4.
     b. The sum is 4, given that the sum is less than 6.
     c. The sum is less than 6, given that the sum is 4.
26. **a.** The sum is 11.
     b. The sum is 11, given that the sum is greater than 10.
     c. The sum is greater than 10, given that the sum is 11.
27. A single die is rolled. Determine which of the following events is least likely and which is most likely. Do so without making any calculations. Explain your reasoning.
     E_1 is the event "rolling a four."
     E_2 is the event "rolling a four, given that the number rolled is even."
     E_3 is the event "rolling a four, given that the number rolled is odd."
28. A pair of dice is rolled. Determine which of the following events is least likely and which is most likely. Do so without making any calculations. Explain your reasoning.
     E_1 is the event "rolling a seven."
     E_2 is the event "rolling a seven, given that the number rolled is even."
     E_3 is the event "rolling a seven, given that the number rolled is odd."

In Exercises 29 and 30, use the following information. To determine what effect the salespeople had on purchases, a department store polled 700 shoppers as to whether or not they had made a purchase and whether or not they were pleased with the service. Of those who had made a purchase, 125 were happy with the service and 111 were not. Of those who had made no purchase, 148 were happy with the service and 316 were not.

29. Find the probability that a shopper who was happy with the service had made a purchase (round off to the nearest hundredth). What can you conclude?
30. Find the probability that a shopper who was unhappy with the service had not made a purchase. (Round off to the nearest hundredth.) What can you conclude?

In Exercises 31–34, five cards are dealt from a full deck. Find the probabilities of the given events. (Round off to four decimal places.)

31. All are spades.
32. The fifth is a spade, given that the first four were spades.

33. The last four are spades, given that the first was a spade.

34. All are the same suit.

In Exercises 35–40, round off to the nearest hundredth.

35. If three cards are dealt from a full deck, use a tree diagram to find the probability that exactly two are spades.

36. If three cards are dealt from a full deck, use a tree diagram to find the probability that exactly one is a spade.

37. If three cards are dealt from a full deck, use a tree diagram to find the probability that exactly one is an ace.

38. If three cards are dealt from a full deck, use a tree diagram to find the probability that exactly two are aces.

39. If a pair of dice is rolled three times, use a tree diagram to find the probability that exactly two throws result in sevens.

40. If a pair of dice is rolled three times, use a tree diagram to find the probability that all three throws result in sevens.

In Exercises 41–44, use the following information: A personal computer manufacturer buys 38% of its chips from Japan and the rest from the United States. 1.7% of the Japanese chips are defective, and 1.1% of the American chips are defective.

41. Find the probability that a chip is defective and made in Japan.

42. Find the probability that a chip is defective and made in the United States.

43. Find the probability that a chip is defective.

44. Find the probability that a chip is defect-free.

45. The 109th U.S. Senate convened on January 5, 2005. Data on the Senate's composition are given in Figure 3.43.

	Democrats	Republicans	Independents
Male	35	50	1
Female	9	5	0

Figure 3.43
U.S. Senate by political party

  a. Find the probability that a senator is female, given that the senator is a Democrat.
  b. Find the probability that a senator is female, given that the senator is a Republican.
  c. What observations can you make?

46. The 109th U.S. Senate convened on January 5, 2005. Data on the Senate's composition are given in Figure 3.43.

  a. Find the probability that a senator is a Democrat, given that the senator is female.
  b. Find the probability that a senator is a Republican, given that the senator is female.
  c. What observations can you make?

47. January 2005 data on the governors of the fifty states are given in Figure 3.44.

	Democrats	Republicans	Independents
Male	16	22	0
Female	6	2	0

Figure 3.44
Governors by political party

  a. Find the probability that a governor is female, given that the governor is a Democrat.
  b. Find the probability that a governor is female, given that the governor is a Republican.
  c. What observations can you make?

48. January 2005 data on the governors of the fifty states are given in Figure 3.44.

  a. Find the probability that a governor is a Democrat, given that the governor is female.
  b. Find the probability that a governor is a Republican, given that the governor is female.
  c. What observations can you make?

49. Figure 3.45 gives the estimated number of diagnoses of AIDS among adults and adolescents in the United States by exposure category through December 2002.

Exposure category	Male	Female	Total
Male-to-male sexual contact	420,790	—	420,790
Injection drug use	172,351	67,917	240,268
Male-to-male sexual contact and injection drug use	59,719	—	59,719
Heterosexual contact	50,793	94,935	145,728
Other*	14,350	6,519	20,869

Figure 3.45
AIDS sources
*includes hemophilia, blood transfusion, perinatal, and risk not reported or not identified
Source: Centers for Disease Control, National Center for HIV, STD and TB Prevention, Divisions of HIV/AIDS Prevention

a. Find the probability that a U.S. adult or adolescent diagnosed with AIDS is male and the probability that a U.S. adult or adolescent diagnosed with AIDS is female.

b. Find the probability that a male U.S. adult or adolescent diagnosed with AIDS was exposed by injection drug use (possibly combined with male-to-male sexual contact) and the probability that a female U.S. adult or adolescent diagnosed with AIDS was exposed by injection drug use.

c. Find the probability that a U.S. adult or adolescent diagnosed with AIDS is male and was exposed by injection drug use (possibly combined with male-to-male sexual contact) and the probability that a U.S. adult or adolescent diagnosed with AIDS is female and was exposed by injection drug use.

d. Find the probability that a male U.S. adult or adolescent diagnosed with AIDS was exposed by heterosexual contact and the probability that a female U.S. adult or adolescent diagnosed with AIDS was exposed by heterosexual contact.

e. Find the probability that a U.S. adult or adolescent diagnosed with AIDS is male and was exposed by heterosexual contact and the probability that a U.S. adult or adolescent diagnosed with AIDS is female and was exposed by heterosexual contact.

f. Explain the difference between parts (b) and (c) and the difference between parts (d) and (e).

50. Figure 3.46 gives the estimated number of diagnoses of AIDS in the United States by age at the time of diagnosis, through December 2002. Also, note that in 2002, there were 287,973,924 residents of the United States, according to the U.S. Census Bureau.

a. Find the probability that a U.S. resident diagnosed with AIDS was 15 to 24 years old at the time of diagnosis.

b. Find the probability that a U.S. resident was diagnosed with AIDS and was 15 to 24 years old at the time of diagnosis.

c. Find the probability that a U.S. resident diagnosed with AIDS was 25 to 34 years old at the time of diagnosis.

d. Find the probability that a U.S. resident was diagnosed with AIDS and was 25 to 34 years old at the time of diagnosis.

e. Explain in words the difference between parts (c) and (d).

Age	Cumulative number of AIDS cases
Under 13	9300
13 to 14	839
15 to 24	35,460
25 to 34	301,278
35 to 44	347,860
45 to 54	138,386
55 to 64	40,548
65 or older	12,868

Figure 3.46
AIDS by age
Source: Centers for Disease Control, National Center for HIV, STD and TB Prevention, Divisions of HIV/AIDS Prevention.

51. In February 1983, the National Center for Health Statistics published a document entitled "Obese and Overweight Adults in the United States." According to that document, 19.4% of adult men and 27.7% of adult women in the United States were obese in 1983. At that time, there were 57.2 million adult men and 63.9 million adult women in the country. *Source:* National Center for Health Statistics, S. Abraham: Obese and overweight adults in the United States. Vital and Health Statistics Series 11, No. 230. DHHS Pub. No. 83–1680. Public Health Service (Washington, D.C.: U.S. Government Printing Office, January 1983).

a. Find the probability that an adult man was obese.

b. Find the probability that an adult was an obese man.

c. Find the probability that an adult woman was obese.

d. Find the probability that an adult was an obese woman.

e. Find the probability that an adult was obese.

f. Explain in words the difference between parts (c) and (d).

52. The document mentioned in Exercise 51 also says that 23.2% of adult men and 29.5% of adult women in the United States were overweight.

a. Find the probability that an adult man was overweight.

b. Find the probability that an adult was an overweight man.

c. Find the probability that an adult woman was overweight.

d. Find the probability that an adult was an overweight woman.

e. Find the probability that an adult was overweight.

53. In 1981 a study on race and the death penalty was released. The data in Figure 3.47 are from that study.

Death penalty imposed?	Victim's race	Defendant's race	Frequency
Yes	White	White	19
Yes	White	Black	11
Yes	Black	White	0
Yes	Black	Black	6
No	White	White	132
No	White	Black	152
No	Black	White	9
No	Black	Black	97

Figure 3.47
Race and the death penalty
Source: M. Radelet (1981) "Racial Characteristics and the Imposition of the Death Penalty." American Sociological Review, 46, 918–927.

a. Find p(death penalty imposed | victim white and defendant white).
b. Find p(death penalty imposed | victim white and defendant black).
c. Find p(death penalty imposed | victim black and defendant white).
d. Find p(death penalty imposed | victim black and defendant black).
e. What can you conclude from parts (a) through (d)?
f. Determine what other conditional probabilities would affect this issue, and calculate those probabilities.
g. Discuss your results.

54. The information in Exercise 53 is rather dated. Use the following information to determine whether things have improved since then. In April 2003, Amnesty International issued a study on race and the death penalty. The following quote is from that study:

"The population of the USA is approximately 75 percent white and 12 per cent black. Since 1976, blacks have been six to seven times more likely to be murdered than whites, with the result that blacks and whites are the victims of murder in about equal numbers. Yet, 80 per cent of the more than 840 people put to death in the USA since 1976 were convicted of crimes involving white victims, compared to the 13 per cent who were convicted of killing blacks. Less than four per cent of the executions carried out since 1977 in the USA were for crimes involving Hispanic victims. Hispanics represent about 12 per cent of the US population. Between 1993 and 1999, the recorded murder rate for Hispanics was more than 40 per cent higher than the national homicide rate."

(*Source:* Amnesty International. "United States of America: Death by discrimination—the continuing role of race in capital cases." AMR 51/046/2003)

a. In the above quote, what conditional probability is 80%? That is, find events A and B such that $p(A | B) = 80\%$.
b. In the above quote, what conditional probability is 13%? That is, find events C and D such that $p(C | D) = 13\%$.
c. Use the information in Exercise 53 to compute $p(A | B)$ for 1981.
d. Use the information in Exercise 53 to compute $p(C | D)$ for 1981.
e. Have things improved since 1981? Justify your answer.

55. A man and a woman have a child. Both parents have sickle-cell trait. They know that their child does not have sickle-cell anemia because he shows no symptoms, but they are concerned that he might be a carrier. Find the probability that he is a carrier.

56. A man and a woman have a child. Both parents are Tay-Sachs carriers. They know that their child does not have Tay-Sachs disease because she shows no symptoms, but they are concerned that she might be a carrier. Find the probability that she is a carrier.

In Exercises 57–60, use the following information. The University of Metropolis requires its students to pass an examination in college-level mathematics before they can graduate. The students are given three chances to pass the exam; 61% pass it on their first attempt, 63% of those that take it a second time pass it then, and 42% of those that take it a third time pass it then.

57. What percent of the students pass the exam?
58. What percent of the students are not allowed to graduate because of their performance on the exam?
59. What percent of the students take the exam at least twice?
60. What percent of the students take the test three times?
61. In the game of blackjack, if the first two cards dealt to a player are an ace and either a ten, jack, queen, or king, then the player has a blackjack, and he or

she wins. Find the probability that a player is dealt a blackjack out of a full deck (no jokers).

62. In the game of blackjack, the dealer's first card is dealt face up. If that card is an ace, then the player has the option of "taking insurance." "Insurance" is a side bet. If the dealer has blackjack, the player wins the insurance bet and is paid 2 to 1 odds. If the dealer does not have a blackjack, the player loses the insurance bet. Find the probability that the dealer is dealt a blackjack if his or her first card is an ace.

63. Use the data in Figure 3.33 to find the following probabilities, where N is the event "saying no," and W is the event "being a woman":
    a. $p(N' \mid W)$ b. $p(N \mid W')$ c. $p(N' \mid W')$
    d. Which event, $N' \mid W$, $N \mid W'$, or $N' \mid W'$, is the complement of the event $N \mid W$? Why?

64. Use the data in Figure 3.33 to find the following probabilities, where Y is the event "saying yes," and M is the event "being a man.":
    a. $p(Y' \mid M)$ b. $p(Y \mid M')$ c. $p(Y' \mid M')$
    d. Which event, $Y' \mid M$, $Y \mid M'$, or $Y' \mid M'$, is the complement of the event $Y \mid M$? Why?

65. If A and B are arbitrary events, what is the complement of the event $A \mid B$?

66. Show that $p(A \mid B) = \dfrac{p(A \cap B)}{p(B)}$.

    Hint: Divide the numerator and denominator of the Conditional Probability Definition by $n(S)$.

67. Use Exercise 66 and appropriate answers from Exercise 9 to find $P(N \mid W)$.

68. Use Exercise 66 and appropriate answers from Exercise 10 to find $P(Y \mid M)$.

━━━━━━━➤ *Answer the following questions using complete sentences and your own words.*

CONCEPT QUESTIONS

69. Which must be true for any events A and B?
    • $P(A \mid B)$ is always greater than or equal to $P(A)$.
    • $P(A \mid B)$ is always less than or equal to $P(A)$.
    • Sometimes $P(A \mid B)$ is greater than or equal to $P(A)$, and sometimes $P(A \mid B)$ is less than or equal to $P(A)$, depending on the events A and B.

Answer this without making any calculations. Explain your reasoning.

70. Compare and contrast the events A, $A \mid B$, $B \mid A$, and $A \cap B$. Be sure to discuss both the similarities and the differences between these events.

WEB PROJECTS

71. There are many different blood type systems, but the ABO and Rh systems are the most important systems for blood donation purposes. These two systems generate eight different blood types: A+, A−, B+, B−, AB+, AB−, O+, and O−.
    a. For each of these eight blood types, use the web to determine the percentage of U.S. residents that have that blood type. Interpret these percentages as probabilities. Use language like "the probability that a randomly-selected U.S. resident . . . ."
    b. You can always give blood to someone with the same blood type. In some cases, you can give blood to someone with a different blood type, or receive blood from someone with a different blood type. This depends on the donor's blood type and the receiver's blood type. Use the web to complete the chart in Figure 3.48.
    c. For each of the eight blood types, use the web to determine the percentage of U.S. residents that can donate blood to a person of that blood type. Interpret these percentages as conditional probabilities.
    d. For each of the eight blood types, use the web to determine the percentage of U.S. residents that can receive blood from a person of that blood type. Interpret these percentages as conditional probabilities.
    e. Are the probabilities in parts (a), (c), and (d) theoretical probabilities or relative frequencies? Why?

Some useful links for this web project are listed on the text web site:
www.thomsonedu.com/mathematics/johnson

72. According to the U.S. Department of Transportation's National Highway Traffic Safety Administration, "rollovers are dangerous incidents and have a higher fatality rate than other kinds of crashes. Of the nearly

Blood group	A+	A−	B+	B−	AB+	AB−	O+	O−
Can donate blood to								
Can receive blood from								

Figure 3.48
Blood types

11 million passenger car, SUV, pickup, and van crashes in 2002, only 3% involved a rollover. However, rollovers accounted for nearly 33% of all deaths from passenger vehicle crashes."
(*Source:* http://www. safercar.gov/Rollover/pages/ RolloCharFat.htm)

Use probabilities and the web to investigate rollovers. How likely is a rollover if you are driving a sedan, an SUV, or a van? Which models have the highest probability of a rollover? Which models have the lowest probability? Wherever possible, give specific conditional probabilities.

Some useful links for this web project are listed on the text web site: www.thomsonedu.com/ mathematics/johnson

PROJECTS

73. In 1973, the University of California at Berkeley admitted 1,494 of 4,321 female applicants for graduate study and 3,738 of 8,442 male applicants.
(*Source:* P.J. Bickel, E.A. Hammel, and J.W. O'Connell, "Sex Bias in Graduate Admissions: Data from Berkeley," *Science,* vol. 187, 7 February 1975.)
   a. Find the probability that:
      • an applicant was admitted
      • an applicant was admitted, given that he was male
      • an applicant was admitted, given that she was female
      Discuss whether these data indicate a possible bias against women.
   b. Berkeley's graduate students are admitted by the department to which they apply rather than by a campuswide admissions panel. When $P(\text{admission} \mid \text{male})$ and $P(\text{admission} \mid \text{female})$ were computed for each of the school's more than 100 departments, it was found that in four departments, $P(\text{admission} \mid \text{male})$ was greater than $P(\text{admission} \mid \text{female})$ by a significant amount and that in six departments, $P(\text{admission} \mid \text{male})$

was less than $P(\text{admission} \mid \text{female})$ by a significant amount. Discuss whether this information indicates a possible bias against women and whether it is consistent with that in part (a).
   c. The authors of "Sex Bias in Graduate Admissions: Data from Berkeley" attempt to explain the paradox by discussing an imaginary school with only two departments: "machismatics" and "social wafare." Machismatics admitted 200 of 400 male applicants for graduate study and 100 of 200 female applicants, while social warfare admitted 50 of 150 male applicants for graduate study and 150 of 450 female applicants. For *the school as a whole and for each of the two departments,* find the probability that:
      • an applicant was admitted
      • an applicant was admitted, given that he was male
      • an applicant was admitted, given that she was female
      Discuss whether these data indicate a possible bias against women.
   d. Explain the paradox illustrated in parts (a)–(c).
   e. What conclusions would you make, and what further information would you obtain, if you were an affirmative action officer at Berkeley?
74. a. Use the data in Figure 3.49 to find the probability that:
      • a New York City resident died from tuberculosis
      • a Caucasian New York City resident died from tuberculosis
      • a non-Caucasian New York City resident died from tuberculosis
      • a Richmond resident died from tuberculosis
      • a Caucasian Richmond resident died from tuberculosis
      • a non-Caucasian Richmond resident died from tuberculosis
   b. What conclusions would you make, and what further information would you obtain, if you were a public health official?

	New York City		Richmond, Virginia	
	Population	TB deaths	Population	TB deaths
Caucasian	4,675,000	8400	81,000	130
Non-Caucasian	92,000	500	47,000	160

Figure 3.49
Tuberculosis deaths by race and location, 1910
Source: Morris R. Cohen and Ernest Nagel, *An Introduction to Logic and Scientific Method* (New York: Harcourt Brace & Co, 1934.)

3.7 Independence; Trees in Genetics

Dependent and Independent Events

Consider the dealing of two cards from a full deck. An observer who saw that the first card was a heart would be better able to predict whether the second card will be a heart than would another observer who did not see the first card. If the first card was a heart, there is one fewer heart in the deck, so it is slightly less likely that the second card will be a heart. In particular,

$$p(\text{2nd heart} \mid \text{1st heart}) = \frac{12}{51} \approx 0.24$$

whereas, as we saw in Example 3 of Section 3.6,

$$p(\text{2nd heart}) = 0.25$$

These two probabilities are different because of the effect the first card drawn has on the second. We say that the two events "first card is a heart" and "second card is a heart" are *dependent*; the result of dealing the second card depends, to some extent, on the result of dealing the first card. In general, two events E and F are **dependent** if $p(E \mid F) \neq p(E)$.

Consider two successive tosses of a single die. An observer who saw that the first toss resulted in a three would be *no better able* to predict whether the second toss will result in a three than another observer who did not observe the first toss. In particular,

$$p(\text{2nd toss is a three}) = \frac{1}{6}$$

and

$$p(\text{2nd toss is a three} \mid \text{1st toss was a three}) = \frac{1}{6}$$

These two probabilities are the same, because the first toss has no effect on the second toss. We say that the two events "first toss is a three" and "second toss is a three" are *independent;* the result of the second toss does *not* depend on the result of the first toss. In general, two events E and F are **independent** if $p(E \mid F) = p(E)$.

Independence/Dependence Definitions

Two events E and F are **independent** if $p(E \mid F) = p(E)$.
(Think: Knowing F does not affect E's probability.)

Two events E and F are **dependent** if $p(E \mid F) \neq p(E)$.
(Think: Knowing F does affect E's probability.)

　　Many people have difficulty distinguishing between *independent* and *mutually exclusive*. (Recall that two events E and F are mutually exclusive if $E \cap F = \varnothing$, that is, if one event excludes the other.) This is probably because the relationship between mutually exclusive events and the relationship between independent events both could be described, in a very loose sort of way, by saying that "the two events have nothing to do with each other." *Never think this way;* mentally replacing "mutually exclusive" or "independent" with "having nothing to do with each other" only obscures the distinction between these two concepts. *E* and *F* are independent if knowing that *F* has occurred *does not* affect the probability that *E* will occur. *E* and *F* are dependent if knowing that *F* has occurred *does* affect the probability that *E* will occur. *E* and *F* are mutually exclusive if *E* and *F* cannot occur simultaneously.

EXAMPLE 1   Let *F* be the event "a person has freckles" and *R* the event "a person has red hair."

  a. Are *F* and *R* independent?
  b. Are *F* and *R* mutually exclusive?

Solution   **a.** *F* and *R* are independent if $p(F \mid R) = p(F)$. With $p(F \mid R)$, we are given that a person has red hair; with $p(F)$, we are not given that information. Does knowing that a person has red hair affect the probability that the person has freckles? Yes, it does; $p(F \mid R) > p(F)$. Therefore, *F* and *R* are not independent; they are dependent.

  b. *F* and *R* are mutually exclusive if $F \cap R = \varnothing$. Many people have both freckles and red hair, so $F \cap R \neq \varnothing$, and *F* and *R* are not mutually exclusive. In other words, having freckles does not exclude the possibility of having red hair; freckles and red hair can occur simultaneously.   ■

EXAMPLE 2   Let *T* be the event "a person is tall" and *R* the event "a person has red hair."

  a. Are *T* and *R* independent?
  b. Are *T* and *R* mutually exclusive?

Solution   **a.** *T* and *R* are independent if $p(T \mid R) = p(T)$. With $p(T \mid R)$, we are given that a person has red hair; with $p(T)$, we are not given that information. Does knowing that a person has red hair affect the probability that the person is tall? No, it does not; $p(T \mid R) = p(T)$, so *T* and *R* are independent.

  b. *T* and *R* are mutually exclusive if $T \cap R = \varnothing$. $T \cap R$ is the event "a person is tall and has red hair." There are tall people who have red hair, so $T \cap R \neq \varnothing$, and *T* and *R* are not mutually exclusive. In other words, being tall does not exclude the possibility of having red hair; being tall and having red hair can occur simultaneously.   ■

In Examples 1 and 2, we had to rely on our personal experience in concluding that knowledge that a person has red hair does affect the probability that he or she has freckles and does not affect the probability that he or she is tall. It may be the case that you have seen only one red-haired person and she was short and without freckles. Independence is better determined by computing the appropriate probabilities than by relying on one's own personal experiences. This is especially crucial in determining the effectiveness of an experimental drug. *Double-blind* experiments, in which neither the patient nor the doctor knows whether the given medication is the experimental drug or an inert substance, are often done to ensure reliable, unbiased results.

Independence is an important tool in determining whether an experimental drug is an effective vaccine. Let *D* be the event that the experimental drug was administered to a patient, and let *R* be the event that the patient recovered. It is hoped that $p(R \mid D) > p(R)$, that is, that the rate of recovery is greater among those who were given the drug. In this case, R and D are dependent. Independence is also an important tool in determining whether an advertisement effectively promotes a product. An ad is effective if p(consumer purchases product | consumer saw ad) $>$ p(consumer purchases product).

EXAMPLE 3   Use probabilities to determine whether the events "thinking there is too much violence in television" and "being a man" in Example 1 of Section 3.6 are independent.

Solution   Two events E and F are independent if $p(E|F) = p(E)$. The events "responding yes to the question on violence in television" and "being a man" are independent if $p(Y|M) = p(Y)$. We need to compute these two probabilities and compare them.

In Example 1 of Section 3.6, we found $p(Y|M) \approx 0.58$. We can use the data from the poll in Figure 3.30 to find $p(Y)$.

$$p(Y) = \frac{418}{600} \approx 0.70$$

$$p(Y|M) \neq p(Y)$$

The events "responding yes to the question on violence in television" and "being a man" are dependent. According to the poll, men are less likely to think that there is too much violence on television.   ■

In Example 3, what should we conclude if we found that $p(Y) = 0.69$ and $p(Y|M) = 0.67$? Should we conclude that $p(Y|M) \neq p(Y)$ and that the events "thinking that there is too much violence on television" and "being a man" are dependent? Or should we conclude that $p(Y | M) \approx p(Y)$ and that the events are (probably) independent? In this particular case, the probabilities are relative frequencies rather than theoretical probabilities, and relative frequencies can vary. A group of 600 people was polled to determine the opinions of the entire viewing public; if the same question was asked of a different group, a somewhat different set of relative frequencies could result. While it would be reasonable to conclude that the events are (probably) independent, it would be more appropriate to include more people in the poll and make a new comparison.

Product Rule for Independent Events

The product rule says that $p(A \cap B) = p(A | B) \cdot p(B)$. If A and B are independent, then $p(A | B) = p(A)$. Combining these two equations, we get the following rule:

Product Rule for Independent Events

If A and B are independent events, then the probability of A and B is

$$p(A \cap B) = p(A) \cdot p(B)$$

A common error that is made in computing probabilities is using the formula $p(A \cap B) = p(A) \cdot p(B)$ without verifying that A and B are independent. In fact, the Federal Aviation Administration (FAA) has stated that this is the most frequently encountered error in probabilistic analysis of airplane component failures. If it is not known that A and B are independent, you must use the Product Rule $p(A \cap B) = p(A | B) \cdot p(B)$.

EXAMPLE 4 If a pair of dice is tossed twice, find the probability that each toss results in a seven.

Solution In Example 4 of Section 3.3, we found that the probability of a seven is $\frac{1}{6}$. The two rolls are independent (one roll has no influence on the next), so the probability of a seven is $\frac{1}{6}$ regardless of what might have happened on an earlier roll; we can use the Product Rule for Independent Events:

$$p(A \cap B) = p(A) \cdot p(B)$$
$$p(\text{1st is 7 and 2nd is 7}) = p(\text{1st is 7}) \cdot p(\text{2nd is 7})$$
$$= \frac{1}{6} \cdot \frac{1}{6}$$
$$= \frac{1}{36}$$

See Figure 3.50. The thicker branch of the tree diagram starts at the event "1st roll is 7" and ends at the event "2nd roll is 7," so it is the conditional probability $p(\text{2nd is 7} \mid \text{1st is 7})$. However, the two rolls are independent, so $p(\text{2nd is 7} \mid \text{1st is 7}) = p(\text{2nd is 7})$. We are free to label this branch as either of these two equivalent probabilities. ∎

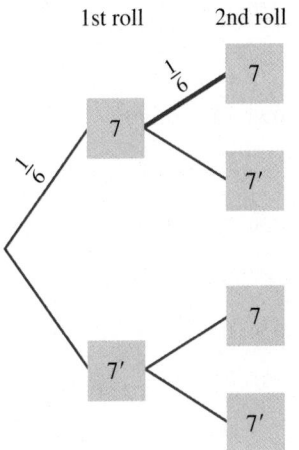

1st roll   2nd roll

Figure 3.50
A tree diagram for Example 4

Trees in Medicine and Genetics

Usually, medical diagnostic tests are not 100% accurate. A test might indicate the presence of a disease when the patient is in fact healthy (this is called a **false positive**), or it might indicate the absence of a disease when the patient does in fact have the disease (a **false negative**). Probability trees can be used to determine the probability that a person whose test results were positive actually has the disease.

EXAMPLE 5 Medical researchers have recently devised a diagnostic test for "white lung" (an imaginary disease caused by the inhalation of chalk dust). Teachers are particularly susceptible to this disease; studies have shown that half of all teachers are afflicted with it. The test correctly diagnoses the presence of white lung in 99% of the people who have it and correctly diagnoses its absence in 98% of the people who do not have it. Find the probability that a teacher whose test results are positive actually has white lung and the probability that a teacher whose test results are negative does not have white lung.

Solution First, we determine which probabilities have been given and find their complements, as shown in Figure 3.51. We use + to denote the event that a person receives a positive diagnosis and − to denote the event that a person receives a negative diagnosis.

Probabilities Given	Complements of Those Probabilities
$p(\text{ill}) = 0.50$	$p(\text{healthy}) = p(\text{not ill}) = 1 - 0.50 = 0.50$
$p(- \mid \text{healthy}) = 98\% = 0.98$	$p(+ \mid \text{healthy}) = 1 - 0.98 = 0.02$
$p(+ \mid \text{ill}) = 99\% = 0.99$	$p(- \mid \text{ill}) = 1 - 0.99 = 0.01$

Figure 3.51
Data from Example 5

The first two of these probabilities [p(ill) and p(healthy)] are not conditional, so they start at the beginning of the tree. The next two probabilities [$p(-\mid\text{healthy})$ and $p(+\mid\text{healthy})$] are conditional, so they start at their condition (healthy). Similarly, the last two probabilities are conditional, so they start at their condition (ill). This placement of the probabilities yields the tree diagram in Figure 3.52.

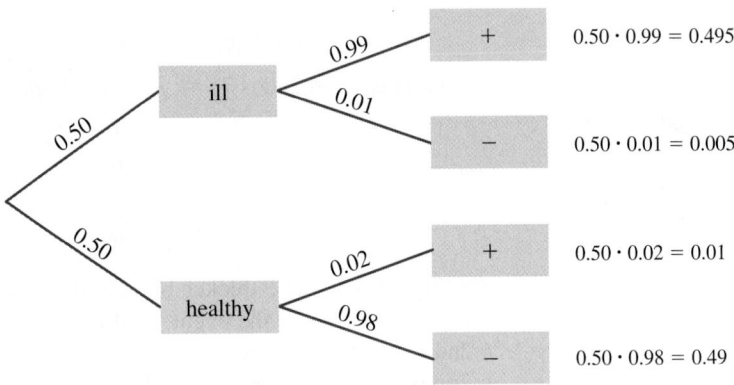

Figure 3.52
A tree diagram for Example 5

The four probabilities to the right of the tree are

p(ill and $+$) = 0.495

p(ill and $-$) = 0.005

p(healthy and $+$) = 0.01

p(healthy and $-$) = 0.49

We are to find the probability that a teacher whose test results are positive actually has white lung; that is, we are to find $p(\text{ill}\mid +)$. Thus, we are given that the test results are positive, and we need only consider branches that involve positive test results: p(ill and $+$) = 0.495 and p(healthy and $+$) = 0.01. The probability that a teacher whose test results are positive actually has white lung is

$$p(\text{ill}\mid +) = \frac{0.495}{0.495 + 0.01} = 0.9801\ldots \approx 98\%$$

The probability that a teacher whose test results are negative does not have white lung is

$$p(\text{healthy}\mid -) = \frac{0.49}{0.49 + 0.005} = 0.98989\ldots \approx 99\%$$

The probabilities show that this diagnostic test works well in determining whether a teacher actually has white lung. In the exercises, we will see how well it would work with schoolchildren.   ■

EXAMPLE 6   Mr. and Mrs. Smith each had a sibling who died of cystic fibrosis. The Smiths are considering having a child. They have not been tested to determine whether they are carriers. What is the probability that their child would have cystic fibrosis?

Solution   In our previous examples involving trees, we were given probabilities; the conditional or nonconditional status of those probabilities helped us to determine the physical layout of the tree. In this example, we are not given any probabilities. To determine the physical layout of the tree, we have to separate what we know to be true from what is only possible or probable. The tree focuses on what is possible or probable. We know that both Mr. and Mrs. Smith had a sibling who died of cystic fibrosis, and it is possible that their child would inherit that disease. The tree's branches will represent the series of possible events that could result in the Smith child inheriting cystic fibrosis.

What events must take place if the Smith child is to inherit the disease? First, the grandparents would have to have had cystic fibrosis genes. Next, the Smiths themselves would have to have inherited those genes from their parents. Finally, the Smith child would have to inherit those genes from his or her parents.

Cystic fibrosis is recessive, which means that a person can inherit it only if he or she receives two cystic fibrosis genes, one from each parent. Mr. and Mrs. Smith each had a sibling who had cystic fibrosis, so each of the four grandparents must have been a carrier. They could not have actually had the disease because the Smiths would have known, and we would have been told.

We now know the physical layout of our tree. The four grandparents were definitely carriers. Mr. and Mrs. Smith were possibly carriers. The Smith child will possibly inherit the disease. The first set of branches will deal with Mr. and Mrs. Smith, and the second set will deal with the child.

The Smith child will not have cystic fibrosis unless Mr. and Mrs. Smith are both carriers. Figure 3.53 shows the Punnett square for Mr. and Mrs. Smith's possible genetic configuration.

	N	**c**
N	NN	Nc
c	Nc	cc

← one grandparent's genes
← possible offspring
← possible offspring

↑
other grandparent's genes

Figure 3.53
A Punnett square for Example 6

Neither Mr. Smith nor Mrs. Smith has the disease, so we can eliminate the cc possibility. Thus, the probability that Mr. Smith is a carrier is $\frac{2}{3}$, as is the probability that Mrs. Smith is a carrier. Furthermore, these two events are independent, since the Smiths are (presumably) unrelated.

Using the Product Rule for Independent Events, we have

p(Mr. S is a carrier and Mrs. S is a carrier)

$$= p(\text{Mr. S is a carrier}) \cdot p(\text{Mrs. S is a carrier}) = \frac{2}{3} \cdot \frac{2}{3} = \frac{4}{9}$$

The same Punnett square tells us that the probability that their child will have cystic fibrosis, given that the Smiths are both carriers, is $\frac{1}{4}$. Letting B be the event that both parents are carriers and F be the event that the child has cystic fibrosis, we obtain the tree diagram in Figure 3.54.

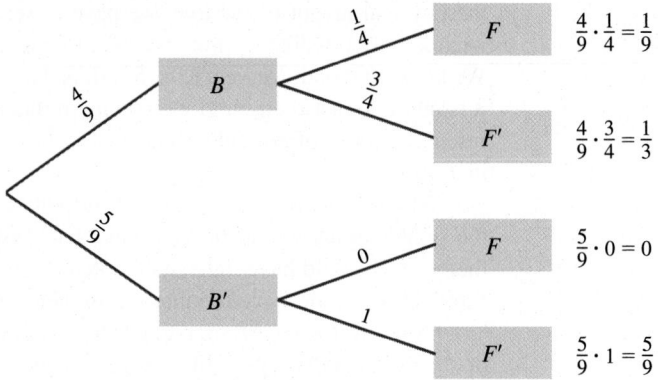

Figure 3.54
A tree diagram for Example 6

The probability that the Smiths' child would have cystic fibrosis is $\frac{1}{9}$.   ■

Notice that the tree in Figure 3.54 could have been drawn differently, as shown in Figure 3.55. It is not necessary to draw a branch going from the B' box to the F box, since the child cannot have cystic fibrosis if both parents are not carriers.

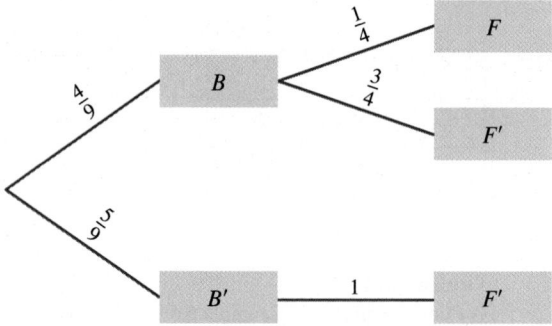

Figure 3.55
An alternative to the tree in Figure 3.54.

Hair Color

Like cystic fibrosis, hair color is inherited, but the method by which it is transmitted is more complicated than the method by which an inherited disease is transmitted. This more complicated method of transmission allows for the possibility that a child might have hair that is colored differently from that of other family members.

Hair color is determined by two pairs of genes: one pair that determines placement on a blond/brown/black spectrum and one pair that determines the presence or absence of red pigment. These two pairs of genes are independent.

Melanin is a brown pigment that affects the color of hair (as well as the colors of eyes and skin). The pair of genes that controls the brown hair colors does so by

Genes	Hair Color
$M^{Bd}M^{Bd}$	blond
$M^{Bd}M^{Bw}$	light brown
$M^{Bd}M^{Bk}$ $M^{Bw}M^{Bw}$	medium brown
$M^{Bw}M^{Bk}$	dark brown
$M^{Bk}M^{Bk}$	black

Figure 3.56
Melanin

determining the amount of melanin in the hair. This gene has three forms, each traditionally labeled with an M (for melanin): M^{Bd}, or blond (a light melanin deposit); M^{Bw}, or brown (a medium melanin deposit); and M^{Bk}, or black (a heavy melanin deposit). Everyone has two of these genes, and their combination determines the brown aspect of hair color, as illustrated in Figure 3.56.

The hair colors in Figure 3.56 are altered by the presence of red pigment, which is determined by another pair of genes. This gene has two forms: R^-, or no red pigment, and R^+, or red pigment. Because everyone has two of these genes, there are three possibilities for the amount of red pigment: R^-R^-, R^+R^-, and R^+R^+. The amount of red pigment in a person's hair is independent of the brownness of his or her hair.

The actual color of a person's hair is determined by the interaction of these two pairs of genes, as shown in Figure 3.57.

Genes	Blond ($M^{Bd}M^{Bd}$)	Light Brown ($M^{Bd}M^{Bw}$)	Medium Brown ($M^{Bd}M^{Bk}$, $M^{Bw}M^{Bw}$)	Dark Brown ($M^{Bw}M^{Bk}$)	Black ($M^{Bk}M^{Bk}$)
R^-R^-	blond	light brown	medium brown	dark brown	black
R^+R^-	strawberry blond	reddish brown	chestnut	shiny dark brown	shiny black
R^+R^+	bright red	dark red	auburn	glossy dark brown	glossy black

Figure 3.57
Melanin and red pigment

ThomsonNOW™
for Liberal Arts Mathematics
www.thomsonedu.com/1pass.

EXAMPLE 7   The Rosses are going to have a child. Mr. Ross has blond hair, and Mrs. Ross has reddish brown hair. Find their child's possible hair colors and the probabilities of each possibility.

Solution   The parent with blond hair has genes $M^{Bd}M^{Bd}$ and R^-R^-. The parent with reddish brown hair has genes $M^{Bd}M^{Bw}$ and R^+R^-. We need to use two Punnett squares, one for the brownness of the hair (Figure 3.58) and one for the presence of red pigment (Figure 3.59).

	M^{Bd}	M^{Bd}
M^{Bd}	$M^{Bd}M^{Bd}$	$M^{Bd}M^{Bd}$
M^{Bw}	$M^{Bd}M^{Bw}$	$M^{Bd}M^{Bw}$

Figure 3.58
Brownness

$$p(M^{Bd}M^{Bd}) = \frac{2}{4} = \frac{1}{2}$$

$$p(M^{Bd}M^{Bw}) = \frac{2}{4} = \frac{1}{2}$$

	R^-	R^-
R^+	R^+R^-	R^+R^-
R^-	R^-R^-	R^-R^-

Figure 3.59
Red pigment

$$p(R^+R^-) = \frac{2}{4} = \frac{1}{2}$$

$$p(R^-R^-) = \frac{2}{4} = \frac{1}{2}$$

We will use a tree diagram to determine the possible hair colors and their probabilities (see Figure 3.60).

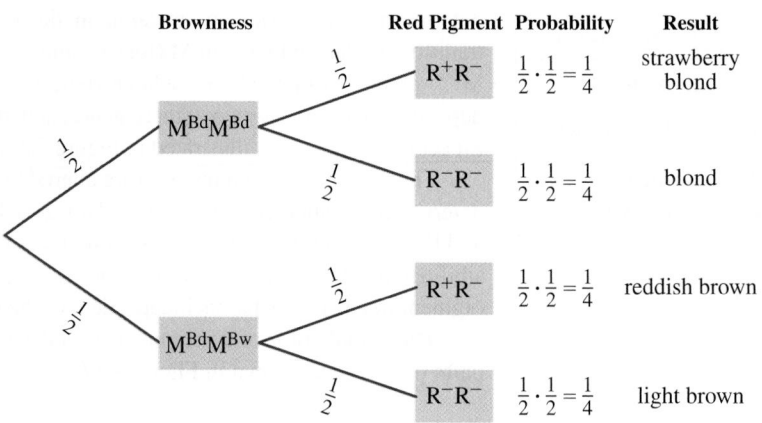

Figure 3.60
A tree diagram for Example 7

The Rosses' child could have strawberry blond, blond, reddish brown, or light brown hair. The probability of each color is $\frac{1}{4}$. Notice that there is a 50% probability that the child will have hair that is colored differently from that of either parent. ■

3.7   Exercises

In Exercises 1–8, use your own personal experience with the events described to determine whether (a) E and F are independent and (b) E and F are mutually exclusive. (Where appropriate, these events are meant to be simultaneous; for example, in Exercise 2, "E and F" would mean "it's simultaneously raining and sunny.") Write a sentence justifying each of your answers.

1. *E* is the event "being a doctor," and *F* is the event "being a woman."

2. *E* is the event "it's raining," and *F* is the event "it's sunny."

3. *E* is the event "being single," and *F* is the event "being married."

4. *E* is the event "having naturally blond hair," and *F* is the event "having naturally black hair."

5. *E* is the event "having brown hair," and *F* is the event "having gray hair."

6. *E* is the event "being a plumber," and *F* is the event "being a stamp collector."

7. *E* is the event "wearing shoes," and *F* is the event "wearing sandals."

8. *E* is the event "wearing shoes," and *F* is the event "wearing socks."

In Exercises 9–10, use probabilities, rather than your own personal experience, to determine whether (a) E and F are independent and (b) E and F are mutually exclusive. (c) Interpret your answers to parts (a) and (b).

9. If a die is rolled once, *E* is the event "getting a 4," and *F* is the event "getting an odd number."

10. If a die is rolled once, *E* is the event "getting a 4," and *F* is the event "getting an even number."

11. Determine whether the events "responding yes to the question on violence in television" and "being a woman" in Example 1 of Section 3.6 are independent.

12. Determine whether the events "having a defect" and "being manufactured in Peoria" in Example 4 of Section 3.6 are independent.

13. A single die is rolled once.
   a. Find the probability of rolling a 5.
   b. Find the probability of rolling a 5 given that the number rolled is even.
   c. Are the events "rolling a 5" and "rolling an even number" independent? Why?
   d. Are the events "rolling a 5" and "rolling an even number" mutually exclusive? Why?
   e. Interpret the results of parts (c) and (d).

14. A pair of dice is rolled once.
  a. Find the probability of rolling a 6.
  b. Find the probability of rolling a 6 given that the number rolled is even.
  c. Are the events "rolling a 6" and "rolling an even number" independent? Why?
  d. Are the events "rolling a 6" and "rolling an even number" mutually exclusive? Why?
  e. Interpret the results of parts (c) and (d).

15. A card is dealt from a full deck (no jokers).
  a. Find the probability of being dealt a jack.
  b. Find the probability of being dealt a jack given that you were dealt a red card.
  c. Are the events "being dealt a jack" and "being dealt a red card" independent? Why?
  d. Are the events "being dealt a jack" and "being dealt a red card" mutually exclusive? Why?
  e. Interpret the results of parts (c) and (d).

16. A card is dealt from a full deck (no jokers).
  a. Find the probability of being dealt a jack.
  b. Find the probability of being dealt a jack given that you were dealt a card above a 7 (count aces high).
  c. Are the events "being dealt a jack" and "being dealt a card above a 7" independent? Why?
  d. Are the events "being dealt a jack" and "being dealt a card above a 7" mutually exclusive? Why?
  e. Interpret the results of parts (c) and (d).

In Exercises 17 and 18, use the following information: To determine what effect salespeople had on purchases, a department store polled 700 shoppers as to whether or not they made a purchase and whether or not they were pleased with the service. Of those who made a purchase, 125 were happy with the service and 111 were not. Of those who made no purchase, 148 were happy with the service and 316 were not.

17. Are the events "being happy with the service" and "making a purchase" independent? What conclusion can you make?

18. Are the events "being unhappy with the service" and "not making a purchase" independent? What conclusion can you make?

19. A personal computer manufacturer buys 38% of its chips from Japan and the rest from the United States. Of the Japanese chips, 1.7% are defective, whereas 1.1% of the U.S. chips are defective. Are the events "defective" and "Japanese-made" independent? What conclusion can you draw? (See Exercises 41–44 in Section 3.6.)

20. A skateboard manufacturer buys 23% of its ball bearings from a supplier in Akron, 38% from one in Atlanta, and the rest from one in Los Angeles. Of the ball bearings from Akron, 4% are defective; 6.5% of those from Atlanta are defective; and 8.1% of those from Los Angeles are defective.
  a. Find the probability that a ball bearing is defective.
  b. Are the events "defective" and "from the Los Angeles supplier" independent?
  c. Are the events "defective" and "from the Atlanta supplier" independent?
  d. What conclusion can you draw?

21. Over the years, a group of nutritionists have observed that their vegetarian clients tend to have fewer health problems. To determine whether their observation is accurate, they collected the following data:
  • They had 365 clients.
  • 281 clients are healthy.
  • Of the healthy clients, 189 are vegetarians.
  • Of the unhealthy clients, 36 are vegetarians.
  a. Use the data to determine whether the events "being a vegetarian" and "being healthy" are independent.
  b. Use the data to determine whether the events "being a vegetarian" and "being healthy" are mutually exclusive.
  c. Interpret the results of parts (a) and (b).

Hint: Start by organizing the data in a chart, similar to that used in the "Violence on Television" poll in Figure 3.33 in Section 3.6.

22. Over the years, a group of exercise physiologists have observed that their clients who exercise solely by running tend to have more ankle problems than do their clients who vary between running and other forms of exercise. To determine whether their observation is accurate, they collected the following data:
  • They had 422 clients.
  • 276 clients are have ankle problems.
  • Of the clients with ankle problems, 191 only run.
  • Of the clients without ankle problems, 22 only run.
  a. Use the data to determine whether the events "running only" and "having ankle problems" are independent.
  b. Use the data to determine whether the events "running only" and "having ankle problems" are mutually exclusive.
  c. Interpret the results of parts (a) and (b).

Hint: Start by organizing the data in a chart, similar to that used in the "Violence on Television" poll in Figure 3.33 in Section 3.6.

23. The Venn diagram in Figure 3.61 contains the results of a survey. Event *A* is "supports proposition 3," and event *B* is "lives in Bishop."

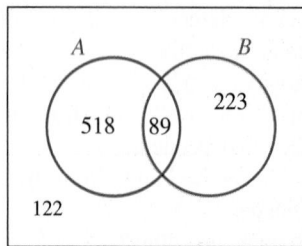

Figure 3.61
Venn diagram for Exercise 23

a. Are the events "supports proposition 3" and "lives in Bishop" independent?
b. Are the events "supports proposition 3" and "lives in Bishop" mutually exclusive?
c. Interpret the results of parts (a) and (b).

24. The Venn diagram in Figure 3.62 contains the results of a survey. Event *A* is "uses Ipana toothpaste," and event *B* is "has good dental checkups."

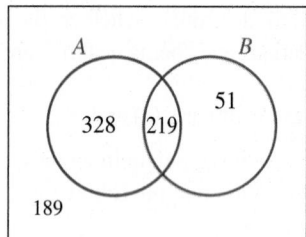

Figure 3.62
Venn diagram for Exercise 24

a. Are the events "uses Ipana toothpaste" and "has good dental checkups" independent?
b. Are the events "uses Ipana toothpaste" and "has good dental checkups" mutually exclusive?
c. Interpret the results of parts (a) and (b).

In Exercises 25–28, you may wish to use Exercise 66 in Section 3.6.

25. Sixty percent of all computers sold last year were HAL computers, 4% of all computer users quit using computers, and 3% of all computer users used HAL computers and then quit using computers. Are the events "using a HAL computer" and "quit using computers" independent? What conclusion can you make?

26. Ten percent of all computers sold last year were Peach computers, 4% of all computer users quit using computers, and 0.3% of all computer users used Peach computers and then quit using computers. Are the events "using a Peach computer" and "quit using computers" independent? What conclusion can you make?

27. Forty percent of the country used SmellSoGood deodorant, 10% of the country quit using deodorant, and 4% of the country used SmellSoGood deodorant and then quit using deodorant. Are the events "used SmellSoGood" and "quit using deodorant" independent? What conclusion can you make?

28. Nationwide, 20% of all TV viewers use VideoLink cable, 8% of all TV viewers switched from cable to a satellite service, and 7% of all TV viewers used VideoLink cable and then switched from cable to a satellite service. Are the events "used VideoLink cable" and "switched from cable to satellite" independent? What conclusion can you make?

29. Suppose that the space shuttle has three separate computer control systems—the main system and two backup duplicates of it. The first backup would monitor the main system and kick in if the main system failed. Similarly, the second backup would monitor the first. We can assume that a failure of one system is independent of a failure of another system, since the systems are separate. The probability of failure for any one system on any one mission is known to be 0.01.
a. Find the probability that the shuttle is left with no computer control system on a mission.
b. How many backup systems does the space shuttle need if the probability that the shuttle is left with no computer control system on a mission must be $\frac{1}{1\text{ billion}}$?

30. Use the information in Example 5 to find $p(\text{healthy} \mid +)$, the probability that a teacher receives a false positive.

31. Use the information in Example 5 to find $p(\text{ill} \mid -)$, the probability that a teacher receives a false negative.

32. Overwhelmed with their success in diagnosing white lung in teachers, public health officials decided to administer the test to all schoolchildren, even though only one child in 1,000 has contracted the disease. Recall from Example 5 that the test correctly diagnoses the presence of white lung in 99% of the people who have it and correctly diagnoses its absence in 98% of the people who do not have it.

a. Find the probability that a schoolchild whose test results are positive actually has white lung.

b. Find the probability that a schoolchild whose test results are negative does not have white lung.

c. Find the probability that a schoolchild whose test results are positive does not have white lung.

d. Find the probability that a schoolchild whose test results are negative actually has white lung.

e. Which of these events is a false positive? Which is a false negative?

f. Which of these probabilities would you be interested in if you or one of your family members tested positive?

g. Discuss the usefulness of this diagnostic test, both for teachers (as in Example 5) and for schoolchildren.

33. In 2004, the Centers for Disease Control estimated that 1,000,000 of the 287,000,000 residents of the United States are HIV-positive. (HIV is the virus that is believed to cause AIDS.) The SUDS diagnostic test correctly diagnoses the presence of AIDS/HIV 99.9% of the time and correctly diagnoses its absence 99.6% of the time.

a. Find the probability that a person whose test results are positive actually has HIV.

b. Find the probability that a person whose test results are negative does not have HIV.

c. Find the probability that a person whose test results are positive does not have HIV.

d. Find the probability that a person whose test results are negative actually has HIV.

e. Which of the probabilities would you be interested in if you or someone close to you tested positive?

f. Which of these events is a false positive? Which is a false negative?

g. Discuss the usefulness of this diagnostic test.

h. It has been proposed that all immigrants to the United States should be tested for HIV before being allowed into the country. Discuss this proposal.

34. Assuming that the SUDS test cannot be made more accurate, what changes in the circumstances described in Exercise 33 would increase the usefulness of the SUDS diagnostic test? Give a specific example of this change in circumstances, and demonstrate how it would increase the test's usefulness by computing appropriate probabilities.

35. Compare and contrast the circumstances and the probabilities in Example 5 and in Exercises 32, 33,

and 34. Discuss the difficulties in using a diagnostic test when that test is not 100% accurate.

36. Use the information in Example 6 to find the following:
a. the probability that the Smiths' child is a cystic fibrosis carrier
b. the probability that the Smiths' child is healthy (i.e., has no symptoms)
c. the probability that the Smiths' child is healthy and not a carrier

Hint: You must consider three possibilities: both of the Smiths are carriers, only one of the Smiths is a carrier, and neither of the Smiths is a carrier.

37. a. Two first cousins marry. Their mutual grandfather's sister died of cystic fibrosis. They have not been tested to determine whether they are carriers. Find the probability that their child will have cystic fibrosis. (Assume that no other grandparents are carriers.)
b. Two unrelated people marry. Each had a grandparent whose sister died of cystic fibrosis. They have not been tested to determine whether they are carriers. Find the probability that their child will have cystic fibrosis. (Assume that no other grandparents are carriers.)

38. It is estimated that one in twenty-five Americans is a cystic fibrosis carrier. Find the probability that a randomly selected American couple's child will have cystic fibrosis (assuming that they are unrelated).

39. In 1989, researchers announced a new carrier-detection test for cystic fibrosis. However, it was discovered in 1990 that the test will detect the presence of the cystic fibrosis gene in only 85% of cystic fibrosis carriers. If two unrelated people are in fact carriers of cystic fibrosis, find the probability that they both test positive. Find the probability that they don't both test positive.

40. Ramon del Rosario's mother's father died of Huntington's disease. His mother died in childbirth, before symptoms of the disease would have appeared in her. Find the probability that Ramon will have the disease. (Assume that Ramon's grandfather had one bad gene, and that there was no other source of Huntington's in the family.) (Huntington's disease is discussed on pages 143–145 in Section 3.2.)

41. Albinism is a recessive disorder that blocks the normal production of pigmentation. The typical albino has white hair, white skin, and pink eyes and is visually impaired. Mr. Jones is an albino, and although Ms. Jones is normally pigmented, her brother is an albino. Neither of Ms. Jones' parents are albinos. Find the probability that their child will be an albino.

42. Find the probability that the Joneses child will not be an albino but will be a carrier. (See Exercise 41.)

43. If the Joneses' first child is an albino, find the probability that their second child will be too. (See Exercise 41.)

44. The Donohues are going to have a child. She has shiny black hair, and he has bright red hair. Find their child's possible hair colors and the probabilities of each possibility.

45. The Yorks are going to have a child. She has black hair, and he has dark red hair. Find their child's possible hair colors and the probabilities of each possibility.

46. The Eastwoods are going to have a child. She has chestnut hair ($M^{Bd}M^{Bk}$), and he has dark brown hair. Find their child's possible hair colors and the probabilities of each possibility.

47. The Wilsons are going to have a child. She has strawberry blond hair, and he has shiny dark brown hair. Find their child's possible hair colors and the probabilities of each possibility.

48. The Breuners are going to have a child. She has blond hair, and he has glossy dark brown hair. Find their child's possible hair colors and the probabilities of each possibility.

49. The Landres are going to have a child. She has chestnut hair ($M^{Bd}M^{Bk}$), and he has shiny dark brown hair. Find their child's possible hair colors and the probabilities of each possibility.

50. The Hills are going to have a child. She has reddish brown hair, and he has strawberry blond hair. Find their child's possible hair colors and the probabilities of each possibility.

51. Recall from Section 3.1 that Antoine Gombauld, the Chevalier de Méré, had made money over the years by betting with even odds that he could roll at least one six in four rolls of a single die. This problem finds the probability of winning that bet, and the expected value of the bet.
  a. Find the probability of rolling a six in one roll of one die.
  b. Find the probability of not rolling a six in one roll of one die.
  c. Find the probability of never rolling a six in four rolls of a die.

  Hint: This would mean that the first roll is not a six *and* the second roll is not a six *and* the third is not a six *and* the fourth is not a six.

  d. Find the probability of rolling at least one six in four rolls of a die.

  Hint: Use complements.

  e. Find the expected value of this bet if $1 is wagered.

52. Recall from Section 3.1 that Antoine Gombauld, the Chevalier de Méré, had lost money over the years by betting with even odds that he could roll at least one pair of sixes in 24 rolls of a pair of dice and that he could not understand why. This problem finds the probability of winning that bet and the expected value of the bet.
  a. Find the probability of rolling a double six in one roll of a pair of dice.
  b. Find the probability of not rolling a double six in one roll of a pair of dice.
  c. Find the probability of never rolling a double six in 24 rolls of a pair of dice.
  d. Find the probability of rolling at least one double six in 24 rolls of a pair of dice.
  e. Find the expected value of this bet if $1 is wagered.

53. Probability theory began when Antoine Gombauld, the Chevalier de Méré, asked his friend Blaise Pascal why he had made money over the years betting that he could roll at least one 6 in four rolls of a single die but he had lost money betting that he could roll at least one pair of 6's in 24 rolls of a pair of dice. Use decision theory and the results of Exercises 51 and 52 to answer Gombauld.

54. Use Exercise 66 of Section 3.6 to explain the calculations of $p(\text{ill} \mid +)$ and $p(\text{healthy} \mid -)$ in Example 5 of this section.

55. Dr. Wellby's patient exhibits symptoms associated with acute neural toxemia (an imaginary disease), but the symptoms can have other, innocuous causes. Studies show that only 25% of those who exhibit the symptoms actually have acute neural toxemia (ANT). A diagnostic test correctly diagnoses the presence of ANT in 88% of the persons who have it and correctly diagnoses its absence in 92% of the persons who do not have it. ANT can be successfully treated, but the treatment causes side effects in 2% of the patients. If left untreated, 90% of those with ANT die; the rest recover fully. Dr. Wellby is considering ordering a diagnostic test for her patient and treating the patient if test results are positive, but she is concerned about the treatment's side effects.
  a. Dr. Wellby could choose to test her patient and administer the treatment if the results are positive.

Find the probability that her patient's good health
will return under this plan.

b. Dr. Wellby could choose to avoid the treatment's
side effects by not administering the treatment.
(This also implies not testing the patient.) Find the
probability that her patient's good health will
return under this plan.

c. Find the probability that the patient's good health
will return if he undergoes treatment regardless of
the test's outcome.

d. On the basis of the probabilities, should Dr. Wellby
order the test and treat the patient if test results are
positive?

━━━━▶ *Answer the following questions using
complete sentences and your own words.*

CONCEPT QUESTIONS

56. In Example 5, we are given that $p(-\mid \text{healthy}) = 98\%$ and $p(+\mid \text{ill}) = 99\%$, and we computed that $p(\text{ill}\mid +) \approx 98\%$ and that $p(\text{healthy}\mid -) \approx 99\%$. Which of these probabilities would be most important to a teacher who was diagnosed as having white lung? Why?

57. Are the melanin hair color genes (M^{Bd}, M^{Bw}, and M^{Bk}) dominant, recessive, or codominant? Are the redness hair color genes (R^+ and R^-) dominant, or recessive, or codominant? Why?

58. Compare and contrast the concepts of independence and mutual exclusivity. Be sure to discuss both the similarities and the differences between these two concepts.

Chapter 3   Review

Thomson NOW for Liberal Arts Mathematics
Preparing for an exam? Test yourself on key material by visiting ThomsonNOW at **www.thomsonedu.com/1pass.**

TERMS

carrier	dominant gene	genes	outcome
certain event	event	impossible event	probability
codominant gene	expected value	independent events	Punnett square
conditional probability	experiment	Law of Large Numbers	recessive gene
decision theory	false negative	mutually exclusive events	relative frequency
dependent events	false positive	odds	sample space

PROBABILITY RULES

1. The probability of the null set is 0: $p(\varnothing) = 0$
2. The probability of the sample space is 1: $p(S) = 1$
3. Probabilities are between 0 and 1 (inclusive):
$0 \leq p(S) \leq 1$
4. $p(E \cup F) = p(E) + p(F) - p(E \cap F)$
5. $p(E \cup F) = p(E) + p(F)$ if E and F are mutually exclusive
6. $p(E) + p(E') = 1$

FORMULAS

If outcomes are equally likely, then:
• the **probability** of an event E is $p(E) = n(E)/n(S)$
• the **odds** of an event E are $o(E) = n(E){:}n(E')$

To find the **expected value** of an experiment, multiply the value of each outcome by its probability and add the results.

The **conditional probability** of A given B is
$p(A\mid B) = n(A \cap B)/n(B)$ if outcomes are equally likely

The **product rule:**
$p(A \cap B) = p(A\mid B) \cdot p(B)$ for any two events A and B
$p(A \cap B) = p(A) \cdot p(B)$ if A and B are independent

REVIEW EXERCISES

In Exercises 1–6, a card is dealt from a well-shuffled deck of fifty-two cards.

1. Describe the experiment and the sample space.
2. Find and interpret the probability and the odds of being dealt a queen.
3. Find and interpret the probability and the odds of being dealt a club.
4. Find and interpret the probability and the odds of being dealt the queen of clubs.
5. Find and interpret the probability and the odds of being dealt a queen or a club.
6. Find and interpret the probability and the odds of being dealt something other than a queen.

In Exercises 7–12, three coins are tossed.

7. Find the experiment and the sample space.
8. Find the event E that exactly two are tails.
9. Find the event F that two or more are tails.
10. Find and interpret the probability of E and the odds of E.
11. Find and interpret the probability of F and the odds of F.
12. Find and interpret the probability of E' and the odds of E'.

In Exercises 13–18, a pair of dice is tossed. Find and interpret the probability of rolling each of the following.

13. a 7
14. an 11
15. a 7, an 11, or doubles
16. a number that is both odd and greater than 8
17. a number that is either odd or greater than 8
18. a number that is neither odd nor greater than 8

In Exercises 19–24, three cards are dealt from a deck of fifty-two cards. Find the probability of each of the following.

19. All three are hearts.
20. Exactly two are hearts.
21. At least two are hearts.
22. The first is an ace of hearts, the second is a two of hearts, and the third is a three of hearts.
23. The second is a heart, given that the first is a heart.
24. The second is a heart, and the first is a heart.

In Exercises 25–30, a pair of dice is rolled three times. Find the probability of each of the following.

25. All three are 7's.
26. Exactly two are 7's.
27. At least two are 7's.
28. None are 7's.
29. The second roll is a 7, given that the first roll is a 7.
30. The second roll is a 7 and the first roll is a 7.

In Exercises 31–32, use the following information. A long-stemmed pea is dominant over a short-stemmed one. A pea with one long-stemmed gene and one short-stemmed gene is crossed with a pea with two short-stemmed genes.

31. Find and interpret the probability that the offspring will be long-stemmed.
32. Find and interpret the probability that the offspring will be short-stemmed.

In Exercises 33–35, use the following information: Cystic fibrosis is caused by a recessive gene. Two cystic fibrosis carriers produce a child.

33. Find the probability that that child will have the disease.
34. Find the probability that that child will be a carrier.
35. Find the probability that that child will neither have the disease nor be a carrier.

In Exercises 36–38, use the following information: Sickle-cell anemia is caused by a codominant gene. A couple, each of whom has sickle-cell trait, produce a child. (Sickle-cell trait involves having one bad gene. Sickle-cell anemia involves having two bad genes.)

36. Find the probability that that child will have the disease.
37. Find the probability that that child will have sickle-cell trait.
38. Find the probability that that child will have neither sickle-cell disease nor sickle-cell trait.

In Exercises 39–41, use the following information: Huntington's disease is caused by a dominant gene. One parent has Huntington's disease. This parent has a single gene for Huntington's disease.

39. Find the probability that that child will have the disease.
40. Find the probability that that child will be a carrier.
41. Find the probability that that child will neither have the disease nor be a carrier.
42. Find the probability of being dealt a pair of tens (and no other tens) when playing five-card draw poker.

43. Find the probability of being dealt a pair of tens and three jacks when playing five-card draw poker.

44. Find the probability of being dealt a pair of tens and a pair jacks (and no other tens or jacks) when playing five-card draw poker.

45. In nine-spot keno, five winning spots breaks even, six winning spots pays $50, seven pays $390, eight pays $6000, and nine pays $25,000.
    a. Find the probability of each of these events.
    b. Find the expected value of a $1 bet.

46. Some $1 bets in craps (specifically, the pass, don't pass, come, and don't come bets) have expected values of −$0.014. Use decision theory to compare these bets with a $1 bet in nine-spot keno. (See Exercise 45.)

47. At a certain office, three people make $6.50 per hour, three make $7, four make $8.50, four make $10, four make $13.50, and two make $25. Find the expected hourly wage at that office.

In Exercises 48–55, use the following information: Jock O'Neill, a sportscaster, and Trudy Bell, a member of the state assembly, are both running for governor of the state of Erehwon. A recent telephone poll asked 800 randomly selected voters whom they planned on voting for. The results of this poll are shown in Figure 3.63.

	Jock O'Neill	Trudy Bell	Undecided
Urban residents	266	184	22
Rural residents	131	181	16

Figure 3.63
Data for Exercises 48–55

48. Find the probability that an urban resident supports O'Neill and the probability that an urban resident supports Bell.

49. Find the probability that a rural resident supports O'Neill and the probability that a rural resident supports Bell.

50. Find the probability that an O'Neill supporter lives in an urban area and the probability that an O'Neill supporter lives in a rural area.

51. Find the probability that a Bell supporter lives in an urban area and the probability that a Bell supporter lives in a rural area.

52. Where are O'Neill supporters more likely to live? Where are Bell supporters more likely to live?

53. Which candidate do the urban residents tend to prefer? Which candidate do the rural residents tend to prefer?

54. Are the events "supporting O'Neill" and "living in an urban area" independent or dependent? What can you conclude?

55. On the basis of the poll, who is ahead in the gubernatorial race?

Are the following events independent or dependent? Are they mutually exclusive?

56. "It is summer" and "it is sunny." (Use your own personal experience.)

57. "It is summer" and "it is Monday." (Use your own personal experience.)

58. "It is summer" and "it is autumn." (Use your own personal experience.)

59. "The first card dealt is an ace" and "the second card dealt is an ace." (Do not use personal experience.)

60. "The first roll of the dice results in a 7" and "the second roll results in a 7" (Do not use personal experience.)

In Exercises 61–66, use the following information: Gregor's Garden Corner buys 40% of their plants from the Green Growery and the balance from Herb's Herbs. Twenty percent of the plants from the Green Growery must be returned, and 10% of those from Herb's Herbs must be returned.

61. What percent of all plants are returned?

62. What percent of all plants are not returned?

63. What percent of all plants are from the Green Growery and are returned?

64. What percent of the returned plants are from the Green Growery?

65. What percent of the returned plants are from Herb's Herbs?

66. Are the events "a plant must be returned" and "a plant was from Herb's Herbs" independent? What conclusion can you make?

In Exercises 67–73, use the following information: The Nissota Automobile Company buys emergency flashers from two different manufacturers, one in Arkansas and one in Nevada. Thirty-nine percent of its turn-signal indicators are purchased from the Arkansas manufacturer, and the rest are purchased from the Nevada manufacturer. Two percent of the Arkansas turn-signal indicators are defective, and 1.7% of the Nevada indicators are defective.

67. What percent of the indicators are defective and made in Arkansas?

68. What percent of the indicators are defective and made in Nevada?
69. What percent of the indicators are defective?
70. What percent of the indicators are not defective?
71. What percent of the defective indicators are made in Arkansas?
72. What percent of the defective indicators are made in Nevada?
73. Are the events "made in Arkansas" and "a defective" independent? What conclusion can you make?

▬▬▶ *Answer the following questions with complete sentences and your own words.*

CONCEPT QUESTIONS

74. What is a conditional probability?
75. Give two events that are independent, and explain why they are independent.
76. Give two events that are dependent, and explain why they are dependent.
77. Give two events that are mutually exclusive, and explain why they are mutually exclusive.
78. Give two events that are not mutually exclusive, and explain why they are not mutually exclusive.
79. Why are probabilities always between 0 and 1, inclusive?
80. Give an example of a permutation and a similar example of a combination.

81. Give an example of two events that are mutually exclusive, and explain why they are mutually exclusive.
82. Give an example of two events that are not mutually exclusive, and explain why they are not mutually exclusive.
83. How is set theory used in probability theory?

HISTORY

84. What two mathematicians invented probability theory? Why?
85. Why was probability theory not considered to be a serious branch of mathematics? What changed that viewpoint? Give a specific example of something that helped change that viewpoint.
86. What role did the following people play in the development of probability theory?
Jacob Bernoulli
Gerolamo Cardano
Pierre de Fermat
Antoine Gombauld, the Chevalier de Méré
John Graunt
Pierre-Simon, the Marquis de Laplace
Gregor Mendel
Blaise Pascal

4 Statistics

© Reuters/Landov

Statistics are everywhere. The news, whether reported in a newspaper, on television, or over the radio, includes statistics of every kind. When shopping for a new car, you will certainly examine the statistics (average miles per gallon, acceleration times, braking distances, and so on) of the various makes and models you are considering. Statistics abound in government studies, and their interpretation affects us all. Industry is driven by statistics; they are essential to the direction of quality control, marketing research, productivity, and many other factors. Sporting events are laden with statistics concerning the past performance of the teams and players.

A person who understands the nature of statistics is equipped to see beyond the short-term and individual perspective. He or she is also better prepared to deal with those who use statistics in misleading ways. To many people, the word *statistics* conjures up an image of an endless list of facts and figures. Where do they come from? What do they mean? In this chapter, you will learn to handle basic statistical problems and expand your knowledge of the meanings, uses, and misuses of statistics.

4.1 Population, Sample, and Data

The field of **statistics** can be defined as the science of collecting, organizing, and summarizing data in such a way that valid conclusions and meaningful predictions can be drawn from them. The first part of this definition, "collecting, organizing, and summarizing data," applies to **descriptive statistics.** The second part, "drawing valid conclusions and making meaningful predictions," describes **inferential statistics.**

Population versus Sample

Who will become the next president of the United States? During election years, political analysts spend a lot of time and money trying to determine what percent of the vote each candidate will receive. However, because there are over 175 million registered voters in the United States, it would be virtually impossible to contact each and every one of them and ask, "Whom do you plan on voting for?" Consequently, analysts select a smaller group of people, determine their intended voting patterns, and project their results onto the entire body of all voters.

Because of time and money constraints, it is very common for researchers to study the characteristics of a small group in order to estimate the characteristics of a larger group. In this context, the set of all objects under study is called the **population,** and any subset of the population is called a **sample** (see Figure 4.1).

When we are studying a large population, we might not be able to collect data from every member of the population, so we collect data from a smaller, more manageable sample. Once we have collected these data, we can summarize by calculating various descriptive statistics, such as the average value. Inferential statistics, then, deals with drawing conclusions (hopefully, valid ones!) about the population, based on the descriptive statistics of the sample data.

Sample data are collected and summarized to help us draw conclusions about the population. A good sample is representative of the population from which it was taken. Obviously, if the sample is not representative, the conclusions concerning the population might not be valid. The most difficult aspect of inferential statistics is obtaining a representative sample. Remember that conclusions are only as reliable as the sampling process and that information will usually change from sample to sample.

Figure 4.1
Population versus sample

What person who lived in the twentieth century do you admire most? The top ten responses in a Gallup poll taken on Dec. 20–21, 1999, were as follows: (1) Mother Teresa, (2) Martin Luther King, Jr., (3) John F. Kennedy, (4) Albert Einstein, (5) Helen Keller, (6) Franklin D. Roosevelt, (7) Billy Graham, (8) Pope John Paul II, (9) Eleanor Roosevelt, and (10) Winston Churchill.

Frequency Distributions

The first phase of any statistical study is the collection of data. Each element in a set of data is referred to as a **data point.** When data are first collected, the data points might show no apparent patterns or trends. To summarize the data and detect any trends, we must organize the data. This is the second phase of descriptive statistics. The most common way to organize raw data is to create a **frequency distribution,** a table that lists each data point along with the number of times it occurs (its **frequency**).

The composition of a frequency distribution is often easier to see if the frequencies are converted to percents, especially if large amounts of data are being summarized. The **relative frequency** of a data point is the frequency of the data point expressed as a percent of the total number of data points (that is, made *relative* to the total). The relative frequency of a data point is found by dividing its frequency by the total number of data points in the data set. Besides listing the frequency of

x = Age	Tally	Frequency	Relative Frequency
$16 \leq x < 24$	⊞ ⊞I	11	$\frac{11}{42} \approx 26\%$
$24 \leq x < 32$	⊞ ⊞III	13	$\frac{13}{42} \approx 31\%$
$32 \leq x < 40$	⊞II	7	$\frac{7}{42} \approx 17\%$
$40 \leq x < 48$	III	3	$\frac{3}{42} \approx 7\%$
$48 \leq x < 56$	II	2	$\frac{2}{42} \approx 5\%$
$56 \leq x < 64$	⊞I	6	$\frac{6}{42} \approx 14\%$
		$n = 42$	total $= 100\%$

Figure 4.3
Frequency distribution of grouped data

When we are working with grouped data, we can choose the groups in any desired fashion. The method used in Example 2 might not be appropriate in all situations. For example, we used the smallest data point as the beginning of the first group, but we could have begun the first group at an even smaller number. The box below gives a general method for constructing a frequency distribution.

Constructing a Frequency Distribution

1. If the raw data consist of many different values, create intervals and work with grouped data. If not, list each distinct data point. [When working with grouped data, choose from four to eight intervals. Divide the range (high minus low) by the desired number of intervals, round off this answer to be consistent with the data, and then add the result to the lowest data point to find the beginning of the second group.]
2. Tally the number of data points in each interval or the number of times each individual data point occurs.
3. List the frequency of each interval or each individual data point.
4. Find the relative frequency by dividing the frequency of each interval or each individual data point by the total number of data points in the distribution. The resulting decimal can be expressed as a percent.

Histograms

When data are grouped in intervals, they can be depicted by a **histogram,** a bar chart that shows how the data are distributed in each interval. To construct a histogram, mark off the class limits on a horizontal axis. If each interval has equal width, we draw two vertical axes; the axis on the left exhibits the frequency of an interval, and the axis on the right gives the corresponding relative frequency. We then draw a rectangle above each interval; the height of the rectangle corresponds to the number of data points contained in the interval. The vertical scale on the right gives the percentage of data contained in each interval. The histogram depicting the distribution of the ages of the people in Keith Reed's acting class (Example 2) is shown in Figure 4.4.

Figure 4.4
Ages of the people in Keith Reed's acting class

What happens if the intervals do not have equal width? For instance, suppose the ages of the people in Keith Reed's acting class are those given in the frequency distribution shown in Figure 4.5.

x = Age	Frequency	Relative Frequency	Class Width
$15 \leq x < 20$	4	$\frac{4}{42} \approx 10\%$	5
$20 \leq x < 25$	11	$\frac{11}{42} \approx 26\%$	5
$25 \leq x < 30$	6	$\frac{6}{42} \approx 14\%$	5
$30 \leq x < 45$	11	$\frac{11}{42} \approx 26\%$	15
$45 \leq x < 65$	10	$\frac{10}{42} \approx 24\%$	20
	$n = 42$	total $= 100\%$	

Figure 4.5
Frequency distribution of age

With frequency and relative frequency as the vertical scales, the histogram depicting this new distribution is given in Figure 4.6.

Does the histogram in Figure 4.6 give a truthful representation of the distribution? No; the rectangle over the interval from 45 to 65 appears to be larger than the rectangle over the interval from 30 to 45, yet the interval from 45 to 65 contains less data than the interval from 30 to 45. This is misleading; rather than comparing the heights of the rectangles, our eyes naturally compare the areas of the rectangles. Therefore, to make an accurate comparison, *the areas of the rectangles must correspond to the relative frequencies of the intervals.* This is accomplished by utilizing the **density** of each interval.

Figure 4.6
Why is this histogram misleading?

Histograms and Relative Frequency Density

Density is a ratio. In science, density is used to determine the concentration of weight in a given volume: Density = weight/volume. For example, the density of water is 62.4 pounds per cubic foot. In statistics, density is used to determine the concentration of data in a given interval: Density = (percent of total data)/(size of an interval). Because relative frequency is a measure of the percentage of data within an interval, we shall calculate the relative frequency density of an interval to determine the concentration of data within the interval.

Definition of Relative Frequency Density

Given a set of n data points, if an interval contains f data points, then the **relative frequency density** (*rfd*) of the interval is

$$rfd = \frac{f/n}{\Delta x}$$

where Δx is the width of the interval.

For example, if the interval $20 \leq x < 25$ contains 11 out of 42 data points, then the relative frequency density of the interval is

$$rfd = \frac{f/n}{\Delta x}$$

$$= \frac{11/42}{5} = 0.052380952\ldots$$

If a histogram is constructed using relative frequency density as the vertical scale, the area of a rectangle will correspond to the relative frequency of the interval, as shown in Figure 4.7.

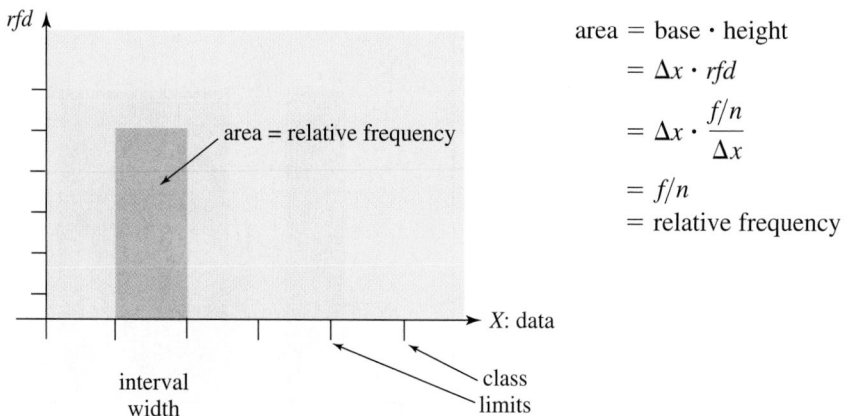

Figure 4.7
Area equals relative frequency

Adding a new column to the frequency distribution given in Figure 4.5, we obtain the relative frequency densities shown in Figure 4.8.

x = Age	Relative Frequency	Class Frequency	Relative Width	Frequency Density
$15 \leq x < 20$	4	$\frac{4}{42} \approx 10\%$	5	$\frac{4}{42} \div 5 \approx 0.020$
$20 \leq x < 25$	11	$\frac{11}{42} \approx 26\%$	5	$\frac{11}{42} \div 5 \approx 0.052$
$25 \leq x < 30$	6	$\frac{6}{42} \approx 14\%$	5	$\frac{6}{42} \div 5 \approx 0.028$
$30 \leq x < 45$	11	$\frac{11}{42} \approx 26\%$	15	$\frac{11}{42} \div 15 \approx 0.017$
$45 \leq x < 65$	10	$\frac{10}{42} \approx 24\%$	20	$\frac{10}{42} \div 20 \approx 0.012$
	$n = 42$	total = 100%		

Figure 4.8
Calculating relative frequency density

We now construct a histogram using relative frequency density as the vertical scale. The histogram depicting the distribution of the ages of the people in Keith Reed's acting class (using the frequency distribution in Figure 4.8) is shown in Figure 4.9.

Figure 4.9
Ages of the people in Keith Reed's acting class

Comparing the histograms in Figures 4.6 and 4.9, we see that using relative frequency density as the vertical scale (rather than frequency) gives a more truthful representation of a distribution when the interval widths are unequal.

EXAMPLE 3   To study the output of a machine that fills bags with corn chips, a quality control engineer randomly selected and weighed a sample of 200 bags of chips. The frequency distribution in Figure 4.10 summarizes the data. Construct a histogram for the weights of the bags of corn chips.

x = Weight (ounces)	f = Number of Bags
$15.3 \leq x < 15.5$	10
$15.5 \leq x < 15.7$	24
$15.7 \leq x < 15.9$	36
$15.9 \leq x < 16.1$	58
$16.1 \leq x < 16.3$	40
$16.3 \leq x < 16.5$	20
$16.5 \leq x < 16.7$	12

Figure 4.10
Weights of bags of corn chips

Solution   Because each interval has the same width ($\Delta x = 0.2$), we construct a combined frequency and relative frequency histogram. The relative frequencies are given in Figure 4.11.

x	f	$rf = f/n$
$15.3 \leq x < 15.5$	10	0.05
$15.5 \leq x < 15.7$	24	0.12
$15.7 \leq x < 15.9$	36	0.18
$15.9 \leq x < 16.1$	58	0.29
$16.1 \leq x < 16.3$	40	0.20
$16.3 \leq x < 16.5$	20	0.10
$16.5 \leq x < 16.7$	12	0.06
	$n = 200$	sum $= 1.00$

Figure 4.11
Relative frequencies

We now draw coordinate axes with appropriate scales and rectangles (Figure 4.12). Notice the (near) symmetry of the histogram. We will study this type of distribution in more detail in Section 4.4.

Figure 4.12
Weights of bags of corn chips

Histograms and Single-Valued Classes

The histograms that we have constructed so far have all utilized intervals of grouped data. For instance, the first group of ages in Keith Reed's acting class was from 16 to 24 years old ($16 \leq x < 24$). However, if a set of data consists of only a few distinct values, it may be advantageous to consider each distinct value to be a "class" of data; that is, we utilize single-valued classes of data.

EXAMPLE 4   A sample of high school seniors was asked, "How many television sets are in your house?" The frequency distribution in Figure 4.13 summarizes the data. Construct a histogram using single-valued classes of data.

Number of Television sets	Frequency
0	2
1	13
2	18
3	11
4	5
5	1

Figure 4.13
Frequency distribution

Solution   Rather than using intervals of grouped data, we use a single value to represent each class. Because each class has the same width ($\Delta x = 1$), we construct a combined frequency and relative frequency histogram. The relative frequencies are given in Figure 4.14.

We now draw coordinate axes with appropriate scales and rectangles. In working with single valued classes of data, it is common to write the single value at the midpoint of the base of each rectangle as shown in Figure 4.15.

Number of Television Sets	Frequency	Relative Frequency
0	2	2/50 = 4%
1	13	13/50 = 26%
2	18	18/50 = 36%
3	11	11/50 = 22%
4	5	5/50 = 10%
5	1	1/50 = 2%
	$n = 50$	total = 100%

Figure 4.14
Calculating relative frequency

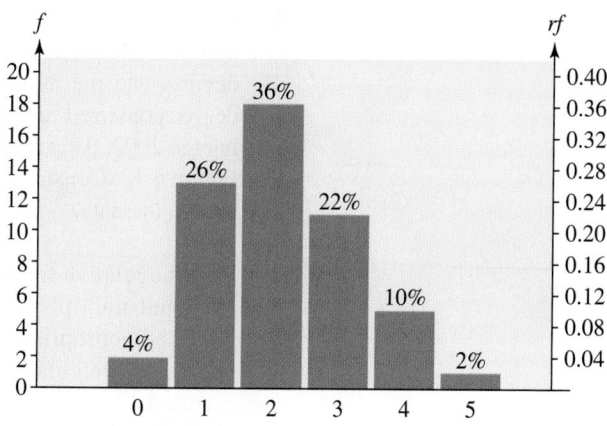

Figure 4.15
Number of television sets

Pie Charts

Many statistical studies involve **categorical data**—that which is grouped according to some common feature or quality. One of the easiest ways to summarize categorical data is through the use of a **pie chart**. A pie chart shows how various categories of a set of data account for certain proportions of the whole. Financial incomes and expenditures are invariably shown as pie charts, as in Figure 4.16.

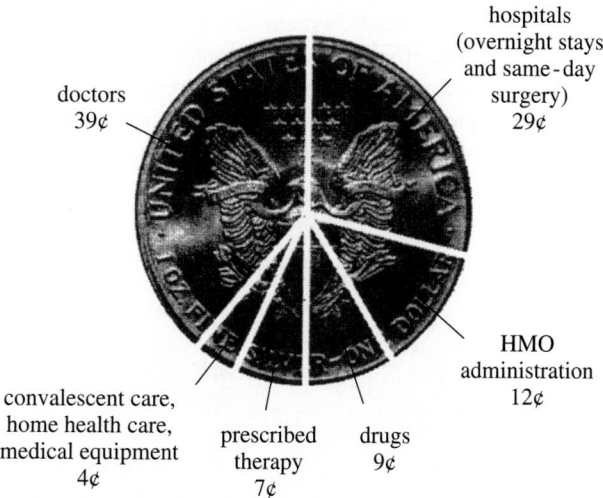

Figure 4.16
How a typical medical dollar is spent

To draw the "slice" of the pie representing the relative frequency (percentage) of the category, the appropriate central angle must be calculated. Since a complete circle comprises 360 degrees, we obtain the required angle by multiplying 360° times the relative frequency of the category.

EXAMPLE 5 What type of academic degree do you hope to earn? The different types of degrees and the number of each type of degree conferred in the United States during the 2002–03 academic year is given in Figure 4.17. Construct a pie chart to summarize the data.

Solution Find the relative frequency of each category and multiply it by 360° to determine the appropriate central angle. The necessary calculations are shown in Figure 4.18.

Type of Degree	Frequency (thousands)
associate's	633
bachelor's	1,349
master's	513
doctorate	46
Total	2,541

Figure 4.17
Academic degrees conferred (2002–03)
Source: National Center for Education Statistics

Type of Degree	Frequency (thousands)	Relative Frequency	Central Angle
associate's	633	$\frac{633}{2,541} \approx 0.249$	$0.249 \times 360° = 89.64°$
bachelor's	1,349	$\frac{1,349}{2,541} \approx 0.531$	$0.531 \times 360° = 191.16°$
master's	513	$\frac{513}{2,541} \approx 0.202$	$0.202 \times 360° = 72.72°$
doctorate	46	$\frac{46}{2,541} \approx 0.018$	$0.018 \times 360° = 6.48°$
	$n = 2,541$	sum = 1.000	total = 360°

Figure 4.18
Calculating relative frequency and central angles

Now use a protractor to lay out the angles and draw the "slices." The name of each category can be written directly on the slice, or, if the names are too long, a legend consisting of various shadings may be used. Each slice of the pie should contain its relative frequency, expressed as a percent. Remember, the whole reason for constructing a pie chart is to convey information visually; pie charts should enable the reader to instantly compare the relative proportions of categorical data. See Figure 4.19.

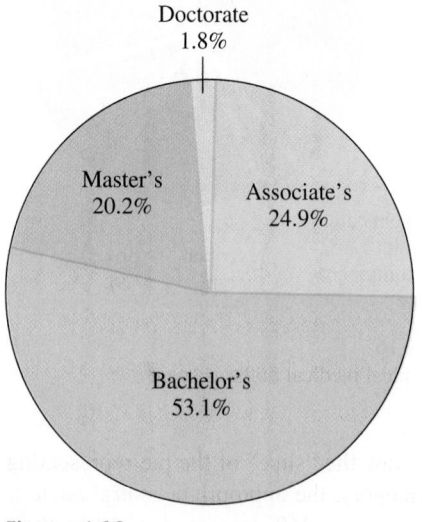

Figure 4.19
Academic degrees conferred (2002–03)

Exercises

1. To study the library habits of students at a local college, thirty randomly selected students were surveyed to determine the number of times they had been to the library during the last week. The following results were obtained:

   1 5 2 1 1 4 2 1 5 4
   5 2 5 1 2 3 4 1 1 2
   3 5 4 1 2 2 4 5 1 2

   a. Organize the given data by creating a frequency distribution.
   b. Construct a pie chart to represent the data.
   c. Construct a histogram using single-valued classes of data.

2. To study the eating habits of students at a local college, thirty randomly selected students were surveyed to determine the number of times they had purchased food at the school cafeteria during the last week. The following results were obtained:

   2 3 1 2 2 3 2 1 2 3
   1 1 5 5 4 3 5 2 2 2
   3 2 3 4 1 4 1 2 3 3

   a. Organize the given data by creating a frequency distribution.
   b. Construct a pie chart to represent the data.
   c. Construct a histogram using single-valued classes of data.

3. To study the composition of families in Manistee, Michigan, forty randomly selected married couples were surveyed to determine the number of children in each family. The following results were obtained:

   2 1 3 0 1 0 2 5 1 2 0 2 2 1
   4 3 1 1 3 4 1 1 0 2 0 0 2 2
   1 0 3 1 1 3 4 2 1 3 0 1

   a. Organize the given data by creating a frequency distribution.
   b. Construct a pie chart to represent the data.
   c. Construct a histogram using single-valued classes of data.

4. To study the spending habits of shoppers in Orlando, Florida, fifty randomly selected shoppers at a mall were surveyed to determine the number of credit cards they carried. The following results were obtained:

   2 5 0 4 2 1 0 6 3 5 4 3 4 0
   5 2 5 2 0 2 5 0 2 5 2 5 4 3
   5 6 1 0 6 3 5 3 4 0 5 2 2 5
   2 0 2 0 4 2 1 0

   a. Organize the given data by creating a frequency distribution.
   b. Construct a pie chart to represent the data.
   c. Construct a histogram using single-valued classes of data.

5. The speeds, in miles per hour, of forty randomly monitored cars on Interstate 40 near Winona, Arizona, were as follows:

   66 71 76 61 73 78 74 67 80 63 69
   78 66 70 77 60 72 58 65 70 64 75
   80 75 62 67 72 59 74 65 54 69 73
   79 64 68 57 51 68 79

   a. Organize the given data by creating a frequency distribution. (Group the data into six intervals.)
   b. Construct a histogram to represent the data.

6. The weights, in pounds, of 35 packages of ground beef at the Cut Above Market were as follows:

   1.0 1.9 2.5 1.2 2.0 0.7 1.3 2.4 1.1
   3.3 2.4 0.8 2.3 1.7 1.0 2.8 1.4 3.0
   0.9 1.1 1.4 2.2 1.5 3.2 2.1 2.7 1.8
   1.6 2.3 2.6 1.3 2.9 1.9 1.2 0.5

   a. Organize the given data by creating a frequency distribution. (Group the data into six intervals.)
   b. Construct a histogram to represent the data.

7. To examine the effects of a new registration system, a campus newspaper asked freshmen how long they had to wait in a registration line. The frequency distribution in Figure 4.20 summarizes the responses. Construct a histogram to represent the data.

x = Time in Minutes	Number of Freshmen
$0 \leq x < 10$	101
$10 \leq x < 20$	237
$20 \leq x < 30$	169
$30 \leq x < 40$	79
$40 \leq x < 50$	51
$50 \leq x < 60$	63

Figure 4.20
Time waiting in line

8. The frequency distribution shown in Figure 4.21 lists the annual salaries of the managers at Universal Manufacturing of Melonville. Construct a histogram to represent the data.

x = Salary (in thousands)	Number of Managers
$\$30 \leq x < 40$	6
$40 \leq x < 50$	12
$50 \leq x < 60$	10
$60 \leq x < 70$	5
$70 \leq x < 80$	7
$80 \leq x < 90$	3

Figure 4.21
Annual salaries

9. The frequency distribution shown in Figure 4.22 lists the hourly wages of the workers at Universal

x = Hourly Wage	Number of Employees
$\$4.00 \leq x < 5.50$	21
$5.50 \leq x < 7.00$	35
$7.00 \leq x < 8.50$	42
$8.50 \leq x < 10.00$	27
$10.00 \leq x < 11.50$	18
$11.50 \leq x < 13.00$	9

Figure 4.22
Hourly wages

Manufacturing of Melonville. Construct a histogram to represent the data.

10. To study the output of a machine that fills boxes with cereal, a quality control engineer weighed 150 boxes of Brand X cereal. The frequency distribution in Figure 4.23 summarizes her findings. Construct a histogram to represent the data.

x = Weight (in ounces)	Number of Boxes
$15.3 \leq x < 15.6$	13
$15.6 \leq x < 15.9$	24
$15.9 \leq x < 16.2$	84
$16.2 \leq x < 16.5$	19
$16.5 \leq x < 16.8$	10

Figure 4.23
Weights of boxes of cereal

11. The ages of the nearly 4 million women who gave birth in the United States in 1997 are given in Figure 4.24. Construct a histogram to represent the data.

Age	Number of Women
$15 \leq x < 20$	486,000
$20 \leq x < 25$	948,000
$25 \leq x < 30$	1,075,000
$30 \leq x < 35$	891,000
$35 \leq x < 40$	410,000
$40 \leq x < 45$	77,000
$45 \leq x < 50$	4,000

Figure 4.24
Ages of women giving birth in 1997
Source: U.S. Bureau of the Census

12. The age composition of the population of the United States in the year 2000 is given in Figure 4.25. Replace the interval "85 and over" with the interval $85 \leq x \leq 100$ and construct a histogram to represent the data.

Age	Number of People (in thousands)
$0 < x < 5$	19,176
$5 \leq x < 10$	20,550
$10 \leq x < 15$	20,528
$15 \leq x < 25$	39,184
$25 \leq x < 35$	39,892
$35 \leq x < 45$	44,149
$45 \leq x < 55$	37,678
$55 \leq x < 65$	24,275
$65 \leq x < 85$	30,752
85 and over	4,240

Figure 4.25
Age composition of the population
of the United States in the year 2000
Source: U.S. Bureau of the Census

In Exercises 13 and 14, use the age composition of the 14,980,000 students enrolled in institutions of higher education in the United States during 2000, as given in Figure 4.26.

13. Using the data in Figure 4.26, replace the interval "35 and over" with the interval $35 \leq x \leq 60$ and construct a histogram to represent the male data.

14. Using the data in Figure 4.26, replace the interval "35 and over" with the interval $35 \leq x \leq 60$ and construct a histogram to represent the female data.

15. The frequency distribution shown in Figure 4.27 lists the ages of 200 randomly selected students who received a bachelor's degree at State University last year. Where possible, determine what percent of the graduates had the following ages:
   a. less than twenty-three
   b. at least thirty-one
   c. at most twenty
   d. not less than nineteen
   e. at least nineteen but less than twenty-seven
   f. not between twenty-three and thirty-five

Age	Number of Students
$10 \leq x < 15$	1
$15 \leq x < 19$	4
$19 \leq x < 23$	52
$23 \leq x < 27$	48
$27 \leq x < 31$	31
$31 \leq x < 35$	16
$35 \leq x < 39$	29
39 and over	19

Figure 4.27
Age of students

Age of Males	Number of Students	Age of Females	Number of Students
$14 \leq x < 18$	94,000	$14 \leq x < 18$	78,000
$18 \leq x < 20$	1,551,000	$18 \leq x < 20$	1,907,000
$20 \leq x < 22$	1,420,000	$20 \leq x < 22$	1,597,000
$22 \leq x < 25$	1,091,000	$22 \leq x < 25$	1,305,000
$25 \leq x < 30$	865,000	$25 \leq x < 30$	1,002,000
$30 \leq x < 35$	521,000	$30 \leq x < 35$	664,000
35 and over	997,000	35 and over	1,888,000
total	6,539,000	total	8,441,000

Figure 4.26
Age composition of students in higher education
Source: U.S. National Center for Education Statistics

16. The frequency distribution shown in Figure 4.28 lists the number of hours per day a randomly selected sample of teenagers spent watching television. Where possible, determine what percent of the teenagers spent the following number of hours watching television:
a. less than 4 hours
b. at least 5 hours
c. at least 1 hour
d. less than 2 hours
e. at least 2 hours but less than 4 hours
f. more than 3.5 hours

Hours per day	Number of Teenagers
$0 \le x < 1$	18
$1 \le x < 2$	31
$2 \le x < 3$	24
$3 \le x < 4$	38
$4 \le x < 5$	27
$5 \le x < 6$	12
$6 \le x < 7$	15

Figure 4.28
Time watching television

17. Figure 4.29 lists the top five reasons given by patients for emergency room visits in 2002. Construct a pie chart to represent the data.

Reason	Number of Patients (thousands)
stomach pain	7,152
chest pain	5,637
fever	5,310
cough	3,016
shortness of breath	2,943

Figure 4.29
Reasons for emergency room visits, 2002
Source: U.S. Department of Health and Human Services

18. Figure 4.30 lists the world's top six countries as tourist destinations in 2003. Construct a pie chart to represent the data.

Country	Number of Arrivals (millions)
France	75.0
Spain	52.5
United States	40.4
Italy	39.6
China	33.0
United Kingdom	24.8

Figure 4.30
World's top tourist destinations, 2003
Source: World Tourism Organization

19. Figure 4.31 lists the race and gender of new AIDS cases in the United States in 2002.
a. Construct a pie chart to represent the male data by race.
b. Construct a pie chart to represent the female data by race.
c. Compare the results of parts (a) and (b). What conclusions can you make?
d. Construct a pie chart to represent the total data.

Race	Male	Female
White	11,221	1,930
Black	14,310	7,339
Hispanic	5,543	1,561
Asian	381	68
Native American	155	42

Figure 4.31
New AIDS cases in the United States, 2002
Source: U.S. Department of Health and Human Services

20. Figure 4.32 lists the types of accidental deaths in the United States in 2003. Construct a pie chart to represent the data.

Type of Accident	Number of Deaths
motor vehicle	44,800
falls	16,200
poison	13,900
choking	4,300
drowning	2,900
fire	2,600
firearms	700

Figure 4.32
Types of accidental deaths in the United States, 2003
Source: National Safety Council

21. Figure 4.33 lists some common specialties of physicians in the United States in 2002.
   a. Construct a pie chart to represent the male data.
   b. Construct a pie chart to represent the female data.
   c. Compare the results of parts (a) and (b). What conclusions can you make?
   d. Construct a pie chart to represent the total data.

Specialty	Male	Female
family practice	63,194	23,317
general surgery	32,678	4,525
internal medicine	101,633	41,658
obstetrics/gynecology	25,606	15,432
pediatrics	33,020	33,351
psychiatry	27,803	12,292

Figure 4.33
Physicians by gender and specialty, 2002
Source: American Medical Association

22. Figure 4.34 lists the major metropolitan areas on intended residence for immigrants admitted to the United States in 2002. Construct a pie chart to represent the data.

Metropolitan Area	Number of Immigrants
Los Angeles/ Long Beach, CA	100,397
New York, NY	86,898
Chicago, IL	41,616
Miami, FL	39,712
Washington DC	36,371

Figure 4.34
Immigrants and areas of residence, 2002
Source: U.S. Immigration and Naturalization Service

Answer the following questions using complete sentences and your own words.

CONCEPT QUESTIONS

23. Explain the meanings of the terms *population* and *sample.*

24. The cholesterol levels of the 800 residents of Land-o-Lakes, Wisconsin, were recently collected and organized in a frequency distribution. Do these data represent a sample or a population? Explain your answer.

25. Explain the difference between frequency, relative frequency, and relative frequency density. What does each measure?

26. When is relative frequency density used as the vertical scale in constructing a histogram? Why?

27. In some frequency distributions, data are grouped in intervals; in others, they are not.
   a. When should data be grouped in intervals?
   b. What are the advantages and disadvantages of using grouped data?

 ## Technology and Statistical Graphs

In Example 2 of this section, we created a frequency distribution and a histogram for the ages of the students in an acting class. Much of this work can be done on a computer or a graphing calculator.

Histograms on a Graphing Calculator

Entering the Data

To enter the data from Example 2, do the following.

Entering the Data on a TI-83/84/86:

- *Put the calculator into statistics mode* by pressing $\boxed{\text{STAT}}$ (TI-86: $\boxed{\text{2nd}}$ $\boxed{\text{STAT}}$).
- *Set the calculator up for entering the data* by selecting "Edit" from the "EDIT" menu (TI-86: select "Edit" by pressing $\boxed{\text{F2}}$), and the "list screen" appears, as shown in Figure 4.35. (A TI-86's list screen says "xStat," "yStat," and "fStat" instead of "L_1," "L_2," "L_3," respectively.) If data already appear in a list (as they do in list L_1 in Figure 4.35), use the arrow buttons to highlight the name of the list (i.e., "L_1" or "xStat") and press $\boxed{\text{CLEAR}}$ $\boxed{\text{ENTER}}$.
- *Enter the students' ages* in list L_1 (TI-86: in list xStat), in any order, using the arrow buttons and the $\boxed{\text{ENTER}}$ button. When completed, your screen should look similar to that in Figure 4.35. Notice the "$L_1(43)=$" at the bottom of the screen; this indicates that 42 entries have been made, and the calculator is ready to receive the 43rd. This allows you to check whether you have left any entries out.

Note: If some data points frequently recur, you can enter the data points in list L_1 and their frequencies in list L_2 (TI-86: in list fStat), rather than reentering a data point each time it recurs.

- Press $\boxed{\text{2nd}}$ $\boxed{\text{QUIT}}$.

Entering the Data on a Casio:

- *Put the calculator into statistics mode* by pressing MENU, highlighting STAT, and pressing EXE.
- *Enter the data* in List 1 in any order, using the arrow buttons and the $\boxed{\text{EXE}}$ button. See Figure 4.36. If data already appear in a list and you want to erase it, use the arrow keys to move to that list, press $\boxed{\text{F6}}$ and then $\boxed{\text{F4}}$ (i.e., $\boxed{\text{DEL-A}}$, which stands for "delete all"), and then press $\boxed{\text{F1}}$ (i.e., $\boxed{\text{YES}}$). Notice that the entries are numbered; this allows you to check whether you have left any entries out.

Drawing a Histogram

Once the data are entered, you can draw a histogram.

Drawing a Histogram on a TI-83/84:

- Press $\boxed{\text{Y=}}$ and clear any functions that may appear.
- *Enter the group boundaries* by pressing $\boxed{\text{WINDOW}}$, entering the left boundary of the first group as xmin (16 for this problem), the right boundary of the last group plus 1 as xmax ($64 + 1 = 65$ for this problem), and the group width as xscl (8 for this problem). (The calculator will create histograms only with equal group widths.) Enter 0 for ymin, and the largest frequency for ymax. (You may guess; it's easy to change it later if you guess wrong.)
- *Set the calculator up to draw a histogram* by pressing $\boxed{\text{2nd}}$ $\boxed{\text{STAT PLOT}}$ and selecting "Plot 1." Turn the plot on and select the histogram icon.
- Tell the calculator to put the data entered in list L_1 on the *x*-axis by selecting "L_1" for "Xlist," and to consider each entered data point as having a frequency of 1 by selecting "1" for "Freq."

Note: If some data points frequently recur and you entered their frequencies in list L_2, then select "L_2" rather than "1" for "Freq" by typing $\boxed{\text{2nd}}$ $\boxed{\text{L}_2}$.

Figure 4.35
A TI-83/84's list screen

Figure 4.36
A Casio's list screen

Figure 4.37
The first bar's boundaries are 16 and 24; its frequency is 11.

- *Draw a histogram* by pressing GRAPH. If some of the bars are too long or too short for the screen, alter ymin accordingly.
- Press TRACE to find out the left and right boundaries and the frequency of the bars, as shown in Figure 4.37. Use the arrow buttons to move from bar to bar.
- Press 2nd STAT PLOT, select "Plot 1" and turn the plot off, or else the histogram will appear on future graphs.

Drawing a Histogram on a TI-86:

- Put the calculator into graphing mode, press y(x) =, and clear any functions that may appear.
- *Enter the group boundaries* by pressing 2nd WIND, entering the left boundary of the first group as xmin (16 for this problem), the right boundary of the last group plus 1 as xmax ($64 + 1 = 65$ for this problem), and the group width as xscl (8 for this problem). (The calculator will create histograms only with equal group widths.) Enter 0 for ymin, and the largest frequency for ymax. (You may guess; it's easy to change it later if you guess wrong.) Quit the graphing mode by pressing 2nd QUIT.
- *Put the calculator back into statistics mode* by pressing 2nd STAT.
- *Set the calculator up to draw a histogram* by pressing PLOT (i.e., F3) and then Plot1 (i.e., F1). Turn the plot on, press the down arrow until the symbol after "Type=" is flashing, and press HIST (i.e., F4).
- Tell the calculator to put the data entered in xStat on the *x*-axis by selecting "xStat" for "Xlist," and to consider each entered data point as having a frequency of 1 by typing "1" for "Freq."

Note: If some data points frequently recur and you entered their frequencies in list fStat, then press fStat (i.e., F3) rather than "1" for "Freq."

- Press 2nd QUIT, and put the calculator back into statistics mode by pressing 2nd STAT.
- *Draw the histogram* by pressing DRAW. If some of the bars are too long or too short for the screen, alter ymax accordingly. If the button labels obscure the histogram, press EXIT once.
- Press PLOT and then Plot1 and turn the plot off, or else the histogram will appear on future graphs.

Drawing a Histogram on a Casio:

- Press GRPH (i.e., F1).
- Press SET (i.e., F6).
- Make the resulting screen, which is labeled "StatGraph1," read as follows:

Graph Type	:Hist
Xlist	:List1
Frequency	: 1

To make the screen read as described above, do the following:
- Use the down arrow button to scroll down to "Graph Type."
- Press F6.
- Press HIST (i.e., F1).
- In a similar manner, change "Xlist" and "Frequency" if necessary.
- Press EXE and return to List1.

Figure 4.38
The first bar's left boundary is 16; its frequency is 11

- Press GPH1 (i.e., F1) and make the resulting screen, which is labeled "Set Interval," read as follows:

  Start: 16   "Start" refers to the beginning of the first group

  Pitch: 8   "Pitch" refers to the width of each group

- Press DRAW (i.e., F6), and the calculator will display the histogram.
- Press SHIFT and then TRCE (i.e., F1) to find out the left boundaries and the frequency of the bars. Use the arrow buttons to move from bar to bar. See Figure 4.38.

Histograms and Pie Charts on a Computerized Spreadsheet

A **spreadsheet** is a large piece of paper marked off in rows and columns. Accountants use spreadsheets to organize numerical data and perform computations. A **computerized spreadsheet,** such as Microsoft Excel, is a computer program that mimics the appearance of a paper spreadsheet. It frees the user from performing any computations; instead, it allows the user merely to give instructions on how to perform those computations. The instructions in this subsection were specifically written for Microsoft Excel; however, all computerized spreadsheets work somewhat similarly.

When you start a computerized spreadsheet, you see something that looks like a table waiting to be filled in. The rows are labeled with numbers and the columns with letters, as shown in Figure 4.39.

◇	A	B	C	D
1				
2				
3				
4				
5				

Figure 4.39
A blank spreadsheet

The individual boxes are called **cells.** The cell in column A row 1 is called cell A1; the cell below it is called cell A2, because it is in column A row 2.

A computerized spreadsheet is an ideal tool to use in creating a histogram or a pie chart. We will illustrate this process by preparing both a histogram and a pie chart for the ages of the students in Keith Reed's acting class, as discussed in Example 2.

Entering the Data

1. *Label the columns.* Use the mouse and/or the arrow buttons to move to cell A1, type in "age of student" and press "return" or "enter." (If there were other data, we could enter it in other columns. For example, if the students' names were included, we could type "name of student" in cell A1, and "age of student" in cell B1.)

2. *Enter the students' ages in column A.* Move to cell A2, type in "26" and press "return" or "enter." Move to cell A3, type in "16" and press "return" or "enter." In a similar manner, enter all of the ages. You can enter the ages in any order. After you

complete this step, your spreadsheet should look like that in Figure 4.40 (except it should go down a lot further).

◇	A	B	C	D
1	age of student			
2	26			
3	16			
4	21			
5	34			

Figure 4.40
The spreadsheet after entering the students' ages

3. *Save the spreadsheet.* Use your mouse to select "File" at the very top of the screen. Then pull your mouse down until "Save As" is highlighted, and let go. Your instructor may give you further instructions on where and how to save your spreadsheet.

Preparing the Data for a Chart

1. *Enter the group boundaries.* Excel uses "bin numbers" rather than group boundaries. A group's bin number is the highest number that should be included in that group. In Example 2, the first group was $16 \leq x < 24$. The highest age that should be included in this group is 23, so the group's bin number is 23. Be careful; this group's bin number is not 24, since people who are 24 years old should be included in the second group.

- If necessary, use the up arrow in the upper-right corner of the spreadsheet to scroll to the top of the spreadsheet.
- Type "bin numbers" in cell B1.
- Determine each group's bin number, and enter them in column B.

After you complete this step, your spreadsheet should look like that in Figure 4.41. You may need to adjust the width of column A. Click on the right edge of the "A" label at the top of the first column, and move it.

◇	A	B
1	age of student	bin numbers
2	26	23
3	16	31
4	21	39
5	34	47
6	45	55
7	18	63
8	41	

Figure 4.41
The spreadsheet after entering the bin numbers

2. *Have Excel determine the frequencies.*
- Use your mouse to select "Tools" at the very top of the screen. Then pull your mouse down until "Data Analysis" is highlighted, and let go.
- If "Data Analysis" is not listed under "Tools," then
  - Select "Tools" at the top of the screen, pull down until "Add-Ins" is highlighted, and let go.
  - Select "Analysis ToolPak-VBA."
  - Use your mouse to press the "OK" button.
  - Select "Tools" at the top of the screen, pull down until "Data Analysis" is highlighted, and let go.
- In the "Data Analysis" box that appears, use your mouse to highlight "Histogram" and press the "OK" button.
- In the "Histogram" box that appears, use your mouse to click on the white rectangle that follows "Input Range," and then use your mouse to draw a box

around all of the ages. (To draw the box, move your mouse to cell A2, press the mouse button, and move the mouse down until all of the entered ages are enclosed in a box.) This should cause "$A$2:$A$43" to appear in the Input Range rectangle.
- Use your mouse to click on the white rectangle that follows "Bin Range," and then use your mouse to draw a box around all of the bin numbers. This should cause "$B$2:$B$7" to appear in the Bin Range rectangle.
- Be certain that there is a dot in the button to the left of "Output Range." Then click on the white rectangle that follows "Output Range," and use your mouse to click on cell C1. This should cause "$C$1" to appear in the Output Range rectangle.
- Use your mouse to press the "OK" button.

After you complete this step, your spreadsheet should look like that in Figure 4.42.

◇	A	B	C	D
1	age of student	bin numbers	*Bin*	*Frequency*
2	26	23	23	11
3	16	31	31	13
4	21	39	39	7
5	34	47	47	3
6	45	55	55	2
7	18	63	63	6
8	41		More	0
9	38			

Figure 4.42
The spreadsheet after Excel determines the frequencies

3. *Prepare labels for the chart.*
- In column E, list the group boundaries.
- In column F, list the frequencies.

After you complete this step, your spreadsheet should look like that in Figure 4.43 (the first few columns are not shown).

C	D	E	F
Bin	*Frequency*	group boundaries	frequency
23	11	$16 \leq x < 24$	11
31	13	$24 \leq x < 32$	13
39	7	$32 \leq x < 40$	7
47	3	$40 \leq x < 48$	3
55	2	$48 \leq x < 56$	2
63	6	$56 \leq x < 64$	6
More	0		

Figure 4.43
The spreadsheet after preparing labels

Drawing a Histogram

1. *Use the Chart Wizard to draw a bar chart.*
   - Use your mouse to press the "Chart Wizard" button at the top of the spreadsheet. (It looks like a histogram and it might have a magic wand.)
   - Select "Column" and then an appropriate style.
   - Press the "Next" button.
   - Use your mouse to click on the white rectangle that follows "Data range," and then use your mouse to draw a box around all of the group boundaries and frequencies from step 3 in "Preparing the data for a chart". This should cause "=Sheet1!\$E\$2:\$F\$7" to appear in the Data range rectangle.
   - Press the "Next" button.
   - Under "Chart Title" type an appropriate title, such as "Acting class ages."
   - After "Category (X)" type an appropriate title for the *x*-axis, such as "Students' ages."
   - After "Value (Y)" type an appropriate title for the *y*-axis, such as "frequencies."
   - Press "Legend" at the top of the chart options box, and then remove the check mark next to "show legend."
   - Press the "Finish" button and the bar chart will appear. See Figure 4.44.

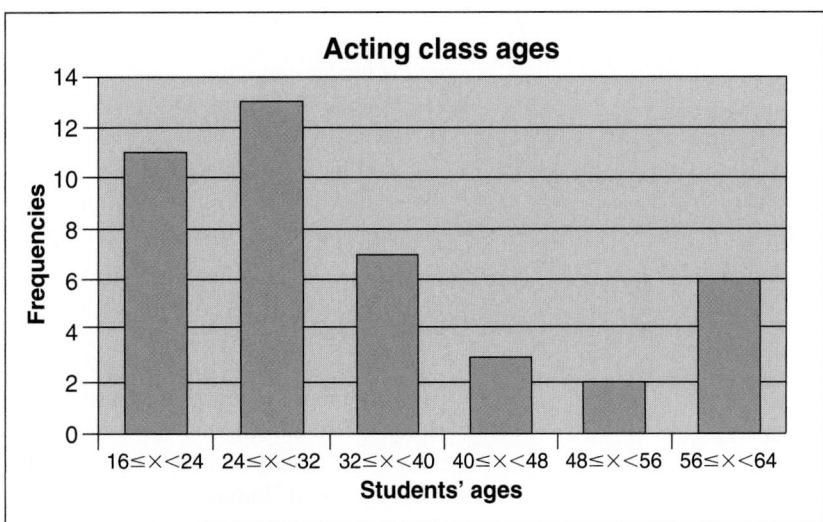

Figure 4.44
The bar chart

2. *Save the spreadsheet.* Use your mouse to select "File" at the very top of the screen. Then pull your mouse down until "Save" is highlighted, and let go.

3. *Convert the bar graph to a histogram.* The graph is not a histogram because of the spaces between the bars.
   - Double click or right click on the bar graph and a "Format Data Series" box will appear.
   - Press the "Options" tab.
   - Remove the spaces between the bars by changing the "Gap width" to 0.
   - Press OK.
   - Save the spreadsheet. See Figure 4.45, on page 242.

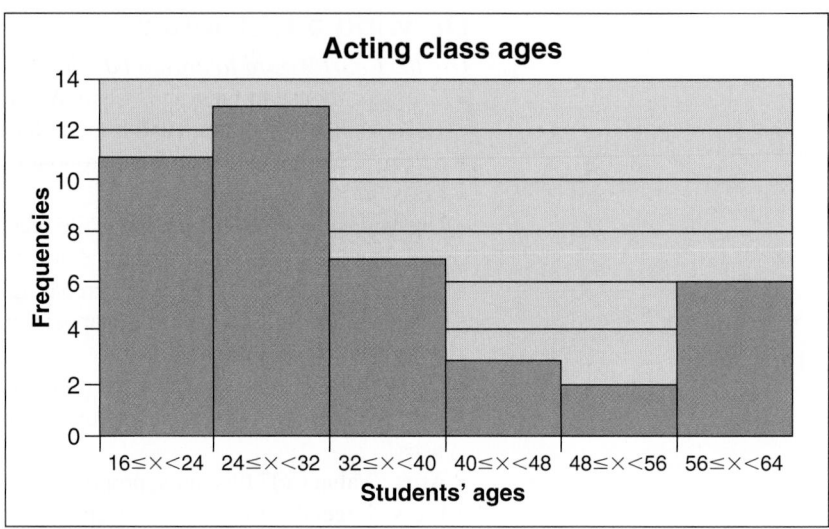

Figure 4.45
The histogram

4. *Print the histogram.*
   - Click on the histogram, and it will be surrounded by a thicker border than before.
   - Use your mouse to select "File" at the very top of the screen. Then pull your mouse down until "Print" is highlighted, and let go.
   - Respond appropriately to the "Print" box that appears.

Drawing a Pie Chart

1. *Use the Chart Wizard to draw the pie chart.*
   - Use your mouse to press the "Chart Wizard" button at the top of the spreadsheet. (It looks like a histogram and it might have a magic wand.)
   - Select "Pie" and then an appropriate style.
   - Press the "Next" button.
   - Use your mouse to click on the white rectangle that follows "Data range," and then use your mouse to draw a box around all of the group boundaries and frequencies from step 3 in "Preparing the data for a chart." This should cause "=Sheet1!$E$2:$F$7" to appear in the Data range rectangle.
   - Press the "Next" button.
   - Under "Chart Title" type an appropriate title.
   - Press the "Finish" button and the pie chart will appear. See Figure 4.46.

2. *Save the spreadsheet.* Use your mouse to select "File" at the very top of the screen. Then pull your mouse down until "Save" is highlighted, and let go.

3. *Print the pie chart.*
   - Click on the chart, and it will be surrounded by a thicker border than before.
   - Use your mouse to select "File" at the very top of the screen. Then pull your mouse down until "Print" is highlighted, and let go.
   - Respond appropriately to the "Print" box that appears.

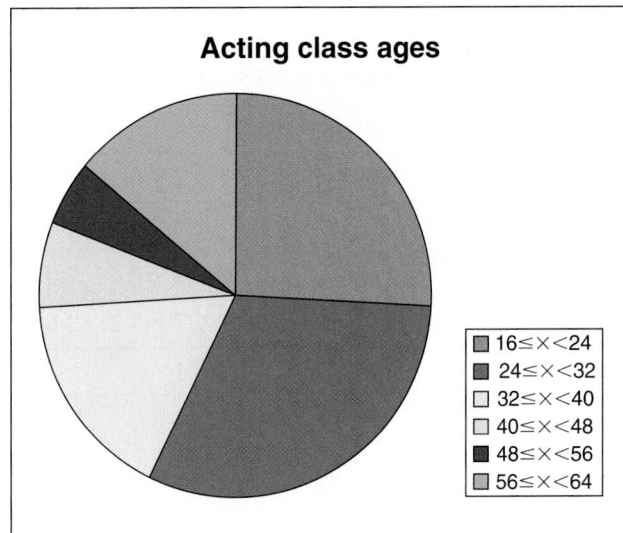

Acting class ages

Legend:
- $16 \leq x < 24$
- $24 \leq x < 32$
- $32 \leq x < 40$
- $40 \leq x < 48$
- $48 \leq x < 56$
- $56 \leq x < 64$

Figure 4.46
The pie chart

Exercises

Use a graphing calculator or Excel on the following exercises. Answers will vary depending on your choice of groups.

Graphing calculator instructions: *By hand, copy a graphing calculator's histogram onto paper for submission.*

Excel instructions: *Print out a histogram for submission.*

28. *Money* magazine claimed that "state regulation (of the insurance industry) is a joke." The article "ranks the states by the percentage of legislators with links to insurers who sit on committees focusing on insurance issues." See Figure 4.47. An unformatted spreadsheet containing these data can be downloaded from the

Mississippi	40%	Louisiana	38%	Arkansas	37%
Virginia	35%	North Carolina	33%	Missouri	32%
Florida	31%	Alabama	26%	Georgia	25%
Texas	25%	West Virginia	24%	Indiana	21%
Ohio	18%	Minnesota	18%	Iowa	17%
North Dakota	17%	Pennsylvania	15%	Arizona	14%
Wyoming	14%	Illinois	14%	Kentucky	13%
Idaho	13%	Delaware	13%	Utah	12%
Kansas	12%	Wisconsin	10%	Hawaii	10%
New Mexico	10%	Rhode Island	9%	Maryland	9%
New Jersey	8%	New York	8%	South Carolina	8%
Washington	8%	Tennessee	8%	Maine	8%
Oklahoma	8%	New Hampshire	7%	Connecticut	5%

(unlisted states were ranked 0%)

Figure 4.47
Percentage of legislators who regulate insurers and have links to the insurance industry
Source: Money, August 1996

Double Stuf Oreos:																		
4.7	6.5	5.5	5.6	5.1	5.3	5.4	5.4	3.5	5.5	6.5	5.9	5.4	4.9	5.6	5.7	5.3	6.9	6.5
6.3	4.8	3.3	6.4	5.0	5.3	5.5	5.0	6.0	5.7	6.3	6.0	6.3	6.1	6.0	5.8	5.8	5.9	6.2
5.9	6.5	6.5	6.1	5.8	6.0	6.2	6.2	6.0	6.8	6.2	5.4	6.6	6.2					

Traditional Oreos:																		
2.9	2.8	2.6	3.5	3.0	2.4	2.7	2.4	2.5	2.2	2.6	2.6	2.9	2.6	2.6	3.1	2.9	2.4	2.8
3.8	3.1	2.9	3.0	2.1	3.8	3.0	3.0	2.8	2.9	2.7	3.2	2.8	3.1	2.7	2.8	2.6	2.6	3.0
2.8	3.5	3.3	3.3	2.8	3.1	2.6	3.5	3.5	3.1	3.1								

Figure 4.48
Amount of Oreo filling
Source: Revak, Marie A. and Jihan G. Williams, "Sharing Teaching Ideas: The Double Stuf Dilemma," The Mathematics Teacher, November 1999, Volume 92, Issue 8, page 674.

text web site: www.thomsonedu.com/mathematics/johnson
a. Create a histogram to represent the data for all 50 states.
b. What do your results say about state regulation of the insurance industry?

29. Do Double Stuf Oreos really contain twice as much "stuf"? In 1999, Marie Revak and Jihan William's "interest was piqued by the unusual spelling of the word stuff and the bold statement 'twice the filling' on the package." So they conducted an experiment in which they weighed the amount of filling in a sample of traditional Oreos and Double Stuf Oreos. The data from that experiment are in Figure 4.48 (weight in grams). An unformatted spreadsheet containing these data can be downloaded from the text web site: www.thomsonedu.com/mathematics/johnson

a. Create one histogram for Double Stuf Oreos and one for Traditional Oreos. Use the same categories for each histogram.
b. Compare the results of part (a). What do they tell you about the two cookies?

30. Figure 4.49 gives EPA fuel efficiency ratings for 2003 compact automobiles with an automatic transmission and the smallest available engine. An unformatted spreadsheet containing this data can be downloaded from the text web site: www.thomsonedu.com/mathematics/johnson
a. For each of the three manufacturing regions, create a histogram for city driving mileage. Use the same categories for the three regions.
b. What conclusions can you make?

	City mpg	Highway mpg
American cars		
Chevrolet Cavalier	24	33
Chrysler Sebring	21	28
Dodge Neon	25	32
Dodge Stratus	21	28
Ford Focus	25	31
Oldsmobile Alero	24	33
Pontiac Grand Am/Sunfire	24	33
Saturn Ion	24	32

Figure 4.49
Fuel efficiency
Source: www.fueleconomy.gov

	City mpg	Highway mpg
Asian cars		
Acura 3.2 CL	20	29
Honda Civic	30	38
Honda Civic Hybrid	48	47
Hyundai Accent	28	36
Hyundai Elantra	24	33
Infiniti G35	19	26
Infiniti M45	17	23
Kia Rio	25	32
Kia Spectra	22	30
Lexus IS300	18	25
Mazda Protege	25	30
Mitsubishi Lancer	24	29
Nissan Sentra	28	35
Subaru Legacy/Outback	22	28
Suzuki Aerio	26	31
Toyota Solara	23	32
Toyota Corolla	29	38
Toyota Echo	33	39
Toyota Prius Hybrid	52	45

	City mpg	Highway mpg
European cars		
Audi A4	23	29
BMW 325i/330i	20	28
BMW 525i/530i	20	28
BMW 540i	18	24
Jaguar X-Type	19	26
Jaguar XJ8	17	24
Mercedes-Benz C230 Kompressor	23	32
Mercedes-Benz C240	19	25
Mercedes-Benz C320	20	26
Saab 9-3	22	31
Volkswagen Golf	34	45
Volkswagen GTI	22	29
Volkswagen Jetta	34	45
Volkswagen Passat	19	26
Volvo S40	22	30
Volvo S60	22	31

Figure 4.49 *Continued*

31. Figure 4.49 gives EPA fuel efficiency ratings for 2003 compact automobiles with an automatic transmission and the smallest available engine. An unformatted spreadsheet containing this data can be downloaded from the text web site:

www.thomsonedu.com/mathematics/johnson

a. For each of the three manufacturing regions, create a histogram for highway driving mileage. Use the same categories for the three regions.
b. What conclusions can you make?

4.2   Measures of Central Tendency

Who is the best running back in professional football? How much does a typical house cost? What is the most popular television program? The answers to questions like these have one thing in common: They are based on averages. To compare the capabilities of athletes, we compute their average performances. This computation usually involves the ratio of two totals, such as (total yards gained)/(total number of carries) = average gain per carry. In real estate, the average-price house is found by listing the prices of all houses for sale (from lowest to highest) and selecting the price in the middle. Television programs are rated by the average number of households tuned in to each particular program.

Rather than listing every data point in a large distribution of numbers, people tend to summarize the data by selecting a representative number, calling it the average. Three figures—the *mean,* the *median,* and the *mode*—describe the "average" or "center" of a distribution of numbers. These averages are known collectively as the **measures of central tendency.**

The Mean

The **mean** is the average people are most familiar with; it can also be the most misleading. Given a collection of n data points, $x_1, x_2, \ldots, x_n$, the mean is found by adding up the data and dividing by the number of data points:

$$\text{the mean of } n \text{ data points} = \frac{x_1 + x_2 + \cdots + x_n}{n}$$

If the data are collected from a sample, then the mean is denoted by $\bar{x}$ (read "x bar"); if the data are collected from an entire population, then the mean is denoted by μ (lowercase Greek letter "mu"). Unless stated otherwise, we will assume that the data represent a sample, so the mean will be symbolized by $\bar{x}$.

Mathematicians have developed a type of shorthand, called **summation notation,** to represent the sum of a collection of numbers. The Greek letter Σ ("sigma") corresponds to the letter S and represents the word *sum.* Given a group of data points $x_1, x_2 \ldots, x_n$, we use the symbol Σx to represent their sum; that is, $\Sigma x = x_1 + x_2 + \cdots + x_n$.

In 2001, Barry Bonds of the San Francisco Giants hit 73 home runs and set an all-time record in professional baseball. On average, how many home runs per season did Bonds hit? See Exercise 13.

> **Definition of the Mean**
>
> Given a sample of n data points, $x_1, x_2, \ldots, x_n$, the **mean,** denoted by $\bar{x}$, is
>
> $$\bar{x} = \frac{\Sigma x}{n} \quad \text{or} \quad \bar{x} = \frac{\text{the sum of the data points}}{\text{the number of data points}}$$

Many scientific calculators have statistical functions built into them. These functions allow you to enter the data points and press the "x-bar" button to obtain the mean. Consult your manual for specific instructions.

EXAMPLE 1 In 2005, Lance Armstrong won his seventh consecutive Tour de France bicycle race. No one in the 100-year history of the race has won so many times. (Miguel Indurain of Spain won five times from 1991 to 1995.) Lance's winning times are given in Figure 4.50. Find the mean winning time of Lance Armstrong's Tour de France victory rides.

Year	1999	2000	2001	2002	2003	2004	2005
Time (h:m:s)	91:32:16	92:33:08	86:17:28	82:05:12	83:41:12	83:36:02	86:15:02

Figure 4.50
Lance Armstrong's Tour de France winning times
Source: San Francisco Chronicle

Solution To find the mean, we must add up the data and divide by the number of data points. However, before we can sum the data, they must be converted to a common unit, say, minutes. Therefore, we multiply the number of hours by 60 (minutes per hour), divide the number of seconds by 60 (seconds per minute), and add each result to the number of minutes. Lance Armstrong's 1999 winning time of 91 hours, 32 minutes, 16 seconds is converted to minutes as follows:

$$\left(91\,\text{hours} \times \frac{60\,\text{minutes}}{1\,\text{hour}}\right) + 32\,\text{minutes} + \left(16\,\text{seconds} \times \frac{1\,\text{minute}}{60\,\text{seconds}}\right)$$

$$= 5,460\,\text{minutes} + 32\,\text{minutes} + 0.266666\ldots\text{minutes}$$

$$= 5,492.267\,\text{minutes} \qquad \text{rounding off to three decimal places}$$

In a similar fashion, all of Lance Armstrong's winning times are converted to minutes and the results are given in Figure 4.51.

Year	1999	2000	2001	2002	2003	2004	2005
Time (min)	5492.267	5553.133	5177.467	4925.200	5021.200	5016.033	5175.033

Figure 4.51
Winning time in minutes

$$\bar{x} = \frac{\Sigma x}{n}$$

$$= \frac{5492.267 + 5553.133 + 5177.467 + 4925.200 + 5021.200 + 5016.033 + 5175.033}{7}$$

$$= \frac{36360.333}{7}$$

$$= 5194.333\,\text{minutes}$$

Converting back to hours, minutes, and seconds, we find that Lance Armstrong's mean winning time is 86 hours, 34 minutes, 20 seconds. ∎

EXAMPLE 2 In 2001, the U.S. Bureau of Labor Statistics tabulated a survey of workers' ages and wages. The frequency distribution in Figure 4.52 summarizes the age distribution of workers who received minimum wage ($5.15 per hour). Find the mean age of a worker receiving minimum wage.

y = Age	Number of Workers
$16 \le y < 20$	640,000
$20 \le y < 25$	660,000
$25 \le y < 35$	372,000
$35 \le y < 45$	276,000
$45 \le y < 55$	171,000
$55 \le y < 65$	111,000
	$n = 2,230,000$

Figure 4.52
Age distribution of workers receiving minimum wage
Source: Bureau of Labor Statistics, U.S. Department of Labor

THE FAR SIDE® By GARY LARSON

"Bob and Ruth! Come on in. ... Have you met Russell and Bill, our 1.5 children?"

If a sample of 460 families have 690 children altogether, then the mean number of children per family is $\bar{x} = 1.5$.

Solution   To find the mean age of the workers, we must add up the ages of all the workers and divide by 2,230,000. However, because we are given grouped data, the ages of the individual workers are unknown to us. In this situation, we use the midpoint of each interval as the representative of the interval; consequently, our answer is an approximation.

To find the midpoint of an interval, add the endpoints and divide by 2. For instance, the midpoint of the interval $16 \leq y < 20$ is

$$\frac{16 + 20}{2} = 18$$

We can then say that each of the 640,000 people in the first interval is approximately eighteen years old. Adding up these workers' ages, we obtain

$$18 + 18 + 18 + \cdots + 18 \quad \text{(six hundred forty thousand times)}$$
$$= (640,000)(18)$$
$$= 11,520,000 \text{ years}$$

If we let f = frequency and x = the midpoint of an interval, the product $f \cdot x$ gives us the total age of the workers in an interval. The results of this procedure for all the workers are shown in Figure 4.53.

y = Age	f = Frequency	x = Midpoint	$f \cdot x$
$16 \le y < 20$	640,000	18	11,520,000
$20 \le y < 25$	660,000	22.5	14,850,000
$25 \le y < 35$	372,000	30	11,160,000
$35 \le y < 45$	276,000	40	11,040,000
$45 \le y < 55$	171,000	50	8,550,000
$55 \le y < 65$	111,000	60	6,660,000
	$n = 2{,}230{,}000$		$\Sigma(f \cdot x) = 63{,}780{,}000$

Figure 4.53
Using midpoints of grouped data

Because $f \cdot x$ gives the sum of the ages of the workers in an interval, the symbol $\Sigma(f \cdot x)$ represents the sum of all the $(f \cdot x)$; that is, $\Sigma(f \cdot x)$ represents the sum of the ages of *all* workers.

The mean age is found by dividing the sum of the ages of all the workers by the number of workers:

$$\bar{x} = \frac{\Sigma(f \cdot x)}{n}$$

$$= \frac{63{,}780{,}000}{2{,}230{,}000}$$

$$= 28.60089 \text{ years}$$

The mean age of the workers earning minimum wage is approximately 28.6 years. ■

> ✓ One common mistake that is made in working with grouped data is to forget to multiply the midpoint of an interval by the frequency of the interval. Another common mistake is to divide by the number of intervals instead of by the total number of data points.

The procedure for calculating the mean when working with grouped data (as illustrated in Example 2) is summarized in the following box.

Calculating the Mean: Grouped Data

Given a frequency distribution containing several groups of data, the mean $\bar{x}$ can be found by using the following formula:

$$\bar{x} = \frac{\Sigma(f \cdot x)}{n}$$

where x = the midpoint of a group, f = the frequency of the group, and $n = \Sigma f$

EXAMPLE 3   Ten college students were comparing their wages earned at part-time jobs. Nine earned $10.00 per hour working at jobs ranging from waiting on tables to working in a bookstore. The tenth student earned $200.00 per hour modeling for a major fashion magazine. Find the mean wage of the ten students.

Solution   The data point $10.00 is repeated nine times, so we multiply it by its frequency. Thus, the mean is as follows:

$$\bar{x} = \frac{\Sigma(f \cdot x)}{n}$$

$$= \frac{(9 \cdot 10) + (1 \cdot 200)}{10}$$

$$= \frac{290}{10}$$

$$= 29$$

The mean wage of the students is $29.00 per hour.   ■

Example 3 seems to indicate that the average wage of the ten students is $29.00 per hour. Is that a reasonable figure? If nine out of ten students earn $10.00 per hour, can we justify saying that their average wage is $29.00? Of course not! Even though the mean wage *is* $29.00, it is not a convincing "average" for this specific group of data. The mean is inflated because one student made $200.00 per hour. This wage is called an **outlier** (or **extreme value**) because it is significantly different from the rest of the data. Whenever a collection of data has extreme values, the mean can be greatly affected and might not be an accurate measure of the average.

The Median

The **median** is the "middle value" of a distribution of numbers. To find it, we first put the data in numerical order. (If a number appears more than once, we include it as many times as it occurs.) If there is an odd number of data points, the median is the middle data point; if there is an even number of data points, the median is defined to be the mean of the two middle values. In either case, the median separates the distribution into two equal parts. Thus, the median can be viewed as an "average." (The word *median* is also used to describe the strip that runs down the middle of a freeway; half the freeway is on one side, and half is on the other. This common usage is in keeping with the statistical meaning.)

EXAMPLE 4   Find the median of the following sets of raw data.

a. 2   8   3   12   6   2   11
b. 2   8   3   12   6   2   11   8

Solution   **a.** First, we put the data in order from smallest to largest. Because there is an odd number of data points, ($n = 7$), we pick the middle one:

2   2   3   6   8   11   12
↑
middle value

The median is 6. (Notice that 6 is the fourth number in the list.)

b. We arrange the data first. Because there is an even number of data points ($n = 8$), we pick the two middle values and find their mean:

$$2 \quad 2 \quad 3 \quad 6 \quad 8 \quad 8 \quad 11 \quad 12$$

$$\frac{(6 + 8)}{2} = 7$$

Therefore, the median is 7. (Notice that 7 is halfway between the fourth and fifth numbers in the list.) ∎

In Example 4, we saw that when $n = 7$, the median was the fourth number in the list, and that when $n = 8$, the median was halfway between the fourth and fifth numbers in the list. Consequently, the *location* of the median depends on n, the number of numbers in the set of data. The formula

$$L = \frac{n + 1}{2}$$

can be used to find the location, L, of the median; when $n = 7$,

$$L = \frac{7 + 1}{2} = 4$$

(the median is the fourth number in the list), and when $n = 8$,

$$L = \frac{8 + 1}{2} = 4.5$$

(the median is halfway between the fourth and fifth numbers in the list).

Location of the Median

Given a sample of n data points, the location of the median can be found by using the following formula:

$$L = \frac{n + 1}{2}$$

Once the data has been arranged from smallest to largest, the median is the Lth number in the list.

EXAMPLE 5   Find the median wage for the ten students in Example 3.

Solution   First, we put the ten wages in order:

$$10 \quad 10 \quad 10 \quad 10 \quad 10 \quad 10 \quad 10 \quad 10 \quad 10 \quad 200$$

Because there are $n = 10$ data points, the location of the median is

$$L = \frac{10 + 1}{2} = 5.5$$

That is, the median is halfway between the fifth and sixth numbers. To find the median, we add the fifth and sixth numbers and divide by 2. Now, the fifth number is 10 and the sixth number is 10, so the median is

$$\frac{10 + 10}{2} = 10$$

Therefore, the median wage is $10.00. This is a much more meaningful "average" than the mean of $29.00.   ■

If a collection of data contains extreme values, the median, rather than the mean, is a better indicator of the "average" value. For instance, in discussions of real estate, the median is usually used to express the "average" price of a house. (Why?) In a similar manner, when the incomes of professionals are compared, the median is a more meaningful representation. The discrepancy between mean (average) income and median income is illustrated in the following news article. Although the article is dated (it's from 1992), it is very informative as to the differences between the mean and the median.

Doctors' average income in U.S. is now $177,000

Associated Press. Reprinted with permission.

WASHINGTON—The average income of the nation's physicians rose to $177,400 in 1992, up 4 percent from the year before, the American Medical Association said yesterday. . . .

The average income figures, compiled annually by the AMA and based on a telephone survey of more than 4,100 physicians, ranged from a low of $111,800 for general practitioners and family practice doctors to a high of $253,300 for radiologists.

Physicians' median income was $148,000 in 1992, or 6.5 percent more than a year earlier. Half the physicians earned more than that and half earned less.

The average is pulled higher than the median by the earnings of the highest paid surgeons, anesthesi-

ologists and other specialists at the top end of the scale. . . .

Here are the AMA's average and median net income figures by specialty for 1992:

General/family practice: $111,800, $100,000.
Internal medicine: $159,300, $130,000.
Surgery: $244,600, $207,000.
Pediatrics: $121,700, $112,000.
Obstetrics/gynecology: $215,100, $190,000.
Radiology: $253,300, $240,000.
Psychiatry: $130,700, $120,000.
Anesthesiology: $228,500, $220,000.
Pathology: $189,800, $170,000.
Other: $165,400, $150,000.

EXAMPLE 6   The students in Ms. Kahlo's art class were asked how many siblings they had. The frequency distribution in Figure 4.54 summarizes the responses. Find the median number of siblings.

Number of Siblings	Number of Responses
0	2
1	8
2	5
3	6

Figure 4.54
Frequency distribution of data

Solution   The frequency distribution indicates that two students had no (0) siblings, eight students had one (1) sibling, five students had two (2) siblings, and six students had three (3) siblings. Therefore, there were $n = 2 + 8 + 5 + 6 = 21$ students in the class; consequently, there are 21 data points. Listing the data in order, we have

$$0, 0, 1, \ldots 1, 2, \ldots 2, 3, \ldots 3$$

Because there are $n = 21$ data points, the location of the median is

$$L = \frac{21 + 1}{2} = 11$$

That is, the median is the eleventh number. Because the first ten numbers are 0s and 1s ($2 + 8 = 10$), the eleventh number is a 2. Consequently, the median number of siblings is 2. ∎

The Mode

The third measure of central tendency is the **mode.** The mode is the most frequent number in a collection of data; that is, it is the data point with the highest frequency. Because it represents the most common number, the mode can be viewed as an average. A distribution of data can have more than one mode or none at all.

ThomsonNOW™
for Liberal Arts Mathematics
www.thomsonedu.com/1pass.

EXAMPLE 7   Find the mode(s) of the following sets of raw data:

a. 4   10   1   8   5   10   5   10
b. 4   9   1   10   1   10   4   9
c. 9   6   1   8   3   10   3   9

Solution   a. The mode is 10, because it has the highest frequency (3).
b. There is no mode, because each number has the same frequency (2).
c. The distribution has two modes—namely, 3 and 9—each with a frequency of 2. A distribution that has two modes is called *bimodal.* ∎

In summarizing a distribution of numbers, it is most informative to list all three measures of central tendency. It helps to avoid any confusion or misunderstanding in situations in which the word *average* is used. In the hands of someone with questionable intentions, numbers can be manipulated to mislead people. In his book *How to Lie with Statistics,* Darrell Huff states, "The secret language of statistics, so appealing in a fact-minded culture, is employed to sensationalize, inflate, confuse, and oversimplify. Statistical methods and statistical terms are necessary in reporting the mass data of social and economic trends, business conditions, 'opinion' polls, the census. But without writers who use the words with honesty and understanding, and readers who know what they mean, the result can only be semantic nonsense." An educated public should always be on the alert for oversimplification of data via statistics. The next time someone mentions an "average," ask "Which one?" Although people might not intentionally try to mislead you, their findings can be misinterpreted if you do not know the meaning of their statistics and the method by which the statistics were calculated.

4.2   Exercises

In Exercises 1–4, find the mean, median, and mode of the given set of raw data.

1.
   9   12   8   10   9   11   12
   15   20   9   14   15   21   10

2.
   20   25   18   30   21   25   32   27
   32   35   19   26   38   31   20   23

3.
   1.2   1.8   0.7   1.5   1.0   0.7   1.9   1.7   1.2
   0.8   1.7   1.3   2.3   0.9   2.0   1.7   1.5   2.2

4.
   0.07   0.02   0.09   0.04   0.10   0.08   0.07   0.13
   0.05   0.04   0.10   0.07   0.04   0.01   0.11   0.08

5. Find the mean, median, and mode of each set of data.
   a. 9   9   10   11   12   15
   b. 9   9   10   11   12   102
   c. How do your answers for parts (a) and (b) differ (or agree)? Why?

6. Find the mean, median, and mode for each set of data.
   a. 80   90   100   110   110   140
   b. 10   90   100   110   110   210
   c. How do your answers for parts (a) and (b) differ (or agree)? Why?

7. Find the mean, median, and mode of each set of data.
   a. 2   4   6   8   10   12
   b. 102   104   106   108   110   112
   c. How are the data in part (b) related to the data in part (a)?
   d. How do your answers for parts (a) and (b) compare?

8. Find the mean, median, and mode of each set of data.
   a. 12   16   20   24   28   32
   b. 600   800   1,000   1,200   1,400   1,600
   c. How are the data in part (b) related to the data in part (a)?
   d. How do your answers for parts (a) and (b) compare?

9. Kaitlin Mowry is a member of the local 4-H club and has six mini Rex rabbits that she enters in regional rabbit competition shows. The weights of the rabbits are given in Figure 4.55. Find the mean, median, and mode of the rabbits' weights.

| Weight (lb:oz) | 3:12 | 4:03 | 3:06 | 3:15 | 3:12 | 4:02 |

Figure 4.55
Weights of rabbits

10. As was stated in Example 1, Lance Armstrong won the Tour de France bicycle race every year from 1999 to 2005. His margins of victory (time difference of the second place finisher) are given in Figure 4.56. Find the mean, median, and mode of Lance Armstrong's victory margins.

Year	1999	2000	2001	2002	2003	2004	2005
Margin (m:s)	7:37	6:02	6:44	7:17	1:01	6:19	4:40

Figure 4.56
Lance Armstrong's Tour de France victory margins
Source: San Francisco Chronicle

11. Jerry Rice holds the all-time record in professional football for scoring touchdowns. The number of touchdown receptions (TDs) for each of his seasons is given in Figure 4.57. Find the mean, median, and mode of the number of touchdown receptions per year by Rice.

Year	TDs	Year	TDs
1985	3	1995	15
1986	15	1996	8
1987	22	1997	1
1988	9	1998	9
1989	17	1999	5
1990	13	2000	7
1991	14	2001	9
1992	10	2002	7
1993	15	2003	2
1994	13	2004	3

Figure 4.57
Touchdown receptions for Jerry Rice
Source: http://sportsillustrated.cnn.com/football/nfl/players/

12. Wayne Gretzky, known as "The Great One," holds the all-time record in professional hockey for scoring goals. The number of goals for each of his seasons is

given in Figure 4.58. Find the mean, median, and mode of the number of goals per season by Gretzky.

Season	Goals	Season	Goals
1979–80	51	1989–90	40
1980–81	55	1990–91	41
1981–82	92	1991–92	31
1982–83	71	1992–93	16
1983–84	87	1993–94	38
1984–85	73	1994–95	11
1985–86	52	1995–96	23
1986–87	62	1996–97	25
1987–88	40	1997–98	23
1988–89	54	1998–99	9

Figure 4.58
Goals made by Wayne Gretzky
Source: The World Almanac

13. Barry Bonds of the San Francisco Giants set an all-time record in professional baseball by hitting 73 home runs in one season (2001). The number of home runs (HR) for each of his seasons in professional baseball is given in Figure 4.59. Find the mean, median, and mode of the number of home runs hit per year by Bonds.

Year	HR	Year	HR
1986	16	1996	42
1987	25	1997	40
1988	24	1998	37
1989	19	1999	34
1990	33	2000	49
1991	25	2001	73
1992	34	2002	46
1993	46	2003	45
1994	37	2004	45
1995	33		

Figure 4.59
Home runs hit by Barry Bonds
Source: http://sportsillustrated.cnn.com/baseball/mlb/players/

14. Michael Jordan has been recognized as an extraordinary player in professional basketball, especially in terms of the number of points per game he has scored. Jordan's average number of points per game (PPG) for each of his seasons is given in Figure 4.60. Find the mean, median, and mode of the average points per game made per season by Jordan.

Season	PPG	Season	PPG
1984–85	28.2	1991–92	30.1
1985–86	22.7	1992–93	32.6
1986–87	37.1	1994–95	26.9
1987–88	35.0	1995–96	30.4
1988–89	32.5	1996–97	29.6
1989–90	33.6	1997–98	28.7
1990–91	31.5	2001–02	22.9
		2002–03	20.0

Figure 4.60
Points per game made by Michael Jordan
Source: http://www.nba.com/playerfile/

15. The frequency distribution in Figure 4.61 lists the results of a quiz given in Professor Gilbert's statistics class. Find the mean, median, and mode of the scores.

Score	Number of Students
10	3
9	10
8	9
7	8
6	10
5	2

Figure 4.61
Quiz scores in Professor Gilbert's statistics class

16. Todd Booth, an avid jogger, kept detailed records of the number of miles he ran per week during the past year. The frequency distribution in Figure 4.62

summarizes his records. Find the mean, median, and mode of the number of miles per week that Todd ran.

Miles Run per Week	Number of Weeks
0	5
1	4
2	10
3	9
4	10
5	7
6	3
7	4

Figure 4.62
Miles run by Todd Booth

17. To study the output of a machine that fills boxes with cereal, a quality control engineer weighed 150 boxes of Brand X cereal. The frequency distribution in Figure 4.63 summarizes his findings. Find the mean weight of the boxes of cereal.

x = Weight (in ounces)	Number of Boxes
$15.3 \leq x < 15.6$	13
$15.6 \leq x < 15.9$	24
$15.9 \leq x < 16.2$	84
$16.2 \leq x < 16.5$	19
$16.5 \leq x \leq 16.8$	10

Figure 4.63
Amount of Brand X cereal per box

18. To study the efficiency of its new price-scanning equipment, a local supermarket monitored the amount of time its customers had to wait in line. The frequency distribution in Figure 4.64 summarizes the findings. Find the mean amount of time spent in line.

x = Time (in minutes)	Number of Customers
$0 \leq x < 1$	79
$1 \leq x < 2$	58
$2 \leq x < 3$	64
$3 \leq x < 4$	40
$4 \leq x \leq 5$	35

Figure 4.64
Time spent waiting in a supermarket checkout line

19. Katrina must take five exams in a math class. If her scores on the first four exams are 71, 69, 85, and 83, what score does she need on the fifth exam for her overall mean to be
   a. at least 70?   b. at least 80?
   c. at least 90?

20. Eugene must take four exams in a geography class. If his scores on the first three exams are 91, 67, and 83, what score does he need on the fourth exam for his overall mean to be
   a. at least 70?   b. at least 80?
   c. at least 90?

21. The mean salary of 12 men is $58,000, and the mean salary of 8 women is $42,000. Find the mean salary of all 20 people.

22. The mean salary of 12 men is $52,000, and the mean salary of 4 women is $84,000. Find the mean salary of all 16 people.

23. Maria drove from Chicago, Illinois, to Milwaukee, Wisconsin, a distance of 90 miles, at a mean speed of 60 miles per hour. On her return trip, the traffic was much heavier, and her mean speed was 45 miles per hour. Find Maria's mean speed for the round trip.

   HINT: Divide the total distance by the total time.

24. Sully drove from Atlanta, Georgia, to Birmingham, Alabama, a distance of 150 miles, at a mean speed of 50 miles per hour. On his return trip, the traffic was much lighter, and his mean speed was 60 miles per hour. Find Sully's mean speed for the round trip.

   HINT: Divide the total distance by the total time.

25. The mean age of a class of 25 students is 23.4 years. How old would a 26th student have to be for the mean age of the class to be 24.0 years?

26. The mean age of a class of 15 students is 18.2 years. How old would a 16th student have to be for the mean age of the class to be 21.0 years?

27. The mean salary of eight employees is $40,000, and the median is $42,000. The highest-paid employee gets a $6,000 raise.
   a. What is the new mean salary of the eight employees?
   b. What is the new median salary of the eight employees?

28. The mean salary of ten employees is $32,000, and the median is $30,000. The highest-paid employee gets a $5,000 raise.
   a. What is the new mean salary of the ten employees?
   b. What is the new median salary of the ten employees?

29. The number of civilians holding various federal government jobs and their mean monthly earnings for March 2004 are given in Figure 4.65.

Department	Number of Civilian Workers	Mean Monthly Earnings
State Department	32,997	$5,578.08
Justice Department	103,318	$5,622.52
Congress	17,154	$4,998.89
Department of the Navy	182,386	$4,047.43
Department of the Air Force	154,411	$4,043.67
Department of the Army	230,941	$2,762.16

Figure 4.65
Monthly earnings for civilian jobs, March 2004
Source: U.S. Office of Personnel Management

   a. Find the mean monthly earnings of all civilians employed by the Navy, Air Force, and Army.
   b. Find the mean monthly earnings of all civilians employed by the State Department, Justice Department, and Congress.

30. The ages of the nearly 4 million women who gave birth in the United States in 1997 are given in Figure 4.66. Find the mean age of these women.

Age	Number of Women
$15 \leq x < 20$	486,000
$20 \leq x < 25$	948,000
$25 \leq x < 30$	1,075,000
$30 \leq x < 35$	891,000
$35 \leq x < 40$	410,000
$40 \leq x < 45$	77,000
$45 \leq x < 50$	4,000

Figure 4.66
Ages of women giving birth in 1997
Source: U.S. Bureau of the Census

31. The age composition of the population of the United States in the year 2000 is given in Figure 4.67.
   a. Find the mean age of all people in the United States under the age of eighty-five.
   b. Replace the interval "85 and over" with the interval $85 \leq X \leq 100$ and find the mean age of all people in the United States.

Age	Number of People (in thousands)
$0 < x < 5$	19,176
$5 \leq x < 10$	20,550
$10 \leq x < 15$	20,528
$15 \leq x < 25$	39,184
$25 \leq x < 35$	39,892
$35 \leq x < 45$	44,149
$45 \leq x < 55$	37,678
$55 \leq x < 65$	24,275
$65 \leq x < 85$	30,752
85 and over	4,240

Figure 4.67
Age composition of the population of the United States in the year 2000
Source: U.S. Bureau of the Census

In Exercises 32 and 33, use the age composition of the 14,980,000 students enrolled in institutions of higher education in the United States during 2000, as given in Figure 4.68.

Age of Males	Number of Students
$14 \leq x < 18$	94,000
$18 \leq x < 20$	1,551,000
$20 \leq x < 22$	1,420,000
$22 \leq x < 25$	1,091,000
$25 \leq x < 30$	865,000
$30 \leq x < 35$	521,000
35 and over	997,000
total	6,539,000

Figure 4.68
Age composition of students in higher education
Source: U.S. National Center for Education Statistics

Age of Females	Number of Students
$14 \leq x < 18$	78,000
$18 \leq x < 20$	1,907,000
$20 \leq x < 22$	1,597,000
$22 \leq x < 25$	1,305,000
$25 \leq x < 30$	1,002,000
$30 \leq x < 35$	664,000
35 and over	1,888,000
total	8,441,000

Figure 4.68 *Continued*

32. a. Find the mean age of all male students in higher education under thirty-five.
  b. Replace the interval "35 and over" with the interval $35 \leq x \leq 60$ and find the mean age of all male students in higher education.

33. a. Find the mean age of all female students in higher education under thirty-five.
  b. Replace the interval "35 and over" with the interval $35 \leq x \leq 60$ and find the mean age of all female students in higher education.

 Answer the following questions using complete sentences and your own words.

CONCEPT QUESTIONS

34. What are the three measures of central tendency? Briefly explain the meaning of each.
35. Suppose the mean of Group I is A and the mean of Group II is B. We combine Groups I and II to form Group III. Is the mean of Group III equal to $\frac{A+B}{2}$? Explain.
36. Why do we use the midpoint of an interval when calculating the mean of grouped data?

WEB PROJECT

37. What were last month's "average" high and low temperatures in your favorite city? Pick a city that interests you and obtain the high and low temperatures for each day last month. Print the data, and submit them as evidence in answering the following.
  a. Find the mean, median, and mode of the daily high temperatures.
  b. How does the mean daily high temperature last month compare to the mean seasonal high temperature for last month? That is, were the high temperatures last month above or below the normal high temperatures for the month?
  c. Find the mean, median, and mode of the daily low temperatures.
  d. How does the mean daily low temperature last month compare to the mean seasonal low temperature for last month? That is, were the low temperatures last month above or below the normal low temperatures for the month?

Some useful links for this web project are listed on the text web site:
www.thomsonedu.com/mathematics/johnson

4.3   Measures of Dispersion

To settle an argument over who was the better bowler, George and Danny agreed to bowl six games, and whoever had the highest "average" would be considered best. Their scores were as shown in Figure 4.69.

George	185	135	200	185	250	155
Danny	182	185	188	185	180	190

Figure 4.69
Bowling scores

Each bowler then arranged his scores from lowest to highest and computed the mean, median, and mode:

George
$$
\begin{cases}
\quad 135 \qquad 155 \qquad 185 \qquad 185 \qquad 200 \qquad 250 \\[2mm]
\text{mean} = \dfrac{\text{sum of scores}}{6} = \dfrac{1{,}110}{6} = 185 \\[3mm]
\text{median} = \text{middle score} = \dfrac{185 + 185}{2} = 185 \\[3mm]
\text{mode} = \text{most common score} = 185
\end{cases}
$$

Danny
$$
\begin{cases}
\quad 180 \qquad 182 \qquad 185 \qquad 185 \qquad 188 \qquad 190 \\[2mm]
\text{mean} = \dfrac{1{,}110}{6} = 185 \\[3mm]
\text{median} = \dfrac{185 + 185}{2} = 185 \\[3mm]
\text{mode} = 185
\end{cases}
$$

Much to their surprise, George's mean, median, and mode were exactly the same as Danny's! Using the measures of central tendency alone to summarize their performances, the bowlers appear identical. Even though their averages were identical, however, their performances were not; George was very erratic, while Danny was very consistent. Who is the better bowler? On the basis of high score, George is better. On the basis of consistency, Danny is better.

George and Danny's situation points out a fundamental weakness in using only the measures of central tendency to summarize data. In addition to finding the averages of a set of data, the consistency, or spread, of the data should also be taken into account. This is accomplished by using **measures of dispersion,** which determine how the data points differ from the average.

Deviations

It is clear from George and Danny's bowling scores that it is sometimes desirable to measure the relative consistency of a set of data. Are the numbers consistently bunched up? Are they erratically spread out? To measure the dispersion of a set of data, we need to identify an average or typical distance between the data points and the mean. The difference between a single data point x and the mean $\bar{x}$ is called the **deviation from the mean** (or simply the **deviation**) and is given by $(x - \bar{x})$. A data point that is close to the mean will have a small deviation, whereas data points far from the mean will have large deviations, as shown in Figure 4.70.

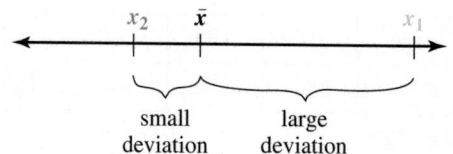

Figure 4.70
Large versus small deviations

To find the typical deviation of the data points, you might be tempted to add up all the deviations and divide by the total number of data points, thus finding the "average" deviation. Unfortunately, this process leads nowhere. To see why, we will find the mean of the deviations of George's bowling scores.

EXAMPLE 1   George bowled six games, and his scores were 185, 135, 200, 185, 250, and 155. Find the mean of the scores, the deviation of each score, and the mean of the deviations.

Solution $\bar{x} = \dfrac{\text{sum of scores}}{6} = \dfrac{1{,}110}{6} = 185$

The mean score is 185.

To find the deviations, subtract the mean from each score, as shown in Figure 4.71.

$$\text{mean of the deviations} = \frac{\text{sum of deviations}}{6} = \frac{0}{6} = 0$$

The mean of the deviations is zero.   ∎

Score (x)	Deviation ($x - 185$)
135	−50
155	−30
185	0
185	0
200	15
250	65
	sum = 0

Figure 4.71
Deviations of George's score

In Example 1, the sum of the deviations of the data is zero. This *always* happens; that is, $\Sigma\,(x - \bar{x}) = 0$ for any set of data. The negative deviations are the "culprits"—they will always cancel out the positive deviations. Therefore, to use deviations to study the spread of the data, we must modify our approach and convert the negatives into positives. We do this by squaring each deviation.

Variance and Standard Deviation

Before proceeding, we must be reminded of the difference between a population and a sample. A population is the universal set of all possible items under study; a sample is any group or subset of items selected from the population. (Samples are used to study populations.) In this context, George's six bowling scores represent a sample, not a population; because we do not know the scores of *all* the games George has ever bowled, we are limited to a sample. Unless otherwise specified, we will consider any given set of data to represent a sample, not an entire population.

To measure the typical deviation contained within a set of data points, we must first find the **variance** of the data. Given a sample of n data points, the variance of the data is found by squaring each deviation, adding the squares, and then dividing the sum by the number $(n - 1)$.*

Because we are working with n data points, you might wonder why we divide by $n - 1$ rather than by n. The answer lies in the study of inferential statistics. Recall that inferential statistics deal with the drawing of conclusions concerning the nature of a population based on observations made within a sample. Hence, the variance of a sample can be viewed as an estimate of the variance of the population. However, because the population will vary more than the sample (a population has more data points),

---

* If the n data points represent the entire population, the population variance, denoted by σ^2, is found by squaring each deviation, adding the squares, and then dividing the sum by n.

Sample Variance Deviation

Given a sample of n data points, $x_1, x_2, \ldots, x_n$, the **variance** of the data, denoted by s^2, is

$$s^2 = \frac{\Sigma(x - \bar{x})^2}{n - 1}$$

The variance of a sample is found by dividing the sum of the squares of the deviations by $n - 1$. The symbol s^2 is a reminder that the deviations have been squared.

dividing the sum of the squares of the sample deviations by n would underestimate the true variance of the entire population. To compensate for this underestimation, statisticians have determined that dividing the sum of the squares of the deviations by $n - 1$ rather than by n produces the best estimate of the true population variance.

Variance is the tool with which we can obtain a measure of the typical deviation contained within a set of data. However, because the deviations have been squared, we must perform one more operation to obtain the desired result: We must take the square root. The square root of variance is called the **standard deviation** of the data.

Standard Deviation Definition

Given a sample of n data points, $x_1, x_2, \ldots, x_n$, the **standard deviation** of the data, denoted by s, is

$$s = \sqrt{\text{variance}}$$

To find the standard deviation of a set of data, first find the variance and then take the square root of the variance.

ThomsonNOW™
for Liberal Arts Mathematics
www.thomsonedu.com/1pass.

EXAMPLE 2

George bowled six games, and his scores were 185, 135, 200, 185, 250, and 155. Find the standard deviation of his scores.

Solution

To find the standard deviation, we must first find the variance. The mean of the six data points is 185. The necessary calculations for finding variance are shown in Figure 4.72.

Data (x)	Deviation $(x - 185)$	Deviation Squared $(x - 185)^2$
135	-50	$(-50)^2 = 2{,}500$
155	-30	$(-30)^2 = 900$
185	0	$(0)^2 = 0$
185	0	$(0)^2 = 0$
200	15	$(15)^2 = 225$
250	65	$(65)^2 = 4{,}225$
		sum $= 7{,}850$

Figure 4.72
Finding variance

$$\text{variance} = \frac{\text{sum of the squares of the deviations}}{n-1}$$

$$s^2 = \frac{7{,}850}{6-1}$$

$$= \frac{7{,}850}{5}$$

$$= 1{,}570$$

The variance is $s^2 = 1{,}570$. Taking the square root, we have

$$s = \sqrt{1{,}570}$$

$$= 39.62322551\ldots$$

It is customary to round off s to one place more than the original data. Hence, the standard deviation of George's bowling scores is $s = 39.6$ points.   ■

Because they give us information concerning the spread of data, variance and standard deviation are called **measures of dispersion.** Standard deviation (and variance) is a relative measure of the dispersion of a set of data; the larger the standard deviation, the more spread out the data. Consider George's standard deviation of 39.6. This appears to be high, but what exactly constitutes a "high" standard deviation? Unfortunately, because it is a relative measure, there is no hard-and-fast distinction between a "high" and a "low" standard deviation.

By itself, the standard deviation of a set of data might not be very informative, but standard deviations are very useful in comparing the relative consistencies of two sets of data. Given two groups of numbers of the same type (for example, two sets of bowling scores, two sets of heights, or two sets of prices), the set with the lower standard deviation contains data that are more consistent, whereas the data with the higher standard deviation are more spread out. Calculating the standard deviation of Danny's six bowling scores, we find $s = 3.7$. Since Danny's standard deviation is less than George's, we infer that Danny is more consistent. If George's standard deviation is less than 39.6 the next time he bowls six games, we would infer that his game has become more consistent (the scores would not be spread out as far).

Alternative Methods for Finding Variance

The procedure for calculating variance is very direct: First find the mean of the data, then find the deviation of each data point, and finally divide the sum of the squares of the deviations by $(n-1)$. However, using the Sample Variance Definition to find the variance can be rather tedious. Fortunately, many scientific calculators are programmed to find the variance (and standard deviation) if you just push a few buttons. Consult your manual to utilize the statistical capabilities of your calculator.

If your calculator does not have built-in statistical functions, you might still be able to take a shortcut in calculating variance. Instead of using the Definition of Variance (as in Example 2), we can use an alternative formula that contains the two sums Σx and Σx^2, where Σx represents the sum of the data and Σx^2 represents the sum of the squares of the data, as shown in the following box.

> **Alternative Formula for Sample Variance**
>
> Given a sample of n data points, $x_1, x_2, \ldots, x_n$, the **variance** of the data, denoted by s^2, can be found by
>
> $$s^2 = \frac{1}{(n-1)}\left[\Sigma x^2 - \frac{(\Sigma x)^2}{n}\right]$$
>
> *Note:* Σx^2 means "square each data point, then add"; $(\Sigma x)^2$ means "add the data points, then square."

Although we will not prove it, this Alternative Formula for Sample Variance is algebraically equivalent to the Sample Variance Definition; given any set of data, either method will produce the same answer. At first glance, the Alternative Formula might appear to be more difficult to use than the Definition. Do not be fooled by its appearance! As we will see, the Alternative Formula is relatively quick and easy to apply.

EXAMPLE 3   Using the Alternative Formula for Sample Variance, find the standard deviation of George's bowling scores as given in Example 2.

Solution   Recall that George's scores were 185, 135, 200, 185, 250, and 155. To find the standard deviation, we must first find the variance.

The Alternative Formula for Sample Variance requires that we find the sum of the data and the sum of the squares of the data. These calculations are shown in Figure 4.73. Applying the Alternative Formula for Sample Variance, we have

$$
\begin{aligned}
s^2 &= \frac{1}{(n-1)}\left[\Sigma x^2 - \frac{(\Sigma x)^2}{n}\right] \\
&= \frac{1}{6-1}\left[213{,}200 - \frac{(1{,}110)^2}{6}\right] \\
&= \frac{1}{5}\left[213{,}200 - 205{,}350\right] \\
&= \frac{7{,}850}{5} = 1{,}570
\end{aligned}
$$

x	x^2
135	18,225
155	24,025
185	34,225
185	34,225
200	40,000
250	62,500
$\Sigma x = 1{,}110$	$\Sigma x^2 = 213{,}200$

Figure 4.73
Data and data squared

The variance is $s^2 = 1{,}570$. (Note that this is the same as the variance calculated in Example 2 using the Definition.)

Taking the square root, we have

$$
\begin{aligned}
s &= \sqrt{1{,}570} \\
&= 39.62322551\ldots
\end{aligned}
$$

Rounded off, the standard deviation of George's bowling scores is $s = 39.6$ points. ∎

When we are working with grouped data, the individual data points are unknown. In such cases, the midpoint of each interval should be used as the representative value of the interval.

EXAMPLE 4   In 2001, the U.S. Bureau of Labor Statistics tabulated a survey of workers' ages and wages. The frequency distribution in Figure 4.74 summarizes the age distribution of workers who received minimum wage ($5.15 per hour). Find the standard deviation of the ages of these workers.

y = Age	Number of Workers
$16 \leq y < 20$	640,000
$20 \leq y < 25$	660,000
$25 \leq y < 35$	372,000
$35 \leq y < 45$	276,000
$45 \leq y < 55$	171,000
$55 \leq y < 65$	111,000
	$n = 2,230,000$

Figure 4.74
Workers receiving minimum wage
Source: Bureau of Labor Statistics,
U.S. Department of Labor

Solution   Because we are given grouped data, the first step is to determine the midpoint of each interval. We do this by adding the endpoints and dividing by 2.

To utilize the Alternate Formula for Sample Variance, we must find the sum of the data and the sum of the squares of the data. The sum of the data is found by multiplying each midpoint by the frequency of the interval and adding the results; that is, $\Sigma(f \cdot x)$. The sum of the squares of the data is found by squaring each midpoint, multiplying by the corresponding frequency, and adding; that is, $\Sigma(f \cdot x^2)$. The calculations are shown in Figure 4.75.

y = Age	f = Frequency	x = Midpoint	$f \cdot x$	$f \cdot x^2$
$16 \leq y < 20$	640,000	18	11,520,000	207,360,000
$20 \leq y < 25$	660,000	22.5	14,850,000	334,125,000
$25 \leq y < 35$	372,000	30	11,160,000	334,800,000
$35 \leq y < 45$	276,000	40	11,040,000	441,600,000
$45 \leq y < 55$	171,000	50	8,550,000	427,500,000
$55 \leq y < 65$	111,000	60	6,660,000	399,600,000
	$n = 2,230,000$		$\Sigma(f \cdot x) = 63,780,000$	$\Sigma(f \cdot x^2) = 2,144,985,000$

Figure 4.75
Finding variance of grouped data

Applying the Alternative Formula for Sample Variance, we have

$$s^2 = \frac{1}{n-1}\left[\Sigma(f \cdot x^2) - \frac{(\Sigma f \cdot x)^2}{n}\right]$$

$$= \frac{1}{2,230,000 - 1}\left[2,144,985,000 - \frac{(63,780,000)^2}{2,230,000}\right]$$

$$= \frac{1}{2,229,999}[2,144,985,000 - 1,824,165,202]$$

$$= \frac{320,819,798}{2,229,999}$$

The variance is $s^2 = 143.8654448$. Taking the square root, we have

$$s = \sqrt{143.8654448}$$

$$= 11.99439222\ldots$$

Rounded off, the standard deviation of the ages of the workers receiving minimum wage is $s = 12.0$ years. ∎

The procedure for calculating variance when working with grouped data (as illustrated in Example 4) is summarized in the following box.

Alternative Formula for Sample Variance: Grouped Data

Given a frequency distribution containing several groups of data, the variance s^2 can be found by

$$s^2 = \frac{1}{(n-1)}\left[\Sigma(f \cdot x^2) - \frac{(\Sigma f \cdot x)^2}{n}\right]$$

where x = the midpoint of a group, f = the frequency of the group, and $n = \Sigma f$.

To obtain the best analysis of a collection of data, we should use the measures of central tendency and the measures of dispersion in conjunction with each other. The most common way to combine these measures is to determine what percent of the data lies within a specified number of standard deviations of the mean. The phrase "one standard deviation of the mean" refers to all numbers within the interval $[\bar{x} - s, \bar{x} + s]$, that is, all numbers that differ from $\bar{x}$ by at most s. Likewise, "two standard deviations of the mean" refers to all numbers within the interval $[\bar{x} - 2s, \bar{x} + 2s]$. One, two, and three standard deviations of the mean are shown in Figure 4.76.

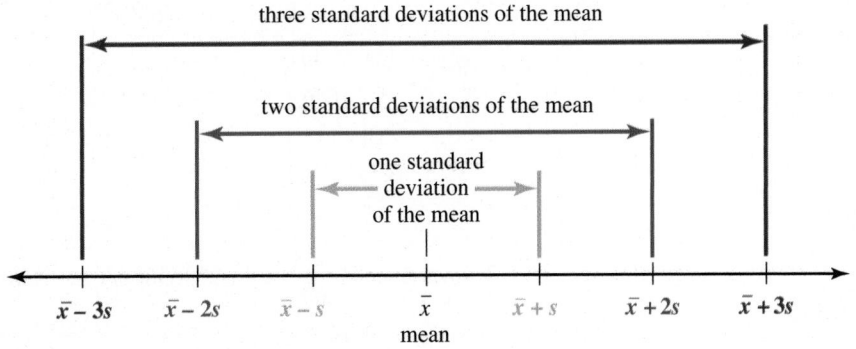

Figure 4.76
One, two, and three standard deviations of the mean

Thomson**NOW**
for Liberal Arts Mathematics
www.thomsonedu.com/1pass.

EXAMPLE 5 Paki Mowry is a rabbit enthusiast and has eleven mini Rex rabbits that she enters in regional rabbit competition shows. The weights of the rabbits are given in Figure 4.77. What percent of the rabbits' weights lie within one standard deviation of the mean?

Weight (lb:oz)	3:10	4:02	3:06	3:15	3:12	4:01
	3:11	4:00	4:03	3:13	3:15	

Figure 4.77
Weights of rabbits in pounds and ounces

Solution   First, the weights must be converted to a common unit, say, ounces. Therefore, we multiply the number of pounds by 16 (ounces per pound) and add the given number of ounces. For instance, 3 pounds, 10 ounces is converted to ounces as follows:

$$3 \text{ pounds, 10 ounces} = \left(3 \text{ pounds} \times \frac{16 \text{ ounces}}{\text{pound}} \right) + 10 \text{ ounces}$$

$$= 48 \text{ ounces} + 10 \text{ ounces}$$

$$= 58 \text{ ounces}$$

The converted weights are given in Figure 4.78.

Weight (oz)	58	66	54	63	60	65
	59	64	67	61	63	

Figure 4.78
Weights of rabbits in ounces

Now, we find the mean and standard deviation.

Summing the eleven data points, we have $\Sigma x = 680$. Summing the squares of the data points, we have $\Sigma x^2 = 42,186$. The mean is

$$\bar{x} = \frac{680}{11} = 61.81818181\ldots = 61.8 \text{ ounces} \text{ (rounded to one decimal place)}$$

Using the Alternative Formula for Sample Variance, we have

$$s^2 = \frac{1}{(n-1)}\left[\Sigma x^2 - \frac{(\Sigma x)^2}{n} \right]$$

$$= \frac{1}{(11-1)}\left[42,186 - \frac{(680)^2}{11} \right]$$

$$= \frac{1}{10}\left[\frac{1646}{11} \right] \qquad \text{subtracting fractions with LCD = 11}$$

$$= \frac{823}{55} \qquad \text{reducing}$$

The variance is $\frac{823}{55} = 14.963636363\ldots$. Taking the square root, we have

$$s = \sqrt{823/55} = 3.868285972\ldots$$

The standard deviation is 3.9 ounces (rounded to one decimal place).

To find one standard deviation of the mean, we add and subtract the standard deviation to and from the mean:

$$[\bar{x} - s, \bar{x} + s] = [61.8 - 3.9, 61.8 + 3.9]$$
$$= [57.9, 65.7]$$

Arranging the data from smallest to largest, we see that eight of the eleven data points are between 57.9 and 65.7:

54   58   59   60   61   63   63   64   65   66   67

$\bar{x} - s$	$\bar{x}$	$\bar{x} + s$
57.9	61.8	65.7

Therefore, $\dfrac{8}{11} = 0.7272727272\ldots$, or 72.7%, of the data lie within one standard deviation of the mean. ∎

4.3 Exercises

1. Perform each task, given the following sample data:

   3   8   5   3   10   13

   a. Use the Sample Variance Definition to find the variance and standard deviation of the data.
   b. Use the Alternate Formula for Sample Variance to find the variance and standard deviation of the data.

2. Perform each task, given the following sample data:

   6   10   12   12   11   17   9

   a. Use the Sample Variance Definition to find the variance and standard deviation of the data.
   b. Use the Alternate Formula for Sample Variance to find the variance and standard deviation of the data.

3. Perform each task, given the following sample data:

   10   10   10   10   10   10

   a. Find the variance of the data.
   b. Find the standard deviation of the data.

4. Find the mean and standard deviation of each set of data.
   a. 2   4   6   8   10   12
   b. 102   104   106   108   110   112
   c. How are the data in (b) related to the data in (a)?
   d. How do your answers for (a) and (b) compare?

5. Find the mean and standard deviation of each set of data.
   a. 12   16   20   24   28   32
   b. 600   800   1,000   1,200   1,400   1,600
   c. How are the data in (b) related to the data in (a)?
   d. How do your answers for (a) and (b) compare?

6. Find the mean and standard deviation of each set of data.
   a. 50   50   50   50   50
   b. 46   50   50   50   54
   c. 5   50   50   50   95
   d. How do your answers for (a), (b), and (c) compare?

7. Joey and Dee Dee bowled five games at the Rock 'n' Bowl Lanes. Their scores are given in Figure 4.79.
   a. Find the mean score of each bowler. Who has the highest mean?
   b. Find the standard deviation of each bowler's scores.
   c. Who is the more consistent bowler? Why?

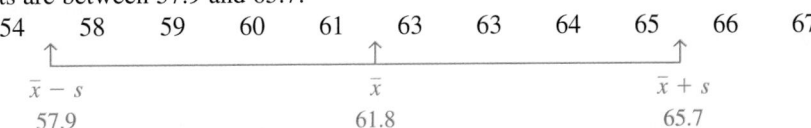

Joey	144	171	220	158	147
Dee Dee	182	165	187	142	159

Figure 4.79
Bowling scores

8. Paki surveyed the price of unleaded gasoline (self-serve) at gas stations in Novato and Lafayette. The raw data, in dollars per gallon, are given in Figure 4.80.
   a. Find the mean price in each city. Which city has the lowest mean?
   b. Find the standard deviation of prices in each city.
   c. Which city has more consistently priced gasoline? Why?

Novato	2.399	2.589	2.929	2.459	2.499	2.599
Lafayette	3.095	2.889	2.699	2.699	3.049	2.849

Figure 4.80
Price (in dollars) of one gallon of unleaded gasoline

9. Kaitlin Mowry is a member of the local 4-H club and has six mini Rex rabbits that she enters in regional rabbit competition shows. The weights of the rabbits are given in Figure 4.81. Find the standard deviation of the rabbits' weights.

| Weight (lb:oz) | 4:12 | 5:03 | 4:06 | 3:15 | 4:12 | 4:08 |

Figure 4.81
Weights of rabbits

10. As was stated in Example 1 of Section 4.2, Lance Armstrong won the Tour de France bicycle race every year from 1999 to 2005. His margins of victory (time difference of the second place finisher) are given in Figure 4.82. Find the standard deviation of Lance Armstrong's victory margins.

Year	1999	2000	2001	2002	2003	2004	2005
Margin	7:37	6:02	6:44	7:17	1:01	6:19	4:40

Figure 4.82
Lance Armstrong's Tour de France victory margins
Source: San Francisco Chronicle

11. Barry Bonds of the San Francisco Giants set an all-time record in professional baseball by hitting 73 home runs in one season (2001). The number of home runs (HR) for each of his seasons in professional baseball is given in Figure 4.83. Find the standard deviation of the number of home runs hit per year by Bonds.

Year	HR	Year	HR
1986	16	1995	33
1987	25	1996	42
1988	24	1997	40
1989	19	1998	37
1990	33	1999	34
1991	25	2000	49
1992	34	2001	73
1993	46	2002	46
1994	37	2003	45
		2004	45

Figure 4.83
Barry Bonds' home runs

12. Michael Jordan has been recognized as an extraordinary player in professional basketball, especially in terms of the number of points per game he has scored. Jordan's average number of points per game (PPG) for each of his seasons is given in Figure 4.84. Find the standard deviation of the average points per game made per season by Jordan.

Season	PPG	Season	PPG
1984–85	28.2	1992–93	32.6
1985–86	22.7	1994–95	26.9
1986–87	37.1	1995–96	30.4
1987–88	35.0	1996–97	29.6
1988–89	32.5	1997–98	28.7
1989–90	33.6	2001–02	22.9
1990–91	31.5	2002–03	20.0
1991–92	30.1		

Figure 4.84
Michael Jordan's points per game

13. The Truly Amazing Dudes are a group of comic acrobats. The heights (in inches) of the ten acrobats are as follows:

68   50   70   67   72   78   69   68   66   67

Is your height or weight "average"? These characteristics can vary considerably within any specific group of people. The mean is used to represent the average, and the standard deviation is used to measure the "spread" of a collection of data.

a. Find the mean and standard deviation of the heights.

b. What percent of the data lies within one standard deviation of the mean?

c. What percent of the data lies within two standard deviations of the mean?

14. The weights (in pounds) of the ten Truly Amazing Dudes are as follows:

152 196 144 139 166 83 186 157 140 138

a. Find the mean and standard deviation of the weights.

b. What percent of the data lies within one standard deviation of the mean?

c. What percent of the data lies within two standard deviations of the mean?

15. The normal monthly rainfall in Seattle, Washington, is given in Figure 4.85.

Month	Jan.	Feb.	Mar.	Apr.	May	June
Inches	5.4	4.0	3.8	2.5	1.8	1.6

Month	July	Aug.	Sept.	Oct.	Nov.	Dec.
Inches	0.9	1.2	1.9	3.3	5.7	6.0

Figure 4.85
Monthly rainfall in Seattle, WA
Source: U.S. Department of Commerce

a. Find the mean and standard deviation of the monthly rainfall in Seattle.

b. What percent of the year will the monthly rainfall be within one standard deviation of the mean?

c. What percent of the year will the monthly rainfall be within two standard deviations of the mean?

16. The normal monthly rainfall in Phoenix, Arizona, is given in Figure 4.86.

Month	Jan.	Feb.	Mar.	Apr.	May	June
Inches	0.7	0.7	0.9	0.2	0.1	0.1

Month	July	Aug.	Sept.	Oct.	Nov.	Dec.
Inches	0.8	1.0	0.9	0.7	0.7	1.0

Figure 4.86
Monthly rainfall in Phoeniz, AZ
Source: U.S. Department of Commerce

a. Find the mean and standard deviation of the monthly rainfall in Phoenix.

b. What percent of the year will the monthly rainfall be within one standard deviation of the mean?

c. What percent of the year will the monthly rainfall be within two standard deviations of the mean?

17. The frequency distribution in Figure 4.87 lists the results of a quiz given in Professor Gilbert's statistics class.

Score	Number of Students	Score	Number of Students
10	5	7	8
9	10	6	3
8	6	5	2

Figure 4.87
Quiz scores in Professor Gilbert's statistics class

a. Find the mean and standard deviation of the scores.

b. What percent of the data lies within one standard deviation of the mean?

c. What percent of the data lies within two standard deviations of the mean?

d. What percent of the data lies within three standard deviations of the mean?

18. Amy surveyed the prices for a quart of a certain brand of motor oil. The sample data, in dollars per quart, is summarized in Figure 4.88.

Price per Quart	Number of Stores
1.99	2
2.09	5
2.19	10
2.29	13
2.39	9
2.49	3

Figure 4.88
Price (in dollars) for a quart of motor oil

a. Find the mean and the standard deviation of the prices.

b. What percent of the data lies within one standard deviation of the mean?

c. What percent of the data lies within two standard deviations of the mean?

d. What percent of the data lies within three standard deviations of the mean?

19. To study the output of a machine that fills boxes with cereal, a quality control engineer weighed 150 boxes of Brand X cereal. The frequency distribution in Figure 4.89 summarizes his findings. Find the standard deviation of the weight of the boxes of cereal.

x = Weight (in ounces)	Number of Boxes
$15.3 \leq x < 15.6$	13
$15.6 \leq x < 15.9$	24
$15.9 \leq x < 16.2$	84
$16.2 \leq x < 16.5$	19
$16.5 \leq x < 16.8$	10

Figure 4.89
Amount of Brand X cereal per box

20. To study the efficiency of its new price-scanning equipment, a local supermarket monitored the amount of time its customers had to wait in line. The frequency distribution in Figure 4.90 summarizes the findings. Find the standard deviation of the amount of time spent in line.

x = Time (in minutes)	Number of Customers
$0 \leq x < 1$	79
$1 \leq x < 2$	58
$2 \leq x < 3$	64
$3 \leq x < 4$	40
$4 \leq x \leq 5$	35

Figure 4.90
Time spent waiting in a supermarket checkout line

21. The ages of the nearly 4 million women who gave birth in the United States in 1997 are given in Figure 4.91. Find the standard deviation of the ages of these women.

Age	Number of Women
$15 \leq x < 20$	486,000
$20 \leq x < 25$	948,000
$25 \leq x < 30$	1,075,000
$30 \leq x < 35$	891,000
$35 \leq x < 40$	410,000
$40 \leq x < 45$	77,000
$45 \leq x < 50$	4,000

Figure 4.91
Ages of women giving birth in 1997
Source: U.S. Bureau of the Census

22. The age composition of the population of the United States in the year 2000 is given in Figure 4.92. Replace the interval "85 and over" with the interval $85 \leq x \leq 100$ and find the standard deviation of the ages of all people in the United States.

Age	Number of People (in thousands)
$0 < x < 5$	19,176
$5 \leq x < 10$	20,550
$10 \leq x < 15$	20,528
$15 \leq x < 25$	39,184
$25 \leq x < 35$	39,892
$35 \leq x < 45$	44,149
$45 \leq x < 55$	37,678
$55 \leq x < 65$	24,275
$65 \leq x < 85$	30,752
85 and over	4,240

Figure 4.92
Age composition of the population of the United States in the year 2000
Source: U.S. Bureau of the Census

Answer the following questions using complete sentences and your own words.

CONCEPT QUESTIONS

23. a. When studying the dispersion of a set of data, why are the deviations from the mean squared?
   b. What effect does squaring have on a deviation that is less than 1?
   c. What effect does squaring have on a deviation that is greater than 1?
   d. What effect does squaring have on the data's units?
   e. Why is it necessary to take a square root when calculating standard deviation?

24. Why do we use the midpoint of an interval when calculating the standard deviation of grouped data?

 WEB PROJECT

25. This project is a continuation of Exercise 34 in Section 4.2. How did last month's daily high and low temperatures vary in your favorite city? Pick a city that interests you, and obtain the high and low temperatures for each day last month. Print the data, and submit them as evidence in answering the following.

a. Find the standard deviation of the daily high temperatures.

b. Find the standard deviation of the daily low temperatures.

c. Comparing your answers to parts (a) and (b), what can you conclude?

Some useful links for this web project are listed on the text web site:

www.thomsonedu.com/mathematics/johnson

PROJECTS

26. The purpose of this project is to explore the variation in the pricing of a common commodity. Go to several different stores that sell food (the more the better) and record the price of one gallon of whole milk.

a. Compute the mean, median, and mode of the data.

b. Compute the standard deviation of the data.

c. What percent of the data lie within one standard deviation of the mean?

d. What percent of the data lie within two standard deviations of the mean?

e. What percent of the data lie within three standard deviations of the mean?

27. The purpose of this project is to explore the variation in the pricing of a common commodity. Go to several different gas stations (the more the better) and record the price of one gallon of premium gasoline (91 octane).

a. Compute the mean, median, and mode of the data.

b. Compute the standard deviation of the data.

c. What percent of the data lie within one standard deviation of the mean?

d. What percent of the data lie within two standard deviations of the mean?

e. What percent of the data lie within three standard deviations of the mean?

 # Technology and Measures of Central Tendency and Dispersion

In Examples 1, 2, and 3 of this section, we found the mean, median, mode, variance, and standard deviation of George's bowling scores. This work can be done quickly and easily on either a graphing calculator or Excel.

 ## Measures of Central Tendency and Dispersion on a Graphing Calculator

Calculating the Mean, the Variance, and the Standard Deviation

On a TI-83/84:
- Enter the data from Example 1 of this section in list L_1 as discussed in Section 4.1.
- Press STAT.
- Select "1-Var Stats" from the "CALC" menu.
- When "1-Var" appears on the screen, press 2nd L_1 ENTER.

On a TI-86:
- Enter the data from Example 1 of this section as discussed in Section 4.1.
- Press 2nd QUIT, and put the calculator back into statistics mode by pressing 2nd STAT.
- Press F1, which is now labeled "CALC," and then F1, now labeled "OneVa", and the screen will read "OneVar".
- Press 2nd LIST and then F3, which is now labeled "NAMES".

Figure 4.93
After computing the mean,
standard deviation, and more
on a TI graphing calculator

- Press [F3], which is now labeled "xStat," and the screen will read "OneVar xStat".
- Press [ENTER].

On a Casio:
- Enter the data from Example 1 of this section as described in Section 4.1.
- Press [CALC] (i.e., [F2]).
- Press [1VAR] (i.e., [F1]).

The above steps will result in the first screen in Figure 4.93. This screen gives the mean, the sample standard deviation (S_x on a TI, $\chi\sigma n\text{-}1$ on a Casio), the population standard deviation (σ_x on a TI, $\chi\sigma n$ on a Casio), and the number of data points (n). The second screen can be obtained by pressing the down arrow. It gives the minimum and maximum data points (minX and maxX, respectively) as well as the median (Med).

Calculating the Sample Variance

The above work does not yield the sample variance. To find it, follow these steps:

- Quit the statistics mode.
- Get S_x on the screen:

TI-83/84:	Press [VARS], select "Statistics," and then select "S_x" from the "X/Y" menu.
TI-86:	Press [2nd] [STAT], press [VARS] (i.e., [F5]) and then [S_x] (i.e., [F3]).

- Once S_x is on the screen, square it by pressing [x^2] [ENTER]. The variance is 1,570.

Calculating the Mean, the Variance, and the Standard Deviation with Grouped Data

To calculate the mean and standard deviation from the frequency distribution that utilizes grouped data, follow the same steps except:

On a TI-83/84:
- Enter the midpoints of the classes in list L_1.
- Enter the frequencies of those classes in list L_2. Each frequency must be less than 100.
- After "1-Var Stats" appears on the screen, press [2nd] [L_1] [,] [2nd] [L_2] [ENTER].

On a TI-86:
- Enter the midpoints of the classes in list xStat.
- Enter the frequencies of those classes in list fStat.
- After "1-Var Stats" appears on the screen, press $\boxed{\text{2nd}}$ $\boxed{\text{LIST}}$, and then $\boxed{\text{NAMES}}$ (i.e., $\boxed{\text{F3}}$).
- Press $\boxed{\text{xStat}}$ $\boxed{\text{,}}$ $\boxed{\text{fStat}}$, and the screen will read "OneVar xStat,fStat."
- Press $\boxed{\text{ENTER}}$.

Measures of Central Tendency and Dispersion on Excel

1. *Enter the data* from Example 1 of this section as discussed in Section 4.1. See column A of Figure 4.94.

2. *Have Excel compute the mean, standard deviation, etc.*
   - Use your mouse to select "Tools" at the very top of the screen. Then pull your mouse down until "Data Analysis" is highlighted, and let go. (If "Data Analysis" is not listed under "Tools," then follow the instructions given in Section 4.1.)
   - In the "Data Analysis" box that appears, use your mouse to highlight "Descriptive Statistics" and press the "OK" button.
   - In the "Descriptive Statistics" box that appears, use your mouse to click on the white rectangle that follows "Input Range," and then use your mouse to draw a box around all of the bowling scores. (To draw the box, move your mouse to cell A2, press the mouse button, and move the mouse down until all of the entered scores are enclosed in a box.) This should cause "$A$2:$A$7" to appear in the Input Range rectangle.
   - Be certain that there is a dot in the button to the left of "Output Range." Then click on the white rectangle that follows "Output Range," and use your mouse to click on cell B1. This should cause "$B$1" to appear in the Output Range rectangle.
   - Select "Summary Statistics."
   - Use your mouse to press the "OK" button.

After you complete this step, your spreadsheet should look like that in Figure 4.94. To compute the population standard deviation, first type "pop std dev" in cell B16, and then click on cell C16. Press the *fx* button on the Excel ribbon at the top of the screen. In the "Paste Function" box that appears, click on "Statistical" under "Function category" and on "STDEVP" under "Function name." Then press the "OK" button. In the "STDEVP" box that appears, click on the white rectangle that follows "Number 1." Then use your mouse to draw a box around all of the bowling scores. Press the "OK" button and the population standard deviation will appear in cell C16.

◇	A	B	C
1	George s scores	*Column 1*	
2	185		
3	135	Mean	185
4	200	Standard Errc	16.1761141
5	185	Median	185
6	250	Mode	185
7	155	Standard Dev	39.6232255
8		Sample Varia	1570
9		Kurtosis	0.838675
10		Skewness	0.60763436
11		Range	115
12		Minimum	135
13		Maximum	250
14		Sum	1110
15		Count	6

Figure 4.94
After computing the mean, standard deviation, and more on Excel

Exercises

28. *Money* magazine claimed that "state regulation (of the insurance industry) is a joke." The article "ranks the states by the percentage of legislators with links to insurers who sit on committees focusing on insurance issues." See Figure 4.95. An unformatted spreadsheet containing these data can be downloaded from the text web site. **www.thomsonedu.com/mathematics/johnson**
   a. Find the mean of these data.
   b. Find the standard deviation.
   c. Did you choose the population or sample standard deviation? Why?
   d. What do your results say about state regulation of the insurance industry? Interpret the results of both parts (a) and (b).

29. Do Double Stuf Oreos really contain twice as much "stuf"? In 1999, Marie Revak and Jihan William's "interest was piqued by the unusual spelling of the word stuff and the bold statement 'twice the filling' on the package." So they conducted an experiment in which they weighed the amount of filling in a sample of traditional Oreos and Double Stuf Oreos.

The data from that experiment are in Figure 4.96 (weight in grams). An unformatted spreadsheet containing these data can be downloaded from the text web site. **www.thomsonedu.com/mathematics/johnson**
   a. Find the mean of these data.
   b. Find the standard deviation.
   c. Did you choose the population or sample standard deviation? Why?
   d. What do your results say about Double Stuf Oreos? Interpret the results of both parts (a) and (b).

Mississippi	40%	Louisiana	38%	Arkansas	37%
Virginia	35%	North Carolina	33%	Missouri	32%
Florida	31%	Alabama	26%	Georgia	25%
Texas	25%	West Virginia	24%	Indiana	21%
Ohio	18%	Minnesota	18%	Iowa	17%
North Dakota	17%	Pennsylvania	15%	Arizona	14%
Wyoming	14%	Illinois	14%	Kentucky	13%
Idaho	13%	Delaware	13%	Utah	12%
Kansas	12%	Wisconsin	10%	Hawaii	10%
New Mexico	10%	Rhode Island	9%	Maryland	9%
New Jersey	8%	New York	8%	South Carolina	8%
Washington	8%	Tennessee	8%	Maine	8%
Oklahoma	8%	New Hampshire	7%	Connecticut	5%

(unlisted states were ranked 0%)

Figure 4.95
Percentage of legislators who regulate insurers and have links to the insurance industry
Source: Money, August 1996

Double Stuf Oreos:																		
4.7	6.5	5.5	5.6	5.1	5.3	5.4	5.4	3.5	5.5	6.5	5.9	5.4	4.9	5.6	5.7	5.3	6.9	6.5
6.3	4.8	3.3	6.4	5.0	5.3	5.5	5.0	6.0	5.7	6.3	6.0	6.3	6.1	6.0	5.8	5.8	5.9	6.2
5.9	6.5	6.5	6.1	5.8	6.0	6.2	6.2	6.0	6.8	6.2	5.4	6.6	6.2					

Traditional Oreos:																		
2.9	2.8	2.6	3.5	3.0	2.4	2.7	2.4	2.5	2.2	2.6	2.6	2.9	2.6	2.6	3.1	2.9	2.4	2.8
3.8	3.1	2.9	3.0	2.1	3.8	3.0	3.0	2.8	2.9	2.7	3.2	2.8	3.1	2.7	2.8	2.6	2.6	3.0
2.8	3.5	3.3	3.3	2.8	3.1	2.6	3.5	3.5	3.1	3.1								

Figure 4.96
Amount of Oreo filling
Source: Revak, Marie A. and Jihan G. Williams, "Sharing Teaching Ideas: The Double Stuf Dilemma," The Mathematics Teacher, November 1999, Volume 92, Issue 8, page 674

30. Figure 4.49 on pages 244–245 gives EPA fuel efficiency ratings for 2003 compact automobiles with an automatic transmission and the smallest available engine. An unformatted spreadsheet containing this data can be downloaded from the text web site. **www.thomsonedu.com/mathematics/johnson**
   a. For each of the three manufacturing regions, find the mean and an appropriate standard deviation for city driving, and defend your choice of standard deviations.
   b. Analyze the results of part (a). What do they tell you about the three manufacturing regions?
   c. For each of the three manufacturing regions, find the median for city driving.
   d. Analyze the results of part (c). What do they tell you about the three manufacturing regions?

31. Figure 4.49 on pages 244–245 gives EPA fuel efficiency ratings for 2003 compact automobiles with an automatic transmission and the smallest available engine. An unformatted spreadsheet containing this data can be downloaded from the text web site.
   a. For each of the three manufacturing regions, find the mean and an appropriate standard deviation for highway driving, and defend your choice of standard deviations.
   b. Analyze the results of part (a). What do they tell you about the three manufacturing regions?
   c. For each of the three manufacturing regions, find the median for highway driving.
   d. Analyze the results of part (c). What do they tell you about the three manufacturing regions?

4.4 The Normal Distribution

Sets of data may exhibit various trends or patterns. Figure 4.97 shows a histogram of the weights of bags of corn chips. Notice that most of the data are near the "center" and that the data taper off at either end. Furthermore, the histogram is nearly symmetric; it is almost the same on both sides. This type of distribution (nearly symmetric, with most of the data in the middle) occurs quite often in many different situations. To study the composition of such distributions, statisticians have created an ideal **bell-shaped curve** describing a **normal distribution,** as shown in Figure 4.98.

Before we can study the characteristics and applications of a normal distribution, we must make a distinction between different types of variables.

Discrete versus Continuous Variables

The number of children in a family is variable, because it varies from family to family. In listing the number of children, only whole numbers (0, 1, 2, and so on) can be used.

Figure 4.97
Weights of bags of corn chips

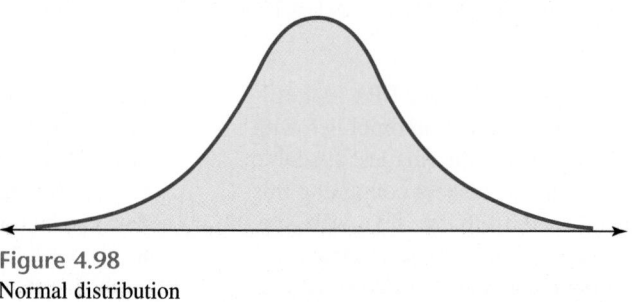

Figure 4.98
Normal distribution

In this respect, we are limited to a collection of discrete, or separate, values. A variable is **discrete** if there are "gaps" between each possible variable value. Consequently, any variable that involves counting is discrete.

On the other hand, a person's height or weight does not have such a restriction. When someone grows, he or she does not instantly go from 67 inches to 68 inches; a person grows continuously from 67 inches to 68 inches, attaining all possible values in between. For this reason, height is called a continuous variable. A variable is **continuous** if it can assume *any value* in an interval of real numbers. Consequently, any variable that involves measurement is continuous; someone might claim to be 67 inches tall and to weigh 152 pounds, but the true values might be 67.13157 inches and 151.87352 pounds. Heights and weights are expressed (discretely) as whole numbers solely for convenience; most people do not have rulers or bathroom scales that allow them to obtain measurements that are accurate to ten or more decimal places!

Normal Distributions

The collection of all possible values that a discrete variable can assume forms a countable set. For instance, we can list all the possible numbers of children in a family. In contrast, a continuous variable will have an uncountable number of possibilities because it can assume any value in an interval. For instance, the weights (a continuous variable) of bags of corn chips could be *any* value x such that $15.3 \leq x \leq 16.7$.

When we sample a continuous variable, some values may occur more often than others. As we can see in Figure 4.97, the weights are "clustered" near the center of the histogram, with relatively few located at either end. If a continuous variable has a symmetric distribution such that the highest concentration of values is at the center and the lowest is at both extremes, the variable is said to have a **normal**

distribution and is represented by a smooth, continuous, bell-shaped curve like that in Figure 4.98.*

The normal distribution, which is found in a wide variety of situations, has two main qualities: (1) the frequencies of the data points nearer the center or "average" are increasingly higher than the frequencies of data points far from the center, and (2) the distribution is symmetric (one side is a mirror image of the other). *Because of these two qualities, the mean, median, and mode of a normal distribution all coincide at the center of the distribution.*

Just like any other collection of numbers, the spread of normal distribution is measured by its standard deviation. It can be shown that for any normal distribution, slightly more than two-thirds of the data (68.26%) will lie within one standard deviation of the mean, 95.44% will lie within two standard deviations, and virtually all the data (99.74%) will lie within three standard deviations of the mean. Recall that μ (the Greek letter "mu") represents the mean of a population and σ (the Greek letter "sigma") represents the standard deviation of the population. The spread of a normal distribution, with μ and σ used to represent the mean and standard deviation, is shown in Figure 4.99.

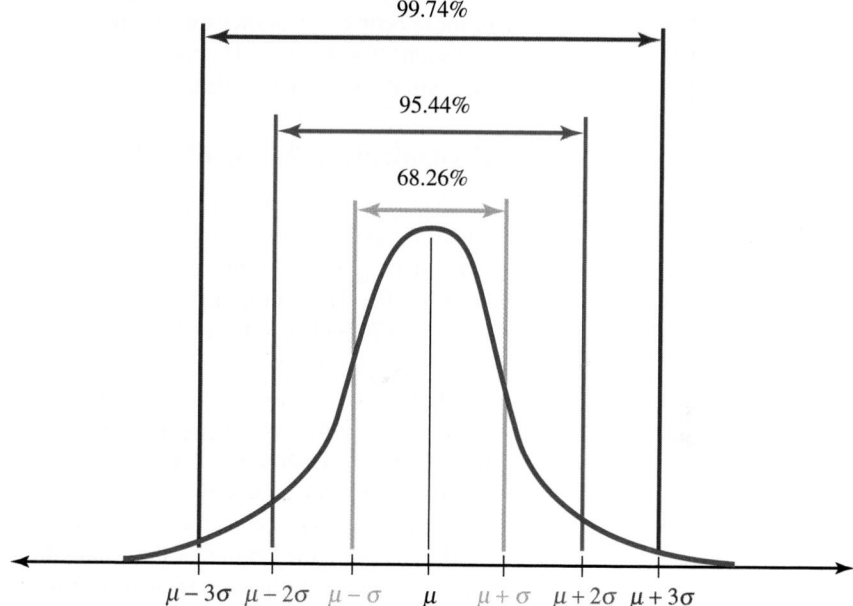

Figure 4.99
The spread of a normal distribution

Thomson NOW
for Liberal Arts Mathematics
www.thomsonedu.com/1pass.

EXAMPLE 1   The heights of a large group of people are assumed to be normally distributed. Their mean height is 66.5 inches, and the standard deviation is 2.4 inches. Find and interpret the intervals representing one, two, and three standard deviations of the mean.

Solution   The mean is $\mu = 66.5$, and the standard deviation is $\sigma = 2.4$.

1. *One standard deviation of the mean:*

$$\mu \pm 1\sigma = 66.5 \pm 1(2.4)$$
$$= 66.5 \pm 2.4$$
$$= [64.1, 68.9]$$

Therefore, approximately 68% of the people are between 64.1 and 68.9 inches tall.

---

* This is an informal definition only. The formal definition of a normal distribution involves the number pi, the natural exponential e^x, and the mean and variance of the distribution.

2. *Two standard deviations of the mean:*

$$\mu \pm 2\sigma = 66.5 \pm 2(2.4)$$
$$= 66.5 \pm 4.8$$
$$= [61.7, 71.3]$$

Therefore, approximately 95% of the people are between 61.7 and 71.3 inches tall.

3. *Three standard deviations of the mean:*

$$\mu \pm 3\sigma = 66.5 \pm 3(2.4)$$
$$= 66.5 \pm 7.2$$
$$= [59.3, 73.7]$$

Nearly all of the people (99.74%) are between 59.3 and 73.7 inches tall.   ■

In Example 1, we found that virtually all the people under study were between 59.3 and 73.7 inches tall. A clothing manufacturer might want to know what percent of these people are shorter than 66 inches or what percent are taller than 73 inches. Questions like these can be answered by using probability and a normal distribution. (We will do this in Example 6.)

Probability, Area, and Normal Distributions

In Chapter 3, we mentioned that relative frequency is really a type of probability. If 3 out of every 100 people have red hair, you could say that the relative frequency of red hair is $\frac{3}{100}$ (or 3%), or you could say that the probability of red hair $p(x = \text{red hair})$ is 0.03. Therefore, to find out what percent of the people in a population are taller than 73 inches, we need to find $p(x > 73)$, the probability that x is greater than 73, where x represents the height of a randomly selected person.

Recall that a sample space is the set S of all possible outcomes of a random experiment. Consequently, the probability of a sample space must always equal 1; that is, $p(S) = 1$ (or 100%). If the sample space S has a normal distribution, its outcomes and their respective probabilities can be represented by a bell curve.

Recall that when constructing a histogram, relative frequency density (*rfd*) was used to measure the heights of the rectangles. Consequently, the *area* of a rectangle gave the relative frequency (percent) of data contained in an interval. In a similar manner, we can imagine a bell curve being a histogram composed of infinitely many "skinny" rectangles, as in Figure 4.100.

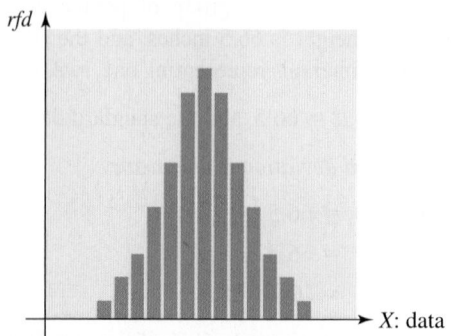

Figure 4.100
Symmetric, bell-shaped histogram

For a normal distribution, the outcomes nearer the center of the distribution occur more frequently than those at either end; the distribution is denser in the middle and sparser at the extremes. This difference in density is taken into account by consideration of the area under the bell curve; the center of the distribution is denser, contains more area, and has a higher probability of occurrence than the extremes. Consequently, we use the area under the bell curve to represent the probability of an outcome. Because $p(S) = 1$, we define the entire area under the bell curve to equal 1.

Because a normal distribution is symmetric, 50% of the data will be greater than the mean, and 50% will be less. (The mean and the median coincide in a symmetric distribution.) Therefore, the probability of randomly selecting a number x greater than the mean is $p(x > \mu) = 0.5$, and that of selecting a number x less than the mean is $p(x < \mu) = 0.5$, as shown in Figure 4.101.

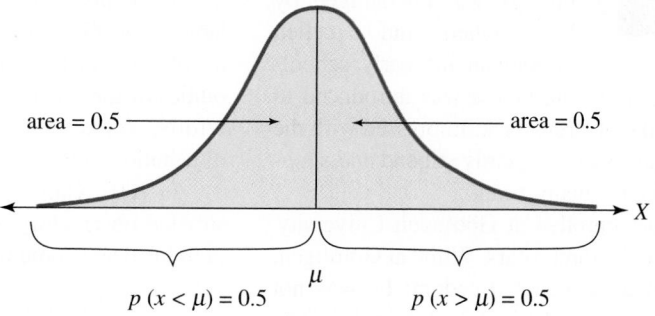

Figure 4.101
Total probability equals 1 (or 100%)

To find the probability that a randomly selected number x is between two values (say a and b), we must determine the area under the curve from a to b; that is, $p(a < x < b) =$ area under the bell curve from $x = a$ to $x = b$, as shown in Figure 4.102(a). Likewise, the probability that x is greater than or less than any specific number is given by the area of the tail, as shown in Figure 4.102(b). To find probabilities involving data that are normally distributed, we must find the area of the appropriate region under the bell curve.

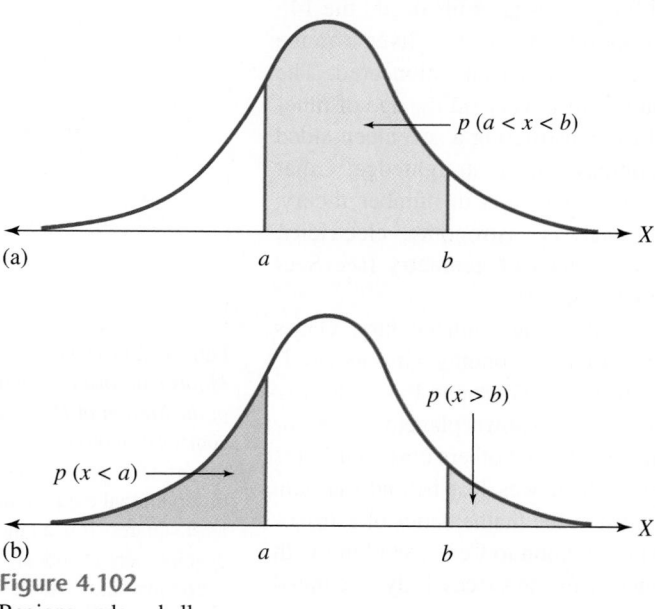

Figure 4.102
Regions under a bell curve

Historical Note   Carl Friedrich Gauss, 1777–1855

© akg-images

Dubbed "the Prince of Mathematics," Carl Gauss is considered by many to be one of the greatest mathematicians of all time. At the age of three, Gauss is said to have discovered an arithmetic error in his father's bookkeeping. The child prodigy was encouraged by his teachers and excelled throughout his early schooling. When he was fourteen, Gauss was introduced to Ferdinand, the Duke of Brunswick. Impressed with the youth, the duke gave Gauss a yearly stipend and sponsored his education for many years.

In 1795, Gauss enrolled at Göttingen University, where he remained for three years. While at Göttingen, Gauss had complete academic freedom; he was not required to attend lectures, he had no required conferences with professors or tutors, and he did not take exams. Much of his time was spent studying independently in the library. For reasons unknown to us, Gauss left the university in 1798 without a diploma. Instead, he sent his dissertation to the University of Helmstedt and in 1799 was awarded his degree without the usual oral examination.

In 1796, Gauss began his famous mathematical diary. Discovered forty years after his death, the 146 sometimes cryptic entries exhibit the diverse range of topics that Gauss pondered and pioneered. The first entry was Gauss's discovery (at the age of nineteen) of a method for constructing a seventeen-sided polygon with a compass and a straightedge. Other entries include important results in number theory, algebra, calculus, analysis, astronomy, electricity, magnetism, the foundations of geometry (see Section 6.7), and probability.

At the dawn of the nineteenth century, Gauss began his lifelong study of astronomy. On January 1, 1801, the Italian astronomer Giuseppe Piazzi discovered Ceres, the first of the known planetoids (minor planets or asteroids). Piazzi and others observed Ceres for forty-one days, until it was lost behind the sun. Because of his interest in the mathematics of astronomy, Gauss turned his attention to Ceres. Working with a minimum amount of data, he successfully calculated the orbit of Ceres. At the end of the year, the planetoid was rediscovered in exactly the spot that Gauss had predicted!

To obtain the orbit of Ceres, Gauss utilized his method of least squares, a technique for dealing with experimental error. Letting x represent the error between an experimentally obtained value and the true value it represents, Gauss's theory involved minimizing x^2—that is, obtaining the least square of the error. Theorizing that the probability of a small error was higher than that of a large error, Gauss subsequently developed the normal distribution, or bell-shaped curve, to explain the probabilities of the random errors. Because of his pioneering efforts, some mathematicians refer to the normal distribution as the Gaussian distribution.

In 1807, Gauss became director of the newly constructed observatory at Göttingen; he held the position until his death some fifty years later.

© Stock Montage

Published in 1809, Gauss's *Theoria Motus Corporum Coelestium (Theory of the Motion of Heavenly Bodies)* contained rigorous methods of determining the orbits of planets and comets from observational data via the method of least squares. It is a landmark in the development of modern mathematical astronomy and statistics.

The Standard Normal Distribution

All normal distributions share the following features: they are symmetric, bell-shaped curves, and virtually all the data (99.74%) lie within three standard deviations of the mean. Depending on whether the standard deviation is large or small, the bell curve will be either flat and spread out or peaked and narrow, as shown in Figure 4.103.

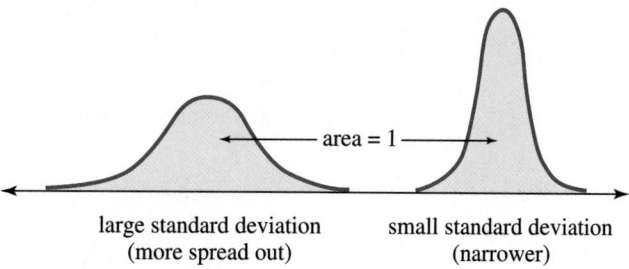

large standard deviation   small standard deviation
(more spread out)   (narrower)

Figure 4.103
Large versus small standard deviation

To find the area under any portion of any bell curve, mathematicians have devised a means of comparing the proportions of any curve with the proportions of a special curve defined as "standard." To find probabilities involving normally distributed data, we utilize the bell curve associated with the standard normal distribution.

The **standard normal distribution** is the normal distribution whose mean is 0 and standard deviation is 1, as shown in Figure 4.104. The standard normal distribution is also called the *z*-**distribution;** we will always use the letter *z* to refer to the standard normal. By convention, we will use the letter *x* to refer to any other normal distribution.

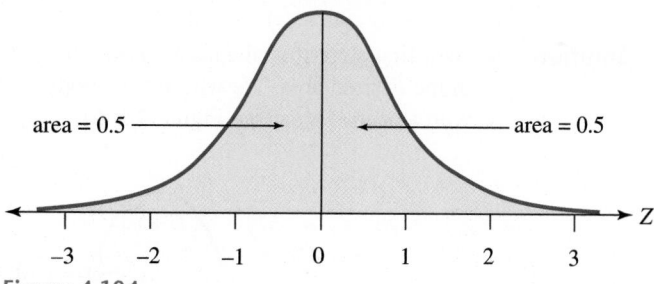

Figure 4.104
The standard normal distribution (mean = 0, standard deviation = 1)

Tables have been developed for finding areas under the standard normal curve using the techniques of calculus. Graphing calculators will also give these areas. We will use the table in Appendix F to find $p(0 < z < z^*)$, the probability that z is between 0 and a positive number z^*, as shown in Figure 4.105(a). The table in Appendix F is known as the **body table** because it gives the probability of an interval located in the middle, or body, of the bell curve.

The tapered end of a bell curve is known as a **tail.** To find the probability of a tail—that is, to find $p(z > z^*)$ or $p(z < z^*)$ where z^* is a positive real number—subtract the probability of the corresponding body from 0.5, as shown in Figure 4.105(b).

ThomsonNOW™
for Liberal Arts Mathematics
www.thomsonedu.com/1pass.

EXAMPLE 2   Find the following probabilities (that is, the areas), where z represents the standard normal distribution.

a. $p(0 < z < 1.25)$ **b.** $p(z > 1.87)$

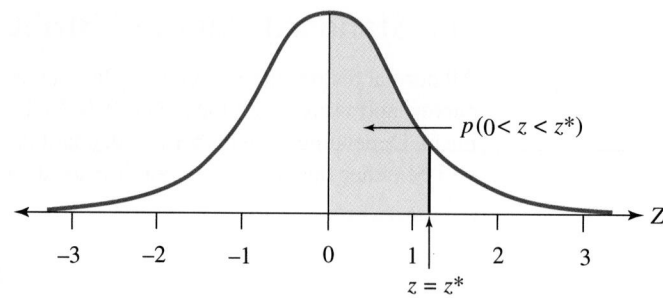

(a) Area found by using the body table (Appendix F)

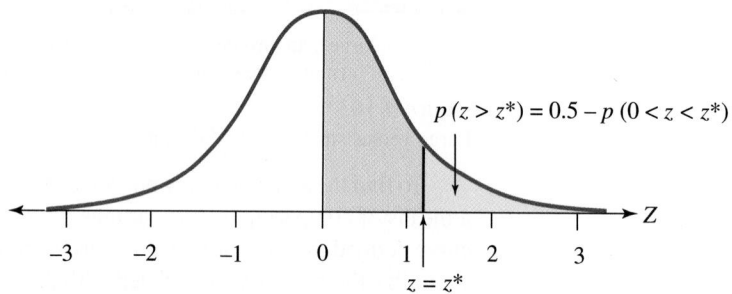

(b) Area of a tail, found by subtracting the corresponding
body area from 0.5

Figure 4.105

Solution   **a.** As a first step, it is always advisable to draw a picture of the z-curve and shade
in the desired area. We will use the body table directly, because we are working
with a central area (see Figure 4.106).

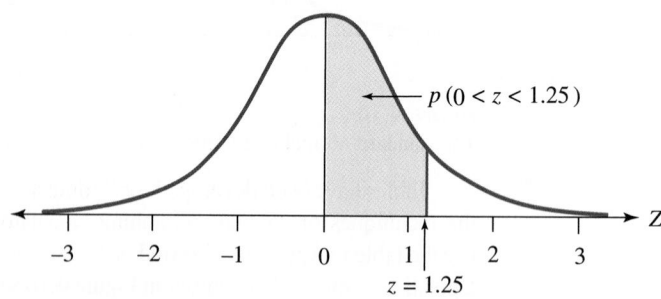

Figure 4.106
A central region, or body

The z-numbers are located along the left edge and the top of the table.
Locate the whole number and the first-decimal-place part of the number (1.2)
along the left edge; then locate the second-decimal-place part of the number
(0.05) along the top. The desired probability (area) is found at the intersection
of the row and column of the two parts of the z-number. Thus, $p(0 < z < 1.25) = 0.3944$, as shown in Figure 4.107.

z^*	0.00	0.01	0.02	0.03	0.04	0.05	0.06	0.07	0.08	0.09
o										
o										
o										
1.1	0.3643	0.3665	0.3686	0.3708	0.3729	0.3749	0.3770	0.3790	0.3810	0.3830
1.2	0.3849	0.3869	0.3888	0.3907	0.3925	0.3944	0.3962	0.3980	0.3997	0.4015
1.3	0.4032	0.4049	0.4066	0.4082	0.4099	0.4115	0.4131	0.4147	0.4162	0.4177
o										
o										

Figure 4.107
A portion of the body table

Hence, we could say that about 39% of the z-distribution lies between $z = 0$ and $z = 1.25$.

b. To find the area of a tail, we subtract the corresponding body area from 0.5, as shown in Figure 4.108. Therefore,

$$p(z > 1.87) = 0.5 - p(0 < z < 1.87)$$
$$= 0.5 - 0.4692$$
$$= 0.0308$$

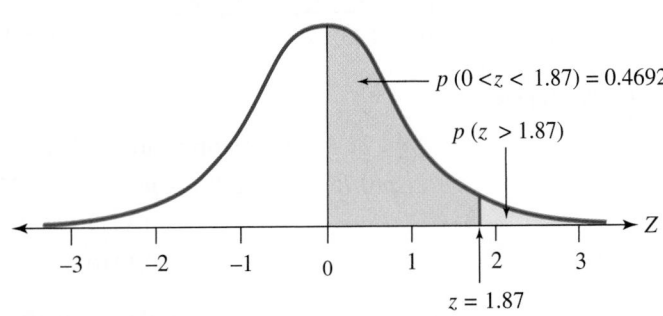

Figure 4.108
Finding the area of a tail ■

The body table can also be used to find areas other than those given explicitly as $p(0 < z < z^*)$ and $p(z > z^*)$ where z^* is a positive number. By adding or subtracting two areas, we can find probabilities of the type $p(a < z < b)$, where a and b are positive or negative numbers, and probabilities of the type $p(z < c)$, where c is a positive or negative number.

EXAMPLE 3   Find the following probabilities (the areas), where z represents the standard normal distribution.

a. $p(0.75 < z < 1.25)$ **b.** $p(-0.75 < z < 1.25)$

Solution   **a.** Because the required region, shown in Figure 4.109, doesn't begin exactly at $z = 0$, we cannot look up the desired area directly in the body table. Whenever z is between two nonzero numbers, we will take an indirect approach to finding the required area.

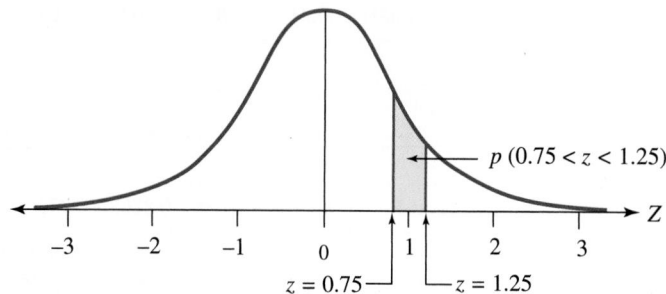

Figure 4.109
A strip

The total area under the curve from $z = 0$ to $z = 1.25$ can be divided into two portions: the area under the curve from 0 to 0.75 and the area under the curve from 0.75 to 1.25.

To find the area of the "strip" between $z = 0.75$ and $z = 1.25$, we *subtract* the area of the smaller body (from $z = 0$ to $z = 0.75$) from that of the larger body (from $z = 0$ to $z = 1.25$), as shown in Figure 4.110.

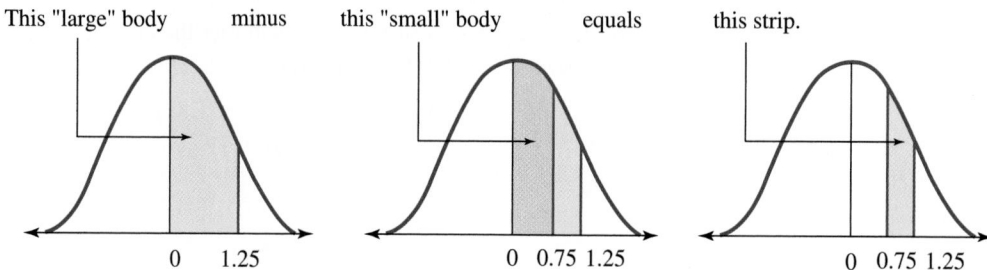

Figure 4.110
Area of a strip

$$\text{area of strip} = \text{area of large body} - \text{area of small body}$$
$$p(0.75 < z < 1.25) = p(0 < z < 1.25) - p(0 < z < 0.75)$$
$$= 0.3944 - 0.2734$$
$$= 0.1210$$

Therefore, $p(0.75 < z < 1.25) = 0.1210$. Hence, we could say that about 12.1% of the z-distribution lies between $z = 0.75$ and $z = 1.25$.

b. The required region, shown in Figure 4.111, can be divided into two regions: the area from $z = -0.75$ to $z = 0$ and the area from $z = 0$ to $z = 1.25$.

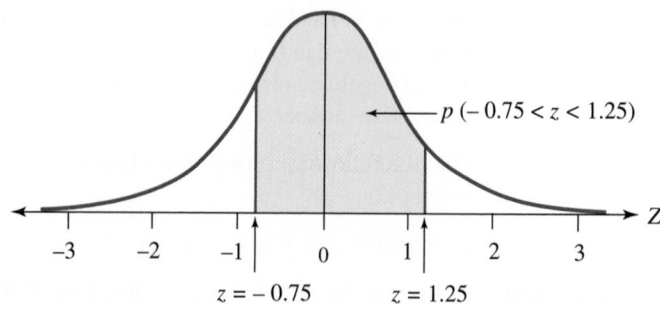

Figure 4.111
A central region

To find the total area of the region between $z = -0.75$ and $z = 1.25$, we *add* the area of the "left" body (from $z = -0.75$ to $z = 0$) to the area of the "right" body (from $z = 0$ to $z = 1.25$), as shown in Figure 4.112.

This total region   equals   this "left" body   plus   this "right" body.

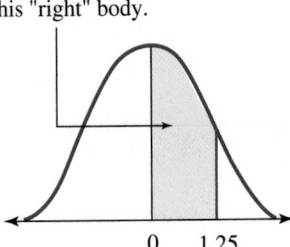

$$-0.75 \quad 0 \quad 1.25 \qquad\qquad -0.75 \quad 0 \qquad\qquad 0 \quad 1.25$$

Figure 4.112
Left body plus right body

This example is different from our previous examples in that it contains a negative z-number. A glance at the tables reveals that negative numbers are not included! However, recall that normal distributions are symmetric. Therefore, the area of the body from $z = -0.75$ to $z = 0$ is the same as that from $z = 0$ to $z = 0.75$; that is, $p(-0.75 < z < 0) = p(0 < z < 0.75)$. Therefore,

$$\text{total area of region} = \text{area of left body} + \text{area of right body}$$
$$p(-0.75 < z < 1.25) = p(-0.75 < z < 0) + p(0 < z < 1.25)$$
$$= p(0 < z < 0.75) + p(0 < z < 1.25)$$
$$= 0.2734 + 0.3944$$
$$= 0.6678$$

Therefore, $p(-0.75 < z < 1.25) = 0.6678$. Hence, we could say that about 66.8% of the z-distribution lies between $z = -0.75$ and $z = 1.25$. ∎

EXAMPLE 4   Find the following probabilities (the areas), where z represents the standard normal distribution.

a. $p(z < 1.25)$ **b.** $p(z < -1.25)$

Solution   **a.** The required region is shown in Figure 4.113. Because 50% of the distribution lies to the left of 0, we can add 0.5 to the area of the body from $z = 0$ to

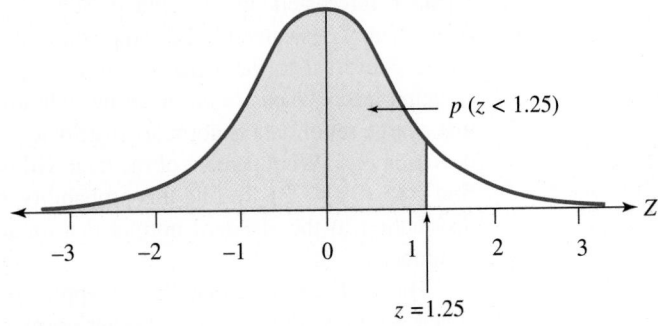

$$p(z < 1.25)$$

$$-3 \quad -2 \quad -1 \quad 0 \quad 1 \quad 2 \quad 3 \qquad Z$$

$$z = 1.25$$

Figure 4.113
A body plus 50%

$z = 1.25$:

$$p(z < 1.25) = p(z < 0) + p(0 < z < 1.25)$$
$$= 0.5 + 0.3944$$
$$= 0.8944$$

Therefore, $p(z < 1.25) = 0.8944$. Hence, we could say that about 89.4% of the z-distribution lies to the left of $z = 1.25$.

b. The required region is shown in Figure 4.114. Because a normal distribution is symmetric, the area of the left tail ($z < -1.25$) is the same as the area of the corresponding right tail ($z > 1.25$).

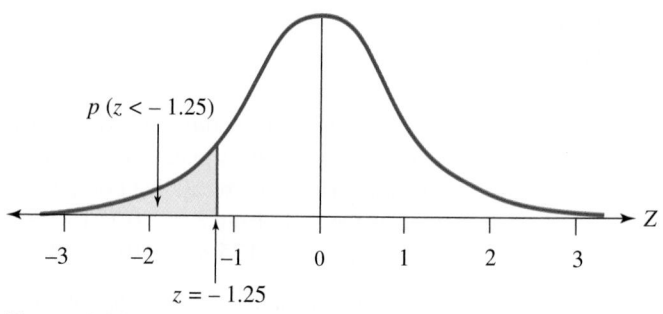

Figure 4.114
A left tail

Therefore,

$$p(z < -1.25) = p(z > 1.25)$$
$$= 0.5 - p(0 < z < 1.25)$$
$$= 0.5 - 0.3944$$
$$= 0.1056$$

Hence, we could say that about 10.6% of the z-distribution lies to the left of $z = -1.25$.   ∎

Converting to the Standard Normal

Weather forecasters in the United States usually report temperatures in degrees Fahrenheit. Consequently, if a temperature is given in degrees Celsius, most people would convert it to Fahrenheit in order to judge how hot or cold it was. A similar situation arises when we are working with a normal distribution. Suppose we know that a large set of data is normally distributed with a mean value of 68 and a standard deviation of 4. What percent of the data will lie between 65 and 73? We are asked to find $p(65 < x < 73)$. To find this probability, we must first convert the given normal distribution to the standard normal distribution and then look up the approximate z-numbers.

The body table (Appendix F) applies to the standard normal z-distribution. When we are working with any other normal distribution (denoted by X), we must first convert the x-distribution into the standard normal z-distribution. This conversion is done with the help of the following rule.

> **Converting a Normal Distribution into the Standard Normal z**
>
> Every number x in a given normal distribution has a corresponding number z in the standard normal distribution. The z-**number** that corresponds to the number x is
>
> $$z = \frac{x - \mu}{\sigma}$$
>
> where μ is the mean and σ the standard deviation of the given normal distribution.

Given a number x, its corresponding z-number counts the number of standard deviations the number lies from the mean. For example, suppose the mean and standard deviation of a normal distribution are $\mu = 68$ and $\sigma = 4$. The z-number corresponding to $x = 78$ is

$$z = \frac{x - \mu}{\sigma} = \frac{78 - 68}{4} = 2.5$$

This implies that $x = 78$ lies two and one-half standard deviations above the mean, 68. Similarly, for $x = 65$,

$$z = \frac{65 - 68}{4} = -0.75$$

Therefore, $x = 65$ lies three-quarters of a standard deviation below the mean, 68.

ThomsonNOW™
for Liberal Arts Mathematics
www.thomsonedu.com/1pass.

EXAMPLE 5   Suppose a population is normally distributed with a mean of 24.6 and a standard deviation of 1.3. What percent of the data will lie between 25.3 and 26.8?

Solution   We are asked to find $p(25.3 < x < 26.8)$, the area of the region shown in Figure 4.115. Because we need to find the area of the strip between 25.3 and 26.8, we must find the body of each and subtract, as in part (a) of Example 3.

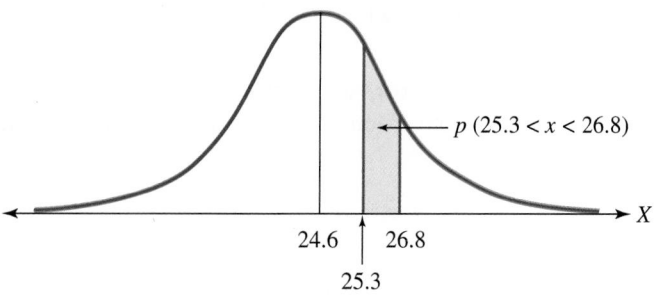

Figure 4.115
A strip

Using the Conversion Formula $z = (x - \mu)/\sigma$ with $\mu = 24.6$ and $\sigma = 1.3$, we first convert $x = 25.3$ and $x = 26.8$ into their corresponding z-numbers.

Converting $x = 25.3$

$$z = \frac{x - \mu}{\sigma}$$

$$= \frac{25.3 - 24.6}{1.3}$$

$$= 0.5384615$$

$$= 0.54$$

Converting $x = 26.8$

$$z = \frac{x - \mu}{\sigma}$$

$$= \frac{26.8 - 24.6}{1.3}$$

$$= 1.6923077$$

$$= 1.69$$

rounding off z-numbers to two decimal places

Therefore,

$$p(25.3 < x < 26.8) = p(0.54 < z < 1.69)$$
$$= p(0 < z < 1.69) - p(0 < z < 0.54)$$
$$= 0.4545 - 0.2054 \quad \text{using the body table}$$
$$= 0.2491$$

Assuming a normal distribution, approximately 24.9% of the data will lie between 25.3 and 26.8. ∎

EXAMPLE 6   The heights of a large group of people are assumed to be normally distributed. Their mean height is 68 inches, and the standard deviation is 4 inches. What percentage of these people are the following heights?

a. taller than 73 inches   **b.** between 60 and 75 inches

Solution   **a.** Let x represent the height of a randomly selected person. We need to find $p(x > 73)$, the area of a tail, as shown in Figure 4.116.

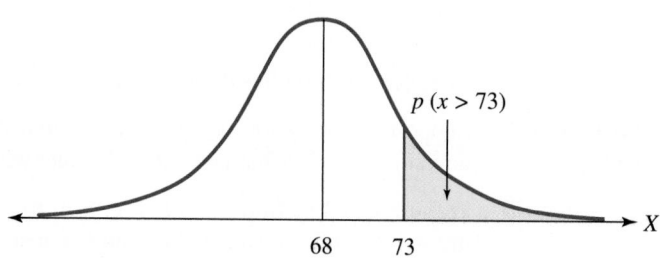

Figure 4.116
A right tail

First, we must convert $x = 73$ to its corresponding z-number. Using the Conversion Formula with $x = 73$, $\mu = 68$, and $\sigma = 4$, we have

$$z = \frac{x - \mu}{\sigma}$$
$$= \frac{73 - 68}{4}$$
$$= 1.25$$

Therefore,

$$p(x > 73) = p(z > 1.25)$$
$$= 0.5 - p(0 < z < 1.25)$$
$$= 0.5 - 0.3944$$
$$= 0.1056$$

Approximately 10.6% of the people will be taller than 73 inches.
b. We need to find $p(60 < x < 75)$, the area of the central region shown in Figure 4.117. Notice that we will be adding the areas of the two bodies.

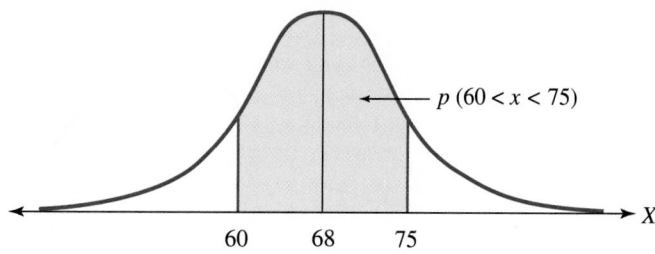

Figure 4.117
A central region

First, we convert $x = 60$ and $x = 75$ to their corresponding z-numbers:

$$p(60 < x < 75) = p\left(\frac{60 - 68}{4} < z < \frac{75 - 68}{4}\right) \qquad \text{using the Conversion Formula}$$

$$= p(-2.00 < z < 1.75)$$

$$= p(-2.00 < z < 0) + p(0 < z < 1.75) \qquad \text{expressing the area as two bodies}$$

$$= p(0 < z < 2.00) + p(0 < z < 1.75) \qquad \text{using symmetry}$$

$$= 0.4772 + 0.4599 \qquad \text{using the body table}$$

$$= 0.9371$$

Approximately 93.7% of the people will be between 60 and 75 inches tall.   ∎

All the preceding examples involved finding probabilities that contained only the strict $<$ or $>$ inequalities, never $\leq$ or $\geq$ inequalities; the endpoints were never included. What if the endpoints are included? How does $p(a < x < b)$ compare with $p(a \leq x \leq b)$? Because probabilities for continuous data are found by determining *area* under a curve, including the endpoints does not affect the probability! The probability of a single point $p(x = a)$ is 0, because there is no "area" over a single point. (We obtain an area only when we are working with an interval of numbers.) Consequently, if x represents continuous data, then $p(a \leq x \leq b) = p(a < x < b)$; it makes no difference whether the endpoints are included.

EXAMPLE 7   Tall Dudes is a clothing store that specializes in fashions for tall men. Its informal motto is "Our customers are taller than 80% of the rest." Assuming the heights of men to be normally distributed with a mean of 67 inches and a standard deviation of 5.5 inches, find the heights of Tall Dudes' clientele.

Solution   Let $c =$ the height of the shortest customer at Tall Dudes, and let x represent the height of a randomly selected man. We are given that the heights of all men are normally distributed with $\mu = 67$ and $\sigma = 5.5$.

Assuming Tall Dudes' clientele to be taller than 80% of all men implies that $x < c$ 80% of the time and $x > c$ 20% of the time. Hence, we can say that the probability of selecting someone shorter than the shortest tall dude is $p(x < c) = 0.80$ and that the probability of selecting a tall dude is $p(x > c) = 0.20$, as shown in Figure 4.118.

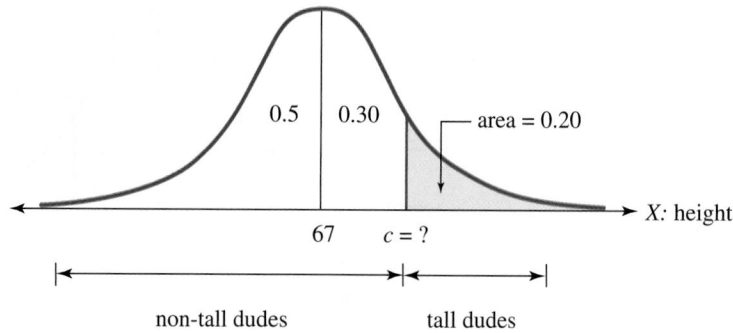

Figure 4.118
Finding a 20% right tail

We are given that the area of the right tail is 0.20 (and the right body is 0.30), and we need to find the appropriate cutoff number c. This is exactly the reverse of all the previous examples, in which we were given the cutoff numbers and asked to find the area. Thus, our goal is to find the z-number that corresponds to a body of area 0.30 and convert it into its corresponding x-number.

When we scan through the *interior* of the body table, the number closest to the desired area of 0.30 is 0.2995, which is the area of the body when $z = 0.84$. This means that $p(0 < z < 0.84) = 0.2995$, as shown in Figure 4.119.

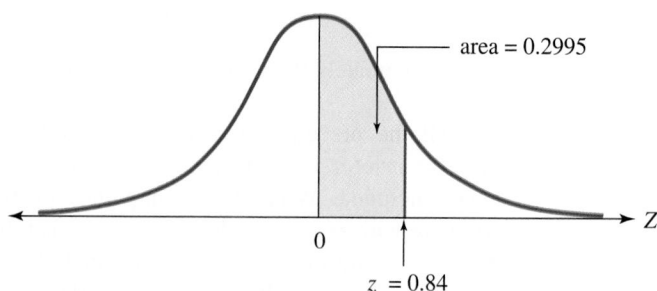

Figure 4.119
A body close to 30%

Therefore, the number c that we are seeking lies 0.84 standard deviations above the mean. All that remains is to convert $z = 0.84$ into its corresponding x-number by substituting $x = c$, $z = 0.84$, $\mu = 67$, and $\sigma = 5.5$ into the Conversion Formula:

$$z = \frac{x - \mu}{\sigma}$$

$$0.84 = \frac{c - 67}{5.5}$$

$$(5.5)(0.84) = c - 67$$

$$4.62 = c - 67$$

$$c = 71.62 \quad (\approx 72 \text{ inches, or 6 feet})$$

Therefore, Tall Dudes caters to men who are at least 71.62 inches, or about 6 feet tall. ∎

4.4   Exercises

1. The weights (in ounces) of several bags of corn chips are given in Figure 4.120. Construct a histogram for the data using the groups $15.40 \leq x < 15.60$, $15.60 \leq x < 15.80, \dots, 16.60 \leq x < 16.80$. Do the data appear to be approximately normally distributed? Explain.

16.08	16.49	15.61	16.66	15.80	15.87
16.02	15.82	16.48	16.08	15.63	16.02
16.00	16.25	15.41	16.22	16.04	15.68
16.45	16.41	16.01	15.82	16.08	15.82
16.29	16.26	16.05	16.25	15.86	

Figure 4.120
Weights (in ounces) of bags of corn chips

2. The weights (in grams) of several bags of chocolate chip cookies are given in Figure 4.121. Construct a histogram for the data using the groups $420 \leq x < 430$, $430 \leq x < 440, \dots, 480 \leq x < 490$. Do the data appear to be approximately normally distributed? Explain.

451	435	482	449	454	451
479	448	432	423	461	475
467	453	448	459	454	444
475	461	450	466	446	458

Figure 4.121
Weights (in grams) of bags of chocolate chip cookies

3. The time (in minutes) spent waiting in line for several students at the campus bookstore are given in Figure 4.122. Construct a histogram for the data using the groups $0 \leq x < 1, 1 \leq x < 2, \dots, 4 \leq x < 5$. Do the data appear to be normally distributed? Explain.

1.75	2	0.5	1.5	0
1	2.5	0.75	2.25	3
1.5	3.5	2.5	1	0.5
1.75	0.25	4.5	0	1.5

Figure 4.122
Time (in minutes) spent waiting in line

4. A die was rolled several times, and the results are given in Figure 4.123. Construct a histogram for the data using the single-values 1, 2, 3, 4, 5, and 6. Do the data appear to be normally distributed? Explain.

5	3	1	1	4	2
6	4	1	2	5	5
4	3	1	6	4	2
3	1	2	5	2	1
4	6	5	1	4	5

Figure 4.123
Rolling a die

5. What percent of the standard normal z-distribution lies between the following values?
    a. $z = 0$ and $z = 1$
    b. $z = -1$ and $z = 0$
    c. $z = -1$ and $z = 1$ (*Note:* This interval represents one standard deviation of the mean.)

6. What percent of the standard normal z-distribution lies between the following values?
    a. $z = 0$ and $z = 2$
    b. $z = -2$ and $z = 0$
    c. $z = -2$ and $z = 2$ (*Note:* This interval represents two standard deviations of the mean.)

7. What percent of the standard normal z-distribution lies between the following values?
    a. $z = 0$ and $z = 3$
    b. $z = -3$ and $z = 0$
    c. $z = -3$ and $z = 3$ (*Note:* This interval represents three standard deviations of the mean.)

8. What percent of the standard normal z-distribution lies between the following values?
    a. $z = 0$ and $z = 1.5$
    b. $z = -1.5$ and $z = 0$
    c. $z = -1.5$ and $z = 1.5$ (*Note:* This interval represents one and one-half standard deviations of the mean.)

9. A population is normally distributed with mean 24.7 and standard deviation 2.3.
    a. Find the intervals representing one, two, and three standard deviations of the mean.
    b. What percentage of the data lies in each of the intervals in part (a)?
    c. Draw a sketch of the bell curve.

10. A population is normally distributed with mean 18.9 and standard deviation 1.8.
    a. Find the intervals representing one, two, and three standard deviations of the mean.
    b. What percentage of the data lies in each of the intervals in part (a)?
    c. Draw a sketch of the bell curve.

11. Find the following probabilities.
    a. $p(0 < z < 1.62)$ b. $p(1.30 < z < 1.84)$
    c. $p(-0.37 < z < 1.59)$ d. $p(z < -1.91)$
    e. $p(-1.32 < z < -0.88)$ f. $p(z < 1.25)$

12. Find the following probabilities.
    a. $p(0 < z < 1.42)$ b. $p(1.03 < z < 1.66)$
    c. $p(-0.87 < z < 1.71)$ d. $p(z < -2.06)$
    e. $p(-2.31 < z < -1.18)$ f. $p(z < 1.52)$

13. Find c such that each of the following is true.
    a. $p(0 < z < c) = 0.1331$
    b. $p(c < z < 0) = 0.4812$
    c. $p(-c < z < c) = 0.4648$
    d. $p(z > c) = 0.6064$
    e. $p(z > c) = 0.0505$
    f. $p(z < c) = 0.1003$

14. Find c such that each of the following is true.
    a. $p(0 < z < c) = 0.3686$
    b. $p(c < z < 0) = 0.4706$
    c. $p(-c < z < c) = 0.2510$
    d. $p(z > c) = 0.7054$
    e. $p(z > c) = 0.0351$
    f. $p(z < c) = 0.2776$

15. A population X is normally distributed with mean 250 and standard deviation 24. For each of the following values of x, find the corresponding z-number. Round off your answers to two decimal places.
    a. $x = 260$ b. $x = 240$
    c. $x = 300$ d. $x = 215$
    e. $x = 321$ f. $x = 197$

16. A population X is normally distributed with mean 72.1 and standard deviation 9.3. For each of the following values of x, find the corresponding z-number. Round off your answers to two decimal places.
    a. $x = 90$ b. $x = 80$
    c. $x = 75$ d. $x = 70$
    e. $x = 60$ f. $x = 50$

17. A population is normally distributed with mean 36.8 and standard deviation 2.5. Find the following probabilities.
    a. $p(36.8 < x < 39.3)$ b. $p(34.2 < x < 38.7)$
    c. $p(x < 40.0)$ d. $p(32.3 < x < 41.3)$
    e. $p(x = 37.9)$ f. $p(x > 37.9)$

18. A population is normally distributed with mean 42.7 and standard deviation 4.7. Find the following probabilities.
    a. $p(42.7 < x < 47.4)$ b. $p(40.9 < x < 44.1)$
    c. $p(x < 50.0)$ d. $p(33.3 < x < 52.1)$
    e. $p(x = 45.3)$ f. $p(x > 45.3)$

19. The mean weight of a box of cereal filled by a machine is 16.0 ounces, with a standard deviation of 0.3 ounce. If the weights of all the boxes filled by the machine are normally distributed, what percent of the boxes will weigh the following amounts?
    a. less than 15.5 ounces
    b. between 15.8 and 16.2 ounces

20. The amount of time required to assemble a component on a factory assembly line is normally distributed with a mean of 3.1 minutes and a standard deviation of 0.6 minute. Find the probability that a randomly selected employee will take the given amount of time to assemble the component.
    a. more than 4.0 minutes
    b. between 2.0 and 2.5 minutes

21. The time it takes an acrylic paint to dry is normally distributed. If the mean is 2 hours 36 minutes with a standard deviation of 24 minutes, find the probability that the drying time will be as follows.
    a. less than 2 hours 15 minutes
    b. between 2 and 3 hours

    HINT: Convert everything to minutes (or to hours).

22. The shrinkage in length of a certain brand of blue jeans is normally distributed with a mean of 1.1 inches and a standard deviation of 0.2 inch. What percent of this brand of jeans will shrink the following amounts?
    a. more than 1.5 inches
    b. between 1.0 and 1.25 inches

23. The mean volume of a carton of milk filled by a machine is 1.0 quart, with a standard deviation of 0.06 quart. If the volumes of all the cartons are normally distributed, what percent of the cartons will contain the following amounts?
    a. at least 0.9 quart
    b. at most 1.05 quarts

24. The amount of time between taking a pain reliever and getting relief is normally distributed with a mean of 23 minutes and a standard deviation of 4 minutes. Find the probability that the time between taking the medication and getting relief is as follows.
    a. at least 30 minutes
    b. at most 20 minutes

25. The results of a statewide exam for assessing the mathematics skills of realtors were normally distributed with a mean score of 72 and a standard deviation of 12. The realtors who scored in the top 10% are to receive a special certificate, while those in the bottom 20% will be required to attend a remedial workshop.
   a. What score does a realtor need in order to receive a certificate?
   b. What score will dictate that the realtor attend the workshop?

26. Professor Harde assumes that exam scores are normally distributed and wants to grade "on the curve." The mean score was 58, with a standard deviation of 16.
   a. If she wants 14% of the students to receive an A, find the minimum score to receive an A.
   b. If she wants 19% of the students to receive a B, find the minimum score to receive a B.

27. The time it takes an employee to package the components of a certain product is normally distributed with $\mu = 8.5$ minutes and $\sigma = 1.5$ minutes. To boost productivity, management has decided to give special training to the 34% of employees who took the greatest amount of time to package the components. Find the amount of time taken to package the components that will indicate that an employee should get special training.

28. The time it takes an employee to package the components of a certain product is normally distributed with $\mu = 8.5$ and $\sigma = 1.5$ minutes. As an incentive, management has decided to give a bonus to the 20% of employees who took the shortest amount of time to package the components. Find the amount of time taken to package the components that will indicate that an employee should get a bonus.

▬▬▬▶ *Answer the following questions using complete sentences and your own words.*

CONCEPT QUESTIONS
29. What are the characteristics of a normal distribution?
30. Are all distributions of data normally distributed? Support your answer with an example.
31. Why is the total area under a bell curve equal to 1?
32. Why are there no negative z-numbers in the body table?
33. When converting an x-number to a z-number, what does a negative z-number tell you about the location of the x-number?
34. Is it logical to assume that the heights of all high school students in the United States are normally distributed? Explain.
35. Is it reasonable to assume that the ages of all high school students in the United States are normally distributed? Explain.

HISTORY QUESTIONS
36. Who is known as "the Prince of Mathematics"? Why?
37. What mathematician was instrumental in the creation of the normal distribution? What application prompted this person to create the normal distribution?

4.5 Polls and Margin of Error

One of the most common applications of statistics is the evaluation of the results of surveys and public opinion polls. Most editions of the daily newspaper contain the results of at least one poll. Headlines announce the attitude of the nation toward a myriad of topics ranging from the actions of politicians to controversial current issues as shown in the newspaper article on the next page. How are these conclusions reached? What do they mean? How valid are they? In this section, we investigate these questions and obtain results concerning the "margin of error" associated with the reporting of "public opinion."

Sampling and Inferential Statistics

The purpose of conducting a survey or poll is to obtain information about a population—for example, adult Americans. Because there are approximately 200 million Americans over the age of eighteen, it would be very difficult, time-consuming, and expensive to

Abortion Views Hold Steady Over Past Year

PUBLIC MAKES SHARP DISTINCTIONS ABOUT ABORTION CIRCUMSTANCES BY LYDIA SAAD

Gallup News Service

Gallup's annual reading of American's basic attitudes on abortion, conducted May 5–7, 2003, finds little change compared with attitudes a year ago. The public appears to be ambivalent about abortion, calling it morally wrong, but widely accepting its use in certain situations. Americans remain closely divided over the leading abortion debate position labels: 48% consider themselves "pro-choice" and 45% call themselves "pro-life."

Most Americans hold one of two middle-of-the road positions on abortion. A large plurality (42%) holds the view that abortion should be legal, but only in a few circumstances. Another 15% of American believe abortion should be legal under most, but not all, circumstances. Then, at the extremes, Gallup finds about a quarter, 23%, believing abortion should be legal under all circumstances. Nearly as many, 19%, take the opposite view, saying it should be illegal in all circumstances.

As Gallup has noted many times before, the strongest differences in perceptions of the morality of abortion are seen along religious lines. In the latest survey, the perception that abortion is morally wrong ranges from 78% among those who attend religious services weekly, to 58% among those who attend often, but not weekly, to only 28% among those who seldom or never attend.

Beyond the religious factor, Gallup finds women slightly more likely to believe abortion is morally wrong than men are (56% of women call it morally wrong vs. 49% of men). Young adults are less likely to consider abortion morally wrong than those over age 30 are. Residents of the East and West are less likely to believe abortion is morally wrong than residents in the Midwest and South are.

Although Democrats tend to be pro-choice, and Republicans tend to be pro-life, the two partisan groups are not all that far apart in their perceptions of the morality of abortion. A solid majority of Republicans consider abortion morally wrong, while Democrats are divided on the question.

Survey Methods

Results are based on telephone interviews with 1,005 national adults, aged 18 and older, conducted May 5–7, 2003. For results based on the total sample of national adults, one can say with 95% confidence that the margin of sampling error is ± 3 percentage points.

contact every one of them. The only realistic alternative is to poll a sample and use the science of inferential statistics to draw conclusions about the population as a whole. Different samples have different characteristics depending on, among other things, the age, sex, education, and locale of the people in the sample. Therefore, it is of the utmost importance that a sample be representative of the population. Obtaining a representative sample is the most difficult aspect of inferential statistics.

Another problem facing pollsters is determining *how many* people should be selected for the sample. Obviously, the larger the sample, the more likely that it will reflect the population. However, larger samples cost more money, so a limited budget will limit the sample size. Conducting surveys can be very costly, even for a small to moderate sample. For example, a survey conducted in 1989 by the Gallup Organization that contacted 1,005 adults and 500 teenagers would have cost $100,000 (the pollsters donated their services for this survey). The results of this poll indicated that Americans thought the "drug crisis" was the nation's top problem (stated by 27% of the adults and 32% of the teenagers).

After a sample has been selected and its data analyzed, information about the sample is generalized to the entire population. Because 27% of the 1,005 adults in a poll stated that the drug crisis was the nation's top problem, we would like to conclude that 27% of *all* adults have the same belief. Is this a valid generalization? That is, how confident is the pollster that the feelings of the people in the sample reflect those of the population?

Sample Proportion versus Population Proportion

If x members (for example, people, automobiles, households) in a sample of size n have a certain characteristic, then the proportion of the sample, or **sample proportion,** having this characteristic is given by $\frac{x}{n}$. For instance, in a sample of $n = 70$ automobiles, if $x = 14$ cars have a defective fan switch, then the proportion of the sample having a defective switch is $\frac{14}{70} = 0.2$, or 20%. The true proportion of the entire population, or **population proportion,** having the characteristic is represented by the letter P. A sample proportion $\frac{x}{n}$ is an estimate of the population proportion P.

Sample proportions $\frac{x}{n}$ vary from sample to sample; some will be larger than P, and some will be smaller. Of the 1,005 adults in the Gallup Poll sample mentioned above, 27% viewed the drug crisis as the nation's top problem. If a different sample of 1,005 had been chosen, 29% might have had this view. If still another 1,005 had been selected, this view might have been shared by only 25%. We will assume that the sample proportions $\frac{x}{n}$ are normally distributed around the population proportion P. The set of all sample proportions, along with their probabilities of occurring, can be represented by a bell curve like the one in Figure 4.124.

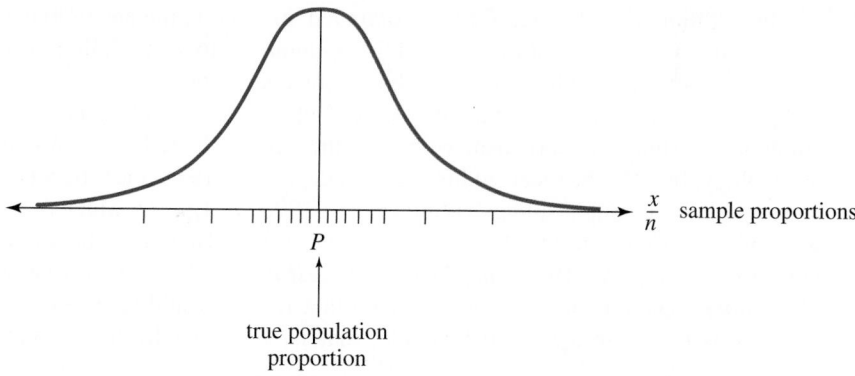

Figure 4.124
Sample proportions normally distributed around the true (population) proportion

In general, a sample estimate is not 100% accurate; although a sample proportion might be close to the true population proportion, it will have an error term associated with it. The difference between a sample estimate and the true (population) value is called the **error of the estimate.** We can use a bell curve (like the one in Figure 4.124) to predict the probable error of a sample estimate.

Before developing this method of predicting the error term, we need to introduce some special notation. The symbol z_α (read "z alpha") will be used to represent the positive z-number that has a right body of area α. That is, z_α is the number such that $p(0 < z < z_\alpha) = \alpha$, as shown in Figure 4.125.

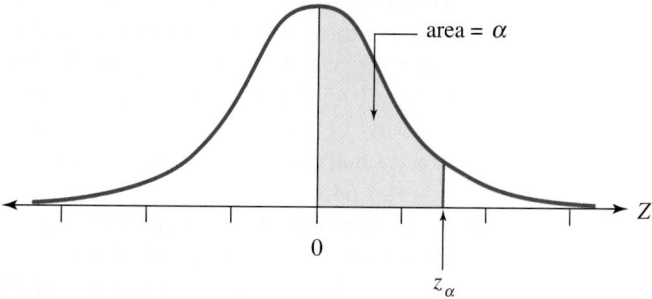

Figure 4.125
$p(0 < z < z_\alpha) = \alpha$

Historical Note   George H. Gallup, 1901–1984

© Corbis/Bettman

To many people, the name *Gallup* is synonymous with opinion polls. George Horace Gallup, the founder of the American Institute of Public Opinion, began his news career while attending the University of Iowa. During his junior year as a student of journalism, Gallup became the editor of his college newspaper, the *Daily Iowan*. After receiving his bachelor's degree in 1923, Gallup remained at the university nine years as an instructor of journalism.

In addition to teaching, Gallup continued his own studies of human nature and public opinion. Interest in how the public reacts to advertisements and perceives various issues of the day led Gallup to combine his study of journalism with the study of psychology. In 1925, he received his master's degree in psychology. Gallup's studies culminated in 1928 with his doctoral thesis, *A New Technique for Objective Methods for Measuring Reader Interest in Newspapers*. Gallup's new technique of polling the public was to utilize a stratified sample, that is, a

sample that closely mirrors the composition of the entire population. Gallup contended that a stratified sample of 1,500 people was sufficient to obtain reliable estimates. For his pioneering work in this new field, Gallup was awarded his Ph.D. in journalism in 1928.

Gallup founded the American Institute of Public Opinion in 1935 with the stated purpose "to impartially measure and report public opinion on political and social issues of the day without regard to the rightness or wisdom of the views expressed." His first triumph was his prediction of the winner of the 1936 presidential election between Franklin D. Roosevelt and Alfred Landon. While many, including the prestigious *Literary Digest,* predicted Landon to win, Gallup correctly predicted Roosevelt as the winner.

Gallup Polls have correctly predicted all presidential elections since, with the exception of the 1948 race between Thomas Dewey and Harry S Truman. Much to his embarrassment, Gallup predicted Dewey as the winner. Truman won the election with 49.9% of the vote, while Gallup had predicted that he would receive only 44.5%. Gallup's explanation was that he had ended his poll too far in advance of

EXAMPLE 1   Use the body table to find the following values:

    a. $z_{0.3925}$ **b.** $z_{0.475}$ **c.** $z_{0.45}$ **d.** $z_{0.49}$

Solution   **a.** $z_{0.3925}$ represents the z-number that has a body of area 0.3925. Looking at the interior of the body table, we find 0.3925 and see that it corresponds to the z-number 1.24—that is, $p(0 < z < 1.24) = 0.3925$. Therefore, $z_{0.3925} = 1.24$.

    b. In a similar manner, we find that a body of area 0.4750 corresponds to $z = 1.96$. Therefore, $z_{0.475} = 1.96$.

    c. Looking through the interior of the table, we cannot find a body of area 0.45. However, we do find a body of 0.4495 (corresponding to $z = 1.64$) and a body of 0.4505 (corresponding to $z = 1.65$). Because the desired body is *exactly halfway* between the two listed bodies, we use a z-number that is exactly halfway between the two listed z-numbers, 1.64 and 1.65. Therefore, $z_{0.45} = 1.645$.

    d. We cannot find a body of the desired area, 0.49, in the interior of the table. The closest areas are 0.4898 (corresponding to $z = 2.32$) and 0.4901 (corresponding to $z = 2.33$). Because the desired area (0.49) is *closer* to 0.4901, we use $z = 2.33$. Therefore, $z_{0.49} = 2.33$. ∎

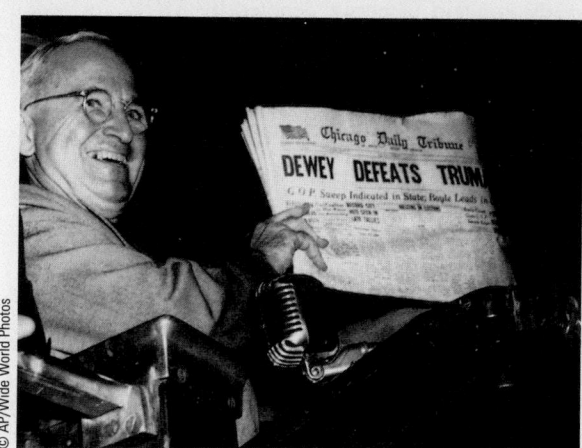

The Gallup Organization predicted that Thomas Dewey would win the 1948 presidential election. Much to Gallup's embarrassment, Harry S. Truman won the election and triumphantly displayed a newspaper containing Gallup's false prediction. With the exception of this election, Gallup Polls have correctly predicted every presidential election since 1936.

Although some people criticize the use of polls, citing their potential influence and misuse, Gallup considered the public opinion poll to be "one of the most useful instruments of democracy ever devised." Answering the charge that he and his polls influenced elections, Gallup retorted, "One might as well insist that a thermometer makes the weather!" In addition, Gallup confessed that he had not voted in a presidential election since 1928. Above all, Gallup wanted to ensure the impartiality of his polls.

Besides polling people regarding their choices in presidential campaigns, Gallup was the first pollster to ask the public to rate a president's performance and popularity. Today, these "presidential report cards" are so common we may take them for granted. In addition to presidential politics, Gallup also dealt with sociological issues, asking questions such as "What is the most important problem facing the country?"

Polling has become a multimillion-dollar business. In 1983, the Gallup Organization had revenues totaling $6.7 million. Today, Gallup Polls are syndicated in newspapers across the country and around the world.

election day and had disregarded the votes of those who were undecided. Of his error, Gallup said, "We are continually experimenting and continually learning."

Margin of Error

Sample proportions $\frac{x}{n}$ vary from sample to sample; some will have a small error, and some will have a large error. Knowing that sample estimates have inherent errors, statisticians make predictions concerning the largest possible error associated with a sample estimate. This error is called the **margin of error** of the estimate and is denoted by **MOE.** Because the margin of error is a prediction, we cannot guarantee that it is absolutely correct; that is, the probability that a prediction is correct might be 0.95, or it might be 0.75.

In the field of inferential statistics, the probability that a prediction is correct is referred to as the **level of confidence** of the prediction. For example, we might say that we are 95% confident that the maximum error of an opinion poll is plus or minus 3 percentage points; that is, if 100 samples were analyzed, 95 of them would have proportions that differ from the true population proportion by an amount less than or equal to 0.03, and 5 of the samples would have an error greater than 0.03.

Assuming that sample proportions are normally distributed around the population proportion (as in Figure 4.124), we can use the z-distribution to determine the margin of error associated with a sample proportion. In general, the margin of error depends on the sample size and the level of confidence of the estimate.

> **Margin of Error Formula**
>
> Given a sample size n, the **margin of error**, denoted by **MOE**, for a poll involving sample proportions is
>
> $$\text{MOE} = \frac{z_{\alpha/2}}{2\sqrt{n}}$$
>
> where α represents the level of confidence of the poll. That is, the probability is α that the sample proportion has an error of at most MOE.

ThomsonNOW
for Liberal Arts Mathematics
www.thomsonedu.com/1pass.

EXAMPLE 2 Assuming a 90% level of confidence, find the margin of error associated with each of the following:

a. sample size $n = 275$ **b.** sample size $n = 750$

Solution

a. The margin of error depends on two things: the sample size and the level of confidence. For a 90% level of confidence, $\alpha = 0.90$. Hence, $\frac{\alpha}{2} = 0.45$, and $z_{\alpha/2} = z_{0.45} = 1.645$.

Substituting this value and $n = 275$ into the MOE formula, we have the following:

$$\text{MOE} = \frac{z_{\alpha/2}}{2\sqrt{n}}$$
$$= \frac{1.645}{2\sqrt{275}}$$
$$= 0.049598616\ldots$$
$$\approx 0.050 \qquad \text{rounding off to three decimal places}$$
$$= 5.0\%$$

When we are polling a sample of 275 people, we can say that we are 90% confident that the maximum possible error in the sample proportion will be plus or minus 5.0 percentage points.

b. For a 90% level of confidence, $\alpha = 0.90$, $\frac{\alpha}{2} = 0.45$, and $z_{\alpha/2} = z_{0.45} = 1.645$. Substituting this value and $n = 750$ into the MOE formula, we have the following:

$$\text{MOE} = \frac{z_{\alpha/2}}{2\sqrt{n}}$$
$$= \frac{1.645}{2\sqrt{750}}$$
$$= 0.030033453\ldots$$
$$\approx 0.030 \qquad \text{rounding off to three decimal places}$$
$$= 3.0\%$$

When we are polling a sample of 750 people, we can say that we are 90% confident that the maximum possible error in the sample proportion will be plus or minus 3.0 percentage points. ∎

If we compare the margins of error in parts (a) and (b) of Example 2, we notice that by increasing the sample size (from 275 to 750), the margin of error was reduced (from 5.0 to 3.0 percentage points). Intuitively, this should make sense; a larger sample gives a better estimate (has a smaller margin of error).

EXAMPLE 3 To obtain an estimate of the proportion of all Americans who think the president is doing a good job, a random sample of 500 Americans is surveyed, and 345 respond, "The president is doing a good job."

a. Determine the sample proportion of Americans who think the president is doing a good job.
b. Assuming a 95% level of confidence, find the margin of error associated with the sample proportion.

Solution

a. $n = 500$ and $x = 345$. The sample proportion is

$$\frac{x}{n} = \frac{345}{500} = 0.69$$

Sixty-nine percent of the sample think the president is doing a good job.

b. We must find MOE when $n = 500$ and $\alpha = 0.95$. Because $\alpha = 0.95$, $\frac{\alpha}{2} = 0.475$ and $z_{\alpha/2} = z_{0.475} = 1.96$.
Therefore,

$$MOE = \frac{z_{\alpha/2}}{2\sqrt{n}}$$

$$= \frac{1.96}{2\sqrt{500}}$$

$$= 0.043826932\ldots$$

$$\approx 0.044 \qquad \text{rounding off to three decimal places}$$

$$= 4.4\%$$

The margin of error associated with the sample proportion is plus or minus 4.4 percentage points. We are 95% confident that 69% (±4.4%) of all Americans think the president is doing a good job. In other words, on the basis of our sample proportion ($\frac{x}{n}$) of 69%, we predict (with 95% certainty) that the true population proportion (P) is somewhere between 64.6% and 73.4%. ∎

EXAMPLE 4 The article shown on the next page was released by the Gallup Organization in June 1999.

a. The poll states that 79% of the Americans questioned said they favored the registration of all firearms. Assuming a 95% level of confidence (the most commonly used level of confidence), find the margin of error associated with the survey.
b. Assuming a 98% level of confidence, find the margin of error associated with the survey.

Solution

a. We must find MOE when $n = 1,022$ and $\alpha = 0.95$. Because $\alpha = 0.95$, $\alpha/2 = 0.475$ and $z_{\alpha/2} = z_{0.475} = 1.96$.

Americans Support Wide Variety of Gun Control Measures
Eight out of ten favor registration of all firearms

Poll Releases, June 16, 1999
by Frank Newport
© 1999 The Gallup Organization.
All rights reserved, used with permission.

PRINCETON, NJ—Motivated in part by the tragic school shootings in Colorado and Georgia this spring, the idea of instituting new and more stringent gun control laws has become one of the most hotly debated issues on the current national agenda. A new Gallup poll shows that Americans strongly support most of the specific types of gun control measures now being debated in Congress, and that support for the idea of a general requirement that all firearms be registered is now higher than it was as recently as last fall.

The "registration of all firearms" proposition now receives somewhat stronger support than in the past. The current higher level of support—79%—may reflect the influence of this spring's tragic school shootings in Colorado and in Georgia.

There are some differences by party in support for the measure, but even among Republicans, support is strong, at 71%, compared to 88% Democrats. Women are also somewhat more likely than men to support registration of all firearms, by an 88% to 69% margin.

An interesting finding: more than sixty years ago, in 1938, Gallup asked Americans about their support for registration of all "pistols and revolvers" and found 84% support, suggesting that the current sentiment in favor of many of these gun measures does not necessarily represent a wholesale change in the attiudes of the public over the ensuing decades.

The results above are based on telephone interviews with a randomly selected national sample of 1,022 adults, 18 years and older, conducted June 11–13, 1999.

Therefore,

$$\text{MOE} = \frac{z_{\alpha/2}}{2\sqrt{n}}$$

$$= \frac{1.96}{2\sqrt{1,022}}$$

$$= 0.0306549511 \ldots$$

$$\approx 0.031 \qquad \text{rounding off to three decimal places}$$

The margin of error associated with the survey is plus or minus 3.1%. We are 95% confident that 79% ± 3.1% of all Americans favor the registration of all firearms.

b. We must find MOE when $n = 1,022$ and $\alpha = 0.98$. Because $\alpha = 0.98$, $\alpha/2 = 0.49$ and $z_{\alpha/2} = z_{0.49} = 2.33$.

Therefore,

$$\text{MOE} = \frac{z_{\alpha/2}}{2\sqrt{n}}$$

$$= \frac{2.33}{2\sqrt{1,022}}$$

$$= 0.0364418551 \ldots$$

$$\approx 0.036 \qquad \text{rounding off to three decimal places}$$

The margin of error associated with the survey is plus or minus 3.6%. We are 98% confident that 79% ± 3.6% of all Americans favor the registration of all firearms. ■

Who Supports Marijuana Legalization?
Support rising; varies most by age and gender

By Joseph Carroll,
Gallup Poll Assistant Editor

Since the late 1960s, Gallup has periodically asked Americans whether the use of marijuana should be made legal in the United States. Although a majority of Americans have consistently opposed the idea of legalizing marijuana, public support has slowly increased over the years. In 1969, just 12% of Americans supported making marijuana legal, but by 1977, roughly one in four endorsed it. Support edged up to 31% in 2000, and now, about a third of Americans say marijuana should be legal.

Support for marijuana legalization varies greatest by gender and age. Overall, younger Americans (aged 18 to 29) are essentially divided, with 47% saying marijuana should be legal and 50% saying it should not be. Support for legalization is much lower among adults aged 30 to 64 (35%) and those aged 65 and older (22%). Men (39%) are somewhat more likely than women (30%) to support the legalization of marijuana in the country.

Americans residing in the western parts of the country are more likely than those living elsewhere to support the legalization of marijuana. These differences perhaps result from the fact that six Western states have, in various ways, already legalized marijuana for medicinal use. Overall, the data show that Westerners are divided about marijuana, with 47% saying it should be legal and 49% saying it should not be. No more than a third of adults living in other parts of the country feel marijuana should be legal.

Support for legalizing marijuana is much lower among Republicans than it is among Democrats or independents. One in five Republicans (21%) say marijuana should be made legal in this country, while 37% of Democrats and 44% of independents share this view.

*Results are based on telephone interviews with 2,034 national adults, aged 18 and older, conducted Aug. 3–5, 2001, Nov. 10–12, 2003, and Oct. 21–23, 2005. For results based on the total sample of national adults, one can say with 95% confidence that the maximum margin of sampling error is ±2 percentage points.

If we compare the margins of error found in parts (a) and (b) of Example 4, we notice that as the level of confidence went up (from 95% to 98%) the margin of error increased (from 3.1% to 3.6%). Intuitively, if we want to be more confident in our predictions, we should give our prediction more leeway (a larger margin of error).

When the results of the polls are printed in a newspaper, the sample size, level of confidence, margin of error, date of survey, and location of survey may be given as a footnote, as shown in the above article.

EXAMPLE 5   Verify the margin of error stated in the article shown above.

Solution   The footnote to the article states that for a sample size of 2,034 and a 95% level of confidence, the margin of error is ±2%, that is, MOE = 0.02 for $n = 2,034$ and $\alpha = 0.95$. Because $\alpha = 0.95$, $\frac{\alpha}{2} = 0.475$ and $z_{\alpha/2} = z_{0.475} = 1.96$. Therefore,

$$\text{MOE} = \frac{z_{\alpha/2}}{2\sqrt{n}}$$

$$= \frac{1.96}{2\sqrt{2,034}}$$

$$= 0.02195127\ldots$$

$$\approx 0.02 \qquad \text{rounding off to two decimal places}$$

Therefore, the stated margin of error of ±2% is correct.   ■

To Many Americans, UFOs Are Real and Have Visited Earth in Some Form

The Roper Poll
UFOs & Extraterrestrial Life
Americans' Beliefs and Personal Experiences
(Prepared for the SCI FI Channel—September 2002)

Most Americans appear comfortable with and even excited about the thought of the discovery of extraterrestrial life. More than half (56 percent) of the American public think that UFOs are something real and not just in people's imagination. Nearly as many (48 percent) believe that UFOs have visited earth in some form. Males are significantly more likely to believe in the reality of UFOs, as are those under the age of 65. A significant drop is witnessed in the percentage of believers among the 65+ age group.

Two-thirds (67 percent) of adults think there are other forms of intelligent life in the universe. This belief tends to be more prevalent among males, adults ages 64 or younger, and residents of the Northeast as opposed to North Central and South.

In the view of many adults (55 percent), the government does not share enough information with the public in general. An even greater proportion (roughly seven in ten) thinks that the government does not tell us everything it knows about extraterrestrial life and UFOs. The younger the age, the stronger the belief that the government is withholding information about these topics.

This study was conducted by RoperASW. The sample consists of 1,021 male and female adults (in approximately equal number), all 18 years of age and over. The telephone interviews were conducted from August 23 through August 25, 2002, using a Random Digit Dialing (RDD) probability sample of all telephone households in the continental United States. The margin of error for the total sample is ±3 percent.

News articles do not always mention the level of confidence of a survey. However, if the sample size and margin of error are given, the level of confidence can be determined, as shown in Example 6.

EXAMPLE 6   Do you think that UFOs are real? The news article shown above presents the results of a Roper poll pertaining to this question. Find the level of confidence of this poll.

Solution   We are given $n = 1,021$ and MOE $= 0.03$. To find α, the level of confidence of the poll, we must first find $z_{\alpha/2}$ and the area of the bodies, as shown in Figure 4.126.

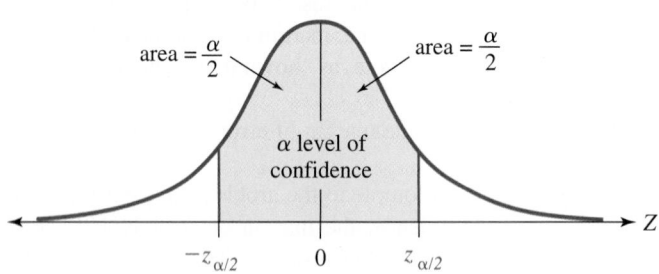

Figure 4.126
Level of confidence equals central region

Substituting the given values into the MOE formula, we have

$$\text{MOE} = \frac{z_{\alpha/2}}{2\sqrt{n}}$$

$$0.03 = \frac{z_{\alpha/2}}{2\sqrt{1,021}}$$

Statistics in the Real World: Random Sampling in Opinion Polls

National opinion polls are pervasive in today's world; hardly a day goes by without the release of yet another glimpse at the American psyche. Topics ranging from presidential performance and political controversy to popular culture and alien abduction vie for our attention in the media: "A recent survey indicates that a majority of Americans. . . ." Where do these results come from? Are they accurate? Can they be taken seriously, or are they mere entertainment and speculation?

Although many people are intrigued by the "findings" of opinion polls, others are skeptical or unsure of the fundamental premise of statistical sampling. How can the opinions of a diverse population of nearly 300 million Americans be measured by a survey of a mere 1,000 to 1,500 individuals? Wouldn't a survey of 10,000 people give better, or more accurate, results than those obtained from the typical sample of about a thousand people? The answer is "no, not necessarily." Although intuition might dictate that a survey's reliability is driven by the size of the sample (bigger is better), in reality, the most important factor in obtaining reliable results is the *method* by which the sample was selected. That is, depending on how it was selected, a sample of 1,000 people can yield far better results than a sample of 10,000, 20,000, or even 50,000 people.

The basic premise in statistical sampling is that the views of a small portion of a population *can* accurately represent the views of the entire population *if* the sample is selected properly, that is, if the sample is selected randomly. So what does it mean to say that a sample is selected "randomly"? The answer is simple: A sample is random if every member of the population has an *equal chance* of being selected. For example, suppose that a remote island has a population of 1,234 people and we wish to select a random sample of 30 inhabitants to interview. We could write each islander's name on a slip of paper, put all of the slips in a large box, shake up the box, close our eyes, and select 30 slips. Each person on the island would have an equal chance of being selected; consequently, we would be highly confident (say, 95%) that the views of the sample (plus or minus a margin or error) would accurately represent the views of the entire population.

The key to reliable sampling is the selection of a random sample; to select a random sample, each person in the population must have an equal chance of being selected. Realistically, how can this be accomplished with such a vast population of people in the United States or elsewhere? Years ago, the most accurate polls (especially Gallup polls) were based on data gathered from knocking on doors and conducting face-to-face interviews. However, in today's world, almost every American adult has a telephone, and telephone surveys have replaced the door-to-door surveys of the past. Although a poll might state that the target population is "all Americans aged 18 and over," it really means "all Americans aged 18 and over who have an accessible telephone number." Who will be excluded from this population? Typically, active members of the military forces and people in prisons, institutions, or hospitals are excluded from the sampling frame of today's opinion polls.

How can a polling organization obtain a list of the telephone numbers of all Americans aged 18 and over? Typically, no such list exists. Telephone directories are not that useful because of the large number of unlisted telephone numbers. However, using high-speed computers and a procedure known as random digit dialing, polling organizations are able to create a list of all possible phone numbers and thus are able to select a random sample. Finally, each telephone number in the sample is called, and "an American aged 18 and over" (or whatever group is being targeted) is interviewed. If no one answers the phone or the appropriate person is not at home, the polling organization makes every effort to establish contact at a later time. This ensures that the random sampling process is accurately applied and therefore that the results are true to the stated level of confidence with an acceptable margin of error.

$$z_{\alpha/2} = 0.03(2\sqrt{1,021}) \qquad \text{multiplying each side by } 2\sqrt{1,021}$$

$$z_{\alpha/2} = 1.917185437\ldots$$

$$z_{\alpha/2} \approx 1.92 \qquad \text{rounding off to two decimal places}$$

Using the body table, we can find the area under the bell curve between $z = 0$ and $z = 1.92$; that is, $p(0 < z < 1.92) = 0.4726$.

Therefore, $\frac{\alpha}{2} = 0.4726$, and multiplying by 2, we have $\alpha = 0.9452$. Thus, the level of confidence is $\alpha = 0.9452$ (or 95%), as shown in Figure 4.127.

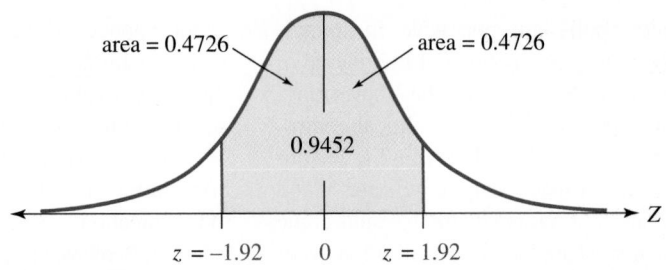

area = 0.4726   area = 0.4726

0.9452

$z = -1.92$ 0 $z = 1.92$

Figure 4.127
Finding the level of confidence   ■

4.5 Exercises

In Exercises 1–4, use the body table to find the specified z-number.

1. **a.** $z_{0.2517}$ **b.** $z_{0.1217}$ **c.** $z_{0.4177}$ **d.** $z_{0.4960}$
2. **a.** $z_{0.0199}$ **b.** $z_{0.2422}$ **c.** $z_{0.4474}$ **d.** $z_{0.4936}$
3. **a.** $z_{0.4250}$ **b.** $z_{0.4000}$ **c.** $z_{0.3750}$ **d.** $z_{0.4950}$
4. **a.** $z_{0.4350}$ **b.** $z_{0.4100}$ **c.** $z_{0.2750}$ **d.** $z_{0.4958}$
5. Find the z-number associated with a 92% level of confidence.
6. Find the z-number associated with a 97% level of confidence.
7. Find the z-number associated with a 75% level of confidence.
8. Find the z-number associated with an 85% level of confidence.

In Exercises 9–22, round off your answers (sample proportions and margins of error) to three decimal places (a tenth of a percent).

9. The Gallup Poll in Example 5 states that one-third (33%) of respondents support general legalization of marijuana. For each of the following levels of confidence, find the margin of error associated with the sample.
    a. a 90% level of confidence
    b. a 98% level of confidence
10. The Roper poll in Example 6 states that 56% of the Americans questioned think that UFOs are real. For each of the following levels of confidence, find the margin of error associated with the sample.
    a. an 80% level of confidence
    b. a 98% level of confidence
11. A survey asked, "How important is it to you to buy products that are made in America?"

Of the 600 Americans surveyed, 450 responded, "It is important." For each of the following levels of confidence, find the sample proportion and the margin of error associated with the poll.
    a. a 90% level of confidence
    b. a 95% level of confidence
12. In the survey in Exercise 11, 150 of the 600 Americans surveyed responded, "It is not important." For each of the following levels of confidence, find the sample proportion and the margin of error associated with the poll.
    a. an 85% level of confidence
    b. a 98% level of confidence
13. A survey asked, "Have you ever bought a lottery ticket?" Of the 2,710 Americans surveyed, 2,141 said yes, and 569 said no.*
    a. Determine the sample proportion of Americans who have purchased a lottery ticket.
    b. Determine the sample proportion of Americans who have not purchased a lottery ticket.
    c. With a 90% level of confidence, find the margin of error associated with the sample proportions.
14. A survey asked, "Which leg do you put into your trousers first?" Of the 2,710 Americans surveyed, 1,138 said left, and 1,572 said right.*
    a. Determine the sample proportion of Americans who put their left leg into their trousers first.
    b. Determine the sample proportion of Americans who put their right leg into their trousers first.
    c. With a 90% level of confidence, find the margin of error associated with the sample proportions.

---

* Data from Poretz and Sinrod, *The First Really Important Survey of American Habits* (Los Angeles: Price Stern Sloan Publishing, 1989).

15. A survey asked, "Do you prefer showering or bathing?" Of the 1,220 American men surveyed, 1,049 preferred showering, and 171 preferred bathing. In contrast, 1,043 of the 1,490 American women surveyed preferred showering, and 447 preferred bathing.*
  a. Determine the sample proportion of American men who prefer showering.
  b. Determine the sample proportion of American women who prefer showering.
  c. With a 95% level of confidence, find the margin of error associated with the sample proportions.

16. A survey asked, "Do you like the way you look in the nude?" Of the 1,220 American men surveyed, 830 said yes, and 390 said no. In contrast, 328 of the 1,490 American women surveyed said yes, and 1,162 said no.*
  a. Determine the sample proportion of American men who like the way they look in the nude.
  b. Determine the sample proportion of American women who like the way they look in the nude.
  c. With a 95% level of confidence, find the margin of error associated with the sample proportions.

Exercises 17–20 are based on a survey (published in April 2000) of 129,593 students in grades 6–12 conducted by USA WEEKEND *magazine.*

17. When asked, "Do you, personally, feel safe from violence in school?" 92,011 said yes, and 37,582 said no.
  a. Determine the sample proportion of students who said yes.
  b. Determine the sample proportion of students who said no.
  c. With a 95% level of confidence, find the margin of error associated with the sample proportions.

18. When asked, "Do kids regularly carry weapons in your school?" 14,255 said yes, and 115,338 said no.
  a. Determine the sample proportion of students who said yes.
  b. Determine the sample proportion of students who said no.
  c. With a 95% level of confidence, find the margin of error associated with the sample proportions.

19. When asked, "Is there a gun in your home?" 58,317 said yes, 60,908 said no, and 10,368 said they did not know.
  a. Determine the sample proportion of students who said yes.
  b. Determine the sample proportion of students who said no.
  c. Determine the sample proportion of students who said they did not know.
  d. With a 90% level of confidence, find the margin of error associated with the sample proportions.

20. When asked, "How likely do you think it is that a major violent incident could occur at your school?" 18,143 said very likely, 64,797 said somewhat likely, and 46,653 said not likely at all.
  a. Determine the sample proportion of students who said "very likely."
  b. Determine the sample proportion of students who said "somewhat likely."
  c. Determine the sample proportion of students who said "not likely at all."
  d. With a 90% level of confidence, find the margin of error associated with the sample proportions.

21. A survey asked, "Can you imagine a situation in which you might become homeless?" Of the 2,503 Americans surveyed, 902 said yes.*
  a. Determine the sample proportion of Americans who can imagine a situation in which they might become homeless.
  b. With a 90% level of confidence, find the margin of error associated with the sample proportion.
  c. With a 98% level of confidence, find the margin of error associated with the sample proportion.
  d. How does your answer to part (c) compare to your answer to part (b)? Why?

22. A survey asked, "Do you think that homeless people are responsible for the situation they are in?" Of the 2,503 Americans surveyed, 1,402 said no.*
  a. Determine the sample proportion of Americans who think that homeless people are not responsible for the situation they are in.
  b. With an 80% level of confidence, find the margin of error associated with the sample proportion.
  c. With a 95% level of confidence, find the margin of error associated with the sample proportion.
  d. How does your answer to part (c) compare to your answer to part (b)? Why?

23. A sample consisting of 430 men and 765 women was asked various questions pertaining to international affairs. With a 95% level of confidence, find the margin of error associated with the following samples.
  a. the male sample
  b. the female sample
  c. the combined sample

* Data from Poretz and Sinrod, *The First Really Important Survey of American Habits* (Los Angeles: Price Stern Sloan Publishing, 1989).

* Data from Mark Clements, "What Americans Say about the Homeless," *Parade Magazine,* Jan. 9, 1994: 4–6.

24. A sample consisting of 942 men and 503 women was asked various questions pertaining to the nation's economy. For a 95% level of confidence, find the margin of error associated with the following samples.
    a. the male sample
    b. the female sample
    c. the combined sample

25. A poll pertaining to environmental concerns had the following footnote: "Based on a sample of 1,763 adults, the margin of error is plus or minus 2.5 percentage points." Find the level of confidence of the poll.

HINT: See Example 6.

26. A poll pertaining to educational goals had the following footnote: "Based on a sample of 2,014 teenagers, the margin of error is plus or minus 2 percentage points." Find the level of confidence of the poll.

HINT: See Example 6.

27. A recent poll pertaining to educational reforms involved 640 men and 820 women. The margin of error for the combined sample is 2.6%. Find the level of confidence for the entire poll.

HINT: See Example 6.

28. A recent poll pertaining to educational reforms involved 640 men and 820 women. The margin of error is 3.9% for the male sample and 3.4% for the female sample. Find the level of confidence for the male portion of the poll and for the female portion of the poll.

HINT: See Example 6.

Answer the following questions using complete sentences and your own words.

CONCEPT QUESTIONS
29. What is a sample proportion? How is it calculated?
30. What is a margin of error? How is it calculated?
31. If the sample size is increased in a survey, would you expect the margin of error to increase or decrease? Why?
32. What is a random sample?
33. What is random digit dailing (RDD)?

HISTORY QUESTIONS
34. Who founded the American Institute of Public Opinion? When? In what two academic fields did this person receive degrees?

 WEB PROJECTS

35. What are the nation's current opinions concerning major issues in the headlines? Pick a current issue (such as abortion, same-sex marriage, gun control, war, or the President's job rating) and find a recent survey regarding the issue. Summarize the results of the survey. Be sure to include the polling organization, date(s) of the survey, sample size, level of confidence, margin of error, and any pertinent information.

36. What is random digit dialing (RDD) and how is it used? Visit the web site of a national polling or news organization, and conduct a search of its FAQs regarding the organization's methods of sampling and the use of RDD. Write a report in which you summarize the use of RDD by the polling or news organization. Some useful links for this web project are listed on the text web site: **www.thomsonedu.com/mathematics/johnson**

PROJECTS
37. The purpose of this project is to conduct an opinion poll. Select a topic that is relevant to you and/or your community. Create a multiple-choice question to gather people's opinions concerning this issue.

    For example:

> If you could vote today, how would you vote on Proposition X?
> i. support   ii. oppose   iii. undecided

    a. Ask 50 people your question, and record their responses. Calculate the sample proportion for each category of response. Use a 95% level of confidence, and calculate the margin of error for your survey.
    b. Ask 100 people your question, and record their responses. Calculate the sample proportion for each category of response. Use a 95% level of confidence, and calculate the margin of error for your survey.
    c. How does the margin of error in part (b) compare to the margin of error in part (a)?

4.6 Linear Regression

When x and y are variables and m and b are constants, the equation $y = mx + b$ has infinitely many solutions of the form (x, y). A specific ordered pair (x_1, y_1) is a solution of the equation if $y_1 = mx_1 + b$. Because every solution of the given equation lies on a straight line, we say that x and y are *linearly related.*

If we are given two ordered pairs (x_1, y_1) and (x_2, y_2), we should be able to "work backwards" and find the equation of the line passing through them; assuming that x and y are linearly related, we can easily find the equation of the line passing through the points (x_1, y_1) and (x_2, y_2). The process of finding the equation of a line passing through given points is known as **linear regression;** the equation thus found is called the **mathematical model** of the linear relationship. Once the model has been constructed, it can be used to make predictions concerning the values of x and y.

EXAMPLE 1 Charlie is planning a family reunion and wants to place an order for custom T-shirts from Prints Alive (the local silk-screen printer) to commemorate the occasion. He has ordered shirts from Prints Alive on two previous occasions; on one occasion, he paid $164 for 24 shirts; on another, he paid $449 for 84. Assuming a linear relationship between the cost of T-shirts and the number ordered, predict the cost of ordering 100 shirts.

Solution Letting $x =$ the number of shirts ordered and $y =$ the total cost of the shirts, the given data can be expressed as two ordered pairs: $(x_1, y_1) = (24, 164)$ and $(x_2, y_2) = (84, 449)$. We must find $y = mx + b$, the equation of the line passing through the two points.

First, we find m, the slope:

$$m = \frac{y_2 - y_1}{x_2 - x_1}$$

$$= \frac{449 - 164}{84 - 24}$$

$$= \frac{285}{60}$$

$$= 4.75$$

Now we use one of the ordered pairs to find b, the y-intercept. Either point will work; we will use $(x_1, y_1) = (24, 164)$.

The slope-intercept form of a line is $y = mx + b$. Solving for b, we obtain

$$b = y - mx$$

$$= 164 - 4.75(24)$$

$$= 164 - 114$$

$$= 50$$

Therefore, the equation of the line is $y = 4.75x + 50$. We use this linear model to predict the cost of ordering $x = 100$ T-shirts.

$$y = 4.75x + 50$$
$$= 4.75(100) + 50$$
$$= 475 + 50$$
$$= 525$$

We predict that it will cost \$525 to order 100 T-shirts.   ■

Linear Trends and Line of Best Fit

Example 1 illustrates the fact that two points determine a unique line. To find the equation of the line, we must find the slope and y-intercept. If we are given more than two points, the points might not be collinear. When collecting real-world data, this is usually the case. However, after plotting the scatter of points on an x-y coordinate system, it may appear that they "almost" fit on a line. If a sample of ordered pairs tend to "go in the same general direction," we say that they exhibit a **linear trend.** See Figure 4.128.

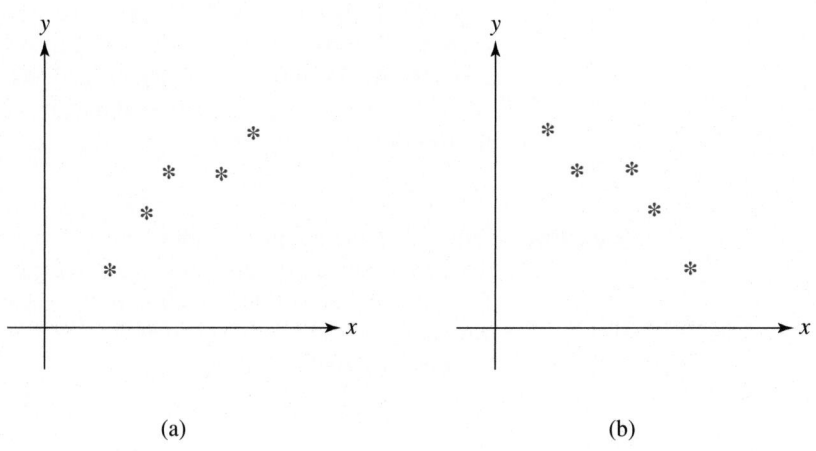

(a)   (b)

Figure 4.128
Ordered pairs that exhibit a linear trend

When a scatter of points exhibits a linear trend, we construct the line that best approximates the trend. This line is called the **line of best fit** and is denoted by $\hat{y} = mx + b$. The "hat" over the y indicates that the calculated value of y is a prediction based on linear regression. See Figure 4.129.

To calculate the slope and y-intercept of the line of best fit, mathematicians have developed formulas based on the method of least squares. (See the Historical Note on Carl Gauss in Section 4.4.)

Recall that the symbol Σ means "sum." Therefore, Σx represents the sum of the

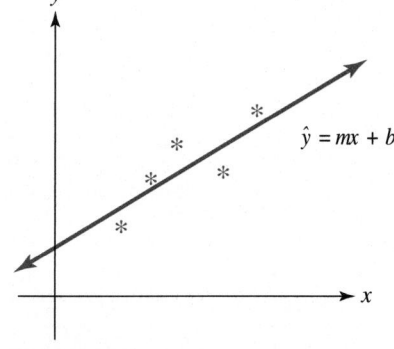

Figure 4.129
The line of best fit $y = mx + b$

x-coordinates of the points and Σy the sum of the y-coordinates. To find Σxy, multiply the x- and y-coordinates of each point and sum the results.

> **Line of Best Fit**
>
> Given a sample of n ordered pairs $(x_1, y_1), (x_2, y_2), \ldots, (x_n, y_n)$.
>   The **line of best fit** (the line that best represents the data) is denoted by $\hat{y} = mx + b$, where the slope m and y-intercept b are given by
>
> $$m = \frac{n(\Sigma xy) - (\Sigma x)(\Sigma y)}{n(\Sigma x^2) - (\Sigma x)^2} \quad \text{and} \quad b = \bar{y} - m\bar{x}$$
>
> $\bar{x}$ and $\bar{y}$ denote the means of the x- and y-coordinates, respectively.

ThomsonNOW
for Liberal Arts Mathematics
www.thomsonedu.com/1pass.

EXAMPLE 2   Given: the ordered pairs (5, 14), (9, 17), (12, 16), (14, 18), and (17, 23).

a. Find the equation of the line of best fit.
b. Plot the given data and sketch the graph of the line of best fit on the same coordinate system.

Solution   a. Organize the data in a table and compute the appropriate sums, as shown in Figure 4.130.

(x, y)	x	x^2	y	xy
(5, 14)	5	25	14	$5 \cdot 14 = 70$
(9, 17)	9	81	17	$9 \cdot 17 = 153$
(12, 16)	12	144	16	$12 \cdot 16 = 192$
(14, 18)	14	196	18	$14 \cdot 18 = 252$
(17, 23)	17	289	23	$17 \cdot 23 = 391$
$n = 5$ ordered pairs	$\Sigma x = 57$	$\Sigma x^2 = 735$	$\Sigma y = 88$	$\Sigma xy = 1{,}058$

Figure 4.130
Table of sums

First, we find the slope:

$$m = \frac{n(\Sigma xy) - (\Sigma x)(\Sigma y)}{n(\Sigma x^2) - (\Sigma x)^2}$$

$$= \frac{5(1{,}058) - (57)(88)}{5(735) - (57)^2}$$

$$= 0.643192488\ldots$$

For graphing calculators, see the instructions on page 317.

Once m has been calculated, we store it in the memory of our calculator. We will need it to calculate b, the y-intercept.

$$b = \bar{y} - m\bar{x}$$

$$= \left(\frac{88}{5}\right) - 0.643192488\left(\frac{57}{5}\right)$$

$$= 10.26760564\ldots$$

Therefore, the line of best fit, $\hat{y} = mx + b$, is

$$\hat{y} = 0.643192488x + 10.26760564$$

Rounding off to one decimal place, we have

$$\hat{y} = 0.6x + 10.3$$

b. To graph the line, we need to plot two points. One point is the y-intercept $(0, b) = (0, 10.3)$. To find another point, we pick an appropriate value for x—say, $x = 18$—and calculate $\hat{y}$:

$$\hat{y} = 0.6x + 10.3$$

$$= 0.6(18) + 10.3$$

$$= 21.1$$

Therefore, the point $(x, \hat{y}) = (18, 21.1)$ is on the line of best fit.

Plotting $(0, 10.3)$ and $(18, 21.1)$, we construct the line of best fit; it is customary to use asterisks (*) to plot the given ordered pairs as shown in Figure 4.131.

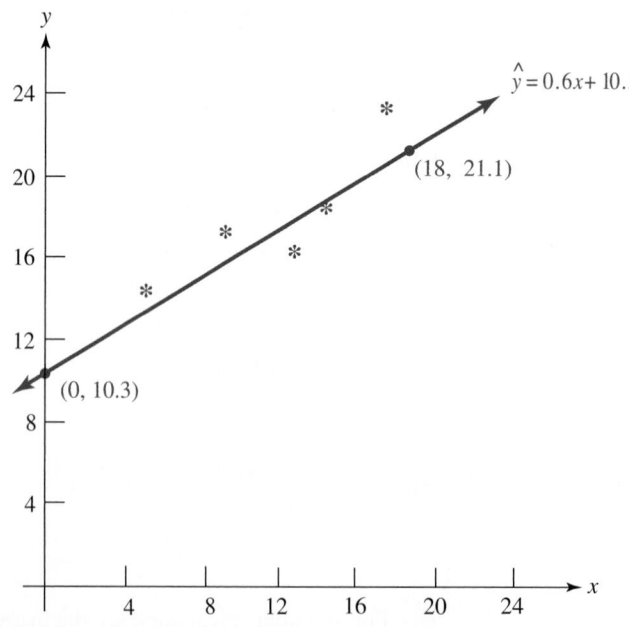

Figure 4.131
Scatter diagram and line of best fit

Coefficient of Linear Correlation

Given a sample of n ordered pairs, we can always find the line of best fit. Does the line accurately portray the data? Will the line give accurate predictions? To answer these questions, we must consider the relative strength of the linear trend exhibited by the given data. If the given points are close to the line of best fit, there is a strong linear relation between x and y; the line will generate good predictions. If the given points are widely scattered about the line of best fit, there is a weak linear relation, and predictions based on it are probably not reliable.

One way to measure the strength of a linear trend is to calculate the **coefficient of linear correlation,** denoted by r. The formula for calculating r is shown in the following box.

Coefficient of Linear Correlation

Given a sample of n ordered pairs, $(x_1, y_1), (x_2, y_2), \ldots, (x_n, y_n)$.
The **coefficient of linear correlation,** denoted by r, is given by

$$r = \frac{n(\Sigma xy) - (\Sigma x)(\Sigma y)}{\sqrt{n(\Sigma x^2) - (\Sigma x)^2}\sqrt{n(\Sigma y^2) - (\Sigma y)^2}}$$

The calculated value of r is always between -1 and 1, inclusive; that is, $-1 \leq r \leq 1$. If the given ordered pairs lie perfectly on a line whose slope is *positive,* then the calculated value of r will equal 1 (think 100% perfect with positive slope). In this case, both variables have the same behavior: As one increases (or decreases), so will the other. On the other hand, if the data points fall perfectly on a line whose slope is *negative,* the calculated value of r will equal -1 (think 100% perfect with negative slope). In this case, the variables have opposite behavior: As one increases, the other decreases, and vice versa. See Figure 4.132.

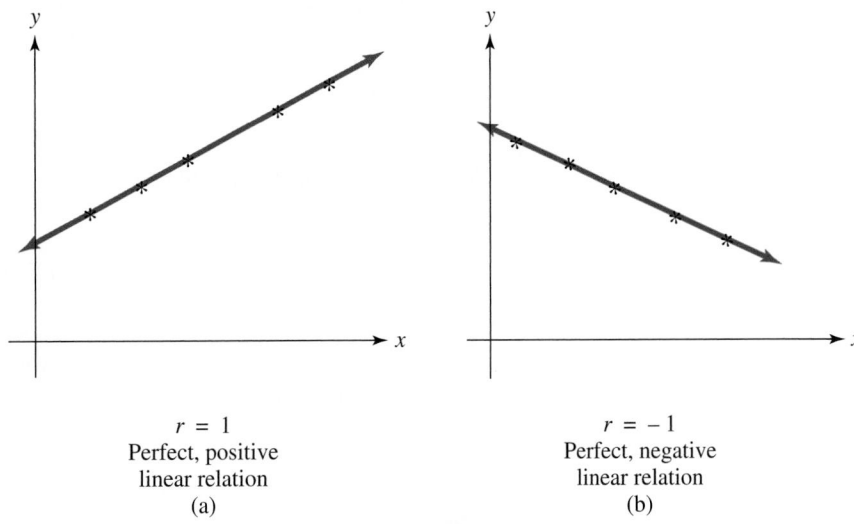

$r = 1$
Perfect, positive
linear relation
(a)

$r = -1$
Perfect, negative
linear relation
(b)

Figure 4.132

If the value of r is close to 0, there is little or no *linear* relation between the variables. This does not mean that the variables are not related. It merely means that no *linear* relation exists; the variables might be related in some nonlinear fashion, as shown in Figure 4.133.

In summary, the closer *r* is to 1 or −1, the stronger is the linear relation between *x* and *y;* the line of best fit will generate reliable predictions. The closer *r* is to 0, the weaker is the linear relation; the line of best fit will generate unreliable predictions. If *r* is positive, the variables have a direct relationship (as one increases, so does the other); if *r* is negative, the variables have an inverse relationship (as one increases, the other decreases).

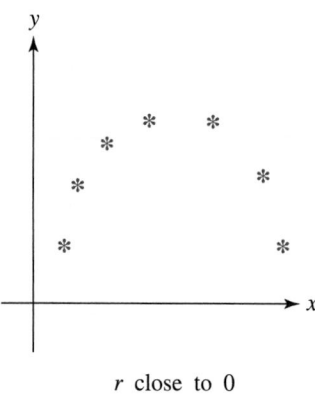

r close to 0

Figure 4.133

ThomsonNOW™
for Liberal Arts Mathematics
www.thomsonedu.com/1pass. **EXAMPLE 3**   Calculate the coefficient of linear correlation for the ordered pairs given in Example 2.

Solution   The ordered pairs are (5, 14), (9, 17), (12, 16), (14, 18), and (17, 23). We add a y^2 column to the table in Example 2, as shown in Figure 4.134.

(*x, y*)	*x*	x^2	*y*	y^2	*xy*
(5, 14)	5	25	14	196	$5 \cdot 14 = 70$
(9, 17)	9	81	17	289	$9 \cdot 17 = 153$
(12, 16)	12	144	16	256	$12 \cdot 16 = 192$
(14, 18)	14	196	18	324	$14 \cdot 18 = 252$
(17, 23)	17	289	23	529	$17 \cdot 23 = 391$
$n = 5$ ordered pairs	$\Sigma x = 57$	$\Sigma x^2 = 735$	$\Sigma y = 88$	$\Sigma y^2 = 1{,}594$	$\Sigma xy = 1{,}058$

Figure 4.134
Table of sums

Now use the formula to calculate *r:*

$$r = \frac{n(\Sigma xy) - (\Sigma x)(\Sigma y)}{\sqrt{n(\Sigma x^2) - (\Sigma x)^2}\sqrt{n(\Sigma y^2) - (\Sigma y)^2}}$$

$$= \frac{5(1{,}058) - (57)(88)}{\sqrt{5(735) - (57)^2}\sqrt{5(1{,}594) - (88)^2}}$$

$$= 0.883062705\dots$$

The coefficient of linear correlation is reasonably close to 1, so the line of best fit will generate reasonably reliable predictions. (Notice that the data points in Figure 4.131 are fairly close to the line of best fit.) ∎

EXAMPLE 4 Unemployment and personal income are undoubtedly related; we would assume that as the national unemployment rate increases, total personal income would decrease. Figure 4.135 gives the unemployment rate and the total personal income for the United States for various years.

a. Use linear regression to predict the total personal income of the United States if the unemployment rate is 5.0%.

b. Use linear regression to predict the unemployment rate if the total personal income of the United States is $10 billion.

c. Are the predictions in parts (a) and (b) reliable? Why or why not?

Year	Unemployment Rate (percent)	Total Personal Income (billions)
1975	8.5	$1.3
1980	7.1	2.3
1985	7.2	3.4
1990	5.6	4.8
1995	5.6	6.1
2000	4.0	8.3

Figure 4.135
Sources: Bureau of Labor Statistics, U.S. Dept. of Labor and Bureau of Economic Analysis, U.S. Dept. of Commerce

Solution **a.** Letting x = unemployment rate, and y = total personal income, we have $n = 6$ ordered pairs, as shown in Figure 4.136.

Using a calculator, we find the following sums:

$$\Sigma x^2 = 253.22 \qquad \Sigma y^2 = 147.68 \qquad \Sigma xy = 146.1$$

First we find the slope:

$$m = \frac{n(\Sigma xy) - (\Sigma x)(\Sigma y)}{n(\Sigma x^2) - (\Sigma x)^2}$$

$$= \frac{6(146.1) - (38)(26.2)}{6(253.22) - (38)^2}$$

$$= -1.579925651 \ldots$$

Now we calculate b, the y-intercept:

$$b = \bar{y} - m\bar{x}$$

$$= \left(\frac{26.2}{6}\right) - (-1.579925651)\left(\frac{38}{6}\right)$$

$$= 14.37286246 \ldots$$

x (percent)	y (billions)
8.5	1.3
7.1	2.3
7.2	3.4
5.6	4.8
5.6	6.1
4.0	8.3
$\Sigma x = 38$	$\Sigma y = 26.2$

Figure 4.136
x = unemployment rate,
y = total personal income

Therefore, the line of best fit, $\hat{y} = mx + b$, is

$$\hat{y} = -1.579925651x + 14.37286246$$

Rounding off to two decimal places (one more than the data), we have

$$\hat{y} = -1.58x + 14.37$$

Now, substituting $x = 5.0$ (5% unemployment) into the equation of the line of best fit, we have

$$\hat{y} = -1.58x + 14.37$$
$$= -1.58(5.0) + 14.37$$
$$= -7.9 + 14.37$$
$$= 6.47$$

If the unemployment rate is 5.0%, we predict that the total personal income of the United States will be approximately $6.47 billion.

b. To predict the unemployment rate when the total personal income is $10 billion, we let $y = 10$, substitute into $\hat{y}$, and solve for x:

$$10 = -1.58x + 14.37$$
$$1.58x + 10 = 14.37 \qquad \text{adding } 1.58x \text{ to both sides}$$
$$1.58x = 14.37 - 10 \qquad \text{subtracting 10 from both sides}$$
$$1.58x = 4.37$$
$$x = \frac{4.37}{1.58} \qquad \text{dividing by 1.58}$$
$$= 2.765822785 \ldots$$

We predict that the unemployment rate will be approximately 2.8% when the total personal income is $10 billion.

c. To investigate the reliability of our predictions (the strength of the linear trend), we must calculate the coefficient of linear correlation:

$$r = \frac{n(\Sigma xy) - (\Sigma x)(\Sigma y)}{\sqrt{n(\Sigma x^2) - (\Sigma x)^2}\sqrt{n(\Sigma y^2) - (\Sigma y)^2}}$$

$$r = \frac{6(146.1) - (38)(26.2)}{\sqrt{6(253.22) - (38)^2}\sqrt{6(147.68) - (26.2)^2}}$$

$$= -0.9704385162 \ldots$$

Because r is extremely close to -1, we conclude that our predictions are very reliable; the linear relationship between x and y is extremely high. Furthermore, since r is negative, we know that y (total personal income) decreases as x (unemployment rate) increases.   ∎

4.6 Exercises

1. A set of $n = 6$ ordered pairs has the following sums:

$$\Sigma x = 64 \quad \Sigma x^2 = 814 \quad \Sigma y = 85$$
$$\Sigma y^2 = 1,351 \quad \Sigma xy = 1,039$$

   a. Find the line of best fit.

   b. Predict the value of y when $x = 11$.
   c. Predict the value of x when $y = 19$.
   d. Find the coefficient of linear correlation.
   e. Are the predictions in parts (b) and (c) reliable? Why or why not?

2. A set of $n = 8$ ordered pairs has the following sums:

$$\Sigma x = 111 \quad \Sigma x^2 = 1{,}869 \quad \Sigma y = 618$$
$$\Sigma y^2 = 49{,}374 \quad \Sigma xy = 7{,}860$$

a. Find the line of best fit.
b. Predict the value of y when $x = 8$.
c. Predict the value of x when $y = 70$.
d. Find the coefficient of linear correlation.
e. Are the predictions in parts (b) and (c) reliable? Why or why not?

3. A set of $n = 5$ ordered pairs has the following sums:

$$\Sigma x = 37 \quad \Sigma x^2 = 299 \quad \Sigma y = 38$$
$$\Sigma y^2 = 310 \quad \Sigma xy = 279$$

a. Find the line of best fit.
b. Predict the value of y when $x = 5$.
c. Predict the value of x when $y = 7$.
d. Find the coefficient of linear correlation.
e. Are the predictions in parts (b) and (c) reliable? Why or why not?

4. Given the ordered pairs (4, 40), (6, 37), (8, 34), and (10, 31):
a. Find and interpret the coefficient of linear correlation.
b. Find the line of best fit.
c. Plot the given ordered pairs and sketch the graph of the line of best fit on the same coordinate system.

5. Given the ordered pairs (5, 5), (7, 10), (8, 11), (10, 15), and (13, 16):
a. Plot the ordered pairs. Do the ordered pairs exhibit a linear trend?
b. Find the line of best fit.
c. Predict the value of y when $x = 9$.
d. Plot the given ordered pairs and sketch the graph of the line of best fit on the same coordinate system.
e. Find the coefficient of linear correlation.
f. Is the prediction in part (c) reliable? Why or why not?

6. Given the ordered pairs (5, 20), (6, 15), (10, 14), (12, 15), and (13, 10):
a. Plot the ordered pairs. Do the ordered pairs exhibit a linear trend?
b. Find the line of best fit.
c. Predict the value of y when $x = 8$.
d. Plot the given ordered pairs and sketch the graph of the line of best fit on the same coordinate system.

e. Find the coefficient of linear correlation.
f. Is the prediction in part (c) reliable? Why or why not?

7. Given the ordered pairs (2, 6), (3, 12), (6, 15), (7, 4), (10, 6), and (11, 12):
a. Plot the ordered pairs. Do the ordered pairs exhibit a linear trend?
b. Find the line of best fit.
c. Predict the value of y when $x = 8$.
d. Plot the given ordered pairs and sketch the graph of the line of best fit on the same coordinate system.
e. Find the coefficient of linear correlation.
f. Is the prediction in part (c) reliable? Why or why not?

8. The average hourly earnings and the unemployment rate in the United States for 1999–2003 are given in Figure 4.137.

Year	Average Hourly Earnings	Unemployment Rate
1999	$13.47	4.2%
2000	14.00	4.0
2001	14.53	4.7
2002	14.95	5.8
2003	15.35	6.0

Figure 4.137
Average hourly earnings and unemployment rates, 1999–2003
Source: Bureau of Labor Statistics

a. Letting $x =$ the average hourly earnings and $y =$ the unemployment rate, plot the data. Do the data exhibit a linear trend?
b. Find the line of best fit.
c. Predict the unemployment rate when the average hourly earning is $15.00.
d. Predict the average hourly earning when the unemployment rate is 5.5%.
e. Find the coefficient of correlation.
f. Are the predictions in parts (a) and (b) reliable? Why or why not?

9. The average hourly earnings and the average tuition at public four-year institutions of higher education in the United States for 1997–2002 are given in Figure 4.138.
a. Letting $x =$ the average hourly earnings and $y =$ the average tuition, plot the data. Do the data exhibit a linear trend?
b. Find the line of best fit.

c. Predict the average tuition when the average hourly earning is $14.75.

d. Predict the average hourly earning when the average tuition is $4,000.

e. Find the coefficient of correlation.

f. Are the predictions in parts (a) and (b) reliable? Why or why not?

Year	Average Hourly Earnings	Average Tuition at 4-year Institutions
1997	$12.49	$3,110
1998	13.00	3,229
1999	13.47	3,349
2000	14.00	3,501
2001	14.53	3,735
2002	14.95	4,059

Figure 4.138
Average hourly earnings and average tuition, 1997–2002
Source: Bureau of Labor Statistics and National Center for Education Statistics

10. The number of domestic and imported retail car sales (in hundred thousands) in the United States for 1999–2003 are given in Figure 4.139.

Year	Domestic Car Sales	Imported Car Sales
1999	69.8	17.2
2000	68.3	20.2
2001	63.2	21.0
2002	58.8	22.3
2003	55.3	20.8

Figure 4.139
Domestic and imported retail car sales (hundred thousands), 1999–2003
Source: Ward's Commission

a. Letting x = the number of domestic car sales and y = the number of imported car sales, plot the data. Do the data exhibit a linear trend?

b. Find the line of best fit.

c. Predict the number of imported car sales when there are 5,600,000 domestic car sales.

d. Predict the number of domestic car sales when there are 2,200,000 imported car sales.

e. Find the coefficient of correlation.

f. Are the predictions in parts (a) and (b) reliable? Why or why not?

11. The numbers of marriages and divorces (in millions) in the United States are given in Figure 4.140.

a. Letting x = the number of marriages and y = the number of divorces in a year, plot the data. Do the data exhibit a linear trend?

b. Find the line of best fit.

c. Predict the number of divorces in a year when there are 2,750,000 marriages.

d. Predict the number of marriages in a year when there are 1,500,000 divorces.

e. Find the coefficient of linear correlation.

f. Are the predictions in parts (c) and (d) reliable? Why or why not?

Year	1965	1970	1975	1980	1985	1990	2000
Marriages	1.800	2.158	2.152	2.413	2.425	2.448	2.329
Divorces	0.479	0.708	1.036	1.182	1.187	1.175	1.135

Figure 4.140
Number of marriages and divorces (millions)
Source: National Center for Health Statistics

12. The median-priced home and average mortgage rate in the United States for 1999–2003 are given in Figure 4.141.

Year	Median Price	Mortgage Rate
1999	$133,300	7.33%
2000	139,000	8.03
2001	147,000	7.03
2002	158,100	6.55
2003	170,000	5.74

Figure 4.141
Median-priced home and average mortgage rate, 1999–2003
Source: National Association of Realtors

a. Letting x = the median price of a home and y = the average mortgage rate, plot the data. Do the data exhibit a linear trend?

b. Find the line of best fit.

c. Predict the average mortgage rate if the median price of a home is $165,000.

d. Predict the median-priced home when if the average mortgage rate is 7.25%.

e. Find the coefficient of correlation.

f. Are the predictions in parts (a) and (b) reliable? Why or why not?

 Answer the following questions using complete sentences and your own words.

CONCEPT QUESTIONS

13. What is a line of best fit? How do you find it?

14. How do you measure the strength of a linear trend?

15. What is a positive linear relation? Give an example.

16. What is a negative linear relation? Give an example.

PROJECTS

17. Measure the heights and weights of 10 people. Let x = height and y = weight.

a. Plot the ordered pairs. Do the ordered pairs exhibit a linear trend?

b. Use the data to find the line of best fit.

c. Find the coefficient of linear correlation.

d. Will your line of best fit produce reliable predictions? Why or why not?

 Technology and Linear Regression

In Example 2 of this section, we computed the slope and y-intercept of the line of best fit for the five ordered pairs (5, 14), (9, 17), (12, 16), (14, 18), and (17, 23). These calculations can be tedious when done by hand, even with only five data points. In the real world, there are always a large number of data points, and the calculations are always done with the aid of technology.

Linear Regression on a Graphing Calculator

Graphing calculators can draw a scatter diagram, compute the slope and y-intercept of the line of best fit, and graph the line.

On a TI-83/84/86:

• *Put the calculator into statistics mode* by pressing STAT (TI-86: 2nd STAT).

• *Set the calculator up for entering the data* from Example 2 by selecting "Edit" from the "EDIT" menu (TI-86: Select "Edit" by pressing F2), and the "List Screen" appears, as shown in Figure 4.142. If data already appear in a list (as they do in Figure 4.142) and you want to clear it, use the arrow buttons to highlight the name of the list and press CLEAR ENTER.

• Use the arrow buttons and the ENTER button to enter the x-coordinates in list L_1 and the corresponding y-coordinates in list L_2. (TI-86: Enter the x-coordinates in list xStat and the corresponding y-coordinates in list yStat. Also enter a frequency of 1 in list fStat, for each of the entries in list xStat.) When done, your screen should look like Figure 4.142 (b).

(a)

(b)

Figure 4.142
(a) A TI-83/84's list screen; (b) A TI-86's list screen.

Figure 4.143
A Casio's list screen

On a Casio:
- *Put the calculator into statistics mode* by pressing MENU, highlighting STAT, and pressing EXE.
- Use the arrow buttons and the EXE button to enter the *x*-coordinates from Example 2 in List 1 and the corresponding *y*-coordinates in List 2. When you are done, your screen should look like Figure 4.143. If data already appears in a list and you want to erase it, use the arrow keys to move to that list, press F6 and then F4 (i.e., DEL-A, which stands for "delete all") and then F1 (i.e., YES).

Finding the Equation of the Line of Best Fit
Once the data have been entered, it is easy to find the equation. Recall from Example 2 that the equation is $y = 0.643192488x + 10.26760564$.

On a TI-83/84:
- Press 2nd QUIT.
- The TI-83/84 does not display the correlation coefficient unless you tell it to. To do so, press 2nd CATALOG, scroll down, and select "DiagnosticOn." When "DiagnosticOn" appears on the screen, press ENTER.
- Press STAT, scroll to the right, and select the "CALC" menu, and select "LinReg(ax + b)" from the "CALC" menu.
- When "LinReg(ax + b)" appears on the screen, press ENTER.
- The slope, the *y*-intercept and the correlation coefficient will appear on the screen, as shown in Figure 4.144. They are not labeled *m, b,* and *r*. Their labels are different and are explained in Figure 4.145.

Figure 4.144
Finding the equation of the line of best fit on a TI-83/84/86

	TI-83/84 labels	TI-86 labels
slope *m*	a	b
y-intercept *b*	b	a
correlation coefficient *r*	r	Corr
number of data points *n*	(not shown)	n

Figure 4.145
Graphing calculator line of best fit labels

On a TI-86:
- Press EXIT.
- Press 2nd STAT and select "CALC" by pressing F1.
- Select linear regression by pressing LINR (that is, F3).
- When "LinR" appears on the screen, press ENTER.
- The slope, the *y*-intercept, the correlation coefficient, and the number of data points will appear on the screen, as shown in Figure 4.144. Their labels are different and are explained in Figure 4.145. (Use the ▼ button to see it all.)

Figure 4.146
A Casio's equation of the line of best fit

On a Casio:
- Press $\boxed{\text{CALC}}$ (i.e., $\boxed{\text{F2}}$) to calculate the equation.
- Press $\boxed{\text{REG}}$ (i.e., $\boxed{\text{F3}}$); "REG" stands for "regression."
- Press $\boxed{\text{X}}$ (i.e., $\boxed{\text{F1}}$) for linear regression. (See Figure 4.146.)

Drawing a Scatter Diagram and the Line of Best Fit

Once the equation has been found, the line can be graphed.

On a TI-83/84:
- Press $\boxed{\text{Y=}}$, and clear any functions that may appear.
- Set the calculator up to draw a scatter diagram by pressing $\boxed{\text{2nd}}$ $\boxed{\text{STAT PLOT}}$ and selecting "Plot 1." Turn the plot on and select the scatter icon.
- Tell the calculator to put the data entered in list L_1 on the x-axis by selecting "L_1" for "Xlist," and to put the data entered in list L_2 on the y-axis by selecting "L_2" for "Ylist."
- Automatically set the range of the window and obtain a scatter diagram by pressing $\boxed{\text{ZOOM}}$ and selecting option 9: "ZoomStat."
- If you don't want the data points displayed on the line of best fit, press $\boxed{\text{2nd}}$ $\boxed{\text{STAT PLOT}}$ and turn off plot 1.
- Quit the statistics mode by pressing $\boxed{\text{2nd}}$ $\boxed{\text{QUIT}}$.

Figure 4.147
A TI-83/84's graph of the line of best fit and the scatter diagram

- Enter the equation of the line of best fit by pressing $\boxed{\text{Y=}}$ $\boxed{\text{VARS}}$, selecting "Statistics," scrolling to the right to select the "EQ" menu, and selecting "RegEQ" (for regression equation).
- Automatically set the range of the window and obtain a scatter diagram by pressing $\boxed{\text{ZOOM}}$ and selecting option 9: "ZoomStat." (See Figure 4.147.)
- Press $\boxed{\text{TRACE}}$ to read off the data points as well as points on the line of best fit. Use the up and down arrows to switch between data points and points on the line of best fit. Use the left and right arrows to move left and right on the graph.

On a TI-86:
- Quit the statistics mode by pressing $\boxed{\text{2nd}}$ $\boxed{\text{QUIT}}$.
- Put the calculator into graphing mode, press $\boxed{\text{y(x)=}}$, and clear any functions that may appear. Press $\boxed{\text{WIND}}$ and enter appropriate values. For this problem, let x and y range from 0 to 25. Quit the graphing mode by pressing $\boxed{\text{2nd}}$ $\boxed{\text{QUIT}}$.
- Put the calculator into statistics mode by pressing $\boxed{\text{2nd}}$ $\boxed{\text{STAT}}$.
- Press $\boxed{\text{DRAW}}$. If this immediately produces a drawing, clear it by pressing $\boxed{\text{CLDRW}}$. (You may need to press $\boxed{\text{MORE}}$ to find $\boxed{\text{CLDRW}}$.)
- Press either $\boxed{\text{SCAT}}$ (for a scatter diagram) or $\boxed{\text{DRREG}}$ (for the regression line) or both. (You may need to press $\boxed{\text{MORE}}$ to find $\boxed{\text{DRREG}}$.) If the button labels obscure the line, press $\boxed{\text{EXIT}}$ once, and then $\boxed{\text{CLEAR}}$.
- Press $\boxed{\text{CLDRW}}$, or else the scatter diagram will appear on future graphs.

On a Casio:
- Press $\boxed{\text{SHIFT}}$ $\boxed{\text{QUIT}}$ to return to List 1 and List 2.
- Press $\boxed{\text{GRPH}}$ (i.e., $\boxed{\text{F1}}$).
- Press $\boxed{\text{SET}}$ (i.e., $\boxed{\text{F6}}$).

- Make the resulting screen, which is labeled "StatGraph1," read as follows:

Graph Type	:Scatter
Xlist	:List1
Ylist	:List2
Frequency	:1

To make the screen read as described above,
- Use the down arrow button to scroll down to "Graph Type."
- Press $\boxed{\text{Scat}}$ (i.e., $\boxed{\text{F1}}$).
- In a similar manner, change "Xlist," "Ylist," and "Frequency" if necessary.
- Press $\boxed{\text{SHIFT}}$ $\boxed{\text{QUIT}}$ to return to List1 and List2.
- Press $\boxed{\text{GRPH}}$ (i.e., $\boxed{\text{F1}}$).
- Press $\boxed{\text{GPH1}}$ (i.e., $\boxed{\text{F1}}$) to obtain the scatter diagram.
- Press $\boxed{\text{X}}$ (i.e., $\boxed{\text{F1}}$).
- Press $\boxed{\text{DRAW}}$ (i.e., $\boxed{\text{F6}}$), and the calculator will display the scatter diagram as well as the line of best fit. (See Figure 4.148.)

Figure 4.148
A Casio's graph of the line of best fit and the scatter diagram

 Linear Regression on Excel

Entering the Data

We will use the data from Example 2. Start by entering the information as shown in Figure 4.149. Then save your spreadsheet.

◇	A	B	C
1	x-coordinate	y-coordinate	
2	5	14	
3	9	17	
4	12	16	
5	14	18	
6	17	23	
7			

Figure 4.149
The given data

Using the Chart Wizard to Draw the Scatter Diagram:
- Press the "Chart Wizard" button at the top of the spreadsheet. (It looks like a bar chart, and it might have a magic wand.)
- Select "XY (Scatter)" and choose the Chart sub-type that displays points without connecting lines.
- Press the "Next" button and a "Chart Source Data" box will appear.
- Click on the triangle to the right of "Data range." This will cause the "Chart Source Data" box to shrink.
- Use your mouse to draw a box around all of the x-coordinates and y-coordinates. This will cause "=Sheet1!$A$2:$B$6" to appear in the "Chart Source Data" box.
- Press the triangle to the right of the "Chart Source Data" box and the box will expand.

- Press the "Next" button.
- Under "Chart Title" type an appropriate title. Our example lacks content, so we will use the title "Example 2."
- After "Category (X)" type an appropriate title for the x-axis. Our example lacks content, so we will use the title "x-axis."
- After "Category (Y)" type an appropriate title for the y-axis.
- Click on the "Gridlines" tab and check the box under "value (X) axis," next to "Major gridlines."
- Click on the "Legend" tab and remove the check from the box next to "show legend."
- Press the "Finish" button and the scatter diagram will appear.
- Save the spreadsheet.

After completing this step your spreadsheet should include the scatter diagram shown in Figure 4.150.

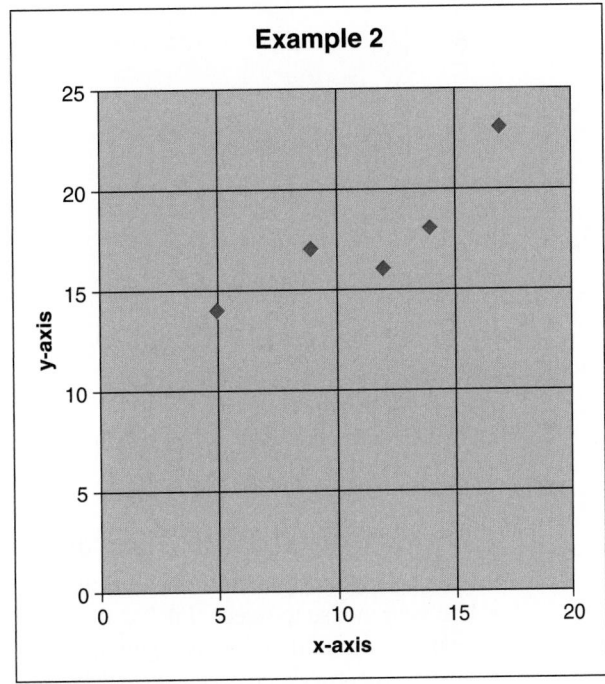

Figure 4.150
An Excel-generated scatter diagram

Finding and Drawing the Line of Best Fit

- Click on the chart until it is surrounded by a thicker border.
- Use your mouse to select "Chart" at the very top of the screen. Then pull your mouse down until "Add Trendline . . ." is highlighted, and let go.
- Press the "Linear" button.
- Click on "Options".
- Select "Display Equation on Chart".
- Select "Display R-Squared value on Chart".
- Press the "OK" button, and the scatter diagram, the line of best fit, the equation of the line of best fit, and r^2 (the square of the correlation coefficient) appear. If the equation is in the way, you can click on it and move it to a better location.

- Use the square root button on your calculator to find r, the correlation coefficient.
- Save the spreadsheet

After completing this step, your spreadsheet should include the scatter diagram, the line of best fit, and the equation of the line of best fit, as shown in Figure 4.151.

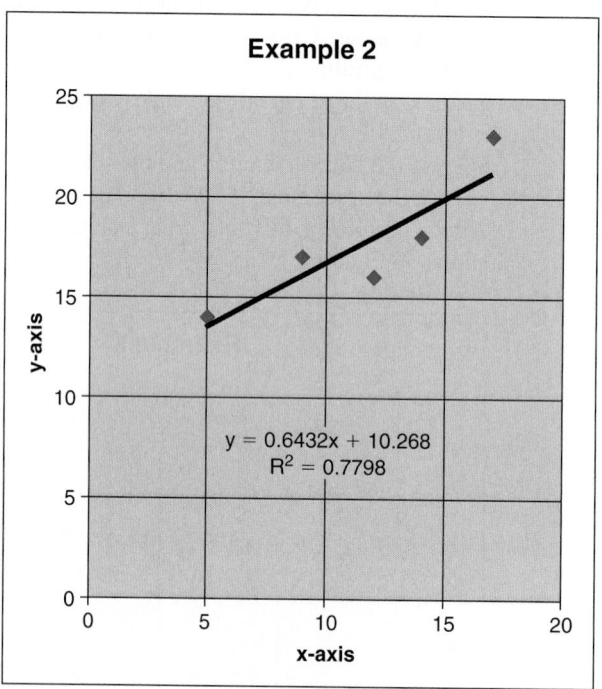

Figure 4.151
The scatter diagram with the line of best fit and its equation

Printing the Scatter Diagram and Line of Best Fit

- Click on the diagram, until it is surrounded by a thicker border.
- Use your mouse to select "File" at the very top of the screen. Then pull your mouse down until "Print" is highlighted, and let go.
- Respond appropriately to the "Print" box that appears. Usually, it is sufficient to press a "Print" button or an "OK" button, but the actual way that you respond to the "Print" box depends on your printer.

Exercises

18. Throughout the twentieth century, the record time for the mile run steadily decreased, from 4 minutes 15.4 seconds in 1911 to 3 minutes 43.1 seconds in 1999. Some of the record times are given in Figure 4.152.
    a. Convert the given data to ordered pairs (x, y), where x is the number of years since 1900 and y is the time in seconds.
    b. Draw a scatter diagram, find the equation of the line of best fit, and find the correlation coefficient.
    c. Predict the record time for the mile run in the year 2020.
    e. Predict when the record time for the mile run will reach 3:30.
    f. Predict when the record time for the mile run will reach 0:30.

Year	1911	1923	1933	1942	1945	1954
Time (min:sec)	4:15.4	4:10.4	4:07.6	4:04.6	4:01.4	3:59.4

Year	1964	1967	1975	1980	1985	1999
Time (min:sec)	3:54.1	3:51.1	3:49.4	3:48.8	3:46.3	3:43.1

Figure 4.152
Record times for the mile run

g. What assumption are these predictions based on? If this assumption is correct, how accurate are these predictions?

19. Heart disease is much less of a problem in France than it is in the United States and other industrialized countries, despite a high consumption of saturated fats in France. One theory is that this lower heart disease rate is due to the fact that the French drink more wine. This so-called French paradox heightened an interest in moderate wine drinking. Figure 4.153 gives data on wine consumption (in liters of alcohol per person) and heart disease (annual deaths, per 100,000 people) for nineteen industrialized countries, including France and the United States.
   a. Draw a scatter diagram, find the equation of line of best fit, and find the correlation for wine consumption versus annual deaths.
   b. Do the data support the theory that wine consumption lowers the heart disease rate?

Country	Wine Consumption	Annual Deaths from Heart Disease
Australia	2.5	211
Austria	3.9	167
Belgium	2.9	131
Canada	2.4	191
Denmark	2.9	220
Finland	0.8	297
France	9.1	71
Iceland	0.8	211
Ireland	0.6	300
Italy	7.9	107
Netherlands	1.8	266
New Zealand	1.9	266
Norway	0.8	227
Spain	6.5	86
Sweden	1.6	207
Switzerland	5.8	115
United Kingdom	1.3	285
United States	1.2	199
West Germany	2.7	172

Figure 4.153
Wine consumption and heart disease
Source: New York Times, December 28, 1994

Chapter 4   Review

ThomsonNOW™ **for Liberal Arts Mathematics**
Preparing for an exam? Test yourself on key material by visiting ThomsonNOW at **www.thomsonedu.com/1pass.**

TERMS

bell-shaped curve	discrete variable	mathematical model	relative frequency density
body table	error of the estimate	mean	sample
categorical data	extreme value	measures of central tendency	sample proportion
cells	frequency	measures of dispersion	spreadsheet
coefficient of linear correlation	frequency distribution	median	standard deviation
	histogram	mode	standard normal distribution
computerized spreadsheet	inferential statistics	normal distribution	statistics
continuous variable	level of confidence	outlier	summation notation
data point	linear regression	pie chart	tail
density	linear trend	population	variance
descriptive statistics	line of best fit	population proportion	z-distribution
deviation from the mean	margin of error (MOE)	relative frequency	z-number

REVIEW EXERCISES

1. Find (a) the mean, (b) the median, (c) the mode, and (d) the standard deviation of the following set of raw data:

   5   8   10   4   8   10   6   8   7   5

2. To study the composition of families in Winslow, Arizona, 40 randomly selected married couples were surveyed to determine the number of children in each family. The following results were obtained:

   3 1 0 4 1 3 2 2 0 2 0 2 2 1
   4 3 1 1 3 4 2 1 3 0 1 0 2 5
   1 2 3 0 0 1 2 3 1 2 0 2

   a. Organize the given data by creating a frequency distribution.
   b. Find the mean number of children per family.
   c. Find the median number of children per family.
   d. Find the mode number of children per family.
   e. Find the standard deviation of the number of children per family.
   f. Construct a histogram using single-valued classes of data.

3. The frequency distribution in Figure 4.154 lists the number of hours per day that a randomly selected sample of teenagers spent watching television. Where possible, determine what percent of the teenagers spent the following number of hours watching television.
   a. less than 4 hours
   b. not less than 6 hours
   c. at least 2 hours
   d. less than 2 hours
   e. at least 4 hours but less than 8 hours
   f. more than 3.5 hours

$x =$ Hours per Day	Frequency
$0 \le x < 2$	23
$2 \le x < 4$	45
$4 \le x < 6$	53
$6 \le x < 8$	31
$8 \le x \le 10$	17

Figure 4.154
Time watching television

4. To study the efficiency of its new oil-changing system, a local service station monitored the amount of time it took to change the oil in customers' cars. The frequency distribution in Figure 4.155 summarizes the findings.

$x =$ Time (in minutes)	Number of Customers
$3 \le x < 6$	18
$6 \le x < 9$	42
$9 \le x < 12$	64
$12 \le x < 15$	35
$15 \le x \le 18$	12

Figure 4.155
Time to change oil

a. Find the mean number of minutes to change the oil in a car.

b. Find the standard deviation of the amount of time to change the oil in a car.

c. Construct a histogram to represent the data.

5. If your scores on the first four exams (in this class) are 74, 65, 85, and 76, what score do you need on the next exam for your overall mean to be at least 80?

6. The mean salary of 12 men is $37,000, and the mean salary of 8 women is $28,000. Find the mean salary of all 20 people.

7. Timo and Henke golfed five times during their vacation. Their scores are given in Figure 4.156.

a. Find the mean score of each golfer. Who has the lowest mean?

b. Find the standard deviation of each golfer's scores.

c. Who is the more consistent golfer? Why?

Timo	103	99	107	93	92
Henke	101	92	83	96	111

Figure 4.156
Golf scores

8. Suzanne surveyed the prices for a quart of a certain brand of motor oil. The sample data, in dollars per quart, are summarized in Figure 4.157.

Price per Quart	Number of Stores
1.99	2
2.09	3
2.19	7
2.29	10
2.39	14
2.49	4

Figure 4.157
Price of motor oil

a. Find the mean and standard deviation of the prices.

b. What percentage of the data lies within one standard deviation of the mean?

c. What percentage of the data lies within two standard deviations of the mean?

d. What percentage of the data lies within three standard deviations of the mean?

9. Classify the following types of data as discrete, continuous, or neither.

a. weights of motorcycles

b. colors of motorcycles

c. number of motorcycles

d. ethnic background of students

e. number of students

f. amounts of time spent studying

10. What percentage of the standard normal z-distribution lies in the following intervals?

a. between $z = 0$ and $z = 1.75$

b. between $z = -1.75$ and $z = 0$

c. between $z = -1.75$ and $z = 1.75$

11. A large group of data is normally distributed with mean 78 and standard deviation 7.

a. Find the intervals that represent one, two, and three standard deviations of the mean.

b. What percentage of the data lies in each interval in part (a)?

c. Draw a sketch of the bell curve.

12. The time it takes a latex paint to dry is normally distributed. If the mean is $3\frac{1}{2}$ hours with a standard deviation of 45 minutes, find the probability that the drying time will be as follows.

a. less than 2 hours 15 minutes

b. between 3 and 4 hours

HINT: Convert everything to hours (or to minutes).

13. All incoming freshmen at a major university are given a diagnostic mathematics exam. The scores are normally distributed with a mean of 420 and a standard deviation of 45. If the student scores less than a certain score, he or she will have to take a review course. Find the cutoff score at which 34% of the students would have to take the review course.

14. Find the specified z-number.

a. $z_{0.4441}$ b. $z_{0.4500}$ c. $z_{0.1950}$ d. $z_{0.4975}$

15. A survey asked, "Do you think that the president is doing a good job?" Of the 1,200 Americans surveyed, 800 responded yes. For each of the following levels of confidence, find the sample proportion and the margin of error associated with the poll.

a. a 90% level of confidence

b. a 95% level of confidence

16. A survey asked, "Do you support capital punishment?" Of the 1,000 Americans surveyed, 750 responded no. For each of the following levels of confidence, find the sample proportion and the margin of error associated with the poll.

a. an 80% level of confidence

b. a 98% level of confidence

17. A sample consisting of 580 men and 970 women was asked various questions pertaining to international affairs. For a 95% level of confidence, find

the margin of error associated with the following samples.

a. the male sample

b. the female sample

c. the combined sample

18. A poll pertaining to environmental concerns had the following footnote: "Based on a sample of 1,098 adults, the margin of error is plus or minus 2 percentage points." Find the level of confidence of the poll.

19. A set of $n = 5$ ordered pairs has the following sums:

$$\Sigma x = 66 \quad \Sigma x^2 = 1{,}094 \quad \Sigma y = 273$$
$$\Sigma y^2 = 16{,}911 \quad \Sigma xy = 4{,}272$$

a. Find the line of best fit.

b. Predict the value of y when $x = 16$.

c. Predict the value of x when $y = 57$.

d. Find the coefficient of linear correlation.

e. Are the predictions in parts (b) and (c) reliable? Why or why not?

20. Given the ordered pairs (5, 38), (10, 30), (20, 33), (21, 25), (24, 18), and (30, 20),

a. Plot the ordered pairs. Do the ordered pairs exhibit a linear trend?

b. Find the line of best fit.

c. Predict the value of y when $x = 15$.

d. Plot the given ordered pairs and sketch the graph of the line of best fit on the same coordinate system.

e. Find the coefficient of linear correlation.

f. Is the prediction in part (c) reliable? Why or why not?

21. The value (in billions of dollars) of agricultural exports and imports in the United States for 1999–2003 are given in Figure 4.158.

a. Letting $x =$ the value of agricultural exports and $y =$ the value of agricultural imports, plot the data. Do the data exhibit a linear trend?

b. Find the line of best fit.

c. Predict the value of agricultural imports when the value of agricultural exports is 58.0 billion dollars.

Year	Agricultural Exports (billion dollars)	Agricultural Imports (billion dollars)
1999	48.4	37.7
2000	51.2	39.0
2001	53.7	39.4
2002	53.1	41.9
2003	59.5	47.3

Figure 4.158

Agricultural exports and imports (billion dollars), 1999–2003

Source: U.S. Department of Agriculture

d. Predict the value of agricultural exports when the value of agricultural imports is 47.0 billion dollars.

e. Find the coefficient of correlation.

f. Are the predictions in parts (a) and (b) reliable? Why or why not?

Answer the following questions using complete sentences and your own words.

CONCEPT QUESTIONS

22. What are the three measures of central tendency? Briefly explain the meaning of each.

23. What does standard deviation measure?

24. What is a normal distribution? What is the standard normal distribution?

25. What is a margin of error? How is it related to a sample proportion?

26. How do you measure the strength of a linear trend?

HISTORY QUESTIONS

27. What role did the following people play in the development of statistics?

• George Gallup

• Carl Friedrich Gauss

5

Finance

© Bob Daemmrich/The Image Works

Today, almost everyone borrows money, whether in the form of a student loan, a car loan, a credit card loan, or a home loan. Almost everyone saves money, through either a savings account at a bank, a money market account, or some form of annuity. To make educated decisions about such matters, you should understand the mathematics of finance.

Webster's Dictionary defines **finance** as "the system that includes the circulation of money, the granting of credit, the making of investments, and the provision of banking facilities." In this chapter, we investigate the mathematics of two aspects of finance: loans and non-speculative investments—such as savings accounts, money market accounts, certificates of deposit, and annuities. (Speculative investments are investments that involve some risk, such as stocks and real estate.)

Loans and investments are very similar financial transactions; both involve the flow of money from one party to another, the return of the money to its source, and the payment of a fee to the source for the use of the money. When you make a deposit in your savings account, you view the transaction as an investment, but the bank views it as a loan; you are lending the bank your money, which they will lend to another customer, perhaps to buy a house. When you borrow money to buy a car, you view the transaction as a loan, but the bank views it as an investment; the bank is investing its money in you in order to make a profit.

In this chapter, we explore some of the different forms of loans and investments, along with the various ways in which they are calculated, so that you can make the best choice when you have to make a financial decision. We investigate the different types of automobile loans available and discuss how home loans work. We also show how you can accumulate an incredibly large amount of money by saving only $50 a month.

5.1   Simple Interest

When an investor puts money into a savings account or buys a certificate of deposit (CD) or a Treasury bill, the investor expects to make a profit. The amount of money that is invested is called the **principal.** The profit is the **interest.** How much interest will be paid depends on the interest **rate** (usually expressed as a percent per year); the **term,** or length of time that the money is invested; and how the interest is calculated.

Some forms of investment and many short-term loans are calculated as simple interest. **Simple interest** means that the amount of interest is computed as a percent-per-year of the principal.

> **Simple Interest Formula**
>
> The **simple interest** I on a principal P at an annual rate of interest r for a term of t years is
>
> $$I = Prt$$

EXAMPLE 1   Tom and Betty buy a two-year CD that pays 5.1% simple interest from their bank for $150,000. (Many banks pay simple interest on larger CDs and compound interest on smaller CDs.) They invest $150,000, so the principal is $P = \$150,000$. The interest rate is $r = 5.1\% = 0.051$, and the term is $t = 2$ years.

a. Find the interest that the investment earns.
b. Find the value of the CD at the end of its term.

Solution   **a.** Using the Simple Interest Formula, we have

$$I = Prt$$
$$= 150,000 \cdot 0.051 \cdot 2$$
$$= \$15,300$$

b. Two years in the future, the bank will pay Tom and Betty

$$\$150,000 \text{ principal} + \$15,300 \text{ interest} = \$165,300$$

This is called the *future value* of the CD, because it is what the CD will be worth in the future. ∎

Future Value

In Example 1, we found the future value by adding the principal to the interest. The **future value** FV is always the sum of the principal P plus the interest I. If we combine this fact with the Simple Interest Formula, we can get a formula for the future value:

$$FV = P + I$$
$$= P + Prt \quad \text{From the Simple Interest Formula}$$
$$= P(1 + rt) \quad \text{Factoring}$$

> **Simple Interest Future Value Formula**
>
> The **future value** FV of a principal P at an annual rate of interest r for t years is
>
> $$FV = P(1 + rt)$$

There is an important distinction between the variables FV, I, P, r, and t: FV, I, and P measure amounts of money, whereas r and t do not. For example, consider the interest rate r and the interest I. Frequently, people confuse these two. However, the interest rate r is a percentage, and the interest I is an amount of money; Tom and Betty's interest rate is $r = 5.1\%$, but their interest is $I = \$15,300$. To emphasize this distinction, we will always use capital letters for variables that measure amounts of money and lowercase letters for other variables. We hope this notation will help you avoid substituting $5.1\% = 0.051$ for I when it should be substituted for r.

One of the more common uses of simple interest is a short-term (such as a year or less) loan that requires a single lump sum payment at the end of the term. Businesses routinely obtain these loans to purchase equipment or inventory, to pay operating expenses, or to pay taxes. A **lump sum** payment is a single payment that pays off an entire loan. Some loans require smaller monthly payments rather than a single lump sum payment. We will investigate that type of loan later in this section and in Section 5.4.

ThomsonNOW™
for Liberal Arts Mathematics
www.thomsonedu.com/1pass.

EXAMPLE 2 Espree Clothing borrowed $185,000 at $7\frac{1}{4}\%$ from January 1 to February 28.

a. Find the future value of the loan.
b. Interpret the future value.

Solution

a. We are given $p = \$185,000$ and $r = 7\frac{1}{4}\% = 0.0725$. There are $31 + 28 - 1 = 58$ days from January 1 to February 28. To understand why we subtract 1, consider this simpler problem. Clearly there is only 1 day from January 1 to January 2. However, if we count the days, we get 2. There are 2 days from January 1 *through* January 2, but there is 1 day from January 1 *to* January 2. Similarly, there are 59 days from January 1 *through* February 28, but there are 58 days from January 1 *to* February 28. The word "to" means "do not count the last day," and the word "through" means "count the last day." Using dimensional analysis, as discussed in Appendix E, we get

$$t = 58 \text{ days} \cdot \frac{1 \text{ year}}{365 \text{ days}} = \frac{58}{365} \text{ years}$$

Using the Simple Interest Future Value Formula, we have

$$FV = P(1 + rt)$$
$$= 185,000 \left(1 + 0.0725 \cdot \frac{58}{365} \right)$$
$$= 187,131.3013\ldots$$
$$\approx \$187,131.30$$

b. This means that Espree has agreed to make a lump sum payment to their lender $187,168.05 at the end of February.

> ✔ As a rough check, notice that the answer is somewhat higher than the original $185,000. This is as it should be—the future value includes the original $185,000 plus interest.

■

Notice that in Example 2, we do not use $t = 2$ months $= \frac{2}{12}$ year. If we did, we would get $187,235.42, an inaccurate answer. Instead, we use the number of days, converted to years.

In Example 2, we naturally used 365 days per year, but some institutions traditionally count a year as 360 days and a month as 30 days (especially if that tradition works in their favor). This is a holdover from the days before calculators and computers—the numbers were simply easier to work with. Also, we used normal round-off rules to round $187,168.0479 \ldots$ to $187,168.05$; some institutions round off some interest calculations in their favor. In this book, we will count a year as 365 days and use normal round-off rules (unless stated otherwise).

The issue of how a financial institution rounds off its interest calculations might seem unimportant. After all, we are talking about a difference of a fraction of a penny. However, consider one classic form of computer crime: the round-down fraud, performed on a computer system that processes a large number of accounts. Frequently, such systems use the normal round-off rules in their calculations and keep track of the difference between the theoretical account balance if no rounding off is done and the actual account balance with rounding off. Whenever that difference reaches or exceeds 1¢, the extra penny is deposited in (or withdrawn from) the account. A fraudulent computer programmer can write the program so that the extra penny is deposited in his or her own account. This fraud is difficult to detect because the accounts appear to be balanced. While each individual gain is small, the total gain can be quite large if a large number of accounts is processed regularly.

A written contract signed by the lender and the borrower is called a **loan agreement** or a **note**. The **maturity value** of the note (or just the **value** of the note) refers to the note's future value. Thus, the value of the note in Example 1 was $187,168.05. This is what the note is worth to the lender in the future—that is, when the note matures.

National Debt

For every year from 1970 through 1998 (and for most years since 1931), the U.S. federal budget called for deficit spending, that is, spending more money than is received. Deficit spending grew tremendously during the Reagan and Bush presidencies, but it ceased during the Clinton years. The 2005 budget (see Figures 5.1(a) and (b)) calls for deficit spending of $364 billion. The total national debt held by the public in October 2005 was $8,003 billion; this amounts to a debt of almost $27,000 per person in the United States. President Bush's 2005 budget calls for spending $178 billion for interest on the national debt.

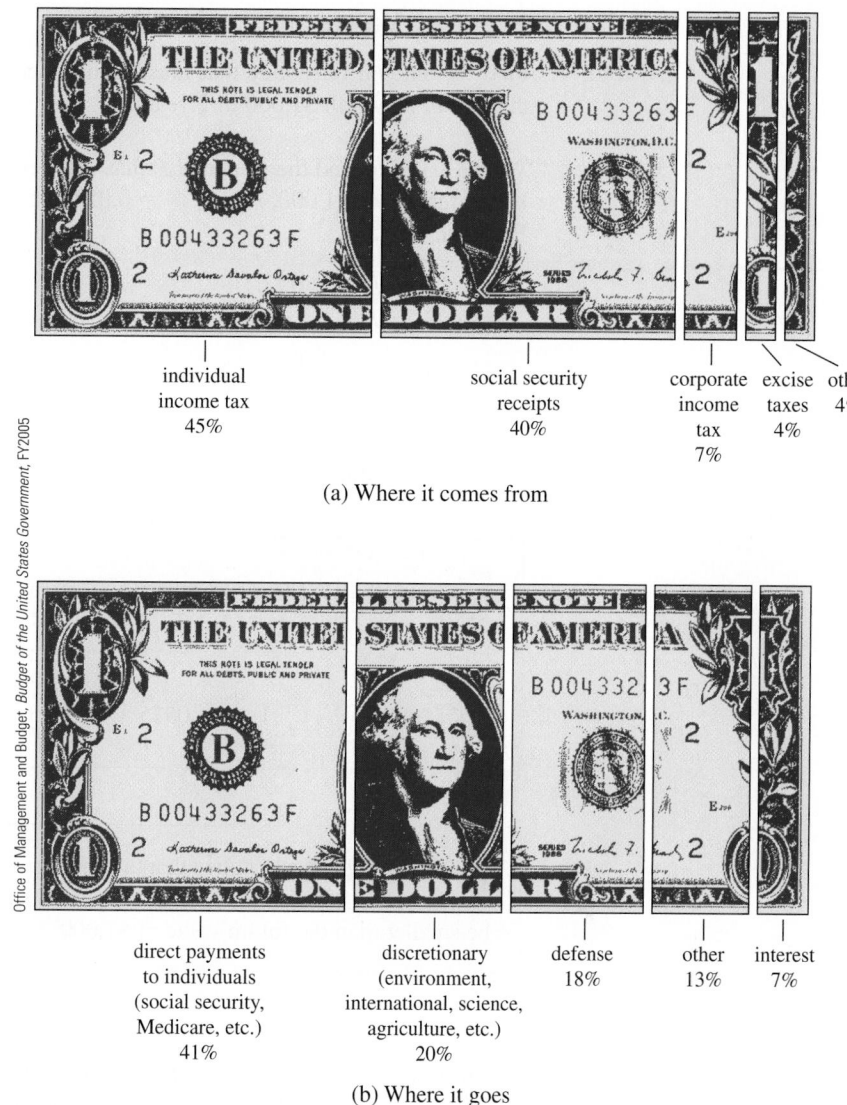

(a) Where it comes from

individual income tax 45% · social security receipts 40% · corporate income tax 7% · excise taxes 4% · other 4%

direct payments to individuals (social security, Medicare, etc.) 41% · discretionary (environment, international, science, agriculture, etc.) 20% · defense 18% · other 13% · interest 7%

(b) Where it goes

Office of Management and Budget, *Budget of the United States Government, FY2005*

Figure 5.1
The federal government dollar (2005)

EXAMPLE 3   Find the simple interest rate that was paid on the national debt in 2005.

Solution   From the information given, we know that I = \$178 billion, P = \$8,003 billion, and t = 1 year. Using the Simple Interest Formula, we have

$$I = Prt$$
$$\$178 \text{ billion} = \$8,003 \text{ billion} \cdot r \cdot 1$$
$$r = \$178 \text{ billion}/\$8,003 \text{ billion} = 0.0224 \ldots \approx 2.2\,\%$$

∎

Present Value

EXAMPLE 4 Find the amount of money that must be invested now at a $5\frac{3}{4}\%$ simple interest so that it will be worth $1,000 in 2 years.

Solution We are asked to find the principal P that will generate a future value of $1,000. We know that $FV = 1,000$, $r = 5\frac{3}{4}\% = 0.0575$, and $t = 2$. Using the Future Value Formula, we have

$$FV = P(1 + rt)$$
$$1,000 = P(1 + 0.0575 \cdot 2)$$
$$P = \frac{1,000}{1 + 0.0575 \cdot 2} \qquad \text{dividing}$$
$$= 896.86099$$
$$\approx \$896.86$$

As a rough check, notice that the answer is somewhat smaller than $1,000. This is as it should be—the principal does not include interest, so it should be smaller than the future value of $1,000.

In Example 4, the investment is worth $1,000 two years *in the future;* that is, $1000 is the investment's *future value.* But the same investment is worth $896.86 *in the present.* For this reason, we say that $896.86 is the investment's **present value.** In this case, this is the same thing as the principal; it is just called the *present value* to emphasize that this is its value in the present.

Add-on Interest

Purchases made at car dealers, appliance stores, and furniture stores can be financed through the store itself. Frequently, this type of loan involves **add-on interest,** where the total amount to be repaid is computed with the simple interest future value formula. The payment is found by dividing that total amount by the number of payments.

EXAMPLE 5 Tom and Betty purchase $1,300 worth of kitchen appliances at Link's Appliance Mart. Link's finances their purchase with a 10% two-year add-on interest loan. Link's requires that Tom and Betty make a $200 down payment, followed by monthly payments. Find the monthly payment.

Solution The loan amount is $P = 1{,}300 - 200 = \$1{,}100$. The total amount due is

$$FV = P(1 + rt)$$
$$= 1{,}100(1 + 0.10 \cdot 2)$$
$$= \$1{,}320$$

The total amount due is spread out over twenty-four monthly payments, so the monthly payment is $\$1{,}320/24 = \$55$. This means that Tom and Betty have to pay Link's $200 when they purchase the appliances and then $55 a month for 24 months. ■

Credit Card Finance Charge

Credit cards have become part of the American way of life. Purchases made with a credit card are subject to a finance charge, but there is frequently a grace period, and no finance charge is assessed if full payment is received by the payment due date. One of the most common methods of calculating credit card interest is the **average daily balance** method. To find the average daily balance, the balance owed on the account is found for each day in the billing period, and the results are averaged. The finance charge consists of simple interest, charged on the result.

EXAMPLE 6 The activity on Tom and Betty's Visa account for one billing is shown below. The billing period is October 15 through November 14, the previous balance was $346.57, and the annual in rate is 21%.

October 21	payment	$50.00
October 23	restaurant	$42.14
November 7	clothing	$18.55

a. Find the average daily balance.
b. Find the finance charge.

Solution **a.** To find the average daily balance, we have to know the balance for each day in the billing period and the number of days at that balance, as shown in Figure 5.2. The average daily balance is then the weighted average of each daily balance, with each balance weighted to reflect the number of days at that balance.

Time Interval	Days	Daily Balance
October 15–20	$20 - 14 = 6$	$346.57
October 21–22	$22 - 20 = 2$	$\$346.57 - \$50 = \$296.57$
October 23–November 6 (October has 31 days)	$31 - 22 = 9$ $9 + 6 = 15$	$\$296.57 + \$42.14 = \$338.71$
November 7–14	$14 - 6 = 8$	$\$338.71 + \$18.55 = \$357.26$

Figure 5.2
Preparing to find the average daily balance

$$\text{Average daily balance} = \frac{6 \cdot 346.57 + 2 \cdot 296.57 + 15 \cdot 338.71 + 8 \cdot 357.26}{6 + 2 + 15 + 8}$$

$$= 342.29967\ldots$$

$$\approx \$342.30$$

b. The finance charge consists of simple interest on the average daily balance. We know that $P = 342.29967\ldots$, $r = 21\% = 0.21$, and $t = 31$ days $= 31/365$ years. Using the simple interest formula, we have

$$I = Prt$$
$$= 342.29967\ldots 0.21 \cdot 31/365$$
$$= 6.1051258\ldots$$
$$\approx \$6.11$$

Credit Card Finance Charge

To find the credit card finance charge with the average daily balance method, do the following:

1. Find the balance for each day in the billing period and the number of days at that balance.
2. The average daily balance is the weighted average of these daily balances, weighted to reflect the number of days at that balance.
3. The finance charge is simple interest applied to the average daily balance.

How Many Days?

Financial calculations usually involve computing the term t either in a whole number of years or in the number of days converted to years. This requires that you know the number of days in each month. As you can see from Figure 5.3, the months alternate between 31 days and 30 days, with two exceptions:

- February has 28 days (29 in leap years).
- The alternation does not happen from July to August.

January	31 days	May	31 days	September	30 days
February	28 days	June	30 days	October	31 days
March	31 days	July	31 days	November	30 days
April	30 days	August	31 days	December	31 days

Figure 5.3
The number of days in a month

Historical Note The History of Credit Cards

Credit cards and charge cards were first used in the United States in 1915, when Western Union issued a metal card to some of its regular customers. Holders of these cards were allowed to defer their payments and were assured of prompt and courteous service. Shortly thereafter, several gasoline companies, hotels, department stores, and railroads issued cards to their preferred customers.

The use of credit and charge cards virtually ceased during World War II, owing to government restrictions on credit. In 1950, a New York lawyer established the Diners' Club after being embarrassed when he lacked sufficient cash to pay a dinner bill. A year later, the club had billed more than $1 million. Carte Blanche and the American Express card soon followed. These "travel and entertainment" cards were and are attractive to the public because they provide a convenient means of paying restaurant, hotel, and airline bills. They eliminate both the possibility of an out-of-town check being refused and the need to carry a large amount of cash.

The first bank card was issued in 1951 by Franklin National Bank in New York. Within a few years, 100 other banks had followed suit. Because these bank cards were issued by individual banks, they were accepted for use only in relatively small geographical areas. In 1965, the California-based Bank of America began licensing other banks (both in the United States and abroad) to issue BankAmericards. Other banks formed similar groups, which allowed customers to use the cards out of state.

In 1970, the Bank of America transferred administration of its bank card program to a new company owned by the banks that issued the card. In 1977, the card was renamed Visa. MasterCard was originally created by several California banks (United California Bank, Wells Fargo, Crocker Bank, and the Bank of California) to compete with BankAmericard. It was then licensed to the First National Bank of Louisville, Kentucky and the Marine Midland Bank of New York.

Credit cards are now a part of the American way of life. In 2004 we started to use credit cards, debit cards and other forms of electronic bill paying more than we use paper checks. Credit card debt has soared, especially among college students who are already in debt with college loans. Credit card issuers have been accused of targeting college students, who tend to make minimum payments and thus incur much more interest. See Exercises 45–48.

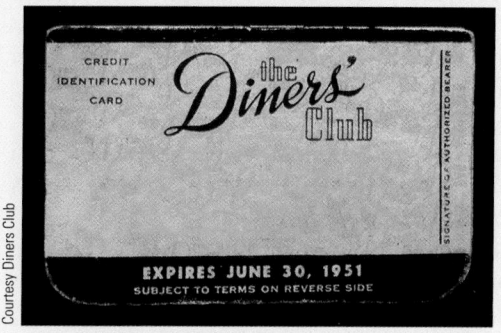

An early Diners' Card

5.1 Exercises

In Exercises 1–4, find the number of days.

1. September 1 through October 31 of the same year
2. July 1 through December 31 of the same year
3. April 1 through July 10 of the same year
4. March 10 through December 20 of the same year

In Exercises 5–10, find the simple interest of the given loan amount.

5. $2,000 borrowed at 8% for 3 years
6. $35,037 borrowed at 6% for 2 years

7. $420 borrowed at $6\frac{3}{4}$% for 325 days
8. $8,950 borrowed at $9\frac{1}{2}$% for 278 days
9. $1,410 borrowed at $12\frac{1}{4}$% from September 1 through October 31 of the same year
10. $5,682 borrowed at $11\frac{3}{4}$% from July 1 through December 31 of the same year

In Exercises 11–14, find the future value of the given present value.

11. Present value of $3,670 at $2\frac{3}{4}$% for 7 years
12. Present value of $4,719 at 14.1% for 11 years

13. Present value of $12,430 at $5\frac{7}{8}$% for 660 days
14. Present value of $172.39 at 6% for 700 days

In Exercises 15–20, find the maturity value of the given loan amount.

15. $1,400 borrowed at $7\frac{1}{8}$% for 3 years
16. $3,250 borrowed at $8\frac{1}{2}$% for 4 years
17. $5,900 borrowed at $14\frac{1}{2}$% for 112 days
18. $2,720 borrowed at $12\frac{3}{4}$% for 275 days
19. $16,500 borrowed at $11\frac{7}{8}$% from April 1 through July 10 of the same year
20. $2,234 borrowed at $12\frac{1}{8}$% from March 10 through December 20 of the same year

In Exercises 21–26 find the present value of the given future value.

21. Future value $8,600 at $9\frac{1}{2}$% simple interest for 3 years
22. Future value $420 at $5\frac{1}{2}$% simple interest for 2 years
23. Future value $1,112 at $3\frac{5}{8}$% simple interest for 512 days
24. Future value $5,750 at $4\frac{7}{8}$% simple interest for 630 days
25. Future value $1,311 at $6\frac{1}{2}$% simple interest from February 10 to October 15 of the same year
26. Future value $4,200 at $6\frac{3}{4}$% simple interest from April 12 to November 28 of the same year

27. If you borrow $1,000 at 8.5% interest and the loan requires a lump sum payment of $1,235.84, what is the term of the loan?
28. If you borrow $1,700 at 5.25% interest and the loan requires a lump sum payment of $1,863.12, what is the term of the loan?

29. The Square Wheel Bicycle store has found that they sell most of their bikes in the spring and early summer. On March 1, they borrowed $226,500 to buy bicycles. They are confident that they can sell most of these bikes by August 1. Their loan is at $6\frac{7}{8}$% interest. What size lump sum payment would they have to make on August 1 to pay off the loan?
30. Ernie Bilko has a business idea. He wants to rent an abandoned gas station for just the months of November and December. He will convert the gas station into a drive-through Christmas wrapping station. Customers will drive in, drop off their gifts, return the next day, and pick up their wrapped gifts. He needs $338,200 to rent the gas station, purchase wrapping paper, hire workers, and advertise. If he borrows this amount at $6\frac{1}{2}$% interest for those two

months, what size lump sum payment will he have to make to pay off the loan?

31. Alice Cohen buys a two-year-old Honda from a car dealer for $19,000. She puts $500 down and finances the rest through the dealer at 13% add-on interest. If she agrees to make 36 monthly payments, find the size of each payment.
32. Sven Lundgren buys a three-year-old Chevrolet from a car dealer for $14,600. He puts $300 down and finances the rest through the dealer at 12.5% add-on interest. If he agrees to make 24 monthly payments, find the size of each payment.
33. Ray and Teresa Martinez buy a bedroom set at Fowler's Furniture for $6,700. They put $500 down and finance the rest through the store at 9.8% add-on interest. If they agree to make 36 monthly payments, find the size of each payment.
34. Helen and Dick Davis buy a refrigerator at Appliance Barn for $1,200. They put $100 down and finance the rest through the store at 11.6% add-on interest. If they agree to make 36 monthly payments, find the size of each payment.

In Exercises 35–36, use the following information. When "trading up," it is preferable to sell your old house before buying your new house because that allows you to use the proceeds from selling your old house to buy your new house. When circumstances do not allow this, the homeowner can take out a bridge loan.

35. Dale and Claudia have sold their house, but they will not get the proceeds from the sale for an estimated $2\frac{1}{2}$ months. The owner of the house they want to buy will not hold the house that long. Dale and Claudia have two choices: let their dream house go or take out a bridge loan. The bridge loan would be for $110,000, at 7.75% simple interest, due in 90 days.
    a. How big of a check would they have to write in 90 days?
    b. How much interest would they pay for this loan?
36. Tina and Mike have sold their house, but they will not get the proceeds from the sale for an estimated 3 months. The owner of the house they want to buy will not hold the house that long. Tina and Mike have two choices: let their dream house go or take out a bridge loan. The bridge loan would be for $85,000, at 8.5% simple interest, due in 120 days.
    a. How big of a check would they have to write in 120 days?
    b. How much interest would they pay for this loan?

37. The activity on Stuart Ratner's Visa account for one billing period is shown below. Find the average daily balance and the finance charge if the billing period is April 11 through May 10, the previous balance was $126.38, and the annual interest rate is 18%.

April 15	payment	$15.00
April 22	DVD store	$25.52
May 1	clothing	$32.18

38. The activity on Marny Zell's MasterCard account for one billing period is shown below. Find the average daily balance and the finance charge if the billing period is June 26 through July 25, the previous balance was $396.68, and the annual interest rate is 19.5%.

June 30	payment	$100.00
July 2	gasoline	$36.19
July 10	restaurant	$53.00

39. The activity on Denise Hellings' Sears account for one billing period is shown below. Find the average daily balance and the finance charge if the billing period is March 1 through March 31, the previous balance was $157.14, and the annual interest rate is 21%.

| March 5 | payment | $25.00 |
| March 17 | tools | $36.12 |

40. The activity on Charlie Wilson's Visa account for one billing period is shown below. Find the average daily balance and the finance charge if the billing period is November 11 through December 10, the previous balance was $642.38, and the annual interest rate is 20%.

November 15	payment	$150
November 28	office supplies	$23.82
December 1	toy store	$312.58

41. Donovan and Pam Hamilton bought a house from Edward Gurney for $162,500. In lieu of a 10% down payment, Mr. Gurney accepted 5% down at the time of the sale and a promissory note from the Hamiltons for the remaining 5%, due in 4 years. The Hamiltons also agreed to make monthly interest payments to Mr. Gurney at 10% interest until the note expires. The Hamiltons obtained a loan from their bank for the remaining 90% of the purchase price. The bank in turn paid the sellers the remaining 90% of the purchase price, less a sales commission of 6% of the purchase price, paid to the sellers' and the buyers' real estate agents.
a. Find the Hamiltons' down payment.
b. Find the amount that the Hamiltons borrowed from their bank.

c. Find the amount that the Hamiltons borrowed from Mr. Gurney.
d. Find the Hamilton's monthly interest-only payment to Mr. Gurney.
e. Find Mr. Gurney's total income from all aspects of the down payment (including the down payment, the amount borrowed under the promissory note, and the monthly payments required by the promissory note).
f. Find Mr. Gurney's net income from the Hamiltons' bank.
g. Find Mr. Gurney's total income from all aspects of the sale.

42. George and Peggy Fulwider bought a house from Sally Sinclair for $233,500. In lieu of a 10% down payment, Ms. Sinclair accepted 5% down at the time of the sale and a promissory note from the Fulwiders for the remaining 5%, due in 4 years. The Fulwiders also agreed to make monthly interest payments to Ms. Sinclair at 10% interest until the note expires. The Fulwiders obtained a loan from their bank for the remaining 90% of the purchase price. The bank in turn paid the sellers the remaining 90% of the purchase price, less a sales commission of 6% of the purchase price, paid to the sellers' and the buyers' real estate agents.
a. Find the Fulwiders' down payment.
b. Find the amount that the Fulwiders borrowed from their bank.
c. Find the amount that the Fulwiders borrowed from Ms. Sinclair.
d. Find the Fulwiders' monthly interest-only payment to Ms. Sinclair.
e. Find Ms. Sinclair's total income from all aspects of the down payment (including the down payment, the amount borrowed under the promissory note, and the monthly payments required by the promissory note).
f. Find Ms. Sinclair's net income from the Fulwiders' bank.
g. Find Ms. Sinclair's total income from all aspects of the sale.

43. The Clintons bought a house from the Bushes for $389,400. In lieu of a 20% down payment, the Bushes accepted a 10% down payment at the time of the sale and a promissory note from the Clintons for the remaining 10%, due in 4 years. The Clintons also agreed to make monthly interest payments to the Bushes at 11% interest until the note expires. The Clintons obtained a loan for the remaining 80% of the purchase price from their bank. The bank in turn paid the sellers the remaining 80% of the purchase price, less a sales commission of 6% of the sales price paid to the sellers' and the buyers' real estate agents.

a. Find the Clintons' down payment.

b. Find the amount that the Clintons borrowed from their bank.

c. Find the amount that the Clintons borrowed from the Bushes.

d. Find the Clintons' monthly interest-only payment to the Bushes.

e. Find the Bushes' total income from all aspects of the down payment (including the down payment, the amount borrowed under the promissory note, and the monthly payments required by the note).

f. Find the Bushes' income from the Clinton's bank.

g. Find the Bushes' total income from all aspects of the sale.

44. Sam Needham bought a house from Sheri Silva for $238,300. In lieu of a 20% down payment, Ms. Silva accepted a 10% down payment at the time of the sale and a promissory note from Mr. Needham for the remaining 10%, due in 4 years. Mr. Needham also agreed to make monthly interest payments to Ms. Silva at 9% interest until the note expires. Mr. Needham obtained a loan for the remaining 80% of the purchase price from his bank. The bank in turn paid Ms. Silva the remaining 80% of the purchase price, less a sales commission (of 6% of the sales price) paid to the sellers' and the buyers' real estate agents.

a. Find Mr. Needham's down payment.

b. Find the amount that Mr. Needham borrowed from his bank.

c. Find the amount that Mr. Needham borrowed from the Ms. Silva.

d. Find Mr. Needham's monthly interest-only payment to Ms. Silva.

e. Find Ms. Silva's total income from all aspects of the down payment (including the down payment, the amount borrowed under the promissory note, and the monthly payments required by the note).

f. Find Ms. Silva's income from Mr. Needham's bank.

g. Find Ms. Silva's total income from all aspects of the sale.

45. Your credit card has a balance of $1,000. Its interest rate is 21%. You have stopped using the card, because you don't want to go any deeper into debt.

Each month, you make the minimum required payment of $20.

a. During the January 10 through February 9 billing period, you pay the minimum required payment on January 25th. Find the average daily balance, the finance charge and the new balance. (The new balance includes the finance charge.)

b. During the February 10 through March 9 billing period, you pay the minimum required payment on February 25th. Find the average daily balance, the finance charge and the new balance. (The new balance includes the finance charge.)

c. During the March 10 through April 9 billing period, you pay the minimum required payment on March 25th. Find the average daily balance, the finance charge and the new balance. (The new balance includes the finance charge.)

d. Discuss the impact of making the minimum required payment, both on yourself and on the credit card issuer.

46. Your credit card has a balance of $1,200. Its interest rate is 20.5%. You have stopped using the card, because you don't want to go any deeper into debt. Each month, you make the minimum required payment of $24.

a. During the September 10 through October 9 billing period, you pay the minimum required payment on September 25th. Find the average daily balance, the finance charge and the new balance. (The new balance includes the finance charge.)

b. During the October 10 through November 9 billing period, you pay the minimum required payment on October 25th. Find the average daily balance, the finance charge and the new balance. (The new balance includes the finance charge.)

c. During the November 10 through December 9 billing period, you pay the minimum required payment on November 25th. Find the average daily balance, the finance charge and the new balance. (The new balance includes the finance charge.)

d. Discuss the impact of making the minimum required payment, both on yourself and on the credit card issuer.

47. Your credit card has a balance of $1,000. Its interest rate is 21%. You have stopped using the card, because you don't want to go any deeper into debt. Each month, you make the minimum required payment. Your credit card issuer recently changed their minimum required payment policy, in response to the Bankruptcy Abuse Prevention and Consumer

Protection Act of 2005. As a result, your minimum required payment is $40.

a. During the January 10 through February 9 billing period, you pay the minimum required payment on January 25ᵗʰ. Find the average daily balance, the finance charge and the new balance. (The new balance includes the finance charge.)

b. During the February 10 through March 9 billing period, you pay the minimum required payment on February 25ᵗʰ. Find the average daily balance, the finance charge and the new balance. (The new balance includes the finance charge.)

c. During the March 10 through April 9 billing period, you pay the minimum required payment on March 25ᵗʰ. Find the average daily balance, the finance charge and the new balance. (The new balance includes the finance charge.)

d. Compare the results of parts (a) through (c) with those of Exercise 45. Discuss the impact of the credit card issuer's change in their minimum required payment policy.

48. Your credit card has a balance of $1,200. Its interest rate is 20.5%. You have stopped using the card, because you don't want to go any deeper into debt. Each month, you make the minimum required payment. Your credit card issuer recently changed their minimum required payment policy, in response to the Bankruptcy Abuse Prevention and Consumer Protection Act of 2005. As a result, your minimum required payment is $48.

a. During the September 10 through October 9 billing period, you pay the minimum required payment on September 25ᵗʰ. Find the average daily balance, the finance charge and the new balance. (The new balance includes the finance charge.)

b. During the October 10 through November 9 billing period, you pay the minimum required payment on October 25ᵗʰ. Find the average daily balance, the finance charge and the new balance. (The new balance includes the finance charge.)

c. During the November 10 through December 9 billing period, you pay the minimum required payment on November 25ᵗʰ. Find the average daily balance, the finance charge and the new balance. (The new balance includes the finance charge.)

d. Compare the results of parts (a) through (c) with those of Exercise 46. Discuss the impact of the credit card issuer's change in their minimum required payment policy.

Answer the following questions using complete sentences and your own words.

CONCEPT QUESTIONS

49. Could Exercises 5–26 all be done with the Simple Interest Formula? If so, how? Could Exercises 5–26 all be done with the Simple Interest Future Value Formula? If so, how? Why do we have both formulas?

50. Which is always higher: future value of principal? Why?

HISTORY QUESTIONS

51. Who offered the first credit card?
52. What was the first post–World War II credit card?
53. Who created the first post–World War II credit card?
54. What event prompted the creation of the first post–World War II credit card?
55. What was the first interstate bank card?

Compound Interest

Many forms of investment (including savings accounts) earn **compound interest,** where interest is periodically paid on the existing account balance, which includes both the original principal and previous interest payments. The **compounding period** is the time period between successive interest payments. Usually, interest is compounded annually, semiannually, quarterly, monthly, or daily.

Earning interest on previous interest payments results in earnings that are significantly higher over a long period of time. It is important that you understand this difference in order to make wise financial decisions.

Compound Interest as Simple Interest, Repeated

EXAMPLE 1   Tom and Betty deposit $1,000 into their new bank account. The account pays 8% interest compounded quarterly. This means that interest is computed and deposited every quarter of a year. Find the account balance after six months, using the Simple Interest Future Value formula to compute the balance at the end of each quarter.

Solution   At the end of the first quarter, $P = \$1,000$, $r = 8\% = 0.08$, and $t =$ one quarter or $\frac{1}{4}$ year.

$$FV = P(1 + rt)$$
$$= 1,000(1 + 0.08 \cdot \tfrac{1}{4})$$
$$= 1,000(1 + 0.02)$$
$$= \$1,020$$

This means that there is $1,020 in Tom and Betty's account at the end of the first quarter. It also means that the second quarter's interest will be paid on this new principal. So at the end of the second quarter, $P = \$1,020$ and $r = 0.08$. Note that $t = \frac{1}{4}$, not $\frac{2}{4}$, because we are computing interest for one quarter.

$$FV = P(1 + rt)$$
$$= 1,020(1 + 0.08 \cdot \tfrac{1}{4})$$
$$= 1,020(1 + 0.02)$$
$$= \$1,040.40$$

At the end of 6 months, the account balance is $1,040.40.

 As a rough check, notice that the future value is slightly higher than the principal, as it should be.

■

The Compound Interest Formula

This process would become tedious if we were computing the balance after 20 years. Because of this, compound interest problems are usually solved with their own formula.

Notice that for each quarter's calculation, we multiplied the annual rate of 8% (0.08) by the time 1 quarter ($\frac{1}{4}$ year) and got 2% (0.02). This 2% is the **quarterly rate** (or more generally, the *periodic rate*). A **periodic rate** is any rate that is pro-rated in this manner from an annual rate.

If i is the periodic interest rate, then the future value at the end of the first period is

$$FV = P(1 + i)$$

Because this is the account balance at the beginning of the second period, it becomes the new principal. The account balance at the end of the second period is

$$FV = P(1 + i) \cdot (1 + i) \quad \text{Substituting } P(1 + i) \text{ for } P$$
$$= P(1 + i)^2$$

This means that $P(1 + i)^2$ is the account balance at the beginning of the third period, and the future value at the end of the third period is

$$FV = \left[P(1 + i)^2\right] \cdot (1 + i) \quad \text{substituting } P(1 + i)^2 \text{ for } P$$
$$= P(1 + i)^3$$

If we generalize these results, we get the Compound Interest Formula.

Compound Interest Formula

If initial principal P earns compound interest at a periodic interest rate i for n periods, the future value is

$$FV = P(1 + i)^n$$

Notice that we have maintained our variables tradition: i and n do not measure amounts of money, so they are not capital letters. We now have three interest-related variables:

- r, the annual interest rate (not an amount of money)
- i, the periodic interest rate (not an amount of money)
- I, the interest itself (an amount of money)

EXAMPLE 2   Use the Compound Interest Formula to recompute Tom and Betty's account balance from Example 1.

Solution   Their 8% interest is compounded quarterly, so each quarter they earn a quarter of $8\% = \frac{1}{4} \cdot 8\% = 2\% = 0.02$. Also, n counts the number of quarters, so $n = 2$.

$$FV = P(1 + i)^n$$
$$= 1,000(1 + 0.02)^2 = \$1,040.40$$

Compound Interest Compared with Simple Interest

As we saw at the beginning of this section, compound interest is just simple interest, repeated. However, there can be profound differences in their results.

EXAMPLE 3   In 1777, it looked as though the Revolutionary War was about to be lost. George Washington's troops were camped at Valley Forge. They had minimal supplies, and the winter was brutal. According to a 1990 class action suit, Jacob DeHaven, a wealthy Pennsylvania merchant, saved Washington's troops and the revolutionary cause by loaning Washington $450,000. The suit, filed by DeHaven's descendants, asked the government to repay the still-outstanding loan plus compound interest at the then-prevailing rate of 6%. (*Source: New York Times*, May 27, 1990.) How much did the government owe on the 1990 anniversary of the loan if the interest is compounded monthly?

Solution   The principal is $P = \$450,000$. If 6% interest is paid each year, then 1/12th of 6% is paid each month, and $i = \frac{1}{12} \cdot 6\% = 0.06/12$. The term is

$$n = 213 \text{ years}$$
$$= 213 \text{ years} \cdot \frac{12 \text{ months}}{1 \text{ year}}$$
$$= 2{,}556 \text{ months}$$

Using the Compound Interest Formula, we get

$$FV = P(1 + i)^n$$
$$= 450{,}000(1 + 0.06/12)^{2556}$$
$$= 1.547627234\ldots \times 10^{11}$$
$$\approx \$154{,}762{,}723{,}400$$

The DeHavens sued for this amount, but they also stated that they were willing to accept a more reasonable payment.

 As a rough check, notice that the future value is much higher than the principal, as it should be because of the long time period.

Some sources have questioned the DeHavens' claim: "There is also no evidence to support the claim of the DeHaven family that their ancestor Jacob DeHaven lent George Washington \$450,000 in cash and supplies while the army was encamped at Valley Forge. This tradition first appeared in print in a history of the DeHaven family penned by Howard DeHaven Ross. Periodically, the descendants of Jacob DeHaven make attempts to get the "loan" repaid with interest. . . . This remarkably persistent tradition has been thoroughly debunked by Judith A. Meier, of the Historical Society of Montgomery County, whose genealogical research revealed that there were no DeHavens living in the immediate area until after 1790 and that Jacob DeHaven had never been rich enough to make such a fabulous loan." (*Source:* Lorett Treese, *Valley Forge: Making and Remaking a National Symbol,* University Park, PA: Pennsylvania University Press, 1995.)

EXAMPLE 4   How much would the government have owed the DeHavens if the interest was simple interest?

Solution   With simple interest, we use r and t, which are *annual* figures, rather than i and n, which are *periodic* figures. So $r = 6\% = 0.06$, and $t = 30$ years.

$$FV = P(1 + rt)$$
$$= 450000(1 + 0.06 \cdot 213)$$
$$= \$6{,}201{,}000$$

Simple interest would have required a payment of only \$6 million. This is a lot, but not in comparison with the \$155 billion payment required by compound interest.   ■

In Examples 3 and 4, the future value with compound interest is almost 25,000 times the future value with simple interest. Over longer periods of time, compound interest is immensely more profitable to the investor than simple interest, because compound interest gives interest on interest. Similarly, compounding more frequently (daily rather than quarterly, for example) is more profitable to the investor.

The effects of the size of the time interval and the compounding period can be seen in Figure 5.4.

Type of interest	After 1 year	Future value of \$1000 at 10% interest	
		After 10 years	After 30 years
Simple interest	\$1100.00	\$2000.00	\$ 4,000.00
Compounded annually	\$1100.00	\$2593.74	\$17,449.40
Compounded quarterly	\$1103.81	\$2685.06	\$19,358.15
Compounded monthly	\$1104.71	\$2707.04	\$19,837.40
Compounded daily	\$1105.16	\$2717.91	\$20,077.29

Figure 5.4
Comparing simple interest and compound interest

Finding the Interest and the Present Value

EXAMPLE 5   Betty's boss paid her an unexpected bonus of \$2,500. Betty and her husband Tom decided to save the money for their daughter's education. They deposited it in an account that pays 10.3% interest compounded daily. Find the amount of interest that they would earn in 15 years by finding the future value and subtracting the principal.

Solution   Compounding daily, we have $P = 2500, i = 1/365$th of $10.3\% = 0.103/365$, and $n = 15$ years $= 15$ years $\cdot 365$ days/year $= 5{,}475$ days.

$$FV = P(1 + i)^n$$
$$= 2{,}500(1 + 0.103/365)^{5475}$$
$$= 11{,}717.374\ldots$$
$$\approx \$11{,}717.37$$

2500 ⊠ ⎛ 1 ⊞ .103 ⊟ 365 ⎞ y^x 5475 🟰

2500 ⎛ 1 ⊞ .103 ⊟ 365 ⎞ ^ 5475 ENTER

Warning: If you compute $\frac{0.103}{365}$ separately, you will get a long decimal. Do not round off that decimal, because the resulting answer will be inaccurate. Doing the calculation all at once (as shown above) avoids this difficulty.

The principal is $2{,}500, and the total of principal and interest is \$11{,}717.37. Thus, the interest is $\$11{,}717.37 - \$2500 = \$9{,}217.37$. ∎

EXAMPLE 6   Find the amount of money that must be invested now at $7\frac{3}{4}\%$ interest compounded annually so that it will be worth \$2,000 in 3 years.

Solution   The question actually asks us to find the present value, or principal P, that will generate a future value of \$2,000. We have $FV = 2{,}000$, $i = 7\frac{3}{4}\% = 0.0775$, and $n = 3$.

$$FV = P(1 + i)^n$$
$$2{,}000 = P(1 + 0.0775)^3$$
$$P = \frac{2{,}000}{1.0775^3}$$
$$= 1{,}598.7412$$
$$\approx \$1{,}598.74$$

 As a rough check, notice that the principal is somewhat lower than the future value, as it should be, because of the short time period.

∎

Annual Yield

Which investment is more profitable: one that pays 5.8% compounded daily or one that pays 5.9% compounded quarterly? It is difficult to tell. Certainly, 5.9% is a better rate than 5.8%, but compounding daily is better than compounding quarterly. The

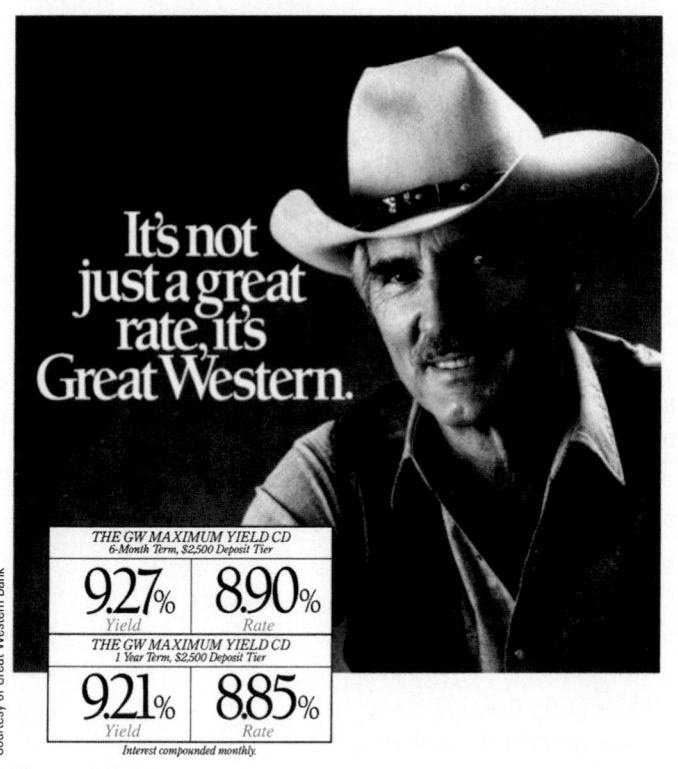

It's not just a great rate, it's Great Western.

THE GW MAXIMUM YIELD CD 6-Month Term, $2,500 Deposit Tier	
9.27%	**8.90**%
Yield	Rate

THE GW MAXIMUM YIELD CD 1 Year Term, $2,500 Deposit Tier	
9.21%	**8.85**%
Yield	Rate

Interest compounded monthly.

Figure 5.5
Most banks advertise their yields as well as their rates

two rates cannot be directly compared because of their different compounding frequencies. The way to tell which is the better investment is to find the annual yield of each.

The **annual yield** (or just **yield**) of a compound interest deposit is the *simple interest rate* that has the same future value as the compound rate would have in one year. The annual yields of two different investments can be compared, because they are both simple interest rates. Annual yield provides the consumer with a uniform basis for comparison. The annual yield should be slightly higher than the compound rate, because compound interest is slightly more profitable than simple interest over a short period of time. The compound rate is sometimes called the **nominal rate** to distinguish it from the yield (here, *nominal* means "named" or "stated").

To find the **annual yield** r of a given compound interest rate, you find the simple interest rate that makes the future value under simple interest the same as the future value under compound interest in one year.

$$FV(\text{simple interest}) = FV(\text{compound interest})$$
$$P(1 + rt) = P(1 + i)^n$$

EXAMPLE 7   Find the annual yield of $2,500 deposited in an account in which it earns 10.3% interest compounded daily for 15 years.

Solution

Simple interest	Compound interest
$P = 2{,}500$	$P = 2{,}500$
$r = $ unknown annual yield	$i = \dfrac{1}{365}$ th of $10.3\% = \dfrac{0.103}{365}$
$t = 1$ year	$n = 1$ year $= 365$ days

(1 year, not 15 years—its *annual* yield)

$$FV \text{ (simple interest)} = FV \text{ (compounded monthly)}$$
$$P(1 + rt) = P(1 + i)^n$$
$$2{,}500(1 + r \cdot 1) = 2{,}500(1 + 0.103/365)^{365}$$
$$(1 + r \cdot 1) = (1 + 0.103/365)^{365} \qquad \text{dividing by } 2{,}500$$
$$r = (1 + 0.103/365)^{365} - 1 \qquad \text{solving for } r$$
$$r = 0.10847\ldots \approx 10.85\%$$

 As a rough check, notice that the annual yield is slightly higher than the compound rate, as it should be.

By tradition, we round this to the nearest hundredth of a percent, so the annual yield is 10.85%. This means that in 1 year's time, 10.3% compounded daily has the same effect as 10.85% simple interest. For any period of time longer than a year, 10.3% compounded daily will yield *more* interest than 10.85% simple interest would, because compound interest gives interest on interest. ∎

Notice that in Example 7, the principal of $2,500 cancelled out. If the principal were two dollars or two million dollars, it would still cancel out, and the annual yield of 10.3% compounded daily would still be 10.85%. The principal does not matter in computing the annual yield. Also notice that the 15 years did not enter into the calculation—*annual* yield is always based on a *one-year* period.

EXAMPLE 8   Find the annual yield corresponding to a nominal rate of 8.4% compounded monthly.

Solution   We are told neither the principal nor the time, but (as discussed above) these variables do not affect the annual yield.

Compounding monthly, we have $i = 1/12$ of $8.4\% = \frac{0.084}{12}$, $n = 1$ year = 12 months, and $t = 1$ year.

$$FV\,(\text{simple interest}) = FV(\text{compounded monthly})$$
$$P(1 + rt) = P(1 + i)^n$$
$$(1 + rt) = (1 + i)^n \qquad \text{dividing by } P$$
$$(1 + r \cdot 1) = (1 + .084/12)^{12} \qquad \text{annual, so } t = 1 \text{ year and } n = 12 \text{ months}$$
$$r = (1 + .084/12)^{12} - 1 \qquad \text{solving for } r$$
$$r = 0.08731\ldots \approx 8.73\%$$

 As a rough check, notice that the annual yield is slightly higher than the compound rate, as it should be.

∎

In Example 8, we found that 8.4% compounded monthly generates an annual yield of 8.73%. This means that 8.4% compounded monthly has the same effect as does 8.73% simple interest in 1 year's time. Furthermore, as Figure 5.6 indicates, 8.4% compounded monthly has the same effect as does 8.73% compounded annually *for any time period*.

For a Principal of $1,000	After 1 Year	After 10 Years
FV at 8.4% compounded monthly	$1,087.31	$2,309.60
FV at 8.73% simple interest	$1,087.30	$1,873.00
FV at 8.73% compounded annually	$1,087.30	$2,309.37

Figure 5.6
What annual yield means

Notice that all three rates have the same future value after 1 year (the 1¢ difference is due to rounding off the annual yield to 8.73%). However, after 10 years, the simple interest has fallen way behind, while the 8.4% compounded monthly and 8.73% compounded annually remain the same (except for the round-off error). This always happens. The annual yield is the simple interest rate that has the same future

Compound Interest in the Real World: Benjamin Franklin's Gift

© Stock Montage/Getty Images

In 1789, when Benjamin Franklin was 83, he added a codicil to his will. That codicil was meant to aid young people, Boston (where Franklin grew up), and Philadelphia (where he had been President of the state of Pennsylvania) over a time span of two hundred years.

In his codicil, Franklin wrote of his experience as an apprentice printer, of the friends who had loaned him the money to set up his own printing business, and of the importance of craftsmen to a city. He then went on to say, "To this End I devote Two thousand Pounds Sterling, which I give, one thousand thereof to the Inhabitants of the Town of Boston, in Massachusetts, and the other thousand to the Inhabitants of the City of Philadelphia, in Trust to and for the Uses, Interests and Purposes hereinafter mentioned and declared." At that time, £1,000 was the equivalent of about $4,500.

Franklin's plan called for each town to use the money as a loan fund for one hundred years. The money would be loaned out to young tradesmen to help them start their own businesses. He calculated that at the end of 100 years, each fund would have grown to £131,000 (or about $582,000 in 1892 dollars). At that point, one-fourth of the money would continue to be used for a loan fund. The remainder would be used for public works.

He calculated that at the end of 200 years, each of the two city's loan funds would have grown to £4,061,000 (or about $7,000,000 in 1992 dollars). At that point, he called for the money to be given to the two cities and states.

Twenty-two tradesmen, including bricklayers, hairdressers, jewelers, and tanners, applied for loans from Franklin's fund in the first month after Franklin's death. The fund remained popular until the onset of the Industrial Revolution in the early 1800s, when young people stopped becoming tradesmen with their own shops. Instead, most went to work as mechanics in factories.

When Boston's fund reached its hundredth anniversary in 1891, its value was approximately $391,000. One-fourth of that money continued to be used for a loan fund, as Franklin wished. The city of Boston decided to use the remainder to build a trade school, because it fit with Franklin's goal of helping young people.

Legal problems delayed the school's founding, but the Benjamin Franklin Institute of Technology was opened in Boston in 1908. At that point, the trade school part of the Boston fund had risen to $432,367, and the loan fund part had grown to $163,971.

The Institute continues to operate today. It has almost 400 students, 90% of whom receive financial aid. It awards Bachelor of Science degrees, Associate in Engineering degrees, and Associate in Science degrees. Ninety-eight percent of its graduates find work in their fields within six months of graduation.

Pennsylvania used its share of the money to fund the Franklin Institute of Philadelphia. Originally, the institute promoted the "mechanical arts." Now it houses a planetarium, an IMAX theater, and a science museum.

Exercises 55–60 explore some aspects of Franklin's bequest.

Source: The Benjamin Franklin Institute of Technology and the Franklin Institute of Philadelphia.

value that the compound rate would have in one year. It is also the annually compounded rate that has the same future value that the nominal rate would have after any amount of time.

An annual yield formula does exist, but the annual yield can be calculated efficiently without it, as was shown above. The formula is developed in the exercises. See Exercise 49.

5.2 Exercises

In Exercises 1–6, find the periodic rate that corresponds to the given compound rate, if the rate is compounded (a) quarterly, (b) monthly, (c) daily, (d) biweekly (every two weeks), and (e) semimonthly (twice a month). Do not round off the periodic rate.

1. 12%
2. 6%
3. 3.1%
4. 6.8%
5. 9.7%
6. 10.1%

In Exercises 7–10, find the number of periods that corresponds to the given time span, if a period is (a) a quarter of a year, (b) a month, and (c) a day. (Ignore leap years.)

7. $8\frac{1}{2}$ years
8. $9\frac{3}{4}$ years
9. 30 years
10. 45 years

In Exercises 11–16, (a) find and (b) interpret the future value of the given amount.

11. $3,000 at 6% compounded annually for 15 years
12. $7,300 at 7% compounded annually for 13 years
13. $5,200 at $6\frac{3}{4}$% compounded quarterly for $8\frac{1}{2}$ years
14. $36,820 at $7\frac{7}{8}$% compounded quarterly for 4 years
15. $1,960 at $4\frac{1}{8}$% compounded daily for 17 years (ignore leap years)
16. $12,350 at 6% compounded daily for 10 years and 182 days (ignore leap years)

In Exercises 17–20, (a) find and (b) interpret the annual yield corresponding to the given nominal rate.

17. 8% compounded monthly
18. $5\frac{1}{2}$% compounded quarterly
19. $4\frac{1}{4}$% compounded daily
20. $12\frac{5}{8}$% compounded daily

In Exercises 21–22, find and interpret the annual yield corresponding to the given nominal rate.

21. 10% compounded (a) quarterly, (b) monthly, and (c) daily
22. $12\frac{1}{2}$% compounded (a) quarterly, (b) monthly, and (c) daily

In Exercises 23–26, (a) find and (b) interpret the present value that will generate the given future value.

23. $1,000 at 8% compounded annually for 7 years
24. $9,280 at $9\frac{3}{4}$% compounded monthly for 2 years and 3 months
25. $3,758 at $11\frac{7}{8}$% compounded monthly for 17 years and 7 months

26. $4,459 at $10\frac{3}{4}$% compounded quarterly for 4 years

27. $10,000 is deposited in an account in which it earns 10% interest compounded monthly. No principal or interest is withdrawn from the account. Instead, both continue to earn interest over time. Find the account balance after six months:
   a. using the simple interest future value formula to compute the balance at the end of each compounding period.
   b. using the compound interest formula.

28. $20,000 is deposited in an account in which it earns 9.5% interest compounded quarterly. No principal or interest is withdrawn from the account. Instead, both continue to earn interest over time. Find the account balance after one year:
   a. using the simple interest future value formula to compute the balance at the end of each compounding period.
   b. using the compound interest formula.

29. $15,000 is deposited in an account in which it earns 6% interest compounded annually. No principal or interest is withdrawn from the account. Instead, both continue to earn interest over time. Find the account balance after three years:
   a. using the simple interest future value formula to compute the balance at the end of each compounding period.
   b. using the compound interest formula.

30. $30,000 is deposited in an account in which it earns 10% interest compounded annually. No principal or interest is withdrawn from the account. Instead, both continue to earn interest over time. Find the account balance after four years:
   a. using the simple interest future value formula to compute the balance at the end of each compounding period.
   b. using the compound interest formula.

31. Donald Trumptobe decided to build his own dynasty. He is considering specifying in his will that at his death, $10,000 would be deposited into a special account that would earn a guaranteed 6% interest compounded daily. This money could not be touched for 100 years, at which point it would be divided among his heirs. Find the future value.

32. How much would Donald Trumptobe in Exercise 31 have to have deposited so that his heirs would have $50,000,000 or more in 100 years if his money earns 7% compounded monthly?

33. Donald Trumptobe in Exercise 31 predicts that in 100 years, he will have four generations of offspring (that is, children, grandchildren, great-grandchildren, and great-great-grandchildren). He estimates that each person will have two children. How much will he have to have deposited so that each of his great-great-grandchildren would have $1,000,000 or more in 100 years if his money earns 7.5% compounded monthly?

34. Donald Trumptobe in Exercise 31 predicts that in 100 years, he will have four generations of offspring (that is, children, grandchildren, great-grandchildren, and great-great-grandchildren). He estimates that each person will have two children. How much will he have to have deposited so that each of his great-great-grandchildren would have $1,000,000 or more in 100 years if his money earns 9.25% compounded daily?

35. When Jason Levy was born, his grandparents deposited $3,000 into a special account for Jason's college education. The account earned $6\frac{1}{2}$% interest compounded daily.
   a. How much will be in the account when Jason is 18?
   b. If, on turning 18, Jason arranges for the monthly interest to be sent to him, how much will he receive each 30-day month?

36. When Alana Cooper was born, her grandparents deposited $5,000 into a special account for Alana's college education. The account earned $7\frac{1}{4}$% interest compounded daily,
   a. How much will be in the account when Alana is 18?
   b. If, on turning 18, Alana arranges for the monthly interest to be sent to her, how much will she receive each 30-day month?

*For Exercises 37–40, note the following information: An **Individual Retirement Account (IRA)** is an account in which the saver does not pay income tax on the amount deposited but is not allowed to withdraw the money until retirement. (The saver pays income tax at that point, but his or her tax bracket is much lower then.)*

37. At age 27, Lauren Johnson deposited $1,000 into an IRA, in which it earns $7\frac{7}{8}$% compounded monthly. What will it be worth when she retires at 65?

38. At age 36, Dick Shoemaker deposited $2,000 into an IRA, in which it earns $8\frac{1}{8}$% compounded semiannually. What will it be worth when he retires at 65?

39. Marlene Silva wishes to have an IRA that will be worth $100,000 when she retires at age 65.
   a. How much must she deposit at age 35 at $8\frac{3}{8}$% compounded daily?
   b. If, at age 65, she arranges for the monthly interest to be sent to her, how much will she receive each 30-day month?

40. David Murtha wishes to have an IRA that will be worth $150,000 when he retires at age 65.
   a. How much must he deposit at age 26 at $6\frac{1}{8}$% compounded daily?
   b. If, at age 65, he arranges for the monthly interest to be sent to him, how much will he receive each 30-day month?

For Exercises 41–46, note the following information: A certificate of deposit (CD) is an agreement between a bank and a saver in which the bank guarantees an interest rate and the saver commits to leaving his or her deposit in the account for an agreed-upon period of time.

41. First National Bank offers 2-year CDs at 9.12% compounded daily, and Citywide Savings offers 2-year CDs at 9.13% compounded quarterly. Compute the annual yield for each institution and determine which is more advantageous for the consumer.

42. National Trust Savings offers 5-year CDs at 8.25% compounded daily, and Bank of the Future offers 5-year CDs at 8.28% compounded annually. Compute the annual yield for each institution, and determine which is more advantageous for the consumer.

43. Verify the annual yield for the 5 year certificate quoted in the bank sign in Figure 5.7, using interest

Figure 5.7
A bank sign for Exercise 43.

that is compounded daily and:

a. 365-day years   **b.** 360-day years

c. Various combinations of 360-day and 365-day years

44. Verify the yield for the 1-year CDs quoted in Figure 5.8.

Worrying what to do with your money? Bury it in Jean Lafitte Savings Bank's 1-year "High Seas" CDs! You'll rest in peace.

| Rate: | 8.7% compounded monthly |
| Yield: | 10.16% |

Figure 5.8
A savings bank advertisement for Exercise 44

45. Verify the yield for the 2-year CDs quoted in Figure 5.9.

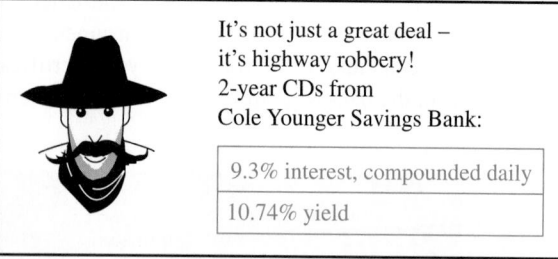

It's not just a great deal – it's highway robbery! 2-year CDs from Cole Younger Savings Bank:

| 9.3% interest, compounded daily |
| 10.74% yield |

Figure 5.9
A savings bank advertisement for Exercise 45

46. Verify the yield for the 6-month CDs quoted in the savings bank advertisement on page 345.

47. Recently, Bank of the West offered 6-month CDs at 5.0% compounded monthly.

a. Find the annual yield of one of these CDs.

b. How much would a $1,000 CD be worth at maturity?

c. How much interest would you earn?

d. What percent of the original $1,000 is this interest?

e. The answer to part (d) is not the same as that of part (a). Why?

f. The answer to part (d) is close to, but not exactly half, that of part (a). Why?

(*Source:* Bank of the West.)

48. Recently, Bank of the West offered 6-month CDs at 5.0% interest compounded monthly and 1-year CDs at 5.20% interest compounded monthly. Maria Ruiz bought a 6-month $2,000 CD, even though she knew

she would not need the money for at least a year, because it was predicted that interest rates would rise.

a. Find the future value of Maria's CD.

b. Six months later, Maria's CD has come to term, and in the intervening time, interest rates have risen. She reinvests the principal and interest from her first CD in a second 6-month CD that pays 5.31% interest compounded monthly. Find the future value of Maria's second CD.

c. Would Maria have been better off if she had bought a 1-year CD in December 1996?

d. If Maria's second CD pays 5.46% interest compounded monthly, rather than 5.31%, would she be better off with the two 6-month CDs or the 1-year CD?

(*Source:* Bank of the West.)

49. Develop a formula for the annual yield of a compound interest rate.

HINT: Follow the procedure given in Example 8, but use the letters i and n in the place of numbers.)

In Exercises 50–54, use the formula found in Exercise 49 to compute the annual yield corresponding to the given nominal rate.

50. $9\frac{1}{2}\%$ compounded monthly

51. $7\frac{1}{4}\%$ compounded quarterly

52. $12\frac{3}{8}\%$ compounded daily

53. $5\frac{5}{8}\%$ compounded (a) semiannually, (b) quarterly, (c) monthly, (d) daily, (e) biweekly, and (f) semimonthly.

54. $10\frac{1}{2}\%$ compounded (a) semiannually, (b) quarterly, (c) monthly, (d) daily, (e) biweekly, and (f) semimonthly.

Exercises 55–60 refer to Benjamin Franklin's gift, discussed on page 347.

55. Would Benjamin Franklin have used simple or compound interest in projecting the 100-year future value of his bequest? Why? What interest rate did he use in calculating the 100-year future value? Use pounds, not dollars, in your calculation.

56. Would Benjamin Franklin have used simple or compound interest in projecting the 200-year future value of his bequest? Why? What interest rate did he use in calculating the 200-year future value? Use pounds, not dollars, in your calculation.

57. What interest rate did Franklin's Boston bequest actually earn in the first hundred years? Use dollars, not pounds, in your calculation.

58. What interest rate did Franklin's Boston bequest actually earn from 1891 to 1908? Use dollars, not pounds, in your calculation.

59. After 100 years, Franklin's total bequest to Boston was worth much more than his original bequest. The future value was what percent of the original bequest? Use dollars, not pounds, in your calculation.

60. In 1908, Franklin's total bequest to Boston was worth much more than his original bequest. The future value was what percent of the original bequest? Use dollars, not pounds, in your calculation.

― *Answer the following questions using complete sentences and your own words.*

CONCEPT QUESTIONS

61. Explain how compound interest is based on simple interest.

62. Why is there no work involved in finding the annual yield of a given simple interest rate?

63. Why is there no work involved in finding the annual yield of a given compound interest rate when that rate is compounded annually?

64. Which should be higher: the annual yield of a given rate compounded quarterly or compounded monthly? Explain why, *without* performing any calculations or referring to any formulas.

65. Why should the annual yield of a given compound interest rate be higher than the compound rate? Why should it be only slightly higher? Explain why, *without* performing any calculations or referring to any formulas.

66. Explain the difference between simple interest and compound interest.

67. If money earns compound interest, why must the future value be *slightly* higher than the principal, after a short amount of time? Why must the future value be *much* higher than the principal after a long amount of time?

68. *Money* magazine and other financial publications regularly list the top-paying money-market funds, the top-paying bond funds, and the top-paying CDs and their yields. Why do they list yields rather than interest rates and compounding periods?

69. Equal amounts are invested in two different accounts. One account pays simple interest, and the other pays compound interest at the same rate. When will the future values of the two accounts be the same?

70. Suppose you invest some money in a new account that pays 5% interest compounded annually, and you do not make any further deposits into or withdrawals from that account. Which of the following must be true?
- The account grows by the same dollar amount in the second year as it did in the first year.
- The account grows by a larger dollar amount in the second year as it did in the first year.
- The account grows by a smaller dollar amount in the second year as it did in the first year. Why?

 WEB PROJECT

71. Go to the web sites of four different banks.
- **a.** For each bank, try to determine the following:
  - The interest rate
  - The compounding frequency
  - The annual yield (also called the *annual percentage yield,* or A.P.Y.)
    for CDs for two different terms. (Use the same terms for each bank.) Some banks might not give all of the above information, but all will give the annual yield, as required by federal law.
- **b.** If a bank omits either the interest rate or the compounding frequency, calculate the omitted information.
- **c.** If a bank omits both the interest rate and the compounding frequency, assume that the compounding frequency is daily and calculate the interest rate.
- **d.** If a bank omits none of the information, verify the annual yield.
- **e.** Which bank offers the best deal? Why?

Some useful links for this web project are listed on the text web site: **www.thomsonedu.com/mathematics/johnson**

PROJECT

72. Suppose you have $1,000 invested at 5% annual interest and you do not make any further deposits into that account. Let x measure years after you invested the money, and let y measure the future value of the account. Draw a graph that shows the relationship between x and y, for $0 \le x \le 5$, if the interest rate is

a. simple interest

b. compounded annually

c. compounded daily

d. Discuss the differences between the three graphs. In your discussion, address the following:
- The difference in their shapes
- Where they coincide
- Which graph is above the others
- Which is below the others

Doubling Time on a Graphing Calculator

Simple interest is a very straightforward concept. If an account earns 5% simple interest, then 5% of the principal is paid for each year that principal is in the account. In one year, the account earns 5% interest; in two years, it earns 10% interest; in three years, it earns 15% interest, and so on.

It is not nearly so easy to get an intuitive grasp of compound interest. If an account earns 5% interest compounded daily, then it does not earn only 5% interest in one year, and it does not earn only 10% interest in two years.

Annual yield is one way of gaining an intuitive grasp of compound interest. If an account earns 5% interest compounded daily, then it will earn 5.13% interest in one year (because the annual yield is 5.13%), but it does not earn merely $2 \cdot 5.13\% = 10.26\%$ interest in two years.

Doubling time is another way of gaining an intuitive grasp of compound interest. **Doubling time** is the amount of time it takes for an account to double in value; that is, it's the amount of time it takes for the future value to become twice the principal. To find the doubling time for an account that earns 5% interest compounded daily, substitute $2P$ for the future value and solve the resulting equation.

$$FV = P(1 + i)^n \qquad \text{Compound Interest Future Value Formula}$$

$$2P = P\left(1 + \frac{0.05}{365}\right)^n \qquad \text{substituting}$$

$$2 = \left(1 + \frac{0.05}{365}\right)^n \qquad \text{dividing by } P$$

Solving this equation for n involves mathematics that will be covered in Section 10.0B. For now, we will use the graphing calculator to approximate n.

EXAMPLE 9   Use a graphing calculator to find the doubling time for an account that earns 5% interest compounded daily.

Solution

1. *Use the calculator to graph two different equations:*

$$y = 2 \qquad \text{and} \qquad y = \left(1 + \frac{0.05}{365}\right)^x$$

Follow the procedure discussed in Appendix C. Fill in the "Y=" screen, as shown in Figure 5.10(a). If you use the "ZoomStandard" command, you will get the graph shown in Figure 5.10(b).

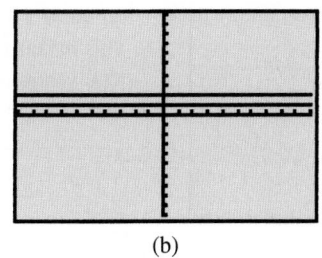

(a)   (b)

Figure 5.10

2. *Adjust the calculator's screen so that it shows the point at which these two equations intersect.* To do this, we must determine appropriate values of xmin, xmax, ymin, and ymax.

   In the equation $y = (1 + \frac{0.05}{365})^x$, x measures time and y measures money. Neither can be negative, so set xmin and ymin to 0.

   We can get a rough idea of how big x should be by finding the doubling time for simple interest (a much easier calculation):

   $$FV = P(1 + rt)\qquad \text{Simple Interest Future Value Formula}$$
   $$2P = P(1 + 0.05t)\qquad \text{substituting } 2P \text{ for } FV$$
   $$2 = 1 + 0.05t\qquad \text{dividing by } P$$
   $$1 = 0.05t$$
   $$t = \frac{1}{0.05} = 20$$

Figure 5.11

   The doubling time for 5% *simple* interest is 20 years. Simple interest is not as productive as compound interest, so the doubling time for 5% *compound* interest will be less than 20 years. In the equation $y = (1 + \frac{0.05}{365})^x$, x measures time in days, and 20 years = 20 · 365 = 7,300 days, so set xmax to 7,300.

   The value of ymax should correspond to xmax = 7,300:

   $$y = \left(1 + \frac{0.05}{365}\right)^{7300}$$
   $$= 2.71809\ldots$$
   $$\approx 2.7$$

   Set ymax to 2.7.

3. *Press the* $\boxed{\text{GRAPH}}$ *button to obtain the screen shown in Figure 5.11. Notice that we can now see the point of intersection.*

4. *Use the calculator to find the point of intersection.* Follow the procedure discussed in Appendix D and summarized below.

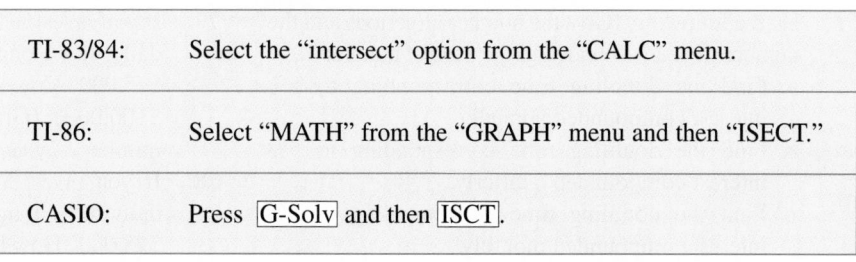

TI-83/84:	Select the "intersect" option from the "CALC" menu.
TI-86:	Select "MATH" from the "GRAPH" menu and then "ISECT."
CASIO:	Press $\boxed{\text{G-Solv}}$ and then $\boxed{\text{ISCT}}$.

Intersection
X=5060.321 ⎐Y=2

Figure 5.12

Figure 5.12 shows the result of this work. The intersection is at (5,060.321, 2). This means that it takes 5,060.321 days (or about 5060.321/365 = 13.86389315 ... ≈ 13.9 years) for money invested at 5% interest compounded daily to double.

 The solution can be checked by substituting 5060.321 for x in the equation

$$2 = \left(1 + \frac{0.05}{365}\right)^x$$

The right side is

$$\left(1 + \frac{0.05}{365}\right)^x = \left(1 + \frac{0.05}{365}\right)^{5060.321}$$

$$= 2.000000005$$

This solution is mathematically (almost) correct, as is shown in the preceding box. Practically speaking, however, the ".321" part of the solution does not make sense. If interest is compounded daily, then at the end of each day, your account is credited with that day's interest. After 5,060 days, your account would contain slightly less than twice the original principal. After 5,060.321 days, your account balance would not have changed, since interest will not be credited until the end of the day. After 5,061 days, your account would contain slightly more than twice the original principal. The doubling time is 5,061 days ≈ 13.9 years. ∎

Exercises

73. The solution in Example 9 didn't quite check; we found that the y-value at the point of intersection is 2.000000005, not 2. Why is there a discrepancy?

74. If $1,000 is deposited into an account that earns 5% interest compounded daily, the doubling time is approximately 5,061 days.
   a. Find the amount in the account after 5,061 days.
   b. Find the amount in the account after 2 · 5,061 days.
   c. Find the amount in the account after 3 · 5,061 days.
   d. Find the amount in the account after 4 · 5,061 days.
   e. What conclusion can you make?

75. Do the following. (Give the number of periods and the number of years, rounded to the nearest hundredth.)
   a. Find the doubling time corresponding to 5% interest compounded annually.
   b. Find the doubling time corresponding to 5% interest compounded quarterly.
   c. Find the doubling time corresponding to 5% interest compounded monthly.

   d. Find the doubling time corresponding to 5% interest compounded daily.
   e. Discuss the effect of the compounding period on doubling time.

76. Do the following. (Give the number of periods and the number of years, rounded to the nearest hundredth.)
   a. Find the doubling time corresponding to 6% interest compounded annually.
   b. Find the doubling time corresponding to 7% interest compounded annually.
   c. Find the doubling time corresponding to 10% interest compounded annually.
   d. Discuss the effect of the interest rate on doubling time.

77. If you invest $10,000 at 8.125% interest compounded daily, how long will it take for you to accumulate $15,000? How long will it take for you to accumulate $100,000? (Give the number of periods and the number of years, rounded to the nearest hundredth.)

78. If you invest $15,000 at $9\frac{3}{8}$% interest compounded daily, how long will it take for you to accumulate $25,000? How long will it take for you to accumulate

$100,000? (Give the number of periods and the number of years, rounded to the nearest hundredth.)

79. If you invest $20,000 at $6\frac{1}{4}\%$ interest compounded daily, how long will it take for you to accumulate

$30,000? How long will it take for you to accumulate $100,000? (Give the number of periods and the number of years, rounded to the nearest hundredth.)

5.3 Annuities

Many people have long-term financial goals and limited means with which to accomplish them. Your goal might be to save $3,000 over the next four years for your college education, to save $10,000 over the next ten years for the down payment on a home, to save $30,000 over the next eighteen years to finance your new baby's college education, or to save $300,000 over the next forty years for your retirement. It seems incredible, but each of these goals can be achieved by saving only $50 a month (if interest rates are favorable)! All you need to do is start an annuity.

An **annuity** is simply a sequence of equal, regular payments into an account in which each payment receives compound interest. Because most annuities involve relatively small periodic payments, they are affordable for the average person. Over longer periods of time, the payments themselves start to amount to a significant sum, but it is really the power of compound interest that makes annuities so amazing. If you pay $50 a month into an annuity for the next forty years, then your total payment is

$$\frac{\$50}{\text{month}} \cdot \frac{12 \text{ months}}{\text{year}} \cdot 40 \text{ years} = \$24,000$$

A **Christmas Club** is an annuity that is set up to save for Christmas shopping. A Christmas Club participant makes regular equal deposits, and the deposits and the resulting interest are released to the participant in December when the money is needed. Christmas Clubs are different from other annuities in that they span a short amount of time—a year at most—and thus earn only a small amount of interest. (People set them up to be sure that they are putting money aside rather than to generate interest.) Our first few examples will deal with Christmas Clubs, because their short time span makes it possible to see how an annuity actually works.

Annuities as Compound Interest, Repeated

EXAMPLE 1   On August 12, Patty Leitner joined a Christmas Club through her bank. For the next three months, she would deposit $200 at the beginning of each month. The money would earn $8\frac{3}{4}\%$ interest compounded monthly, and on December 1, she could withdraw her money for shopping. Use the compound interest formula to find the future value of the account.

Solution   We are given $P = 200$ and $r = \frac{1}{12}$th of $8\frac{3}{4}\% = \frac{0.0875}{12}$.

First, calculate the future value of the first payment (made on September 1). Use $n = 3$ because it will receive interest during September, October, and November.

$$FV = P(1 + i)^n$$
$$= 200\left(1 + \frac{0.0875}{12}\right)^3$$
$$= 204.40698$$
$$\approx \$204.41$$

Next, calculate the future value of the second payment (made on October 1). Use $n = 2$ because it will receive interest during October and November.

$$FV = P(1 + i)^n$$
$$= 200\left(1 + \frac{0.0875}{12}\right)^2$$
$$= 202.9273$$
$$\approx \$202.93$$

Next, calculate the future value of the third payment (made on November 1). Use $n = 1$ because it will receive interest during November.

$$FV = P(1 + i)^n$$
$$= 200\left(1 + \frac{0.0875}{12}\right)^1$$
$$= 201.45833$$
$$\approx \$201.46$$

The payment schedule and interest earned are illustrated in Figure 5.13.

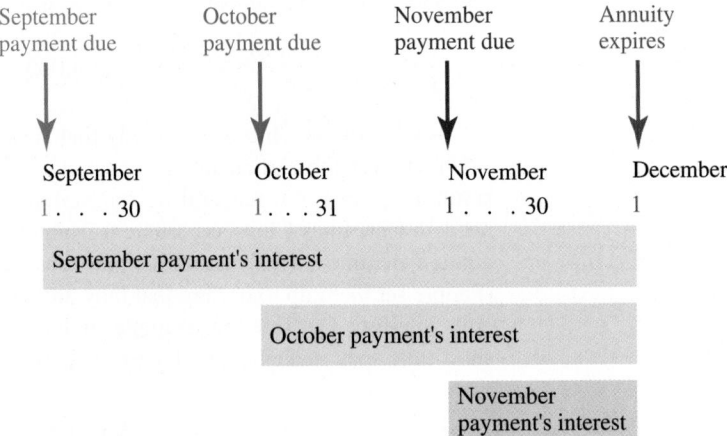

Figure 5.13
Patty's payment schedule and interest earning periods

The future value of Patty's annuity is the sum of the future values of each payment:

$$FV \approx \$204.41 + \$202.93 + \$201.46$$
$$= \$608.80$$

Patty's deposits will total $600.00. She will earn $8.80 interest on her deposits.   ∎

The **payment period** of an annuity is the time between payments; in Example 1, the payment period was one month. The **term** is the time from the beginning of the first payment period to the end of the last; the term of Patty's Christmas Club was three months. When an annuity has **expired** (that is, when its term is over), the entire account or any portion of it may be withdrawn. Most annuities are **simple,** that is, their compounding period is the same as their payment period (for example, if payments are made monthly, then interest is compounded monthly). In this book, we will work only with simple annuities.

Ordinary Annuities and Annuities Due

An **annuity due** is one in which each payment is due at the beginning of its time period. Patty's annuity in Example 1 was an annuity due, because the payments were due at the *beginning* of each month. An **ordinary annuity** is an annuity for which each payment is due at the end of its time period. As the name implies, this form of annuity is more typical. As we will see in the next example, the difference is one of timing.

EXAMPLE 2   Dan Bach also joined a Christmas Club through his bank. His was just like Patty's except that his payments were due at the end of each month, and his first payment was due September 30. Use the Compound Interest Formula to find the future value of the account.

Solution   This is an *ordinary* annuity because payments are due at the *end* of each month. Interest is compounded monthly. From Example 1, we know that $P = 200$, $i = \frac{1}{12}$ of $8\frac{3}{4}\% = \frac{0.0875}{12}$.

To calculate the future value of the first payment (made on September 30), use $n = 2$. This payment will receive interest during October and November.

$$FV = P(1 + i)^n$$
$$= 200\left(1 + \frac{0.0875}{12}\right)^2$$
$$= 202.9273$$
$$\approx \$202.93$$

To calculate the future value of the second payment (made on October 31), use $n = 1$. This payment will receive interest during November.

$$FV = P(1 + i)^n$$
$$= 200\left(1 + \frac{0.0875}{12}\right)^1$$
$$= 201.45833$$
$$\approx \$201.46$$

To calculate the future value of the second payment (made on November 30), note that no interest is earned, because the payment is due November 30 and the annuity expires December 1. Therefore,

$$FV = \$200$$

Dan's payment schedule and interest payments are illustrated in Figure 5.14 on page 358.
The future value of Dan's annuity is the sum of the future values of each payment:

$$FV \approx \$200 + \$201.46 + \$202.93$$
$$= \$604.39$$

Dan earned $4.39 interest on his deposits.

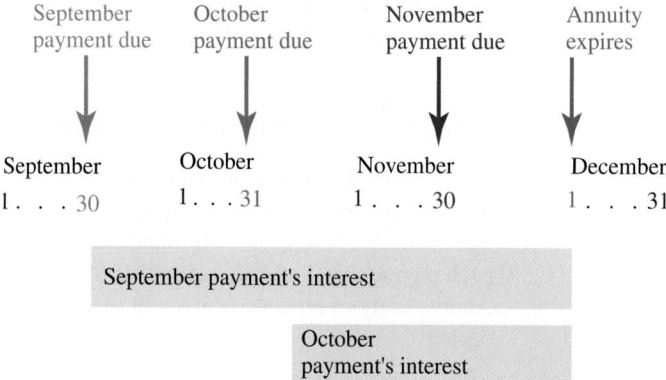

Figure 5.14
Dan's payment schedule and interest earning periods   ■

In Examples 1 and 2, why did Patty earn more interest than Dan? The reason is that each of her payments was made a month earlier and therefore received an extra month's interest. In fact, we could find the future value of Patty's account by giving Dan's future value one more month's interest.

$$\text{Patty's } FV = \text{Dan's } FV \cdot (1 + i)^1$$

$$\$608.80 = \$604.39 \cdot \left(1 + \frac{0.0875}{12}\right)^1$$

More generally, we can find the future value of an annuity due by giving an ordinary annuity's future value one more month's interest.

$$FV(\text{due}) = FV(\text{ordinary}) \cdot (1 + i)^1$$

The difference between an ordinary annuity and an annuity due is strictly a timing difference, because any ordinary annuity in effect will become an annuity due if you leave all funds in the account for one extra period.

The Annuity Formulas

The procedure followed in Examples 1 and 2 reflects what actually happens with annuities, and it works fine for a small number of payments. However, most annuities are long-term, and the procedure would become tedious if we were computing the future value after forty years. Because of this, long-term annuities are calculated with their own formula.

For an *ordinary* annuity with payment *pymt,* a periodic rate *i,* and a term of *n* payments, the first payment receives interest for $n - 1$ periods. The payment is made at the end of the first period, so it received no interest for that one period. Its future value is

$$FV(\text{first } pymt) = pymt(1 + i)^{n - 1}$$

The last payment receives no interest (under the annuity), because it is due at the end of the last period and it expires the next day. Its future value is

$$FV(\text{last } pymt) = pymt$$

The next-to-last payment receives one period's interest, so its future value is

$$FV(\text{next-to-last } pymt) = pymt(1 + i)^1$$

The future value of the annuity is the sum of all of these future values of individual payments:

$$FV = pymt + pymt(1 + i)^1 + pymt(1 + i)^2 + \cdots + pymt(1 + i)^{n-1}$$

To get a short-cut formula from all this, we will multiply each side of this equation by $(1 + i)$ and then subtract the original equation from the result. This leads to a lot of cancelling.

$$FV(1 + i) = \cancel{pymt(1 + i)} + \cancel{pymt(1 + i)^2} + \cdots + \cancel{pymt(1 + i)^{n-1}} + pymt(1 + i)^n$$

minus: $$FV = pymt + \cancel{pymt(1 + i)^1} + \cancel{pymt(1 + i)^2} + \cdots + \cancel{pymt(1 + i)^{n-1}}$$

equals: $$FV(1 + i) - FV = pymt(1 + i)^n - pymt \qquad \text{subtracting}$$
$$FV(1 + i - 1) = pymt[(1 + i)^n - 1] \qquad \text{factoring}$$
$$FV(i) = pymt[(1 + i)^n - 1]$$
$$FV = pymt\frac{(1 + i)^n - 1}{i} \qquad \text{dividing}$$

This is the future value of the ordinary annuity.

Ordinary Annuity Formula

The future value FV of an ordinary annuity with payment size *pymt*, a periodic rate i, and a term of n payments is

$$FV(\text{ord}) = pymt\frac{(1 + i)^n - 1}{i}$$

As we saw in Examples 1 and 2, the future value of an annuity due is the future value of an ordinary annuity plus one more period's interest.

$$FV(\text{due}) = FV(\text{ord}) \cdot (1 + i)$$

This gives us a formula for the future value of an annuity due.

Annuity Due Formula

The future value FV of an annuity due with payment size *pymt*, a periodic rate i, and a term of n payments is

$$FV(\text{due}) = FV(\text{ord}) \cdot (1 + i)$$
$$= pymt\frac{(1 + i)^n - 1}{i}(1 + i)$$

Tax-Deferred Annuities

A **tax-deferred annuity (TDA)** is an annuity that is set up to save for retirement. Money is automatically deducted from the participant's paychecks until retirement, and the federal (and perhaps state) tax deduction is computed *after* the annuity payment has been deducted, resulting in significant tax savings. In some cases, the employer also makes a regular contribution to the annuity.

The following example involves a long-term annuity. Usually, the interest rate of a long-term annuity varies somewhat from year to year. In this case, calculations must be viewed as predictions, not guarantees.

EXAMPLE 3   Tom and Betty decided that they should start saving for retirement, so they set up a tax-deferred annuity. They arranged to have $200 taken out of each of Tom's monthly checks, which will earn $8\frac{3}{4}\%$ interest. Because of the tax-deferring effect of the TDA, Tom's take-home pay went down by only $115. Tom just had his thirtieth birthday, and his ordinary annuity will come to term when he is 65.

a. Find the future value of the annuity.
b. Find Tom's contribution and the interest portion.

Solution   **a.** This is an ordinary annuity, with *pymt* = 200, $i = \frac{1}{12}$th of $8\frac{3}{4}\%$ = 0.0875/12, and n = 35 years = 35 years · 12 months/year = 420 monthly payments.

$$FV(\text{ord}) = pymt\frac{(1 + i)^n - 1}{i}$$

$$= 200\frac{(1 + 0.0875/12)^{420} - 1}{0.0875/12}$$

$$\approx \$552,539.96$$

Because 0.0875/12 occurs twice in the calculation, compute it first and put it into your calculator's memory. Then type

(1 + RCL) y^x 420 − 1 = ÷ RCL × 200 =

In the above, "RCL" refers to recalling the stored number. The way to do this varies with different calculators.

Because 0.0875/12 occurs twice in the calculation, compute it and store it by typing

with a TI-83/84: .0875 ÷ 12 STO▶ ALPHA I
with a TI-86: .0875 ÷ 12 STO▶ I

Then type

200 × ((1 + ALPHA I) ^ 420 − 1) ÷ ALPHA
I ENTER

---

b. The principal part of this $552,539.96 is Tom's contribution, and the rest is interest.

- *Tom's contribution* is 420 payments of $200 each = 420 · $200 = $84,000.
- The *interest portion* is then $552,539.96 − $84,000 = $468,539.96. ■

In Example 3, the interest portion is almost six times as large as Tom's contribution! The magnitude of the earnings illustrates the amazing power of annuities and the effect of compound interest over a long period of time.

Sinking Funds

A **sinking fund** is an annuity in which the future value is a specific amount of money that will be used for a certain purpose, such as a child's education or the down payment on a home.

ThomsonNOW™
for Liberal Arts Mathematics
www.thomsonedu.com/1pass.

EXAMPLE 4

Tom and Betty have a new baby. They agreed that they would need $30,000 in 18 years for the baby's college education. They decided to set up a sinking fund and have money deducted from each of Betty's biweekly paychecks. That money will earn $9\frac{1}{4}\%$ interest in Betty's ordinary annuity. Find their monthly payment.

Solution

This is an ordinary annuity, with $i = \frac{1}{12}$th of $9\frac{1}{4}\% = 0.0925/26$, and $n = 18$ years · 26 periods/year = 468 periods, and $FV = \$30,000$.

$$FV(\text{ord}) = pymt\frac{(1 + i)^n - 1}{i}$$

$$\$30,000 = pymt\frac{(1 + 0.0925/26)^{468} - 1}{0.0925/26}$$

To find *pymt*, we must divide 30,000 by the fraction on the right side of the equation. Because the fraction is so complicated, it is best to first calculate the fraction and then multiply its reciprocal by 30,000.

This gives *pymt* = 24.995038. . . . Betty would need to have only $25 taken out of each of her biweekly paychecks to save $30,000 in 18 years. Notice that she will not have exactly $30,000 saved, because she cannot have exactly $24.995048. . . deducted from each paycheck.   ∎

Present Value of an Annuity

The **present value of an annuity** is the lump sum that can be deposited at the beginning of the annuity's term, at the same interest rate and with the same compounding period, that would yield the same amount as the annuity. This value can help the saver to understand his or her options; it refers to an alternative way of saving the same amount of money in the same time. It is called the *present value* because it refers to the single action that the saver can take *in the present* (i.e., at the beginning of the annuity's term) that would have the same effect as would the annuity.

EXAMPLE 5   Find the present value of Tom and Betty's annuity.

Solution

$$FV = P(1 + i)^n$$

$$30{,}005.95588 = P\left(1 + \frac{0.0925}{26}\right)^{468}$$

$$P = \frac{30{,}005.95588}{\left(1 + \dfrac{0.0925}{26}\right)^{468}}$$

$$= 5{,}693.6451\ldots \approx \$5693.65$$

This means that Tom and Betty would have to deposit $5,693.55 as a lump sum to save as much money as the annuity would yield. They chose an annuity over a lump sum deposit because they could not afford to tie up $5,700 for 18 years, but they could afford to deduct $25 out of each paycheck.   ∎

EXAMPLE 6   Find the present value of an ordinary annuity that has $200 monthly payments for 25 years, where the account receives $10\frac{1}{2}\%$ interest.

Solution   We could find the future value of the annuity and then find the lump sum deposit whose future value matches it, as we did in Example 5. However, it is simpler to do the calculation all at once. The key is to realize that the future value of the lump sum must equal the future value of the annuity:

Future value of lump sum = future value of annuity

$$P(1 + i)^n = pymt\frac{(1 + i)^n - 1}{i}$$

For both the lump sum and the annuity, $i = \frac{1}{12}$ of $10\frac{1}{2}\% = \frac{0.105}{12}$ and $n = 25$ years = 300 months. The annuity's payment is *pymt* = $200.

$$P(1 + i)^n = pymt\frac{(1 + i)^n - 1}{i}$$

$$P(1 + 0.105/12)^{300} = 200\frac{(1 + 0.105/12)^{300} - 1}{0.105/12}$$

First, calculate the right side, as with any annuity calculation. Then divide by $(1 + 0.105/12)^{300}$ to find P.

$$P = 21182.363\ldots$$
$$P \approx \$21{,}182.36$$

This means that one would have to make a lump sum deposit of more than $21,000 to have as much money after 25 years as with monthly $200 annuity payments. ■

Present Value of Annuity Formula

$$FV(\text{lump sum}) = FV(\text{annuity})$$
$$P(1 + i)^n = pymt\frac{(1 + i)^n - 1}{i}$$

The present value is the lump sum P.

5.3 Exercises

In Exercises 1–12, find the future value of the given annuity.

1. ordinary annuity, $120 monthly payment, $5\frac{3}{4}\%$ interest, 1 year
2. ordinary annuity, $175 monthly payment, $6\frac{1}{8}\%$ interest, 11 years
3. annuity due, $100 monthly payment, $5\frac{7}{8}\%$ interest, 4 years
4. annuity due, $150 monthly payment, $6\frac{1}{4}\%$ interest, 13 years

5. On September 8, Bert Sarkis joined a Christmas club. His bank will automatically deduct $75 from his checking account at the end of each month and deposit it into his Christmas club account, where it will earn 7% interest. The account comes to term on December 1. Find the following:
   a. The future value of the account, using an annuity formula
   b. The future value of the account, using the compound interest formula
   c. Bert's total contribution to the account
   d. The total interest
6. On August 19, Rachael Westlake joined a Christmas club. Her bank will automatically deduct $110 from her checking account at the end of each month and deposit it into her Christmas club account, where it will earn $6\frac{7}{8}\%$ interest. The account comes to term on December 1. Find the

following:
   a. The future value of the account, using an annuity formula
   b. The future value of the account, using the compound interest formula
   c. Rachael's total contribution to the account
   d. The total interest
7. On August 23, Ginny Deus joined a Christmas club. Her bank will automatically deduct $150 from her checking account at the beginning of each month and deposit it into her Christmas club account, where it will earn $7\frac{1}{4}\%$ interest. The account comes to term on December 1. Find the following:
   a. The future value of the account, using an annuity formula
   b. The future value of the account, using the compound interest formula
   c. Ginny's total contribution to the account
   d. The total interest
8. On September 19, Lynn Knight joined a Christmas club. Her bank will automatically deduct $100 from her checking account at the beginning of each month and deposit it into her Christmas club account, where it will earn 6% interest. The account comes to term on December 1. Find the following:
   a. The future value of the account, using an annuity formula
   b. The future value of the account, using the compound interest formula

c. Lynn's total contribution to the account
d. The total interest

9. Pat Gilbert recently set up a TDA to save for her retirement. She arranged to have $175 taken out of each of her monthly checks; it will earn $10\frac{1}{2}$% interest. She just had her thirty-ninth birthday, and her ordinary annuity comes to term when she is 65. Find the following:
   a. The future value of the account
   b. Pat's total contribution to the account
   c. The total interest

10. Dick Eckel recently set up a TDA to save for his retirement. He arranged to have $110 taken out of each of his biweekly checks; it will earn $9\frac{7}{8}$% interest. He just had his twenty-ninth birthday, and his ordinary annuity comes to term when he is 65. Find the following:
    a. The future value of the account
    b. Dick's total contribution to the account
    c. The total interest

11. Sam Whitney recently set up a TDA to save for his retirement. He arranged to have $290 taken out of each of his monthly checks; it will earn 11% interest. He just had his forty-fifth birthday, and his ordinary annuity comes to term when he is 65. Find the following:
    a. The future value of the account
    b. Sam's total contribution to the account
    c. The total interest

12. Art Dull recently set up a TDA to save for his retirement. He arranged to have $50 taken out of each of his biweekly checks; it will earn it will earn $9\frac{1}{8}$% interest. He just had his 30th birthday, and his ordinary annuity comes to term when he is 65. Find the following:
    a. The future value of the account
    b. Art's total contribution to the account
    c. The total interest

In Exercises 13–18, find and interpret the present value of the given annuity.

13. The annuity in Exercise 1
14. The annuity in Exercise 2
15. The annuity in Exercise 5
16. The annuity in Exercise 6
17. The annuity in Exercise 9
18. The annuity in Exercise 10

In Exercises 19–24, find the monthly payment that will yield the given future value.

19. $100,000 at $9\frac{1}{4}$% interest for 30 years; ordinary annuity

20. $45,000 at $8\frac{7}{8}$% interest for 20 years; ordinary annuity
21. $250,000 at $10\frac{1}{2}$% interest for 40 years, ordinary annuity
22. $183,000 at $8\frac{1}{4}$% interest for 25 years, ordinary annuity
23. $250,000 at $10\frac{1}{2}$% interest for 40 years, annuity due
24. $183,000 at $8\frac{1}{4}$% interest for 25 years, annuity due

25. Mr. and Mrs. Gonzales set up a TDA to save for their retirement. They agreed to have $100 deducted from each of Mrs. Gonzales's biweekly paychecks, which will earn $8\frac{1}{8}$% interest.
    a. Find the future value of their ordinary annuity if it comes to term after they retire in $35\frac{1}{2}$ years.
    b. After retiring, the Gonzales family convert their annuity to a savings account, which earns 6.1% interest compounded monthly. At the end of each month, they withdraw $650 for living expenses. Complete the chart in Figure 5.15 for their postretirement account.

Month Number	Account Balance at Beginning of the Month	Interest for the Month	With-drawal	Account Balance at End of the Month
1				
2				
3				
4				
5				

Figure 5.15
Chart for Exercise 25

26. Mr. and Mrs. Jackson set up a TDA to save for their retirement. They agreed to have $125 deducted from each of Mrs. Jackson's biweekly paycheck, which will earn $7\frac{5}{8}$% interest.
    a. Find the future value of their ordinary annuity, if it comes to term after they retire in $32\frac{1}{2}$ years.
    b. After retiring, the Jacksons convert their annuity to a savings account, which earns 6.3% interest compounded monthly. At the end of each month, they withdraw $700 for living expenses. Complete the chart in Figure 5.16 for their postretirement account.

Month Number	Account Balance at Beginning of the Month	Interest for the Month	With-drawal	Account Balance at End of the Month
1				
2				
3				
4				
5				

Figure 5.16
Chart for Exercise 26

27. Jeanne and Harold Kimura want to set up a TDA that will generate sufficient interest at maturity to meet their living expenses, which they project to be $950 per month.

  a. Find the amount needed at maturity to generate $950 per month interest if they can get $6\frac{1}{2}\%$ interest compounded monthly.

  b. Find the monthly payment that they would have to put into an ordinary annuity to obtain the future value found in part (a) if their money earns $8\frac{1}{4}\%$ and the term is 30 years.

28. Susan and Bill Stamp want to set up a TDA that will generate sufficient interest at maturity to meet their living expenses, which they project to be $1,200 per month.

  a. Find the amount needed at maturity to generate $1,200 per month interest, if they can get $7\frac{1}{4}\%$ interest compounded monthly.

  b. Find the monthly payment that they would have to make into an ordinary annuity to obtain the future value found in part (a) if their money earns $9\frac{3}{4}\%$ and the term is 25 years.

In Exercises 29–32, use the following information. An Individual Retirement Account (IRA) is an annuity that is set up to save for retirement. IRAs differ from TDAs in that an IRA allows the participant to contribute money whenever he or she wants, whereas a TDA requires the participant to have a specific amount deducted from each of his or her paychecks.

29. When Shannon Pegnim was 14, she got an after-school job at a local pet shop. Her parents told her that if she put some of her earnings into an IRA, they would contribute an equal amount to her IRA. That year and every year thereafter, she deposited $1,000 into her IRA. When she became 25 years old, her parents stopped contributing, but Shannon increased her annual deposit to $2,000 and continued depositing that amount annually until she retired at age 65. Her IRA paid 8.5% interest. Find the following:

  a. The future value of the account
  b. Shannon's and her parents' total contributions to the account
  c. The total interest
  d. The future value of the account if Shannon waited until she was 19 before she started her IRA
  e. The future value of the account if Shannon waited until she was 24 before she started her IRA

30. When Bo McSwine was 16, he got an after-school job at his parents' barbecue restaurant. His parents told him that if he put some of his earnings into an IRA, they would contribute an equal amount to his IRA. That year and every year thereafter, he deposited $900 into his IRA. When he became 21 years old, his parents stopped contributing, but Bo increased his annual deposit to $1,800 and continued depositing that amount annually until he retired at age 65. His IRA paid 7.75% interest. Find the following:

  a. The future value of the account
  b. Bo's and his parents' total contributions to the account
  c. The total interest
  d. The future value of the account if Bo waited until he was 18 before he started his IRA
  e. The future value of the account if Bo waited until he was 25 before he started his IRA

31. If Shannon Pegnim from Exercise 29 started her IRA at age 35 rather than age 14, how big of an annual contribution would she have had to have made to have the same amount saved at age 65?

32. If Bo McSwine from Exercise 30 started his IRA at age 35 rather than age 16, how big of an annual contribution would he have had to have made to have the same amount saved at age 65?

33. Toni Torres wants to save $1,200 in the next two years to use as a down payment on a new car. If her bank offers her 9% interest, what monthly payment would she need to make into an ordinary annuity to reach her goal?

34. Fred and Melissa Furth's daughter Sally will be a freshman in college in six years. To help cover their extra expenses, the Furths decide to set up a sinking fund of $12,000. If the account pays 7.2% interest and they wish to make quarterly payments, find the size of each payment.

35. Anne Geyer buys some land in Utah. She agrees to pay the seller a lump sum of $65,000 in five years. Until then, she will make monthly simple interest payments to the seller at 11% interest.
   a. Find the amount of each interest payment.
   b. Anne sets up a sinking fund to save the $65,000. Find the size of her semiannual payments if her payments are due at the end of every six-month period and her money earns $8\frac{3}{8}\%$ interest.
   c. Prepare a table showing the amount in the sinking fund after each deposit.

36. Chrissy Fields buys some land in Oregon. She agrees to pay the seller a lump sum of $120,000 in six years. Until then, she will make monthly simple interest payments to the seller at 12% interest.
   a. Find the amount of each interest payment.
   b. Chrissy sets up a sinking fund to save the $120,000. Find the size of her semiannual payments if her money earns $10\frac{3}{4}\%$ interest.
   c. Prepare a table showing the amount in the sinking fund after each deposit.

37. Develop a new formula for the present value of an ordinary annuity by solving the Present Value of Annuity Formula for P and simplifying.
38. Use the formula developed in Exercise 37 to find the present value of the annuity in Exercise 2.
39. Use the formula developed in Exercise 37 to find the present value of the annuity in Exercise 1.
40. Use the formula developed in Exercise 37 to find the present value of the annuity in Exercise 6.
41. Use the formula developed in Exercise 37 to find the present value of the annuity in Exercise 5.

Answer the following questions using complete sentences and your own words.

CONCEPT QUESTIONS
42. Explain the difference between compound interest and an annuity.
43. Explain how an annuity is based on compound interest.
44. Describe the difference between an ordinary annuity and an annuity due.
45. Compare and contrast an annuity with a lump sum investment that receives compound interest. Be sure to discuss both the similarity and the difference between these two concepts, as well as the advantages and disadvantages of each.
46. Which is always greater: the present value of an annuity or the future value? Why?

47. If you want to retire on $2,000 a month for 25 years, do you need to save $2,000 a month for 25 years before retiring? Why or why not?
48. *For those who have completed Section 1.1:* Is the logic used in deriving the Ordinary Annuity Formula inductive or deductive? Why? Is the logic used in deriving the relationship

$$FV(\text{due}) = FV(\text{ordinary}) \cdot (1 + i)$$

inductive or deductive? Why?

 WEB PROJECTS

49. Think of something that you would like to be able to afford but cannot—perhaps a car, boat, or motorcycle. In this exercise, you will explore how to make that unaffordable dream a realistic goal.
   a. Just what is it that you would like to be able to afford?
   b. Determine an appropriate but realistic date for making the purchase. Justify your date.
   c. Do some research and determine how much your goal would cost currently. Cite your sources.
   d. Do some web research and determine the current rate of inflation. Inflation rate information is readily available on the web. Cite your sources.
   e. Use the results of parts (b), (c), and (d), as well as either the simple or compound interest formula, to predict how much your goal will cost in the future.
   f. Discuss why you chose to use the simple or compound interest formula in part (e).
   g. Go to the web sites of four different banks and determine the current interest rate available for an annuity of an appropriate term.
   h. Determine the necessary annuity payment that will allow you to meet your goal. If necessary, alter the date from part (b).

50. Suppose you had a baby in 2005 and you decided that it would be wise to start saving for his or her college education. In this exercise, you will explore how to go about doing that.
   a. According to the College Board, four-year public schools cost an average of $5,491 per year in 2005–2006. Four-year private schools cost an average of $21,235 per year. These costs include tuition and fees, room and board, books and supplies, and personal expenses. Furthermore, four-year public schools increased 7.1% from 2004–2005, and four-year private schools increased 5.9% from 2004–2005. Use this information, as well as either the simple or compound interest formula, to predict the cost of your child's college education in his or

her freshman, sophomore, junior, and senior years, for both public and private institutions. Discuss your assumptions and justify your work.
(*Source:* **www.collegeboard.com/pay**.)

b. Will you save for a public or a private institution? Why?

c. Discuss why you chose to use the simple or compound interest formula in part (a).

d. Go to the web sites of four different banks and determine the current interest rate available for an annuity of an appropriate term.

e. Determine the necessary annuity payment that will allow you to save the total amount from part (a) in time to meet your goal.

Some useful links for these web projects are listed on the text web site:
www.thomsonedu.com/mathematics/johnson

 Annuities on a Graphing Calculator

Exercises

51. **a.** Use the method discussed in Section 5.2 on doubling time to find how long it takes for an annuity to have a balance of $500,000, if the ordinary annuity requires monthly $200 payments that earn 5% interest. Notice that payments are made on a monthly basis, so the answer must be a whole number of months.

b. State what values you used for *x*min, *x*max, *x*scl, *y*min, *y*max, and *y*scl. Explain how you arrived at those values.

c. Draw a freehand sketch of the graph obtained on your calculator.

52. Analyze the effect of the interest rate on annuities by recalculating Exercise 51 with interest rates of 6%, 8%, and 10%. Discuss the impact of the increased rate.

53. Analyze the effect of the payment period on annuities by recalculating Exercise 40 with twice-monthly payments of $100 each that earn 5% interest. Discuss the impact of the altered period.

5.4 Amortized Loans

An **amortized loan** is a loan for which the loan amount, plus interest, is paid off in a series of regular equal payments. In Section 5.1, we looked at one type of amortized loan: the add-on interest loan. A second type of amortized loan is the *simple interest amortized loan.* There is an important difference between these two types of loans that any potential borrower should be aware of: *the payments are smaller with a simple interest amortized loan than they are with an add-on interest loan* (assuming, naturally, that the loan amounts, interest rates, and number of payments are the same).

A **simple interest amortized loan** is an ordinary annuity whose future value is the same as the loan amount's future value, under compound interest. Realize, though, that a simple interest amortized loan's payments are used to pay off a loan, where other annuities' payments are used to generate savings.

Simple Interest Amortized Loan Formula

Future value of annuity = future value of loan amount

$$pymt \frac{(1 + i)^n - 1}{i} = P(1 + i)^n$$

where *pymt* is the loan payment, *i* is the periodic interest rate, *n* is the number of periods, and *P* is the present value or loan amount

---

Algebraically, this formula could be used to determine any one of the four unknowns (*pymt*, *i*, *n*, and *P*) if the other three are known. We will use it to find the payment when the annual interest rate, number of periods, and loan amount are known.

EXAMPLE 1   Tom and Betty decide that they need a more dependable car, now that they have a baby. They buy one for $13,518.77. They make a $1,000 down payment and finance the balance through a four-year simple interest amortized loan from their bank. They are charged 12% interest.

a. Find their monthly payment.
b. Find the total interest they will pay over the life of the loan.

Solution   **a.** Use the simple interest amortized loan formula, with a loan amount of $P = \$13,518.77 - \$1,000 = \$12,518.77$, a monthly interest rate of $12\%/12 = 1\% = 0.01$, and a term of $n = 4$ years $= 48$ months.

$$pymt\frac{(1 + i)^n - 1}{i} = P(1 + i)^n$$

$$pymt\frac{(1 + 0.01)^{48} - 1}{0.01} = 12{,}518.77(1 + 0.01)^{48}$$

To find *pymt*, we need to divide the right side by the fraction on the left side or, equivalently, multiply by its reciprocal. First, find the fraction on the left side, as with any annuity calculation. Then multiply the right side by the fraction's reciprocal.

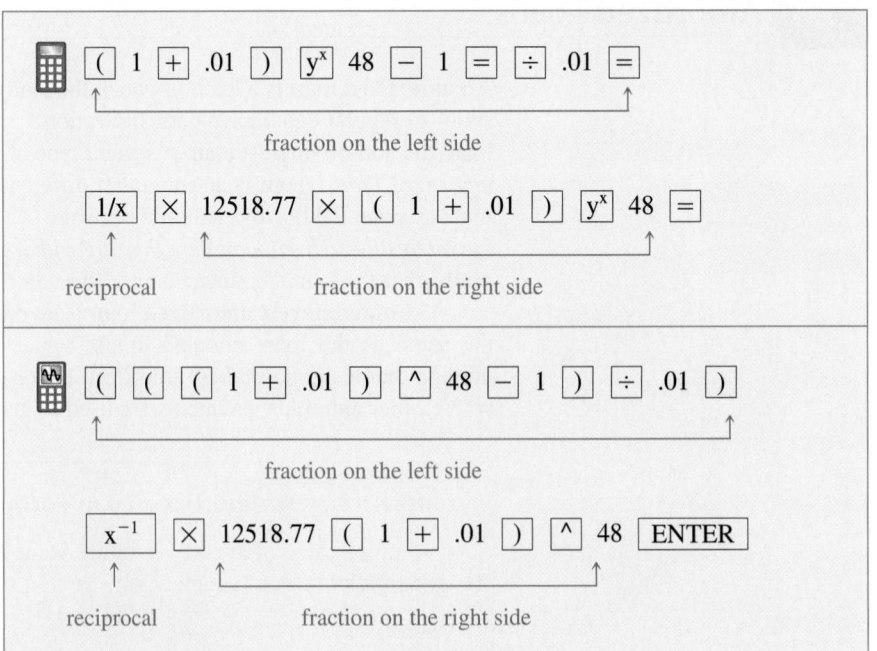

We get *pymt* = 329.667... ≈ $329.67.

> ✓ As a rough check, note that Tom and Betty borrowed $12,518.77. If they paid no interest, each payment would be $12,518.77/48 ≈ $260.81. Since they must pay interest, each payment must be larger than $260.81. Our calculation checks, because $329.67 > $260.81.

b. The total of their payments is 48 · $329.67 = $15,824.16. This includes both principal and interest. Of this, $12,518.77 is principal, so $15,824.16 − $12,518.77 = $3,305.39 is interest. Over the life of the loan, Tom and Betty are paying $3,305.39 in interest.   ∎

Amortization Schedules

The simple interest amortized loan formula is an equation with the ordinary annuity formula on one side of the equal symbol and the compound interest formula on the other side. However, with such a loan, you do not pay compound interest. Instead, the interest portion of each payment is simple interest on the outstanding principal.

An **amortization schedule** is a list of several periods of payments, the principal and interest portions of those payments, and the **outstanding principal** (or **balance**) after each of those payments is made.

The data on the amortization schedule are important to the borrower for two reasons. The borrower needs to know the total interest paid, for tax purposes. Interest paid on a home loan is usually deductible from the borrower's income tax, and interest paid on a loan by a business is usually deductible. The borrower would also need the data on an amortization schedule if he or she is considering paying off the loan early. Such prepayment could save money because an advance payment would be all principal and would not include any interest; however, the lending institution may charge a **prepayment penalty** which would absorb some of the interest savings.

EXAMPLE 2   Prepare an amortization schedule for the first two months of Tom and Betty's loan.

Solution   For any simple interest loan, the interest portion of each payment is simple interest on the outstanding principal, so use the Simple Interest Formula, $I = Prt$, to compute the interest. Recall that r and t are annual figures. For each payment, $r = 12\% = 0.12$ and $t = 1$ month $= \frac{1}{12}$ year. For payment number 1, $P = \$12,418.77$ (the amount borrowed). The interest portion of payment number 1 is

$$I = Prt$$

$$= 12,518.77 \cdot 0.12 \cdot \frac{1}{12}$$

$$\approx \$125.19$$

The principal portion of payment number 1 is

payment − interest portion

$$= \$329.67 - \$125.19$$

$$= \$204.48$$

The outstanding principal or balance is

Previous principal − principal portion
$$= \$12{,}518.77 - \$204.48$$
$$= \$12{,}314.29$$

All of the above information goes into the amortization schedule. See Figure 5.17.

Payment number	Principal portion	Interest portion	Total payment	Balance
0	—	—	—	$12,518.77
1	$204.48	$125.19	$329.67	$12,314.29
2	$206.53	$123.14	$329.67	$12,107.76

Figure 5.17
An amortization schedule for the first two months of Tom and Betty's loan

When it is time to make payment number 2, less money is owed. The outstanding principal is $12,314.29, so this is the new value of P. The interest portion of payment number 2 is

$$I = Prt$$
$$= 12{,}314.29 \cdot 0.12 \cdot \tfrac{1}{12}$$
$$\approx \$123.14$$

The principal portion is

Payment − interest portion
$$= \$329.67 - \$123.14$$
$$= \$206.53$$

The outstanding principal or balance is

Previous principal − principal portion
$$= \$12{,}314.29 - \$206.53$$
$$= \$12{,}107.76$$

All of the above information is shown in the amortization schedule in Figure 5.17.   ■

Notice how in Example 2, the principal portion increases and the interest portion decreases. This continues throughout the life of the loan, and the final payment is mostly principal. This happens because after each payment, the amount due is somewhat smaller, so the interest on the amount due is somewhat smaller also.

EXAMPLE 3   Comp-U-Rent needs to borrow $60,000 to increase their inventory of rental computers. The company is confident that their expanded inventory will generate sufficient extra income to allow them to pay off the loan in a short amount of time, so they wish to borrow the money for only three months. First National Bank offered them a simple interest amortized loan at $8\tfrac{3}{4}\%$ interest.

a. Find what their monthly payment would be with First National.
b. Prepare an amortization schedule for the entire term of the loan.

Solution   **a.** $P = \$60,000$, $i = \frac{1}{12}$ of $8\frac{3}{4}\% = 0.0875/12$, and $n = 3$ months.

Future value of annuity = future value of loan amount

$$pymt\frac{(1 + i)^n - 1}{i} = P(1 + i)^n$$

$$pymt\frac{\left(1 + \dfrac{0.0875}{12}\right)^3 - 1}{\dfrac{0.0875}{12}} = 60000\left(1 + \frac{0.0875}{12}\right)^3$$

We get

$$pymt = 20{,}292.375\ldots$$
$$\approx \$20{,}292.38$$

> ✓ As a rough check, notice that Comp-U-Rent borrowed \$60,000, and each payment includes principal and interest, so the payment must be larger than $\frac{\$60{,}000}{3} = \$20{,}000$. Our calculation checks, because \$20,292.38 > \$20,000.

b. For each payment, $r = 8\frac{3}{4}\% = 0.0875$ and $t = 1$ month $= \frac{1}{12}$ year.

For payment number 1, $P = \$60,000.00$ (the amount borrowed). The interest portion of payment number 1 is

$$I = Prt$$

$$= 60,000.00 \cdot 0.0875 \cdot \frac{1}{12}$$

$$= \$437.50$$

The principal portion of payment number 1 is

Payment $-$ interest portion

$$= \$20,292.38 - \$437.50 = \$19,854.88$$

The outstanding principal or balance is

Previous principal $-$ principal portion

$$= \$60,000.00 - \$19,854.88 = \$40,152.12$$

For payment number 2, $P = \$40,152.12$ (the outstanding principal). The interest portion of payment number 2 is

$$I = Prt$$

$$= 40,152.12 \cdot 0.0875 \cdot \frac{1}{12}$$

$$\approx \$146.89$$

The principal portion is

Payment $-$ interest portion $= \$20,292.38 - \$146.89 = \$20,145.49$

The outstanding principal or balance is

Previous principal $-$ principal portion

$$= \$20,145.46 - \$20,145.49 = -\$0.03$$

Negative three cents cannot be correct. After the final payment is made, the amount due *must be* $0.00. The discrepancy arises from the fact that we rounded off the payment size from $20,292.375 ... to $20,292.38. If there were some way that the borrower could make a monthly payment that is not rounded off, then the calculation above would have yielded an amount due of $0.00. To repay the exact amount owed, we must compute the last payment differently.

The principal portion of payment number 3 *must be* $20,145.46, because that is the balance, and this payment is the only chance to pay it. The payment must also include $146.89 interest, as calculated above. The last payment is the sum of the principal due and the interest on that principal:

$$\$20,145.46 + \$146.89 = \$20,292.35$$

The closing balance is then $20,145.46 $-$ $20,145.46 = $0.00, as it should be. The amortization schedule is given in Figure 5.18.

Payment number	Principal portion	Interest portion	Total payment	Balance
0	—	—	—	$60,000.00
1	$19,854.88	$437.50	$20,292.38	$40,145.12
2	$19,999.66	$292.72	$20,292.38	$20,145.46
3	$20,145.46	$146.89	$20,292.35	$0.00

Figure 5.18
The amortization schedule for Comp-U-Rent's loan ∎

> ## Amortization Schedule Steps
>
> For each payment, list the payment number, principal portion, interest portion, total payment and balance.
>
> *For each payment:*
> 1. Find the interest on the balance, using the simple interest formula.
>
> *For each payment except the last:*
> 2. The principal portion is the payment minus the interest portion.
> 3. The new balance is the previous balance minus the principal portion.
>
> *For the last payment:*
> 4. The principal portion is the previous balance.
> 5. The total payment is the sum of the principal portion and interest portion.
>
> Figure 5.19 shows these steps in chart form.

Payment number	Principal portion	Interest portion	Total payment	Balance
0	—	—	—	Loan amount
First through next-to-last	Total payment minus interest portion	Simple interest on previous balance; use $I = Prt$	Use simple interest amortized loan formula	Previous balance minus this payment's principal portion
Last	Previous balance	Simple interest on previous balance; use $I = Prt$	Principal portion plus interest portion	$0.00

Figure 5.19
Amortization schedule steps

In preparing an amortization schedule, the new balance is the previous balance minus the principal portion. This means that only the principal portion of a payment goes toward paying off the loan. The interest portion is the lender's profit.

Prepaying a Loan

Sometimes a borrower needs to pay his or her loan off early, before the loan's term is over. This is called **prepaying a loan.** This often happens when people sell their houses. It also happens when interest rates have fallen since the borrower obtained his or her loan and the borrower decides to refinance the loan at a lower rate.

When you prepay a loan, you pay off the unpaid balance. You can find that unpaid balance by looking at an amortization schedule. If an amortization schedule has not already been prepared, the borrower can easily approximate the **unpaid balance** by subtracting the current value of the annuity from the current value of the loan. This approximation is usually off by at most a few pennies.

Unpaid Balance Formula

Unpaid balance = current value of loan amount − current value of annuity

$$\approx P(1 + i)^n - pymt \frac{(1 + i)^n - 1}{i}$$

where *pymt* is the loan payment, i is the periodic interest rate, n is the number of periods *from the beginning of the loan to the present*, and P is the loan amount.

A common error in using this formula is to let n equal the number of periods in the entire life of the loan, rather than the number of periods from the beginning of the loan until the time of prepayment. This results in an answer of 0. After all of the payments have been made, the unpaid balance should be 0.

ThomsonNOW™
for Liberal Arts Mathematics
www.thomsonedu.com/1pass.

EXAMPLE 4 Ten years ago, Tom and Betty bought a house for $140,000. They paid the sellers a 20% down payment and obtained a simple interest amortized loan for $112,000 from their bank at $10\frac{3}{4}\%$ for 30 years. Their monthly payment is $1,045.50.

Their old home was fine when there were just two of them. But with their new baby, they need more room. They are considering selling their existing home and buying a new home. This would involve paying off their home loan. They would use the income from the house's sale to do this. Find the cost of paying off the existing loan.

Solution The loan is a 30-year loan, so if we were computing the payment, we would use $n = 30$ years $\cdot$ 12 months/year $= 360$ months. However, we are computing the unpaid balance, not the payment, so n is not 360. Tom and Betty have made payments on their loan for 10 years $= 120$ months, so for this calculation, $n = 120$.

Unpaid balance = current value of loan amount − current value of annuity

$$\approx P(1 + i)^n - pymt \frac{(1 + i)^n - 1}{i}$$

$$= 112{,}000 \left(1 + \frac{0.1075}{12} \right)^{120} - 1{,}045.50 \frac{\left(1 + \dfrac{0.1075}{12} \right)^{120} - 1}{\dfrac{0.1075}{12}}$$

$$= \$102{,}981.42$$

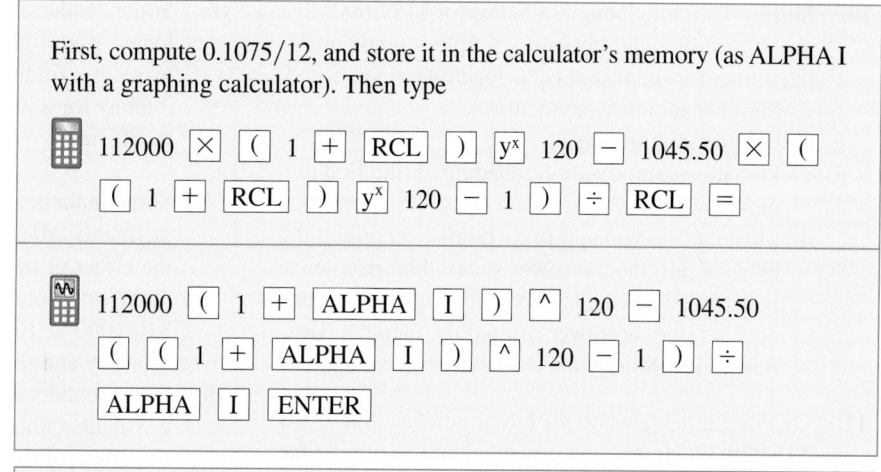

First, compute 0.1075/12, and store it in the calculator's memory (as ALPHA I with a graphing calculator). Then type

112000 × (1 + RCL) yˣ 120 − 1045.50 × (
(1 + RCL) yˣ 120 − 1) ÷ RCL =

112000 (1 + ALPHA I) ^ 120 − 1045.50
((1 + ALPHA I) ^ 120 − 1) ÷
ALPHA I ENTER

✓ As a rough check, note that Tom and Betty borrowed $112,000, so their unpaid balance must be less. Our calculation checks, because $102,981.42 < $112,000.

The result of Example 4 means that after 10 years of payments of over $1000 a month on a loan of $112,000, Tom and Betty still owe approximately $102,981.42! They were shocked. It is extremely depressing for first-time home purchasers to discover how little of their beginning payments actually goes toward paying off the loan. At the beginning of the loan, you owe a lot of money, so most of each payment is interest, and very little is principal. Later, you do not owe as much, so most of each payment is principal and very little is interest.

5.4 Exercises

In the following exercises, all loans are simple interest amortized loans with monthly payments unless labeled otherwise.

In Exercises 1–6, find (a) the monthly payment and (b) the total interest for the given simple interest amortized loan.

1. $5,000, $9\frac{1}{2}$%, 4 years
2. $8,200, $10\frac{1}{4}$%, 6 years
3. $10,000, $6\frac{1}{8}$%, 5 years
4. $20,000, $7\frac{3}{8}$%, $5\frac{1}{2}$ years
5. $155,000, $9\frac{1}{2}$%, 30 years
6. $289,000, $10\frac{3}{4}$%, 35 years

7. Wade Ellis buys a new car for $16,113.82. He puts 10% down and obtains a simple interest amortized loan for the rest at $11\frac{1}{2}$% interest for four years.
   a. Find his monthly payment.
   b. Find the total interest.
   c. Prepare an amortization schedule for the first two months of the loan.

8. Guy dePrimo buys a new car for $9,837.91. He puts 10% down and obtains a simple interest amortized loan for the rest at $8\frac{7}{8}$% interest for four years.
   a. Find his monthly payment.
   b. Find the total interest.
   c. Prepare an amortization schedule for the first two months of the loan.

9. Chris Burditt bought a house for $212,500. He put 20% down and obtains a simple interest amortized loan for the rest at $10\frac{7}{8}$% for thirty years.
   a. Find his monthly payment.
   b. Find the total interest.
   c. Prepare an amortization schedule for the first two months of the loan.
   d. Most lenders will approve a home loan only if the total of all the borrower's monthly payments, including the home loan payment, is no more than 38% of the borrower's monthly income. How much must Chris make to qualify for the loan?

10. Shirley Trembley bought a house for $187,600. She put 20% down and obtains a simple interest amortized loan for the rest at $6\frac{3}{8}\%$ for thirty years.
    a. Find her monthly payment.
    b. Find the total interest.
    c. Prepare an amortization schedule for the first two months of the loan.
    d. Most lenders will approve a home loan only if the total of all the borrower's monthly payments, including the home loan payment, is no more than 38% of the borrower's monthly income. How much must Shirley make to qualify for the loan?

11. Dennis Lamenti wants to buy a new car that costs $15,829.32. He has two possible loans in mind. One loan is through the car dealer; it is a four-year add-on interest loan at $7\frac{3}{4}\%$ and requires a down payment of $1,000. The second is through his bank; it is a four-year simple interest amortized loan at at $7\frac{3}{4}\%$ and requires a down payment of $1,000.
    a. Find the monthly payment for each loan.
    b. Find the total interest paid for each loan.
    c. Which loan should Dennis choose? Why?

12. Barry Wood wants to buy a used car that costs $4,000. He has two possible loans in mind. One loan is through the car dealer; it is a three-year add-on interest loan at 6% and requires a down payment of $300. The second is through his credit union; it is a three-year simple interest amortized loan at 9.5% and requires a 10% down payment.
    a. Find the monthly payment for each loan.
    b. Find the total interest paid for each loan.
    c. Which loan should Barry choose? Why?

13. Investigate the effect of the term on simple interest amortized auto loans by finding the monthly payment and the total interest for a loan of $11,000 at $9\frac{7}{8}\%$ interest if the term is
    a. 3 years   b. 4 years
    c. 5 years.

14. Investigate the effect of the interest rate on simple interest amortized auto loans by finding the monthly payment and the total interest for a four-year loan of $12,000 at
    a. 8.5%   b. 8.75%
    c. 9%   d. 10%.

15. Investigate the effect of the interest rate on home loans by finding the monthly payment and the total interest for a thirty-year loan of $100,000 at
    a. 6%   b. 7%
    c. 8%   d. 9%
    e. 10%   f. 11%

16. Some lenders are now offering fifteen-year home loans. Investigate the effect of the term on home loans by finding the monthly payment and total interest for a loan of $100,000 at 10% if the term is
    a. 30 years   b. 15 years

17. Some lenders are now offering loans with biweekly payments rather than monthly payments. Investigate the effect of this on home loans by finding the payment and total interest on a thirty-year loan of $100,000 at 10% interest if payments are made (a) monthly and (b) biweekly.

18. Some lenders are now offering loans with biweekly payments rather than monthly payments. Investigate the effect of this on home loans by finding the payment and total interest on a thirty-year loan of $150,000 at 8% interest if payments are made (a) monthly and (b) biweekly.

19. Verify (a) the monthly payments and (b) the interest savings in the savings and loan advertisement in Figure 5.20.

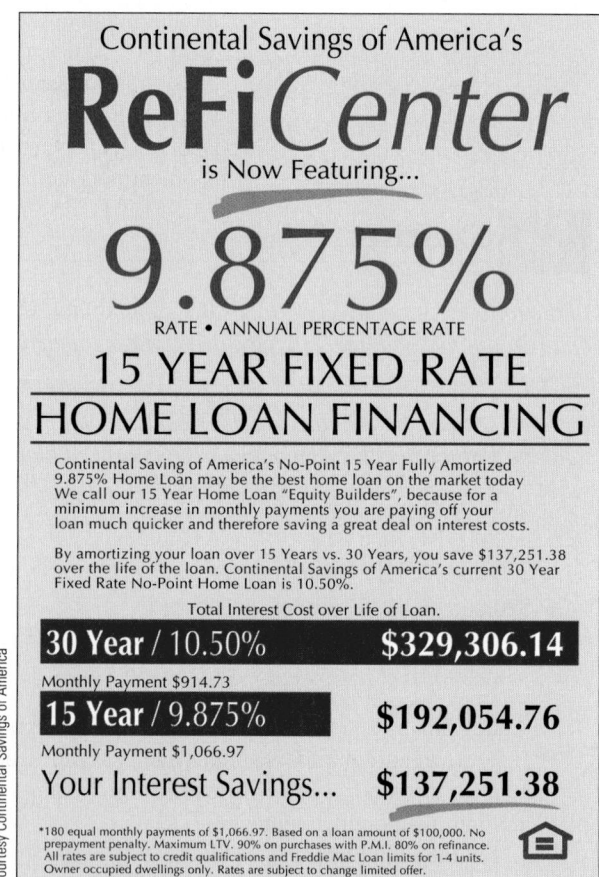

Figure 5.20
Loan terms for Exercise 19

20. The home loan in Exercise 19 presented two options. The thirty-year option required a smaller monthly payment. A consumer who chooses the thirty-year option could take the savings in the monthly payment that this option generates and invest that savings in an annuity. At the end of fifteen years, the annuity might be large enough to pay off the thirty-year loan. Determine whether this is a wise plan if the annuity's interest rate is 7.875%. (Disregard the tax ramifications of this approach.)

21. Pool-N-Patio World needs to borrow $75,000 to increase its inventory for the upcoming summer season. The owner is confident that he will sell most, if not all, of the new inventory during the summer, so he wishes to borrow the money for only four months. His bank has offered him a simple interest amortized loan at $7\frac{3}{4}$% interest.
a. Find the size of the monthly bank payment.
b. Prepare an amortization schedule for all four months of the loan.

22. Slopes R Us needs to borrow $120,000 to increase its inventory of ski equipment for the upcoming season. The owner is confident that she will sell most, if not all, of the new inventory during the winter, so she wishes to borrow the money for only five months. Her bank has offered her a simple interest amortized loan at $8\frac{7}{8}$% interest.
a. Find the size of the monthly bank payment.
b. Prepare an amortization schedule for all five months of the loan.

23. The owner of Video Extravaganza is opening a second store and needs to borrow $93,000. Her success with her first store has made her confident that she will be able to pay off her loan quickly, so she wishes to borrow the money for only four months. Her bank has offered her a simple interest amortized loan at $9\frac{1}{8}$% interest.
a. Find the size of the monthly bank payment.
b. Prepare an amortization schedule for all four months of the loan.

24. The Green Growery Nursery needs to borrow $48,000 to increase its inventory for the upcoming summer season. The owner is confident that he will sell most, if not all, of the new plants during the summer, so he wishes to borrow the money for only four months. His bank has offered him a simple interest amortized loan at $9\frac{1}{4}$% interest.
a. Find the size of the monthly bank payment.
b. Prepare an amortization schedule for all four months of the loan.

*For Exercises 25–28, note the following information. A **line of credit** is an agreement between a bank and a borrower by which the borrower can borrow any amount of money (up to a mutually agreed-upon maximum) at any time, simply by writing a check. Typically, monthly simple interest payments are required, and the borrower is free to make principal payments as frequently or infrequently as he or she wants. Usually, a line of credit is secured by the title to the borrower's house, and the interest paid to the bank by the borrower is deductible from the borrower's income taxes.*

25. Kevin and Roxanne Gahagan did not have sufficient cash to pay their income taxes. However, they had previously set up a line of credit with their bank. On April 15, they wrote a check to the Internal Revenue Service on their line of credit for $6,243. The line's interest rate is 5.75%.
a. Find the size of the required monthly interest payment.
b. The Gahagans decided that it would be in their best interests to get this loan paid off in eight months. Find the size of the monthly principal-plus-interest payment that would accomplish this. (*Hint:* In effect, the Gahagans are converting the loan to an amortized loan.)
c. Prepare an amortization schedule for all eight months of the loan.
d. Find the amount of line of credit interest that the Gahagans could deduct from their taxes next year.

26. James and Danna Wright did not have sufficient cash to pay their income taxes. However, they had previously set up a line of credit with their bank. On April 15, they wrote a check to the Internal Revenue Service on their line of credit for $10,288. The line's interest rate is 4.125%.
a. Find the size of the required monthly interest payment.
b. The Wrights decided that it would be in their best interests to get this loan paid off in six months. Find the size of the monthly principal-plus-interest payment that would accomplish this. (*Hint:* In effect, the Wrights are converting the loan to an amortized loan.)
c. Prepare an amortization schedule for all six months of the loan.
d. Find the amount of line of credit interest that the Wrights could deduct from their taxes next year.

27. Homer Simpson had his bathroom remodeled. He did not have sufficient cash to pay for it. However, he had previously set up a line of credit with his bank. On

July 12, he wrote a check to his contractor on his line of credit for $12,982. The line's interest rate is 8.25%.

a. Find the size of the required monthly interest payment.

b. Homer decided that it would be in his best interests to get this loan paid off in seven months. Find the size of the monthly principal-plus-interest payment that would accomplish this. (*Hint:* In effect, Simpson is converting the loan to an amortized loan.)

c. Prepare an amortization schedule for all seven months of the loan.

d. Find the amount of line of credit interest that Simpson could deduct from his taxes next year.

28. Harry Trask had his kitchen remodeled. He did not have sufficient cash to pay for it. However, he had previously set up a line of credit with his bank. On July 12, he wrote a check to his contractor on his line of credit for $33,519. The line's interest rate is 7.625%.

a. Find the size of the required monthly interest payment.

b. Harry decided that it would be in his best interests to get this loan paid off in seven months. Find the size of the monthly principal-plus-interest payment that would accomplish this. (*Hint:* In effect, Trask is converting the loan to an amortized loan.)

c. Prepare an amortization schedule for all seven months of the loan.

d. Find the amount of line of credit interest that Trask could deduct from his taxes next year.

29. Wade Ellis buys a car for $16,113.82. He puts 10% down and obtains a simple interest amortized loan for the balance at $11\frac{1}{2}\%$ interest for four years. Three years and two months later, he sells his car. Find the unpaid balance on his loan.

30. Guy de Primo buys a car for $9837.91. He puts 10% down and obtains a simple interest amortized loan for the balance at $10\frac{7}{8}\%$ interest for four years. Two years and six months later, he sells his car. Find the unpaid balance on his loan.

31. Gary Kersting buys a house for $212,500. He puts 20% down and obtains a simple interest amortized loan for the balance at $10\frac{7}{8}\%$ interest for thirty years. Eight years and two months later, he sells his house. Find the unpaid balance on his loan.

32. Shirley Trembley buys a house for $187,600. She puts 20% down and obtains a simple interest amortized loan for the balance at $11\frac{3}{8}\%$ interest for thirty years. Ten years and six months later, she sells her house. Find the unpaid balance on her loan.

33. Harry and Natalie Wolf have a 3-year old loan with which they purchased their house. Their interest rate is $13\frac{3}{8}\%$. Since they obtained this loan, interest rates have dropped, and they can now get a loan for $8\frac{7}{8}\%$ through their credit union. Because of this, the Wolfs are considering refinancing their home. Each loan is a thirty-year simple interest amortized loan, and neither has a prepayment penalty. The existing loan is for $152,850, and the new loan would be for the current amount due on the old loan.

a. Find their monthly payment with the existing loan.

b. Find the loan amount for their new loan.

c. Find the monthly payment with their new loan.

d. Find the total interest they will pay if they do *not* get a new loan.

e. Find the total interest they will pay if they *do* get a new loan.

f. Should the Wolfs refinance their home? Why or why not?

34. Russ and Roz Rosow have a ten-year old loan with which they purchased their house. Their interest rate is $10\frac{5}{8}\%$. Since they obtained this loan, interest rates have dropped, and they can now get a loan for $9\frac{1}{4}\%$ through their credit union. Because of this, the Rosows are considering refinancing their home. Each loan is a 30-year simple interest amortized loan, and neither has a prepayment penalty. The existing loan is for $112,000, and the new loan would be for the current amount due on the old loan.

a. Find their monthly payment with the existing loan.

b. Find the loan amount for their new loan.

c. Find the monthly payment with their new loan.

d. Find the total interest they will pay if they do *not* get a new loan.

e. Find the total interest they will pay if they *do* get a new loan.

f. Should the Rosows refinance their home? Why or why not?

35. Michael and Lynn Sullivan have a 10-year old loan for $187,900 with which they purchased their house.

They just sold their highly profitable import-export business and are considering paying off their home loan. Their loan is a thirty-year simple interest amortized loan at 10.5% interest and has no prepayment penalty.

a. Find their monthly payment.

b. Find the unpaid balance of the loan.

c. Find the amount of interest they will save by prepaying.

d. The Sullivans decided that if they paid off their loan, they would deposit the equivalent of half their monthly payment into an annuity. If the ordinary annuity pays 9% interest, find its future value after 20 years.

e. The Sullivans decided that if they do not pay off their loan, they would deposit an amount equivalent to their unpaid balance into an account that pays $9\frac{3}{4}$% interest compounded monthly. Find the future value of this account after 20 years.

f. Should the Sullivans prepay their loan? Why or why not?

36. Charlie and Ellen Wilson have a 25-year-old loan for $47,000 with which they purchased their house. The Wilsons are retired and are living on a fixed income, so they are contemplating paying off their home loan. Their loan is a thirty-year simple interest amortized loan at $4\frac{1}{2}$% and has no prepayment penalty. They also have savings of $73,000, which they have invested in a certificate of deposit currently paying $8\frac{1}{4}$% interest compounded monthly. Should they pay off their home loan? Why or why not?

37. Ray and Helen Lee bought a house for $189,500. They put 10% down, borrowed 80% from their bank for thirty years at $11\frac{1}{2}$%, and convinced the owner to take a second mortgage for the remaining 10%. That 10% is due in full in five years (this is called a *balloon payment*), and the Lees agree to make monthly interest-only payments to the seller at 12% simple interest in the interim.

a. Find the Lees' down payment

b. Find the amount that the Lees borrowed from their bank.

c. Find the amount that the Lees borrowed from the seller.

d. Find the Lees' monthly payment to the bank.

e. Find the Lees' monthly interest payment to the seller.

38. Jack and Laurie Worthington bought a house for $163,700. They put 10% down, borrowed 80% from their bank for thirty years at 12% interest, and convinced the owner to take a second mortgage for the remaining 10%. That 10% is due in full in five years, and the Worthingtons agree to make monthly interest-only payments to the seller at 12% simple interest in the interim.

a. Find the Worthingtons' down payment

b. Find the amount that the Worthingtons borrowed from their bank.

c. Find the amount that the Worthingtons borrowed from the seller.

d. Find the Worthingtons' monthly payment to the bank.

e. Find the Worthingtons' monthly interest payment to the seller.

39. a. If the Lees in Exercise 37 save for their balloon payment with a sinking fund, find the size of the necessary monthly payment into that fund if their money earns 6% interest.

b. Find the Lees' total monthly payment for the first five years.

c. Find the Lees' total monthly payment for the last twenty-five years.

40. a. If the Worthingtons in Exercise 38 save for their balloon payment with a sinking fund, find the size of the necessary monthly payment into that fund if their money earns 7% interest.

b. Find the Worthingtons' total monthly payment for the first five years.

c. Find the Worthingtons' total monthly payment for the last twenty-five years.

> Answer the following questions using complete sentences and your own words.

CONCEPT QUESTIONS

41. In Exercise 11, an add-on interest amortized loan required larger payments than did a simple interest amortized loan at the same interest rate. What is there about the structure of an add-on interest loan that makes its payments larger than those of a simple interest amortized loan? (*Hint:* The interest portion is a percent of what quantity?)

42. Why are the computations for the last period of an amortization schedule different from those for all preceding periods?

43. Give two different situations in which a borrower would need the information contained in an amortization schedule.

44. A borrower would need to know the unpaid balance of a loan if the borrower were to prepay the loan. Give three different situations in which it might be in a borrower's best interests to prepay a loan. (*Hint:* See some of the preceding exercises.)

45. If you double the period of an amortized loan, does your monthly payment halve? Why or why not?

46. If you double the loan amount of an amortized loan, does your monthly payment double? Why or why not?

 WEB PROJECTS

47. This is an exercise in buying a car. It involves choosing a car and selecting the car's financing. Write a paper describing all of the following points.

    a. You might not be in a position to buy a car now. If that is the case, fantasize about your future. What job do you have? How long have you had that job? What is your salary? If you are married, does your spouse work? Do you have a family? What needs will your car fulfill? Make your fantasy realistic. Briefly describe what has happened between the present and your future fantasy. (If you are in a position to buy a car now, discuss these points on a more realistic level.)

    b. Go shopping for a car. Look at new cars, used cars, or both. Read newspaper and magazine articles about your choices (see, for example, *Consumer Reports, Motor Trend,* and *Road and Track*). Discuss in detail the car you selected and why you did so. How will your selection fulfill your (projected) needs? What do newspapers and magazines say about your selection? Why did you select a new or a used car?

    c. Go to the web sites of four different banks. Get all of the information you need about a car loan. Perform all appropriate computations yourself— do not have the lenders tell you the payment size, and do not use web calculators. Summarize the appropriate data (including the down payment, payment size, interest rate, duration of loan, type of loan, and loan fees) in your paper, and discuss which loan you would choose. Explain how you would be able to afford your purchase.

48. This is an exercise in buying a home. It involves choosing a home and selecting the home's financing. Write a paper describing all of the following points.

    a. You might not be in a position to buy a home now. If this is the case, fantasize about your future. What job do you have? How long have you had that job? What is your salary? If you are married, does your spouse work? Do you have a family? What needs will your home fulfill? Make your future fantasy realistic. Briefly describe what has happened between the present and your future fantasy. (If you are in a position to buy a home now, discuss these points on a more realistic level.)

    b. Go shopping for a home. Look at houses, condominiums, or both. Look at new homes, used homes, or both. Used homes can easily be visited by going to an "open house," where the owners are gone and the real estate agent allows interested parties to inspect the home. Open houses are probably listed in your local newspaper. Read appropriate newspaper and magazine articles (for example, in the real estate section of your local newspaper). Discuss in detail the home you selected and why you did so. How will your selection fulfill your (projected) needs? Why did you select a house or a condominium? Why did you select a new or a used home? Explain your choice of location, house size, and features of the home.

    c. Go to the web sites of four different banks. Get all of the information you need about a home loan. Perform all appropriate computations yourself—do not have the lenders tell you the payment size, and do not use web calculators. Summarize the appropriate data in your paper and discuss which loan you would choose. Include in your discussion the down payment, the duration of the loan, the interest rate, the payment size, and other terms of the loan.

    d. Also discuss the real estate taxes (your instructor will provide you with information on the local tax rate) and the effect of your purchase on your income taxes (interest paid on a home loan is deductible from the borrower's income taxes).

e. Most lenders will approve a home loan only if the total of all the borrower's monthly payments, including the home loan payment, real estate taxes, credit card payments, and car loan payments, is no more than 38% of the borrower's monthly income. Discuss your ability to qualify for the loan.

Some useful links for the above web projects are listed on the text web site: www.thomsonedu.com/mathematics/johnson

PROJECTS

*For Exercises 49–51, note the following information. An **adjustable-rate mortgage** (or ARM) is, as the name implies, a mortgage in which the interest rate is allowed to change. As a result, the payment changes too. At first, an ARM costs less than a fixed-rate mortgage—its initial interest rate is usually two or three percentage points lower. As time goes by, the rate is adjusted. As a result, it might or might not continue to hold this advantage.*

49. Trustworthy Savings offers a thirty-year adjustable-rate mortgage with an initial rate of 5.375%. The rate and the required payment are adjusted annually. Future rates are set at 2.875 percentage points above the 11th District Federal Home Loan Bank's cost of funds. Currently, that cost of funds is 4.839%. The loan's rate is not allowed to rise more than two percentage points in any one adjustment, nor is it allowed to rise above 11.875%. Trustworthy Savings also offers a thirty-year fixed-rate mortgage with an interest rate of 7.5%.
  a. Find the monthly payment for the fixed-rate mortgage on a loan amount of $100,000.
  b. Find the monthly payment for the ARM's first year on a loan amount of $100,000.
  c. How much would the borrower save in the mortgage's first year by choosing the adjustable rather than the fixed-rate mortgage?
  d. Find the unpaid balance at the end of the ARM's first year.
  e. Find the interest rate and the value of *n* for the ARM's second year if the 11th District Federal Home Loan Bank's cost of funds does not change during the loan's first year.
  f. Find the monthly payment for the ARM's second year if the 11th District Federal Home Loan

Bank's cost of funds does not change during the loan's first year.
  g. How much would the borrower save in the mortgage's first two years by choosing the adjustable-rate mortgage rather than the fixed-rate mortgage if the cost of funds does not change?
  h. Discuss the advantages and disadvantages of an adjustable-rate mortgage.

50. American Dream Savings Bank offers a thirty-year adjustable-rate mortgage with an initial rate of 4.25%. The rate and the required payment are adjusted annually. Future rates are set at three percentage points above the one-year Treasury bill rate, which is currently 5.42%. The loan's rate is not allowed to rise more than two percentage points in any one adjustment, nor is it allowed to rise above 10.25%. American Dream Savings Bank also offers a thirty-year fixed-rate mortgage with an interest rate of 7.5%.
  a. Find the monthly payment for the fixed-rate mortgage on a loan amount of $100,000.
  b. Find the monthly payment for the ARM's first year on a loan amount of $100,000.
  c. How much would the borrower save in the mortgage's first year by choosing the adjustable rather than the fixed-rate mortgage?
  d. Find the unpaid balance at the end of the ARM's first year.
  e. Find the interest rate and the value of *n* for the ARM's second year if the one-year Treasury bill rate does not change during the loan's first year.
  f. Find the monthly payment for the ARM's second year if the Treasury bill rate does not change during the loan's first year.
  g. How much would the borrower save in the mortgage's first two years by choosing the adjustable-rate mortgage rather than the fixed-rate mortgage if the Treasury bill rate does not change?
  h. Discuss the advantages and disadvantages of an adjustable-rate mortgage.

51. Bank Two offers a thirty-year adjustable-rate mortgage with an initial rate of 3.95%. This initial rate is in effect for the first six months of the loan, after which it is adjusted on a monthly basis. The monthly payment is adjusted annually. Future rates are set at 2.45 percentage points above the 11th District

Federal Home Loan Bank's cost of funds. Currently, that cost of funds is 4.839%.

a. Find the monthly payment for the ARM's first year on a loan amount of $100,000.

b. Find the unpaid balance at the end of the ARM's first six months.

c. Find the interest portion of the seventh payment if the cost of funds does not change.

d. Usually, the interest portion is smaller than the monthly payment, and the difference is subtracted from the unpaid balance. However, the interest portion found in part (c) is larger than the monthly payment found in part (a), and the difference is added to the unpaid balance found in part (b). Why would this difference be added to the unpaid balance? What effect will this have on the loan?

e. The situation described in part (d) is called **negative amortization.** Why?

f. What is there about the structure of Bank Two's loan that allows negative amortization?

Amortization Schedules on a Computer

Interest paid on a home loan is deductible from the borrower's income taxes, and interest paid on a loan by a business is usually deductible. A borrower with either of these types of loans needs to know the total interest paid on the loan during the final year. The way to determine the total interest paid during a given year is to prepare an amortization schedule for that year. Typically, the lender provides the borrower with an amortization schedule, but it is not uncommon for this schedule to arrive after taxes are due. In this case, the borrower must either do the calculation personally or pay taxes without the benefit of the mortgage deduction and then file an amended set of tax forms after the amortization schedule has arrived.

Computing a year's amortization schedule is rather tedious, and neither a scientific calculator nor a graphing calculator offers relief. The best tool for the job is a computer, combined either with the Amortrix computer program (available with this book) or with a computerized spreadsheet. Each is discussed below.

Amortization Schedules and Amortrix

Amortrix is one of the features of the text web site (**www.thomsonedu.com/ mathematics/johnson**). This software will enable you to quickly and easily compute an amortization schedule for any time period. We'll illustrate this process by using Amortrix to prepare an amortization schedule for a 15 year $175,000 loan at 7.5% interest with monthly payments.

When you start Amortrix, a main menu appears. Click on the "Amortization Schedule" option. Once you're at the page labeled "Amortization Schedule," enter "175000" (without commas or dollar signs) for the Loan Amount (P), "7.5" (without a percent sign) for the Annual Interest Rate, and "180" for the Total Number of Payment Periods (n). See the top part of Figure 5.21. Click on "Calculate" and the software will create an amortization schedule. See the bottom part of Figure 5.21.

The software will *not* correctly compute the last loan payment; it will compute the last payment in the same way it computes all other payments rather than in the way shown earlier in this section. You will have to correct this last payment if you use the computer to prepare an amortization schedule for a time period that includes the last payment.

Figure 5.21
Part of an amortization schedule prepared with Amortrix

Amortization Schedules and Excel

A **spreadsheet** is a large piece of paper marked off in rows and columns. Accountants use spreadsheets to organize numerical data and perform computations. A **computerized spreadsheet** such as Microsoft Excel is a computer program that mimics the appearance of a paper spreadsheet. It frees the user from performing any computations.

When you start a computerized spreadsheet, you see something that looks like a table waiting to be filled in. The rows are labeled with numbers and the columns are labeled with letters, as shown in Figure 5.22. The individual boxes are called **cells.** The cell in column A, row 1 is called cell A1.

◇	A	B	C	D	E	F	G
1	payment number	principal portion	interest portion	total payment	balance		
2	0				$175,000.00	rate	7.50%
3						years	15
4						payments/yr	12
5							

Figure 5.22
The spreadsheet after step 1

Excel is an ideal tool to use in creating an amortization schedule. We will illustrate this process by preparing the amortization schedule for a fifteen-year $175,000 loan at 7.5% interest with monthly payments.

1. *Set up the spreadsheet.* Start by entering the information shown in Figure 5.22.

   - Adjust the columns' widths. To make column A wider, place your cursor on top of the line dividing the "A" and "B" labels. Then use your cursor to move that dividing line.
   - Format cell E2 as currency by highlighting that cell and pressing the "$" button on the Excel ribbon at the top of the screen.
   - Be certain that you include the % symbol in cell G2.
   - Save the spreadsheet.

   If you have difficulty creating a spreadsheet that looks like this, you may download it from our web site at www.thomsonedu.com/mathematics/johnson. See the file 5.4.xls and select the "step 1" sheet. Other sheets show later steps.

2. *Compute the monthly payment.* While you can do this with your calculator, as we discussed earlier in this section, you can also use the Excel "PMT" function to compute the monthly payment. This allows us to change the rate (in cell G2) or the loan amount (in cell E2) and Excel will automatically compute the new payment and change the entire spreadsheet accordingly. To use the "PMT" function:

   - Click on cell D3, where we will put the total payment.
   - Click on the *fx* button at the top of the screen.
   - In the resulting "Paste Function" box, select "Financial" under "Function Category", select "PMT" under "Function name", and press "OK." A "PMT" box will appear.
   - Cell G2 holds 7.5%, the annual interest rate, and cell G4 holds 12, the number of periods/year. The periodic interest rate is $7.5\%/12 = G2/G4$. In the PMT box after "Rate", type "$G$2/$G$4".
   - The number of periods is $15 \cdot 12 = G3 \cdot G4$. In the PMT box after "Nper", type "$G$3*$G$4."
   - After "PV", type "$E$2". This is where we have stored the present value of the loan.
   - Do not type anything after "Fv" or "type".
   - Press "OK" at the bottom of the "PMT" box, and "($1,622.27)" should appear in cell D2. The parentheses mean that the number is negative.
   - Save the spreadsheet.

   Doing the above creates two problems that we must fix. The payment is negative. We can fix that with an extra minus sign. Also, the payment is not rounded to the nearest penny. We can fix that with the ROUND function.

   In cell D2, Excel displays ($1622.27), but it stores $-1622.27163\ldots$, and all calculations involving cell D2 will be done by using $-1622.27163\ldots$. This will make the amortization schedule incorrect. To fix this, click on cell D3 and the long box at the top of the screen will have

   $$=\text{PMT}(\$G\$2/\$G\$4,\$G\$3*\$G\$4,\$E\$2)$$

   in it. Change this to

   $$=\text{ROUND}\,(-\text{PMT}(\$G\$2/\$G\$4,\$G\$3*\$G\$4,\$E\$2),2)$$

   the round fuction   a new minus sign   part of the round function

   The new parts here are "ROUND(−" at the beginning and ",2)" at the end. The "ROUND (,2)" part tells Excel to round to two decimal places, as a bank would. The minus sign after "ROUND(" makes the payment positive. Place your cursor where the new parts go, and add them.

3. *Fill in row 3.*
   - We need to add a year each time we go down one row. Type "= A2 + 1" in cell A3.
   - Cell C3 should contain instructions on computing simple interest on the previous balance.

   $$I = P \cdot r \cdot t$$
   $$= \text{the previous balance} \cdot 0.075 \cdot 1/12$$
   $$= E2*\$G\$2/\$G\$4$$

   Move to cell C3 and type in "=ROUND(E2*\$G\$2/\$G\$4,2)" and press "return". The "ROUND (,2)" part rounds the interest to two decimal places, as a bank would.
   - Cell B3 should contain instructions on computing the payment's principal portion.

   $$\text{Principal portion} = \text{payment} - \text{interest portion}$$
   $$= D3 - C3$$

   In cell B3, type "= D3 − C3".
   - Cell E3 should contain instructions on computing the new balance.

   $$\text{New balance} = \text{previous balance} - \text{principal portion}$$
   $$= E2 - B3$$

   In cell E3, type "= E2 − B3".

   See Figure 5.23.

◇	A	B	C	D	E	F	G
1	payment number	principal portion	Interest portion	total payment	balance		
2	0				$175,000.00	rate	7.50%
3	1	$ 528.52	$1,093.75	$1,622.27	$174,471.48	years	15
4						payments/yr	12
5							

Figure 5.23
The spreadsheet after completing step 3

4. *Fill in rows 4 and beyond.* All of the remaining payments' computations are just like payment 1 computations (except for the last payment), so all we have to do is copy the payment 1 instructions in row 3 and paste them in rows 4 and beyond. After completion of this step, the top and bottom of your spreadsheet should look like that in Figure 5.24 on page 386.

5. *Fix the last payment.* Clearly, the ending balance of $0.56 in cell E182 is not correct—we cannot owe $0.56 or any other amount after making our last payment. See Exercise 52.

6. *Find the interest paid.* Typing "SUM(C3:C6)" in a cell will result in the sum of cells C4, C4, C5, and C6 appearing in that cell. Using this function makes it easy to find the total interest paid for any time period.

◇	A	B	C	D	E	F	G
1	payment number	principal portion	interest portion	total payment	balance		
2	0				$175,000.00	rate	7.50%
3	1 $	528.52	$1,093.75	$1,622.27	$174,471.48	years	15
4	2 $	531.82	$1,090.45	$1,622.27	$173,939.66	payments/yr	12
5	3 $	535.15	$1,087.12	$1,622.27	$173,404.51		
6	4 $	538.49	$1,083.78	$1,622.27	$172,866.02		

The top of the spreadsheet, and ...

178	176 $	$1,572.51	$49.76	$1,622.27	$	6,389.50
179	177 $	$1,582.34	$39.93	$1,622.27	$	4,807.16
180	178 $	$1,592.23	$30.04	$1,622.27	$	3,214.93
181	179 $	$1,602.18	$20.09	$1,622.27	$	1,612.75
182	180 $	$1,612.19	$10.08	$1,622.27	$	0.56
183						

... the bottom of the spreadsheet, after completing step 4

Figure 5.24

Exercises

In Exercises 52–58, first use Amortrix or Excel to create an amortization schedule for a fifteen-year $175,000 loan at 7.5% interest with monthly payments. (Figure 5.24 shows the first and last lines of this amortization schedule, created with Excel.) You will alter parts of this amortization schedule to answer Exercises 52–58.

52. **a.** The very last line of your schedule is incorrect, because the calculations for the last line of any amortization schedule are done differently from those of all other lines. What should the last line be?

    b. If you use Excel, state exactly what you should type in row 182 to fix the last payment.

53. Use Excel or Amortrix to find the total interest paid in

    a. the loan's first year and

    b. the loan's last year if the first month of the loan is January.

    c. If you use Excel, state exactly what you type in parts (a) and (b) to find the total interest paid.

54. Right before you signed your loan papers, the interest rate dropped from 7.5% to 7%.

    a. How does this affect the total payment?

    b. How does this affect the total interest paid in the loan's first year, if the first month of the loan is January?

    c. How does this affect the total interest paid in the loan's last year?

55. Right before you signed your loan papers, you convinced the seller to accept a smaller price. As a result, your loan amount dropped from $175,000 to $165,000. (The interest rate remains at 7.5%)

    a. How does this affect the total payment?

    b. How does this affect the total interest paid in the loan's first year, if the first month of the loan is January?

    c. How does this affect the total interest paid in the loan's last year?

56. You are considering a twenty-year loan. (The interest rate remains at 7.5% and the loan amount remains at $175,000.)

    a. How does this affect the total payment?

    b. How does this affect the total interest paid in the loan's first year, if the first month of the loan is January?

    c. How does this affect the total interest paid in the loan's fifteenth year?

    d. Explain, without performing any calculations, why the total payment should go down and the total interest paid during the first year should go up.

387 Annual Percentage Rate on a Graphing Calculator 387

57. You are considering a thirty-year loan. (The interest rate remains at 7.5%, and the loan amount remains at $175,000.)
  a. How does this affect the total payment?
  b. How does this affect the total interest paid in the loan's first year if the first month of the loan is January?
  c. How does this affect the total interest paid in the loan's fifteenth year?
  d. Explain, without performing any calculations, why the total payment should go down and the total interest paid during the first year should go up.

58. You are considering a loan with payments every two weeks. (The interest rate remains at 7.5%, the loan amount remains at $175,000, and the term remains at fifteen years.)
  a. How does this affect the total payment?
  b. How does this affect the total interest paid in the loan's first year, if the first month of the loan is January?
  c. How does this affect the total interest paid in the loan's last year?
  d. Explain, without performing any calculations, why the total payment and the total interest paid during the first year should both go down.

In Exercises 59–62:

  a. *Use Amortrix or Excel to prepare an amortization schedule for the given loan.*
  b. *Find the amount that could be deducted from the borrower's taxable income (that is, find the total interest paid) in the loan's first year if the first payment was made in January. (Interest on a car loan is deductible only in some circumstances. For the purpose of these exercises, assume that this interest is deductible.)*
  c. *Find the amount that could be deducted from the borrower's taxable income (that is, find the total interest paid) in the loan's last year.*

  d. *If you sell your home or car after three years, how much will you still owe on the mortgage?*

59. A five-year simple interest amortized car loan for $32,600 at 12.25% interest

60. A four-year simple interest amortized car loan for $26,200 at 5.75% interest

61. A fifteen-year simple interest amortized home loan for $220,000 at 6.25% interest

62. A thirty-year simple interest amortized home loan for $350,000 at 14.5% interest

63. Use Amortrix or Excel to do Exercise 29 on page 378. Is your answer the same as that of Exercise 29? If not, why not? Which answer is more accurate? Why?

64. Use Amortrix or Excel to do Exercise 30 on page 378. Is your answer the same as that of Exercise 30? If not, why not? Which answer is more accurate? Why?

65. Use Amortrix or Excel to do Exercise 31 on page 378. Is your answer the same as that of Exercise 31? If not, why not? Which answer is more accurate? Why?

66. Use Amortrix or Excel to do Exercise 32 on page 378. Is your answer the same as that of Exercise 32? If not, why not? Which answer is more accurate? Why?

67. *For Excel users only:* When interest rates are high, some lenders offer the following type of home loan. You borrow $200,000 for thirty years. For the first five years, the interest rate is 5.75%, and the payments are calculated as if the interest rate were going to remain unchanged for the life of the loan. For the last twenty-five years, the interest rate is 9.25%.
  a. What are the monthly payments for the first five years?
  b. What are the monthly payments for the last twenty-five years?
  c. What is the total interest paid, during the loan's fifth year if the first month of the loan is January?
  d. What is the total interest paid during the loan's sixth year?
  e. Why would a lender offer such a loan?

5.5 Annual Percentage Rate on a Graphing Calculator

A **simple interest loan** is any loan for which the interest portion of each payment is simple interest on the outstanding principal. A *simple interest amortized loan* fulfills this requirement; in fact, we compute an amortization schedule for a simple interest amortized loan by finding the simple interest on the outstanding principal.

Thomson NOW™
for Liberal Arts Mathematics
www.thomsonedu.com/1pass.

EXAMPLE 1   Chris Dant paid $18,327 for a new car. Her dealer offered her a five-year add-on interest loan at 8.5% interest.

a. Find the size of her monthly payment.
b. Find the principal portion and the interest portion of each payment.
c. Determine if this loan is a simple interest loan.

Solution   a. *Finding the monthly payment.* We are given $P = 18{,}327$, $r = 0.085$, and $t = 5$ years. The total interest charge is simple interest on the loan amount:

$$I = Prt$$
$$= 18{,}327 \cdot 0.085 \cdot 5$$
$$= 7{,}788.975$$
$$\approx \$7{,}788.98$$

The total of principal and interest is then

$$P + I = 18{,}327 + 7{,}788.98 = \$26{,}115.98$$

This amount is equally distributed over 60 monthly payments, so the monthly payment is

$$\frac{26{,}115.98}{60} = 435.26633 \ldots \approx \$435.27$$

b. *Finding the principal and interest portions.* The total amount due consists of $18,327 in principal and $7,788.98 in interest. The total amount due is equally distributed over 60 monthly payments, so the principal and interest are also equally distributed over 60 monthly payments. The principal portion of each payment is then

$$\frac{18{,}327}{60} = \$305.45$$

and the interest portion is

$$\frac{7{,}788.98}{60} = 129.81633 \ldots \approx \$129.82$$

c. *Determining whether the loan is a simple interest loan.* The loan is a simple interest loan if the interest portion of each payment is simple interest on the outstanding principal. For the first payment, the outstanding principal is $18,327; simple interest on this amount is

$$I = Prt$$
$$= 18{,}327 \cdot 0.085 \cdot \frac{1}{12}$$
$$= 129.81625 \approx \$129.82$$

In part (b), we found that the interest portion of each payment is also $129.82; thus, for the first payment, the interest portion is simple interest on the outstanding principal.

For the second payment, the outstanding principal is

$$18{,}327 - 305.45 = 18{,}021.55$$

Simple interest on this amount is

$$I = Prt$$

$$= 18{,}021.55 \cdot 0.085 \cdot \frac{1}{12}$$

$$= 127.65264 \ldots \approx \$127.65$$

However, the interest portion of each payment is $129.82; thus, for the second payment, the interest portion is *more than* simple interest on the outstanding principal. The loan is *not* a simple interest loan; it requires higher interest payments than would a simple interest loan at the same rate. ∎

The APR of an Add-on Interest Loan

Whenever a loan is not a simple interest loan, the Truth in Lending Act requires the lender to disclose the annual percentage rate to the borrower. The **annual percentage rate (APR) of an add-on interest loan** is the simple interest rate that makes the dollar amounts the same if the loan is recomputed as a simple interest amortized loan.

EXAMPLE 2   Find the APR of the add-on interest loan in Example 1.

Solution   Substitute the dollar amounts into the Simple Interest Amortized Loan Formula and solve for the interest rate i.

$$pymt\frac{(1 + i)^n - 1}{i} = P(1 + i)^n$$

$$435.27\frac{(1 + i)^{60} - 1}{i} = 18{,}327(1 + i)^{60}$$

Solving this equation without the aid of a graphing calculator would be difficult if not impossible. To solve it with a graphing calculator:

- Enter $435.27\dfrac{(1 + x)^{60} - 1}{x}$ for Y_1
- Enter $18{,}327(1 + x)^{60}$ for Y_2
- Graph the two functions
- Find the point at which the two functions intersect, as discussed in Appendix D and summarized here.

TI-83/84:	select the "intersect" option from the "CALC" menu
TI-86:	select "MATH" from the "GRAPH" menu, and then "ISECT" from the "GRAPH MATH" menu
Casio:	press G-Solv and then ISCT

Figure 5.25
Finding the intersection

The graph is shown in Figure 5.25. The intersection is at $i = 0.0124369$. This is a monthly rate; the corresponding annual rate is $12 \cdot 0.0124369 = 0.1492428 \approx$ 14.92%. The 8.5% add-on interest loan in Example 1 has an APR of 14.92%. This means that the add-on interest loan requires the same monthly payment that a 14.92% simple interest amortized loan would have.

Courtesy of LBJ Library

The **Truth in Lending Act** was signed by President Lyndon Johnson in 1968. Its original intent was to promote credit shopping by requiring lenders to use uniform language and calculations and to make full disclosure of credit charges so that consumers could shop for the most favorable credit terms. The Truth in Lending Act is interpreted by the Federal Reserve Board's Regulation Z.

During the 1970s, Congress amended the act many times, and corresponding changes were made in Regulation Z. These changes resulted in a huge increase in the length and complexity of the law. Its scope now goes beyond disclosure to grant significant legal rights to the borrower.

The Truth in Lending Act, which requires a bank to divulge its loan's APR, allows a tolerance of one-eighth of 1% (= 0.125%) in the claimed APR. Thus, the dealership would be legally correct if it stated that the APR was between 14.92428% − 0.125% = 14.79928% and 14.92428% + 0.125% = 15.04928%. ■

The APR of a Simple Interest Loan

Sometimes **finance charges** other than the interest portion of the monthly payment are associated with a loan; these charges must be paid when the loan agreement is signed. For example, a **point** is a finance charge that is equal to 1% of the loan amount; a **credit report fee** pays for a report on the borrower's credit history, including any late or missing payments and the size of all outstanding debts; an **appraisal fee** pays for the determination of the current market value of the property (the auto, boat, or home) to be purchased with the loan. The Truth in Lending Act requires the lender to inform the borrower of the total finance charge, which includes the interest, the points, and some of the fees (see Figure 5.26). Most of the finance charges must be paid before the loan is awarded, so in essence the borrower must pay money now in order to get more money later. The law says that this means the lender is not really borrowing as much as he or she thinks. According to the law, the actual amount loaned is the loan amount minus all points and those fees included in the finance charge. The **APR of a simple interest amortized loan** is the rate that reconciles the payment and this actual loan amount.

Costs Included in the Prepaid Finance Charges		Other Costs Not Included in the Finance Charge	
2 points	$2,415.32	appraisal fee	$ 70.00
prorated interest	$1,090.10	credit report	$ 60.00
prepaid mortgage		closing fee	$ 670.00
insurance	$ 434.50	title insurance	$ 202.50
loan fee	$1,242.00	recording fee	$ 20.00
document		notary fee	$ 20.00
preparation fee	$ 80.00	tax and insurance	
tax service fee	$ 22.50	escrow	$ 631.30
processing fee	$ 42.75		
subtotal	$5,327.17	subtotal	$1,673.80

Figure 5.26
Sample portion of a federal truth-in-lending disclosure statement
(loan amount = $120,765.90)

EXAMPLE 3   Glen and Tanya Hansen bought a home for $140,000. They paid the sellers a 20% down payment and obtained a simple interest amortized loan for the balance from their bank at $10\frac{3}{4}\%$ for 30 years. The bank in turn paid the sellers the remaining 80% of the purchase price, less a 6% sales commission paid to the sellers' and the buyers' real estate agents. (The transaction is illustrated in Figure 5.27.) The bank charged the Hansens two points plus fees totaling $3,247.60; of these fees, $1,012.00 were included in the finance charge.

a. Find the size of the Hansens' monthly payment.
b. Find the total interest paid.
c. Compute the total finance charge.
d. Find the APR

Figure 5.27

Solution   **a.** *Finding their monthly payment*   We are given down payment = 20% of $140,000 = $28,000, P = loan amount = $140,000 - $28,000 = $112,000, $i = \frac{1}{12}$ of $10\frac{3}{4}\%$ = $\frac{0.1075}{12}$, and n = 30 years = 30 years · (12 months)/(1 year) = 360 months.

$$\text{future value of annuity} = \text{future value of loan amount}$$

$$pymt\frac{(1+i)^n - 1}{i} = P(1+i)^n$$

$$pymt\frac{\left(1 + \dfrac{0.1075}{12}\right)^{360} - 1}{\dfrac{0.1075}{12}} = 112{,}000\left(1 + \dfrac{0.1075}{12}\right)^{360}$$

Computing the fraction on the left side and multiplying its reciprocal by the right side, we get

$$pymt = 1,045.4991 \approx \$1,045.50$$

b. *Finding the total interest paid*   The total interest paid is the total amount paid minus the amount borrowed. The Hansens agreed to make 360 monthly payments of $1045.50 each, for a total of $360 \cdot \$1,045.50 = \$376,380.00$. Of this, $112,000 is principal; therefore, the total interest is

$$376,380 - 112,000 = \$264,380$$

c. *Computing their total finance charge*

$$
\begin{aligned}
2 \text{ points} = 2\% \text{ of } \$112,000 &= \$\quad 2,240 \\
\text{included fees} &= \$\quad 1,012 \\
\underline{\text{total interest paid}} &= \underline{\$264,380} \quad \text{from part (b)} \\
\text{total finance charge} &= \$267,632
\end{aligned}
$$

d. *Finding the APR*   The APR is the simple interest rate that makes the dollar amounts the same if the loan is recomputed using the legal loan amount (loan amount less points and fees) in place of the actual loan amount. The legal loan amount is

$$P = \$112,000 - \$3252 = \$108,748$$

Substitute the dollar amounts into the simple interest amortized loan formula and solve for the interest rate i.

$$pymt\frac{(1 + i)^n - 1}{i} = P(1 + i)^n \qquad \begin{array}{l}\text{Simple Interest}\\ \text{Amortized Loan Formula}\end{array}$$

$$1,045.50\frac{(1 + i)^{360} - 1}{i} = 108,748(1 + i)^{360} \qquad \text{substituting}$$

To solve this with a graphing calculator:

- Enter $1,045.50 \dfrac{(1 + x)^{360} - 1}{x}$ for Y_1.
- Enter $108,748(1 + x)^{360}$ for Y_2.
- Graph the two functions.
- Find the point at which the two functions intersect.

Figure 5.28
Finding the intersection

The graph is shown in Figure 5.28. The intersection is at $i = 0.00926661$. This is a monthly rate; the corresponding annual rate is $12 \cdot 0.00926661 = 0.11119932 \approx 11.12\%$.

The APR is 11.12%. This means that the $10\frac{3}{4}\%$ loan requires the same monthly payment as a 11.12% loan with no points or fees would have.   ■

If you need to obtain a loan, it is not necessarily true that the lender with the lowest interest rate will give you the least expensive loan. One lender may charge more points or higher fees than does another lender. One lender may offer an add-on interest loan, while another lender offers a simple interest amortized loan. These differences can have a significant impact on the cost of a loan. If two lenders offer loans at the same interest rate but differ in any of these ways, that difference will be reflected in the APR. "Lowest interest rate" does *not* mean "least expensive loan," but "lowest APR." *does* mean "least expensive loan."

APR Steps

1. *Compute the payment.*
   a. Use the methods of Section 5.1 with an add-on interest loan.
   b. Use the methods of Section 5.4 with a simple interest amortized loan.
2. *Use the Simple Interest Amortized Loan Formula to compute the APR.*
   a. Substitute the payment from step 1 for pymt.
   b. If there are points and fees, substitute the legal loan amount (legal loan amount = loan amount less points and those fees included in the finance charge) for *P*.
   c. Use your graphing calculator to find the periodic interest rate.
   d. Convert the periodic interest rate to an annual rate. This annual rate is the APR.
3. *Determine whether the claimed APR is legally correct.* If the lender claims a certain APR, the Truth in Lending Act allows a tolerance of one-eighth of 1% (= 0.125%).

Estimating Prepaid Finance Charges

As we have seen, a borrower usually must pay an assortment of fees when obtaining a home loan. These fees can be quite substantial, and they can vary significantly from lender to lender. For example, the total fees in Figure 5.26 were $7,000.97; the Hansens' total fees in Example 3 were $5,487.60. When shopping for a home loan, a borrower should take these fees into consideration.

All fees must be disclosed to the borrower when he or she signs the loan papers. Furthermore, the borrower must be given an estimate of the fees when he or she applies for the loan. Before an application is made, the borrower can use a loan's APR to obtain a reasonable approximation of those fees that are included in the finance charge (for a fixed rate loan). This would allow the borrower to make a more educated decision in selecting a lender.

ThomsonNOW
for Liberal Arts Mathematics
www.thomsonedu.com/1pass.

EXAMPLE 4   Felipe and Delores Lopez are thinking of buying a home for $152,000. A potential lender advertises an 80%, 30-year simple interest amortized loan at $7\frac{1}{2}\%$ interest with an APR of 7.95%.

   a. Find the size of the Lopezes' monthly payment.
   b. Use the APR to approximate the fees included in the finance charge.

Solution   a. *Finding the monthly payment*   The loan amount is 80% of $152,000 = $121,600, $i = \frac{1}{12}$ of 7.5% $= \frac{0.075}{12}$, and $n = 30$ years = 360 months.

$$\text{Future value of annuity} = \text{future value of loan amount}$$

$$pymt\frac{(1 + i)^n - 1}{i} = P(1 + i)^n$$

$$pymt\frac{\left(1 + \dfrac{0.075}{12}\right)^{360} - 1}{\dfrac{0.075}{12}} = 121{,}600\left(1 + \dfrac{0.075}{12}\right)^{360}$$

Computing the fraction on the left side and multiplying its reciprocal by the right side, we get

$$pymt = 850.2448 \approx \$850.25$$

b. *Approximating the fees included in the finance charge*   We approximate the fees by computing the legal loan amount (loan amount less points and fees), using the APR as the interest rate and the payment computed in part (a) as *pymt*. Therefore, $i = \frac{1}{12}$ of $7.95\% = \frac{0.0795}{12}$, *pymt* = 850.25.

$$\text{Future value of annuity} = \text{future value of loan amount}$$

$$pymt \frac{(1 + i)^n - 1}{i} = P(1 + i)^n$$

$$850.25 \frac{\left(1 + \dfrac{0.0795}{12}\right)^{360} - 1}{\dfrac{0.0795}{12}} = P\left(1 + \dfrac{0.0795}{12}\right)^{360}$$

Computing the left side, and dividing by $(1 + \frac{0.0795}{12})^{360}$, we get

$$P = 116{,}427.6304 \approx \$116{,}427.63$$

This is the legal loan amount, that is, the loan amount less points and fees. The Lopezes were borrowing $121,600, so that leaves $121,600 − $116,427.63 = $5,172.37 in points and fees. This is an estimate of the points and fees included in the finance charge, such as a loan fee, a document preparation fee, and a processing fee; it does not include fees such as an appraisal fee, a credit report fee, and a title insurance fee.   ■

5.5   Exercises

1. Wade Ellis buys a new car for $16,113.82. He puts 10% down and obtains a simple interest amortized loan for the balance at $11\frac{1}{2}\%$ interest for four years. If loan fees included in the finance charge total $814.14, find the APR.

2. Guy de Primo buys a new car for $9,837.91. He puts 10% down and obtains a simple interest amortized loan for the balance at $10\frac{7}{8}\%$ interest for four years. If loan fees included in the finance charge total $633.87, find the APR.

3. Chris Burditt bought a house for $212,500. He put 20% down and obtained a simple interest amortized loan for the balance at $10\frac{7}{8}\%$ interest for 30 years. If Chris paid 2 points and $4,728.60 in fees, $1,318.10 of which are included in the finance charge, find the APR.

4. Shirley Trembley bought a house for $187,600. She put 20% down and obtained a simple interest amor-

tized loan for the balance at $11\frac{3}{8}\%$ for 30 years. If Shirley paid 2 points and $3,427.00 in fees, $1,102.70 of which are included in the finance charge, find the APR.

5. Jennifer Tonda wants to buy a used car that costs $4,600. The used car dealer has offered her a four-year add-on interest loan that requires no down payment at 8% annual interest, with an APR of $14\frac{1}{4}\%$.
   a. Find the monthly payment.
   b. Verify the APR.

6. Melody Shepherd wants to buy a used car that costs $5,300. The used car dealer has offered her a four-year add-on interest loan that requires a $200 down payment at 7% annual interest, with an APR of 10%.
   a. Find the monthly payment.
   b. Verify the APR.

7. Anne Scanlan is buying a used car that costs $10,340. The used car dealer has offered her a five-year add-on

interest loan at 9.5% interest, with an APR of 9.9%. The loan requires a 10% down payment.
a. Find the monthly payment.
b. Verify the APR.

8. Stan Loll bought a used car for $9,800. The used car dealer offered him a four-year add-on interest loan at 7.8% interest, with an APR of 8.0%. The loan requires a 10% down payment.
a. Find the monthly payment.
b. Verify the APR.

9. Susan Chin is shopping for a car loan. Her savings and loan offers her a simple interest amortized loan for four years at 9% interest. Her bank offers her a simple interest amortized loan for four years at 9.1% interest. Which is the less expensive loan?

10. Stephen Tamchin is shopping for a car loan. His credit union offers him a simple interest amortized loan for 4 years at 7.1% interest. His bank offers him a simple interest amortized loan for four years at 7.3% interest. Which is the less expensive loan?

11. Ruben Lopez is shopping for a home loan. Really Friendly Savings and Loan offers him a 30-year simple interest amortized loan at 9.2% interest, with an APR of 9.87%. The Solid and Dependable Bank offers him a 30-year simple interest amortized loan at 9.3% interest, with an APR of 9.80%. Which loan would have the lower payments? Which loan would be the least expensive, taking into consideration monthly payments, points, and fees? Justify your answers.

12. Keith Moon is shopping for a home loan. Sincerity Savings offers him a 30-year simple interest amortized loan at 8.7% interest, with an APR of 9.12%. Pinstripe National Bank offers him a 30-year simple interest amortized loan at 8.9% interest, with an APR of 8.9%. Which loan would have the lower payments? Which loan would be the least expensive, taking into consideration monthly payments, points, and fees? Justify your answers.

13. The Nguyens are thinking of buying a home for $119,000. A potential lender advertises an 80%, 30-year simple interest amortized loan at $8\frac{1}{4}$% interest, with an APR of 9.23%. Use the APR to approximate the fees included in the finance charge.

14. Ellen Taylor is thinking of buying a home for $126,000. A potential lender advertises an 80%, 30-year simple interest amortized loan at $10\frac{3}{4}$% interest, with an APR of 11.57%. Use the APR to approximate the fees included in the finance charge.

15. James Magee is thinking of buying a home for $124,500. Bank of the Future advertises an 80%, 30-year simple interest amortized loan at $9\frac{1}{4}$% interest, with an APR of 10.23%. R.T.C. Savings and Loan advertises an 80%, 30-year simple interest amortized loan at 9% interest with an APR of 10.16%.
a. Find James's monthly payment if he borrows through Bank of the Future.
b. Find James's monthly payment if he borrows through R.T.C. Savings and Loan.
c. Use the APR to approximate the fees included in the finance charge by Bank of the Future.
d. Use the APR to approximate the fees included in the finance charge by R.T.C. Savings and Loan.
e. Discuss the advantages of each of the two loans. Who would be better off with the Bank of the Future loan? Who would be better off with the R.T.C. loan?

16. Holly Kresch is thinking of buying a home for $263,800. State Bank advertises an 80%, 30-year simple interest amortized loan at $6\frac{1}{4}$% interest, with an APR of 7.13%. Boonville Savings and Loan advertises an 80%, 30-year simple interest amortized loan at $6\frac{1}{2}$% interest with an APR of 7.27%.
a. Find Holly's monthly payment if she borrows through State Bank.
b. Find Holly's monthly payment if she borrows through Boonville Savings and Loan.
c. Use the APR to approximate the fees included in the finance charge by State Bank.
d. Use the APR to approximate the fees included in the finance charge by Boonville Savings and Loan.
e. Discuss the advantages of the two loans. Who would be better off with the State Bank loan? Who would be better off with the Boonville loan?

Answer the following questions using complete sentences and your own words.

CONCEPT QUESTIONS

17. If the APR of a simple interest amortized home loan is equal to the loan's interest rate, what conclusions could you make about the loan's required fees and points?

18. Compare and contrast the annual percentage rate of a loan with the annual yield of a compound interest rate. Be sure to discuss both the similarity and the difference between these two concepts.

5.6 Payout Annuities

The annuities that we have discussed are all savings instruments. In Section 5.3, we defined an annuity as a sequence of equal, regular payments into an account in which each payment receives compound interest. A saver who utilizes such an annuity can accumulate a sizable sum.

Annuities can be payout instruments rather than savings instruments. After you retire, you might wish to have part of your savings sent to you each month for living expenses. You might wish to receive equal, regular payments from an account where each payment has earned compound interest. Such an annuity is called a **payout annuity.** Payout annuities are also used to pay for a child's college education.

Calculating Short-Term Payout Annuities

EXAMPLE 1   On November 1, Debra Landre will make a deposit at her bank that will be used for a payout annuity. For the next three months, commencing on December 1, she will receive a payout of $1,000 per month. Use the Compound Interest Formula to find how much money she must deposit on November 1 if her money earns 10% compounded monthly.

Solution   First, calculate the principal necessary to receive $1,000 on December 1. Use $FV = 1,000$ and $n = 1$ (interest is earned for one month).

$$FV = P(1 + i)^n$$

$$1,000 = P\left(1 + \frac{0.10}{12}\right)^1$$

$$P = 1,000 \div \left(1 + \frac{0.10}{12}\right)^1 = 991.7355 \ldots \approx \$991.74$$

Next, calculate the principal necessary to receive $1,000 on January 1. Use $n = 2$ (interest is earned for two months).

$$FV = P(1 + i)^n$$

$$1,000 = P\left(1 + \frac{0.10}{12}\right)^2$$

$$P = 1,000 \div \left(1 + \frac{0.10}{12}\right)^2 = 983.5393 \ldots \approx \$983.54$$

Now calculate the principal necessary to receive $1,000 on February 1. Use $n = 3$ (interest is earned for three months).

$$FV = P(1 + i)^n$$

$$1,000 = P\left(1 + \frac{0.10}{12}\right)^3$$

$$P = 1,000 \div \left(1 + \frac{0.10}{12}\right)^3 = 975.4109 \ldots \approx \$975.41$$

Debra must deposit the sum of the above three amounts if she is to receive three monthly payouts of $1,000 each. Her total principal must be

$$\$991.74 + \$983.54 + \$975.41 = \$2,950.69$$

> ✓ If Debra's principal received no interest, then she would need $3 \cdot \$1,000 = \$3,000$. Since her principal does receive interest, she needs slightly less than $3,000.

■

Comparing Payout Annuities and Savings Annuities

Payout annuities and savings annuities are similar but not identical. It is important to understand their differences before we proceed. The following example is the "savings annuity" version of Example 1; that is, it is the savings annuity most similar to the payout annuity discussed in Example 1.

EXAMPLE 2   On November 1, Debra Landre set up an ordinary annuity with her bank. For the next three months she will make a payment of $1,000 per month. Each of those payments will receive compound interest. At the end of the three months (i.e., on February 1), she can withdraw her three $1,000 payments plus the interest that they will have earned. Use the Compound Interest Formula to find how much money she can withdraw.

Solution   Her first payment of $1,000 would be due on November 30. It would earn interest for two months (December and January).

$$FV = P(1 + i)^n$$
$$FV = 1,000\left(1 + \frac{0.10}{12}\right)^2 \approx \$1,016.74$$

Her second payment of $1,000 would be due on December 31. It would earn interest for one month (January).

$$FV = P(1 + i)^n$$
$$FV = 1,000\left(1 + \frac{0.10}{12}\right)^1 \approx \$1,008.33$$

Her third payment would be due on January 31. It would earn no interest (since the annuity expires February 1), so its future value is $1,000. On February 1, Debra could withdraw

$$\$1,016.74 + \$1,008.33 + \$1,000 = \$3,025.07$$

■

Naturally, we could have found the future value of Debra's ordinary annuity more easily with the Ordinary Annuity Formula.

$$FV(\text{ordinary}) = pymt\frac{(1 + i)^n - 1}{i}$$
$$= 1,000\frac{\left(1 + \frac{0.10}{12}\right)^3 - 1}{\frac{0.10}{12}} \approx \$3,025.07$$

The point of the method shown in Example 2 is to illustrate the differences between a savings annuity and a payout annuity.

Calculating Long-Term Payout Annuities

The procedure used in Example 1 reflects what actually happens with payout annuities, and it works fine for a small number of payments. However, most annuities are long-term. In the case of a savings annuity, we do not need to calculate the future value of each individual payment, as we did in Example 2; instead, we can use the Ordinary Annuity Formula. We need such a formula for payout annuities. We can find a formula if we look more closely at how the payout annuity from Example 1 compares with the savings annuity from Example 2.

In Example 2, we found that the future value of an ordinary annuity with three $1,000 payments is

$$FV = 1{,}000\left(1 + \frac{0.10}{12}\right)^2 + 1{,}000\left(1 + \frac{0.10}{12}\right)^1 + 1{,}000$$

In Example 1, we found that the total principal necessary to generate three $1,000 payouts is

$$P = 1{,}000 \div \left(1 + \frac{0.10}{12}\right)^1 + 1{,}000 \div \left(1 + \frac{0.10}{12}\right)^2 + 1{,}000 \div \left(1 + \frac{0.10}{12}\right)^3$$

This can be rewritten, using exponent laws, as

$$P = 1{,}000\left(1 + \frac{0.10}{12}\right)^{-1} + 1{,}000\left(1 + \frac{0.10}{12}\right)^{-2} + 1{,}000\left(1 + \frac{0.10}{12}\right)^{-3}$$

This is quite similar to the future value of the ordinary annuity—the only difference is the exponents. If we multiply each side by $(1 + \frac{0.10}{12})^3$, even the exponents will match.

$$P\left(1 + \frac{0.10}{12}\right)^3 = 1{,}000\left(1 + \frac{0.10}{12}\right)^{-1}\left(1 + \frac{0.10}{12}\right)^3$$

$$+ 1{,}000\left(1 + \frac{0.10}{12}\right)^{-2}\left(1 + \frac{0.10}{12}\right)^3$$

$$+ 1{,}000\left(1 + \frac{0.10}{12}\right)^{-3}\left(1 + \frac{0.10}{12}\right)^3$$

$$P\left(1 + \frac{0.10}{12}\right)^3 = 1{,}000\left(1 + \frac{0.10}{12}\right)^2 + 1{,}000\left(1 + \frac{0.10}{12}\right)^1 + 1{,}000\left(1 + \frac{0.10}{12}\right)^0$$

The right side of the above equation is the future value of an ordinary annuity, so we can use the Ordinary Annuity Formula to rewrite it.

$$P\left(1 + \frac{0.10}{12}\right)^3 = 1{,}000\,\frac{\left(1 + \frac{0.10}{12}\right)^3 - 1}{\frac{0.10}{12}}$$

If we generalize by replacing $\frac{0.10}{12}$ with i, 3 with n, and 1,000 with *pymt*, we have our Payout Annuity Formula (see page 399).

We have seen this formula before. In Section 5.3, we used it to find the present value P of a savings annuity. In Section 5.4, we used it to find the payment of a simple interest amortized loan. In this section, we use it to find the required principal for a payout annuity. This is a versatile formula.

> **Payout Annuity Formula**
>
> $$P(1 + i)^n = pymt\frac{(1 + i)^n - 1}{i}$$
>
> where P is the total principal necessary to generate n payouts, $pymt$ is the size of the payout, and i is the periodic interest rate.

The following example involves a long-term annuity. Usually, the interest rate of a long-term annuity varies somewhat from year to year. In this case, calculations must be viewed as predictions, not guarantees.

ThomsonNOW™
for Liberal Arts Mathematics
www.thomsonedu.com/1pass.

EXAMPLE 3   Fabiola Macias is about to retire, so she is setting up a payout annuity with her bank. She wishes to receive a payout of $1,000 per month for the next twenty-five years. Use the Payout Annuity Formula to find how much money she must deposit if her money earns 10% compounded monthly.

Solution   We are given that $pymt = 1,000$, $i = \frac{1}{12}$ of $10\% = \frac{0.10}{12}$, and $n = 25 \cdot 12 = 300$.

$$P(1 + i)^n = pymt\frac{(1 + i)^n - 1}{i}$$

$$P\left(1 + \frac{0.10}{12}\right)^{300} = 1,000\frac{\left(1 + \frac{0.10}{12}\right)^{300} - 1}{\frac{0.10}{12}}$$

To find P, we need to calculate the right side and then divide by the $\left(1 + \frac{0.10}{12}\right)^{300}$ from the left side.

We get $110,047.23005 \approx \$110,047.23$. This means that if Fabiola deposits $110,047.23, she will receive monthly payouts of $1,000 each for 25 years, or a total of $25 \cdot 12 \cdot 1,000 = \$300,000$.

> ✓ If Fabiola's principal received no interest, then she would need $300 \cdot \$1,000 = \$300,000$. Since her principal does receive compound interest for a long time, she needs significantly less than $300,000.

∎

If Fabiola Macias in Example 3 were like most people, she would not have $110,047.23 in savings when she retires, so she wouldn't be able to set up a payout annuity for herself. However, if she had set up a savings annuity thirty years before retirement, she could have saved that amount by making monthly payments of only $48.68. This is something that almost anyone can afford, and it's a wonderful deal. Thirty years of monthly payments of $48.68, while you are working, can generate twenty-five years of monthly payments of $1,000 when you are retired. In the exercises, we will explore this combination of a savings annuity and a payout annuity.

Payout Annuities with Inflation

The only trouble with Fabiola's retirement payout annuity in Example 3 is that she is ignoring inflation. In twenty-five years, she will still be receiving $1,000 a month, but her money won't buy as much as it does today. Fabiola would be better off if she allowed herself an annual **cost-of-living adjustment (COLA).**

ThomsonNOW™
for Liberal Arts Mathematics
www.thomsonedu.com/1pass.

EXAMPLE 4   After retiring, Fabiola Macias set up a payout annuity with her bank. For the next twenty-five years, she will receive payouts that start at $1,000 per month and then receive an annual COLA of 3%. Find the size of her monthly payout for

a. the first year
b. the second year
c. the third year
d. the 25th year

Solution   **a.** During the first year, no adjustment is made, so she will receive $1,000 per month.
b. During the second year, her monthly payout of $1,000 will increase 3%, so her new monthly payout will be

$$1,000 \cdot (1 + .03) = \$1,030$$

c. During the third year, her monthly payout of $1,030 will increase 3%, so her new monthly payout will be

$$1,030 \cdot (1 + .03) = \$1,060.90$$

Since the 1,030 in the above calculation came from computing $1,000 \cdot (1 + .03)$, we could rewrite this calculation as

$$1000 \cdot (1 + .03)^2 = \$1,060.90$$

d. By the twenty-fifth year, she will have received twenty-four 3% increases, so her monthly payout will be

$$1,000 \cdot (1 + .03)^{24} = 2,032.7941 \ldots \approx \$2,032.79$$

If a payout annuity is to have automatic annual cost-of-living adjustments, the following formula should be used to find the principal that must be deposited. The COLA is an annual one, so all other figures must also be annual figures; in particular, r is the *annual* interest rate, t is the duration of the annuity in *years,* and we use an *annual* payout.

Annual Payout Annuity with COLA Formula

A payout annuity of t years, where the payouts receive an annual COLA, requires a principal of

$$P = (pymt)\frac{1 - \left(\dfrac{1 + c}{1 + r}\right)^t}{r - c}$$

where *pymt* is the annual payout for the first year, c is the annual COLA rate, and r is the annual rate at which interest is earned on the principal.

EXAMPLE 5   After retiring, Sam Needham set up a payout annuity with his bank. For the next twenty-five years, he will receive annual payouts that start at $12,000 and then receive an annual COLA of 3%. Use the Annual Payout Annuity with COLA Formula to find how much money he must deposit if his money earns 10% interest per year.

Solution   We are given that the annual payout is $pymt = 12,000$, $r = 10\% = 0.10$, $c = 3\% = 0.03$, and $t = 25$.

$$P = (pymt)\frac{1 - \left(\dfrac{1 + c}{1 + r}\right)^t}{r - c}$$

$$= (12,000)\frac{1 - \left(\dfrac{1 + 0.03}{1 + 0.10}\right)^{25}}{0.10 - 0.03}$$

$$= (12,000)\frac{1 - \left(\dfrac{1.03}{1.10}\right)^{25}}{0.10 - 0.03}$$

$$= 138,300.4587 \approx \$138,300.46$$

---

1 $\boxed{-}$ $\boxed{(}$ 1.03 $\boxed{\div}$ 1.10 $\boxed{)}$ $\boxed{y^x}$ 25 $\boxed{=}$ $\boxed{\div}$ $\boxed{(}$ .10 $\boxed{-}$ .03 $\boxed{)}$ $\boxed{\times}$ 12000 $\boxed{=}$

With a graphing calculator, press $\boxed{\char`\^}$ rather than $\boxed{y^x}$ and $\boxed{\text{ENTER}}$ rather than $\boxed{=}$.

This means that if Sam deposits $138,300.46 now, he will receive

$12,000 in one year

$12,000 \cdot (1 + .03)^1 = \$12,360$ in two years

$12,000 \cdot (1 + .03)^2 = \$12,730.80$ in three years

$12,000 \cdot (1 + .03)^{24} = \$24,393.53$ in 25 years

> ✓ If Sam's principal received no interest, he would need $25 \cdot \$12,000 = \$300,000$. Since his principal does receive compound interest for a long time, he needs significantly less than $300,000.

■

Compare Sam's payout annuity in Example 5 with Fabiola's in Example 3. Sam receives annual payouts that start at $12,000 and slowly increase to $24,393. Fabiola receives exactly $1,000 per month (or $12,000 per year) for the same amount of time. Sam was required to deposit $138,300, and Fabiola was required to deposit $110,047.23.

5.6 Exercises

1. Suzanne Miller is planning for her retirement, so she is setting up a payout annuity with her bank. She wishes to receive a payout of $1,200 per month for twenty years.
   a. How much money must she deposit if her money earns 8% interest compounded monthly?
   b. Find the total amount that Suzanne will receive from her payout annuity.

2. James Magee is planning for his retirement, so he is setting up a payout annuity with his bank. He wishes to receive a payout of $1,100 per month for twenty-five years.
   a. How much money must he deposit if his money earns 9% interest compounded monthly?
   b. Find the total amount that James will receive from his payout annuity.

3. Dean Gooch is planning for his retirement, so he is setting up a payout annuity with his bank. He wishes to receive a payout of $1,300 per month for twenty-five years.
   a. How much money must he deposit if his money earns 7.3% interest compounded monthly?
   b. Find the total amount that Dean will receive from his payout annuity.

4. Holly Krech is planning for her retirement, so she is setting up a payout annuity with her bank. She wishes to receive a payout of $1,800 per month for twenty years.
   a. How much money must she deposit if her money earns 7.8% interest compounded monthly?
   b. Find the total amount that Holly will receive from her payout annuity.

5. a. How large a monthly payment must Suzanne Miller (from Exercise 1) make if she saves for her payout annuity with an ordinary annuity, which she sets up thirty years before her retirement? (The two annuities pay the same interest rate.)
   b. Find the total amount that Suzanne will pay into her ordinary annuity, and compare it with the total amount that she will receive from her payout annuity.

6. a. How large a monthly payment must James Magee (from Exercise 2) make if he saves for his payout annuity with an ordinary annuity, which he sets up twenty-five years before his retirement? (The two annuities pay the same interest rate.)
   b. Find the total amount that James will pay into his ordinary annuity, and compare it with the total amount that he will receive from his payout annuity.

7. a. How large a monthly payment must Dean Gooch (from Exercise 3) make if he saves for his payout annuity with an ordinary annuity, which he sets up thirty years before his retirement? (The two annuities pay the same interest rate.)
   b. How large a monthly payment must he make if he sets the ordinary annuity up twenty years before his retirement?

8. a. How large a monthly payment must Holly Krech (from Exercise 4) make if she saves for her payout annuity with an ordinary annuity, which she sets up thirty years before her retirement? (The two annuities pay the same interest rate.)
   b. How large a monthly payment must she make if she sets the ordinary annuity up twenty years before her retirement?

9. Lily Chang is planning for her retirement, so she is setting up a payout annuity with her bank. For twenty years, she wishes to receive annual payouts that start at $14,000 and then receive an annual COLA of 4%.
   a. How much money must she deposit if her money earns 8% interest per year?
   b. How large will Lily's first annual payout be?
   c. How large will Lily's second annual payout be?
   d. How large will Lily's last annual payout be?

10. Wally Brown is planning for his retirement, so he is setting up a payout annuity with his bank. For twenty-five years, he wishes to receive annual payouts that start at $16,000 and then receive an annual COLA of 3.5%.
    a. How much money must he deposit if his money earns 8.3% interest per year?
    b. How large will Wally's first annual payout be?
    c. How large will Wally's second annual payout be?
    d. How large will Wally's last annual payout be?

11. Oshri Karmon is planning for his retirement, so he is setting up a payout annuity with his bank. He is now thirty years old, and he will retire when he is sixty. He wants to receive annual payouts for twenty-five years, and he wants those payouts to receive an annual COLA of 3.5%.
    a. He wants his first payout to have the same purchasing power as does $13,000 today. How big should that payout be if he assumes inflation of 3.5% per year?

b. How much money must he deposit when he is sixty if his money earns 7.2% interest per year?

c. How large a monthly payment must he make if he saves for his payout annuity with an ordinary annuity? (The two annuities pay the same interest rate.)

d. How large a monthly payment would he make if he waits until he is forty before starting his ordinary annuity?

12. Shelly Franks is planning for her retirement, so she is setting up a payout annuity with her bank. She is now thirty-five years old, and she will retire when she is sixty-five. She wants to receive annual payouts for twenty years, and she wants those payouts to receive an annual COLA of 4%.

a. She wants her first payout to have the same purchasing power as does $15,000 today. How big should that payout be if she assumes inflation of 4% per year?

b. How much money must she deposit when she is sixty-five if her money earns 8.3% interest per year?

c. How large a monthly payment must she make if she saves for her payout annuity with an ordinary annuity? (The two annuities pay the same interest rate.)

d. How large a monthly payment would she make if she waits until she is forty before starting her ordinary annuity?

In Exercises 13–16, use the Annual Payout Annuity with COLA Formula to find the deposit necessary to receive monthly payouts with an annual cost-of-living adjustment. To use the formula, all figures must be annual figures, including the payout and the annual rate. You can adapt the formula for monthly payouts by using

- *the future value of a one-year ordinary annuity in place of the annual payout, where pymt is the monthly payout, and*
- *the annual yield of the given compound interest rate in place of the annual rate r.*

13. Fabiola Macias is about to retire, so she is setting up a payout annuity with her bank. She wishes to receive a monthly payout for the next twenty-five years, where the payout starts at $1,000 per month and receives an annual COLA of 3%. Her money will earn 10% compounded monthly.

a. The annual payout is the future value of a one-year ordinary annuity. Find this future value.

b. The annual rate r is the annual yield of 10% interest compounded monthly. Find this annual yield. (*Do not* round it off.)

c. Use the Annual Payout Annuity with COLA Formula to find how much money she must deposit.

d. Fabiola could have saved for her payout annuity with an ordinary annuity. If she had started doing so thirty years ago, what would the required monthly payments have been? (The two annuities pay the same interest rate.)

14. Gary Kersting is about to retire, so he is setting up a payout annuity with his bank. He wishes to receive a monthly payout for the next twenty years, where the payout starts at $1,300 per month and receives an annual COLA of 4%. His money will earn 8.7% compounded monthly.

a. The annual payout is the future value of a one-year ordinary annuity. Find this future value.

b. The annual rate r is the annual yield of 8.7% interest compounded monthly. Find this annual yield. (*Do not* round it off.)

c. Use the Annual Payout Annuity with COLA Formula to find how much money he must deposit.

d. Gary could have saved for his payout annuity with an ordinary annuity. If he had started doing so twenty-five years ago, what would the required monthly payments have been? (The two annuities pay the same interest rate.)

15. Conrad von Schtup is about to retire, so he is setting up a payout annuity with his bank. He wishes to receive a monthly payout for the next twenty-three years, where the payout starts at $1,400 per month and receives an annual COLA of 5%. His money will earn 8.9% compounded monthly.

a. The annual payout is the future value of a one-year ordinary annuity. Find this future value.

b. The annual rate r is the annual yield of 8.9% interest compounded monthly. Find this annual yield. (*Do not* round it off.)

c. Use the Annual Payout Annuity with COLA Formula to find how much money he must deposit.

d. Conrad could have saved for his payout annuity with an ordinary annuity. If he had started doing so twenty years ago, what would the required monthly payments have been? (The two annuities pay the same interest rate.)

16. Mitch Martinez is about to retire, so he is setting up a payout annuity with his bank. He wishes to receive a monthly payout for the next thirty years, where the payout starts at $1,250 per month and receives an annual COLA of 4%. His money will earn 7.8% compounded monthly.

a. The annual payout is the future value of a one-year ordinary annuity. Find this future value.

b. The annual rate r is the annual yield of 7.8% interest compounded monthly. Find this annual yield.

c. Use the Annual Payout Annuity with COLA Formula to find how much money he must deposit.

d. Mitch could have saved for his payout annuity with an ordinary annuity. If he had started doing so twenty years ago, what would the required monthly payments have been? (The two annuities pay the same interest rate.)

17. Bob Pirtle won $1 million in a state lottery. He was surprised to learn that he will not receive a check for $1 million. Rather, for twenty years, he will receive an annual check from the state for $50,000. The state finances this series of checks by buying Bob a payout annuity. Find what the state pays for Bob's payout annuity if the interest rate is 8%.

18. John-Paul Ramin won $2.3 million in a state lottery. He was surprised to learn that he will not receive a check for $2.3 million. Rather, for twenty years, he will receive an annual check from the state for $\frac{1}{20}$ of his winnings. The state finances this series of checks by buying John-Paul a payout annuity. Find what the state pays for John-Paul's payout annuity if the interest rate is 7.2%.

──── Answer the following questions using complete sentences and your own words.

CONCEPT QUESTIONS

19. Compare and contrast a savings annuity with a payout annuity. How do they differ in purpose? How do they differ in structure? How do their definitions differ?

HINT: Compare Examples 1 and 2.

20. Under what circumstances would a savings annuity and a payout annuity be combined?

 WEB PROJECTS

21. This is an exercise in saving for your retirement. Write a paper describing all of the following points.

a. Go on the Web and find out what the annual rate of inflation has been for each of the last ten years. Use the average of these figures as a prediction of the future annual rate of inflation.

b. Estimate the total monthly expenses you would have if you were retired now. Include housing, food, and utilities.

c. Use parts (a) and (b) to predict your total monthly expenses when you retire, assuming that you retire at age sixty-five.

d. Plan on financing your monthly expenses with a payout annuity. How much money must you deposit when you are sixty-five, if your money earns 7.5% interest per year?

e. How large a monthly payment must you make if you save for your payout annuity with an ordinary annuity, starting now? (The two annuities pay the same interest rate.)

f. How large a monthly payment must you make if you save for your payout annuity with an ordinary annuity, starting ten years from now?

g. How large a monthly payment must you make if you save for your payout annuity with an ordinary annuity, starting 20 years from now?

Some useful links for this web project are listed on the text web site:
www.thomsonedu.com/mathematics/johnson

Chapter 5   Review

ThomsonNOW for Liberal Arts Mathematics
Preparing for an exam? Test yourself on key material by visiting ThomsonNOW at **www.thomsonedu.com/1pass**.

TERMS

add-on interest loan	annual yield	expire	lump sum
adjustable rate mortgage	annuity	finance	maturity value
amortization schedule	annuity due	finance charges	negative amortization
amortized loan	average daily balance	future value	nominal rate
annual percentage rate (APR) of an add-on interest loan	balance	Individual Retirement Account (IRA)	note
	compound interest		ordinary annuity
	compounding period	interest	outstanding principal
annual percentage rate (APR) of a simple interest amortized loan	cost-of-living adjustment (COLA)	interest rate	payment period
		line of credit	payout annuity
	doubling time	loan agreement	periodic rate

point	principal	sinking fund	unpaid balance
prepaying a loan	simple interest	term	yield
prepayment penalty	simple interest amortized	tax-deferred annuity	
present value	loan	(TDA)	
present value of an annuity	simple interest loan	Truth in Lending Act	

FORMULAS

The **simple interest** I on a principal P at a interest rate r for t years is $I = Prt$

and the **future value** is $FV = P(1 + rt)$

After n compounding periods the future value FV of an initial principal P earning **compound interest** at a periodic interest rate i for n periods is

$$FV = P(1 + i)^n$$

The **annual yield** of a given compound rate is the simple interest rate r that has the same future value as the compound rate in 1 year. To find it, solve

$$FV(\text{compound interest}) = FV(\text{simple interest})$$
$$P(1 + i)^n = P(1 + rt)$$

for the simple interest rate r after making appropriate substitutions.

The future value FV of an **ordinary annuity** is

$$FV(\text{ordinary}) = pymt\frac{(1 + i)^n - 1}{i}$$

and the future value of an **annuity due** is

$$FV(\text{due}) = FV(\text{ordinary}) \cdot (1 + i)$$
$$= pymt\frac{(1 + i)^n - 1}{i}(1 + i)$$

The **present value of an ordinary annuity** is the lump sum P such that

$$FV(\text{lump sum}) = FV(\text{annuity})$$
$$P(1 + i)^n = pymt\frac{(1 + i)^n - 1}{i}$$

Simple interest amortized loan:

$$FV(\text{annuity}) = FV(\text{loan amount})$$
$$pymt\frac{(1 + i)^n - 1}{i} = P(1 + i)^n$$

Unpaid balance:

$$\text{Unpaid balance} = \text{current value of loan amount}$$
$$- \text{current value of annuity}$$
$$\approx P(1 + i)^n - pymt\frac{(1 + i)^n - 1}{i}$$

where n is the number of periods *from the beginning of the loan to the present.*

Payout annuity: The total principal necessary to generate n payouts of size *pymt* is P, where

$$P(1 + i)^n = pymt\frac{(1 + i)^n - 1}{i}$$

Annual payout annuity with COLA: A payout annuity with a term of t years, a first-year payout of *pymt,* and an annual COLA rate c, requires a principal of

$$P = (pymt)\frac{1 - \left(\dfrac{1 + c}{1 + r}\right)^t}{r - c}$$

STEPS

To find the **credit card finance charge** with the **average daily balance method:**
1. Find the balance for each day in the billing period and the number of days at that balance.
2. The average daily balance is the weighted average of these daily balances, weighted to reflect the number of days at that balance.
3. The finance charge is simple interest applied to the average daily balance.

To create an **amortization schedule:**
For each payment, list the payment number, principal portion, interest portion, total payment and balance.

For each payment:
1. Find the interest on the balance—use the simple interest formula.

For each payment except the last:
2. The principal portion is the payment minus the interest portion.

3. The new balance is the previous balance minus the principal portion.

For the last payment:
4. The principal portion is the previous balance.
5. The total payment is the sum of the principal portion and interest portion.

To find the **APR** of a loan:
1. *Compute the payment.*
2. *Use the simple interest amortized loan formula to compute the APR.*
   - Substitute the payment from step 1 for *pymt*

- If there are points and fees, substitute the legal loan amount (i.e., loan amount less points and those fees included in the finance charge) for *P*.
- Use technology to solve the resulting equation for the periodic interest rate.
- Convert the periodic interest rate to an annual rate. This annual rate is the APR.

3. *Determine whether the claimed APR is legally correct* by seeing if the claimed APR is within one-eighth of 1% of the actual APR found in step 2.

Exercises

1. Find the interest earned by a deposit of $8,140 at $9\frac{3}{4}$% simple interest for eleven years.
2. Find the interest earned by a deposit of $10,620 at $8\frac{1}{2}$% simple interest for twenty-five years.
3. Find the future value of a deposit of $12,288 at $4\frac{1}{4}$% simple interest for fifteen years.
4. Find the future value of a deposit of $22,880 at $5\frac{3}{4}$% simple interest for thirty years.
5. Find the maturity value of a loan of $3,550 borrowed at $12\frac{1}{2}$% simple interest for one year and two months.
6. Find the maturity value of a loan of $12,250 borrowed at $5\frac{1}{2}$% simple interest for two years.
7. Find the present value of a future value of $84,120 at $7\frac{1}{4}$% simple interest for twenty-five years.
8. Find the present value of a future value of $10,250 at $5\frac{3}{4}$% simple interest for twenty years.
9. Find the future value of a deposit of $8,140 at $9\frac{3}{4}$% interest compounded monthly for eleven years.
10. Find the future value of a deposit of $7,250 at $5\frac{1}{4}$% interest compounded monthly for twenty years.
11. Find the interest earned by a deposit of $7,990 at $4\frac{3}{4}$% interest compounded monthly for eleven years.
12. Find the interest earned by a deposit of $22,250 at $9\frac{1}{4}$% interest compounded monthly for twenty years.
13. Find the present value of a future value of $33,120 at $6\frac{1}{4}$% interest compounded daily for twenty-five years.
14. Find the present value of a future value of $10,600 at $7\frac{7}{8}$% interest compounded daily for four years.
15. Find the annual yield corresponding to a nominal rate of 7% compounded daily.
16. Find the annual yield corresponding to a nominal rate of $8\frac{1}{2}$% compounded monthly.
17. Find the future value of a twenty-year ordinary annuity with monthly payments of $230 at 6.25% interest.

18. Find the future value of a thirty-year ordinary annuity with biweekly payments of $130 at 7.25% interest.
19. Find the future value of a twenty-year annuity due with monthly payments of $450 at 8.25% interest.
20. Find the future value of a thirty-year annuity due with biweekly payments of $240 at 6.75% interest.
21. Find (a) the monthly payment and (b) the total interest for a simple interest amortized loan of $25,000 for five years at $9\frac{1}{2}$% interest.
22. Find (a) the monthly payment and (b) the total interest for a simple interest amortized loan of $130,000 for twenty years at $8\frac{1}{4}$% interest.
23. The Square Wheel Bicycle store has found that they sell most of their bikes in the spring and early summer. On February 15, they borrowed $351,500 to buy bicycles. They are confident that they can sell most of these bikes by September 1. Their loan is at $5\frac{7}{8}$% simple interest. What size lump sum payment would they have to make on September 1 to pay off the loan?
24. Mike Taylor buys a four-year old Ford from a car dealer for $16,825. He puts 10% down and finances the rest through the dealer at 10.5% add-on interest. If he agrees to make thirty-six monthly payments, find the size of each payment.
25. The activity on Sue Washburn's MasterCard account for one billing period is shown below. Find the average daily balance and the finance charge if the billing period is August 26 through September 25, the previous balance was $3,472.38, and the annual interest rate is $19\frac{1}{2}$%.

August 30	payment	$100.00
September 2	gasoline	$34.12
September 10	restaurant	$62.00

Exercises 407

26. George and Martha Simpson bought a house from Sue Sanchez for $205,500. In lieu of a 20% down payment, Ms. Sanchez accepted 5% down at the time of the sale and a promissory note from the Simpsons for the remaining 15%, due in eight years. The Simpsons also agreed to make monthly interest payments to Ms. Sanchez at 12% simple interest until the note expires. The Simpsons borrowed the remaining 80% of the purchase price from their bank. The bank paid that amount, less a commission of 6% of the purchase price, to Ms. Sanchez.
   a. Find the Simpsons' monthly interest-only payment to Ms. Sanchez.
   b. Find Ms. Sanchez's total income from all aspects of the down payment.
   c. Find Ms. Sanchez's total income from all aspects of the sale of the house, including the down payment.

27. Tien Ren Chiang wants to have an IRA that will be worth $250,000 when he retires at age sixty-five.
   a. How much must he deposit at age twenty-five at $8\frac{1}{8}\%$ compounded quarterly?
   b. If he arranges for the monthly interest to be sent to him starting at age sixty-five, how much would he receive each month? (Assume that he will continue to receive $8\frac{1}{8}\%$ interest, compounded monthly.)

28. Extremely Trustworthy Savings offers five-year CDs at 7.63% compounded annually, and Bank of the South offers five-year CDs at 7.59% compounded daily. Compute the annual yield for each institution and determine which offering is more advantageous for the consumer.

29. You are 32, and you have just set up an ordinary annuity to save for retirement. You make monthly payments of $200 that earn $6\frac{1}{8}\%$ interest. Find the future value when you reach age 65.

30. Find and interpret the present value of the annuity in Exercise 29.

31. Find the future value of an annuity due with monthly payments of $200 that earns $6\frac{1}{8}\%$ interest, after 11 years.

32. Matt and Leslie Silver want to set up a TDA that will generate sufficient interest on maturity to meet their living expenses, which they project to be $1,300 per month.
   a. Find the amount needed at maturity to generate $1,300 per month interest if they can get $8\frac{1}{4}\%$ interest compounded monthly.
   b. Find the monthly payment they would have to make into an ordinary annuity to obtain the future value found in part (a) if their money earns $9\frac{3}{4}\%$ and the term is 30 years.

33. Mr. and Mrs. Liberatore set up a TDA to save for their retirement. They agreed to have $100 deducted from each of Mrs. Liberatore's monthly paychecks, which will earn $6\frac{1}{8}\%$ interest.
   a. Find the future value of their ordinary annuity, if it comes to term after they retire in thirty years
   b. After retiring, the Liberatores convert their annuity to a savings account, which earns 5.75% interest compounded monthly. At the end of each month, they withdraw $1,000 for living expenses. Complete the chart in Figure 5.29 for their postretirement account.

Month Number	Account Balance at Beginning of the Month	Interest for the Month	With-drawal	Account Balance at End of the Month
1				
2				
3				
4				
5				

Figure 5.29
Chart for Exercise 33

34. Delores Lopez buys some land in Nevada. She agrees to pay the seller a lump sum of $235,000 in five years. Until then, she will make monthly simple interest payments to the seller at 10% interest.
   a. Find the amount of each interest payment.
   b. Delores sets up a sinking fund to save the $235,000. Find the size of her monthly payments if her payments are due at the end of every month and her money earns $9\frac{3}{8}\%$ interest.
   c. Prepare a table showing the amount in the sinking fund after each of the first two deposits.

35. Maude Frickett bought a house for $225,600. She put 20% down and obtains a simple interest amortized loan for the rest at $7\frac{3}{8}\%$ for thirty years.
   a. Find her monthly payment.
   b. Find the total interest.
   c. Prepare an amortization schedule for the first two months of the loan.

36. Navlet's Nursery needs to borrow $228,000 to increase its inventory for the upcoming spring season. The owner is confident that he will sell most if not all of the new plants during the summer, so he wishes to borrow the money for only six months.

His bank has offered him a simple interest amortized loan at $8\frac{1}{4}\%$ interest.

a. Find the size of the monthly bank payment.

b. Prepare an amortization schedule for all six months of the loan.

37. Harry Carry had his kitchen remodeled. He did not have sufficient cash to pay for it. However, he had previously set up a line of credit with his bank. On May 16, he wrote a check to his contractor on his line of credit for $41,519. The line's interest rate is $7\frac{3}{4}\%$.

a. Find the size of the required monthly interest payment.

b. Harry decided that it would be in his best interests to get this loan paid off in seven months. Find the size of the monthly principal-plus-interest payment that would accomplish this. (*Hint:* In effect, Carry is converting the loan to an amortized loan.)

c. Prepare an amortization schedule for all seven months of the loan.

d. Find the amount of line of credit interest that Carry could deduct from his taxes next year.

38. Ben Suico buys a car for $13,487.31. He puts 10% down and obtains a simple interest amortized loan for the rest at $10\frac{7}{8}\%$ interest for five years.

a. Find his monthly payment.

b. Find the total interest.

c. Prepare an amortization schedule for the first two months of the loan.

d. Mr. Suico decides to sell his car two years and six months after he bought it. Find the unpaid balance on his loan.

39. Scott Frei wants to buy a used car that costs $6,200. The used car dealer has offered him a four-year add-on interest loan that requires a $200 down payment at 9.9% annual interest with an APR of 10%.

a. Find the monthly payment.

b. Verify the APR.

40. Miles Archer bought a house for $112,660. He put 20% down and obtains a simple interest amortized loan for the rest at $9\frac{7}{8}\%$ for thirty years. If Miles paid two points and $5,738.22 in fees, $1,419.23 of which are included in the finance charge, find the APR.

41. Susan and Steven Tamchin are thinking of buying a home for $198,000. A potential lender advertises an 80%, thirty-year simple interest amortized loan at $8\frac{1}{2}\%$ interest, with and APR of 9.02%.

a. Find the size of the Tamchin's monthly payment.

b. Use the APR to approximate the fees included in the finance charge.

42. Fred Rodgers is planning for his retirement, so he is setting up a payout annuity with his bank. He wishes to receive a payout of $1,700 per month for twenty-five years.

a. How much money must he deposit if his money earns 6.1% interest compounded monthly?

b. How large a monthly payment would Fred have made if he had saved for his payout annuity with an ordinary annuity, set up thirty years before his retirement? (The two annuities pay the same interest rate.)

c. Find the total amount that Fred will pay into his ordinary annuity and the total amount that he will receive from his payout annuity.

43. Sue West is planning for her retirement, so she is setting up a payout annuity with her bank. She is now thirty years old, and she will retire when she is sixty. She wants to receive annual payouts for twenty-five years, and she wants those payouts to have an annual COLA of 4.2%.

a. She wants her first payout to have the same purchasing power as does $17,000 today. How big should that payout be if she assumes inflation of 4.2% per year?

b. How much money must she deposit when she is sixty if her money earns 8.3% interest per year?

c. How large a monthly payment must she make if she saves for her payout annuity with an ordinary annuity? (The two annuities pay the same interest rate.)

Answer the following questions using complete sentences and your own words.

CONCEPT QUESTIONS

44. What is the difference between simple interest and compound interest?

45. Describe a situation in which simple interest, rather than compound interest, would be expected.

46. Describe a situation in which compound interest, rather than simple interest, would be expected.

47. What is the difference between an account that earns compound interest and an annuity that earns compound interest?

48. What is the difference between a simple interest amortized loan and an add-on interest loan?

HISTORY QUESTIONS

49. What does the Truth In Lending Act do for borrowers?

50. Who offered the first credit card?

51. How did the first credit card differ from the first post–World War II credit card?

6

Voting and Apportionment

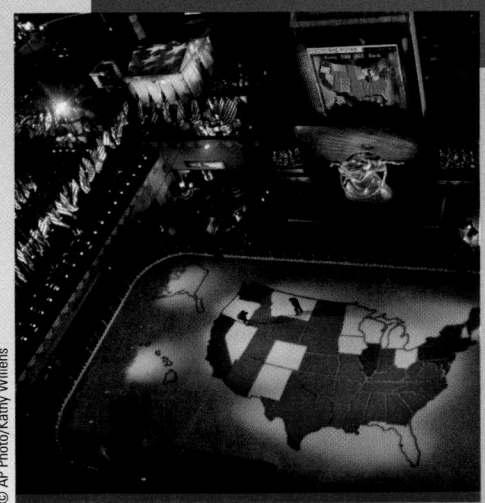

© AP Photo/Kathy Willens

6.1 Voting Systems

6.2 Methods of Apportionment

6.3 Flaws of Apportionment

The right to vote for one's governing officials is the cornerstone of any democracy. In the United States, individual citizens have a voice in the selection of local mayors and city councils, county officials, state legislators, governors, members of the House of Representatives and the Senate, and even the ultimate office: the President of the United States. Elections of all types are commonplace, and their outcomes range from overwhelming landslides to razor-thin victories, legal suits, and tedious recounts.

When we vote to fill a political office, the outcome obviously depends on the number of votes cast for each candidate; when we vote to create laws, the numbers for and against a proposition determine its fate. However, determining the outcome of an election might not be as simple as some may think. The premises of "one person, one vote" and "the candidate with the most votes wins" are neither universal nor absolute. In fact, several different voting systems have been used in democratic societies today and throughout history.

Elected governmental bodies range in scope from local to regional to national levels. That is, elected officials represent their constituencies in matters that affect others in a broader sense. For example, precincts within a city elect officials to represent the precinct on citywide matters, districts within a state elect officials to represent the district on statewide matters, and states elect officials to represent the individual states on nationwide matters. How many officials will a region receive to represent it in these matters? In some cases, each region may receive the same number of representative seats; in others, different regions may receive different numbers of seats. The determination of the number of representatives a region receives is known as apportionment.

In this chapter, we will study various voting systems and methods of apportionment. In addition, we will examine problems, deficiencies, and flaws associated with the methods. Democracies ultimately strive for fairness, but exactly what is meant by the word *fair*?

6.1   Voting Systems

Voting is an essential element in any democratic form of government. Whether it is the selection of leaders, the creation of laws and regulations, or deciding the outcome of issues ranging from mundane to volatile, decisions are made on the basis of the results of elections.

Voting is an essential element in any democratic form of government.

Once votes have been cast, they must be counted, and a winner must be declared. Many people assume that "the candidate with the most votes wins"; however, that is not always the case. Deciding who wins an election is ultimately determined by the type of voting system that is used. "The person with the most votes wins" is a commonly used system and is called the plurality method. However, the plurality method is not the only method that is used. In this section, we will examine several common voting systems: (1) the plurality method, (2) the plurality with elimination method, (3) the ranked-choice or instant run-off method (sometimes called the Australian method), (4) the Borda count method, and (5) the pairwise comparison method.

The Plurality Method

A common way to determine the outcome of an election is to declare the candidate with the most votes the winner. This is the basis of the **plurality method of voting.**

Plurality Method of Voting

Each person votes for his or her favorite candidate (or choice). The candidate (or choice) who receives the most votes is declared the winner. (In case of a tie, a special run-off election might be held.)

EXAMPLE 1 City Cab is a new taxi service that will soon begin operation in a major metropolitan area. The board of directors for City Cab must purchase a fleet of new vehicles, and the vehicles must all be the same color. Four colors are available: black, white, red, and green. The five directors vote for their choice of color and the results are given in Figure 6.1. Using the plurality method of voting, which color choice wins?

	Director 1	Director 2	Director 3	Director 4	Director 5
Choice	green	red	red	green	red

Figure 6.1
Results of voting for the color of City Cab vehicles

Solution Tallying the results, we see that green received two votes and red received three. Because red received the most votes (3), red is declared the winner. ■

Referring to Example 1, it should be noted that in addition to receiving the most votes, red also received a majority of the votes, that is, more than 50% of the votes were for red. As we can see, three out of five, or 60%, of the votes were for red, while 40% were not for red.

Although the plurality method is easy to apply, this method can produce an unusual winner, as we shall see in the next example.

EXAMPLE 2 Referring to City Cab in Example 1, suppose the results of the election are those given in Figure 6.2. Using the plurality method of voting, which color choice wins?

	Director 1	Director 2	Director 3	Director 4	Director 5
Choice	green	red	red	black	white

Figure 6.2
Results of voting for the color of City Cab vehicles

Solution Once again, because red received the most votes (two), red would be declared the winner. Although red is the winner, it should be noted that red did *not* receive a *majority* of the votes. Fewer than half of the votes, that is 40%, were for red, while significantly more than half (60%) did not want the color red. This type of "winner" can lead to very difficult situations in political elections. Imagine yourself as the "winner" of a local election when more than half of the voters do *not* support you. ■

The Plurality with Elimination Method

As shown in Example 2, the plurality method of voting may produce a "winner" even though a majority of voters did not vote for the "winner." This dilemma can be avoided in several ways. One common way is to eliminate the candidate with the fewest votes and then hold another election. This is the basis of the **plurality with elimination method.** Consequently, the voters who supported the eliminated candidate now vote for someone else; that is, they must vote for their second choice. Typically, this second election will produce a winner with a majority of support. However, if a majority is still not attained, the process is repeated until a candidate obtains a majority of the votes.

> **Plurality with Elimination Method of Voting**
>
> Each person votes for his or her favorite candidate. If a candidate receives a majority of votes, that candidate is declared the winner. If no candidate receives a majority, then the candidate with the fewest votes is eliminated and a new election is held. This process continues until a candidate receives a majority of the votes.

ThomsonNOW™
for Liberal Arts Mathematics
www.thomsonedu.com/1pass.

EXAMPLE 3   The managers of We Deliver (a local shipping business) are planning a party for their 60 employees. There are three possible locations for the party: the warehouse, the park, and the beach. The employees voted on which location they preferred, and the results are given in Figure 6.3.

	Warehouse	Park	Beach
Number of votes	23	19	18

Figure 6.3
Results of voting for party location

a. Using the plurality method of voting, which location wins?
b. Using the plurality with elimination method of voting, which location wins?

Solution   **a.** Because the warehouse received the most votes (23), the party will be held at the warehouse.
   b. Although the warehouse received the most votes, it did not receive a majority; the warehouse received only 23/60, or approximately 38.3%, of the vote. Because the beach received the fewest number of votes (18), it is eliminated, and a new election is held. The results of the second election are given in Figure 6.4.

	Warehouse	Park	Beach
Number of votes	25	35	0

Figure 6.4
Results of the second election

Because the park received a majority of the votes ($35/60 \approx 58.3\%$), the party will be held at the park.   ■

As is shown in Example 3, the outcome of an election may change depending on the type of voting system used. Therefore, it is imperative that the voters know ahead of time which system is being used!

As we have seen, the elimination of one candidate forces some people to shift their vote to their second choice in the new election. In the realm of real-world political elections, this method of elimination can be very expensive both for the local office of elections and for the candidates themselves, and it is overly time-consuming. Rather than holding a new election after eliminating the candidate with the fewest votes, a popular alternative is to have each voter rank each candidate during the first election. This is the basis of the ranked-choice or instant runoff method.

The Ranked-Choice or Instant Runoff Method

In Example 3, the employees of We Deliver had three choices for the location of their party. If each of the 60 employees was asked to rank these choices in order of preference, in how many different ways could the locations be ranked? Because the choices are being put in an order (that is, a person must make a first choice, a second choice, and a third choice), we conclude that *permutations* can be used to determine the number of different rankings (see Section 2.4). Specifically, there are

$$_3P_3 = \frac{3!}{(3-3)!} = \frac{3!}{0!} = \frac{3 \cdot 2 \cdot 1}{1} = 6$$

different rankings (permutations) of the three locations. One such ranking is "warehouse, park, beach" or, more conveniently, WPB. Another such ranking is BPW (beach, park, warehouse). Figure 6.5 lists all possible permutations (or rankings) of the three locations.

1st choice	W	W	P	P	B	B
2nd choice	P	B	B	W	W	P
3rd choice	B	P	W	B	P	W

Figure 6.5
All permutations (or rankings) of "Warehouse, Park, Beach"

After each employee ranked the locations and the ballots were tallied, it was determined that 9 employees chose the WPB ranking, 14 chose WBP, 15 chose PBW, 4 chose PWB, 2 chose BWP, and 16 chose BPW. These results are portrayed in the table shown in Figure 6.6.

		Number of Ballots Cast					
		9	14	15	4	2	16
Ranked Ballot	1st choice	W	W	P	P	B	B
	2nd choice	P	B	B	W	W	P
	3rd choice	B	P	W	B	P	W

Figure 6.6
Voter preference table

Tables like that shown in Figure 6.6 are often called **voter preference tables,** that is, tables that list different rankings of the candidates along with the number of voters who chose each specific ranking. Tables of this nature are often used to depict the voting patterns in a ranked election.

Referring to the voter preference table in Figure 6.6, notice that the sum of the row of numbers is 60, that is, the sum of the row equals the number of people voting. In addition, to determine the number of first-choice votes for each candidate, go to the row labeled "1st choice" and observe that $9 + 14 = 23$ people had W (the warehouse) as their first choice, whereas $15 + 4 = 19$ people had P (the park) as their first choice, and $2 + 16 = 18$ had B (the beach) as their first choice. These subtotals agree with the values given in Example 3.

We now proceed as we did when using the plurality with elimination method; that is, we eliminate the candidate with the fewest first-choice votes (the beach). However, rather than holding a new election, we simply modify the original table by shifting the votes in the voter preference table upwards to fill the voids created by eliminating the beach, as shown in Figure 6.7.

		Number of Ballots Cast					
		9	14	15	4	2	16
Ranked Ballot	1st choice	W	W	P	P	~~B~~	~~B~~
	2nd choice	P	~~B~~	~~B~~	W	W↑	P↑
	3rd choice	~~B~~	P↑	W↑	~~B~~	P↑	W↑

Figure 6.7
Shifting votes in the voter preference table

The modified voter preference table in shown is Figure 6.8.

	Number of Ballots Cast					
	9	14	15	4	2	16
1st choice	W	W	P	P	W	P
2nd choice	P	P	W	W	P	W

Figure 6.8
Modified voter preference table

Referring to Figure 6.8, we now see that the Warehouse received $9 + 14 + 2 = 25$ first choice votes, whereas the Park received $15 + 4 + 16 = 35$ votes. Of course, these values agree with those given in Example 3.

The process described above is called the **ranked-choice or instant runoff method.** This voting system is commonly used in Australia. In March 2002, the voters of San Francisco, California, adopted the instant runoff method for the election of most political posts, including mayor, sheriff, district attorney, and many other high-visibility positions. It has been estimated that this voting method will save the City of San Francisco $2 million per election, and voters will not have to wait several weeks for a second election. The instant runoff method of voting is summarized as follows.

Instant Runoff Method of Voting

Each voter ranks all of the candidates; that is, each voter selects his or her first choice, second choice, third choice, and so on. If a candidate receives a majority of first-choice votes, that candidate is declared the winner. If no candidate receives a majority, then the candidate with the fewest first-choice votes is eliminated, and those votes are given to the next preferred candidate. If a candidate now has a majority of first-choice votes, that candidate is declared the winner. If no candidate receives a majority, this process continues until a candidate receives a majority.

Thomson**NOW**™
for Liberal Arts Mathematics
www.thomsonedu.com/1pass.

EXAMPLE 4   Townsville is electing a new mayor. The candidates are Alturas (A), Bellum (B), Chan (C), and Dushay (D). The ranked ballots are tallied, and the results are summarized as shown in the voter preference table given in Figure 6.9.

		Number of Ballots Cast					
		15,527	15,287	10,023	9,105	7,978	5,660
Ranked Ballot	**1st choice**	C	B	C	B	A	D
	2nd choice	B	C	D	A	B	A
	3rd choice	A	D	A	D	C	C
	4th choice	D	A	B	C	D	B

Figure 6.9
Voter preference table for Mayor of Townsville

a. How many voters participated in the election?
b. Use the plurality method to determine the winner.
c. Use the instant runoff method to determine the winner.

Solution   **a.** To determine the number of voters, we find the sum of the row of numbers representing the number of ballots cast for each ranking:

$$15{,}527 + 15{,}287 + 10{,}023 + 9{,}105 + 7{,}978 + 5{,}660 = 63{,}580$$

Therefore, 63,580 voters participated in the election.

b. Looking at the row of first-choice votes, we see that

A received 7,978 votes,

B received $15{,}287 + 9{,}105 = 24{,}392$ votes,

C received $15{,}527 + 10{,}023 = 25{,}550$ votes, and

D received 5,660 votes.

Because Chan (C) received the most votes (25,550), Mr. Chan becomes the new mayor of Townsville. However, it should be noted that Mr. Chan did not receive a majority of the votes; he received only $25{,}550/63{,}580 \approx 40.2\%$.

c. First, we determine the number of votes needed to obtain a majority. Because 63,580 votes were cast, a candidate needs more than half of 63,580, that is, more than $63{,}580 \div 2 = 31{,}790$ votes; in other words, a candidate needs 31,791 votes to win. Because D received the fewest number of votes (5,660), Dushay is eliminated, and we obtain the modified voter preference table shown in Figure 6.10.

	Number of Ballots Cast					
	15,527	15,287	10,023	9,105	7,978	5,660
1st choice	C	B	C	B	A	A
2nd choice	B	C	A	A	B	C
3rd choice	A	A	B	C	C	B

Figure 6.10
Modified voter preference table for Mayor of Townsville

Examining the new row of first-choice votes, we see that

A received $7,978 + 5,660 = 13,638$ votes,

B received $15,287 + 9,105 = 24,392$ votes, and

C received $15,527 + 10,023 = 25,550$ votes.

Because no candidate has a majority, we eliminate the candidate with the fewest votes, that is, we eliminate Alturas and obtain the modified voter preference table shown in Figure 6.11.

	Number of Ballots Cast					
	15,527	15,287	10,023	9,105	7,978	5,660
1st choice	C	B	C	B	B	C
2nd choice	B	C	B	C	C	B

Figure 6.11
Modified voter preference table for Mayor of Townsville

Looking at the new row of first-choice votes, we see that

B received $15,287 + 9,105 + 7,978 = 32,370$ votes, and

C received $15,527 + 10,023 + 5,660 = 31,210$ votes.

Consequently, Ms. Bellum becomes the new mayor with $32,370/63,580 \approx 50.9\%$ of the vote. ∎

Whereas the plurality method utilizes only a voter's first choice, the instant runoff method is designed to accommodate a voter's alternative choices. Another popular method that requires voters to rank their choices is the Borda count method.

The Borda Count Method

To win an election using the instant runoff method, a candidate must capture a majority of first-choice votes. Rather than tallying the first-choice votes only, we might want to tally the first-choice votes, the second-choice votes, the third-choice votes, and so on. This is the basis of the **Borda count method,** which is summarized in the following box.

Borda Count Method of Voting

Each voter ranks all of the candidates; that is, each voter selects his or her first choice, second choice, third choice, and so on. If there are k candidates, each candidate receives k points for each first-choice vote, $(k - 1)$ points for each second-choice vote, $(k - 2)$ points for each third-choice vote, and so on. The candidate with the most total points is declared the winner.

EXAMPLE 5 Use the Borda count method to determine the location of the party for the We Deliver employees referenced in Example 3.

Solution Recall that the employees were asked to rank the three locations (warehouse, park, and beach); the results are summarized in the voter preference table given in Figure 6.12.

		Number of Ballots Cast					
		9	14	15	4	2	16
Ranked Ballot	**1st choice**	W	W	P	P	B	B
	2nd choice	P	B	B	W	W	P
	3rd choice	B	P	W	B	P	W

Figure 6.12
Voter preference table

First, we tally the votes for each location as shown in Figure 6.13. Because there were $k = 3$ candidates (warehouse, park, and beach), each first-choice vote is worth 3 points, each second-choice vote is worth 2 points, and each third-choice vote is worth 1 point.

	Warehouse	**Park**	**Beach**
1st choice	$9 + 14 = 23$ votes	$15 + 4 = 19$ votes	$2 + 16 = 18$ votes
2nd choice	$4 + 2 = 6$ votes	$9 + 16 = 25$ votes	$14 + 15 = 29$ votes
3rd choice	$15 + 16 = 31$ votes	$14 + 2 = 16$ votes	$9 + 4 = 13$ votes

Figure 6.13
Tally of votes

Now determine the score for each candidate by multiplying the number of votes times the appropriate number of points as shown in Figure 6.14.

	Warehouse	**Park**	**Beach**
1st choice votes (3 points each)	23×3 points $= 69$ points	19×3 points $= 57$ points	18×3 points $= 54$ points
2nd choice votes (2 points each)	6×2 points $= 12$ points	25×2 points $= 50$ points	29×2 points $= 58$ points
3rd choice votes (1 point each)	31×1 points $= 31$ points	16×1 points $= 16$ points	13×1 points $= 13$ points

Figure 6.14
Points

Summing each column, we obtain the results of the election:

$$\text{Warehouse} = 69 + 12 + 31 = 112 \text{ points}$$
$$\text{Park} = 57 + 50 + 16 = 123 \text{ points}$$
$$\text{Beach} = 54 + 58 + 13 = 125 \text{ points}$$

When the Borda count method is used, the beach is declared the winner because it has the most points (125). ∎

As is illustrated by the examples referring to the We Deliver employees' party location, the outcome of the election depends on the voting system that is used. Specifically, the plurality method produced the warehouse as the winner, whereas the plurality with elimination and instant runoff methods declared the park to be the winner, and the Borda count method chose the beach. Obviously, it is very important to inform voters *before* the election as to which system will be used!

ThomsonNOW
for Liberal Arts Mathematics
www.thomsonedu.com/1pass.

EXAMPLE 6   Referring to Example 4, use the Borda count method to determine the next Mayor of Townsville.

Solution   Referring to the voter preference table given in Figure 6.9, we tally the votes as shown in Figure 6.15.

	Alturas	Bellum	Chan	Dushay
1st choice	7,978 votes	15,287 + 9,105 = 24,392 votes	15,527 + 10,023 = 25,550 votes	5,660 votes
2nd choice	9,105 + 5,660 = 14,765 votes	15,527 + 7,978 = 23,505 votes	15,287 votes	10,023 votes
3rd choice	15,527 + 10,023 = 25,550 votes	0 votes	7,978 + 5,660 = 13,638 votes	15,287 + 9,105 = 24,392 votes
4th choice	15,287 votes	10,023 + 5,660 = 15,683 votes	9,105 votes	15,527 + 7,978 = 23,505 votes

Figure 6.15
Tally of votes

Because there were $k = 4$ candidates (Alturas, Bellum, Chan, and Dushay), each first-choice vote is worth 4 points. Consequently, the second-, third-, and fourth-choice votes are worth 3, 2, and 1 points each, respectively. Now determine the score for each candidate by multiplying the number of votes times the appropriate number of points as shown in Figure 6.16.

	Alturas	Bellum	Chan	Dushay
1st choice (4 points each)	7,978 × 4 = 31,912 points	24,392 × 4 = 97,568 points	25,550 × 4 = 102,200 points	5,660 × 4 = 22,640 points
2nd choice (3 points each)	14,765 × 3 = 44,295 points	23,505 × 3 = 70,515 points	15,287 × 3 = 45,861 points	10,023 × 3 = 30,069 points
3rd choice (2 points each)	25,550 × 2 = 51,100 votes	0 × 2 = 0 points	13,638 × 2 = 27,276 points	24,392 × 2 = 48,784 points
4th choice (1 point each)	15,287 × 1 = 15,287 votes	15,683 × 1 = 15,683 points	9,105 × 1 = 9,105 points	23,505 × 1 = 23,505 points

Figure 6.16
Points

Summing each column, we obtain the results of the election:

Alturas = 31,912 + 44,295 + 51,100 + 15,287 = 142,594 points

Bellum = 97,568 + 70,515 + 15,683 = 183,766 points

Chan = 102,200 + 45,861 + 27,276 + 9,105 = 184,442 points

Dushay = 22,640 + 30,069 + 48,784 + 23,505 = 124,998 points

When the Borda count method is used, Mr. Chan becomes the new mayor because he has the most points (184,442). ∎

The Pairwise Comparison Method

An election that has more than two candidates can be viewed as several mini-elections in which each possible pair of candidates compete against each other. That is, if there are three candidates, X, Y, and Z, we may consider the mini-elections or pairwise comparisons of X versus Y, X versus Z, and Y versus Z. We then determine the winner of each of these pairwise comparisons, and the candidate who wins the most of these is declared the winner of the overall election. This process is the basis of the **pairwise comparison method,** which is summarized in the following box.

Pairwise Comparison Method of Voting

Each voter ranks all of the candidates; that is, each voter selects his or her first choice, second choice, third choice, and so on. For each possible pairing of candidates, the candidate with the most votes receives 1 point; if there is a tie, each candidate receives $1/2$ point. The candidate who receives the most points is declared the winner.

How many possible pairwise comparisons are there? That is, if there are k candidates and we must select two for a mini-election, how many different mini-elections are possible? The answer lies in using *combinations,* as illustrated in Section 2.4; that is, the order in which we select two candidates to compete does not matter (X versus Y is the same as Y versus X). Consequently, if there are k candidates, we must examine

$$_kC_2 = \frac{k!}{(k-2)!2!} = \frac{k(k-1)}{2}$$

possible pairwise comparisons.

EXAMPLE 7 Use the pairwise comparison method to determine the location of the party for the We Deliver employees referenced in Example 3.

Solution Recall that the employees were asked to rank the three locations (warehouse, park, and beach) and the results are summarized in the voter preference table in Figure 6.17.

		Number of Ballots Cast					
		9	14	15	4	2	16
Ranked Ballot	1st choice	W	W	P	P	B	B
	2nd choice	P	B	B	W	W	P
	3rd choice	B	P	W	B	P	W

Figure 6.17
Voter preference table

Because there are $k = 3$ "candidates," we must examine

$$_3C_2 = \frac{3!}{(3-2)!2!} = \frac{3!}{1!2!} = \frac{3 \cdot 2 \cdot 1}{1 \cdot 2 \cdot 1} = 3$$

pairwise comparisons; specifically, we investigate W versus P, W versus B, and P versus B. For each comparison, we examine each column of the voter preference table to determine which candidate is preferred.

W versus P   In the first column of Figure 6.17, we see that W is preferred over P; therefore, W receives the 9 votes listed in column 1. In the second column, W is again ranked over P and receives 14 votes. In the third column, P is preferred over W, so P receives 15 votes. In the fourth column, P is again preferred over W and receives 4 votes. In the fifth column, W is preferred over P and receives 2 votes. Finally, P is preferred over W in the sixth column and receives 16 votes. Tallying these results, we obtain the following totals:

votes for W $= 9 + 14 + 2 = 25$

votes for P $= 15 + 4 + 16 = 35$

For the comparison of W versus P, voters preferred P over W by a vote of 35 to 25. Consequently, P receives 1 point.

W versus B   Examining Figure 6.17, we see that the voters in the first, second, and fourth columns all preferred W over B. Consequently, W receives the 9, 14, and 4 votes listed in those columns. We also see that the voters in the third, fifth, and sixth columns all preferred B over W, and so B receives the 15, 2, and 16 votes listed in those columns. Tallying these results, we obtain the following totals:

votes for W $= 9 + 14 + 4 = 27$

votes for B $= 15 + 2 + 16 = 33$

For the comparison of W versus B, voters preferred B over W by a vote of 33 to 27. Consequently, B receives 1 point.

P versus B   Figure 6.17 indicates that the voters in the first, third, and fourth columns all ranked P over B. Therefore, P receives the 9, 15, and 4 votes listed in those columns. The voters in the remaining columns (second, fifth, and sixth) preferred B over P, and B receives 14, 2, and 16 votes accordingly. Tallying these results, we obtain the following totals:

votes for P $= 9 + 15 + 4 = 28$

votes for B $= 14 + 2 + 16 = 32$

For the comparison of P versus B, voters preferred B over P by a vote of 32 to 28. Consequently, B receives 1 point.

Tallying the points, we see that B receives 2 points (B was preferred in two of the comparisons), P receives 1 point (P was preferred in one comparison), and W receives 0 points (W was not preferred in any of the comparisons). Consequently, B is declared the winner; the party will be held at the beach. ∎

EXAMPLE 8   Use the pairwise comparison method to determine the new mayor of Townsville referenced in Example 4.

Solution   Recall that the candidates are Alturas (A), Bellum (B), Chan (C), and Dushay (D) and that 63,580 votes were cast. The ranked ballots are tallied, and the results are summarized as shown in the voter preference table in Figure 6.18.

		Number of Ballots Cast					
		15,527	15,287	10,023	9,105	7,978	5,660
Ranked Ballot	1st choice	C	B	C	B	A	D
	2nd choice	B	C	D	A	B	A
	3rd choice	A	D	A	D	C	C
	4th choice	D	A	B	C	D	B

Figure 6.18
Voter preference table for Mayor of Townsville

Because there are $k = 4$ candidates, we must examine

$$_4C_2 = \frac{4!}{(4-2)!2!} = \frac{4!}{2! \cdot 2!} = \frac{4 \cdot 3 \cdot 2 \cdot 1}{2 \cdot 1 \cdot 2 \cdot 1} = 6$$

pairwise comparisons; specifically, we investigate A versus B, A versus C, A versus D, B versus C, B versus D, and C versus D. For each comparison, we examine each column of the voter preference table to determine which candidate is preferred.

A versus B   Figure 6.18 indicates that the voters in the third, fifth, and sixth columns all chose A over B. Therefore, A receives the $10,023 + 7,978 + 5,660 = 23,661$ votes listed in those columns. To determine the number of voters who prefer B over A, we could sum the votes listed in the remaining columns, or using a simpler approach, we subtract the 23,661 votes for A from the total number of votes to obtain the number of votes for B. Therefore, B receives $63,580 - 23,661 = 39,919$ votes. Consequently, B is preferred and receives 1 point.

A versus C   Figure 6.18 indicates that the voters in the fourth, fifth, and sixth columns all preferred A over C. Therefore, the votes are tallied as follows:

A = $9,105 + 7,978 + 5,660 = 22,743$ votes
C = $63,580 - 22,743 = 40,837$ votes

Consequently, C is preferred and receives 1 point.

A versus D   Figure 6.18 indicates that the voters in the first, fourth, and fifth columns all preferred A over D. Therefore, the votes are tallied as follows:

A = $15,527 + 9,105 + 7,978 = 32,610$ votes
D = $63,580 - 32,610 = 30,970$ votes

Consequently, A is preferred and receives 1 point.

B versus C   Figure 6.18 indicates that the voters in the second, fourth, and fifth columns all preferred B over C. Therefore, the votes are tallied as follows:

B = 15,287 + 9,105 + 7,978 = 32,370 votes

C = 63,580 − 32,370 = 31,210 votes

Consequently, B is preferred and receives 1 point.

B versus D   Figure 6.18 indicates that the voters in the first, second, fourth, and fifth columns all preferred B over D. Therefore, the votes are tallied as follows:

B = 15,527 + 15,287 + 9,105 + 7,978 = 47,897 votes

D = 63,580 − 47,897 = 15,683 votes

Consequently, B is preferred and receives 1 point.

C versus D   Figure 6.18 indicates that the voters in the first, second, third, and fifth columns all preferred C over D. Therefore, the votes are tallied as follows:

C = 15,527 + 15,287 + 10,023 + 7,978 = 48,815 votes

D = 63,580 − 48,815 = 14,765 votes

Consequently, C is preferred and receives 1 point.

Tallying the points, we see that A receives 1 point, B receives 3 points, C receives 2 points, and D receives 0 points. B has the most points (3) and so is declared the winner; Ms. Bellum is the new mayor of Townsville.   ■

Flaws of Voting Systems

As is illustrated by the examples involving the We Deliver employees' party location and the election of the mayor of Townsville, the outcome of an election may depend on the voting system that is used. Some (especially the losing candidates) might claim that this is not fair. But what exactly is meant by the word *fair*? Political scientists and mathematicians have created a list of criteria that any "fair" voting system should meet. Specifically, the four fairness criteria are known as the *majority criterion,* the *head-to-head criterion,* the *monotonicity criterion,* and the *irrelevant alternatives criterion.*

If you were a candidate in an election and you received a **majority** of the votes (that is, you received more than half of the votes), you would expect to be declared the winner. This seems to be a fair and logical conclusion; most would agree that if a candidate received more than half of the votes, that candidate should be declared the winner. This conclusion is known as the **majority criterion of fairness.**

The Majority Criterion

If candidate X receives a majority of the votes, then candidate X should be declared the winner.

Unfortunately, not all systems of voting satisfy this criterion; that is, in some circumstances, a candidate might receive a majority of the votes, yet another candidate might be declared the winner! Specifically, Exercise 23 will illustrate that the Borda count method can at times violate the majority criterion.

If you were a candidate in an election running against two other candidates, say, A and B, and you received more votes when compared to either A or B individually, you would expect to be declared the winner. In other words, suppose that you were running only against candidate A and you beat candidate A; then suppose that you were running only against candidate B and you beat candidate B. The fair and logical conclusion would be that you triumphed over each candidate, A and B, and therefore, you should be declared the winner. This conclusion is known as the **head-to-head criterion of fairness.**

The Head-to-Head Criterion

If candidate X is favored when compared head-to-head (individually) with each of the other candidates, then candidate X should be declared the winner.

Unfortunately, not all systems of voting satisfy this criterion; that is, in some circumstances, a candidate might triumph in every possible head-to-head comparison, yet another candidate might be declared the overall winner! Specifically, Exercise 24 will illustrate that the plurality method can in fact violate the head-to-head criterion.

Many times, regular elections are preceded by preliminary, nonbinding elections called *straw votes*. Suppose you were the winning candidate in a straw vote. Then, the regular election was held, and the only changes in the votes were changes in your favor; that is, some people now voted for you instead of their original candidate. Under these conditions, a fair and logical conclusion would be that you should be declared the winner. This conclusion is known as the **monotonicity criterion of fairness.**

The Monotonicity Criterion

If candidate X wins an election, and in a subsequent election, the only changes are changes in favor of candidate X, then candidate X should be declared the winner.

Unfortunately, not all systems of voting satisfy this criterion; that is, in some circumstances, a candidate might win an election (or straw vote) and gather more votes in a second election but lose out to another candidate in the second election. Specifically, Exercise 25 will illustrate that the instant run-off method can in fact violate the monotonicity criterion.

Suppose you win an election. However, one or more of the (losing) candidates are subsequently determined to be ineligible to run in the election; consequently, they are removed from the ballot, and a recount is initiated. Under these conditions, a fair and logical conclusion would be that you should still be declared the winner. This conclusion is known as the **irrelevant alternatives criterion of fairness.**

The Irrelevant Alternatives Criterion

If candidate X wins an election, and in a recount, the only changes are that one or more of the losing candidates are removed from the ballot, then candidate X should still be declared the winner.

Unfortunately, not all systems of voting satisfy this criterion; that is, in some circumstances, a candidate might win an election, but on the removal of other candidates from the ballot, another candidate might be declared the winner! Specifically, Exercise 26 will illustrate that the pairwise comparison method can in fact violate the irrelevant alternatives criterion.

Each of the common systems of voting studied in this section can be shown to violate at least one of the four fairness criteria. Is it possible to create an ultimate system that satisfies all four criteria? The answer is "no"; it is mathematically impossible to create a system of voting that satisfies all four fairness criteria. This result was proven in 1951 by Kenneth Arrow and is known as **Arrow's Impossibility Theorem.**

Arrow's Impossibility Theorem

It is mathematically impossible to create any system of voting (involving three or more candidates) that satisfies all four fairness criteria.

The systems of voting studied in this section are summarized in Figure 6.19.

System of Voting	Description
Plurality	Each person votes for his or her favorite candidate. The candidate who receives the most votes is declared the winner.
Instant Run-Off	Each voter ranks all of the candidates. If a candidate receives a majority (more than half) of first-choice votes, that candidate is declared the winner. If no candidate receives a majority, then the candidate with the fewest first-choice votes is eliminated and those votes are given to the next preferred candidate. If a candidate now has a majority of first-choice votes, that candidate is declared the winner. If no candidate receives a majority, this process continues until a candidate receives a majority.
Borda count	Each voter ranks all of the candidates. If there are k candidates, each candidate receives k points for each first-choice vote, $(k-1)$ points for each second-choice vote, $(k-2)$ points for each third-choice vote, and so on. The candidate with the most total points is declared the winner.
Pairwise comparison	Each voter ranks all of the candidates. For each possible pairing of candidates, the candidate with the most votes receives 1 point; if there is a tie, each candidate receives $1/2$ point. The candidate who receives the most points is declared the winner.

Figure 6.19
Systems of voting

The systems of voting studied in this section each satisfy (or violate) the fairness criteria as summarized in Figure 6.20.

Fairness criteria	System of Voting			
	Plurality Method	Instant Runoff Method	Borda Count Method	Pairwise Comparison Method
Majority Criterion	always satisfies	always satisfies	may not satisfy	always satisfies
Head-to-Head Criterion	may not satisfy	may not satisfy	may not satisfy	always satisfies
Monotonicity Criterion	always satisfies	may not satisfy	always satisfies	always satisfies
Irrelevant Alternatives Criterion	may not satisfy	may not satisfy	may not satisfy	may not satisfy

Figure 6.20
Fairness and flaws of the systems of voting

6.1 Exercises

1. Four candidates, Alliotti, Baker, Cruz, and Daud, are running for president of the student government. After the polls close, votes are tallied, and the results in Figure 6.21 are obtained.

Candidate	Alliotti	Baker	Cruz	Daud
Number of votes	314	155	1,052	479

Figure 6.21
Table for Exercise 1

   a. How many votes were cast?
   b. Using the plurality method of voting, which candidate wins?
   c. Did the winner receive a majority of the votes?

2. Five candidates, Edwards, Fischer, Gelinas, Horner, and Inclan, are running for president of the faculty senate. After the polls close, votes are tallied, and the results in Figure 6.22 are obtained.

Candidate	Edwards	Fischer	Gelinas	Horner	Inclan
Number of votes	32	25	17	102	24

Figure 6.22
Table for Exercise 2

   a. How many votes were cast?
   b. Using the plurality method of voting, which candidate wins?
   c. Did the winner receive a majority of the votes?

3. Three candidates, Arce, Edelstein, and Spence, are running for president of academic affairs at a local college. After the polls close, votes are tallied, and the results in Figure 6.23 are obtained.

Candidate	Arce	Edelstein	Spence
Number of votes	3,021	4,198	3,132

Figure 6.23
Table for Exercise 3

   a. How many votes were cast?
   b. Using the plurality method of voting, which candidate wins?
   c. Did the winner receive a majority of the votes?

4. Four candidates, Brecha, Parks, Wilcox, and Willett, are running for district supervisor. After the polls close, votes are tallied, and the results in Figure 6.24 are obtained.

Candidate	Brecha	Parks	Wilcox	Willett
Number of votes	3,007	2,957	10,541	2,851

Figure 6.24
Table for Exercise 4

   a. How many votes were cast?
   b. Using the plurality method of voting, which candidate wins?
   c. Did the winner receive a majority of the votes?

5. The managers of Prints Alive (a local silk-screening business) are planning a party for their 30 employees.

There are three possible locations for the party: the warehouse, the park, or the beach. The employees are asked to rank these choices in order of preference, and the results are summarized in Figure 6.25.

Number of Ballots Cast			
6	8	11	5
1st choice P	P	B	W
2nd choice B	W	W	B
3rd choice W	B	P	P

Figure 6.25
Voter preference table for Exercise 5

a. How many votes were cast?
b. Use the plurality method of voting to determine the winner.
c. What percent of the votes did the winner in part (b) receive?
d. Use the instant runoff method to determine the winner.
e. What percent of the votes did the winner in part (d) receive?
f. Use the Borda count method to determine the winner.
g. How many points did the winner in part (f) receive?
h. Use the pairwise comparison method to determine the winner.
i. How many points did the winner in part (h) receive?

6. The members of a local service club are volunteering to clean up and modernize the playground at one of the elementary schools in town. There are three schools: Hidden Lakes (H), Strandwood (S), and Valhalla (V). The members are asked to rank these choices in order of preference, and the results are summarized in Figure 6.26.

Number of Ballots Cast			
13	8	10	19
1st choice S	V	V	H
2nd choice H	S	H	S
3rd choice V	H	S	V

Figure 6.26
Voter preference table for Exercise 6

a. How many votes were cast?
b. Use the plurality method of voting to determine the winner.
c. What percent of the votes did the winner in part (b) receive?
d. Use the instant runoff method to determine the winner.
e. What percent of the votes did the winner in part (d) receive?
f. Use the Borda count method to determine the winner.
g. How many points did the winner in part (f) receive?
h. Use the pairwise comparison method to determine the winner.
i. How many points did the winner in part (h) receive?

7. The members of a youth club are raising money so that they can attend a summer camp. There are three camps in the area: Coastline (C), Pinewood (P), and The Ranch (R). The members are asked to rank these choices in order of preference, and the results are summarized in Figure 6.27.

Number of Ballots Cast				
19	12	10	11	13
1st choice C	C	P	R	R
2nd choice P	R	C	P	C
3rd choice R	P	R	C	P

Figure 6.27
Voter preference table for Exercise 7

a. How many votes were cast?
b. Use the plurality method of voting to determine the winner.
c. What percent of the votes did the winner in part (b) receive?
d. Use the instant runoff method to determine the winner.
e. What percent of the votes did the winner in part (d) receive?
f. Use the Borda count method to determine the winner.
g. How many points did the winner in part (f) receive?
h. Use the pairwise comparison method to determine the winner.
i. How many points did the winner in part (h) receive?

8. The members of a local charitable group are raising money to send a group of neighborhood children to a special event. There are three events to choose from: the circus (C), the ice show (I), and the symphony (S). The members are asked to rank these choices in order of preference, and the results are summarized in Figure 6.28.

		Number of Ballots Cast			
	6	8	7	10	15
1st choice	C	C	I	I	S
2nd choice	I	S	C	S	C
3rd choice	S	I	S	C	I

Figure 6.28
Voter preference table for Exercise 8

a. How many votes were cast?
b. Use the plurality method of voting to determine the winner.
c. What percent of the votes did the winner in part (b) receive?
d. Use the instant runoff method to determine the winner.
e. What percent of the votes did the winner in part (d) receive?
f. Use the Borda count method to determine the winner.
g. How many points did the winner in part (f) receive?
h. Use the pairwise comparison method to determine the winner.
i. How many points did the winner in part (h) receive?

9. Three candidates, Budd (B), Nirgiotis (N), and Shattuck (S), are running for union president. After the polls close, ranked ballots are tallied, and the results are summarized in Figure 6.29.

		Number of Ballots Cast			
	25	13	19	27	30	26
1st choice	B	B	N	N	S	S
2nd choice	N	S	S	B	B	N
3rd choice	S	N	B	S	N	B

Figure 6.29
Voter preference table for Exercise 9

a. How many votes were cast?
b. Use the plurality method of voting to determine the winner.
c. What percent of the votes did the winner in part (b) receive?
d. Use the instant runoff method to determine the winner.
e. What percent of the votes did the winner in part (d) receive?
f. Use the Borda count method to determine the winner.
g. How many points did the winner in part (f) receive?
h. Use the pairwise comparison method to determine the winner.
i. How many points did the winner in part (h) receive?

10. Three candidates, Maruyama (M), Peters (P), and Vilas (V), are running for district representative. After the polls close, ranked ballots are tallied, and the results are summarized in Figure 6.30.

		Number of Ballots Cast			
	675	354	451	387	601	297
1st choice	M	M	P	P	V	V
2nd choice	P	V	V	M	M	P
3rd choice	V	P	M	V	P	M

Figure 6.30
Voter preference table for Exercise 10

a. How many votes were cast?
b. Use the plurality method of voting to determine the winner.
c. What percent of the votes did the winner in part (b) receive?
d. Use the instant runoff method to determine the winner.
e. What percent of the votes did the winner in part (d) receive?
f. Use the Borda count method to determine the winner.
g. How many points did the winner in part (f) receive?
h. Use the pairwise comparison method to determine the winner.
i. How many points did the winner in part (h) receive?

11. Four candidates, Dolenz (D), Jones (J), Nesmith (N), and Tork (T), are running for director of public relations. After the polls close, ranked ballots are tallied, and the results are summarized in Figure 6.31.

Number of Ballots Cast						
225	134	382	214	81	197	109
1st choice D	D	J	J	N	T	T
2nd choice J	N	D	T	J	D	J
3rd choice T	T	T	D	T	N	D
4th choice N	J	N	N	D	J	N

Figure 6.31
Voter preference table for Exercise 11

   a. How many votes were cast?
   b. Use the plurality method of voting to determine the winner.
   c. What percent of the votes did the winner in part (b) receive?
   d. Use the instant runoff method to determine the winner.
   e. What percent of the votes did the winner in part (d) receive?
   f. Use the Borda count method to determine the winner.
   g. How many points did the winner in part (f) receive?
   h. Use the pairwise comparison method to determine the winner.
   i. How many points did the winner in part (h) receive?

12. Four candidates, Harrison (H), Lennon (L), McCartney (M), and Starr (S), are running for regional manager. After the polls close, ranked ballots are tallied, and the results are summarized in Figure 6.32.

Number of Ballots Cast						
23	98	45	82	32	17	21
1st choice H	L	L	M	M	S	S
2nd choice L	M	S	L	L	H	M
3rd choice S	H	M	H	S	L	H
4th choice M	S	H	S	H	M	L

Figure 6.32
Voter preference table for Exercise 12

   a. How many votes were cast?
   b. Use the plurality method of voting to determine the winner.
   c. What percent of the votes did the winner in part (b) receive?
   d. Use the instant runoff method to determine the winner.
   e. What percent of the votes did the winner in part (d) receive?
   f. Use the Borda count method to determine the winner.
   g. How many points did the winner in part (f) receive?
   h. Use the pairwise comparison method to determine the winner.
   i. How many points did the winner in part (h) receive?

13. Five candidates, Addley (A), Burke (B), Ciento (C), Darter (D), and Epp (E), are running for mayor. After the polls close, ranked ballots are tallied, and the results are summarized in Figure 6.33.

Number of Ballots Cast						
1,897	1,025	4,368	2,790	6,897	9,571	5,206
1st choice A	A	B	C	D	D	E
2nd choice C	D	D	E	B	E	D
3rd choice D	C	C	D	C	B	C
4th choice B	E	E	B	E	A	B
5th choice E	B	A	A	A	C	A

Figure 6.33
Voter preference table for Exercise 13

   a. How many votes were cast?
   b. Use the plurality method of voting to determine the winner.
   c. What percent of the votes did the winner in part (b) receive?
   d. Use the instant runoff method to determine the winner.
   e. What percent of the votes did the winner in part (d) receive?
   f. Use the Borda count method to determine the winner.
   g. How many points did the winner in part (f) receive?
   h. Use the pairwise comparison method to determine the winner.
   i. How many points did the winner in part (h) receive?

14. Five candidates, Fino (F), Gempler (G), Holloway (H), Isho (I), and James (J), are running for president of the Polar Bear Swim Club. After the polls close, ranked ballots are tallied, and the results are summarized in Figure 6.34.

	Number of Ballots Cast						
	12	18	29	11	21	19	18
1st choice	F	F	G	H	I	J	J
2nd choice	G	I	F	G	J	H	F
3rd choice	I	G	H	F	G	I	G
4th choice	J	H	J	I	F	G	I
5th choice	H	J	I	J	H	F	H

Figure 6.34
Voter preference table for Exercise 14

  a. How many votes were cast?
  b. Use the plurality method of voting to determine the winner.
  c. What percent of the votes did the winner in part (b) receive?
  d. Use the instant runoff method to determine the winner.
  e. What percent of the votes did the winner in part (d) receive?
  f. Use the Borda count method to determine the winner.
  g. How many points did the winner in part (f) receive?
  h. Use the pairwise comparison method to determine the winner.
  i. How many points did the winner in part (h) receive?

15. If there are six candidates in an election and voters are asked to rank all of the candidates, how many different rankings are possible?

16. If there are eight candidates in an election and voters are asked to rank all of the candidates, how many different rankings are possible?

17. If there are six candidates in an election and voters are asked to rank all of the candidates, how many different pairwise comparisons are there?

18. If there are eight candidates in an election and voters are asked to rank all of the candidates, how many different pairwise comparisons are there?

19. In an election, there are three candidates and 25 voters.

  a. What is the maximum number of points that a candidate can receive using the Borda count method?
  b. What is the minimum number of points that a candidate can receive using the Borda count method?

20. In an election, there are four candidates and 75 voters.
  a. What is the maximum number of points that a candidate can receive using the Borda count method?
  b. What is the minimum number of points that a candidate can receive using the Borda count method?

21. In an election, there are seven candidates.
  a. What is the maximum number of points that a candidate can receive using the pairwise comparison method?
  b. What is the minimum number of points that a candidate can receive using the pairwise comparison method?

22. In an election, there are nine candidates.
  a. What is the maximum number of points that a candidate can receive using the pairwise comparison method?
  b. What is the minimum number of points that a candidate can receive using the pairwise comparison method?

23. Candidates A, B, and C are being considered as supervisor of a local school district. There are thirteen directors on the school board, and they have ranked their choices as shown in Figure 6.35.

	Number of Ballots Cast		
	7	4	2
1st choice	A	B	C
2nd choice	B	C	B
3rd choice	C	A	A

Figure 6.35
Voter preference table for Exercise 23

  a. Does any candidate have a majority of first-choice votes? Who should win?
  b. Use the Borda count method to determine the winner.
  c. Does the Borda count method violate the majority criterion of fairness? Explain why or why not.

24. Candidates A, B, and C are being considered as chancellor of a local college district. There are twenty-one

directors on the college board, and they have ranked their choices as shown in Figure 6.36.

Number of Ballots Cast				
	9	2	4	6
1st choice	A	C	B	B
2nd choice	B	A	A	C
3rd choice	C	B	C	A

Figure 6.36
Voter preference table for Exercise 24

  a. Use the plurality method to determine the winner.
  b. Who wins when A is compared to B?
  c. Who wins when A is compared to C?
  d. Who wins when B is compared to C?
  e. Does the plurality method violate the head-to-head criterion of fairness? Explain why or why not.

25. Candidates A, B, and C are being considered as director of a public service agency. There are twenty-seven trustees on the executive board, and after the initial discussion and casting of straw votes, the trustees have ranked their choices as shown in Figure 6.37.

Number of Ballots Cast				
	6	8	4	9
1st choice	A	B	A	C
2nd choice	B	C	C	A
3rd choice	C	A	B	B

Figure 6.37
Original voter preference table for Exercise 25

  a. Use the instant runoff method to determine the winner of the straw vote.
  b. After discussing the results of the straw vote, four trustees changed their votes to support candidate C as shown in Figure 6.38. Use the instant runoff method to determine the winner of the subsequent election.

Number of Ballots Cast			
	6	8	13
1st choice	A	B	C
2nd choice	B	C	A
3rd choice	C	A	B

Figure 6.38
Revised voter preference table for Exercise 25

  c. Does the instant runoff method violate the monotonicity criterion of fairness? Explain why or why not.

26. Candidates A, B, C, and D are being considered as director of an endowment for the arts trust fund. There are twenty-eight trustees on the executive board, and they have ranked their choices as shown in Figure 6.39.

Number of Ballots Cast				
	10	8	6	4
1st choice	A	C	D	D
2nd choice	B	B	A	A
3rd choice	C	D	C	B
4th choice	D	A	B	C

Figure 6.39
Original voter preference table for Exercise 26

  a. Use the pairwise comparison method to determine the winner.
  b. After the trustees had ranked the candidates, it was discovered that candidates B and C did not meet the minimum qualifications and hence were ineligible for the position. Consequently, candidates B and C were removed, and the voter preference table in Figure 6.40 was established. Use the pairwise comparison method to determine the new winner.

	10	8	6	4
1st	A	D	D	D
2nd	D	A	A	A

Figure 6.40
Revised voter preference table for Exercise 26

  c. Does the pairwise comparison method violate the irrelevant alternatives criterion of fairness? Explain why or why not.

———— *Answer the following questions using complete sentences and your own words.*

CONCEPT QUESTIONS

27. What is a majority?
28. What is a ranked ballot?
29. What is a voter preference table?
30. What are the four fairness criteria? Explain the meaning of each.
31. What is Arrow's Impossibility Theorem?

6.2 Methods of Apportionment

In the constitution, the "founding fathers" created two legislative bodies: the Senate (two members per state) and the House of Representatives (apportioned to each state's population).

After the thirteen former British colonies won their independence, the "founding fathers" had to come up with a constitution to govern the newly formed nation. Needless to say, it was an enormous task to create the framework of a government rooted in revolution and driven by the underlying principle of "power to the people." The framers of the Constitution believed that the former colonies, because of their diversity, should be relatively independent states; however, there should also be some sort of central or federal government to serve as a cohesive binding element to strengthen and secure the sovereignty of the nation. To that effect, the Preamble of the Constitution reads, "We, the People of the United States, in Order to form a more perfect Union, establish Justice, insure domestic Tranquility, provide for the common defense, promote the general Welfare, and secure the Blessings of Liberty to ourselves and our Posterity, do ordain and establish this Constitution for the United States of America."

An important feature of any government is the method for creating the nation's laws; this is commonly called the legislative branch of government. When delegates of the original thirteen states met in Philadelphia in 1787, a common debate was the manner in which the states would be represented in the federal legislature. Some states, in particular the smaller ones, advocated that each state receive the same number of representatives. On the other hand, many of the larger states wanted representation in proportion to the number of people residing in the individual states.

The result of this great debate was the creation of two distinct legislative bodies. Article I, Section 1 of the Constitution states, "All legislative Powers herein granted shall be vested in a Congress of the United States, which shall consist of a Senate and House of Representatives." Article I, Section 3 created the Senate and states "The Senate of the United States shall be composed of two Senators from each state." In contrast, Article I, Section 2 created the House of Representatives and states, "Representatives and direct taxes shall be apportioned among the several States which may be included within this Union, according to their respective Numbers, which shall be determined by adding to the whole Number of free Persons, including those bound to Service for a Term of Years, and excluding Indians not taxed, three-fifths of all other persons. The actual Enumeration shall be made within three Years after the first Meeting of the Congress of the United States, and within every subsequent Term of ten Years, in such Manner as they shall by Law direct. The Number of Representatives shall not exceed one for every thirty Thousand, but each State shall have at Least one Representative."

It is very easy to interpret the phrase "The Senate of the United States shall be composed of two Senators from each state." However, the phrase "Representatives shall be *apportioned* among the several states according to their respective numbers" is nebulous at best. When we consult a standard dictionary, we find that the word **apportion** is defined as "to divide and distribute in shares according to a plan." Although the Constitution specifies that Representatives be apportioned or distributed according to a plan, it does not specify the plan. Consequently, several methods of apportionment have been proposed, and several different methods have actually been used since the first apportionment in 1790.

In this section, we will study the following methods of apportionment:

Hamilton's method was proposed by Alexander Hamilton (1757–1804) and was the first plan to be approved by Congress in 1790. However, President George Washington vetoed the plan (the first use of a presidential veto in the history of the United States). Eventually, Hamilton's method was used following each federal government census (every ten years) from 1850 to 1900.

Jefferson's method was proposed by Thomas Jefferson (1743–1826) and was the first method to actually be used; it was used following each census from 1790 to 1830.

Adams's method was proposed by John Quincy Adams (1767–1848) and has never been used. The method merits investigation, as it is viewed as the exact opposite of Jefferson's Method.

Webster's method was proposed by Daniel Webster (1782–1852). It was used following the census of 1840 and then again in 1910 and 1930. (There was no apportionment in 1920.)

Hill-Huntington method was proposed by Joseph Hill (1860–1938) and Edward Huntington (1874–1952), and it has been used following every census from 1940 to the present. During the early twentieth century, Hill was the chief statistician at the U.S. Census Bureau, and Huntington was a professor of mathematics and mechanics at Harvard.

Basic Terminology

Before we discuss any specific method of apportionment, we must introduce some basic concepts and define some basic terms.

Standard Divisor

The **standard divisor,** denoted by d, is the ratio of the total population to the total number of seats to be allocated; d is found by dividing the total population by the total number of seats to be allocated.

$$d = \text{standard divisor} = \frac{\text{total population}}{\text{total number of seats}}$$

It is customary to round the standard divisor to 2 decimal places.

For example, if the total population is 201,000 people and 10 seats are to be allocated, the standard divisor will be

$$\frac{201{,}000 \text{ people}}{10 \text{ seats}} = 20{,}100$$

Therefore, $d = 20{,}100$ people per seat, or alternatively, each seat will represent 20,100 people. Once the standard divisor has been calculated, each state's **standard quota** must be determined

---

Standard Quota

The **standard quota** (of a specific state), denoted by *q,* is the ratio of a state's population to the standard divisor; q is found by dividing the state's population by the standard divisor.

$$q = \text{standard quota} = \frac{\text{state's population}}{\text{standard divisor}} = \frac{\text{state's population}}{d}$$

It is customary to round the standard quota to two decimal places.

For example, if the population of state A was 94,700 and the standard divisor was $d = 20{,}100$, then state A's standard quota would be

$$\frac{94{,}700}{20{,}100} = 4.711442786\ldots$$

Therefore, $q = 4.71$ (rounded to two decimal places). Subscripts may be used to identify a specific state's standard quota, that is, we can use q_A to represent the standard quota of state A; hence, $q_A = 4.71$. Similarly, we use q_B to represent the standard quota of state B, q_C to represent the standard quota of state C, and so on. Rounding to two decimal places provides satisfactory results in most cases. However, if rounding to two decimal places creates an integer (when the standard quota is not an exact integer), then round to three (or more) decimal places so that an integer is not obtained. That is, if we round $q = \frac{94{,}700}{18{,}950} = 4.997361478\ldots$ to two decimal places, we obtain $q = 5.00$, an integer. Instead, we round to three decimal places and obtain $q = 4.997$.

A standard quota can be interpreted as the number of seats allocated to a specific state; therefore, if $q_A = 4.71$, state A should receive 4.71 seats. Because a state cannot receive part of a seat, we conclude that state A should receive at least four seats. The "whole number" part of a standard quota is called the **lower quota** of the state.

Lower Quota

The **lower quota** (of a specific state) is the standard quota of a state truncated (rounded down) to a whole number. That is, the lower quota is the whole number part of a standard quota. (Use the complete, original standard quota, not the rounded version, when finding the lower quota.) If the standard quota is an exact integer, then the lower quota equals the standard quota.

Hamilton's Method

We begin our study of the methods of apportionment with Hamilton's Method. Hamilton's Method is the easiest and most direct method to implement; it was the first plan to be approved by Congress in 1790, although it was not officially used until 1850.

Hamilton's Method of Apportionment

1. Using the standard divisor *d,* calculate the standard quotas and the lower quotas of each state. Initially, each state receives a number of seats equal to its lower quota.

2. If the sum of the lower quotas equals the total number of seats to be apportioned, the apportionment process is complete.

3. If the sum of the lower quotas is less than the total number of seats to be apportioned, then assign a seat to the state that has the highest decimal part in its standard quota.

4. Repeat step 3 (using the next highest decimal part) until the total number of seats has been apportioned.

Inhabitants of Middle Earth.

EXAMPLE 1   Middle Earth is a fantastic world created by the literary genius of J. R. R. Tolkien; Middle Earth's principle inhabitants are hobbits, dwarves, elves, wizards, men, and many mutant forms of evil. Let us suppose that the good inhabitants of the realms of Gondor, Eriador, and Rohan have agreed to form a federation to foster mutual protection, trade, and cultural exchange. The (estimated) populations of these realms (or states) are given in Figure 6.41. The high council of the federation is to have ten seats. Use Hamilton's Method to apportion the seats among these three realms (states).

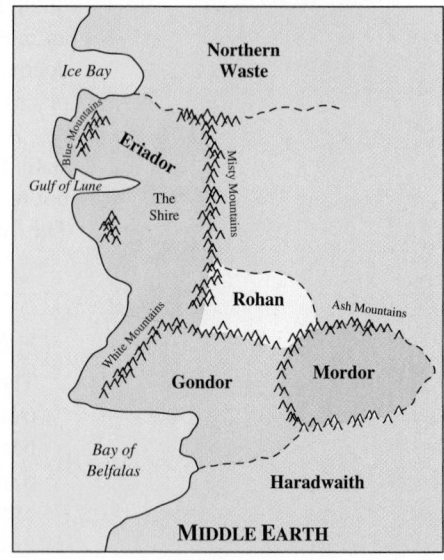

Realm (State)	Gondor	Eriador	Rohan	Total
Population (estimated)	94,700	72,600	33,700	201,000

Figure 6.41
Realms of Middle Earth

Solution
It is often advantageous to express large numbers that have several zeroes at the end in terms of simpler multiples. That is, rather than working with the numeral 94,700, we can express it as 94.7 thousand (or 94.7 × 1,000), and then use only the numeral 94.7. This will consistently simplify our calculations. Figure 6.42 shows the simplified population data in terms of thousands.

Realm (State)	Gondor	Eriador	Rohan	Total
Population (thousands)	94.7	72.6	33.7	201

Figure 6.42
Realms of Middle Earth

First, we must find the standard divisor.

d = standard divisor

$$= \frac{\text{total population}}{\text{total number of seats}}$$

$$= \frac{201}{10} = 20.1$$

Therefore, $d = 20.1$.

Now calculate each realm's standard quota, starting with Gondor (G):

q_G = Gondor's standard quota

$$= \frac{\text{Gondor's population}}{d}$$

$$= \frac{94.7}{20.1}$$

$$= 4.711442786\ldots$$

Therefore $q_G = 4.71$.

In a similar fashion, we calculate the standard quotas for Eriador (E) and Rohan (R).

q_E = Eriador's standard quota q_R = state C's standard quota

$$= \frac{\text{Eriador's population}}{d}$$ $$= \frac{\text{Rohan's population}}{d}$$

$$= \frac{72.6}{20.1}$$ $$= \frac{33.7}{20.1}$$

$q_E = 3.61$ rounded to two decimal places $q_R = 1.68$ rounded to two decimal places

The standard quotas and lower quotas for Gondor, Eriador, and Rohan are summarized in Figure 6.43.

Realm (State)	Gondor	Eriador	Rohan	Total
Population (thousands)	94.7	72.6	33.7	201
Standard Quota (using $d = 20.1$)	4.71	3.61	1.68	10
Lower Quota	4	3	1	8

Figure 6.43
Standard and lower quotas of the realms

Because the sum of the lower quotas (8) is less than the total number of seats (10), we must decide who receives the two $(10 - 8 = 2)$ "surplus" seats. According to Hamilton's method, Gondor will receive the first additional seat (Figure 6.44) because Gondor has the highest decimal part in its standard quota (0.71 is larger than either 0.61 or 0.68), and Rohan will receive the next seat (because $0.68 > 0.61$).

Realm (State)	Gondor	Eriador	Rohan	Total
Population (thousands)	94.7	72.6	33.7	201
Standard Quota ($d = 20.1$)	4.71	3.61	1.68	10
Lower Quota	4	3	1	8
Additional Seats	1	0	1	2
Number of Seats	5	3	2	10

Figure 6.44
Hamilton's method applied to the realms of Middle Earth

By using Hamilton's Method, the ten seats will be apportioned as shown in Figure 6.45.

Realm (State)	Gondor	Eriador	Rohan	Total
Hamilton's Apportionment	5	3	2	10

Figure 6.45
Final apportionment of the seats of the high council ∎

Jefferson's Method

As we have seen, Hamilton's method utilizes the standard divisor $(d = \frac{\text{state's population}}{\text{total number of seats}})$, and this method may result in surplus seats that must be distributed accordingly. Consequently, some states receive preferential treatment by obtaining extra seats. Thomas Jefferson proposed a method in which no surplus seats are created. The essence of his method is this: If the standard divisor creates surplus seats, then find a new divisor (called a modified divisor) that will result in all of the seats being allocated. As we shall see, Jefferson's method requires a bit of trial and error.

A **modified divisor** is a number that is close to the standard divisor, and it is denoted by d_m. A modified divisor may be less than or greater than the standard divisor. Jefferson's method calls for a modified divisor that is *less than* the standard divisor, that is, $d_m < d$. Now, rather than calculating the standard quota, we calculate the **modified quota.**

Modified Quota

The **modified quota** (of a specific state), denoted by q_m, is the ratio of a state's population to the modified divisor; q_m is found by dividing the state's population by the modified divisor.

$$q_m = \text{modified quota} = \frac{\text{state's population}}{\text{modified divisor}} = \frac{\text{state's population}}{d_m}$$

It is customary to round off the modified quota to two decimal places.

As a consequence, the modified quota (in Jefferson's method) will always be larger than the standard quota (in Hamilton's method) because we are dividing the state's population by a smaller number. (Dividing by a smaller number produces a larger result.) Because the modified quotas are larger than the standard quotas, our goal is to find a modified divisor that will automatically "use up" any surplus seats.

Jefferson's Method of Apportionment

1. Using the standard divisor d, calculate the standard quotas and the lower quotas of each state.
2. If the sum of the lower quotas equals the total number of seats to be apportioned, the apportionment process is complete, that is, each state receives a number of seats equal to its lower quota.
3. If the sum of the lower quotas does not equal the total number of seats to be apportioned, choose a modified divisor d_m less than the standard divisor, that is, $d_m < d$, and calculate the modified quotas and lower modified quotas.
4. Repeat step 3 until you find a modified divisor such that the sum of the lower modified quotas equals the total number of seats to be apportioned. Each state receives a number of seats equal to its lower modified quota and the apportionment process is complete.

EXAMPLE 2 As in Example 1, the populations of Gondor, Eriador, and Rohan are given in Figure 6.46. Use Jefferson's method to apportion the ten seats of the high council (legislature) among these three realms.

Realm (State)	Gondor	Eriador	Rohan	Total
Population	94,700	72,600	33,700	201,000

Figure 6.46
Realms of Middle Earth

Solution   As we saw in Example 1, Hamilton's Method and the standard divisor $d = 20.1$ results in two surplus seats, as shown in Figure 6.47.

Realm (State)	Gondor	Eriador	Rohan	Total
Population (thousands)	94.7	72.6	33.7	201
Standard Quota ($d = 20.1$)	4.71	3.61	1.68	10
Lower Quota	4	3	1	8

Figure 6.47
Standard and lower quotas of the realms

Therefore, to apply Jefferson's method, we must find a modified divisor, $d_m < d = 20.1$, such that no surplus is created. As a first guess, we try $d_m = 19$ and calculate each realm's modified quota and lower modified quota; that is, divide each population by $d_m = 19$, round to two decimal places, and then truncate (round down) to a whole number as shown in Figure 6.48.

Realm (State)	Gondor	Eriador	Rohan	Total
Population (thousands)	94.7	72.6	33.7	201
Standard Quota ($d = 20.1$)	4.71	3.61	1.68	10
Modified Quota ($d_m = 19$)	4.98	3.82	1.77	(not needed)
Lower Modified Quota	4	3	1	8

Figure 6.48
Modified and lower modified quotas using $d_m = 19$

The modified divisor $d_m = 19$ still creates two surplus seats, so we must try an even smaller divisor, say, $d_m = 18$, and repeat the process; that is, divide each population by $d_m = 18$, round to two decimal places, and then truncate to a whole number as shown in Figure 6.49.

Realm (State)	Gondor	Eriador	Rohan	Total
Population (thousands)	94.7	72.6	33.7	201
Standard Quota ($d = 20.1$)	4.71	3.61	1.68	10
Modified Quota ($d_m = 19$)	4.98	3.82	1.77	(not needed)
Modified Quota ($d_m = 18$)	5.26	4.03	1.87	(not needed)
Lower Modified Quota	5	4	1	10

Figure 6.49
Jefferson's method using $d_m = 18$

The modified divisor $d_m = 18$ has properly apportioned the ten seats, and the final results shown in Figure 6.50.

Realm (State)	Gondor	Eriador	Rohan	Total
Jefferson's Apportionment	5	4	1	10
Hamilton's Apportionment	5	3	2	10

Figure 6.50
Comparison of final apportionments   ■

Figure 6.50 shows that different methods of apportionment can lead to different allocations of legislative seats. Observe that Jefferson's method favors larger states while Hamilton's method favors smaller states. That is, under Jefferson's method, Eriador received an extra seat (and the population of Eriador is greater than that of Rohan), while under Hamilton's method, Rohan received an extra seat (and the population of Rohan is less than that of Eriador). We will explore these "flaws" in more detail in Section 6.3.

Adams's Method

As we have seen, Jefferson's method utilizes a modified divisor that is less than the standard divisor, and the modified quotas are truncated (rounded down) to a whole number (lower modified quota). Like Jefferson, John Quincy Adams proposed using a modified divisor; however, his approach was the opposite of Jefferson's: Adams proposed using a modified divisor *greater than* the standard divisor, and he rounded his modified quotas *up* to the **upper quota.**

Upper Quota

The **upper quota** (of a specific state) is the standard (or modified) quota of a state rounded up to the next whole number. (Use the complete, original standard quota, not the rounded version, when finding the upper quota.) If the standard quota is an exact integer, then the upper quota equals the standard quota plus 1.

Adams's Method of Apportionment

1. Using the standard divisor d, calculate the standard quotas and the upper quotas of each state.
2. If the sum of the upper quotas equals the total number of seats to be apportioned, the apportionment process is complete, that is, each state receives a number of seats equal to its upper quota.
3. If the sum of the upper quotas does not equal the total number of seats to be apportioned, choose a modified divisor d_m greater than the standard divisor, that is, $d_m > d$ and calculate the modified quotas and upper modified quotas.
4. Repeat step 3 until you find a modified divisor such that the sum of the upper modified quotas equals the total number of seats to be apportioned. Each state receives a number of seats equal to its upper modified quota and the apportionment process is complete.

In Adams's method, the modified quota will always be smaller than the standard quota (in Hamilton's method) because we are dividing the state's population by a larger number. (Dividing by a larger number produces a smaller result.) Because the modified quotas are smaller than the standard quotas (and we will use upper modified quotas), our goal is to find a modified divisor that will automatically allocate the proper number of seats.

EXAMPLE 3   As in Example 1, the populations of Gondor, Eriador, and Rohan are given in Figure 6.51. Use Adams's method to apportion the ten seats of the high council among these three realms.

Realm (State)	Gondor	Eriador	Rohan	Total
Population	94,700	72,600	33,700	201,000

Figure 6.51
Realms of Middle Earth

Solution   As we saw in Example 1, Hamilton's method and the standard divisor $d = 20.1$ results in two unused seats as shown in Figure 6.52. Figure 6.52 also shows the upper quotas (standard quotas rounded up) for each realm.

Realm (State)	Gondor	Eriador	Rohan	Total
Population (thousands)	94.7	72.6	33.7	201
Standard Quota ($d = 20.1$)	4.71	3.61	1.68	10
Lower Quota	4	3	1	8
Upper Quota	5	4	2	11

Figure 6.52
Standard, lower, and upper quotas of the realms

Because the sum of the upper quotas (11) does not equal the total number of seats to be allocated (10), the apportionment is not complete (we have allocated one seat too many). Therefore, to apply Adams's method, we must find a modified divisor, $d_m > d = 20.1$, such that the seats will be properly allocated. As a first guess, we try $d_m = 23$ and calculate each realm's modified quota and upper modified quota, that is, divide each population by $d_m = 23$, round to two decimal places, and then round up to a whole number as shown in Figure 6.53.

Realm (State)	Gondor	Eriador	Rohan	Total
Population (thousands)	94.7	72.6	33.7	201
Standard Quota ($d = 20.1$)	4.71	3.61	1.68	10
Modified Quota ($d_m = 23$)	4.12	3.16	1.47	(not needed)
Upper Modified Quota	5	4	2	11

Figure 6.53
Modified and upper modified quotas using $d_m = 23$

The modified divisor $d_m = 23$ does not properly allocate the ten seats, so we must try an even larger divisor. Divisors need not equal whole numbers. For example, we can choose $d_m = 23.8$ and repeat the process; that is, we divide each population by $d_m = 23.8$, round to two decimal places, and find the upper quotas as shown in Figure 6.54.

Realm (State)	Gondor	Eriador	Rohan	Total
Population (thousands)	94.7	72.6	33.7	201
Standard Quota ($d = 20.1$)	4.71	3.61	1.68	10
Modified Quota ($d_m = 23$)	4.12	3.16	1.47	(not needed)
Modified Quota ($d_m = 23.8$)	3.98	3.05	1.42	(not needed)
Upper Modified Quota	4	4	2	10

Figure 6.54
Adams's method using $d_m = 23.8$

Because the sum of the upper quotas (10) equals the total number of seats to be allocated (10), the apportionment process is complete. By using Adams's method, the ten seats will be apportioned as shown in Figure 6.55.

Realm (State)	Gondor	Eriador	Rohan	Total
Adams's Apportionment	4	4	2	10
Jefferson's Apportionment	5	4	1	10
Hamilton's Apportionment	5	3	2	10

Figure 6.55
Comparison of final apportionments

Once again, we see (Figure 6.55) that different methods of apportionment can lead to significantly different allocations of seats in a legislature (high council). However, different methods of apportionment do not always lead to different allocations; different methods may lead to the same allocation of seats. Note that Adams's method favors smaller states; that is, under Adams's method, Eriador "took away" a seat from Gondor (and the population of Eriador is less than that of Gondor). Different methods of apportionment have "flaws," which will be examined in Section 6.3.

Webster's Method

As we have seen, both Hamilton's method and Jefferson's method consistently round *down* the quotas and thereby utilize *lower* quotas, whereas Adams's method consistently rounds *up* the quotas and thereby utilizes *upper* quotas. Webster's method differs from the previous methods in that this method utilizes *regular* rounding rules, that is, round down if a number is less than 5 and round up if it is 5 or more.

> ## Webster's Method of Apportionment
>
> 1. Using the standard divisor d, calculate the standard quotas of each state and use the regular rules of rounding to round each standard quota to a whole number.
> 2. If the sum of the rounded standard quotas equals the total number of seats to be apportioned, the apportionment process is complete, that is, each state receives a number of seats equal to its rounded standard quota.
> 3. If the sum of the rounded standard quotas does not equal the total number of seats to be apportioned, choose a modified divisor d_m that is different from the standard divisor (either less than or greater than d), and calculate the modified quotas and rounded modified quotas.
> 4. Repeat step 3 until you find a modified divisor such that the sum of the rounded modified quotas equals the total number of seats to be apportioned. Each state receives a number of seats equal to its rounded modified quota and the apportionment process is complete.

ThomsonNOW™ for Liberal Arts Mathematics www.thomsonedu.com/1pass.

EXAMPLE 4 As in Example 1, the populations of Gondor, Eriador, and Rohan are given in Figure 6.56. Use Webster's Method to apportion the ten seats of the high council among these three realms.

Realm (State)	Gondor	Eriador	Rohan	Total
Population	94,700	72,600	33,700	201,000

Figure 6.56
Realms of Middle Earth

Solution As we saw in Example 1, the standard divisor $d = 20.1$ produces the standard quotas shown in Figure 6.57. Figure 6.57 also shows the rounded quotas (using the rules of regular rounding) for each realm.

Realm (State)	Gondor	Eriador	Rohan	Total
Population (thousands)	94.7	72.6	33.7	201
Standard Quota ($d = 20.1$)	4.71	3.61	1.68	10
Rounded Quota	5	4	2	11

Figure 6.57
Standard and (regular) rounded quotas of the realms

The sum of the rounded quotas (11) is too large, so we must choose a modified divisor, $d_m > d = 20.1$, because dividing by a larger number creates a smaller result. As a first guess, we try $d_m = 21$ and calculate each realm's modified quota, that is, divide each population by $d_m = 21$, round to two decimal places, and then round to a whole number as shown in Figure 6.58.

Realm (State)	Gondor	Eriador	Rohan	Total
Population (thousands)	94.7	72.6	33.7	201
Standard Quota ($d = 20.1$)	4.71	3.61	1.68	10
Modified Quota ($d_m = 21$)	4.51	3.46	1.60	(not needed)
Rounded Quota	5	3	2	10

Figure 6.58
Webster's method using $d_m = 21$

Because the sum of the rounded quotas (10) equals the total number of seats to be allocated (10), the apportionment process is complete. By using Webster's method, the ten seats will be apportioned as shown in Figure 6.59.

Realm (State)	Gondor	Eriador	Rohan	Total
Webster's Apportionment	5	3	2	10
Adams's Apportionment	4	4	2	10
Jefferson's Apportionment	5	4	1	10
Hamilton's Apportionment	5	3	2	10

Figure 6.59
Comparison of final apportionments   ∎

In Example 4, the apportionment due to Webster's method is the same as that due to Hamilton's method. This does not always occur. Sometimes different methods lead to the same allocation of seats; other times, different methods may lead to different allocation of seats. It should also be noted that Webster's method tends to favor smaller states.

Hill-Huntington Method

As we have seen, the methods of Hamilton and Jefferson consistently round *down* the quotas, the method advocated by Adams consistently rounds *up* the quotas, and quotas are subjected to the rules of *regular* rounding when Webster's method is applied. The Hill-Huntington method is similar to Webster's method in that sometimes the quotas are rounded up and sometimes they are rounded down. Before this method is discussed, we must first explore the geometric mean.

When you study algebra, you undoubtedly encounter the calculation of an "average," that is, many people recall that the "average" of two numbers is the sum of the numbers divided by 2. (See Section 4.2.) Technically, this type of average is called the **arithmetic mean,** that is, the arithmetic mean of a group of numbers equals the sum of the numbers divided by the number of numbers. There are other types of means. In particular, the Hill-Huntington Method utilizes the geometric mean.

Geometric Mean

Given two numbers x and y, the **geometric mean** of x and y, denoted by **gm,** is the square root of the product of x and y. That is,

$$gm = \text{geometric mean} = \sqrt{xy}$$

For example, the geometric mean of 2 and 18 is $gm = \sqrt{2 \times 18} = \sqrt{36} = 6$, whereas the arithmetic mean is $\frac{2+18}{2} = \frac{20}{2} = 10$. We now describe the Hill-Huntington method of apportionment.

Hill-Huntington Method of Apportionment

1. Using the standard divisor d, calculate the standard quotas, lower quotas, and upper quotas of each state.
2. For each state, calculate the geometric mean (rounded to two decimal places) of its lower quota and upper quota. If the standard quota is less than the geometric mean, round the quota down; if the standard quota is greater than or equal to the geometric mean, round the quota up.
3. If the sum of the rounded standard quotas equals the total number of seats to be apportioned, the apportionment process is complete, that is, each state receives a number of seats equal to its rounded standard quota.
4. If the sum of the rounded standard quotas does not equal the total number of seats to be apportioned, choose a modified divisor d_m different from the standard divisor (either less than or greater than d), and calculate the modified quotas and rounded modified quotas.
5. Repeat step 4 until you find a modified divisor such that the sum of the rounded modified quotas equals the total number of seats to be apportioned. Each state receives a number of seats equal to its rounded modified quota and the apportionment process is complete.

ThomsonNOW for Liberal Arts Mathematics www.thomsonedu.com/1pass.

EXAMPLE 5 As in Example 1, the populations of Gondor, Eriador, and Rohan are given in Figure 6.60. Use the Hill-Huntington Method to apportion the ten seats of the high council among these three realms.

Realm (State)	Gondor	Eriador	Rohan	Total
Population	94,700	72,600	33,700	201,000

Figure 6.60
Realms of Middle Earth

Solution As we saw in Example 3, the standard divisor $d = 20.1$ produces the standard, lower, and upper quotas shown in Figure 6.61.

Beginning with Gondor, we now calculate the geometric mean of the lower and upper quotas for each state.

$$gm_G = \sqrt{4 \times 5} = \sqrt{20} = 4.472135955\ldots$$

Therefore, $gm_G = 4.47$

Realm (State)	Gondor	Eriador	Rohan	Total
Population (thousands)	94.7	72.6	33.7	201
Standard Quota ($d = 20.1$)	4.71	3.61	1.68	10
Lower Quota	4	3	1	8
Upper Quota	5	4	2	11

Figure 6.61
Standard, lower, and upper quotas of the realms

In a similar fashion, we calculate the geometric means for Eriador (E) and Rohan (R).

gm_E = Eriador's geometric mean
$= \sqrt{3 \times 4}$
$= \sqrt{12}$
$gm_E = 3.46$ rounded to two decimal places

gm_R = Rohan's geometric mean
$= \sqrt{1 \times 2}$
$= \sqrt{2}$
$gm_R = 1.41$ rounded to two decimal places

The standard quotas and geometric means for Gondor, Eriador, and Rohan are summarized in Figure 6.62.

Realm (State)	Gondor	Eriador	Rohan	Total
Population (thousands)	94.7	72.6	33.7	201
Standard Quota ($d = 20.1$)	4.71	3.61	1.68	10
Geometric Mean	4.47	3.46	1.41	(not needed)

Figure 6.62
Geometric means of the realms

For each state, the standard quota is greater than the geometric mean; consequently, each standard quota is rounded up as shown in Figure 6.63.

Realm (State)	Gondor	Eriador	Rohan	Total
Population (thousands)	94.7	72.6	33.7	201
Standard Quota ($d = 20.1$)	4.71	3.61	1.68	10
Geometric Mean	4.47	3.46	1.41	(not needed)
Rounded Quota	5	4	2	11

Figure 6.63
Rounded quotas of the realms

The sum of the rounded quotas (11) is too large, so we choose a modified divisor, $d_m > d = 20.1$, because dividing by a larger number creates a smaller result. As a first guess, we try $d_m = 21$ and calculate each realm's modified quota; that is, we divide each population by $d_m = 21$, round to two decimal places, and then round to a whole number (using the geometric mean) as shown in Figure 6.64.

Realm (State)	Gondor	Eriador	Rohan	Total
Population (thousands)	94.7	72.6	33.7	201
Standard Quota ($d = 20.1$)	4.71	3.61	1.68	10
Geometric Mean	4.47	3.46	1.41	(not needed)
Modified Quota ($d_m = 21.2$)	4.51	3.46	1.60	(not needed)
Rounded Quota	5	4	2	11

Figure 6.64
The Hill-Huntington method using $d_m = 21$

The sum of the rounded quotas (11) is still too large, so we must choose an even larger modified divisor, say, $d_m = 21.2$, and repeat the process. That is, we divide each population by $d_m = 21.2$, round to two decimal places, and then round to a whole number (using the geometric mean) as shown in Figure 6.65.

Realm (State)	Gondor	Eriador	Rohan	Total
Population (thousands)	94.7	72.6	33.7	201
Standard Quota ($d = 20.1$)	4.71	3.61	1.68	10
Geometric Mean	4.47	3.46	1.41	(not needed)
Modified Quota ($d = 21.2$)	4.47	3.42	1.59	(not needed)
Rounded Quota	5	3	2	10

Figure 6.65
The Hill-Huntington method using $d_m = 21.2$

Because the sum of the rounded quotas (10) equals the total number of seats to be allocated (10), the apportionment process is complete. Using the Hill-Huntington method, the ten seats will be apportioned as shown in Figure 6.66.

Realm (State)	Gondor	Eriador	Rohan	Total
Hill-Huntington's Apportionment	5	3	2	10
Webster's Apportionment	5	3	2	10
Adams's Apportionment	4	4	2	10
Jefferson's Apportionment	5	4	1	10
Hamilton's Apportionment	5	3	2	10

Figure 6.66
Comparison of final apportionments

Additional Seats

Once the seats of a legislature have been allocated, it might be decided that the size of the legislature should be increased; that is, new seats might be added to an existing apportionment. Who gets the new seats? The answer lies in the calculation of **Hill-Huntington numbers.**

Hill-Huntington Number

The **Hill-Huntington number** for a state, denoted **HHN,** is the square of the state's population divided by the product of its current number of seats n and $n + 1$. That is,

$$HHN = \frac{(\text{state's population})^2}{n(n+1)}$$

where n = state's current number of seats.

In allocating a new seat to an existing legislature, the state with the highest *HHN* should receive the seat.

EXAMPLE 6 As we saw in Example 5, the populations of Gondor, Eriador, and Rohan and the apportionment of the ten seats via the Hill-Huntington Method are given in Figure 6.67. Suppose the high council decides to add an additional seat; that is, it is decided that the council should now consist of eleven seats. Use Hill-Huntington numbers to determine which realm should receive the new seat.

Realm (State)	Gondor	Eriador	Rohan	Total
Population (thousands)	94.7	72.6	33.7	201
Number of Seats	5	3	2	10

Figure 6.67
Apportionment of the high council

Solution Beginning with Gondor, we calculate the Hill-Huntington number for each realm.

$$HHN_G = \frac{94.7^2}{5(5+1)} = \frac{94.7^2}{30} = 298.936333\ldots$$

Therefore, $HHN_G = 298.94$.

In a similar fashion, we calculate the Hill-Huntington numbers for Eriador (E) and Rohan (R).

HHN_E = Eriador's Hill-Huntington number

$$= \frac{72.6^2}{3(3+1)}$$

$$= \frac{72.6^2}{12}$$

$HHN_E = 439.23$ rounded to two decimal places

HHN_R = Rohan's Hill-Huntington number

$$= \frac{33.7^2}{2(2+1)}$$

$$= \frac{33.7^2}{6}$$

$HHN_R = 189.28$ rounded to two decimal places

The Hill-Huntington numbers for the three realms are summarized in Figure 6.68.

Realm (State)	Gondor	Eriador	Rohan
Hill-Huntington Number	298.94	439.23	189.28

Figure 6.68
Hill-Huntington numbers for each realm

Because Eriador has the highest *HHN* (439.23 is greater than 298.23 or 189.28), Eriador should receive the new seat.   ∎

Why does this method work? Why does the state with the highest Hill-Huntington number deserve the additional seat? The answer lies in what is called *relative unfairness* and in the application of algebra. Consider the following scenario: A community college has a main campus and a satellite campus that is located several miles from the main campus. The main campus has an enrollment of 23,000 students with a faculty of 460; the satellite has an enrollment of 2,000 with a faculty of 40. If the college hires one new instructor, which campus is more deserving of receiving the new hire?

Using intuition, it might seem appropriate to calculate student-to-teacher ratios; that is, the campus that has more students per teacher would appear to be more deserving. However, in our scenario, both campuses have the same ratio, as shown in the following calculations:

Main Campus

$$\frac{\text{number of students}}{\text{number of teachers}}$$

$$= \frac{23,000}{460}$$

$$= 50 \text{ students per teacher}$$

Satellite Campus

$$\frac{\text{number of students}}{\text{number of teachers}}$$

$$= \frac{2,000}{40}$$

$$= 50 \text{ students per teacher}$$

To continue, we must investigate the meaning of *relative unfairness*.

Suppose two groups, *A* and *B*, are vying for an additional representative. Let p_A and p_B denote the populations of groups *A* and *B*, respectively, and let *a* and *b* denote the current number of representatives of groups *A* and *B*, respectively. The mean representations of *A* and *B* are given by $\frac{p_A}{a}$ and $\frac{p_B}{b}$, respectively (the student-to-teacher ratios shown above). Now, if *A* receives one new representative, its mean representation becomes $\frac{p_A}{a+1}$, whereas if *B* receives the new representative, its mean representation becomes $\frac{p_B}{b+1}$.

Our goal is to allocate the new representative in such a way as to minimize the relative unfairness of the allocation. Relative unfairness can be interpreted as the difference (subtraction) in mean representations compared to (divided by) the new mean representation. So if *A* receives one new representative, the difference in mean representations is $\frac{p_B}{b} - \frac{p_A}{a+1}$, and the relative unfairness to *A* may be expressed as

$$\frac{\frac{p_B}{b} - \frac{p_A}{a+1}}{\frac{p_A}{a+1}}$$

In a similar fashion, if B receives the new representative, the relative unfairness to B may be expressed as

$$\frac{\dfrac{p_A}{a} - \dfrac{p_B}{b+1}}{\dfrac{p_B}{b+1}}$$

For the sake of argument, let us now suppose that the relative unfairness to A is smaller than the relative unfairness to B; consequently, A deserves the new representative. We can express this analytically as follows.

Relative unfairness to A < relative unfairness to B

$$\left[\frac{\dfrac{p_B}{b} - \dfrac{p_A}{a+1}}{\dfrac{p_A}{a+1}}\right] < \left[\frac{\dfrac{p_A}{a} - \dfrac{p_B}{b+1}}{\dfrac{p_B}{b+1}}\right]$$

Because $\frac{p_A}{a+1}$ and $\frac{p_B}{b+1}$ are both positive, we multiply each side of the inequality by the expression $\left(\frac{p_A}{a+1}\right)\left(\frac{p_B}{b+1}\right)$ and obtain the following:

$$\left(\frac{p_A}{a+1}\right)\left(\frac{p_B}{b+1}\right)\left[\frac{\dfrac{p_B}{b} - \dfrac{p_A}{a+1}}{\dfrac{p_A}{a+1}}\right] < \left(\frac{p_A}{a+1}\right)\left(\frac{p_B}{b+1}\right)\left[\frac{\dfrac{p_A}{a} - \dfrac{p_B}{b+1}}{\dfrac{p_B}{b+1}}\right]$$

multiplying each side by the expression $\left(\dfrac{p_A}{a+1}\right)\left(\dfrac{p_B}{b+1}\right)$

$$\left(\frac{p_B}{b+1}\right)\left[\frac{p_B}{b} - \frac{p_A}{a+1}\right] < \left(\frac{p_A}{a+1}\right)\left[\frac{p_A}{a} - \frac{p_B}{b+1}\right]$$

canceling $\dfrac{p_A}{a+1}$ on the left side and $\dfrac{p_B}{b+1}$ on the right side

$$\frac{p_B^2}{b(b+1)} - \frac{p_A p_B}{(a+1)(b+1)} < \frac{p_A^2}{a(a+1)} - \frac{p_A p_B}{(a+1)(b+1)} \qquad \text{distributing}$$

$$\frac{p_B^2}{b(b+1)} < \frac{p_A^2}{a(a+1)} \qquad \text{adding } \frac{p_A p_B}{(a+1)(b+1)} \text{ to each side}$$

$$HHN_B < HHN_A \qquad \text{definition of Hill-Huntington numbers}$$

Therefore, if the relative unfairness to A is less than the relative unfairness to B, we see that $HHN_B < HHN_A$, and consequently, A deserves the new seat; that is, the group with the highest HHN should receive an additional seat. Regarding our scenario involving a community college and its satellite, we conclude that the main campus should receive the new instructor because $HHN_{main} > HHN_{satellite}$. (Verify this!)

We conclude this section with an example that recaps the five methods of apportionment.

EXAMPLE 7   Suppose that the governors of six New England region states have agreed to form an interstate bureau to foster historical awareness, tourism, and commerce. The bureau will have sixteen seats, and the populations of the states are given in Figure 6.69.

State	Connecticut (CT)	Maine (ME)	Massachusetts (MA)	New Hampshire (NH)	Rhode Island (RI)	Vermont (VT)
Population	3,405,584	1,274,923	6,349,097	1,235,786	1,048,319	608,827

Figure 6.69
New England region states
Source: Bureau of the Census, 2000

a. Use Hamilton's method to apportion the seats.
b. Use Jefferson's method to apportion the seats.
c. Use Adams's method to apportion the seats.
d. Use Webster's method to apportion the seats.
e. Use the Hill-Huntington method to apportion the seats.

Solution To simplify calculations, we express the populations in millions, rounded to three decimal places, as shown in Figure 6.70.

State	CT	ME	MA	NH	RI	VT	Total
Population (millions)	3.406	1.275	6.349	1.236	1.048	0.609	13.923

Figure 6.70
New England region states

a. The standard divisor is $d =$ total population/total number of seats $= 13.923/16 = 0.87$, rounded to two decimal places. Now, divide each state's population by 0.87 and round to two decimal places to obtain the standard quotas and lower quotas shown in Figure 6.71.

State	CT	ME	MA	NH	RI	VT	Total
Population (millions)	3.406	1.275	6.349	1.236	1.048	0.609	13.923
Standard Quota ($d = 0.87$)	3.91	1.47	7.30	1.42	1.20	0.70	16
Lower Quota	3	1	7	1	1	0	13

Figure 6.71
Standard and lower quotas of New England region states

Finally, we allocate three additional seats to the three states (CT, VT, ME) that have the highest decimal parts (0.91, 0.70, 0.47) in their standard quotas and obtain the final apportionment shown in Figure 6.72.

State	CT	ME	MA	NH	RI	VT	Total
Lower Quota	3	1	7	1	1	0	13
Additional Seats	1	1	0	0	0	1	3
Hamilton's Apportionment	4	2	7	1	1	1	16

Figure 6.72
Final apportionment using Hamilton's method

b. Using the modified divisor $d_m = 0.7$ (less than d), we obtain the modified quotas, lower modified quotas, and final apportionment shown in Figure 6.73.

State	CT	ME	MA	NH	RI	VT	Total
Population (millions)	3.406	1.275	6.349	1.236	1.048	0.609	13.923
Modified Quota $(d_m = 0.7)$	4.87	1.82	9.07	1.77	1.50	0.87	(not needed)
Lower Modified Quota	4	1	9	1	1	0	16
Jefferson's Apportionment	4	1	9	1	1	0	16

Figure 6.73
Final apportionment using Jefferson's method

c. Using the modified divisor $d_m = 1.1$ (greater than d), we obtain the modified quotas, upper modified quotas, and final apportionment shown in Figure 6.74.

State	CT	ME	MA	NH	RI	VT	Total
Population (millions)	3.406	1.275	6.349	1.236	1.048	0.609	13.923
Modified Quota $(d_m = 1.1)$	3.10	1.16	5.77	1.12	0.95	0.55	(not needed)
Upper Modified Quota	4	2	6	2	1	1	16
Adams' Apportionment	4	2	6	2	1	1	16

Figure 6.74
Final apportionment using Adams's method

d. Using the modified divisor $d_m = 0.85$ (different from d), we obtain the modified quotas, rounded modified quotas, and final apportionment shown in Figure 6.75.

State	CT	ME	MA	NH	RI	VT	Total
Population (millions)	3.406	1.275	6.349	1.236	1.048	0.609	13.923
Modified Quota $(d_m = 0.85)$	4.01	1.50	7.47	1.45	1.23	0.72	(not needed)
Rounded Modified Quota	4	2	7	1	1	1	16
Webster's Apportionment	4	2	6	2	1	1	16

Figure 6.75
Final apportionment using Webster's method

e. Using the modified divisor $d_m = 0.9$ (different from d), we obtain the modified quotas, lower modified quotas, geometric means, rounded quotas, and final apportionment shown in Figure 6.76.

State	CT	ME	MA	NH	RI	VT	Total
Population (millions)	3.406	1.275	6.349	1.236	1.048	0.609	13.923
Modified Quota $(d_m = 0.9)$	3.78	1.42	7.05	1.37	1.16	0.68	16
Lower Modified Quota	3	1	7	1	1	0	(not needed)
Geometric Mean	$\sqrt{3 \times 4}$ $= 3.46$	$\sqrt{1 \times 2}$ $= 1.41$	$\sqrt{7 \times 8}$ $= 7.48$	$\sqrt{1 \times 2}$ $= 1.41$	$\sqrt{1 \times 2}$ $= 1.41$	$\sqrt{0 \times 1}$ $= 0$	(not needed)
Rounded Quota	4	2	7	1	1	1	16
Hill-Huntington Apportionment	4	2	7	1	1	1	16

Figure 6.76
Final apportionment using the Hill-Huntington method

Voting in the Real World: Apportionment in the U.S. House of Representatives

After the adoption of the Constitution, the United States of America took its first census in the year 1790. This census, along with Jefferson's method of apportionment, was used to distribute the 105 seats of the first House of Representatives. The population (in descending magnitude) and apportionment of each of the states are listed in Figure 6.77.

It could be argued that in 1790, Virginia was the state with the most power or influence because it had the greatest number of representatives (nineteen). Notice that on the basis of the first apportionment, New York and North Carolina had equal power (each had ten representatives), while North Carolina had twice as many representatives as New Jersey (five).

Over time, the relative rankings of the "power" of these states has dramatically changed. For instance, on the basis of the 2000 census, New York now has more than twice as many representatives as North Carolina has (twenty-nine compared to thirteen), while North Carolina and New Jersey are now equal (thirteen each). Go to www.census.gov/population/www/censusdata/apportionment.html to see the current apportionment of the House of Representatives.

As we have seen, different methods of apportionment can lead to different allocations of seats. In the following example and in the exercises, we will explore the distribution of seats of the first House of Representatives, using several different methods of apportionment.

State	Population	Representatives
Virginia	630,560	19
Massachusetts	475,327	14
Pennsylvania	432,879	13
North Carolina	353,523	10
New York	331,589	10
Maryland	278,514	8
Connecticut	236,841	7
South Carolina	206,236	6
New Jersey	179,570	5
New Hampshire	141,822	4
Vermont	85,533	2
Georgia	70,835	2
Kentucky	68,705	2
Rhode Island	68,446	2
Delaware	55,540	1
Total	3,615,920	105

Figure 6.77
The first census and apportionment of the United States (1790)

For comparison, we have included all five apportionments in Figure 6.78.

State	CT	ME	MA	NH	RI	VT	Total
Hamilton's Apportionment	4	2	7	1	1	1	16
Jefferson's Apportionment	4	1	9	1	1	0	16
Adams's Apportionment	4	2	6	2	1	1	16
Webster's Apportionment	4	2	6	2	1	1	16
Hill-Huntington Apportionment	4	2	7	1	1	1	16

Figure 6.78
Comparison of final apportionments

EXAMPLE 8   Use Jefferson's method of apportionment and the population data in Figure 6.77 to verify the apportionment of the 105 seats in the 1790 House of Representatives (as shown in Figure 6.77).

Solution   To simplify calculations, as express the populations in hundred thousands, rounded to two decimal places, as shown in Figure 6.79.

State	Population (hundred thousands)
Virginia	6.31
Massachusetts	4.75
Pennsylvania	4.33
North Carolina	3.54
New York	3.32
Maryland	2.79
Connecticut	2.37
South Carolina	2.06
New Jersey	1.80
New Hampshire	1.42
Vermont	0.86
Georgia	0.71
Kentucky	0.69
Rhode Island	0.68
Delaware	0.56
Total	36.19

Figure 6.79
The first United States

First, we must find the standard divisor.

$$d = \frac{\text{total population}}{\text{total number of seats}} = \frac{36.19}{105} = 0.3446666667$$

Therefore, $d = 0.34$.

Now, calculate each state's standard quota, that is, divide each population by $d = 0.34$, round to two decimal places, and then truncate to a whole number to obtain the lower quotas, all as shown in Figure 6.80.

State	Population (hundred thousands)	Standard Quota ($d = 0.34$)	Lower Quota
Virginia	6.31	$6.31/0.34 = 18.56$	18
Massachusetts	4.75	$4.75/0.34 = 13.97$	13
Pennsylvania	4.33	$4.33/0.34 = 12.74$	12
North Carolina	3.54	$3.54/0.34 = 10.41$	10
New York	3.32	$3.32/0.34 = 9.76$	9
Maryland	2.79	$2.79/0.34 = 8.21$	8
Connecticut	2.37	$2.37/0.34 = 6.97$	6
South Carolina	2.06	$2.06/0.34 = 6.06$	6
New Jersey	1.80	$1.80/0.34 = 5.29$	5
New Hampshire	1.42	$1.42/0.34 = 4.18$	4
Vermont	0.86	$0.86/0.34 = 2.53$	2
Georgia	0.71	$0.71/0.34 = 2.09$	2
Kentucky	0.69	$0.69/0.34 = 2.03$	2
Rhode Island	0.68	$0.68/0.34 = 2.00$	2
Delaware	0.56	$0.56/.34 = 1.65$	1
Total	36.19		100

Figure 6.80
Standard and lower quotas of the first states

Because the sum of the lower quotas (100) is less than the number of seats to be allocated (105), the standard quota creates five surplus seats. Therefore, to apply Jefferson's method, we must find a modified divisor, $d_m < d = 0.34$ such that no surplus is created. After trial and error, we find that the modified divisor $d_m = 0.33$ will properly apportion the 105 seats, and the final results are shown in Figure 6.81.

State	Population (hundred thousands)	Modified Quota ($d_m = 0.33$)	Lower Modified Quota
Virginia	6.31	$6.31/0.33 = 19.12$	19
Massachusetts	4.75	$4.75/0.33 = 14.39$	14
Pennsylvania	4.33	$4.33/0.33 = 13.12$	13
North Carolina	3.54	$3.54/0.33 = 10.73$	10
New York	3.32	$3.32/0.33 = 10.06$	10
Maryland	2.79	$2.79/0.33 = 8.45$	8
Connecticut	2.37	$2.37/0.33 = 7.18$	7
South Carolina	2.06	$2.06/0.33 = 6.24$	6
New Jersey	1.80	$1.80/0.33 = 5.45$	5
New Hampshire	1.42	$1.42/0.33 = 4.30$	4
Vermont	0.86	$0.86/0.33 = 2.61$	2
Georgia	0.71	$0.71/0.33 = 2.15$	2
Kentucky	0.69	$0.69/0.33 = 2.09$	2
Rhode Island	0.68	$0.68/0.33 = 2.06$	2
Delaware	0.56	$0.56/.33 = 1.70$	1
Total	36.19		105

Figure 6.81
Jefferson's method applied to the first states ■

6.2 Exercises

1. A small country consists of three states, *A, B,* and *C*; the population of each state is given in Figure 6.82. The country's legislature is to have forty seats.

State	A	B	C
Population	900,000	700,000	400,000

Figure 6.82
Table for Exercise 1

   a. Express each state's population (and the total population) in terms of thousands.
   b. Find the standard divisor.
   c. Find each state's standard, lower, and upper quotas.

2. A small country consists of three states, *A, B,* and *C*; the population of each state is given in Figure 6.83. The country's legislature is to have fifty seats.

State	A	B	C
Population	810,000	720,000	510,000

Figure 6.83
Table for Exercise 2

   a. Express each state's population (and the total population) in terms of thousands.
   b. Find the standard divisor.
   c. Find each state's standard, lower and upper quotas.

3. Suppose that the governors of three Middle Atlantic region states have agreed to form an interstate bureau to foster historical awareness, tourism, and commerce. The bureau will have fifteen seats, and the populations of the states are given in Figure 6.84.

State	New York	Pennsylvania	New Jersey
Population	18,976,821	12,281,054	8,414,347

Figure 6.84
Table for Exercise 3
Source: Bureau of the Census, 2000

    a. Express each state's population (and the total population) in terms of millions, rounded to three decimal places.
    b. Find the standard divisor.
    c. Find each state's standard, lower, and upper quotas.

4. Suppose that the governors of three midwestern region states have agreed to form an interstate bureau to foster historical awareness, tourism, and commerce. The bureau will have ten seats, and the populations of the states are given in Figure 6.85.

State	Michigan	Wisconsin	Minnesota
Population	9,938,480	5,363,704	4,919,485

Figure 6.85
Table for Exercise 4
Source: Bureau of the Census, 2000

    a. Express each state's population (and the total population) in terms of millions, rounded to three decimal places.
    b. Find the standard divisor.
    c. Find each state's standard, lower, and upper quotas.

5. Use Hamilton's method to apportion the legislative seats in Exercise 1.

6. Use Hamilton's method to apportion the legislative seats in Exercise 2.

7. Use Hamilton's method to apportion the bureau seats in Exercise 3.

8. Use Hamilton's method to apportion the bureau seats in Exercise 4.

9. A local school district contains four middle schools: Applewood, Boatwright, Castlerock, and Dunsmuir. The number of students attending each school is given in Figure 6.86. The school district has received a generous donation of 200 graphing calculators for student use. Use Hamilton's method to determine how the calculators should be divided among the schools.

School	Applewood	Boatwright	Castlerock	Dunsmuir
Students	1,768	1,357	1,091	893

Figure 6.86
Table for Exercise 9

10. A local school district contains four elementary schools: Elmhurst, Fernwood, Greenbriar, and Hawthorne. The number of students attending each school is given in Figure 6.87. The school district has received a generous donation of sixty digital cameras for student use. Use Hamilton's method to determine how the cameras should be divided among the schools.

School	Elmhurst	Fernwood	Greenbriar	Hawthorne
Students	1,214	1,008	901	766

Figure 6.87
Table for Exercise 10

11. In J.R.R. Tolkien's Middle Earth, a region known as the Shire is homeland to numerous clans of hobbits. The Shire is divided into four regions: North Farthing, South Farthing, East Farthing, and West Farthing. Suppose the hobbits decide to form an association to foster the preservation of their cultural and culinary traditions. If the association is to have twenty-four seats, use Hamilton's method and the (estimated) population data given in Figure 6.88 to apportion the seats.

Region	North Farthing	South Farthing	East Farthing	West Farthing
Population (estimated)	2,680	6,550	2,995	8,475

Figure 6.88
Table for Exercise 11

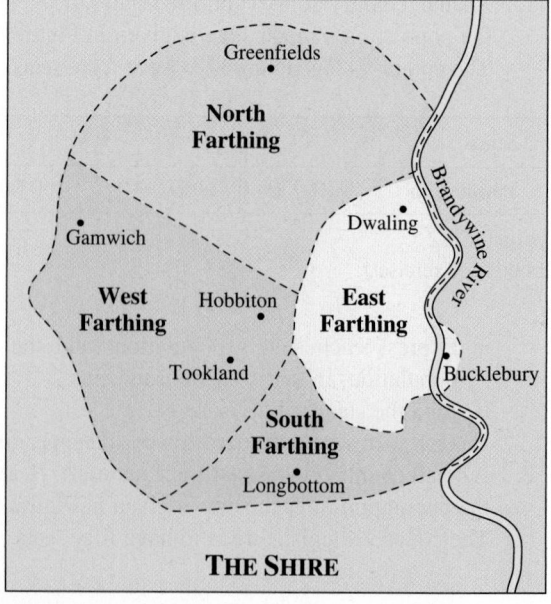

THE SHIRE

12. In J.R.R. Tolkien's Middle Earth, the regions in and around the Misty Mountains are inhabited by elves, dwarves, and Ents (tree shepherds). Elves reside in Rivendell and Lothlórien, dwarves dwell in Moria, and Ents roam the forests of Fangorn. Suppose that these inhabitants decide to form a federation to foster mutual protection, historical preservation, and cultural exchange. If the federation is to have twenty-one seats, use Hamilton's method and the (estimated) population data given in Figure 6.89 to apportion the seats.

Region	Rivendell	Lothlórien	Moria	Fangorn
Population (estimated)	5,424	4,967	6,821	587

Figure 6.89
Table for Exercise 12

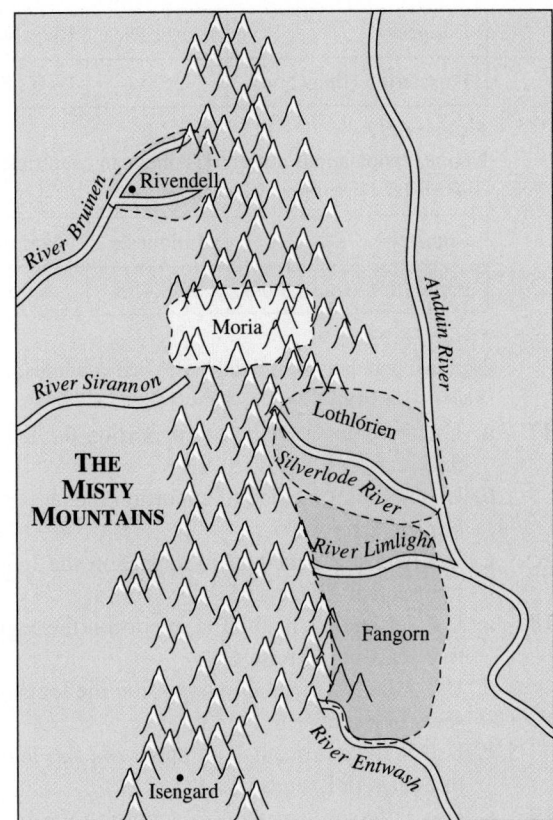

13. Suppose that the governments of several Scandinavian countries have agreed to form an international bureau to foster tourism, commerce, and cultural awareness. The bureau will have twenty seats and the populations of the countries are given in Figure 6.90. Use Hamilton's method to apportion the bureaucratic seats.

14. Suppose that the governments of several North African countries have agreed to form an international bureau to foster tourism, commerce, and education. The bureau will have twenty-five seats, and the populations of the countries are given in Figure 6.91. Use Hamilton's method to apportion the bureau seats.

15. Suppose that the governments of several Central American countries have agreed to form an international bureau to foster tourism, commerce, and education. The bureau will have twenty-five seats, and the populations of the countries are given in Figure 6.92

on the next page. Use Hamilton's method to apportion the bureau seats.

16. Suppose that the governments of several Southeast Asian countries have agreed to form an international bureau to foster tourism, commerce, and education. The bureau will have thirty seats, and the populations of the countries are given in Figure 6.93 on the next page. Use Hamilton's method to apportion the bureau seats.

Country	Denmark	Finland	Iceland	Norway	Sweden
Population (thousands)	5,413	5,215	294	4,575	8,986

Figure 6.90
National populations of Scandinavia, 2004
Source: U.S. Department of Commerce

Country	Algeria	Egypt	Libya	Morocco	Tunisia
Population (thousands)	32,129	76,117	5,632	32,209	9,975

Figure 6.91
National populations of North African countries, 2004
Source: U.S. Department of Commerce

Country	Costa Rica	El Salvador	Guatemala	Honduras	Nicaragua	Panama
Population (thousands)	3,957	6,588	14,281	6,824	5,360	3,000

Figure 6.92

National populations of Central American countries, 2004
Source: U.S. Department of Commerce

Country	Indonesia	Malaysia	Philippines	Taiwan	Thailand	Vietnam
Population (thousands)	238,453	23,522	86,242	22,750	64,866	82,690

Figure 6.93

National populations of Southeast Asian countries, 2004
Source: U.S. Department of Commerce

17. a. Use Jefferson's method to apportion the legislative seats in Exercise 5.

b. Use Adams's method to apportion the legislative seats in Exercise 5.

c. Use Webster's method to apportion the legislative seats in Exercise 5.

18. a. Use Jefferson's method to apportion the legislative seats in Exercise 6.

b. Use Adams's method to apportion the legislative seats in Exercise 6.

c. Use Webster's method to apportion the legislative seats in Exercise 6.

19. a. Use Jefferson's method to apportion the bureau seats in Exercise 7.

b. Use Adams's method to apportion the bureau seats in Exercise 7.

c. Use Webster's method to apportion the bureau seats in Exercise 7.

20. a. Use Jefferson's method to apportion the bureau seats in Exercise 8.

b. Use Adams's method to apportion the bureau seats in Exercise 8.

c. Use Webster's method to apportion the bureau seats in Exercise 8.

21. a. Use Jefferson's method to determine how the calculators should be divided among the school in Exercise 9.

b. Use Adams's method to determine how the calculators should be divided among the school in Exercise 9.

c. Use Webster's method to determine how the calculators should be divided among the school in Exercise 9.

22. a. Use Jefferson's method to determine how the cameras should be divided among the school in Exercise 10.

b. Use Adams's method to determine how the cameras should be divided among the schools in Exercise 10.

c. Use Webster's method to determine how the cameras should be divided among the schools in Exercise 10.

23. a. Use Jefferson's method to apportion the associations seats in Exercise 11.

b. Use Adams's method to apportion the association seats in Exercise 11.

c. Use Webster's method to apportion the association seats in Exercise 11.

24. a. Use Jefferson's method to apportion the federation seats in Exercise 12.

b. Use Adams's method to apportion the federation seats in Exercise 12.

c. Use Webster's method to apportion the federation seats in Exercise 12.

25. a. Use Jefferson's method to apportion the bureau seats in Exercise 13.

b. Use Adams's method to apportion the bureau seats in Exercise 13.

c. Use Webster's method to apportion the bureau seats in Exercise 13.

26. a. Use Jefferson's method to apportion the bureau seats in Exercise 14.

b. Use Adams's method to apportion the bureau seats in Exercise 14.

c. Use Webster's method to apportion the bureau seats in Exercise 14.

27. a. Use Jefferson's method to apportion the bureau seats in Exercise 15.

b. Use Adams's method to apportion the bureau seats in Exercise 15.

c. Use Webster's method to apportion the bureau seats in Exercise 15.

28. a. Use Jefferson's method to apportion the bureau seats in Exercise 16.

b. Use Adams's method to apportion the bureau seats in Exercise 16.

c. Use Webster's method to apportion the bureau seats in Exercise 16.

29. Use the Hill-Huntington method to apportion the legislative seats in Exercise 5.

30. Use the Hill-Huntington method to apportion the legislative seats in Exercise 6.

31. Use the Hill-Huntington method to apportion the bureau seats in Exercise 7.

32. Use the Hill-Huntington method to apportion the bureau seats in Exercise 8.

33. Use the Hill-Huntington method to determine how the calculators should be divided among the school in Exercise 9.

34. Use the Hill-Huntington method to determine how the cameras should be divided among the schools in Exercise 10.

35. Use the Hill-Huntington method to apportion the association seats in Exercise 11.

36. Use the Hill-Huntington method to apportion the federation seats in Exercise 12.

37. Use the Hill-Huntington method to apportion the bureau seats in Exercise 13.

38. Use the Hill-Huntington method to apportion the bureau seats in Exercise 14.

39. Use the Hill-Huntington method to apportion the bureau seats in Exercise 15.

40. Use the Hill-Huntington method to apportion the bureau seats in Exercise 16.

41. A community college has a main campus and a satellite campus that is located several miles away from the main campus. The main campus has an enrollment of 25,000 students with a faculty of 465; the satellite has an enrollment of 2,600 with a faculty of 47. The college decides to hire one new instructor. Use Hill-Huntington numbers to determine which location should receive the new instructor.

42. A waste management district operates two recycling locations: Alpha and Beta. Alpha services 20,000 households and has a staff of 23 workers; Beta's staff of 34 services 30,000 households. The district decides to hire one new worker. Use Hill-Huntington numbers to determine which location should receive the new staff member.

43. A school district has three elementary schools: Agnesi, Banach, and Cantor. If the district decides to hire one new teacher, use Hill-Huntington numbers to determine which school should receive the new instructor. The current enrollments and faculty of the schools are given in Figure 6.94.

44. A school district has three middle schools: Descartes, Euclid, and Fermat. If the district decides to hire one new teacher, use Hill-Huntington numbers to determine which school should receive the

new instructor. The current enrollments and faculty of the schools are given in Figure 6.95.

	Agnesi	**Banach**	**Cantor**
Students	567	871	666
Teachers	21	32	25

Figure 6.94
Table for Exercise 43

	Descartes	**Euclid**	**Fermat**
Students	752	984	883
Teachers	25	33	29

Figure 6.95
Table for Exercise 44

45. Use Hamilton's method of apportionment and the population data in Figure 6.77 to apportion the 105 seats in the 1790 House of Representatives. Does this hypothetical apportionment differ from the actual apportionment (shown in Figure 6.77)? If so, how?

46. Use Adams's method of apportionment and the population data in Figure 6.77 to apportion the 105 seats in the 1790 House of Representatives. Does this hypothetical apportionment differ from the actual apportionment (shown in Figure 6.77)? If so, how?

47. Use Webster's method of apportionment and the population data in Figure 6.77 to apportion the 105 seats in the 1790 House of Representatives. Does this hypothetical apportionment differ from the actual apportionment (shown in Figure 6.77)? If so, how?

48. Use the Hill-Huntington method of apportionment and the population data in Figure 6.77 to apportion the 105 seats in the 1790 House of Representatives. Does this hypothetical apportionment differ from the actual apportionment (shown in Figure 6.77)? If so, how?

▬▬▬▶ *Answer the following questions using complete sentences and your own words.*

CONCEPT QUESTIONS

49. What is apportionment?

50. What is a standard divisor?

51. What is a modified divisor?

52. What is a standard quota?

53. What is a modified quota?

54. What are lower and upper quotas?

HISTORY QUESTIONS

55. What was the first method of apportionment of the House of Representatives to be approved by Congress?

56. What was the first method of apportionment of the House of Representatives to actually be used?

57. Why are the answers to Exercises 55 and 56 different?

58. Since the founding of the United States, what methods of apportionment of the House of Representatives have been used? What is their chronology?

59. What method of apportionment of the House of Representatives is currently being used?

 WEB PROJECT

60. In 1790, the United States had 3,615,920 people, and the House of Representatives had 105 seats. Therefore, on average, each seat represented approximately 34,437 people.

a. What was the population of the United States in 1850? How many seats did the House have? On average, how many people did each seat represent?

b. What was the population of the United States in 1900? How many seats did the House have? On average, how many people did each seat represent?

c. What was the population of the United States in 1950? How many seats did the House have? On average, how many people did each seat represent?

d. What was the population of the United States in 2000? How many seats did the House have? On average, how many people did each seat represent?

e. When you compare the answers to parts (a) through (d), what conclusion(s) can you make about the average representation in the House of Representatives?

Some useful links for this web project are listed on the text website: **www.thomsonedu.com/ mathematics/johnson**

6.3　Flaws of Apportionment

In the previous section, we studied several different methods of apportionment. Quite often, different methods of apportionment produce different allotments of the items being apportioned. Are the methods "fair"? Is one method "better" than the rest? We will attempt to answer these questions in this section.

The Quota Rule

The first step in any method of apportionment is the calculation of the standard divisor and the standard quota. That is,

$$d = \text{standard divisor} = \frac{\text{total population}}{\text{total number of seats}}$$

$$q = \text{standard quota} = \frac{\text{state's population}}{\text{standard divisor}} = \frac{\text{state's population}}{d}$$

Most people would agree that in any fair method of apportionment, each state should receive either its lower quota (truncated standard quota) of seats or its upper quota (rounded up standard quota) of seats. After all, if a state received too few seats, the state would appear to have been penalized, whereas if it received too many, favoritism could be alleged. Consequently, this fundamental criterion of fairness is referred to as the **Quota Rule.**

> **The Quota Rule**
>
> The apportionment of a group should equal either the lower quota of the group or the upper quota of the group.

EXAMPLE 1 A small nation of 18 million people is composed of four states as shown in Figure 6.96. The national legislature has 120 seats.

State	A	B	C	D	Total
Population	1,548,000	2,776,000	3,929,000	9,747,000	18,000,000

Figure 6.96
National population figures

a. Use Jefferson's method to apportion the 120 legislative seats.
b. Compare the final apportionment with the lower and upper quotas of each state.

Solution **a.** After each state's population is expressed in terms of millions, we find the standard divisor.

$$d = \text{standard divisor} = \frac{\text{total population}}{\text{total number of seats}} = \frac{18.0}{120} = 0.15$$

However, the standard divisor $d = 0.15$ results in two surplus seats, as shown in Figure 6.97.

State	A	B	C	D	Total
Population (millions)	1.548	2.776	3.929	9.747	18.000
Standard Quota ($d = 0.15$)	10.32	18.51	26.19	64.98	120
Lower Quota	10	18	26	64	118

Figure 6.97
Standard and lower quotas of the states

Therefore, to apply Jefferson's method, we must find a modified divisor, $d_m < d = 0.15$, such that no surplus is created. As a first guess, we try $d_m = 0.147$ and calculate each state's modified quota and lower modified quota, that is, divide each state's population by $d_m = 0.147$, round to two decimal places, and then truncate (round down) to a whole number, as shown in Figure 6.98.

State	A	B	C	D	Total
Population (millions)	1.548	2.776	3.929	9.747	18.000
Standard Quota ($d = 0.15$)	10.32	18.51	26.19	64.98	120
Modified Quota ($d_m = 0.147$)	10.53	18.88	26.73	66.31	122.45
Lower Modified Quota	10	18	26	66	120

Figure 6.98
Modified and lower modified quotas using $d_m = 0.147$

The modified divisor $d_m = 0.147$ has properly apportioned the 120 seats, and the final results are shown in Figure 6.99.

State	A	B	C	D	Total
Jefferson's Apportionment	10	18	26	66	120

Figure 6.99
Final apportionment using Jefferson's method

b. The standard quota, lower quota, upper quota, and final apportionment for each state are shown in Figure 6.100.

State	A	B	C	D	Total
Standard Quota ($d = 0.15$)	10.32	18.51	26.19	64.98	120
Lower Quota	10	18	26	64	118
Upper Quota	11	19	27	65	122
Jefferson's Apportionment	10	18	26	66	120

Figure 6.100
Comparison of apportionment with lower and upper quotas

Notice that the apportionments for A, B, and C are "fair" in that they equal either the lower quota or the upper quota of the state. However, the apportionment for D (66 seats) is "unfair" because it is greater than D's upper quota of 65. ■

Example 1 illustrates the fact that Jefferson's method can violate the Quota Rule. In a similar fashion, it can be shown that Adams's method, Webster's method, and the Hill-Huntington method can all produce apportionments that violate the Quota Rule. In contrast, Hamilton's method always produces a fair apportionment with respect to the Quota Rule; it always allocates either the lower quota or the upper quota. However, even though it satisfies the Quota Rule, Hamilton's method gives rise to several other problems. In particular, we shall investigate the Alabama Paradox, the New States Paradox, and the Population Paradox.

The Alabama Paradox

In 1870, the population of the United States was 38,558,371, and the House of Representatives had 292 seats. Ten years later, the 1880 census recorded a population of 50,189,209—a 30% increase. Subsequently, the number of seats in the House of Representatives was to be increased. A discussion on whether to increase to 299 or 300 seats ensued. At the time, there were thirty-eight states in the Union, and Hamilton's method was the official means of apportionment. The U.S. Census Office calculated the apportionment for each state using both 299 and 300 seats. It was then discovered that with 299 seats, Alabama would receive eight seats, whereas with 300, Alabama's share would be only seven. That is, an increase of one seat overall would result in a reduction in a state's apportionment. Consequently, this scenario has become known as the **Alabama Paradox.**

> **The Alabama Paradox**
>
> Adding one new seat to the total number of seats being allocated causes one of the states to lose one of its seats (even though the population has not changed).

ThomsonNOW™
for Liberal Arts Mathematics
www.thomsonedu.com/1pass.

EXAMPLE 2 A small nation of 12 million people is composed of three states as shown in Figure 6.101. The national legislature has 150 seats.

State	A	B	C	Total
Population (thousands)	595	5,615	5,790	12,000

Figure 6.101
National population figures

a. Use Hamilton's method to apportion the 150 legislative seats.
b. Suppose the total number of seats increases by one. Reapportion the 151 legislative seats.
c. Compare the apportionments in parts (a) and (b).

Solution a. First, we must find the standard divisor.

$$d = \text{standard divisor} = \frac{\text{total population}}{\text{total number of seats}} = \frac{12{,}000}{150} = 80$$

Therefore, $d = 80$.

Now, calculate each state's standard quota (divide each state's population by $d = 80$) and lower quota (truncate the standard quotas to whole numbers), as shown in Figure 6.102. By using Hamilton's method, state A receives the one additional seat because the decimal part of state A's standard quota is highest (0.44); the final apportionment is shown in the last row of Figure 6.102.

State	A	B	C	Total
Population (thousands)	595	5,615	5,790	12,000
Standard Quota ($d = 80$)	7.44	70.19	72.38	150.01
Lower Quota	7	70	72	149
Additional Seats	1	0	0	1
Number of Seats	8	70	72	150

Figure 6.102
Hamilton's apportionment of 150 seats

b. The new standard divisor is

$$d = \frac{\text{total population}}{\text{total number of seats}} = \frac{12{,}000}{151} = 79.47019868\ldots$$

Therefore, $d = 79.47$ (rounded to two decimal places).

Now, calculate each state's standard quota (divide each state's population by $d = 79.47$) and lower quota (truncate the standard quotas to whole numbers), as shown in Figure 6.103. By using Hamilton's Method, state C and state B each receive an additional seat because the decimal parts of their standard quotas are highest (0.86 and 0.66); the final apportionment is shown in the last row of Figure 6.103.

State	A	B	C	Total
Population (thousands)	595	5,615	5,790	12,000
Standard Quota ($d = 79.47$)	7.49	70.66	72.86	151.01
Lower Quota	7	70	72	149
Additional Seats	0	1	1	2
Number of Seats	7	71	73	151

Figure 6.103
Hamilton's apportionment of 151 seats

c. Figure 6.104 summarizes the two apportionments found in parts (a) and (b). Although the states' populations did not change, state A lost one of its seats when one new seat was added to the legislature; this is an illustration of the Alabama Paradox. Notice that the state that lost a seat is also the smallest state. This is typical of the so-called Alabama Paradox; large states benefit at the expense of small states.

State	Apportionment of 150 seats	Apportionment of 151 seats
A	8	7
B	70	71
C	72	73

Figure 6.104
Comparison of apportionments

The New States Paradox

Oklahoma was admitted to the United States in 1907 as the forty-sixth state. On the basis of its populations, it was determined that the new state should have five seats in the House of Representatives. Therefore, the House was increased from 386 seats to 391 with the intention of not altering the apportionment of the other forty-five states. However, when Hamilton's method was applied to the new House of 391 seats, it was discovered that the apportionments for two other states would be affected; Maine would gain one seat at the expense of New York losing a seat! Therefore, after the 1910 census, Congress jettisoned Hamilton's method, and Webster's method was reinstated. Consequently, this scenario has become known as the **New States Paradox.**

> **The New States Paradox**
>
> Adding a new state, and the corresponding number of seats based on its population, alters the apportionments for some of the other states.

ThomsonNOW™
for Liberal Arts Mathematics
www.thomsonedu.com/1pass.

EXAMPLE 3 A small nation of 10 million people is composed of two states, as shown in Figure 6.105. The national legislature has 100 seats.

State	A	B	Total
Population	3,848,000	6,152,000	10,000,000

Figure 6.105
National population figures

a. Use Hamilton's method to apportion the 100 legislative seats.
b. Suppose state C has a population of 4,332,000 and joins the union. How many news seats should be added to the legislature?
c. Apportion the new total number of seats.
d. Compare the apportionments in parts (a) and (c).

Solution a. After expressing each state's population in terms of thousands (an arbitrary decision), we must find the standard divisor.

$$d = \text{standard divisor} = \frac{\text{total population}}{\text{total number of seats}} = \frac{10,000}{100} = 100$$

Therefore, $d = 100$.

Now, calculate each state's standard quota (divide each state's population by $d = 100$) and lower quota (truncate the standard quotas to whole numbers), as shown in Figure 6.106. By using Hamilton's method, state B receives the one additional seat because the decimal part of state B's standard quota is highest (0.52); the final apportionment is shown in the last row of Figure 6.106.

State	A	B	Total
Population (thousands)	3,848	6,152	10,000
Standard Quota ($d = 100$)	38.48	61.52	100
Lower Quota	38	61	99
Additional Seats	0	1	1
Number of Seats	38	62	100

Figure 6.106
Hamilton's apportionment of the original 100 seats

b. In terms of thousands, state C's population of 4,332,000 would be expressed as 4,332 thousand. Therefore, state C's quota of (new) seats is $\frac{4,332}{100} = 43.32$ or 43 seats.

That is, when state C is admitted to the union, forty-three seats should be added to the legislature; the total number of seats to be allocated is now $100 + 43 = 143$ seats.

c. We must now find the new standard divisor.

$$d = \text{standard divisor} = \frac{\text{total population}}{\text{total number of seats}} = \frac{10,000 + 4,332}{100 + 43}$$

$$= \frac{14,332}{143} = 100.2237762\ldots$$

Therefore, $d = 100.22$.

Now, calculate each state's standard quota (divide each population by $d = 100.22$) and lower quota (truncate the standard quotas to whole numbers), as shown in Figure 6.107. Using the lower quotas, all 143 seats are allocated as shown in the last row of Figure 6.107.

State	A	B	C	Total
Population (thousands)	3,848	6,152	4,332	14,332
Standard Quota ($d = 100.22$)	39.40	61.38	43.22	144
Lower Quota	39	61	43	143
Number of Seats	39	61	43	143

Figure 6.107
Hamilton's apportionment of the new 143 seats

d. Figure 6.108 summarizes the apportionments from parts (a) and (c). On comparison, we see that when state C joined the union, state A gained 1 seat and state B lost one seat. That is, the addition of state C altered the apportionments for the other states; this is an illustration of the New States Paradox.

State	A	B	C	Total
(Original) Number of Seats	38	62	0	100
(New) Number of Seats	39	61	43	143

Figure 6.108
Comparison of apportionments ∎

The Population Paradox

Throughout history, the populations of the various states have been in flux; changes in population subsequently affect each state's apportionment of seats in the House of Representatives. Hypothetically, if one state grew at a fast rate while another grew at a slow rate, it might seem more appropriate that the state with the larger growth rate be entitled to an "extra" seat, if any became available. However, Hamilton's method does not adhere to this premise. During the late 1800s and early 1900s, Virginia was growing at rates much faster than Maine; for example, from 1890 to 1910, Virginia's population increased by 24.5%, while Maine's growth was only 12.3%. To everyone's surprise, Virginia lost one of its seats to Maine in the early 1900s. Consequently, this scenario has become known as the **Population Paradox.**

> **The Population Paradox**
>
> One state loses a seat to another state even though it is growing at a faster rate;
> state A loses a seat to state B even though state A is growing at a faster rate than
> state B.

EXAMPLE 4 A school district of slightly more than 6,000 students is composed of three campuses, as shown in Figure 6.109. The district has twelve educational specialists to assign to three locations.

Campus	A	B	C	Total
Enrollment (2005)	777	1,792	3,542	6,111
Enrollment (2006)	820	1,880	3,645	6,345

Figure 6.109
School district enrollments

a. Use Hamilton's method and the 2005 enrollments to apportion the twelve specialists.
b. Use Hamilton's method and the 2006 enrollments to reapportion the twelve specialists.
c. Find the differences, if any, in the number of specialists apportioned to each campus in 2005 versus 2006.
d. Which campus had the highest growth rate (percentagewise) from 2005 to 2006? Do the growth rates of the campuses conform with the differences found in part (c)?

Solution a. First, we must find the standard divisor for 2005.

$$d = \text{standard divisor} = \frac{\text{total enrollment}}{\text{total number of specialists}} = \frac{6,111}{12} = 509.25$$

Therefore, $d = 509.25$.

Now, calculate each campus's standard quota (divide each enrollment by $d = 509.25$) and lower quota (truncate the standard quotas to whole numbers), as shown in Figure 6.110. By using Hamilton's method, campus C and campus A each receive one additional specialist because the decimal parts of their standard quotas are highest (0.96 and 0.53); the 2005 apportionment is shown in the last row of Figure 6.110.

Campus	A	B	C	Total
Enrollment	777	1,792	3,542	6,111
Standard Quota ($d = 509.25$)	1.53	3.52	6.96	12.01
Lower Quota	1	3	6	10
Additional Specialists	1	0	1	2
Number of Specialists (in 2005)	2	3	7	12

Figure 6.110
Hamilton's apportionment in 2005

b. Now, we find the standard divisor for 2006.

$$d = \text{standard divisor} = \frac{\text{total enrollment}}{\text{total number of specialists}} = \frac{6{,}345}{12} = 528.75$$

Therefore, $d = 528.75$.

Now, calculate each campus's standard quota (divide each enrollment by $d = 528.75$) and lower quota (truncate the standard quotas to whole numbers), as shown in Figure 6.111. By using Hamilton's method, campus C and campus B each receive one additional specialist because the decimal parts of their standard quotas are highest (0.89 and 0.56); the 2006 apportionment is shown in the last row of Figure 6.101.

Campus	A	B	C	Total
Enrollment	820	1,880	3,645	6,345
Standard Quota ($d = 509.25$)	1.55	3.56	6.89	12
Lower Quota	1	3	6	10
Additional Specialists	0	1	1	2
Number of Specialists (in 2006)	1	4	7	12

Figure 6.111
Hamilton's apportionment in 2006

c. Comparing the number of specialists apportioned in 2005 and 2006 (Figure 6.112), we see that campus A lost one specialist, campus B gained one specialist, and the apportionment of campus C was unchanged.

Campus	A	B	C
Number of Specialists (in 2005)	2	3	7
Number of Specialists (in 2006)	1	4	7
Difference	-1	$+1$	0

Figure 6.112
Comparison of apportionment in 2005 and 2006

d. To calculate the growth rate of each campus (expressed as a percentage), we must find the difference in the enrollment (2006 minus 2005), divide the difference by the original (2005) enrollment, and then multiply by 100. These calculations are summarized in Figure 6.113.

Consequently, campus A had the highest growth rate (5.5%) from 2005 to 2006. However, the growth rates do not conform with the differences in apportionment of specialists (Figure 6.114).

Notice that campus A grew at a faster rate than campus B (5.5% > 4.9%), but campus A lost one of its specialists to campus B! This is an illustration of the Population Paradox; one group loses an apportioned position to another group even though the first group is growing at a faster rate.

Campus	A	B	C
2005 Enrollment	777	1,792	3,542
2006 Enrollment	820	1,880	3,645
Difference (increase)	$820 - 777 = 43$	$1,880 - 1,792 = 88$	$3,645 - 3,542 = 103$
Proportion	$43/777$ $= 0.05534\ldots$	$88/1,792$ $= 0.04910\ldots$	$103/3,542$ $= 0.02907\ldots$
Percentage growth	5.5%	4.9%	2.9%

Figure 6.113
Growth rates from 2005 and 2006

Campus	A	B	C
Enrollment growth rate	5.5%	4.9%	2.9%
Change in apportionment	lost 1 specialist	gained 1 specialist	no change

Figure 6.114
Comparison of growth rates and changes in apportionment.   ■

The Balinski-Young Impossibility Theorem

We have studied several methods of apportionment, specifically the methods of Hamilton, Jefferson, Adams, Webster, and Hill-Huntington, We have also seen various problems associated with these methods of apportionment. Although Hamilton's method does not violate the Quota Rule, it has several flaws in that it gives rise to the Alabama, New States, and Population Paradoxes. In contrast, it can be shown that the methods of Jefferson, Adams, Webster, and Hill-Huntington are all immune to these paradoxes; however, they may all violate the Quota Rule. In addition, some methods favor large states, whereas others favor small states. In other words, if a method of apportionment does not have one type of problem, it has another type of problem. In the quest of searching for the ideal method of apportionment, can we ever hope to find a method that will satisfy the Quota Rule, be immune from paradoxes, and favor neither large nor small states? Unfortunately, the answer is "no." In 1980, mathematicians Michel Balinski and H. Payton Young proved that it is mathematically impossible for any method of apportionment to satisfy the Quota Rule and not produce any paradoxes. This is known as the **Balinski-Young Impossibility Theorem.**

The Balinski-Young Impossibility Theorem

There is no perfect method of apportionment: If a method satisfies the Quota Rule, it must give rise to paradoxes; and if a method does not give rise to paradoxes, it must violate the Quota Rule.

Just as Arrow's Impossibility Theorem (Section 6.1) ruled out the possibility of the existence of a perfect system of voting, the Balinski-Young Impossibility Theorem rules out the possibility of creating a perfect method of apportionment.

Figure 6.115 lists several methods of apportionment and summarizes the flaws discussed in this section.

Flaw	Method of Apportionment				
	Hamilton	Jefferson	Adams	Webster	Hill-Huntington
May violate the Quota Rule.	No	Yes	Yes	Yes	Yes
May produce the Alabama Paradox.	Yes	No	No	No	No
May produce the New States Paradox.	Yes	No	No	No	No
May produce the Population Paradox.	Yes	No	No	No	No

Figure 6.115
Flaws of the methods of apportionment

6.3 Exercises

1. A small country consists of three states: A, B, and C. The population of each state is given in Figure 6.116. The country's legislature is to have thirty-two seats.

State	A	B	C
Population	3,500,000	4,200,000	16,800,000

Figure 6.116
Table for Exercise 1

a. Express each state's population (and the total population) in terms of millions.
b. Find the standard divisor, and round it off to two decimal places.
c. Find each state's standard, lower, and upper quotas.
d. Use Jefferson's method to apportion the thirty-two seats.
e. Is the Quota Rule violated? Explain.

2. A small country consists of three states, A, B, C. The population of each state is given in Figure 6.117. The country's legislature is to have ninety-six seats.

State	A	B	C
Population	1,200,000	3,400,000	15,400,000

Figure 6.117
Table for Exercise 2

a. Express each state's population (and the total population) in terms of millions.
b. Find the standard divisor, and round it off to two decimal places.

c. Find each state's standard, lower, and upper quotas.
d. Use Jefferson's method to apportion the ninety-six seats.
e. Is the Quota Rule violated? Explain.

3. A small country consists of three states: A, B, and C. The population of each state is given in Figure 6.118. The country's legislature is to have seventy-nine seats.

State	A	B	C
Population	3,500,000	4,200,000	16,800,000

Figure 6.118
Table for Exercise 3

a. Express each state's population (and the total population) in terms of millions.
b. Find the standard divisor, and round it off to two decimal places.
c. Find each state's standard, lower, and upper quotas.
d. Use Adams's method to apportion the seventy-nine seats.
e. Is the Quota Rule violated? Explain.

4. A small country consists of three states: A, B, and C. The population of each state is given in Figure 6.119. The country's legislature is to have seventy-three seats.

State	A	B	C
Population	1,200,000	3,400,000	15,400,000

Figure 6.119
Table for Exercise 4

a. Express each state's population (and the total population) in terms of millions.

b. Find the standard divisor, and round it off to two decimal places.

c. Find each state's standard, lower, and upper quotas.

d. Use Adams's method to apportion the seventy-three seats.

e. Is the Quota Rule violated? Explain.

5. A small country consists of four states: A, B, C, and D. The population of each state is given in Figure 6.120. The country's legislature is to have 200 seats.

State	A	B	C	D
Population	1,200,000	3,400,000	17,500,000	19,400,000

Figure 6.120
Table for Exercise 5

a. Express each state's population (and the total population) in terms of millions.

b. Find the standard divisor, and round it off to two decimal places.

c. Find each state's standard, lower, and upper quotas.

d. Use Webster's method to apportion the 200 seats.

e. Is the Quota Rule violated? Explain.

6. A small country consists of four states: A, B, C, and D. The population of each state is given in Figure 6.121. The country's legislature is to have 177 seats.

State	A	B	C	D
Population	1,100,000	3,200,000	17,300,000	19,400,000

Figure 6.121
Table for Exercise 6

a. Express each state's population (and the total population) in terms of millions.

b. Find the standard divisor, and round it off to two decimal places.

c. Find each state's standard, lower, and upper quotas.

d. Use Webster's method to apportion the 177 seats.

e. Is the Quota Rule violated? Explain.

7. A small country consists of four states: A, B, C, and D. The population of each state is given in Figure 6.122. The country's legislature is to have 200 seats.

State	A	B	C	D
Population	1,200,000	3,400,000	17,500,000	19,400,000

Figure 6.122
Table for Exercise 7

a. Express each state's population (and the total population) in terms of millions.

b. Find the standard divisor, and round it off to two decimal places.

c. Find each state's standard, lower, and upper quotas.

d. Use the Hill-Hunington method to apportion the 200 seats.

e. Is the Quota Rule violated? Explain.

8. A small country consists of four states: A, B, C, and D. The population of each state is given in Figure 6.123. The country's legislature is to have 177 seats.

State	A	B	C	D
Population	1,100,000	3,200,000	17,300,000	19,400,000

Figure 6.123
Table for Exercise 8

a. Express each state's population (and the total population) in terms of millions.

b. Find the standard divisor and round it off to two decimal places.

c. Find each state's standard, lower, and upper quotas.

d. Use the Hill-Huntington method to apportion the 177 seats.

e. Is the Quota Rule violated? Explain.

9. A small country consists of three states: A, B, and C. The population of each state is given in Figure 6.124. The country's legislature is to have 120 seats.

State	A	B	C
Population	690,000	5,700,000	6,410,000

Figure 6.124
Table for Exercise 9

a. Express each state's population (and the total population) in terms of thousands.

b. Find the standard divisor and round it off to two decimal places.

c. Use Hamilton's method to apportion the 120 legislative seats.

d. Suppose the total number of seats increases by one. Reapportion the 121 legislative seats.

e. Is the Alabama Paradox exhibited? Explain.

10. A small country consists of three states: A, B, C. The population of each state is given in Figure 6.125. The country's legislature is to have 140 seats.

a. Express each state's population (and the total population) in terms of thousands.

State	A	B	C
Population	700,000	5,790,000	6,510,000

Figure 6.125
Table for Exercise 10

   b. Find the standard divisor and round it off to two decimal places.

   c. Use Hamilton's method to apportion the 140 legislative seats.

   d. Suppose the total number of seats increases by one. Reapportion the 141 legislative seats.

   e. Is the Alabama Paradox exhibited? Explain.

11. A small country consists of three states: A, B, and C. The population of each state is given in Figure 6.126. The country's legislature is to have 110 seats.

State	A	B	C
Population	1,056,000	1,844,000	2,100,000

Figure 6.126
Table for Exercise 11

   a. Express each state's population (and the total population) in terms of thousands.

   b. Find the standard divisor and round it off to two decimal places.

   c. Use Hamilton's method to apportion the 110 legislative seats.

   d. Suppose state D has a population of 2,440,000 and joins the union. How many new seats should be added to the legislature?

   e. Apportion the new total number of seats.

   f. Is the New States Paradox exhibited? Explain

12. A small country consists of three states: A, B, and C. The population of each state is given in Figure 6.127. The country's legislature is to have 150 seats.

State	A	B	C
Population	2,056,000	1,844,000	3,100,000

Figure 6.127
Table for Exercise 12

   a. Express each state's population (and the total population) in terms of thousands.

   b. Find the standard divisor and round it off to two decimal places.

   c. Use Hamilton's method to apportion the 150 legislative seats.

   d. Suppose state D has a population of 2,500,000 and joins the union. How many new seats should be added to the legislature?

   e. Apportion the new total number of seats.

   f. Is the New States Paradox exhibited? Explain.

13. A school district has three campuses with enrollments given in Figure 6.128. The district has eleven educational specialists to assign to the locations.

Campus	A	B	C
Enrollment (2005)	780	1,700	3,520
Enrollment (2006)	820	1,880	3,740

Figure 6.128
Table for Exercise 13

   a. Use Hamilton's method and the 2005 enrollments to apportion the eleven specialists.

   b. Use Hamilton's method and the 2006 enrollments to reapportion the eleven specialists.

   c. Find the growth rate (from 2005 to 2006) for each campus.

   d. Is the Population Paradox exhibited? Explain.

14. A school district has three campuses with enrollments given in Figure 6.129. The district has twelve educational specialists to assign to the locations.

Campus	A	B	C
Enrollment (2005)	750	1,680	3,980
Enrollment (2006)	788	1,858	4,210

Figure 6.129
Table for Exercise 14

   a. Use Hamilton's method and the 2005 enrollments to apportion the twelve specialists.

   b. Use Hamilton's method and the 2006 enrollments to reapportion the twelve specialists.

   c. Find the growth rate (from 2005 to 2006) for each campus.

   d. Is the Population Paradox exhibited? Explain.

Answer the following questions using complete sentences and your own words.

CONCEPT QUESTIONS

15. What is the Quota Rule? Explain its meaning.

16. What is the Alabama Paradox? Explain its meaning.

17. What is the New States Paradox? Explain its meaning.

18. What is the Population Paradox? Explain its meaning.

19. What is the Balinski-Young Impossibility Theorem? Explain its meaning.

Chapter 6   Review

ThomsonNOW™ for Liberal Arts Mathematics
Preparing for an exam? Test yourself on key material by visiting ThomsonNOW at **www.thomsonedu.com/1pass.**

TERMS

Adams's method
Alabama paradox
apportion
arithmetic mean
Arrow's Impossibility
  Theorem
Balinski-Young
  Impossibility Theorem
Borda count method
geometric mean
Hamilton's method

head-to-head criterion of
  fairness
Hill-Huntington method
Hill-Huntington number
irrelevant alternatives
  criterion of fairness
Jefferson's method
lower quota
majority
majority criterion of
  fairness

modified divisor
modified quota
monotonicity criterion of
  fairness
New States paradox
pairwise comparison
  method
plurality method of voting
plurality with elimination
  method
population paradox

ranked-choice or instant
  runoff method
Quota Rule
standard divisor
standard quota
upper quota
voter preference table
Webster's method

REVIEW EXERCISES

1. The Metropolitan Symphonic Orchestra has volunteered to play a benefit concert, with proceeds to support music programs at local high schools. The organizers need to know what music will be featured at the concert. The conductor has asked each member of the orchestra to rank the following composers: Beethoven (B), Mozart (M), Tchaikovsky (T), and Vivaldi (V). The ranked ballots have been tallied, and the results are summarized in Figure 6.130.
   a. How many votes were cast?
   b. Use the plurality method of voting to determine the winner.
   c. What percent of the votes did the winner in part (b) receive?
   d. Use the instant runoff method to determine the winner.
   e. What percent of the votes did the winner in part (d) receive?
   f. Use the Borda count method to determine the winner.

   g. How many points did the winner in part (f) receive?
   h. Use the pairwise comparison method to determine the winner.
   i. How many points did the winner in part (h) receive?
2. A small country consists of three states: A, B, and C. The population of each state is given in Figure 6.131. The country's legislature is to have seventy-six seats.

State	A	B	C
Population	1,200,000	2,300,000	2,500,000

Figure 6.131
Table for Exercise 2

   a. Express each state's population (and the total population) in terms of thousands.
   b. Find the standard divisor and round it off to two decimal places.
   c. Find each state's standard, lower, and upper quotas.

	Number of Ballots Cast						
	5	7	6	12	13	17	14
1st choice	M	M	M	B	B	V	T
2nd choice	B	V	T	T	M	M	V
3rd choice	V	B	V	V	T	B	B
4th choice	T	T	B	M	V	T	M

Figure 6.130
Voter preference table for Exercise 1

3. Use Hamilton's method to apportion the legislative seats in Exercise 2.
4. Use Jefferson's method to apportion the legislative seats in Exercise 2.
5. Use Adams's method to apportion the legislative seats in Exercise 2.
6. Use Webster's method to apportion the legislative seats in Exercise 2.
7. Use the Hill-Huntington method to apportion the legislative seats in Exercise 2.
8. Suppose that the governments of several South American countries have agreed to form an international bureau to foster tourism, commerce, and education. The bureau will have twenty-five seats, and the populations of the countries are given in Figure 6.132. Use Hamilton's method to apportion the bureau seats.

Country	Argentina	Bolivia	Chile	Paraguay	Uruguay
Population (thousands)	39,145	8,724	15,824	6,191	3,399

Figure 6.132
National population figures, 2004
Source: U.S. Department of Commerce

9. Use Jefferson's method to apportion the bureau seats in Exercise 8.
10. Use Adams's method to apportion the bureau seats in Exercise 8.
11. Use Webster's method to apportion the bureau seats in Exercise 8.
12. Use the Hill-Huntington method to apportion the bureau seats in Exercise 8.
13. A school district has three high schools: Leibniz, Maclaurin, and Napier. If the district decides to hire one new teacher, use Hill-Huntington numbers to determine which school gets the new instructor. The current enrollments and faculty of the schools are given in Figure 6.133.

	Leibniz	Maclaurin	Napier
Students	987	1,242	1,763
Teachers	25	31	44

Figure 6.133
Table for Exercise 13

14. A small country consists of three states: A, B, and C. The population of each state is given in Figure 6.134. The country's legislature is to have thirty-one seats.

State	A	B	C
Population	3,500,000	4,100,000	16,800,000

Figure 6.134
Table for Exercise 14

a. Express each state's population (and the total population) in terms of millions.
b. Find the standard divisor, and round it off to two decimal places.
c. Find each state's standard, lower, and upper quotas.
d. Use Jefferson's method to apportion the thirty-one seats.
e. Is the Quota Rule violated? Explain.

15. A small country consists of three states: A, B, and C. The population of each state is given in Figure 6.135. The country's legislature is to have seventy-two seats.

State	A	B	C
Population	1,600,000	3,500,000	15,300,000

Figure 6.135
Table for Exercise 15

a. Express each state's population (and the total population) in terms of millions.
b. Find the standard divisor, and round it off to two decimal places.
c. Find each state's standard, lower, and upper quotas.
d. Use Adams's method to apportion the seventy-two seats.
e. Is the Quota Rule violated? Explain.

16. A small country consists of four states: A, B, C, and D. The population of each state is given in Figure 6.136. The country's legislature is to have 201 seats.

State	A	B	C	D
Population	1,350,000	3,430,000	17,600,000	19,650,000

Figure 6.136
Figure for Exercise 16

a. Express each state's population (and the total population) in terms of millions.
b. Find the standard divisor, and round it off to two decimal places.
c. Find each state's standard, lower, and upper quotas.
d. Use Webster's method to apportion the 201 seats.
e. Is the Quota rule violated? Explain.

17. A small country consists of four states: A, B, C, and D. The population of each state is given in Figure 6.137. The country's legislature is to have 201 seats.

State	A	B	C	D
Population	1,100,000	3,500,000	17,600,000	19,400,000

Figure 6.137
Figure for Exercise 17

   a. Express each state's population (and the total population) in terms of millions.
   b. Find the standard divisor, and round it off to two decimal places.
   c. Find each state's standard, lower, and upper quotas.
   d. Use the Hill-Huntington method to apportion the 201 seats.
   e. Is the Quota Rule violated? Explain.

18. A small country consists of three states: A, B, and C. The population of each state is given in Figure 6.138. The country's legislature is to have 140 seats.

State	A	B	C
Population	700,000	5,790,000	6,520,000

Figure 6.138
Figure for Exercise 18

   a. Express each state's population (and the total population) in terms of thousands.
   b. Find the standard divisor and round it off to two decimal places.
   c. Use Hamilton's method to apportion the 140 legislative seats.
   d. Suppose the total number of seats increases by one. Reapportion the 141 legislative seats.
   e. Is the Alabama Paradox exhibited? Explain.

19. A small country consists of three states: A, B, and C. The population of each state is given in Figure 6.139. The country's legislature is to have 110 seats.
   a. Express each state's population (and the total population) in terms of thousands.
   b. Find the standard divisor and round it off to two decimal places.
   c. Use Hamilton's method to apportion the 110 legislative seats.

State	A	B	C
Population	1,057,000	1,942,000	2,001,000

Figure 6.139
Figure for Exercise 19

   d. Suppose state D has a population of 2,450,000 and joins the union. How many new seats should be added to the legislature?
   e. Apportion the new total number of seats.
   f. Is the New States Paradox exhibited? Explain.

20. A school district has three campuses with enrollments given in Figure 6.140. The district has twelve educational specialists to assign to the locations.

Campus	A	B	C
Enrollment (2005)	760	1,670	3,980
Enrollment (2006)	788	1,858	4,210

Figure 6.140
Figure for Exercise 20

   a. Use Hamilton's method and the 2005 enrollments to apportion the twelve specialists.
   b. Use Hamilton's method and the 2006 enrollments to reapportion the twelve specialists.
   c. Find the growth rate (from 2005 to 2006) for each campus.
   d. Is the Population Paradox exhibited? Explain.

▬▬▬▬▬▶ *Answer the following questions using complete sentences and your own words.*

CONCEPT QUESTIONS

21. What is a majority?
22. What is a ranked ballot?
23. What is a voter preference table?
24. What is a standard divisor?
25. What is a standard quota?
26. What is apportionment?
27. What is the Quota Rule? Explain its meaning.
28. What is the Alabama Paradox? Explain its meaning.
29. What is the New States Paradox? Explain its meaning.
30. What is the Population Paradox? Explain its meaning.
31. What is the Balinski-Young Impossibility Theorem? Explain its meaning.
32. What are the four fairness criteria? Explain their meanings.
33. What is Arrow's Impossibility Theorem?

HISTORY QUESTIONS

34. What methods of apportionment for the House of Representatives have actually been used? What is their chronology?
35. What method of apportionment for the House of Representatives is currently being used?

7 Number Systems and Number Theory

7.1 Place Systems

7.2 Arithmetic in Different Bases

7.3 Prime Numbers and Perfect Numbers

7.4 Fibonacci Numbers and the Golden Ratio

Have you ever wondered why our timekeeping system uses such weird units? Why are there *sixty* minutes in an hour and *sixty* seconds in a minute? Why not break hours and minutes down into tenths and hundredths, as we do in so many other cases? It turns out that we use sixty here because of the ancient Babylonian number system. This number system not only gave us our timekeeping system, it is also a forerunner of our own decimal system.

In this chapter, we will discuss the decimal system, the system that you use every day. Because of their increasing prominence in technological areas, we will discuss some computer number systems—the binary system, the octal system, and the hexadecimal system. And because we still use it today, we will discuss the Babylonian system.

In the second half of this chapter, we will discuss some topics from number theory. **Number theory** is the branch of mathematics involving the study of the properties and relationships of numbers. We will discuss prime numbers and Fibonacci numbers. When you buy something over the Web, prime numbers are used to keep your credit card number secret. The next time you look at a flower, you might find that the number of petals is a Fibonacci number. The flower in the photograph on this page is loaded with Fibonacci numbers. Fibonacci numbers occur frequently in nature, for reasons that we are only beginning to understand.

This chapter's topics are quite varied. The first half of the chapter, on number systems, might seem to be totally unrelated to the second half, on prime numbers and Fibonacci numbers. There is, however, an important connection. Each topic in this chapter focuses on numbers. Of course, numbers are a part of every topic in mathematics. Usually, though, numbers are a tool rather than the principal focus. In this chapter, numbers are the principal focus.

7.1 Place Systems

We use a system to write numbers. For example, we write the number "two hundred and thirty-four" as 234. We are so familiar with this system that it feels natural, and we are not even aware that we are using a system. To understand the system and how it works, we need to talk briefly about how this system came to be. It all started with the abacus.

The Abacus

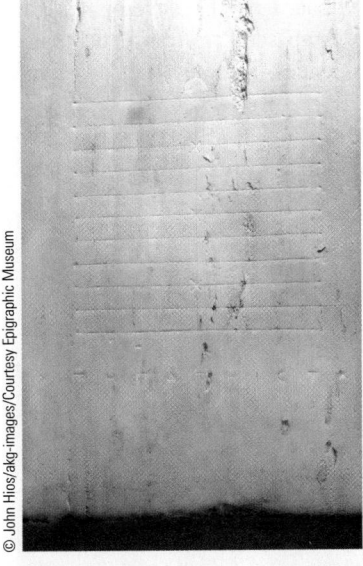

The oldest surviving counting board is made of marble and was used by the Babylonians in about 300 B.C.

The **abacus** was one of the first counting devices. The earliest abacus was probably a flat surface with sand spread evenly across it. It was used as a counting tool. In fact, the word *abacus* points to this early version—it is derived from either the Phoenician word *abak* ("sand") or the Hebrew word *avak* ("dust"). Merchants used abacuses to count how many items they bought or sold and to calculate the cost of those items. This type of abacus is also called a **counting board.**

Later, counting boards did not use sand. Instead, they had lines carved into them. Users placed stones in the areas between the lines. Every time an abacus user got ten stones in the first area, he would put one stone in the new second area and remove ten stones from the first area. And every time he got ten stones on the second area, he would put one stone in the new third area and remove ten stones from the second area.

How much is one stone in the third area worth? It is the same as ten stones in the second area, and each of those is itself worth ten stones in the first area. So one stone in the third area is worth $10 \cdot 10 = 100$ stones in the first area.

Later abacuses had beads and rods instead of stones and lines. The European abacus consists of several rods, each with ten beads, as shown in Figure 7.1. Some European abacuses had five beads per rod, with one or two extra beads for carrying. Many different cultures used abacuses, including the Chinese, Japanese, Phoenicians, and Egyptians. Each culture had either ten beads on each rod or five beads per rod and some carrying beads. It is thought that the abacus represents a refinement of counting on one's fingers and that each wire has ten beads because we have ten fingers.

Each bead on the rightmost wire counts as 1, each bead on the next wire counts as 10, and each bead on the third wire counts as $10^2 = 100$. The number 234 is indicated by raising four beads on the rightmost wire, 3 beads on the next wire, and two beads on the third wire. See Figure 7.1.

The Hindu Number System

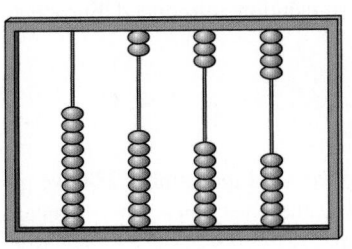

Figure 7.1
The number 234 on an abacus

Hindu mathematicians in India developed the number system that we use today. They were the first people to write "234" to represent the number on the abacus in Figure 7.1.

The Hindu number system is a place system. A **place system** is a system in which the value of a digit is determined by its place. For example, 31 and 13 are different numbers because of the digits' places. Roman numerals do not comprise a place system; V means "five" regardless of where the V is placed. It is probable that the place system aspect of the Hindu number system came from the abacus.

The place system made arithmetic calculations much easier. Imagine what multiplication or long division would be like with a nonplace system such as Roman numerals! In fact, the Hindus invented the multiplication algorithm and the division algorithm that we use today. An **algorithm** is a logical procedure for solving a certain type of problem.

Arabic Mathematics

Mathematics and the sciences entered an extremely long period of stagnation when the Greek civilization fell and was replaced by the Roman Empire. This inactivity was uninterrupted until after the Islamic religion and the resulting Islamic culture were founded by the prophet Muhammad in A.D. 622. Within a century, the Islamic Empire stretched from Spain, Sicily, and Northern Africa to India.

Unlike the Roman Empire, Islamic culture encouraged the development of mathematics and the sciences as well as the arts. Arab scholars translated many Greek and Hindu works in mathematics and the sciences. In doing so, they came upon Hindu numerals.

The Arab mathematician Mohammed ibn Musa al-Khowarizmi wrote two important books around A.D. 830, which were translated into Latin and published in Europe in the twelfth century. Much of the mathematical knowledge of medieval Europe was derived from the Latin translations of al-Khowarizmi's two works.

A calculation competition. One competitor uses an abacus, while the other uses Hindu numerals.

Hindu numerals (top two rows), Arabic numerals (middle row), and early European numerals.

Al-Khowarizmi's first book, on Hindu numerals, was titled *Algorithmi de numero Indorum* (or *al-Khowarizmi on Indian Numbers*). The Latin translation of this book introduced to Europe the Hindu number system. Before this book came out, Europeans used Roman numerals, so a European wrote CCXXXIV instead of 234.

Al-Khowarizmi's book also introduced Europe to the simpler calculation techniques of the Hindu system, including the multiplication algorithm and the division algorithm that we use today. In fact, the title *Algorithmi de numero Indorum* is the origin of the word *algorithm*.

Al-Khowarizmi's second book, *Al-Jabr w'al Muqabalah,* is on algebra rather than Hindu numbers, multiplication, and division. In fact, this book marks the beginning of algebra, and the word *algebra* comes from *Al-Jabr* in its title. The book discusses linear and quadratic equations. Its Latin translation introduced Europe to algebra.

Place Values and Expanded Form

The Hindu number system is a *place* system. When we write the number 234, we put 2 in the hundreds *place,* 3 in the tens *place,* and 4 in the ones *place.* So 234 means 2 hundreds, 3 tens and 4 ones. In symbols, it means

$$234 = 2 \cdot 100 + 3 \cdot 10 + 4 \cdot 1$$

The **place values** are $100 = 10^2$, $10 = 10^1$, and $1 = 10^0$. Using these powers of ten, we get

$$234 = 2 \cdot 10^2 + 3 \cdot 10^1 + 4 \cdot 10^0$$

When a number is written in this way, in terms of its place values, the number is said to be written **expanded form.**

The Decimal System

Recall that in the expression 10^2, 2 is called the exponent and 10 is called the base. The Hindu number system is called a **base ten** system because the system uses powers of 10; that is, it involves exponents with a base of 10.

Our base ten system uses the digits 1, 2, 3, 4, 5, 6, 7, 8, 9, and 0 and nothing else. We create bigger numbers by putting these digits in different places. Notice that there are ten digits in the above list. A base ten system uses ten digits.

The base ten system is also called the **decimal system.** The word *decimal* is derived from the Latin word *decim,* which means "ten."

Base Ten Facts

- It uses ten digits: 1, 2, 3, 4, 5, 6, 7, 8, 9, and 0.
- The place values are powers of ten: $100 = 10^2$, $10 = 10^1$, and $1 = 10^0$.
- It is called the decimal system.

Base Eight

Base ten systems are the most common, because most cultures count on their fingers. But they are not the only systems. Some historic cultures used different bases, and some modern applications use different bases. The Mayans used a base twenty system, perhaps because they counted on their fingers and toes. The Yuki tribe of Native Americans used a base eight system because they counted using the spaces between their fingers rather than the fingers themselves. And some computers use base eight internally, even though they display everything in base ten.

The base eight system is also called the **octal system.** The word *octal* is derived from the Greek word for "eight."

Base eight differs from base ten in several ways. It uses eight digits: 1, 2, 3, 4, 5, 6, 7, and 0. It does not use the digits 8 and 9. Also, the place values are not $10^2 = 100$, $10^1 = 10$, and $10^0 = 1$. Instead, they are $8^2 = 64$, $8^1 = 8$, and $8^0 = 1$. So the columns are sixty-fours, eights, and ones.

When we write 10_8, we mean something totally different than when we write 10. The subscript 8 indicates that the number is in base eight. (When a number is in base ten, no subscript is used.) There is a 1 in the eights column and a 0 in the ones column, so 10_8 means 1 eight and 0 ones—or, using expanded form:

$$10_8 = 1 \cdot 8^1 + 0 \cdot 8^0 = 8$$

We cannot write eight in base eight as "8" because that is not one of the digits that base eight uses.

Similarly, 11_8 has a 1 in the eights column and a 1 in the ones column. So 11_8 means 1 eight and 1 one.

$$11_8 = 1 \cdot 8^1 + 1 \cdot 8^0 = 9$$

We cannot write nine in base eight as "9" because that is not one of the digits that base eight uses.

> **Base Eight Facts**
>
> - It uses eight digits: 1, 2, 3, 4, 5, 6, 7, and 0.
> - The place values are powers of eight: $8^0 = 1$, $8^1 = 8$, and $8^2 = 64$, etc.
> - It is called the octal system.

Reading Numbers in Different Bases

We do not read the number 234_8 as "two hundred thirty four, base 8" because "two hundred" means "$2 \cdot 10^2$" and "thirty" means "$3 \cdot 10^1$." Instead, we read it as "two three four, base eight."

Converting from Base Eight to Base Ten

EXAMPLE 1   **a.** Rewrite 234_8 in expanded form.
b. Use part (a) to convert 234_8 to base ten.

Solution   **a.**

Expanded form for 234_8 is $2 \cdot 8^2 + 3 \cdot 8^1 + 4 \cdot 8^0$.
b. The computation described by the above expanded form gives 156, so $234_8 = 156$.

Converting from Base Ten to Base Eight

EXAMPLE 2   Convert 234 to base eight.

Solution   Some of the place values are $8^0 = 1$, $8^1 = 8$, and $8^2 = 64$. To convert 234 to base eight is to find how many sixty-fours, eights and ones there are in 234. We will start by finding how many sixty-fours there are in 234.

$$\begin{array}{r} 3 \\ 64\overline{)234} \\ -192 \\ \hline 42 \end{array}$$

Rewriting this result with multiplication, we get

$$234 = 3 \cdot 64 + 42$$

This tells us that there are three sixty-fours in 234.

Next, we need to find out how many eights there are in 42.

$$\begin{array}{r} 5 \\ 8\overline{)42} \\ -40 \\ \hline 2 \end{array}$$

Rewriting this result with multiplication, we get

$$42 = 5 \cdot 8 + 2$$

Substituting this into our previous result, we get

$$234 = 3 \cdot 64 + 42 \qquad \text{our previous result}$$
$$234 = 3 \cdot 64 + 5 \cdot 8 + 2 \quad \text{substituting for 42}$$
$$= 352_8 \qquad\qquad\qquad ■$$

Converting to and from Base Ten

To convert a given number (in base b) to base ten:

Use base b's place values to rewrite the given number in expanded form. Then do the computation described by the expanded form.

To convert a given number from base ten to base b:

1. Divide the given number by base b place values, and rewrite the results with multiplication.
2. Use the result of step 1 to rewrite the given number in expanded form, using base b digits.

Base Sixteen

Some computers use base sixteen internally, even though they display everything in base ten. The base sixteen system is also called the **hexadecimal system.** The current hexadecimal system was created by IBM in 1963. IBM decided that the term *sexidecimal,* from the Latin words for "six" and "ten," was too risqué. To avoid it, they welded together *hex,* the Greek word for "six," and *decim,* the Latin word for "ten."

Base sixteen uses sixteen digits, which is more than we base ten users have at our disposal. So we use some letters as digits. The sixteen digits are shown in Figure 7.2.

Base Sixteen Digit	1	2	3	4	5	6	7	8	9	A	B	C	D	E	F	0
Base Ten Equivalent	1	2	3	4	5	6	7	8	9	10	11	12	13	14	15	0

Figure 7.2
The base sixteen numerals

Base Sixteen Facts

• It uses sixteen digits: 1, 2, 3, 4, 5, 6, 7, 8, 9, A, B, C, D, E, F, and 0.
• The place values are powers of sixteen: $16^0 = 1$, $16^1 = 16$, and $16^2 = 256$, etc.
• It is called the hexadecimal system.

Converting Between Base Sixteen and Base Ten

ThomsonNOW for Liberal Arts Mathematics www.thomsonedu.com/1pass.

EXAMPLE 3 Convert 234_{16} to base ten.

Solution We will use base sixteen's place values to rewrite 234 in expanded form, as described in the section "Converting To and From Base Ten" above.

| Place value is 16^2. | Place value is 16^1. | Place value is 16^0. |

$$2 \quad 3 \quad 4_{16}$$

$$234_{16} = 2 \cdot 16^2 + 3 \cdot 16^1 + 4 \cdot 16^0 = 564$$

We have found that the number 234_{16} has 2 two hundred fifty-sixes ($16^2 = 256$), 3 sixteens, and 4 ones. This amounts to the same as the base ten number 564, which has 5 hundreds, 6 tens, and 4 ones. ∎

ThomsonNOW for Liberal Arts Mathematics www.thomsonedu.com/1pass.

EXAMPLE 4 Convert 234 to base sixteen.

Solution We will use the steps given in "Converting To and From Base Ten" above.

1. *Divide the given number by base sixteen place values, and rewrite the results with multiplication.*

   Some of the place values are $16^2 = 256$, $16^1 = 16$, and $16^0 = 1$. The highest of these that will divide into 256 is 16.

   $$\begin{array}{r} 14 \\ 16\overline{)234} \\ -16 \\ \hline 74 \\ -64 \\ \hline 10 \end{array} \qquad 234 = 14 \cdot 16 + 10$$

2. *Use the result of step 1 to rewrite the given number in expanded form, using base sixteen digits.*

   $$234 = 14 \cdot 16 + 10$$
   $$= 14 \cdot 16^1 + 10 \cdot 16^0$$

   We have now rewritten in expanded form, but we have not used base sixteen digits. The base sixteen digit for "14" is "E," and the base sixteen digit for "10" is "A," as shown in Figure 7.2. So, using base sixteen digits, 234 has E sixteens and A ones. Putting the E in the sixteens place and the A in the ones place, we get $234 = EA_{16}$. ∎

The result of Example 4 is rather strange. Remember that in base sixteen, E and A are digits. They are just digits that are not used in base ten.

Base Two

All computers use *machine language,* which is a system of codes that are directly understandable by a computer's central processor. Machine language is composed only of the numerals 0 and 1. So at their deepest level, computers use base two. For certain applications, computers use base eight or base sixteen rather than base two. This is because programmers prefer to program in base eight or sixteen, because these systems are more compact. It takes a lot of space to write a big number in base two. It takes far less space to write the same number in base eight or base sixteen. Base ten is also more compact than base two. However, it is much easier to convert between base two and base eight or sixteen than it is to convert between base two and base ten.

The base two system is also called the **binary system.** The word *bi* is a Latin word that means "two."

Base Two Facts

- It uses two digits: 1 and 0.
- The place values are powers of two: $2^0 = 1$, $2^1 = 2$, $2^2 = 4$, $2^3 = 8$, $2^4 = 16$, $2^5 = 32$, etc.
- It is called the binary system.

Converting Between Base Two and Base Ten

EXAMPLE 5   Convert 234_2 to base ten.

Solution   This cannot be done, because there is no such thing as 234_2. Base two uses only two numerals: 0 and 1. It does not use 2, 3, or 4. ∎

EXAMPLE 6   Convert 10010_2 to base ten.

Solution

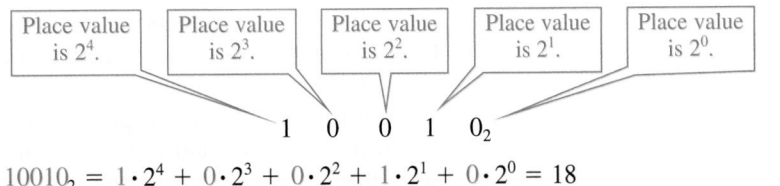

$$10010_2 = 1 \cdot 2^4 + 0 \cdot 2^3 + 0 \cdot 2^2 + 1 \cdot 2^1 + 0 \cdot 2^0 = 18$$ ∎

Example 6 shows that $18 = 10010_2$. It takes a lot of space to write 18 in base two. This bulkiness is one reason computer programmers prefer working in a base other than base two.

ThomsonNOW
for Liberal Arts Mathematics
www.thomsonedu.com/1pass.

EXAMPLE 7   Convert 25 to base two.

Solution

1.   *Divide the given number by base two place values, and rewrite the results with multiplication.*

Some of the place values are $2^5 = 32$, $2^4 = 16$, $2^3 = 8$, $2^2 = 4$, $2^1 = 2$, and 1.

$$\begin{array}{r} 1 \\ 16\overline{)25} \\ -16 \\ \hline 9 \end{array}$$ $25 = 1 \cdot 16 + 9$

$$\begin{array}{r} 1 \\ 8\overline{)9} \\ -8 \\ \hline 1 \end{array}$$ $9 = 1 \cdot 8 + 1$

2. *Use the result of step 1 to rewrite the given number in expanded form, using base two digits.* Combining the above results, we get

$$25 = 1 \cdot 16 + 9 \qquad \text{the first result}$$
$$= 1 \cdot 16 + 1 \cdot 8 + 1 \qquad \text{using the second result}$$
$$= 1 \cdot 16 + 1 \cdot 8 + 0 \cdot 4 + 0 \cdot 2 + 1 \qquad \text{inserting the missing place values}$$
$$= 1 \cdot 2^4 + 1 \cdot 2^3 + 0 \cdot 2^2 + 0 \cdot 2^1 + 1 \cdot 2^0 \qquad \text{This is expanded form.}$$
$$= 11001_2$$

Base Sixty

We have talked about bases 2, 8, and 16. Each of these bases is important in our high-tech world, and you might well have encountered them. So why talk about base sixty? Believe it or not, you have already encountered base sixty, and you use it every day.

The ancient Babylonians counted in base sixty. There are many theories about why they used such a large base, but it is really not known why. One possibility is the fact that a base of sixty makes division and multiplication easier because sixty has so many factors (1, 2, 3, 4, 5, 6, 10, 12, 15, 20, 30, and 60).

The Babylonians used base sixty when they originated the system by which the day is divided into 24 hours, an hour is divided into 60 minutes, and a minute is divided into 60 seconds. They also originated the system by which a degree is divided into 60 minutes and a minute into 60 seconds.

This similarity of timekeeping and angle-measuring systems apparently started with Babylonian and Sumerian astronomers. Because of their perspective from the earth, they observed that the sun and planets seemed to move in relation to the fixed background of stars and constellations in a circular path, now called the *ecliptic*. (Of course, we now know that the earth moves, rather than the sun.) They determined that if we could see the stars during the daytime, the sun would appear to pass through a different constellation about once every thirty days. These twelve constellations are called the *zodiac*. It takes the sun about $12 \cdot 30 = 360$ days to complete its path through the zodiac. They divided this circular path into 360 degrees to track each day's portion of the sun's journey. As a result, we now divide a circle into 360 degrees, and we now define a degree as one-three hundred sixtieth of a circle.

When we talk about 2 hours, 18 minutes, and 43 seconds, we are using the Babylonian base sixty system to measure time. In discussing base sixty, we will not

Courtesy of NASA

A small part of the ecliptic is easily seen in this photo from the 1994 Clementine spacecraft. We see, from right to left, the moon, the sun rising behind the moon, Saturn, Mars, and Mercury. Connect these dots, and you have drawn part of the ecliptic.

use sixty different numerals. If we used all ten of our base ten numerals and all twenty-six letters in the alphabet, we would not have enough symbols. Instead, we will just use our base ten numerals and separate the places with commas. We will write 2 hours, 18 minutes, and 43 seconds in base sixty as

$$2,18,43_{60}$$

EXAMPLE 8   Convert $2,18,43_{60}$ to base ten.

Solution $$2,18,43_{60} = 2 \cdot 60^2 + 18 \cdot 60^1 + 43 \cdot 60^0 = 8323$$

This means that $2,18,43_{60} = 8323$. It also means that 2 hours, 18 minutes, and 43 seconds = 8323 seconds. ∎

Converting Between Computer Bases

When you convert between the computer bases (that is, base two, eight, and sixteen), *do not think about tens,* and *do not convert to base ten.* Doing either of these things makes things much harder. Instead, *think about twos, eights, and sixteens.*

EXAMPLE 9   Convert 562_{16} to base eight.

Solution   Base eight place values are $8^2 = 64$, $8^1 = 8$, and $8^0 = 1$. Base sixteen place values are $16^2 = 256$, $16^1 = 16$, and $16^0 = 1$. To convert between these place values, use the following conversions:

$$16^2 = (2 \cdot 8)^2 = 2^2 \cdot 8^2 = 4 \cdot 8^2$$
$$16^1 = 2 \cdot 8^1$$

We are going to rewrite 562_{16} using its base sixteen place values. Then we are going to use the above conversions to convert base sixteen place values to base eight place values.

$$562_{16} = 5 \cdot 16^2 + 6 \cdot 16^1 + 2 \qquad \text{using base sixteen place values}$$
$$= 5 \cdot 4 \cdot 8^2 + 6 \cdot 2 \cdot 8^1 + 2 \qquad \text{using the conversions}$$
$$= 20 \cdot 8^2 + 12 \cdot 8^1 + 2 \qquad \text{simplifying}$$

The Binary System in the Real World: Computer Imaging

The word *bit* is formed by combining the words *binary* and *digit*. A **bit** is a binary digit. That is, it is either of the digits 0 and 1.

A **bitmapped image** is a computer-generated image that is made up of **pixels,** which are tiny square dots. These pixels are lined up next to each other in a grid, forming a complete image. A computer forms a bitmapped image by associating a number, made up of bits, with each pixel. The number associated with one particular pixel determines the pixel's color.

The simplest of all bitmapped images is a **one-bit image,** in which each pixel is associated with one bit. If a pixel's bit is 0, then the pixel is colored white. If the pixel's bit is 1, then the pixel is black. A one-bit image is made up of blacks and whites, with no grays.

With a **grayscale image,** each pixel is typically associated with eight bits (also called one **byte**). That is, each pixel is associated with a number that has eight places when written in binary. The smallest eight-bit binary number is

$$00000000_2 = 0$$

A pixel with this number is colored white. The largest eight-bit binary number is

$$11111111_2 = 1 \cdot 2^7 + 1 \cdot 2^6 + 1 \cdot 2^5 + 1 \cdot 2^4$$
$$+ 1 \cdot 2^3 + 1 \cdot 2^2 + 1 \cdot 2^1 + 1 \cdot 2^0$$
$$= 255$$

A pixel with this number is colored black. A pixel with a number between 00000000_2 and 11111111_2 is colored a shade of gray. There are 256 numbers between 0 and 255, so each pixel in a grayscale image is colored one of 256 different shades of gray.

With an **RGB image,** each pixel is typically associated with three different one-byte numbers. One of these three numbers determines the amount of red (R) in the pixel, one determines the amount of green (G),

Cyan, magenta, yellow and black—
the CMYK colors.

and one determines the amount of blue (B). Each number has one byte (that is, eight bits), so an RGB image can contain 256 different shades of red, 256 different shades of green, and 256 different shades of blue. Computer monitors and TVs display RGB images. RGB color is also called *three-channel color.*

With a **CMYK image,** each pixel is typically associated with four different one-byte numbers. The four numbers determine the amount of cyan (C), magenta (M), yellow (Y), and black (K) in the pixel. Each number has one byte, so a CMYK image can produce 256 shades of the colors cyan, magenta, yellow, and black. Offset printing and color photography use the CMYK system. CMYK color is called *four-channel color,* and CMYK printing is called *four-color printing.*

Digital photographers and Adobe Photoshop users need to understand *bit depth.* **Bit depth** is the number of bits used to store color information about each pixel. The bit depth of a one-bit image is one. The bit depth of a grayscale image is 8, the bit depth of an RGB image is $3 \cdot 8 = 24$, and the bit depth of a CMYK image is $4 \cdot 8 = 32$.

See Exercise 95.

CMYK

Grayscale

One-bit image

© PhotoDisc/Getty Images

Now the first part of this, $20 \cdot 8^2$, is more than $8 \cdot 8^2 = 8^3$. And 8^3 is one of the base eight place values.

$$\begin{array}{r} 2 \\ 8)\overline{20} \\ -16 \\ \hline 4 \end{array} \qquad 20 = 2 \cdot 8 + 4$$

$$\begin{aligned} 20 \cdot 8^2 &= (2 \cdot 8 + 4) \cdot 8^2 & \text{substituting for 20} \\ &= 2 \cdot 8 \cdot 8^2 + 4 \cdot 8^2 & \text{distributing} \\ &= 2 \cdot 8^3 + 4 \cdot 8^2 & \text{simplifying} \end{aligned}$$

If we combine this with our previous work, we get

$$\begin{aligned} 562_{16} &= 20 \cdot 8^2 + 12 \cdot 8^1 + 2 & \text{our previous work} \\ &= 2 \cdot 8^3 + 4 \cdot 8^2 + 12 \cdot 8^1 + 2 & \text{substituting for } 20 \cdot 8^2 \end{aligned}$$

The only thing that we have left to deal with is the $12 \cdot 8^1$. This is more than $8 \cdot 8^1 = 8^2$. So we will do the same type of thing with it that we did with $20 \cdot 8^2$.

$$\begin{array}{r} 1 \\ 8)\overline{12} \\ -8 \\ \hline 4 \end{array} \qquad 12 = 1 \cdot 8 + 4$$

$$\begin{aligned} 562_{16} &= 2 \cdot 8^3 + 4 \cdot 8^2 + 12 \cdot 8^1 + 2 & \text{our previous result} \\ &= 2 \cdot 8^3 + 4 \cdot 8^2 + (1 \cdot 8^1 + 4) \cdot 8^1 + 2 & \text{substituting} \\ &= 2 \cdot 8^3 + 4 \cdot 8^2 + 1 \cdot 8^1 \cdot 8^1 + 4 \cdot 8^1 + 2 & \text{distributing} \\ &= 2 \cdot 8^3 + 4 \cdot 8^2 + 1 \cdot 8^2 + 4 \cdot 8^1 + 2 & \text{simplifying} \\ &= 2 \cdot 8^3 + 5 \cdot 8^2 + 4 \cdot 8^1 + 2 & \text{This is expanded form.} \\ &= 2542_8 & \blacksquare \end{aligned}$$

EXAMPLE 10 Convert 110101_2 to base eight.

Solution Base eight place values are $8^2 = 64$, $8^1 = 8$, and 1. Base two place values are $2^5 = 32$, $2^4 = 16$, $2^3 = 8$, and $2^2 = 4$, $2^1 = 2$, and 1. To convert between these place values, use the following conversions:

$$\begin{aligned} 2^5 &= 2^2 \cdot 2^3 = 4 \cdot 8^1 \\ 2^4 &= 2^1 \cdot 2^3 = 2 \cdot 8^1 \\ 2^3 &= 8 = 1 \cdot 8^1 \end{aligned}$$

We are going to rewrite 110101_2 using its base two place values. Then we are going to use the above conversions to convert base two place values to base eight place values.

$$\begin{aligned} 110101_2 &= 1 \cdot 2^5 + 1 \cdot 2^4 + 0 \cdot 2^3 + 1 \cdot 2^2 + 0 \cdot 2^1 + 1 \\ &= 1 \cdot 4 \cdot 8^1 + 1 \cdot 2 \cdot 8^1 + 1 \cdot 4 + 1 & \text{using the conversions} \\ &= 4 \cdot 8^1 + 2 \cdot 8^1 + 4 + 1 & \text{simplifying} \\ &= 6 \cdot 8^1 + 5 & \text{This is expanded form.} \\ &= 65_8 & \blacksquare \end{aligned}$$

Converting Between Computer Bases

1. Rewrite the given number in expanded form.
2. Use the conversions given below to convert the given base place values to the new base place values.
3. Use the expanded form that results from Step 2 to rewrite the number in the new base.

Converting from base sixteen to base eight:

- $16^3 = (2 \cdot 8)^3 = 2^3 \cdot 8^3 = 8 \cdot 8^3 = 1 \cdot 8^4$
- $16^2 = (2 \cdot 8)^2 = 2^2 \cdot 8^2 = 4 \cdot 64 = 4 \cdot 8^2$
- $16^1 = 2 \cdot 8^1$

Converting from base eight to base sixteen:

- $8^2 = 8 \cdot 8 = 2 \cdot 2 \cdot 2 \cdot 8 = 4 \cdot 16$
- $8^3 = 8 \cdot 8 \cdot 8 = 2 \cdot 2 \cdot 2 \cdot 8 \cdot 8 = 2 \cdot 2 \cdot 8 \cdot 2 \cdot 8 = 2 \cdot 16 \cdot 16 = 2 \cdot 16^2$

Converting from base two to base eight:

- $2^5 = 2^2 \cdot 2^3 = 4 \cdot 8^1$
- $2^4 = 2^1 \cdot 2^3 = 2 \cdot 8^1$
- $2^3 = 8 = 1 \cdot 8^1$

Converting from base eight to base two:

- $8 = 2^3$
- $8^2 = (2^3)^2 = 2^6$
- $8^3 = (2^3)^3 = 2^9$

Conversions for converting between bases two and sixteen will be developed in Exercises 65 and 66.

7.1 Exercises

In Exercises 1–20, write the given number in expanded form, or explain why there is no such number.

1. 891
2. 458
3. 3,258
4. 86,420
5. 372_8
6. 543_8
7. 3592_{16}
8. 8403_{16}
9. $ABCDE0_{16}$
10. $FA3B02_{16}$
11. 1011001_2
12. 11001110_2
13. 1324_2
14. 2103_2
15. $5,32,85_{60}$
16. $27,19,01_{60}$
17. 4312_5
18. 643_7
19. 123_4
20. 8034_9

In Exercises 21–48, convert the given number to base ten.

21. 372_8 (See Exercise 5)
22. 543_8 (See Exercise 6)
23. 3592_{16} (See Exercise 7)
24. 8403_{16} (See Exercise 8)
25. $ABCDE0_{16}$ (See Exercise 9)
26. $FA3B02_{16}$ (See Exercise 10)
27. 1011001_2 (See Exercise 11)
28. 11001110_2 (See Exercise 12)
29. $5,32,85_{60}$ (See Exercise 15)
30. $27,19,01_{60}$ (See Exercise 16)
31. 4312_5 (See Exercise 17)
32. 643_7 (See Exercise 18)
33. 123_4 (See Exercise 19)
34. 8034_9 (See Exercise 20)
35. 1705_8
36. 2264_8
37. 11011011_2
38. 1011001_2
39. $55,28,33,59_{60}$
40. $43,56,19,20_{60}$
41. 5798_{16}
42. $AB839_{16}$
43. 7224_8
44. 11011101_2
45. $23,44,14_{50}$
46. $340GD1_{17}$
47. 253_7
48. 2412_5

49. Convert 452_8 to base two.

50. Convert 1502_8 to base two.
51. Convert 5260_8 to base two.
52. Convert 31570_8 to base two.
53. Convert 1010_2 to base eight.
54. Convert 11011_2 to base eight.
55. Convert 10110110_2 to base eight.
56. Convert 101001_2 to base eight.
57. Convert 42_8 to base sixteen.
58. Convert 352_8 to base sixteen.
59. Convert 540_8 to base sixteen.
60. Convert 3670_8 to base sixteen.
61. Convert $AB8_{16}$ to base eight.
62. Convert $10C_{16}$ to base eight.
63. Convert $801D_{16}$ to base eight.
64. Convert $986B_{16}$ to base eight.

65. In this exercise, you will develop conversions from base sixteen to base two, using the fact that $2^4 = 16$.
    a. Rewrite $16^3 = (2^4)^3$ as 2 to a power.
    b. Rewrite 16^2 as 2 to a power.
    c. Rewrite 16^1 as 2 to a power.
66. In this exercise, you will develop conversions from base two to base sixteen, using the fact that $2^4 = 16$.
    a. Rewrite $2^5 = 2 \cdot 2^4$ as a number times 16 to a power.
    b. Rewrite $2^6 = 2^2 \cdot 2^4$ as a number times 16 to a power.
    c. Rewrite 2^7 as a number times 16 to a power.
    d. Rewrite $2^8 = (2^4)^2$ as a number times 16 to a power.

In Exercises 67–74, use the conversions developed in Exercises 65–66.

67. Convert $53A2_{16}$ to base two.
68. Convert $FB0_{16}$ to base two.
69. Convert $3AB0_{16}$ to base two.
70. Convert $30AB_{16}$ to base two.
71. Convert 10100_2 to base sixteen.
72. Convert 110011_2 to base sixteen.
73. Convert 10111010_2 to base sixteen.
74. Convert 10100_2 to base sixteen.

75. Convert $22,12_{60}$ to base ten.
76. Convert $57,02_{60}$ to base ten.
77. Convert $33,14,25_{60}$ to base ten.
78. Convert $52,58,12_{60}$ to base ten.
79. Convert 364 to base sixty.
80. Convert 9481 to base sixty.
81. Convert 129,845 to base sixty.
82. Convert 5,383,221 to base sixty.
83. Convert 534_9 to base seven.

84. Convert 512_6 to base four.
85. Convert 331_4 to base five.
86. Convert 58_9 to base eleven.

87. a. If $121_x = 16$, find base x.
    b. If $121_x = 36$, find base x.
88. a. If $243_x = 129$, find base x.
    b. If $243_x = 73$, find base x.
89. a. If $370_x = 248$, find base x.
    b. If $370_x = 306$, find base x.
90. a. If $504_x = 409$, find base x.
    b. If $504_x = 184$, find base x.

91. a. What digits does base five use?
    b. What are the first four place values in base five?
    c. Write the first twenty-five whole numbers in base five.
92. a. What digits does base four use?
    b. What are the first four place values in base four?
    c. Write the first twenty-five whole numbers in base four.
93. a. What digits does base eleven use?
    b. What are the first four place values in base eleven?
    c. Write the first twenty-five whole numbers in base eleven.
94. a. What digits does base twelve use?
    b. What are the first four place values in base twelve?
    c. Write the first twenty-five whole numbers in base twelve.

95. With 8-bit color, each of the basic colors (red, green, and blue in the RGB system; cyan, magenta, yellow, and black in the CMYK system) has 256 different shades. How many different shades are there with 16-bit color? How many different shades are there with 24-bit color? Justify your answers.

▬▬▬ *Answer the following using complete sentences and your own words.*

CONCEPT QUESTIONS
96. Why is there no base one?
97. Throughout history, most cultures have used either base ten or base twenty. Why are those bases the most popular ones?
98. Why did the digit 0 have to be invented before a written place system could be invented?
99. Describe how the number 531 would appear on a base seven abacus.

100. The words *calculate* and *calculator* are derived from the word *calculus,* which means "pebble" or "stone." Why?

101. What do place systems have in common with the abacus?

HISTORY QUESTIONS

102. What did the Hindus do to further the decimal system?

103. What did the Arabs do to further the decimal system?

104. What was the subject matter of Mohammed ibn Musa al-Khowarizmi's two books?

105. What two English words come from the titles of Mohammed ibn Musa al-Khowarizmi's two books?

 WEB PROJECTS

106. Write an essay on one of the following topics:
- History of the abacus
- Calculations using the abacus
- The abacus versus the calculator
- The Roman abacus
- The Chinese suan pan
- The Mesoamerican abacus
- The Mayan base 20

Some useful links for this web project are listed on the text web site: **www.thomsonedu.com/mathematics/johnson**

7.2 Arithmetic in Different Bases

Arithmetic in different bases works just like arithmetic in base ten. The algorithms that the Hindus invented are the same in base eight or base two as they are in base ten. What is different is how you write numbers. For example, we write "eight" as

- 8 in base ten
- 10_8 in base eight
- 1000_2 in base two

Addition

Base eight uses the digits 0 through 7. Because of this, whole numbers between zero and seven are written the same way in base eight as they are in base ten. For example, $7_8 = 7$. But numbers larger than 7_8 are written with multiple digits, using the place system. For example, $8 = 10_8$.

Compare the following two additions:

- $3 + 4 = 7$ Adding in base ten
- $3_8 + 4_8 = 7_8$ Adding in base eight

Everything is the same, except for the base. We are adding the same two numbers, we get the same answer, and we write the answers the same way (except for the subscript indicating base eight).

Compare another pair of additions:

- $4 + 4 = 8$ Adding in base ten
- $4_8 + 4_8 = 10_8$ Adding in base eight

The answers have the same meaning, but they are written differently. This happens whenever the sum is greater than or equal to eight, the base.

When we add numbers with multiple digits, such as $27 + 36$, we use the Hindu algorithm and add column by column. That is, we do not really add 27 and 36; instead, we add 7 and 6. Then we add 2 and 3 (and anything carried from the right column).

When we add in other bases, we also add column by column. But when a column's sum is greater than or equal to the base, we convert the sum from base ten to that base. This happened when we added $4_8 + 4_8$ above.

EXAMPLE 1   **a.** Compute:

$$27_8$$
$$+ \ 36_8$$

b. Check your answer by converting both the problem and the answer to base ten.

Solution   **a.** As always, we start by adding the numbers in the right column. The column's sum is $7_8 + 6_8 = 13$, and $13 \geq 8$, the base. So we convert 13 from base ten to base eight.

$$7_8 + 6_8 = 13$$
$$= 1 \cdot 8^1 + 5 \cdot 8^0 = 15_8$$

As always, we put the 5 in the right column, and we carry the 1 to the second column.

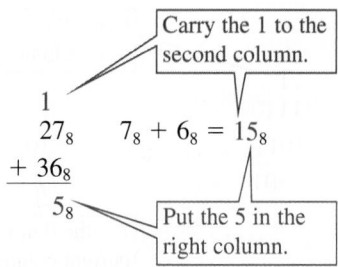

Now we add the digits in the second column.

$$1_8 + 2_8 + 3_8 = 6_8$$

The column's sum is 6, and $6 < 8$, the base. There is no need to convert 6 to base 8, because $6 = 6_8$. Instead, we put the 6 in the second column, and we are done.

$$\begin{array}{r} 1 \\ 27_8 \\ + \ 36_8 \\ \hline 65_8 \end{array}$$

The sum is $27_8 + 36_8 = 65_8$.

b. We will check this answer by converting both the problem and the answer to base ten.

$$27_8 = 2 \cdot 8^1 + 7 \cdot 8^0 = 23$$
$$36_8 = 3 \cdot 8^1 + 6 \cdot 8^0 = 30$$
$$65_8 = 6 \cdot 8^1 + 5 \cdot 8^0 = 53$$

Our work checks because $23 + 30 = 53$.   ∎

EXAMPLE 2   **a.** Compute:

$$11110_2$$
$$+ \ 1011_2$$

b. Check your answer by converting both the problem and the answer to base ten.

Solution **a.** Start by adding the numbers in the right column. We get $0_2 + 1_2 = 1_2$. The column's sum is 1, and $1 < 2$, the base. There is no need to convert 1 to base 2, because $1 = 1_2$. We put 1 in the right column and move on to the second column.

$$\begin{array}{r} 11110_2 \\ + \ 1011_2 \\ \hline 1_2 \end{array}$$

In the second column, $1_2 + 1_2 = 2$. The column's sum is 2, and $2 \geq 2$, the base. So we convert 2 from base ten to base two.

$$1_2 + 1_2 = 2$$
$$= 1 \cdot 2^1 + 0 \cdot 2^0 = 10_2$$

As always, we put the 0 in the current column, and we carry the 1 to the next column.

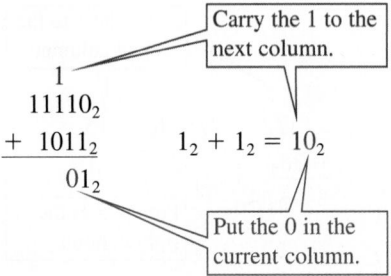

Now we add the digits in the third column.

$$1_2 + 1_2 + 0_2 = 2 = 10_2$$

Put the 0 in the current column, and carry the 1 to the next column.

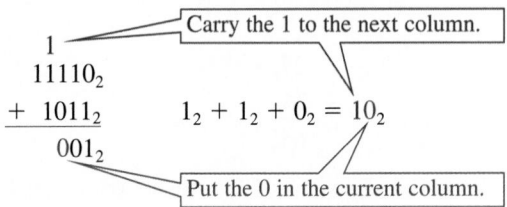

Now we add the digits in the fourth column.

$$1_2 + 1_2 + 1_2 = 3$$
$$= 1 \cdot 2^1 + 1 \cdot 2^0 = 11_2$$

Put the 1 in the current column, and carry the 1 to the next column.

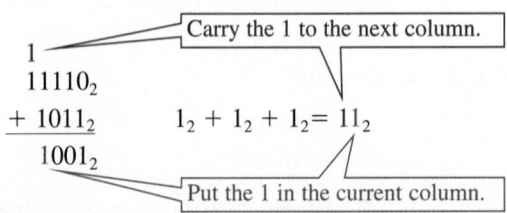

Now we add the digits in the fifth column, and we are done.

$$1_2 + 1_2 = 2$$
$$= 1 \cdot 2^1 + 0 \cdot 2^0 = 10_2$$

$$
\begin{array}{r}
1 \\
11110_2 \\
+\ 1011_2 \\
\hline
101001_2
\end{array}
$$

The sum is 101001_2.

b. $11110_2 = 1 \cdot 2^4 + 1 \cdot 2^3 + 1 \cdot 2^2 + 1 \cdot 2^1 + 0 \cdot 2^0 = 30$
$1011_2 = 1 \cdot 2^3 + 0 \cdot 2^2 + 1 \cdot 2^1 + 1 \cdot 2^0 = 11$
$101001_2 = 1 \cdot 2^5 + 0 \cdot 2^4 + 1 \cdot 2^3 + 0 \cdot 2^2 + 0 \cdot 2^1 + 1 \cdot 2^0 = 41$

Our work checks because $30 + 11 = 41$.   ■

Subtraction

EXAMPLE 3   **a.** Compute:

$$
\begin{array}{r}
41_8 \\
-27_8
\end{array}
$$

b. Check your answer by converting both the problem and the answer to base ten.

Solution   **a.** In the right column, we cannot subtract $1_8 - 7_8$ without borrowing 1 from the 4 in the second column. We are working in base eight, so we borrow one group of 8. Combining the borrowed group of 8 with the 1 in the second column, we get $8 + 1 = 9$. We can now subtract in the right column.

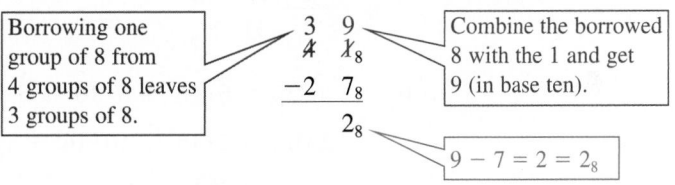

Borrowing one group of 8 from 4 groups of 8 leaves 3 groups of 8.

Combine the borrowed 8 with the 1 and get 9 (in base ten).

$9 - 7 = 2 = 2_8$

In the second column, $3_8 - 2_8 = 1_8$.

$$
\begin{array}{r}
3\ \ 9 \\
\not{4}\ \ \not{1}_8 \\
-2\ \ 7_8 \\
\hline
1\ \ 2_8
\end{array}
$$

The difference is 12_8.

b. $41_8 = 4 \cdot 8^1 + 1 \cdot 8^0 = 33$
$27_8 = 2 \cdot 8^1 + 7 \cdot 8^0 = 23$
$12_8 = 1 \cdot 8^1 + 2 \cdot 8^0 = 10$

Our work checks because $33 - 23 = 10$.   ■

EXAMPLE 4   **a.** Compute:

$$
\begin{array}{r}
683_{16} \\
-2AB_{16}
\end{array}
$$

b. Check your answer by converting both the problem and the answer to base ten.

Solution **a.** In the right column, we cannot subtract $3_{16} - B_{16}$ without borrowing 1 from the 8 in the second column. (Remember, $B_{16} = 11$.) We are working in base sixteen, so we borrow one group of 16. Combining the borrowed 16 with the 3 in the right column, we get $16 + 3 = 19$. We can now subtract $B_{16} = 11$ from 19 in the right column.

In the second column, we cannot subtract $7_{16} - A_{16}$ without borrowing. (Remember, $A_{16} = 10$.)

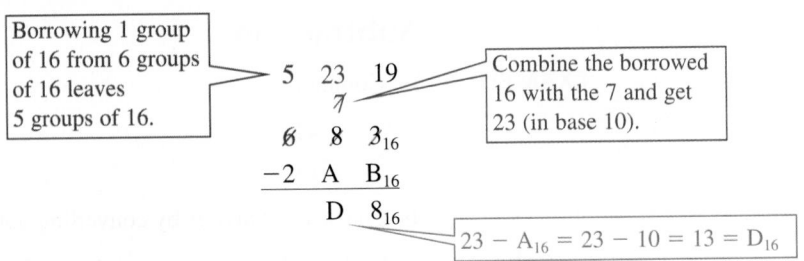

In the third column, $5_{16} - 2_{16} = 3_{16}$

$$
\begin{array}{ccc}
5 & 23 & 19 \\
\cancel{6} & \cancel{8} & \cancel{3}_{16} \\
-2 & A & B_{16} \\
\hline
3 & D & 8_{16}
\end{array}
$$

The difference is $3D8_{16}$.

b. $683_{16} = 6 \cdot 16^2 + 8 \cdot 16^1 + 3 \cdot 16^0 = 1667$

$2AB_{16} = 2 \cdot 16^2 + 10 \cdot 16^1 + 11 \cdot 16^0$ Remember, $A_{16} = 10$ and $B_{16} = 11$.

$\qquad = 683$

$3D8_{16} = 3 \cdot 16^2 + 13 \cdot 16^1 + 8 \cdot 16^0$ Remember, $D_{16} = 13$ and $B_{16} = 11$.

$\qquad = 984$

Our work checks because $1667 - 683 = 984$.   ■

Multiplication

ThomsonNOW™
for Liberal Arts Mathematics
www.thomsonedu.com/1pass.

EXAMPLE 5   **a.** Compute:

$$
\begin{array}{r}
32_8 \\
\times\ 5_8 \\
\hline
\end{array}
$$

b. Check your answer by converting both the problem and the answer to base ten.

Solution   **a.** We start in the right column and multiply the 5 by the 2 above it.

$$5_8 \cdot 2_8 = 5_{10} \cdot 2_{10} = 10_{10}$$
$$= 1 \cdot 8^1 + 2 \cdot 8^0 = 12_8$$

Write the 2 in the right column and carry the 1 to the next column.

Next, we multiply the 5 by the 3 in the second column, and we add the carried 1.

$$(5_8 \cdot 3_8) + 1_8 = (5_{10} \cdot 3_{10}) + 1_{10} = 16_{10}$$
$$= 2 \cdot 8^1 + 0 \cdot 8^0 = 20_8$$

$$
\begin{array}{r}
1 \\
32_8 \\
\times\ 5_8 \\
\hline
202_8
\end{array}
$$

The answer is 202_8.

b. $32_8 = 3 \cdot 8^1 + 2 = 26$

$5_8 = 5$

$202_8 = 2 \cdot 8^2 + 0 \cdot 8^1 + 2 = 130$

Our work checks because $26 \times 5 = 130$.   ∎

EXAMPLE 6   **a.** Compute:

$$
\begin{array}{r}
32_8 \\
\times\ 45_8
\end{array}
$$

b. Check your answer by converting both the problem and the answer to base ten.

Solution   **a.** In Example 5, we did half of this problem—the half involving multiplication by 5. What remains is to do the half involving multiplication by 4.

$$
\left.
\begin{array}{r}
32_8 \\
\times\ 45_8 \\
\hline
202_8
\end{array}
\right\} \quad \text{from Example 5}
$$

Applying the 4 to the right column, we have

$$4_8 \cdot 2_8 = 4_{10} \cdot 2_{10} = 8_{10}$$
$$= 1 \cdot 8^1 + 0 \cdot 8^0 = 10_8$$

We carry the 1, and we write the 0 under the 202_8 that we computed in Example 5, but one column to the left.

In the next column, we have

$$(4_8 \cdot 3_8) + 1_8 = (4_{10} \cdot 3_{10}) + 1_{10} = 13_{10}$$
$$= 1 \cdot 8^1 + 5 \cdot 8^0 = 15_8$$

We write this next to the 0, at the bottom line. Then we add.

$$32_8$$
$$\underline{\times\ 45_8}$$
$$202_8$$
$$\underline{150_8}$$
$$1702_8$$

b. $32_8 = 3 \cdot 8^1 + 2 \cdot 8^0 = 26$

$45_8 = 4 \cdot 8^1 + 5 \cdot 8^0 = 37$

$1702_8 = 1 \cdot 8^3 + 7 \cdot 8^2 + 0 \cdot 8^1 + 2 \cdot 8^0 = 962$

Our work checks because $26 \cdot 37 = 962$.   ∎

Division

EXAMPLE 7   **a.** Compute:

$$3_8 \overline{)237_8}$$

b. Check your answer by converting both the problem and the answer to base ten.

Solution   **a.** Start by dividing 3_8 into 23_8, just as we would if we were in base ten. Asking how many times 3_8 goes into 23_8 is the same as asking what we can multiply by 3_8 to get 23_8. To answer this question, we will create a base eight multiplication table for a multiplier of 3_8, as shown in Figure 7.3. The table's entries are found by using the multiplication process illustrated in Examples 5 and 6.

$3_8 \cdot$	1_8	2_8	3_8	4_8	5_8	6_8	7_8
$=$	3_8	6_8	11_8	14_8	17_8	22_8	25_8

Figure 7.3
A base eight multiplication table for a multiplier of 3_8

Figure 7.3 shows that $3_8 \cdot 6_8 = 22_8$ (which is a bit too small) and $3_8 \cdot 7_8 = 25_8$ (which is a bit too big). We use the "bit too small" answer, as we would if we were in base ten.

$$3_8 \overline{)237_8}^{\,6}$$

Use Figure 7.3 to multiply $6_8 \cdot 3_8 = 22_8$, and put the result underneath. Subtract and bring down the 7_8.

$$3_8 \overline{)237_8}^{\,6}$$

$23 - 22 = 1$ \\ -22 / Bring down the 7.
17

Now use Figure 7.3 to divide 3_8 into 17_8 and get 5_8.

$$3_8 \overline{)237_8}^{\,65}$$
$$\underline{-22}$$
$$17$$

Use Figure 7.3 to multiply $5_8 \cdot 3_8$, get 17_8, put the result underneath, and subtract.

$$
\begin{array}{r}
65_8 \\
3_8)\overline{237_8} \\
-22 \\
\hline
17 \\
-17 \\
\hline
0
\end{array}
$$

There is no remainder, so we are done. The quotient is 65_8.

b. $237_8 = 2 \cdot 8^2 + 3 \cdot 8^1 + 7 \cdot 8^0 = 159$

$3_8 = 3$

$65_8 = 6 \cdot 8^1 + 5 \cdot 8^0 = 53$

Our work checks because $159 \div 3 = 53$.   ∎

EXAMPLE 8   **a.** Compute: $3671_8 \div 12_8$, working solely in base eight.
b. Check your answer by multiplying, in base eight.

Solution   **a.** We are dividing by 12_8, so we need a multiplication table for a multiplier of 12_8, as shown in Figure 7.4. The table's entries are found by using the multiplication process illustrated in Examples 5 and 6.

$12_8 \cdot$	1_8	2_8	3_8	4_8	5_8	6_8	7_8
=	12_8	24_8	36_8	50_8	62_8	74_8	106_8

Figure 7.4
A base eight multiplication table for a multiplier of 12_8

Use Figure 7.4 to divide 12_8 into 36_8 and get 3_8.

$$
\begin{array}{r}
3 \\
12_8)\overline{3671_8}
\end{array}
$$

Use Figure 7.4 to multiply $3_8 \cdot 12_8 = 36_8$, and put the result underneath. Subtract and bring down the 7_8.

$$
\begin{array}{r}
3 \\
12_8)\overline{3671_8} \\
-36 \\
\hline
07
\end{array}
$$

$36 - 36 = 0$ Bring down the 7.

Now divide 12_8 into 7_8, get 0, and bring down the 1.

$$
\begin{array}{r}
30 \\
12_8)\overline{3671_8} \\
-36 \\
\hline
071
\end{array}
$$

Bring down the 1.

Use Figure 7.4 to divide 12_8 into 71_8 and get 5_8.

$$
\begin{array}{r}
305 \\
12_8)\overline{3671_8}
\end{array}
$$

Use Figure 7.4 to multiply $5_8 \cdot 12_8 = 62_8$, put the result underneath, and subtract.

$$
\begin{array}{r}
305_8 \\
12_8 \overline{)\,3671_8} \\
-36 \\
\hline
071 \\
-62 \\
\hline
7_8
\end{array}
$$

$71_8 - 62_8 = 7_8$

We are done, because 12_8 will not divide into 7_8. The quotient is 305_8, with a remainder of 7_8.

b. We will not check the answer by converting both the problem and the answer to base ten, as we have in the past, because the remainder means that we would have to work with fractions in base eight. It is less work to check by multiplying.

$$
\begin{array}{r}
305_8 \\
\times\ 12_8 \\
\hline
612 \\
305 \\
\hline
3662_8
\end{array}
$$

Add the remainder of 7_8 to this and get $3662_8 + 7_8 = 3671_8$. Our answer to part (a) checks.   ∎

7.2   Exercises

In Exercises 1–12, perform the given computation, working solely in the given base.

1. $3_8 + 7_8$
2. $5_8 + 6_8$
3. $4_8 + 5_8$
4. $7_8 + 7_8$
5. $9_{16} + 7_{16}$
6. $A_{16} + 2_{16}$
7. $B_{16} + 9_{16}$
8. $D_{16} + 6_{16}$
9. $1_2 + 1_2$
10. $10_2 + 11_2$
11. $11_2 + 11_2$
12. $101_2 + 10_2$

In Exercises 13–24, do the following.

a. *Perform the given computation, working solely in the given base.*
b. *Check your answer by converting both the problem and the answer to base ten. (Answers are not given in the back of the book.)*

13. $13_8 + 72_8$
14. $54_8 + 63_8$
15. $475_8 + 254_8$
16. $765_8 + 117_8$
17. $A9_{16} + 7B_{16}$
18. $A01_{16} + 236_{16}$
19. $BBC_{16} + CCD_{16}$
20. $1094_{16} + 2399_{16}$
21. $110_2 + 101_2$
22. $10110_2 + 11001_2$
23. $1100110_2 + 1101101_2$
24. $10110_2 + 11111_2$

In Exercises 25–30, perform the given computation, working solely in the given base.

25. $13_8 - 7_8$
26. $51_8 - 16_8$
27. $194_{16} - 52_{16}$
28. $A5_{16} - 61_{16}$
29. $1101_2 - 1010_2$
30. $10110_2 - 1010_2$

In Exercises 31–52, do the following.

a. *Perform the given computation, working solely in the given base.*
b. *Check your answer by converting both the problem and the answer to base ten. (Answers are not given in the back of the book.)*

31. $15_8 - 6_8$
32. $55_8 - 26_8$
33. $1B3_{16} - 105_{16}$
34. $A4_{16} - 31_{16}$
35. $11011_2 - 10110_2$
36. $101110_2 - 10010_2$
37. $341_5 - 243_5$
38. $4632_7 - 326_7$
39. $25_8 \cdot 5_8$
40. $53_8 \cdot 16_8$
41. $1BA_{16} \cdot 15_{16}$
42. $A41_{16} \cdot 11_{16}$
43. $1101_2 \cdot 10110_2$
44. $10110_2 \cdot 1010_2$
45. $31_5 \cdot 23_5$
46. $462_7 \cdot 26_7$
47. $46_8 \div 2_8$
48. $54_8 \div 4_8$

49. $14D_{16} \div 3_{16}$ **50.** $A5_{16} \div 5_{16}$

51. $144_8 \div 31_8$ **52.** $234_8 \div 14_8$

In Exercises 53–64, do the following.

a. *Perform the given computation, working solely in the given base.*

b. *Check your answer by multiplying, in the given base. (Answers are not given in the back of the book.)*

53. $153_8 \div 15_8$ **54.** $227_8 \div 25_8$

55. $277_8 \div 152_8$ **56.** $7541_8 \div 370_8$

57. $79AB_{16} \div 38_{16}$ **58.** $A104_{16} \div 21_{16}$

59. $790B_{16} \div 218_{16}$ **60.** $A14B_{16} \div 2B1_{16}$

61. $1100111_2 \div 110_2$ **62.** $1010110_2 \div 1010_2$

63. $1101111_2 \div 1100_2$ **64.** $1010010_2 \div 10010_2$

65. Create a base five multiplication table, with entries ranging from $1_5 \cdot 1_5$ through $11_5 \cdot 11_5$.

66. Create a base six multiplication table, with entries ranging from $1_6 \cdot 1_6$ through $11_6 \cdot 11_6$.

67. Create a base fourteen multiplication table, with entries ranging from $1_{14} \cdot 1_{14}$ through $11_{14} \cdot 11_{14}$.

68. Create a base thirteen multiplication table, with entries ranging from $1_{13} \cdot 1_{13}$ through $11_{13} \cdot 11_{13}$.

69. Use the result of Exercise 65 to compute $4230_5 \div 23_5$.

70. Use the result of Exercise 66 to compute $5312_6 \div 35_6$.

71. Use the result of Exercise 67 to compute $BADC_{14} \div 102_{14}$.

72. Use the result of Exercise 68 to compute $C9AB_{13} \div B3_{13}$.

— *Answer the following using complete sentences and your own words.*

CONCEPT QUESTION

73. Describe in detail how to add $38 + 99$ on a base ten abacus.

74. Describe in detail a base eight abacus and how to add $27_8 + 54_8$ on a base eight abacus.

75. Describe in detail how to subtract $251 - 89$ on a base ten abacus.

76. Describe in detail how to subtract $332_8 - 67_8$ on a base eight abacus.

7.3   Prime Numbers and Perfect Numbers

In this section, we discuss some interesting patterns that occur within the counting numbers. **Counting numbers** are the numbers that you count with: 1, 2, 3, 4, ...

Factorizations and Primes

To **factor** a counting number is to rewrite the number as a product of counting numbers. For example, the number 40 factors into $4 \cdot 10$. We say that 4 and 10 are **factors** of 40.

But $40 = 4 \cdot 10$ is not factored completely, because 4 and 10 can each be factored further.

$$40 = 4 \cdot 10$$
$$= 2 \cdot 2 \cdot 2 \cdot 5$$

This is **factored completely,** because you cannot factor it further.

Numbers like 2 and 5 that cannot be factored are called *primes*. Of course, $2 = 1 \cdot 2$, and $5 = 1 \cdot 5$. Any number can be written as the product of 1 and itself. We do not include these 1s when we factor completely.

A **prime number** is a counting number that has exactly two factors: 1 and itself. A complete factorization, such as $40 = 2 \cdot 2 \cdot 2 \cdot 5$, is also called a **prime factorization,** because it is a factorization into primes. A counting number that has factors other than 1 and itself is called a **composite number.**

In Figure 7.5, we list the numbers 2 through 20 and their prime factorizations. Prime numbers do not have prime factorizations, so when there is no prime factorization listed, the number is prime.

Number	Prime Factorization	Prime or Composite?
2		Prime
3		Prime
4	$2 \cdot 2$	Composite
5		Prime
6	$2 \cdot 3$	Composite
7		Prime
8	$2 \cdot 2 \cdot 2$	Composite
9	$3 \cdot 3$	Composite
10	$2 \cdot 5$	Composite
11		Prime
12	$2 \cdot 2 \cdot 3$	Composite
13		Prime
14	$2 \cdot 7$	Composite
15	$3 \cdot 5$	Composite
16	$2 \cdot 2 \cdot 2 \cdot 2$	Composite
18	$2 \cdot 3 \cdot 3$	Composite
19		Prime
20	$2 \cdot 2 \cdot 2 \cdot 5$	Composite

Figure 7.5
Prime factorizations and primes

Is 1 Prime or Composite?

Notice that the number 1 is not listed in Figure 7.5. Is 1 a prime number or a composite number? To see whether 1 is prime, we must determine whether 1 has exactly two factors: 1 and itself. The only way to factor 1 is to write $1 = 1 \cdot 1$. One does not have two factors; it has only one factor. So according to the definition, 1 is not a prime number.

Furthermore, 1 is not composite, because it does not have factors other than 1 and itself.

It can be hard to understand why 1 is not prime, because you cannot break 1 down into prime factors the way you can with $40 = 2 \cdot 2 \cdot 2 \cdot 5$. But the whole idea of factoring does not apply to 1. How do we factor 1? The best we can do is

$$1 = 1 \cdot 1 = 1 \cdot 1 \cdot 1 = 1 \cdot 1 \cdot 1 \cdot 1 = \cdots$$

The whole distinction of prime versus composite does not apply to 1.

Prime Numbers in the Real World: Periodic Cicadas

Cicadas live underground and suck sap from tree roots. They are common in the United States. Annual cicadas reach maturity after living about seven years underground. Each year, some annual cicadas in a particular area will reach maturity. These mature cicadas come out from under the ground to mate. If you are around when this happens, you might notice a few flying around.

Periodical cicadas do it differently. These insects reach maturity after seventeen years. Unlike annual cicadas, all periodical cicadas mature in the same year. So once every seventeen years, an onslaught of periodical cicadas emerges from under the ground. They

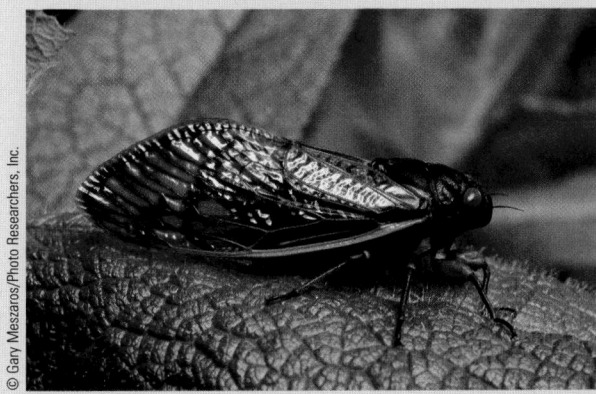

© Gary Meszaros/Photo Researchers, Inc.

A cicada

cover sidewalks, cars, tree branches, and houses in Washington, D.C.; Baltimore; Long Island; Detroit; and other parts of the Northeast and Midwest. This last happened in 2004. If you are around in 2021 when this happens next, you will be overwhelmed. A related species of periodical cicada reaches maturity after thirteen years.

Glenn Webb, a mathematician at Vanderbilt University, argues that it is not a coincidence that the two types of periodical cicadas' lifespans are prime numbers. He has proposed that these prime numbers offer evolutionary benefits.

If cicadas emerged from under the ground once every sixteen years, then predators with two-, four-, and eight-year life cycles would be there to eat them, potentially wiping out an entire population. And eating them would be very easy, because when cicadas are present, they are present in big numbers. But because they cycle at a larger prime number, cicadas do not have predators that specialize in eating them. Instead, there are only opportunistic predators. As a result, more cicadas live to mate and lay eggs.

Source: Webb, G. F. 2001. The prime number periodical cicada problem. *Discrete and Continuous Dynamical Systems,* Series B, Volume 1, Number 3, page 387–399.

How to Tell Whether a Number Is Prime or Composite

ThomsonNOW™
for Liberal Arts Mathematics
www.thomsonedu.com/1pass.

EXAMPLE 1

a. Is 516 prime or composite?
b. Is 517 prime or composite?

Solution

a. 516 is an even number, so it is divisible by 2. This means that 2 is a factor, so 516 is composite: $516 = 2 \cdot 258$.

b. 517 is not an even number, so it is not divisible by 2. We will divide 517 by counting numbers until we get an answer with no remainder. This might seem like a daunting task, but there are two shortcuts: Only divide by primes, and stop once you reach $\sqrt{517} = 22.7\ldots$ (We will discuss the first shortcut in this example and the second shortcut later.)

- $517/3 = 172.3\ldots$ This has a remainder.
- $517/4$: We already found that 517 is not divisible by 2. This tells us that 517 is not divisible by 4 or any other even number such as 2, 4, 6, or 8.
- $517/5 = 103.4\ldots$ This has a remainder.

- $517/7 = 73.8. \ldots$ This has a remainder.
- $517/9$: We already found that 517 is not divisible by 3. This tells us that 517 is not divisible by 9.
- $517/11 = 47$. This has no remainder.

The last result tells us that $517 = 11 \cdot 47$, so 517 is composite. Notice that we did not have to check for divisibility by composites, such as $4 = 2 \cdot 2$, $6 = 2 \cdot 3$, and $9 = 3 \cdot 3$. We only had to check for divisibility by primes. ∎

How Many Prime Numbers Are There?

If we were to continue the list in Figure 7.5, would we continue to find primes, or would we reach a point at which they just stopped occurring? It is easy to think that they would stop—a large number has so many more *possible* factors than does a small number. For example, 4, which is a small number, has only 1, 2, 3, and 4 as possible factors (1, 2, and 4 are the actual factors). But 1,000, which is a large number, has possible factors of 1, 2, 3, 4, . . . , 999, and 1000. We might assume that as the list of factors gets larger, we would have to find a factor sooner or later.

But the ancient Greek mathematician Euclid proved in about 300 B.C. that this is not true. He proved that there is an infinite number of finite numbers. So no matter how many you list, they never stop.

Prime Numbers and Eratosthenes

In about 200 B.C., another ancient Greek mathematician named Eratosthenes developed a way to list prime numbers that used multiplication rather than division. Of course, calculators had not been invented, and division is much more difficult by hand than is multiplication. We will apply his method to the numbers 2 through 25.

First, go through and eliminate every multiple of 2 (except 2 itself, because 2 is prime). We will show eliminated numbers in black.

2	3	4	5	6	7	8	9	10	11	12	13
14	15	16	17	18	19	20	21	22	23	24	25

Next, go through and eliminate every multiple of 3 (except 3 itself, because 3 is prime).

2	3	4	5	6	7	8	9	10	11	12	13
14	15	16	17	18	19	20	21	22	23	24	25

There is no need to eliminate multiples of 4 because they are all multiples of 2 and were eliminated earlier. So eliminate every multiple of 5 (except for 5 itself). The only one that has not already been eliminated is 25.

2	3	4	5	6	7	8	9	10	11	12	13
14	15	16	17	18	19	20	21	22	23	24	25

Continue this process through the square root of the highest number on the list. The highest number on our list is 25, and $\sqrt{25} = 5$. So we are done, because we just eliminated multiples of 5. Of the numbers 2–25, the prime numbers are 2, 3, 5, 7, 11, 13, 17, 19, and 23.

This method is called the **Sieve of Eratosthenes,** because a *sieve* is a type of filter, and this method filters out the composites.

The above process allows us to stop after we reach $\sqrt{25} = 5$. We will continue the process anyway, with multiples of 7, to show why we can stop after we reach $\sqrt{25} = 5$. All of the smaller multiples of 7, such as $7 \cdot 2$ and $7 \cdot 6$, have already been eliminated because they are multiples of 2 and 6, respectively. The first multiple of 7 that has not already been eliminated is $7 \cdot 7 = 49$, and that is not on the list because $49 > 25$. Stopping after we reach $\sqrt{25} = 5$ avoids this needless checking.

Why Search for Primes?

Eratosthenes was not the last person to search for primes. Many people are still actively searching for primes today. Why do people do this? Here are some reasons.

It is a tradition. Euclid started the search in approximately 300 B.C. Since then, many famous mathematicians have continued the search.

In the search for prime numbers, unforeseen benefits are discovered along the way. For example, prime numbers are used to encrypt credit card numbers when they are transmitted over the web. Also, software that searches for prime numbers is used by Intel to test Pentium chips and by Cray Research, the manufacturer of some of the most powerful computers ever built, to test their computers. This finding of unforeseen benefits is a common occurrence in mathematical and the scientific research.

Some people search for primes because they enjoy the glory of discovery. There is a group of on-line prime searchers, called GIMPS—the Great Internet Prime Search (www.mersenne.org/prime.htm). This group has found eight previously unknown primes. In February 2005, Martin Novak, a GIMPS member from Germany, found a previously unknown prime with 7,816,230 decimal places, using his home computer. In December 2005, a team at Central Missouri State University found an even larger previously unknown prime with 9.1 million decimal places. The team is affiliated with GIMPS.

The Search for Prime Numbers and Fermat

Many mathematicians tried to find a formula that could be used to generate primes. Because they are divisible by 2, it was clear that none of the even numbers are prime. Therefore, the focus should be on odd numbers. Some mathematicians studied the formula $2^n + 1$. The number 2^n is even, and adding 1 to an even number gives an odd number, so $2^n + 1$ must be odd. This formula fails to generate primes rather quickly. See Figure 7.6.

n	1	2	3
$2^n + 1$	$2^1 + 1 = 2 + 1$ $= 3$	$2^2 + 1 = 4 + 1$ $= 5$	$2^3 + 1 = 8 + 1$ $= 9$
Result	Prime	Prime	Composite

Figure 7.6
Is $2^n + 1$ always prime?

In the early seventeenth century, French mathematician Pierre de Fermat used the same formula but used only powers of 2 for n. See Figure 7.7 on page 504. Fermat checked these numbers through $2^{16} + 1 = 65,537$, dividing by hand, and found that each outcome is prime. He conjectured that this formula always works, but he was unable to prove it. More than 100 years later, the Swiss mathematician

Leonhard Euler showed that $2^{32} + 1 = 4,294,967,297$ is not prime, because it has 641 as a factor. (Euler comes up a lot in this book. He was instrumental in creating graph theory, studied in Chapter 9, and in exponential and logarithmic functions, studied in Chapter 10. There is an historical note about him in Section 10.0.A.)

n	$2^1 = 2$	$2^2 = 4$	$2^3 = 8$
$2^n + 1$	$2^2 + 1 = 4 + 1$ $= 5$	$2^4 + 1 = 16 + 1$ $= 17$	$2^8 + 1 = 256 + 1$ $= 257$
Result	Prime	Prime	Prime

n	$2^4 = 16$	$2^5 = 32$	$2^6 = 64$
$2^n + 1$	$2^{16} + 1$ $= 65,537$	$2^{32} + 1$ $= 4,294,967,297$	$2^{64} + 1 = ?$
Result	Prime	Composite	?

Figure 7.7
Is $2^n + 1$ always prime, if n is a power of 2?

The Search for Prime Numbers and Mersenne

In the 1600s, the French monk Marin Mersenne studied the formula $2^n - 1$. The number 2^n is even, so $2^n - 1$ must be odd. It was proven that if n is composite, then $2^n - 1$ is also composite. So only prime numbers were used for n. See Figure 7.8.

n	2	3	5
$2^n - 1$	$2^2 - 1 = 4 - 1$ $= 3$	$2^3 - 1 = 8 - 1$ $= 7$	$2^5 - 1 = 32 - 1$ $= 31$
Result	Prime	Prime	Prime

n	7	11	13
$2^n - 1$	$2^7 - 1 = 128 - 1$ $= 127$	$2^{11} - 1 = 2047$	$2^{13} - 1 = 8191$
Result	Prime	?	?

Figure 7.8
Is $2^n + 1$ always prime if n is prime?

Unfortunately, this formula fails to generate *only* prime numbers, because $2^{11} - 1 = 2047 = 23 \cdot 89$ is not prime. However, it generates *many* prime numbers, and these Mersenne numbers continue to be one of the largest sources of prime numbers. In the late 1700s, Euler showed that $2^{31} - 1$ is a Mersenne prime. A **Mersenne number** is a number of the form $2^n - 1$, where n is prime. A **Mersenne prime** is a Mersenne number that is itself prime.

In the 1950s, mathematicians started to use computers to check Mersenne numbers to find out whether they are prime. In 1952, a computer was used to find that

$2^{2281} - 1$ is prime. The primes discovered by GIMPS—the Great Internet Prime Search—are all Mersenne primes. And the prime discovered in February 2005 by Martin Novak of Germany is the forty-second Mersenne prime, $2^{25,964,951} - 1$.

Composite Numbers and Their Factors

In ancient times, numbers were thought to have mystical properties. One such property had to do with how many factors a number has. It turns out that this rather odd detail might help mathematicians with their search for prime numbers.

Look at the number 20. It has an abundance of factors, because there are many ways to factor it.

$20 = 1 \cdot 20$ 1 and 20 are factors.

$20 = 2 \cdot 10$ 2 and 10 are factors.

$20 = 4 \cdot 5$ 4 and 5 are factors.

The factors of 20 are 1, 2, 4, 5, 10, and 20. But the *proper* factors of 20 are 1, 2, 4, 5, and 10. The **proper factors** of a number are that number's factors except for the number itself.

Notice that in each of the above factorizations, one factor is less than $\sqrt{20} = 4.47\ldots$ and the other is greater than $\sqrt{20} = 4.47\ldots$:

- Of the factors 1 and 20, $1 < 4.47\ldots$ and $20 > 4.47\ldots$.
- Of the factors 2 and 10, $2 < 4.47\ldots$ and $10 > 4.47\ldots$.
- Of the factors 4 and 5, $4 < 4.47\ldots$ and $5 > 4.47\ldots$.

This always happens. As a result, it is unnecessary to search above the square root of a number in searching for that number's factors.

Abundant, Deficient, and Perfect Numbers

The sum of 20's proper factors is

$1 + 2 + 4 + 5 + 10 = 22$

Since 22 (the sum of the proper factors) is greater than 20 (the number itself), we say that 22 is an *abundant* number. A counting number is **abundant** if the sum of its proper factors is greater than the number itself.

Now consider the number 49:

$49 = 1 \cdot 49$ 1 and 49 are factors.

$49 = 7 \cdot 7$ 7 is a factor.

The proper factors of 49 are 1 and 7. The number 49 is rather deficient in its factors.

The sum of 49's proper factors is $1 + 7 = 8$. Since 8 (the sum of the proper factors) is less than 49 (the number itself), we say that 49 is a *deficient* number. A counting number is **deficient** if the sum of its proper factors is less than the number itself.

Finally, consider the number 6.

$6 = 1 \cdot 6$ 1 and 6 are factors.

$6 = 2 \cdot 3$ 2 and 3 are factors.

The proper factors of 6 are 1, 2 and 3. The sum of 6's proper factors is $1 + 2 + 3 = 6$. Since 6 (the sum of the proper factors) is equal to 6 (the number itself), we say

Historical Note   Pythagoras (approximately 560–480 B.C.)

North Wind Picture Archives

Pythagoras founded a philosophical and religious school in Crotone, Italy, that had many followers. They lived permanently at the school, had no personal possessions, and were vegetarians. They worshiped Pythagoras as a demigod. All Pythagoreans were required to practice obedience, loyalty, secrecy, silence, ritualistic fasting, simplicity in dress and possessions, and the habit of frequent self-examination. In fact, the Pythagoreans were so loyal and secretive that the contributions of Pythagoras cannot be separated from those of his followers.

The famous Pythagorean Theorem was not discovered by Pythagoras; in fact, it was known by the Babylonians 1000 years earlier. However, Pythagoras was probably the first to prove it.

The Pythagoreans believed that geometry is the highest form of knowledge and that mathematics is the best, most direct approach to reality. In fact, they believed that mathematics constitutes the true nature of things.

The Pythagoreans observed that vibrating strings produce pleasant tones if one string's length divided by the other string's length is a rational number (that is, a fraction made of whole numbers, such as 1/2, 4/3, etc.). To them, this meant that even sound is reducible to mathematics.

The Pythagoreans saw a mystical significance in numbers.

- "One" is the generator of all numbers and therefore the number of reason.
- "Two" is the number of women.
- "Three" is the number of men.
- "Four" is the number of justice.
- "Five" is the number of marriage; it is the sum of woman and man.
- "Six" is the number of creation; it is the product of woman and man. Six is a perfect number, so creation is perfect.
- "Seven" is the number of opportunity.
- "Ten" is the number of the universe. Ten is the sum of the numbers one through four, so ten represents reason, women, men, and justice, combined.
- Numbers have special mystical properties if they are perfect, deficient, or abundant.

that 6 is a *perfect* number. A counting number is **perfect** if the sum of its proper factors is equal to the number itself.

Euclid wrote about perfect numbers in about 300 B.C., but many ancient cultures studied perfect numbers before then. Pythagoras also studied them and their mystical properties. See the Historical Note for more on Pythagoras and his view of numbers.

More Perfect Numbers

ThomsonNOW™
for Liberal Arts Mathematics
www.thomsonedu.com/1pass.

EXAMPLE 2   **a.** List all of 28's proper factors.
b. Is 28 an abundant, perfect, or deficient number?

Solution   **a.** First, note that $\sqrt{28} = 5.2 \ldots$, so in searching for 28's factors, it will be unnecessary to search above 5.

- $28/2 = 14$ $\Rightarrow$ $28 = 2 \cdot 14$ $\Rightarrow$ 2 and 14 are factors of 28
- $28/3 = 9.3 \ldots$ $\Rightarrow$ 3 is not a factor of 28
- $28/4 = 7$ $\Rightarrow$ $28 = 4 \cdot 7$ $\Rightarrow$ 4 and 7 are factors of 28
- $28/5 = 5.6$ $\Rightarrow$ 5 is not a factor of 28

- It is unnecessary to search above 5, so we will not try 28/6.
- Clearly, $28 = 1 \cdot 28$, so 1 and 28 are factors of 28, but 28 is not a proper factor.

28's proper factors are 1, 2, 4, 7, and 14.

b. $1 + 2 + 4 + 7 + 14 = 28$, so 28 is a perfect number.   ■

The first four perfect numbers are 6, 28, 496, and 8,128:

- 6's proper factors are 1, 2, and 3.
  Also, $6 = 1 + 2 + 3$.
- 28's proper factors are 1, 2, 4, 7, and 14.
  Also, $28 = 1 + 2 + 4 + 7 + 14$.
- 496's proper factors are 1, 2, 4, 8, 16, 31, 62, 124, and 248.
  Also, $496 = 1 + 2 + 4 + 8 + 16 + 31 + 62 + 124 + 248$.
- 8128's proper factors are 1, 2, 4, 8, 16, 32, 64, 127, 254, 508, 1016, 2032, and 4064.

Also, $8128 = 1 + 2 + 4 + 8 + 16 + 32 + 64 + 127 + 254 + 508 + 1,016 + 2,032 + 4,064$.

These perfect numbers have been known since prehistoric times. They were discovered so long ago that there is no record of their discovery.

Perfect Numbers and Mersenne Primes

Perfect numbers contain an abundance of patterns. Let's look at some factorizations of some perfect numbers.

- $6 = 2 \cdot 3$
- $28 = 4 \cdot 7$
- $496 = 16 \cdot 31$
- $8,128 = 64 \cdot 127$

In each case, the first factor is a power of 2.

- 6's first factor is $2^1 = 2$
- 28's first factor is $2^2 = 4$
- 496's first factor is $2^4 = 16$
- 8,128's first factor is $2^6 = 64$

Furthermore, the powers are all 1 less than a prime number.

- 2^1's power is 1, 1 is one less than 2, and 2 is a prime number.
- 2^2's power is 2, 2 is one less than 3, and 3 is a prime number.
- 2^4's power is 4, 4 is one less than 5, and 5 is a prime number.
- 2^6's power is 6, 6 is one less than 7, and 7 is a prime number.

So the pattern, so far, is that the first factors are all of the form 2^{n-1}, where n is a prime number.

Now let's look at the second factors of the perfect numbers. Each is a Mersenne prime. That is, each is of the form $2^n - 1$, where n is a prime number.

- 6's second factor is $3 = 2^2 - 1$.
- 28's second factor is $7 = 2^3 - 1$.
- 496's second factor is $31 = 2^5 - 1$.
- 8,128's second factor is $127 = 2^7 - 1$.

Prime Numbers in the Real World: RSA Encryption and Web Purchases

When you make an Internet purchase, you usually have to give your credit card number. There are people who eavesdrop electronically on transactions like this, hoping to steal your number. So for your protection, the credit card number is usually **encrypted.** That is, it is converted into another number that is useless to anyone who does not know how to **decrypt** it back to the actual credit card number. One common method is **RSA encryption,** which involves prime numbers and factoring.

Here's how RSA encryption works. Suppose you are buying something from Amazon.com. When you get to the web page at which you enter your credit card number, that page's address will begin with "https://www.amazon.com." The "s" in "https" indicates that you are now using a secure server and that your credit card number will be encrypted. Amazon.com uses RSA encryption, so their secure server has stored a pair of two huge prime numbers, which we will call *p* and *q*. These two numbers are called the *private key.* They are kept private, because they are used to decrypt your credit card number.

Amazon.com's secure server transmits the product of *p* and *q* to your computer, which in turn uses it to encrypt your credit card number. This number, *pq*, cannot be used to decrypt. It is called the *public key.*

An electronic eavesdropper could easily find out the public key *pq*. But the eavesdropper would have to factor it into *p* and *q* to be able to decrypt. It is very easy to program a computer to factor a number, but it is also very time-consuming for the computer to suc-

ceed if the number is large. This is because the computer has to try dividing the number to be factored by almost every prime up to the number's square root.

Typically, the number *pq* has about 400 digits, so there are an incredibly large number of primes to try. According to an article published by the *Berkeley Math Circle,* "the computer would have to run for 10^{176} times the life of the universe to factor" such a number. So it is essentially not factorable.

Of course, all of the encoding and decoding, as well as the creation of the public key and the private key, happen behind the scenes. The Internet user is never aware of it.

In summary:

- The public key is *pq*, which is the product of two huge prime numbers. It is used only to encrypt, so it can be made public without risk.
- The private key is *p* and *q*, the two prime factors of the public key. It is used only to decrypt, so it is never made public.
- Using a computer to try to factor *pq* would take far longer than all of human history.

Exercises 41–44 and 46–47 involve RSA encryption.

(See *Amazon.com Licenses RSA Security's RSA BSAFE® Crypto-C Software.* Press release from RSA Security, http://www.rsasecurity.com/press_release.asp?doc_id= 189 and *RSA Encryption,* by Tom Davis, http:// mathcircle.berkeley.edu/BMC3/rsa/rsa.html.)

In fact, every single number of the form $(2^{n-1})(2^n - 1)$, where n and $2^n - 1$ are prime numbers, is a perfect number. Euclid proved this in about 300 B.C.

This relationship between perfect numbers and Mersenne primes is an example of why people—mystics and mathematicians alike—have a sense that there are some secrets waiting to be discovered here. Perhaps there is a formula that will generate all of the primes. Perhaps there is some theory that makes the existence, location, and importance of the primes clearer.

We have discussed $(2^{n-1})(2^n - 1)$, where n is one of the first prime numbers: 2, 3, 5, and 7. What about the next prime, $n = 11$?

$$(2^{n-1})(2^n - 1) = (2^{11-1})(2^{11} - 1) = 1{,}024 \cdot 2{,}047 = 2{,}096{,}128$$

It was not known for quite a while whether this is a perfect number, because of the difficulty in factoring a number this large without the aid of technology. In 1536, it was discovered that this is not a perfect number and that $2^{11} - 1 = 2,047$ is not a Mersenne prime, because $2,047 = 23 \cdot 89$.

The fifth perfect number was not discovered by Europeans until 1461, but an Arab mathematician found it in about 1200. It comes from using the next prime number, 13, for n.

7.3 Exercises

In Exercises 1–6, do the following.

a. Give the prime factorization of the number, if it has one.

b. State whether the number is prime or composite.

1. 42
2. 66
3. 23
4. 52
5. 54
6. 29

In Exercises 7–12, determine whether the number is prime or composite.

7. 299
8. 301
9. 207
10. 501
11. 233
12. 283

13. Use the Sieve of Eratosthenes to determine which of the numbers from 26 to 49 are primes.
14. Use the Sieve of Eratosthenes to determine which of the numbers from 50 to 75 are primes.

15. Evaluate the formula $2^n + 1$ for $n = 5$. Is the result prime or composite? Why?
16. Evaluate the formula $2^n + 1$ for $n = 6$. Is the result prime or composite?
17. Evaluate the formula $2^n + 1$ for $n = 7$. Is the result prime or composite? Why?
18. Evaluate the formula $2^n + 1$ for $n = 9$. Is the result prime or composite?

19. Evaluate the formula $2^n - 1$ for $n = 4$. Is the result prime or composite? Why?
20. Evaluate the formula $2^n - 1$ for $n = 6$. Is the result prime or composite? Why?
21. Evaluate the formula $2^n - 1$ for $n = 8$. Is the result prime or composite? Why?
22. Evaluate the formula $2^n - 1$ for $n = 9$. Is the result prime or composite? Why?

23. List all of the proper factors of 42.
24. List all of the proper factors of 43.
25. List all of the proper factors of 54.

26. List all of the proper factors of 55.
27. Is 42 abundant, perfect, or deficient? Why? (See Exercise 23.)
28. Is 43 abundant, perfect, or deficient? Why? (See Exercise 24.)
29. Is 54 abundant, perfect, or deficient? Why? (See Exercise 25.)
30. Is 55 abundant, perfect, or deficient? Why? (See Exercise 26.)
31. Is 61 abundant, perfect, or deficient? Why?
32. Is 60 abundant, perfect, or deficient? Why?
33. Is 62 abundant, perfect, or deficient? Why?
34. Is 65 abundant, perfect, or deficient? Why?

35. Find the fifth perfect number.
36. What is the fifth perfect number's first factor? Use the factoring method described in the text.
37. What is the fifth perfect number's second factor? Use the factoring method described in the text.
38. Which factor of the fifth perfect number is a Mersenne prime? Why?

39. $2^{127} - 1$ is a Mersenne prime. Find the perfect number associated with this number. Write your answer (a) using exponents with a base of 2 and (b) in scientific notation.
40. $2^{521} - 1$ is a Mersenne prime. Find the perfect number associated with this number. Write your answer (a) using exponents with a base of 2 and (b) in scientific notation.

41. In RSA encryption, if the public key is 35, what is the private key?
42. In RSA encryption, if the public key is 51, what is the private key?
43. In RSA encryption, if the public key is 143, what is the private key?
44. In RSA encryption, if the public key is 221, what is the private key?

45. **a.** Rewrite the first four perfect numbers with the binary system.
b. Describe the pattern or patterns you see in the result of part (a).

➤ *Answer the following, using complete sentences and your own words.*

CONCEPT QUESTIONS

46. What do encryption and decryption have to do with prime numbers?

47. What do encryption and decryption have to do with factoring?

HISTORY QUESTIONS

48. What is odd about Pythagoras' most famous result?

49. What strange feature did the Pythagoreans see in numbers?

 WEB PROJECTS

50. Write an essay on one of the following topics:
- Mersenne primes
- GIMPS—the Great Internet Mersenne Prime Search
- Large prime numbers
- Prime numbers and perfect numbers
- History of prime numbers
- Mysticism and numbers

Some useful links for this web project are listed on the text web site: **www.thomsonedu.com/mathematics/ johnson**

7.4 Fibonacci Numbers and the Golden Ratio

Fibonacci Numbers

In 1202, the Italian mathematician Fibonacci tried to determine how quickly animals would breed in ideal circumstances. In discussing this issue, Fibonacci used rabbits as an example. However, the logic in Fibonacci's rabbit example was unnecessarily complicated. Henry Dudeny, an English puzzle book author in the early 1900s, came up with the following simplified version of Fibonacci's logic.

If a cow produces its first female calf at age two years and after that produces another female calf every year, how many female calves are there after five years if we start with one newborn female calf and no calves die?

- At the beginning, there is one female calf.
- After one year, there is still only one female calf.
- After two years, the one female has produced a calf, so there is now 1 new cow, and $1 + 1 = 2$ cows in total. This new cow will start to produce in two years.
- After three years, the original female has produced a second calf, so there is 1 new cow, and there are $1 + 2 = 3$ cows in total.
- After four years, the original female has produced yet another calf, and the calf born two years ago has also produced a calf, so there are now 2 new cows, and there are $2 + 3 = 5$ cows in total.

Notice that each year, the number of new cows is the same as the number of cows that were alive two years ago, because each of those cows has produced a calf.

- After five years, there are three new cows: one born to the original female, one to the second female, and one to the third female. There are now $3 + 5 = 8$ cows in total.

See Figure 7.9, in which the cows are color-coded. The original female is red, and the first-born calf is blue.

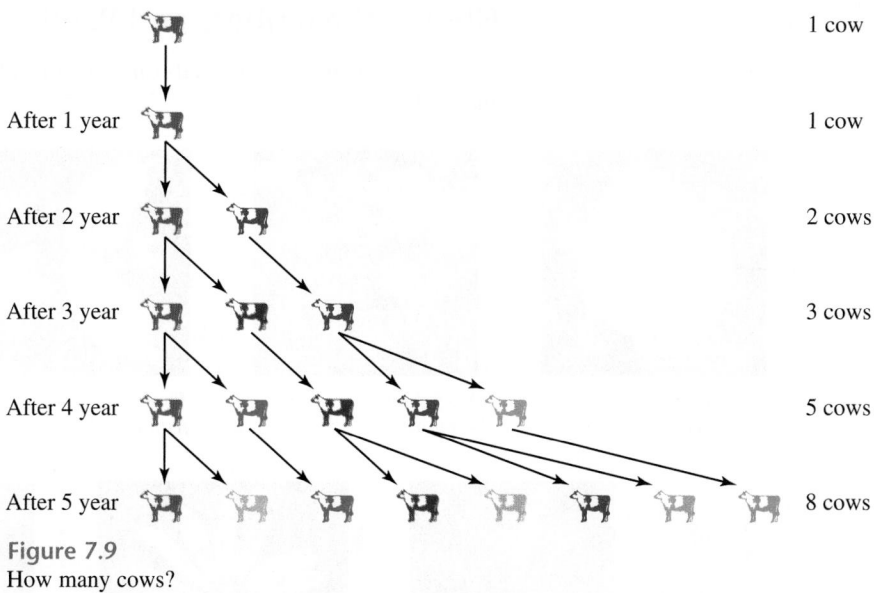

After 1 year   1 cow

After 2 year   2 cows

After 3 year   3 cows

After 4 year   5 cows

After 5 year   8 cows

Figure 7.9
How many cows?

How many cows will there be in one more year? Two years earlier there were five cows, and each will produce a calf. So there will be 5 + 8 = 13 cows.

The number of cows in any given year is the sum of those alive two years ago (because they will each produce a calf) and those alive last year (because they're still alive).

The pattern is as follows: 1, 1, 1 + 1 = 2, 1 + 2 = 3, 2 + 3 = 5, 3 + 5 = 8, 5 + 8 = 13, ... Each number in the sequence is the sum of the previous two numbers. This sequence of numbers is called the **Fibonacci sequence.** And the numbers themselves are called **Fibonacci numbers.**

The Fibonacci Sequence and Honeybees

You might be thinking that the cow story is rather unrealistic. If so, you are right. But there are many realistic occurrences of Fibonacci numbers in nature, one of which has to do with honeybees.

There are two types of female honeybees: workers, which produce no eggs, and the queen, which produces eggs. Drone bees are males that are produced from the queen's unfertilized eggs. A drone's mother is the queen that laid the egg. But a drone *has no father,* because the egg was never fertilized.

Worker females and queen females are produced from the queen's fertilized eggs. So each type of female has a father as well as a mother, because the female eggs were fertilized.

Let's look at a drone's ancestors. A drone has:

- 1 parent—his mother
- 2 grandparents, who are his mother's parents—a male and a female
- 3 great-grandparents: his grandmother has two parents (his great-grandmother and great-grandfather), but his grandfather, like all males, has only one parent (his great-grandmother)
- 5 great-great-grandparents: each of his two great-grandmothers had two parents, and his one great-grandfather had one parent

The numbers of ancestors of a drone honeybee are the Fibonacci numbers.

Fibonacci Numbers and Plants

Frequently, the number of petals on a plant is a Fibonacci number. See Figures 7.10 and 7.11.

One petal:
calla lily

Two petals:
euphorbia

Three petals:
trillium

Five petals:
hibiscus

Eight petals:
clematis

Thirteen petals:
black-eyed susan

Twenty-one petals:
shasta daisy

Figure 7.10
Fibonacci numbers and flowers

Some plants have a spiral structure. For example, pinecones, pineapples, and cauliflowers all have a spiral structure. Frequently, the number of spirals is a Fibonacci number. And although you cannot see it in the photograph, the number of scales along any one spiral is usually a Fibonacci number. See Figures 7.12 and 7.13.

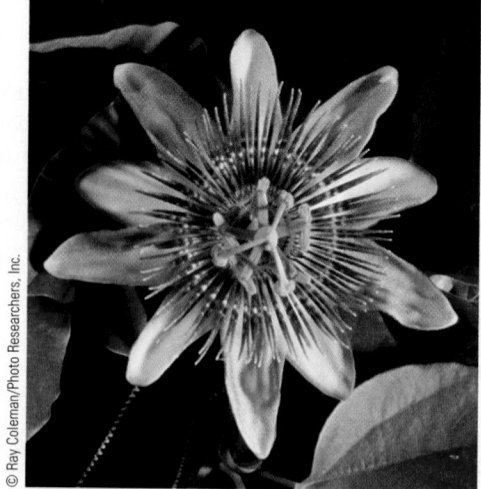

Figure 7.11
This passionflower has three inner T-shaped petals, five white petals behind it, and then a set of five purple petals, behind which is another set of five light purple petals

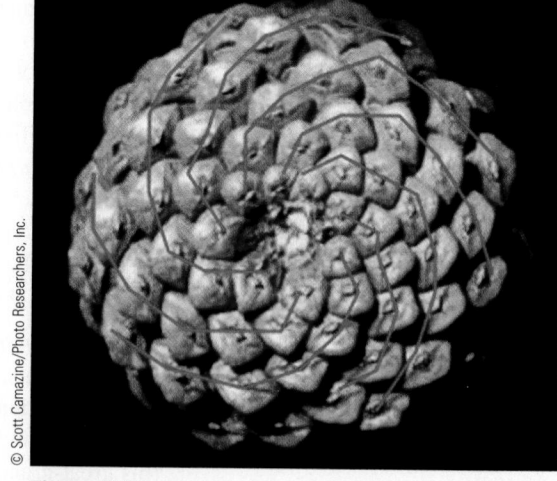

Figure 7.12
This pinecone has eight clockwise spirals. How many counterclockwise spirals does it have?

The number of petals on some plants is a Fibonacci number, and the number of spirals on some plants is a Fibonacci number. But these things do not always happen. You can find plants for which the number of petals is not a Fibonacci number. Nonetheless, it happens frequently. It is not a universal law, but rather a tendency. Nature tends to use the Fibonacci numbers.

Figure 7.13
This sunflower has fifty-five clockwise spirals. How many counterclockwise spirals does it have?

The Fibonacci Spiral

Here is another example of how nature tends to use Fibonacci numbers. The first two Fibonacci numbers are 1 and 1, so draw two adjacent squares, each with a side of length 1. See Figure 7.14.

The left edge of those two squares is of length 2, and 2 is the next Fibonnaci number. Draw a square with a side of length 2 along this edge. See Figure 7.15.

The bottom edge is of length 3, and 3 is the next Fibonnaci number. Draw a square with a side of length 3 along this edge. See Figure 7.16.

Figure 7.14
Two squares with sides of length 1

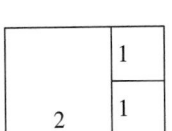

Figure 7.15
Squares with sides of length 1, 1, and 2

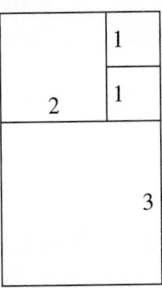

Figure 7.16
Squares with sides of length 1, 1, 2, and 3

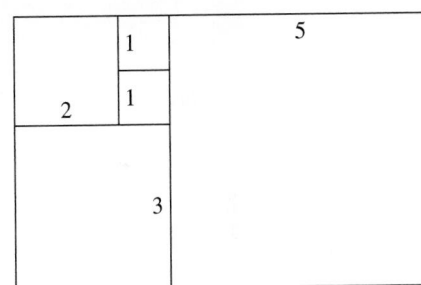

Figure 7.17
Squares with sides of length 1, 1, 2, 3, and 5

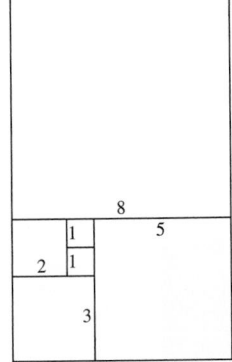

Figure 7.18
Squares with sides of length 1, 1, 2, 3, 5, and 8

The right edge is of length 5, and 5 is the next Fibonnaci number. Draw a square with a side of length 5 along this edge. See Figure 7.17.

The top edge is of length 8, and 8 is the next Fibonnaci number. Draw a square with a side of length 8 along this edge. See Figure 7.18.

We could continue this process indefinitely, and the next square would have an edge whose length is the next Fibonacci number. However, the result does not resemble anything found in nature. At least it doesn't until you insert quarter circles in the squares so that they form a spirallike curve. See Figure 7.19 on page 514.

This **Fibonacci spiral** is a close approximation of the spirals found in nautilus shells, snail shells, and chameleon tails. See Figure 7.20 on page 514.

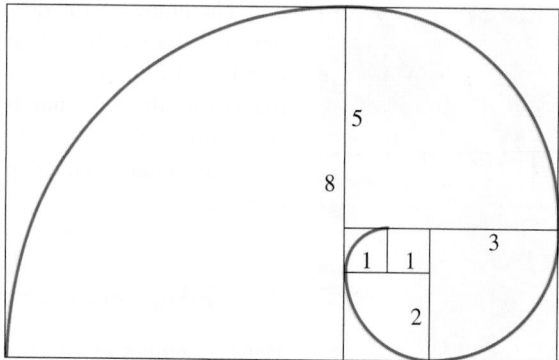

Figure 7.19
The Fibonacci spiral

Figure 7.20
Nautilus shell, chameleon tails, and the Fibonacci spiral

The Golden Rectangle

Which of the rectangles in Figure 7.21 is most esthetically pleasing?

Since early in human history, some artists and architects have claimed that the second rectangle is the most esthetically pleasing. Because of this, it is called a **golden rectangle.** This rectangle's length is about 1.6 times its width, so the ratio of its length to its width is

$$\frac{\text{length}}{\text{width}} \approx \frac{1.6 \cdot \text{width}}{\text{width}} = 1.6$$

This ratio is called the **golden ratio.**

Figure 7.21
Three rectangles

The larger rectangle in Figure 7.22 is twice as long and twice as wide as is the smaller rectangle. However, its ratio is

$$3.2/2 = 1.6$$

and the smaller rectangle's ratio is

$$1.6/1 = 1.6$$

Each is a golden rectangle.

Figure 7.22
Two golden rectangles

The Golden Rectangle in Art and Architecture

The ancient Greeks considered the golden rectangle to be very esthetically pleasing, and they used it extensively in their art and architecture. The Parthenon in Athens, Greece, forms a rectangle with this ratio. See Figure 7.23. The golden ratio is present also—the ratio of the lengths of the two purple lines is the golden ratio.

Some artists have favored the golden rectangle. They appear in the art of Leonardo da Vinci. See Figure 7.24. It has been claimed that Leonardo da Vinci

Figure 7.23
The Parthenon—a golden rectangle

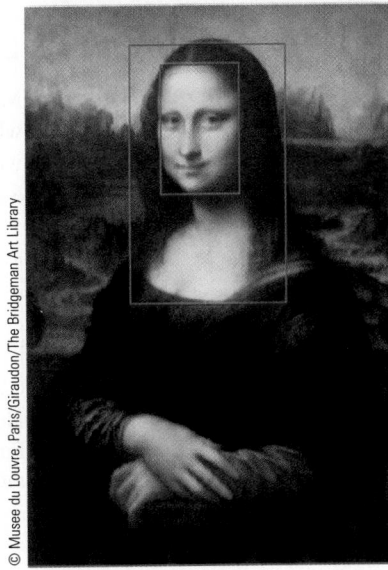

Figure 7.24
Leonardo da Vinci's *Mona Lisa* and golden rectangles

deliberately used golden rectangles in his art. However, we do not know whether this is a fact. We do know that Leonardo was a friend of Luca Pacioli, the Italian friar who wrote *The Divine Proportion,* a book on the golden ratio. And we know that Pacioli collaborated with Leonardo in other areas. So it is quite possible that Leonardo deliberately used golden rectangles.

We do know that several artists deliberately used the golden rectangle in their art, including Piet Mondrian and Georges Seurat. See Figures 7.25 and 7.26.

Figure 7.25
Piet Mondrian, *Composition with Red, Blue and Yellow,* 1930. Oil on Canvas, 18 1/8 × 18 1/8 inches. © 2006 Mondrian/Holtzman Trust c/o HCR International, Warrenton, VA, USA.

Figure 7.26
Georges-Pierre Seurat, *Circus Sideshow,* 1887–88. Oil on canvas, 39 1/4 × 59 inches. The Metropolitan Museum of Art, Bequest of Stephen C. Clark, 1960 (61.101.17). Photograph © 1989 The Metropolitan Museum of Art.

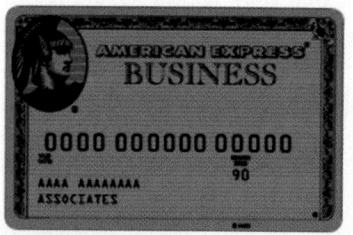

Figure 7.27
Your own personal golden rectangle

As a last example of the ubiquity of golden rectangles, it must be pointed out that you probably have a golden rectangle in your pocket or purse. See Figure 7.27.

Deriving the Golden Ratio

Mathematically, where does the golden ratio come from? Consider a rectangle whose base has two parts, a larger part of length a and a smaller part of length b, and whose height is a. See Figure 7.28. This rectangle is the combination of a square and a smaller rectangle. See Figure 7.29.

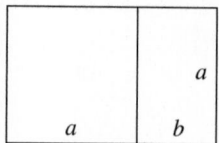

Figure 7.28
Finding the golden ratio

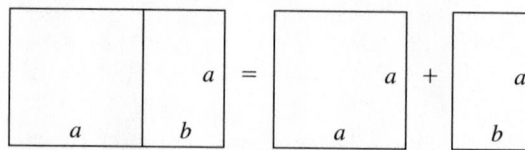

Figure 7.29
Continuing to find the golden ratio

In a golden rectangle, the smaller rectangle and the larger rectangle have the same shape. In other words, the two rectangles are similar. And the golden ratio comes from finding the lengths a and b that make these two rectangles similar. Each rectangle has its own ratio of length to width, as shown in Figure 7.30.

Rectangle	Length	Width	Length/Width
Larger	$a + b$	a	$(a + b)/a$
Smaller	a	b	a/b

Figure 7.30
The length/width ratios

If we make these ratios the same and solve the resulting equation for a/b, we get

$$a/b = \frac{1 + \sqrt{5}}{2} = 1.61803398\ldots$$

See Exercise 13. This number, which is called ϕ, is the golden ratio. (The symbol ϕ is the Greek letter phi.)

$$\phi = \frac{a}{b} = \frac{1 + \sqrt{5}}{2}$$

In its design, the golden rectangle is composed of two rectangles with the same ratio of length to width: a larger rectangle whose length is $a + b$ and whose width is a and a smaller rectangle whose length is a and whose width is b. See Figure 7.31.

Sometimes, artists use both golden rectangles. See Figure 7.32.

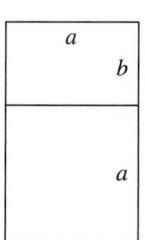

Figure 7.31
A golden rectangle within a golden rectangle

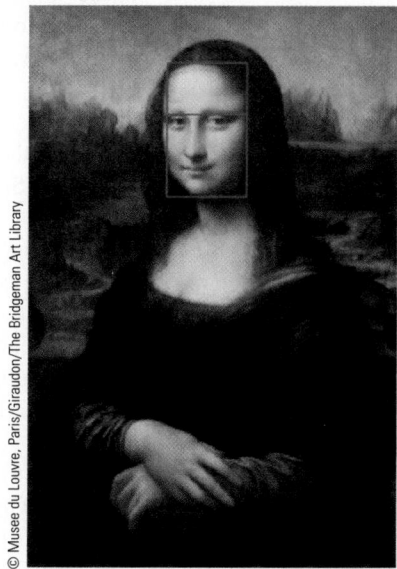

© Musee du Louvre, Paris/Giraudon/The Bridgeman Art Library

Figure 7.32
A golden rectangle within a golden rectangle

7.4 Exercises

Exercises 1 and 2 refer to Henry Dudeny's version of Fibonacci's puzzle.

1. How many female calves are there after six years, assuming that none die? Of these calves, how many are newly born?

2. How many female calves are there after seven years, assuming that none die? Of these calves, how many are newly born?

3. Give the ten Fibonacci numbers after 13.

4. Give the five Fibonacci numbers after those listed in Exercise 3.

5. How many great-great-great-grandparents does a drone bee have? How is each related to the drone?

6. How many great-great-great-great-grandparents does a drone bee have? How is each related to the drone?

7. How many counterclockwise spirals does the pinecone in Figure 7.12 have?

8. How many counterclockwise spirals does the sunflower in Figure 7.13 have?

9. Carefully draw the Fibonacci spiral composed of squares with sides of lengths 1, 1, 2, 3, 5, 8, and 13.

10. Carefully draw the Fibonacci spiral composed of squares with sides of lengths 1, 1, 2, 3, 5, 8, 13, and 21.

11. *Binet's Formula* states that the nth Fibonacci number is

$$\frac{1}{\sqrt{5}}\left[\left(\frac{1 + \sqrt{5}}{2}\right)^n - \left(\frac{1 - \sqrt{5}}{2}\right)^n\right]$$

a. Use Binet's Formula to find the thirtieth and fortieth Fibonacci numbers.

b. What advantage does Binet's Formula hold over the method of generating Fibonacci numbers discussed in this section?

c. What disadvantage does Binet's Formula hold over the method of generating Fibonacci numbers discussed in this section?

12. a. Use Binet's Formula (see Exercise 11) to find the 50th and 60th Fibonacci numbers.

b. What would you have to do to find the 50th and 60th Fibonacci numbers, without Binet's Formula?

13. Complete the derivation of the golden ratio that was started on page 516. Specifically, do the following:

a. Make an equation by setting the larger rectangle's length to width ratio equal to that of the smaller rectangle.

b. Find the lowest common denominator of the two fractions in part (a), and multiply each side of the equation in part (a) by it. This will eliminate denominators.

c. Simplify the result of part (b) by distributing.

d. Divide the result of part (c) by b^2.

e. Simplify the result of part (d) by canceling.

f. Substitute x for a/b in part (e). This should eliminate all a's and b's from the equation.

g. Use the quadratic equation to solve the result of part (f). This generates two different answers. The positive one is the golden ratio.

h. Why does a negative answer not make sense in this context?

14. The golden ratio is the only number whose square can be produced simply by adding 1.

a. Use your calculator to approximate the square of the golden ratio.

b. Use your calculator to approximate the sum of 1 and the golden ratio. What do you observe?

c. Use the result of Exercise 13 part (f) to verify that the square of the golden ratio is exactly equal to the sum of 1 and the golden ratio?

15. Find a work of art that uses the golden ratio. Provide a sketch or photograph of the work of art, and discuss its use of the golden ratio. Do not use anything mentioned in the text.

16. Find a building that uses a golden rectangle. Provide a sketch or photograph of the building, and discuss its use of the golden rectangle. Do not use anything mentioned in the text.

17. Find something in nature that involves Fibonacci numbers. Provide a sketch, photograph, or sample of the thing itself. Describe its involvement with Fibonacci numbers. Do not use anything discussed in detail in the text.

18. Complete the chart in Figure 7.33. Do not round off the decimals in the third column. As you read down the third column, the numbers get closer to a certain number discussed in this section. What number is that?

Fibonacci number	Dividing consecutive Fibonacci numbers	Decimal approximation of the result
1		
1	1/1	1
2	2/1	2
3	3/2	1.5
5		
8		
13		
21		
34		
55		

Figure 7.33
Chart for Exercise 18

19. The twenty-fourth and twenty-fifth Fibonacci numbers are 46,368 and 75,025, respectively. Divide the larger of these numbers by the smaller. What do you observe?

20. Exercises 18 and 19 indicate a relationship between Fibonacci numbers and the golden ratio. There is another relationship between them that has to do with Binet's Formula in Exercise 11 and the result of Exercise 13. Describe that relationship.

21. Carefully inspect the photograph of the Parthenon in Figure 7.23. Describe other occurrences of the golden rectangle or the golden ratio.

22. Carefully inspect *Composition in Red, Yellow and Blue* in Figure 7.25. Which rectangles are golden rectangles?

23. Carefully inspect *The Circus Sideshow* in Figure 7.26. Describe occurrences of golden rectangles and golden ratios.

PROJECTS

24. It has been claimed that Fibonacci numbers are nature's numbers. Investigate this claim in some or all of the following situations.
- Get several pineapples. Pineapples have three sets of spirals, as shown in Figure 7.34. Count the number of spirals in each direction.

Figure 7.34

- Get several cauliflowers. Count the number of clockwise spirals and counter clockwise spirals on each. This can be difficult due to the wartiness of the flowerlets.
- Get several sunflowers. Count the number of clockwise spirals on the flower, the number of counterclockwise spirals on the flower, and the number of seeds on each spiral.
- Count the number of petals on many flowers.
- Usually, leaves are located on the stem of a plant in such a way that the leaves actually spiral around the stem (Figure 7.35). If you start at one leaf and follow one complete spiral, you might come to another leaf that points in the same

Figure 7.35

direction as does the first leaf. With some plants, one complete spiral is not enough, and you need to follow two complete spirals before you come to another leaf that points in the same direction as does the first leaf. For several different plants, count the number of spirals before you come to another leaf that points in the same direction as does the first leaf. Also, count the number of leaves from the first leaf to the next leaf that points in the same direction (not counting the first leaf).

Write an essay in which you describe your research, and discuss your results.

25. Obtain good copies of the following works of art:
- Leonardo da Vinci's *Mona Lisa* and *Vitruvian Man*
- Boticelli's *The Birth of Venus*

Search for golden rectangles in these works. Then determine whether or not those rectangles are actually golden by measuring their lengths and widths and dividing. You might also investigate works by Paul Serusier, Giro Severini, and Juan Gris, as well as other works by Leonardo, Seurat, and Boticelli.

 ## WEB PROJECTS

26. Write an essay on one of the following topics:
- Fibonacci numbers in nature
- Fibonacci puzzles
- Fibonacci numbers and the golden rectangle
- The mathematics of Fibonacci numbers
- Fibonacci numbers and continued fractions
- Fibonacci's biography
- The golden rectangle and art
- Leonardo da Vinci's friendship and collaborative efforts with Fibonacci
- Fibonacci numbers and the stock market
- Fibonacci numbers and music
- The claim that the Egyptian pyramids were constructed using the golden ratio
- The claim that Debussy, Mozart, and Bartok used the golden ratio in their music and that some Gregorian chants are based on the golden ratio

Some useful links for this web project are listed on the text web site: **www.thomsonedu.com/mathematics/johnson**

Chapter 7 Review

ThomsonNOW **for Liberal Arts Mathematics**
Preparing for an exam? Test yourself on key material by visiting ThomsonNOW at **www.thomsonedu.com/1pass.**

TERMS

abacus	counting numbers	Fibonacci spiral	perfect numbers
abundant numbers	decimal system	golden ratio	place system
algorithm	deficient numbers	golden rectangle	place values
base ten	expanded form	hexidecimal system	prime factorization
binary system	factor	Mersenne number	prime number
bit	factor completely	Mersenne prime	proper factors
byte	Fibonacci numbers	number theory	sieve of Eratosthenes
composite number	Fibonacci sequence	octal system	

Exercises

In Exercises 1–18, write the given number in expanded form, or explain why there is no such number.

1. 5372
2. 1208
3. 325_8
4. 8642_8
5. 390_8
6. 2205_8
7. 905_{16}
8. 3265_{16}
9. ABC_{16}
10. $D013_{16}$
11. 11011001_2
12. 2103_2
13. 101112_2
14. 1011001_2
15. $39,22,54_{60}$
16. $12,14,75_{60}$
17. $25,44,34_{60}$
18. $12,18,05_{60}$

19. Convert 514_8 to base ten.
20. Convert $229A_{16}$ to base ten.
21. Convert 110101_2 to base ten.
22. Convert $32,55,49_{60}$ to base ten.
23. Convert $BC9A_{16}$ to base ten.
24. Convert $41,39,18_{60}$ to base ten.
25. Convert 111000_2 to base ten.
26. Convert 37292_8 to base ten.
27. Convert 1254_9 to base ten.
28. Convert 3241_5 to base ten.

29. Convert 514 to base eight.
30. Convert 3,922 to base two.
31. Convert 8,430 to base sixteen.
32. Convert 22,876 to base sixty.
33. Convert 39,287 to base sixty.
34. Convert 30,281 to base eight.
35. Convert 12,003 to base two.
36. Convert 6,438 to base sixteen.
37. Convert 59,375 to base seven.
38. Convert 44,886 to base three.

39. If $543_x = 207$, find base x.
40. If $123_x = 38$, find base x.
41. If $421_x = 343$, find base x.
42. If $111_x = 57$, find base x.

In Exercises 43–66, do the following.

a. *Perform the given computation, working solely in the given base.*
b. *Check your answer by converting it to base ten.*

43. $52_8 + 62_8$
44. $39_{16} + AB_{16}$
45. $101_2 + 110_2$
46. $22,12,53_{60} + 18,49,03_{60}$
47. $A9_{16} + 9B_{16}$
48. $1101_2 + 101_2$
49. $22,33,44_{60} + 55,44,33_{60}$
50. $32_8 + 157_8$
51. $201_8 - 112_8$
52. $50C_{16} - BA_{16}$
53. $1101_2 - 1010_2$
54. $9,22,32_{60} - 1,53,26_{60}$
55. $A08_{16} - 7B_{16}$
56. $1111_2 - 1001_2$
57. $22,33,44_{60} - 10,44,33_{60}$
58. $540_8 - 157_8$
59. $22_8 \cdot 72_8$
60. $3A_{16} \cdot A4_{16}$
61. $101_2 \cdot 110_2$
62. $42_7 \cdot 13_7$
63. $B3_{16} \cdot 92_{16}$
64. $1001_2 \cdot 101_2$
65. $34_5 \cdot 41_5$
66. $32_8 \cdot 57_8$

In Exercises 67–74, do the following:

a. *Perform the given computation, working solely in the given base.*
b. *Check your answer by multiplying in the given base.*

67. $301_8 \div 12_8$
68. $50C_{16} \div B1_{16}$
69. $1101_2 \div 101_2$
70. $147_8 \div 17_8$
71. $A08_{16} \div 1B_{16}$
72. $1111_2 \div 11_2$
73. $110110_2 \div 10_2$
74. $547_8 \div 17_8$

In Exercises 75–78, do the following:

a. *Give the prime factorization of the number, if it has one.*
b. *State whether the number is prime or composite.*

75. 33
76. 37
77. 41
78. 44

79. Use the Sieve of Eratosthenes to determine which of the numbers from 50 to 59 are primes.
80. Use the Sieve of Eratosthenes to determine which of the numbers from 60 to 69 are primes.

81. Evaluate the formula $2^n + 1$ for $n = 2$ and 3. Determine which of the results are prime and which are composites.
82. Evaluate the formula $2^n + 1$ for $n = 4$ and 5. Determine which of the results are prime and which are composites.
83. Evaluate the formula $2^n - 1$ for $n = 3$. Is the result prime or composite? Why?
84. Evaluate the formula $2^n - 1$ for $n = 4$. Is the result prime or composite? Why?

85. List all of the proper factors of 70.
86. List all of the proper factors of 66.

87. Is 20 abundant, perfect or deficient? Why?
88. Is 21 abundant, perfect or deficient? Why?
89. Is 49 abundant, perfect or deficient? Why?
90. Is 48 abundant, perfect or deficient? Why?
91. $2^{17} - 1$ is a Mersenne prime. Find the perfect number associated with this number. Write your answer (a) using exponents with a base of 2 and (b) in scientific notation.
92. $2^{19} - 1$ is a Mersenne prime. Find the perfect number associated with this number. Write your answer (a) using exponents with a base of 2 and (b) in scientific notation.

93. Give the first twenty Fibonacci numbers.
94. Referring to Henry Dudeny's version of Fibonacci's puzzle, how many female calves are there after five years, assuming that none die? Of these calves, how many are newly born?
95. How many great-great-grandparents does a drone bee have? How is each related to the drone?
96. How many great-grandparents does a drone bee have? How is each related to the drone?
97. Referring to Henry Dudeny's version of Fibonacci's puzzle, how many female calves are there after four years, assuming that none die? Of these calves, how many are newly born?
98. Carefully draw the Fibonacci spiral composed of squares with sides of lengths 1, 1, 2, 3, 5, and 8.
99. *Binet's Formula* states that the nth Fibonacci number is

$$\frac{1}{\sqrt{5}}\left[\left(\frac{1 + \sqrt{5}}{2}\right)^n - \left(\frac{1 - \sqrt{5}}{2}\right)^n\right]$$

a. Use Binet's Formula to find the twenty-fifth and twenty-sixth Fibonacci numbers.
b. Use the results of part (a) to find the twenty-seventh Fibonacci number.
100. Give six different examples of the presence of Fibonacci numbers in nature.
101. What is the golden rectangle?
102. What is the golden ratio?
103. Give an example of a work of art that uses a golden ratio.
104. Give an example of a building that uses a golden rectangle.
105. Describe two different relationships between the golden ratio and Fibonacci numbers.

➤ *Answer the following, using complete sentences and your own words.*

CONCEPT QUESTIONS
106. Why is there no base zero?
107. How does the place system use the number 0?
108. Do Roman numerals have a symbol for zero? Why or why not?
109. Would the Babylonian base sixty system have had some way of representing the number zero? Why or why not?
110. Describe in detail how to add $14_8 + 76_8$ on a base eight abacus.
111. Describe in detail how to subtract $321 - 74$ on a base ten abacus.

HISTORY QUESTIONS
112. Describe the origins of Hindu-Arabic numerals.
113. What was the subject matter of Mohammed ibn Musa al-Khowarizmi's two books?
114. Describe the Pythagorean's beliefs regarding numbers.
115. Why were people originally interested in perfect numbers?
116. Is the interest in prime numbers relatively new or old part of the human pursuit of knowledge?
117. What was Fibonacci trying to do when he invented Fibonacci numbers?
118. Describe one of the earlier occurrences of the golden rectangle or golden ratio.

8 Geometry

© Pasieka/Photo Researchers, Inc.

Geometry is one of the oldest branches of mathematics; its roots include two very practical human endeavors: surveying and astronomy. From the beginning of recorded time, people have attempted to measure the land and the heavens. The word *geometry* literally means "earth measurement"; it is derived from the Greek words *geos,* meaning "earth," and *metros,* meaning "measure."

Ancient geometry consisted of a fragmented assortment of examples, guesses, tricks, and rule-of-thumb procedures. It was an empirical subject in which approximate answers were usually good enough for practical purposes. A typical problem might have been the determination of the area of a triangular field of grain. The Babylonians and Egyptians were masters at using this collection of "how to's."

As a result of commerce, travel, and warfare, the empirical geometry of the Egyptians reached the ancient Greeks; however, the Greeks' pursuit of absolute truth through abstract thought set them apart from their predecessors. Beginning with Thales of Miletus, the Greeks insisted that geometric statements be established by *deductive reasoning* rather than by trial and error. The orderly development of theorems by *proof* was characteristic of Greek mathematics. This was an entirely new concept in the realm of mathematics. Consequently, the Greeks no longer concerned themselves with triangular fields of grain but with "triangles" and the characteristics associated with "triangularity." To this end, Greek scholars set out to systematically develop the relationships between all known mathematics of the day. The culmination of this effort was the writing of the monumental *Elements* by Euclid in the fourth century B.C.

Conceptually, geometry underwent little change over the next 2,000 years. The seventeenth century witnessed the marriage of geometry and algebra, the so-called *analytic geometry* of Descartes and Fermat (see Chapter 13). Euclid's iron grip on the bases of geometry was finally challenged in the nineteenth century, when non-Euclidean geometries were established. In the twentieth century, there was a proliferation of many types of new geometries. Just as the union of geometry and algebra heralded the new analytic geometry, the modern-day union of geometry and computers led to the new fractal geometry of Benoit Mandelbrot.

In this chapter, we review the standard concepts and calculations of length, area, and volume and present a brief history of the major accomplishments of ancient, analytic, and modern geometries.

ThomsonNOW™ **for Liberal Arts Mathematics**
Throughout this chapter, the ThomsonNOW logo indicates an opportunity for online self-study, linking you to interactive tutorials and videos based on your level of understanding.

8.1 Perimeter and Area

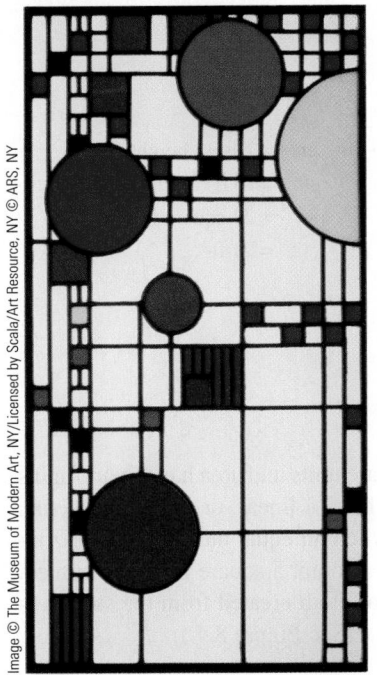

Architect Frank Lloyd Wright used simple geometric shapes to design this stained glass window in 1911.

Geometric shapes have intrigued people throughout history. Art, religion, science, engineering, architecture, psychology, and advertising are just a few of the many areas that make use of triangles, rectangles, squares, circles, cubes, pyramids, cones, spheres, and a host of other forms. Having seen many of these shapes in nature, geometers have constructed ideal representations of them and have developed various formulas for measuring their lengths (one-dimensional), areas (two-dimensional), and volumes (three-dimensional). In this section, we examine some features of the most commonly encountered two-dimensional figures.

Polygons

Two-dimensional figures can be classified by the number of sides they have. A **polygon** is a many-sided figure. A pentagon is a five-sided figure, a hexagon is a six-sided figure, and an octagon is an eight-sided figure. However, the names of polygons do not necessarily end with -*gon*. Although a three-sided figure could be called a *trigon,* we prefer *triangle*. Likewise, a four-sided figure is referred to as a quadrilateral rather than a quadragon.

Our study of polygons focuses on finding the distance around a figure and the amount of space enclosed within the figure. Some of the polygons we will examine are shown in Figure 8.1. The symbol ⌐ represents an angle of 90° (90 degrees = a square corner).

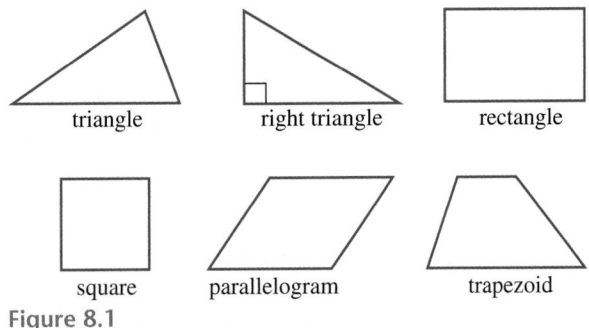

Figure 8.1
Common polygons

The **perimeter** of (or distance around) a two-dimensional figure is the sum of the lengths of its sides. As shown in Figure 8.2, the perimeter of a rectangular scarf 18 inches wide and 2 feet long is 7 feet. (We must first convert 18 inches into 1.5 feet.)

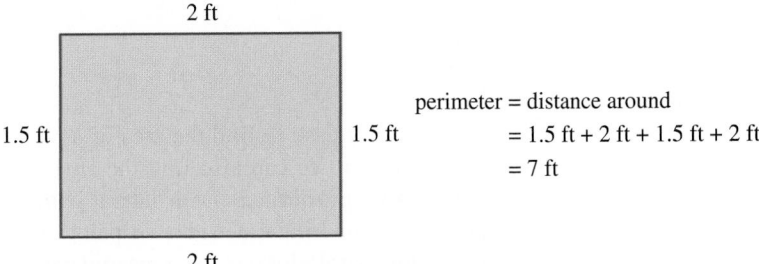

perimeter = distance around
= 1.5 ft + 2 ft + 1.5 ft + 2 ft
= 7 ft

Figure 8.2
Finding the perimeter of a rectangular scarf

The **area** of a two-dimensional figure is the number of square units (for example, square inches, square miles) it takes to fill the interior of that figure. As shown in Figure 8.3, a rectangular rug 6 feet long and 3 feet 6 inches wide has an area of 21 square feet (or 21 ft²).

area = base × height
= (6 ft)(3.5 ft)
= 21 square ft
= 21 ft²

3.5 ft

6 ft

Figure 8.3
Finding the area of a rectangular scarf

It has been suggested that the concepts of square units and area have their origins in the weaving of fabric. Single strands of yarn have a linear, or one-dimensional, measurement, such as inches or feet. However, when an equal number of "horizontal" and "vertical" strands are woven together on a loom, a square figure is formed. Therefore, a natural way to measure the amount of cloth created from the strands is to employ units consisting of squares (square feet). (See Figure 8.4.)

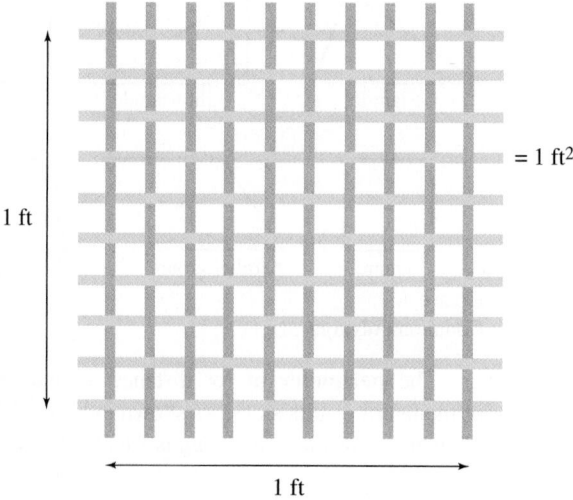

= 1 ft²

1 ft

1 ft

Figure 8.4
Woven cloth

It is very easy to find the area of a rectangle or square. On the basis of these quadrilaterals, we can also find the areas of triangles, trapezoids, and parallelograms. A **trapezoid** is a quadrilateral with one pair of parallel sides; a **parallelogram** is a quadrilateral with two pairs of parallel sides. The area of a **triangle, rectangle,** parallelogram, or trapezoid can be found by use of the appropriate formula given in Figure 8.5, with A = area, b = base, h = height, and b_1 and b_2 = bases.

Figure 8.5
Area formulas for common polygons

(a) **triangle** $A = \frac{1}{2} bh$

(b) **rectangle** $A = bh$

(c) **parallelogram** $A = bh$

(d) **trapezoid** $A = \frac{1}{2}(b_1+b_2)h$

Why does the area of a triangle equal one-half the product of the base times the height? The answer lies in the formula for the area of a rectangle. A triangle can be divided into two smaller triangles. If copies of these smaller triangles are then "added on" to the original triangle, a rectangle of area $b \cdot h$ can be formed, as shown in Figure 8.6. Because the area of the original triangle is half that of the rectangle, we have the desired result.

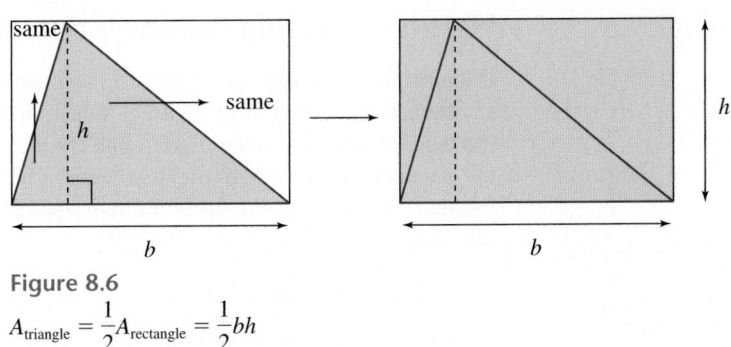

Figure 8.6
$A_{\text{triangle}} = \frac{1}{2}A_{\text{rectangle}} = \frac{1}{2}bh$

Why does the area of a parallelogram equal the product of the base times the height? Once again, the answer lies in the formula for the area of a rectangle. A parallelogram can be rearranged to form a rectangle of area $b \cdot h$, as shown in Figure 8.7. That is, the area of the parallelogram is the same as the area of the rectangle, and we have the desired result.

Why does the area of a trapezoid equal one-half the product of the sum of the bases times the height? Yet again, the answer lies in the formula for the area of a rectangle! The two triangular "tips" of the trapezoid can be cut off and rearranged to form a rectangle, as shown in Figure 8.8. The base of the rectangle equals the average of the two bases of the trapezoid; that is, $b_{\text{rectangle}} = (b_1 + b_2)/2$. Thus, we have the desired result.

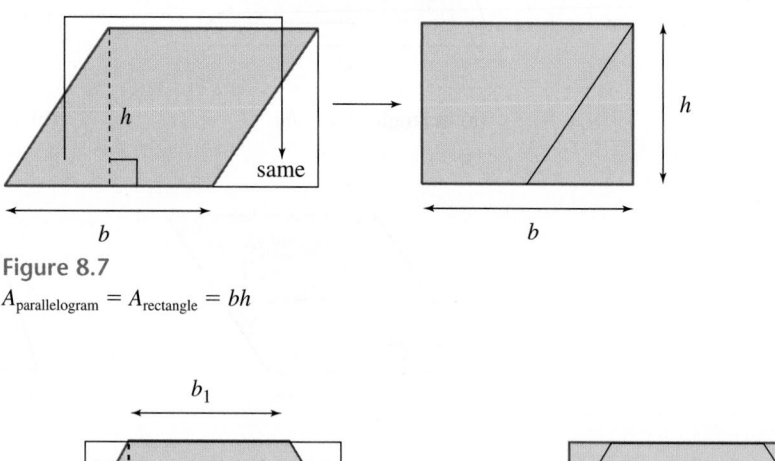

Figure 8.7

$A_{\text{parallelogram}} = A_{\text{rectangle}} = bh$

Figure 8.8

$A_{\text{trapezoid}} = A_{\text{rectangle}} = \left(\dfrac{b_1 + b_2}{2}\right)(h)$

Heron's Formula for the Area of a Triangle

If the height of triangle is not known, we cannot use the common formula $A = \frac{1}{2}bh$; an alternative method must be used. When the lengths of all three sides are known, the "semiperimeter" can be used to find the area of the triangle. As the name implies, the semiperimeter is half the perimeter. The formula is referred to as Heron's Formula, in honor of the ancient Greek mathematician Heron of Alexandria (circa A.D. 75).

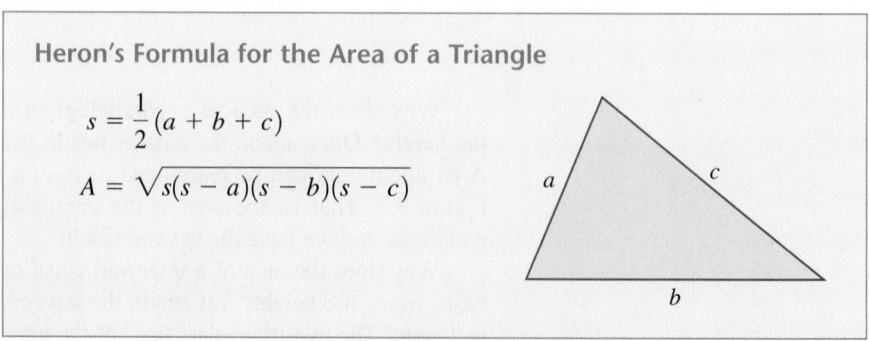

Heron's Formula for the Area of a Triangle

$$s = \frac{1}{2}(a + b + c)$$

$$A = \sqrt{s(s - a)(s - b)(s - c)}$$

EXAMPLE 1   Assuming the same growing conditions, which of the fields shown in Figures 8.9 and 8.10 would produce more grain?

a.

b.

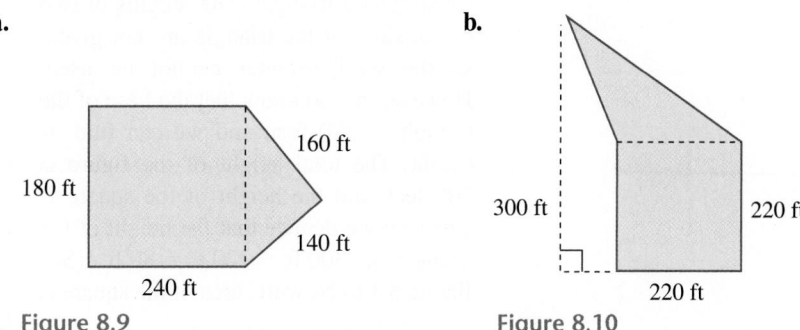

Figure 8.9   Figure 8.10

Solution   The field with the larger area would produce more grain.

Finding the Area of the Field in Part (a)   The field consists of a rectangle and a triangle, as shown in Figure 8.11. To find the total area, we must find the area of each separate shape and add the results together.

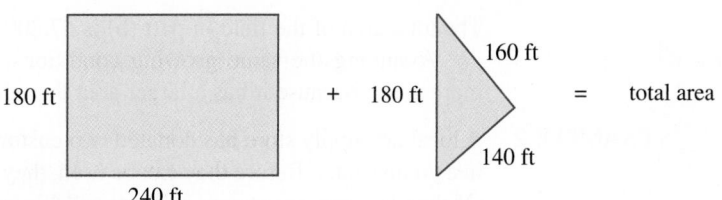

Figure 8.11
Finding the area of a field.

$$A_{\text{rectangle}} + A_{\text{triangle}} = A_{\text{total}}$$

The area of the rectangle is

$$A_{\text{rectangle}} = bh = (240 \text{ ft})(180 \text{ ft}) = 43{,}200 \text{ ft}^2$$

To find the area of the triangle, we use the semiperimeter, because the lengths of all three sides are known:

$$s = \frac{1}{2}(a + b + c) = \frac{1}{2}(180 \text{ ft} + 160 \text{ ft} + 140 \text{ ft}) = 240 \text{ ft}$$

Using Heron's Formula, we have

$$\begin{aligned}
A_{\text{triangle}} &= \sqrt{240(240 - 180)(240 - 160)(240 - 140)} \\
&= \sqrt{(240 \text{ ft})(60 \text{ ft})(80 \text{ ft})(100 \text{ ft})} \\
&= \sqrt{115{,}200{,}000 \text{ ft}^4} \\
&= 10{,}733.12629 \ldots \text{ ft}^2 \\
&\approx 10{,}733 \text{ ft}^2
\end{aligned}$$

$$\begin{aligned}
A_{\text{total}} &= A_{\text{rectangle}} + A_{\text{triangle}} = 43{,}200 \text{ ft}^2 + 10{,}733 \text{ ft}^2 \\
&= 53{,}933 \text{ ft}^2
\end{aligned}$$

The total area of the field in part (a) is 53,933 square feet.

Finding the Area of the Field in Part (b)

The field in part (b) is composed of a square and a triangle. The lengths of two of the sides of the triangle are not given, so the semiperimeter cannot be used. However, we do know that the base of the triangle is 220 feet, and we can find its height. The total height of the figure is 300 feet, and the height of the square is 220 feet; we deduce that the height of the triangle is 300 ft − 220 ft = 80 ft. (See Figure 8.12.) Now the area of the square is

$$A_{\text{square}} = b^2 = (220 \text{ ft})^2 = 48,400 \text{ ft}^2$$

The area of the triangle is

$$A_{\text{triangle}} = \frac{1}{2}bh = \frac{1}{2}(220 \text{ ft})(80 \text{ ft}) = 8,800 \text{ ft}^2$$

$$A_{\text{total}} = A_{\text{square}} + A_{\text{triangle}} = 48,400 \text{ ft}^2 + 8,800 \text{ ft}^2$$
$$= 57,200 \text{ ft}^2$$

The total area of the field in part (b) is 57,200 square feet.

Assuming the same growing conditions, the field in part (b) would produce more grain, because it has a larger area than the field in part (a). ∎

Figure 8.12
Finding the area of a field

EXAMPLE 2 A local art supply store has donated two custom-size canvases to be used for murals in a youth center. Before they can be used, they must be treated with a special primer. Of the two canvases shown in Figures 8.13 and 8.14, which requires more primer?

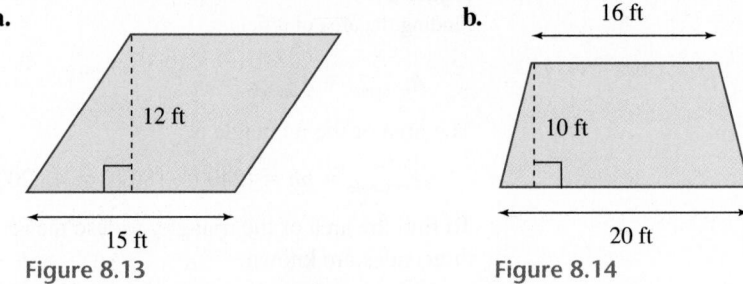

a.

12 ft

15 ft

Figure 8.13

b.

16 ft

10 ft

20 ft

Figure 8.14

Solution The canvas with the larger area requires more primer.

Finding the Area of the Parallelogram in Part (a)

$$A_{\text{parallelogram}} = bh = (15 \text{ ft})(12 \text{ ft}) = 180 \text{ ft}^2$$

The canvas in part (a) contains 180 square feet of material to be primed.

Finding the Area of the Trapezoid in Part (b)

$$A_{\text{trapezoid}} = \frac{1}{2}(b_1 + b_2)h$$

$$= \frac{1}{2}(20 \text{ ft} + 16 \text{ ft})(10 \text{ ft})$$

$$= \frac{1}{2}(36 \text{ ft})(10 \text{ ft})$$

$$= 180 \text{ ft}^2$$

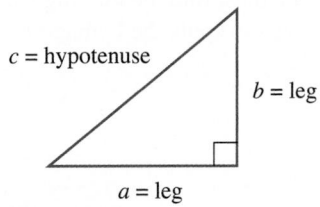

Figure 8.15
Right triangle

The canvas in part (b) also contains 180 square feet of material to be primed. Because the figures have the same area, each would require the same amount of primer. ∎

Right Triangles

If one of the angles of a triangle is a right angle (a square corner, or 90°), the triangle is called a **right triangle.** The side opposite the right angle is called the **hypotenuse** and is labeled c, while the remaining two sides are called the **legs** and are labeled a and b, as shown in Figure 8.15.

An early military handbook utilized the Pythagorean Theorem. Because $18^2 + 24^2 = 30^2$, a 30-foot ladder is required to reach the top of a 24-foot wall from the opposite side of an 18-foot moat.

A special relationship exists between the hypotenuse and the legs of a right triangle. Over 2,000 years ago, early geometers observed that if the longest side of a right triangle (the hypotenuse) was squared, the number obtained was always the same as the sum of the squares of the two other sides (the legs). This observation was proved to be true for *all* right triangles. It is referred to as the **Pythagorean Theorem,** in honor of the ancient Greek mathematician Pythagoras of Samos (circa 580 B.C.), although the result was known to earlier peoples.

Pythagorean Theorem

For any right triangle, the square of the hypotenuse equals the sum of the squares of the legs.

$$c^2 = a^2 + b^2$$

c = hypotenuse
b = leg
a = leg

The converse of this theorem is also true; if the sides of a triangle satisfy the relationship $c^2 = a^2 + b^2$, then the triangle is a right triangle.

Figure 8.16
A right triangle

EXAMPLE 3   The length of the hypotenuse of a right triangle is 25 feet, and the length of one of the legs is 7 feet. (See Figure 8.16.) Find the area of the triangle.

Solution   To find the area of a triangle, we must know either a base and its corresponding (perpendicular) height, or all three sides. In this case, we must find the missing side of the given triangle. Because we have a right triangle, we can apply the Pythagorean Theorem to find the missing leg, b:

$$a^2 + b^2 = c^2$$
$$b^2 = c^2 - a^2$$
$$b^2 = (25)^2 - (7)^2$$
$$b^2 = 625 - 49 = 576$$
$$b = \sqrt{576} = 24 \text{ ft}$$

Now we find the area of the triangle:

$$A = \frac{1}{2}\,bh$$
$$= \frac{1}{2}\,(24 \text{ ft})(7 \text{ ft})$$
$$= 84 \text{ ft}^2$$

The area of the triangle is 84 square feet. (We obtain the same answer if we use Heron's Formula.)   ∎

Circles

Many people consider the circle a perfect geometric figure; it has no beginning or end, it has total symmetry, and it motivated the invention of the famous irrational (nonfraction) number π **(pi).** The early Greeks defined a **circle** as the set of all points in a plane equidistant from a fixed point. The fixed point is called the **center** of the circle; a line segment from the center to any point on the circle is called a **radius;** and any line segment connecting two points on the circle and passing through the center is called a **diameter.** (The length of a diameter is twice the length of a radius.) See Figure 8.17.

Recall that the distance around a figure is called its *perimeter*; however, as a special case, the distance around a circle is called its **circumference.** Thousands of years ago, geometers observed a curious relationship: If they measured the circumference C of *any* circle (probably by using a string or a rope), they found that it was always a little bit longer than 3 times the diameter d of the circle, as shown in Figure 8.18. In other words, the ratio of circumference to diameter is constant: circumference/diameter = constant. This constant number, which is a little larger than 3, is represented by the Greek letter π (read "pi"; rhymes with *sly*). Hence,

$$\frac{\text{circumference}}{\text{diameter}} = \pi \qquad \text{or} \qquad \text{circumference} = \pi \cdot \text{diameter}$$

radius — center — diameter

Figure 8.17
A circle

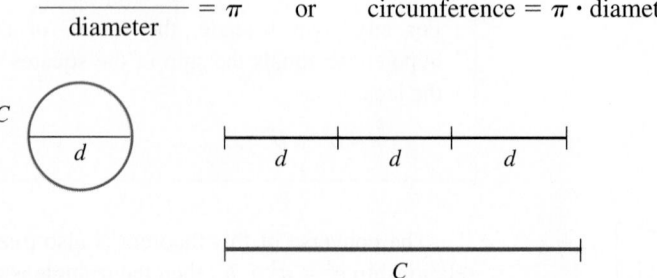

Figure 8.18
Circumference of a circle

Mathematicians have shown that π is irrational; that is, it cannot be written exactly as a fraction, and its decimal expansion never repeats or terminates. Computers have calculated π to hundreds of thousands of decimal places, and we have

$$\pi \approx 3.14159265358979323846264338327 9 \ldots$$

Various approximations of π have been used throughout history (some of them are investigated in Sections 8.3 and 8.4). Today, the most commonly used classroom estimates are 3.1416 and $\frac{22}{7}$. When π is needed in calculations, simply press the appropriate button on your calculator. If your calculator doesn't have a $\boxed{\pi}$ button, use $\pi = 3.1416$.

Pi is used to calculate the circumference and the area of a circle. We will investigate the origins of the area formula in Section 8.3.

Circumference and Area of a Circle

The **circumference** C of a circle of radius r is

$$C = 2\pi r$$

The **area** A of a circle of radius r is

$$A = \pi r^2$$

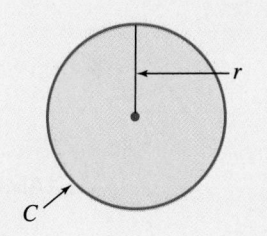

EXAMPLE 4 A Norman window consists of a rectangle with a semicircle mounted on top. For the window shown in Figure 8.19, find the following.

a. the area **b.** the perimeter

Solution **a.** The area of the entire window is the sum of the areas of the rectangle and the semicircle. A semicircle is half a circle, so the semicircular region has area equal to one-half the area of a complete circle. The base of the rectangle is the same as the diameter of the semicircle. Hence, the diameter is 4 feet, and the radius is 2 feet (see Figure 8.20).

$$A_{\text{total}} = A_{\text{rectangle}} + A_{\text{semicircle}}$$
$$= bh + \frac{1}{2}(\pi r^2)$$
$$= (4 \text{ ft})(6 \text{ ft}) + \frac{1}{2}\pi(2 \text{ ft})^2$$
$$= 24 \text{ ft}^2 + 2\pi \text{ ft}^2$$
$$= (24 + 2\pi) \text{ ft}^2$$
$$= 30.28318531 \ldots \text{ ft}^2$$

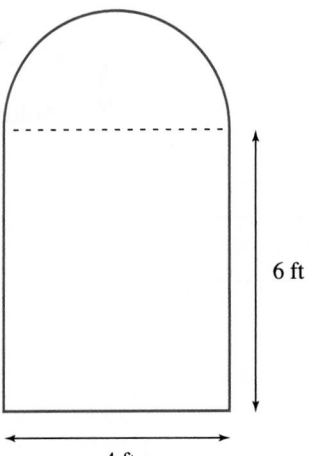

6 ft

4 ft

Figure 8.19
A Norman window

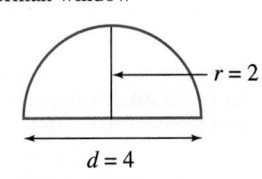

$r = 2$

$d = 4$

Figure 8.20
Half a circle

Thus, the area of the Norman window is approximately 30.3 square feet.
b. The perimeter of the window consists of a semicircle (one-half the circumference of a circle), one horizontal line segment, and two vertical line segments.

$$P = \frac{1}{2}C + b + 2h$$

$$= \frac{1}{2}(\pi \cdot 4 \text{ ft}) + 4 \text{ ft} + 2(6 \text{ ft})$$

$$= 2\pi \text{ ft} + 16 \text{ ft}$$

$$= 22.28318531\ldots \text{ ft}$$

Thus, the perimeter of the Norman window is approximately 22.3 feet. ∎

EXAMPLE 5   Like all the planets, Uranus revolves around the sun in an elliptical orbit. Consequently, the distance between Uranus and the sun varies from approximately 1,703 million to 1,866 million miles.

a. What is the (approximate) average distance from Uranus to the sun?
b. On average, how far from the sun is Uranus in AU?
c. Interpret the answer to part (b) in terms of the earth's distance from the sun.

Solution   **a.** To find the (approximate) average distance, we add the minimum and maximum distances and divide by 2 to obtain

$$\frac{1,703 + 1,866}{2} = 1,784.5 \text{ million miles}$$

b. To convert miles to astronomical units, we divide 1,784.5 million miles by 93 million miles (= 1 AU) to obtain

$$1,784.5 \text{ million miles} \times \frac{1 \text{ AU}}{93 \text{ million miles}} = 19.18817204\ldots \text{ AU}$$

Therefore, the average distance from Uranus to the sun is approximately 19.2 AU.
c. Because 1 AU represents the earth's average distance from the sun, a distance of 19.2 AU indicates that Uranus is slightly more than 19 times farther from the sun than is the earth. ∎

ThomsonNOW
for Liberal Arts Mathematics
www.thomsonedu.com/1pass.

EXAMPLE 6   The star closest to earth (other than the sun) is Proxima Centauri. Scientists believe that this star is approximately 40 trillion kilometers from the earth.

a. How far, in astronomical units, is Proxima Centauri from the earth?
b. How far, in light-years, is Proxima Centauri from the earth?
c. How far, in parsecs, is Proxima Centauri from the earth?

Solution   **a.** To convert kilometers to astronomical units, we divide 40 trillion km by 150 million km (= 1 AU) and obtain

$$40{,}000{,}000{,}000{,}000 \text{ km} \times \frac{1 \text{ AU}}{150{,}000{,}000 \text{ km}} = 266{,}667 \text{ AU}$$

Therefore, the distance from Proxima Centauri to the earth is approximately 266,667 AU; the distance between Proxima Centauri and the earth is 266,667 times the distance between our sun and the earth.

Geometry in the Real World (and Universe): Astronomical Measurement

What is the diameter of a tennis ball? How tall is the Statue of Liberty? What is the distance between London and Paris or between San Francisco and Singapore? When measuring objects or distances, most Americans use the familiar units of inches, feet, and miles; elsewhere in the world, the metric units of millimeters, centimeters, meters, and kilometers are prevalent. The answers to the previous questions can be found by browsing through an almanac; the diameter of a tennis ball is $2\frac{1}{2}$ inches (63.5 millimeters), Ms. Liberty is 151 feet (46 meters) tall (from base to torch), London and Paris are 210 miles (338 kilometers) apart, and there are 8,440 miles (13,583 kilometers) between San Francisco and Singapore.

How far is it from the earth is the sun? What is the distance between planets, say, Mercury and Pluto? How far must one travel to reach the next galaxy? Obviously, these distances are great; miles and kilometers pale in magnitude when astronomical distances are involved. Consequently, scientists have designed various standard units to express and calculate distances within our solar system and beyond. Three common units, from smallest to largest, are the **astronomical unit (AU), light-year (ly),** and **parsec (pc).**

The earth travels around the sun in an elliptical orbit. (See Section 8.6.) Consequently, the distance between the earth and the sun varies from approximately 91.4 million to 94.5 million miles. Scientists have made the following definition: 1 astronomical unit (AU) = the mean (or average) distance between the earth and the sun. Consequently, 1 AU ≈ (91.4 + 94.5)/2 = 92.95 million miles. It is common to round off this calculation to a whole number and say that

1 AU is about 93 million miles, or about 150 million kilometers. When distances within our solar system are measured, AU are the commonly used units.

Objects outside of our solar system are very, very far away, and AU are considered to be relatively small units. The next larger unit in measuring distance is the light-year. Scientists have made the following definition: 1 light-year (ly) = the distance that light can travel in one year through a vacuum (empty space). Because light travels at a speed of about 186,282.34 miles per second, 1 light-year equals approximately

$$\frac{186{,}282.34 \text{ miles}}{\text{second}} \times \frac{3{,}600 \text{ seconds}}{\text{hour}}$$

$$\times \frac{24 \text{ hours}}{\text{day}} \times \frac{365.25 \text{ days}}{\text{year}} \times 1 \text{ year}$$

$$= 5{,}878{,}623{,}573{,}000 \text{ miles}$$

Most references state that 1 light-year is about 5.88 trillion miles (mi), or about 9.46 trillion kilometers (km), or about 63,240 astronomical units (AU).

After the light-year, the next larger unit in measuring distance is the parsec. The exact definition of a parsec is based on a right triangle and an acute angle of measure: 1/3600th of a degree. This definition will be examined in Section 8.5. For our purposes here, we define 1 parsec (pc) = 3.26 light-years (ly).

Summarizing the commonly used units in astronomy, we have the following:

1 AU = 93 million mi = 150 million km

1 ly = 5.88 trillion mi = 9.46 trillion km

     = 63,240 AU

1 pc = 3.26 ly

b. To convert astronomical units to light-years, we divide 266,677 AU by 63,240 AU (= 1 ly) and obtain

$$266,667 \text{ AU} \times \frac{1 \text{ ly}}{63,240 \text{ AU}} = 4.216745731 \ldots \text{ly}$$

Therefore, the distance from Proxima Centauri to the earth is approximately 4.22 ly; light emitted from Proxima Centauri takes a little more than 4 years to reach the earth.

c. To convert light-years to parsecs, we divide 4.22 ly by 3.26 ly (= 1 pc) and obtain

$$4.22 \text{ ly} \times \frac{1 \text{ pc}}{3.26 \text{ ly}} = 1.294478528 \ldots \text{pc}$$

Therefore, the distance from Proxima Centauri to the earth is approximately 1.29 pc. ∎

8.1 Exercises

When necessary, round off answers to one decimal place. In Exercises 1–8, find the area of each figure.

1.

2.

3.

4.

5.

6.

7.

8.

In Exercises 9 and 10, find (a) the area and (b) the circumference of the circle.

9.

6.5 in.

10.

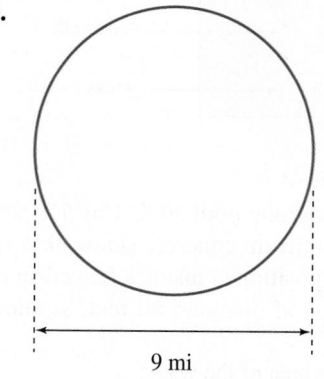

9 mi

In Exercises 11–20, find (a) the area and (b) the perimeter of each figure.

11.

5 m

13 m

12.

4 km

5 km

13.

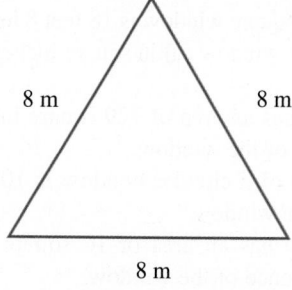

8 m   8 m

8 m

14.

9 cm   9 cm

6 cm

15.

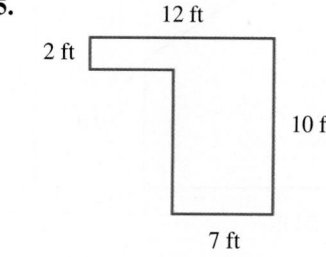

12 ft

2 ft

10 ft

7 ft

16.

11 m

14 m

5 m

8 m

17.

100 yd

18.

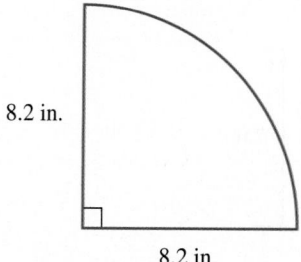

8.2 in.

8.2 in.

19.

20.

In Exercises 21 and 22, find (a) the area and (b) the perimeter of each Norman window.

HINT: See Example 4.

21.

22.

23. A circular swimming pool has diameter 50 feet and is centered in a fenced-in square region measuring 80 feet by 80 feet. A concrete sidewalk 5 feet wide encircles the pool, and the rest of the region is grass, as shown in Figure 8.21.
   a. Find the surface area of the water.
   b. Find the area of the concrete sidewalk.
   c. Find the area of the grass.

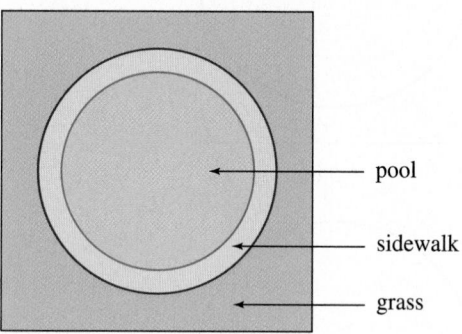

Figure 8.21
Diagram for Exercise 23

24. A rectangular swimming pool 30 feet by 15 feet is surrounded by a uniform concrete sidewalk 5 feet wide. The pool is positioned inside a fenced-in circular grassy region of diameter 60 feet, as shown in Figure 8.22.
   a. Find the surface area of the water.
   b. Find the area of the concrete sidewalk.
   c. Find the area of the grass.

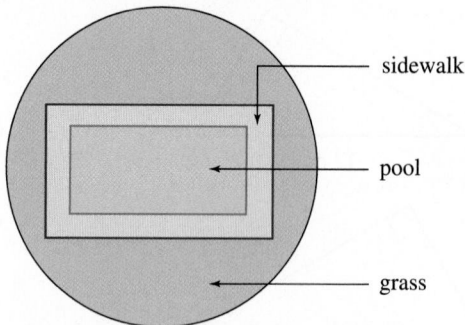

Figure 8.22
Diagram for Exercise 24

25. The perimeter of a square window is 18 feet 8 inches. Find the area of the window (a) in square inches and (b) in square feet.

26. A square window has an area of 729 square inches. Find the perimeter of the window.

27. The circumference of a circular window is 10 feet. Find the area of the window.

28. A circular window has an area of 10 square feet. Find the circumference of the window.

29. You jog $\frac{3}{4}$ mile due north, then jog $1\frac{1}{2}$ miles due east, and then return to your starting point via a straight-line path. How many miles have you jogged?

30. You walk 100 yards due south, then 120 yards due west, and then 30 yards due north. How far are you from your starting point?

31. A 10-foot ladder leans against a wall. If the base of the ladder is 6 feet from the wall, how far up the wall does the ladder reach?

32. A ladder is leaning against a building. If the bottom of the ladder is $7\frac{1}{2}$ feet from the wall and the top of the ladder is 10 feet above the ground, how long is the ladder?

33. An oval athletic field is the union of a rectangle and semicircles at opposite ends, as shown in Figure 8.23.
    a. Find the total area of the field.
    b. Find the distance around the field.

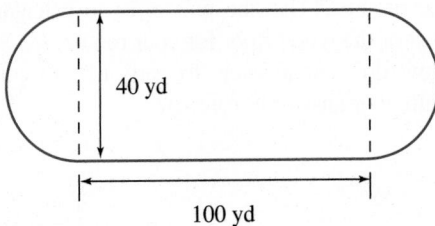

Figure 8.23
Diagram for Exercise 33

34. An oval athletic field is the union of a square and semicircles at opposite ends, as shown in Figure 8.24. If the total area of the field is 1,300 square yards, find the dimensions of the square.

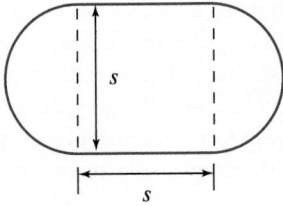

Figure 8.24
Diagram for Exercise 34

35. The lengths of the edges of a triangular canvas are 100 feet, 140 feet, and 180 feet. You want to waterproof the tarp. If one can of waterproofing will treat 1,000 square feet, how many cans do you need?

36. Steve Loi wants to fertilize his backyard. The yard is a triangle whose sides are 110 feet, 120 feet, and

150 feet. If one bag of fertilizer will cover 1,200 square feet, how many bags does he need?

37. Reddie's Pizza Bash is famous for its Loads-o-Meat Deluxe Pizza. It comes in three sizes: small (13 inches) for $11.75, large (16 inches) for $14.75, and super (19 inches) for $22.75. Which size is the best deal?

   HINT: Find the price per square inch.

38. The best-selling pizza at Magic Mushroom Pizza Deli is the Vegetarian Surprise. It comes in three sizes: small (12 inches) for $11.25, large (14 inches) for $14.75, and super (18 inches) for $20.75. Which size is the best deal?

   HINT: Find the price per square inch.

39. Mercury (the closest planet to the sun) revolves around the sun in an elliptical orbit. Consequently, the distance between Mercury and the sun varies from approximately 28.6 million to 43.4 million miles.
    a. What is the (approximate) average distance from Mercury to the sun?
    b. On average, how far from the sun is Mercury in AU?
    c. Interpret the answer to part (b) in terms of the earth's distance from the sun.

40. Pluto (the farthest planet to the sun) revolves around the sun in an elliptical orbit. Consequently, the distance between Pluto and the sun varies from approximately 2,756 million to 4,539 million miles.
    a. What is the (approximate) average distance from Pluto to the sun?
    b. On average, how far from the sun is Pluto in AU?
    c. Interpret the answer to part (b) in terms of the earth's distance from the sun.

41. Scientists believe that the star Polaris (known as the North Star) is approximately 4,080 trillion kilometers from the earth.
    a. How far, in astronomical units, is Polaris from the earth?
    b. How far, in light-years, is Polaris from the earth?
    c. How far, in parsecs, is Polaris from the earth?

42. Scientists believe that the star Sirius (known as the Dog Star) is approximately 81.5 trillion kilometers from the earth.
    a. How far, in astronomical units, is Sirius from the earth?
    b. How far, in light-years, is Sirius from the earth?
    c. How far, in parsecs, is Sirius from the earth?

Answer the following questions using complete sentences and your own words.

PROJECTS

43. The purpose of this project is to verify the Pythagorean Theorem.

   a. On a sheet of paper that has a square corner, measure 3 inches from the corner along the edge of the paper and put a dot. Now measure 4 inches from the same corner along the other edge of the paper and put a dot. Use the Pythagorean Theorem to calculate the distance between the dots. Now, measure the distance between the dots. How does your measurement compare to your calculation?

   b. Repeat part (a) using 6 and 8 inches.

   c. Repeat part (a) using $7\frac{1}{2}$ and 10 inches.

   d. Repeat part (a) using 3 and 4 centimeters.

   e. Repeat part (a) using 5 and 12 centimeters.

   f. Repeat part (a) using 7 and 24 centimeters.

44. The purpose of this project is to calculate the cost of materials to resurface a floor.

   a. Pick a rectangular room in your apartment or house. Measure the dimensions of the floor, and calculate its area in square feet.

   b. Convert your answer from part (a) to square yards.

   (*HINT:* How many square feet are in 1 square yard?)

   c. Suppose you want to install a new floor covering in your selected room. Of the three materials, carpet, vinyl, and wood, which do you think would be most expensive? Least expensive?

   d. Go to a local carpet and floor covering store, and select a style of carpet for your room. What is the unit price of the product? Use the unit price to calculate the cost of the carpet for your room.

   e. Instead of carpet, select a vinyl product for the floor of your room. What is the unit price of the product? Use the unit price to calculate the cost of the vinyl for the floor of your room.

   f. Instead of carpet or vinyl, select a wood floor covering for your room. What is the unit price of the product? Use the unit price to calculate the cost of the wood floor for your room.

   g. How do your answers to parts (d)–(f) compare with your answer to part (c)?

8.2   Volume and Surface Area

Having examined the perimeter (one-dimensional measurement) and area (two-dimensional measurement) of common geometric figures in Section 6.1, we now explore the geometry of three-dimensional figures. In particular, we investigate the calculation of volume and surface area and apply the results to a variety of situations.

Problem Solving

Outdoors Unlimited, a camping supply store, is having a sale on its demonstration models, discontinued lines, and irregular items. Two water containers are on sale—one cylindrical and the other rectangular, as shown in Figure 8.25. You want to purchase the container that holds the most water, but unfortunately the original information on the capacity of each container has been lost. Which container would you choose? To compare the capacities of the two vessels, you must find their volumes.

Volume is a measure of the amount of space occupied by a three-dimensional object. Volume was originally investigated in the course of bartering and commerce. When grain, wine, and oil were traded, various vessels or containers were used as standards. However, different cultures used different standard vessels, so elaborate conversion systems were necessary. Pints, quarts, gallons, hogsheads, cords, pecks, and bushels are the modern-day legacy of early vessel measurement. Today, the fundamental figure used in the calculation of volume is the cube; volume is expressed in terms of cubic units such as cubic feet (ft^3) and cubic centimeters (cc). (See Figure 8.26.)

Figure 8.25
Containers

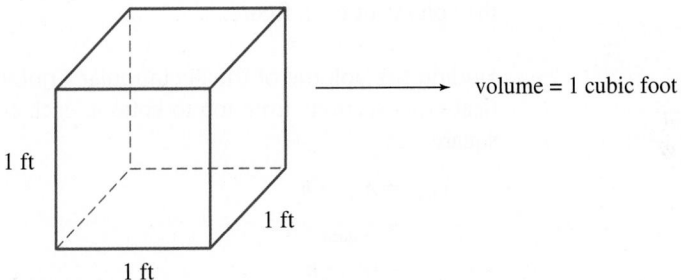

Figure 8.26
1 cubic foot

The volume of a rectangular box is found by multiplying its length times its width times its height; that is, $V = L \cdot w \cdot h$. Notice that $(L \cdot w)$ equals the area of the (rectangular) base of the box. Consequently, the volume of the box can be found by multiplying the area of the base times the height of the box, as shown in Figure 8.27.

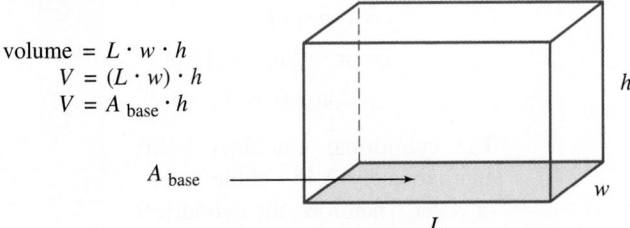

volume $= L \cdot w \cdot h$
$V = (L \cdot w) \cdot h$
$V = A_{base} \cdot h$

Figure 8.27
A rectangular box

A rectangular box has the property that every cross section, or slice, taken parallel to the base produces an identical rectangle. It is precisely this property that allows us to calculate the volume of the box simply by multiplying the area of the base times the height. In general, the volume of any figure that has *identical* cross sections (same shape and size) from top to bottom is found by multiplying the area of the base times the height of the figure.

Volume of a Figure Having Identical Cross Sections

If a three-dimensional figure has **height** h and identical cross sections from top to bottom, each of **area** A, the **volume** V of the figure is

$$V = A \cdot h$$

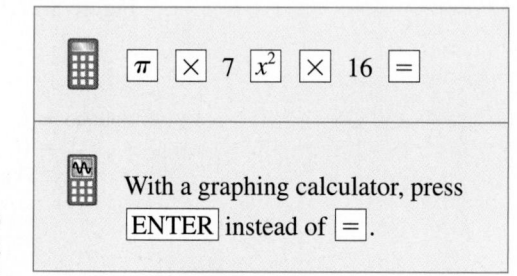

identical cross sections (top to bottom)

ThomsonNOW™
for Liberal Arts Mathematics
www.thomsonedu.com/1pass.

EXAMPLE 1   Referring to the cylindrical and rectangular water containers shown in Figure 8.25, determine which holds the most water.

Solution   The holding capacity of each container is determined by its volume, so we must find the volume of each figure.

Finding the Volume of the Rectangular Container   The rectangular box has identical cross sections from top to bottom; each cross section is a 12-inch-by-12-inch square.

$$
\begin{aligned}
V_{\text{box}} &= A_{\text{base}} \cdot h \\
&= A_{\text{square}} \cdot h \\
&= (b^2) \cdot h \\
&= (12 \text{ in.})^2 \cdot (17 \text{ in.}) \\
&= 2{,}448 \text{ in.}^3
\end{aligned}
$$

The rectangular container holds 2,448 cubic inches of water.

Finding the Volume of the Cylindrical Container   The **cylinder** has identical cross sections from top to bottom; each is a circle of radius 7 inches (half the diameter).

$$
\begin{aligned}
V_{\text{cylinder}} &= A_{\text{base}} \cdot h \\
&= A_{\text{circle}} \cdot h \\
&= (\pi r^2) \cdot h \\
&= \pi \cdot (7 \text{ in.})^2 \cdot (16 \text{ in.}) \\
&= 2{,}463.00864 \ldots \text{ in.}^3
\end{aligned}
$$

| π | $\times$ | 7 | x^2 | $\times$ | 16 | $=$ |

With a graphing calculator, press ENTER instead of $=$.

The cylindrical container holds approximately 2,463 cubic inches of water. Therefore, the cylindrical container holds about 15 cubic inches more than the rectangular container. ∎

Surface Area

The **surface area** of a three-dimensional figure is the sum total of the areas of all the surfaces that compose the figure. Although a formula for the surface area of a rectangular box could be given, it is not necessary because we already know how to find the area of each rectangular face. Nonrectangular objects, however, deserve more attention. Cylinders and spheres have specific formulas requiring more analysis.

EXAMPLE 2   Referring to the cylindrical and rectangular water containers shown in Figure 8.25, determine which has the greater surface area.

Solution   Finding the Surface Area of the Rectangular Container   The surface area of a box is composed of six pieces: the top, the bottom, and the four sides. In this case, the top and bottom are identical squares; each has area b^2. The sides are identical rectangles; each has area $b \cdot h$.

$$A_{box} = 2 \cdot A_{square} + 4 \cdot A_{rectangle}$$
$$= 2(b^2) + 4(b \cdot h)$$
$$= 2(12 \text{ in.})^2 + 4(12 \text{ in.})(17 \text{ in.})$$
$$= 288 \text{ in.}^2 + 816 \text{ in.}^2$$
$$= 1{,}104 \text{ in.}^2$$

The surface area of the rectangular container is 1,104 square inches.

Finding the Surface Area of the Cylindrical Container   The surface area of a cylinder is composed of three pieces: the top, the bottom, and the curved side. The top and bottom are circles, each with an area of πr^2. We find the area of the curved side by cutting it from top to bottom and rolling it out flat. The result is a rectangle with length equal to the circumference of the cylinder, as shown in Figure 8.28.

Figure 8.28
The curved side of a cylinder is a rectangle.

$$A_{cylinder} = 2 \cdot A_{circle} + A_{side}$$
$$= 2(\pi r^2) + (2\pi r) \cdot h$$
$$= 2\pi \cdot (7 \text{ in.})^2 + 2\pi \cdot (7 \text{ in.})(16 \text{ in.})$$
$$= 98\pi \text{ in.}^2 + 224\pi \text{ in.}^2$$
$$= 322\pi \text{ in.}^2$$
$$= 1{,}011.592834 \ldots \text{ in.}^2$$

The surface area of the cylindrical container is approximately 1,012 square inches. Therefore, the surface area of the rectangular container is about 92 square inches more than that of the cylindrical container.   ∎

Comparing Examples 1 and 2, we see that even though the volume of the cylinder is greater than that of the box, the surface area of the cylinder is less than that of the box. In essence, the cylinder is more "efficient" because of its curved surface.

Spheres

How much larger is a basketball than a soccer ball? Depending on which measurement you use for the comparison, different answers are possible. The "size" of a ball can be given in terms of its diameter (one-dimensional measurement), its surface area (two-dimensional measurement), or its volume (three-dimensional measurement).

A **sphere** is the three-dimensional counterpart of a circle; it is the set of all points *in space* equidistant from a fixed point. The fixed point is called the **center** of the sphere, a line segment connecting the center and any point on the sphere is a **radius,** and a line segment connecting any two points of the sphere *and* passing through the center is a **diameter.** The length of a diameter is twice that of a radius. (See Figure 8.29.)

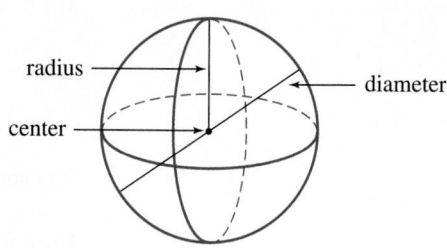

Figure 8.29
A sphere

The ancient Greeks knew the following formulas for determining the volume and surface area of a sphere.

Volume and Surface Area of a Sphere

The **volume** V of a sphere of radius r is

$$V = \frac{4}{3}\pi r^3$$

The **surface area** A of a sphere of radius r is

$$A = 4\pi r^2$$

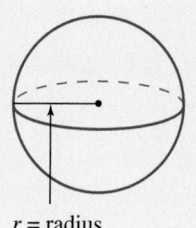

r = radius

EXAMPLE 3   A regulation basketball has a circumference of 30 inches; the circumference of a standard soccer ball is 27 inches. (Each ball is a sphere.)

a. Find and compare the volumes of the two types of balls.
b. Find and compare the surface areas of the two types of balls.

Solution   **a.** To find the volume of a sphere, we must know its radius. We are given the circumference, so we can find the radius as follows:

$$C = 2\pi r$$

$$r = \frac{C}{2\pi} \quad \text{dividing by } 2\pi$$

Finding the Volume of the Basketball   The circumference of a basketball is $C = 30$ inches, so we have

$$r_{\text{basketball}} = \frac{C}{2\pi} = \frac{30 \text{ in.}}{2\pi} = \frac{15}{\pi} \text{ in.}$$

Rather than using a calculator at this point, we substitute this exact value directly into the volume formula and then use a calculator:

$$V_{\text{basketball}} = \frac{4}{3}\pi r^3$$

$$= \frac{4}{3}\pi \left(\frac{15}{\pi} \text{ in.}\right)^3$$

$$= 455.9453264 \ldots \text{ in.}^3$$

The volume of the basketball is approximately 456 cubic inches.

Finding the Volume of the Soccer Ball   The circumference of a soccer ball is $C = 27$ inches; thus,

$$r_{\text{soccer ball}} = \frac{C}{2\pi} = \frac{27 \text{ in.}}{2\pi} = \frac{13.5 \text{ in.}}{\pi}$$

Therefore,

$$V_{\text{soccer ball}} = \frac{4}{3}\pi r^3$$

$$= \frac{4}{3}\pi \left(\frac{13.5 \text{ in.}}{\pi}\right)^3$$

$$= 332.3841429 \ldots \text{ in.}^3$$

The volume of the soccer ball is approximately 332 cubic inches.

To compare the volumes, consider the ratio of the larger ball to the smaller ball:

$$\frac{V_{\text{basketball}}}{V_{\text{soccer ball}}} = \frac{456 \text{ in.}^3}{332 \text{ in.}^3} \approx 1.37$$

$$V_{\text{basketball}} \approx 1.37 V_{\text{soccer ball}}$$

$$V_{\text{basketball}} \approx 137\% \text{ of } V_{\text{soccer ball}}$$

Therefore, the volume of the basketball is 37% more than the volume of the soccer ball.

b. To find the surface area of each ball, substitute the radii found in part (a) into the surface area formula.

Finding the Surface Area of the Basketball   Recall that $r_{\text{basketball}} = \frac{15}{\pi}$ in.

$$A = 4\pi r^2$$

$$= 4\pi \left(\frac{15}{\pi} \text{ in.}\right)^2$$

$$= 286.4788976 \ldots \text{ in.}^2$$

$$4 \times \pi \times (\; 15 \div \pi \;) \; x^2 \; =$$

With a graphing calculator, press $\boxed{\text{ENTER}}$ instead of $\boxed{=}$.

The surface area of the basketball is approximately 286 square inches.

Finding the Surface Area of the Soccer Ball   Recall that $r_{\text{soccer ball}} = \frac{13.5}{\pi}$ in.

$$A = 4\pi r^2$$
$$= 4\pi \left(\frac{13.5}{\pi} \text{ in.} \right)^2$$
$$= 232.047907 \ldots \text{ in.}^2$$

The surface area of the soccer ball is approximately 232 square inches.

As with the volume comparison, use the ratio of the larger ball to the smaller ball to compare their surface areas:

$$\frac{A_{\text{basketball}}}{A_{\text{soccer ball}}} = \frac{286 \text{ in.}^2}{232 \text{ in.}^2} \approx 1.23$$
$$A_{\text{basketball}} \approx 1.23 \, A_{\text{soccer ball}}$$
$$A_{\text{basketball}} \approx 123\% \text{ of } A_{\text{soccer ball}}$$

Therefore, the surface area of the basketball is 23% more than the surface area of the soccer ball.   ■

Even though the circumference of a basketball is 11% more than that of a soccer ball ($C_{\text{basketball}}/C_{\text{soccer ball}} = 30/27 = 1.11$), the surface area is 23% more, and the volume is 37% more. These comparisons differ because of the dimensions of the measurements; C is one-dimensional, A is two-dimensional, and V is three-dimensional. Each measurement involves the multiplication of an additional value of r.

Cones and Pyramids

Two other common three-dimensional objects are the cone and the pyramid. A **cone** has a circular base, whereas the base of a **pyramid** is a polygon. In either case, cross sections parallel to the base all have the same shape as the base, but they differ in size; starting at the base, the cross sections get progressively smaller until they reach a single point at the top of the figure. For both a cone and a pyramid, the volume is one-third the product of the area of the base times the height.

Volume of a Cone or a Pyramid

If a cone or pyramid has height h and if the area of the base is A, the **volume** V of the figure is

$$v = \frac{1}{3} A \cdot h$$

ThomsonNOW
for Liberal Arts Mathematics
www.thomsonedu.com/1pass.

EXAMPLE 4   Find and compare the volumes of the cone and pyramid shown in Figure 8.30.

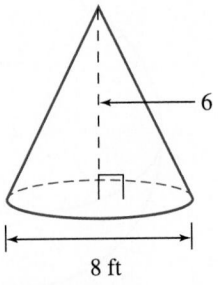

Figure 8.30
A cone and a pyramid

Solution   Finding the Volume of the Cone   We have $r = 8$ ft$/2 = 4$ ft and $h = 6$ ft.

$$V = \frac{1}{3} A \cdot h$$

$$= \frac{1}{3} (\pi r^2) h$$

$$= \frac{1}{3} \pi (4 \text{ ft})^2 (6 \text{ ft})$$

$$= 32 \pi \text{ ft}^3$$

$$= 100.5309649 \ldots \text{ ft}^3$$

The volume of the cone is approximately 101 cubic feet.

Finding the Volume of the Pyramid

$$V = \frac{1}{3} A \cdot h$$

$$= \frac{1}{3} (b^2) h$$

$$= \frac{1}{3} (8 \text{ ft})^2 (6 \text{ ft})$$

$$= 128 \text{ ft}^3$$

The volume of the pyramid is 128 cubic feet.

Comparing the Larger Volume to the Smaller Volume   We have

$$\frac{V_{\text{pyramid}}}{V_{\text{cone}}} = \frac{128 \text{ ft}^3}{101 \text{ ft}^3} \approx 1.27$$

$$V_{\text{pyramid}} \approx 1.27 \, V_{\text{cone}}$$

$$V_{\text{pyramid}} \approx 127\% \text{ of } V_{\text{cone}}$$

Therefore, the volume of the pyramid is 27% larger than the volume of the cone. ∎

8.2 Exercises

When necessary, round off answers to two decimal places.

In Exercises 1–6, find (a) the volume and (b) the surface area of each figure.

1.

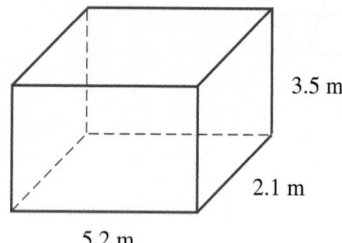

3.5 m
2.1 m
5.2 m

2.

1.5 m
0.6 m
0.6 m

3.

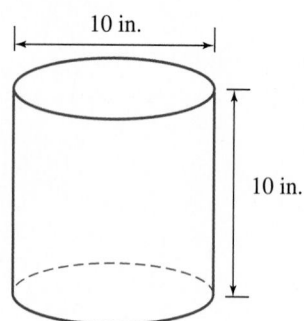

10 in.
10 in.

4.

2 cm
12 cm

5.

$1\frac{3}{4}$ in.

6.

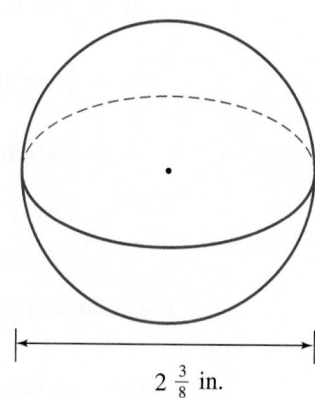

$2\frac{3}{8}$ in.

In Exercises 7–14, find the volume of each figure. All dimensions are given in feet.

7.

4
4

8.

9.

10.

11.

12.

13.

14.

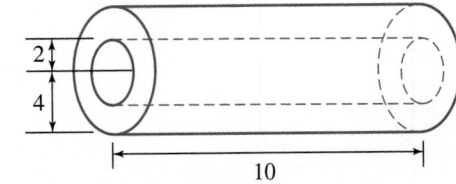

15. From a 10-inch-by-16-inch piece of cardboard, 1.5-inch-square corners are cut out, as shown in Figure 8.31, and the resulting flaps are folded up to form an open box. Find (a) the volume and (b) the external surface area of the box.

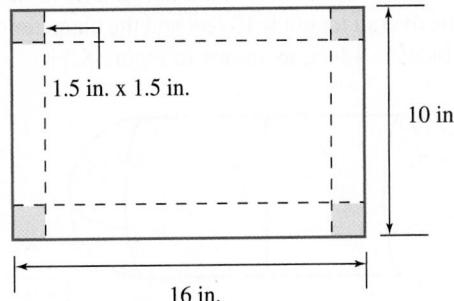

Figure 8.31
Diagram for Exercise 15

16. From a 24-inch-square piece of cardboard, square corners are to be cut out as shown in Figure 8.32, and the resulting flaps folded up to form an open box. Find the volume of the resulting box if (a) 4-inch-square corners are cut out and (b) 8-inch-square corners are cut out.

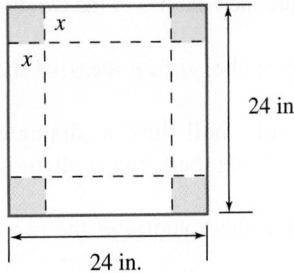

Figure 8.32
Diagram for Exercise 16

17. A grain silo consists of a cylinder with a hemisphere on top. Find the volume of a silo in which the cylindrical part is 50 feet tall and has a diameter of 20 feet, as shown in Figure 8.33.

50 ft

20 ft

Figure 8.33
Diagram for Exercise 17

18. A propane gas tank consists of a cylinder with a hemisphere at each end. Find the volume of the tank if the overall length is 10 feet and the diameter of the cylinder is 4 feet, as shown in Figure 8.34.

4 ft

10 ft

Figure 8.34
Diagram for Exercise 18

19. A regulation baseball (hardball) has a circumference of 9 inches; a regulation softball has a circumference of 12 inches.
   a. Find and compare the volumes of the two types of balls.
   b. Find and compare the surface areas of the two types of balls.

20. A regulation tennis ball has a diameter of $2\frac{1}{2}$ inches; a Ping-Pong ball has a diameter of $1\frac{1}{2}$ inches.
   a. Find and compare the volumes of the two types of balls.
   b. Find and compare the surface areas of the two types of balls.

21. The diameter of the earth is approximately 7,920 miles; the diameter of the moon is approximately 2,160 miles. How many moons could fit inside the earth?

HINT: Compare their volumes.

22. The diameter of the earth is approximately 7,920 miles; the diameter of the planet Jupiter (the largest planet in our solar system) is approximately 88,640 miles. How many earths could fit inside Jupiter?

HINT: Compare their volumes.

23. The diameter of the earth is approximately 7,920 miles; the diameter of the planet Pluto (the smallest) is approximately 1,500 miles. How many Plutos could fit inside the earth?

HINT: Compare their volumes.

24. The diameter of the planet Jupiter (the largest planet in our solar system) is approximately 88,640 miles; the diameter of the planet Pluto (the smallest) is approximately 1,500 miles. How many Plutos could fit inside Jupiter?

HINT: Compare their volumes.

In Exercises 25 and 26, use the following information: Firewood is measured and sold by the cord. A cord of wood is a rectangular pile of wood 4 feet wide, 4 feet high, and 8 feet long, as shown in Figure 8.35. Therefore, one cord = 128 cubic feet.

= 1 cord

4 ft

8 ft

4 ft

Figure 8.35
1 cord of wood

25. A cord of seasoned almond wood costs $190. You paid $190 for a pile that was 4 feet wide, 2 feet high, and 10 feet long. Did you get an honest deal? If not, what should the cost have been?

26. A cord of seasoned oak wood costs $160. You paid $640 for a pile that was 4 feet wide, 6 feet high, and 20 feet long. Did you get an honest deal? If not, what should the cost have been?

27. Tennis balls are packaged three to a cylindrical can. If the diameter of a tennis ball is $2\frac{1}{2}$ inches, find the volume of the can.

28. Golf balls are packaged three to a rectangular box. If the diameter of a golf ball is $1\frac{3}{4}$ inches, find the volume of the box.

29. Ron Thiele bought an older house and wants to put in a new concrete driveway. The driveway will be 36 feet long, 9 feet wide, and 6 inches thick. Concrete (a mixture of sand, gravel, and cement) is measured by the cubic yard. One sack of dry cement mix costs $7.30, and it takes four sacks to mix up 1 cubic yard of concrete. How much will it cost Ron to buy the cement?

30. Marcus Robinson bought an older house and wants to put in a new concrete patio. The patio will be 18 feet long, 12 feet wide, and 3 inches thick. Concrete is measured by the cubic yard. One sack of dry cement mix costs $7.30, and it takes four sacks to mix up 1 cubic yard of concrete. How much will it cost Marcus to buy the cement?

31. The Great Pyramid of Cheops (erected about 2600 B.C.) originally had a square base measuring 756 feet by 756 feet and was 480 feet high. (It has since been eroded and damaged.) Find the volume of the original Great Pyramid.

32. The student union at the University of Utopia (U^2) is a pyramid with a 150-foot square base and a height of 40 feet. Find the volume of the student union at U^2.

33. A water storage tank is an upside-down cone, as shown in Figure 8.36. If the diameter of the circular top is 12 feet and the length of the sloping side is 10 feet, how much water will the tank hold?

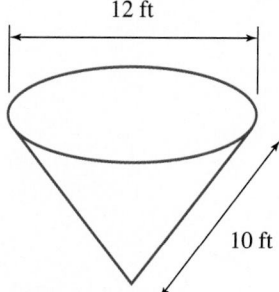

12 ft

10 ft

Figure 8.36
Diagram for Exercise 33

34. The diameter of a conical paper cup is 3.5 inches, and the length of the sloping side is 4.55 inches, as shown in Figure 8.37. How much water will the cup hold?

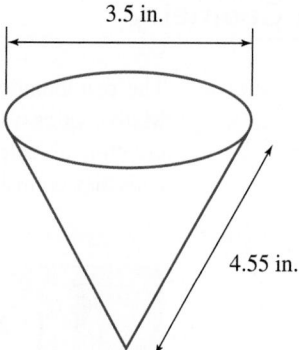

3.5 in.

4.55 in.

Figure 8.37
Diagram for Exercise 34

➤ *Answer the following questions using complete sentences and your own words.*

PROJECTS

35. How do you convert fluid ounces to cubic inches or vice versa? The purpose of this project is to obtain appropriate conversion factors.

   a. Select a canned food product or beverage such as a can of soup, soda, juice, or vegetable. According to the label, how many fluid ounces of product does the can contain?

   b. Measure the height and diameter of the can in inches. Use these measurements to calculate the volume of the can in cubic inches.

   c. Divide the volume calculated in part (b) by the amount given in part (a). This ratio will give the number of cubic inches per fluid ounce.

   d. Your answer to part (c) should be close to 1.8; that is, a commonly used conversion factor is 1 fluid ounce = 1.8047 cubic inches. How accurate is your calculation?

   e. If a can contains 32 fluid ounces of juice, how large is the can in terms of cubic inches?

   f. Now measure the height and diameter of the can in centimeters. Use these measurements to calculate the volume of the can in cubic centimeters.

   g. Divide the volume calculated in part (f) by the amount given in part (a). This ratio will give the number of cubic centimeters per fluid ounce.

   h. Your answer to part (g) should be close to 29.6; that is, a commonly used conversion factor is 1 fluid ounce = 29.5735 cubic centimeters. How accurate is your calculation?

   i. If a can contains 32 fluid ounces of juice, how large is the can in terms of cubic centimeters?

8.3 Egyptian Geometry

The mathematics of the early Egyptians was very practical in its content and use. Mathematics was required in the surveying of land, the construction of buildings, the creation of calendars, and the recordkeeping necessary for commerce. All these activities contributed to the advancement of Egyptian civilization.

Harvest Scene, ca. 1420 B.C. The Metropolitan Museum of Art, Rogers Fund, 1930 (30.4.44)

Found in the Tomb of Menna, this Egyptian wall painting depicts a harvest scene. Notice the use of knotted ropes to measure a field of grain.

Repeated surveying was a necessity for the Egyptians because of the periodic flooding of the Nile. The floodbanks were very fertile, so this land was most valuable and was taxed accordingly. However, when the river flooded its banks, all landmarks and boundaries would be washed away, making it imperative to be able to lay out geometric figures of the correct size and shape.

The Egyptian surveyors used ropes with equally spaced knots to measure distance. In fact, surveyors were referred to as "rope stretchers." These knotted ropes were also used to construct right angles. A right triangle is formed when sides of lengths 3, 4, and 5 are used. A rope with twelve equally spaced knots can easily be stretched to form such a triangle, as shown in Figure 8.38.

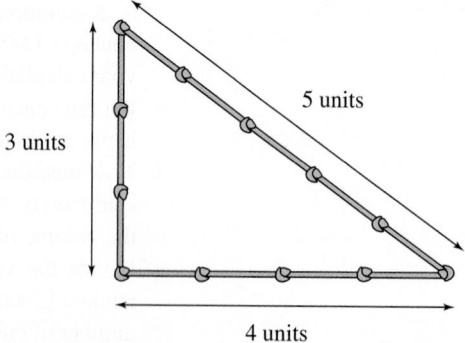

Figure 8.38
A knotted rope

ThomsonNOW
for Liberal Arts Mathematics
www.thomsonedu.com/1pass.

EXAMPLE 1   Which of the configurations of knotted ropes shown in Figures 8.39 and 8.40 would form a right triangle?

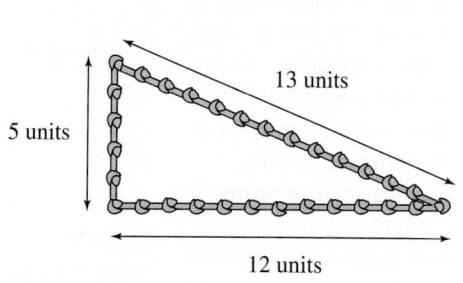

Figure 8.39
A knotted rope

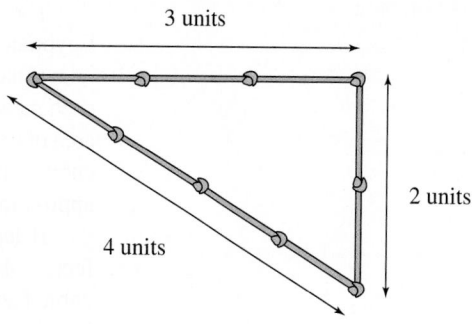

Figure 8.40
A knotted rope

Solution   **a.** Counting the number of segments on each side, we find the sides to be of lengths 5, 12, and 13 units. If the triangle is a right triangle, the sides will satisfy the Pythagorean theorem, $a^2 + b^2 = c^2$. Because 13 is the largest value, we let $c = 13$ and check the theorem:

$$a^2 + b^2 = 5^2 + 12^2$$
$$= 25 + 144$$
$$= 169$$
$$= 13^2 = c^2$$

The sides satisfy the Pythagorean Theorem; therefore, a right angle is formed at the vertex opposite the side of length 13.

b. The sides of the triangle have lengths 2, 3, and 4. Let $a = 2$ and $b = 3$ (the two shortest sides) and check the Pythagorean Theorem:

$$a^2 + b^2 = 2^2 + 3^2$$
$$= 4 + 9$$
$$= 13$$
$$\neq 4^2 = c^2$$

The lengths do not satisfy the Pythagorean Theorem; no right angle is formed. ■

Units of Measurement

The basic unit used by the ancient Egyptians for measuring length was the cubit. A *cubit* (*cubitum* is Latin for "elbow") is the distance from a person's elbow to the end of the middle finger. Just as a yard can be subdivided into smaller units of feet and inches, a cubit can be subdivided into smaller units of *palms* and *fingers*. One "royal" cubit (the unit used in official land measurement) equals seven palms, whereas one "common" cubit equals six palms; a royal cubit is longer than a common cubit. (In this book, we will take each cubit to be a royal cubit.) In either case, one palm equals four fingers. Since a cubit is a relatively small length, it is an inconvenient unit to use when measuring large distances. Consequently, the Egyptians defined a *khet* to equal 100 cubits; khets were used when land was surveyed.

The basic unit of area used by the Egyptians was the setat. A *setat* is equal to one square khet; a square whose sides each measure one khet (100 cubits) has an area of exactly one setat (or 10,000 square cubits). (See Figure 8.41.) One setat is approximately two-thirds of an acre.

If lengths are measured in terms of feet, volume is calculated in terms of cubic feet. Since the Egyptians measured lengths in terms of cubits, we would expect volume to be expressed in terms of cubic cubits. However, the basic unit

one khet (100 cubits)

Figure 8.41
1 setat = 1 square khet
= 10,000 square cubits

of volume used by the Egyptians was the *khar;* one khar equals two-thirds of a cubic cubit (or 1 cubic cubit = $\frac{3}{2}$ khar). The Egyptian units of measurement are summarized in the following box.

Egyptian Units of Measurement

1 cubit = 7 palms   1 setat = 1 square khet = 10,000 square cubits
1 palm = 4 fingers   1 khar = $\frac{2}{3}$ cubic cubit
1 khet = 100 cubits

Empirical Geometry (If It Works, Use It)

The body of mathematical knowledge possessed by the Egyptians was organized and presented quite differently from ours today. Whereas contemporary mathematics deals with general formulas, theorems, and proofs, most of the surviving written records of the Egyptians consist of single problems without mention of a general formula. The Egyptians lacked a formal system of variables and algebra; they were unable to write a formula such as $A = Lw$. Consequently, their work resembles a verbal narration on how to solve a specific problem or puzzle.

A specific example of this single-problem approach comes from what is now called the Moscow Papyrus, an Egyptian papyrus dating from about 1850 B.C. and containing 25 mathematical problems. It now resides in the Museum of Fine Arts in Moscow. Problem 14 of the Moscow Papyrus deals with finding the volume of a truncated pyramid (that is, a pyramid with its top cut off). The problem translates as follows: "Example of calculating a truncated pyramid. If you are told: a truncated pyramid of 6 for the vertical height by 4 on the base by 2 on the top: You are to square this 4; result 16. You are to double 4; result 8. You are to square this 2; result 4. You are to add the 16 and the 8 and the 4; result 28. You are to take $\frac{1}{3}$ of 6; result 2. You are to take 28 twice; the result 56. See, it is of 56. You will find it right."

The corresponding modern-day formula for finding the volume of a truncated pyramid like the one shown in Figure 8.42 is $V = \frac{h}{3}(a^2 + ab + b^2)$, where h is the height and a and b are the lengths of the sides of the square top and square base. The following line-by-line comparison shows the relationship between the Egyptians' work and the modern-day formula.

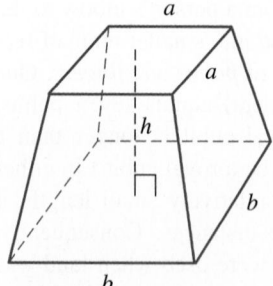

Figure 8.42
A truncated pyramid

Problem from Moscow Papyrus	Modern Counterparts
Example of calculating a truncated pyramid.	$V = \frac{h}{3}(a^2 + ab + b^2)$
If you are told:	Given
a truncated pyramid of 6 for the vertical height	$h = 6$,
by 4 on the base	$b = 4$, and
by 2 on the top:	$a = 2$.
You are to square this 4; result 16.	$b^2 = 16$
You are to double 4; result 8.	$ab = 8$
You are to square this 2; result 4.	$a^2 = 4$
You are to add the 16 and the 8 and the 4; result 28.	$a^2 + ab + b^2 = 28$
You are to take $\frac{1}{3}$ of 6; result 2.	$\frac{h}{3} = 2$
You are to take 28 twice;	$\frac{h}{3}(a^2 + ab + b^2)$
the result 56.	$V = 56$

ThomsonNOW™
for Liberal Arts Mathematics
www.thomsonedu.com/1pass.

EXAMPLE 2   Which of the following structures has the greater storage capacity?

a. a truncated pyramid 15 cubits high with a 6-cubit-by-6-cubit top and an 18-cubit-by-18-cubit bottom (Figure 8.43)

b. a regular pyramid 20 cubits tall with an 18-cubit-by-18-cubit base (Figure 8.44)

Solution   Because volume is a measure of the amount of space within a three-dimensional figure, it is used to compare the storage capacities of structures.

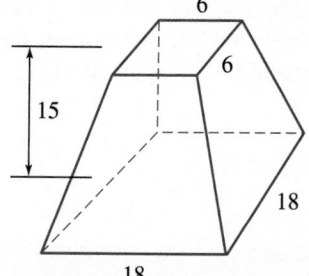

Figure 8.43
A truncated pyramid

a. To find the volume of the truncated pyramid, we use $h = 15$ cubits, $a = 6$ cubits, and $b = 18$ cubits:

$$V = \frac{h}{3}(a^2 + ab + b^2)$$

$$= \frac{15 \text{ cubits}}{3}[(6 \text{ cubits})^2 + (6 \text{ cubits})(18 \text{ cubits}) + (18 \text{ cubits})^2]$$

$$= (5 \text{ cubits})(36 \text{ cubits}^2 + 108 \text{ cubits}^2 + 324 \text{ cubits}^2)$$

$$= (5 \text{ cubits})(468 \text{ cubits}^2)$$

$$= 2{,}340 \text{ cubits}^3$$

$$= (2{,}340 \text{ cubits}^3)\left(\frac{\frac{3}{2}\text{ khar}}{1 \text{ cubit}^3}\right) \quad \text{converting cubic cubits to khar}$$

$$= 3{,}510 \text{ khar}$$

The volume of the truncated pyramid is 3,510 khar.

b. To find the volume of the pyramid, we use $h = 20$ cubits and $b = 18$ cubits. The pyramid is not truncated, so $a = 0$:

$$V = \frac{h}{3}(b^2)$$

$$= \frac{20 \text{ cubits}}{3}(18 \text{ cubits})^2$$

$$= 2{,}160 \text{ cubits}^3$$

$$= (2{,}160 \text{ cubits}^3)\left(\frac{\frac{3}{2}\text{ khar}}{1 \text{ cubit}^3}\right) \quad \text{converting cubic cubits to khar}$$

$$= 3{,}240 \text{ khar}$$

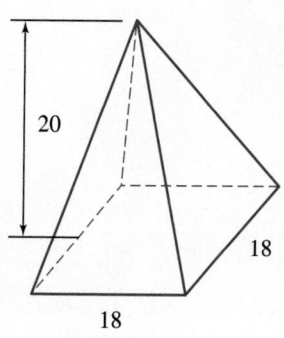

Figure 8.44
A regular pyramid

The volume of the regular pyramid is 3,240 khar.

Even though the two figures have the same base and the regular pyramid is taller, the truncated pyramid has a larger volume and thus a greater storage capacity. This result is due to the fact that the faces of the truncated pyramid have a steeper slope than those of the regular pyramid.   ■

Because of its single-problem approach, Egyptian geometry consisted of a fragmented assortment of guesses, tricks, and rule-of-thumb procedures. It was an empirical subject in which approximate answers were usually good enough for practical purposes. This does not mean that the Egyptians weren't concerned with accuracy. The Great Pyramid of Cheops is a testament to the Egyptians' ability to calculate, measure, and construct with a high degree of precision.

The Great Pyramid of Cheops

Built around 2600 B.C. with an original height of 481.2 feet, the Great Pyramid is the most awesome of all ancient structures. The blocks of limestone composing the pyramid have a combined weight of over 5 million tons. The base is almost a perfect square; the average length of the four sides is 755.78 feet, with a maximum discrepancy of only 4.5 inches (a relative error of only 0.05%)! In addition, the sides of the pyramid are oriented to correspond to the four major compass headings of north, south, east, and west. This positioning has an error of less than 1 degree.

Because of its near perfection, some people believe that the pyramid contains mystic puzzles and the answers to the riddles of the universe. For instance, if you divide the semiperimeter of the base (the distance halfway around, or two times the length of a side) by the height of the pyramid, an interesting result is obtained:

$$\frac{\text{semiperimeter}}{\text{height}} = \frac{2(755.78)}{481.2} = 3.141230258\ldots$$

© Rolf Richardson/Alamy

The precision of the Great Pyramid of Cheops is a testament to the ancient Egyptians' geometric ability. Some people believe that it also contains mystic puzzles.

The comparison of this number to the modern-day approximation of π is remarkable (π is approximately 3.141592654 . . .). Was this planned, or is it merely coincidental? The debate continues.

As surveyors, astronomers, and architects, the early Egyptians were undoubtedly concerned with the mathematics of a circle. By taking careful measurements, they knew that the circumference of a circle was proportional to its diameter—that is, that circumference/diameter = constant. Today, we call this constant π. On the basis of a contemporary interpretation of an ancient document known as the Rhind Papyrus, it appears that the Egyptians would have concluded that $\pi = \frac{256}{81}$ (≈ 3.16).

The Rhind Papyrus

Most of our knowledge of early Egyptian mathematics was obtained from two famous papyri: the Moscow (or Golenischev) Papyrus and the Rhind Papyrus. In 1858, the Scottish lawyer and amateur archeologist Henry Rhind was vacationing in Luxor, Egypt. While investigating the buildings and tombs of Thebes, he came across an old rolled-up papyrus. The papyrus was written by the scribe Ahmes around 1650 B.C. and contained eighty-four mathematical problems and their solutions. We know that the papyrus is a copy of an older original, for it begins: "The entrance into the knowledge of all existing things and all obscure secrets. This book was copied in the year 33, the fourth month of the inundation season, under the King of Upper and Lower Egypt 'A-user-Re,' endowed with life, in likeness to writings made of old in the time of the King of Upper and Lower Egypt Ne-mat'et-Re. It is the scribe Ahmes who copies this writing." The reference to the ruler Ne-mat'et-Re places the original script around 2000 B.C. Of particular interest is Problem 48, in which the area of a circle (and hence π) is investigated.

Problem 48 of the Rhind Papyrus concludes that "*the area of a circle of diameter 9 is the same as the area of a square of side 8.*" Today, we know that this is not exactly true; a circle of diameter 9 has area $A = \pi(4.5)^2 \approx 63.6$, whereas a square of side 8 has area $A = 8^2 = 64$. As we can see, the geometry of the early Egyptians was not perfect. But in many practical applications, the answers were close enough for their intended use.

The reasoning behind the conclusion of Problem 48 went like this: Construct a square whose sides are 9 units each, and inscribe a circle in the square. Thus, the diameter of the circle is 9 units. Now divide the sides of the square into three equal segments of 3 units each. Remove the triangular corners, as shown in Figure 8.45, and an irregular octagon is formed. The area of this octagon is then used as an approximation of the area of the circle: $A_{\text{circle}} \approx A_{\text{octagon}}$.

Notice that the area of the octagon is equal to the area of the large square minus the area of the four triangular corners. Rearranging these triangles, we can see that the area of two triangular corners is the same as the area of one square of side 3 units, as shown in Figure 8.46.

Therefore,

$$
\begin{aligned}
A_{\text{circle}} \approx A_{\text{octagon}} &= A_{\text{square of side 9}} - 4A_{\text{triangle}} \\
&= A_{\text{square of side 9}} - 2A_{\text{square of side 3}} \\
&= (9)^2 - 2(3)^2 \\
&= 81 - 18 \\
&= 63
\end{aligned}
$$

Therefore, $A_{\text{circle}} \approx 63$ square units. The Egyptians realized that the area of the circle was not 63 but close to it. They mistakenly thought that the area of the circle must be

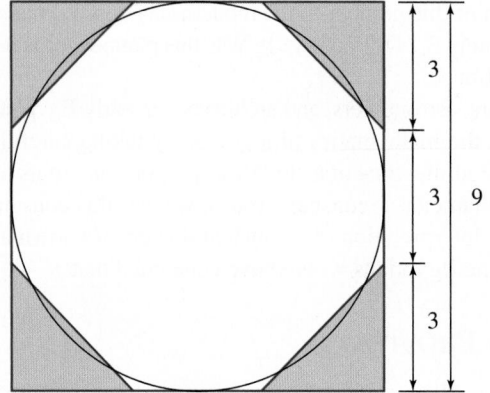

Figure 8.45
A circle is approximated by an irregular octagon.

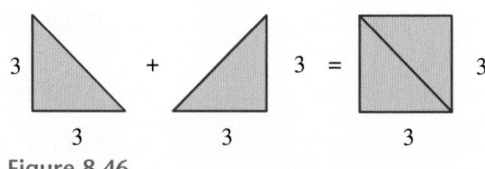

Figure 8.46
Two triangles = a square

exactly 64, because 64 is close to 63 and 64 ($= 8^2$) is the area of a square with whole-number sides. (This 4,000-year-old "logic" doesn't seem reasonable today.) Thus, the ancient mathematicians reached their conclusion that

$$A_{\text{circle of diameter 9}} = A_{\text{square of side 8}}$$

Two approximations were used in Problem 48:

$$A_{\text{circle of diameter 9}} \approx A_{\text{octagon}}$$
$$A_{\text{octagon}} = 63 \approx 64 = A_{\text{square of side 8}}$$

Of these two estimates, the first (Figure 8.45) was an underestimate, and the second an overestimate; thus, the net "canceling" of errors contributed to the accuracy of their overall calcula-

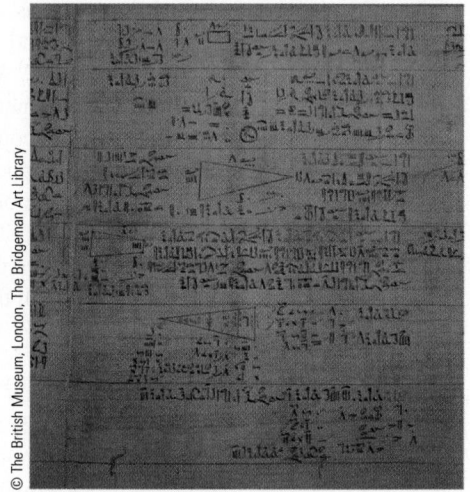

The Rhind Papyrus is one of the world's oldest known mathematics textbooks. Notice the triangles in this portion of the 18-foot-long papyrus.

tion. The Egyptians' geometry was not perfect, but it did give reasonable answers.

Pi and the Area of a Circle

The Egyptians obtained the "formula" $C = \pi d$ by observing that the ratio circumference/diameter is a constant. How did they obtain the "formula" for the area of a circle? (The Egyptians did not work with actual formulas; instead, they used verbal examples, as we saw with truncated pyramids.) The Egyptians might have used a method of rearrangement to find the area of any given circle. The method of rearrangement says that if a region is cut up into smaller pieces and rearranged to form a new shape, the new shape will have the same area as the original figure. We will apply this method to a circle, as the Egyptians might have done, and see what happens.

Suppose a circle of radius r is cut in half, and each semicircular region is then cut up into an equal number of uniform slices (like a pie). The circumference of the entire circle is known to be $2\pi r$, so the length of each semicircle is πr, as shown in Figure 8.47.

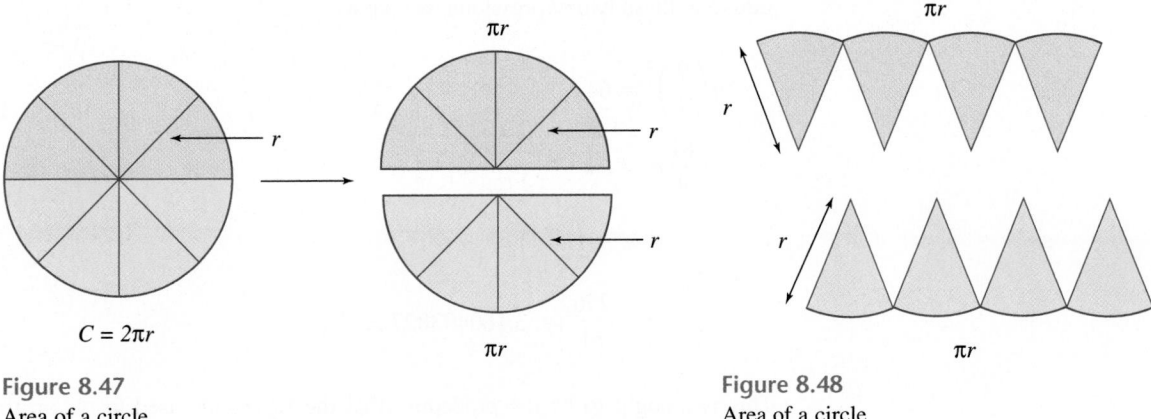

Figure 8.47
Area of a circle

Figure 8.48
Area of a circle

Now rearrange the slices as shown in Figure 8.48. The total area of these slices is still the same as the area of the circle. These pieces are then joined together as in Figure 8.49.

If we continue this process, cutting up the circle into more and more slices that get thinner and thinner, the rearranged figure becomes quite rectangular, as shown in Figure 8.50.

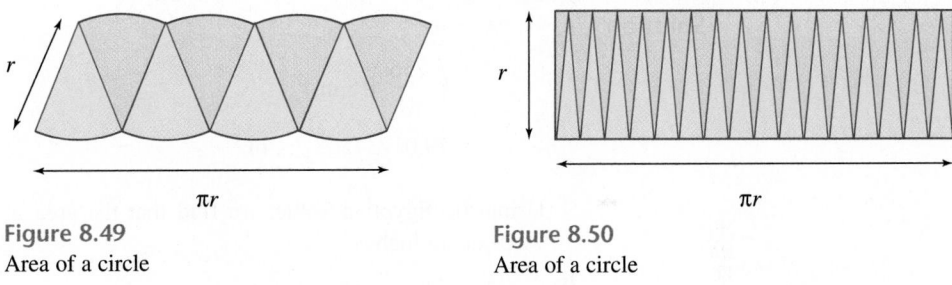

Figure 8.49
Area of a circle

Figure 8.50
Area of a circle

Rather than having a bumpy top and bottom, as in Figure 8.49, the top and bottom of this new figure are fairly straight, owing to the arcs of the circle becoming so short that they appear to be straight line segments.

Using the formula for the area of a rectangle (the rearranged figure), we have $A = bh = (\pi r)(r) = \pi r^2$. Because the area of the rearranged figure is πr^2, the area of the original circle is also πr^2.

EXAMPLE 3   Combining the result of Problem 48 of the Rhind Papyrus with the method of rearrangement of area, find the Egyptian approximation of π.

Solution   Problem 48 of the Rhind Papyrus states that the area of a circle of diameter 9 (or radius $\frac{9}{2}$) is equal to the area of a square of side 8; that is,

$$A_{\text{circle of diameter 9}} = A_{\text{square of side 8}} = 64$$

Using rearrangement of area, we also have

$$A_{\text{circle of diameter 9}} = \pi \left(\frac{9}{2}\right)^2$$

Equating these two expressions, we have

$$\pi\left(\frac{9}{2}\right)^2 = 64$$

$$\frac{81}{4}\pi = 64$$

$$\pi = 64\left(\frac{4}{81}\right)$$

$$\pi = \frac{256}{81} \ (= 3.160493827\ldots)$$

This is thought to be the procedure that the Egyptians used to obtain their value of π. This approximation is remarkably close to our modern-day value!   ■

EXAMPLE 4   For a circle of radius 5 inches, do the following:

a. Use $\pi = \frac{256}{81}$ (the Egyptian approximation of pi) to find the area of the circle.
b. Use the value of π contained in a scientific calculator to find the area of the circle.
c. Find the error of the Egyptian calculation relative to the calculator value.

Solution   a. $A = \pi r^2$

$$= \left(\frac{256}{81}\right)(5 \text{ in.})^2$$

$$= 79.01234568\ldots \text{ in.}^2$$

Using the Egyptian value, we find that the area of the circle is approximately 79.0 square inches.

b. $A = \pi r^2$

$$= \pi(5 \text{ in.})^2$$

$$= 25\pi \text{ in.}^2$$

$$= 78.53981634\ldots \text{ in.}^2$$

Using the calculator value, we find that the area of the circle is approximately 78.5 square inches. Note that if both decimals were rounded off to the nearest whole unit, they would yield the same answer.

c. To find the error of the Egyptian calculation relative to the calculator value, we must find the difference of the calculations and divide this difference by the calculator value:

$$A_{\text{Egyptian}} - A_{\text{calculator}} = 79.01234568 \text{ in.}^2 - 78.53981634 \text{ in.}^2$$

$$= 0.47252934 \text{ in.}^2$$

$$\text{Error relative to calculator} = \frac{0.47252934 \text{ in.}^2}{78.53981634 \text{ in.}^2}$$

$$= 0.00601643\ldots$$

$$= 0.6\%$$

The error of the Egyptian calculation relative to the calculator value is 0.6%.   ■

8.3 Exercises

In Exercises 1–4, determine whether each configuration of knotted ropes would form a right triangle.

1.

2.

3.

4.

5. Which of the following structures has the larger storage capacity?
   a. a truncated pyramid that is 12 cubits high with a 2-cubit-by-2-cubit top and a 15-cubit-by-15-cubit bottom
   b. a regular pyramid that is 18 cubits tall with a 15-cubit-by-15-cubit base

6. Which of the following structures has the larger storage capacity?
   a. a truncated pyramid that is 33 cubits high with a 6-cubit-by-6-cubit top and a 30-cubit-by-30-cubit bottom

   b. a regular pyramid that is 50 cubits tall with a 30-cubit-by-30-cubit base

In Exercises 7 and 8, fill in the blanks as in the problem from the Moscow Papyrus (the Egyptian method for finding the volume of a truncated pyramid, immediately preceding Example 2).

7. "If you are told: a truncated pyramid of 9 for the vertical height by 6 on the base by 3 on the top: You are to square this 6; result _____. You are to triple 6; result _____. You are to square this 3; result _____. You are to add the _____ and the _____ and the _____; result _____. You are to take $\frac{1}{3}$ of _____; result _____. You are to take _____ three times; the result _____. See, it is of _____. You will find it right."

8. "If you are told: a truncated pyramid of 12 for the vertical height by 10 on the base by 4 on the top: You are to square this 10; result _____. You are to quadruple 10; result _____. You are to square this 4; result _____. You are to add the _____ and the _____ and the _____; result _____. You are to take $\frac{1}{3}$ of _____; result _____. You are to take _____ four times; the result _____. See, it is of _____. You will find it right."

9. Following the method of Problem 48 of the Rhind Papyrus, find a square, with a whole number as the length of its side, that has the same (approximate) area as a circle of diameter 24 units.

   HINT: Divide the sides of a 24-by-24 square into three equal segments.

10. Following the method of Problem 48 of the Rhind Papyrus, find a square, with a whole number as the length of its side, that has the same (approximate) area as a circle of diameter 42 units.

    HINT: Divide the sides of a 42-by-42 square into three equal segments.

11. Using the result of Exercise 9 and the method of Example 3, obtain an approximation of π.

12. Using the result of Exercise 10 and the method of Example 3, obtain an approximation of π.

13. For a circle of radius 3 palms, do the following.
    a. Use $\pi = \frac{256}{81}$ (the Egyptian approximation of pi) to find the area of the circle.

b. Use the value of π contained in a scientific calculator to find the area of the circle.

c. Find the error of the Egyptian calculation relative to the calculator value.

14. For a circle of radius 6 palms, do the following.
  a. Use $\pi = \frac{256}{81}$ (the Egyptian approximation of pi) to find the area of the circle.
  b. Use the value of π contained in a scientific calculator to find the area of the circle.
  c. Find the error of the Egyptian calculation relative to the calculator value.

15. For a sphere of radius 3 palms, do the following.
  a. Use $\pi = \frac{256}{81}$ (the Egyptian approximation of pi) to find the volume of the sphere.
  b. Use the value of π contained in a scientific calculator to find the volume of the sphere.
  c. Find the error of the Egyptian calculation relative to the calculator value.

16. For a sphere of radius 6 palms, do the following.
  a. Use $\pi = \frac{256}{81}$ (the Egyptian approximation of pi) to find the volume of the sphere.
  b. Use the value of π contained in a scientific calculator to find the volume of the sphere.
  c. Find the error of the Egyptian calculation relative to the calculator value.

17. Find the perimeter of a triangle having sides with the following measurements: 2 cubits, 5 palms, 3 fingers; 3 cubits, 3 palms, 2 fingers; 4 cubits, 4 palms, 3 fingers.

18. It has been suggested that one palm equals 7.5 centimeters. Using a standard equivalence, one centimeter equals 0.3937 inch.
  a. How many inches are in one royal cubit?
  b. How many inches are in one common cubit?
  c. Measure your own personal cubit. How does it compare to the royal and common cubits?

19. Use part (a) of Exercise 18 to determine which is larger: one khar or one cubic foot.

20. A rectangular field measures 50 cubits by 200 cubits. Find the area of this field in setats.

21. A triangular field has sides that measure 150 cubits, 200 cubits, and 250 cubits. Find the area of this field in setats.

22. A rectangular room is 20 cubits long, 15 cubits wide, and 8 cubits high. Find the volume of this room in khar.

23. The area of a square is 4 setats. Find the length of a side of this square. Express your answer in the following units.
  a. khet   **b.** cubits

24. The volume of a cube is 12 khar. Find the length, in cubits, of an edge of this cube.

25. Problem 41 of the Rhind Papyrus pertains to finding the volume of a cylindrical granary. The diameter of the granary is 9 cubits, and its height is 10 cubits.
  a. Calculate the volume of this granary, in khar, using the Egyptian method (Problem 48 of the Rhind Papyrus) to calculate the area of a circle.
  b. Calculate the volume of this granary, in khar, using the conventional formula $A = \pi r^2$ to calculate the area of a circle.
  c. Find the error of the Egyptian calculation of volume relative to the conventional calculation.

━━━━━━ *Answer the following questions using complete sentences and your own words.*

HISTORY QUESTIONS

26. List four activities in which the ancient Egyptians used mathematics and, in so doing, contributed to the advancement of their civilization.

27. How did the Egyptians construct right angles?

28. Who wrote the Rhind Papyrus?

29. What does Problem 48 of the Rhind Papyrus state?

30. Write a research paper on a historical topic referred to in this section or on a related topic. Following is a list of possible topics:
  • Ancient agriculture and the Nile River (How did agriculture lead to mathematical and scientific advancement?)
  • Early astronomy (What motivated the early astronomers? What is the connection between early astronomy and mathematics?)
  • Early calendars (What purpose did calendars serve? Who created them?)
  • The Great Pyramid of Cheops (How was it constructed? What was its purpose?)
  • The history of π (Trace its history and applications.)
  • The Moscow Papyrus (What does it contain? When was it translated?)
  • The Rhind Papyrus (Is it intact? What does it contain? When was it translated?)
  • The Rosetta Stone (How was it used in mathematics? In politics? In science? In archeology?)

The primary question was not "What do we know," but
"How do we know it."
—Aristotle to Thales

Many people accept common knowledge without questioning it, especially if they have firsthand experience of it. However, as Aristotle's comment to Thales points out, the ancient Greek scholars were not satisfied with mere facts; they were forever asking "why" in their search for absolute truth in the world around them. Whereas the Egyptians constructed a right angle by stretching a knotted rope to form a triangle with sides of three, four, and five units, the Greeks wanted to know *why* this method worked. The Greeks were not content to accept a claim simply because experience indicated that it worked. They wanted proof, and they obtained proof through the systematic application of logic and deductive reasoning.

Thales of Miletus

The empirical geometry of the Egyptians reached the ancient Greeks through commerce, travel, and warfare. One of the first Greek scholars to study this geometry was Thales of Miletus (625–547 B.C.). Thales is known as the first of the Seven Sages of Greece. (The Seven Sages of Greece were famous for their practical knowledge. In addition to Thales, the sages were Solon of Athens, Bias of Priene, Chilo of Sparta, Cleobulus of Rhodes, Periander of Corinth, and Pittacus of Mitylene.)

A successful businessman, statesman, and philosopher, Thales had occasion to travel to Egypt and consequently learned of Egyptian geometry. While in Egypt, he won the respect of the pharaoh by calculating the height of the Great Pyramid of Cheops. He did this by measuring the shadows of the pyramid and of his walking staff. Knowing the height of his staff, he used a proportion to calculate the height of the pyramid.

In the history of mathematics, Thales is credited with being the first to prove that corresponding sides of similar triangles are proportional. If the three angles of one triangle are equal to the three angles of another triangle, the triangles are **similar.** Consequently, similar triangles have the same shape but may differ in size.

The Sides of Similar Triangles Are Proportional

If the three angles of one triangle equal the three angles of another triangle, then

$$\frac{a}{d} = \frac{b}{e} = \frac{c}{f}$$

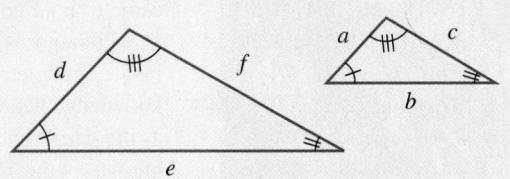

EXAMPLE 1 For the two triangles in Figure 8.51, find the lengths of the unknown sides.

Figure 8.51
Similar triangles

Solution Because the triangles are similar (they have equal angles), their sides are proportional.

Comparing the small triangle to the large triangle, we have

$$\frac{x}{180} = \frac{12}{216}$$

$$x = \frac{12 \cdot 180}{216} \qquad \text{multiplying both sides by 180}$$

$$x = 10$$

Comparing the large triangle to the small triangle, we have

$$\frac{y}{5} = \frac{216}{12}$$

$$y = \frac{216 \cdot 5}{12} \qquad \text{multiplying both sides by 5}$$

$$y = 90 \qquad\qquad\qquad\qquad\qquad\qquad\qquad ■$$

Pythagoras of Samos

The name Pythagoras is synonymous with the famous formula that relates the lengths of the sides of a right triangle, $a^2 + b^2 = c^2$. However, the life and teachings of this famous Greek mathematician, philosopher, and mystic are shrouded in mystery, legend, and conjecture.

Born on the Aegean island of Samos about 500 B.C., Pythagoras traveled and studied extensively as a young man. His travels took him to Egypt, Phoenicia, and Babylonia. Returning to Samos when he was fifty, Pythagoras found his homeland under the rule of the tyrant Polycrates. For an unknown reason, Pythagoras was banned from Samos and migrated to Croton, in the south of present-day Italy, where he founded a school. His students and disciples, known as Pythagoreans, eventually came to hold considerable social power.

Pythagoras had his students concentrate on four subjects: the theory of numbers, music, geometry, and astronomy. To him, these subjects constituted the core of knowledge necessary for an educated person. This core of four subjects, later known as the *quadrivium*, persisted until the Middle Ages. At that time, three other subjects, known as the *trivium*, were added to the list: logic, grammar, and rhetoric. These seven areas of study came to be known as the seven liberal arts and formed the proper course of study for all educated people.

Pythagoras of Samos

The Pythagoreans were a secretive sect. Their motto, "All is number," indicates their belief that numbers have a mystical quality and are the essence of the universe. Numbers had a mystic aura about them, and all things had a numerical representation. For instance, the number 1 stood for reason (the number of absolute truth), 2 stood for woman (the number of opinion), 3 stood for man (the number of harmony), 4 stood for justice (the product of equal terms, $4 = 2 \times 2 = 2 + 2$), 5 stood for marriage (the sum of man and woman, $5 = 2 + 3$), and 6 stood for creation (from which man and woman came, $6 = 2 \times 3$). In general, odd numbers were masculine, and even numbers were feminine. Indeed, the Pythagorean doctrine was a strange mixture of cosmic philosophy and number mysticism. The Pythagoreans' beliefs were augmented by many rites and taboos, which included refusing to eat beans (they were sacred), refusing to pick up a fallen object, and refusing to stir a fire with an iron.

In the realm of astronomy, Pythagoras had a curious theory. Being geocentric, he believed the earth to be the center of the universe. The sun, moon, and five known planets circled the earth, each traveling on its own crystal sphere. Because of the friction of these gigantic bodies whirling about, each body would produce a unique tone based on its distance from the earth. The combined effect of the seven bodies was the harmonious celestial music of the gods, which Pythagoras was the only mortal blessed to hear!

Near the end of his life, Pythagoras and his followers became more political. Their power and influence were felt in Croton and the other Greek cities of southern Italy. Legend has it that this power and the Pythagoreans' sense of autocratic supremacy initiated a conflict with the local peoples and governments that ultimately led to his death.

Historians do not agree on the exact circumstances of Pythagoras's death, but it is known that during a violent revolt against the Pythagoreans, the locals set fire to the school. Some say that Pythagoras died in the flames. Others say that he escaped the inferno but was chased to the edge of a bean field. Rather than trample the sacred beans, Pythagoras allowed the crowd to overtake and kill him. Some people believe that Pythagoras was murdered by a disgruntled disciple. In death, as in life, Pythagoras is shrouded in mystery.

Although Pythagoras is given credit for proving the theorem that bears his name, there is no hard evidence that he did so. All his teachings were verbal; there are no written records of his actual work. Furthermore, his disciples had to swear not to reveal any of his doctrines to outsiders.

Since the days of Pythagoras, many proofs of "his" theorem have been put forth. Most involve the method of rearrangement of area. A few years before being elected the twentieth president of the United States, James Garfield produced an original proof. His method, shown below, involved the creation of a trapezoid that consisted of three right triangles.

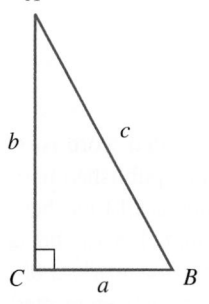

Figure 8.52
A right triangle

James Garfield's Proof of the Pythagorean Theorem

Given: the right triangle ABC, as shown in Figure 8.52.

Extend segment CA a distance of a, and call this point D.

Construct a perpendicular segment of length b at D.

The endpoint of this segment is E. (See Figure 8.53.)

Triangles ABC and EAD are congruent, so $\angle EAB$ is a right angle. (Why?)

Figure $BCDE$ is a trapezoid with parallel bases CB and DE.

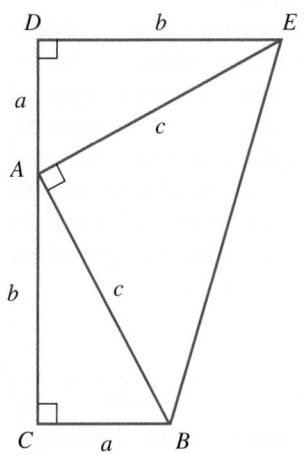

Figure 8.53
A trapezoid

The area of *BCDE* can be found by using the formula for the area of a trapezoid, or it can be found by adding the areas of the three right triangles. Apply both methods and set their results equal as follows:

$$\text{Area of trapezoid} = (\text{average base})(\text{height}) = \left(\frac{a+b}{2}\right)(a+b)$$

$$\text{Area of } \Delta\, ABC = \frac{1}{2}(\text{base})(\text{height}) \qquad = \frac{1}{2}ab$$

$$\text{Area of } \Delta\, EAD = \frac{1}{2}(\text{base})(\text{height}) \qquad = \frac{1}{2}ab$$

$$\text{Area of } \Delta\, EAB = \frac{1}{2}(\text{base})(\text{height}) \qquad = \frac{1}{2}c^2$$

Area of trapezoid = sum of the areas of the three triangles

$$\left(\frac{a+b}{2}\right)(a+b) = \left(\frac{1}{2}ab\right) + \left(\frac{1}{2}ab\right) + \left(\frac{1}{2}c^2\right)$$

$$(a+b)(a+b) = ab + ab + c^2 \qquad \text{\small multiplying each side by 2}$$

$$a^2 + ab + ab + b^2 = ab + ab + c^2 \qquad \text{\small multiplying } (a+b)(a+b)$$

$$a^2 + b^2 = c^2 \qquad \text{\small subtracting } 2ab \text{ from each side}$$

Thus, given any right triangle with legs a and b and hypotenuse c, we have shown that $a^2 + b^2 = c^2$.

Euclid of Alexandria

Euclid's proof of the Pythagorean Theorem as it appeared in an Arabic translation of *Elements*.

The most widely published and read book in the history of the printed word is, of course, the Bible. And what do you suppose the second most widely published book is? Many people are quite surprised to discover that the number 2 all-time best-seller is a mathematics textbook! Written by the Greek mathematician Euclid of Alexandria around 300 B.C., *Elements* was hand-copied by scribes for nearly 2,000 years. Since the coming of the printing press, over 1,000 editions have been printed, the first in 1482. *Elements* consists of 13 "books," or units. Although most people associate Euclid with geometry, *Elements* also contains number theory and elementary algebra, all the mathematics known to the scholars of the day.

Euclid's monumental contribution to the world of mathematics was his method of organization, not his discovery of new theorems. He took all the mathematical knowledge that had been compiled since the days of Thales (300 years earlier) and organized it in such a way that every result followed logically from its predecessors. To begin this chain of proof, Euclid had to start with a handful of assumptions, or things that cannot be proved. These assumptions are called *axioms*, or *postulates*, and are accepted without proof. By carefully choosing five geometric postulates, Euclid proceeded to prove 465 results, many of which were quite complicated and not at all intuitively obvious. The beauty of his work is that so much was proved from so few assumptions.

Euclid's Postulates of Geometry

1. A straight line segment can be drawn from any point to any other point.

2. A (finite) straight line segment can be extended continuously into an (infinite) straight line.

3. A circle may be described with any center and any radius.

4. All right angles are equal to one another.

5. Given a line and a point not on that line, there is one and only one line through the point parallel to the original line.

The only "controversial" item on this list is postulate 5, the so-called Parallel Postulate. This proved to be the unraveling of Euclidean geometry, and 2,000 years later non-Euclidean geometries were born. Non- Euclidean geometries will be discussed in Section 8.6.

The rigor of Euclid's work was unmatched in the ancient world. *Elements* was written for mature, sophisticated thinkers. Even today, the real significance of Euclid's work lies in the superb training it gives in logical thinking. Abraham Lincoln knew of the rigor of Euclid's *Elements*. At the age of forty, while still a struggling lawyer, Lincoln mastered

Euclid by Joos van Ghent and Pedro Berruguete. Galleria Nazionale della Marche, Urbino, Italy.

the first six books in the *Elements* on his own, solely as training for his mind.

Teachers of mathematics invariably encounter resistance from students when abstract mathematics with little apparent application is presented. Even in the days of Euclid, students were dismayed at the complexity of the proofs and asked the question so often echoed by today's students: "Why do we have to learn this stuff?" According to legend, one of Euclid's beginning students asked, "What shall I gain by learning these things?" Euclid quickly summoned his servant, responding, "Give him a coin, since he must make gain out of what he learns!" A second legend handed down through the years claims that King Ptolemy once asked Euclid if there was a shorter way of learning geometry than the study of his *Elements*, to which Euclid replied, "There is no royal road to geometry." Sorry, no shortcuts today!

Deductive Proof

To investigate the Greeks' method of geometric deduction, we will examine congruent triangles. **Congruent triangles** are triangles that have exactly the same shape and size; their corresponding angles and sides are equal, as shown in Figure 8.54.

Figure 8.54
Congruent triangles

Historical Note Archimedes of Syracuse (287–212 B.C.)

Archimedes demonstrated that a very heavy object could be raised with relatively little effort by using a series of pulleys (a compound pulley) or by using levers. He was so confident of his principles of levers and pulleys that he boasted that he could move anything. One of Archimedes' many famous quotes is "Give me a place to stand and I will move the earth!"

Aware of his genius, King Hieron turned to Archimedes on many occasions. One amusing legend concerns the problem of the king's gold crown. After giving a goldsmith a quantity of gold for the creation of an elaborate crown, King Hieron suspected the smith of keeping some of the gold and replacing it with an equal weight of silver. Since he had no proof, Hieron turned to Archimedes. While pondering the king's prob-lem, Archimedes submerged himself in a full tub of water to bathe. As he climbed in, he noticed the water overflowing the tub. In a flash of brilliance, he discov-ered the solution to the king's problem. Overjoyed with his idea, Archimedes is rumored to have run down the street, totally naked, shouting, "Eureka! Eureka!" ("I have found it! I have found it!")

What Archimedes had found was that because sil-ver is less dense than gold, an equal weight of silver would have a greater volume than an equal weight of gold and consequently would displace more water. When the crown and a piece of pure gold of the same weight were immersed in water, the smith's fate was sealed. Even though the crown and the pure gold had the same weight, the crown displaced more water, thus proving that it had been adulterated.

A seventeenth-century engraving depicting Archimedes' invention of a system of mirrors designed to focus the sun's rays on attacking ships to set them afire. The focusing properties of these shapes will be discussed in Section 8.6.

"Triangle *ABC* is congruent to triangle *DEF*" (denoted by $\triangle ABC \cong \triangle DEF$) means that all six of the following statements are true:

$$\angle A = \angle D \qquad \angle B = \angle E \qquad \angle C = \angle F$$
$$AB = DE \qquad BC = EF \qquad CA = FD$$

In 215 B.C., the Romans laid siege to Syracuse on the island of Sicily. The Roman commander Marcellus had no idea of the fierce resistance his troops were to face. The attack was from the sea. A native of Syracuse, Archimedes (then seventy-two years old) designed and helped to build a number of ingenious weapons to defend his home. He designed huge catapults that hurled immense boulders at the Roman ships. These catapults were set to throw the projectiles at different ranges so that no matter where the ships were, they were always under fire. When a ship managed to come in close, the defenders lowered large hooks over the city walls, grabbed the ship, lifted it into the air with pulleys, and tossed it back into the sea.

Needless to say, the Greek defenders put up a good fight. They fought so well that the siege lasted nearly three years. However, one day, while the people of Syracuse were feasting and celebrating a religious fes-tival, Roman sympathizers within the city informed the attackers of weaknesses in the city's defense, and a bloody sack began.

Marcellus had given strict orders to take Archimedes alive; no harm was to come to him or his house. At the time, Archimedes was studying the drawings of circles he had made in the sand. Preoccupied, he did not notice the Roman soldier standing next to him until the soldier cast a shadow on his drawings in the sand. The agitated mathematician called out, "Don't disturb my circles!" At that, the insulted soldier used his sword and killed Archimedes of Syracuse.

The Roman commander Marcellus so grieved at the loss of Archimedes that when he learned of the mathematician's wish for the design of his tombstone, he fulfilled it.

North Wind Picture Archives

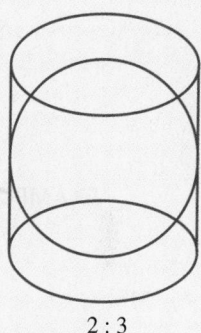

2 : 3

Archimedes' tombstone

Euclid proved that two triangles are congruent—that is, $\triangle ABC \cong \triangle DEF$—in any of the following three circumstances:

1. SSS: If the three sides of one triangle equal the three sides of the other triangle, then the triangles are congruent. (The corresponding angles will automatically be equal.) See Figure 8.55.

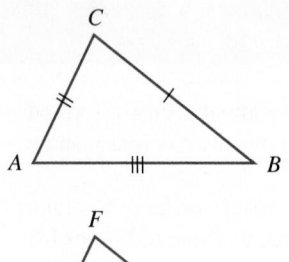

Figure 8.55
SSS implies $\triangle ABC \cong \triangle DEF$.

2. SAS: If two sides and the included angle of one triangle equal two sides and the included angle of the other triangle, then the triangles are congruent. (The other side and angles will automatically be equal.) See Figure 8.56.

3. ASA: If two angles and the included side of one triangle equal two angles and the included side of the other triangle, then the triangles are congruent. (The other angle and sides will automatically be equal.) See Figure 8.57.

Figure 8.56
SAS implies $\triangle ABC \cong \triangle DEF$.

Figure 8.57
ASA implies $\triangle ABC \cong \triangle DEF$.

When proving a result in geometry, we use a two-column method. The first column is a list of each valid statement used in the proof, starting with the given statements and progressing to the conclusion. The second column gives the reason for or justification of each statement in the first column.

EXAMPLE 2 Given: $CA = DB$
$\angle CAB = \angle DBA$
Prove: $CB = DA$
(See Figure 8.58.)

Figure 8.58
Hypotheses

Solution

Statements	Reasons
1. $CA = DB$	1. Given
2. $\angle CAB = \angle DBA$	2. Given
3. $AB = AB$	3. Anything equals itself.
4. $\triangle CAB \cong \triangle DBA$	4. SAS
5. $CB = DA$	5. Corresponding parts of congruent triangles are equal. ∎

Archimedes of Syracuse

Many scholars consider Archimedes the father of physics; he was the first scientific engineer, for he combined mathematics and the deductive logic of Euclid with the scientific method of experimentation. Born around 287 B.C. in the Greek city of Syracuse on the island of Sicily, Archimedes gave the world many labor-saving devices, such as pulleys, levers, and a water pump that utilized a screw-shaped cylinder. In addition, he created formidable weapons that were used in the defense of Syracuse, developed the fundamental concept of buoyancy, and wrote many treatises

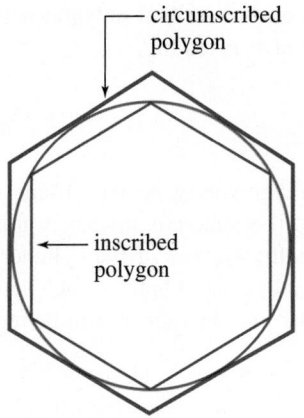

Figure 8.59
Inscribed and circum-
scribed polygons

ThomsonNOW
for Liberal Arts Mathematics
www.thomsonedu.com/1pass.

on mathematics and geometry, including ones on the calculation of π and on the surface area and volume of spheres, cylinders, and cones.

To calculate the value of π to any desired level of accuracy, Archimedes approximated a circle by using a regular polygon—a polygon whose sides all have the same length. His method was to inscribe a regular polygon of n sides inside a circle and determine the perimeter of the polygon. Because it was inside the circle, the perimeter of the polygon was smaller than the circumference of the circle. Then he would circumscribe an n-sided polygon around the circle and find its perimeter, which of course was larger than the circumference, as shown in Figure 8.59. By increasing n (the number of sides of the polygon) from 6 to 12 to 24 and so on, he got polygons whose perimeters were progressively closer and closer to the circumference of the circle. Using a polygon of 96 sides, he obtained

$$3\frac{10}{71} < \pi < 3\frac{1}{7}$$

This upper limit of $\frac{22}{7}$ is still used today as a common approximation for π.

EXAMPLE 3   A regular hexagon is inscribed in a circle of radius r, while another regular hexagon is circumscribed around the same circle.

a. Using the perimeter of the inscribed hexagon as an approximation of the circumference of the circle, obtain an estimate of π.

b. If the perimeter of the circumscribed hexagon has sides of length $s = (2\sqrt{3}/3)r$, obtain an estimate of π.

Solution   a. Let s = length of a side of the inscribed hexagon; hence, $P_{\text{hexagon}} = 6s$. A regular hexagon is composed of six equilateral triangles; that is, triangles in which all sides are equal. Therefore, $s = r$, as shown in Figure 8.60.

$$C_{\text{circle}} \approx P_{\text{inscribed hexagon}}$$
$$2\pi r \approx 6s$$
$$2\pi r \approx 6r \qquad \text{substituting } s \text{ for } r$$
$$\pi \approx \frac{6r}{2r} \qquad \text{dividing by } 2r$$
$$\pi \approx 3$$

Therefore, π is approximately 3. Note that an **inscribed polygon** will underestimate the circumference, so our approximation of π is too small.

b. Let s = length of a side of the circumscribed hexagon; hence, $P = 6s$, as shown in Figure 8.61. We are given that $s = (2\sqrt{3}/3)r$.

$$C_{\text{circle}} \approx P_{\text{circumscribed hexagon}}$$
$$2\pi r \approx 6s$$
$$2\pi r \approx 6\left(\frac{2\sqrt{3}}{3}r\right) \qquad \text{substituting } s = (2\sqrt{3}/3)r$$
$$2\pi r \approx 4\sqrt{3}r$$
$$\pi \approx \frac{4\sqrt{3}r}{2r} \qquad \text{dividing by } 2r$$
$$\pi \approx 2\sqrt{3}$$
$$\pi \approx 3.4641016$$

Figure 8.60
Inscribed hexagon

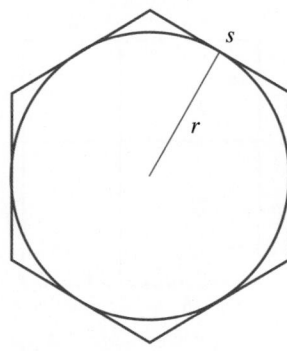

Figure 8.61
Circumscribed hexagon

Therefore, π is approximately 3.4641016. Note that a **circumscribed polygon** will overestimate the circumference, so our approximation of π is too big.

Combining (b) with (a), we have

$$3 < \pi < 3.4641016$$

Of all the work Archimedes did in the field of solid geometry, he was proudest of one particular achievement. Archimedes found that if a sphere is inscribed in a cylinder, the volume of the sphere will be two-thirds the volume of the cylinder. In other words, the ratio of their volumes is $2:3$. The result pleased him so much that he wanted a diagram of a sphere inscribed in a cylinder and the ratio $2:3$ to be the only markings on his tombstone.

8.4 Exercises

In Exercises 1–4, the given triangles are similar. Find the lengths of the missing sides.

1.

2.

3.

4.

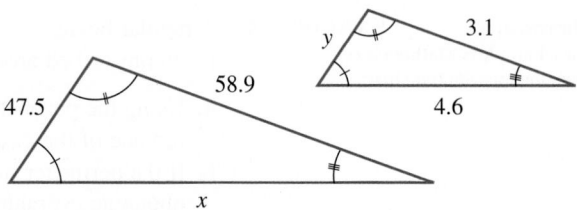

5. A 6-foot-tall man casts a shadow of 3.5 feet at the same instant that a tree casts a shadow of 21 feet. How tall is the tree?

6. A 5.4-foot-tall woman casts a shadow of 2 feet at the same instant that a telephone pole casts a shadow of 9 feet. How tall is the pole?

7. Use the diagrams in Figure 8.62 to show that $a^2 + b^2 = c^2$.

HINT: The area of square #1 = the area of square #2. (Why?)

Square #1

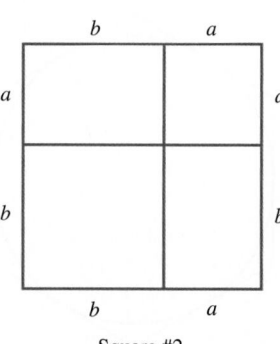
Square #2

Figure 8.62
Diagrams for Exercise 7

8. Use Figure 8.63 and the method of rearrangement to show that $a^2 + b^2 = c^2$.

HINT: Find the area of the figure in two ways, as was done in Garfield's proof.

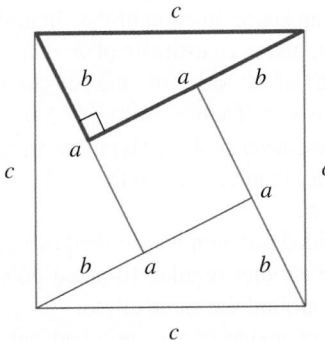

Figure 8.63
Diagram for Exercise 8

9. Find the length of the longest object that will fit inside a rectangular box 4 feet long, 3 feet wide, and 2 feet high.

HINT: Draw a picture and use the Pythagorean Theorem.

10. Find the length of the longest object that will fit inside a cube 2 feet long, 2 feet wide, and 2 feet high.

HINT: Draw a picture and use the Pythagorean Theorem.

11. Find the length of the longest object that will fit inside a cylinder that has a radius of $2\frac{1}{4}$ inches and is 6 inches high.

HINT: Draw a picture and use the Pythagorean Theorem.

12. Find the length of the longest object that will fit inside a cylinder that has a radius of 1 inch and is 2 inches high.

HINT: Draw a picture and use the Pythagorean Theorem.

In Exercises 13–18, use the two-column method to prove the indicated result.

13. Given: $AD = CD$
             $AB = CB$
Prove: $\angle DBA = \angle DBC$

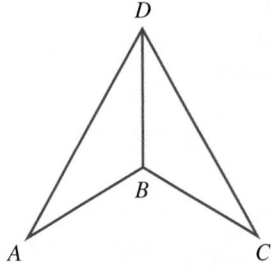

14. Given: $CA = DB$
             $CB = DA$
Prove: $\angle ACB = \angle ADB$

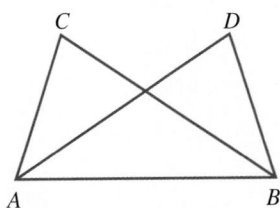

15. Given: $AD = BD$
             $\angle ADC = \angle BDC$
Prove: $AC = BC$

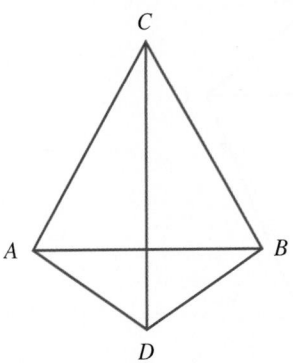

16. Given: $\angle ACD = \angle BCD$
             $\angle ADC = \angle BDC$
Prove: $AC = BC$

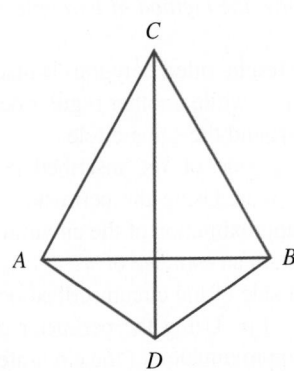

17. Given: $AE = CE$
    $AB = CB$
Prove: $\angle ADB = \angle CDB$

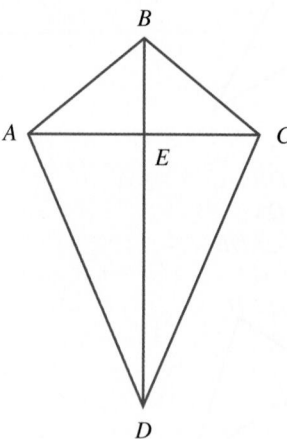

18. Given: $AE = CE$
    $\angle AEB = \angle CEB$
Prove: $AD = CD$

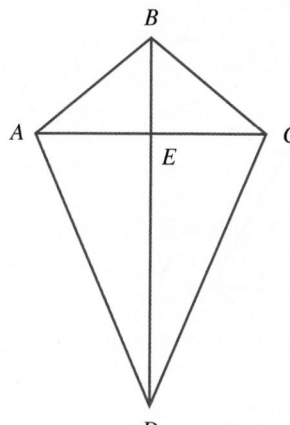

In Exercises 19–22, follow the method of Example 3.

19. A regular octagon (eight-sided polygon) is inscribed in a circle of radius r, while another regular octagon is circumscribed around the same circle.
  a. The length of a side of the inscribed octagon is $s = \sqrt{2 - \sqrt{2}}\,r$. Using the perimeter of this polygon as an approximation of the circumference of the circle, obtain an estimate of π.
  b. The length of a side of the circumscribed octagon is $s = 2(\sqrt{2} - 1)r$. Using the perimeter of this polygon as an approximation of the circumference of the circle, obtain an estimate of π.

20. A regular dodecagon (12-sided polygon) is inscribed in a circle of radius r, while another regular dodecagon is circumscribed around the same circle.
  a. The length of a side of the inscribed dodecagon is $s = [(\sqrt{6} - \sqrt{2})/2]r$. Using the perimeter of this polygon as an approximation of the circumference of the circle, obtain an estimate of π.
  b. The length of a side of the circumscribed dodecagon is $s = [(\sqrt{6} - \sqrt{2})/(\sqrt{2} + \sqrt{3})]r$. Using the perimeter of this polygon as an approximation of the circumference of the circle, obtain an estimate of π.

21. A regular 16-sided polygon is inscribed in a circle of radius r, while another regular 16-sided polygon is circumscribed around the same circle.
  a. The length of a side of the inscribed polygon is $s = \sqrt{2 - \sqrt{2 + \sqrt{2}}}\,r$. Using the perimeter of this polygon as an approximation of the circumference of the circle, obtain an estimate of π.
  b. The length of a side of the circumscribed polygon is

$$s = \left(\frac{2\sqrt{2 - \sqrt{2}}}{2 + \sqrt{2 + \sqrt{2}}}\right)r$$

Using the perimeter of this polygon as an approximation of the circumference of the circle, obtain an estimate of π.

22. A regular 24-sided polygon is inscribed in a circle of radius r, while another 24-sided polygon is circumscribed around the same circle.
  a. The length of a side of the inscribed polygon is $s = [\sqrt{8 - \sqrt{24 - \sqrt{8}}}/2]r$. Using the perimeter of this polygon as an approximation of the circumference of the circle, obtain an estimate of π.
  b. The length of a side of the circumscribed polygon is

$$s = \left(\frac{2\sqrt{8 - \sqrt{24 - \sqrt{8}}}}{\sqrt{8 + \sqrt{24 + \sqrt{8}}}}\right)r$$

Using the perimeter of this polygon as an approximation of the circumference of the circle, obtain an estimate of π.

23. If a sphere is inscribed in a cylinder, the diameter of the sphere equals the height of the cylinder. Using this relationship, show Archimedes' favorite result, that is, that the volumes of the sphere and the cylinder are in the ratio $2 : 3$.

Answer the following questions using complete sentences and your own words.

HISTORY QUESTIONS

24. What is the connection between Thales and Egypt?

25. What was the motto of the Pythagoreans?

26. Which president of the United States developed an original proof of the Pythagorean Theorem?

27. What is the Parallel Postulate?

28. How did the geometry of the Greeks differ from that of the Egyptians?

29. Write a research paper on a historical topic referred to in this section or on a related topic. Following is a list of possible topics:

- Alexandria as a center of knowledge (Why and when did Alexandria flourish as a center of knowledge in the ancient world?)
- Ancient war machines (How were mathematics and physics used in the design and construction of ancient weapons of war?)
- Archimedes
- Classic construction problems (What do the problems "trisecting an angle," "squaring a circle," and "doubling a cube" have in common? Why are these problems special in the history of geometry?)
- Eratosthenes of Cyrene (How did he calculate the earth's circumference? What did he contribute to the making of maps?)
- Euclid
- Eudoxus of Cnidos (What is the "Golden Section" or "Golden Ratio"?)
- Heron of Alexandria (What is Heron's Formula?)
- Hippocrates of Chios (What are the lunes of Hippocrates?)
- Pappus of Alexandria (What did Pappus's Mathematical Collection contain?)
- Pulleys and levers (How do pulleys and levers reduce the force required to lift an object?)
- Pythagoras and the Pythagoreans
- Thales of Miletus (How did he calculate the height of the Great Pyramid of Cheops?)
- Zeno of Elea (What are Zeno's paradoxes?)

PROJECT

30. The purpose of this project is to calculate the height of a tall building by using similar triangles. Choose a tall building and measure the length of its shadow. Now have an assistant measure the length of your shadow. The top of the building, the base of the building, and the tip of the shadow of the building form a triangle, say, $\triangle ABC$. Likewise, the top of your head, your heels, and the tip of your shadow form a triangle, say, $\triangle DEF$. (See Figure 8.64.) Assuming that rays of sunlight are parallel, we conclude that $\triangle ABC$ and $\triangle DEF$ are similar. Use the ratio of your height to the length of your shadow to calculate the height of the building.

 ## WEB PROJECTS

31. There are numerous web sites devoted to the works of Euclid; many are animated and interactive. Investigate some of these web sites and write an essay and/or make a presentation about your findings.

32. There are numerous web sites devoted to the works of Pythagoras; many are animated and interactive. Investigate some of these web sites and write an essay and/or make a presentation about your findings.

Some useful links for these web projects are listed on the text web site: **www.thomsonedu.com/mathematics/johnson**

Figure 8.64
Using shadows to calculate the height of a building

8.5 | Right Triangle Trigonometry

Plane geometry includes the study of many simple figures, such as circles, triangles, rectangles, squares, parallelograms, and trapezoids. Triangles are of special interest; they are the simplest figures that can be drawn with straight sides. **Trigonometry** is the branch of mathematics that details the study of triangles. The word *trigonometry* literally means "triangle measurement"; it is derived from the Greek words *trigon*, meaning "triangle," and *metros*, meaning "measure." Although modern trigonometry has numerous applications that do not involve triangles, trigonometry's roots lie in the study of right triangles; for example, we will use trigonometry to calculate the length of a cable that runs from the ground to the top of a warehouse, as shown in Figure 8.65.

Figure 8.65
How long is a cable?

Angle Measurement

A triangle has three sides and three angles. The sides can be measured in terms of any linear unit, such as inches, feet, centimeters, or meters. How are angles measured? The two most common units are *degrees* and *radians*. We will consider only degrees in this textbook.

A **right angle** is an angle that makes a square corner. By definition, a right angle is said to have a measure of **90 degrees,** denoted 90°. See Figure 8.66(a). Consequently, a straight line can be viewed as an angle of measure 180°, as shown in Figure 8.66(b).

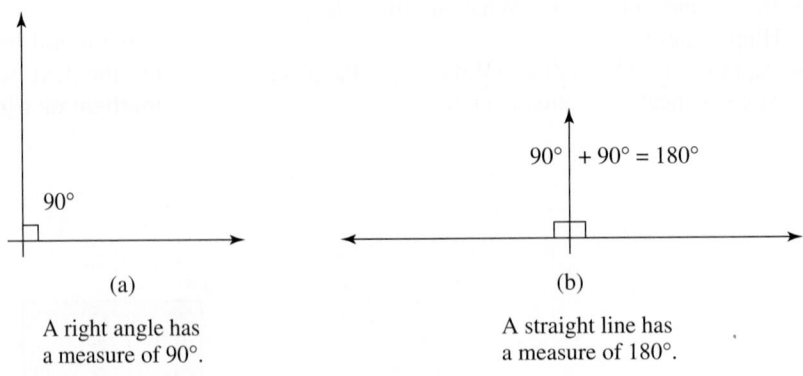

(a)

A right angle has
a measure of 90°.

(b)

A straight line has
a measure of 180°.

Figure 8.66

If a right angle is "cut up" into 90 equal pieces, each "slice" would represent an angle of measure 1°. From a previous math course, you may recall that for any triangle, the sum of the measures of the three angles is 180°. Why is this true? Suppose the angles of a triangle are *A, B,* and *C,* as shown in Figure 8.67.

Now construct two copies of the triangle and place them next to the original, as shown in Figure 8.68.

Figure 8.67
An arbitrary triangle

Figure 8.68
Multiple copies of a triangle

If the triangle on the right is flipped backward and upside down, it will fit perfectly into the "notch" between the other two triangles, as shown in Figure 8.69.

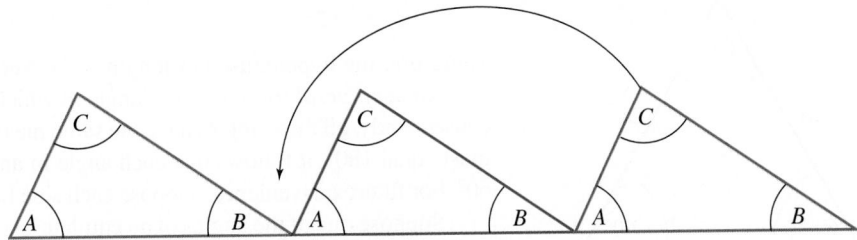

Figure 8.69
Moving a triangle

Notice that the three angles *A*, *B*, and *C* now form a straight line, as shown on the bottom edge of Figure 8.70; consequently, the sum of the angles equals 180°.

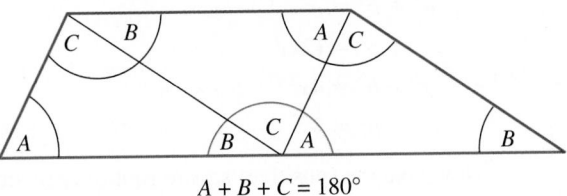

$$A + B + C = 180°$$

Figure 8.70
The sum of the angles in a triangle $= 180°$

Special Triangles

There are two special triangles that frequently arise in trigonometry: the 30°-60°-90° triangle and the 45°-45°-90° triangle. The sides of each of these triangles have special relationships.

Each angle of a square has a measure of 90°. If a square with sides of length *x* is cut in two along one of its diagonals, each resulting right triangle will have two equal sides (of length *x*) and two 45° angles. See Figure 8.71.

Figure 8.71
A square and its diagonal

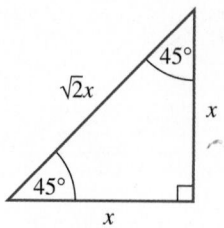

Figure 8.72
Isosceles right triangle

Figure 8.73
Equilateral triangle

Figure 8.74
Bisected equilateral triangle

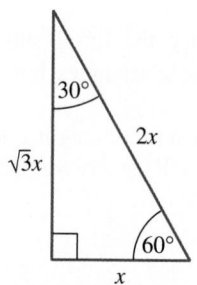

Figure 8.75
30°-60°-90° triangle

A triangle of this type is called an *isosceles right triangle*. (An isosceles triangle is a triangle that has two equal sides.) Applying the Pythagorean theorem, we find the length of the hypotenuse as follows:

$$c^2 = a^2 + b^2 \quad c = \text{hypotenuse, } a \text{ and } b = \text{legs}$$
$$c^2 = x^2 + x^2 \quad \text{substituting } a = x \text{ and } b = x$$
$$c^2 = 2x^2 \quad \text{combining like terms}$$

Now take the (positive) square root of each side and simplify:

$$\sqrt{c^2} = \sqrt{2x^2}$$
$$c = \sqrt{2}x$$

Therefore, the hypotenuse has length $\sqrt{2}x$. See Figure 8.72.

An *equilateral triangle* is a triangle in which all three sides have the same length; consequently, all three angles have the same measure. Since the sum of the three angles must equal 180°, it follows that each angle in an equilateral triangle has a measure of 60°. For future convenience, suppose each side has length $2x$, as shown in Figure 8.73.

Suppose one of the angles of an equilateral triangle is bisected (cut into two equal pieces); the bisected 60° angle will generate two 30° angles. Consequently, two right triangles are created. In addition, the side opposite the bisected angle will be cut into two equal segments of length x. See Figure 8.74.

Applying the Pythagorean theorem to one of these right triangles (called a *30°-60°-90° triangle*), we find the length of the missing (vertical) leg as follows:

$$a^2 + b^2 = c^2 \quad a \text{ and } b = \text{legs, } c = \text{hypotenuse}$$
$$x^2 + b^2 = (2x)^2 \quad \text{substituting } a = x \text{ and } c = 2x$$
$$x^2 + b^2 = 4x^2 \quad \text{multiplying}$$
$$b^2 = 3x^2 \quad \text{subtracting } x^2 \text{ from both sides}$$

Now take the (positive) square root of each side and simplify:

$$\sqrt{b^2} = \sqrt{3x^2}$$
$$b = \sqrt{3}x$$

Therefore, the leg has length $\sqrt{3}x$. See Figure 8.75.

Trigonometric Ratios

An *acute angle* is an angle whose measure is between 0° and 90°. Let θ (the Greek letter theta) equal the measure of one of the acute angles of a right triangle. Using the abbreviations "opp," "adj," and "hyp" to represent, respectively, the side opposite angle θ, the side adjacent to angle θ, and the hypotenuse, a right triangle can be labeled, as in Figure 8.76.

Figure 8.76
The sides of a right triangle

In Section 8.4, we saw that for similar triangles (triangles that have the same shape but different size), corresponding sides are proportional; that is, ratios of corresponding sides are constant. Consequently, no matter how large or how small a right triangle is, the ratio of its sides is determined solely by the size of the acute angles. Three specific ratios of sides form the basis of trigonometry; these ratios are called the *sine, cosine*, and *tangent* and are defined as follows.

Trigonometric Ratios

Let θ equal the measure of one of the acute angles of a right triangle.

1. The **sine** of θ, denoted **sinθ**, is the ratio of the side opposite θ compared to the hypotenuse: $\sin\theta = \dfrac{\text{opp}}{\text{hyp}}$

2. The **cosine** of θ, denoted **cosθ**, is the ratio of the side adjacent θ compared to the hypotenuse: $\cos\theta = \dfrac{\text{adj}}{\text{hyp}}$

3. The **tangent** of θ, denoted **tanθ**, is the ratio of the side opposite θ compared to the side adjacent θ: $\tan\theta = \dfrac{\text{opp}}{\text{adj}}$

The following example examines the special isosceles right triangle.

EXAMPLE 1   Find the sine, cosine, and tangent of 45°.

Solution   Recall the general isosceles right triangle shown in Figure 8.72 and reproduced in Figure 8.77. The length of the side opposite 45° is x, as is the length of the side adjacent 45°; the length of the hypotenuse is $\sqrt{2}x$. Applying the trigonometric ratios to the angle of measure 45°, we have the following:

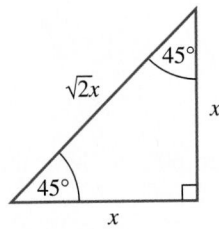

Figure 8.77
Isosceles right triangle

$$\sin 45° = \frac{\text{opp}}{\text{hyp}} = \frac{x}{\sqrt{2}x} = \frac{1}{\sqrt{2}}$$ To rationalize the denominator, multiply by $\dfrac{\sqrt{2}}{\sqrt{2}}$.

Therefore, $\sin 45° = \dfrac{\sqrt{2}}{2}$.

$$\cos 45° = \frac{\text{adj}}{\text{hyp}} = \frac{x}{\sqrt{2}x} = \frac{1}{\sqrt{2}}$$ To rationalize the denominator, multiply by $\dfrac{\sqrt{2}}{\sqrt{2}}$.

Therefore, $\cos 45° = \dfrac{\sqrt{2}}{2}$.

$$\tan 45° = \frac{\text{opp}}{\text{adj}} = \frac{x}{x} = 1$$

Therefore, $\tan 45° = 1$.   ∎

Trigonometric Ratios for 45°

$$\sin 45° = \frac{1}{\sqrt{2}} = \frac{\sqrt{2}}{2}$$

$$\cos 45° = \frac{1}{\sqrt{2}} = \frac{\sqrt{2}}{2}$$

$$\tan 45° = \frac{1}{1} = 1$$

Notice that in Example 1, the value of x has no influence on the trigonometric ratios; x always "cancels" in both the numerator and denominator. In working with a 45° angle, an "easy" triangle to remember is the triangle in which $x = 1$. The results are summarized below.

The following example examines the special 30°-60°-90° triangle.

EXAMPLE 2   **a.** Find the sine, cosine, and tangent of 30°.
b. Find the sine, cosine, and tangent of 60°.

Solution   **a.** Recall the general 30°-60°-90° triangle shown in Figure 8.75 and reproduced in Figure 8.78. The length of the side opposite 30° is x, the length of the side adjacent 30° is $\sqrt{3}x$, and the length of the hypotenuse is $2x$. Applying the trigonometric ratios to the angle of measure 30°, we have the following:

$$\sin 30° = \frac{\text{opp}}{\text{hyp}} = \frac{x}{2x} = \frac{1}{2}$$

Therefore, $\sin 30° = \dfrac{1}{2}$.

$$\cos 30° = \frac{\text{adj}}{\text{hyp}} = \frac{\sqrt{3}x}{2x} = \frac{\sqrt{3}}{2}$$

Therefore, $\cos 30° = \dfrac{\sqrt{3}}{2}$.

$$\tan 30° = \frac{\text{opp}}{\text{adj}} = \frac{x}{\sqrt{3}x} = \frac{1}{\sqrt{3}}$$ To rationalize the denominator, multiply by $\dfrac{\sqrt{3}}{\sqrt{3}}$.

Therefore, $\tan 30° = \dfrac{\sqrt{3}}{3}$.

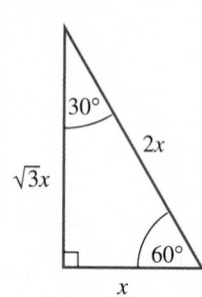

Figure 8.78
30°-60°-90° triangle

b. Applying the trigonometric ratios to the angle of measure 60°, we have the following:

$$\sin 60° = \frac{\text{opp}}{\text{hyp}} = \frac{\sqrt{3}x}{2x} = \frac{\sqrt{3}}{2}$$

Therefore, $\sin 60° = \dfrac{\sqrt{3}}{2}$.

$$\cos 60° = \frac{\text{adj}}{\text{hyp}} = \frac{x}{2x} = \frac{1}{2}$$

Therefore, $\cos 60° = \dfrac{1}{2}$.

$$\tan 60° = \frac{\text{opp}}{\text{adj}} = \frac{\sqrt{3}x}{x} = \frac{\sqrt{3}}{1}$$

Therefore, $\tan 60° = \sqrt{3}$.   ∎

Notice that in Example 2, the value of x has no influence on the trigonometric ratios; x always "cancels" in both the numerator and denominator. In working with a 30° or 60° angle, an "easy" triangle to remember is the triangle in which $x = 1$. The results are summarized below.

Trigonometric Ratios for 30° and 60°

$$\sin 30° = \frac{1}{2} \qquad\qquad \sin 60° = \frac{\sqrt{3}}{2}$$

$$\cos 30° = \frac{\sqrt{3}}{2} \qquad\qquad \cos 60° = \frac{1}{2}$$

$$\tan 30° = \frac{1}{\sqrt{3}} = \frac{\sqrt{3}}{3} \qquad \tan 60° = \sqrt{3}$$

Before we investigate acute angles other than the special ones (30°, 45°, and 60°), we first consider an application of trigonometric ratios.

EXAMPLE 3   A cable runs from the top of a warehouse to a point on the ground 54.0 feet from the base of the building. If the cable makes an angle of 30° with the ground (as shown in Figure 8.79), calculate

a. the height of the warehouse
b. the length of the cable

Figure 8.79
A cable attached to a warehouse

Solution   Let h = height of the warehouse and c = length of the cable. Assuming that the building makes a right angle with the ground, we have the right triangle shown in Figure 8.80.

a. The unknown height of the building is the side opposite the given angle of 30°; it is labeled *opp*. Similarly, the known side of 54.0 feet is adjacent to the angle; it is labeled *adj*. By comparing *opp* to *adj*, we use the tangent ratio to find the height of the building:

Figure 8.80
A right triangle

$$\tan \theta = \frac{\text{opp}}{\text{adj}} \qquad \text{definition of the tangent ratio}$$

$$\tan 30° = \frac{h}{54.0} \qquad \text{substituting } \theta = 30°, \text{ opp} = h, \text{ adj} = 54.0$$

$$h = 54.0 \tan 30° \qquad \text{multiplying both sides by 54.0}$$

However, we know that $\tan 30° = \dfrac{\sqrt{3}}{3}$. Substituting, we obtain

$$h = 54.0 \left(\frac{\sqrt{3}}{3} \right) = 18\sqrt{3} = 31.1769\ldots$$

The height of the warehouse is approximately 31.2 feet.

b. The unknown length of the cable is the hypotenuse; it is labeled *hyp*. By comparing *adj* (which is known) to *hyp*, we use the cosine ratio to find the length of the cable:

$$\cos \theta = \frac{adj}{hyp} \qquad \text{definition of the cosine ratio}$$

$$\cos 30° = \frac{54.0}{c} \qquad \text{substituting } \theta = 30°, adj = 54.0, hyp = c$$

$$c(\cos 30°) = 54.0 \qquad \text{multiplying both sides by } c$$

$$c = \frac{54.0}{\cos 30°} \qquad \text{dividing both sides by } \cos 30°$$

However, we know that $\cos 30° = \dfrac{\sqrt{3}}{2}$. Substituting, we obtain

$$c = \frac{54.0}{\dfrac{\sqrt{3}}{2}}$$

$$c = \left(\frac{54.0}{1}\right)\left(\frac{2}{\sqrt{3}}\right) \qquad \text{dividing by a fraction: invert and multiply}$$

$$c = \frac{108}{\sqrt{3}} \cdot \frac{\sqrt{3}}{\sqrt{3}} \qquad \text{rationalizing the denominator}$$

$$c = \frac{108\sqrt{3}}{3} = 36\sqrt{3} = 62.3538 \ldots$$

The length of the cable is approximately 62.4 feet. ∎

Using a Calculator

So far, we have found trigonometric ratios of only three angles: 30°, 45°, and 60°. Historically, these were the "easiest" owing to their association with special triangles. How do we find the trigonometric ratios for *any* acute angle? In the past, mathematicians developed extensive tables that gave the desired values; today, we use calculators.

When using a calculator, the proper unit of angle measure must be selected. In this textbook, we measure angles in terms of degrees; the calculator must be in the "degree mode." If your calculator has a $\boxed{\text{DRG}}$ button, press it as many times as necessary so that the display reads "D" or "DEG." (*DRG* signifies *Degrees-Radians-Gradiens*.) Some scientific calculators have a reference chart printed near the display window; press $\boxed{\text{MODE}}$ and the appropriate symbol to select "degrees." If you have a graphing calculator, press the $\boxed{\text{MODE}}$ button and use the arrow buttons to select "degrees."

EXAMPLE 4   Use a calculator to find the following values.

a. sin 17°   **b.** cos 25.3°   **c.** tan 83.45°

Solution   **a.** Locate the SIN button. With most scientific calculators, you enter the measure of the angle first, then press SIN; with most graphing calculators, you press SIN first, type the measure of the angle, and then press ENTER. In any case, we find that sin 17° = 0.292371704723. . . .

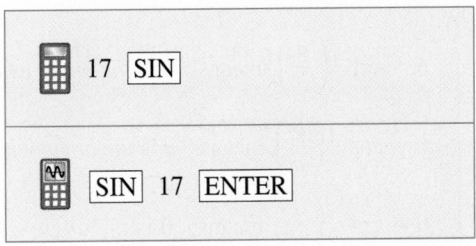

b. Using the COS button, we find that cos 25.3° = 0.9040825497. . . .

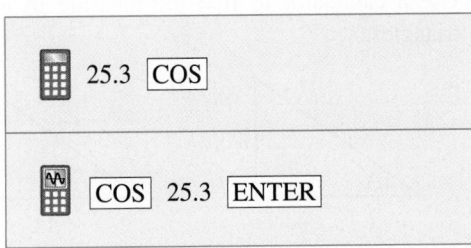

c. Using the TAN button, we find that tan 83.45° = 8.70930765678. . . .

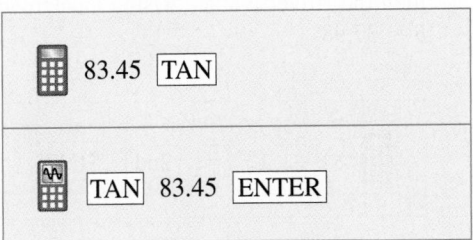

Finding an Acute Angle

If you know the measure of an acute angle, you can use a calculator to find the angle's trigonometric ratios. A calculator can also be used to work a problem in reverse; if you know a trigonometric ratio, a calculator can find the measure of the desired acute angle.

Specifically, let $a = opp$, $b = adj$, and $c = hyp$ be the lengths of the sides of a right triangle, where θ equals the measure of the acute angle opposite side a, as shown in Figure 8.81.

To find θ, note that $\sin \theta = opp/hyp = a/c$; alternatively, we can say that θ is *the angle whose sine is a/c*. This description of θ can be symbolized as $\theta = \sin^{-1}(a/c)$; that is, θ is the "inverse sine" of the ratio (a/c). Similar statements can be made concerning the cosine and tangent. These results are summarized below.

Figure 8.81
A right triangle

Inverse Trigonometric Ratios

If a, b, and c are the lengths of the sides of a right triangle and θ is the acute angle shown in the figure to the right, then

$\theta = \sin^{-1}\left(\dfrac{a}{c}\right)$ means θ is *the angle whose sine is* $\dfrac{a}{c}$;   that is, $\sin\theta = \dfrac{a}{c}$.

$\theta = \cos^{-1}\left(\dfrac{b}{c}\right)$ means θ is *the angle whose cosine is* $\dfrac{b}{c}$;   that is, $\cos\theta = \dfrac{b}{c}$.

$\theta = \tan^{-1}\left(\dfrac{a}{b}\right)$ means θ is *the angle whose tangent is* $\dfrac{a}{b}$,   that is, $\tan\theta = \dfrac{a}{b}$.

EXAMPLE 5   Use a calculator to find the measure of angle θ in each of the following right triangles:

a.   **b.**   **c.**

Solution   **a.** Applying the sine ratio, we have $\sin\theta = \frac{8}{13}$ or, equivalently, $\theta = \sin^{-1}\left(\frac{8}{13}\right)$.
On a calculator, pressing $\boxed{\text{INV}}$ $\boxed{\text{SIN}}$ or $\boxed{\text{2nd}}$ $\boxed{\text{SIN}}$ or $\boxed{\text{shift}}$ $\boxed{\text{SIN}}$ will summon the "inverse sine." Using a calculator, we obtain $\theta = 37.97987244\ldots^{\circ}$, or $\theta \approx 37.98^{\circ}$.

b. Applying the cosine ratio, we have $\cos\theta = \frac{7}{11}$ or, equivalently, $\theta = \cos^{-1}\left(\frac{7}{11}\right)$.
Using a calculator, we obtain $\theta = 50.47880364\ldots^{\circ}$, or $\theta \approx 50.48^{\circ}$.

c. Applying the tangent ratio, we have $\tan \theta = \frac{3}{4}$ or, equivalently, $\theta = \tan^{-1}(\frac{3}{4})$. Using a calculator, we obtain $\theta = 36.86989765\ldots°$, or $\theta \approx 36.87°$.

Angles of Elevation and Depression

If an object is above your eye level, you must raise your eyes to focus on the object. The angle measured from a horizontal line to the object is called the **angle of elevation.** If an object is below your eye level, you must lower your eyes to focus on the object. The angle measured from a horizontal line to the object is called the **angle of depression.** (See Figure 8.82.)

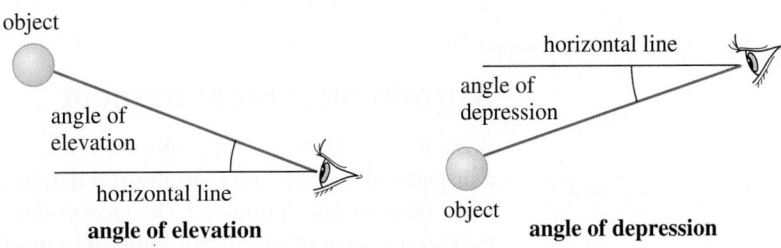

Figure 8.82

EXAMPLE 6   A communications antenna is on top of a building. You are 100.0 feet from the building. Using a transit (a surveyor's instrument), you measure the angle of elevation of the bottom of the antenna as 20.6° and the angle of elevation of the top of the antenna as 27.3°. How tall is the antenna?

Solution   Let h = height of the antenna, and x = height of the building. The given information is depicted by the two (superimposed) right triangles, shown in Figure 8.83

Figure 8.83
Finding the height of an antenna

Referring to the large right triangle, the side opposite the 27.3° angle is $(x + h)$, and the side adjacent is 100.0 feet. Using the tangent ratio (*opp/adj*), we have

$$\tan 27.3° = \frac{x + h}{100.0}$$

Solving for h, we obtain

$$h = 100.0 \, (\tan 27.3°) - x \qquad \qquad \textbf{(1)}$$

Using the tangent ratio for the small right triangle, we have

$$\tan 20.6° = \frac{x}{100.0}$$

Hence,

$$x = 100.0 \, (\tan 20.6°) \qquad \qquad \textbf{(2)}$$

Substituting equation (2) into equation (1), we obtain

$$h = 100.0 \, (\tan 27.3°) - 100.0 \, (\tan 20.6°)$$

or

$$h = 100.0 \, (\tan 27.3° - \tan 20.6°) \quad \text{factoring 100.0 from each term}$$

Using a calculator, we have

$$h = 14.0263198 \ldots$$

The antenna is approximately 14.0 feet tall.   ■

Astronomical Measurement

In the feature "Geometry in the Real World (and Universe)" presented in Section 8.1, a distance of 1 parsec was (arbitrarily) defined to equal 3.26 light-years (ly). What is the basis of this definition? The answer lies in right triangle trigonometry and in the measurement of very small angles. For most purposes, measuring an angle to the nearest degree is sufficient. However, if a finer measurement is required, 1 degree can be subdivided into smaller units called **minutes** (or arcminutes); by definition, 1 degree = 60 minutes. If still finer measurements are required, 1 minute can be subdivided into smaller units called **seconds** (or arcseconds); by definition, 1 minute = 60 seconds. Consequently, 1 degree = 3,600 seconds, or 1 second = 1/3600th of a degree.

Now, the unit of parsec can be defined precisely; 1 parsec (pc) is the length of the longer leg of a right triangle in which the shorter leg is exactly 1 AU (astronomical unit, or the mean distance between the earth and the sun) and the angle between the sun and the earth, as seen from an object in space (typically a star) is one arcsecond, as shown in Figure 8.84.

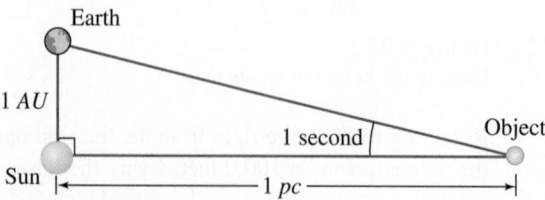

Figure 8.84
Definition of 1 parsec

Given that 1 AU = 93 million miles and 1 second = 1/3600th of a degree, we can now calculate the length of 1 parsec by using the tangent ratio.

$$\tan(1 \text{ second}) = \frac{1 \text{ AU}}{1 \text{ pc}} \qquad \tan = \frac{opp}{adj}$$

$$1 \text{ pc} = \frac{1 \text{ AU}}{\tan(1 \text{ second})} \qquad \text{multiplying by 1 pc and dividing by } \tan(1 \text{ second})$$

$$= \frac{93{,}000{,}000 \text{ mi}}{\tan(1/3600)^{\circ}} \qquad \text{substituting}$$

$$= \frac{93{,}000{,}000 \text{ mi}}{\tan(1/3600)^{\circ}} \times \frac{1 \text{ ly}}{5{,}880{,}000{,}000{,}000 \text{ mi}} \qquad 1 \text{ ly} = 5.88 \text{ trillion mi}$$

$$= \frac{93}{\tan(1/3600)^{\circ}} \times \frac{1 \text{ ly}}{5{,}880{,}000} \qquad \text{cancelling } 1{,}000{,}000 \text{ mi}$$

$$= 3.262351527 \ldots \text{ ly} \qquad \text{using a calculator}$$

Therefore, 1 parsec = 3.26 light-years, as was stated in Section 8.1.

8.5 Exercises

In Exercises 1–8, use trigonometric ratios to find the unknown sides and angles in each right triangle. Do not use a calculator.

1.

2.

3.

4.

5.

6.

7.

8.
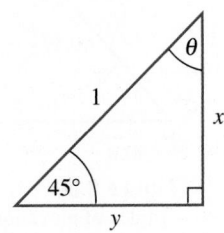

In Exercises 9–26, use the given information to draw a right triangle labeled like the one shown in Figure 8.85. Use trigonometric ratios and a calculator to find the unknown sides and angles. Round off sides and angles to the same number of decimal places as the given sides and angles; in Exercises 21–26, round off angles to one decimal place.

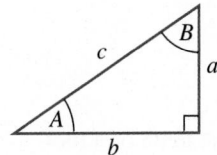

Figure 8.85
Standard triangle

9. $a = 12.0$ and $A = 37°$
10. $a = 23.0$ and $A = 74°$
11. $b = 5.6$ and $A = 54.3°$
12. $b = 19.5$ and $A = 20.1°$
13. $c = 0.92$ and $B = 49.9°$
14. $c = 0.086$ and $B = 18.7°$
15. $a = 1{,}546$ and $B = 9.15°$
16. $a = 2{,}666$ and $B = 81.07°$
17. $c = 54.40$ and $A = 53.125°$
18. $c = 81.60$ and $A = 32.375°$
19. $b = 1.002$ and $B = 5.00°$
20. $b = 1.321$ and $B = 3.00°$
21. $a = 15.0$ and $c = 23.0$
22. $a = 6.0$ and $c = 17.0$

23. $a = 6.0$ and $b = 7.0$

24. $a = 52.1$ and $b = 29.6$

25. $b = 0.123$ and $c = 0.456$

26. $b = 0.024$ and $c = 0.067$

27. A cable runs from the top of a building to a point on the ground 66.5 feet from the base of the building. If the cable makes an angle of 43.9° with the ground (as shown in Figure 8.86), find

Figure 8.86
Cable and building

    a. the height of the building
    b. the length of the cable
Round off answers to the nearest tenth of a foot.

28. A support cable runs from the top of a telephone pole to a point on the ground 42.7 feet from its base. If the cable makes an angle of 29.6° with the ground (as shown in Figure 8.87), find

Figure 8.87
Cable and pole

    a. the height of the pole
    b. the length of the cable
Round off answers to the nearest tenth of a foot.

29. A 20-foot-high diving tower is built on the edge of a large swimming pool. From the tower, you measure the angle of depression of the far edge of the pool. If the angle is 7.6°, how wide (to the nearest foot) is the pool?

30. You are at the top of a lighthouse and measure the angle of depression of an approaching ship. If the lighthouse is 90 feet tall and the angle of depression of the ship is 4.12°, how far from the lighthouse is the ship? Round your answer to the nearest multiple of ten.

31. The "pitch" of a roof refers to the vertical rise measured against a standard horizontal distance of 12 inches. If the pitch of a roof is 4 in 12 (the roof rises 4 inches for every 12 horizontal inches), find the acute angle the roof makes with a horizontal line. Round your answer to one decimal place.

32. If the pitch of a roof is 3 in 12, find the acute angle the roof makes with a horizontal line. Round your answer to one decimal place. See Exercise 31.

33. A bell tower is known to be 48.5 feet tall. You measure the angle of elevation of the top of the tower. How far from the tower (to the nearest tenth of a foot) are you if the angle is
    a. 15.4°?   **b.** 61.2°?

34. You are standing on your seat in the top row of an outdoor sports stadium, 200 feet above ground level. You see your car in the parking lot below and measure its angle of depression. How far (to the nearest foot) from the stadium is your car if the angle is
    a. 73.5°?   **b.** 22.9°?

35. A billboard is on top of a building. You are 125.0 feet from the building. You measure the angle of elevation of the bottom of the billboard as 33.4° and the angle of elevation of the top of the billboard as 41.0°. How tall is the billboard? Round your answer to one decimal place.

36. The distance between two buildings is 180 feet. You are standing on the roof of the taller building. You measure the angle of depression of the top and bottom of the shorter building. If the angles are 36.7° and 71.1°, respectively, find
    a. the height of the taller building
    b. the height of the shorter building
Round your answers to the nearest foot.

37. You are hiking along a river and see a tall tree on the opposite bank. You measure the angle of elevation of the top of the tree and find it to be 61.0°. You then walk 50 feet directly away from the tree and measure the angle of elevation. If the second measurement is 49.5° (see Figure 8.88), how tall is the tree? Round your answer to the nearest foot.

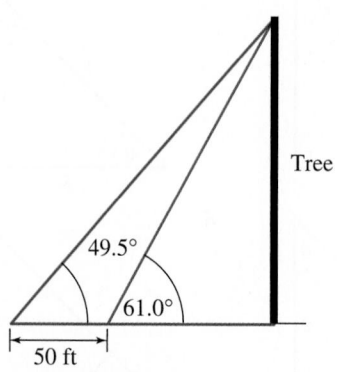

Figure 8.88
Angles of elevation

38. While sightseeing in Washington, D.C., you visit the Washington Monument. From an unknown distance, you measure the angle of elevation of the top of the monument. You then move 100 feet backward (directly away from the monument) and measure the angle of elevation of the top of the monument. If the angles are 61.6° and 54.2°, respectively, how tall (to the nearest foot) is the Washington Monument?

39. While sightseeing in St. Louis, you visit the Gateway Arch. Standing near the base of the structure, you measure the angle of elevation of the top of the arch. You then move 120 feet backward and measure the angle of elevation of the top of the arch. If the angles are 73.72° and 64.24°, respectively, how tall (to the nearest foot) is the Gateway Arch?

© Brian Yarvin/The Image Works

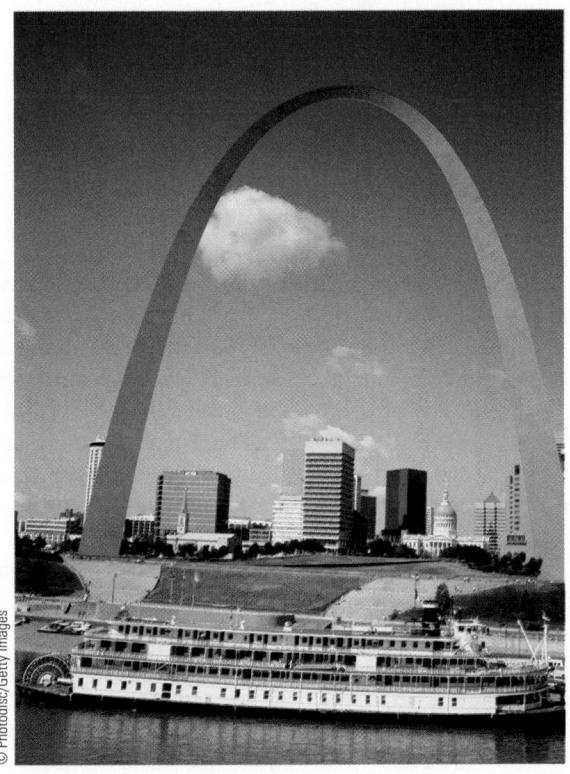

© Photodisc/Getty Images

40. While sightseeing in Seattle, you visit the Space Needle. From a point 175 feet from the structure, you measure the angle of elevation of the top of the structure. If the angle is 73.87°, how tall (to the nearest foot) is the Space Needle?

41. The Sears Tower (in Chicago) is the tallest building in the United States. From a point 900 feet from the building, you measure the angle of elevation of the top of the building. If the angle is 58.24°, how tall (to the nearest foot) is the Sears Tower?

42. A tree is growing on a hillside. From a point 100 feet downhill from the base of the tree, the angle of elevation to the base of the tree is 28.3° with an additional 12.5° to the top of the tree. How tall is the tree?

43. A tree is growing on a hillside. From a point 100 feet uphill from the base of the tree, the angle of depression to the top of the tree is 28.3° with an additional 12.5° to the base of the tree. How tall is the tree?

44. Two observers are 500 feet apart. Each measures the angle of elevation of a hot air balloon that lies in a vertical plane passing through their locations; the angles are 53° and 42°. Find the height of the balloon if the observers are on

a. opposite sides of the balloon. See Figure 8.89.

b. the same side of the balloon. See Figure 8.90.

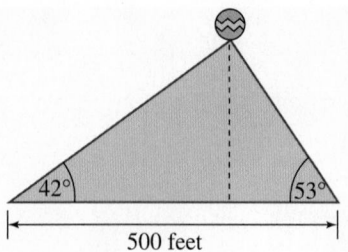

Figure 8.89
Observers on opposite sides

Figure 8.90
Observers on same side

45. Calculate the distance (in parsecs) from the sun to an object in space if the angle between the sun and the earth (as seen from the object) is 2 seconds and the distance between the earth and the sun is 1 AU. (See Figure 8.84.)

46. Calculate the distance (in parsecs) from the sun to an object in space if the angle between the sun and the earth (as seen from the object) is 1/5 second and the distance between the earth and the sun is 1 AU. (See Figure 8.84.)

47. Suppose the distance from the sun to an object in space is 0.9 parsec and the distance between the earth and the sun is 1 AU (as in Figure 8.84).
  a. How far from the sun is the object in terms of AU?
  b. Calculate the angle between the sun and the earth, as seen from the object. Express your answer in terms of seconds, rounded off to the nearest hundredth of a second.

48. Suppose the distance from the sun to an object in space is 0.8 parsec and the distance between the earth and the sun is 1 AU (as in Figure 8.84).
  a. How far from the sun is the object in terms of AU?
  b. Calculate the angle between the sun and the earth, as seen from the object. Express your answer in terms of seconds, rounded off to the nearest hundredth of a second.

PROJECT

49. The purpose of this project is to calculate the height of a tall building by measuring an angle of elevation.

a. First, we must construct a device to measure an angle of elevation. You will need a protractor, a plastic drinking straw, a bead (or button), and a piece of thread. Attach the straw to the straight edge of the protractor so that the straw lines up with the 0° marks on the protractor. Next, tie the thread to the bead and attach the thread to the center mark of the protractor as shown in Figure 8.91.

Figure 8.91
Home made device

Note that when the straw is horizontal, the thread is positioned over the 90° mark on the scale. To measure an angle of elevation, look through the straw and focus on the top of the object being measured. Allow the thread to hang vertically, and have an assistant read the larger of the two angles indicated by the thread. (One scale will give an angle greater than 90°, the other less than 90°.) Subtract 90° from the larger angle, and you will obtain the angle of elevation. The angle of elevation indicated by the device shown in Figure 8.91 is 115° − 90° = 25°.

b. Choose a tall building. Pick an accessible point away from the building, and measure the angle of elevation of the top of the building. Now use a tape measure to determine the distance from your location to the base of the building. Set up the appropriate trigonometric ratio, and calculate the height of the building.

c. Pick a second location, and repeat part (b).

d. Your answers to parts (b) and (c) should be roughly the same. Are they? If not, give possible reasons for the discrepancy.

Conic Sections and Analytic Geometry

In order to seek truth it is necessary once in the course of our life to doubt as far as possible all things.
—René Descartes

It is easy to take our system of mathematical notation and algebraic manipulation for granted or even to regard the study of mathematics as an inconvenience and doubt its relevance to our individual lives and goals. René Descartes, one of the great philosophers and mathematicians of the seventeenth century, felt the same way. In doubting the "system" of mathematics of his predecessors, Descartes contributed to the creation of one of the major foundations of modern mathematics: analytic geometry. Simply put, **analytic geometry** is the marriage of algebra and geometry.

Most mathematics from ancient times through the Middle Ages consisted of a verbal description of a geometric method for solving a single problem. (The empirical geometry of the Egyptians is a case in point.) In the Europe of the Middle Ages, "advanced" mathematics was available only to those who knew Latin, as all major works were written in this "language of scholars." Rather than relying on language (which differs from culture to culture), Descartes pioneered the modern notion of using single letters to represent known and unknown quantities. He used the last letters of the alphabet—x, y, and z—to represent variables and the first letters—a, b, and c—to represent constants.

The Greeks verbally defined a parabola to be "the set of all points in a plane equidistant from a given line and a point not on the line"; modern mathematicians, by contrast, favor the equation $y = ax^2 + bx + c$. When we say that the graph of the equation $y = ax^2 + bx + c$ is a parabola, we are operating in the realm of analytic geometry. No doubt your initial exposure to this field was in your beginning or intermediate algebra course, when you were asked to graph an equation such as $y = 2x + 1$ or $y = x^2$.

Analytic geometry was not created overnight. Many individuals from diverse cultures made significant contributions. Some historians take the easy way out and say that Descartes invented it, but the story of the creation of analytic geometry is much more complicated than that. Because of its intimate association with the creation of calculus, the development of analytic geometry is discussed in Chapter 13.

Conic Sections

A **conic section** is the figure formed when a plane intersects a cone. What do the cross sections of a cone look like? The ancient Greeks studied this problem extensively. Recall that their approach was verbal, involving statements such as "a parabola is the set of all points in a plane equidistant from a given line and a point not on the line" or "a circle is the set of all points in a plane equidistant from a given point." When we study the various conic sections, we use a "double cone" like the one shown in Figure 8.92. Depending on the inclination of the plane of intersection, a circle, a parabola, ellipse, or hyperbola is formed. (See Figure 8.93.)

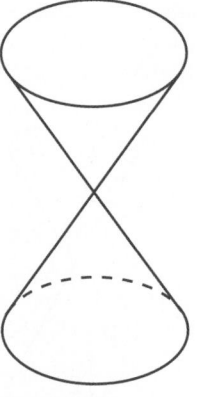

Figure 8.92
A double cone

The Circle

When a cross section is taken parallel to the base of the cone, as in Figure 8.93(a), a **circle** is obtained. The Greeks defined a circle as "the set of all points in a plane equidistant from a given fixed point." That is, no matter which point P on the circle

Historical Note Hypatia, A.D. 370–415

© CORBIS/Bettman

Hypatia, one of the first women to be recognized for her mathematical accomplishments, lived in Alexandria, one of the largest and most academically prominent cities on the Mediterranean Sea. This Hellenistic city was famed for its university and its library, said to be the largest of its day.

Hypatia's early environment was filled with intellectual challenge and stimulation. Her father, Theon, was a professor of mathematics and director of the museum and library at the University of Alexandria. He gave his daughter a classic education in the arts, literature, mathematics, science, and philosophy. In addition to her studies at the University of Alexandria, Hypatia traveled the Mediterranean. While in Athens, she attended a school conducted by the famed writer Plutarch.

On her return to Alexandria, Hypatia continued in her father's footsteps: She lectured on mathematics and philosophy at the university and directed the museum and library. Her lectures were well received by enthusiastic students and scholars alike; many considered Hypatia to be an oracle. Although none of her writings remains intact, historians attribute several mathematical treatises to Hypatia, including commentaries on the astronomical works of Diophantus and Ptolemy, on the conics of Apollonius, and on the geometry of Euclid. In addition to her insightful lectures and mathematical works, letters written by Hypatia's contemporaries credit her with the invention of devices used in the study of astronomy.

At this time, Alexandria was part of the Roman Empire and was undergoing a power struggle between Christians and pagans (worshippers of Greek and Roman gods). Bishop Cyril was using his position in the Christian church to usurp the power of the Alexandrian government, which was under the rule of the Roman prefect Orestes. Orestes was known to have attended many of Hypatia's lectures and was believed to have been Hypatia's lover. Since Hypatia was a symbol of classic Greek culture, Bishop Cyril associated her with paganism and viewed her as a threat to his quest for Christian power.

In a frenzied attempt to eradicate the pagan influences of Alexandria, a Christian mob incited by Bishop Cyril attacked Hypatia while she was riding in her chariot. She was stripped naked and dragged through the streets, then tortured and murdered.

Hypatia was the last symbol of the ancient culture of Alexandria. She preserved and carried forward the knowledge and wisdom of the Golden Age of Greek civilization. It is tragic that her intelligence and devotion led to her violent death.

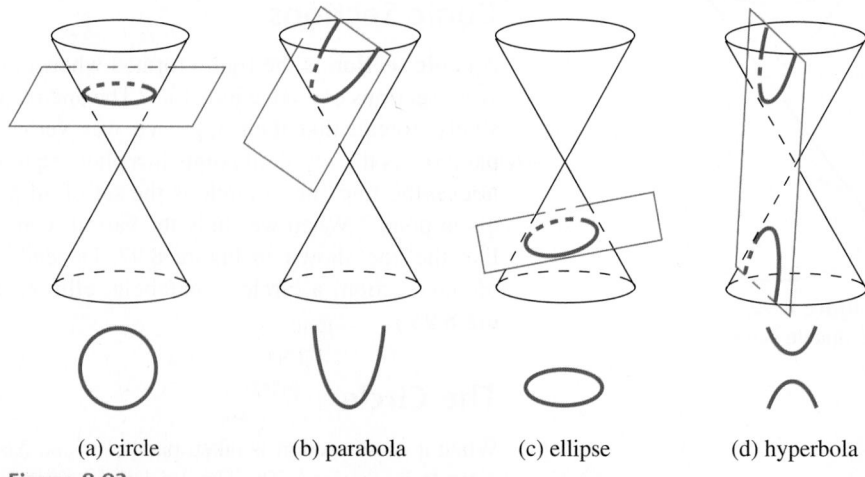

(a) circle   (b) parabola   (c) ellipse   (d) hyperbola

Figure 8.93
Conic sections

you select, its distance from the fixed point (the center) is always the same, as shown in Figure 8.94.

Using modern algebra, a circle can also be described as the graph of an algebraic equation, as shown in Figure 8.95.

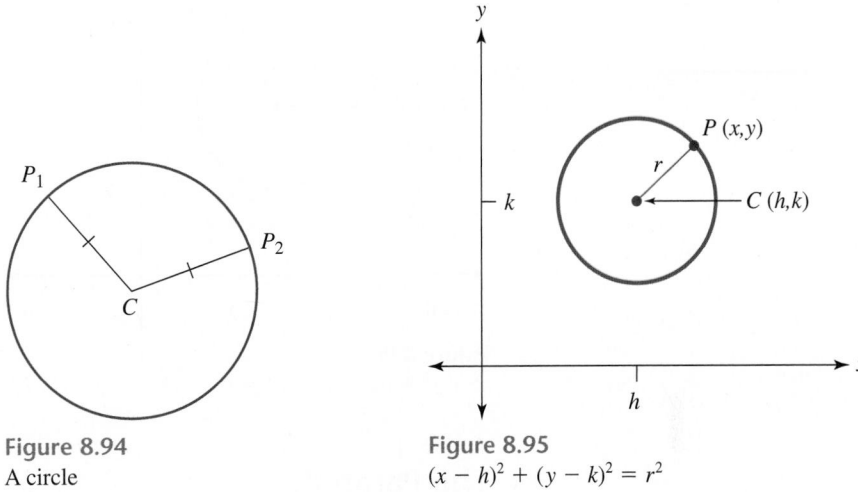

Figure 8.94
A circle

Figure 8.95
$(x - h)^2 + (y - k)^2 = r^2$

Equation of a Circle

A **circle** of radius r, centered at the point (h, k), is the set of all points satisfying the equation

$$(x - h)^2 + (y - k)^2 = r^2$$

ThomsonNOW™
for Liberal Arts Mathematics
www.thomsonedu.com/1pass.

EXAMPLE 1   Find the center and radius of the circle $x^2 + y^2 + 6x - 8y + 21 = 0$.

Solution   To find the coordinates of the center (that is, to find h and k), we must put the given equation in the form $(x - h)^2 + (y - k)^2 = r^2$. The expressions $(x - h)^2$ and $(y - k)^2$ are called *perfect squares;* each is the square of a binomial.

First, group the terms containing x and those containing y and put the constant on the right side of the equation:

$$x^2 + y^2 + 6x - 8y + 21 = 0$$
$$(x^2 + 6x) + (y^2 - 8y) = -21$$

The given expression $(x^2 + 6x)$ is *not* a perfect square; we must add the appropriate term to complete the square. To find this missing term, take half the coefficient of x and square it; we must add $(\frac{6}{2})^2 = 9$. Likewise, we must add $(\frac{-8}{2})^2 = 16$ to complete the square of y.

$$(x^2 + 6x + 9) + (y^2 - 8y + 16) = -21 + 9 + 16$$
$$(x + 3)^2 + (y - 4)^2 = 4$$
$$[x - (-3)]^2 + (y - 4)^2 = 2^2$$

Therefore, the center is the point $C(-3, 4)$, and the radius is $r = 2$, as shown in Figure 8.96.

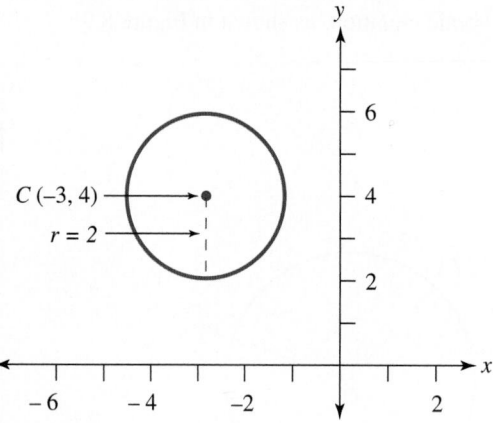

Figure 8.96
$x^2 + y^2 + 6x - 8y + 21 = 0$

The Parabola

When a cross section passes through the base of the cone and intersects only one portion, as in Figure 8.93(b), the resulting figure is a **parabola.** The Greeks defined a parabola as "the set of all points in a plane equidistant from a given line and a given point not on the line." No matter which point P on the parabola you select, its distance from the given line (called the *directrix*) is the same as its distance from the given point (called the *focus*), as shown in Figure 8.97. Every parabola can be divided into two equal pieces. The line that cuts the parabola into symmetric halves is called the **line of symmetry.** The point at which the line of symmetry intersects the parabola is called the **vertex** of the parabola.

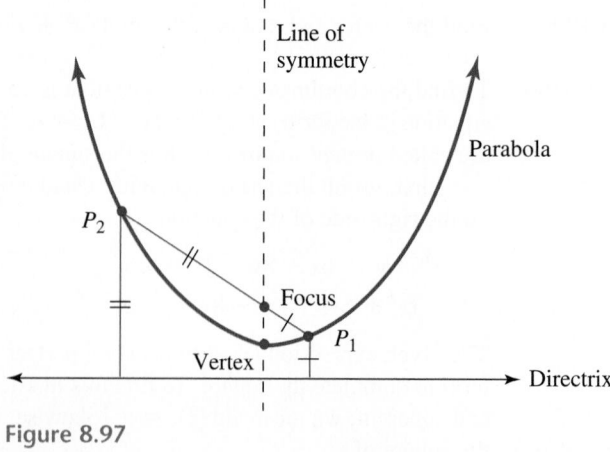

Figure 8.97
A parabola

Parabolas have many applications. In particular, satellite dish antennas and the dishes used with sports microphones are designed so that their cross sections are parabolas; that is, they are parabolic reflectors. When incoming radio waves, microwaves, or sound waves hit the dish, they are reflected to the focus, where they are collected and converted to electronic pulses. The focusing ability of a parabola is also applied to the construction of mirrors used in telescopes and in solar heat collection. (See Figure 8.98.)

Our study of parabolas concentrates on locating the focus. In this respect, we will consider only a parabola whose vertex is at the origin (0, 0) of a rectangular coordinate system.

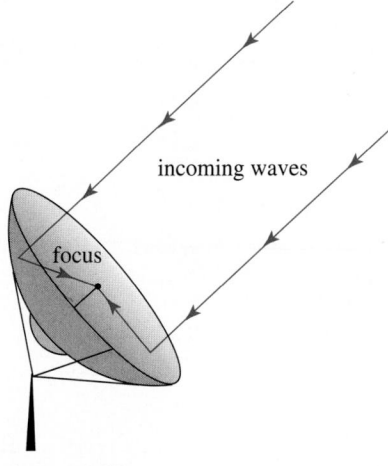

Figure 8.98
Satellite dish antenna

A Parabola and Its Focus

A **parabola,** with vertex at the point (0, 0) and **focus** at the point (0, p), is the set of all points (x, y) satisfying the equation

$$4py = x^2$$

A parabola and its focus, located on a rectangular coordinate system, are shown in Figure 8.99.

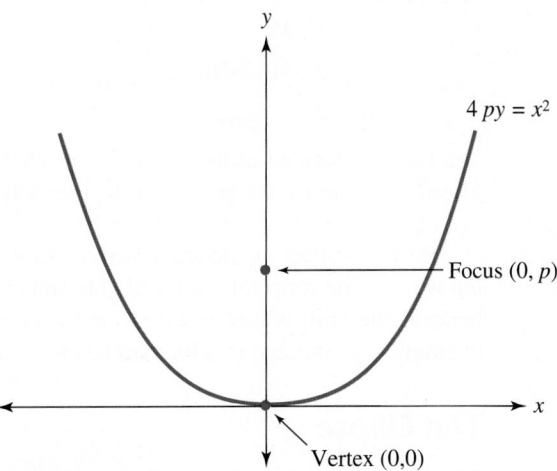

Figure 8.99
A parabola and its focus

EXAMPLE 2   A Peace Corps worker has found a way for villagers to obtain Mylar-coated parabolic dishes that will concentrate the sun's rays for the solar heating of water. If a dish is 7 feet wide and 1.5 feet deep, where should the water container be placed?

Solution   The water will be heated most rapidly if the container is placed at the focus. (The incoming solar radiation will be reflected by the parabolic dish and concentrated at the focus of the parabola.)

Draw a parabola with its vertex at the origin. As Figure 8.100 demonstrates, the given dimensions tell us that the point $Q(3.5, 1.5)$ is on the parabola.

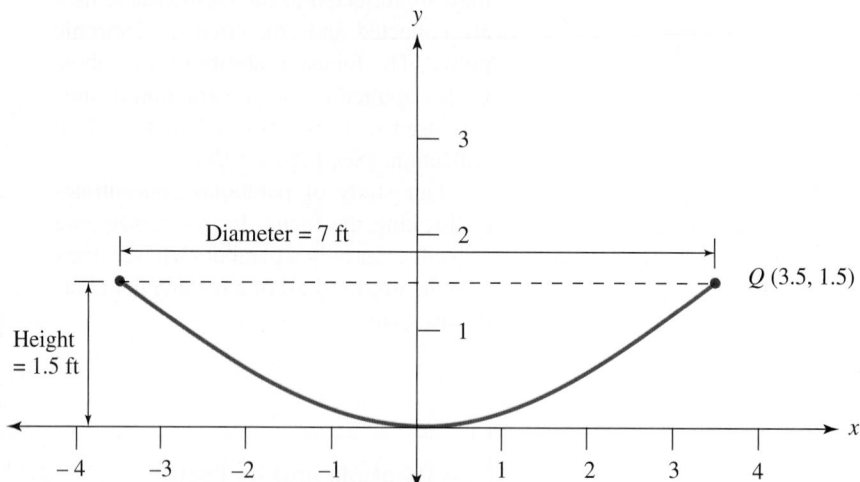

Figure 8.100
Solar parabolic dish

Substituting $x = 3.5$ and $y = 1.5$ into the general equation $4py = x^2$, we can find p, the location of the focus:

$$4py = x^2$$
$$4p(1.5) = (3.5)^2$$
$$6p = 12.25$$
$$p = 2.04166666\ldots$$
$$p \approx 2$$

The focus is located at the point $(0, 2)$. Therefore, the water container should be placed 2 feet above the bottom of the parabolic dish.   ■

Besides collecting incoming waves, parabolic reflectors also concentrate outgoing waves. The reflectors in flashlights and classic automotive headlights are parabolic. If the bulb is positioned at the focus, then the outgoing light will be more intensely concentrated in a forward beam.

The Ellipse

When a cross section of a cone is not parallel to the base and does not pass through the base, as in Figure 8.93(c), the resulting figure is an **ellipse.** An ellipse is a symmetric oval. The Greeks defined an ellipse as "the set of all points in a plane the *sum* of whose distances from two given points is constant." That is, if C_1 and C_2 represent the given points (called the **foci**), then no matter which point P on the ellipse you select, the distance from P to C_1 plus the distance from P to C_2 is always the same, as shown in Figure 8.101.

Ellipses are commonly used in astronomy. The planets revolve around the sun, each following its own elliptical path with the sun at one of the foci. The elliptical motion of the planets was first hypothesized by Johann Kepler in 1609.

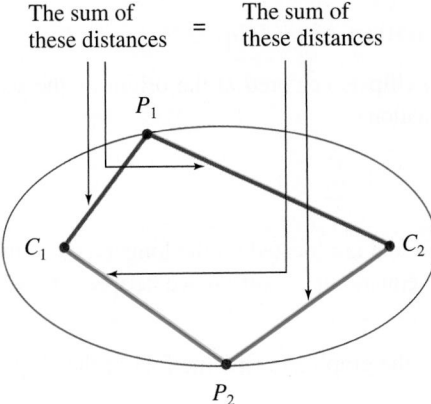

The sum of these distances = The sum of these distances

Figure 8.101
Ellipse

An ellipse has an interesting reflective property. Anything that is emitted from one focus and strikes the ellipse will be directed to the other focus. For example, suppose you had an elliptical billiard table with a hole located at one of the foci. A cue ball located at the other focus would always go in the hole, regardless of the direction in which it was shot. This principle is used in the design of whispering galleries. If the walls and ceiling of a room are elliptical, a person standing at one focus will be able to hear the whispering of someone standing at the other focus; the sound waves are reflected and directed from one focus to the other. The world's most famous whispering gallery is located in St. Paul's Cathedral in London, England, which was completed in 1710. Other galleries are located in the U.S. Capitol in Washington, D.C.; in the Mormon Tabernacle in Salt Lake City, Utah; and in Gloucester Cathedral (built in the eleventh century) in Gloucester, England.

Although they can be located anywhere in a plane, we will consider only ellipses that are centered at the origin of a rectangular coordinate system. (See Figure 8.102.) (If the center is not at the origin, the method of completing the square can be used to locate the center, as in Example 1.)

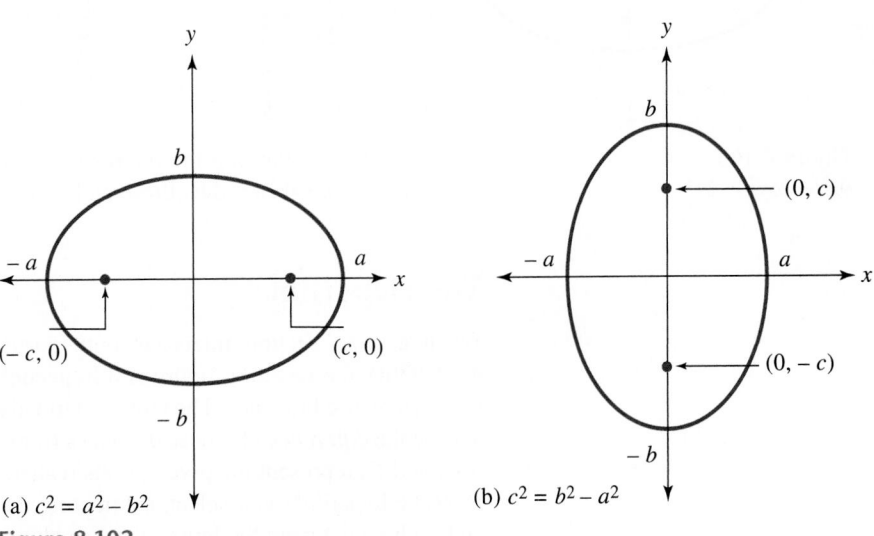

(a) $c^2 = a^2 - b^2$ (b) $c^2 = b^2 - a^2$

Figure 8.102
Standard ellipses

> **Equation of an Ellipse**
>
> An **ellipse,** centered at the origin, is the set of all points (x, y) satisfying the equation
>
> $$\frac{x^2}{a^2} + \frac{y^2}{b^2} = 1 \qquad a \text{ and } b \text{ are positive constants.}$$
>
> The **foci** are located on the longer axis of the ellipse and are found by solving the equation $c^2 = $ difference between a^2 and b^2.

ThomsonNOW™
for Liberal Arts Mathematics
www.thomsonedu.com/1pass.

EXAMPLE 3   Sketch the graph and find the foci of the ellipse $9x^2 + 25y^2 = 225$.

Solution   First, we put the equation in the general formula $x^2/a^2 + y^2/b^2 = 1$. To obtain this form, we divide each side of the equation by 225 so that the right side will equal 1:

$$9x^2 + 25y^2 = 225$$

$$\frac{9x^2}{225} + \frac{25y^2}{225} = \frac{225}{225} \qquad \text{dividing by 225}$$

$$\frac{x^2}{25} + \frac{y^2}{9} = 1 \qquad \text{simplifying the fractions}$$

(Note that $a^2 = 25$ and $b^2 = 9$.) We can now sketch the ellipse by finding the x- and y-intercepts.

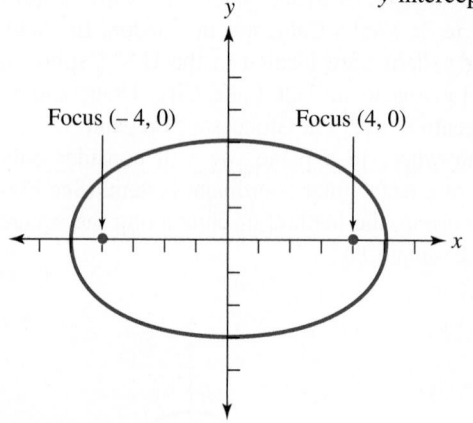

Finding the x-Intercepts:	Finding the y-Intercepts:	Finding the Foci:
(Substitute $y = 0$ and solve for x.)	(Substitute $x = 0$ and solve for y.)	$c^2 = a^2 - b^2$
$\dfrac{x^2}{25} + \dfrac{0^2}{9} = 1$	$\dfrac{0^2}{25} + \dfrac{y^2}{9} = 1$	$= 25 - 9$
$\dfrac{x^2}{25} = 1$	$\dfrac{y^2}{9} = 1$	$= 16$
$x^2 = 25$	$y^2 = 9$	Therefore, $c = \pm 4$.
$x = \pm 5$	$y = \pm 3$	

Figure 8.103
$9x^2 + 25y^2 = 225$

Plotting the intercepts and connecting them with a smooth oval, we sketch the ellipse and its foci as shown in Figure 8.103. ∎

The Hyperbola

When a cross section intersects both portions of the double cone, as in Figure 8.93(d), the resulting figure is a **hyperbola.** A hyperbola consists of two separate, symmetric branches. The Greeks defined a hyperbola as "the set of all points in a plane the *difference* of whose distances from two given points is constant." That is, if C_1 and C_2 represent the given points (called the **foci**), then no matter which point P on the hyperbola you select, subtracting the smaller distance (between the point and each focus) from the larger distance always results in the same value, as shown in Figure 8.104.

Figure 8.104
Hyperbola

As with the ellipse, we will consider only hyperbolas centered at the origin of a rectangular coordinate system, as shown in Figure 8.105.

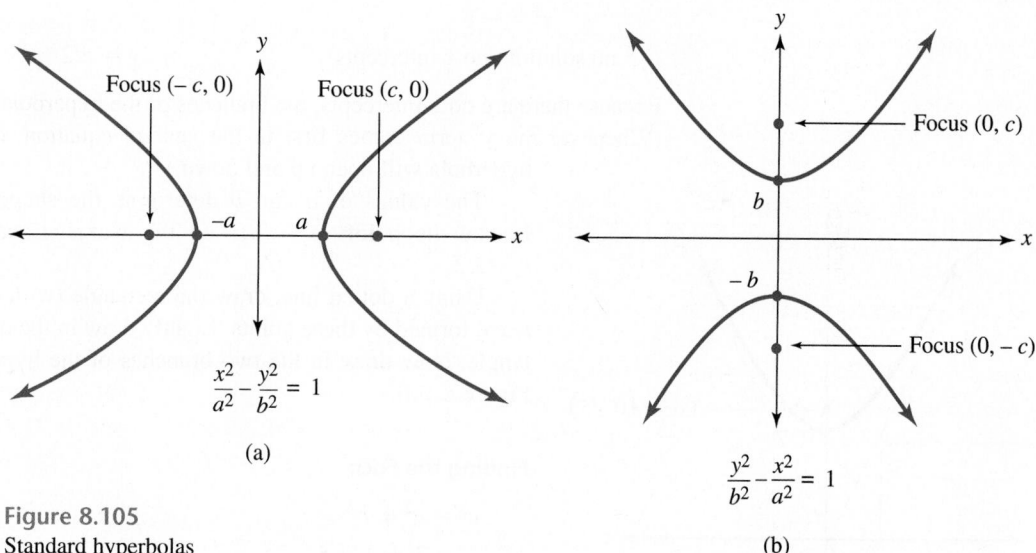

Figure 8.105
Standard hyperbolas

Equation of a Hyperbola

A **hyperbola,** centered at the origin, is the set of all points (x, y) satisfying either the equation

$$\frac{x^2}{a^2} - \frac{y^2}{b^2} = 1 \quad \text{or} \quad \frac{y^2}{b^2} - \frac{x^2}{a^2} = 1 \qquad a \text{ and } b \text{ are positive constants.}$$

The **foci** are found by solving the equation $c^2 = a^2 + b^2$.

EXAMPLE 4   Sketch the graph and find the foci of the hyperbola $y^2 - 4x^2 = 4$.

Solution   First, put the equation in the general form, with the right side of the equation equal to 1:

$$y^2 - 4x^2 = 4$$

$$\frac{y^2}{4} - \frac{4x^2}{4} = \frac{4}{4} \quad \text{dividing by 4}$$

$$\frac{y^2}{4} - \frac{x^2}{1} = 1 \quad \text{simplifying the fractions}$$

(Note that $a^2 = 1$ and $b^2 = 4$.)

Finding the x-Intercepts:
(Substitute $y = 0$ and solve for x.)

$$\frac{0^2}{4} - \frac{x^2}{1} = 1$$

$$-\frac{x^2}{1} = 1$$

$$x^2 = -1$$

no solution; no x-intercepts

Finding the y-Intercepts:
(Substitute $x = 0$ and solve for y.)

$$\frac{y^2}{4} - \frac{0^2}{1} = 1$$

$$\frac{y^2}{4} = 1$$

$$y^2 = 4$$

$$y = \pm 2$$

Because there are no x-intercepts, the branches of the hyperbola open up and down. (Whenever the y^2 term comes first in the general equation, the branches of the hyperbola will open up and down.)

The values of a and b determine the shape of the hyperbola. Locate the points $a = \pm 1$ on the x-axis. Locate the y-intercepts $b = \pm 2$.

Using a dotted line, draw the rectangle (with sides parallel to the axes) formed by these points. Lightly draw in the diagonals of this rectangle. Now draw in the two branches of the hyperbola, as shown in Figure 8.106.

Finding the Foci:

$$c^2 = a^2 + b^2$$

$$= 1 + 4 = 5$$

Therefore, $c = \pm\sqrt{5}$ (≈ 2.24). The foci are the points $(0, \sqrt{5})$ and $(0, -\sqrt{5})$.   ∎

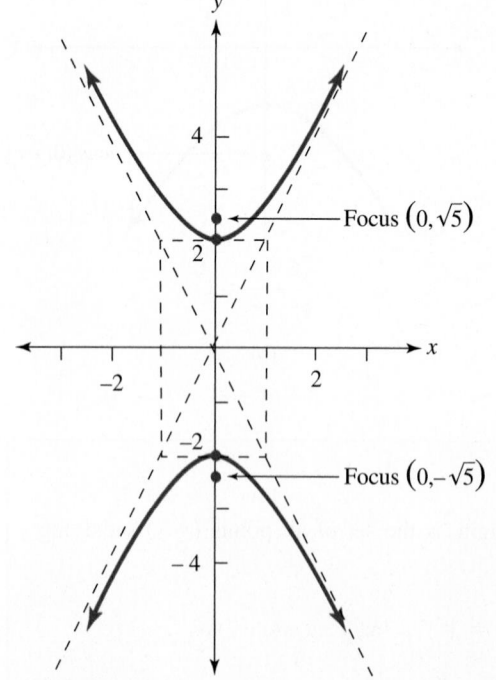

Figure 8.106
$y^2 - 4x^2 = 4$

The most common application of the hyperbola is in radio-assisted-navigation, or LORAN (LOng-RAnge Navigation). Various national and international transmitters continually emit radio signals at regular time intervals. By measuring the *difference* in reception times of a pair of these signals, the navigator of a ship at sea knows that the vessel is somewhere on the hyperbola that has the transmitters as foci. Using a second pair of transmitters, the navigator knows that the ship is somewhere on a second hyperbola. By plotting both hyperbolas on a map, the navigator knows that the ship is located at the intersection of the two hyperbolas, as shown in Figure 8.107.

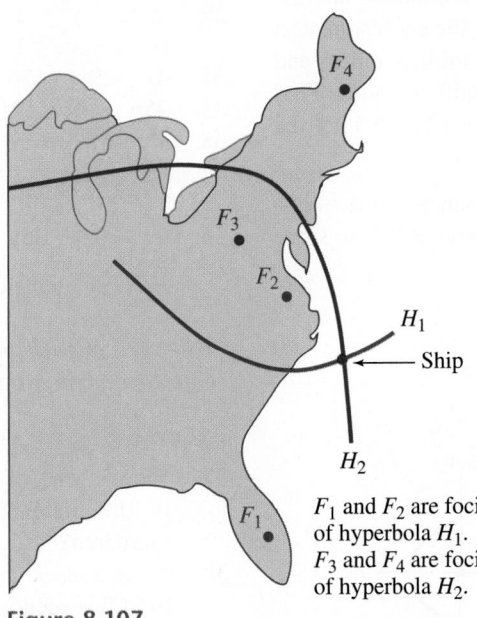

Figure 8.107
Hyperbolas are used in radio-assisted navigation.

8.6 Exercises

In Exercises 1–8, find the center and radius of the given circle and sketch its graph.

1. $x^2 + y^2 = 1$
2. $x^2 + y^2 = 4$
3. $x^2 + y^2 - 4x - 5 = 0$
4. $x^2 + y^2 + 8y + 12 = 0$
5. $x^2 + y^2 - 10x + 4y + 13 = 0$
6. $x^2 + y^2 + 2x - 6y - 15 = 0$
7. $x^2 + y^2 - 10x - 10y + 25 = 0$
8. $x^2 + y^2 + 8x + 8y + 16 = 0$

In Exercises 9–12, sketch the graph of the given parabola and find its focus.

9. $y = x^2$
10. $y = 2x^2$
11. $y = \frac{1}{2}x^2$
12. $y = \frac{1}{4}x^2$

13. A Mylar-coated parabolic reflector dish (used for the solar heating of water) is 9 feet wide and $1\frac{3}{4}$ feet deep. Where should the water container be placed to heat the water most rapidly?

14. A satellite dish antenna has a parabolic reflector dish that is 18 feet wide and 4 feet deep. Where is the focus located?

15. A flashlight has a diameter of 3 inches. If the reflector is 1 inch deep, where should the light bulb be located to concentrate the light in a forward beam?

16. An automotive headlight has a diameter of 7 inches. If the reflector is 3 inches deep, where should the light bulb be located to concentrate the light in a forward beam?

In Exercises 17–24, sketch the graph and find the foci of the given ellipse.

17. $4x^2 + 9y^2 = 36$
18. $x^2 + 4y^2 = 16$
19. $25x^2 + 4y^2 = 100$
20. $x^2 + 9y^2 = 9$
21. $16x^2 + 9y^2 = 144$
22. $4x^2 + y^2 = 36$
23. $4x^2 + y^2 - 8x - 4y + 4 = 0$

HINT: Complete the square for x and for y.

24. $25x^2 + 9y^2 + 150x - 90y + 225 = 0$

HINT: Complete the square for x and for y.

25. An elliptical room is designed to function as a whispering gallery. If the room is 30 feet long and 24 feet wide, where should two people stand to optimize the whispering effect?

26. An elliptical billiard table is 8 feet long and 5 feet wide. Where are the foci located?

27. During the earth's elliptical orbit around the sun (the sun is located at one of the foci), the earth's greatest distance from the sun is 94.51 million miles, and the shortest distance is 91.40 million miles. Find the equation (in the form $x^2/a^2 + y^2/b^2 = 1$) of the earth's elliptical orbit.

HINT: Use $a + c$ = greatest distance and $a - c$ = shortest distance, as shown in Figure 8.108, to solve for a and then c; then find b.

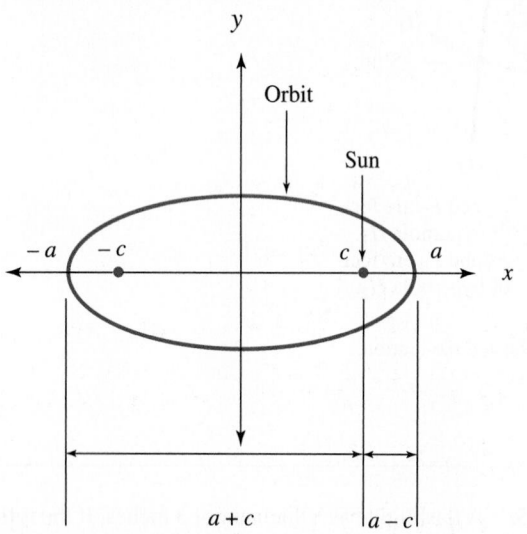

Figure 8.108
Elliptical orbit of the earth around the Sun

28. During Mercury's elliptical orbit around the sun (the sun is located at one of the foci), the planet's greatest distance from the sun is 43.38 million miles, and the shortest distance is 28.58 million miles. Find the equation (in the form $x^2/a^2 + y^2/b^2 = 1$) of Mercury's elliptical orbit.

HINT: Use $a + c$ = greatest distance and $a - c$ = shortest distance, as shown in Figure 6.93, to solve for a and c; then find b.

In Exercises 29–36, sketch the graph and find the foci of the given hyperbola.

29. a. $x^2 - y^2 = 1$
    b. $y^2 - x^2 = 1$

30. a. $x^2 - y^2 = 4$
    b. $y^2 - x^2 = 4$

31. $4x^2 - 9y^2 = 36$

32. $x^2 - 4y^2 = 16$

33. $25y^2 - 4x^2 = 100$

34. $y^2 - 9x^2 = 9$

35. $x^2 - 4y^2 - 6x + 5 = 0$

HINT: Complete the square for x and for y.

36. $9y^2 - x^2 - 36y - 2x + 26 = 0$

HINT: Complete the square for x and for y.

———— *Answer the following questions using complete sentences and your own words.*

HISTORY QUESTIONS

37. Who is one of the first women to be mentioned in the history of mathematics? Why was she murdered?

38. Write a research paper on a historical topic referred to in this section or on a related topic. Following is a list of possible topics:
- Apollonius of Perga (What contributions did he make to the study of conic sections?)
- Hypatia (Discuss her accomplishments.)
- René Descartes (What contributions did he make to the study of conic sections?)
- Edmond Halley (What is Halley's comet? How did Halley predict its arrival?)
- Johann Kepler (How did he develop his theory of elliptical orbits?)
- LORAN (How is LORAN used in navigation? When and where did it originate?)
- Parabolic reflectors and antennas (How have parabolic reflectors changed the dissemination of information around the world?)

8.7   Non-Euclidean Geometry

Mathematics is that subject in which we do not know what we are talking about, or whether what we are saying is true.
—Bertrand Russell

From the time of Thales of Miletus, mathematics has been based on the axiomatic system of deduction and proof. Euclid's monumental work, *Elements,* was the epitome of this effort. By the careful choice of five geometric postulates (things that are assumed to be true), Euclid proceeded to build the entire structure of all geometric knowledge of the day. The ancient Greeks' attitude was "the truth of these five axioms is obvious; therefore, everything that follows from them is true also." Over 2,000 years and many headaches, heartbreaks, and humiliations later, the attitude of a current mathematician is more "*if* we *assume* these axioms to be valid, then everything that follows from them is valid also." While the Greeks considered geometry (and all of mathematics) to be a system of absolute truth, Bertrand Russell's twentieth-century comment implies that results that are based on *assumptions* cannot be considered true or false; the most one can hope for is *consistency*.

The Parallel Postulate

Euclid's empire was based on his perception of reality. This reality in turn consisted of five axiomatic concepts so simple and so obvious that no one could deny that they were in fact true. After all, reality is what is confirmed by experience, and experience had never shown the following statements to be false:

1. A straight line can be drawn from any point to any other point.
2. A (finite) straight line segment can be extended continuously in an (infinite) straight line.
3. A circle may be described with any center and any radius.
4. All right angles are equal to one another.
5. Given a line and a point not on that line, there is one and only one line through the point parallel to the original line.

Taking the axiomatic method to heart, scholars scrutinized even the five "obvious" postulates for any possible overlap or dependence. Could one of the five be deduced logically from the others? The search was on for the smallest possible set of postulates that could serve as the basis of all geometry.

The most likely candidate for a dependent postulate was number 5, the so-called **Parallel Postulate** (see Figure 8.109). From the beginning, it aroused the curiosity and inquiry of mathematicians the world over. Many "proofs" of this postulate were put forth, but each was dismissed upon further scrutiny by later investigations. Each "proof" had its flaw, its logical demise. Even Euclid himself had tried to prove it but couldn't, and he consequently called it a postulate (it was obviously true).

The more mathematicians studied the postulate, the more elusive it became. Indeed, the competition to prove the Parallel Postulate became so intense that the French mathematician d'Alembert called it "the scandal of elementary geometry." On one occasion, the highly admired Joseph-Louis Lagrange presented a paper on the postulate to the French Academy. However, halfway through his presentation, he stopped, with the comment "I must meditate further on this." He put the paper away and never spoke of it again.

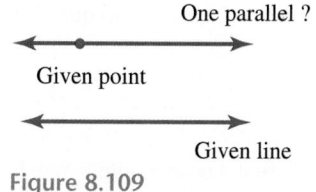

One parallel ?

Given point

Given line

Figure 8.109
The Parallel Postulate

Girolamo Saccheri

In all the furor to prove the Parallel Postulate, one particular attempt is worth mentioning. In 1733, a Jesuit priest named Girolamo Saccheri published *Euclides ab Omni Naevo Vindicatus* (*Euclid Vindicated of Every Blemish*). This work contained Saccheri's "proof" of the Parallel Postulate. Saccheri was a professor of mathematics, logic, and philosophy at the University of Pavia, Italy. He is said to have had an incredible memory; commonly mentioned was his ability to play three games of chess simultaneously, without looking at the chessboards!

Saccheri was the first to examine the consequences of assuming the Parallel Postulate to be false; that is, he attempted a proof by contradiction. Proof by contradiction is an indirect means of proving a statement. Simply put, to prove something, you assume it to be false. If you can show that this assumption leads to a contradiction, then the assumed falsity of the statement is itself false, implying that the statement is true. This is an acceptable line of reasoning that has been used since the days of classic Greek logic. Euclid himself used proof by contradiction in *Elements,* though not in reference to the Parallel Postulate.

In Saccheri's quest to prove the Parallel Postulate, he assumed it to be false, thereby hoping to obtain some sort of logical contradiction. The denial of the Parallel Postulate consists of two alternatives:

1. Given a line and a point not on the line, there are no lines through the point parallel to the original line. (Parallels don't exist.)

2. Given a line and a point not on the line, there are at least two lines through the point parallel to the original line.

Surely, each of these assumptions would lead to a contradiction. To his surprise, Saccheri couldn't find one! How could this be? Were the foundations of Euclid's world built on rock or on sand? Rather than accepting the implications of his work, Saccheri manufactured a spurious contradiction in order to be able to proclaim support of the Parallel Postulate.

Saccheri's work had little impact at the time. The implication that the Parallel Postulate was independent of the other four postulates (that is, that it could not be proved) was too far out of line for the thinking of the day. Mathematicians were so convinced of the dependence of the Parallel Postulate that little interest was given to any other alternative. However, less than 100 years later, three men (Carl Gauss, Janos Bolyai, and Nikolai Lobachevsky) independently resurrected the denial of the Parallel Postulate and took roads similar to the one that had led Saccheri to question the "truth" of Euclidean geometry.

Carl Friedrich Gauss

Carl Friedrich Gauss, the "Prince of Mathematicians," first became interested in the study of parallels in 1792, at the age of fifteen (see Historical Note on page 280). Although he never published any theory that went counter to Euclid's Parallel Postulate, his diaries and letters to close friends over the years affirmed his realization that geometries based on axioms different from Euclid's could exist.

Having failed to prove the Parallel Postulate, Gauss began to ponder its sanctity. According to his diaries, he put forth an axiom that contradicted Euclid's; specifically, he assumed that more than one parallel could be drawn through a point not on a line. Rather than trying to obtain a contradiction as Saccheri had attempted (and thus prove the Parallel Postulate via reductio ad absurdum), Gauss began to see that a geometry that was at odds with Euclid's but internally consistent could be developed. However, Gauss never went public; he didn't want to face the inevitable public controversy.

Janos Bolyai

Janos Bolyai was a Hungarian army officer and the son of a respected mathematician. From an early age, Janos was exposed to the world of mathematics. Janos's father, Wolfgang, had studied with Gauss at Göttingen, and their sporadic correspondence lasted a lifetime. In fact, in their correspondence, Gauss had pointed out the fallacy in the elder Bolyai's "proof" of the Parallel Postulate.

Carl Friedrich Gauss

In 1817, Janos entered the Imperial Engineering Academy in Vienna at the age of fifteen. After completing his studies in 1823, he embarked upon a military career. An expert fencer with numerous successful duels to his credit, Bolyai was also an accomplished violinist.

The early teachings of his father prompted Bolyai to continue his study of mathematics. However, when Janos informed his father of his interest in the study of parallels, the elder Bolyai wrote, "Do not waste an hour's time on that problem. It does not lead to any result; instead it will come to poison all your life." Ignoring his father's decree, the younger Bolyai pursued the elusive Parallel Postulate.

After several failed attempts to prove the Parallel Postulate, Bolyai, too, proceeded down the path of denying Euclid's assumption of the existence of a unique parallel. In a manner not unlike that of Gauss, even though he was unaware of Gauss's work, Bolyai assumed that more than one parallel existed through a point not on a line. He developed several theorems in this new, consistent geometry. He was so excited over his discovery that he wrote, "Out of nothing I have created a strange new world."

Hastening to publish his work, the younger Bolyai published his theory as a twenty-four-page appendix to the elder Bolyai's two-volume *Tentamen Juventutem Studiosam in Elementa Matheseos Purae* (*An Attempt to Introduce Studious Youth to the Elements of Pure Mathematics*). Although the book is dated 1829, it was actually printed in 1832.

When Wolfgang Bolyai sent a copy to his old friend Carl Gauss, the reply was less than heartening to Janos. Gauss responded, "If I begin by saying that I dare not praise this work, you will of course be surprised for a moment; but I cannot do otherwise. To praise it would amount to praising myself. For the entire content of the work, the approach which your son has taken, and the results to which he is led, coincide almost exactly with my own meditations which have occupied my mind for the past thirty or thirty-five years. It was my plan to put it all down on paper eventually, so that at least it would not perish with me. So I am greatly surprised to be spared this effort, and am overjoyed that it happens to be the son of my old friend who outstrips me in such a remarkable way."

Not seeing the true compliment that the great Gauss had bestowed on him, Bolyai feared that his work was being stolen. His feelings were compounded by the total lack of interest on the part of other mathematicians, and deep periods of depression set in. Janos Bolyai never published again.

Nikolai Lobachevsky

Labeling it "imaginary geometry," Nikolai Ivanovitch Lobachevsky published the first complete text on non-Euclidean geometry in 1829. Many have since heralded him as the Copernicus of geometry. Just as Copernicus had challenged the long-established theory that the earth was the center of the universe, Lobachevsky's alternative view of geometry was in direct contradiction to the long-revered work of Euclid. Referring to Lobachevsky, Einstein said, "He dared to challenge an axiom."

Lobachevsky spent most of his life at the University of Kazan, near Siberia. Founded by Czar Alexander I in 1804, Kazan was Europe's easternmost center of higher education. Among its first students, Lobachevsky received his master's degree in mathematics and physics in 1811 at the age of eighteen. He remained at Kazan teaching courses for civil servants until 1816, when he was promoted to full professorship.

Geometry received Nikolai's special attention. Being inquisitive, Lobachevsky made several attempts to prove the Parallel Postulate; he failed at each. He then proceeded to examine the consequences of substituting an alternative to Euclid's postulate of a unique parallel; Lobachevsky assumed that more than one parallel could

Nikolai Lobachevsky

Lobachevsky's *On the Foundations of Geometry* was published in this 1829 issue of the academic journal of the University of Kazan.

be drawn through a point. Living in distant isolation from the learning capitals of Europe, Lobachevsky was unaware of the similar approaches taken by Gauss and Bolyai.

Lobachevsky's idea of a geometry based on an axiom in opposition to Euclid began to take form in 1823, when he drew up an outline for a geometry course he was teaching. In 1826, he gave a lecture and presented a paper incorporating his belief in the feasibility of a geometry based on axioms different from Euclid's. Although this paper has been lost (as have so many in the history of mathematics), it was the first recorded attempt to breach Euclid's bastion. In 1829, the monthly academic journal of the University of Kazan printed a series of his works titled *On the Foundations of Geometry;* this publication is considered by many to be the official birth of the radically new, non-Euclidean geometry.

As with anything new and at odds with the status quo, Lobachevsky's work was not embraced with open arms. The St. Petersburg Academy rejected it for publication in its scholarly journal and printed an uncomplimentary review. In contrast to Gauss, who did not have the courage to print, and Bolyai, who did not have the fortitude to face his opponents, Lobachevsky remained undaunted. He proceeded in his work, expanding *Foundations* into *New Elements of Geometry, with a Complete Theory of Parallels,* which was also published in Kazan's academic journal of 1835. Shortly thereafter, his work began to be recognized outside of Kazan; he was published in Moscow, Paris, and Berlin.

In 1846, when Gauss received a copy of Lobachevsky's latest book, *Geometrical Investigations on the Theory of Parallels* (which contained only sixty-one pages), he wrote to a colleague, "I have had occasion to look through again that little volume by Lobachevsky. You know that for fifty-four years now I have held the same conviction. I have found in Lobachevsky's work nothing that is new to me, but the development is made in a way different from that which I have followed, and certainly by Lobachevsky in a skillful way and a truly geometrical spirit." However, Gauss did not give public approval to Lobachevsky's work.

As is the case with those who pioneer ideas and art forms that are incomprehensible to the world at large, the radicals of geometry, Lobachevsky and Bolyai, never received full recognition of the value of their work during their lifetimes. However, they opened the door for a host of new ideas in geometry and in the axiomatic system in general.

Bernhard Riemann

The groundbreaking work of Lobachevsky and Bolyai was ignored for many years. The mathematician who finally convinced the academic world of the merits of non-Euclidean geometry (that is, geometries based on the denial of Euclid's Parallel Postulate) was born in 1826, at the time when Lobachevsky and Bolyai were initially presenting their ideas. Although his life was cut short (he died of tuberculosis at the age of thirty-nine), Georg Friedrich Bernhard Riemann's contributions to the world of modern mathematics were monumental.

At the age of nineteen, Bernhard Riemann enrolled at the University of Göttingen with the intention of pursuing the study of theology and philosophy. However, he became so interested in the study of mathematics that he devoted his life entirely to that field. As a graduate student at Göttingen, Riemann studied under Gauss, who was unusually impressed with his protégé's abilities. In his report to the faculty, Gauss said that Riemann's dissertation came from "a creative, active, truly mathematical mind, and of a gloriously fertile originality." Coming from the Prince of Mathematicians, that was a compliment indeed!

Bernhard Riemann

After receiving his degree, Riemann wanted to stay at Göttingen as a member of the faculty. To prove himself, he first had to deliver a "probationary" lecture to his former teachers. When he submitted three possible topics for his oration, Gauss chose the third, "On the Hypotheses That Underlie the Foundation of Geometry." Delivered in 1854 and published in 1868 (two years after his death), Riemann's lecture is considered by many to be one of the highlights of modern mathematical history.

Unlike Gauss, Lobachevsky, and Bolyai (each of whom assumed that through a point not on a line more than one parallel existed), Riemann denied the existence of all parallel lines! He philosophized that we could just as well assume that all lines eventually intersect as assume that some (parallel lines) do not intersect.

Geometric Models

One reason for the initial resistance to non-Euclidean geometry was practical experience. In our perception of the "flatness" of the immediate world in which we live, we naturally accept the Parallel Postulate as true. It is easy to envision a rectangular grid of city blocks with straight streets that do not intersect. However, our world is not flat; the earth is spherical. An age-old problem associated with map making has been the projection of the curved earth (a globe) into a plane (a flat map) and vice versa. (What happens if we try to peel the "skin" off of a globe or stretch a piece of rectangular graph paper around a sphere?)

Granted, non-Euclidean geometries run counter to common sense. To understand them, we need to draw pictures; we must create models. Each model will have its own unique definition of a line and a plane. In all cases, **parallel lines** are lines in a plane that do not intersect one another. Geometries can be categorized by three types, depending on what is assumed concerning parallel lines:

1. *Euclidean geometry:* Through a point not on a given line, there exists exactly one line parallel to the given line (Euclid's Parallel Postulate).

2. *Lobachevskian geometry:* Through a point not on a given line, there exist more than one line parallel to the given line.

3. *Riemannian geometry:* Through a point not on a given line, there exists no line parallel to the given line.

Because they are based on postulates that contradict Euclid's Parallel Postulate, Lobachevskian and Riemannian geometries are called non-Euclidean geometries.

The model that is used to express Euclidean geometry is the plane. Lying in a flat plane, lines extend indefinitely into space and have infinite length; they do not "wrap" around the earth. See Figure 8.110.

In contrast, the model used to express Riemannian geometry is the sphere. Using this model, we define lines to be the great circles that encompass the

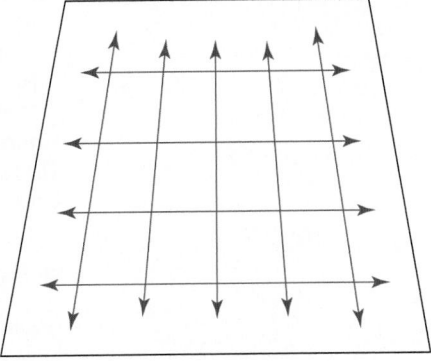

Figure 8.110
The Euclidean planar model

sphere. A **great circle** is a circle whose center lies at the center of the sphere, as shown in Figure 8.111. No matter how they are drawn, each pair of great circles will always intersect in two points. Consequently, parallel lines do not exist! Riemannian geometry is important in navigation, because the shortest distance between two points on a sphere is the path along a great circle.

A model that is used to express Lobachevskian geometry is the interior of a circle. This model was described by the Frenchman Henri Poincaré (1854–1912). The Poincaré model defines a plane to be all points "inside" a circle. (The points *on* the circle are excluded.) This region is also called a **disk;** it is shown in Figure 8.112.

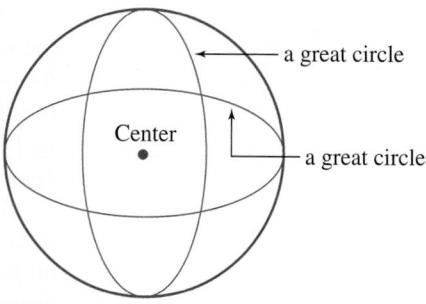

Figure 8.111
The Riemannian spherical model

Poincaré's model of a Lobachevskian geometry defines a line to be either of the following:

1. a diameter of the disk

2. a circular arc connecting two points on the boundary of the disk*

The two types of Poincaré lines are shown in Figure 8.113. Because the boundary of the disk is excluded, a (Poincaré) line has no endpoints.

Using Poincaré's disk model, we can construct more than one parallel line through a given point, as shown in Figure 8.114. Notice that lines *CD* and *EF* pass through the point *Q* and that each line is parallel to line *AB*. (*CD* and *AB* are parallel because they do not intersect. Likewise, *EF* and *AB* are parallel because they do not intersect.)

Figure 8.112
A Poincaré plane (disk)

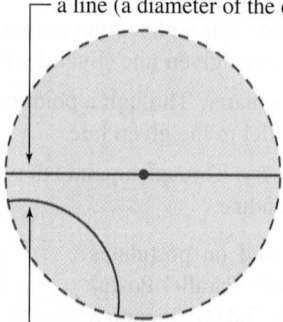

Figure 8.113
The Poincaré disk model for Lobachevskian geometry

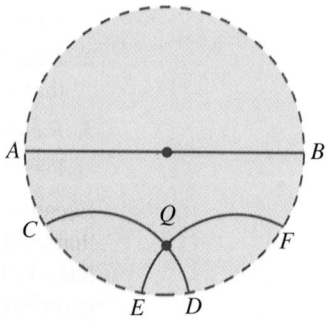

Figure 8.114
More than one parallel line through *Q*

A Comparison of Triangles

Owing to the difference in the number of parallel lines, non-Euclidean geometries have properties that differ from those of Euclidean geometry. As you know, the sum of the angles of a triangle always *equals* 180° in Euclidean geometry, as shown in Figure 8.115(a). In contrast, when a triangle is drawn on a sphere, the sum of the

---

* Technically, the arc is the intersection of the disk with an orthogonal circle. Orthogonal circles are circles whose radii are perpendicular at the points where the circles intersect.

angles is *greater* than 180°, as shown in Figure 8.115(b). Finally, the sum of the angles of a triangle drawn on a Poincaré disk is *less* than 180°, as shown in Figure 8.115(c).

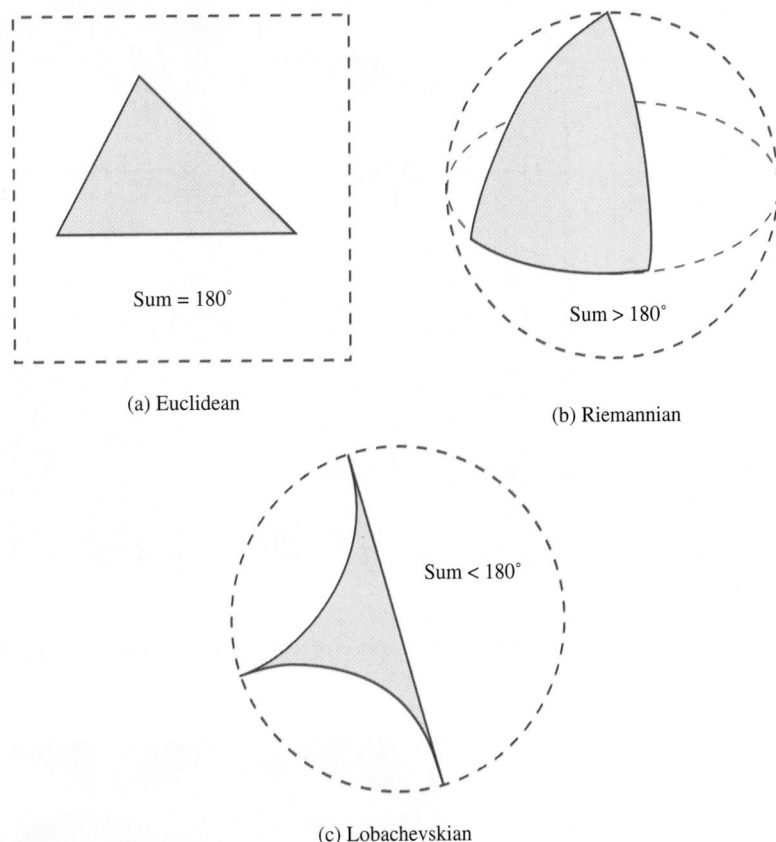

(a) Euclidean

(b) Riemannian

(c) Lobachevskian

Figure 8.115
The angles of a triangle

Geometry in Art

The study of geometry has always been an integral part of an education in art. Whether an artist leans toward realism or surrealism, a fundamental knowledge of geometric relationships is essential. No artist has been more successful in combining art and geometry than the Dutch artist Maurits Cornelis Escher (1898–1972). Escher is known for his repetitious plane-filling patterns, which often depict the metamorphosis of one figure into another.

The precise geometric appearance of Escher's work is not accidental; he possessed the knowledge, the talent, and the creativity to manipulate perspective and dimension in both Euclidean and non-Euclidean geometry. In fact, several of Escher's images were based on the non-Euclidean geometry of a Poincaré disk. A sketch from one of his many notebooks (Figure 8.116) shows Escher's study of a disk in his preparation to work in non-Euclidean space. Notice the triangles formed by the circular arcs.

Escher often created the same image on different geometric models. For instance, after he developed a pattern of complementary angels and devils on a "flat" Euclidean plane (Figure 8.117), Escher then employed Lobachevskian geometry to map the pattern onto a disk (Figure 8.118) and Riemannian geometry to transpose the pattern onto a sphere (Figure 8.119).

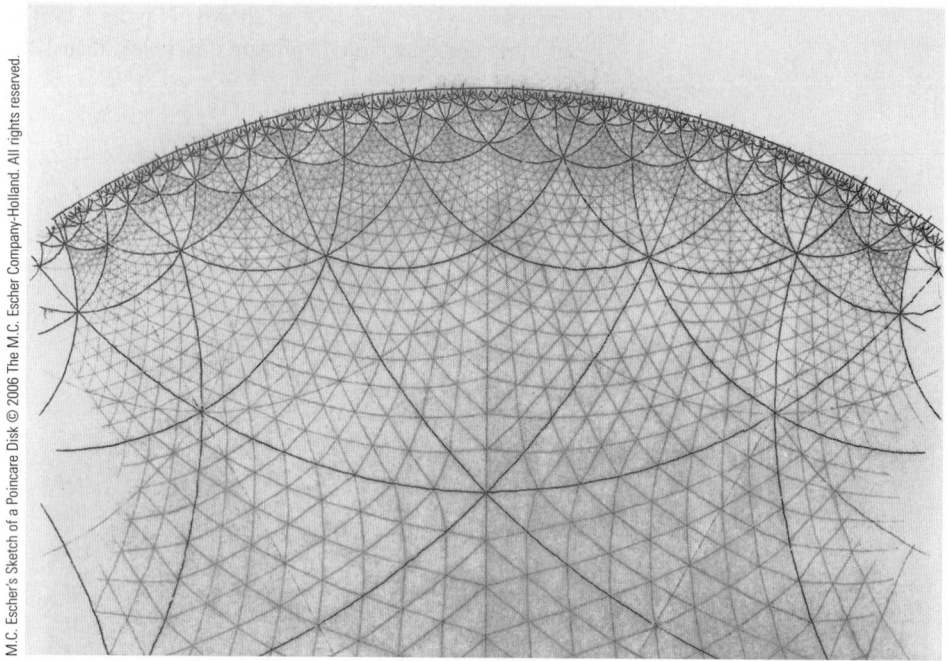

Figure 8.116
Escher drew this sketch of a Poincaré disk before creating the work of art shown in Figure 8.118.

Figure 8.117
A repetitive pattern that completely fills a plane is known as a tiling; these complementary angels and devils form a tiling of the Euclidian plane.

Figure 8.118
Utilizing Lobachevskian geometry, Escher tiled a Poincaré disk with
the angels and devils of Figure 8.117.

Figure 8.119
Utilizing Riemannian geometry, Escher transposed the angels and devils
of Figure 8.117 onto a sphere; the design is carved in solid maple.

8.7   Exercises

In Exercises 1–6, answer the questions for the planar model of Euclidean geometry.

1. Through a point not on a given line, how many parallels to the given line are there?
2. What can be said about the sum of the angles of a triangle?
3. How many right angles can a triangle have?
4. If two different lines are each parallel to a third line, are they necessarily parallel to each other?
5. In how many points can a pair of distinct lines intersect?
6. Do lines have finite or infinite length?

In Exercises 7–12, answer the questions for the spherical model of Riemannian geometry.

7. Through a point not on a given line, how many parallels to the given line are there?
8. What can be said about the sum of the angles of a triangle?
9. How many right angles can a triangle have?
10. Do lines have finite or infinite length?
11. In how many points can a pair of distinct lines intersect?
12. What is the minimum number of sides required to form a polygon?

In Exercises 13–18, answer the questions for Poincaré's model of Lobachevskian geometry.

13. Through a point not on a given line, how many parallels to the given line are there?
14. What can be said about the sum of the angles of a triangle?
15. How many right angles can a triangle have?
16. If two different lines are each parallel to a third line, are they necessarily parallel to each other?
17. In how many points can a pair of distinct lines intersect?
18. Do lines have finite or infinite length?

Answer the following questions using complete sentences and your own words.

HISTORY QUESTIONS

19. What is Euclid's Parallel Postulate?
20. What was "the scandal of elementary geometry"?

21. Who was the first mathematician to examine the consequences of assuming the Parallel Postulate to be false? What were his findings? What was his reaction?
22. Who was the first mathematician to affirm the existence of a non-Euclidean geometry? Did he publish his work? Why or why not?
23. Who stopped publishing his mathematical ideas after he had published his work on non-Euclidean geometry? Why did he stop?
24. Who is called the Copernicus of geometry? Why?
25. Who introduced a geometry in which parallel lines do not exist?
26. Write a research paper on a historical topic referred to in this section or on a related topic. Following is a list of possible topics:
   - Janos Bolyai (What influence did he have on the development of non-Euclidean geometry?)
   - M. C. Escher (How did he combine art and geometry?)
   - Carl Friedrich Gauss (What influence did he have on the development of non-Euclidean geometry?)
   - Immanuel Kant (How did his philosophy affect the development of non-Euclidean geometry?)
   - Felix Klein (What model did he create to represent Lobachevskian geometry?)
   - Adrien-Marie Legendre (How did he attempt to prove the Parallel Postulate?)
   - Nikolai Lobachevsky (Why is he given most of the credit for the creation of non-Euclidean geometry?)
   - Henri Poincaré (What did he contribute to the study of geometry?)
   - Georg Friedrich Bernhard Riemann (What did he contribute to the study of geometry?)
   - Girolamo Saccheri (What is the connection between Saccheri quadrilaterals and non-Euclidean geometry?)

 WEB PROJECT

27. There are numerous web sites devoted to the works of M. C. Escher; many are animated and interactive. Investigate some of these web sites and write an essay and/or make a presentation about your findings. Some useful links for this web project are listed on the text web site: **www.thomsonedu.com/mathematics/johnson**

8.8 | **Fractal Geometry**

The classic geometry of the Greeks is not the only geometry. In Section 8.7, we learned of the non-Euclidean geometries created by Gauss, Bolyai, Lobachevsky, and Riemann. Intuitively, Euclid's geometry seems most natural to us, probably because we seem to live on a plane. However, it could be argued that Riemannian geometry is more natural because we actually live on a sphere, not a plane.

The shapes of the classic geometry of the Greeks—the triangle, square, rectangle, circle, ellipse, parabola and hyperbola—do not seem natural to us. Although these idealized shapes occasionally exist in nature, we more commonly see complex, less perfect shapes such as those of mountains and plants, rather than the simple, Euclidean shapes of circles and rectangles. As mathematician Benoit Mandelbrot said, "Clouds are not spheres, mountains are not cones, coastlines are not circles, and bark is not smooth, nor does lightning travel in a straight line" (Benoit Mandelbrot, *The Fractal Geometry of Nature*).

Amazingly, the fern and the mountain in Figure 8.120 are neither photographs nor drawings; they are shapes formed with fractal geometry, a new alternative

A fractal forgery of a landscape

Figure 8.120
A fractal forgery of a fern

geometry. In many respects, fractal geometry has more in common with nature than does the classic geometry of the Greeks.

The Sierpinski Gasket

The **Sierpinski gasket** is one of the older examples of fractal geometry. It was first studied by Polish mathematician Waclaw Sierpinski in 1915. To understand this shape, we will create one. Take the following steps, which are illustrated in Figure 8.121:

1. Draw a triangle and fill in its interior.

2. Put a hole in the triangle:
   - Find the midpoints of the sides of the triangle.
   - Connect the midpoints and form a new triangle.
   - Form a hole by removing the new triangle.

   This leaves three filled in triangles, each a smaller version of the original triangle from step 1.

3. Put a hole in each of the three triangles from step 2 by applying the procedure from step 2 to each triangle.

4. Put a hole in each of the triangles from step 3 by applying the procedure from step 2 to each triangle.

5 (and beyond). Put a hole in each of the triangles from the previous step by applying the procedure from step 2 to each triangle.

step 1   step 2   step 3   step 4

Figure 8.121
Creating a Sierpinski gasket

The Sierpinski gasket (Figure 8.122) is the result of continuing the process described above through steps 6 and 7 and 8 and beyond without stopping.

After several steps, it becomes impossible to visually tell the difference between the result of one step and the result of the next step. That is, you probably couldn't tell the difference between the result of stopping at step 10 and the result of stopping at step 11 unless you used a magnifying glass. However, neither of these shapes would be a Sierpinski gasket; the only way to get a Sierpinski gasket is to continue the process without stopping, forming more and more tinier triangles, each with holes and even tinier triangles inside it.

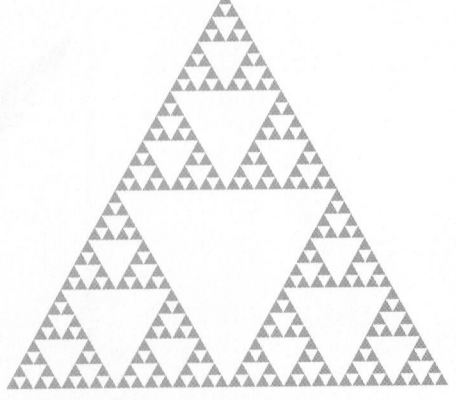

Figure 8.122
Step 5 of a Sierpinski gasket

Self-Similarity

If you did use a magnifying glass to enlarge a portion of a Sierpinski gasket, what would you see? The enlarged portion would look exactly the same as the non-enlarged portion, as shown in Figure 8.123. This is called *self-similarity*. A shape has **self-similarity** if parts of that shape appear within itself at different scales. Fractal geometry utilizes shapes that have self-similarity. Some shapes have **exact self-similarity** (in which the enlarged portion is exactly the same as the nonenlarged portion), and some shapes have **approximate self-similarity.**

Figure 8.123
The self-similarity of a Sierpinski gasket

Notice that the fern in Figure 8.120 has self-similarity. One branch of the fern looks just like the fern itself. And that one branch consists of many smaller branches, each of which looks just like the fern itself. Approximate self-similarity is a common feature of shapes that exist in nature. For example, cauliflower has a self-similar structure (see Figures 8.124 and 8.125), as do some trees. The fact that fractal geometry utilizes self-similarity is one reason why fractal geometry generates shapes that could exist in the real world.

Figure 8.124
The approximate self-similarity of a cauliflower. One of these pictures is a blow-up of the other. Which is which?

Figure 8.125

Self-similarity can be found in art, too. For example, consider M. C. Escher's *Smaller and Smaller* (Figure 8.126), and Jonathan Swift's *On Poetry. A Rhapsody*. These two works certainly exhibit self-similarity.

Figure 8.126
M. C. Escher's *Smaller and Smaller*

So Nat'ralists observe, a Flea
Hath smaller fleas that on him prey;
And these have smaller Fleas to bite 'em
And so proceed ad infinitum.

—JONATHAN SWIFT

ON POETRY. A RHAPSODY

Recursive Processes

Step 2 of building a Sierpinski gasket gives a procedure for putting a hole in a triangle. Steps 3 and beyond instruct us to apply that procedure to the result, over and over again. Such a process, in which a procedure is applied and the same procedure is applied to the result, over and over again, is called a **recursive process.**

Two of the key elements of fractal geometry are self-similarity and recursive processes. We will not attempt a more formal definition of fractal geometry, since it goes beyond the scope of this text. Benoit Mandelbrot, who coined the term **fractal,** described a fractal as a mathematical set or object whose form is extremely irregular and/or fragmented at all scales. Mandelbrot derived the term *fractal* from the Latin *fractus* (to break, to create irregular fragments).

The Skewed Sierpinski Gasket

The **skewed Sierpinski gasket** is a simple variation of the Sierpinski gasket. The steps are the same as before, except for a twist.

1. Draw a triangle and fill in its interior (as before).

2. Put a twisted hole in the triangle: (See Figure 8.127.)

   - Find the midpoints of the sides of the triangle (as before).
   - For each midpoint, find a nearby point (this is the new twist).
   - Connect the nearby points and the vertices of the original triangle.
   - Form a twisted hole by removing the center triangle.

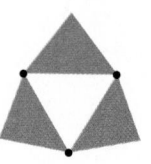

The midpoints and nearby points in step 2

Finishing step 2

Figure 8.127
Creating a skewed Sierpinski gasket

3. Put a twisted hole in each of the three triangles from step 2 by applying the procedure from step 2 to each triangle.

4. Put a twisted hole in each of the triangles from step 3 by applying the procedure from step 2 to each triangle.

5 (and beyond). Put a twisted hole in each of the triangles from the previous step by applying the procedure from step 2 to each triangle. Continue this process indefinitely.

Figure 8.128 on page 616 shows a skewed Sierpinski gasket. Surprisingly, this fractal looks somewhat like a mountain! It has both the shape and the texture of a real mountain. Fractal geometry generates shapes that could exist in the real world.

The Sierpinski Gasket and the Chaos Game

There is an interesting connection between probability (as studied in Chapter 3) and the Sierpinski gasket. This connection involves the **chaos game,** which is similar to a game of chance.

Chaos Game Equipment:

- A sketch of a triangle with the vertices labeled *A, B,* and *C.* (See Figure 8.129 on page 616.)
- A die
  - If you roll a 1 or a 2, vertex *A* "wins."
  - If you roll a 3 or a 4, vertex *B* "wins."
  - If you roll a 5 or a 6, vertex *C* "wins."

Chaos Game Procedure:

1. Roll the die. Put a dot at the winning vertex. For example, if you roll a 3, put a dot at vertex B. (See Figure 8.129 on page 616.)

2. Roll the die. Put a dot halfway between the last dot and the winning vertex. For example, if you roll a 2, put a dot halfway between vertex *B* (the last dot) and vertex *A* (the winning vertex). (See Figure 8.129 on page 616.)

3. Roll the die. Put a dot halfway between the last dot and the winning vertex, as determined by the die. For example, if you roll a 6, put a dot halfway between the last dot and vertex *C* (the winning vertex). (See Figure 8.129 on page 616.)

Figure 8.128
A skewed Sierpinski gasket and a real mountain

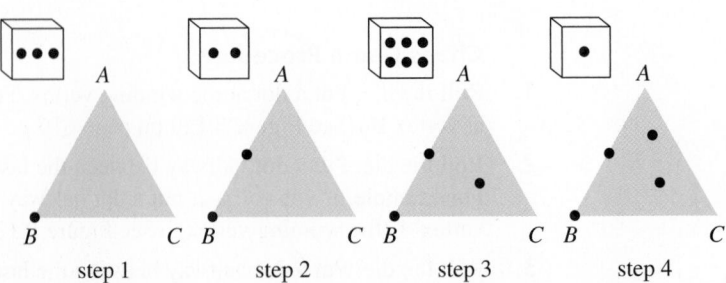

Figure 8.129
The beginnings of the chaos game

4 (and beyond).   Continue rolling the die. Each time, put a dot halfway between the last dot and the winning vertex, as determined by the die. For example, if you roll a 1, put a dot halfway between the last dot and vertex A (the winning vertex). (See Figure 8.129 on page 616.) Then roll again.

Figure 8.130(a) shows the chaos game after 100 rolls. Not surprisingly, it's just a scattering of dots. Figure 8.130(d) shows the chaos game after 10,000 rolls. Quite surprisingly, it looks like the Sierpinski gasket! After all, the chaos game is a sequence of random events, so you would think that you would get a random result, and if you play the game twice, you would think that you'd get two different results. But this is not the case; no matter how many times you play the chaos game, if you roll the die enough, you get the same result: the Sierpinski gasket.

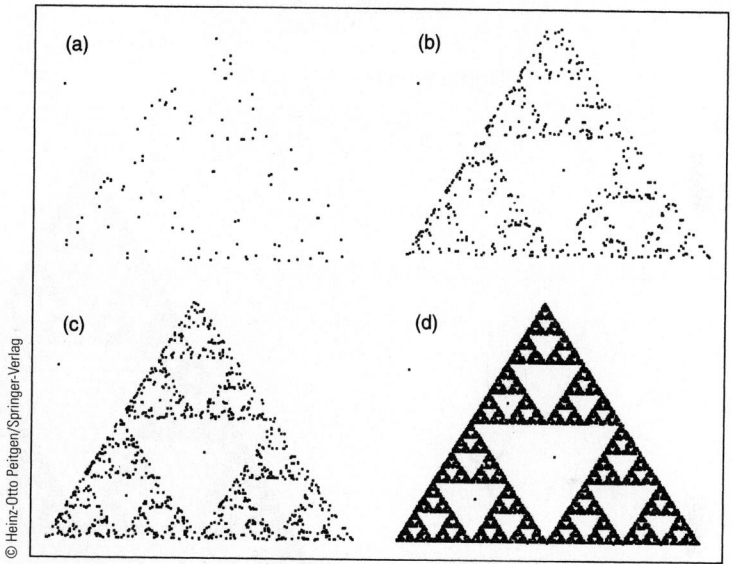

Figure 8.130
The chaos game after (a) 100 rolls, (b) 500 rolls, (c) 1,000 rolls, and (d) 10,000 rolls

Pascal's Triangle and the Sierpinski Gasket
(for those who have read Section 2.4: Permutations and Combinations.)

There is an interesting connection between Pascal's triangle (as studied in Section 2.4) and the Sierpinski gasket. If you color the odd numbers of the first eight rows of Pascal's triangle black and the even numbers white, the result looks like the result of step 3 in creating the Sierpinski gasket. This is shown in Figure 8.131 on page 618.

Figure 8.132 shows the result of coloring the first 64 rows of Pascal's triangle. Is this a Sierpinski gasket? (See Exercise 20.)

The Koch Snowflake

The **Koch snowflake** is another example of fractal geometry. It was first studied by Swedish mathematician Helge von Koch in 1904. To create this shape, follow the following steps, which are illustrated in Figure 8.133 on page 618.

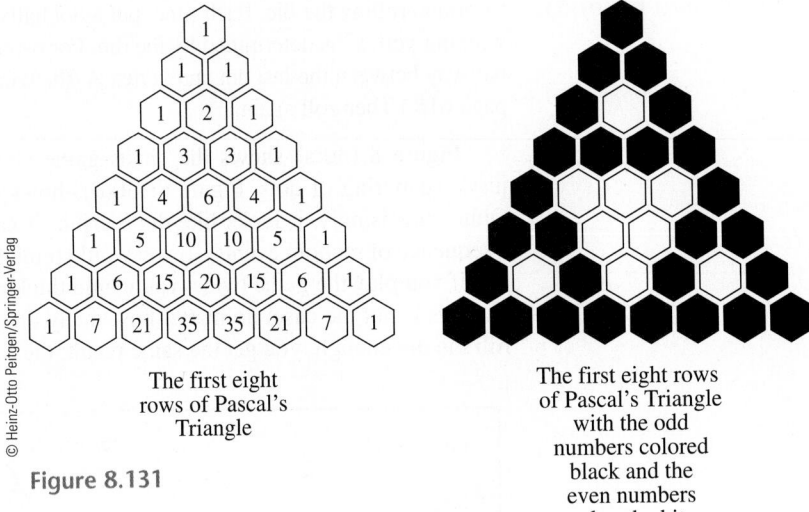

Figure 8.131

The first eight rows of Pascal's Triangle

The first eight rows of Pascal's Triangle with the odd numbers colored black and the even numbers colored white

Figure 8.132
The first 64 rows of Pascal's Triangle with the odd numbers colored black and the even numbers colored white

The Koch Snowflake *Continued*

1. Draw an equilateral triangle (that is, a triangle whose sides are equal and whose angles are equal).

2. In the middle of each side of the equilateral triangle, attach a smaller equilateral triangle (pointing outwards). The smaller equilateral triangle's sides should be

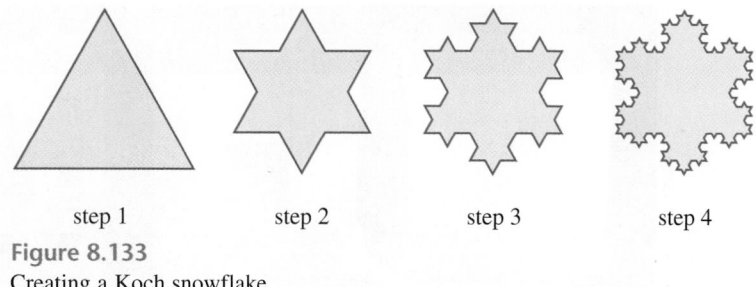

step 1   step 2   step 3   step 4

Figure 8.133
Creating a Koch snowflake

one third as long as the original larger equilateral triangle's sides. This forms a Star of David, with 12 sides.

3. In the middle of each side of the Star of David, attach an equilateral triangle (pointing outward). This equilateral triangle's sides should be one third as long as the Star of David's sides.

4 (and beyond). Alter the result of the previous step by attaching an equilateral triangle to the middle of each side. Continue this process indefinitely. The resulting curve, shown in Figure 8.134, is called a Koch snowflake. Notice its similarity to a real snowflake, such as that in Figure 8.135.

© Phototake, Inc./Alamy

Figure 8.134
A Koch snowflake

Figure 8.135
Notice the similarity between the Koch snowflake and the real snowflake shown here.

The Menger Sponge

The **Menger sponge** is closely related to the Sierpinski gasket. To create this shape, follow the following steps, which are illustrated in Figure 8.136 on page 620.

1. Draw a cube.

2. Remove some of the cube:
    * Subdivide each of the cube's faces into nine equal squares. This subdivides the cube itself into 27 small cubes (nine in the front, nine in the middle, and nine in the back).
    * Remove the small cube at the center of each face. There are six such small cubes.
    * Remove the small cube at the very center of the original cube.

3. A number of small cubes remain at the end of step 2. Remove some of each of these cubes by applying the procedure from step 2 to them.

Courtesy of www.wikipedia.org.

Step 1.   step 2.   step 3.   step 4.

Figure 8.136
Creating a Menger sponge

4 (and beyond).   A number of cubes remain at the end of the previous step. Remove some of each of these cubes by applying the procedure from step 2 to them. Continue this process indefinitely. The resulting shape is called a Menger sponge.

Some Applications of Fractals

Biologists use diffusion fractals to analyze how bacteria cultures grow. Biologists also use l-systems, a method of generating fractals, to study the structures of plants. Chemists use strange attractors, a type of fractal, to study chaotic behavior in chemical reactions. Anatomists use fractal canopies to study the structure of lungs. Computer scientists use fractal image compression to squeeze 7,000 photographs onto the Microsoft Encarta encyclopedia CD. And moviemakers use fractals to create special effects in movies.

Dimension

You're probably familiar with the term *dimension*. A box has three dimensions: up/down, back/forth, and left/right. A rectangle has two dimensions: up/down and left/right. A line has one dimension: left/right. A circle (one that's not filled in) also has one dimension: clockwise/counterclockwise. See Figure 8.137.

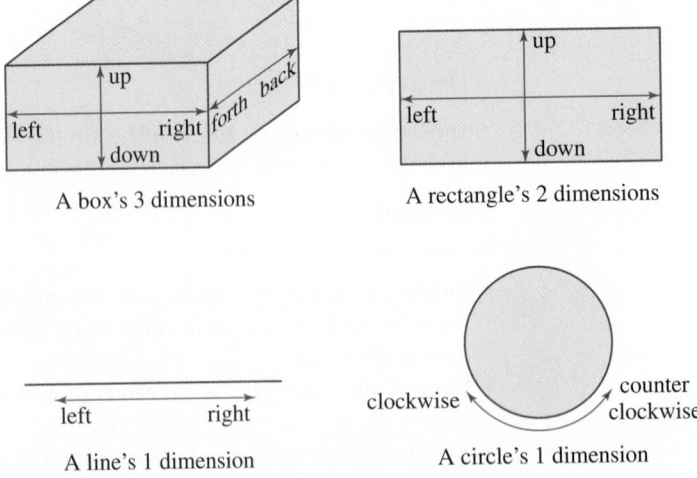

A box's 3 dimensions   A rectangle's 2 dimensions

A line's 1 dimension   A circle's 1 dimension

Figure 8.137

Fractals in the Real World: Computer Graphics in Movies

In the late 1970s, Loren Carpenter was using computer-aided design to create images of airplanes at Boeing. His personal goal was to use computers to help make science fiction movies. He worked at Boeing to gain access to the computers that might help him accomplish that goal.

Carpenter read Benoit Mandelbrot's *The Fractal Geometry of Nature* (see the historical note on Mandelbrot on page 626 of this section). The book's fractal landscapes appealed to Carpenter. In a recent interview, he said, "I wanted to be able to create landscapes for my airplanes. But the method described (in Mandelbrot's book) did not allow you to stand *in* the landscape and have a level of detail in the background and foreground that was consistent with reality, to have a situation where the magnification factor can vary by 1000 in the same object. And that is what I needed." After some time, he figured out how to write computer algorithms "to make lightning bolts, landscapes, clouds, and a host of other things that had an infinite variety of detail and scale and would also animate, because they had a consistent geometry."*

Carpenter wanted to work at Lucasfilm's new computer graphics department, but he had to find a way to be noticed. So he made a movie to demonstrate his fractal discoveries. He used Boeing's computers at night and on weekends for four months to make the two-minute film *Vol Libre* ("Free Flight"), in which the viewer flies through a range of mountains. He used fractals to create the landscapes. These landscapes were different from those in Mandelbrot's book in that the viewer could move through them and see them from different perspectives and different distances.

SIGGRAPH is a professional organization of computer scientists, artists and filmmakers who are interested in computer graphics. Carpenter showed his 16 mm film at their annual conference in 1980. He reported that "the audience erupted. And Ed Catmull, then head of the computer division at Lucasfilm, and Alvy Ray Smith, then director of the computer graphics group, were in the front row. They made me a job offer on the spot."

At Lucasfilm, Carpenter used fractals in the Genesis Sequence of *Star Trek II: The Wrath of Khan*. The result was overwhelmingly successful. The Genesis Sequence was used in the next three Star Trek movies.

"Hot Planet" from the Star Wars Genesis Sequence

Carpenter's fractal mountains in *Vol Libre*

In 1986, Steve Jobs bought Lucasfilm's computer division and renamed it Pixar. Carpenter became the company's first Senior Scientist, and he still holds that position. In 1993, he received a technical Academy Award for his invention of RenderMan image synthesis software system. RenderMan was used in rendering the dinosaurs in *Jurassic Park* and in making many other movies. See Exercise 35.

In 2001, he was awarded an Oscar for his fundamental contributions to the technology of motion pictures.

Thanks to Loren Carpenter for the information in this article, and for the *Vol Libre* and *Star Wars* images.

*Source for all quotes: Perry, Tekla S. "And The Oscar Goes To . . . ," *IEEE Spectrum,* April 2001.

What is the dimension of a piece of paper? Many would say that it has two dimensions: up/down and left/right. Actually, a piece of paper has a third dimension, which is thickness. Tissue paper is very thin, typing paper is somewhat thicker, and card stock is thicker still. It's just that we tend to forget about this third dimension, the thickness, because it's so small relative to the paper's length and width. The only way a piece of paper would have two dimensions would be if it were infinitely thin. And paper isn't infinitely thin.

The Dimension of a Fractal

What is the dimension of a Koch snowflake? Many would say that, like a circle, it has one dimension: clockwise/counterclockwise. But a Koch snowflake wiggles around a lot; it's filled with little tiny crinkles. And those crinkles seem to make it somewhat thicker than a circle. The fact that a piece of paper has some thickness makes the paper's dimension 3 rather than 2. Does the fact that the Koch snowflake has some thickness affect its dimension?

The dimension of a fractal is certainly less obvious than the dimension of a box, a rectangle, or a line. Because of this, it must be determined with a new approach. We'll start by applying this new approach to a square.

If we take a square and double both its length and its width, we get four new squares, each equal to the original square. (See Figure 8.138.) We say that the scale factor is 2, because we multiplied the length and width by 2.

What happens if we apply a scale factor of 3 to the original square? If we multiply both the length and the width by 3, we get nine new squares, each equal to the original square. (See Figure 8.139.)

The original
square

Double its length
and width and
get 4 new squares,
each equal to the
original square.

Figure 8.138

The original
square

Triple its length
and width and get
9 new squares,
each equal to the
original square.

Figure 8.139

In each case, the scale factor, raised to the second power, gives the number of new squares:

- When the scale factor is 2, we get $2^2 = 4$ new squares.
- When the scale factor is 3, we get $3^2 = 9$ new squares.

What happens with cubes? If we apply a scale factor of 2 to a cube, then we get eight new cubes, each equal to the original cube. And if we apply a scale factor of 3, then we get twenty-seven new cubes, each equal in size to the original cube. See Figure 8.140.

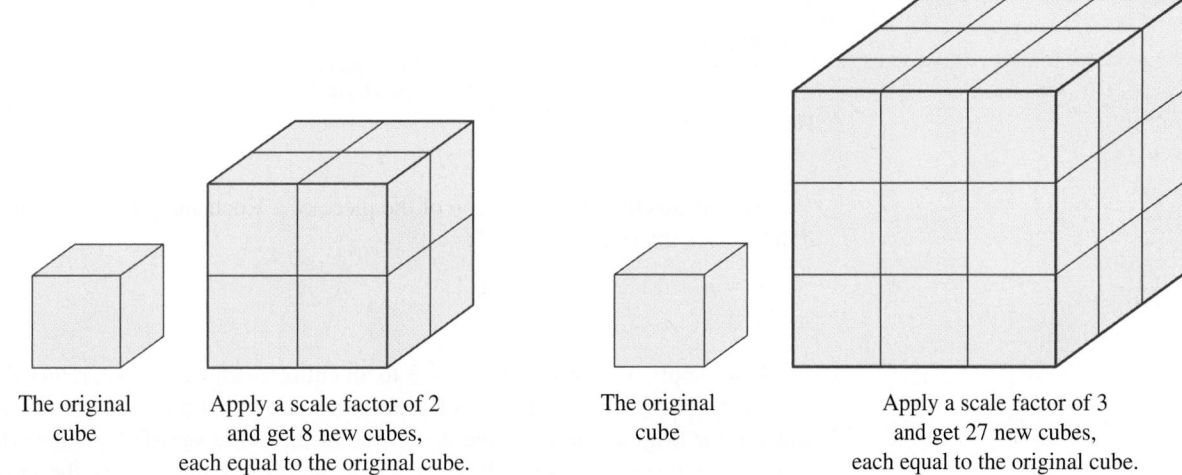

The original cube   Apply a scale factor of 2 and get 8 new cubes, each equal to the original cube.   The original cube   Apply a scale factor of 3 and get 27 new cubes, each equal to the original cube.

Figure 8.140

With the square, the scale factor raised to the *second* power gives the number of new squares. But with the cube, the scale factor raised to the *third* power gives the number of new cubes:

- When the scale factor is 2, we get $2^3 = 8$ new cubes.
- When the scale factor is 3, we get $3^3 = 27$ new cubes.

Clearly, there is a relationship between the scale factor, the number of new objects, and the dimension. The scale factor, raised to the power of the dimension, equals the number of new objects. It turns out that this relationship can be applied to fractals just as well as it can be applied to squares and cubes.

> ### Dimension Formula
>
> The dimension of an object is the number d that satisfies the equation
>
> $$s^d = n$$
>
> where s is the scale factor and n is the number of new objects that result, where each new object is equal in size to the original object.

ThomsonNOW™
for Liberal Arts Mathematics
www.thomsonedu.com/1pass.

EXAMPLE 1   Find the dimension of a Koch snowflake.

Solution   We will start by looking at one piece of a Koch snowflake. If we apply a scale factor of $s = 3$ to this piece, we get a larger piece of a Koch snowflake. This larger piece has $n = 4$ parts, each of which is equal in size to the original piece. See Figure 8.141 on page 624.

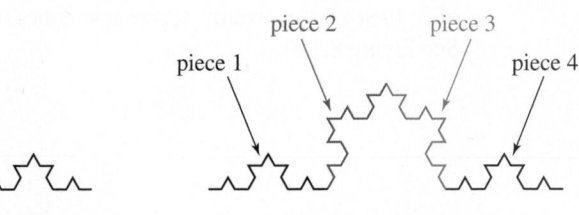

The original
piece

Apply a scale factor of 3
and get 4 new pieces,
each equal to the original piece.

Figure 8.141

This means that the dimension of the piece of a Koch snowflake is the number d that solves the equation $3^d = 4$:

$s^d = n$ the dimension formula

$3^d = 4$ substituting

If we apply a scale factor of $s = 3$ to an entire Koch snowflake, rather than to just one piece of the snowflake, each piece turns into $n = 4$ pieces, each of which is equal to the original piece. So the dimension of the entire snowflake is the same as that of the piece of a snowflake. It is the number d that solves the equation $3^d = 4$.

What number would d be? Solving this equation involves mathematics that will be covered in Chapter 10. Instead, we'll approximate d to the nearest tenth, using a trial and error process.

Clearly, $d \neq 1$, because $3^1 = 3 \neq 4$. Also, $d \neq 2$, because $3^2 = 9 \neq 4$. Substituting 1 for d gives a result that's too small, and substituting 2 for d gives a result that's too big. Instead, d is a number between 1 and 2. We'll approximate d to the nearest tenth, using a trial and error process. Using a calculator, we get the following results:

$3^{1.1} = 3.348\ldots$

$3^{1.2} = 3.737\ldots$

$3^{1.3} = 4.171\ldots$

Substituting 1.2 for d gives a result that's a little too small, and substituting 1.3 gives a result that's a little too big. This means that d is between 1.2 and 1.3. And since $4.171\ldots$ is closer to 4 than is $3.737\ldots$, we can tell that d is closer to 1.3 than to 1.2. Thus, $d \approx 1.3$.

So the dimension of a Koch snowflake is a little bigger than the dimension of a circle. The Koch snowflake has dimension $d \approx 1.3$, while the circle has dimension

$d = 1$. The little tiny crinkles do make the snowflake somewhat thicker than a circle. And this extra thickness makes the snowflake's dimension somewhat bigger than that of a circle. But it's quite strange that a fractal can have a dimension that is not a whole number. ∎

In discussing squares and cubes, it didn't matter what scale factor we used. We got the same result for the dimension regardless of whether we used $s = 2$ or $s = 3$. In fact, we would get the same result for any scale factor. But in Example 1, we deliberately used a scale factor of 3 because that scale factor gave us a larger piece of a Koch snowflake with four parts, *each of which is equal to the original piece.* If we used a scale factor of 2, we would still get a larger piece of a Koch snowflake with four parts, *but those parts would not be equal to the original piece.* They would be too small to equal the original piece. See Figure 8.142. With fractals, it's important to choose a scale factor carefully so that each of the new pieces is equal in size to the original piece.

piece 1   piece 2   piece 3   piece 4

The original piece

Apply a scale factor of 2 and get 4 new pieces, but they are <u>not</u> equal in size to the original piece.

Figure 8.142

Some Applications of Fractal Dimension

A country's coastline has self-similarity. If you took two photographs of an uninhabited coastline, one from an elevation of 1,000 feet and the other from an elevation of 5,000 feet, you would have difficulty determining which photo was which. This self-similarity means that fractal geometry can be used to analyze the complexity of coastlines. In fact, geographers classify coastlines according to their fractal dimension. South Africa's coastline is quite smooth, with a fractal dimension close to 1. On the other hand, Norway's coastline is heavily indented by glacier-carved fjords; its fractal dimension is about 1.5.

It's interesting to note that a Koch snowflake is sometimes called a *Koch island,* because the snowflake's outline looks somewhat like the coastline of an island. (See Exercise 34.)

Geologists have found that the fractal dimension of geologic features such as mountains, coastlines, faults, valleys, and river courses are related to the conditions under which they were formed and the processes to which they have been subjected. The fractal dimension of a mountain's contour line will be higher if the mountain is made of easily eroded rock or if it is in an area of high rainfall.

Scientists have also found that the fractal dimension of the surfaces of small particles such as soot and pharmaceuticals is related to their physical characteristics. For example, the tendency of soot to adhere to lung surfaces and the rate of absorption of ingested pharmaceuticals are both related to their fractal dimensions.

Historical Note   Benoit Mandelbrot 1924–

Benoit Mandelbrot was born in Warsaw, Poland. His father made his living buying and selling clothes. His mother was a doctor. When Mandelbrot was 11, he emigrated with his family to France. There, his uncle, who was a professor of mathematics at the Collège de France, took responsibility for his education. After World War II started, poverty and the need to survive kept Mandelbrot away from school. As a result, he educated himself. He now attributes much of his success to this self-education. It allowed him to think in different ways and to avoid the standard modes of thought encouraged by a conventional education.

After obtaining a master's degree in aeronautics from the California Institute of Technology and a Ph.D. at the Université de Paris, Mandelbrot went to the Institute for Advanced Study at Princeton. Later, he became an IBM Fellow at their Watson Research Center in New York. IBM allowed him

great latitude in choosing the direction of his research. At IBM, he started his famous work on fractals, which led to his being called "the father of fractal geometry." Also at IBM, he developed one particularly important fractal called the Mandelbrot set, and he developed some of the first computer programs that print graphics.

In part because of his unique education, he decided early on to make contributions to many different branches of science. His academic posts indicate this; they include professor of mathematics at Yale, Harvard, and the École Polytechnique and visiting professor at Harvard (first in economics and later in applied mathematics), Yale (engineering), and the Einstein College of Medicine (physiology). He has received many honors and prizes, including the Barnard Medal for Meritorious Service to Science, the Franklin Medal, the Alexander von Humbolt Prize, the Steinmetz Medal, the Science for Art Prize, the Harvey Prize for Science and Technology, the Nevada Medal, the Wolf Prize in physics, and the Honda Prize.

This is the Mandelbrot set, shown in black and white. It can be intriguingly beautiful along it's edge. The result of zooming in on the boxed portion is shown to the right.

We'll zoom in again. The result is shown on page 627.

The edge has many fascinating details. We'll zoom in on the spiral.

Now we'll zoom in on a detail of the spiral.

Another zoom.

One more zoom, and what do we see?

We see the Mandelbrot set again, surrounded by beautiful lace filigree. What an extreme example of self-similarity! And, if we zoom in one final time . . .

. . . we find another Mandelbrot set, slightly to the left and below the center, surrounded by ornate details. Even more self-similarity.

8.8 Exercises

1. *The Sierpinski gasket*
   a. Using a straightedge, carefully and accurately draw five equilateral triangles, each on a separate piece of paper. Make each side 1 foot in length (or, optionally, make your triangle as large as the paper will allow and label each side as being 1 foot in length).
   b. With the first equilateral triangle, complete the first step of a Sierpinski gasket. With the second equilateral triangle, complete the first two steps of a Sierpinski gasket. Continue in this fashion until you have completed the first five steps with the fifth equilateral triangle. With each triangle, shade in the appropriate regions.

2. *The Sierpinski carpet.* The Sierpinski carpet is closely related to the Sierpinski gasket and the Menger sponge. It is the square version of the Sierpinski gasket and the flat version of the Menger sponge. To create this shape, use the following steps. (See Figure 8.143.)

step 1   step 2   step 3

Figure 8.143
Creating a Sierpinski carpet

   1. Draw a square.
   2. Subdivide the square into nine equal squares. Remove the central square.
   3. The original square now has a square hole in its center, surrounded by eight other squares. Subdivide each of these eight squares into nine smaller squares. Remove each of the central squares.
   4 (**and beyond**). Continue this process indefinitely.
   a. Using a straightedge, carefully and accurately draw four squares, each on a separate piece of paper. Make each side 1 foot in length (or, optionally, make your triangle as large as the paper will allow and label each side as being 1 foot in length).

   b. With the first square, complete the first step of a Sierpinski carpet. With the second square, complete the first two steps of a Sierpinski carpet. Continue in this fashion until you have completed the first four steps with the fourth square. With each square, shade in the appropriate regions.

3. *The Mitsubishi gasket.* Carefully and accurately complete the first four steps of a Mitsubishi gasket. To create this shape, use the following steps. (See Figure 8.144.)

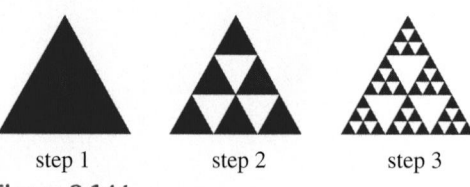

step 1   step 2   step 3

Figure 8.144
Creating a Mitsubishi gasket

   1. Draw a triangle and fill in its interior.
   2. Put three holes in the triangle:
      • Find the points that divide the sides of the triangle into thirds.
      • Form nine new triangles by connecting those points with lines parallel to the sides.
      • Form three holes by removing three of these new triangles, as shown in Figure 8.144.

   This leaves six filled in triangles, each a smaller version of the original triangle from step 1.
   3. Put three holes in each of the six triangles from step 2 by applying the procedure from step 2 to each triangle.
   4. Put three holes in each of the triangles from step 3 by applying the procedure from step 2 to each triangle.
   5 (**and beyond**). Put three holes in each of the triangles from the previous step by applying the procedure from step 2 to each triangle.

4. *The square snowflake.* Carefully and accurately complete the first four steps of a square snowflake. To create this shape, use the following steps. (See Figure 8.145.)
   1. Draw a square.
   2. Divide each of the square's straight lines into fourths. In a clockwise order:
      • Leave the first fourth alone.
      • Replace the second fourth with a square that's on the outer side of the line.

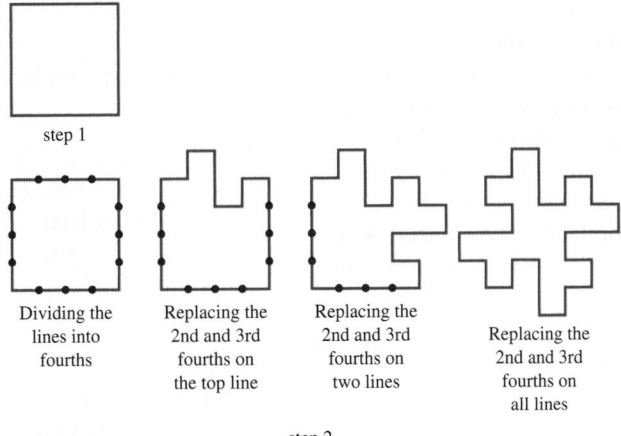

step 1

Dividing the lines into fourths

Replacing the 2nd and 3rd fourths on the top line

Replacing the 2nd and 3rd fourths on two lines

Replacing the 2nd and 3rd fourths on all lines

step 2

Figure 8.145
Creating a square snowflake

- Replace the third fourth with a square that's on the inner side of the line.
- Leave the last fourth alone.

Apply the above procedure to each of the square's straight lines. *Be sure to do the above in a clockwise order!*

3. Apply the procedure from step 2 to each of the straight lines in step 2.

4 (and beyond). Apply the procedure from step 2 to each of the previous step's straight lines.

5. Find the exact dimension of a fractal that has $n = 25$ and $s = 5$. Do not approximate.

6. Find the exact dimension of a fractal that has $n = 5$ and $s = 25$. Do not approximate.

7. Find the exact dimension of a fractal that has $n = 3$ and $s = 27$. Do not approximate.

8. Find the exact dimension of a fractal that has $n = 2$ and $s = 32$. Do not approximate.

9. Find the approximate dimension of the Sierpinski gasket. Round off to the nearest tenth. What does this number say about the fractal?

10. Find the approximate dimension of the Sierpinski carpet from Exercise 2. Round off to the nearest tenth. What does this number say about the fractal?

11. Find the approximate dimension of the Mitsubishi gasket from Exercise 3. Round off to the nearest tenth. What does this number say about the fractal?

12. Find the approximate dimension of the Menger sponge. Round off to the nearest tenth. What does this number say about the fractal?

13. Find the approximate dimension of the square snowflake from Exercise 4. Round off to the nearest tenth. What does this number say about the fractal?

14. Of the assigned problems from 9 through 13, which fractal is the least complex? Which is the most complex? Why?

15. Use a scale factor of 4 to find the dimension of a cube. What do you observe?

16. Use a scale factor of 5 to find the dimension of a square. What do you observe?

17. Use scale factors to find the dimension of a rectangle.

18. Play the chaos game through 40 rolls of a die. Label the points "P_1," "P_2," "P_3," etc. Discuss your results.

▬▬▬▬ *Answer the following using complete sentences and your own words.*

CONCEPTS QUESTIONS

19. When playing the chaos game, why is it impossible to ever land in the hole formed in step 2 of the Sierpinski gasket process? Why is this an important consideration in seeing that the chaos game yields the Sierpinski gasket?

20. Figure 8.130(d) shows the result of the chaos game after 10,000 rolls. Is this a Sierpinski gasket, or is it the result of one of the steps in forming a Sierpinski gasket or is it merely similar to the Sierpinski gasket or one of the steps in forming the gasket? Explain your answer.

21. *For those who have read Section 2.4: Permutations and Combinations.* Figure 8.132 shows the result of coloring the first 64 rows of Pascal's triangle. Is this a Sierpinski gasket, or is it the result of one of the steps in forming a Sierpinski gasket? Explain your answer. If it is the result of one of the steps in forming a Sierpinski gasket, determine which step.

22. Describe the self-similarity of the given fractal. Also determine whether the self-similarity is exact or approximate.
   a. the Sierpinski gasket
   b. the skewed Sierpinski gasket
   c. the Koch snowflake
   d. the Menger sponge

23. Describe the self-similarity for the fractal described in each of the assigned exercises (see Exercises 2–4). In each case, describe the self-similarity. Also determine whether the self-similarity is exact or approximate.

24. Discuss the recursive nature of:
   a. the Sierpinski gasket
   b. the skewed Sierpinski gasket
   c. the Koch snowflake
   d. the Menger sponge

25. Discuss the recursive nature of the fractal described in each of the assigned exercises (see Exercises 2–4).

26. Name three naturally occurring objects that have self-similarity. Do these objects have exact or approximate self-similarity?

27. Which type of self-similarity is most typically found in nature: exact or approximate? Which type is most typically found in fractal geometry?

28. Name a naturally occurring object that has a recursive nature. Describe its recursive nature.

29. In fractal geometry, self-similarity occurs at ever-smaller scales. That is, parts of a shape appear within itself at infinitely many different scales. Is this the case with the self-similarity found in nature? Why?

30. Describe two applications of fractal dimension.

HISTORY QUESTIONS

31. Who is known as the father of fractal geometry?

32. What was different about the education of the father of fractal geometry? What impact did this difference have on him?

PROJECTS

33. *Design your own fractal.* Discuss its self-similarity and its recursive nature. Find its dimension

34. *Create your own coastline.* Draw a line segment. Roll a die. If either a 1, 2, or 3 comes up, apply step 2 of the Koch snowflake procedure to that line segment, with the triangle(s) pointing upwards. If a 4, 5, or 6 comes up, apply step 2 of the Koch snowflake procedure to that line segment, with the triangle(s) pointing downwards. Continue this process until your fractal resembles a coastline. Include in your work the final sketch and a description of exactly how you created that shape. In your description, list each roll of the die and describe how that roll changed your image.

  WEB PROJECTS

35. Write an essay on one of the use of fractals by the film industry in creating computer graphics.

36. Investigate some fractal-related web sites and write an essay about your findings.

Some useful links for these web projects are listed on the text companion web site: Go to www.thomsonedu.com/mathematics/johnson to access the web site.

8.9   The Perimeter and Area of a Fractal

There is something both puzzling and useful about the perimeter and area of a fractal. We're going to find the perimeter of a Koch snowflake. To do so, we'll first find the perimeter of the first few steps of the process of creating a Koch snowflake. Then we'll use that work to find the perimeter of a Koch snowflake.

The Perimeter of the Koch Snowflake

EXAMPLE 1   Find the perimeter P_1 of the equilateral triangle formed in step 1 of the process of creating a Koch snowflake, if each side is of length 1 foot. (See Figure 8.146.)

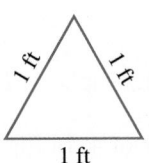

Figure 8.146
Finding the perimeter P_1

Solution The perimeter is just the sum of the lengths of the sides, so it is

$$P_1 = 1 \text{ ft} + 1 \text{ ft} + 1 \text{ ft} = 3 \text{ ft}$$

Alternatively, there are three sides, each of length 1 foot, so the perimeter is

$$P_1 = (3 \text{ sides}) \cdot (1 \text{ foot/side}) = 3 \text{ ft}$$

(Notice that we are using dimensional analysis, as discussed in Appendix E.) We call this P_1 as a memory device; P stands for perimeter, and the subscript 1 indicates that it is the perimeter of the triangle formed in step one. ∎

Figure 8.147
Finding the perimeter P_2

EXAMPLE 2 Find the perimeter P_2 of the Star of David that is formed in step 2 of the process of creating a Koch snowflake. (See Figure 8.147.)

Solution The sides of the new smaller triangles are one third as long as the sides of the larger equilateral triangle from Example 1, so each side is

$$1/3 \cdot 1 \text{ ft} = 1/3 \text{ ft}$$

There are 12 sides, so the perimeter is

$$P_2 = (12 \text{ sides}) \cdot (1/3 \text{ ft/side}) = 4 \text{ ft} \quad \text{using dimensional analysis} \quad ∎$$

EXAMPLE 3 Find the perimeter P_3 of the shape that is formed in step 3 of the process of creating a Koch snowflake. (See Figure 8.148.)

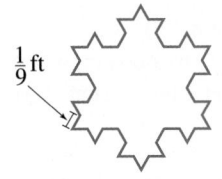

Figure 8.148
Finding the perimeter P_3

Solution The sides of the new smaller triangles are one-third as long as the sides of the triangle from Example 2, so each side is

$$1/3 \cdot 1/3 \text{ ft} = 1/9 \text{ ft}$$

There are 48 sides (count them!), so the perimeter is

$$P_3 = (48 \text{ sides}) \cdot (1/9 \text{ ft/side})$$
$$= 48/9 \text{ ft} = 16/3 \text{ ft} \quad \text{using dimensional analysis}$$

The results of Examples 1, 2, and 3 are summarized in Figure 8.149.

Step	Picture	Number of Sides	Length of Each Side	Perimeter
1		3 sides	1 ft/side	$P_1 = 3 \text{ sides} \cdot 1 \text{ ft/side}$ $= 3 \text{ ft}$
2		12 sides	1/3 ft/side	$P_2 = 12 \text{ sides} \cdot 1/3 \text{ ft/side}$ $= 4 \text{ ft}$
3		48 sides	1/9 ft/side	$P_3 = 48 \text{ sides} \cdot 1/9 \text{ ft/side}$ $= 16/3 \text{ ft}$

Figure 8.149
Finding the perimeter P ∎

EXAMPLE 4 Find the perimeter P of the Koch snowflake.

Solution To find the perimeter P, we have to generalize from the solutions of Examples 1, 2, and 3. There are several patterns to observe.

First notice that the number of sides is increasing by a factor of 4. In Figure 8.149, under "Number of Sides," each entry is 4 times the previous entry. In particular, $4 \cdot 3$ sides $= 12$ sides and $4 \cdot 12$ sides $= 48$ sides. This should make sense, because in each step in the Koch snowflake procedure, you replace every ⁀‾‾‾‾ with a four-sided ＿／\＿.

This means that every single side is replaced with four sides. So naturally, the number of sides increases by a factor of 4. (The word *factor* means "multiplier," so saying that the number of sides increases by a factor of 4 means that the number of sides is multiplied by 4.)

The second pattern to observe is that the length of each side is decreasing by a factor of $1/3$. In Figure 8.149, under "Length of Each Side," each entry is $1/3$ times the previous entry. In particular, $1/3 \cdot 1$ ft $= 1/3$ ft and $1/3 \cdot 1/3$ ft $= 1/9$ ft. This too should make sense, because each step in the Koch snowflake procedure makes each new side $1/3$ as long as each side from the previous step.

The last pattern to observe is that the perimeter is increasing by a factor of $4/3$. In Figure 8.149, under "Perimeter," each entry is $4/3$ times the previous entry. In particular, $4/3 \cdot 3$ ft $= 4$ ft and $4/3 \cdot 4$ ft $= 16/3$ ft. This should also make sense, because there are four times as many sides, and each side is $1/3$ as long, so the perimeter increases by a factor of $4 \cdot 1/3 = 4/3$.

These observations are summarized and extended in Figure 8.150.

Notice that for each step number, the perimeter is 3 times $4/3$ to a power, where the power is 1 less than the step number. In other words, if we let n be the step number, then the perimeter at step n is

$$P_n = 3 \cdot (4/3)^{n-1}$$

Let's check this formula by using it to find P_5:

$$P_5 = 3 \cdot (4/3)^{5-1} = 3 \cdot (4/3)^4$$

This checks with the last entry in Figure 8.150.

The Koch snowflake is the result of continuing the process illustrated in Figure 8.150 indefinitely. That means that the perimeter of the Koch snowflake is the result of multiplying previous perimeters by $4/3$, indefinitely. At step 100, the perimeter is huge:

$$P_{100} = 3 \cdot (4/3)^{100-1} = 3 \cdot (4/3)^{99} \text{ ft}$$

With a calculator, we find that this is approximately 7×10^{12} ft $= 7,000,000,000,000$ ft. At step 1,000, the perimeter is incredibly huge:

$$P_{1000} = 3 \cdot (4/3)^{1000-1} = 3 \cdot (4/3)^{999} \text{ ft}$$

On the calculator, this is approximately 2×10^{125} ft. (Remember, 2×10^{125} is a 2 followed by 125 zeros!)

Step	Picture	Number of Sides	Length of Each Side	Perimeter
1		3 sides	1 ft/side	$P_1 = 3$ ft
2		$4 \cdot 3$ sides $= 12$ sides	$1/3 \cdot 1$ ft/side $= 1/3$ ft/side	$P_2 = 4/3 \cdot P_1$ $= 4/3 \cdot 3$ ft
3		$4 \cdot 12$ sides $= 48$ sides	$1/3 \cdot 1/3$ ft/side $= 1/9$ ft/side	$P_3 = 4/3 \cdot P_2$ $= 4/3 \cdot 4/3 \cdot 3$ ft $= (4/3)^2 \cdot 3$ ft
4		$4 \cdot 48$ sides $= 192$ sides	$1/3 \cdot 1/9$ ft/side $= 1/27$ ft/side	$P_4 = 4/3 \cdot P_3$ $= 4/3 \cdot (4/3)^2 \cdot 3$ ft $= (4/3)^3 \cdot 3$ ft
5		$4 \cdot 192$ sides $= 768$ sides	$1/3 \cdot 1/27$ ft/side $= 1/81$ ft/side	$P_5 = 4/3 \cdot P_4$ $= 4/3 \cdot (4/3)^3 \cdot 3$ ft $= (4/3)^4 \cdot 3$ ft

Figure 8.150
Continuing to find the perimeter P

We must conclude that the perimeter of the Koch snowflake is infinite! This is hard to make sense of, because the perimeter just doesn't look that long. But remember, after several steps, it becomes impossible to visually tell the difference between the result of one step and the result of the next step. So you can't visually tell the difference between the result of step 5 (which has perimeter $P_5 \approx 9.48$ ft) and a Koch snowflake (which has infinite perimeter). Most of the perimeter is in all the little tiny crinkles.   ∎

The Area of the Koch Snowflake

EXAMPLE 5   Find the area A_1 of the equilateral triangle formed in step 1 of the process of creating a Koch snowflake, if each side is of length 1 foot. (See Figure 8.151.)

Solution   To find the area, we need the base and the height. It's easy to see that the base is 1 foot, but we have to calculate the height h. The height h splits the equilateral triangle into two identical triangles, one on the each side of the height. Because the two triangles are identical, their bases are equal, so each must be $1/2$ ft. We can apply the Pythagorean Theorem to either of these triangles, because they are right triangles. (See Figure 8.152.)

Figure 8.151
Finding the area A_1

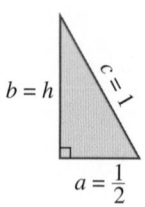

Figure 8.152
Finding the height h

$$a^2 + b^2 = c^2 \quad \text{Pythagorean Theorem}$$
$$(1/2)^2 + h^2 = 1^2 \quad \text{substituting}$$
$$1/4 + h^2 = 1$$
$$h^2 = 1 - 1/4 = 3/4$$
$$h = \sqrt{\frac{3}{4}} = \frac{\sqrt{3}}{\sqrt{4}} = \frac{\sqrt{3}}{2} \text{ ft}$$

This is the height of the right triangle, but it is also the height of the triangle formed in step 1 of the Koch snowflake process. This means that the area of that triangle is

$$A_1 = \frac{1}{2}bh = \frac{1}{2} \cdot 1 \text{ ft} \cdot \frac{\sqrt{3}}{2} \text{ ft} = \frac{\sqrt{3}}{4} \text{ sq ft}$$ ∎

EXAMPLE 6   Find the area A_2 of the Star of David formed in step 2 of the process of creating a Koch snowflake. (See Figure 8.153.)

Solution   The area is the sum of the area of the original equilateral triangle plus the areas of the three new smaller triangles. The original equilateral triangle's area was found in Example 5. It is

$$A_1 = \frac{\sqrt{3}}{4} \text{ sq ft}$$

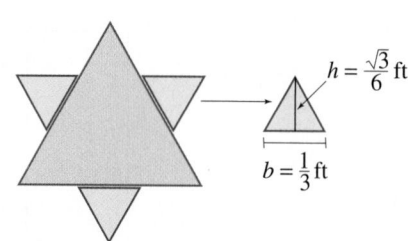

Figure 8.153
Finding the area A_2

The new smaller triangles are identical, so they have the same area. In step 2, all sides are 1/3 of what they were in step 1, so the base is $b = 1/3 \cdot 1$ ft $= 1/3$ ft, and the height is $h = 1/3 \cdot \sqrt{3}/2 = \sqrt{3}/6$ ft. This means that the area of each smaller triangle is

$$A = \frac{1}{2}bh = \frac{1}{2} \cdot \frac{1}{3} \text{ ft} \cdot \frac{\sqrt{3}}{6} \text{ ft} = \frac{\sqrt{3}}{36} \text{ sq ft}$$

There are three of these new smaller triangles. Thus, the area A_2 of the Star of David is

$$A_2 = \text{(area of original equilateral triangle)} + 3 \cdot \text{(area of new smaller triangle)}$$
$$= \frac{\sqrt{3}}{4} + 3 \cdot \frac{\sqrt{3}}{36} \text{ sq ft}$$
$$= \frac{\sqrt{3}}{4} + \frac{\sqrt{3}}{12} \text{ sq ft} \quad \text{canceling}$$ ∎

EXAMPLE 7   Find the area A_3 of the shape formed in step 3 of the process of creating a Koch snowflake. (See Figure 8.154.)

Solution   The area is the sum of the area of the Star of David from Example 6 plus the areas of the new smaller triangles. The Star of David's area is

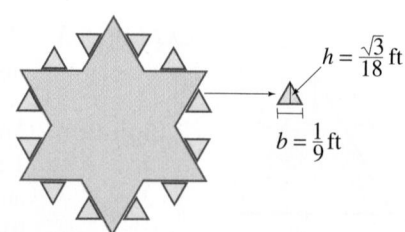

Figure 8.154
Finding the area A_3

$$A_2 = \frac{\sqrt{3}}{4} + \frac{\sqrt{3}}{12} \text{ sq ft}$$

The new smaller triangles are identical, so they have the same area. The base is $b = 1/3 \cdot 1/3$ ft $= 1/9$ ft, and the height is $h = 1/3 \cdot \sqrt{3}/6 = \sqrt{3}/18$ ft. This means that the area of each smaller triangle is

$$A = \frac{1}{2}bh = \frac{1}{2} \cdot \frac{1}{9} \text{ ft} \cdot \frac{\sqrt{3}}{18} \text{ ft} = \frac{\sqrt{3}}{324} \text{ sq ft}$$

Also, there are 12 new smaller triangles (count them!). Thus, the area A_3 is

$$A_3 = \text{(area of Star of David)} + 12 \cdot \text{(area of new smaller triangle)}$$

$$= \frac{\sqrt{3}}{4} + \frac{\sqrt{3}}{12} + 12 \cdot \frac{\sqrt{3}}{324} \text{ sq ft}$$

$$= \frac{\sqrt{3}}{4} + \frac{\sqrt{3}}{12} + \frac{\sqrt{3}}{27} \text{ sq ft} \qquad \text{canceling} \qquad ■$$

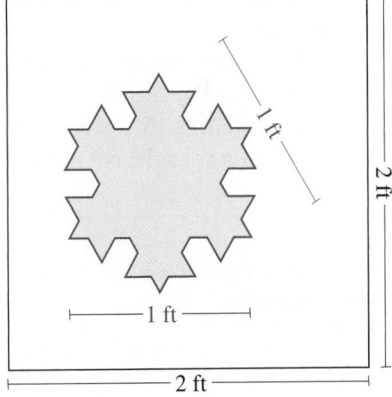

While there are a number of patterns that we could observe, these patterns lead to a more complicated result than do the perimeter patterns discussed in Example 4. Utilizing these patterns to compute the exact area of the Koch snowflake is beyond the scope of this text. In Exercise 11, we will use Figure 8.154 to approximate the area.

Notice that the result of each step of the process of creating a Koch snowflake can be completely enclosed in a square that is 2 feet on each side, with plenty of room left over, as shown in Figure 8.155. This means that the area of the Koch snowflake is significantly less than the area of the square, which is 4 square feet.

Figure 8.155
Approximating the area A of a koch snowflake

The Perimeter and Area of a Fractal

At the beginning of this section, we said, "There is something both puzzling and useful about the perimeter and area of a fractal." Now that we've found the perimeter and area of the Koch snowflake, we can discuss the puzzling and useful aspect of this topic.

The Koch snowflake has an infinite perimeter but an area that is significantly less than 4 square feet. This means that the Koch snowflake fits an infinitely long curve (the perimeter) inside a small space. The Sierpinski gasket also has an infinite perimeter but an area of zero! (The area is zero because of all of the holes—they just take over the Sierpinski gasket.) So the Sierpinski gasket also fits an infinitely long curve inside a small space. The Menger sponge has an infinite surface area but a volume of zero! This means that the Menger sponge fits an infinitely large surface inside a small space.

An Application of Fractals

It is certainly puzzling that a Koch snowflake has an infinite perimeter but a small area, and it's certainly puzzling that a Menger sponge has an infinite surface area but a volume of zero. But why are these facts useful?

Consider the human body's system of veins, arteries, and capillaries (see Figure 8.156). This system takes up a small part of the body's space, but its length is

Figure 8.156
The human body's system of veins, arteries and capillaries

incredibly large; laid end to end, the system of veins and arteries would be more than 40,000 miles long. Like a Koch snowflake, the length is huge, but it fits inside a relatively small space.

Also, consider the human kidney. Its surface area is practically infinite (and for good reason, since the kidney's surface area determines how much blood it can cleanse), but its volume is small—the kidney is only 4 inches long.

Scientists use fractal geometry to analyze these various parts of the body. It may well be that fractal geometry will help us better understand the functioning of the human body.

8.9 Exercises

1. *Finding the perimeter of a Sierpinski gasket.* (By "perimeter," we mean the total distance around all of the filled in regions, not the distance around the outside only.)

   a. Use your drawings from Exercise 1 of Section 8.8 to explain each of the calculations in the chart in Figure 8.157, where "triangle" refers to a filled-in triangle, not a triangular hole.

Step	Number of Triangles	Length of Each Side	Perimeter of One Triangle	Total Perimeter of All Triangles
1	1	1 ft	3 · 1 ft = 3 ft	1 · 3 ft = 3 ft
2	3*	1/2 · 1 ft = 1/2 ft	3 · 1/2 ft = 3/2 ft	3 · 3/2 ft = 9/2 ft
3				
4				
5				
6				

*3 not 4, because only three triangles are filled iin.

Figure 8.157
For Exercise 1

   b. Use your drawings from Exercise 1 of Section 8.8 to complete the chart in Figure 8.157. Include units.
   c. By what factor does the number of triangles increase? Explain why your answer makes sense.
   d. By what factor does the length of each side decrease? Explain why your answer makes sense.
   e. By what factor does the perimeter of one triangle change? Explain why your answer makes sense. Is the perimeter of one triangle increasing or decreasing? Why?

   f. By what factor does the total perimeter of all triangles change? Explain why your answer makes sense. Is the total perimeter of one triangle increasing or decreasing? Why?
   g. Use your answers to parts (a)–(f) to find a formula for the total perimeter of all triangles at step n.
   h. Use your answer to part (g) and your calculator to find the perimeter of the Sierpinski gasket. Explain your work. (By "perimeter," we mean the total distance around all of the filled in regions, not the distance around the outside only.)

2. *Finding the area of a Sierpinski gasket.* (By "area," we mean the total area of all the filled in regions.)
   a. Use your drawings in Exercise 1 of Section 8.8 to complete the chart in Figure 8.158, where "triangle" refers to a filled in triangle, not to a triangular hole. Include units.
   b. By what factor does the number of triangles increase? Explain why your answer makes sense.
   c. By what factor does the length of the base decrease? Explain why your answer makes sense.
   d. By what factor does the length of the height decrease? Explain why your answer makes sense.
   e. By what factor does the area of each triangle decrease? Explain why your answer makes sense.
   f. By what factor does the total area of all triangles change? Explain why your answer makes sense. Is the total area increasing or decreasing? Why?
   g. Use your answers to parts (a)–(f) to find a formula for the total area of all triangles at step n.
   h. Use your answer to part (g) and your calculator to find the area of the Sierpinski gasket. (By "area," we mean the total area of all of the filled in regions.) Explain your work.

3. *Finding the perimeter of a Sierpinski carpet.* (See Exercise 2 in Section 8.8 for a description of this fractal.) (By "perimeter," we mean the total distance around all of the filled in regions.)

Step	Number of Triangles	Length of the Base	Length of the Height	Area of one Triangle	Total area of all Triangles
1	1				
2	3*				
3					
4					
5					
6					

*3 not 4, because only three triangles are filled iin.

Figure 8.158
For Exercise 2

a. Use your drawings in Exercise 2 of Section 8.8 to explain each of the calculations in the chart in Figure 8.159. These calculations are illustrated in Figure 8.160 on page 640.

b. Use your drawings from Exercise 2 of Section 8.8 to complete the chart in Figure 8.159. Include units.

c. By what factor does the number of squares increase (starting at step 2)? Explain why your answer makes sense.

d. By what factor does the length of each side decrease? Explain why your answer makes sense.

e. By what factor does the perimeter of one new square change? Explain why your answer makes sense. Is the perimeter of one new square increasing or decreasing? Why?

f. By what factor does the total perimeter of all new squares change? Explain why your answer makes sense. Is the total perimeter of all new squares increasing or decreasing? Why?

g. Use your answers to parts (a)–(f) to find a formula for the total perimeter of all new squares at step n.

Step	Number of New Squares*	Length of Each Side	Perimeter of One New Square*	Total Perimeter of All New Squares*	Total Perimeter of All Squares*
1	1	1 ft	4 · 1 ft = 4 ft	1 · 4 ft = 4 ft	4 ft
2	1	1/3 · 1 ft = 1/3 ft	4 · 1/3 ft = 4/3 ft	1 · 4/3 ft = 4/3 ft	4 ft + 4/3 ft
3	8				
4					
5					
6					

*"New square" refers to the squares that are both new to this step and that contribute to the total perimeter. "All squares" refers to the squares that are new to this step as well as those from previous steps.

Figure 8.159
For Exercise 3

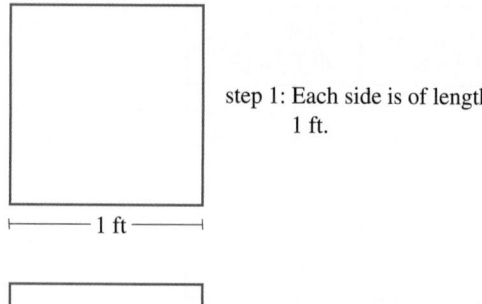

step 1: Each side is of length 1 ft.

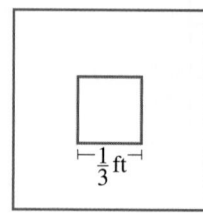

step 2: There is one new square. Each of its sides is of length $\frac{1}{3}$ ft.

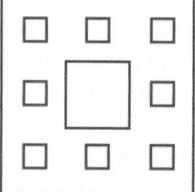

step 3: There are eight new squares. What is the length of each of their sides?

Finding the perimeter and area of a Sierpinski carpet.

Figure 8.160
For Exercise 3

h. Use your answer to part (g) and your calculator to determine what happens to the total perimeter of all new squares as we move from step to step.

i. Use your answer to part (h) to find the perimeter of the Sierpinski carpet. Explain your work. (By "perimeter," we mean the total distance around all of the filled in regions, not the distance around the outside only.)

(*HINT:* How must the total perimeter of all squares compare with the total perimeter of all new squares?)

4. *Finding the area of a Sierpinski carpet.* (See Exercise 2 in Section 8.8 for a description of this fractal.) (By "area," we mean the total area of all the filled in regions.)

a. Use your drawings in Exercise 2 of Section 8.8 to explain each of the calculations in the chart in Figure 8.161. These calculations are illustrated in Figure 8.160.

b. Use your drawings from Exercise 2 of Section 8.8 to complete the chart in Figure 8.161.

c. By what factor does the number of squares increase (starting at step 2)? Explain why your answer makes sense.

d. By what factor does the length of each side decrease? Explain why your answer makes sense.

e. By what factor does the area of one square change? Explain why your answer makes sense. Is the area of one square increasing or decreasing? Why?

Step	Number of New Squares*	Length of Each Side	Area of One New Square	Total Area of All New Squares*	Total area after Removing All New Squares*
1	1	1 ft	$(1 \text{ ft})^2$ = 1 sq ft	$1 \cdot 1$ sq ft = 1 sq ft	1 sq ft − 0 = 1 sq ft (Nothing is removed in step 1.)
2	1	$1/3 \cdot 1$ ft = 1/3 ft	$(1/3 \text{ ft})^2$ = 1/9 sq ft	$1 \cdot 1/9$ sq ft = 1/9 sq ft	1 sq ft − 1/9 sq ft = 8/9 sq ft
3	8				
4					
5					
6					

*"New square" refers to the squares that are both new to this step and that contribute to the area.

Figure 8.161
For Exercise 4

f. By what factor does the total area of all new squares change? Explain why your answer makes sense. Is the total area of all new squares increasing or decreasing? Why?

g. By what factor is the total area after removing all new squares changing? Is the total area after removing all new squares increasing or decreasing? Why?

h. Use your answers to parts (a)–(g) to find a formula for the total area after removing all new squares at step n.

i. Use your answer to part (h) and your calculator to find the area of the Sierpinski carpet. Explain your answer.

5. *The Mitsubishi Gasket.* (See Exercise 3 in Section 8.8 for a description of this fractal.)

a. Find a formula for the total perimeter of step n of the process described in Exercise 3, Section 8.8.

b. Use the formula from part (a) to find the perimeter of the Mitsubishi gasket.

c. Find a formula for the total area of step n of the above process.

d. Use the formula from part (c) to find the area of the Mitsubishi gasket.

6. *A square snowflake.* (See Exercise 4 in Section 8.8 for a description of this fractal.)

a. Find a formula for the total perimeter of step n of the process described in Exercise 4, Section 8.8.

b. Use the formula from part (a) to find the perimeter of a square snowflake.

c. Find a formula for the total area of step n of the above process.

(*HINT:* It's an incredibly easy formula. Don't make it hard.)

d. Use the formula from part (c) to find the area of a square snowflake.

7. *For those who have read Section 1.1: Deductive versus Inductive Reasoning.* In Example 4, we discussed the perimeter of the results of the first few steps of the process of creating a Koch snowflake, and we concluded that $P_4 = 64/9$ and $P_5 = 256/27$. In arriving at these conclusions, some of the reasoning was inductive and some was deductive. Which reasoning was inductive? Which was deductive? Why?

8. *For those who have read Section 8.5: Right Triangle Trigonometry.* In part (b) of Example 1, we found the base and height of a triangle. Discuss how to use the information developed in Section 8.5, under *Special Triangles,* to find this base and height.

━━━━━━▶ Answer the questions 9 and 10 using complete sentences and your own words.

CONCEPT QUESTIONS

9. Describe an application of fractal area, perimeter, and volume.

PROJECTS

10. *Design your own fractal.* Discuss its self-similarity and its recursive nature. Find its perimeter and area (or its surface area and volume, if appropriate).

11. *Approximating the area of a Koch snowflake.*

a. Use Example 7 to complete the chart in Figure 8.162. Include units.

b. Expand the chart to include steps 4–8.

c. Use your calculator and the "Total Area" entries in the chart you completed in part (b) to compute decimal approximations of A_1 through A_8. Round off to four decimal places.

d. So far, we haven't found the area of the Koch snowflake. Instead, we've found the areas of the first few steps of creating a Koch snowflake. Would the area of the Koch snowflake be larger or smaller than A_8? Why? Use the result of part (c) to estimate the area of the Koch snowflake.

Step	Number of New Triangles	Base of New Triangle	Height of New Triangle	Area of each New Triangle	Area of all New Triangles	Total Area
1	1	1 ft	$\frac{\sqrt{3}}{2}$ ft	$A_1 = \frac{1}{2} \cdot 1 \cdot \frac{\sqrt{3}}{2}$ $= \frac{\sqrt{3}}{4}$ sq ft	$1 \cdot \frac{\sqrt{3}}{4}$ sq ft $= \frac{\sqrt{3}}{4}$ sq ft	$A_1 = \frac{\sqrt{3}}{4}$ sq ft
2	3	$\frac{1}{3} \cdot 1$ ft $= \frac{1}{3}$ ft	$\frac{1}{3} \cdot \frac{\sqrt{3}}{2}$ ft $= \frac{\sqrt{3}}{6}$ ft	$\frac{1}{2} \cdot \frac{1}{3} \cdot \frac{\sqrt{3}}{6}$ $= \frac{\sqrt{3}}{36}$ sq ft	$3 \cdot \frac{\sqrt{3}}{36}$ sq ft $= \frac{\sqrt{3}}{12}$ sq ft	$A_2 = \frac{\sqrt{3}}{4}$ $+ \frac{\sqrt{3}}{12}$ sq ft
3						

Figure 8.162
For Exercise 11

Chapter 8 Review

TERMS

analytic geometry	degree	light-year (ly)	recursive process
angle of depression	diameter	line of symmetry	right triangle
angle of elevation	dimension	Menger sponge	seconds
approximate self-similarity	disk	minutes	self-similarity
area	ellipse	parabola	Sierpinski gasket
astronomical unit (AU)	exact self-similarity	parallel lines	similar triangles
center	foci	Parallel Postulate	sine
chaos game	focus	parallelogram	skewed Sierpinski gasket
circle	fractal	parsec (pc)	sphere
circumference	great circle	perimeter	surface area
circumscribed polygon	height	π (pi)	tangent
cone	hyperbola	polygon	trapezoid
congruent triangles	hypotenuse	pyramid	triangle
conic section	inscribed polygon	Pythagorean Theorem	trigonometry
cosine	Koch snowflake	radius	vertex
cylinder	leg	rectangle	volume

REVIEW EXERCISES

1. Find the area and perimeter of Figure 8.163.

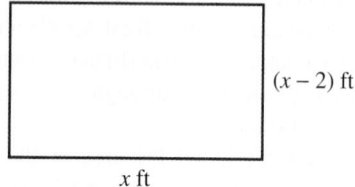

Figure 8.163
Diagram for Exercise 1

2. Find the area and perimeter of Figure 8.164.

Figure 8.164
Diagram for Exercise 2

3. Find the area and perimeter of Figure 8.165.

Figure 8.165
Diagram for Exercise 3

4. Find the area and perimeter of Figure 8.166.

Figure 8.166
Diagram for Exercise 4

5. Find the area and perimeter of Figure 8.167.

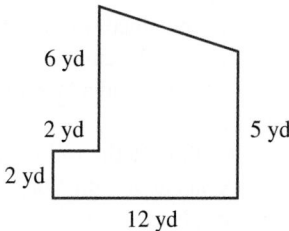

Figure 8.167
Diagram for Exercise 5

6. Saturn (the ringed planet) revolves around the sun in an elliptical orbit. Consequently, the distance between Saturn and the sun varies from approximately 840.4 million to 941.1 million miles.
   a. What is the (approximate) average distance from Saturn to the sun?
   b. On average, how far from the sun is Saturn in AU?
   c. Interpret the answer to part (b) in terms of the earth's distance from the sun.

7. Scientists believe that the star Vega (the brightest star in the constellation Lyra) is approximately 239.7 trillion kilometers from the earth.
   a. How far, in astronomical units, is Vega from the earth?
   b. How far, in light-years, is Vega from the earth?
   c. How far, in parsecs, is Vega from the earth?

8. Find the volume of Figure 8.168.

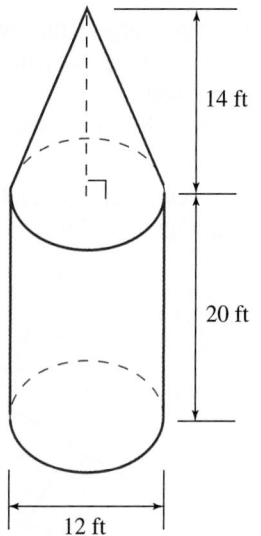

Figure 8.168
Diagram for Exercise 8

9. Find the volume and surface area of Figure 8.169.

Figure 8.169
Diagram for Exercise 9

10. Find the volume and surface area of Figure 8.170.

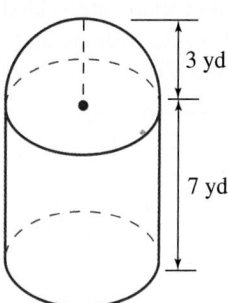

Figure 8.170
Diagram for Exercise 10

11. Find the volume and surface area of Figure 8.171.

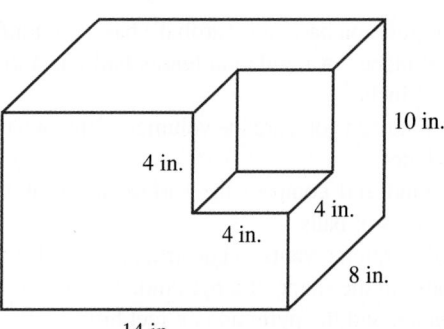

Figure 8.171
Diagram for Exercise 11

12. You jog $1\frac{3}{4}$ miles due south, then jog $\frac{1}{2}$ mile due west, and then return to your starting point via a straight-line path. How many miles have you run?

13. Tony Endres wants to fertilize his backyard. The yard is a triangle whose sides are 100 feet, 130 feet, and 160 feet. If one bag of fertilizer will cover 800 square feet, how many bags does he need?

14. An oval athletic field is the combination of a square and semicircles at opposite ends, as shown in Figure 8.172. If the total area of the field is 1,200 square yards, find the area of the square.

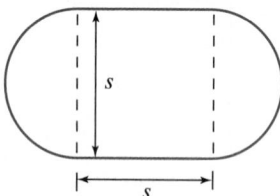

Figure 8.172
Diagram for Exercise 14

15. From a 12-inch-by-18-inch piece of cardboard, 2-inch-square corners are cut out as shown in Figure 8.173, and the resulting flaps are folded up to form an open box. Find (a) the volume and (b) the external surface area of the box.

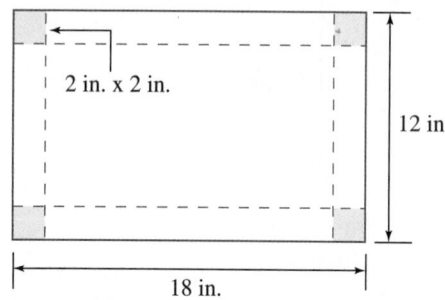

Figure 8.173
Diagram for Exercise 15

16. A regulation baseball (hardball) has a circumference of 9 inches. A regulation tennis ball has a diameter of $2\frac{1}{2}$ inches.
  a. Find and compare the volumes of the two types of balls.
  b. Find and compare the surface areas of the two types of balls.

17. Ted Nirgiotis wants to construct a plexiglass greenhouse in the shape of a pyramid. The base is 13 feet square, and the pyramid is 8 feet high.
  a. Find the volume of the greenhouse.
  b. Find the surface area of the greenhouse. (The floor is not included.)

18. For a circle of radius 1 cubit, do the following:
  a. Use $\pi = \frac{256}{81}$ (the Egyptian approximation of pi) to find the area of the circle.
  b. Use the value of π contained in a scientific calculator to find the area of the circle.
  c. Find the error of the Egyptian calculation relative to the calculator value.

19. For a sphere of radius 1 cubit, do the following.
  a. Use $\pi = \frac{256}{81}$ (the Egyptian approximation of pi) to find the volume of the sphere.
  b. Use the value of π contained in a scientific calculator to find the volume of the sphere.
  c. Find the error of the Egyptian calculation relative to the calculator value.

20. Which of the following structures has the larger storage capacity?
  a. a truncated pyramid 30 cubits high with a 6-cubit-by-6-cubit top and a 40-cubit-by-40-cubit bottom
  b. a regular pyramid 50 cubits tall with a 40-cubit-by-40-cubit base

21. Following the method of Problem 48 of the Rhind Papyrus, find a square, with a whole number as the length of its side, that has the same (approximate) area as a circle of diameter 18 units.

  HINT: Divide the sides of an 18-by-18 square into three equal segments.

22. Use the result of Exercise 21 to obtain an approximation of π.

23. Find the length of the longest object that will fit inside a rectangular trunk 5 feet long, 3.5 feet wide, and 2.5 feet high.

24. Find the length of the longest object that will fit inside a garbage can (a cylinder) that has a diameter of 2.5 feet and a height of 4 feet.

25. Use the diagrams in Figure 8.174 to prove the Pythagorean Theorem.

  HINT: Use the method of rearrangement and label the sides of each square in Figure 8.174(b).

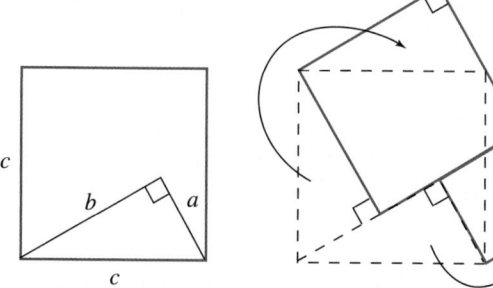

Figure 8.174
Diagram for Exercise 25

In Exercises 26 and 27, use the two-column method to prove the indicated result.

26. Given: $AD = CD$
$\qquad\quad AB = CB$
$\quad$ Prove: $AE = CE$

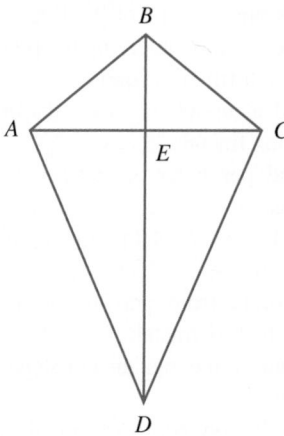

27. Given: $\angle CBA = \angle DAB$
$\qquad\quad BC = AD$
$\quad$ Prove: $\angle CAD = \angle DBC$

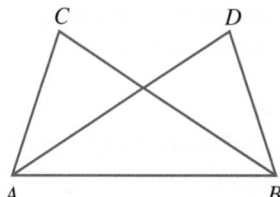

28. A regular octagon (eight-sided polygon) is circumscribed around a circle of radius r. The length of a side of the octagon is $s = 2(\sqrt{2} - 1)r$. Using the perimeter of the octagon as an approximation of the circumference of the circle, obtain an estimate of π.

29. A regular dodecagon (12-sided polygon) is inscribed in a circle of radius r. The length of a side of the dodecagon is $s = [(\sqrt{6} - \sqrt{2})/2]r$. Using the perimeter of the dodecagon as an approximation of the circumference of the circle, obtain an estimate of π.

30. Find the center and radius and sketch the graph of the circle given by

$$x^2 + y^2 - 2x - 2y + 1 = 0$$

31. Find the foci and sketch the graph of the hyperbola given by each of the following equations.
$\quad$ **a.** $9x^2 - 4y^2 = 36$ $\qquad$ **b.** $4y^2 - 9x^2 = 36$

32. A Mylar-coated parabolic reflector is used to solar-heat water. The disk is 8 feet wide and 2 feet deep. Where should the water container be placed to heat the water most rapidly?

33. Find the foci and sketch the graph of the ellipse given by

$$9x^2 + 4y^2 = 36$$

34. An elliptical room is designed to function as a whispering gallery. If the room is 40 feet long and 20 feet wide, where should two people stand to optimize the whispering effect?

In Exercises 35 and 36, draw and label a right triangle like the one shown in Figure 8.175, and use trigonometric ratios and a calculator to find the unknown sides and angles. Round off your answers to one decimal place.

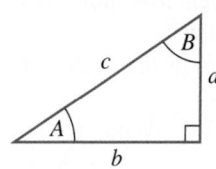

Figure 8.175
Diagram for Exercise 35 and 36

35. $A = 25.4°$ and $c = 56.1$ ft.

36. $a = 45.2$ cm and $b = 67.8$ cm

37. An electrician is at the top of a 50.0-foot utility pole. She measures the angle of depression of the base of another utility pole on the opposite side of a river. If the angle is $13.2°$, how far apart are the poles?

38. While sightseeing in Paris, you visit the Eiffel Tower. From an unknown distance, you measure the angle of elevation of the top of the tower. You then move 120 feet backwards (directly away from the tower) and measure the angle of elevation of the top of the tower. If the angles are $79.7°$ and $73.1°$, respectively, how tall (to the nearest foot) is the Eiffel tower?

39. Calculate the distance (in parsecs) from the sun to an object in space if the angle between the sun and the earth (as seen from the object) is 3 seconds and the distance between the earth and the sun is 1 AU.

40. Calculate the distance (in parsecs) from the sun to an object in space if the angle between the sun and the earth (as seen from the object) is $\frac{2}{3}$ second and the distance between the earth and the sun is 1 AU.

41. Suppose the distance from the sun to an object in space is 0.85 parsec and the distance between the earth and the sun is 1 AU.

a. How far from the sun is the object in terms of AU?

b. Calculate the angle between the sun and the earth, as seen from the object. Express your answer in terms of seconds, rounded off to the nearest hundredth of a second.

42. a. What is the Parallel Postulate?

b. What are the two alternatives if we assume the Parallel Postulate to be false?

c. What geometric models can be used to describe a geometry based on each of the alternatives listed in part (b)?

d. Draw a triangle using each of the geometric models listed in part (c).

In Exercises 43 and 44, answer each question for (a) Euclidean, (b) Lobachevskian, and (c) Riemannian geometry.

43. Through a point not on a given line, how many parallels to the given line are there?

44. What can be said about the sum of the angles of a triangle?

45. *The box fractal.* To create this shape, use the following steps:

1. Draw a square and fill in its interior.

2. Remove some of the square:
   - Subdivide the square into nine equal squares.
   - Remove the four small squares at the middle of each edge.

This leaves five filled in squares, one at each corner and one at the very center. See Figure 8.176.

Step 1   Subdivide the square into 9 equal squares   Remove the small squares at the middle of the edges

Figure 8.176
Creating a box fractal

3. Remove some of the five squares from step 2 by applying the procedure from step 2 to each filled in square.

4. Remove some of the squares from step 3 by applying the procedure from step 2 to each filled in square.

5 (and beyond). Remove some of the five squares from the previous step by applying the procedure from step 2 to each filled in square.

a. Carefully and accurately complete the first five steps of the box fractal.

b. Find the total perimeter of steps 1–5 of the above process.

c. Find a formula for the total perimeter of step n of the above process.

d. Use the formula from part (c) to find the perimeter of the box fractal.

e. Find a formula for the total area of step n of the above process.

f. Use the formula from part (e) to find the area of the box fractal.

g. Find the dimension of the box fractal.

h. Discuss the self-similarity of the box fractal.

i. Discuss the recursive nature of the box fractal.

Answer the following questions using complete sentences and your own words.

HISTORY QUESTIONS

46. What roles did the following people play in the development and application of geometry?
   - Archimedes of Syracuse
   - Euclid of Alexandria
   - Hypatia
   - Johann Kepler
   - Pythagoras of Samos
   - Girolamo Saccheri
   - Janos Bolyai
   - Carl Friedrich Gauss
   - Nikolai Lobachevsky
   - Bernhard Riemann
   - Thales of Miletus
   - Benoit Mandelbrot

9

Graph Theory

© Mark L. Stephenson/Corbis

In algebra you learned that the graph of $y = 3x + 2$ is a line with slope $m = 3$ and intercept $b = 2$. That graph is a diagram used to indicate the relationship between the variables x and y. This chapter is about a different type of graph, one that doesn't involve variables and equations.

In this chapter, we will use a graph to diagram a community and determine efficient postal routes. We will diagram airplane flight options and determine efficient routes for traveling. We will diagram office spaces and determine how to connect the cubicles with a computer network. Graph theory is used to schedule complex projects such as the Apollo moon landings. We will use it to schedule less complex projects, problems that we would encounter in everyday life.

Graph theory arose from the humblest of origins. Over 250 years ago, in the medieval European city of Konigsberg, residents tried to find a way to take an evening stroll, crossing each of the city's seven bridges once. They never succeeded, but graph theory was born in their failure.

9.1 | A Walk Through Konigsberg

A City's Dilemma

There used to be a country in Eastern Europe called Prussia. Its capitol was Konigsberg, where the Prussian king lived in a castle on the Pregel River. People often strolled through the city, enjoying its beauty.

Konigsberg, the Pregel River, and the king's castle

Konigsberg residents tried to find a way to tour the city, walking across each of the city's seven bridges only once. A map of the city and a diagram of an unsuccessful attempt are shown in Figure 9.1.

Figure 9.1
Konigsberg's seven bridges and a diagram of an unsuccessful walking tour

A walker who followed the route shown in Figure 9.1, starting at the upper left corner, would end up on the island. The walker has already crossed each of the island's bridges, so there is no way to get off the island and cross the remaining bridge in the lower right corner.

The City's Solution

The problem was solved by Swiss mathematician Leonhard Euler (pronounced "Oiler"). He started by simplifying the picture. He thought of the separate pieces of land as points and the bridges as lines or curves connecting the points. See Figure 9.2.

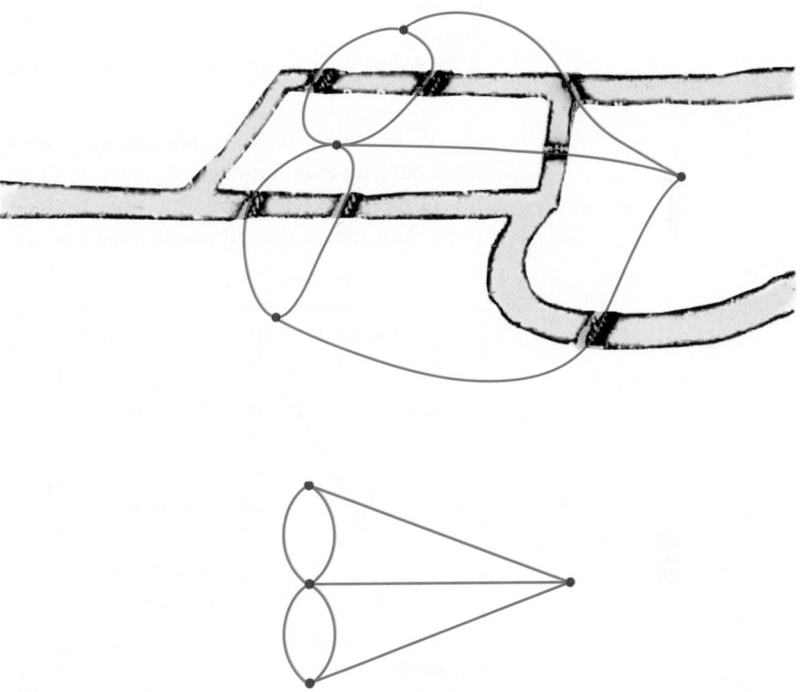

Figure 9.2
Euler's points and connecting lines, both with (above) and without (below) the river and bridges

Euler realized that when a walker starts the walk and crosses the first bridge, the walker uses up one connecting line. When the walker visits a point (or walks across a piece of land), the walker uses up two connecting lines (that is, two bridges)—one in approaching the point and one in departing the point. And when the walker crosses the last bridge, the walker uses up one connecting line.

This means that if the walker were to start and stop the walk at different places, the starting point must have an odd number of lines coming out of it: one line to start the walk and two lines (or four or six . . .) to revisit the point when crossing other bridges. And the ending point must have an odd number of lines coming out of it for the same reason. But all other points must have an even number of lines so that the walker can both approach the point and depart the point. See Figure 9.3 on page 650.

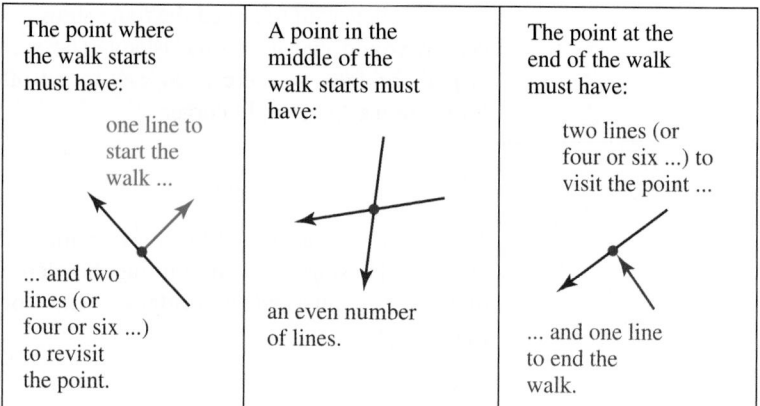

Figure 9.3
If the walker were to start and stop the walk at different places

If the walker were to start and stop the walk at the same point, then that point must have an even number of lines: one line to start the walk, an even number to revisit the point when crossing other bridges, and one line to return to the starting point. This means that all points must have an even number of lines. See Figure 9.4.

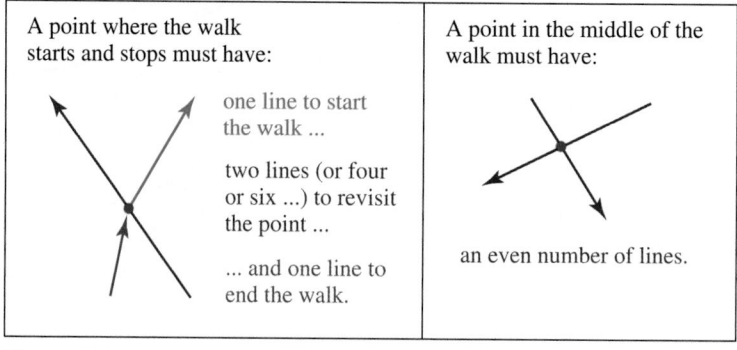

Figure 9.4
If the walker were to start and stop the walk at different places

In the graph of Konigsberg, every point has an odd number of lines. See Figure 9.5. This means that each point could be used as a starting point or as a stopping point. But none of the points could be visited without starting or stopping there. So there is no way to walk across each of Konigsberg's seven bridges once, much to the chagrin of its residents.

Graph Theory

In solving the Konigsberg puzzle, Euler started a new type of geometry called *graph theory*. Despite its lighthearted beginnings, graph theory has important applications in business, psychology, sociology, computer science, chemistry, and biology.

A **graph** is a diagram that consists of points, called **vertices** (the singular is **vertex**) and connecting lines, called **edges.** An edge that connects a vertex with itself is called a **loop.** Euler's simplified picture of Konigsberg is a graph that has edges and vertices but no loops. Figure 9.6 shows that graph, with an added loop.

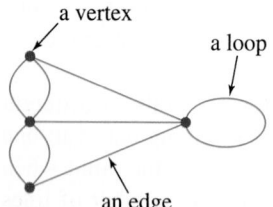

Figure 9.5
Each of Konigsberg's points has an odd number of lines

Figure 9.6
The Konigsberg graph with an added loop

Euler's graph representing Konigsberg and its bridges was rather novel at the time. Now, it's quite common to use a graph to represent a subway system or a rapid transit system. In such a graph, the vertices represent stations and the edges represent subway or rapid transit lines. See Figure 9.7.

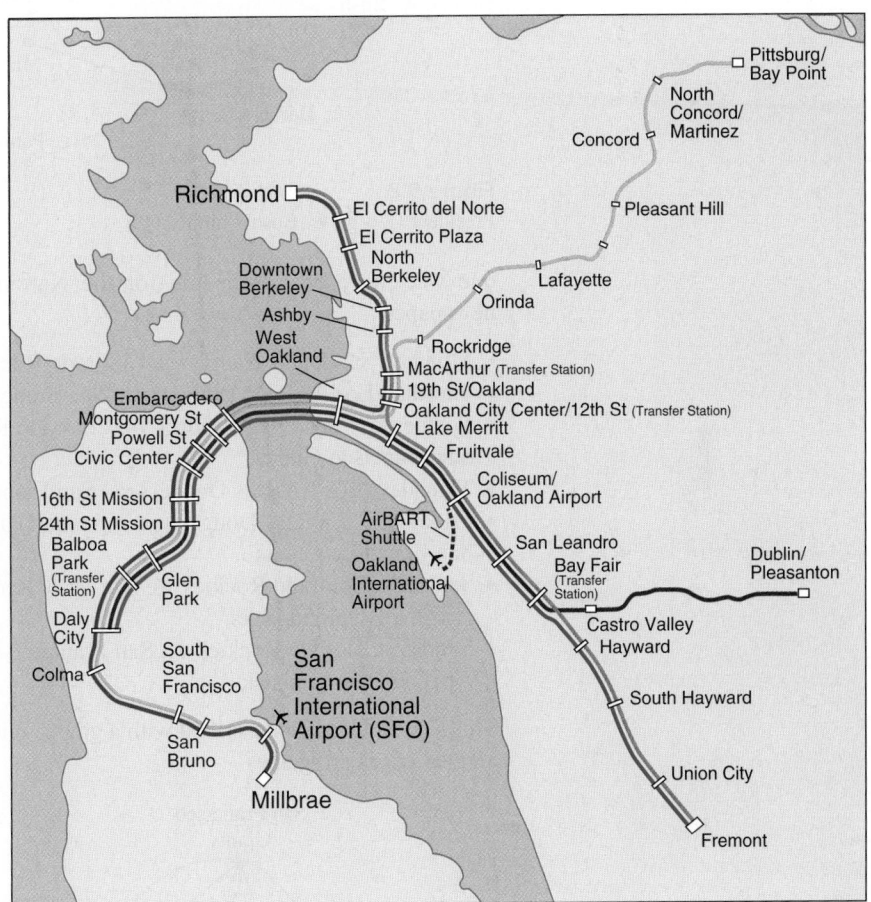

Figure 9.7
A graph of San Francisco's Bay Area Rapid Transit System (BART)

A graph lacks details that are common to most maps. For example, you can't look at the Konigsberg graph and tell how far it is from one bridge to another or what streets to follow to get to a bridge. And the shape of an edge does not necessarily resemble the shape of the connection that it represents. But these details don't matter to subway riders.

Graphs Describe Relationships

A family tree is a graph. Figure 9.8 on page 652 shows a part of President George W. Bush's family tree. Its vertices are family members, and its edges are parent/child relationships.

Graphs can be used to describe relationships *of any type*. The Konigsberg graph and the Bay Area Rapid Transit graph show geographic relationships. The Bush family tree graph shows familial relationships. Graphs can even show relationships such as which baseball teams are scheduled to play each other. Here is

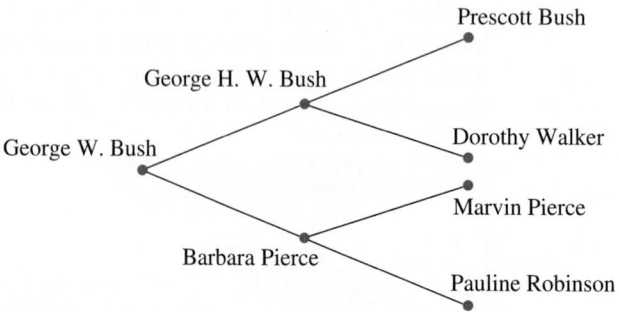

Figure 9.8
President George W. Bush's family tree

one week's baseball schedule for the National League West, right out of the newspaper:

- Monday: San Diego Padres vs. Colorado Rockies
- Tuesday: Los Angeles Dodgers vs. San Francisco Giants
- Wednesday: San Diego Padres vs. Colorado Rockies, Los Angeles Dodgers vs. San Francisco Giants
- Thursday: Los Angeles Dodgers vs. San Francisco Giants
- Friday: Los Angeles Dodgers vs. Arizona Diamondbacks, Colorado Rockies vs. San Francisco Giants
- Saturday: Colorado Rockies vs. San Francisco Giants, Los Angeles Dodgers vs. Arizona Diamondbacks
- Sunday: Colorado Rockies vs. San Francisco Giants, Los Angeles Dodgers vs. Arizona Diamondbacks

This schedule can be described with a graph. With this graph, the vertices are teams, and the edges are games.

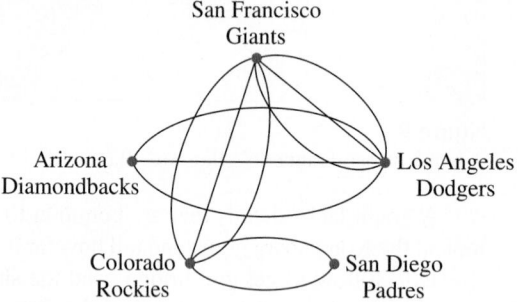

Figure 9.9
One week's baseball schedule

Notice that depending on what you want to know, Figure 9.9 can be easier to read than is the written schedule. For example, a glance at Figure 9.9 is sufficient to determine that the Giants play the Dodgers three times and the Dodgers don't play the Padres at all. You would have to inspect the written schedule rather closely to determine these things.

Also notice that in Figure 9.9 the edges cross each other at many places, only some of which are vertices. Some graphs have this quality. Others, like the Konigsberg graph, don't. It's important to understand that "vertex" isn't the same as "where two edges cross."

9.1 Exercises

In Exercises 1–3, use the following information: After much discussion, the Konigsberg city council decided that it would be a good thing if the city's residents succeeded in finding a bridge walk, that is, a walk that would take the walker across each of the city's bridges once. Council members reasoned that a bridge walk would foster civic pride and attract tourists. They voted to close one bridge if doing so would allow the creation of a bridge walk. (Note: The Konigsberg story presented earlier in this section is true. However, the details presented in the exercises are not.)

1. Is there a bridge whose closure would create a bridge walk if the walk's starting point had to be different from its stopping point? If your answer is "yes," specify the bridge, draw a graph of the new six-bridge system, and describe the walk. If your answer is "no," explain why.

2. Is there a bridge whose closure would create a bridge walk if the walk's starting point had to be the same as its stopping point? If your answer is "yes," specify the bridge, draw a graph of the new six-bridge system, and describe the walk. If your answer is "no," explain why.

3. Is there a bridge whose closure would not create a bridge walk? If so, specify the bridge and explain why a walk would not be created.

In Exercises 4–6, use the following information: After closing a bridge, as discussed in Exercises 1–3, a number of Konigsberg merchants complained to the city council that there was insufficient access to their stores. The council voted to reopen the closed bridge and to build a new bridge if doing so would allow the creation of a bridge walk.

4. Is there a site for a new bridge whose opening would create a bridge walk if the walk's starting point had to be the same as its stopping point? If your answer is "yes," specify the site, draw a graph of the new eight-bridge system, and describe the walk. If your answer is "no," explain why.

5. Is there a site for a new bridge whose opening would create a bridge walk if the walk's starting point had to be different from its stopping point? If your answer is "yes," specify the site, draw a graph of the new eight-bridge system, and describe the walk. If your answer is "no," explain why.

6. Is there a site for a new bridge whose opening would not create a bridge walk? If so, specify the site and explain why a walk would not be created.

In Exercises 7–10, determine the number of vertices, edges, and loops in the given graph.

7.

8.

9.

10.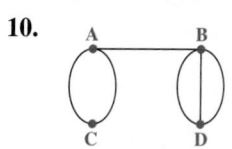

11. **a.** Determine the number of vertices, edges and loops in the Bush family tree graph in Figure 9.8.
    b. Could a family tree graph ever have a loop? Why or why not?

12. **a.** Determine the number of vertices, edges and loops in the baseball schedule graph in Figure 9.9.
    b. Could a baseball schedule graph ever have a loop? Why or why not?

13. Draw a graph with three vertices and six edges, one of which is a loop.

14. Draw a graph with four vertices and eight edges, two of which are loops.

15. **a.** Why do the two diagrams in Figure 9.10 represent the same graph?
    b. Draw another representation of the same graph that looks significantly different from either of these.

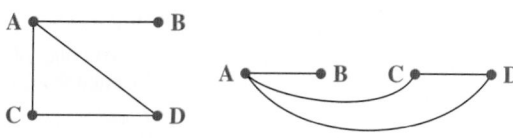

Figure 9.10
Graphs for Exercise 15

16. **a.** Why do the two diagrams in Figure 9.11 on page 654 represent the same graph?
    b. Draw another representation of the same graph that looks significantly different from either of these.

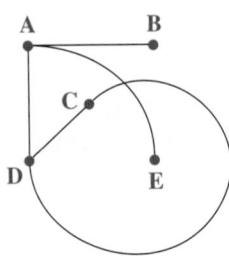

Figure 9.11
Graphs for Exercise 16

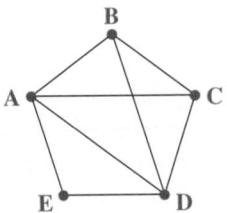

Figure 9.13
A graph for Exercise 18

17. Draw another representation of the graph in Figure 9.12 that looks significantly different.

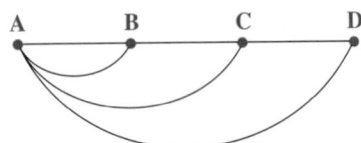

Figure 9.12
A graph for Exercise 17

18. Draw another representation of the graph in Figure 9.13 that looks significantly different.

CONCEPT QUESTIONS

19. The Konigsberg graph shows the relationship between bridges and landmasses. The Bay Area Rapid Transit graph shows the relationship between transit stations and transit lines. The Bush family tree graph shows familial relationships. The baseball schedule graph shows which baseball teams are scheduled to play each other. Give an example of another type of relationship that could be displayed with a graph, and describe how that relationship could be displayed.

PROJECTS

20. Find some information in a newspaper, magazine, or other similar source, and present that information as a graph.

9.2 Graphs and Euler Trails

Terminology

Before going any further, we need to discuss some of the terms that are used in studying graphs.

Two vertices are **adjacent** if they are joined by an edge. In Figure 9.14, the Arizona Diamondbacks and Los Angeles Dodgers vertices are adjacent vertices, because they are joined by an edge. Notice that the Arizona Diamondbacks and San Francisco Giants vertices are not adjacent, even though they are next to each other on the graph.

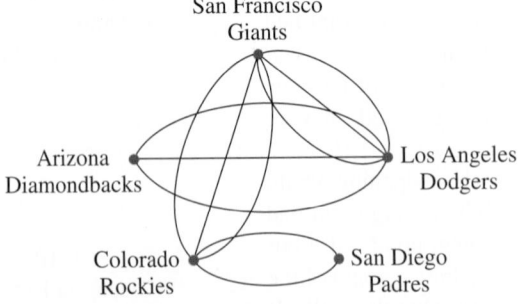

Figure 9.14
One week's baseball schedule

Two edges are **adjacent** if they have a vertex in common. In Figure 9.14, the two edges coming out of the San Diego Padres vertex are adjacent edges, because they have the San Diego Padres vertex in common.

The graph in Figure 9.15 has a different appearance than does the graph in Figure 9.14. However, it describes the same schedule of games. Edges that are adjacent in Figure 9.14 are adjacent in Figure 9.15, and vertices that are adjacent in Figure 9.14 are adjacent in Figure 9.15. For this reason, the graphs in Figures 9.14 and 9.15 are said to be **identical.**

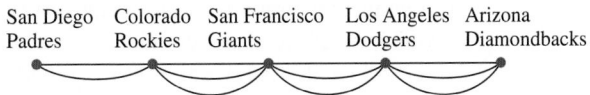

San Diego   Colorado   San Francisco   Los Angeles   Arizona
Padres   Rockies   Giants   Dodgers   Diamondbacks

Figure 9.15
A different description of the same schedule, with an identical graph

A **trail** is a sequence of adjacent vertices and the distinct edges that connect them. Trails are most easily described by listing the distinct edges in order. Figure 9.16 shows the Konigsberg graph, with the edges (that is, the bridges) labeled. One trail is a, d. That trail describes a walk that starts at the upper bank, crosses a bridge (edge a) to the island, and then crosses another bridge (edge d) to the lower bank.

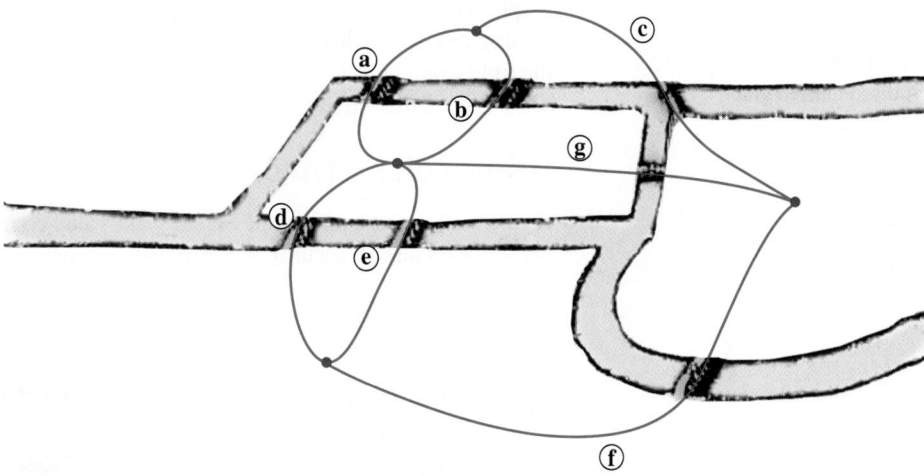

Figure 9.16
The Konigsberg graph with the edges labeled

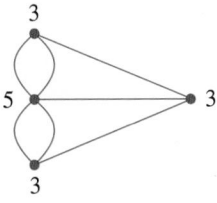

Figure 9.17
The vertices' degrees

A **circuit** is a trail that begins and ends at the same vertex. One circuit is a, d, f, c.

The **degree of a vertex** is the number of edges that connect to that vertex. A loop connects to a vertex twice, so a loop contributes to the degree twice. An **odd vertex** is a vertex with an odd degree, and an **even vertex** is a vertex with an even degree. Figure 9.17 shows each vertex's degree in the Konigsberg graph.

An **Euler trail** is a trail that travels through *every* edge of a graph *exactly once.* An **Euler circuit** is an Euler trail that is a circuit.

Euler's Theorem

In Section 9.1, we stated, "In the graph of Konigsberg, every point has an odd number of lines . . . [so] each point could be used as a starting point or as a stopping point. But none of the points could be visited without starting or stopping there. So there is no way to walk across each of Konigsberg's seven bridges once." The same statement, phrased with the terminology of graph theory, is "The Konigsberg graph has more than two odd vertices, so that graph has no Euler trails and no Euler circuits."

If we generalize from the above statement, we obtain Euler's Theorem. This was Euler's conclusion, after studying the Konigsberg Bridge problem.

Euler's Theorem

- A connected graph with *only even vertices* has at least one Euler trail, which is also an Euler circuit.
- A connected graph with *exactly two odd vertices* and any number of even vertices has at least one Euler trail. Each of these trails will start at one odd vertex and end at the other odd vertex.
- A graph with *more than two odd vertices* has no Euler trails and no Euler circuits.
- It's impossible for a connected graph to have only one odd vertex.

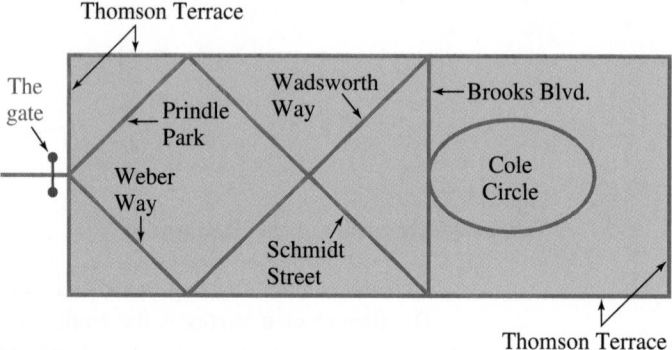

Figure 9.18
A disconnected graph

Euler's Theorem refers to *connected graphs*. A graph is **connected** if every pair of vertices is connected by a trail. Figure 9.18 shows a **disconnected graph,** or a graph that is not connected. This graph is disconnected because there is no trail connecting any of the three vertices on the left triangle with any of the three vertices on the right triangle. It's easy to see that disconnected graphs never have Euler trails and never have Euler circuits.

Using Euler's Theorem

EXAMPLE 1 Figure 9.19 shows a map of the Serene Oaks gated community. After several burglaries, the residents hired a security guard to walk through the subdivision once every night.

Figure 9.19
The Serene Oaks gated community

a. Draw a graph of the neighborhood, where vertices are corners or intersections and edges are streets. In that graph, show the degree of each vertex.
b. The guard, who is not paid by the hour, would prefer to park his car, walk a route that allows him to walk through every block just once, and then return to his car. Is this possible?

Solution a. The map itself is almost a graph. It lacks only vertices. The graph is shown in Figure 9.20. Notice that the vertex that is the intersection of Brooks Blvd. and Cole Circle is of degree 4. A loop connects to a vertex twice, so a loop contributes to the degree twice.

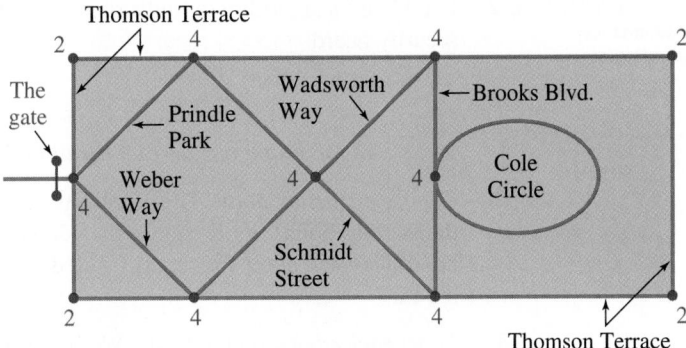

Figure 9.20
A graph of the Serene Oaks neighborhood, showing vertices and their degrees

b. This is a connected graph with only even vertices, so it has at least one Euler trail, which is also an Euler circuit. This means that it is possible for the guard to walk a route that allows him to walk through every block just once.   ■

Using Fleury's Algorithm to Find Euler Trails and Euler Circuits

With some graphs, it's easy to find an Euler trail or an Euler circuit, once you have used Euler's Theorem to verify that the graph has such a trail or circuit. However, there are some traps that can be easily avoided, if you know what to look for. *Fleury's algorithm* describes those traps and specifies how to avoid them. An **algorithm** is a logical step-by-step procedure, or recipe, for solving a problem. Fleury's algorithm is a logical step-by-step procedure for finding Euler circuits and Euler trails.

Fleury's Algorithm for finding Euler Trails and Euler Circuits

1. *Verify that the graph has an Euler trail or Euler circuit,* using Euler's Theorem.

2. *Choose a starting point.*
   - If the graph has two odd vertices, we can find an Euler trail. Pick either of the odd vertices as the starting point.
   - If the graph has no odd vertices, we can find an Euler circuit. Pick any point as the starting point.

3. *Label each edge alphabetically* as you travel that edge.

4. When choosing edges:
   - *Never choose an edge that would make the yet-to-be-traveled part of the graph disconnected,* because you won't be able to get from one portion of the graph to the other.
   - *Never choose an edge that has already been followed,* since you can't trace any edges twice in Euler trails and Euler circuits.
   - *Never choose an edge that leads to a vertex that has no other yet-to-be-traveled edges,* because you won't be able to leave that vertex.

ThomsonNOW™
for Liberal Arts Mathematics
www.thomsonedu.com/1pass.

EXAMPLE 2 Use Fleury's algorithm to find a route through the Serene Oaks neighborhood for the security guard.

Solution

1. *Verify that the graph has an Euler trail or Euler circuit.* We did this in Example 1.

2. *Choose a starting point.* The graph has no odd vertices, so we can pick any point as the starting point. We'll choose the corner at the lower left corner of the neighborhood. This will allow the security guard to drive through the gate, turn right and park his car.

3. *Label each edge alphabetically.* We'll start by following Thomson Terrace along the perimeter of the neighborhood, from edge a to b to c, and so on. See Figure 9.21.

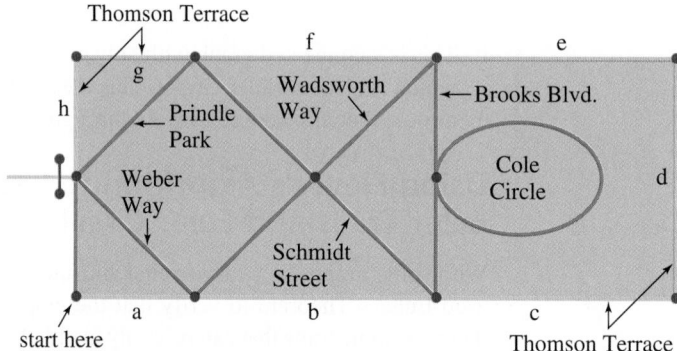

Figure 9.21
Edges a through h

It's only when we get through edge h and arrive at the vertex that is the intersection of Thomson Terrace, Prindle Park, and Weber Way that we have to choose an edge with step 4 in mind. If we continue along Thomson Terrace, we will arrive at a vertex (our starting part, actually) that *has no yet-to-be traveled edges.* That route is prohibited by Fleury's algorithm, so we will take Weber Way instead.

At the next vertex, we must take Wadsworth Way rather than Thomson Terrace, because *we cannot choose an edge that has already been traveled.* For the same reason, after turning onto Schmidt Street, we must choose Brooks Blvd. rather than Thomson Terrace. See Figure 9.22.

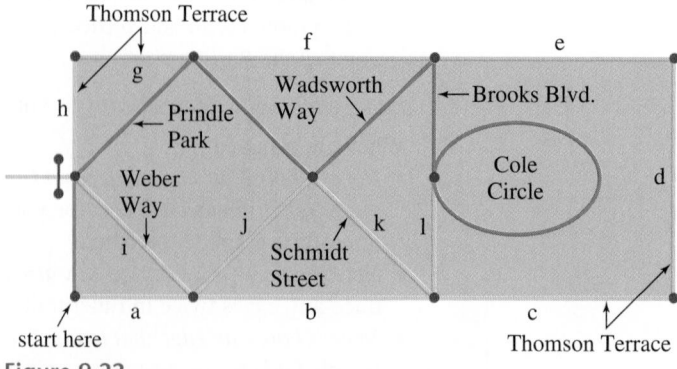

Figure 9.22
Edges a through l

Now we're at the intersection of Brooks Blvd. and Cole Circle. If we continue on Brooks Blvd., we would be choosing an edge that would *make the yet-to-be-traveled part of the graph disconnected.* Specifically, Cole Circle would be disconnected from the rest of the yet-to-be-traveled part of the graph. See Figure 9.23.

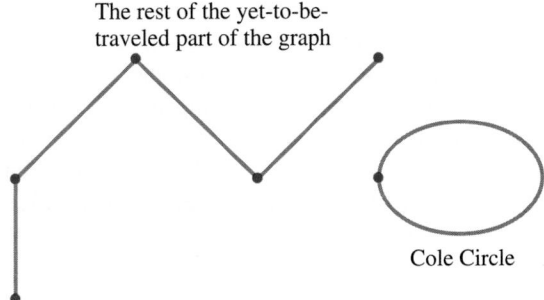

Figure 9.23
Choosing Brooks Blvd. would make the yet-to-be-traveled part of the graph disconnected.

Continuing on Brooks Blvd. is prohibited by Fleury's algorithm, so we will take Cole Circle instead. See Figure 9.24.

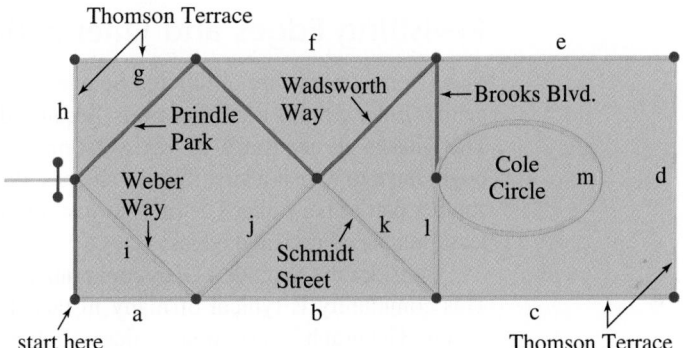

Figure 9.24
Edges a through m

The remaining edge choices are straightforward. The finished Euler circuit is shown in Figure 9.25.

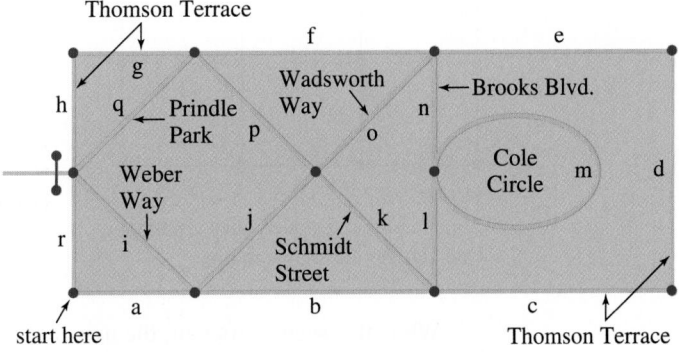

Figure 9.25
The complete Euler circuit

Applications of Euler Trails and Circuits

Euler trails and circuits have many practical applications. Any time services must be delivered along streets, Euler's theorem and Fleury's algorithm can make the job more efficient. The post office must deliver mail six days a week. Utility companies must read meters every month. Garbage companies must pick up garbage every week. Inefficient routes create unnecessary expenses for these firms—expenses that must be passed on to their customers: you and me.

Notice that a utility company must visit each side of the street, because each house has its own meter. This creates a different type of graph. Figure 9.26 shows the effect of this on the lower left part of the Serene Oaks gated community. The right-hand graph has **multiple edges,** which are edges that connect the same two vertices.

If each street must be visited once.

If each *side* of the street must be visited once.

Figure 9.26

Revisiting Edges and Eulerization

Many communities are unlike the Serene Oaks gated community in that they have neither Euler circuits nor Euler trails, because they have more than two odd vertices. This makes it necessary to revisit some edges. For example, a postal carrier might have to revisit a street where she has already delivered the mail, just to get to another part of her route. The most efficient trails and circuits are those that have the least number of revisited edges.

Consider the Secluded Glen community, whose graph is shown in Figure 9.27. This community is typical of many in that it is laid out like a grid, with parallel streets. The graph has six odd vertices, shown in blue, so it has no Euler circuit.

The goal here is to create an efficient route, so we want to add as few new edges as possible. The most efficient thing to do would be to add three new edges, each connecting a pair of odd vertices. We have done this in Figure 9.28.

   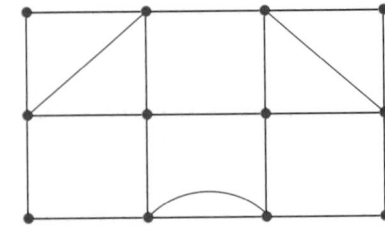

Figure 9.27
A graph of the Secluded Glen community. Odd vertices are shown in blue.

Figure 9.28
A graph of the Secluded Glen community, with three new edges

While this seems efficient, the new diagonal edges require the postal carrier to leave the street and cut through people's backyards, which is not feasible. When we add edges, we can add only edges that are duplicates of existing edges.

One possible graph is shown in Figure 9.29. The added edges (shown in blue) turn all of the odd vertices into even vertices, so the resulting graph has an Euler circuit. The added edges look as though we're requiring the postal carrier to leave the street and cut through somebody's front yard. Instead, they mean that the postal carrier is to revisit that particular edge. That is, she is to walk that block a second time.

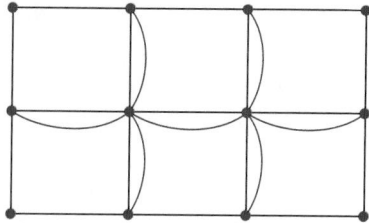

Figure 9.29
A graph of the Secluded Glen community with some added edges

This approach is acceptable in that all added edges are duplicates of existing edges. However, it has seven added edges. Wherever the postal carrier starts and stops, this plan would require her to walk seven blocks a second time. Let's try to find a more efficient plan.

The process of adding duplicate edges so that all odd vertices are turned into even vertices is called **Eulerization** (or **Eulerizing** a graph) because it results in a graph that has an Euler circuit. Figure 9.29 shows one Eulerization of the graph shown in Figure 9.27.

Whenever a graph is laid out like a grid, with parallel edges, the odd vertices are always along the outer perimeter of the graph. The following Eulerization algorithm results in efficient Eulerizations because it recognizes that fact and adds duplicate edges along the outer perimeter only.

Eulerization Algorithm

To efficiently Eulerize a graph that is laid out like a grid:

1. Choose a vertex along the outer perimeter of the graph.
2. If a vertex is an even vertex, add no edges and move on to the next vertex along the outer perimeter.
3. If a vertex is an odd vertex, add a duplicate edge that connects it to the next adjacent vertex along the outer perimeter, and move on to the next vertex along the outer perimeter.
4. Repeat these steps until you return to the vertex in Step 1.

EXAMPLE 3   **a.** Apply the Eulerization algorithm to the Secluded Glen community.
b. Use the result of part (a) to find an efficient route through the community for the postal carrier. All mailboxes are on one side of the street.

Solution   **a.** *Applying the Eulerization algorithm.*

1. *Choose a vertex along the outer perimeter of the graph.* We'll choose the vertex in the upper left corner.

2. *If a vertex is an even vertex, add no edges and move on to the next vertex along the outer perimeter.* The vertex in the upper left corner is indeed an even vertex, so we add no edges and move to the right to the next vertex.

3. *If a vertex is an odd vertex, add a duplicate edge that connects it to the next adjacent vertex along the outer perimeter.* The second vertex is an odd vertex, so we add a duplicate edge that connects it to the next vertex. See Figure 9.30 on page 662.

4. *Repeat these steps until you return to the vertex in Step 1.* This results in the graph shown in Figure 9.31.

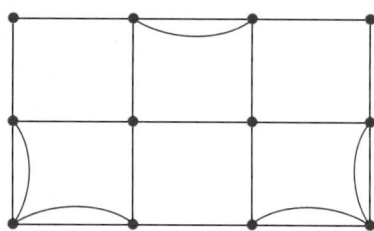

Figure 9.30
Starting the Eulerizing of the Secluded Glen community

Figure 9.31
After Eulerizing the Secluded Glen community

This scheme has five added edges. Wherever the postal carrier starts and stops, this scheme would require her to walk five blocks a second time. This is the most efficient scheme.

b. *Using the result to find an efficient route through the community for the postal carrier.* We'll do this with Fleury's algorithm. The result is shown in Figure 9.32.

Remember that the added edges are supposed to be duplicates of existing edges. For example, edge q is supposed to be a duplicate of edge b. We draw them this way so that we can tell when we're traveling an edge for the second time, and so we can apply Euler's theorem and Fleury's algorithm.

In Figure 9.33 we show a map of the Secluded Glen community without the added edges but with street names.

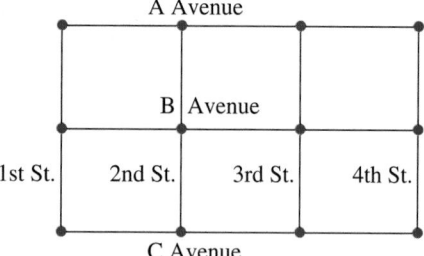

Figure 9.32
Applying Fleury's algorithm to Figure 9.31

Figure 9.33
Applying Fleury's algorithm to Figure 9.32

The beginning of the route described in Figure 9.32 with the letters a, b, c, and so on translates like this:

- Start at the corner of A Avenue and 1st Street.
- Go along A Avenue until you come to 4th Street.
- Turn onto 4th Street.
- Go along 4th Street until you come to C Avenue.
- Turn onto C Avenue.
- Go along C Avenue until you come to 3rd Street.
- Turn onto 3rd Street.
- Go along 3rd Street until you come to B Avenue.
- Turn onto B Avenue and go towards 4th Street.
- Turn onto 4th Street and go towards C Avenue. You are traveling this block for the second time.
- Turn onto C Avenue. You are traveling this block for the second time. ∎

9.2 Exercises

In Exercises 1–6, do the following:

a. *Find two adjacent edges.*
b. *Find two adjacent vertices.*
c. *Find the degree of each vertex.*
d. *Determine whether the graph is connected.*

1.

2.

3.

4.

5.

6.

7. Draw a graph with four vertices in which:
   a. each vertex is of degree 2.
   b. each vertex is of degree 3.
   c. two vertices are of degree 2 and two are of degree 3.

8. Draw a graph with five vertices in which:
   a. each vertex is of degree 2.
   b. each vertex is of degree 4.
   c. one vertex is of degree 2 and four are of degree 3.

9. Draw a graph with six vertices in which:
   a. each vertex is of degree 1.
   b. each vertex is of degree 0.

10. Draw a graph with eight vertices in which each vertex is of degree 3.

11. Draw a graph with six vertices in which each is of degree 3 and that has:
   a. loops but no multiple edges.
   b. multiple edges but no loops.
   c. neither loops nor multiple edges.
   d. both loops and multiple edges.

12. Draw a graph with five vertices in which each is of degree 4 and that has:
   a. loops but no multiple edges.
   b. multiple edges but no loops.
   c. neither loops nor multiple edges.
   d. both loops and multiple edges.

In Exercises 13–18, do the following.

a. *Use Euler's theorem to determine whether the graph has an Euler trail or an Euler circuit.*
b. *If it does, find it. If it does not, say why. Then Eulerize the graph if possible.*

13. The graph in Exercise 1
14. The graph in Exercise 2
15. The graph in Exercise 3
16. The graph in Exercise 4
17. The graph in Exercise 5
18. The graph in Exercise 6
19. Find an Euler circuit for the Serene Oaks gated community other than that created in Example 2.
20. Find an Eulerization for the Secluded Glen community other than that created in Example 3.
21. Once a month, the utility company sends an agent through the Serene Oaks gated community (see Figure 9.19) to read the meters. Each house has its own meter. Find a route for the meter reader:
   a. if there are houses on both sides of each street.

b. if there are houses on both sides of each street except that there is a greenbelt consisting of lawns and playgrounds along the outer edge of the community, outside of Thomson Terrace.

c. there are houses on both sides of each street except that the inside of Cole Circle is a park.

22. Once a month, the utility company sends an agent through the Secluded Glen community (see Figure 9.33) to read the meters. Each house has its own meter. Find a route for the meter reader:

a. if there are houses on both sides of each street.

b. if there are houses on both sides of each street except that there is a greenbelt consisting of lawns and playgrounds along the outer edge of the community, outside of A Avenue, C Avenue, First Street, and Fourth Street.

c. if there are houses on both sides of each street except that the area bounded by A Avenue, B Avenue, First Street, and Second Street is a park.

23. Find an efficient route through the community shown in Figure 9.34 for the postal carrier if mailboxes are all on one side of each street.

A Avenue

B Avenue

1st St.　2nd St.　3rd St.　4th St.　5th St.

C Avenue

Figure 9.34
A map for Exercises 23 and 24

24. Once a month, the utility company sends an agent through the community shown in Figure 9.34 to read the meters. Each house has its own meter. Find a route for the meter reader:

a. if there are houses on both sides of each street.

b. if there are houses on both sides of each street except that there is a greenbelt consisting of lawns and playgrounds along the outer edge of the community, outside of A Avenue, C Avenue, First Street, and Fifth Street.

c. if there are houses on both sides of each street except the area bounded by A Avenue, B Avenue, First Street, and Second Street is a park.

25. Figure 9.35 shows a map of New York City's bridges and tunnels.

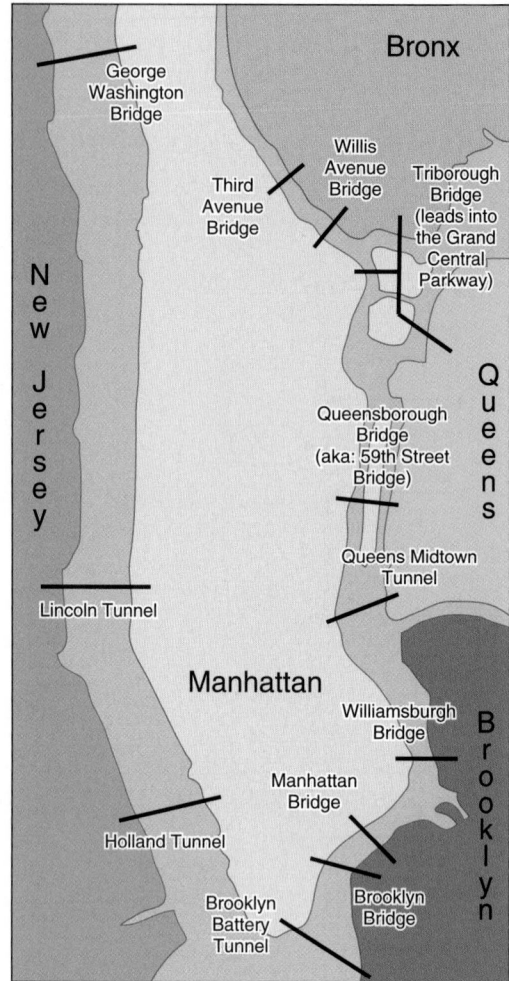

Figure 9.35
New York City's Bridges and Tunnels

a. Draw a graph of this. Use "Manhattan," "the Bronx," "Queens," "Brooklyn," and "New Jersey" as vertices. Consider the Triborough bridge to be three separate bridges: one from Manhattan to the Bronx, one from Manhattan to Queens, and one from Queens to the Bronx.

b. Does New York have a bridge and tunnel drive, that is, a drive that would traverse each of the city's bridges and tunnels once if the drive's starting point must be the same as its stopping point? If your answer is "yes," describe the drive. If your answer is "no," explain why. Then Eulerize the graph if possible.

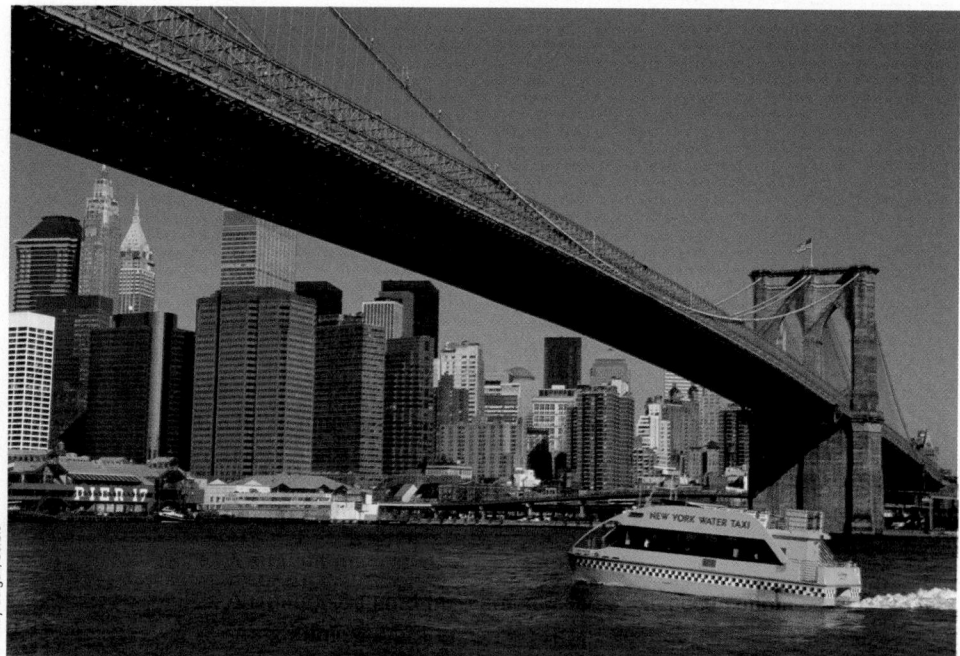

New York's Brooklyn Bridge

c. Does New York have a bridge and tunnel drive, that is, a drive that would traverse each of the city's bridges and tunnels once if the drive's starting point must be different from its stopping point? If your answer is "yes," describe the drive. If your answer is "no," explain why. Then Eulerize the graph if possible.

26. Figure 9.36 shows a map of the San Francisco Bay Area's bridges.

a. Draw a graph of this. Use "the north bay," "the east bay," and "the peninsula" as vertices.

b. Does the Bay Area have a bridge drive, that is, a drive that would traverse each of the area's bridges once if the drive's starting point must be the same as its stopping point? If your answer is "yes," describe the drive. If your answer is "no," explain why. Then Eulerize the graph if possible.

c. Does the Bay Area have a bridge drive, that is, a drive that would traverse each of the area's bridges once if the drive's starting point must be different from its stopping point? If your answer is "yes," describe the drive. If your answer is "no," explain why. Then Eulerize the graph if possible.

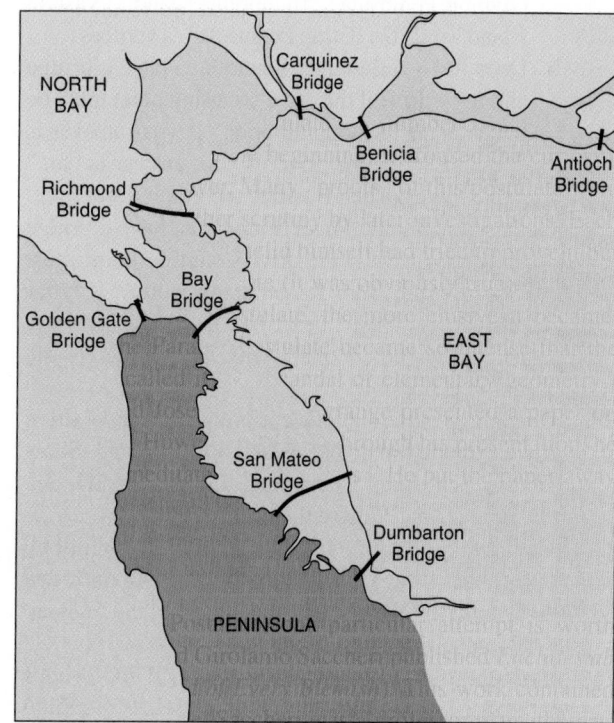

Figure 9.36
The San Francisco Bay Area's bridges

San Francisco's Golden Gate Bridge

27. Figure 9.37 shows a map of the bridges and ferry routes in the Seattle/Puget Sound area. These ferries take cars as well as passengers.
   a. Draw a graph of this. Use "mainland," "Whidbey Island," "Vancouver Island," Orcas Island," "Shaw Island," "San Juan Island," "Lopez Island," "the Olympic Peninsula," "Bainbridge Island," "Vashon Island,"and "the Kitsap Peninsula" as vertices.
   b. Does the Seattle/Puget Sound area have a bridge and ferry drive if the drive's starting point must be the same as its stopping point? If your answer is "yes," describe the drive. If your answer is "no," explain why. Then Eulerize the graph if possible.
   c. Does the Seattle/Puget Sound area have a bridge and ferry drive if the drive's starting point must be different from its stopping point? If your answer is "yes," describe the drive. If your answer is "no," explain why. Then Eulerize the graph if possible.
28. Does the Seattle area have a bridge and ferry drive if we exclude Vancouver Island, Orcas Island, Shaw Island, San Juan Island, and Lopez Island? See Exercise 27 and Figure 9.37.
   a. Draw a graph. Use "mainland," "Whidbey Island," "the Olympic Peninsula," "Bainbridge Island," "Vashon Island," and "the Kitsap Peninsula" as vertices.
   b. Is there a bridge and ferry drive, if the drive's starting point must be the same as its stopping point? If your answer is "yes," describe the drive. If your answer is "no," explain why. Then Eulerize the graph, if possible.

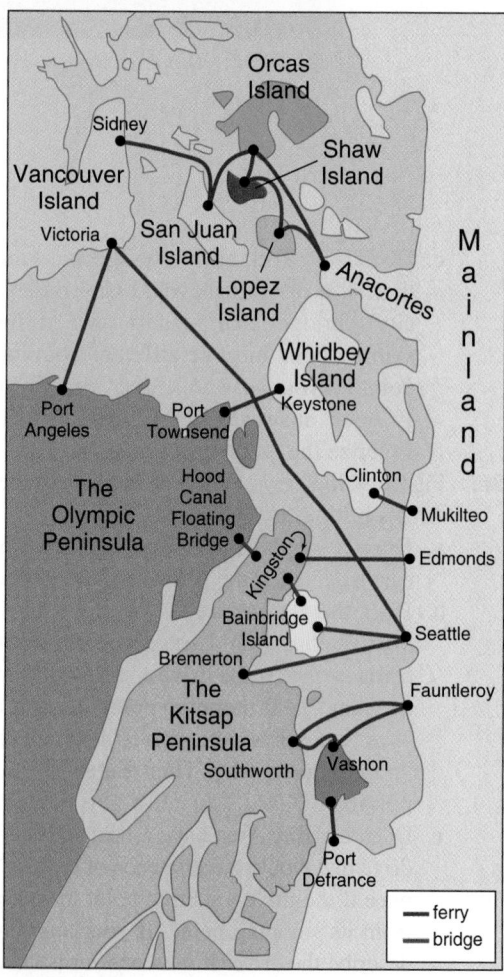

Figure 9.37
The Seattle/Puget Sound area

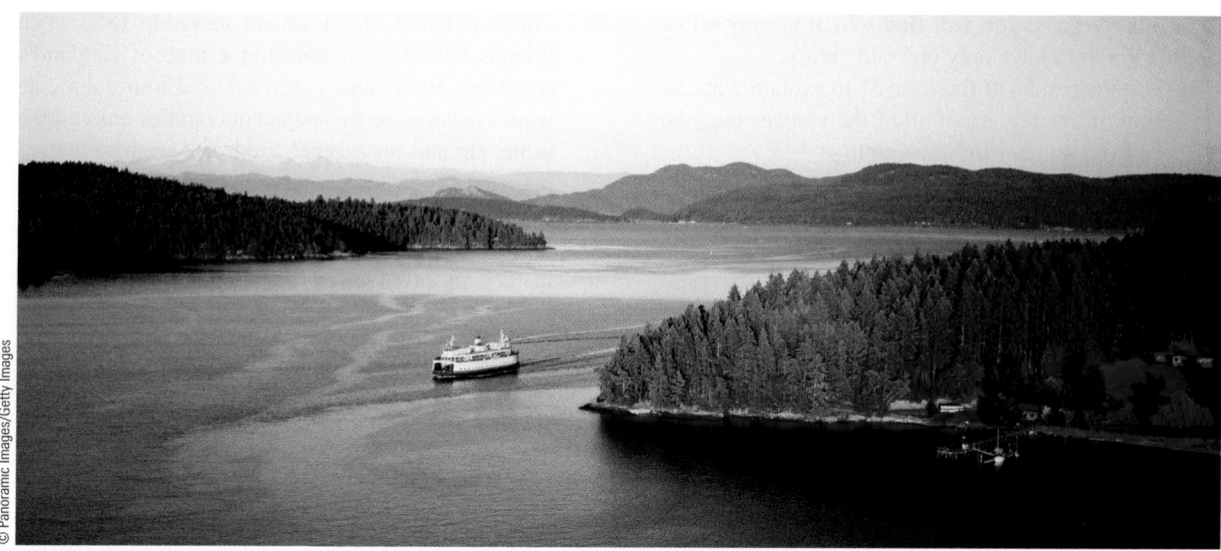

A ferry in the Puget Sound

c. Is there a bridge and ferry drive if the drive's starting point must be different from its stopping point? If your answer is "yes," describe the drive. If your answer is "no," explain why. Then Eulerize the graph, if possible.

29. Figure 9.7 on page 651 shows a map of the BART (Bay Area Rapid Transit) system. On that map, different rapid transit lines are colored differently. For example, the rapid transit line between Millbrae and Pittsburg/Bay Point is colored yellow.

a. Draw a graph of this. If two adjacent stations have three different lines connecting them, then those stations' vertices should have three different edges connecting them.

b. Is there a BART ride if the drive's starting point must be the same as its stopping point? If your answer is "yes," describe the drive. If your answer is "no," explain why. Then Eulerize the graph if possible.

c. Is there a BART ride if the drive's starting point must be different from its stopping point? If your answer is "yes," describe the drive. If your answer is "no," explain why. Then Eulerize the graph if possible.

d. Explain why the existence of a BART ride would be important to BART police, who regularly ride BART to provide security.

e. Which type of BART ride would be more useful to BART police: the type described in part (b) or the type described in part (c)? Why?

30. Is there a BART ride (see Exercise 29) if we disregard the difference between different rapid transit lines (that is, if adjacent stations have only one edge connecting them)? If your answer is "yes," describe the drive. If your answer is "no," explain why.

31. In this exercise, you will find the relationship between the number of edges in a graph and the sum of the degrees of all of the vertices.

a. Complete the chart in Figure 9.38.

Graph	Number of edges in that graph	Sum of the degrees of all of the vertices
Figure 9.15	11	2 + 5 + 6 + 6 + 3 = 22
Figure 9.17		
Figure 9.20		

Figure 9.38
Chart for Exercise 31

b. Using the result of part (a), make a guess as to the relationship between the number of edges in a graph and the sum of the degrees of all of the vertices.

c. How many vertices must an edge connect?

d. How much does an edge contribute to the degree of each of the vertices it connects?

e. Use the results of parts (c) and (d) to explain why your guess in part (a) must be correct.

32. In this exercise, you will find why it is impossible for a graph to have only one odd vertex.
   a. Use the results of Exercise 31 to explain why the sum of the degrees of all of the vertices must be an even number for any graph.
   b. Use the results of part (a) to explain why a graph must have an even number of odd vertices.

WEB PROJECTS

33. There are many interesting problems in graph theory. Some of these problems have been solved, and some remain unsolved. Many of these problems are discussed on the Web. Visit several web sites and choose a specific problem. Describe the problem, its history, and its applications. If it has been solved, describe the method of solution if possible.

34. The Four Color Map Problem started in 1852 when Francis Guthrie was coloring a map of England's countries. He noticed that if he used four colors, he would never have two adjacent counties colored the same. He and his brother Frederick went to mathematician Augustus DeMorgan and asked whether this was always true. Investigate this problem on the Web. Describe its connection with graph theory, its history, its proof in 1976 by Wolfgang Haken and Kenneth Appel, and the nature of the proof.

Some useful links for these web projects are listed on the text companion web site. Go to www.thomsonedu.com/mathematics/johnson to access the web site.

9.3 Hamilton Circuits

Finding the Best Route

Kim Blum owns her own business in New York City. One month, she had to fly to Los Angeles and Chicago to meet with clients. Her firm was a recent startup, and money was tight, so she made a point of finding the least expensive flights. She logged onto a travel web site and found the following costs of one-way flights between the cities:

New York to Chicago	$204
Chicago to New York	$219
New York to Los Angeles	$177
Los Angeles to New York	$199
Los Angeles to Chicago	$314
Chicago to Los Angeles	$314

Source: www.travelocity.com

There are two ways to make this trip: She could go to Chicago first and then Los Angeles, or she could go to Los Angeles first and then Chicago. See Figure 9.39.

Chicago first:

Los Angeles first:

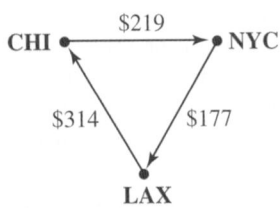

Total cost: $717

Total cost: $710

Figure 9.39
Two ways to make the trip

Kim flew to Los Angeles first and saved $7.

Weighted Graphs and Digraphs

We described each of Kim's possible routes with a graph. However, these graphs are different from those in Sections 9.1 and 9.2 in that each edge has a number associated with it. For example, the edge from CHI to NYC in the "Los Angeles route first" graph above has the cost $219 associated with it. These numbers are called **weights.** A graph that includes weights is called a **weighted graph.**

Kim's circuits are different from those in Sections 9.1 and 9.2 in another way: Each edge has a direction associated with it. Here, it is necessary to distinguish between directions, because the edge from CHI to NYC has a cost of $219 associated with it, while the edge from NYC and CHI has a cost of $204 associated with it. A graph whose edges have directions is called a **directed graph,** or a **digraph.**

So each of Kim's possible routes is illustrated above with a particular type of graph—a *weighted digraph.*

A Weighted Digraph with Four Vertices

A month later, Kim had to visit the same clients and a new client in Miami. She wasn't impressed with her $7 savings a month earlier, so she priced out only two routes. To her surprise, she found that the total cost of her two routes varied quite a bit, so she checked out all the possibilities. The old airfares hadn't changed. She found the following new fares:

New York to Miami	$98
Miami to New York	$286
Los Angeles to Miami	$179
Miami to Los Angeles	$514
Chicago to Miami	$124
Miami to Chicago	$244

Source: www.travelocity.com

Then Kim started listing routes. She quickly realized that there were several. She figured that she was going to three different cities, so there were three different choices for her trip's first leg. That left two choices for the trip's second leg and only one possibility for the third leg. All together, there are $3 \cdot 2 \cdot 1 = 3! = 6$ different routes. See Figure 9.40 on page 670.

Kim flew route 1 and paid $819. This saved $326.

Hamilton Circuits

Kim's route through four cities forms a circuit, because it is a trail that begins and ends at the same vertex. Kim's circuit is different from the Euler circuits that we studied in Sections 9.1 and 9.2 in that her goal is not to pass through every edge exactly once. In fact, none of her possible circuits passed through every edge. For example, route 6 in Figure 9.40 did not pass through the edge joining NYC and LAX, and it did not pass through the edge joining CHI and MIA. (The illustrations show only the edges that were used.)

Instead, Kim's goal is to pass through every *vertex* exactly one time. This type of circuit is called a **Hamilton circuit.**

Route 1:

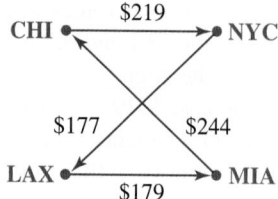

Total trip cost: $819

Route 2:

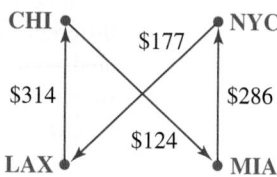

Total trip cost: $901

Route 3:

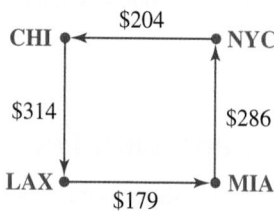

Total trip cost: $983

Route 4:

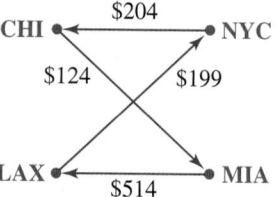

Total trip cost: $1041

Route 5:

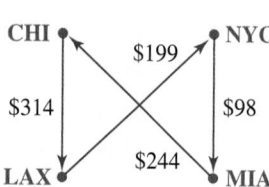

Total trip cost: $855

Route 6:

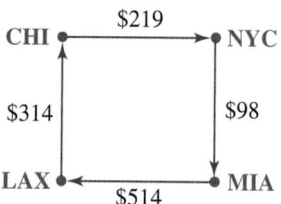

Total trip cost: $1145

Figure 9.40
Six different routes

Many Possibilities Create a Need for Strategies

A few months later, Kim had to visit the same clients and two new clients, one in Seattle and one in Houston. Kim instructed her newly hired assistant to start pricing all the possibilities. After some time, the assistant told her that there were just too many possibilities to try.

Kim calculated that she was going to five different cities, so there were five different choices for her trip's first leg. That left four choices for the trip's second leg, three choices for the third leg, two for the fourth leg, and only one possibility for the fifth leg. All together, there are $5 \cdot 4 \cdot 3 \cdot 2 \cdot 1 = 5! = 120$ different routes. Pricing out every route could end up costing more money (in the form of the assistant's wages) than it would save.

Kim's original approach was to list every possible circuit, add up each circuit's total weight, and choose the circuit with the least total weight. This *brute force algorithm* will find the minimum Hamilton circuit, but most real-world problems have too many circuits for this approach to be realistic.

Figure 9.41
The complete graph

Nonetheless, past experience had taught Kim that different routes could have wildly different prices. And her booming business was going to require a significant amount of traveling. She needed some strategies.

Figure 9.41 shows a complete graph in that it shows every possible leg. In fact, a graph in which every pair of vertices is joined with one edge is called a **complete graph.**

Figure 9.42 shows the various costs.

		From:					
		CHI	HOU	LAX	MIA	NYC	SEA
To:	**CHI**		$305	$314	$244	$204	$179
	HOU	$288		$324	$214	$336	$319
	LAX	$314	$284		$514	$177	$242
	MIA	$124	$190	$179		$98	$307
	NYC	$219	$339	$199	$286		$169
	SEA	$179	$315	$242	$307	$132	

Figure 9.42
Flight costs
Source: www.travelocity.com

The Nearest Neighbor Algorithm

Kim decided that it would be a good beginning to choose the cheapest flight out of New York. That was a $98 flight to Miami. From Miami, the cheapest flight is a $214 flight to Houston. Figure 9.43 continues to choose cheap flights, avoiding flights that take her to a city she has already visited, and avoiding flights that take her home before she's visited all the necessary cities. This method of finding a cheap circuit is called the *nearest neighbor algorithm.*

From:	The cheapest and most usable flight is:
New York City	NYC to MIA at $98
Miami	MIA to HOU at $214
Houston	HOU to MIA at $190—but she'd already been to Miami so use the second cheapest which is HOU to LAX at $284
Los Angeles	LAX to MIA—no, been to Miami LAX to NYC—no, can't go home yet LAX to SEA at $242
Seattle	SEA to NYC—no, can't go home yet SEA to CHI at $179
CHI	Time to go home so CHI to NYC at $219

Figure 9.43
Using the nearest neighbor algorithm

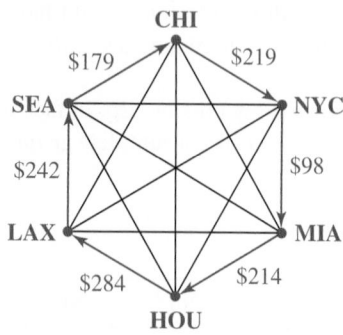

Figure 9.44
The result of the nearest neighbor algorithm

The total cost is $98 + $214 + $284 + $242 + $179 + $219 = $1,236. See Figure 9.44.

The Nearest Neighbor Algorithm for Finding a Minimum Hamilton Circuit

1. From the starting vertex, go to the "nearest neighbor"—that is, use the edge with the smallest weight. (If there is a tie, choose at random. Different choices will result in different circuits, which may have different totals.)

2. From all other vertices, use the edge that:
   • has the smallest weight and
   • leads to a vertex that hasn't been visited yet and
   • does not lead to the starting vertex.

   (If there is a tie, choose at random. Different choices will result in different circuits, which may have different totals.)

3. The last edge must lead to the starting vertex.

Applications of Hamilton Circuits

Finding the cheapest route for Kim's trip is called a **traveling salesman problem.** The goal of any traveling salesman problem is to find a **minimum Hamilton circuit.** That is, the goal is to find a circuit that visits every vertex exactly once and that does so *with the least total weight.*

Finding cheap airplane routes is not the only application of Hamilton circuits. Here are two more:

• *Delivery services.* Companies such as Fed Ex (Federal Express), UPS (United Parcel Service), and DHL deliver hundreds of thousands of packages every day. Using less efficient routes can be prohibitively expensive. Using Hamilton circuits can make a huge impact on a firm's profitability. Here, a vertex is a place where a package must be delivered. An edge's weight is the time it takes to drive between that edge's vertices. Typically, each truck makes between 100 and 200 stops, so one truck's graph would have between 100 and 200 vertices. And a graph with 200 vertices has $200! \approx 8 \times 10^{374}$ different Hamilton circuits. (Remember, 8×10^{374} means an 8 followed by 374 zeros.) So finding an efficient route by listing them all would be impossibly time consuming, even for the fastest of computers.

• *Circuit boards.* Computers and other electronic gadgets have circuit boards inside of them. A circuit board has thousands of small holes drilled at precise locations so that components can be inserted into the holes. The holes are drilled robotically. The cost of producing a circuit board is least if the order in which the holes are drilled is such that the entire drilling sequence is completed in the least amount of time. Here, a vertex is a hole, and an edge's weight is the distance between the edge's vertices.

Delivery services such as FedEx and circuit board manufacturers benefit if they use minimum Hamilton circuits.

Approximation Algorithms

The nearest neighbor algorithm is not guaranteed to find the minimum Hamilton circuit. Instead, it only approximates the minimum Hamilton circuit. For this reason it is called an **approximation algorithm.**

There is no known algorithm that is guaranteed to find the minimum Hamilton circuit in an acceptable amount of time. Instead, there are only approximation algorithms. The brute force algorithm is usually unacceptable, because most real-world problems have too many circuits to do this. Recall that the brute force algorithm involves listing every possible circuit, adding up each circuit's total weight, and choosing the circuit with the least total weight. We used this algorithm earlier in this section when Kim visited three cities and when she visited four cities.

The Repetitive Nearest Neighbor Algorithm

Kim regretted not being able to take advantage of the $190 flight out of Houston to Miami. The nearest neighbor algorithm required that she instead take the $284 flight out of Houston to Los Angeles, because she had already been to Miami. For a moment, she entertained the absurd idea that she should move her business to Houston when she had a flash of insight: Once you have a circuit, it doesn't matter where you start because the circuit goes through all of the vertices once. So she could use the nearest neighbor algorithm with Houston as a starting vertex. Then, once the complete circuit is built, she could use that circuit, with New York City as the starting vertex. This process is illustrated in Figure 9.45.

From:	*The cheapest and most usable flight is:*
Houston	HOU to MIA at $190
Miami	MIA to HOU—no, been to MIA MIA to CHI at $244
Chicago	CHI to MIA—no, been to MIA CHI to SEA at $179
Seattle	SEA to NYC—no, can't go home yet SEA to CHI—no, been to Chicago SEA to LAX at $242
Los Angeles	LAX to MIA—no, been to Miami LAX to NYC at $199
New York City	NYC to HOU at $336

Figure 9.45
Starting to use the repetitive nearest neighbor algorithm

Figure 9.46
You can start this circuit at Houston or at New York City for a total cost of $1390.

The total cost is $1,390. See Figure 9.46.

Kim was disappointed because this is worse than the $1,236 that the nearest neighbor algorithm gave. But this approach could still yield a better price. We have yet to try the other cities as starting vertices, and at least one of those other starting vertices yields a very good total price. We'll save the details for the exercises, but one starting city yields a total price of $1,132. See Exercise 31.

The *repetitive nearest neighbor algorithm* involves applying the nearest neighbor algorithm repeatedly, with each vertex as a starting point. Kim's trip from New York City to five other cities involves six separate applications of the nearest neighbor algorithm—one with New York City as the starting vertex, one with Houston as the starting vertex, one with Chicago as the starting vertex, and so on. This seems inefficient, but remember, there are 5! = 120 different possible circuits. We would be checking only six circuits, not all 120 circuits. This is efficient.

> **The Repetitive Nearest Neighbor Algorithm for Finding a Minimum Hamilton Circuit**
>
> 1. Apply the nearest neighbor algorithm, using any particular vertex as the starting point, and compute the total cost of the circuit obtained.
> 2. Apply the nearest neighbor algorithm, using another vertex as the starting point, and compute the total cost of the circuit obtained.
> 3. Repeat the above process until you have applied the nearest neighbor algorithm, using each vertex as the starting point. If there are any ties, explore each option if the number of options is acceptably small.
> 4. Use the best of the Hamilton circuits obtained in steps 1–3. If this circuit starts at an inappropriate vertex, resequence the circuit so that it starts at the appropriate vertex.

The Cheapest Edge Algorithm

The key idea behind the repetitive nearest neighbor algorithm is that you can build a circuit with any city as the starting vertex, regardless of the fact that you really need to start at one particular city. You just keep adding cheap flights. And you add them in order so that they connect up and make a circuit.

The key idea behind the cheapest edge algorithm is that you can just keep adding cheap flights so that they connect up and make a circuit, but you don't have to add them in order. Instead, you can connect up cheapest flights at the end.

We'll start by listing flights in order of cost. For the time being, we'll list only the flights that cost under $300. See Figure 9.47. With any luck, we won't have to use flights that cost more than $300.

Next, we'll start building a circuit, using the cheapest flights first. The first flight we can use is the $98 special from NYC to MIA. See Figure 9.48.

Cost	Flight
$98	NYC to MIA
$124	CHI to MIA
$132	NYC to SEA
$169	SEA to NYC
$177	NYC to LAX
$179	LAX to MIA
$179	SEA to CHI
$179	CHI to SEA
$190	HOU to MIA
$199	LAX to NYC
$204	NYC to CHI
$214	MIA to HOU
$219	CHI to NYC
$242	SEA to LAX
$242	LAX to SEA
$244	MIA to CHI
$284	HOU to LAX
$286	MIA to NYC
$288	CHI to HOU

Cost	Flight	Comment
$98	NYC to MIA	Use it!
$124	CHI to MIA	Can't use—already flew in to MIA.
$132	NYC to SEA	Can't use—already flew out of NYC.

Figure 9.47
Flights in order of cost

Figure 9.48
Starting the cheapest edge algorithm

The second flight we can use is the $169 SEA to NYC flight. See Figure 9.49.

Cost	Flight	Comment
$169	SEA to NYC	Use it!
$177	NYC to LAX	Can't use—already flew out of NYC.
$179	LAX to MIA	Can't use—already flew in to MIA.
$179	SEA to CHI	Can't use—already flew out of SEA.

Figure 9.49
Continuing the cheapest edge
algorithm

The third flight we can use is the $179 CHI to SEA flight. See Figure 9.50.

Cost	Flight	Comment
$179	CHI to SEA	Use it!
$190	HOU to MIA	Can't use—already flew in to MIA.
$199	LAX to NYC	Can't use—already flew in to NYC.
$204	NYC to CHI	Can't use—already flew out of NYC.

Figure 9.50
Continuing the cheapest edge
algorithm

The fourth flight we can use is the $214 MIA to HOU flight. See Figure 9.51.

Cost	Flight	Comment
$214	MIA to HOU	Use it!
$219	CHI to NYC	Can't use—already flew out of CHI.
$242	SEA to LAX	Can't use—already flew out of SEA.
$242	LAX to SEA	Can't use—already flew in to SEA.
$244	MIA to CHI	Can't use—already flew out of MIA.

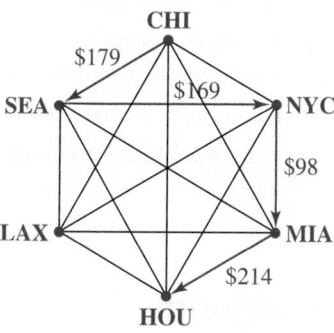

Figure 9.51
Continuing the cheapest edge algorithm

The fifth flight we can use is the $284 HOU to LAX flight. See Figure 9.52. Now we have to complete the circuit with the $314 LAX to CHI flight. See Figure 9.53.

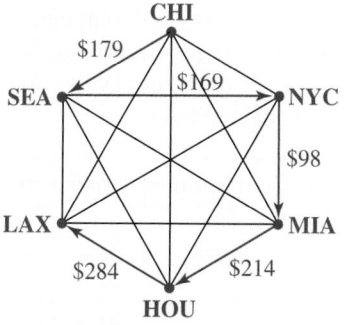

Figure 9.52
Continuing the cheapest edge algorithm

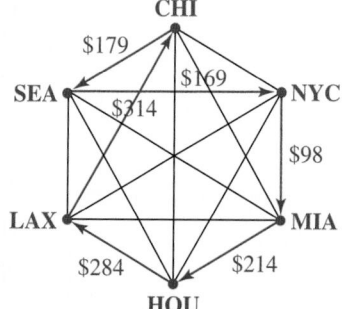

Figure 9.53
Finishing the cheapest edge algorithm

The total cost of this route is $1,258. For this particular problem, the repetitive nearest neighbor algorithm gives the cheapest route at $1,132. You never know which algorithm will work best until you try each one. And you never know if there is a better route unless you use the brute force algorithm and actually find the total weight for every possible route.

> ### The Cheapest Edge Algorithm for Finding a Minimum Hamilton Circuit
>
> 1. List the weights, in order, from smallest to largest.
> 2. Draw a complete graph for the problem.
> 3. Pick the edge with the smallest weight. (In case of a tie, pick one at random. Different choices will result in different circuits, which may have different totals.) Mark that edge on the graph from step 2.
> 4. Continue picking edges, in order, from the list from step 1. Do not pick an edge if:
>   - it leads to a vertex that you've already "flown in to."
>   - it departs from a vertex that you've already "flown out of."
>   - it closes a circuit that shouldn't be closed yet.
>
>   (If there is a tie, choose at random. Different choices will result in different circuits, which may have different totals.
> 5. When all of the vertices are joined, close the circuit.
>
> If more than one edge will close the circuit, choose the one with the smallest weight.

Choosing an Algorithm

We have discussed four different methods: the brute force method, the nearest neighbor algorithm, the repetitive nearest neighbor algorithm, and the cheapest edge algorithm. The last three are approximation algorithms—that is, they're not guaranteed to find the minimum Hamilton circuit. Instead, they approximate the minimum Hamilton circuit.

How do you tell which method to use? It depends on how many different circuits there are. If there aren't too many, use the brute force method. This ensures that you have found the best option. Otherwise, use one or all of the other methods, and be satisfied with something that's close to the best option.

It's easy to tell how many different circuits there are. Think of our experience with Kim Blum's various flights. See Figure 9.54. If Kim had to fly between n cities, there would be $(n - 1)!$ different circuits. And a complete digraph with n vertices has $(n - 1)!$ different Hamilton circuits.

When Kim had to fly between:	There were:
3 cities	$2! = 2 \cdot 1 = 2$ different circuits
4 cities	$3! = 3 \cdot 2 \cdot 1 = 6$ different circuits
6 cities	$5! = 5 \cdot 4 \cdot 3 \cdot 2 \cdot 1 = 120$ different circuits

Figure 9.54
Kim's travels

When we graphed Kim's flight options, we gave each edge a direction. This was necessary because a flight's direction can affect its cost. For example, a flight from Miami to Chicago would cost Kim $244, but a flight from Chicago to Miami would cost $124. Recall that a graph with directed edges is called a *directed graph* or a *digraph*.

If the edges' weights were flight times rather than flight costs, we wouldn't give the edges directions, because a flight from Miami to Chicago takes the same amount of time as does a flight from Chicago to Miami. This means that there would be half as many different Hamilton circuits. And a complete graph with n vertices has $\frac{(n-1)!}{2}$ different Hamilton circuits.

Choosing an Algorithm

1. Find how many different circuits your graph has.
   - A complete directed graph with n vertices has $(n-1)!$ different Hamilton circuits.
   - A complete nondirected graph with n vertices has $\frac{(n-1)!}{2}$ different Hamilton circuits.

2. Determine the appropriate method(s).
   - If the number of different circuits is acceptably small, then use the brute force algorithm and be certain that you've found the Hamilton circuit with the least total weight.
   - If the number of different circuits is not acceptably small, then use any (or, better yet, all) of the approximation algorithms:
     - the nearest neighbor algorithm
     - the repetitive nearest neighbor algorithm
     - the cheapest edge algorithm

9.3 Exercises

Note: In the following exercises, answers will vary if a tie is encountered.

Exercises 1–4 use the flight costs shown in Figure 9.55.

1. You live in Atlanta, and you need to visit Boston, Denver, and Phoenix.
   a. How many different circuits does your graph have?
   b. Approximate the cheapest route, using the nearest neighbor algorithm. Draw this route's graph.

		From:						
		ATL	BOS	DEN	PHX	PORT	SFO	WASH
To:	ATL		$104	$144	$357	$467	$154	$74
	BOS	$104		$312	$446	$179	$134	$54
	DEN	$144	$310		$156	$122	$192	$175
	PHX	$447	$444	$156		$104	$136	$182
	PORT	$216	$177	$122	$104		$84	$492
	SFO	$154	$132	$192	$136	$84		$124
	WASH	$74	$52	$175	$182	$492	$144	

Figure 9.55
Flight costs between Atlanta, Boston, Denver, Phoenix, Portland, San Francisco, and Washington, D.C.
Source: www.travelocity.com

c. Approximate the cheapest route, using the repetitive nearest neighbor algorithm. Draw this route's graph.

d. Approximate the cheapest route, using the cheapest edge algorithm. Draw this route's graph.

e. Find the cheapest route, using the brute force method. Draw this route's graph.

2. You live in Phoenix, and you need to visit Portland, San Francisco, and Washington, D.C.

a. How many different circuits does your graph have?

b. Approximate the cheapest route, using the nearest neighbor algorithm. Draw this route's graph.

c. Approximate the cheapest route, using the repetitive nearest neighbor algorithm. Draw this route's graph.

d. Approximate the cheapest route, using the cheapest edge algorithm. Draw this route's graph.

e. Find the cheapest route, using the brute force method. Draw this route's graph.

3. You live in Atlanta, and you need to visit Boston, Denver, Phoenix, and Portland.

a. How many different circuits does your graph have?

b. Approximate the cheapest route, using the nearest neighbor algorithm. Draw this route's graph.

c. Approximate the cheapest route, using the repetitive nearest neighbor algorithm. Draw this route's graph.

d. Approximate the cheapest route, using the cheapest edge algorithm. Draw this route's graph.

e. Why would it not be appropriate to find the cheapest route, using the brute force method?

4. You live in Denver, and you need to visit Phoenix, Portland, San Francisco, and Washington D.C.

a. How many different circuits does your graph have?

b. Approximate the cheapest route, using the nearest neighbor algorithm. Draw this route's graph.

c. Approximate the cheapest route, using the repetitive nearest neighbor algorithm. Draw this route's graph.

d. Approximate the cheapest route, using the cheapest edge algorithm. Draw this route's graph.

e. Why would it not be appropriate to find the cheapest route, using the brute force method?

Exercises 5–12 use the FedEx travel times shown in Figure 9.56.

Exercises 5–8 involve deliveries to Kinko's, City Hall, the insurance office, the lawyer's office, and the bank.

5. You must make deliveries from the FedEx warehouse to Kinko's, City Hall, the insurance office, the lawyer's office, and the bank. Use the nearest neighbor algorithm to determine the best sequence for those deliveries.

6. You must make deliveries from the FedEx warehouse to Kinko's, City Hall, the insurance office, the lawyer's office, and the bank. Use the repetitive nearest neighbor algorithm to determine the best sequence for those deliveries.

7. You must make deliveries from the FedEx warehouse to Kinko's, City Hall, the insurance office, the

	Kinko's	City hall	Insurance office	Lawyer's office	Bank	Real estate office
FedEx warehouse	14	22	10	35	27	11
Kinko's		17	8	32	15	8
City hall			14	37	18	11
Insurance office				15	25	17
Lawyer's office					19	30
Bank						29

Figure 9.56
FedEx travel times in minutes

lawyer's office, and the bank. Use the cheapest edge algorithm to determine the best sequence for those deliveries.

8. If you must make deliveries from the FedEx warehouse to Kinko's, City Hall, the insurance office, the lawyer's office, and the bank, how many different routes are possible? Of Exercises 5, 6, and 7, which assigned exercise generates the best delivery sequence?

Exercises 9–12 involve deliveries to City Hall, the insurance office, the lawyer's office, the real estate office, and the bank.

9. You must make deliveries from the FedEx warehouse to City Hall, the insurance office, the lawyer's office, the real estate office, and the bank. Use the nearest neighbor algorithm to determine the best sequence for those deliveries.

10. You must make deliveries from the FedEx warehouse to City Hall, the insurance office, the lawyer's office, the real estate office, and the bank. Use the repetitive nearest neighbor algorithm to determine the best sequence for those deliveries.

11. You must make deliveries from the FedEx warehouse to City Hall, the insurance office, the lawyer's office, the real estate office, and the bank. Use the cheapest edge algorithm to determine the best sequence for those deliveries.

12. If you must make deliveries from the FedEx warehouse to City Hall, the insurance office, the lawyer's office, the real estate office, and the bank, how many different routes are possible? Of Exercises 9, 10, and 11, which assigned exercise generates the best delivery sequence?

Use the following information in Exercises 13–22.

You are programming a robotic drill to drill holes in a circuit board. After the holes have been drilled, electronic components will be inserted into them. You use an ordered pair to specify a hole's location. For example, the ordered pair (1, 2) specifies a location that is 1 mm to the right and 2 mm above the board's lower left corner, as shown in Figure 9.57. The drill is programmed to return to (0, 0) after completing a circuit board. The cost of drilling a board is determined by the distance the drill must travel, and the drill moves only horizontally and vertically. For example, if the drill starts at (0, 0) and then travels to (1, 2) to drill a hole, it travels horizontally 1 mm and vertically 2 mm, for a total of 3 mm.

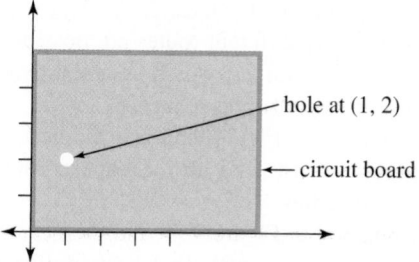

Figure 9.57
A hole in a circuit board at (1, 2)

13. The machine drills holes at (4, 1) and then (3, 7).
   a. How far does the drill travel when it moves from (0, 0) to (4, 1)?
   b. How far does the drill travel when it moves from (4, 1) to (3, 7)?
   c. How far does the drill travel when it moves from (3, 7) to (0, 0)?
   d. How far does the drill travel in total?

14. The machine drills holes at (5, 2) and then (1, 1).
   a. How far does the drill travel when it moves from (0, 0) to (5, 2)?
   b. How far does the drill travel when it moves from (5, 2) to (1, 1)?
   c. How far does the drill travel when it moves from (1, 1) to (0, 0)?
   d. How far does the drill travel in total?

15. A board needs holes at (4, 5), (3, 1), (6, 4), (7, 5), and (2, 2). Use the nearest neighbor algorithm to determine a sequence for the drilling of those holes.

16. A board needs holes at (4, 5), (3, 1), (6, 4), (7, 5), and (2, 2). Use the repetitive nearest neighbor algorithm to determine a sequence for the drilling of those holes.

17. A board needs holes at (4, 5), (3, 1), (6, 4), (7, 5), and (2, 2). Use the cheapest edge algorithm to determine a sequence for the drilling of those holes.

18. If a board needs holes at (4, 5), (3, 1), (6, 4), (7, 5), and (2, 2), how many different routes are possible? Of Exercises 15, 16, and 17, which assigned exercise generates the best drilling sequence?

19. A board needs holes at (1, 8), (2, 7), (5, 1), (3, 4), and (4, 3). Use the nearest neighbor algorithm to determine a sequence for the drilling of those holes.

20. A board needs holes at (1, 8), (2, 7), (5, 1), (3, 4), and (4, 3). Use the repetitive nearest neighbor algorithm to determine a sequence for the drilling of those holes.

21. A board needs holes at (1, 8), (2, 7), (5, 1), (3, 4), and (4, 3). Use the cheapest edge algorithm to determine a sequence for the drilling of those holes.

22. If a board needs holes at (1, 8), (2, 7), (5, 1), (3, 4), and (4, 3), how many different routes are possible? Of Exercises 19, 20, and 21, which assigned exercise generates the best drilling sequence?

23. Use the information in Figure 9.55 and the nearest neighbor algorithm to find the cheapest route between all seven cities.

24. Use the information in Figure 9.56 and the nearest neighbor algorithm to find the best delivery sequence from the FedEx warehouse to Kinko's, City Hall, the insurance office, the lawyer's office, the bank, and the real estate office.

25. Use the information in Figure 9.55 and the repetitive nearest neighbor algorithm to find the cheapest route between all seven cities.

26. Use the information in Figure 9.56 and the repetitive nearest neighbor algorithm to find the best delivery sequence from the FedEx warehouse to Kinko's, City Hall, the insurance office, the lawyer's office, the bank, and the real estate office.

27. Use the information in Figure 9.55 and the cheapest edge algorithm to find the cheapest route between all seven cities.

28. Use the information in Figure 9.56 and the cheapest edge algorithm to find the best delivery sequence from the FedEx warehouse to Kinko's, City Hall, the insurance office, the lawyer's office, the bank, and the real estate office.

29. Of Exercises 23, 25, and 27, which assigned exercise generates the cheapest route?

30. Of Exercises 24, 26, and 28, which assigned exercise generates the best delivery sequence?

31. Find the result of applying the repetitive nearest neighbor algorithm to Kim Blum's trip from New York to all five cities: Chicago, Miami, Houston, Los Angeles, and Seattle.

WEB PROJECTS

32. Investigate traveling salesman problems on the Web. What do they have to do with Hamilton circuits? Summarize their history. Describe specific problems that led to progress.

Some useful links for this web project are listed on the text companion web site. Go to **www.thomsonedu.com/mathematics/johnson** to access the web site.

9.4   Networks

Installing a Network

Sarah and Miles are joining the growing ranks of telecommuters. Each has an office in their home. They are going to install a network so that they can share one fast Internet connection and one color laser printer. For security reasons, their employers have insisted that they have a hardwired network rather than a wireless network.

Their daughter Liz and their son Pete are working on degrees at State University. Each has a laptop computer, which they use in their bedrooms. So altogether, the network should connect four computers.

Sarah and Miles looked into having the network installed by a professional. However, the cost was rather excessive—labor ran over $100 an hour. After buying a book on networks, they realized that they could do the work themselves.

They measured the various distances between the rooms. They found that they would need 60 feet of wire to form a link between Sarah's and Miles's offices. These and other measurements are shown in the complete weighted graph in Figure 9.58.

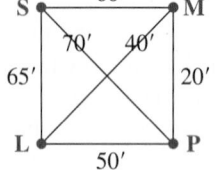

Figure 9.58
Amounts of wire needed

Avoiding Unnecessary Connections

Sarah and Miles quickly realized that it is not necessary to wire up every edge of the graph in Figure 9.58. In particular, there is no reason to form a circuit. For example, if they wired up the 60' edge joining Sarah's and Miles's office, and the 40' edge joining Miles's office and Liz's room, then Liz's room and Sarah's office would be connected through Miles's office. In this case, there would be no need to wire up the 65' edge joining Liz's room and Sarah's office. See Figure 9.59. And avoiding extra connections means saving money.

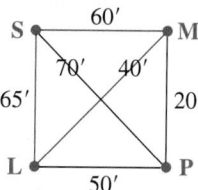

There is no need to form this blue circuit . . .

Figure 9.59

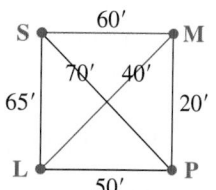

because these two blue edges connect S, M and L.

Choose the Shortest Edges First

Sarah and Miles decided that their network should definitely involve wiring up the 20' edge joining Miles's office and Pete's room, because that's the shortest (and therefore cheapest) connection possible.

They also decided that their network should include the second-shortest connection, which is the 40' edge joining Miles's office and Liz's room. These two decisions gave them the partially completed network shown in Figure 9.60. The network isn't complete yet, so we'll continue to show all the edges.

The third-shortest connection is the 50' edge joining Pete's room and Liz's room. But there's no need to wire up this edge because Pete's room and Liz's room are already connected through Miles's office.

The fourth-shortest connection is the 60' edge joining Sarah's office and Miles's office. This connection completes the network, because all four rooms are now connected. See Figure 9.61.

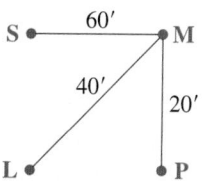

Figure 9.60
A partially completed network.

Trees

The network in Figure 9.61 is an example of a *tree*. A **tree** is a connected graph that has no circuits. The graph of Sarah's and Miles's network must be connected because if it weren't, someone's computer wouldn't have access to the fast Internet connection. And as was discussed previously, it's unnecessary to form a circuit.

There are many other trees that would describe a functional network for Sarah and Miles. For example, the tree in Figure 9.62 would work. However, it involves 60' + 70' + 65' = 195' of wire, so it would be more expensive and probably more time consuming to install.

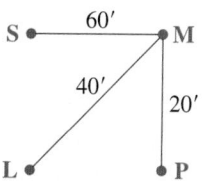

Figure 9.61
Sarah's and Miles's network

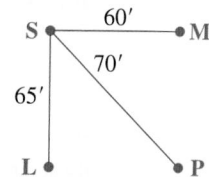

Figure 9.62
A tree describing a more expensive network

The trees in Figure 9.61 and 9.62 are both **spanning trees** because they both include all of the vertices in the original graph. (Here, *span* means "to reach over or across something.") However, the spanning tree in Figure 9.61 has less total weight than does the spanning tree in Figure 9.62. In fact, it has the least possible weight of all spanning trees. For this reason, it is called the **minimum spanning tree.**

Kruskal's Algorithm

While it seems to be just common sense, the method that Sarah and Miles used to design their network guarantees a minimum spanning tree. It is called *Kruskal's algorithm,* named after the Bell Labs mathematician who described it in 1956. Kruskal's algorithm is very similar to the cheapest edge algorithm discussed in Section 9.3. However the cheapest edge algorithm is an approximation algorithm, so it does not guarantee a minimum Hamilton circuit.

Kruskal's Algorithm for Finding a Minimum Spanning Tree

1. List the weights, in order, from smallest to largest.
2. Draw a complete graph for the problem.
3. Pick the edge with the smallest weight. (In case of a tie, pick one at random.) Mark that edge on the graph from step 2.
4. Continue picking edges, in order, from the list from step 1. Do not pick an edge if it closes a circuit.
5. Stop adding edges when all the vertices are joined.

Adding a Vertex

With Sarah's and Miles's network, all lines come together at Miles's office, as shown in Figure 9.61. This means that they would install a *router* in Miles's office. A router is a small, specialized computer that controls the various computer and Internet connections. The four computers, the color laser printer, and the Internet all plug into the router.

Sarah suggested to Miles that if they put the router in the attic, they might end up with shorter connections. They measured the appropriate distances and came up with a new design. That new design is shown in Figure 9.63.

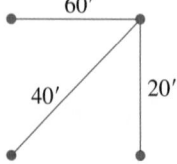

By adding a vertex they could replace this . . .

with an even shorter minimum spanning tree.

Figure 9.63

Using the language of graph theory, putting the router in the attic means adding a vertex to the graph. And adding a vertex might allow Sarah and Miles to replace their minimum spanning tree with a new minimum spanning tree that has less total weight.

Where to Add a Vertex

Suppose you were designing a telephone network that would connect three desert towns: Badwater, Dry Springs, and Hotspot. Each of these towns is 300 miles from each of the other two towns. See Figure 9.64.

A minimum spanning tree consists of any two of the three edges, with a total length of 600 miles. See Figure 9.65.

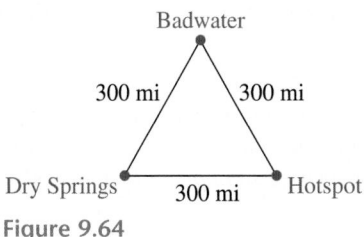

Figure 9.64
Three desert towns

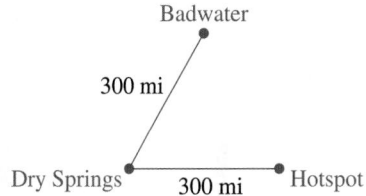

Figure 9.65
A minimum spanning tree

There is nothing but desert between the towns, so we are free to run the telephone cables anywhere we want. In particular, we could shorten the edge that connects Hotspot to the rest of the network by adding a vertex somewhere along the Dry Springs/Badwater edge and connecting there instead of at Dry Springs. See Figure 9.66.

A 90° angle makes for the shortest Hotspot edge. See Figure 9.67.

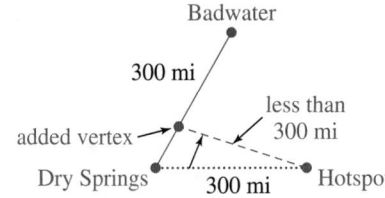

Figure 9.66
Making the network shorter by adding a vertex and shortening an edge

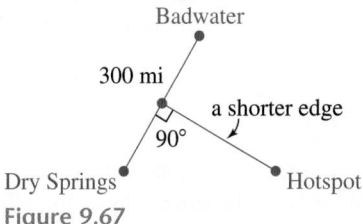

Figure 9.67
A 90° angle means a shorter edge.

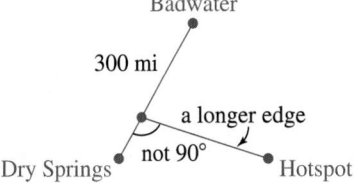

An angle other than 90° means a longer edge.

Example 1   Find the total length of the network with a 90° angle shown in Figure 9.67.

Solution   The three sides of the original triangle have the same length, so the three angles must be the same size. The three angles of any triangle always add to 180°, so our triangle must have three 60° angles. That is, the triangle is equilateral. See Figure 9.68.

The new, shorter edge divides this triangle into two right triangles. We'll use trigonometry, as studied in Section 6.5, to find the needed lengths. See Figure 9.69.

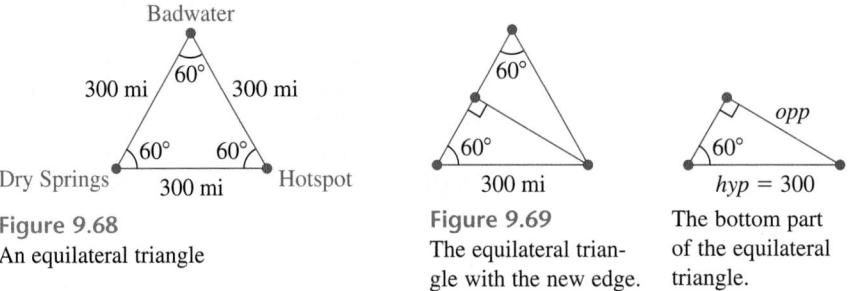

Figure 9.68
An equilateral triangle

Figure 9.69
The equilateral triangle with the new edge.

The bottom part of the equilateral triangle.

Looking at the rightmost triangle, we need to find the new edge's length. This edge is the side opposite the 60° angle, so we need *opp*. And we know that *hyp* = 300. We're interested in *opp* and *hyp*, so we'll use the sine ratio.

$$\sin\theta = \frac{opp}{hyp} \qquad \text{definition of the sine ratio}$$

$$\sin 60° = \frac{opp}{300} \qquad \text{substituting}$$

$$opp = 300 \cdot \sin 60° \qquad \text{multiplying by 300}$$

$$opp = 300 \cdot \frac{\sqrt{3}}{2} = 150\sqrt{3} \approx 259.8$$

The new edge is of length 259.8. And the total length of the network is 300 + 259.8 = 559.8. This is a real improvement over the 600-mile-long minimum spanning tree in Figure 9.65. See Figure 9.70.

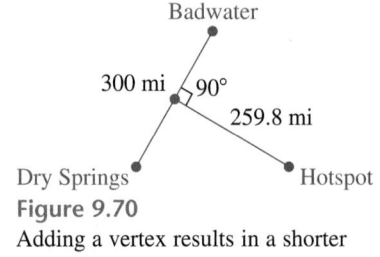

Figure 9.70
Adding a vertex results in a shorter minimum spanning tree. ∎

In Example 1, we used the fact that the three angles of any triangle always add to 180°. This and other triangle facts are given below.

Triangle Facts

- The sum of the three angles is 180°.
- If two sides have the same length, then the two angles across from those sides have the same size.
- If the two angles have the same size, then the two sides across from those angles have the same length.
- If the three sides have the same length, then the three angles have the same size.
- If the three angles have the same size, then the three sides have the same length.

By not connecting Hotspot directly to Dry Springs, we shortened the network by 40.2 miles. Perhaps we could do even better by not connecting Dry Springs directly to Badwater. Let's move the newly added vertex closer to Hotspot and see what happens. See Figure 9.71.

One possibility is shown in Figure 9.72. Here, we moved the added vertex so that the edges form three 120° angles.

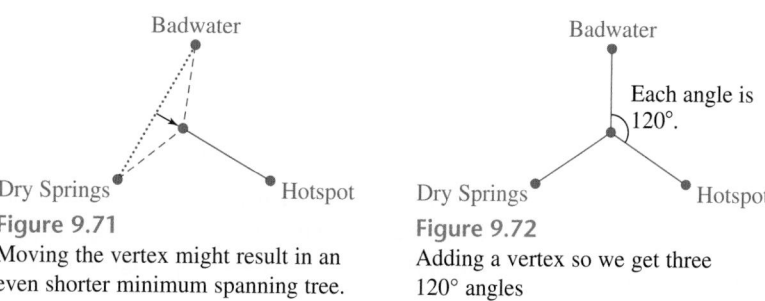

Figure 9.71
Moving the vertex might result in an even shorter minimum spanning tree.

Figure 9.72
Adding a vertex so we get three 120° angles

Example 2   Find the total length of the network with three 120° angles, as shown in Figure 9.72.

Solution   To use trigonometry, we need to find a triangle with a right angle. Reinserting the Dry Springs/Hotspot edge gives us a triangle. Continuing the Badwater edge straight down gives us what appears to be a right triangle. See Figure 9.73.

The fact that it *appears* to be a right triangle doesn't mean that it *is* a right triangle. Let's see if it is. The minimum spanning tree in Figure 9.73 has left/right symmetry. That is, the left side and the right side are identical. This means that the two angles at the bottom of the triangle are identical. They form a straight line, so they must add to 180°. This means that each must be $1/2 \cdot 180° = 90°$. So we do have a right triangle. Also, the two angles at the top of the triangle are equal in size. They add to 120°, so each must be $1/2 \cdot 120° = 60°$. See Figure 9.74.

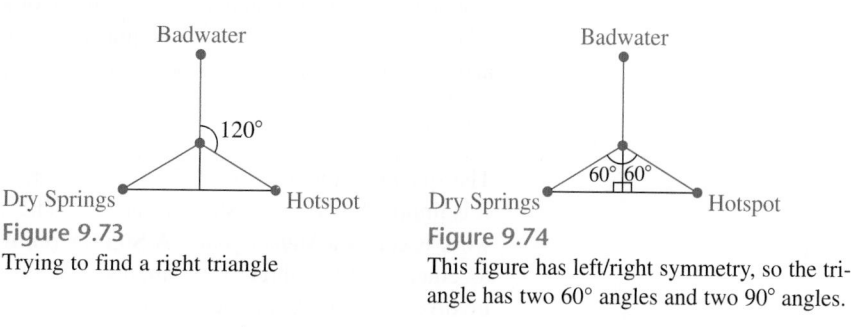

Figure 9.73
Trying to find a right triangle

Figure 9.74
This figure has left/right symmetry, so the triangle has two 60° angles and two 90° angles.

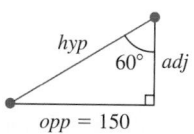

Figure 9.75
A right triangle

We already know that the bottom edge's length is 300, so the left and right halves must each be $\frac{1}{2} \cdot 300 = 150$. This gives us the right triangle in Figure 9.75.

We need to know the length of the Dry Springs edge, which is the hypotenuse. And we know that the opposite is 150. So we'll use the sine ratio.

$$\sin \theta = \frac{opp}{hyp} \qquad \text{definition of the sine ratio}$$

$$\sin 60° = \frac{150}{hyp} \qquad \text{substituting}$$

$$hyp \cdot \sin \theta = 150 \qquad \text{multiplying by } hyp$$

$$hyp \cdot \frac{\sqrt{3}}{2} = 150$$

$$hyp \cdot \frac{\sqrt{3}}{2} \cdot \frac{2}{\sqrt{3}} = 150 \cdot \frac{2}{\sqrt{3}} \qquad \text{solving for } hyp$$

$$hyp = \frac{300}{\sqrt{3}} \approx 173.2$$

This makes each edge 173.2 miles, and the network has a total length of $3 \cdot 173.2 = 519.6$ miles, which is significantly shorter than the earlier networks of 559.8 miles and 600 miles. See Figure 9.76.

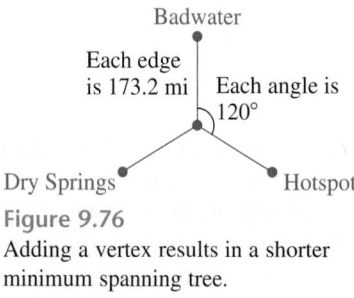

Figure 9.76
Adding a vertex results in a shorter minimum spanning tree.

■

When we studied Euler paths and Hamilton circuits in Sections 9.1–9.3, the size of an angle between two edges didn't matter. In the context of finding an Euler path, the graph in Figure 9.70 that shows the solution to Example 1 and the graph in Figure 9.76 are *identical*. That is, edges that are adjacent in Figure 9.70 are adjacent in Figure 9.76, and vertices that are adjacent in Figure 9.70 are adjacent in Figure 9.76. However, in the context of finding the shortest network connecting the three cities, the graphs in Figures 9.70 and 9.76 have an important difference: One is shorter than the other. The size of an angle matters quite a bit when we try to find the shortest network, and the term *identical,* as defined in Section 9.2, is inappropriate in this context.

Steiner Points

The spanning tree in Figure 9.76 is the shortest network connecting the three cities. It is impossible to find a shorter network. The added vertex at the center of this network is called a *Steiner point.* A **Steiner point** is a vertex where three edges come together, forming three 120° angles. Usually, but not always, shortest networks involve one or more Steiner points.

Example 3   Determine whether adding a Steiner point forms the shortest network connecting the points in Figure 9.77.

Solution   If we don't add a Steiner point, we get a spanning tree of length 30 mi + 40 mi = 70 mi.
As we see in Figure 9.78, it is impossible to add a Steiner point vertex. The 170° angle is just too big for a Steiner point to work.

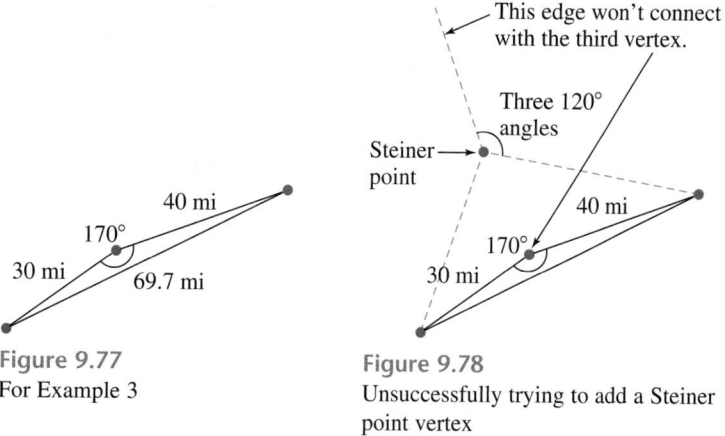

Figure 9.77
For Example 3

Figure 9.78
Unsuccessfully trying to add a Steiner
point vertex

In Example 3 we were unsuccessful in adding a Steiner point vertex, because of the 170° angle. It's easy to see that we would have the same difficulty if the angle were 160° or 150°. In fact, this difficulty arises whenever a triangle has an angle of 120° or more.

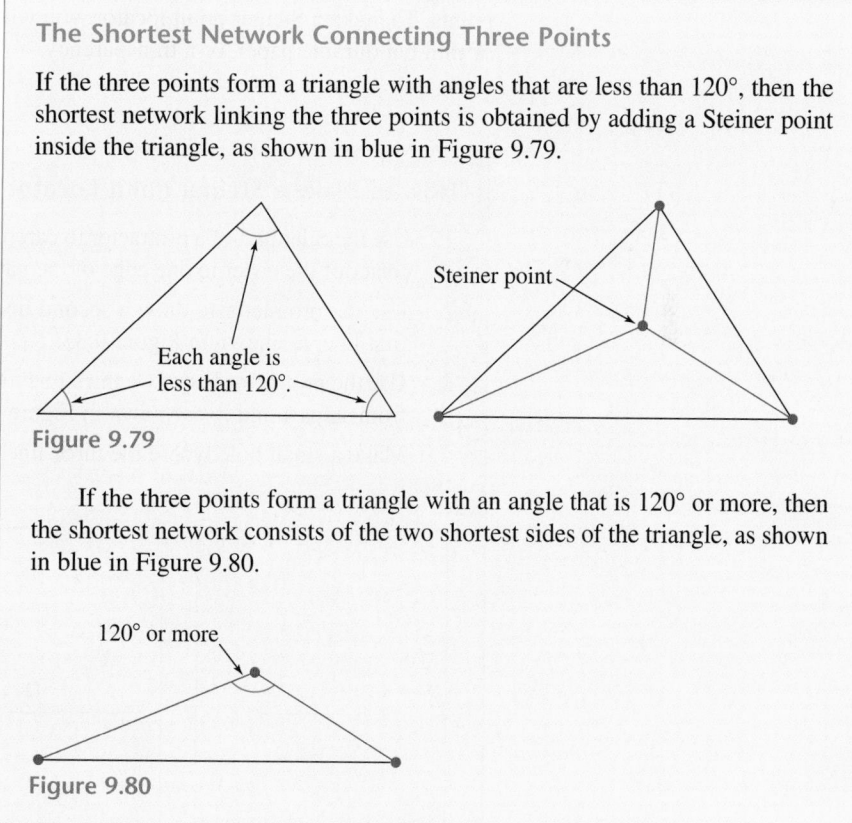

The Shortest Network Connecting Three Points

If the three points form a triangle with angles that are less than 120°, then the shortest network linking the three points is obtained by adding a Steiner point inside the triangle, as shown in blue in Figure 9.79.

Figure 9.79

If the three points form a triangle with an angle that is 120° or more, then the shortest network consists of the two shortest sides of the triangle, as shown in blue in Figure 9.80.

Figure 9.80

Finding Steiner Points Without Symmetry

In Examples 1 and 2, we used right triangles, symmetry and trigonometry to find lengths of edges. Sometimes those tools aren't all available.

Consider the triangle with edges of length 5 km, 6 km, and 7 km, as shown in Figure 9.81. This triangle has no symmetry, so when we insert a Steiner point, we can't use symmetry to determine distances.

If we draw a new edge from the Steiner point that makes a right angle with the 7 km side, the two parts of the 7 km edge are clearly of different lengths. This means that we can't use symmetry to find lengths. See Figure 9.82.

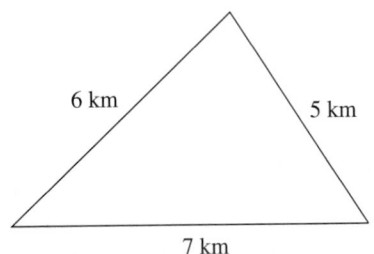

Figure 9.81
A triangle without symmetry

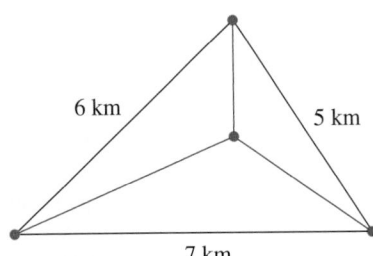

The same triangle with a Steiner point

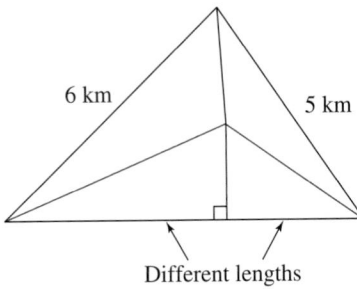

Figure 9.82
We can't use symmetry to find lengths.

Instead, we'll make a *Steiner point locator,* and we'll use it to find Steiner points. To make a Steiner point locator, you will need a protractor and either a piece of thin but durable paper, or a transparency.

How to Make a Steiner Point Locator

1. Use the ruler part of a protractor to carefully draw a horizontal line from the center of the paper to one edge of the paper, as shown in Figure 9.83.

2. Use the protractor to draw a second line that forms a 120° angle with the first line, as shown in Figure 9.83.

3. Use the protractor to draw a third line that forms 120° angles with the lines from steps 1 and 2, as shown in Figure 9.83.

4. Make a small hole where the three lines come together.

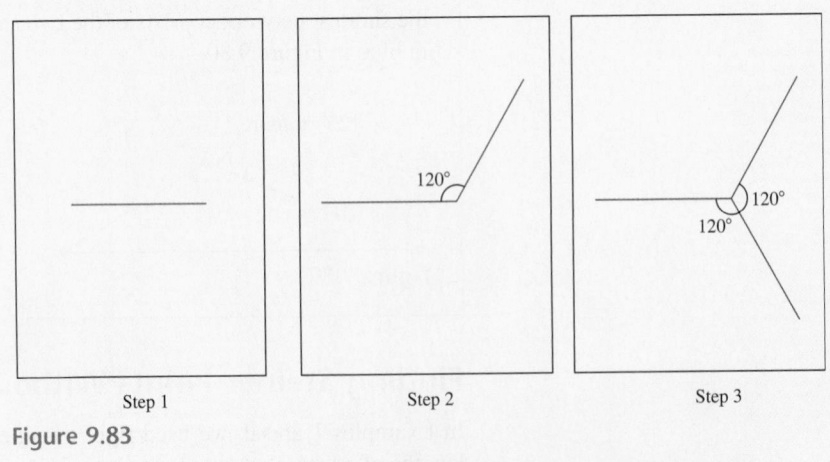

Step 1   Step 2   Step 3

Figure 9.83

How to Use a Steiner Point Locator

Place a Steiner point locator on top of a drawing of a triangle. Move the Steiner point locator around until each of its edges goes through one of the triangle's vertices. Then use your pencil to put make a mark through the small hole in the Steiner point locator.

ThomsonNOW™
for Liberal Arts Mathematics
www.thomsonedu.com/1pass.

Example 4

Use a Steiner point locator to locate the Steiner point for the triangle shown in Figure 9.84.

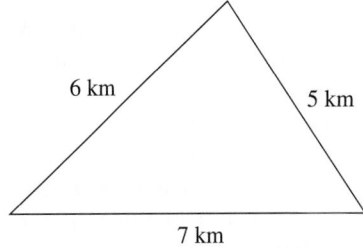

Figure 9.84
The triangle for Example 4

Solution

First, make an accurate drawing of the triangle. Let 1 km correspond to 1 in. A compass can be helpful in doing this. Next, place your Steiner point locator on top of your drawing, as shown in Figure 9.85.

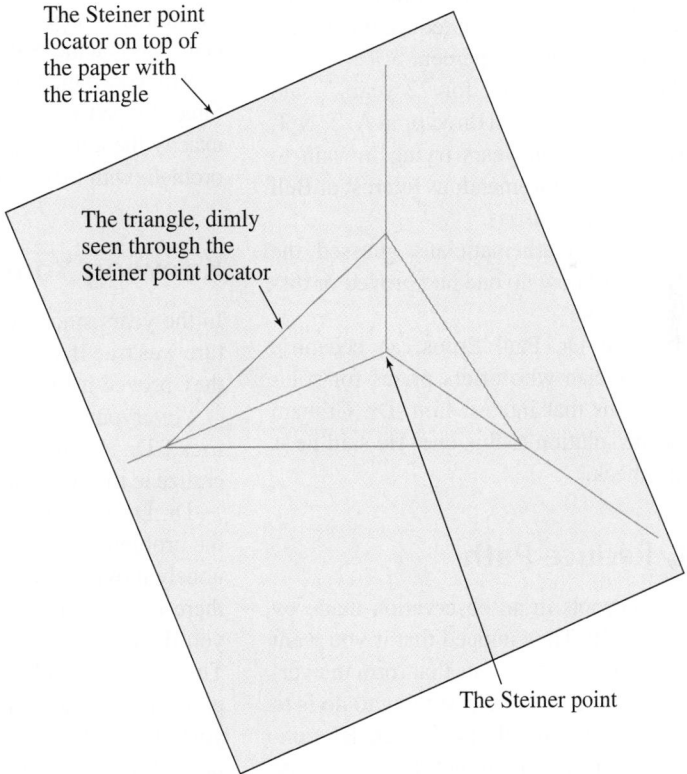

The Steiner point locator on top of the paper with the triangle

The triangle, dimly seen through the Steiner point locator

The Steiner point

Figure 9.85
Using a Steiner point locator

Once the Steiner point has been found, it's easy to use a ruler to measure the distances. ∎

Solution to Old Puzzle: How Short a Shortcut?

By GINA KOLATA
Source: New York Times, 1990

Two mathematicians have solved an old problem in the design of networks that has enormous practical importance but has baffled some of the sharpest minds in the business.

Dr. Frank Hwang of A.T. & T. Bell Laboratories in Murray Hill, N. J., and Dr. Ding Xhu Du, a postdoctoral student at Princeton University, announced at a meeting of theoretical computer scientists last week that they had found a precise limit to the design of paths connecting three or more points.

Designers of things like computer circuits, long-distance telephone lines, or mail routings place great stake in finding the shortest path to connect a set of points, like the cities an airline will travel to or the switching stations for long-distance telephone lines. Dr. Hwang and Dr. Du proved, without using any calculations, that an old trick of adding extra points to a network to make the path shorter can reduce the length of the path by no more than 13.4 percent.

"This problem has been open for 22 years," said Dr. Ronald L. Graham, a research director at A. T. & T. Bell Laboratories who spent years trying in vain to solve it. "The problem is of tremendous interest at Bell Laboratories, for obvious reasons."

In 1968, two other mathematicians guessed the proper answer, but until now no one had proved or disapproved their conjecture.

In the tradition of Dr. Paul Erdos, an eccentric Hungarian mathematician who offers prizes for solutions to hard problems that interest him, Dr. Graham offered $500 for the solution to this one. He said he is about to mail his check.

Add Point, Reduce Path

The problem has its roots in an observation made by mathematicians in 1640. They noticed that if you want to find a path connecting three points that form the vertices of an equilateral triangle, the best thing to do is to add an extra point in the middle of the set. By connecting these four points, you can get a shorter route around the original three than if you had simply drawn lines between those three points in the first place.

Later, mathematicians discovered many more examples of this paradoxical phenomenon. It turned out to be generally true that if you want to find the shortest path connecting a set of points, you can shorten the path by adding an extra point or two.

But it was not always clear where to add the extra points, and even more important, it was not always clear how much you would gain. In 1977, Dr. Graham and his colleagues, Dr. Michael Garey and Dr. David Johnson of Bell Laboratories, proved that there was no feasible way to find where to place the extra points, and since there are so many possibilities, trial and error is out of the question.

The Shorter Route

The length of string needed to join the three corners of a triangle is shorter if a point is added in the middle, as in the figure at right.

Source: F. K. Hwang

These researchers and others found ways to guess where an extra point or so might be beneficial, but they were always left with the nagging question of whether this was the best they could do.

The conjecture, made in 1968 by Dr. Henry Pollack, who was formerly a research director at Bell Laboratories, and Dr. Edger Gilbert, a Bell mathematician, was that the best you could ever do by adding points was to reduce the length of the path by the square root of three divided by two, or about 13.4 percent. This was exactly the amount the path was reduced in the original problem with three vertices of an equilateral triangle.

Knowing Your Limits

In the years since, researchers proved that the conjecture was true if you wanted to connect four points, then they proved it for five.

"Everyone who contributed devised some special trick," Dr. Hwang said. "But still, if you wanted to generalize to the next number of points, you couldn't do it."

Dr. Du and Dr. Hwang's proof relied on converting the problem into one involving mathematics of continuously moving variables. The idea was to suppose that there is an arrangement for which added points make you do worse than the square root of three over two. Then, Dr. Du and Dr. Hwang showed, you can slide the points in your network around and see what happens to your attempt to make the connecting path shorter. They proved that there was no way you could slide points around to do any better than a reduction in the path length of the square root of three over two.

Dr. Graham said the proof should allow network designers to relax as they search for the shortest routes. "If you can never save more than the square route of three over two, it may not be worth all the effort it takes to look for the best solution" he said.

9.4 | Exercises

In Exercises 1–12, determine whether the given graph is a tree. If not, say why.

1.

2.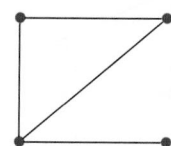

In Exercises 13–18, find a spanning tree for the given graph. Note: Answers will vary, so answers are not given in the back of the book.

13.

14.

3.

4.

15.

5.

6.

16.

7.

8.

9.

10.

17.

11.

12.

18.

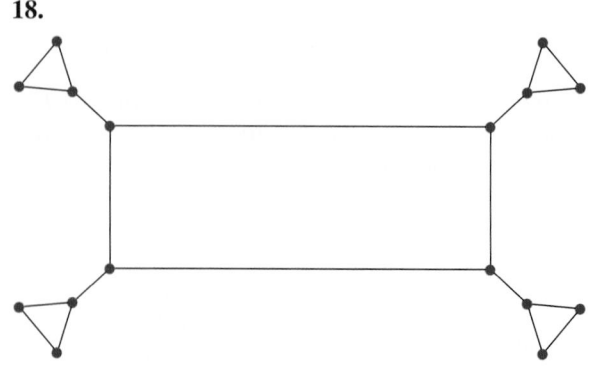

In Exercises 19–24, use Kruskal's algorithm to find the minimum spanning tree for the given weighted graph. Give the total weight of the minimum spanning tree.

19.   **20.**

21.   **22.**

23.

24.

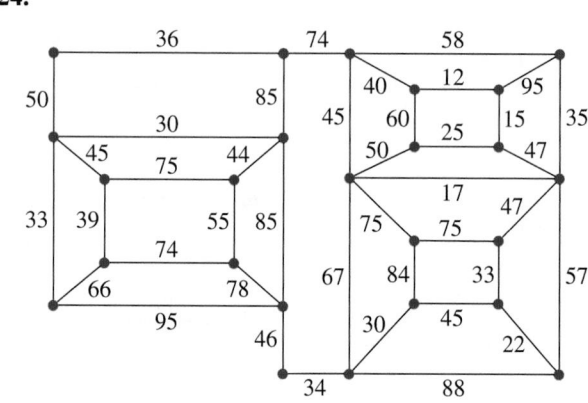

In Exercises 25–30, use the information in Figure 9.86. The distances given are not along roads. Rather, they are "as the crow flies" distances.

	Boston	Chicago	Denver	Kansas City MO	New York City	Philadelphia	Portland OR	San Franciso
Atlanta	932	585	1213	678	744	658	2170	2138
Boston		851	1767	1247	187	273	2536	2696
Chicago			917	407	714	664	1755	1855
Denver				558	1630	1574	979	947
Kansas City MO					1094	1032	1493	1500
New York						86	2442	2570
Philadelphia							2407	2518
Portland OR								538

Figure 9.86
Distances between cities, in miles

25. You have been hired to design a network linking New York City, Chicago, Boston, Kansas City, and San Francisco. Using only those cities as vertices, find the minimum spanning tree connecting the cities and its total length.

26. You have been hired to design a network linking Atlanta, Denver, Philadelphia, Portland, and San Francisco. Using only those cities as vertices, find the minimum spanning tree connecting the cities and its total length.

27. You have been hired to design a network linking Atlanta, Denver, Philadelphia, Portland, New York City, and San Francisco. Using only those cities as vertices, find the minimum spanning tree connecting the cities and its total length.

28. You have been hired to design a network linking New York City, Chicago, Boston, Kansas City, Philadelphia, and San Francisco. Using only those cities as vertices, find the minimum spanning tree connecting the cities and its total length.

29. You have been hired to design a network linking Atlanta, New York City, Chicago, Boston, Kansas City, Philadelphia, Portland, and San Francisco. Using only those cities as vertices, find the minimum spanning tree connecting those cities and its total length.

30. You have been hired to design a network linking Atlanta, New York City, Boston, Denver, Kansas City, Philadelphia, Portland, and San Francisco. Using only those cities as vertices, find the minimum spanning tree connecting those cities and its total length.

31. Figure 9.87 shows the floor plan of Pinstripe National Bank's new information technologies office. Each vertex represents a cubicle, and the edges' weights represent the distances between the cubicles. The cubicles are laid out in a grid pattern, so edges are parallel. You have been asked to design a computer network linking the cubicles. Without adding a vertex, find the minimum spanning tree connecting the cubicles and its total length.

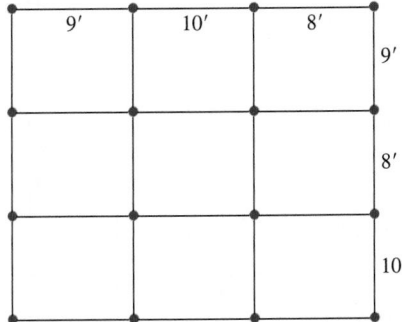

Figure 9.87
The information technologies office

32. Figure 9.88 shows the floor plan of Second National Bank's new web development office. Each vertex represents a cubicle, and the edges' weights represent the distances between the cubicles. The cubicles are laid out in a grid pattern, so edges are parallel. You have been asked to design a computer network linking the cubicles. Without adding a vertex, find the minimum spanning tree connecting the cubicles and its total length.

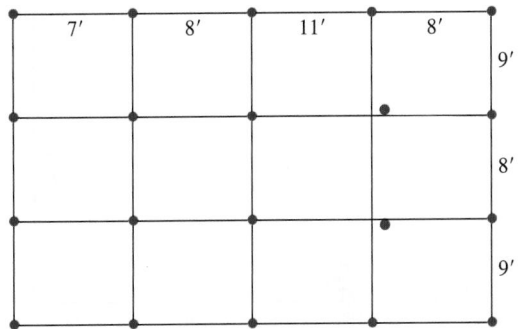

Figure 9.88
The web development office

33. A triangle has three vertices: X, Y, and Z. Inside the triangle are three points: A, B, and C, one of which is a Steiner point. The distances between various points are given in Figure 9.89. Which point, A, B, or C, is a Steiner point? Why?

	A	B	C
X	30	40	50
Y	40	30	50
Z	80	70	60

Figure 9.89
Data for Exercise 33

34. A triangle has three vertices: X, Y, and Z. Inside the triangle are three points: A, B, and C, one of which is a Steiner point. The distances between various points are given in Figure 9.90. Which point, A, B, or C, is a Steiner point? Why?

	A	B	C
X	50	60	70
Y	50	70	40
Z	60	40	60

Figure 9.90
Data for Exercise 34

35. a. Use trigonometry and the combined 20° + 20° = 40° angle at the left of the big triangle in Figure 9.91 to find the size of x + y.
b. Use trigonometry and the 20° angle at the left of the lower triangle to find the size of x.
c. Use the answers to parts (a) and (b) to find the size of y.

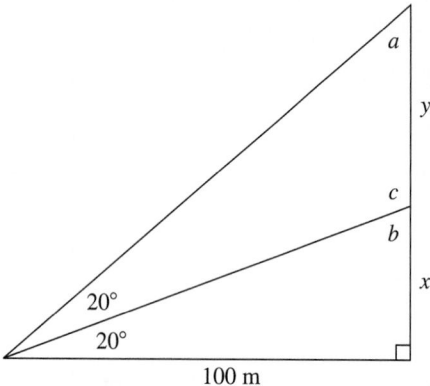

Figure 9.91
A triangle for Exercises 35 and 37

36. a. Use trigonometry and the combined 10° + 15° = 25° angle at the left of the big triangle in Figure 9.92 to find the size of x + y.
b. Use trigonometry and the 15° angle at the left of the lower triangle to find the size of x.
c. Use the answers to parts (a) and (b) to find the size of y.

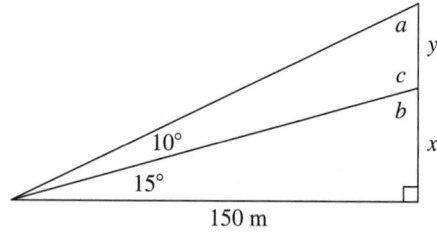

Figure 9.92
A triangle for Exercises 36 and 38

37. a. Use the combined 20° + 20° = 40° angle at the left of the big triangle in Figure 9.91 and the fact that the three angles of a triangle always add to 180° to find the size of angle a.
b. Use the 20° angle at the left of the lower triangle and the fact that the three angles of a triangle always add to 180° to find the size of angle b.
c. Use the answer to part (b) to find the size of angle c.
38. a. Use the combined 10° + 15° = 25° angle at the left of the big triangle in Figure 9.92 and the fact

that the three angles of a triangle always add to 180° to find the size of angle a.
b. Use the 15° angle at the left of the lower triangle and the fact that the three angles of a triangle always add to 180° to find the size of angle b.
c. Use the answer to part (b) to find the size of angle c.

39. a. Use trigonometry and the combined 20° + 25° = 45° angle at the left of the big triangle in Figure 9.93 to find the size of x + y.
b. Use trigonometry and the 25° angle at the left of the lower triangle to find the size of x.
c. Use the answers to parts (a) and (b) to find the size of y.

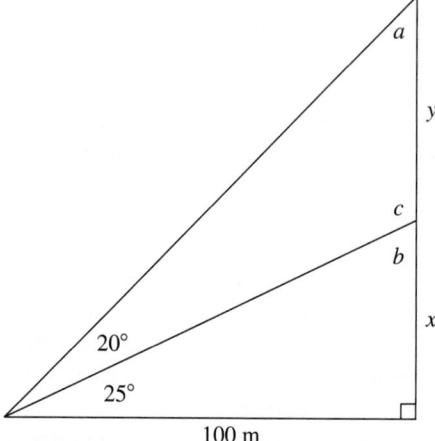

Figure 9.93
A triangle for Exercises 39 and 41

40. a. Use trigonometry and the combined 20° + 15° = 35° angle at the left of the big triangle in Figure 9.94 to find the size of x + y.
b. Use trigonometry and the 15° angle at the left of the lower triangle to find the size of x.
c. Use the answers to parts (a) and (b) to find the size of y.

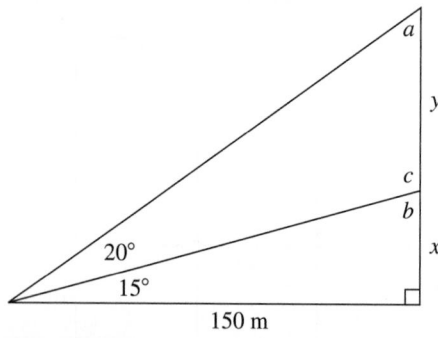

Figure 9.94
A triangle for Exercises 40 and 42

41. a. Use the combined 20° + 25° = 45° angle at the left of the big triangle in Figure 9.93 and the fact that the three angles of a triangle always add to 180° to find the size of angle *a*.
  b. Use the 25° angle at the left of the lower triangle and the fact that the three angles of a triangle always add to 180° to find the size of angle *b*.
  c. Use the answer to part (b) to find the size of angle *c*.

42. a. Use the combined 20° + 15° = 35° angle at the left of the big triangle in Figure 9.94 and the fact that the three angles of a triangle always add to 180° to find the size of angle *a*.
  b. Use the 15° angle at the left of the lower triangle and the fact that the three angles of a triangle always add to 180° to find the size of angle *b*.
  c. Use the answer to part (b) to find the size of angle *c*.

In Exercises 43–46, use symmetry, trigonometry, and Steiner points to find the shortest network connecting the three cities, and the length of that network.

43.

44.

45.

46.

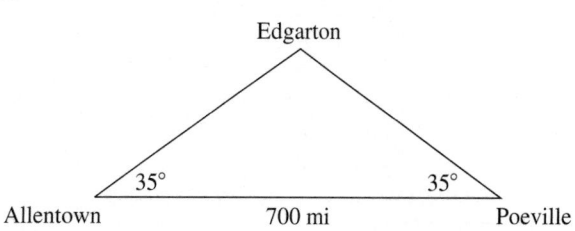

47. Which is the shortest network? What is its length? All interior points are Steiner points, and the object is a rectangle.

a. 600 mi, 500 mi   **b.** 600 mi, 500 mi
c. 600 mi, 500 mi   **d.** 600 mi, 500 mi

48. Which is the shortest network? What is its length? All interior points are Steiner points, and the object is a rectangle.

a. 800 mi, 600 mi

b. 800 mi, 600 mi

c.

d.

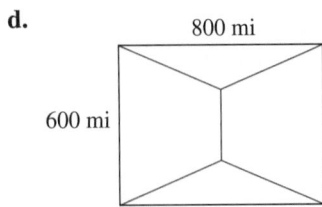

In Exercises 49–52, use symmetry, trigonometry, and Steiner points to find the shortest network and the length of that network.

49. Find the shortest network, and its length, for a rectangle that is 700 miles wide and 600 miles long.

50. Find the shortest network, and its length, for a rectangle that is 700 miles wide and 800 miles long.

51. Which is the shortest network? What is its length? The interior points in (c) are Steiner points, and the object is a square.

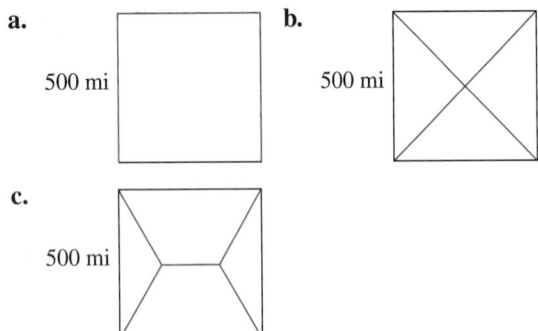

52. Find the shortest network and its length for a square that is 800 miles wide.

53. Find the length of the network in Example 4.

54. Use a Steiner point locator to find the shortest network for a triangle with sides 7 mi, 8 mi, and 10 mi and its length.

55. Use a Steiner point locator to find the shortest network for a triangle with sides 8 mi, 9 mi, and 10 mi and its length.

56. Use a Steiner point locator to find the shortest network for a triangle with sides 9 mi, 10 mi, and 12 mi and its length.

57. Use a Steiner point locator and the map of Nevada in Figure 9.95 on page 699 to find the shortest network joining Reno, Las Vegas, and Wells and its length.

58. Use a Steiner point locator and the map of Arizona in Figure 9.96 on page 700 to find the shortest network joining Phoenix, Winslow, and Prescott and its length.

59. Use a Steiner point locator and the map of Florida in Figure 9.97 on page 701 to find the shortest network joining Tampa, Orlando, and Miami and its length.

60. Use a Steiner point locator and the map of New York in Figure 9.98 on page 702 to find the shortest network joining Poughkeepsie, Rochester, and Troy and its length.

Answer the following questions using complete sentences and your own words.

CONCEPT QUESTIONS

61. In the *New York Times* article on page 692, Dr. Ronald Graham is quoted as saying "The problem is of tremendous interest . . . for obvious reasons." What are the obvious reasons?

62. Figure 9.99 shows a Japanese stamp issued in honor of TPC-3 ("Trans-Pacific Cable number 3"). This was the first fiber-optic telephone cable across the Pacific. It links Japan, Guam, and Hawaii. It was constructed by a consortium of what were then the worlds largest telephone companies: AT&T, Sprint, MCI, and British Telephone.

Figure 9.99
A Japanese stamp in honor of TPC-3

a. TPC-3, the network colored purple on the stamp, links Hawaii, Japan, and Guam. What type of network is this? Why was this type of network used?

b. HAW-4, the network colored blue on the stamp, links Hawaii and the west coast of the United States. What type of network is this? Why was this type of network used?

c. Japan is the purple island on the stamp. It is 3,910 miles from Hawaii and 1,620 miles from Guam. Guam is the blue dot directly below Japan. It is 3,820 miles from Hawaii. Use this information and a Steiner point locator to approximate the length of the network linking Hawaii, Japan, and Guam.

Figure 9.95
Nevada

Figure 9.96
Arizona

Figure 9.97
Florida

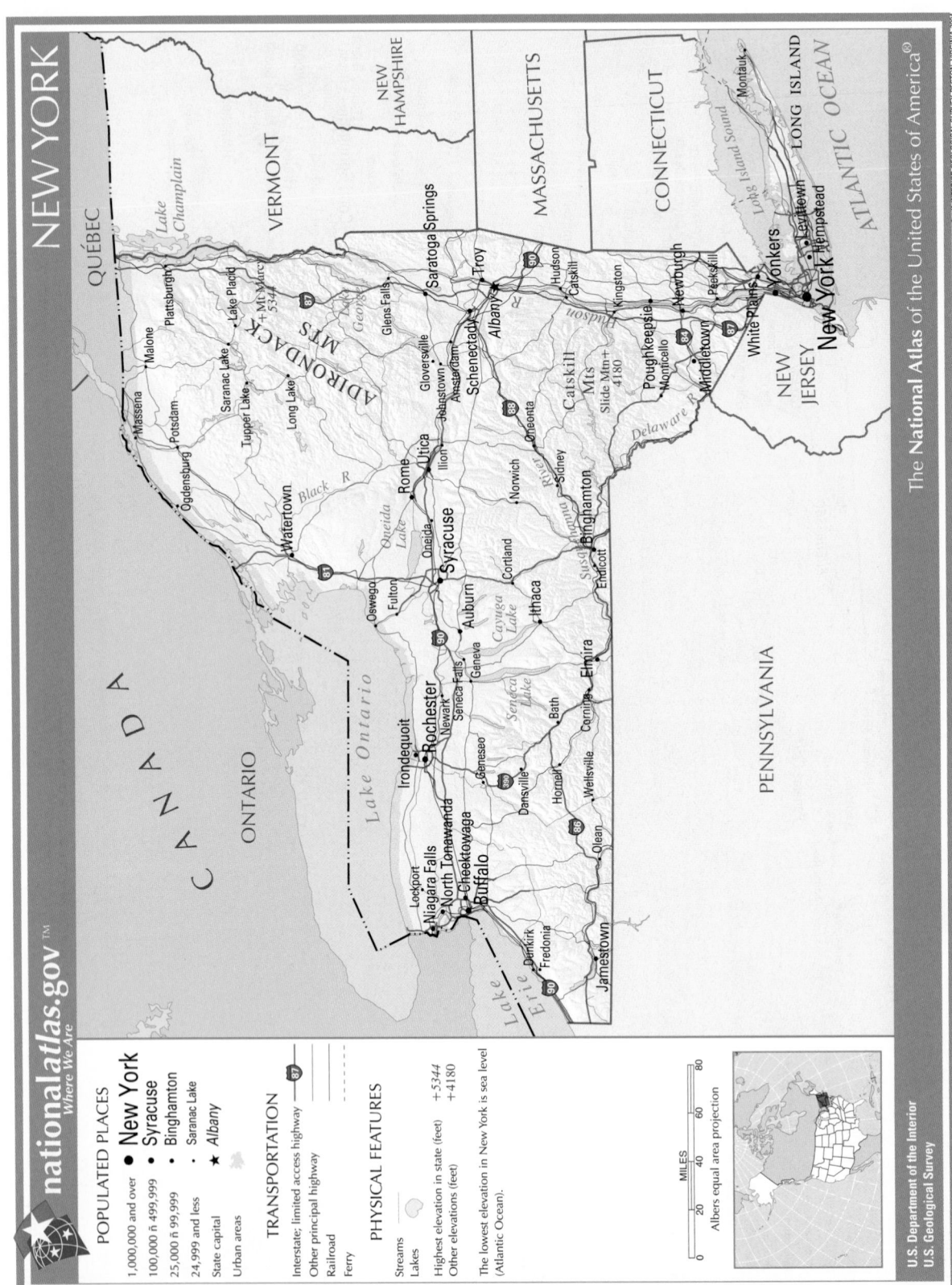

The National Atlas of the United States of America®

nationalatlas.gov™
Where We Are

POPULATED PLACES

● New York — 1,000,000 and over
● Syracuse — 100,000 ñ 499,999
● Binghamton — 25,000 ñ 99,999
· Saranac Lake — 24,999 and less
★ Albany — State capital
— Urban areas

TRANSPORTATION

Interstate; limited access highway
Other principal highway
Railroad
Ferry

PHYSICAL FEATURES

Streams
Lakes
Highest elevation in state (feet)   +5344
Other elevations (feet)   +4180

The lowest elevation in New York is sea level (Atlantic Ocean).

U.S. Department of the Interior
U.S. Geological Survey

MILES
0   20   40   60   80
Albers equal area projection

Figure 9.98
New York

Project managers and other people involved in scheduling projects typically use PERT charts and Gantt charts in producing and maintaining schedules. These charts make it easier to schedule a project's tasks efficiently. They can have a huge impact on the project's cost and completion time.

A Group Project

You have been assigned to work with three other people to write a large research paper. You have been given five weeks to complete the task.

At the group's first planning meeting, one group member pointed out that the assigned topic naturally breaks down to two separate parts. That person went on to suggest that two group members could work on one topic, and two could work on the other. Furthermore, of the two people working on one topic, one could do the research, and one could do the writing. So this plan involved four distinct jobs, one for each of the group's members:

> *Topic 1:*
>
> Al would research.
>
> Betty would write.
>
> *Topic 2:*
>
> Carla would research.
>
> Dave would write.

Dependencies, Milestones, and PERT Charts

Another group member suggested that the group wait four weeks before starting. He reasoned that in the last week, Betty and Dave could do their writing while Al and Carla did their research. The others politely pointed out that the writing couldn't happen until after the research was done. In scheduling, this is called a **dependency.**

The group decided that Al's research would take two weeks, but Carla's research would take only one week. They allotted two weeks for Betty and Dave to write, but they decided that the writing couldn't start until all the research was finished. They also realized that someone should make sure that the two topics were put together well when the writing was over. Carla agreed to do this, since her research task was less time consuming.

The weighted digraph in Figure 9.100 on page 704 is an easy way of describing the proposed schedule. This type of graph is called a **PERT chart.** *PERT* stands for "program evaluation and review technique." PERT was developed in the 1950s by the U.S. Navy to aide in developing and managing the Polaris submarine project. It is used frequently today by project managers.

A PERT chart's vertices are called **milestones.** They are numbered, and the numbers are used to list each milestone and its date.

A PERT chart's edges are the project's tasks. The tasks "Al research" and "Betty write" are called **sequential tasks** because they must happen in sequence, one after the other. The tasks "Al research" and "Carla research" are called **parallel tasks** because there is no dependency between them and they can happen simultaneously.

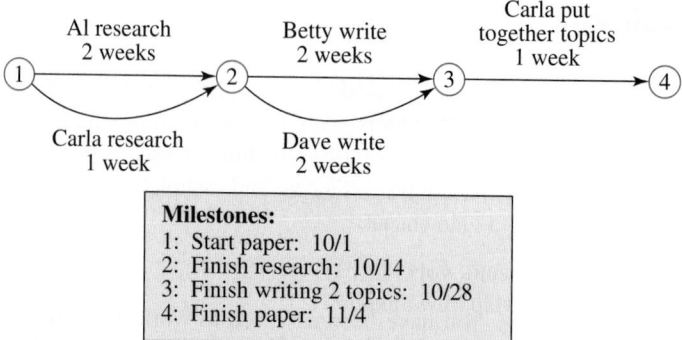

Figure 9.100
The group's proposed schedule, displayed on a PERT chart

Limiting Tasks and Critical Paths

Of the two parallel research tasks, Al's research takes two weeks, and Carla's only takes one week. This means that the task "Al research" limits the total time to complete the project, and "Carla research" does not limit the total time. For this reason, "Al research" is called a **limiting task.**

The project's **critical path** is the path that includes all of the limiting tasks. The length of the critical path is the project's **completion time.** Figure 9.101 uses heavier arrows to show the research paper's critical path. The paper's completion time is

$$2 \text{ weeks} + 2 \text{ weeks} + 1 \text{ week} = 5 \text{ weeks}$$

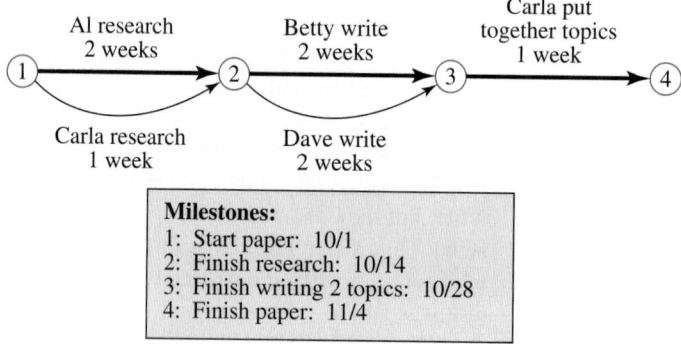

Figure 9.101
The research paper's critical path

Gantt Charts

Gantt charts are also used in scheduling. A Gantt chart describing the proposed schedule is shown in Figure 9.102. Tasks are described along the side, and dates are given along the top. There is a horizontal bar for each task. Each bar shows the task's start, duration, and completion. The dependencies are shown with arrows.

Figure 9.102
The group's proposed schedule, displayed on a Gantt chart

Gantt charts were invented by Henry Gantt (1861–1919), a mechanical engineer and management consultant. The Gantt chart was an important innovation in the 1920s. They were used on large construction projects such as the Hoover Dam in 1931 and the interstate highway network in 1956. Today, they are frequently used by project managers, often in conjunction with PERT charts.

A More Extensive Project

Projects that are more extensive than the research paper require a more organized approach. We will use the following steps.

PERT Steps

1. *List all tasks.* Give each task's estimated time length and dependencies. Usually, this step is done by someone who is very familiar with the project. In this book, this step will be done for you.

2. *Draw an edge for each task that has no dependencies.* These edges should all begin at one vertex. Each edge should be weighted with the task's time length and labeled with a brief description of the task. If space doesn't allow this, list the tasks in a key, as is done with milestones.

3. *Draw an edge for each task that depends on the tasks in step 2.* Position these edges so that dependencies are clear.

4. *Draw the rest of the edges.* A task's edge can be drawn after the tasks that it depends on are drawn.

5. *Create a milestone key* and possibly a task key. Each vertex (or milestone) should be numbered, and a key should be included that describes the vertex's (or milestone's) date and the event that is to occur on that date.

6. *Determine the project's critical path and its completion time.*

EXAMPLE 1 Producing a film is a very complex project. A general description of such a project, starting from when a script is obtained, is given in Figure 9.103 on page 706.

a. Draw a PERT chart for the project.
b. Determine the project's critical path and its completion time.

Task	Length	Dependency
a. Create budget, obtain funds	8 weeks	
b. Hire director	4 weeks	
c. Hire actors	4 weeks	b
d. Hire production staff	2 weeks	a
e. Advertise and contact agents	3 weeks	
f. Scout locations	2 weeks	c, d, e
g. Build sets	4 weeks	f
h. Film non-set-dependent scenes	12 weeks	f
i. Film set-dependent scenes	9 weeks	g
j. Pick conductor and music	3 weeks	h, i
k. Edit film	4 weeks	h, i
l. Write press releases, create preview	2 weeks	h, i
m. Prescreen with audiences	2 weeks	k
n. Create soundtrack CD	1 weeks	j

Figure 9.103
A film project

Solution   **a.** We'll start with tasks a, b, and e, because they have no dependencies. See Figure 9.104.

Now we'll insert tasks c and d because they depend on a and b. See Figure 9.105.

Next, we'll insert task f, because it depends on tasks c, d, and e, and we'll insert the tasks that depend on task f. See Figure 9.106.

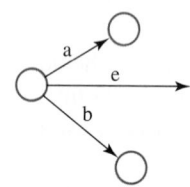

Figure 9.104
Tasks a, b, and e

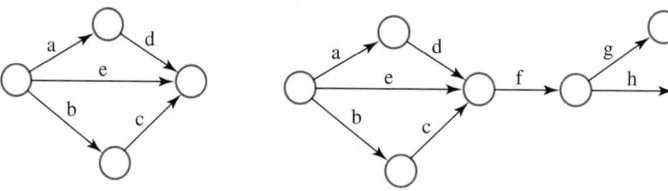

Figure 9.105
Tasks a through e

Figure 9.106
Tasks a through h

Now we'll insert task i, because it depends on task g, and we'll insert tasks j, k, and l, because they depend on h and i. See Figure 9.107.

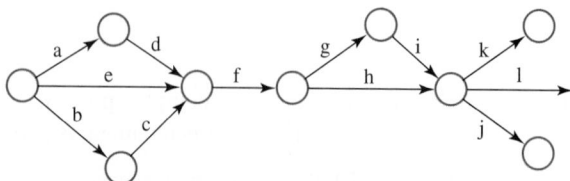

Figure 9.107
Tasks a through l

Finally, we'll insert the remaining tasks. See Figure 9.108.

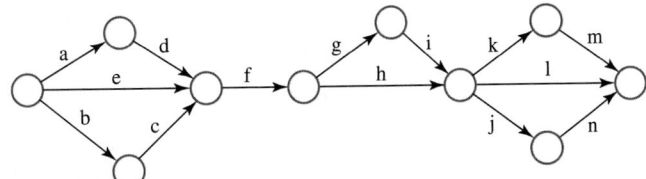

Figure 9.108
All tasks are inserted

The completed PERT chart, including tasks and milestones, is shown in Figure 9.109.

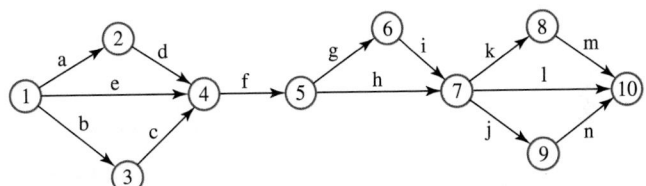

Tasks:	**Milestones:**
a. Create budget, obtain funds: 8 wks	1. Script obtained
b. Hire director: 4 wks	2. Funding acquired
c. Hire actors: 4 wks	3. Director signed
d. Hire production staff: 2 wks	4. Actors and production staff signed
e. Advertise and contact agents: 3 wks	5. Locations picked
f. Scout locations: 2 wks	6. Sets finished
g. Build sets: 4 wks	7. Filming completed
h. Film non-set-dependent scenes: 12 wks	8. Film edited
i. Film set-dependent scenes: 9 wks	9. Soundtrack completed
j. Pick conductor, music: 3 wks	10. Film released
k. Edit film: 4 wks	
l. Write press releases, create preview: 2 wks	
m. Prescreen with audiences: 2 wks	
n. Create soundtrack CD: 1 wk	

Figure 9.109
The complete PERT chart

b. To find the critical path, we must first find the limiting tasks.
Tasks a and d, e, and b and c are parallel tasks.

Tasks a and d take 8 wks + 2 wks = 10 wks

Task e takes 3 wks

Tasks b and c take 4 wks + 4 wks = 8 wks

So tasks a and d are limiting tasks.
Task f has no parallel tasks so it is a limiting task.
Tasks g and i, and h are parallel.

Tasks g and i take 4 wks + 9 wks = 13 wks

Task h takes 12 wks

So tasks g and i are limiting tasks.
   Tasks k and m, l, and j and n are parallel tasks.

Tasks *k* and *m* take 4 wks + 2 wks = 6 wks

Task *l* takes 2 wks

Tasks *j* and *n* take 3 wks + 1 wk = 4 wks

So tasks k and m are limiting tasks.
   The critical path is shown in Figure 9.110.

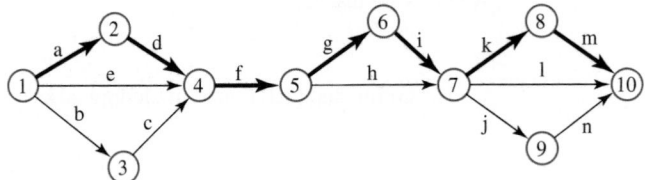

Figure 9.110
The critical path

The completion time is

8 wks + 2 wks + 2 wks + 4 wks + 9 wks + 4 wks + 2 wks = 31 wks ∎

Once a PERT chart has been made, it's easy to make a Gantt chart.

Gantt Steps

1. *Draw a PERT chart.*
2. *Draw a bar for each task that has no dependencies.* These bars should all begin at the same time.
3. *Draw a bar for each task that depends on the tasks in step 2.* These bars should start at the same time that their dependencies stop. Use arrows to make dependencies clear.
4. *Draw the rest of the bars.* A task's bars can be drawn after the tasks that it depends on are drawn.

EXAMPLE 2 **a.** Draw a Gantt chart for the project described in Example 1.
b. If the film proceeds on schedule, how much time should have elapsed before the actors are hired?

Solution **a.** The Gantt chart is shown in Figure 9.111. This Gantt chart differs from the one in Figure 9.102 in that week numbers are given rather than dates, and tasks are described to the right of the bars because of space considerations.

b. The Gantt chart indicates that the actors could be hired as early as week 4. However, the only task that is immediately dependent on this is task f, "scout location." This task cannot start until week 10, because of other dependencies. So the actors should be hired between weeks 4 and 10.

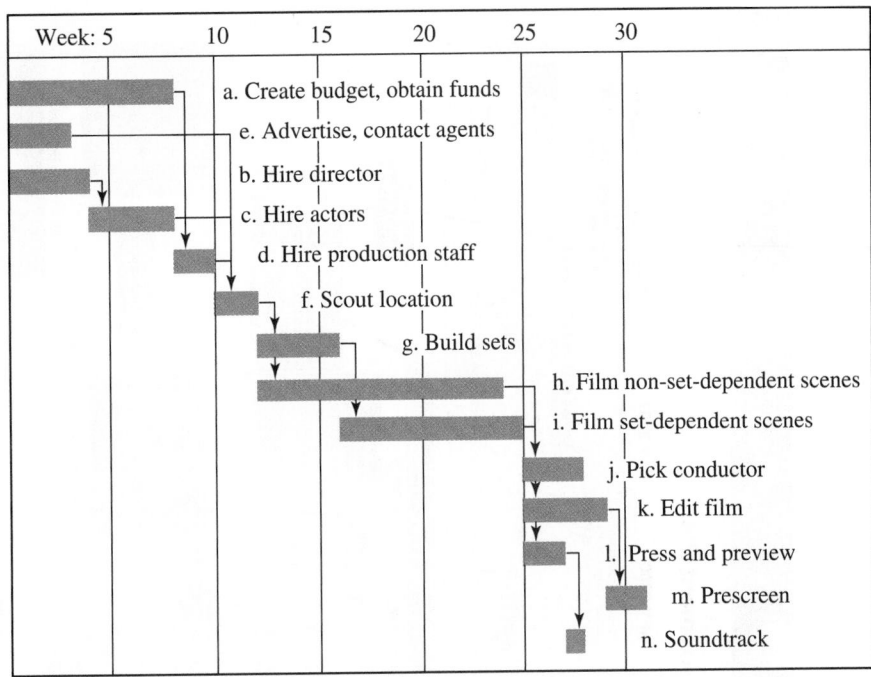

Figure 9.111
A Gantt chart for Example 2.   ■

Advantages and Disadvantages

A PERT chart shows a project's critical path. This in turn makes it easier to use a PERT chart to determine which tasks must be finished on time and which can be delayed for a while. It also makes it easier to use a PERT chart to determine which tasks would be good candidates for acceleration if the project's completion time must be shortened.

It's easier to use a Gantt chart to determine whether the project is on schedule. If the project is not on schedule, it's easier to use a Gantt chart to find remedial actions that would allow it to return to schedule.

The Helios aircraft

PERT Charts and the Real World: NASA

The following is excerpted from a NASA article: "Most program/project and task managers use the Gantt chart format for their graphic display of project plans and actual accomplishments. It is a simple tool to use, and displays a lot of information on a computer screen. . . . From the standpoint of communicating the overall picture of what needs to be done, when and why, to both the project team and our customers, however, I've found the PERT chart to be better. . . . In our solar aircraft development program, we used two types of precedence charts extensively for communication of program/project plans. . . . A top-level program chart, spanning eight years, is shown in Figure 1 [Figure 9.112 in this text].

"The chart was much more than window dressing, as we often referred back to it in team meetings to help redefine the importance of a current task and to see how it fit into 'the big picture.'" This became a very valuable tool for the team.

"With pride, we saw blocks filled in with actual pictures of our accomplishments (as well as programmatic re-adjustments when necessitated by problems). Enthusiasm for accomplishing the next goal was reborn each time we looked at the graphics on our wall."

See Exercises 28–34.

(Box continued on next page)

PERT Charts and the Real World: NASA (Continued)

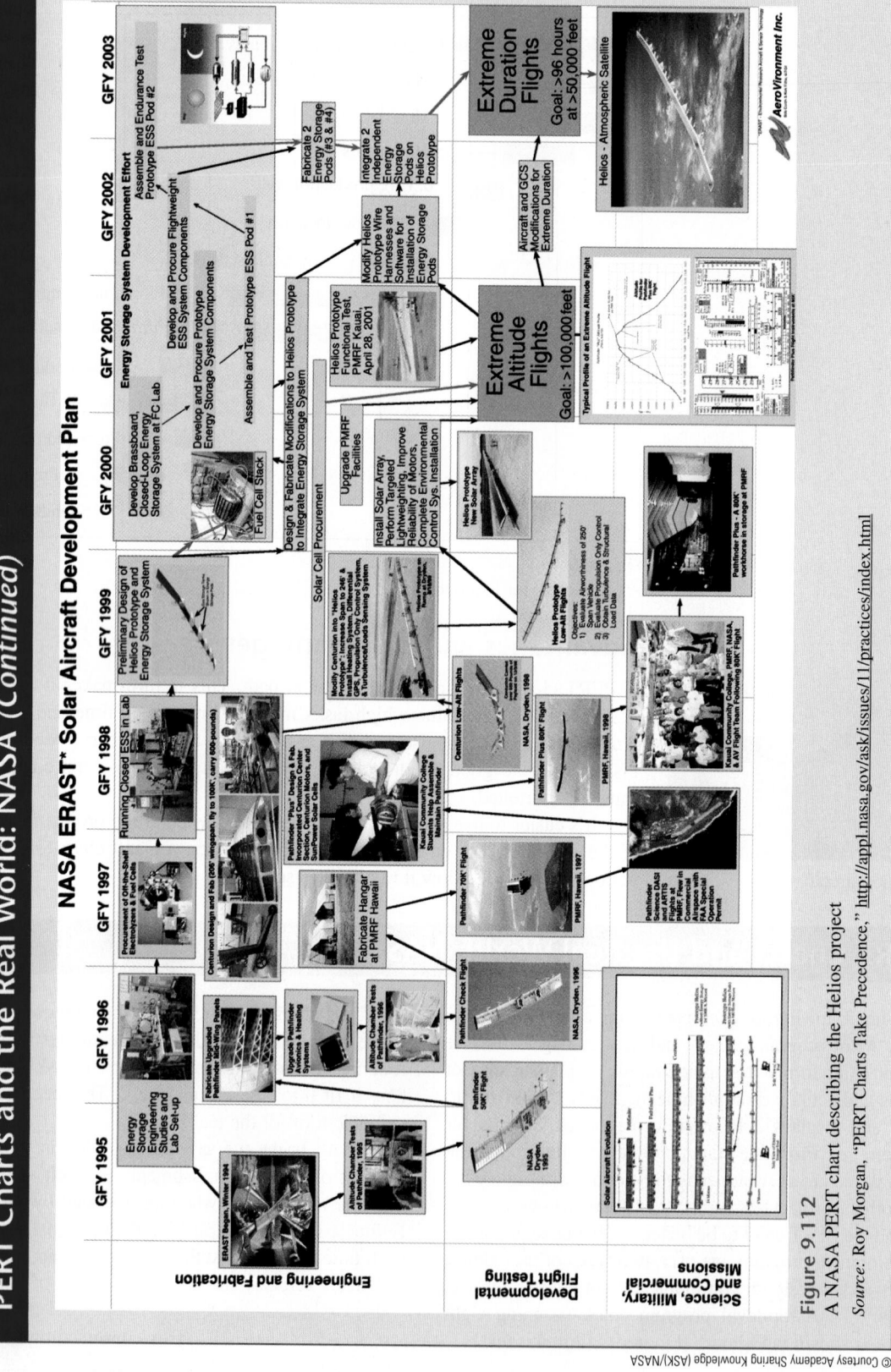

Figure 9.112

A NASA PERT chart describing the Helios project

Source: Roy Morgan, "PERT Charts Take Precedence," http://appl.nasa.gov/ask/issues/11/practices/index.html

9.5 Exercises

Exercises 1–7 involve the information in Figure 9.113 on painting a room.

1. Create a PERT chart for painting a room. Show the critical path. List milestones.
2. What is the completion time?
3. How many workers would it take to finish the painting job within the completion time?
4. How long would it take to finish the painting job if there were only one worker?
5. Create a Gantt chart for painting a room.
6. If the painting job proceeds on schedule, how much time should have elapsed before the trim is painted? If appropriate, give a range of time.
7. If the painting job proceeds on schedule, how much time should have elapsed before the wall paint is dry and the room is inhabitable? If appropriate, give a range of time.

Task	Length	Dependencies
a. Select paint color	2 days	
b. Calculate square footage of area to be painted.	30 minutes	
c. Buy paint	1 hour	a, b
d. Buy drop cloths, brushes, rollers, TSP, tape	1 hour	
e. Wash walls with TSP	2 hours	d
f. Remove lights, plug covers and switch covers	30 minutes	
g. Tape walls and windows alongside trim	1 hour	d
h. Paint trim	3 hours	c, d, e, g
i. Let trim paint dry	2 hours	h
j. Remove wall tape	15 minutes	i
k. Tape trim alongside walls	1 hour	d, i
l. Paint walls	3 hours	c, d, e, f, g, h, k
m. Let wall paint dry	2 hours	l
n. Remove trim tape and window tape	15 minutes	m
o. Replace lights, plug covers and switch covers	30 minutes	m
p. Clean brushes, rollers	30 minutes	l
q. Dispose of drop cloths, tape	15 minutes	n

Figure 9.113
Painting a room

Exercises 8–14 involve the information in Figure 9.114 on installing a drip irrigation system.

8. Create a PERT chart for installing a drip irrigation system. Show the critical path. List milestones.
9. What is the completion time?
10. How many workers would it take to finish the installation within that completion time?
11. How long would it take to finish the installation if there were only one worker?
12. Create a Gantt chart for installing a drip irrigation system.
13. If the installation proceeds on schedule, how much time should have elapsed before all necessary purchases are made? If appropriate, give a range of time.
14. If the installation proceeds on schedule, how much time should have elapsed before the valves are wired to the control panel? If appropriate, give a range of time.

Excercises 15–21 involve the information in Figure 9.115 on having a wedding reception or anniversary celebration.

15. Create a PERT chart for having a wedding reception or anniversary celebration. Show the critical path. List milestones.
16. What is the completion time?
17. How many workers would it take to finish the preparations within that completion time?
18. How long would it take to finish the preparations if there were only one worker?
19. Create a Gantt chart for having a wedding reception or anniversary party.
20. If the project proceeds on schedule, how much time should have elapsed before the caterer is hired? If appropriate, give a range of time.
21. If the project proceeds on schedule, how much time should have elapsed before needed clothing is purchased? If appropriate, give a range of time.

Task	Length	Dependencies
a. Determine water needs of plants	3 hours	
b. Lay out system	1 hour	a
c. Determine length of tubing needed	15 minutes	a, b
d. Determine number and type of emitters needed	15 minutes	a, b
e. Determine location of control panel	15 minutes	
f. Determine number of separately controlled lines	15 minutes	a, b, d
g. Buy emitters, tubing, valves, control panel, wire	1 hour	b, c, d, e, f
h. Get wire from control panel to valves	3 hours	e, f, g
i. Put tubing in place	1 hours	b, c, f, g
j. Cut tubing	30 minutes	g, i
k. Connect cut sections of tubing	30 minutes	g, j
l. Install emitters	1 hour	d, g, i, j, k
m. Test system	15 minutes	k, l

Figure 9.114
Installing a drip irrigation system

Task	Length	Dependencies
a. Create guest list	1 week	
b. Check out different room options	6 hours	
c. Select invitations and place cards	2 hours	
d. Wait for invitations to return from printer	1 week	c
e. Mail invitations	1 hour	d
f. Wait for RSVP's to return from invited guests	2 weeks	d
g. Rent a room	1 hour	a
h. Check out different hotel options for out of town guests	4 hours	
i. Make hotel arrangements	1 hour	h
j. Check out catering options	6 hours	
k. Hire caterer	1 hour	j
l. Determine menu	2 hours	k
m. Give caterer info on number of guests and their food choices	1 hour	f, k
n. Give printer list of attendees for place cards	30 mins	f
o. Wait for place cards to return from printer	1 week	n
p. Check out band options	3 hours	
q. Hire band	1 hour	g, p
r. Check out table decoration options	4 hours	
s. Hire florist	1 hour	r
t. Determine bar needs	2 hours	f
u. Arrange for bar	1 hour	t
v. Determine clothing needs of immediate family	1 hour	
w. Purchase needed clothing	8 hours	v

Figure 9.115
Having a wedding reception or anniversary celebration

Exercises 22–27 involve the information in Figure 9.116 on constructing a home.

22. Create a PERT chart for constructing a home. Show the critical path. List milestones.

23. What is the completion time?
24. Create a Gantt chart for constructing a home.
25. If the project proceeds on schedule, how much time should have elapsed before the roof is built? If appropriate, give a range of time.

Task	Length	Dependency
a. Write and agree on contracts	4 weeks	
b. Secure financing	4 weeks	a
c. Obtain permits	6 weeks	a, b
d. Site work	4 weeks	a, b, c
e. Build foundation	3 weeks	b, c, d
f. Erect walls	3 weeks	d, e
g. Build roof	4 weeks	f
h. Rough in plumbing	2 weeks	f
i. Rough in electrical	2 weeks	f
j. Rough in furnace and air conditioning	1 week	f
k. Finish exterior	6 weeks	f, g, i
l. Finish interior	6 weeks	f, g, h, i, j
m. Finish pluming	2 weeks	h, i
n. Finish electrical	1 week	i, l
o. Finish furnace and air conditioning	1 day	j, l
p. Wait for inspection	2 weeks	k, l, m, n, o

Figure 9.116
Constructing a home

26. If the roof is late, what other tasks could be accelerated to keep the job on schedule?

27. If the project proceeds on schedule, how much time should have elapsed before the inspection? If appropriate, give a range of time.

Use the PERT chart in Figure 9.112 to answer Exercises 28–34.

28. When did the Helios program begin? What were the first tasks?

29. The task "Preliminary design of Helios prototype" was immediately dependent on what other tasks? What is the task's approximate length? What is the task's approximate completion date?

30. The task "Centurion low altitude flights" was immediately dependent on what other tasks? What is the task's approximate length? What is the task's approximate completion date?

31. The task "Extreme altitude flights" was immediately dependent on what other tasks? What is the task's approximate length? What is the task's approximate completion date?

32. Of all the tasks, which has the largest number of immediate dependencies?

33. What is the project's approximate completion time? What is its approximate completion date?

34. How does the NASA PERT chart differ from those in the text? Why do you think it has this difference?

PROJECTS

35. Identify a project (such as painting a room, installing a drip irrigation system, or planning a wedding reception) that you are familiar with. Do not use the projects discussed in the text. Create a detailed PERT chart and Gantt chart for the project.

Chapter 9   Review

TERMS

adjacent edges	digraph	identical graphs	spanning tree
adjacent vertices	directed graph	limiting task	Steiner point
algorithm	disconnected graph	loop	trail
approximation algorithm	edge	milestones	traveling salesman
circuit	Euler circuit	minimum Hamilton circuit	problem
complete graph	Euler trail	minimum spanning tree	tree
completion time	Eulerization	multiple edges	vertex
connected graph	even vertex	odd vertex	weighted graph
critical path	Gantt chart	parallel tasks	weights
degree of a vertex	graph	PERT chart	
dependency	Hamilton circuit	sequential tasks	

THEOREMS AND ALGORITHMS
Euler's theorem

- A connected graph with only even vertices has at least one Euler trail, which is also an Euler circuit.
- A connected graph with exactly two odd vertices and any number of even vertices has at least one Euler trail.

- A graph with more than two odd vertices has no Euler trails and no Euler circuits.
- It's impossible for a connected graph to have only one odd vertex.

Fleury's algorithm for finding Euler circuits

1. Verify that the graph has an Euler trail or Euler circuit, using Euler's theorem.
2. Choose a starting point.
   - If the graph has two odd vertices, pick either of the odd vertices as the starting point.
   - If the graph has no odd vertices, pick any point as the starting point.
3. Label each edge alphabetically as you travel that edge.

4. When choosing edges:
   - Never choose an edge that would make the yet-to-be-traveled part of the graph disconnected.
   - Never choose an edge that has already been followed.
   - Never choose an edge that leads to a vertex which has no other yet-to-be-traveled edges.

Eulerization algorithm

To efficiently Eulerize a graph that is laid out like a grid:
1. Choose a vertex along the outer perimeter of the graph.
2. If a vertex is an even vertex, add no edges and move on to the next vertex along the outer perimeter.

3. If a vertex is an odd vertex, add a duplicate edge that connects it to the next adjacent vertex along the outer perimeter, and move on to the next vertex along the outer perimeter.
4. Repeat these steps until you return to the vertex in Step 1.

The nearest neighbor algorithm for finding a minimum Hamilton circuit

1. From the starting vertex, go to the "nearest neighbor"
2. From all other vertices, use the edge that:
   - has the smallest weight and
   - leads to vertex that hasn't been visited yet and
   - does not lead to the starting vertex.

(If there are any ties in steps 1 or 2, choose at random. Different choices will result in different circuits, which may have different totals.)
3. The last edge must lead to the starting vertex.

The repetitive nearest neighbor algorithm for finding a minimum Hamilton circuit

1. Apply the nearest neighbor algorithm repeatedly, using each vertex as the starting point, and compute the total cost of the circuit obtained.

2. Choose the best of the Hamilton circuits obtained in step 1. If this circuit starts at an inappropriate vertex, resequence the circuit so that it starts at the appropriate vertex.
   (If there are any ties in steps 1 or 2, explore each option if the number of options is acceptably small.)

The cheapest edge algorithm for finding a minimum Hamilton circuit

1. List the weights, in order, from smallest to largest.
2. Draw a complete graph for the problem.
3. Pick edges, in order, from the list from step 1. Do not pick an edge if:
   - it leads to a vertex that you've already "flown in to."

   - it departs from a vertex that you've already "flown out of."
   - it closes a circuit that shouldn't be closed yet.
   (If there is a tie, choose at random. Different choices will result in different circuits, which may result in different totals.)
4. When all of the vertices are joined, close the circuit.

Choosing a Hamilton circuit algorithm

1. Find how many different circuits your graph has.
   - A complete directed graph with n vertices has $(n - 1)!$ different Hamilton circuits.
   - A complete nondirected graph with n vertices $\frac{(n - 1)!}{2}$ different Hamilton circuits.

2. Determine the appropriate method(s).
   - If the number of different circuits is acceptably small, then use the brute force method and be certain that you've found the Hamilton circuit with the least total weight.
   - If the number of different circuits is not acceptably small, then use any (or, better yet, all) of the approximation algorithms: the nearest neighbor algorithm, the repetitive nearest neighbor algorithm, or the cheapest edge algorithm.

Kruskal's algorithm for finding a minimum spanning tree

1. List the weights, in order, from smallest to largest.
2. Draw a complete graph for the problem.

3. Pick edges, in order, from the list from step 1. Do not pick an edge if it closes a circuit.
4. Stop adding edges when all the vertices are joined.

Triangle facts
- The sum of the three angles is 180°.
- If two sides have the same length, then the two angles across from those sides have the same size.
- If the two angles have the same size, then the two sides across from those angles have the same length.

- If the three sides have the same length, then the three angles have the same size.
- If the three angles have the same size, then the three sides have the same length.

REVIEW EXERCISES

In Exercises 1–6, do the following:

a. *Determine the number of vertices, edges, and loops in the given graph.*

b. *Draw another representation of the graph that looks significantly different.*

c. *Find two adjacent edges.*

d. *Find two adjacent vertices.*

e. *Find the degree of each vertex.*

f. *Determine if the graph is connected.*

1.

2.

3.

4.

5.

6.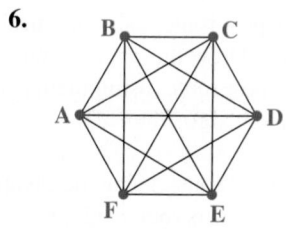

7. Draw a graph with five vertices: three of degree 2 and two of degree 3.

8. Draw a graph with six vertices: five of degree 1 and one of degree 3.

9. Draw a graph with four vertices: two of degree 3 and two of degree 4.

10. Draw a graph with four vertices: all of degree 3.

11. Draw a graph with four vertices in which each is of degree 3 and that has:

   a. loops but no multiple edges.

   b. multiple edges but no loops.

   c. neither loops nor multiple edges.

   d. both loops and multiple edges.

12. Draw a graph with six vertices in which each is of degree 4 and that has:

   a. loops but no multiple edges.

   b. multiple edges but no loops.

   c. both loops and multiple edges.

In Exercises 13–20, do the following.

a. *Use Euler's theorem to determine whether the graph has an Euler trail or an Euler circuit.*

b. *If it does, find it. If it does not, say why. Then Eulerize the graph, if possible.*

13. The graph in Exercise 1

14. The graph in Exercise 2

15. The graph in Exercise 3

16. The graph in Exercise 4

17. The graph in Exercise 5

18. The graph in Exercise 6

19.

20.

21. Figure 9.117 shows a map of the Hidden Circles gated community. After several burglaries, the residents hired a security guard to walk through the subdivision once every night.

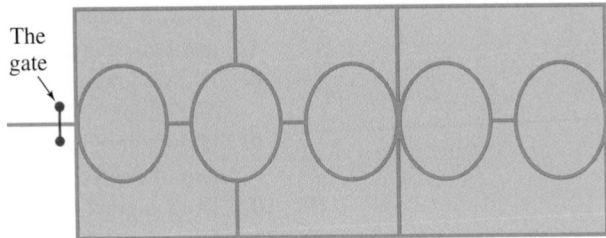

Figure 9.117
The Hidden Circles gated community

a. The guard would prefer to park his car, walk a route that allows him to walk through every block just once, and then return to his car. Is this possible? Explain why or why not.

b. If such a route is possible, find it. If such a route is not possible, find an efficient route.

22. Once a month, the utility company sends an agent through the Hidden Circles gated community to read the meters. (See Exercise 21 and Figure 9.117.) Each house has its own meter. Find a route for the meter reader if there are houses on both sides of each street.

23. Find an efficient route through the community shown in Figure 9.118 for the postal carrier it mailboxes are all on one side of each street.

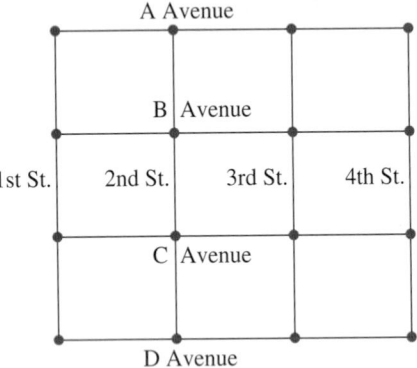

Figure 9.118
A graph for Exercise 23

24. Once a month, the utility company sends an agent through the community shown in Figure 9.118 to read the meters. Each house has its own meter.

a. Find a route for the meter reader if there are houses on both sides of each street.

b. Find a route for the meter reader if there are houses on both sides of each street except that the area bounded by A Avenue, B Avenue, First Street, and Second Street is a park.

Exercises 25–31 use the flight costs shown in Figure 9.119.

25. You live in Chicago, and your need to visit New York, Los Angeles, and Miami.

a. How many different circuits does your graph have?

b. Approximate the cheapest route, using the nearest neighbor algorithm. Draw this route's graph.

c. Approximate the cheapest route, using the repetitive nearest neighbor algorithm. Draw this route's graph.

d. Approximate the cheapest route, using the cheapest edge algorithm. Draw this route's graph.

e. Find the cheapest route, using the brute force method. Draw this route's graph.

26. You live in Los Angeles, and you need to visit Miami, Seattle, and Houston.

a. How many different circuits does your graph have?

b. Approximate the cheapest route, using the nearest neighbor algorithm. Draw this route's graph.

c. Approximate the cheapest route, using the repetitive nearest neighbor algorithm. Draw this route's graph.

d. Approximate the cheapest route, using the cheapest edge algorithm. Draw this route's graph.

e. Find the cheapest route, using the brute force method. Draw this route's graph.

27. You live in Chicago, and you need to visit New York, Los Angeles, Miami, and Seattle.

a. How many different circuits does your graph have?

b. Approximate the cheapest route, using the nearest neighbor algorithm. Draw this route's graph.

c. Approximate the cheapest route, using the repetitive nearest neighbor algorithm. Draw this route's graph.

d. Approximate the cheapest route, using the cheapest edge algorithm. Draw this route's graph.

		From:				
		CHI	NYC	LAX	MIA	SEA	HOU
	CHI		$194	$299	$74	$129	$133
	NYC	$243		$119	$79	$109	$148
To:	LAX	$299	$119		$124	$89	$154
	MIA	$74	$79	$124		$169	$94
	SEA	$129	$109	$89	$169		$294
	HOU	$133	$148	$154	$94	$294	

Figure 9.119
Flight costs between Chicago, New York, Los Angeles, Miami, Seattle, and Houston
Source: www.travelocity.com

e. Why would it not be appropriate to find the cheapest route, using the brute force method?

28. You live in New York, and you need to visit Los Angeles, Miami, Seattle, and Houston.
    a. How many different circuits does your graph have?
    b. Approximate the cheapest route, using the nearest neighbor algorithm. Draw this route's graph.
    c. Approximate the cheapest route, using the repetitive nearest neighbor algorithm. Draw this route's graph.
    d. Approximate the cheapest route, using the cheapest edge algorithm. Draw this route's graph.
    e. Why would it not be appropriate to find the cheapest route, using the brute force method?

29. You live in Chicago, and you need to visit New York, Los Angeles, Miami, Seattle, and Houston. Approximate the cheapest route, using the nearest neighbor algorithm. Draw this route's graph.

30. You live in Chicago, and you need to visit New York, Los Angeles, Miami, Seattle, and Houston. Approximate the cheapest route, using the repetitive nearest neighbor algorithm. Draw this route's graph.

31. You live in Chicago, and you need to visit New York, Los Angeles, Miami, Seattle, and Houston. Approximate the cheapest route, using the cheapest edge algorithm. Draw this route's graph.

Use the following information in Exercises 32–34.

You are programming a robotic drill to drill holes in a circuit board. The cost of drilling a board is determined by the distance the drill must travel, and the drill moves only horizontally and vertically.

32. A board needs holes at (5, 7), (1, 1), (6, 3), (7, 2), and (2, 4). Use the nearest neighbor algorithm to determine a sequence for the drilling of those holes.

33. A board needs holes at (5, 7), (1, 1), (6, 3), (7, 2), and (2, 4). Use the repetitive nearest neighbor algorithm to determine a sequence for the drilling of those holes.

34. A board needs holes at (5, 7), (1, 1), (6, 3), (7, 2), and (2, 4). Use the cheapest edge algorithm to determine a sequence for the drilling of those holes.

In Exercises 35–42, find a spanning tree for the given graph. Note: Answers will vary.

35. The graph in Exercise 1
36. The graph in Exercise 2
37. The graph in Exercise 3
38. The graph in Exercise 4
39. The graph in Exercise 5
40. The graph in Exercise 6
41. The graph in Exercise 19
42. The graph in Exercise 20

In Exercises 43–48, use Kruskal's algorithm to find the minimum spanning tree for the given weighted graph. Give the total weight of the minimum spanning tree.

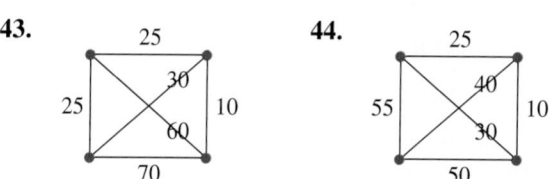

43.
25
25   30   10
      60
   70

44.
25
55   40   10
      30
   50

45.

46.

47.

48.

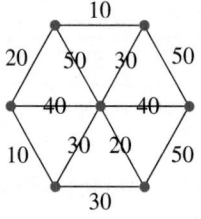

In Exercises 49–52, use the information in Figure 9.120.

49. You have been hired to design a network linking Boston, Chicago, Denver, Kansas City, and New York. Using only those cities as vertices, find the minimum spanning tree connecting the cities, and its total length.

50. You have been hired to design a network linking Kansas City, New York, Philadelphia, Portland, and San Francisco. Using only those cities as vertices, find the minimum spanning tree connecting the cities and its total length.

51. You have been hired to design a network linking Boston, Chicago, Denver, Kansas City, New York, Philadelphia, and San Francisco. Using only those cities as vertices, find the minimum spanning tree connecting the cities and its total length.

52. You have been hired to design a network linking Boston, Chicago, Denver, Kansas City, New York, Philadelphia, Portland, and San Francisco. Using only those cities as vertices, find the minimum spanning tree connecting the cities and its total length.

In Exercises 53–55, use symmetry, trigonometry, and Steiner points to find the shortest network connecting the three cities and the length of that network.

53.

54.

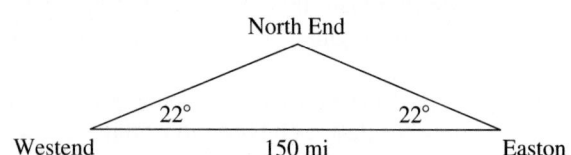

	Boston	Chicago	Denver	Kansas City MO	New York City	Philadelphia	Portland OR	San Francisco
Atlanta	932	585	1213	678	744	658	2170	2138
Boston		851	1767	1247	187	273	2536	2696
Chicago			917	407	714	664	1755	1855
Denver				558	1630	1574	979	947
Kansas City MO					1094	1032	1493	1500
New York						86	2442	2570
Philadelphia							2407	2518
Portland OR								538

Figure 9.120
Distances between cities, in miles

55.

Pleasant Hill

28° 28°

Concord 300 mi Four Corners

56. In Figure 9.121, four different networks are drawn for the same rectangle. Which is the shortest network: a, b, c, or d? What is its length? All interior points are Steiner points.

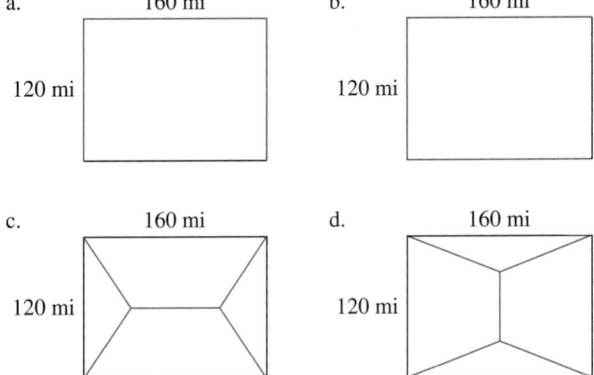

a. 160 mi   b. 160 mi

120 mi   120 mi

c. 160 mi   d. 160 mi

120 mi   120 mi

Figure 9.121
Graphs for Exercise 56

57. Use symmetry, trigonometry, and Steiner points to find the shortest network and its length for a rectangle that is 700 miles wide and 900 miles long.

58. Use symmetry, trigonometry, and Steiner points to find the shortest network and its length for a rectangle that is 700 miles wide and 1,000 miles long.

Exercises 59–64 involve the information in Figure 9.122 on staining a deck.

59. Create a PERT chart for staining a deck. Show the critical path. List milestones.

60. What is the completion time?

61. How many workers would it take to finish the painting job within the completion time?

62. How long would it take to finish the painting job if there were only one worker?

63. Create a Gantt chart for staining a deck.

64. If the staining job proceeds on schedule, how much time should have elapsed before the stain is purchased? It appropriate, give a range of time.

Task	Length	Dependency
a. Rent power washer	1 hour	
b. Remove patio furniture from deck	15 minutes	
c. Power wash deck	16 hours	a, b
d. Select type of stain (opaque or semitransparent) and stain color	2 days	
e. Calculate square footage of area to be stained	15 minutes	
f. Buy stain	1 hour	d, e
g. Buy drop clothes, brushes, rollers	1 hour	
h. Allow deck to dry	2 days	c
i. Stain deck	24 hours	f, h
j. Allow deck to dry	2 days	i

Figure 9.122
Staining a deck

10 Exponential and Logarithmic Functions

© Frank Whitney/Photographer's Choice/Getty Images

In this chapter, we will study a wide assortment of seemingly unrelated topics:

- the prediction of the size of human and animal populations
- the effect of inflation on home prices
- the decay of radioactive materials
- radiocarbon dating of archeological artifacts
- the Richter scale (which rates earthquakes)
- the decibel scale (which rates volumes of sounds)

We will answer questions such as the following:

- When will the world's population exceed the earth's ability to support it?
- How long will it take for my home to double its value?
- How long must the radioactive waste from a nuclear plant be stored?
- What does radiocarbon dating say about the authenticity of the Dead Sea Scrolls and the Shroud of Turin?
- How does a 7.0 earthquake compare with an 8.0 earthquake?

These diverse topics and questions are all related mathematically. Each involves the use of exponential and logarithmic functions.

In Sections 10.0A and 10.0B, we will discuss the algebra of exponential and logarithmic functions and the related use of calculators. In Sections 10.1 through 10.3, we will discuss the topics and questions listed above.

ThomsonNOW™ **for Liberal Arts Mathematics**
Throughout this chapter, the ThomsonNOW logo indicates an opportunity for online self-study, linking you to interactive tutorials and videos based on your level of understanding.

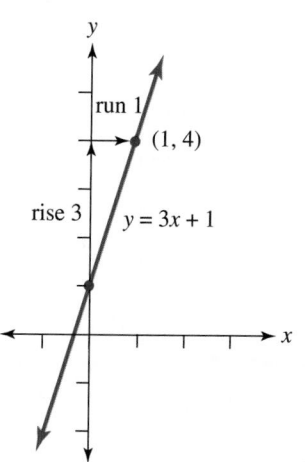

10.0A Review of Exponentials and Logarithms

Functions

An equation is said to be a **function** if to each value of x there corresponds one and only one value of y. For example, consider the equation $y = 3x + 1$; $x = 1$ corresponds to one and only one value of y (in particular, to $y = 4$). Other values of x also correspond to one and only one value of y, so the equation $y = 3x + 1$ is a function. When the value of y depends in this manner on the value of x, we say that y is a function of x; x is called the **independent variable,** and y is called the **dependent variable.**

Our equation $y = 3x + 1$ is called a **linear function,** because it is a function whose graph is a line. The graph of the equation $y = 3x + 1$ is shown in Figure 10.1. The slope of this line is 3; for every one-unit increase in x, the value of y increases by three units [slope = rise/run = (change in y)/(change in x)].

Figure 10.1
A linear function

Exponential Functions

A function in which x appears only in the exponent is called an **exponential function.** For example, $y = 2^x$ is an exponential function. Again, y is the dependent variable and x is the independent variable.

> **Exponential Function**
>
> An equation of the form $y = b^x$, where b is a positive constant, $(b \neq 1)$ is called an **exponential function.** The positive constant b is called the **base.**

As Example 1 shows, the graph of an exponential function is quite different from the graph of a linear function.

ThomsonNOW™
for Liberal Arts Mathematics
www.thomsonedu.com/1pass.

EXAMPLE 1   Sketch the graph of $y = 2^x$.

Solution   We can graph this exponential function by finding several ordered pairs, as in Figure 10.2.

The graph of all ordered pairs (x, y) satisfying the equation $y = 2^x$ is the curve shown in Figure 10.3.   ■

The graph of $y = 2^x$ is not a straight line. Typically, the graph of an exponential function is nearly horizontal over a large portion of the x-axis and then turns upward rather abruptly, increasing without bound. Because of this curvature, the slope is not constant, as it is with a linear function; the vertical change is minimal where the graph is nearly horizontal, whereas the vertical change is quite extreme where the graph is steep.

In this chapter, we will study various applications of exponential functions. These applications include the prediction of the size of human and animal populations, inflation, the decay of radioactive materials, the dating of archeological artifacts, the Richter scale (which rates earthquakes), and the decibel scale (which rates volumes of sounds).

x	$y = 2^x$	Ordered Pair (x, y)
3	$2^3 = 8$	$(3, 8)$
2	$2^2 = 4$	$(2, 4)$
1	$2^1 = 2$	$(1, 2)$
0	$2^0 = 1$	$(0, 1)$
-1	$2^{-1} = \dfrac{1}{2}$	$\left(-1, \dfrac{1}{2}\right)$
-2	$2^{-2} = \dfrac{1}{4}$	$\left(-2, \dfrac{1}{4}\right)$
-3	$2^{-3} = \dfrac{1}{8}$	$\left(-3, \dfrac{1}{8}\right)$

Figure 10.2
Finding ordered pairs

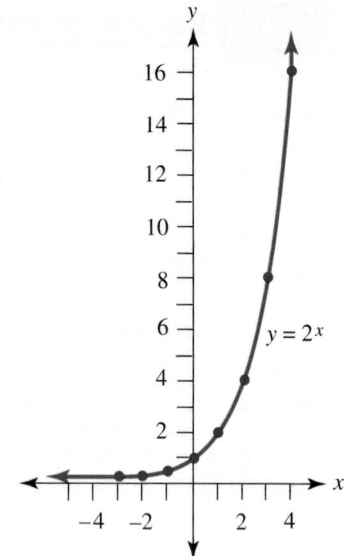

Figure 10.3
An exponential function

Rational and Irrational Numbers

Real numbers are either rational or irrational. A real number that is a terminating decimal, such as $\frac{3}{4} = 0.75$, or a repeating decimal, such as $\frac{2}{3} = 0.6666\ldots$, is called a *rational* **number** because it can be written as the *ratio* of two integers—that is, as a fraction. A real number that is neither a terminating decimal nor a repeating decimal is called an **irrational number.** The irrational numbers with which you are probably most familiar are square roots. For example, $\sqrt{2} = 1.414213562\ldots$ and $\sqrt{23} = 4.795831523\ldots$ are irrational.

One important irrational number is π. Whenever the circumference of any circle is divided by its diameter, the resulting number is *always* $\pi = 3.141592654.\ldots$ Another "famous" irrational number is $e = 2.71828182.\ \ldots$ The number e is used in most of the applications of mathematics in this chapter, including population prediction, inflation, radioactive decay, and the dating of archeological artifacts.

The Natural Exponential Function

The **natural exponential function** is $y = e^x$. Calculations involving e^x will be performed on a calculator. Some scientific calculators have a button that is labeled "e^x" on the button itself and "ln" or "ln x" above the button; to calculate e^1 with such a calculator, press

    1 $\boxed{e^x}$

Other scientific calculators have a button that is labeled "ln" or "ln x" on the button itself and "e^x" above the button; to calculate e^1 with such a calculator, press either

    1 $\boxed{\text{2nd}}$ $\boxed{e^x}$ or   1 $\boxed{\text{shift}}$ $\boxed{e^x}$ or   1 $\boxed{\text{INV}}$ $\boxed{e^x}$

In the future, we will simply write $\boxed{e^x}$ to refer to each of these sequences of keystrokes.

Most graphing calculators have a button that is labeled "LN" on the button itself and "e^x" above the button; to calculate e^1 with such a calculator, press

$$\boxed{\text{2nd}} \ \boxed{e^x} \ 1 \ \boxed{\text{ENTER}}$$

EXAMPLE 2   Use a calculator to find the following values:

a. e **b.** $1/e$ **c.** e^2 **d.** $1/e^2$ **e.** e^3
f. Using the above values, sketch the graph of $y = e^x$.

Solution   *Finding the Values.*

a. $e = e^1 = 2.718281828 \ldots$

b. $1/e = 0.367879441 \ldots$

An alternative way to find $1/e$ is to realize that $1/e = e^{-1}$.

c. $e^2 = 7.389056099 \ldots$

d. $1/e^2 = 0.135335283\ldots$

e. $e^3 = 20.08553692\ldots$

f. *Sketching the graph of* $y = e^x$. First, express the above values as ordered pairs, rounding off y-coordinates to the nearest tenth (which is sufficient for plot plotting), as shown in Figure 10.4. Then plot the ordered pairs and connect them with a smooth curve, as shown in Figure 10.5.

x	$y = e^x$	Ordered Pair (x, y)
3	$e^3 \approx 20.1$	$(3, 20.1)$
2	$e^2 \approx 7.4$	$(2, 7.4)$
1	$e^1 \approx 2.7$	$(1, 2.7)$
0	$e^0 = 1$	$(0, 1)$
-1	$e^{-1} \approx 0.4$	$(-1, 0.4)$
-2	$e^{-2} \approx 0.1$	$(-2, 0.1)$

Figure 10.4
Finding ordered pairs

The graph of $y = e^x$ has the same shape as the graph of $y = 2^x$. Both are nearly horizontal over a portion of the x-axis and then turn upward, increasing without bound; both graphs are entirely above the x-axis. Exponential functions (functions of the form $y = b^x$) always have these qualities. The independent variable x can have any value, but the dependent variable y can have only a positive value.

Because $y = e^x$ is an exponential function, its graph is not a straight line, and the slope is not constant; going from $x = -2$ to $x = -1$, the slope is

$$m \approx \frac{0.4 - 0.1}{-1 - (-2)} = 0.3 \qquad \text{using the ordered pairs generated in Figure 10.4}$$

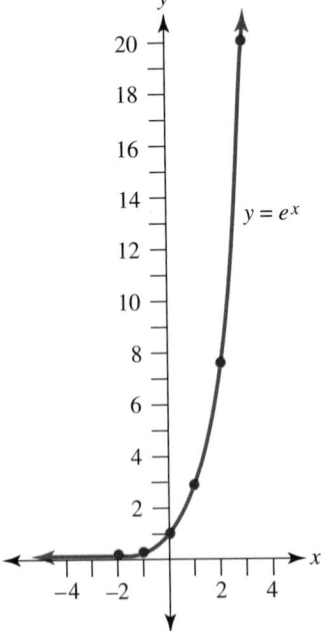

Figure 10.5
The natural exponential function

However, going from $x = 2$ to $x = 3$, the slope is

$$m \approx \frac{20.1 - 7.4}{3 - 2} = 12.7$$

This means that e^x grows much more quickly when x is a larger number. When we are dealing with population growth, this trait can become quite disturbing; the larger the population, the faster it grows. The use of exponential functions to analyze population growth is explored in Section 10.1.

Logarithms

Ten raised to what power gives 100? 1,000? 346? The answers to the first and second questions are easily found; the solution to $10^x = 100$ is $x = 2$, and the solution to $10^x = 1,000$ is $x = 3$. However, the solution to $10^x = 346$ is not so obvious! We could safely say that because $10^2 = 100$ and $10^3 = 1,000$, and because 346 is between 100 and 1,000, the solution to $10^x = 346$ must be between 2 and 3. To be more accurate would be difficult.

The idea of finding the exponent to which a number must be raised in order to get some particular number is the central concept of a **logarithm.** The question "3 raised to what power gives 9?" is the same question as "What is the logarithm (base 3) of 9?" (The answer to either question is 2.) Algebraically, the question "3 raised to what power gives 9?" is written as "$3^x = 9$," and the question "What is the logarithm (base 3) of 9?" is written "$\log_3 9 = x$." When we say $x = \log_3 9$, we are saying that x is the exponent to which we must raise 3 in order to get 9.

Logarithm Definition

$$\log_b u = v \text{ (or the \textbf{logarithm} of } u)$$

means the same as

$$b^v = u$$

In either equation, b is called the **base** and must be a positive number.

This definition of a logarithm allows us to rewrite a logarithmic equation as an exponential equation. The next example uses this fact.

EXAMPLE 3   **a.** Find v if $v = \log_2 8$.   **b.** Find u if $\log_5 u = -2$.   **c.** Find b if $\log_b 9 = 2$.

Solution   **a.** $v = \log_2 8$ can be rewritten as $2^v = 8$. By inspection, $v = 3$.
b. $\log_5 u = -2$ can be rewritten as $5^{-2} = u$. Therefore, $u = 1/5^2 = 1/25$.
c. $\log_b 9 = 2$ can be rewritten as $b^2 = 9$. By inspection, $b = \pm 3$. However, the base b of a logarithm is required to be positive. Thus, $b = 3$. ∎

Common Logarithms

$\text{Log}_{10} x$, in which 10 is the base, is called the **common logarithm.** Base 10 logarithms are used so commonly that the base is understood to be 10 when no base is written. Thus, $\log 346$ is an abbreviation for $\log_{10} 346$; $\log 346$ is the exponent to which 10 must be raised in order to get 346.

Historical Note Leonhard Euler, 1707–1783

North Wind Picture Archives

Leonhard Euler (pronounced "Oiler") was probably the most versatile and prolific writer in the history of mathematics. Euler produced over 700 books and papers, many of which were created during the last seventeen years of his life while he was totally blind. He wrote an average of 800 pages of mathematics each year. Euler's writings spanned pure mathematics, astronomy, annuities, life expectancy, lotteries, and music.

Euler was the son of a Swiss minister and mathematician who studied under Jacob Bernoulli. Euler was schooled in theology and mathematics and was to become a minister. However, his mathematical curiosity and ability prevailed, and his father eventually allowed him to concentrate on mathematics. He was awarded his master's degree at the age of sixteen! At the age of twenty-six, at the invitation of Catherine I, he was appointed to the Academy of St. Petersburg, a major center of scientific research. Later he was invited to Berlin by Frederick the Great.

Euler's phenomenal memory has been the subject of many legends. Having memorized Virgil's *Aeneid,* Euler could recite the first and last lines on any page of his copy. When confronted by two students who asked him to settle a disputed mathematical calculation in the fiftieth place, Euler successfully calculated the true result in his head!

One of Euler's skills was the creation of notations that were useful and compact. His writings were very popular, and the notations he introduced have endured. He introduced the notation $f(x)$ for a function, and he was the first to use π for $3.14159\ldots$, i for $\sqrt{-1}$, and e for $2.718281828.\ldots$

The Granger Collection, NY

METHODUS
INVENIENDI
LINEAS CURVAS
Maximi Minimive proprietate gaudentes,
SIVE
SOLUTIO
PROBLEMATIS ISOPERIMETRICI
LATISSIMO SENSU ACCEPTI
AUCTORE
LEONHARDO EULERO,
*Profeſſore Regio , & Academiæ Imperialis Scientia-
rum* PETROPOLITANÆ *Socio.*

LAUSANNÆ & GENEVÆ,
Apud MARCUM MICHAELEM BOUSQUET & Socios.
MDCCXLIV.

Common Logarithm Definition

$$y = \log x \qquad (\text{or} \qquad y = \log_{10} x)$$

means the same as

$$10^y = x$$

EXAMPLE 4 Find the following values by using a calculator:

    a. $\log 346$ **b.** $\log(0.82)$ **c.** $\log(-10)$ **d.** $\log(10^5)$

Solution   **a.** The $\boxed{\log}$ button on your calculator is used to find the common logarithm of a number. Simply enter the number you want to find the logarithm of and press $\boxed{\log}$. Some calculators require that you press $\boxed{\log}$ first and then enter the desired number.

$$\log 346 = 2.539076099 \ldots$$

Therefore, $10^{2.539076099} \approx 346$. (This is the answer to the earlier question "Ten raised to what power gives 346?" We had estimated that the power must be between 2 and 3; now we know that the power 2.539076099 gives 346.)

b. $\log(0.82) = -0.086186147 \ldots$

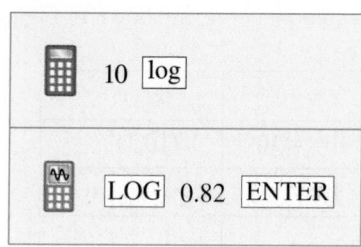

Therefore, $10^{-0.086186147} \approx 0.82$

c. $\log(-10) = \text{ERROR}$

Your calculator cannot find the common logarithm of -10 because negative numbers do not have logarithms! To see why, let $x = \log(-10)$. The corresponding exponential equation would be $10^x = -10$. This equation has no solution, because 10 raised to *any* power *always* gives a positive result; it is impossible for 10^x to equal -10. Consequently, $\log(-10)$ is undefined.

d. $\log(10^5) = \log 100{,}000 = 5$

■

The values of the independent variable x in the common logarithm function $y = \log x$ must always be positive, as was discussed in part (c) above, and the values of the dependent variable y can be either positive or negative, as was shown in parts (a) and (b). This situation is exactly the reverse of that for an exponential function, in which the independent variable can have any value and the dependent variable must be positive.

It should be noted that the explanation given in part (c) of Example 4 is an oversimplification; technically, it should be stated that negative numbers do not have logarithms that are *real numbers*. Negative numbers *do* have logarithms; however, the logarithm of a negative number is a complex (or imaginary) number. Some graphing calculators will compute the (complex) logarithm of a negative number; however, we will not study complex numbers in this textbook.

EXAMPLE 5   Sketch the graph of $y = \log x$.

Solution   Rather than computing the value of y for specific values of x, recall that $y = \log x$ means the same as $x = 10^y$. This exponential form allows for easier computations than does the original logarithmic form. Therefore, we compute the value of x for specific integer values of y. See Figure 10.6. Now plot the ordered pairs and connect them with a smooth curve, as shown in Figure 10.7.

y	$x = 10^y$	(x, y)
1	$10^1 = 10$	$(10,1)$
0	$10^0 = 1$	$(1,0)$
-1	$10^{-1} = \dfrac{1}{10}$	$(\dfrac{1}{10}, -1)$
-2	$10^{-2} = \dfrac{1}{100}$	$(\dfrac{1}{100}, -2)$

Figure 10.6
Finding ordered pairs

Figure 10.7
A logarithmic function   ∎

Notice that when x is between 0 and 1, the graph rises very steeply and the slope is large. In this interval, small changes in x produce large changes in y; for example, as x changes from $x = \frac{1}{100}$ to $x = \frac{1}{10}$, y changes from $y = -2$ to $y = -1$. On the other hand, when x is greater than 1, the graph rises very slowly and the slope is small. Here, large changes in x produce very small changes in y; for example, as x changes from $x = 1$ to $x = 10$, y changes only from $y = 0$ to $y = 1$. Consequently, logarithms can be used to expand small variations and compress large ones (see Figure 10.8). This characteristic will be important in our study of the Richter scale and the decibel scale in Section 10.3.

The Natural Logarithm Function

The number e is often used as the base of a logarithm. $\text{Log}_e\, x$ is called the **natural logarithm function** and is abbreviated $\ln x$. Thus, $\ln 2$ is an abbreviation for $\log_e 2$; $\ln 2$ is the exponent to which e must be raised in order to get 2.

Natural Logarithm Definition

$$y = \ln x \quad (\text{or} \quad y = \log_e x)$$

means the same as

$$e^y = x$$

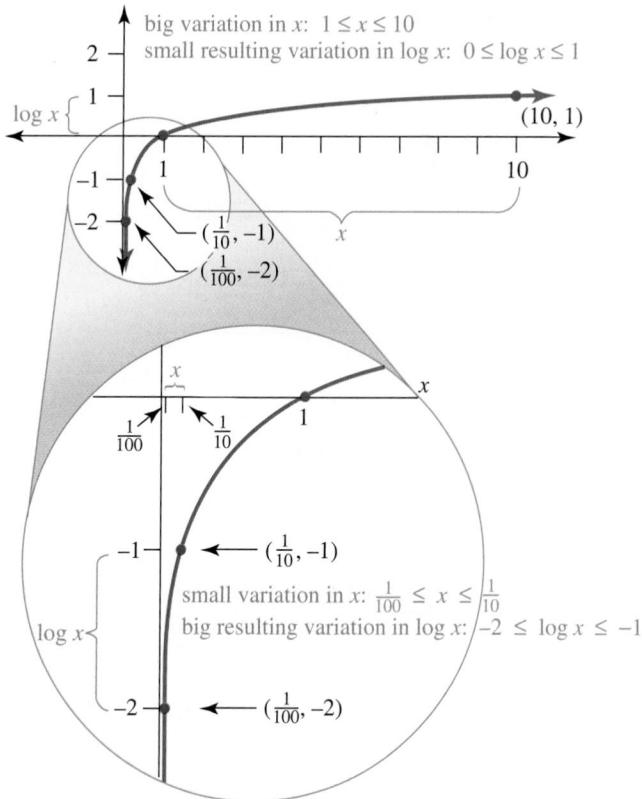

Figure 10.8
Logarithms expand small variations and compress large ones

The common logarithm (base 10) and the natural logarithm (base e) are the most frequently used types of logarithms.

ThomsonNOW™ **EXAMPLE 6**
for Liberal Arts Mathematics
www.thomsonedu.com/1pass.

Find the following values by using a calculator:

a. $\ln 5.2$ **b.** $\ln 0.4$ **c.** $\ln(-1.2)$ **d.** $\ln(e^2)$

Solution   **a.** The $\boxed{\ln x}$ button on your calculator is used to find the natural logarithm of a number. Simply enter the number you want to find the logarithm of and press $\boxed{\ln x}$. On a graphing calculator, you must press $\boxed{\text{LN}}$ first and then enter the desired number.

$$\ln 5.2 = 1.648658626\ldots$$

Therefore, $e^{1.648658626} \approx 5.2$.

b. ln 0.4 = −0.916290731 . . .

Therefore, $e^{-0.916290731} \approx 0.4$.

c. ln(−1.2) = ERROR

Your scientific calculator cannot find the natural logarithm of −1.2 because negative numbers do not have logarithms (in the real number system). If we let $y = \ln(-1.2)$, then the corresponding exponential equation would be $e^y = -1.2$. However, this equation has no solution (in the real number system) because e raised to any power always gives a positive result. Because we are working with real numbers only, $\ln(-1.2)$ is undefined.

d. To find $\ln(e^2)$, you must first calculate e^2: $e^2 = 7.389056099. \ldots$ Now find the natural logarithm of this number.

$$\ln(e^2) = \ln(7.389056099 \ldots) = 2$$

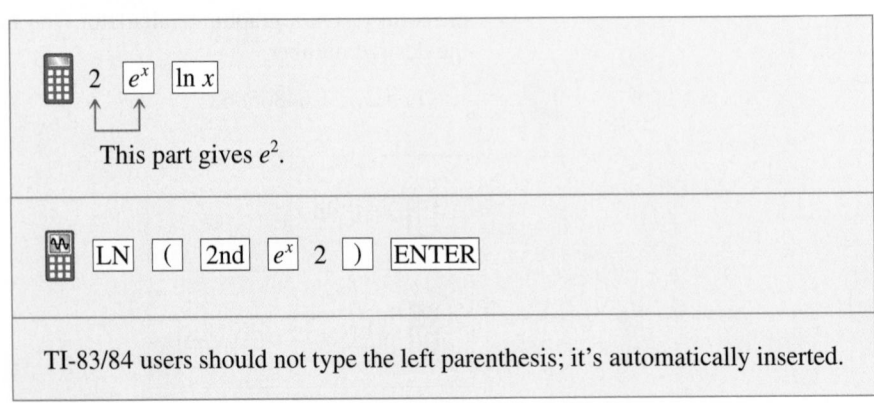

10.0A Exercises

In Exercises 1–12, find the value of u, v, or b.

1. $v = \log_2 4$

2. $v = \log_2 32$

3. $v = \log_2\left(\dfrac{1}{16}\right)$

4. $v = \log_2\left(\dfrac{1}{2}\right)$

5. $\log_5 u = 2$

6. $\log_5 u = 3$

7. $\log_3 u = 0$

8. $\log_3 u = 1$

9. $\log_b 16 = 2$

10. $\log_b 36 = 2$

11. $\log_b 8 = -3$

12. $\log_b 27 = -3$

In Exercises 13–18, rewrite the logarithm as an exponential equation.

13. $P = \log_b Q$

14. $S = \log_b T$

15. $M = \log_b(N + T)$

16. $J = \log_b(K - L)$

17. $M + R = \log_b(N + T)$

18. $J + E = \log_b(K - L)$

In Exercises 19–26, rewrite the exponential equation as a logarithm.

19. $b^F = G$

20. $b^w = Y$

21. $b^{F+2} = G$

22. $b^{w-1} = Y$

23. $b^{CD} = E - F$

24. $b^{EF} = C - D$

25. $b^{2-H} = Z + 3$

26. $b^{2+K} = Q - 9$

In Exercises 27–56, use a calculator to find each value. Give the entire display of your calculator.

27. **a.** $e^{1.4}$ **b.** $10^{1.4}$

28. **a.** $e^{3.2}$ **b.** $10^{3.2}$

29. **a.** $2e^{0.07}$ **b.** $2(10^{0.07})$

30. **a.** $3e^{0.09}$ **b.** $3(10^{0.09})$

31. **a.** $\dfrac{1}{e^{1.2}}$ **b.** $\dfrac{1}{10^{1.2}}$

32. **a.** $\dfrac{1}{e^{2.1}}$ **b.** $\dfrac{1}{10^{2.1}}$

33. **a.** $\dfrac{5}{e^{0.24}}$ **b.** $\dfrac{5}{10^{0.24}}$

34. **a.** $\dfrac{13}{e^{0.68}}$ **b.** $\dfrac{13}{10^{0.68}}$

35. **a.** $\dfrac{e^{5.6}}{2e^{3.4}}$ **b.** $\dfrac{10^{5.6}}{2(10^{3.4})}$

36. **a.** $\dfrac{e^{0.078}}{3e^{0.065}}$ **b.** $\dfrac{10^{0.078}}{3(10^{0.065})}$

37. **a.** $\dfrac{2e^{5.6}}{e^{3.4}}$ **b.** $\dfrac{2(10^{5.6})}{10^{3.4}}$

38. **a.** $\dfrac{3e^{0.078}}{e^{0.065}}$ **b.** $\dfrac{3(10^{0.078})}{10^{0.065}}$

39. **a.** $\dfrac{e^{4.5}}{10^{1.2}}$ **b.** $\dfrac{10^{1.2}}{e^{4.5}}$

40. **a.** $\dfrac{e^{3.9}}{10^{1.5}}$ **b.** $\dfrac{10^{1.5}}{e^{3.9}}$

41. **a.** $\ln 2.67$ **b.** $\log 2.67$

42. **a.** $\ln 8.76$ **b.** $\log 8.76$

43. **a.** $\ln 0.85$ **b.** $\log 0.85$

44. **a.** $\ln 0.33$ **b.** $\log 0.33$

45. **a.** $\ln(e^{4.1})$ **b.** $\log(10^{4.1})$

46. **a.** $\ln(e^{3.8})$ **b.** $\log(10^{3.8})$

47. **a.** $\ln(2e^{4.1})$ **b.** $\log[2(10^{4.1})]$

48. **a.** $\ln(2e^{3.8})$ **b.** $\log[2(10^{3.8})]$

49. **a.** $e^{\ln 2.3}$ **b.** $10^{\log 2.3}$

50. **a.** $e^{\ln 3.2}$ **b.** $10^{\log 3.2}$

51. **a.** $e^{3\ln 2}$ **b.** $10^{3\log 2}$

52. **a.** $e^{2\ln 2}$ **b.** $10^{2\log 2}$

53. **a.** $\ln 10$ **b.** $\ln(10^2)$ **c.** $\ln(10^3)$

54. **a.** $\ln 5$ **b.** $\ln(5^2)$ **c.** $\ln(5^3)$

55. **a.** $\log e$ **b.** $\log(e^2)$ **c.** $\log(e^3)$

56. **a.** $\log 5$ **b.** $\log(5^2)$ **c.** $\log(5^3)$

57. Put the following in numerical order, from smallest to largest: $e^{2.9}$, $\ln 2.9$, $10^{2.9}$, $\log 2.9$.

58. Put the following in numerical order, from smallest to largest: $e^{1.6}$, $\ln 1.6$, $10^{1.6}$, $\log 1.6$.

Answer the following questions using complete sentences and your own words.

CONCEPT QUESTIONS

59. What is an exponential function?

60. What is a logarithmic function?

61. If $x > 0$, which value is larger, 10^x or $\log x$? Why?

62. If $x > 0$, which value is larger, e^x or $\ln x$? Why?

63. If $x > 1$, which value is larger, $\log x$ or $\ln x$? Why?

64. If $0 < x < 1$, which value is larger, $\log x$ or $\ln x$? Why?

65. If $x > 0$, which value is larger, e^x or 10^x? Why?

66. If $x < 0$, which value is larger, e^x or 10^x? Why?

HISTORY QUESTION

67. Who invented the symbol e? What other symbols did this person introduce?

10.0B Review of Properties of Logarithms

Logarithms have several properties that can be used in simplifying complicated expressions and solving equations. We will use these properties extensively in the applications of logarithms and exponentials in this chapter. Perhaps the most important of these properties are the *Inverse Properties*.

The Inverse Properties

In Section 10.0A, we used the calculator to compute $\log(10^5) = 5$ and $\ln(e^2) = 2$ (Examples 4d and 6d). Note that in either case, the base of the logarithm is the same as the base of the exponential and that the logarithm appears to have "canceled" the exponential. Does this "canceling" always work? Does $\log_b(b^v) = v$ for all bases b and all numbers v?

Recall that by the definition of a logarithm, $\log_b u = v$ means the same as $b^v = u$. Since b^v is equal to u, we can substitute b^v for u in the equation $\log_b u = v$. This substitution results in $\log_b(b^v) = v$, which indicates that a base b logarithm will always "cancel" a base b exponential. *Cancel* is not really the right word here; *canceling* refers to what happens when you reduce a fraction. Instead, we say that a base b logarithm is the **inverse** of a base b exponential; whatever an exponential function does to a number, a logarithm undoes it and gives back the number.

Because $\log_b(b^x) = x$ for *any* base b, it is true for bases $b = 10$ and $b = e$. In other words, $\log_{10}(10^x) = x$ and $\log_e(e^x) = x$. Using the alternate notations, $\log(10^x) = x$ and $\ln(e^x) = x$.

Inverse Properties

$$\log(10^x) = x \quad \text{(or} \quad \log_{10}(10^x) = x)$$

$$\ln(e^x) = x \quad \text{(or} \quad \log_e(e^x) = x)$$

EXAMPLE 1 Simplify the following by using the Inverse Properties:

a. $\log(10^{3x})$ **b.** $\ln(e^{-0.012x})$ **c.** $\ln(10^{5x})$

Solution **a.** We are applying a base 10 logarithm to a base 10 exponential, so we can apply the Inverse Property $\log(10^x) = x$ and obtain

$$\log(10^{3x}) = \log_{10}(10^{3x})$$
$$= 3x$$

b. $\ln(e^{-0.012x}) = \log_e(e^{-0.012x})$
$$= -0.012x \qquad \text{using the Inverse Property } \ln(e^x) = x$$

c. The Inverse Properties do *not* apply to $\ln(10^{5x}) = \log_e(10^{5x})$, because the logarithm and the exponential are different bases (base e and base 10, respectively). Thus, $\ln(10^{5x})$ cannot be simplified by using the Inverse Properties. ∎

We have just seen that a logarithm "undoes" an exponential when the logarithm is applied to the exponential and the bases are the same. Will the same thing happen when an exponential is applied to a logarithm? Does $b^{(\log_b x)}$ simplify to just x?

Recall that by the definition of a logarithm, $b^v = u$ means the same as $\log_b u = v$. Because $\log_b u$ is equal to v, we can substitute it for v in the equation $b^v = u$. This substitution results in $b^{(\log_b u)} = u$. This means that a base b exponential will always "undo" a base b logarithm; an exponential (of the same base) is the **inverse** of a logarithm.

Because $b^{(\log_b x)} = x$ for *any* base b, it is true for bases $b = 10$ and $b = e$. In other words, $10^{(\log_{10} x)} = x$ and $e^{(\log_e x)} = x$. Using the alternate notations, $10^{(\log x)} = x$ and $e^{(\ln x)} = x$.

Inverse Properties

$$10^{\log x} = x \quad \text{(or} \quad 10^{\log_{10} x} = x)$$
$$e^{\ln x} = x \quad \text{(or} \quad e^{\log_e x} = x)$$

EXAMPLE 2 Simplify the following.

a. $10^{\log(2x-3)}$ **b.** $e^{\ln(0.023x)}$

Solution **a.** We are raising 10 to a common log (base 10) power, so we can apply an Inverse Property.

$$10^{\log(2x-3)} = 2x - 3 \quad \text{using the Inverse Property } 10^{\log x} = x$$

b. We are raising e to a natural log (base e) power, so we can apply an Inverse Property.

$$e^{\ln(0.023x)} = 0.023x \quad \text{using the Inverse Property } e^{\ln x} = x \quad \blacksquare$$

Solving Exponential Equations

Recall that an exponential equation is one in which x appears only in the exponent. If the variable can be removed from the exponent, then the resulting equation can be solved in a manner that is familiar to you from algebra. To solve an exponential equation, apply a logarithm and use the properties of logarithms to remove the variable from the exponent.

ThomsonNOW™
for Liberal Arts Mathematics
www.thomsonedu.com/1pass. **EXAMPLE 3** Solve $10^x = 0.47$.

Solution We have a base 10 exponential, so we apply the common logarithm to both sides and simplify.

$$10^x = 0.47$$
$$\log(10^x) = \log(0.47)$$
$$x = \log 0.47 \quad \text{using the Inverse Property } \log(10^x) = x$$
$$x = -0.327902142\ldots$$

As a rough check, we know that $x = -0.33$ is between 0 and -1, so we should expect $10^x = 10^{-0.33}$ to be between $10^0 = 1$ and $10^{-1} = \frac{1}{10} = 0.1$. As a more accurate check, use your calculator to verify that $10^{-0.327902142} \approx 0.47$.

ThomsonNOW™
for Liberal Arts Mathematics
www.thomsonedu.com/1pass.

EXAMPLE 4   Solve $5e^{0.01x} = 8$.

Solution   Before we apply a logarithm, we must first "isolate" the exponential by dividing by 5.

$$5e^{0.01x} = 8$$

$$e^{0.01x} = \frac{8}{5}$$

$$\ln(e^{0.01x}) = \ln\frac{8}{5} \qquad \text{taking } \ln \text{ of each side}$$

$$0.01x = \ln\frac{8}{5} \qquad \text{using the Inverse Property } \ln(e^x) = x$$

$$x = \frac{\ln(8/5)}{0.01}$$

$$x = 47.00036292\ldots$$

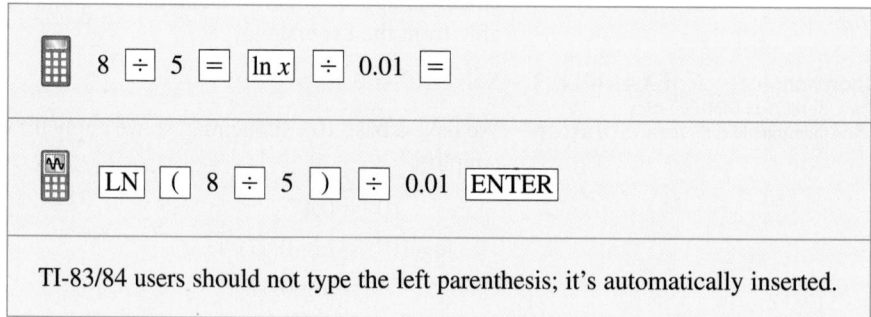

TI-83/84 users should not type the left parenthesis; it's automatically inserted.

Note: A common mistake is to improperly apply the natural logarithm at the first step; that is, if $5e^{0.01x} = 8$, then $5 \ln(e^{0.01x}) \neq \ln 8$. You should always isolate the exponential as your first step.

 Use your calculator to verify that $5e^{(0.01)(47.00036292)}$ is 8.

TI-83/84 users should not type the left parenthesis; it's automatically inserted.

Steps for Solving Exponential Equations

1. Isolate the exponential; that is, rewrite the problem in the form $e^A = B$.
2. Take the natural logarithm of each side.
3. Use the Inverse Property $\ln(e^x) = x$ to simplify.
4. Solve.
5. Use your calculator to check the answer.

(These steps also apply to solving a base 10 exponential equation. However, instead of applying the natural logarithm to each side, apply the common log.)

The Exponent-Becomes-Multiplier Property

A property of logarithms that is useful in solving exponential equations is the *Exponent-Becomes-Multiplier* Property.

Exponent-Becomes-Multiplier Property

$$\log(A^n) = n \cdot \log A$$

$$\ln(A^n) = n \cdot \ln A$$

When we take the logarithm of a base raised to a power, the exponent can be brought down in front of the logarithm, thus becoming a multiplier of the logarithm. Of course, these equations can be "reversed" to obtain $n \cdot \log A = \log(A^n)$ and $n \ln A = \ln (A^n)$; a multiplier of a logarithm can become an exponent.

This property seems rather strange—after all, exponents don't normally turn into multipliers. Or do they? Consider the Exponent Property:

$$(x^m)^n = x^{m \cdot n}$$

In the left-hand side of the equation, n is an exponent; in the right-hand side, n is a multiplier. This property is actually the basis of the Exponent-Becomes-Multiplier Property.

To see why $\log(A^n) = n \cdot \log A$, let $\log A = a$ and rewrite this logarithm as an exponential.

$$\log A = a$$

$$A = 10^a \qquad \text{using the Common Logarithm Definition}$$

Historical Note John Napier, 1550–1617

The Granger Collection, NY

John Napier was a Scottish landowner and member of the upper class. As such, he had a great deal of leisure time, much of which he devoted to mathematics, politics, and religion.

As is the case today, scientists of Napier's time frequently had to multiply and divide large numbers. Of course, at that time no calculators (or even slide rules) existed; a scientist had to make all calculations by hand. Such work was tedious and prone to errors. Napier invented logarithms as a system that would allow the relatively easy calculation of products and quotients, as well as powers and roots.

With this system, the product of two numbers was calculated by finding the logarithms of those numbers in a book of tables and adding the results. Napier spent *twenty years* creating these tables. This procedure represented a real shortcut: Addition replaced multiplication, and addition is much simpler than multiplication when done by hand. This method of multiplying by adding was an especially useful application of what we know as the Multiplication-Becomes-Addition Property, $\log(A \cdot B) = \log A + \log B$. Napier did not invent the notation $\log_b x$; this notation was invented much later by Leonhard Euler.

Napier did not use any notation in his writings on logarithms; he wrote everything out verbally.

Although Napier did not invent the decimal point, he is responsible for its widespread use. Napier's calculating system was very popular, and the logarithms in his tables were decimal numbers written with a decimal point.

A similar system was developed independently by Joost Bürgi, a Swiss mathematician and watchmaker. Napier is generally credited with the invention of logarithms, because he published his work before Bürgi.

As a member of the Scottish aristocracy, Napier was an active participant in local and national affairs. He was also a very religious man and belonged to the Church of Scotland. These two interests led to an interesting story about Napier.

In sixteenth-century Scotland, politics and religion were inexorably entwined. Mary, Queen of Scots, was a Roman Catholic; Elizabeth, Queen of England, was a Protestant; and both the Church of Scotland and the Church of England were Protestant churches. Furthermore, Mary had a strong claim to the throne of England. Catholic factions wanted Scotland to form an alliance with France; Protestant factions wanted an alliance with England. After a series of power struggles, Mary was forced to abdicate her throne to her son, who then became

Raising each side to the nth power, we get

$$A^n = (10^a)^n$$
$$= 10^{an} \qquad \text{using the exponent law } (x^m)^n = x^{mn}$$

Taking the log of each side and simplifying, we have

$$\log(A^n) = \log(10^{an})$$
$$= an \qquad \text{using the Inverse Property } \log(10^x) = x.$$

Finally, we substitute the original expression $\log A = a$ to get the desired result.

$$\log(A^n) = n \cdot \log A$$

A similar method can be used to obtain the natural logarithm version of the Exponent-Becomes-Multiplier Property. (See Exercise 56.)

EXAMPLE 5 **a.** Rewrite $\log(1.0125^x)$ so that the exponent is eliminated.
b. Rewrite $\ln(1 + 0.09/12)^x$ so that the exponent is eliminated.

The logarithm of a quotient can be rewritten as the logarithm of the numerator *minus* the logarithm of the denominator, hence the title *Division-Becomes-Subtraction*. Of course, these equations can be "reversed" to obtain

$$\log A - \log B = \log \frac{A}{B} \quad \text{and} \quad \ln A - \ln B = \ln \frac{A}{B}$$

These properties seem rather strange; after all, division doesn't normally turn into subtraction. Or does it? Consider the exponent property

$$\frac{x^m}{x^n} = x^{m-n}$$

When we use exponents, division *does* become subtraction. Furthermore, logarithms are closely related to exponents, so it shouldn't be surprising that both logarithms and exponents have a Division-Becomes-Subtraction Property.

To see why $\log A - \log B = \log \frac{A}{B}$, let $\log A = a$ and $\log B = b$. Now rewrite each expression as an exponential.

$$A = 10^a \quad \text{and} \quad B = 10^b \quad \text{using the Common Logarithm Definition}$$

Dividing the first equation by the second yields the following:

$$\frac{A}{B} = \frac{10^a}{10^b}$$

$$\frac{A}{B} = 10^{a-b} \quad \text{using the exponent law } \frac{x^m}{x^n} = x^{m-n}$$

$$\log \frac{A}{B} = \log(10^{a-b}) \quad \text{taking the log of each side}$$

$$\log \frac{A}{B} = a - b \quad \text{using the Inverse Property } \log(10^x) = x$$

Finally, substitute the original expressions for a and b to get the desired result:

$$\log \frac{A}{B} = \log A - \log B$$

A similar method can be used to obtain the natural logarithm version of the Division-Becomes-Subtraction Property. (See Exercise 55.)

EXAMPLE 7 **a.** Rewrite $\log \frac{x}{3}$ so that the fraction is eliminated.
b. Rewrite $\log(2x) - \log 8$ as one logarithm.

Solution **a.** We have the log of a fraction, so we can apply the Division-Becomes-Subtraction Property.

$$\log \frac{A}{B} = \log A - \log B \quad \text{division-Becomes-Subtraction Property}$$

$$\log \frac{x}{3} = \log x - \log 3 \quad \text{substituting } x \text{ for } A \text{ and 3 for } B$$

b. We are subtracting two logs, so we can apply the reverse of the Division-Becomes-Subtraction Property.

$$\log A - \log B = \log \frac{A}{B} \qquad \text{division-Becomes-Subtraction Property}$$

$$\log 2x - \log 8 = \log \frac{2x}{8} \qquad \text{substituting } 2x \text{ for } A \text{ and } 8 \text{ for } B$$

$$= \log \frac{x}{4} \qquad \text{canceling} \qquad \blacksquare$$

In the last step of Example 6, we had

$$x = \frac{\log 2}{\log 1.03}$$

Frequently, students will attempt to rewrite this as

$$x = \log(2 - 1.03)$$

This is incorrect; it is the result of misremembering the Division-Becomes-Subtraction Property. That property says that $\log \frac{A}{B} = \log A - \log B$; it does *not* refer to $\frac{\log A}{\log B}$ or to $\log(A - B)$.

Common Errors

You might be tempted to "simplify" the expression $\frac{\log A}{\log B}$. However,

1. $\frac{\log A}{\log B} \neq \log A - \log B$

(Subtracting two logs is not the same as dividing two logs.)

2. $\frac{\log A}{\log B} \neq \frac{A}{B}$

(You cannot cancel the logs in the numerator and denominator of a fraction.)

3. $\frac{\log A}{\log B} \neq \log(A - B)$

(The two logs on the left cannot be reduced into one log.)

The Multiplication-Becomes-Addition Property

We have seen that when logarithms are subtracted, they can be combined into one logarithm via the Division-Becomes-Subtraction Property. There is a similar property when logarithms are added.

Multiplication-Becomes-Addition Property

$$\log(A \cdot B) = \log A + \log B$$
$$\ln(A \cdot B) = \ln A + \ln B$$

The logarithm of a product can be rewritten as the sum of two logarithms, hence the title *Multiplication-Becomes-Addition*. Of course, these equations can be "reversed" to obtain $\log A + \log B = \log(A \cdot B)$ and $\ln A + \ln B = \ln(A \cdot B)$.

EXAMPLE 8   **a.** Rewrite log(3x) as two logarithms.
    b. Rewrite ln(2x) + ln 8 as one simplified logarithm.

Solution   **a.** We have the log of a product, so we can apply the Multiplication-Becomes-Addition Property.

$$\log(A \cdot B) = \log A + \log B \qquad \text{multiplication-Becomes-Addition Property}$$

$$\log(3x) = \log 3 + \log x \qquad \text{substituting 3 for } A \text{ and } x \text{ for } B$$

b. We are adding two logs, so we can apply the reverse of the Multiplication-Becomes-Addition Property.

$$\ln A + \ln B = \ln(A \cdot B) \qquad \text{multiplication-Becomes-Addition Property}$$

$$\ln(2x) + \ln 8 = \ln(2x \cdot 8) \qquad \text{substituting } 2x \text{ for } A \text{ and 8 for } B$$

$$= \ln(16x) \qquad \text{multiplying} \qquad ■$$

Solving Logarithmic Equations

An equation in which x appears "inside" a logarithm is called a **logarithmic equation.** To solve an equation of this type, we apply an exponential function to each side and use the various properties that we have developed.

EXAMPLE 9   Solve log $x = 5 + \log 3$.

Solution   To solve for x, we have to eliminate the common logarithm. We can accomplish this by using the Inverse Property $10^{\log x} = x$. However, for this property to apply, the equation must contain only one logarithm. Therefore, our first step is to get all the log terms on one side and then combine them into one logarithm.

$$\log x = 5 + \log 3$$

$$\log x - \log 3 = 5$$

$$\log \left(\frac{x}{3} \right) = 5 \qquad \text{using the Division-Becomes-Subtraction Property}$$

We can now apply the base 10 exponential function to each side—that is, exponentiate each side (base 10), and simplify.

$$10^{\log(x/3)} = 10^5$$

$$\frac{x}{3} = 100{,}000 \qquad \text{using the Inverse Property } 10^{\log x} = x$$

$$x = 300{,}000$$

 Use your calculator to verify that log 300,000 = 5 + log 3 by computing the right and left sides separately and verifying that they are equal.

The left side: 300000 [log]
The right side: 5 [+] 3 [log] [=]

The left side: [LOG] 300000 [ENTER]
The right side: 5 [+] [LOG] 3 [ENTER]

■

pH: An Application of Logarithms

Chemists define pH by the formula $pH = -\log[H^+]$, where $[H^+]$ is the hydrogen ion concentration measured in moles per liter. Solutions with a pH of 7 are said to be neutral, whereas a pH less than 7 is classified as an acid and a pH greater than 7 is classified as a base.

EXAMPLE 10   **a.** An unknown substance has a hydrogen ion concentration of $[H^+] = 1.7 \times 10^{-5}$ moles per liter. Determine the pH and classify the substance as an acid or a base.

b. If a solution has a pH of 6.2, find the hydrogen ion concentration of the solution.

Solution   **a.** Because we know the hydrogen ion concentration, we substitute it into the pH formula and use a calculator.

$$pH = -\log[H^+] \qquad \text{pH formula}$$
$$= -\log(1.7 \times 10^{-5}) \quad \text{substituting } 1.7 \times 10^{-5} \text{ for } [H^+]$$
$$= 4.769551079 \qquad \text{using a calculator}$$
$$= 4.8 \qquad \text{rounding to one decimal place}$$

The pH of the substance is 4.8, so the substance is classified as an acid.

b. Because we know the pH, we substitute it into the pH formula and solve for $[H^+]$, the hydrogen ion concentration.

$$pH = -\log[H^+] \qquad \text{pH formula}$$
$$6.2 = -\log[H^+] \qquad \text{substituting 6.2 for pH}$$
$$-6.2 = \log[H^+] \qquad \text{multiplying by } -1$$
$$10^{-6.2} = 10^{\log[H^+]} \qquad \text{exponentiating with base 10}$$
$$10^{-6.2} = [H^+] \qquad \text{using the Inverse Property } 10^{\log x} = x$$
$$[H^+] = 6.309573445 \times 10^{-7} \quad \text{using a calculator}$$

The hydrogen ion concentration is 6.3×10^{-7} moles per liter.   ∎

Steps for Solving Logarithmic Equations

1. Get all the log terms on one side and all the nonlog terms on the other.

2. Combine the log terms into one term, using the **Division-Becomes-Subtraction Property**

$$\log\left(\frac{A}{B}\right) = \log A - \log B$$

and the **Multiplication-Becomes-Addition Property**

$$\log(A \cdot B) = \log A + \log B$$

3. Exponentiate each side (base 10).

4. Use the **Inverse Property** $10^{\log x} = x$ and simplify.

5. Solve.

(These steps also apply to solving an equation that contains natural logarithms. However, instead of exponentiating base 10, use the base e exponential and the inverse property $e^{\ln x} = x$ to simplify.)

For Those Who Have Read Section 8.8, "Fractal Geometry"

EXAMPLE 11   In Example 1 of Section 8.8 we found that the dimension of a Koch snowflake is the number d that satisfies the equation $3^d = 4$. We used a trial-and-error process to approximate the value of d because we were unable to solve this equation.

a. Use logarithms to solve the equation and find the exact value of d.
b. Use the result of part (a) to approximate d to the nearest hundredth.
c. Use your calculator to check your answer.

Solution   **a.** *Finding the exact value of d.*

$$3^d = 4$$
$$\log (3^d) = \log 4 \qquad \text{taking the log of each side}$$
$$d \cdot \log 3 = \log 4 \qquad \text{using the Exponent-Becomes-Multiplier Property}$$
$$d = \frac{\log 4}{\log 3} \qquad \text{dividing by log 3}$$

The exact value of d is

$$d = \frac{\log 4}{\log 3}$$

b. *Approximating d.*   Using a calculator, we get

$$d = 1.26185950 \ldots \approx 1.26$$

Notice that this result agrees with that found in Example 1 of Section 8.8.

c. *Checking the answer.*   You can check your answer by verifying that $3^{1.26185950\cdots}$ is 4.

10.0B Exercises

In Exercises 1–12, simplify by using the Inverse Properties.

1. $\log(10^{6x})$
2. $\log(10^{-2x})$
3. $\ln(e^{-0.036x})$
4. $\ln(e^{0.114x})$
5. $10^{\log(2x+5)}$
6. $10^{\log(4-3x)}$
7. $e^{\ln(1-x)}$
8. $e^{\ln(1.75x)}$
9. **a.** $e^{2\ln x}$ **b.** $10^{2\log x}$
10. **a.** $e^{3\ln x}$ **b.** $10^{3\log x}$
11. **a.** $e^{2\ln(3x)}$ **b.** $10^{2\log(3x)}$
12. **a.** $e^{3\ln(2x)}$ **b.** $10^{3\log(2x)}$

In Exercises 13–22, rewrite the given logarithm so that all products, quotients, and exponents are eliminated.

13. $\log\left(\dfrac{x}{4}\right)$
14. $\ln\left(\dfrac{x}{7}\right)$
15. $\ln(1.8x)$
16. $\log(12x)$
17. $\log(1.225^x)$
18. $\ln(1.005^x)$
19. $\ln(3x^4)$
20. $\log(6x^5)$
21. $\log\left(\dfrac{5x^2}{7}\right)$
22. $\ln\left(\dfrac{7x^3}{11}\right)$

In Exercises 23–32, rewrite as one simplified logarithm.

23. $\ln(4x) + \ln 5$
24. $\log(6x) + \log 2$
25. $\log(6x) - \log 2$
26. $\ln(5x) - \ln 5$
27. $\ln x - \ln 3 + \ln 6$
28. $\log x - \log 4 + \log 22$
29. $3\log(2x) - \log 8$
30. $2\ln(5x) - \ln 5$
31. $\ln(9x) + \ln(4x) - 2\ln(6x)$
32. $2\log(3x) + 4\log(2x) - 2\log(12x)$

In Exercises 33–54, solve the given equation. (The solution of every other odd-numbered exercise is given in the back of the book; check the other exercises as shown in the text.)

33. **a.** $e^x = 0.35$ **b.** $10^x = 0.35$
34. **a.** $e^x = 0.75$ **b.** $10^x = 0.75$
35. **a.** $145e^{0.024x} = 290$ **b.** $145(10)^{0.024x} = 290$
36. **a.** $72e^{0.068x} = 144$ **b.** $72(10)^{0.068x} = 144$
37. **a.** $2,000e^{0.004x} = 8,500$ **b.** $2,000(10)^{0.004x} = 8,500$
38. **a.** $5,000e^{0.15x} = 18,000$
    b. $5,000(10)^{0.15x} = 18,000$
39. **a.** $50e^{-0.035x} = 25$ **b.** $50(10)^{-0.035x} = 25$
40. **a.** $13e^{-0.001x} = 6.5$ **b.** $13(10)^{-0.001x} = 6.5$
41. **a.** $80e^{-0.0073x} = 65$ **b.** $80(10)^{-0.0073x} = 65$
42. **a.** $42e^{-0.0037x} = 35$ **b.** $42(10)^{-0.0037x} = 35$
43. **a.** $\ln x = 0.66$ **b.** $\log x = 0.66$
44. **a.** $\ln x = 0.86$ **b.** $\log x = 0.86$
45. **a.** $\ln x = 3.66$ **b.** $\log x = 3.66$
46. **a.** $\ln x = 2.86$ **b.** $\log x = 2.86$

47. **a.** $\ln x + \ln 6 = 2$ **b.** $\log x + \log 6 = 2$
48. **a.** $\ln x + \ln 4 = 3$ **b.** $\log x + \log 4 = 3$
49. **a.** $\ln x - \ln 6 = 2$ **b.** $\log x - \log 6 = 2$
50. **a.** $\ln x - \ln 4 = 3$ **b.** $\log x - \log 4 = 3$
51. **a.** $\ln x = 4.8 + \ln 6.9$ **b.** $\log x = 4.8 - \log 6.9$
52. **a.** $\ln x = 2.7 + \ln 3.5$ **b.** $\log x = 2.7 + \log 3.5$
53. **a.** $\ln 0.9 = 3.1 - \ln(4x)$ **b.** $\log 0.9 = 3.1 - \log(4x)$
54. **a.** $\ln 0.8 = 2.4 - \ln(3x)$ **b.** $\log 0.8 = 2.4 - \log(3x)$
55. Show that $\ln \frac{A}{B} = \ln A - \ln B$.

   HINT: Use the same method as in the proof of $\log \frac{A}{B} = \log A - \log B$.

56. Show that $\ln(A^n) = n \cdot \ln A$.

   HINT: Use the same method as in the proof of $\log(A^n) = n \cdot \log A$.

57. Show that $\log(A \cdot B) = \log A + \log B$.

   HINT: Use a method similar to that used in Exercise 49.

58. Show that $\ln(A \cdot B) = \ln A + \ln B$.

Exercises 59–70 refer to Example 10.

59. For a certain fruit juice, $[H^+] = 3.0 \times 10^{-4}$. Determine the pH and classify the juice as acid or base.
60. For milk, $[H^+] = 1.6 \times 10^{-7}$. Determine the pH and classify milk as acid or base.
61. An unknown substance has a hydrogen ion concentration of 3.7×10^{-8} moles per liter. Determine the pH and classify the substance as acid or base.
62. An unknown substance has a hydrogen ion concentration of 2.4×10^{-5} moles per liter. Determine the pH and classify the substance as acid or base.
63. Fresh-brewed coffee has a hydrogen ion concentration of about 1.3×10^{-5} moles per liter. Determine the pH of fresh-brewed coffee.
64. Normally, human blood has a hydrogen ion concentration of about 3.98×10^{-8} moles per liter. Determine the normal pH of human blood.
65. When the pH of a person's blood drops below 7.4, a condition called *acidosis* sets in. Acidosis can result in death if pH reaches 7.0. What would the hydrogen ion concentration of a person's blood be at that point?
66. When the pH of a person's blood rises above 7.4, a condition called *alkalosis* sets in. Alkalosis can result in death if pH reaches 7.8. What would the hydrogen ion concentration of a person's blood be at that point?

67. You want to plant tomatoes in your backyard. Tomatoes prefer soil that has a pH range of 5.5 to 7.5. Using a testing kit, you determine the hydrogen ion concentration of your soil to be
   a. 3.5×10^{-7} moles per liter. Should you plant tomatoes? Why or why not?
   b. 3.5×10^{-4} moles per liter. Should you plant tomatoes? Why or why not?

68. You want to plant potatoes in your backyard. Potatoes prefer soil that has a pH range of 4.5 to 6.0. Using a testing kit, you determine the hydrogen ion concentration of your soil to be
   a. 2.1×10^{-7} moles per liter. Should you plant potatoes? Why or why not?
   b. 2.1×10^{-5} moles per liter. Should you plant potatoes? Why or why not?

69. Paprika prefers soil that has a pH range of 7.0 to 8.5. What range of hydrogen ion concentration does paprika prefer?

70. Spruce prefers soil that has a pH range of 4.0 to 5.0. What range of hydrogen ion concentration does spruce prefer?

For those who have covered Section 5.2: Compound Interest

71. Use logarithms to solve Exercise 75 on page 354 in Section 5.2.

72. Use logarithms to solve Exercise 76 on page 354 in Section 5.2.

73. Use logarithms to solve Exercise 77 on page 354 in Section 5.2.

74. Use logarithms to solve Exercise 78 on page 354 in Section 5.2.

75. Use logarithms to solve Exercise 79 on page 355 in Section 5.2.

76. Use logarithms to solve Example 9 on page 352 in Section 5.2.

For those who have read Section 8.8, "Fractal Geometry." In Exercises 77–81, (a) Use logarithms to find the exact value of the dimension of the given fractal. (b) Use the result of part (a) to approximate d to the nearest hundredth. (The solution to part (b) is not given in the back of the book; instead, check as shown in the text.)

77. The Sierpinski gasket (See Exercise 9 in Section 8.8.)
78. The Sierpinski carpet (See Exercises 2 and 10 in Section 8.8.)
79. The Mitsubishi gasket (See Exercises 3 and 11 in Section 8.8.)
80. The Menger sponge (See Exercise 12 in Section 8.8.)
81. The square snowflake (See Exercises 4 and 13 in Section 8.8.)

⸻ *Answer the following questions using complete sentences and your own words.*

HISTORY QUESTIONS
82. Who invented logarithms?
83. What motivated the invention of logarithms?
84. Historically, how were logarithms used to calculate quotients?

   HINT: Alter the discussion of how logarithms were used to calculate products.

85. Was Napier in favor of or opposed to the attempt of James VI to succeed Elizabeth to the English throne? Why?
86. What possible catastrophe did Napier's political pursuits save him from?

10.1 Exponential Growth

In Sections 10.0A and 10.0B, we reviewed the algebra that is used in connection with the exponential function $y = ae^{bx}$. In this and the following section, we will see how this function is used in topics as diverse as population growth, inflation, the decay of nuclear wastes, and radiocarbon dating. Each of these quantities grows (or decays) at a rate that reflects its current size: A larger population grows faster than a smaller one, and a larger amount of uranium decays faster than a smaller amount. This common relationship yields a powerful model that allows us to predict what the world population will be in the year 2100, how long it takes a home to double in value, and how long nuclear wastes must be stored. Or, looking back in time, we can use the model to determine the age of the Dead Sea Scrolls.

The first thing we need to do is determine why some things grow at a rate that reflects their current size.

EXAMPLE 1 A farmer has been studying the aphids in her alfalfa so that she can determine the best time to spray or release natural predators. In each of her two fields, she marked off one square foot of alfalfa and counted the number of aphids in that square foot. In her first field, she counted 100 aphids. One week later, the population had increased to 140; that is, its average growth rate is 40 aphids per week. In her second field, she counted 200 aphids. If the two fields provide the aphids with similar living conditions (favorable temperature, an abundance of food and space, and so on), what would be the most likely average growth rate for the second field?

Solution Forty aphids per week would not be realistic. Certainly, the larger population would have more births per week than the smaller one. The most likely average growth rate for the second field would be 80 aphids per week; this reflects the difference in size. Either population, then, would grow by the same percent; that is, the growth rate would be the same percentage of the population (see Figure 10.9).

	Average Growth Rate	Population	Growth as a Percent of (or Relative to) the Population
First Field	40 aphids/week	100 aphids	$\frac{40}{100} = 0.40 = 40\%$
Second Field	80 aphids/week	200 aphids	$\frac{80}{200} = 0.40 = 40\%$

Figure 10.9
Aphid growth

Delta Notation

Delta (Δ) notation is frequently used to describe changes in quantities, with the symbol Δ meaning "change in." Thus, in Example 1, the change in population p could be written Δp, and the change in time t could be written Δt. Expressed in delta notation, the **average growth rate** of 40 aphids per week would be written as

$$\frac{\Delta p}{\Delta t} = \frac{40 \text{ aphids}}{1 \text{ week}}$$

while the **relative growth rate** (that is, the growth rate relative to, or as a percent of, the population) would be written as

$$\frac{\Delta p/\Delta t}{p} = \frac{40 \text{ aphids}/1 \text{ week}}{100 \text{ aphids}} = 40\% \text{ per week}$$ using dimensional analysis, as discussed in Appendix E, to cancel "aphids" with "aphids"

A **rate of change** always means one change divided by another change; the average growth rate

$$\frac{\Delta p}{\Delta t} = \frac{40 \text{ aphids}}{1 \text{ week}} = 40 \text{ aphids/week}$$

is an example of a rate of change.

EXAMPLE 2 Two years ago, Anytown, U.S.A., had a population of 30,000. Last year, there were 900 births and 300 deaths, for a net growth of 600. There was no immigration to or emigration from the town, so births and deaths were the only sources of population change. The average growth rate was therefore

$$\frac{\Delta p}{\Delta t} = \frac{600 \text{ people}}{1 \text{ year}}$$

What is the most likely average growth rate for Anytown this year, assuming no change in living conditions?

Solution   Anytown should grow by the same percent each year, but by a percent of a changing amount. Because Anytown's population has increased slightly, the average growth rate would be a percent of a larger amount and would therefore be slightly higher. During the first year, the relative growth rate was

$$\frac{\Delta p/\Delta t}{p} = \frac{600 \text{ people}/1 \text{ year}}{30{,}000 \text{ people}} = 2\% \text{ per year}$$ using dimensional analysis to cancel "people" with "people"

In the following year, $\dfrac{\Delta p/\Delta t}{p}$ should also be 2%, but p is now 30,600

$$\frac{\Delta p/\Delta t}{p} = 2\%$$

$$\frac{\Delta p/\Delta t}{30{,}600} = 2\%$$

$$\Delta p/\Delta t = 2\% \text{ of } 30{,}600 = 612$$

This means that the most likely average growth rate for Anytown this year would be 612 people/year (see Figure 10.10). It is important to realize that this is only a prediction; it might be a good prediction, but it is not a guarantee. For example, there might be fewer residents in their childbearing years this year than previously. This would have a significant effect on the average growth rate, as would changes in living conditions.

	Average Growth Rate $\Delta p/\Delta t$	Population p	Relative Growth Rate $\dfrac{\Delta p/\Delta t}{p}$
First Year	600 people/year	30,000 people	$\dfrac{600}{30{,}000} = 0.02 = 2\%$
Second Year	612 people/year	30,600 people	$\dfrac{612}{30{,}600} = 0.02 = 2\%$

Figure 10.10
Anytown's growth   ■

In Examples 1 and 2, we saw that populations tend to grow at rates that reflect their size; it is most likely that a population will have an average growth rate that is a fixed percent of the current population size. In the next example, we will see that populations are not the only things that grow in this manner.

EXAMPLE 3   A house is purchased for $140,000 in January 2005. A year later, the house next door is sold for $149,800. The two houses are of the same style and size and are in similar condition, so they should have equal value. What would either house be worth in January 2007?

Solution Another increase of $9,800, for a total value of $159,600, would not be realistic. Experience has shown that more expensive homes tend to increase in value by a larger amount than do less expensive homes. The value would be more likely to increase by the same percent each year than by the same dollar amount. During 2005, the relative growth rate was

$$\frac{\Delta v / \Delta t}{v} = \frac{\$9,800/1 \text{ year}}{\$140,000} = 7\% \text{ per year}$$

In the following year, $\dfrac{\Delta v / \Delta t}{v}$ should also be 7% per year, but v is now $149,800.

$$\frac{\Delta v / \Delta t}{v} = 7\% \text{ per year}$$

$$\frac{\Delta v / \Delta t}{149,800} = 7\% \text{ per year}$$

$$\Delta v / \Delta t = 7\% \text{ of } \$149,800 \text{ per year} = \$10,486 \text{ per year}$$

Therefore, the average growth rate in 2006 should be

$$\frac{\Delta v}{\Delta t} = \frac{\$10,486}{1 \text{ year}} = \$10,486 \text{ per year}$$

and the best guess of the value in January 2007 would be

$$\$149,800 + \$10,486 = \$160,286$$

Remember that this is only a prediction, based on the assumption that the pattern set in 2005 will continue in 2006. This is a common assumption, unless economic forecasts indicate otherwise. ∎

The Exponential Model

In the examples on populations, the average growth rate $\Delta p / \Delta t$ was a constant percent of the current population. In the example on inflation, the average growth rate $\Delta v / \Delta t$ was a constant percent of the current value. More generally, we are looking at situations in which *the average growth rate $\Delta y / \Delta t$ is a constant percent of the current value of y*, that is, in which

$$\frac{\Delta y}{\Delta t} = k \cdot y$$

In the examples above, we used p for population and v for value in place of y, as a memory device, and k was a specific number, such as $7\% = 0.07$. In calculus, it is determined that if a quantity y behaves in such a way that the average growth rate $\Delta y / \Delta t$ is a constant percent of the current value of y, then the size of y at some later time can be predicted using the equation

$$y = ae^{bt}$$

where t is time and a and b are constants. We used equations like this in Sections 10.0A and 10.0B, with specific constants such as 3 and 5 in place of a and b.

A **mathematical model** is an equation that is used to describe a real-world subject. The exponential equation

$$y = ae^{bt}$$

is called the **exponential model** because it is an exponential equation and is used to describe subjects such as populations and real estate appreciation. This model is extremely powerful—it allows us to predict the value of y, and its growth rate, with just a few data. Because many different quantities have a growth rate proportional to their size, the applications of this model are wide and varied. The model's use is illustrated in Examples 4–6, which are continuations of Examples 1–3.

EXAMPLE 4 One square foot of alfalfa has a population of 100 aphids. In one week, the population increases to 140; that is, its average growth rate is 40 aphids per week.

  a. Develop the model that represents the population of aphids.
  b. Predict the aphid population after three weeks.

Solution **a.** *Developing the model.* As we discovered in Examples 1 and 2, the rate at which a population changes is a constant percent of the size of the population (as long as the population's growth is not limited by a lack of food, space, or other constraints). Thus, we can use the model $y = ae^{bt}$ or, with a more appropriate letter, $p = ae^{bt}$. To develop this model for our situation, we need to find the constants a and b.

Step 1 *Write the given information as two ordered pairs (t, p).* At the beginning of the experiment (at "time 0"), there were 100 aphids; that is, when $t = 0$, $p = 100$.

$$(t, p) = (0, 100)$$

After one week, there were 140 aphids, that is, when $t = 1$, $p = 140$:

$$(t, p) = (1, 140)$$

Step 2 *Substitute the first ordered pair into the model $p = ae^{bt}$ and simplify.*

$(t, p) = (0, 100)$	the first ordered pair
$100 = ae^{b \cdot 0}$	substituting
$100 = ae^0$	simplifying
$100 = a \cdot 1$	
$a = 100$	

Because a is a constant, we can rewrite our model $p = ae^{bt}$ as $p = 100\, e^{bt}$. To totally develop this model to fit our situation, we need to determine the value of the remaining constant b.

Step 3 *Substitute the second ordered pair into the model $p = 100\, e^{bt}$, simplify, and solve for b.*

$(t, p) = (1, 140)$	the second ordered pair
$140 = 100\, e^{b \cdot 1}$	substituting
$140 = 100\, e^b$	simplifying
$\dfrac{140}{100} = e^b$	solving for b
$1.40 = e^b$	
$\ln 1.40 = \ln e^b$	
$\ln 1.40 = b$	using the Inverse Property $\ln(e^x) = x$
$b \approx 0.3364722366$	

We have now developed the model $p = ae^{bt}$ to fit our situation. After substituting $a = 100$ and $b = 0.3364722366$, we get

$$p = 100 \; e^{0.3364722366t}$$

where t is the number of weeks after the beginning of the experiment. Notice that a is the initial size of the population and that b is somewhat close to the relative growth rate of $40\% = 0.40$. The values of a and b always have these characteristics. The graph of the model is shown in Figure 10.11.

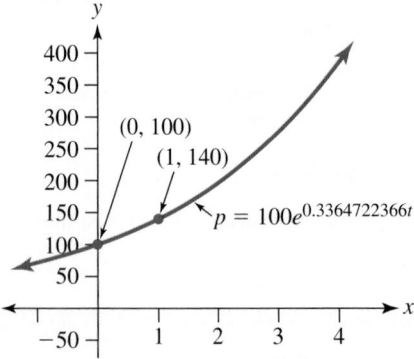

Figure 10.11
A graph showing the population of aphids

Our model will be most accurate if we do not round off $b = 0.3364722366$ at this point—that is, if we use all the decimal places that the calculator displays. (Your calculator might display more or fewer decimal places than shown here.) One way to do this is to store the number in your calculator's memory. To do so, press

when the number is on your calculator's screen.

b. *Predicting the population after three weeks.* Now that the model has been developed, predictions can be made. At the end of three weeks, the aphid population would probably be as follows:

$$p = 100\, e^{(0.3364722366)(3)} \quad \text{substituting 3 for } t$$

$$= 274.3999\ldots$$

$$\approx 274 \qquad \text{rounding off to the nearest whole number of aphids}$$

Therefore, at the end of three weeks, we would expect 274 aphids.

TI-83/84 users should not type the left parenthesis—it's automatically inserted.

ThomsonNOW™
for Liberal Arts Mathematics
www.thomsonedu.com/1pass.

EXAMPLE 5   In the year 2005, Anytown, U.S.A., had a population of 30,000. In the following year, there were 900 births and 300 deaths, for a net growth of 600. There was no immigration to or emigration from the town, so births and deaths were the only sources of population change.

a. Develop the model that represents Anytown's population.
b. Predict Anytown's population in the year 2011.

Solution   **a.** *Developing the model.*   As we discovered, the rate at which a population changes is a constant percent of the size of the population. Thus, we can use the model $p = ae^{bt}$. To develop this model for our situation, we need to find the constants a and b.

Step 1   *Write the given information as two ordered pairs (t, p).* For ease in calculating, let 2005 be $t = 0$; then 2006 is $t = 1$ and 2011 is $t = 6$. Also, we will count the population in thousands. The ordered pairs are $(0, 30)$ and $(1, 30.6)$.

Step 2   *Substitute the first ordered pair into the model $p = ae^{bt}$ and simplify.*

$$(t, p) = (0, 30) \quad \text{the first ordered pair}$$

$$30 = ae^{b \cdot 0} \quad \text{substituting}$$

$$30 = ae^{0} \quad \text{simplifying}$$

$$30 = a \cdot 1$$

$$a = 30$$

As a shortcut, recall that a is always the initial population.

Step 3   *Substitute the second ordered pair into the model $p = 30e^{bt}$, simplify, and solve for b.*

$$(t, p) = (1, 30.6) \quad \text{the second ordered pair}$$

$$30.6 = 30\, e^{b \cdot 1} \quad \text{substituting}$$

$$\frac{30.6}{30} = e^b \qquad\qquad \text{simplifying}$$

$$1.02 = e^b \qquad\qquad \text{solving for } b$$

$$\ln 1.02 = \ln e^b$$

$$\ln 1.02 = b \qquad\qquad \text{using the Inverse Property } \ln(e^x) = x$$

$$b \approx 0.0198026273$$

Store this value of b in your calculator's memory.

 Notice that the value of b is close to the relative growth rate of $2\% = 0.02$, as it should be.

Thus, our model for Anytown's growth is

$$p = 30\, e^{0.0198026273t}$$

where t is years beyond 2005 and p is population in thousands.

b. *Predicting Anytown's population in the year 2011.* Now that the model has been determined, predictions can be made. In the year 2011, $t = 6$ and

$$p = 30\, e^{(0.0198026273)(6)} \qquad \text{substituting 6 for } t$$

$$\approx 33.78487258$$

TI-83/84 users should not type the left parenthesis—it's automatically inserted.

The population would probably be 33,785. ■

At this point, you might ask, "Why can't I just substitute the relative growth rate for b rather than use the second ordered pair to actually calculate b? After all, b is close to the relative growth rate." You can, but the resulting calculation will *not* necessarily be close to the true figure. In Example 5, if we use the relative growth rate for b, the resulting calculation is off by 40 people. Using the relative growth rate for b is equivalent to rounding off b; whenever you round off early, you lose accuracy.

Furthermore, while b is always somewhat close to the relative growth rate, it is not necessarily extremely close to that rate. In Example 4 on aphids, b was 0.3364722, and the relative growth rate was somewhat close at $40\% = 0.40$. In Example 5 on Anytown, b was 0.0198026273, and the relative growth rate was extremely close at $2\% = 0.02$. In general, b is always close to the relative growth rate, and the smaller these two numbers are, the closer they are.

In the previous two examples, we saw how to use the exponential model to predict the size of a population at some future date. The model can also be used to determine the time at which a quantity will have grown to any specified value.

EXAMPLE 6   A house is purchased for $140,000 in January 2005. A year later, the house next door is sold for $149,800. The two houses are of the same style and size and are in similar condition, so they should have equal value.

    a. Develop the mathematical model that represents the home's value.
    b. Find when the house would be worth $200,000 (assuming that the rate of appreciation for houses continues unchanged).

Solution   **a.** *Developing the model.*

Step 1 *Write the given information as two ordered pairs (t, v). If we let t measure years after January 2005 and v measure value in thousands of dollars, then the ordered pairs are $(t, v) = (0, 140)$ and $(t, v) = (1, 149.8)$.*

Step 2 *Substitute the first ordered pair into the model $y = ae^{bt}$ and simplify to find a. (Or remember that a is always the initial value of y.)* Since a is the initial value of y (or here, v), $a = 140$. Alternatively, you could substitute $(0, 140)$ into $v = ae^{bt}$.

$$140 = ae^{b \cdot 0}$$
$$140 = ae^0$$
$$140 = a \cdot 1$$
$$a = 140$$

The model is now

$$v = 140e^{bt}$$

Step 3 *Substitute the second ordered pair into the model, simplify, and solve for b.*

$$(t, v) = (1, 149.8)$$
$$149.8 = 140\, e^{b \cdot 1}$$
$$149.8 = 140\, e^b$$
$$\frac{149.8}{140} = e^b$$
$$1.07 = e^b$$
$$\ln 1.07 = \ln e^b$$
$$\ln 1.07 = b$$
$$b \approx 0.0676586485$$

Store this value of b in your calculator's memory.

 Notice that b is close to the relative growth rate of $7\% = 0.07$.

The model is

$$v = 140\, e^{0.0676586485t}$$

where t is the number of years after January 2005 and v is the value in thousands of dollars.

    b. *Finding when the house will be worth $200,000.*   The problem is to find t when $v = 200$, so we substitute 200 for v in our model:

$$200 = 140\, e^{0.0676586485t} \qquad \text{substituting}$$
$$\frac{200}{140} = e^{0.0676586485t} \qquad \text{solving for } t$$

$$\ln \frac{200}{140} = \ln(e^{0.0676586485t}) \qquad \text{taking the natural log of each side}$$

$$\ln \frac{200}{140} = 0.0676586485t \qquad \text{using the Inverse Property } \ln(e^x) = x$$

$$t = \frac{\ln \dfrac{200}{140}}{0.0676586485} \qquad \text{dividing}$$

$$t = 5.27168 \ldots$$

TI-83/84 users should not type the left parenthesis; it's automatically inserted.

Therefore, our prediction is that the house will be worth $200,000 about $5\frac{1}{4}$ years after January 2005—that is, in April 2010. ∎

Does the prediction above mean that the house definitely will be worth $200,000 in April 2010? No; this calculation is based on the assumption that real estate appreciation continues at 7%. However, real estate values are affected by many things, and the rate changes frequently. The model $y = ae^{bt}$ allows us to make predictions, and it is important to remember that they are *only* predictions.

> ### Steps in Developing an Exponential Model
>
> 1. *Write the information given as to the value of the quantity at two points in time as two ordered pairs (t, y). (You might want to use another letter in place of y as a memory device.)*
> 2. *Substitute the first ordered pair into the model $y = ae^{bt}$ and simplify to find a. (Or remember that a is always the initial value of y.)*
> 3. *Substitute the second ordered pair into the model, simplify and solve for b. (As a check, recall that b is close to the relative growth rate.)*

Exponential Growth and Compound Interest (For Those Who Have Read Chapter 5: Finance)

You might be wondering whether there is a relationship between the inflation of real estate, discussed in Examples 3 and 6, and compound interest, as discussed in Section 5.2. In Example 3, we found that a $140,000 home increased in value by 7% from

January 2005 to January 2006 (that is, its relative growth rate is 7%). Is this equivalent to depositing $140,000 in an account that earns 7% interest compounded annually?

ThomsonNOW™
for Liberal Arts Mathematics
www.thomsonedu.com/1pass.

EXAMPLE 7   In January 2005, $140,000 is deposited in an account that earns 7% interest compounded annually.

a. Find the future value in January 2006.
b. Find when the account would hold $200,000.
c. Develop a compound interest model that can be used to answer questions involving the future value of the account.

Solution   **a.** $FV = P(1 + i)^n$ the Compound Interest Formula

$\quad\quad = 140{,}000(1 + 0.07)^1$ substituting

$\quad\quad = \$149{,}800$

b. $FV = P(1 + i)^n$ the Compound Interest Formula

$200{,}000 = 140{,}000(1 + 0.07)^n$ substituting

We will solve this exponential equation by following the steps developed in Section 10.0B. First, isolate the exponential by dividing by 140,000.

$$\frac{200000}{140000} = (1.07)^n$$

$$\frac{20}{14} = (1.07)^n \quad\quad \text{reducing}$$

$$\ln(20/14) = \ln 1.07^n \quad\quad \text{taking ln of each side}$$

$$\ln(20/14) = n \ln 1.07 \quad\quad \text{using the Exponent-Becomes-Multiplier Property}$$
$$\quad\quad\quad\quad\quad\quad\quad\quad\quad \ln(A^n) = n \cdot \ln A$$

$$\frac{\ln(20/14)}{\ln 1.07} = n \quad\quad \text{dividing by ln 1.07}$$

$$n = 5.27168 \ldots$$

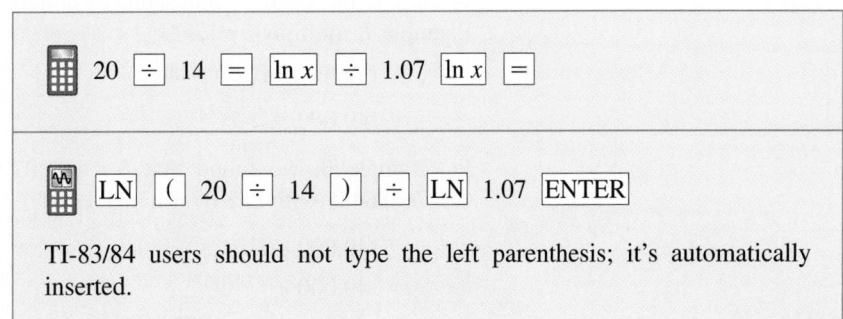

TI-83/84 users should not type the left parenthesis; it's automatically inserted.

Since n is the number of compounding periods and interest is compounded annually, $n = 5.27168 \ldots$ means that it will take $5.27168 \ldots$ years for the account to hold $200,000.

This solution is mathematically correct. Practically speaking, however, the ".27168 . . ." part of the solution doesn't make sense. If interest is compounded annually, then at the end of each year, your account is credited with that year's interest. After five years, your account balance would be $140{,}000(1 + 0.07)^5 \approx$ $196,357.24. After $5.27168 \ldots$ years, your account balance would not have changed, since interest won't be credited until the end of the year. After six years, your account balance would be $140,000(1 + 0.07)^6 \approx \$210,102.25$. Thus, the best answer to the question is that the account will never hold exactly $200,000, but after six years (i.e., in January of 2011) the account will hold more than $200,000.

c. To develop a compound interest model that can be used to answer questions involving the future value of the account, substitute 140,000 for P and $7\% = 0.07$ for i into the Compound Interest Formula.

$$FV = P(1 + i)^n \qquad \text{the Compound Interest Formula}$$
$$FV = 140,000(1.07)^n \qquad \text{substituting}$$

In this model, n must be a whole number of years.   ■

Notice that the future value in part (a) of Example 7 matches the future value in Example 3. That is, a $140,000 home that appreciates 7% in 1 year has the same future value as does a $140,000 deposit that earns 7% interest in 1 year.

Notice also that the mathematically correct (but practically incorrect) answer to part (b) of Example 7 matches that of part (b) of Example 6. That is, the home and the bank account will each be worth $200,000 in 5.27168 years.

These two observations imply that $v = 140\,e^{0.0676586485t}$, the exponential model developed in Example 6, and $FV = 140,000(1.07)^n$, the compound interest model developed in Example 7, are mathematically interchangeable. Regardless of which model is used, only a whole number of years makes sense in a compound interest problem, if the interest is compound annually.

EXAMPLE 8   Show that the exponential model $v = 140\,e^{0.0676586485t}$ from Example 6 and the compound interest model $v = 140,000(1.07)^n$ from Example 7 are mathematically interchangeable.

Solution   We will use the properties of logarithms to convert the exponential model to the compound interest model. In Example 6, we modeled the value of a house with

$$v = 140\,e^{0.0676586485t}$$

where v is the value in thousands of dollars. To convert this to dollars, as used in Example 7, multiply by 1,000:

$$v = 1,000 \cdot 140\,e^{0.067586485t}$$
$$= 140,000\,e^{0.0676586485t}$$

In Example 6, we found that $b = \ln 1.07 \approx 0.0676586485$, so we can replace 0.0676586485 with $\ln 1.07$. This gives us

$$v = 140,000\,e^{0.0676586485t}$$
$$= 140,000\,e^{(\ln 1.07)(t)}$$

For the moment, focus on e's exponent, $(\ln 1.07)(t)$:

$$(\ln 1.07)(t) = t \cdot \ln 1.07$$
$$= \ln(1.07^t) \qquad \text{using the Exponent-Becomes-Multiplier Property}$$
$$\qquad\qquad n \cdot \ln A = \ln(A^n)$$

Thus, we can replace e's exponent, $(\ln 1.07)(t)$, with $\ln(1.07^t)$:

$$v = 140,000\,e^{(\ln 1.07)(t)}$$
$$= 140,000\,e^{\ln(1.07^t)}$$

$$= 140{,}000(1.07^t) \quad \text{using the Inverse Property } e^{\ln x} = x$$

$$= 140{,}000(1.07^n) \quad t \text{ and } n \text{ both measure number of years}$$

Since we were able to convert the exponential model $v = 140\, e^{0.0676586485t}$ to the compound interest model $v = 140{,}000(1.07)^n$, we know that they are algebraically equivalent and therefore are mathematically interchangeable. Regardless of which model is used, only a whole number of years makes sense in a compound interest problem (if the interest is compounded annually). ∎

10.1 Exercises

1. Use the model $p = 30\, e^{0.0198026273t}$ developed in Example 5 to predict the population of Anytown in the year 2009.

2. Use the model $p = 30\, e^{0.0198026273t}$ developed in Example 5 to predict the population of Anytown in the year 2028.

3. Use the model $v = 140\, e^{0.0676586485t}$ developed in Example 6 to predict when the house would be worth $250,000.

4. Use the model $v = 140\, e^{0.0676586485t}$ developed in Example 6 to predict when the house would be worth $300,000.

Exercises 5–12 deal with data from the U.S. Bureau of the Census on populations of metropolitan areas.* These data allow us to find how fast the population is growing and when it will reach certain levels. Such calculations are very important, because they indicate the future needs of the population for goods and services and how well the area can support the population.

5. The third largest metropolitan area in the United States is the Chicago/Naperville/Joliet metropolitan area. Its population in 2000 was 9,098 (in thousands); in 2004, it was 9,392.
   a. Convert this information to two ordered pairs (t, p), where t measures years since 2000 and p measures population in thousands.
   b. Find Δt, the change in time.
   c. Find Δp, the change in population.
   d. Find $\Delta p / \Delta t$, the average growth rate.
   e. Find $(\Delta p / \Delta t)/p$, the relative growth rate.

6. The second largest metropolitan area in the United States is the Los Angeles/Long Beach/Santa Ana metropolitan area. Its population in 2000 was 12,366 (in thousands); in 2004, it was 12,925.

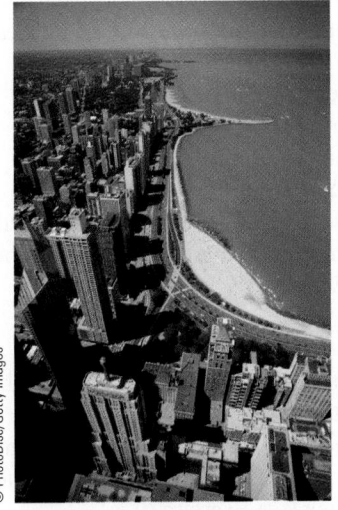

Chicago is the third largest metropolitan area in the United States.

   a. Convert this information to two ordered pairs (t, p), where t measures years since 2000 and p measures population in thousands.
   b. Find Δt, the change in time.
   c. Find Δp, the change in population.
   d. Find $\Delta p / \Delta t$, the average growth rate.
   e. Find $(\Delta p / \Delta t)/p$, the relative growth rate.

7. The largest metropolitan area in the United States is the New York/northern New Jersey/Long Island metropolitan area. Its population in 2000 was 18,323 (in thousands); in 2004, it was 18,710.

New York is the largest metropolitan area in the United States.

a. Convert this information to two ordered pairs (t, p), where t measures years since 2000 and p measures population in thousands.
b. Find Δt, the change in time.
c. Find Δp, the change in population.
d. Find $\Delta p / \Delta t$, the average growth rate.
e. Find $(\Delta p / \Delta t)/p$, the relative growth rate.

8. The twelfth largest metropolitan area in the United States is the San Francisco/Oakland/Fremont metropolitan area. Its population in 2000 was 4,124 (in thousands); in 2004, it was 4,154.
a. Convert this information to two ordered pairs (t, p), where t measures years since 2000 and p measures population in thousands.
b. Find Δt, the change in time.
c. Find Δp, the change in population.
d. Find $\Delta p / \Delta t$, the average growth rate.
e. Find $(\Delta p / \Delta t)/p$, the relative growth rate.

9. a. Develop the model that represents the population of the Chicago/Naperville/Joliet metropolitan area (see Exercise 5).
b. Predict the population in 2009.
c. Predict the population in 2014.
d. Predict when the population will be double its 2000 population.

10. a. Develop the model that represents the population of the Los Angeles/Long Beach/Santa Ana metropolitan area (see Exercise 6).
b. Predict the population in 2010.
c. Predict the population in 2014.
d. Predict when the population will be 50% more than its 2000 population.

11. a. Develop the model that represents the population of the New York/northern New Jersey/Long Island metropolitan area (see Exercise 7).
b. Predict the population in 2007.
c. Predict the population in 2014.
d. Predict when the population will be 50% more than it was in 2000.

12. a. Develop the model that represents the population of the San Francisco/Oakland/Fremont metropolitan area (see Exercise 8).
b. Predict the population in 2008.
c. Predict the population in 2024.
d. Predict when the population will be double what it was in 2000.

13. A biologist is conducting an experiment that involves a colony of fruit flies. (Biologists frequently study fruit flies because their short life span allows the experimenters to easily study several generations.) One day, there were 2,510 flies in the colony. Three days later, there were 5,380.

a. Develop the mathematical model that represents the population of flies.
b. Use the model to predict the population after one week.
c. Use the model to predict when the population will be double its initial size.

14. A university keeps a number of mice for psychology experiments. One day, there were 89 mice. Three weeks later, there were 127.
a. Develop the mathematical model that represents the population of mice.
b. Use the model to predict the population after two months.
c. Use the model to predict when the population will be double its initial size.

15. In August 2005, Buck Meadows bought a house for $230,000. In February 2006, a nearby house with the same floor plan was sold for $310,000.
a. Develop the mathematical model that represents the value of the house.
b. Use the model to predict when the house will double its 2005 value.
c. Use the model to determine the value of the house one year after Mr. Meadows purchased it.
d. Use part (c) to determine $\Delta v / \Delta t$, that is, the rate at which the value of the house grew.

16. In July 2004, Alvarado Niles bought a house for $189,000. In September 2005, a nearby house with the same floor plan was sold for $207,000.
a. Develop the mathematical model that represents the value of the house.
b. Use the model to predict when the house will double its 2004 value.
c. Use the model to determine the value of the house one year after Mr. Niles purchased it.
d. Use part (c) to determine $\Delta v / \Delta t$, that is, the rate at which the value of the house grew.

Exercises 17–20 deal with data from the U.S. Bureau of the Census* on energy consumption and the Gross Domestic Product (the total value of the nation's output of goods and services). These data allow us to predict future energy needs and to monitor economic growth.

17. In 1995, 91.22 quadrillion British Thermal Units (Btu's) of energy were consumed in the United States. In 2000, 98.90 quadrillion Btu's were consumed.
a. Develop the exponential model that represents the amount of energy consumed.

_____

* *Source:* U.S. Department of Commerce, Bureau of the Census, *Statistical Abstract of the United States,* 2006.

b. Use the model to predict the amount of energy consumed in 2002. (The actual 2002 consumption was 98.03 quadrillion Btu's.)

c. Discuss whether an exponential model might be appropriate to this situation and why.

18. In 2002, 3,628.7 billion kilowatt-hours (kwh) of electricity were consumed in the United States. In 2003, 3,656.5 billion kwh were consumed.

a. Develop the exponential model that represents the amount of electricity consumed.

b. Use the model to predict the amount of electricity consumed in 2004. (The actual 2004 consumption was 3,716.5 billion kwh.)

c. Discuss whether an exponential model might be appropriate to this situation and why.

19. In 1995, the Gross Domestic Product (GDP) of the United States was $7,398 billion. In 2000, it was $9,817 billion.

a. Develop the exponential model that represents the nation's GDP.

b. Use the model to predict the GDP in 2004. (The actual 2004 GDP was $11,735 billion.)

c. Discuss whether an exponential model might be appropriate to this situation and why.

20. In 2000, there were 109,478 thousand cellular telephone subscriptions in the United States. In 2002, there were 140,766.

a. Develop the exponential model that represents the nation's cellular telephone subscriptions.

b. Use the model to predict the number of subscriptions in 2004. (The actual number in 2004 was 182,140.)

c. Use the model to predict the number of subscriptions in 2010.

21. The first census of the United States was taken on August 2, 1790; the population then was 3,929,214. Since then, a census has been taken every ten years. The next census gave a U.S. population of 5,308,483.*

a. Use these data to develop the mathematical model that represents the population of the United States.

b. Which prediction would be more accurate, U.S. population in 1810 or in 1990? Why?

c. Use the model to predict the U.S. population in 1810 and in 2000. (The actual populations were 7,239,881, and 281,421,906.)

22. In 1969, the National Academy of Sciences published a study titled "Resources and Man." That study places "the earth's ultimate carrying capacity at about 30 billion people, at a level of chronic near-starvation for the great majority (and with massive immigration to the now less-densely populated lands)!" The study goes on to state that 10 billion people is "close to (if not above) the maximum that an intensively managed world might hope to support with some degree of comfort and individual choice." The world population in 1990 was 5.283 billion; in 2000, it was 6.082 billion.

a. Develop the mathematical model that represents world population.

b. Use the model to predict when world population would reach the "somewhat comfortable" level of 10 billion.

c. Use the model to predict when world population would reach "the earth's ultimate carrying capacity" of 30 billion.

23. Perhaps the most intuitive way to compare how fast different populations are growing is to calculate their "doubling times," that is, how long it takes for the populations to double. Find the doubling times for Africa, North America, and Europe, using the data in Figure 10.12.

	Africa	North America	Europe
1990 population	643 million	277 million	509 million
2000 population	794 million	314 million	727 million

Figure 10.12

24. China's population in 1980 was 983 million; in 1990, it was 1,154 million.

a. Develop the mathematical model that represents China's population.

b. Use the model developed in Exercise 23 that represents Africa's population and the model in part (a) to predict when Africa's population will exceed China's.

25. Use the data in Exercise 22 to compute the doubling time for the world's population.

26. Use the data in Exercise 24 to compute the doubling time for China's population.

27. A real estate investor bought a house for $125,000. He estimated that houses in that area increase their value by 10% per year.

a. If that estimate is accurate, why would it not take five years for the house to increase its value by 50%?

Historical Statistics of the United States—Colonial Times to 1970, U.S. Department of Commerce, Bureau of the Census.

b. How long would it take for the house to increase its value by 50%?

28. The average home in Metropolis in 2005 was $235,600. In 2006 it was $257,400. The average home price in Smallville in 2005 was $112,100. In 2006, it was $137,600. If these trends continue, when would Smallville's average home price exceed that of Metropolis?

Exercises 29–37 are only for those who have read Chapter 5.

29. Develop a compound interest model that represents the population of the New York/northern New Jersey/Long Island metropolitan area. Use the information in Exercise 7 and the method of Example 7.

30. Develop a compound interest model that represents the population of the San Francisco/Oakland/Fremont metropolitan area. Use the information in Exercise 8 and the method of Example 7.

31. Develop a compound interest model that represents the population of the Chicago/Naperville/Joliet metropolitan area. Use the information in Exercise 5 and the method of Example 7.

32. Develop a compound interest model that represents the population of the Los Angeles/Long Beach/Santa Ana metropolitan area. Use the information in Exercise 6 and the method of Example 7.

33. Show that the exponential model developed in Exercise 9 and the compound interest model developed in Exercise 31 are mathematically interchangeable.

34. Show that the exponential model developed in Exercise 10 and the compound interest model developed in Exercise 32 are mathematically interchangeable.

35. In Section 5.2, we found the doubling time of an account that pays 5% interest compounded daily by using a graphing calculator to solve the equation.

$$2P = P\left(1 + \frac{0.05}{365}\right)^n$$

Use logarithms to solve the same equation.

36. Compare and contrast the method used in Exercise 35 with the method used in Section 5.2 to find doubling time. Which method do you prefer? Why?

37. Use logarithms to solve Exercise 51a in Section 5.3.

Answer the following questions using complete sentences and your own words.

CONCEPT QUESTIONS

38. The exponential model is a powerful predictor of population growth, but it is based on the assumption that the two ordered pairs used to find a and b describe a steady tendency in the population's growth. In fact, that tendency may not be so steady, and the model's prediction may not be a good one.
a. What information were you given in Example 1 that implied a steady tendency in the aphid population's growth?
b. What information were you given in Example 2 that implied a steady tendency in Anytown's growth?
c. Discuss the factors of an animal population's growth that affect the accuracy of a prediction based on the exponential model. What other factors would be involved if the population were human?

39. In which of the following four situations would the exponential model give the most accurate prediction? The least accurate? Why?
- p is the population of a specific country; the two ordered pairs used to find a and b cover a time span of five years; you are asked to predict the population five years later.
- p is the population of a specific country; the two ordered pairs used to find a and b cover a time span of 100 years; you are asked to predict the population five years later.
- p is the population of the world; the two ordered pairs used to find a and b cover a time span of five years; you are asked to predict the population five years later.
- p is the population of the world; the two ordered pairs used to find a and b cover a time span of 100 years; you are asked to predict the population five years later.

PROJECTS

40. Do some research and find out the population of your community at two fairly recent points in time. Use that data to:
a. Find $\Delta p/\Delta t$, the average growth rate.
b. Find $(\Delta p/\Delta t)/p$, the relative growth rate.
c. Develop the model that represents the population of your community.
d. Predict the population of your community in 10 years.

e. Determine your community's doubling time.

f. Also, discuss the accuracy of your model. Be certain to include in your discussion the prominence of emigration and immigration in your community.

 WEB PROJECT

41. This web project will take approximately two weeks to complete. The U.S. Census Bureau has two population clocks on its web page (www.census.gov). One gives the estimated current U.S. population, and the other gives the estimated current world population.

a. Visit this site at two different times approximately one week apart, and record the time and the population (either U.S. or world population). Use these two different ordered pairs to create an exponential model, where t is time in hours.

b. Use the exponential model to predict the population approximately one week into the future.

c. Revisit the Census Bureau's web site at the appropriate time to compare your prediction from part (b) with the actual population. What is the error in your prediction? That is, what is the difference between your prediction and the actual population?

d. What is the percent error in your prediction? To find this percent error, take the error from part (c), divide it by the actual population, and convert the result to a percent.

10.2 Exponential Decay

What does the first article below mean when it says that tritium gas "decays at a rate of 5.5% a year"? Why do scientists claim that the frozen corpse found in 1991 in the Austrian Alps "is 5,000 to 5,500 years old"? (See the second article on the next page.) The answers have to do with radioactivity. Radioactive materials have become more and more prevalent and useful since Marie Curie introduced them to the world in 1898.

A radioactive substance is not stable; over time, it transforms itself into another substance. This is called **radioactive decay,** and it is due to the interaction between nuclear particles (protons, neutrons, and electrons) in the radioactive substance. Because a larger quantity of a radioactive substance has more nuclear particles (and hence more interactions), we might guess that it decays faster, just as a larger population produces more offspring. This, in fact, is the case; a larger amount of radioactive material does experience more decay. That is, *the rate of decay is proportional to the amount of radioactive substance present,* just as the rate of growth of a population is proportional to the size of the population. Consequently, we can use the exponential model $y = ae^{bt}$, developed in Section 10.1. As a memory device, we will use the variable Q instead of y to represent "quantity"; therefore, our model will be $Q = ae^{bt}$.

Radioactive Gas Used in A-bomb May Be Missing

WASHINGTON — The Energy Department and the Nuclear Regulatory Commission are investigating the possible loss of enough tritium to help make a nuclear bomb, and authorities have suspended sales of the radioactive gas during the probe.

Tritium is used in nuclear weapons to increase their power. Since it decays at the rate of 5.5 percent a year, the gas must be replenished if weapons are to maintain their full explosive potential.

The federal government also sells 200 to 300 grams of tritium a year to American and foreign companies for use in biological and energy research and to make self-luminous lights, signs and dials.

The incident has led to fears in Congress that some of the gas is missing and has fallen into unfriendly hands. But federal officials say that there are no signs of a diversion and that the discrepancy may have resulted from errors in measuring the tritium.

Tests on Iceman Make Him Part of Stone Age

By John Noble Wilford
Copyright © 1992 by the New York Times Co.
Reprinted with permission.

NEW YORK — Carbon dating tests show that the well-preserved body of a prehistoric human hunter found in an Alpine glacier last year is 5,000 to 5,500 years old, scientists reported yesterday.

The first scientifically established age for the frozen corpse is more than 1,000 years older than original estimates. It means the man lived and presumably froze to death well before the Bronze Age replaced the late Stone Age in Europe.

"Now we know he's not from the Bronze Age, but much older," said Dr. Werner Platzer, head of the anatomy department at Innsbruck University in Austria, who is directing research on the mummified corpse. "It means, I believe, that this is the only corpse we have from the Stone Age."

The tests on bones and skin tissue were conducted by scientists at Oxford University in England and a Swiss physics institute in Zurich. The body is being kept in cold storage at Innsbruck University.

The corpse was found in September in a glacier 10,500 feet up in the Austrian Alps, close to the Italian border. The body was mummified, and its leather and fur clothing was badly deteriorated. But scholars of prehistoric people were fascinated by a leather quiver with 14 arrows and an ax found by the iceman's side.

On first examination, the ax was thought to be made of bronze, which led to the assumption that this was an early Bronze Age man who lived about 4,000 years ago. More careful study showed the ax to be made of copper, thus a product of a simpler technology that seemed to place the iceman in the late Stone Age. This seemed to be confirmed by previous tests on samples of grasses taken from the hunter's clothing, which indicated that the body might be 5,000 years old.

ThomsonNOW™
for Liberal Arts Mathematics
www.thomsonedu.com/1pass.

EXAMPLE 1   Hospitals use the radioactive substance iodine-131 in research. It is effective in locating brain tumors and in measuring heart, liver, and thyroid activity. A hospital purchased 20 grams of the substance. Eight days later, when a doctor wanted to use some of the iodine-131, he observed that only 10 grams remained (the rest had decayed).

a. Develop a mathematical model that represents the amount of iodine-131 present.
b. Predict the amount remaining two weeks after purchase.

Solution   **a.** *Developing the model.*   Because we're using the exponential model $Q = ae^{bt}$, we follow the steps from Section 10.1.

1. *Write the given information as two ordered pairs* (t, Q). Originally there were 20 grams, so we have $(t, Q) = (0, 20)$. Eight days later, there were 10 grams, so we have $(t, Q) = (8, 10)$.

2. Recall that *a is always the initial value of Q, so* a = 20. Alternatively, we could *substitute* (0, 20) *into the model* $Q = ae^{bt}$ *and simplify*.

$$Q = ae^{bt}$$
$$20 = ae^{b(0)}$$
$$20 = ae^0$$
$$20 = a \cdot 1$$
$$a = 20$$

The model is now $Q = 20e^{bt}$.

3. *Substitute the second ordered pair* (8, 10) *into this new model and simplify.*

$$Q = 20\,e^{bt}$$

$$10 = 20\,e^{b(8)}$$

$$\frac{10}{20} = e^{8b}$$

$$0.5 = e^{8b}$$

$\ln 0.5 = \ln e^{8b}$ taking ln of each side

$\ln 0.5 = 8b$ using the Inverse Property $\ln(e^x) = x$

$$b = \frac{\ln 0.5}{8}$$

$$b \approx -0.086643397$$

Store this value of b in your calculator's memory.

Thus, our model for the amount Q of iodine-131 remaining t days after purchase is $Q = 20\,e^{-0.086643397t}$.

b. *Predictions.*   Now that the model has been determined, predictions can be made. Two weeks after purchasing the iodine-131, $t = 14$ days, and

$$Q = 20\,e^{-0.086643397t}$$

$$= 20\,e^{-0.086643397(14)}$$

$$= 5.946035575\ldots$$

$$\approx 5.9$$

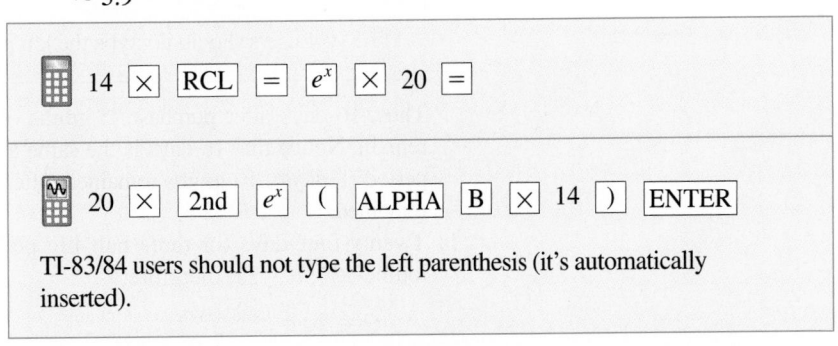

After two weeks (14 days), we would expect approximately 5.9 grams to remain. ∎

Notice that b is negative in Example 1. This indicates that the amount of iodine-131 is *decreasing*. In contrast, b was positive in Section 10.1, because the quantities under study were *increasing*.

Exponential Growth or Exponential Decay

In general, any quantity for which *the rate of change is proportional to the amount present* can be modeled by the formula $y = ae^{bt}$.

1. If $b > 0$, then y is growing exponentially.
2. If $b < 0$, then y is decaying exponentially.

Half-Life

A radioactive substance is not stable. One way to measure its instability is to determine the **half-life** of the substance, that is, the amount of time required for a quantity to reduce to one-half its initial size. In the previous example, the half-life of iodine-131 was 8 days; it took 8 days for 20 grams of the substance to decay into 10 grams. As we will see, it will take an additional 8 days for those 10 grams to decay into 5 grams.

EXAMPLE 2   Use the model $Q = 20e^{-0.086643397t}$ from Example 1 to calculate the amount of iodine-131 remaining at the following times.

a. 16 days after purchase   **b.** 24 days after purchase

Solution   **a.** Sixteen days after purchasing the 20 grams of iodine-131 (or eight days after observing that half of it had decayed), $t = 16$; therefore,

$$Q = 20\, e^{-0.086643397t}$$
$$= 20\, e^{-0.086643397(16)}$$
$$= 5$$

TI-83/84 users should not type the left parenthesis (it's automatically inserted).

Thus, 16 days after purchase, 5 grams of the initial 20 grams of iodine-131 will remain. Notice that 16 days is the same as two half-life periods. After one half-life period (8 days), 10 grams remained; after two half-life periods (16 days), 5 grams remained.

b. Twenty-four days (or three half-life periods) after purchasing the 20 grams of iodine-131, $t = 24$; therefore,

$$Q = 20\, e^{-0.086643397t}$$
$$= 20\, e^{-0.086643397(24)}$$
$$= 2.5$$

Thus, 24 days after purchase, 2.5 grams of the initial 20 grams of iodine-131 will remain. After another 8 days (or a total of 32 days), only 1.25 grams will remain. Every 8 days, half of the material decays.   ■

We can obtain a graph showing the exponential decay of the initial 20 grams of iodine-131 from the example above by plotting the ordered pairs (0, 20), (8, 10), (16, 5), and (24, 2.5) and continuing in this manner (after each 8-day period, only half the previous quantity remains). This graph is shown in Figure 10.13.

The half-life of a radioactive substance does not depend on the amount of substance present. For any given radioactive substance, the half-life is intrinsic to the substance itself; each radioactive substance has its own half-life. Figure 10.14 lists the half-life for various substances. Notice that half-lives have an incredibly wide range of values—from a few seconds to well over 20,000 years.

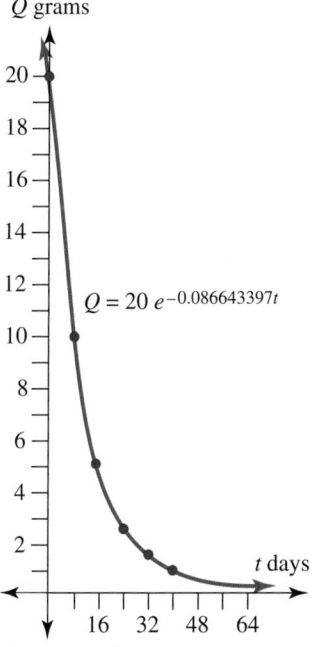

$Q = 20\,e^{-0.086643397t}$

Figure 10.13
A graph showing the amount of iodine-131 remaining t days after purchasing 20 grams

Radioactive Substance	Half-life
krypton-91	10 seconds
silicon-31	2.6 hours
cobalt-55	18.2 hours
magnesium-28	21.0 hours
iodine-124	4.5 days
iodine-131	8.0 days
cobalt-60	5.3 years
plutonium-241	13 years
plutonium-238	86 years
carbon-14	5,730 years
plutonium-239	24,400 years

Figure 10.14
Selected half-lives

EXAMPLE 3   Professor Frank Stein received 8.2 grams of plutonium-241 and stored it in his laboratory for future use in experiments. If he does not use any of the substance for experimental purposes, how much will remain one year later?

Solution   Rather than focusing on the *specific* initial amount of 8.2 grams, we will determine the *general* model $Q = ae^{bt}$ for *any* initial amount and substitute our given values later. We do this here to emphasize the general approach of developing a model, rather than dwell on the specific numerical calculations.

1. *Express the information as ordered pairs* (t, Q). Because a is the initial amount of plutonium-241, $(t, Q) = (0, a)$. To obtain a second ordered pair, we find the half-life of plutonium-241 in Figure 10.14 (the half-life is $t = 13$ years). In 13 years, half of a, or $\frac{a}{2}$, will remain, and we obtain $(t, Q) = (13, \frac{a}{2})$.

2. *Substitute the first ordered pair into the model, and simplify to find* a. Because we are not using a specific value for the initial amount a, we can omit this step. (Our model is still $Q = ae^{bt}$.)

3. *Substitute* $(t, Q) = (13, \frac{a}{2})$ *into the model and simplify.*

$$Q = ae^{bt}$$

$$\frac{a}{2} = ae^{b(13)}$$

$$\frac{1}{2} = e^{13b} \qquad \text{dividing by } a$$

$$0.5 = e^{13b}$$

$$\ln 0.5 = \ln e^{13b} \qquad \text{taking the natural log}$$

$$\ln 0.5 = 13b \qquad \text{using the Inverse Property } \ln(e^x) = x$$

$$b = \frac{\ln 0.5}{13}$$

$$b \approx -0.053319013$$

Store this value of b in your calculator's memory.

Thus, our model for the amount Q of plutonium-241 remaining t years after receiving an initial quantity of a grams is $Q = ae^{-0.053319013t}$. To predict the amount of plutonium-241 that will remain one year after Professor Stein receives his 8.2 grams, we substitute 8.2 for a and 1 for t.

$$Q = a\,e^{-0.053319013t}$$
$$= 8.2 \cdot e^{-0.053319013(1)}$$
$$= 7.774235618\ldots$$
$$\approx 7.8$$

$$\boxed{\text{RCL}}\ \boxed{e^x}\ \boxed{\times}\ 8.2\ \boxed{=}$$

$$8.2\ \boxed{\times}\ \boxed{\text{2nd}}\ \boxed{e^x}\ \boxed{\text{ALPHA}}\ \boxed{\text{B}}\ \boxed{\text{ENTER}}$$

On the basis of the model $Q = 8.2\ e^{-0.053319013t}$, Professor Stein should expect approximately 7.8 grams of plutonium-241 to remain one year after receiving the 8.2 grams.   ■

Relative Decay Rate

In Section 10.1, we discussed the average growth rate $\Delta p/\Delta t$, which indicates how fast a quantity is growing, and the relative growth rate $(\Delta p/\Delta t)/p$, which indicates how fast the quantity is growing *as a percent*. When calculating the constant b in the model $p = ae^{bt}$, we were able to check our work by seeing whether b was close to the relative growth rate.

This same concept can be applied to exponential decay. When using the exponential decay model $Q = ae^{bt}$, we can compute the **average decay rate** $\Delta Q/\Delta t$ by dividing the change in the amount of radioactive material by the corresponding change in time. We can also compute the **relative decay rate** by calculating $(\Delta Q/\Delta t)/Q$. Because Q is decreasing, ΔQ will be negative, as will the relative decay rate.

EXAMPLE 4   Find (a) the average decay rate and (b) the relative decay rate for the data given in Example 3.

Solution   **a.** *Finding the average decay rate.* The initial quantity was 8.2 grams, and 7.774235625 grams remained after $t = 1$ year. (Using the rounded-off figure of 7.8 in place of 7.774235625 would result in a less accurate answer.)

$$\Delta Q = \text{final quantity} - \text{initial quantity}$$
$$= 7.774235625 - 8.2 = -0.425764375 \text{ gram}$$
$$\Delta t = 1 \text{ year}$$

Therefore, the average decay rate is

$$\frac{\Delta Q}{\Delta t} = \frac{-0.425764375 \text{ gram}}{1 \text{ year}} \approx -0.4 \text{ gram per year}$$

The following year, there will not be as much material left, and what remains will decay at a slower rate. The average decay rate will not remain constant.

b. *Finding the relative decay rate.*

$$\frac{\Delta Q/\Delta t}{Q} = \frac{-0.425764375 \text{ gram per year}}{8.2 \text{ grams}}$$
$$= -0.051922484 \text{ per year} = -5.1922484\% \text{ per year}$$

This implies that the amount of radioactive material is decreasing by about 5.2% per year. While the average decay rate will change from year to year (because the material decays more slowly as the amount decreases), the relative decay rate will remain constant. Notice that the relative decay rate is close to b, which was -0.053319013.   ■

ThomsonNOW™
for Liberal Arts Mathematics
www.thomsonedu.com/1pass.

EXAMPLE 5   Plutonium-239 is a waste product of nuclear reactors. How long will it take for this waste to lose 99.9% of its radioactivity and therefore be considered relatively harmless to the biosphere?*

Solution   Regardless of the initial amount a of plutonium-239, we need to determine the time t required for 0.1% of the radioactivity to *remain*. (Remember, the model $Q = ae^{bt}$ determines the amount Q remaining after a time period t.) We first need to determine the model for plutonium-239.

1. *Express the initial data.* Let a = the initial amount of plutonium-239, and note that the half-life is 24,400 years (Figure 9.13). *Express this information as ordered pairs* (t, Q). We have $(0, a)$ and $(24{,}400, \frac{a}{2})$.

2. *Substitute the first ordered pair into the model, and simplify to find a.* Because we are not using a specific value for the initial amount a, we can omit this step. (Our model is still $Q = ae^{bt}$.)

3. *Substitute $(t, Q) = (24{,}400, \frac{a}{2})$ into the model and simplify.*

$$Q = ae^{bt}$$
$$\frac{a}{2} = ae^{b(24{,}400)}$$
$$\frac{1}{2} = e^{24{,}400b}$$
$$0.5 = e^{24{,}400b}$$
$$\ln 0.5 = \ln(e^{24{,}400b}) \quad \text{taking ln of each side}$$
$$\ln 0.5 = 24{,}400b \quad \text{using an Inverse Property}$$
$$b = \frac{\ln 0.5}{24{,}400}$$
$$b \approx -0.000028407$$

Store this value of b in your calculator's memory.

For plutonium-239, therefore, our model is $Q = ae^{-0.000028407t}$. Because the initial amount of plutonium-239 is a, we need to determine the time t when 0.1% of a remains—that is, when the amount Q of plutonium-239 will equal $0.001a$. We substitute $0.001a$ for Q in the model $Q = ae^{-0.000028407t}$ and solve.

$$Q = ae^{-0.000028407t}$$
$$0.001a = ae^{-0.000028407t}$$
$$0.001 = e^{-0.000028407t} \quad \text{dividing by } a$$
$$\ln 0.001 = \ln(e^{-0.000028407t}) \quad \text{taking ln of each side}$$
$$\ln 0.001 = -0.000028407t \quad \text{using an Inverse Property}$$
$$t = \frac{\ln 0.001}{-0.000028407}$$
$$= 243{,}165.1365\ldots$$
$$\approx 240{,}000$$

*H. A. Bethe, "The Necessity of Fission Power," *Scientific American*, January 1976.

It will take approximately 240,000 years for any given quantity of plutonium-239 to lose 99.9% of its radioactivity and thus be considered relatively harmless to the biosphere. How long a time span is this, relative to all of human history?　■

The United States has assembled a vast stockpile of about 50 tons of plutonium. Because plutonium is so deadly, its storage and disposal are of major concern, as indicated by the 1993 news article shown below. However, not every country in the world shares the United States' concern over the perils of plutonium. The Japanese Nuclear Agency has been promoting the use of plutonium in its power plants, as indicated by the 1994 news article.

Plutonium Stockpile Could Last Forever
U.S. has 50 tons to think about disposing

By Matthew L. Wald
Copyright © 1993 by the New York Times Co.
Reprinted with permission.

AMARILLO, TEXAS — In 16 unremarkable concrete bunkers built by the Army for a war with Hirohito and Hitler, the United States has begun assembling about 50 tons of plutonium, a vast stockpile of one of the most expensive materials ever produced and perhaps the most important to safeguard.

The Energy Department says the bunkers, each about the size of a two-car garage, are going to be used for interim storage, meaning six or seven years.

But plutonium, which was invented by the Energy Department's predecessor, the Manhattan Project, may turn out to be the hardest thing on Earth to dispose of. And at the Energy Department, "interim" can have an elastic meaning. "Immediate" tends to mean several years. "Several years" can mean never.

President Clinton announced the formation of an inter-agency task force in September to consider how much plutonium the nation needs in the post-Soviet era and how to dispose of what is surplus. The Energy Department is also drawing up a plan for its weapons production complex for the next century.

But officials are choosing among a short list of unattractive options.

At the Pantex plant near Amarillo, where the plutonium is piling up, general manager Rich Loghry said when asked what would happen next, "I don't think people have a really good answer for what is going to happen to the plutonium."

Storage may be the leading option, but even this is tricky. Plutonium loses half its radioactivity every 24,000 years, so it will reach background levels of radioactivity in 10 "half-lives," or 240,000 years; the U.S. political system is focused largely on problems that can be solved in less than four.

Japanese Make Plutonium Cute
Video claims it's safe to drink

By Seth Sutel
ASSOCIATED PRESS
Reprinted with permission.

TOKYO—Meet Mr. Pluto, the Japanese nuclear agency's round-faced, rosy-cheeked, animated answer to the public's concern about its plan to import 30 tons of plutonium as fuel for power plants.

In the country that best knows the dark side of atomic energy, not everyone is charmed by Mr. Pluto, who is featured in a promotional videotape prepared by the Power Reactor and Nuclear Fuel Development Corp.

Anti-nuclear groups said yesterday that they will campaign against distribution of the video, entitled "The Story of Plutonium: That Dependable Fellow, Mr. Pluto." They contend that it irresponsibly plays down the dangers plutonium poses.

Perky and pint-sized, Mr. Pluto is childlike, with cute red boots and a green helmet with antennae. On the front of the helmet is the chemical symbol for plutonium, Pu.

In one scene in the video, he shakes the hand of a cheerful youngster who is drinking a mug of plutonium-laced soda. The narration says that if plutonium were ingested, most of it would pass through the body without harm.

"The most fundamental lie in this video is the idea that plutonium is not dangerous," said Jinzaburo Takagi,

BY ASSOCIATED PRESS

The agency's video shows 'Mr. Pluto' greeting a boy drinking soda dosed with plutonium.

a former nuclear chemist who heads the Citizens' Nuclear Information Center.

"Of course, it's very dangerous to drink plutonium," Takagi said. "To say otherwise, as they do in this video, is completely outrageous."

The highly radioactive, silvery metal is toxic to humans because it is absorbed by bone marrow. The inhalation of .0001 of a gram can induce lung cancer.

Radiocarbon Dating

Radioactive substances are used to determine the age of fossils and artifacts. The procedure is based on the fact that two types of carbon occur naturally: carbon-12, which is stable, and carbon-14, which is radioactive. The total amount of carbon-14 in the environment is rather small; there is only one atom of carbon-14 for every 1 trillion atoms of carbon-12! Living organisms maintain this ratio due to their intake of water, air, and nutrients. However, when an organism dies, the amount of carbon-14 decreases exponentially due to radioactive decay. By making extremely delicate measurements of the amounts of carbon-14 and carbon-12 in a fossil or artifact, scientists are able to estimate the age of the item under investigation.

EXAMPLE 6   Determine the model representing the amount Q of carbon-14 remaining t years after the death of an organism.

Solution

1.   Let a represent the original quantity of carbon-14 present in a living organism. The half-life of carbon-14 is 5,730 years (from Figure 10.16), so at time $t = 5,730$,

Historical Note Marie Curie, 1867–1934

Born Marie Sklodowska in Warsaw, Poland, Marie Curie moved to Paris and enrolled as a student of physics at the Sorbonne in 1891. While researching the magnetic properties of various steel alloys, she met Pierre Curie, and they married in 1895. In the following year, Antoine Henri Becquerel discovered radioactivity in uranium. As a team, the Curies further investigated uranium and discovered the elements radium and polonium (named after Marie's native country). They also discovered that diseased, tumor-forming cells were destroyed faster than healthy cells when exposed to radium, laying the groundwork for modern radiation therapy. The word *radioactivity* was coined by Madame Curie.

In addition to publishing many important scientific papers and books, Madame Curie received many honors. She was the first person to win two Nobel Prizes, one in 1903 for the discovery of radioactivity (which she shared with her husband and Becquerel) and one in 1911 in chemistry. In 1908, at the University of Paris, she taught the first course on radioactivity ever offered.

The Sorbonne created a special chair in physics for Pierre Curie; Marie was appointed his successor after he died in a street accident.

During World War I, Madame Curie devoted much of her time to providing radiological services to hospitals. On her death, Albert Einstein said, "Marie Curie is, of all celebrated beings, the only one whom fame has not corrupted."

the quantity of carbon-14 present would be $\frac{a}{2}$. Expressed as ordered pairs (t, Q), we have $(0, a)$ and $(5730, \frac{a}{2})$.

2. Because we are not using a specific value for the initial amount a, we can omit this step. (Our model is still $Q = ae^{bt}$.)

3. *We substitute* $(t, Q) = (5730, \frac{a}{2})$ *into the model and simplify.*

$$Q = ae^{bt}$$

$$\frac{a}{2} = ae^{b(5,730)}$$

$$\frac{1}{2} = e^{5,730b} \qquad \text{dividing by } a$$

$$0.5 = e^{5,730b}$$

$$\ln 0.5 = \ln(e^{5,730b}) \qquad \text{taking ln of each side}$$

$$\ln 0.5 = 5,730b \qquad \text{using an Inverse Property}$$

$$b = \frac{\ln 0.5}{5,730}$$

$$b \approx -0.000120968$$

Historical Note   Willard Frank Libby, 1908–1980

Willard Frank Libby developed the radiocarbon dating technique in the mid-1940s. Carbon-14 was known to exist in nature, but little was known of its origins and properties. In 1939, Libby discovered that cosmic rays interacting with nitrogen at high altitudes produced a rapid formation of carbon-14. This high-altitude formation is the basis of the claim that the current ratio of carbon-14 to carbon-12 has been constant throughout history.

While working at the Enrico Fermi Institute of Nuclear Studies in Chicago, Libby was able to artificially produce carbon-14 and accurately determine its half-life. In addition, he devised a relatively simple device that measures the amount of carbon-14 in an organic sample. Before the creation of this device, measuring carbon-14 was a very expensive and difficult process. Libby's method made radiocarbon dating a practical possibility and revolutionized the fields of archeology and geology.

Libby received his doctorate degree in chemistry from the University of California at Berkeley in 1933 and taught there until 1945. During World War II, Libby also worked on the Manhattan Project, which developed the atomic bomb. During the years 1955–1959, he served on the U.S. Atomic Energy Commission, where he was instrumental in the formulation of many aspects of the commission.

After many years of dedicated research and numerous discoveries, Libby was awarded the Nobel Prize in chemistry in 1960 "for his method of using carbon-14 as a measurer of time in archeology, geology, geophysics, and other sciences."

Store this value of b in your calculator's memory.
We now have our model, $Q = ae^{-0.000120968t}$ ∎

The calculations in Example 6 produce the following general model.

Radiocarbon Dating Model

The quantity Q of carbon-14 remaining t years after the death of an organism (that had an initial amount a) is

$$Q = ae^{-0.000120968t} \qquad \text{or} \qquad Q = ae^{bt} \quad \text{where } b = \frac{\ln 0.5}{5{,}730}$$

ThomsonNOW
for Liberal Arts Mathematics
www.thomsonedu.com/1pass.

EXAMPLE 7   The first of the Dead Sea Scrolls were discovered in 1947 in caves near the northwestern shore of the Dead Sea in the Middle East. The parchment scrolls, which were wrapped in linen and leather, contained all the books of the Old Testament (except Esther). An analysis showed that the scrolls contained approximately 79% of the expected amount of carbon-14 found in a living organism. Determine the age of the scrolls.

Solution   Using the model $Q = ae^{-0.000120968t}$, we need to determine the amount of time t required so that 79% of a remains. To do this, we substitute $0.79a$ for Q in the model and solve.

THE FAR SIDE® By GARY LARSON

Early archaeologists

People have always been curious as to the age of artifacts.

A fragment of the Habbakuk Commentary of the Old Testament

$$Q = ae^{-0.000120968t}$$
$$0.79a = ae^{-0.000120968t} \quad \text{substituting } Q = 0.79a$$
$$0.79 = e^{-0.000120968t} \quad \text{dividing by } a$$
$$\ln 0.79 = \ln(e^{-0.000120968t}) \quad \text{taking ln of both sides}$$
$$\ln 0.79 = -0.000120968t \quad \text{using an Inverse Property}$$
$$t = \frac{\ln 0.79}{-0.000120968} \quad \text{dividing by } -0.000120968$$
$$t = 1,948.22 \ldots$$
$$t \approx 1,950$$

Thus, the Dead Sea Scrolls were approximately 1,950 years old when they were discovered and were therefore apparently created around the time when Jesus was alive. This result has been used to support the authenticity of the scrolls. ■

Shroud of Turin Was Created in 14th Century, Official Says

ASSOCIATED PRESS
Reprinted with permission.

ROME — Laboratory tests show that the Shroud of Turin was made in the 14th century and could not be the burial cloth of Christ, the scientific adviser to the archbishop of Turin said he learned yesterday.

Professor Luigi Gonella said he has not yet seen the official report from the three laboratories that conducted the carbon-14 dating tests, but that all the leaks to the press dated it to the 14th century, and "somebody let me understand that the rumors were right."

He refused to identify who had told him about the results of the tests at Oxford University, the University of Arizona and the Swiss Federal Institute of Technology at the University of Zurich.

© Alberto Pizzoli/Corbis Sygma

It should be noted that an underlying assumption in radiocarbon dating is that the current ratio of carbon-14 to carbon-12 in the biosphere remains constant over time and location. There have been disagreements within the scientific community over the validity of this assumption. Therefore, the dates obtained from radiocarbon dating are not guaranteed to be 100% reliable. Some of the results of radiocarbon dating have been controversial, as the article above indicates.

EXAMPLE 8   Suppose that a dose of a mCu of Tc-99m is administered to a patient so that a blood loss imaging scan can be conducted.

a. Develop a mathematical model that represents the amount of Tc-99m remaining in the patient t hours after the injection.

b. What portion of the initial dose will remain in the patient 1 hour after injection?

c. What portion of the initial dose will remain in the patient 10 hours after injection?

d. How long will it take so that 99% of the initial amount of Tc-99m has decayed?

Solution   **a.** To develop the mathematical model $Q = ae^{bt}$, we use the same procedure as in Example 3. That is, we first express the information as ordered pairs.

$$(t, Q) = (0, a) \text{ and } (t, Q) = \left(6.02, \frac{a}{2} \right) \qquad \text{the initial amount is } a, \text{ and the half-life is 6.02 hours.}$$

Substituting $(t, Q) = (6.02, \frac{a}{2})$ into the model, we have

$$Q = ae^{bt}$$

$$\frac{a}{2} = ae^{b(6.02)}$$

$$\frac{1}{2} = e^{b(6.02)} \qquad \text{dividing by } a$$

$$0.5 = e^{6.02b}$$

Radioactivity in the Real World: Nuclear Medicine

If a person has the misfortune to experience internal bleeding, how can a doctor determine the source and severity of the problem? A common investigative procedure (known as blood loss imaging) that involves the introduction of a radioactive material into the patient's bloodstream can assist in the diagnosis. Typically, a small amount of blood is drawn from the patient, a radioactive material (called a *tracer*) is combined with the blood sample, and the mixture is injected back into the bloodstream. Then the person's body is scanned with highly sensitive electronic sensors. These sensors detect the presence of gamma rays. Gamma rays result from the decay of radioactive elements. By monitoring the location and concentration of the gamma rays, doctors can pinpoint internal pools of blood and their sources. Once the bleeding sources have been located, an appropriate course of medical action or surgery can be initiated. The use of radioactive materials in the medical field is known as *nuclear medicine*.

One of the radioactive elements that is frequently used in blood loss imaging is technetium, commonly referred to as Tc-99m. Technetium was discovered in the early twentieth century. However, there is a debate as to when and where it was discovered; some say that German scientists discovered technetium in 1925, while others dispute this "finding" (their experimental data could not be reproduced) and say that technetium was "officially" discovered in 1937 in Italy.

Tc-99m is used in nuclear medicine because it has a relatively short half-life of 6.02 hours and is easily detectable in small amounts. When measuring the amount of radioactivity in a substance, technicians use units called *curies* (named after Pierre Curie) and *millicuries;* one curie equals 1000 millicuries. Curies and millicuries are abbreviated as "Cu" and "mCu," respectively; hence, 1 Cu = 1000 mCu. Unlike a gram or milligram, a curie does not measure weight (or mass). Instead, a curie measures the actual amount of radioactivity in a substance.

The definition of a curie is based on the element radium (discovered by Pierre and Marie Curie); 1 curie was originally defined as the amount of radioactivity in 1 gram of pure radium. However, radioactivity is now measured in terms of the number of atomic disintegrations per second; in 1953, scientists agreed that 1 curie would be defined as 37 billion (3.7×10^{10}) atomic disintegrations per second.

When administering Tc-99m for blood loss imaging, the suggested dose ranges from 10 to 20 mCu. Therefore, if a patient receives 10 mCu of Tc-99m, then 5 mCu will remain in the patient approximately 6 hours after being injected, 2.5 mCu will remain approximately 12 hours after being injected, approximately 1.25 mCu will remain 18 hours after being injected, and so on (owing to Tc-99m's half-life of 6.02 hours). See Example 8.

$$\ln 0.5 = \ln e^{6.02b} \qquad \text{taking the natural log}$$

$$\ln 0.5 = 6.02b \qquad \text{using the Inverse Property } \ln(e^x) = x$$

$$b = \frac{\ln 0.5}{6.02}$$

$$b \approx -0.1151407277 \qquad \text{store this value in your calculator's memory.}$$

Thus, our model for the amount of Tc-99m remaining t hours after an injection of a mCu is $Q = ae^{-0.1151407277t}$.

b. To calculate what portion of the initial dose remains 1 hour after injection, we substitute $t = 1$ and obtain

$$Q = ae^{-0.1151407277t}$$

$$Q = ae^{-0.1151407277(1)}$$

$$Q = ae^{-0.1151407277}$$

$$Q = a(0.8912407129)$$

$$Q = 0.891a \qquad \text{rounding off to three decimal places}$$

Therefore, 1 hour after injection, 0.891 (or 89.1%) of initial amount of Tc-99m will remain.

c. To calculate what portion of the initial dose remains 10 hours after injection, we substitute $t = 10$ and obtain

$$Q = ae^{-0.1151407277t}$$
$$Q = ae^{-0.1151407277(10)}$$
$$Q = ae^{-1.151407277}$$
$$Q = a(0.3161914872)$$
$$Q = 0.316a \qquad \text{rounding off to three decimal places}$$

Therefore, 10 hours after injection, 0.316 (or 31.6%) of the initial amount of Tc-99m will remain.

d. If 99% of the Tc-99m has decayed, 1% will still remain in the patient. To determine how long it will take so that only 1% remains, we substitute $Q = 0.01a$ into the model and solve for t.

$$Q = ae^{-0.1151407277t}$$
$$0.01a = ae^{-0.1151407277t}$$
$$0.01 = e^{-0.1151407277t} \qquad \text{dividing by } a$$
$$\ln 0.01 = \ln e^{-0.1151407277t} \qquad \text{taking the natural log}$$
$$\ln 0.01 = -0.1151407277t \qquad \text{using the Inverse Property } \ln(e^x) = x$$
$$t = \frac{\ln 0.01}{-0.1151407277}$$
$$t = 39.99601426$$

Therefore, 40.0 hours after injection, 1% of the initial amount of Tc-99m remains in the patient, that is, 99% has decayed.   ∎

10.2 Exercises

1. Using the model $Q = 20\,e^{-0.086643397t}$ developed in Example 1, predict how much iodine-131 will remain after three weeks.

2. Using the model $Q = 20\,e^{-0.086643397t}$ developed in Example 1, predict how much iodine-131 will remain after thirty days.

3. Using the model $Q = 8.2\,e^{-0.053319013t}$ developed in Example 3, predict how much plutonium-241 will remain after two years.

4. Using the model $Q = 8.2\,e^{-0.053319013t}$ developed in Example 3, predict how much plutonium-241 will remain after ten years.

In Exercises 5–16, use Figure 10.14.

5. Silicon-31 is used to diagnose certain medical ailments. Suppose a patient is given 50 milligrams.

a. Develop the mathematical model that represents the amount of silicon-31 present at time t.
b. Predict the amount of silicon-31 remaining after one hour.
c. Predict the amount of silicon-31 remaining after one day.
d. Find $\Delta Q/\Delta t$, the average decay rate, for the first hour.
e. Find $(\Delta Q/\Delta t)/Q$, the relative decay rate, for the first hour.
f. Find $\Delta Q/\Delta t$, the average decay rate, for the first day.
g. Find $(\Delta Q/\Delta t)/Q$, the relative decay rate, for the first day.
h. Why is the absolute value of the answer to part (e) greater than the answer to part (g)?

6. Plutonium-238 is used as a compact source of electrical power in many applications, ranging from pacemakers to spacecraft. Suppose a power cell initially contains 1.6 grams of plutonium-238.
   a. Develop the mathematical model that represents the amount of plutonium-238 present at time t.
   b. Predict the amount of plutonium-238 remaining after one year.
   c. Predict the amount of plutonium-238 remaining after 20 years.
   d. Find $\Delta Q/\Delta t$, the average decay rate, for the first year.
   e. Find $(\Delta Q/\Delta t)/Q$, the relative decay rate, for the first year.
   f. Find $\Delta Q/\Delta t$, the average decay rate, for the first 20 years.
   g. Find $(\Delta Q/\Delta t)/Q$, the relative decay rate, for the first 20 years.
   h. Why is the absolute value of the answer to part (e) greater than the answer to part (g)?

7. How long will it take 64 grams of magnesium-28 to decay into the following amounts?
   a. 32 grams
   b. 16 grams
   c. 8 grams

8. How long will it take 56 milligrams of cobalt-55 to decay into the following amounts?
   a. 28 milligrams
   b. 14 milligrams
   c. 7 milligrams

9. How long will it take 500 grams of plutonium-241 to decay into 100 grams?

10. How long will it take 300 grams of cobalt-60 to decay into 10 grams?

11. How long will it take 30 milligrams of plutonium-238 to decay into 20 milligrams?

12. How long will it take 900 milligrams of silicon-31 to decay into 700 milligrams?

13. How long will it take a given quantity of plutonium-239 to lose 90% of its radioactivity?

14. How long will it take a given quantity of plutonium-239 to lose 95% of its radioactivity?

15. How long will it take a given quantity of krypton-91 to lose 99.9% of its radioactivity?

16. How long will it take a given quantity of carbon-14 to lose 99.9% of its radioactivity?

17. The article on the "Iceman" in this section stated that the frozen corpse was 5,000 to 5,500 years old. Assuming the corpse was 5,250 years old, how much carbon-14 should it contain?

18. The article on radioactive gas in this section at the beginning of this section stated that tritium decays at the rate of 5.5% a year; that is, 94.5% of any given quantity will remain at the end of 1 year.
   a. Find the value of b in the mathematical model for the exponential decay of tritium.
   b. Determine the half-life of tritium gas.

19. A lab technician had 58 grams of a radioactive substance. Ten days later, only 52 grams remained. Find the half-life of the radioactive substance.

20. A lab technician had 32 grams of a radioactive substance. Eight hours later, only 30 grams remained. Find the half-life of the radioactive substance.

21. In 1989, a Mayan codex (a remnant of ancient written records) was found in the remains of a thatched hut that had been buried under 15 feet of volcanic ash after a prehistoric eruption of Laguna Caldera Volcano, just north of San Salvador. An analysis of the roof material concluded that the material contained 84% of the expected amount of carbon-14 found in a living organism. Determine the age of the roof material and hence the age of the codex.

22. Two Ohlone Indian skeletons, along with burial goods such as quartz crystals, red ocher, mica ornaments, and olivella shell beads, were dug up at a construction site in San Francisco in 1989. An analysis of the skeletons revealed that they contained 88% of the expected amount of carbon-14 found in a living person. Determine the age of the skeletons.

23. In 1940, beautiful prehistoric cave paintings of animals, hunters, and abstract designs were found in the Lascaux cave near Montignac, France. Analyses revealed that charcoal found in the cave had lost 83% of the expected amount of carbon-14 found in living plant material. Determine the age of the charcoal and the paintings.

24. To determine the onset of the last Ice Age, scientists analyzed fossilized wood found in Two Creeks Forest, Wisconsin. They theorized that the trees from which the wood was taken had been killed by the glacial advance. If the wood had lost 75% of its carbon-14, determine its age.

25. An ancient parchment contained 70% of the expected amount of carbon-14 found in living matter. Estimate the age of the parchment.

26. When analyzing wood found in an Egyptian tomb, scientists determined that the wood contained 55% of the expected amount of carbon-14 found in living matter. Estimate the age of the wood.

27. As noted in the article on the Shroud of Turin in this section, radiocarbon dating analysis concluded that the shroud was made in the fourteenth century. Assuming the shroud was made in 1350 A.D., how much carbon-14 did it contain when it was tested in 1988? If the shroud was made at the time of Christ's death in 33 A.D., how much carbon-14 should it have contained when it was tested in 1988?

28. In 1993, archeologists digging in southern Turkey found an ancient piece of cloth. Radiocarbon dating determined that the cloth was 9,000 years old. How much carbon-14 did the cloth have when it was tested?

29. How much carbon-14 would you expect to find in a 5,730-year-old relic?

30. How much carbon-14 would you expect to find in an 11,460-year-old relic?

31. A museum claims that one of its mummies is 5,000 years old. An analysis reveals that the mummy contains 62% of the expected amount of carbon-14 found in living organisms. Is the museum's claim justified?

32. A museum claims that one of its skeletons is 9,000 years old. An analysis reveals that the skeleton contains 42% of the expected amount of carbon-14 found in living organisms. Is the museum's claim justified?

33. A flute carved from the wing bone of a crane was discovered in Jiahu, an excavation site of Stone Age artifacts in China's Yellow River Valley in the 1980s. Scientists estimate that the artifact has lost 63.5% of its carbon-14. Estimate the age of the flute.

34. Suppose a patient receives 20.0 mCu of Tc-99m.
  a. Complete the chart in Figure 10.15 by determining the amount of Tc-99m remaining in the patient 5, 10, 15, 20, 25, and 30 hours after the injection. (The half-life of Tc-99m is 6.02 hours.)

Hours after injection	0	5	10	15	20	25	30
Amount remaining (mCu)	20.0						

Figure 10.15
Chart for Exercise 34

  b. Sketch a graph of the results obtained in part (a).

35. In Example 8, we determined that 0.891 (or 89.1%) of an initial dose of Tc-99m would remain in a patient 1 hour after injection. In a similar fashion, complete the chart in Figure 10.16.

Hours after injection	1	2	3	4	5
Portion remaining	0.891				

Figure 10.16
Chart for Exercise 35

36. In Example 8, we determined that 0.316 (or 31.6%) of an initial dose of Tc-99m would remain in a patient 10 hours after injection. In a similar fashion, complete the chart in Figure 10.17.

Hours after injection	6	7	8	9	10
Portion remaining					0.316

Figure 10.17
Chart for Exercise 36

37. How long will it take so that 99.5% of an initial dose of Tc-99m has decayed?

38. How long will it take so that 99.9% of an initial dose of Tc-99m has decayed?

39. **a.** How many atomic disintegrations per second occur in 1 mCu of radioactivity?
  b. How many atomic disintegrations per second occur in 10 mCu of radioactivity?
  c. How many atomic disintegrations per second occur in 20 mCu of radioactivity?

▬▬▬▶ *Answer the following questions using complete sentences and your own words.*

CONCEPT QUESTIONS

40. What does the phrase *exponential decay* mean?

41. What does the phrase *half-life* mean?

42. In terms of mathematical models, what is the basic difference between exponential growth and exponential decay?

43. In terms of units, what is the difference between the average decay rate and the relative decay rate?

44. How does radiocarbon dating work?

45. What aspect of radiocarbon dating is controversial?

46. What is nuclear medicine?

47. What basic unit is used in measuring radioactivity? How is it defined?

HISTORY QUESTIONS

48. Who invented radiocarbon dating? When? What other famous project did this person work on?

49. Who received a Nobel Prize for the discovery of radioactivity? When?

50. Who was the first person to receive two Nobel Prizes?

 WEB PROJECT

51. Although it is widely used, radiocarbon dating is not the only method employed when determining the age of artifacts. Write a report, or make a presentation, on an alternative method used in archaeometry (the science of determining the age of artifacts).

Some useful links for this web project are listed on the text web site: **www.thomsonedu.com/ mathematics/johnson**

10.3 Logarithmic Scales

Until now, we have been studying exponential models that are based on the function $y = ae^{bt}$. Now let's look at another topic closely related to exponential models: logarithmic scales. A **logarithmic scale** is a scale in which logarithms serve to make data more manageable by expanding small variations and compressing large ones, as shown in Figure 10.18. We will look at the Richter scale,

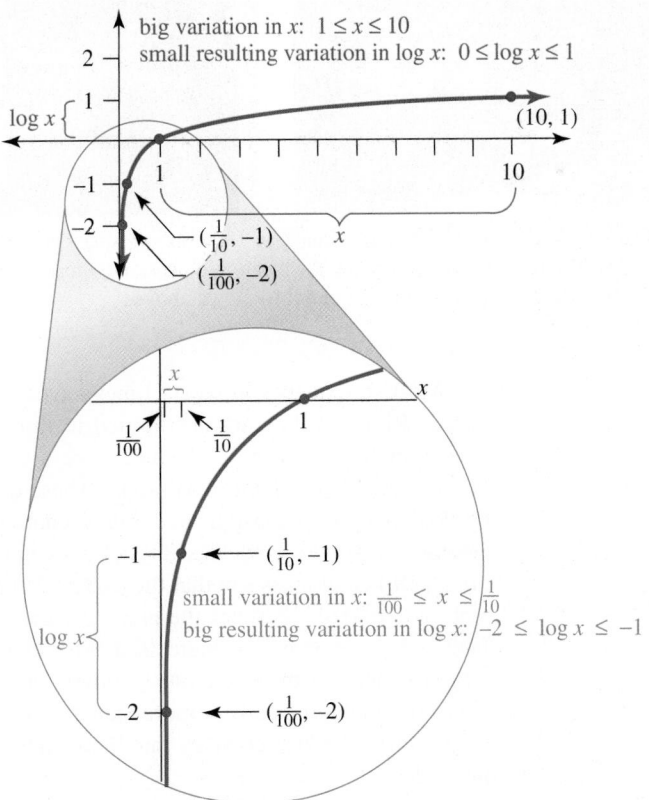

Figure 10.18
With logarithmic scales, logarithms expand small variations and compress large ones.

which is used to rate earthquakes, and the decibel scale, which is used to rate the loudness of sounds.

Earthquakes

Most earthquakes are mild and cause little or no damage, but some are incredibly devastating. San Francisco was almost totally destroyed in 1906 by a large earthquake (and the fires it caused). Over 200,000 people were killed by a large quake in the Indian Ocean in 2004. However, engineers have made great progress in designing structures that can withstand major earthquakes. San Francisco survived its 1989 quake with very limited damage; although there was some major damage, most of the city survived without a scratch (contrary to many sensationalistic news reports).

San Francisco was almost totally destroyed by the great quake of 1906 (above left). Only limited damage was sustained in the quake of 1989 (above right). The few buildings that collapsed were old ones, and most were made of unreinforced masonry or stucco or were built over first-floor garages, as was the building shown above.

Most earthquakes in recent times have been in the Middle East and along the Pacific Rim, but they are not limited to those regions. The most powerful earthquakes in North America's history were a series of three temblors in 1811–1812 on the New Madrid fault in Missouri. These quakes rerouted the Mississippi River, cracked plaster in Boston, and made church bells ring in Montreal. The U.S. Geological Survey now lists 39 states as having earthquake potential. (See Figure 10.19.) Geologists say that the eastern United States is certain to receive a violent jolt by 2010.* Perhaps the greatest current threat is posed by the New Madrid fault, where experts say there is a 90% chance of a serious quake in the next 50 years. Because of the region's geology, such a quake would cause damage over a huge area, including parts of more than a dozen states. (See Figure 10.20.) And it has only been in the last ten years that these states have started to adopt seismic building codes.

Newsweek, 30 October 1989, "East of the Rockies: A Lot of Shaking Going On."

Seismicity of the United States: 1990 - 2000

Figure 10.19
California is not the only state with tectonic stresses and thus prone to earthquakes. Each dot represents an earthquake between 1990 and 2000.

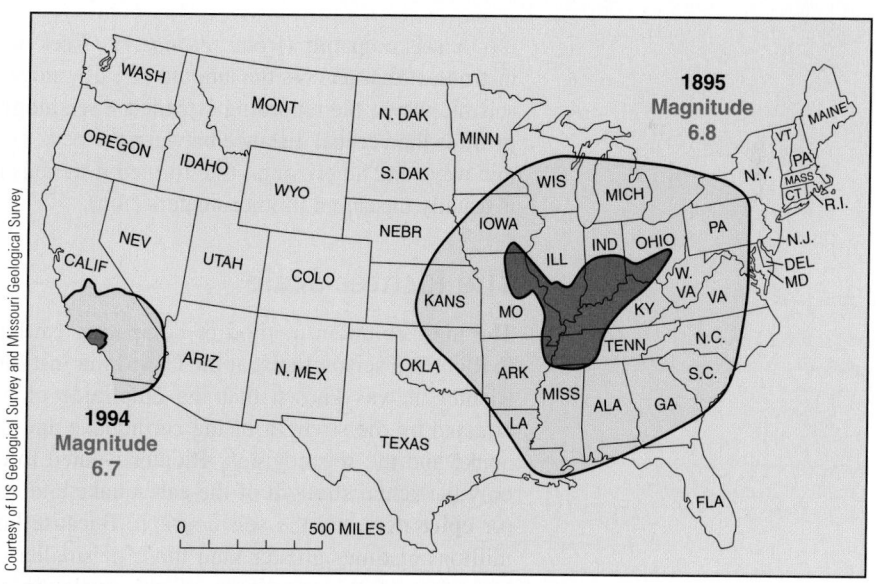

Figure 10.20
Comparing the extent of two earthquakes of similar magnitudes, one in Southern California and one in Missouri.

Most earthquakes are related to **tectonic stress** (that is, stress in the earth's crust). A group of more than 30 scientists from 19 different countries is compiling a global data base of stresses. A result of their efforts is the *World Stress Map,* shown in Figure 10.21, which indicates the presence of tectonic stresses throughout most of North America as well as much of Asia and Europe.

Figure 10.21
The *World Stress Map* shows the pressure of tectonic stresses in most of North America, Europe, and Asia.

A **seismograph** (from *seismos*, a Greek word meaning "earthquake") is an instrument that records the amount of earth movement generated by an earthquake's seismic wave; the recording is called a **seismogram.** The **amplitude of a seismogram** is the vertical distance between the peak or valley of the recording of the seismic wave and a horizontal line formed if there is no earth movement; the amplitude is usually measured in micrometers (μm).

The Richter Scale

The most common method of comparing earthquakes was developed by Charles F. Richter, a seismologist at the California Institute of Technology, in 1935. Prior to Richter, it was known that the amplitude of a recording of a seismic wave is affected by the strength of the earthquake and by the distance between the earthquake and the seismograph. Richter wanted to develop a scale that would reflect only the actual strength of the earthquake and not the distance between the source (or **epicenter**) and the seismograph. Because larger earthquakes have amplitudes millions of times greater than those of smaller quakes, Richter used the common logarithm of the amplitude of the quake in developing his scale in order to compress the enormous variation inherent in earth movement down to a more manageable range of numbers.

Examining data from many earthquakes, Richter discovered an interesting pattern: If A_{10} and A_{20} are the amplitudes of one earthquake measured 10 and 20 kilometers, respectively, from the epicenter and if B_{10} and B_{20} are similar 10- and 20-kilometer measurements for a second earthquake, as shown in Figure 10.22, then

$$\log A_{10} - \log B_{10} = \log A_{20} - \log B_{20}$$

Because the epicenter of the 1994 Northridge quake was within Los Angeles itself, the city suffered more extensive damage than did San Francisco in 1989.

Figure 10.22
The concept behind the Richter scale

or, equivalently, by the Division-Becomes-Subtraction Property,

$$\log \frac{A_{10}}{B_{10}} = \log \frac{A_{20}}{B_{20}}$$

This pattern continues *regardless of the distance*. That is, if A_{95} and B_{95} are the amplitudes of two quakes measured 95 kilometers from their epicenters, then

$$\log A_{10} - \log B_{10} = \log A_{20} - \log B_{20} = \log A_{95} - \log B_{95}$$

Richter's goal was to develop a scale that would measure the actual strength of an earthquake *without being biased by the distance between the quake's epicenter and the seismograph*. This pattern went a long way toward fulfilling Richter's goal; the only difficulty was that the pattern is a comparison of two quakes, not a measurement of a single quake. Richter remedied that problem by creating a "standard earthquake" of a certain fixed strength; this standard earthquake would be a yardstick against which all future earthquakes would be measured. Richter's standard earthquake is an average of a large number of extremely small southern California earthquakes, and the magnitude of an earthquake is a measure of how much stronger than the standard a given earthquake is. Figure 10.23 gives logs of amplitudes at different distances for the standard quake of "magnitude 0."

Distance (km)	20	60	100	140	180	220	260	300	340	380	420	460	500	540	580
Log of amplitude (πm)	−1.7	−2.8	−3.0	−3.2	−3.4	−3.6	−3.8	−4.0	−4.2	−4.4	−4.5	−4.6	−4.7	−4.8	−4.9

Figure 10.23
Richter's standard earthquake

Richter's Definition of Earthquake Magnitude

The magnitude M of an earthquake of amplitude A is

$$M = \log A - \log A_0$$

where A_0 is the amplitude of the "standard earthquake" measured at the same distance.

EXAMPLE 1   A seismograph 20 kilometers from an earthquake's epicenter recorded a maximum amplitude of 5.0×10^6 micrometers. Find the earthquake's magnitude.

Solution

$$M = \log A - \log A_0$$
$$= \log(5.0 \times 10^6) - \log A_0 \quad \text{substituting for } A$$
$$= \log(5.0 \times 10^6) - (-1.7) \quad \text{using Figure 10.23 to find } \log A_0$$
$$\approx 6.699 + 1.7$$
$$= 8.399$$
$$\approx 8.4$$

The earthquake was 8.4 on the Richter scale. (By tradition, earthquake magnitudes are rounded to the nearest tenth.)   ■

An earthquake is usually measured by seismographs in at least three different locations in order to accurately determine its Richter scale rating and the location of its epicenter. The Richter scale ratings of some major earthquakes are given in Figure 10.24.

In addition to comparing the amplitude of an earthquake with that of Richter's artificial "standard earthquake," we can also use Richter's definition to compare the amplitudes of two actual earthquakes.

1811–1812	New Madrid, MO	8.7	1990	Philippines	7.7
1906	San Francisco	8.3	1991	India and Nepal	7.7
1933	Japan	8.9	1992	Yucca Valley, CA	7.4
1950	India	8.7	1994	Los Angeles (Northridge)	6.8
1960	Chile (the largest earthquake ever recorded)	9.5	1995	Kobe, Japan	7.2
1964	Alaska	8.5	1994	Kuril Islands (between Russia and Japan)	8.2
1968	22 midwestern states	5.5	1997	Western Pakistan	7.3
1976	China	8.0	1999	Turkey	7.8
1983	Coalinga, CA	6.5	1999	Taiwan	7.6
1983	Hilo, Hawaii	6.3	1999	Oaxaca, Mexico	7.6
1985	Mexico City	8.1	2001	El Salvador	7.7
1988	Armenia	6.9	2001	India	7.7
1989	San Francisco/Loma Prieta (the "world series quake")	7.1	2001	Peru	8.4
1990	Iran	7.7	2004	Indian Ocean (the second largest ever recorded)	9.2

Figure 10.24
Magnitudes of some major earthquakes

Magnitude Comparison Formula

If M_1 and M_2 are the magnitudes of two earthquakes, and if A_1 and A_2 are their amplitudes measured at equal distances, then

$$M_1 - M_2 = \log\left(\frac{A_1}{A_2}\right)$$

The Magnitude Comparison Formula is true because

$$M_1 - M_2 = (\log A_1 - \log A_0) - (\log A_2 - \log A_0) \quad \text{using Richter's definition to rewrite } M_1 \text{ and } M_2$$

$$= \log A_1 - \log A_2 \quad \text{simplifying}$$

$$= \log\left(\frac{A_1}{A_2}\right) \quad \text{using the Division-Becomes-Subtraction Property}$$

Notice that if the second earthquake in the formula above is Richter's standard earthquake, $M_2 = 0$ (Richter's standard quake is one of magnitude zero), and A_2 is A_0, the magnitude comparison formula turns into Richter's definition of earthquake magnitude.

ThomsonNOW
for Liberal Arts Mathematics
www.thomsonedu.com/1pass.

EXAMPLE 2   If one earthquake has magnitude 6 and another has magnitude 3, it does not mean that the first caused twice as much earth movement, as many people believe. Find how the stronger earthquake actually compares to the weaker one.

Solution   We use the Magnitude Comparison Formula:

$$M_1 - M_2 = \log\left(\frac{A_1}{A_2}\right)$$

$$6 - 3 = \log\left(\frac{A_1}{A_2}\right)$$

$$3 = \log\left(\frac{A_1}{A_2}\right)$$

$$10^3 = 10^{\log(A_1/A_2)}$$

$$10^3 = \frac{A_1}{A_2} \qquad \text{using the Inverse Property } 10^{\log x} = x$$

$$A_1 = 10^3 A_2$$

The stronger earthquake's amplitude is actually $10^3 = 1{,}000$ times that of the weaker earthquake; thus, the stronger earthquake would cause about 1,000 times as much earth movement (not twice as much)!   ■

Earthquake Magnitude and Energy

The magnitude of an earthquake is determined by the amplitude of the earthquake's seismic wave and thus by the amount of ground movement caused by the quake. The amount of ground movement does not depend exclusively on the amount of energy radiated by the earthquake, because different geological compositions will transmit the energy in different ways. However, ground movement and energy are closely related.

Energy Formula

The energy E (in ergs) released by an earthquake of magnitude M is approximated by

$$\log E \approx 11.8 + 1.45M$$

EXAMPLE 3   Find approximately how much more energy is released by an earthquake of magnitude 6 than by an earthquake of magnitude 3.

Solution

$$M_1 = 6 \qquad\qquad M_2 = 3$$

$$\log E_1 \approx 11.8 + 1.45M_1 \qquad \log E_2 \approx 11.8 + 1.45M_2$$

$$= 11.8 + 1.45 \cdot 6 \qquad\qquad = 11.8 + 1.45 \cdot 3$$

$$= 20.5 \qquad\qquad\qquad = 16.15$$

$$10^{\log E_1} \approx 10^{20.5} \qquad\qquad 10^{\log E_2} \approx 10^{16.15}$$

$$E_1 \approx 10^{20.5} \qquad\qquad\qquad E_2 \approx 10^{16.15}$$

$$\frac{E_1}{E_2} \approx \frac{10^{20.5}}{10^{16.15}}$$

$$= 10^{20.5 - 16.15}$$

$$= 10^{4.35}$$

$$= 22{,}387.21139 \ldots$$

$$\approx 22{,}000$$

$$E_1 \approx 22{,}000 \cdot E_2$$

The earthquake of magnitude 6 releases approximately 22,000 times as much energy as the earthquake of magnitude 3. (Recall from Example 2 that the magnitude 6 quake causes 1,000 times as much earth movement as the magnitude 3 quake.) ■

The Decibel Scale

A sound is a vibration received by the ear and processed by the brain. The **intensity of a sound** is a measure of the "strength" of the vibration; it is determined by placing a surface in the path of the sound and measuring the amount of energy in that surface per unit of area per second. This surface acts like an eardrum.

Intensity is not a good measure of loudness as perceived by the human brain, for two reasons. First, experiments have shown that humans perceive loudness on the basis of the ratio of intensities of two different sounds. For example, in order for the human ear and brain to distinguish an increase in loudness, one sound must typically have 25% more energy than another. If two sounds have intensities I_1 and I_2 and if I_1 is 25% greater than I_2, then

$$I_1 = 125\% \text{ of } I_2$$
$$= 1.25 I_2$$

and the ratio of I_1 to I_2 is

$$\frac{I_1}{I_2} = 1.25$$

and the first sound would be perceived as slightly louder than the second. To determine loudness (as humans perceive it), we must look at the ratio of sound intensities.

The second reason why intensity is not a good measure of loudness has to do with how the brain processes sound. The human ear registers an amazing range of sound intensities. The intensity of a painfully loud sound is 100 trillion ($10^{14} = 100{,}000{,}000{,}000{,}000$) times that of a barely audible sound. More precisely, if I_1 is the intensity of a painfully loud sound and I_2 is the intensity of a barely audible sound, then $I_1 = 10^{14} \cdot I_2$, or $I_1/I_2 = 10^{14}$. The ear registers this range of sound intensities, but the brain condenses it to a smaller, more manageable range. The brain processes sound in a roughly logarithmic fashion.

The **decibel scale** approximates loudness as perceived by the human brain. It is based on the ratio of intensities of sounds, and its range is roughly on a par with the range in loudness that a human perceives rather than with the range in energy intensities that the ear or a mechanical device would actually measure.

Decibel Rating Definition

If a sound has intensity I (in watts per square centimeter, measured at a standard distance), then its decibel rating is

$$D = 10 \log\left(\frac{I}{I_0}\right)$$

where I_0 is a "standard intensity" ($I_0 \approx 10^{-16}$ watts/cm^2, the intensity of a barely audible sound).

Historical Note   Alexander Graham Bell, 1847–1922

The Granger Collection, NY

Alexander Melville Bell lectured at the University of Edinburgh (Scotland), at the University of London, and in Boston. He developed a system of visible speech for the deaf, with symbols for every sound of the human voice.

Alexander Graham Bell, the son of Alexander Melville Bell and the husband of a deaf woman, had his own school of vocal physiology in Boston and was very active in issues related to education of the deaf. In 1875, he conceived the idea of the telephone. On March 10, 1876, he used his experimental apparatus to transmit the now famous "Watson, come here; I want you" to his assistant. Later that same year, the telephone was introduced to the world at the Philadelphia Centennial Exposition. The Bell Telephone Company was organized a year later.

Bell also established the Volta Laboratory in Washington, D.C., where the first successful phonograph record was produced. He invented both the flat and cylindrical wax recorders for phonographs, as well as the photophone, which transmits speech by light rays, and the audiometer, which measures a person's hearing ability. He investigated the nature and causes of deafness and studied its heredity. He helped found the magazine *Science,* was president of the National Geographic Society, and was a regent of the Smithsonian Institution.

The Granger Collection, NY

Bell's telephone, as described in an 1877 newspaper. The first two figures are the transmitter and receiver that were shown at the Philadelphia Exposition.

The word *decibel* is abbreviated *dB*. The prefix *deci* is the metric system's prefix meaning "tenth." One decibel, therefore, is one-tenth of a bel. The *bel* is named for Alexander Graham Bell, the inventor of the telephone.

EXAMPLE 4   The background noise of a quiet library has a measured intensity of 10^{-12} watts/cm^2. Find the decibel rating of the sound.

Solution

$$I_1 = 10^{-12} \qquad \text{the given intensity}$$

$$D_1 = 10 \log\left(\frac{I_1}{I_0}\right) \qquad \text{the Decibel Rating definition}$$

$$= 10 \log\left(\frac{10^{-12}}{10^{-16}}\right) \qquad \text{substituting}$$

$$= 10 \log\left(10^{-12 - (-16)}\right) \qquad \text{using an Exponent Property}$$

$$= 10 \log 10^4$$

$$= 10 \cdot 4 \qquad \text{using the Inverse Property } \log 10^x = x$$

$$= 40$$

The decibel rating of the background noise is 40 dB. ∎

EXAMPLE 5 When two students started a conversation in the library (see Example 4), the intensity shot up to 10^{-10} watts/cm². Find the increase in decibels.

Solution

$$I_2 = 10^{-10} \qquad \text{the given intensity}$$

$$D_2 = 10 \log\left(\frac{I_2}{I_0}\right) \qquad \text{the Decibel rating definition}$$

$$= 10 \log\left(\frac{10^{-10}}{10^{-16}}\right) \qquad \text{substituting}$$

$$= 10 \log(10^{-10 - (-16)}) \qquad \text{using an Exponent Property}$$

$$= 10 \log 10^6$$

$$= 10 \cdot 6 \qquad \text{using the Inverse Property } \log 10^x = x$$

$$= 60$$

$$D_2 - D_1 = 60 - 40 \qquad D_1 = 40, \text{ from Example 4}$$

$$= 20$$

The increase in decibels, or **dB gain,** is 20 dB. ∎

Recording engineers monitor the music's decibel level so as to get a high-quality recording.

An alternative approach to finding the dB gain is to use a slightly altered version of the Decibel Formula.

dB Gain Formula

If I_1 and I_2 are the intensities of two sounds, then the dB gain is

$$D_1 - D_2 = 10 \log\left(\frac{I_1}{I_2}\right)$$

The dB Gain Formula is true because

$$D_1 - D_2 = 10 \log\left(\frac{I_1}{I_0}\right) - 10 \log\left(\frac{I_2}{I_0}\right) \qquad \text{using the Decibel Rating Definition to rewrite } D_1 \text{ and } D_2$$

$$= 10\left[\log\left(\frac{I_1}{I_0}\right) - \log\left(\frac{I_2}{I_0}\right)\right] \qquad \text{factoring}$$

$$= 10[(\log I_1 - \log I_0) - (\log I_2 - \log I_0)] \qquad \text{using the Division-Becomes-Subtraction Property}$$

$$= 10[\log I_1 - \log I_2]$$

$$= 10 \log\left(\frac{I_1}{I_2}\right) \qquad \text{using the Division-Becomes-Subtraction Property}$$

EXAMPLE 6   Use the dB Gain Formula to find the increase in decibels when the quiet library in Example 4 became somewhat less quiet due to the students' conversation in Example 5.

Solution

$$I_1 = 10^{-10} \text{ watts/cm}^2 \qquad \text{the conversation's intensity}$$
$$I_2 = 10^{-12} \text{ watts/cm}^2 \qquad \text{the background noise's intensity}$$

$$D_1 - D_2 = 10 \log\left(\frac{I_1}{I_2}\right) \qquad \text{the dB Gain Formula}$$

$$= 10 \log\left(\frac{10^{-10}}{10^{-12}}\right) \qquad \text{substituting}$$

$$= 10 \log(10^{-10-(-12)}) \qquad \text{using an Exponent Property}$$

$$= 10 \log 10^2$$

$$= 10 \cdot 2 = 20 \text{ dB gain}$$

Notice that $I_1/I_2 = 10^{-10}/10^{-12} = 10^2$, so the conversation had a sound intensity $10^2 = 100$ times that of the background noise. This means that there was 100 times as much vibration. ∎

EXAMPLE 7   The noise on a busy freeway varies from 81 dB to 92 dB. Find the corresponding variation in intensities.

Solution   We are given $D_1 = 92$ dB and $D_2 = 81$ dB, and we are asked to compare I_1 and I_2.

$$D_1 - D_2 = 10 \log\left(\frac{I_1}{I_2}\right) \qquad \text{the dB Gain Formula}$$

$$92 - 81 = 10 \log\left(\frac{I_1}{I_2}\right) \qquad \text{substituting}$$

$$11 = 10 \log\left(\frac{I_1}{I_2}\right)$$

$$1.1 = \log\left(\frac{I_1}{I_2}\right) \qquad \text{solving for } I_1$$

$$10^{1.1} = 10^{\log (I_1/I_2)}$$

$$10^{1.1} = \frac{I_1}{I_2} \qquad \text{using the Inverse Property } 10^{\log x} = x$$

$$I_1 = 10^{1.1} I_2 \approx 12.5893 I_2$$

The louder freeway noise has almost 13 times the sound intensity of the quieter freeway noise. This means that the louder freeway noise has almost 13 times as much vibration as does the quieter freeway noise. ∎

Music Making Fans Deaf?
How the iPod generation may be losing its hearing without even knowing it

Jonathan Ringen
Rollingstone.com, posted Nov. 18, 2005

Hearing loss is one of the dirty secrets of the music business, and everyone involved—from musicians onstage to fans who crank MP3s through headphones—is at risk. "We turn it up without realizing that we're doing damage," says Brian Fligor, an audiologist at Boston Children's Hospital. "Noise-induced hearing loss develops so slowly and insidiously that we don't know it's happened until it's too late."

In 1989, Pete Townshend admitted that he had sustained "very severe hearing damage." Since then, Neil Young, Beatles producer George Martin, Sting, Ted Nugent and Jeff Beck have all discussed their hearing problems.

Twenty-two million American adults own an iPod or other digital-music player, and studies show that sustained listening, even at moderate volume, can cause serious harm.

In 2001, Fligor began a study to determine how loud—and for how long—you can safely listen to a portable music player through headphones. He found that the kind of headphones you use greatly affects the risk. "The closer to the eardrum, the higher the sound levels the system is capable of producing," Fligor says. On average, Fligor found that you can safely listen to over-the-ear headphones with a player set at level six (out of ten) for an hour a day. For most in-the-ear headphones, like the earbuds that come with most MP3 players, the acceptable time at that level is less—around thirty minutes for some models before you've exceeded your safe daily dose. Fligor's preliminary findings indicate that the iPod is comparable to a Sony CD Walkman with earbuds, which can go as high as 130 decibels—equivalent to a jackhammer. European iPods, in contrast, are capped at 100 decibels by law.

"Every other musician I know has some form of problem with their hearing," says Garbage singer Shirley Manson. "All the guys in my band have weird schisms that come from being around really loud music." Pearl Jam guitarist Stone Gossard says he isn't affected but that bassist Jeff Ament, guitarist Mike McCready and drummer Matt Cameron all have hearing loss or tinnitus. "It's hard for them to deal with it, so we've made adjustments," Gossard says. "We're trying to keep the volume at a level that allows everyone to keep playing music."

Five Ways to Save Your Ears

1. Wear earplugs: Coldplay and Dave Matthews Band wear ear protection. You should too.
2. Turn it down: Don't crank up your portable music player too loud, especially to compensate for other noise around you.
3. Get better headphones: Those that shut out external noise allow you to turn down the tunes.
4. Give your ears a rest: "There's nothing wrong with going to a rock concert on Friday night," says Marshall Chasin. "Just don't mow your lawn on Saturday." Your ears need about eighteen hours after exposure to sustained high volumes before they return to normal.
5. Quit smoking: It doubles the risk of noise-induced hearing loss.

Logarithms and the Real World: The Negative Impact of Sounds

Sound has an incredibly positive impact on our lives. It provides us with music, and it provides us with oral communication. However, sound can have a negative impact as well. Loud concerts and iPods are damaging our hearing. iPods and other personal listening devices can be especially damaging, for several reasons:

- The earbud is placed in the ear, close to the eardrum. This proximity makes loud music more damaging than it would be if it were produced by a band at a concert or even over-the-ear headphones.
- Some people listen to their iPods for hours without interruption, because of the iPod's portability and its high quality sound. According to federal government safety standards, workers should not be exposed to noise above 90 decibels for more than eight hours. For every 5-decibel increase, the exposure time is cut in half. But iPod users listen at 130 decibels for hours.
- Hearing loss comes on very slowly. If it is due to loud music, the loss starts in the high frequencies, above the pitch of most conversations, and people don't notice. Instead, they continue to abuse their ears with overly loud music until it's too late.

See the *Rolling Stone* article on page 793, Figure 10.25, and Exercise 33.

Loud music isn't the only cause of harmful sounds. The U.S. Navy's sonar system is possibly harming marine mammals, as could its proposed East Coast undersea warfare training range. See the articles on these topics on page 797 and Exercises 35–36.

© William Thomas Cain/Getty Images

Decibels	Example of sound at this level	OSHA maximum exposure time
0 dB	Threshold of hearing	
10 dB	Soundproof room	
20 dB	Radio, television or recording studio	
30–40 dB	Quiet room	
50 dB	Private office	Safe
60 dB	Normal conversation	Listening
70 dB	Street noise	Zone
80 dB	Telephone dial tone	
90 dB	Train whistle at 500 ft	
95 dB	Subway train at 200 ft	4 hours
100 dB	Chain saw	2 hours
105 dB	Power mower	1 hour
110 dB	Dance clubs	30 minutes
120 dB	Band practice	7.5 minutes
130 dB	Threshold of pain	Less than 2 minutes
140 dB	Jet engine at 100 ft	

Figure 10.25
When it's safe to listen

10.3 Exercises

In Exercises 1–4, find the magnitude of the given earthquake.

1. An earthquake was measured by a seismograph 100 kilometers from the epicenter that recorded a maximum amplitude of 3.9×10^4 μm.

2. An earthquake was measured by a seismograph 60 kilometers from the epicenter that recorded a maximum amplitude of 6.3×10^5 μm.

3. An earthquake was measured by three seismographs. The closest, 60 kilometers from the epicenter, recorded a maximum amplitude of 250 μm. A second seismograph 220 kilometers from the epicenter recorded a maximum amplitude of 40 μm. The third seismograph, 460 kilometers from the epicenter, recorded a maximum amplitude of 4 μm.

4. An earthquake was measured by three seismographs. The closest, 20 kilometers from the epicenter, recorded a maximum amplitude of 25 μm. A second seismograph 60 kilometers from the epicenter recorded a maximum amplitude of 2 μm. The third seismograph, 180 kilometers from the epicenter, recorded a maximum amplitude of 0.50 μm.

In Exercises 5–12, use the information in Figure 10.24 (page 787) to compare the two earthquakes given by (a) finding how much more earth movement was caused by the larger and (b) finding how much more energy was released by the larger.

5. San Francisco (1989) and San Francisco (1906)
6. New Madrid and Los Angeles
7. San Francisco (1906) and Los Angeles
8. San Francisco (1989) and Los Angeles
9. Indian Ocean and San Francisco (1989)
10. Turkey and Taiwan
11. New Madrid and Coalinga
12. Los Angeles and Indian Ocean
13. Shortly after the 1989 San Francisco quake, it was announced that the quake was of magnitude 7.0. Later, after the data from more seismographs had been analyzed, the rating was increased to 7.1.
    a. How large an increase of earth movement corresponds to this increase in magnitude?
    b. How large an increase of energy released corresponds to this increase in magnitude?
14. Find a rule of thumb for the increase in earth movement that corresponds to an increase in Richter magnitude of one unit.

HINT: Follow Example 2, using magnitudes $x + 1$ and x.

15. Find a rule of thumb for the increase in energy released that corresponds to an increase in Richter magnitude of one unit.

HINT: Follow Example 3, using magnitudes $x + 1$ and x.

In Exercises 16–20, find the decibel rating of the given sound.

16. a very faint whisper at 10^{-14} watts/cm^2
17. a television at 10^{-9} watts/cm^2
18. a quiet residence at 3.2×10^{-12} watts/cm^2
19. a vacuum cleaner at 2.5×10^{-9} watts/cm^2
20. an outboard motor at 1.6×10^{-6} watts/cm^2

In Exercises 21–24, find the dB gain for the given sound.

21. a whisper increasing from 10^{-14} watts/cm^2 to 10^{-13} watts/cm^2
22. noise in a dormitory increasing from 3.2×10^{-12} watts/cm^2 to 2.1×10^{-11} watts/cm^2
23. a stereo increasing from 3.9×10^{-9} watts/cm^2 to 3.2×10^{-8} watts/cm^2
24. a motorcycle increasing from 6.3×10^{-8} watts/cm^2 to 3.1×10^{-6} watts/cm^2

25. If a single singer is singing at 74 dB, how many singers have joined him if the level increases to 81 dB and each singer is equally loud?
26. If a single singer is singing at 74 dB, how many singers have joined him if the level increases to 83 dB and each singer is equally loud?
27. If a single singer is singing at 74 dB, how many singers have joined her if the level increases to 77 dB and each singer is equally loud?
28. If a single singer is singing at 74 dB, how many singers have joined her if the level increases to 82.5 dB and each singer is equally loud?
29. If a single trumpet is playing at 78 dB, how many trumpets have joined in if the level increases to 85.8 dB and each trumpet is equally loud?
30. If a single trumpet is playing at 78 dB, how many trumpets have joined in if the level increases to 88 dB and each trumpet is equally loud?

31. Find a rule of thumb for the dB gain if the number of sound sources doubles (where each source produces sounds at the same level).

32. Find a rule of thumb for the dB gain if the number of sound sources increases tenfold (where each source produces sounds at the same level).

33. According to the article from *Rolling Stone,* an American iPod can be turned up louder than a European iPod.

    a. How much more impact does an American iPod have on the eardrum than does a European iPod if they are both turned up all the way?

    b. Use the information in the box on page 793 and Figure 10.25 to determine the maximum exposure time for an American iPod and for a European iPod if they are both turned up all the way.

34. According to the following articles, the rock group Pink Floyd reduced its volume to protect Venice's buildings.

    a. How much effect would the stated decrease in volume have?

    b. Use the information in Figure 10.25 to determine the maximum exposure time for the concert at both the proposed and actual decibel levels.

Venice Balks at Concert by Floating Pink Floyd

UNITED PRESS INTERNATIONAL
Reprinted with permission.

VENICE — The British rock band Pink Floyd arrived yesterday to prepare for a weekend world television concert as promoters and city leaders argued over whether the group's decibels would damage ancient monuments.

Pink Floyd originally was scheduled to perform tomorrow on a giant floating platform moored alongside the lagoon city's world famous St. Mark's Square.

But the city's superintendent for architectural and environmental property vetoed the idea on grounds that vibrations from the group's amplified music could damage the structures of such historic buildings as St. Mark's Basilica and the Ducal Palace.

Yesterday morning, the organizers and city officials reached agreement that the concert could go ahead provided the floating platform is moored far enough out in the Venice lagoon to restrict the sound hitting the famed buildings to 60 decibels.

Scientists calculated that could be achieved if the huge platform—which measures 295 feet by 88 feet—is anchored 150 yards away from the St. Mark's Square shore.

Pink Floyd's Venice Concert Draws 200,000

ASSOCIATED PRESS
Reprinted with permission.

VENICE — Fans of the rock group Pink Floyd, including Woody Allen and Tom Cruise, more than doubled the population of this canal city to see the band perform on a floating stage near St. Mark's Square.

100 Million Listeners

It was broadcast live in Italy and 23 other countries, reaching a total audience of 100 million people, according to the state-run RAI network. It was not broadcast in the United States.

The concert set an attendance record as Italy's largest rock event and attracted more visitors to the historic city than did the annual Carnival festival.

City preservation officials had threatened to deny permission for the concert. But on Thursday they dropped their objections when concert organizers agreed to measures to reduce potential damage to the surrounding buildings.

A Quieter Sound

The stage was pulled farther from the shore than previously planned and the rock group reduced its volume to a maximum of 60 decibels, instead of 105 as planned.

Italy's vice premier, Gianni De Michelis, spoke out in favor of the concert.

"Venice must give space to all cultural languages, including that of rock music," he said.

35. According to the article "Suit Filed to Halt Navy Sonar System" from the *San Francisco Chronicle,* the Navy will restrict its sonar operators to a lower sound level to protect marine mammals. How much effect would the stated decrease in volume have?

36. According to the article "Whales at Risk" from *National Geographic News,* mid-frequency sonar can emit continuous sound at 235 dB or higher. How many times as intense is this sound than an American iPod turned up all the way to 130 dB?

Suit Filed to Halt Navy Sonar System

SAN FRANCISCO CHRONICLE,
8 AUGUST 2002

Environmental groups sued . . . to stop the U.S. Navy from deploying a new anti-submarine sonar system that they say could harm sea creatures wherever it's used in the world's oceans. The federal lawsuit seeks an injunction against using the low-frequency active sonar system, which critics fear will threaten entire populations of whales, dolphins, seals and other marine mammals. . . . The Navy says active sonar arrays are capable of generating 215 decibels of sound—dangerous for marine mammals—but that operators will be restricted to 180 decibels, a sound level emitted by some whales. . . . Naomi Rose, marine mammal scientist for the New York-based Humane Society of the United States, . . . called a safety exposure level of 180 decibels "completely unacceptable and not substantiated by data." . . . "It's not precautionary, as the Marine Mammal Protection Act demands," Rose said, "and is out of step with the standards that other countries are setting."

Whales at Risk From New U.S. Navy Sonar Range, Activists Say

Maryann Mott
for *National Geographic News*
November 3, 2005

The U.S. Navy is moving ahead with plans to build an undersea warfare training range on the U.S. East Coast despite fierce opposition from conservation and animal welfare organizations. Groups opposed to the military project say endangered North Atlantic right whales, dolphins, and sea turtles could potentially be injured or killed from powerful sonar blasts emitted during training exercises. "Protecting whales and preserving national security are not mutually exclusive," said Fred O'Regan, president of the International Fund for Animal Welfare.

Mid-frequency sonar used during training exercises can emit continuous sound well above 235 decibels—an intensity roughly comparable to a rocket blastoff, according to the Natural Resources Defense Council (NRDC), a conservation nonprofit group. Experts say that's a problem for marine mammals and other aquatic animals. Sound is their primary means of learning about their environment, communicating, and navigating. "Military sonar needlessly threatens whole populations of whales and other marine animals," said Joel Reynolds, an attorney for NRDC. "The Navy refuses to take basic precautions that could spare these majestic creatures." The Washington, D.C.-based nonprofit, along with several animal welfare and environmental groups, sued the Navy last week, citing harm to whales caused by mid-frequency sonar.

The technology is associated with strandings—when marine animals swim or float onto shore and become beached or stuck in shallow water. In 2000 17 whales from three species beached themselves in the Bahamas after Navy ships conducted mid-range sonar exercises. Of those stranded, seven animals died. A federal investigation concluded sonar was the most likely cause of the whales' stranding. Adding to the controversy, the military's preferred training site in North Carolina is near an area where a mass whale stranding took place in January, not long after the Navy conducted sonar exercises in the region.

Answer the following questions using complete sentences and your own words.

CONCEPT QUESTION

37. Compare and contrast the Richter scale and the decibel scale. Why are the two formulas so similar? Why does each formula involve logarithms?

HISTORY QUESTIONS

38. Name three of Bell's other inventions besides the telephone.

39. The inventor of the Richter scale performed his research at what school?

40. In addition to electronics, what field of study interested Alexander Graham Bell?

Chapter 10   Review

ThomsonNOW™ for Liberal Arts Mathematics
Preparing for an exam? Test yourself on key material by visiting ThomsonNOW at **www.thomsonedu.com/1pass.**

TERMS

amplitude of a seismogram
average decay rate
average growth rate
base of a logarithmic
  function
base of an exponential
  function
common logarithm
dB gain

decibel scale
delta notation
dependent variable
epicenter
exponential function
exponential model
function
half-life
independent variable

intensity of a sound (versus
  its perceived loudness)
irrational number
logarithm
logarithmic equation
logarithmic scale
mathematical model
natural exponential
  function

natural logarithm
radioactive decay
rate of change
rational number
relative decay rate
relative growth rate
seismogram
seismograph
tectonic stress

PROCEDURES

Solving exponential equations

1. Isolate the exponential.
2. Take the natural logarithm of each side.
3. Use an Inverse Property to simplify.
4. Solve.

Solving logarithmic equations

1. Get all the log terms on one side and all the nonlog terms on the other.
2. Combine the log terms into one terms using the Division-Becomes-Subtraction Property and the Multiplication-Becomes-Addition Property.
3. Exponentiate each side.
4. Use an Inverse Property to simplify.
5. Solve.

Developing an exponential model

1. Write the information given as to the value of the quantity at two points in time as two ordered pairs (t, y).
2. Substitute the first ordered pair into the model $y = ae^{bt}$ and simplify to find a.
3. Substitute the second ordered pair into the model and simplify to find b.

APPLICATION DEFINITIONS AND FORMULAS

Richter's definition of earthquake magnitude: $M = \log A - \log A_0$ (get $\log A_0$ from the table on page 786)

Magnitude comparison formula: $M_1 - M_2 = \log\left(\dfrac{A_1}{A_2}\right)$

Energy formula: $\log E \approx 11.8 + 1.45M$

Decibel rating definition: $D = 10 \log\left(\dfrac{I}{I_0}\right)$

dB gain formula: $D_1 - D_2 = 10 \log\left(\dfrac{I_1}{I_2}\right)$

LOGARITHM DEFINITIONS AND FORMULAS

For any base b, $\log_b u = v$ means the same as $b^v = u$. For base 10 and base e:

$y = \log x$ means the same as $10^y = x$ \qquad $y = \ln x$ means the same as $e^y = x$

Division-Becomes-Subtraction Properties

$\log\dfrac{A}{B} = \log A - \log B$ \qquad\qquad $\ln\dfrac{A}{B} = \ln A - \ln B$

Multiplication-Becomes-Addition Property

$\log(A \cdot B) = \log A + \log B$ \qquad\qquad $\ln(A \cdot B) = \ln A + \ln B$

Exponent-Becomes-Multiplier Properties

$\log(A^n) = n \cdot \log A$ \qquad\qquad $\ln(A^n) = n \cdot \ln A$

Inverse Properties

$\log(10^x) = x$ \qquad\qquad $\ln(e^x) = x$

$10^{\log x} = x$ \qquad\qquad $e^{\ln x} = x$

REVIEW EXERCISES

In Exercises 1–3, find the value of x.

1. $x = \log_3 81$
2. $2 = \log_x 16$
3. $5 = \log_5 x$
4. State the Inverse Properties of the common logarithm.
5. State the Inverse Properties of the natural logarithm.
6. State the Division-Becomes-Subtraction Property of the common logarithm.
7. State the Division-Becomes-Subtraction Property of the natural logarithm.
8. State the Exponent-Becomes-Multiplier Property of the common logarithm.
9. State the Exponent-Becomes-Multiplier Property of the natural logarithm.
10. State the Multiplication-Becomes-Addition Property of the common logarithm.
11. State the Multiplication-Becomes-Addition Property of the natural logarithm.

In Exercises 12–14, rewrite each of the expressions (if possible) using properties of logarithms.

12. $\log 3x - \log 4 + \log 7x^2$
13. $\log(x + 2)$
14. $\log(\frac{x}{2})$
15. Solve $520\, e^{0.03x} = 730$.
16. Solve $\ln x - \ln 4 = \ln 2$.
17. Solve $\log 5x + \log x^2 - \log x = 12$.
18. A bacteria culture had a population of about 10,000 at 12 noon. At 2 P.M., the population had grown to 25,000.

a. Develop the mathematical model that represents the population of the bacteria culture.

b. Use the model to predict the population at 6 A.M. the next day.

c. Use the model to predict when the population will double its original size.

d. On what assumptions is the model based?

19. Cobalt-60 is used by hospitals in radiation therapy. Suppose a hospital purchases 300 grams.

a. Develop the mathematical model that represents the amount of cobalt-60 at time t.

b. Predict the amount *lost* if delivery takes two weeks.

c. Predict the amount *remaining* after one year (if none is used).

d. Determine how long it will take for 90% of the cobalt-60 to decay.

20. Archeologists found a wooden bowl in an Indian burial mound. The bowl contained 73% of the amount of carbon-14 found in living matter. Estimate the age of the burial mound.

21. An earthquake was measured by two seismographs. The closest, 20 kilometers from the epicenter, recorded a maximum amplitude of 25 μm. The second seismograph, 60 km from the epicenter, recorded a maximum amplitude of 2 μm. Find the magnitude of the earthquake.

22. A 1989 earthquake in Newcastle, Australia measured 5.5 on the Richter scale. The 1989 San Francisco earthquake measured 7.1.

a. How much more earth movement was caused by the San Francisco quake?

b. How much more energy was released by the San Francisco quake?

23. Find the decibel rating of a scooter at 1.6×10^{-6} watts/cm^2.

24. The scooter in Exercise 23 increases to 1.3×10^{-5} watts/cm^2. Find the dB gain.

25. If a single trumpet is playing at 78 dB, how many trumpets have joined in if the level increases to 84 dB?

Answer the following questions using complete sentences and your own words.

CONCEPT QUESTIONS

26. What is an exponential function?

27. What is a logarithmic function?

28. In terms of mathematical models, what is the basic difference between exponential growth and exponential decay?

29. How does radiocarbon dating work? What aspect of radiocarbon dating is controversial?

HISTORY QUESTIONS

30. What underlying theme unites all of Alexander Graham Bell's professional interests?

31. Name three of Bell's other inventions besides the telephone.

32. Who introduced the symbol e? What other symbols did this person invent?

33. Who invented logarithms? What motivated their invention?

34. Who invented radiocarbon dating?

35. Who was the first person to receive two Nobel Prizes? For what was the first prize awarded?

11 Matrices and Markov Chains

© Directphoto.org/Alamy

Markov chains are used to analyze trends and predict the future. They have many applications in business, sociology, the physical sciences, and biology. For example, in business, Markov chains can be used to predict the future success of a product. In sociology, Markov chains can be used to predict the eventual outcome of trends, such as the shrinking of the American middle class and the growth of the suburbs at the expense of central cities and rural areas. In this chapter, we investigate these two sociological problems and also see how businesses use Markov chains.

The method of Markov chains consists of using probabilistic information in the analysis of current trends in order to predict their outcomes. Probability trees are used to analyze trends, and matrices are used to summarize the probability trees and simplify the calculations. As an alternative to probability trees, matrices make the work easier. (A **matrix** is a rectangular arrangement of numbers enclosed by brackets, such as $\begin{bmatrix} 3 & -2 \\ 0 & 1 \end{bmatrix}$.)

The method of Markov chains also involves solving systems of equations. We will mainly use the elimination method, from intermediate algebra, to solve them, but we will also discuss how to use technology to solve systems.

Matrices will be used throughout this chapter. In Section 11.0, we will review matrix terminology and notation and matrix multiplication. In Section 11.1, we will introduce Markov chains and discuss the application of matrix multiplication to Markov chains. In Section 11.2, we will review the elimination method of solving systems of equations, and we will discuss how to use technology to solve systems. In Section 11.3, we will further explore Markov chains and discuss the use of systems of equations in Markov chains. In Section 11.4, we will discuss the use of the elimination method in solving larger systems of equations, and we will discuss how to use technology to solve larger systems. In Section 11.5, we will continue to explore Markov chains and discuss the use of systems of equations in Markov chains.

11.0 Review of Matrices

Terminology and Notation

A **matrix** (plural *matrices*) is a rectangular arrangement of numbers enclosed by brackets. For example,

$$\begin{bmatrix} 3 & -2 & 0 \\ 1 & -27 & 5 \end{bmatrix}$$

is a matrix. Each number in the matrix is called an **element** or an **entry** of the matrix; 3 is an element of the above matrix, as is -27. The horizontal listings of elements are called **rows;** the vertical listings are called **columns.** The first row of the matrix

$$\begin{bmatrix} 3 & -2 & 0 \\ 1 & -27 & 5 \end{bmatrix} \leftarrow \text{first row}$$

$$\uparrow$$
last column

is "3 -2 0," and the last column is "0   5." The **dimensions of a matrix** with m rows and n columns are $m \times n$ (read "m by n"). The matrix above has two rows and three columns, so its dimensions are 2×3; that is, it is a 2×3 matrix.

The dimensions of a matrix are like the dimensions of a room. If a room is 7 feet on one side and 10 feet on the other, then the dimensions of that room are 7×10. We do not actually multiply 7 times 10 to determine the dimensions (although we would to find the area). Likewise, the dimensions of the above matrix are 2×3; we do not actually multiply 2 times 3 and get 6. If we did, we would have the *number of elements* in the matrix rather than the *dimensions* of the matrix.

A matrix with only one row is called a **row matrix.** A matrix with only one column is called a **column matrix.** A matrix is **square** if it has as many rows as columns. For example,

$\begin{bmatrix} 3 & -2 & 0 \end{bmatrix}$ is a row matrix.

$\begin{bmatrix} 26 \\ 89 \end{bmatrix}$ is a column matrix.

$\begin{bmatrix} 0 & -1 \\ 5 & 3 \end{bmatrix}$ is a square matrix.

$\begin{bmatrix} 3 & -4 & 5 \\ 0 & 18 & 2 \end{bmatrix}$ is neither a row matrix, a column matrix, nor a square matrix.

A matrix is usually labeled with a capital letter, and the entries of that matrix are labeled with the same letter in lowercase with a double subscript. The first subscript refers to the entry's row; the second subscript refers to its column. For example, b_{23} refers to the element of matrix B that is in the second row and the third column. Similarly, if

$$A = \begin{bmatrix} 5 & 7 \\ 9 & 11 \end{bmatrix}$$

then $a_{11} = 5$, $a_{12} = 7$, $a_{21} = 9$, and $a_{22} = 11$.

Matrix Multiplication

In your intermediate algebra class, you might have learned to add, subtract, and multiply matrices and to take the determinant of a matrix. In this chapter, we will be studying applications that involve the multiplication of matrices but not their addition or subtraction or the taking of determinants. Thus, we limit our discussion to matrix multiplication.

Matrix multiplication is a process that at first seems rather strange and obscure. Its usefulness will become apparent when we apply it to probability trees in the method of Markov chains later in this chapter.

EXAMPLE 1 If $A = \begin{bmatrix} 3 & -2 \end{bmatrix}$ and $B = \begin{bmatrix} 1 \\ 0 \end{bmatrix}$, find the product AB.

Solution $AB = \begin{bmatrix} 3 & -2 \end{bmatrix} \begin{bmatrix} 1 \\ 0 \end{bmatrix} = ?$

To find AB,

- multiply the first element of A by the first element of B,
- then multiply the second element of A by the second element of B,
- and then add the results:

$$AB = \begin{bmatrix} 3 & -2 \end{bmatrix} \begin{bmatrix} 1 \\ 0 \end{bmatrix} = [3 \cdot 1 + -2 \cdot 0] = [3]$$ ∎

EXAMPLE 2 If $A = \begin{bmatrix} 3 & -2 \\ 4 & 1 \end{bmatrix}$ and $B = \begin{bmatrix} 1 & 5 \\ 0 & -7 \end{bmatrix}$, find the product AB.

Solution $AB = \begin{bmatrix} 3 & -2 \\ 4 & 1 \end{bmatrix} \begin{bmatrix} 1 & 5 \\ 0 & -7 \end{bmatrix} = ?$

We start with row 1 of A and column 1 of B. We multiply corresponding elements together, add the products, and put the sum in row 1, column 1 of the product matrix, as we did in Example 1.

$$AB = \begin{bmatrix} 3 & -2 \\ 4 & 1 \end{bmatrix} \begin{bmatrix} 1 & 5 \\ 0 & -7 \end{bmatrix} = \begin{bmatrix} 3 \cdot 1 + -2 \cdot 0 & ? \\ ? & ? \end{bmatrix} = \begin{bmatrix} 3 & ? \\ ? & ? \end{bmatrix}$$

We continue in this manner until each row of A has been multiplied by each column of B. Thus, we now multiply row 1 of A and column 2 of B and put the result in row 1, column 2 of the product matrix:

$$AB = \begin{bmatrix} 3 & -2 \\ 4 & 1 \end{bmatrix} \begin{bmatrix} 1 & 5 \\ 0 & -7 \end{bmatrix} = \begin{bmatrix} 3 & 3 \cdot 5 + -2 \cdot -7 \\ ? & ? \end{bmatrix} = \begin{bmatrix} 3 & 29 \\ ? & ? \end{bmatrix}$$

We're finished with row 1; what remains is to multiply row 2 of A first by column 1 of B and then by column 2. Multiplying row 2 of A by column 1 of B gives

$$AB = \begin{bmatrix} 3 & -2 \\ 4 & 1 \end{bmatrix} \begin{bmatrix} 1 & 5 \\ 0 & -7 \end{bmatrix} = \begin{bmatrix} 3 & 29 \\ 4 \cdot 1 + 1 \cdot 0 & ? \end{bmatrix} = \begin{bmatrix} 3 & 29 \\ 4 & ? \end{bmatrix}$$

Finally, we multiply row 2 of A by column 2 of B and put the result in row 2, column 2 of the product matrix:

$$AB = \begin{bmatrix} 3 & -2 \\ 4 & 1 \end{bmatrix} \begin{bmatrix} 1 & 5 \\ 0 & -7 \end{bmatrix} = \begin{bmatrix} 3 & 29 \\ 4 & 4 \cdot 5 + 1 \cdot -7 \end{bmatrix} = \begin{bmatrix} 3 & 29 \\ 4 & 13 \end{bmatrix}$$

The product of A and B is

$$AB = \begin{bmatrix} 3 & 29 \\ 4 & 13 \end{bmatrix} \qquad \blacksquare$$

A matrix is really just a table. Consider the table in Figure 11.1, which gives prices of CDs and DVDs.

The table could be rewritten as the matrix

	Sale Price	Regular Price
CDs	$14	$19
DVDs	$20	$25

Figure 11.1
Disc prices

$$\begin{bmatrix} 14 & 19 \\ 20 & 25 \end{bmatrix}$$

Matrix multiplication is a natural way to manipulate data in tables. If we were to purchase three CDs and two DVDs, we would calculate the sale price as

$$3 \cdot \$14 + 2 \cdot \$20 = \$82$$

and the regular price as

$$3 \cdot \$19 + 2 \cdot \$25 = \$107$$

This calculation is matrix multiplication; it is the same as

$$\begin{bmatrix} 3 & 2 \end{bmatrix} \begin{bmatrix} 14 & 19 \\ 20 & 25 \end{bmatrix} = \begin{bmatrix} 3 \cdot 14 + 2 \cdot 20 & 3 \cdot 19 + 2 \cdot 25 \end{bmatrix}$$

$$= \begin{bmatrix} 82 & 107 \end{bmatrix}$$

It's not always possible to multiply two matrices together. In the previous matrix multiplication problem, if the first matrix had one extra element, we wouldn't be able to multiply.

$$\begin{bmatrix} 3 & 2 & 77 \end{bmatrix} \begin{bmatrix} 14 & 19 \\ 20 & 25 \end{bmatrix} = \begin{bmatrix} 3 \cdot 14 + 2 \cdot 20 + 77 \cdot ? & 3 \cdot 19 + 2 \cdot 25 + 77 \cdot ? \end{bmatrix}$$

It's easy to tell when two matrices cannot be multiplied: If you run out of elements to pair together when you're multiplying the first row by the first column, then the two matrices cannot be multiplied. In the above problem, we paired 3 and 14, and 2 and 20, but we had nothing to pair with 77. This tells us that the two matrices cannot be multiplied together.

Some people prefer to determine whether it's possible to multiply two matrices before they actually start multiplying. In the above problem, we could not multiply a 1×3 matrix by a 2×2 matrix.

$$\begin{bmatrix} 3 & 2 & 77 \end{bmatrix} \begin{bmatrix} 14 & 19 \\ 20 & 25 \end{bmatrix}$$

$$\quad 1 \times 3 \qquad 2 \times 2$$

$$\qquad \uparrow \quad \uparrow$$

Different, so we can't multiply.

Previously, when 77 wasn't there, we *were* able to multiply.

$$\begin{bmatrix} 3 & 2 \end{bmatrix} \begin{bmatrix} 14 & 19 \\ 20 & 25 \end{bmatrix}$$

$$1 \times 2 \quad 2 \times 2$$

$$\uparrow \quad \uparrow$$

The same, so we can multiply.

EXAMPLE 3 $A = \begin{bmatrix} 3 & -2 \\ 4 & 1 \end{bmatrix}$, $B = \begin{bmatrix} 1 & 0 \end{bmatrix}$, and $C = \begin{bmatrix} 1 \\ 0 \end{bmatrix}$.

a. Find AB, if it exists.

b. Find AC, if it exists.

Solution   **a.** $AB = \begin{bmatrix} 3 & -2 \\ 4 & 1 \end{bmatrix} \begin{bmatrix} 1 & 0 \end{bmatrix} = \begin{bmatrix} 3 \cdot 1 + -2 \cdot ? & 3 \cdot 0 + -2 \cdot ? \\ 4 \cdot 1 + 1 \cdot ? & 4 \cdot 0 + 1 \cdot ? \end{bmatrix}$

The product AB does not exist, because we ran out of elements to pair together. Alternatively, A is a 2×2 matrix, and B is a 1×2 matrix.

$$\begin{bmatrix} 3 & -2 \\ 4 & 1 \end{bmatrix} \begin{bmatrix} 1 & 0 \end{bmatrix}$$

$$2 \times 2 \quad 1 \times 2$$

$$\uparrow \quad \uparrow$$

Different, so we can't multiply.

b. $AC = \begin{bmatrix} 3 & -2 \\ 4 & 1 \end{bmatrix} \begin{bmatrix} 1 \\ 0 \end{bmatrix} = \begin{bmatrix} 3 \cdot 1 + -2 \cdot 0 \\ 4 \cdot 1 + 1 \cdot 0 \end{bmatrix} = \begin{bmatrix} 3 \\ 4 \end{bmatrix}$

The product of A and C is

$$\begin{bmatrix} 3 \\ 4 \end{bmatrix}$$

If we had checked the dimensions in advance, we would have found that A and C can be multiplied:

$$\begin{bmatrix} 3 & -2 \\ 4 & 1 \end{bmatrix} \begin{bmatrix} 1 \\ 0 \end{bmatrix}$$

$$2 \times 2 \quad 2 \times 1$$

$$\uparrow \quad \uparrow$$

The same, so we can multiply.

This dimension check also gives the dimensions of the product matrix. The matching twos "cancel" and leave a 2×1 matrix.   ∎

Matrix Multiplication

If matrix A is an $m \times n$ matrix and matrix B is an $n \times p$ matrix, then the product AB exists and is an $m \times p$ matrix.

   (*Think*: $m \times n \cdot n \times p$ yields $m \times p$.)

To find the entry in the ith row and the jth column of the product matrix AB, multiply each element of A's ith row by the corresponding element of B's jth column and add the results.

Properties of Matrix Multiplication

When you multiply numbers together, you use certain properties so automatically that you don't even realize you are using them. In particular, multiplication of real numbers is *commutative;* that is, $ab = ba$ (the order in which you multiply doesn't matter). Multiplication of real numbers is also *associative;* that is, $a(bc) = (ab)c$ (if you multiply three numbers together, it doesn't matter which two you multiply first).

Matrix multiplication is certainly different from real number multiplication. It's important to know whether matrix multiplication is commutative and associative, because people tend to use those properties without being aware of doing so.

ThomsonNOW™
for Liberal Arts Mathematics
www.thomsonedu.com/1pass.

EXAMPLE 4   Determine whether matrix multiplication is commutative by finding the product BA for the matrices B and A given in Example 2 and comparing BA with AB.

Solution $BA = \begin{bmatrix} 1 & 5 \\ 0 & -7 \end{bmatrix} \begin{bmatrix} 3 & -2 \\ 4 & 1 \end{bmatrix} = \begin{bmatrix} 1\cdot3 + 5\cdot4 & 1\cdot-2 + 5\cdot1 \\ 0\cdot3 + -7\cdot4 & 0\cdot-2 + -7\cdot1 \end{bmatrix}$

$= \begin{bmatrix} 23 & 3 \\ -28 & -7 \end{bmatrix}$

The product of B and A is $BA = \begin{bmatrix} 23 & 3 \\ -28 & -7 \end{bmatrix}$.

In Example 2, we found that $AB = \begin{bmatrix} 3 & 29 \\ 4 & 13 \end{bmatrix}$.

Clearly, AB is not the same as BA; that is, $AB \neq BA$. ∎

Matrix multiplication is not commutative; in general, $AB \neq BA$. This means that you have to be careful to multiply in the right order—it's easy to be careless and find BA when you're asked to find AB.

ThomsonNOW™
for Liberal Arts Mathematics
www.thomsonedu.com/1pass.

EXAMPLE 5   Check the associative property for matrix multiplication by finding $A(BC)$ and $(AB)C$ for matrices A, B, and C given below.

$$A = \begin{bmatrix} 2 & -3 \end{bmatrix} \qquad B = \begin{bmatrix} 4 & 0 & -3 \\ 2 & -1 & 5 \end{bmatrix} \qquad C = \begin{bmatrix} -3 & 5 \\ 2 & 0 \\ 1 & -1 \end{bmatrix}$$

Solution   To find $A(BC)$, we first find BC and then multiply it by A. Notice that B's dimensions are 2×3 and C's dimensions are 3×2. This tells us that B and C can be multiplied and that the product will be a 2×2 matrix.

$$BC = \begin{bmatrix} 4 & 0 & -3 \\ 2 & -1 & 5 \end{bmatrix} \begin{bmatrix} -3 & 5 \\ 2 & 0 \\ 1 & -1 \end{bmatrix}$$

$$= \begin{bmatrix} 4\cdot-3 + 0\cdot2 + -3\cdot1 & 4\cdot5 + 0\cdot0 + -3\cdot-1 \\ 2\cdot-3 + -1\cdot2 + 5\cdot1 & 2\cdot5 + -1\cdot0 + 5\cdot-1 \end{bmatrix}$$

$$= \begin{bmatrix} -15 & 23 \\ -3 & 5 \end{bmatrix}$$

To find $A(BC)$, we place A on the *left* of BC and multiply. Notice that A's dimensions are 1×2 and BC's dimensions are 2×2, so A and BC can be multiplied, and the product will be a 1×2 matrix.

$$A(BC) = \begin{bmatrix} 2 & -3 \end{bmatrix} \begin{bmatrix} -15 & 23 \\ -3 & 5 \end{bmatrix}$$
$$= \begin{bmatrix} 2 \cdot -15 + -3 \cdot -3 & 2 \cdot 23 + -3 \cdot 5 \end{bmatrix}$$
$$= \begin{bmatrix} -21 & 31 \end{bmatrix}$$

Now that we've found $A(BC)$, we need to find $(AB)C$ and then see whether the products are the same. To find $(AB)C$, we first find AB and then multiply it by C. Notice that A's dimensions are 1×2 and B's dimensions are 2×3, so A and B can be multiplied.

$$AB = \begin{bmatrix} 2 & -3 \end{bmatrix} \begin{bmatrix} 4 & 0 & -3 \\ 2 & -1 & 5 \end{bmatrix}$$
$$= \begin{bmatrix} 2 \cdot 4 + -3 \cdot 2 & 2 \cdot 0 + -3 \cdot -1 & 2 \cdot -3 + -3 \cdot 5 \end{bmatrix}$$
$$= \begin{bmatrix} 2 & 3 & -21 \end{bmatrix}$$

To find $(AB)C$, we place AB on the *left* of C and multiply.

$$(AB)C = \begin{bmatrix} 2 & 3 & -21 \end{bmatrix} \begin{bmatrix} -3 & 5 \\ 2 & 0 \\ 1 & -1 \end{bmatrix}$$
$$= \begin{bmatrix} 2 \cdot -3 + 3 \cdot 2 + -21 \cdot 1 & 2 \cdot 5 + 3 \cdot 0 + -21 \cdot -1 \end{bmatrix}$$
$$= \begin{bmatrix} -21 & 31 \end{bmatrix}$$

Therefore, $A(BC) = (AB)C$. ∎

In Example 5, $A(BC) = (AB)C$, even though the work involved in finding $A(BC)$ was different from the work involved in finding $(AB)C$. This always happens; $A(BC) = (AB)C$, provided that the dimensions of A, B, and C are such that they can be multiplied together. *Matrix multiplication is associative.*

Identity Matrices

EXAMPLE 6 If $I = \begin{bmatrix} 1 & 0 \\ 0 & 1 \end{bmatrix}$ and $A = \begin{bmatrix} 2 & 3 \\ 4 & 5 \end{bmatrix}$, find IA.

Solution
$$IA = \begin{bmatrix} 1 & 0 \\ 0 & 1 \end{bmatrix} \begin{bmatrix} 2 & 3 \\ 4 & 5 \end{bmatrix}$$
$$= \begin{bmatrix} 1 \cdot 2 + 0 \cdot 4 & 1 \cdot 3 + 0 \cdot 5 \\ 0 \cdot 2 + 1 \cdot 4 & 0 \cdot 3 + 1 \cdot 5 \end{bmatrix}$$
$$= \begin{bmatrix} 2 & 3 \\ 4 & 5 \end{bmatrix}$$
∎

In Example 6, notice that $IA = A$. If you were to multiply in the opposite order, you would find that $AI = A$. The matrix I is similar to the number 1, because $I \cdot A = A \cdot I = A$, just as $1 \cdot a = a \cdot 1 = a$.

The matrix $\begin{bmatrix} 1 & 0 \\ 0 & 1 \end{bmatrix}$ is called an identity matrix, because multiplying this matrix by any other 2×2 matrix A yields a product *identical* to A. The matrix $\begin{bmatrix} 1 & 0 & 0 \\ 0 & 1 & 0 \\ 0 & 0 & 1 \end{bmatrix}$ is also called an identity matrix, because multiplying this 3×3 matrix by any other 3×3 matrix A yields a product identical to A. An **identity matrix** is a square matrix I that has ones for each entry in the **diagonal** (the diagonal that starts in the upper left corner) and zeros for all other entries. The product of any matrix A and an identity matrix is always A, as long as the dimensions of the two matrices are such that they can be multiplied.

Properties of Matrix Multiplication

1. *There is no commutative property:* In general, $AB \neq BA$. You must be careful about the order in which you multiply.
2. **Associative property:** $A(BC) = (AB)C$, provided the dimensions of A, B, and C are such that they can be multiplied together.
3. **Identity property:** An **identity matrix** is a square matrix I that has ones for each entry in the diagonal (the diagonal that starts in the upper left corner) and zeros for all other entries. If I and A have the same dimensions, then $IA = AI = A$.

1814–1897

his interest in mathematics only through tutoring. Florence Nightingale was one of his private pupils. When Sylvester became thoroughly bored with insurance work, he studied for a legal career and met Cayley.

Cayley and Sylvester revived and intensified each other's interest in mathematics, and each started to write mathematics again. During his fourteen years spent practicing law, Cayley wrote almost 300 papers. Each frequently expressed gratitude to the other for assistance and inspiration. In one of his papers, Sylvester wrote that "the theorem above enunciated was in part suggested in the course of a conversation with Mr. Cayley (to whom I am indebted for my restoration to the enjoyment of mathematical life)." In another, he said, "Mr. Cayley habitually discourses pearls and rubies."

Cayley joyfully departed from the legal profession when Cambridge offered him a professorship in mathematics, even though his income suffered as a result. He was finally able to spend his life studying, teaching, and writing mathematics. He became quite famous as a mathematician, writing almost 1,000 papers in algebra and geometry, often in collaboration with Sylvester. Many of these papers are pioneering works of scholarship. Cayley also played an important role in changing Cambridge's policy that had prohibited the admission of women as students.

Sylvester was repeatedly honored for his pioneering work in algebra. He left the law but was unable to obtain a professorship in mathematics at a prominent institution until late in his life. At the age of sixty-two, he accepted a position at the newly founded Johns Hopkins University in Baltimore as its first professor of mathematics. While there, he founded the *American Journal of Mathematics,* introduced graduate work in mathematics into American universities, and generally stimulated the development of mathematics in America. He also arranged for Cayley to spend a semester at Johns Hopkins as guest lecturer. At the age of seventy, Sylvester returned to England to become Savilian Professor of Geometry at Oxford University.

Cayley and Sylvester were responsible for the theory of matrices, including the operation of matrix multiplication. Sixty-seven years after the invention of matrix theory, Heisenberg recognized it as the perfect tool for his revolutionary work in quantum mechanics. The work of Cayley and Sylvester in algebra became quite important for modern physics, particularly in the theory of relativity. Cayley also wrote on non-Euclidean geometry (see Chapter 8 for information on non-Euclidean geometry).

11.0 Exercises

In Exercises 1–10, (a) find the dimensions of the given matrix and (b) determine whether the matrix is a row matrix, a column matrix, a square matrix, or none of these.

1. $A = \begin{bmatrix} 5 & 0 \\ 22 & -3 \\ 18 & 9 \end{bmatrix}$

2. $B = \begin{bmatrix} 1 & 13 & 207 \\ -4 & 8 & 100 \\ 0 & 1 & 5 \end{bmatrix}$

3. $C = \begin{bmatrix} 23 \\ 41 \end{bmatrix}$

4. $D = \begin{bmatrix} 2 & 0 & 19 & -3 \\ 62 & 13 & 44 & 1 \\ 5 & 5 & 30 & 12 \\ 0 & 0 & 0 & 0 \end{bmatrix}$

5. $E = \begin{bmatrix} 3 & 0 \end{bmatrix}$

6. $F = \begin{bmatrix} -2 & 10 \\ 4 & -3 \end{bmatrix}$

7. $G = \begin{bmatrix} 12 & -11 & 5 \\ -9 & 4 & 0 \\ 1 & 9 & 5 \end{bmatrix}$

8. $H = \begin{bmatrix} 1 & 0 \\ 0 & -1 \end{bmatrix}$

9. $J = \begin{bmatrix} 5 \\ -3 \\ 11 \end{bmatrix}$

10. $K = \begin{bmatrix} 2 & -5 & 13 & 0 \\ -1 & 4 & 3 & 6 \\ 8 & -10 & 4 & 0 \end{bmatrix}$

In Exercises 11–20, find the indicated elements of the matrices given in Exercises 1–10.

11. a_{21}

(*HINT:* You're asked to find the entry in row 2, column 1 of the matrix A, which is given in Exercise 1.)

12. b_{23} **13.** c_{21} **14.** d_{34}
15. e_{11} **16.** f_{22} **17.** g_{12}
18. h_{21} **19.** j_{21} **20.** k_{23}

In Exercises 21–30, find the indicated products (if they exist) of the matrices given in Exercises 1–10.

21. **a.** *AC*   **b.** *CA*   **22.** **a.** *AE*   **b.** *EA*
23. **a.** *AD*   **b.** *DA*   **24.** **a.** *GH*   **b.** *HG*
25. **a.** *CG*   **b.** *GC*   **26.** **a.** *DK*   **b.** *KD*
27. **a.** *JB*   **b.** *BJ*   **28.** **a.** *HK*   **b.** *KH*
29. **a.** *AF*   **b.** *FA*   **30.** **a.** *AB*   **b.** *BA*

31. Use matrix multiplication and Figure 11.2 to find the sale price and the regular price of purchasing five CDs and three tapes.

	Sale Price	Regular Price
CDs	$14	$19
DVDs	$20	$25

Figure 11.2
Disc data for Exercises 31 and 32

32. Use matrix multiplication and Figure 11.2 to find the sale price and the regular price of purchasing four CDs and one tape.

33. Use matrix multiplication and Figure 11.3 to find the price of purchasing two slices of pizza and one cola at Blondie's and at SliceMan's.

	Blondie's Price	SliceMan's Price
Pizza	$1.25	$1.30
Cola	$ .95	$1.10

Figure 11.3
Pizza data for Exercises 33 and 34

34. Use matrix multiplication and Figure 11.3 to find the price of purchasing six slices of pizza and two colas at Blondie's and at SliceMan's.

35. Jim, Eloise, and Sylvie are enrolled in the same four courses: English, Math, History, and Business. English and Math are four-unit courses, while History and Business are three-unit courses. Their grades are given in Figure 11.4.

	English	Math	History	Business
Jim	A	B	B	C
Eloise	C	A	A	C
Sylvie	B	A	B	A

Figure 11.4
Grade data for Exercise 35

Counting A's as four grade points, B's as three grade points, and C's as two grade points, use matrix multiplication to compute Jim's, Eloise's, and Sylvie's grade point averages (GPAs).

HINT: Start by computing Jim's total grade points without using matrices. Then set up two matrices, one that is a matrix version of Figure 11.4 with grade points rather than letter grades and another that gives the four courses' units. Set up these two matrices so that their product mirrors your by-hand calculation of Jim's total grade points. Finally, multiply by 1 over the number of units taken so that the result is GPA, not total grade points.

36. Dave, Jay, Johnny, and Larry are each enrolled in the same four courses: Communications, Broadcasting, Marketing, and Contracts. Communications and Contracts are four-unit courses, while Broadcasting and Marketing are three-unit courses. Their grades are given in Figure 11.5.

	Communi-cations	Broad-casting	Marketing	Contracts
Dave	A	B	C	A
Jay	A	B	C	B
Johnny	A	A	A	A
Larry	B	A	B	C

Figure 11.5
Grade data for Exercise 36

Counting A's as 4 grade points, B's as 3 grade points, and C's as 2 grade points, use matrix multiplication to compute Dave's, Jay's, Johnny's, and Larry's grade point averages. (See *HINT,* Exercise 35.)

37. **Tourism Sales.** Figure 11.6 gives the total sales in the United States for two tourism-related industries: hotels and restaurants.

	2002	2003
Hotels	$105.8 billion	$105.5 billion
Restaurants	$ 61.9 billion	$ 66.3 billion

Figure 11.6

Tourism data for Exercise 37

Source: Bureau of Economic Analysis, News Release: Sales of Tourism Industries, www.bea.gov.

  a. Convert this chart into a 2×2 matrix S, and find the product $S\begin{bmatrix} -1 \\ 1 \end{bmatrix}$.

  b. Interpret the solution to part (a) in the context of the data.

38. **Tourism Sales.** Figure 11.7 gives the total sales in the United States for two tourism-related industries: air transportation and auto rental.

	2001	2002
Air Transportation	$103.1 billion	$93.3 billion
Auto Rental	$ 21.6 billion	$23.5 billion

Figure 11.7

Tourism data for Exercise 38

Source: Bureau of Economic Analysis, News Release: Sales of Tourism Industries, www.bea.gov.

  a. Convert this chart into a 2×2 matrix S, and find the product $S\begin{bmatrix} -1 \\ 1 \end{bmatrix}$.

  b. Interpret the solution to part (a) in the context of the data.

39. **Fence Construction.** Borg Fences builds wood fences. They purchase their materials from two suppliers: Piedmont Lumber and Truitt and White. Figure 11.8 shows current prices for their materials and the materials needed to build one 6-foot section of fence.

  a. Make two matrices out of the given data, one showing prices and the other showing materials needed.

  b. Use matrix multiplication and the two matrices from part (a) to determine the materials cost of one 6-foot section of fence at each of the two suppliers.

40. **Fence Construction.** Piedmont Lumber increased all their prices by 10%, and Truitt and White increased theirs by 9%. Use matrix multiplication and the solutions to Exercise 39 to find the materials cost of one 6-foot section of fence at each of the two suppliers after the price increase.

41. **U.S. Population Movement.**

  a. In 2000, the regional population of the United States was 53,594 in the Northeast, 64,393 in the Midwest, 100,237 in the South, and 63,198 in the West. Make a 1×4 matrix A out of these data.

  b. Figure 11.9 on page 812 shows the population movement in the United States during the period from 2000 to 2001. For example, 98.85% = 0.9885 of the residents of the Northeast stayed in the Northeast, and 0.15% = 0.0015 of the residents of the Northeast moved to the Midwest. Make a 4×4 matrix B out of these data.

  c. Compute AB. What do its entries measure?

42. **U.S. Population Movement.** Assume that the data in Figure 11.9 also describe the population movement from 2001 to 2002. Compute $(AB)B$, using the matrices A and B from Exercise 41. What do its entries measure?

43. Verify the associative property by finding $A(BC)$ and $(AB)C$.

$$A = \begin{bmatrix} -4 & 5 \\ 2 & 3 \end{bmatrix} \qquad B = \begin{bmatrix} -3 & 0 & -1 \\ 1 & 4 & -2 \end{bmatrix}$$

$$C = \begin{bmatrix} 0 \\ 5 \\ -1 \end{bmatrix}$$

	Piedmont Lumber	Truitt and White	Materials Needed for a 6′ Section
50 lb. bag of concrete	$6.25	$6.10	2
pressure treated 4″ × 4″ × 8′	$8.97	$8.75	1
redwood 2″ × 4″ × 6′	$4.97	$5.25	2
framed redwood lattice 2′ × 6′	$24.85	$22.12	1
redwood 5/8″ × 5 1/2″ × 8′	$6.98	$6.98	9
redwood 1″ × 1″ × 6′	$3.88	$3.75	4

Figure 11.8

Fence data for Exercise 39

Region Moved From	Region Moved To			
	Northeast	Midwest	South	West
Northeast	0.9885	0.0015	0.0076	0.0024
Midwest	0.0011	0.9901	0.0057	0.0032
South	0.0018	0.0042	0.9897	0.0043
West	0.0017	0.0035	0.0077	0.9870

Figure 11.9
Population data for Exercise 41
Source: U.S. Census Bureau, Statistical Abstract of the United States, 2000 and Current Population Reports, P20–538: Geographical Mobility.

44. Verify the associative property by finding $A(BC)$ and $(AB)C$.

$$A = \begin{bmatrix} 2 & 0 \\ -1 & 1 \\ 3 & 2 \end{bmatrix} \quad B = \begin{bmatrix} 5 & 3 & -2 \\ 2 & 0 & 1 \end{bmatrix}$$

$$C = \begin{bmatrix} 1 \\ 2 \\ 0 \end{bmatrix}$$

In Exercises 45–50, find the product (if it exists).

45. $\begin{bmatrix} 1 & 0 \\ 0 & 1 \end{bmatrix} \cdot \begin{bmatrix} 3 & -2 \\ 4 & 0 \end{bmatrix}$

46. $\begin{bmatrix} 1 & 0 & 0 \\ 0 & 1 & 0 \\ 0 & 0 & 1 \end{bmatrix} \cdot \begin{bmatrix} -4 & 5 & 2 \\ 8 & -1 & 9 \\ -2 & 27 & 4 \end{bmatrix}$

47. $\begin{bmatrix} 1 & 0 & 0 \\ 0 & 1 & 0 \\ 0 & 0 & 1 \end{bmatrix} \cdot \begin{bmatrix} 4 & 1 & -1 \\ 5 & 12 & 3 \end{bmatrix}$

48. $\begin{bmatrix} 27 & 19 \\ 42 & 25 \end{bmatrix} \cdot \begin{bmatrix} 1 & 0 \\ 0 & 1 \end{bmatrix}$

49. $\begin{bmatrix} 19 & 7 & 34 \\ 74 & 0 & -11 \\ 13 & -2 & 44 \end{bmatrix} \cdot \begin{bmatrix} 1 & 0 & 0 \\ 0 & 1 & 0 \\ 0 & 0 & 1 \end{bmatrix}$

50. $\begin{bmatrix} 1 & 0 \\ 0 & 1 \end{bmatrix} \cdot \begin{bmatrix} 6 & 18 & -3 \\ 0 & 1 & 5 \\ -14 & 5 & -2 \end{bmatrix}$

▬▬▬▬ *Answer the following questions using complete sentences and your own words.*

CONCEPT QUESTIONS

51. Describe the two ways of determining whether two matrices can be multiplied.
52. Why is an identity matrix so named?
53. Give an example of two matrices A and B such that $AB = BA$.
54. Give an example of two matrices A and B such that the product AB exists but the product BA does not exist.
55. What must be true about matrices A and B if the products AB and BA both exist?
56. Make up a realistic word problem that involves matrix multiplication, as Exercises 31–42 do.

HISTORY QUESTIONS

57. Compare and contrast Arthur Cayley and James Joseph Sylvester. Be sure to discuss both the similarities and the differences between these two people. In particular, compare and contrast why each turned his back on mathematics, why each returned to mathematics, their relative success in schooling, and their relative success in employment.
58. Why did Cambridge University not award Sylvester his degrees?
59. Why was no position as a mathematician open to Cayley early in his career?

Matrix Multiplication on a Graphing Calculator

The TI-83, TI-84, TI-86, and Casio graphing calculators can multiply matrices. Consider the matrices

$$A = \begin{bmatrix} 1 & 2 \\ 3 & 4 \end{bmatrix} \quad \text{and} \quad B = \begin{bmatrix} 1 & 5 \\ 0 & -7 \end{bmatrix}$$

The TI-83 and TI-84 require that you name these matrices "[A]" and "[B]" with brackets; the TI-86 requires that you name them "A" and "B" without brackets. The Casio requires that you name them "Mat A" and "Mat B."

TI-86 Users
When your calculator is in "ALPHA" mode, some buttons will generate a letter of the alphabet. If your calculator has a flashing "A" on the screen, then it is in "ALPHA" mode. One way to put your calculator in "ALPHA" mode is to press the ALPHA button. However, some other commands will automatically put your calculator in "ALPHA" mode. Don't confuse the 2nd and ALPHA buttons. The 2nd button refers to the mathematical labels above and to the *left* of the buttons.

Entering a Matrix
To enter matrix A on a TI, select "EDIT" from the "MATRIX" menu and then matrix A from the "MATRX EDIT" menu:

TI-83/84	• Press MATRX • Use the ▶ button to highlight "EDIT" • Highlight option 1 "[A]"* • Press ENTER
TI-86	• Press 2nd MATRX • Press EDIT , (i.e. F2) • Press A MATRX†
Casio	• Press MENU • Use the arrow buttons to highlight "MAT" • Press EXE • Use the arrow buttons to highlight "Mat A"

Then enter the dimensions of our 2×2 matrix A by typing

2   ENTER   2   ENTER

---

*Option 1 is automatically highlighted. If we were selecting some other option, we would use the ▲ and ▼ buttons to highlight it.

†The LOG button becomes the A button when the calculator is in "ALPHA" mode. Usually this requires preceding the A button with the ALPHA button. However, the TI-85/86 was automatically placed in "ALPHA" mode when you pressed EDIT, as shown by the flashing "A" on the screen.

To enter the elements, type

1 ENTER 2 ENTER 3 ENTER 4 ENTER

In a similar manner, enter matrix B. When you're done entering, type 2nd QUIT.

Note that the calculator uses double subscript notation. When we entered the "4", the calculator showed that this was entered as a_{22}.

Viewing a Matrix

To view matrix A, type the following:

Figure 11.10
Matrix A.

📟 TI-83/84	• Type MATRX • Highlight option 1: "[A]" • Press ENTER, and "[A]" will appear on the screen • Press ENTER, and the matrix itself will appear on the screen, as shown in Figure 11.10.	
TI-86	Type ALPHA A ENTER and matrix A will appear on the screen, as shown in Figure 11.10	
Casio	• Press MENU • Use the arrow buttons to highlight "MAT" and press EXE • Use the arrow buttons to highlight "Mat A" • Press EXE, and the matrix will appear on the screen, similar to Figure 11.10	

Multiplying Two Matrices

To calculate AB, make the screen read "A * B" or "[A] * [B]" or "MatA × MatB" by typing the following:

📟 TI-83/84	MATRX 1 × MATRX 2 ENTER
TI-86	ALPHA A × ALPHA B ENTER
Casio	• Press MENU • Use the arrow buttons to highlight "RUN" and press EXE • Press OPTN • Press MAT (i.e., F2) • Make the screen read "Mat A × Mat B" by typing MAT ALPHA A × MAT ALPHA B • Press EXE

Figure 11.11
Multiplying two matrices

The result on a TI-83/84 is shown in Figure 11.11; other calculators' results are similar.

Exercises

In Exercises 60–68, use the following matrices.

$$A = \begin{bmatrix} 5 & -7 \\ 3 & 9 \end{bmatrix} \qquad B = \begin{bmatrix} 7 & 9 \\ -8 & 0 \end{bmatrix}$$

$$C = \begin{bmatrix} 8 & -12 & 13 \\ 52 & 17 & -31 \\ 72 & 28 & -15 \end{bmatrix}$$

$$D = \begin{bmatrix} 14 & -22 & 53 \\ 94 & -15 & -35 \\ 83 & 0 & 7 \end{bmatrix} \qquad E = \begin{bmatrix} 5 & -31 \\ 83 & -33 \\ 60 & 0 \end{bmatrix}$$

$$F = \begin{bmatrix} 93 & -11 & 39 \\ 53 & 66 & 83 \end{bmatrix}$$

In Exercises 60–61, find the indicated product (a) by hand and (b) with a graphing calculator. Check your work by comparing the solutions to parts (a) and (b); answers are not given in the back of the book.

60. *AB*   **61.** *BA*

In Exercises 62–68, find the indicated product with a graphing calculator. (When a matrix is too big for the screen, use the arrow keys to view it.)

62. a. *CD*   **b.** *DC*   **63. a.** *EF*   **b.** *FE*
64. a. *AB*   **b.** *BA*   **65. a.** *(BF)C*   **b.** *B(FC)*
66. a. *(DE)A*   **b.** *D(EA)*   **67. a.** C^2 **b.** C^5
68. a. D^2 **b.** D^6

11.1   Introduction to Markov Chains

Suppose a college student observes her own class standing at the beginning of each school year. These observations vary among elements of the finite set {freshman, sophomore, junior, senior, quit, graduated}. If the student is a freshman at the beginning of the 2006–2007 school year, then she could be a freshman or a sophomore or quit at the beginning of the next school year, each with a certain probability. Her class standing at the beginning of any one year is dependent on her class standing in the preceding school year and not on her class standing in any earlier year.

We're talking about repeatedly observing a certain changeable quality at successive points in time (such as class standing, observed at the beginning of the school year). If the probability of making a certain observation at a certain time depends only on the immediately preceding observation and if the observations' outcomes are elements of a finite set, then these observations and their probabilities form a **Markov chain.** Our college student's observations form a Markov chain. The probability of having a certain class standing at any one year depends only on her class standing in the preceding year. And her possible class standings are elements of the finite set {freshman, sophomore, junior, senior, quit, graduated}.

Markov chains were developed in the early 1900s by Andrei Markov, a Russian mathematician. They have many applications in the physical sciences, business, sociology, and biology. In business, Markov chains are used to analyze data on customer satisfaction with a product and the effect of the product's advertising and to predict what portion of the market the product will eventually command. In sociology, Markov chains are used to analyze sociological trends such as the shrinking of the U.S. middle class and the growth of the suburbs at the expense of the central cities and the rural areas and to predict the eventual outcome of such trends. Markov chains are also used to predict the weather, to analyze genetic inheritance, and to predict the daily fluctuation in a stock's price.

EXAMPLE 1   Brute is a nutrition bar for men. Manfood, the company that makes Brute, has just launched a new advertising campaign. A market analysis indicates that as a result of this campaign, 40% of the men who currently *do not* use Brute will buy it the next time they buy a nutrition bar. Another market analysis has studied customer loyalty toward Brute. It indicates that 80% of the consumers who currently *do* use Brute will buy it again the next time they buy a nutrition bar. Rewrite these data in probability form and find the complements of these events.

Solution   This is probabilistic data. Saying that "40% of the men who currently do not use Brute will buy it the next time they buy a nutrition bar" is the same as saying

$$p(\text{next purchase is Brute} \mid \text{current purchase is not Brute}) = 40\% = 0.4$$

And saying that "80% of the consumers who currently do use Brute will buy it again the next time they buy a nutrition bar" is the same as saying

$$p(\text{next purchase is Brute} \mid \text{current purchase is Brute}) = 80\% = 0.8$$

Each of these two events has a complement. If 40% of the men who currently do not use Brute *will* buy it the next time they buy a nutrition bar, then the other 60% of the men who currently do not use Brute *will not* buy it the next time they buy a nutrition bar. That is,

$$p(\text{next purchase is not Brute} \mid \text{current purchase is not Brute})$$
$$= 100\% - 40\% = 60\% = 0.6$$

Similar reasoning gives us

$$p(\text{next purchase is not Brute} \mid \text{current purchase is Brute}) = 1 - 0.8 = 0.2$$

These relationships are summarized in Figure 11.12.

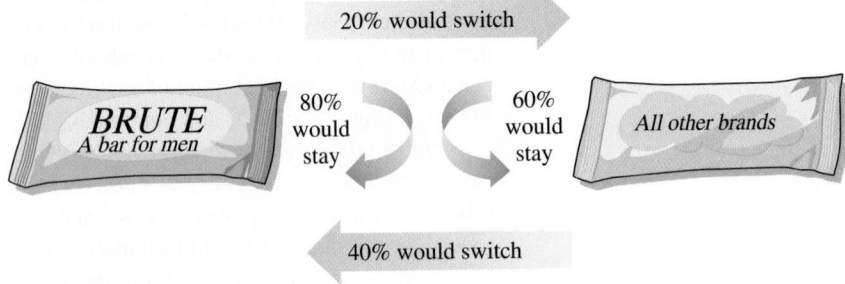

Figure 11.12
Comparing Brute with the competition

Notice that Example 1 describes a Markov chain because:

- the observations are made at successive points in time (shopping trips during which a nutrition bar is purchased);
- the observations' outcomes are elements of the set {Brute is purchased, something other than Brute is purchased}; and
- the probability that the next purchase is Brute depends only on the current purchase and not on any previous purchases—or, more generally, the probability of making a certain observation at a certain time depends only on the outcome of the immediately preceding observation.

The observations' outcomes are called *states*. A **state** is a condition, form, or stage. The states in Example 1 are "a consumer purchases Brute ("Brute," or B for short) and "a consumer purchases something other than Brute ("not Brute," or B′ for short). All states manifest themselves as both **current states** and **following states.**

Our current states are "a consumer currently purchases Brute" and "a consumer currently purchases something other than Brute," and the next following states are "a consumer will purchase Brute the next time he purchases a nutrition bar" and "a consumer will purchase something other than Brute the next time he purchases a nutrition bar."

Transition Matrices

A **transition matrix** is a matrix whose entries are the probabilities of passing from current states to following states. A transition matrix has a row and a column for each state. The rows refer to current manifestations of those states, and the columns refer to later (or following) manifestations of those states. A transition matrix is read like a chart. An entry in a certain row and column represents the probability of making a transition from the current state represented by that row to the following state represented by that column.

EXAMPLE 2   Write the transition matrix T for the data in Example 1.

Solution   The states are "Brute" and "not Brute" (or just B and B′). Therefore our transition matrix T will be the following 2×2 matrix:

a column for each following state

$$
\begin{array}{cc}
& \downarrow \quad\ \downarrow \\
& \text{B} \quad\ \text{B}' \\
T = & \begin{bmatrix} 0.8 & 0.2 \\ 0.4 & 0.6 \end{bmatrix} \begin{array}{l} \text{B} \\ \text{B}' \end{array} \quad \begin{array}{l} \longleftarrow \\ \longleftarrow \end{array} \ \text{a row for each current state}
\end{array}
$$

The entry 0.4 is in the B′ row and the B column, so it is the probability that a consumer makes a transition from buying something other than Brute (current state is B′) to buying Brute (first following state is B). ∎

We can make several observations about transition matrices.

1. *A transition matrix must be square,* because there is one row and one column for each state. If you create a transition matrix that is not square, go back and see which state you did not list as both a current state and a following state.

2. *Each entry in a transition matrix must be between 0 and 1 (inclusive),* because the entries are probabilities. If you create a transition matrix that has an entry that is less than 0 or greater than 1, go back and find your error.

3. *The sum of the entries of any row must be 1,* because the entries of a row are the probabilities of changing from the state represented by that row to *all* of the possible following states. If you create a transition matrix that has a row that does not add to 1, go back and find your error.

Transition Matrix Observations

1. A transition matrix must be square.
2. Each entry in a transition matrix must be between 0 and 1 (inclusive).
3. The sum of the entries of any row must be 1.

The transition matrix in Example 2 shows the transitional trends in the consumers' selection of their next nutrition bar purchases. What effect will these trends have on Brute's success as a product? Will Brute's market share increase or decrease?

Probability Matrices

EXAMPLE 3 A marketing analysis for Brute shows that Brute currently commands 25% of the market. Write this information and its complement in probability form.

Solution We are given

$$p(\text{current purchase is Brute}) = 0.25$$

Its complement is

$$p(\text{current purchase is not Brute}) = 1 - 0.25 = 0.75 \qquad ■$$

A **probability matrix** is a row matrix, with a column for each state. The entries are the probabilities of the different states. The columns of a probability matrix must be labeled in the same way, as are the rows and columns of the transition matrices.

EXAMPLE 4 Write the probability matrix P for the data in Example 3.

Solution The probability matrix is a row matrix, with a column for B and a column for B'.

$$\begin{array}{cc} \text{B} & \text{B}' \end{array}$$
$$P = \begin{bmatrix} 0.25 & 0.75 \end{bmatrix} \qquad ■$$

We can make several observations about probability matrices that parallel the observations made earlier about transition matrices.

> **Probability Matrix Observations**
>
> 1. A probability matrix is a row matrix.
> 2. Each entry in a probability matrix must be between 0 and 1 (inclusive).
> 3. The sum of the entries of the row must be 1.

Using Markov Chains to Predict the Future

ThomsonNOW™
for Liberal Arts Mathematics
www.thomsonedu.com/1pass.

EXAMPLE 5 Use a probability tree to predict Brute's market share after the first following purchase. In other words, find $p(\text{1st following purchase is Brute})$ and $p(\text{1st following purchase is not Brute})$.

Solution We will use the tree in Figure 11.13.

We want $p(\text{1st following purchase is Brute})$, which the sum of the probabilities of the limbs that stop at Brute. Similarly, $p(\text{1st following purchase is not Brute})$ is the sum of the probabilities of the limbs that stop at not Brute.

$$p(\text{1st following purchase is Brute}) = (0.25 \cdot 0.8) + (0.75 \cdot 0.4)$$
$$= 0.2 + 0.3 = 0.5$$

$$p(\text{1st following purchase is not Brute}) = (0.25 \cdot 0.2) + (0.75 \cdot 0.6)$$
$$= 0.05 + 0.45 = 0.5$$

Figure 11.13
Predicting Brute's future success

In other words, after the first following purchase, Brute will command 50% of the market (up from a previous 25%). Remember, though, that this is only a prediction, not a guarantee. And it's based on the assumption that current trends continue unchanged. ∎

Notice that the calculations done in Example 5 are the same ones that are done in computing the product of the probability matrix P and the transition matrix T:

$$PT = \begin{bmatrix} 0.25 & 0.75 \end{bmatrix} \begin{bmatrix} 0.8 & 0.2 \\ 0.4 & 0.6 \end{bmatrix}$$
$$= \begin{bmatrix} 0.25 \cdot 0.8 + 0.75 \cdot 0.4 & 0.25 \cdot 0.2 + 0.75 \cdot 0.6 \end{bmatrix}$$
$$= \begin{bmatrix} 0.2 + 0.3 & 0.05 + 0.45 \end{bmatrix}$$
$$= \begin{bmatrix} 0.5 & 0.5 \end{bmatrix}$$

This parallelism between trees and matrices is illustrated in Figure 11.14.

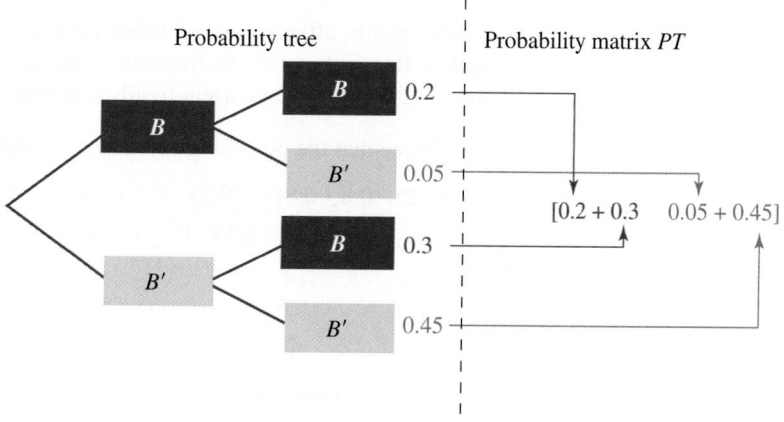

Figure 11.14
Comparing tree results and matrix results

EXAMPLE 6   Use a probability tree to predict Brute's market share after the second following purchase.

Solution   We'll use the tree in Figure 11.15. We want p(2nd following purchase is Brute), which is the sum of the probabilities of the limbs that stop at Brute under "2nd following purchase."

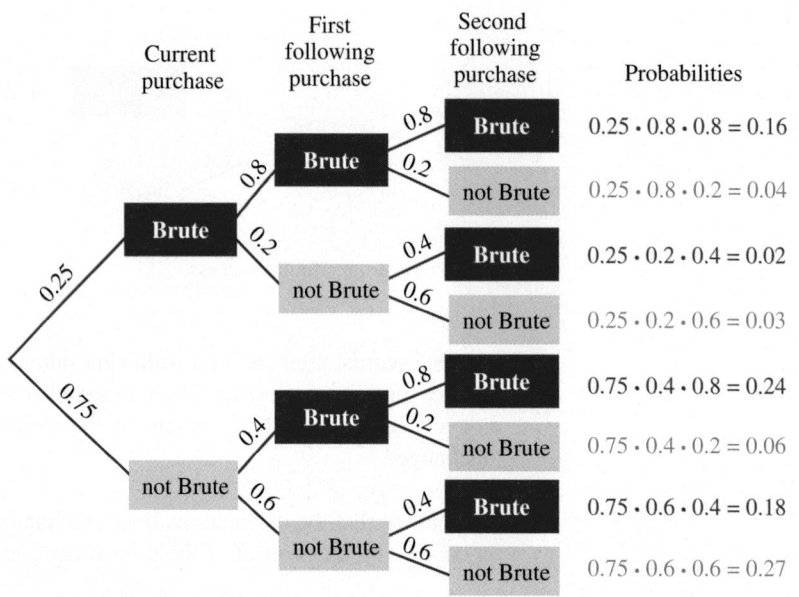

Figure 11.15
Predicting Brute's future success after the second following purchase

This tree gives us

$$p(\text{2nd following purchase is B}) = (0.25 \cdot 0.8 \cdot 0.8) + (0.25 \cdot 0.2 \cdot 0.4)$$
$$+ (0.75 \cdot 0.4 \cdot 0.8) + (0.75 + 0.6 + 0.4)$$
$$= 0.6$$

In other words, after the second following purchase, Brute will command 60% of the market. Remember, though, that this is only a prediction, not a guarantee. It is based on the assumption that current trends continue unchanged. ∎

Notice that the preceding calculation could be rewritten in a factored form.

$$(0.25 \cdot 0.8 \cdot 0.8) + (0.25 \cdot 0.2 \cdot 0.4) + (0.75 \cdot 0.4 \cdot 0.8) + (0.75 \cdot 0.6 \cdot 0.4)$$
$$= (0.25 \cdot 0.8 + 0.75 \cdot 0.4) \cdot 0.08 \quad + (0.75 \cdot 0.4 + 0.75 \cdot 0.6) \cdot 0.4$$

factoring

$$= (\qquad 0.5 \qquad) \cdot 0.08 \quad + (\qquad 0.5 \qquad) \quad \cdot 0.4$$
$$\uparrow \qquad\qquad\qquad\qquad \uparrow$$
$$\text{from } PT \qquad\qquad\qquad \text{from } PT$$

Part of this $(0.25 \cdot 0.8 + 0.75 \cdot 0.4 = 0.5$ and $0.75 \cdot 0.4 + 0.75 \cdot 0.6 = 0.5)$ is the same as the calculation that we did in finding PT. And the whole thing is the same calculation that is done in computing row one, column one of the product of the probability matrix PT and the transition matrix T.

$$PT^2 = PT \cdot T$$
$$= [0.5 \quad 0.5] \cdot \begin{bmatrix} 0.8 & 0.2 \\ 0.4 & 0.6 \end{bmatrix}$$
$$= [0.5 \cdot 0.8 + 0.5 \cdot 0.4 \quad 0.5 \cdot 0.2 + 0.5 \cdot 0.6]$$
$$= [0.6 \quad 0.4]$$

If this parallelism between probability trees and matrices continues, then P(2nd following purchase is not Brute), when computed from the above tree, should match the second entry in PT^2. From the tree,

P(2nd following purchase is B$'$)

$$(0.25 \cdot 0.8 \cdot 0.2) + (0.25 \cdot 0.2 \cdot 0.6) + (0.75 \cdot 0.4 \cdot 0.2) + (0.75 \cdot 0.6 \cdot 0.6)$$
$$= (0.25 \cdot 0.8 + 0.75 \cdot 0.4) \cdot 0.02 \qquad + (0.75 \cdot 0.4 + 0.75 \cdot 0.6) \cdot 0.6 \qquad \text{factoring}$$
$$= (\qquad 0.5 \qquad) \cdot 0.02 \quad + (\qquad 0.5 \qquad) \cdot 0.6$$

This does indeed match the second entry of PT^2. Predictions of Brute's future market shares can be made either with trees or with matrices. The matrix work is certainly simpler, both to set up and to calculate.

EXAMPLE 7   Use matrices to predict Brute's market share after the third following purchase.

Solution   Brute's market share will be an entry in PT^3. One way to do this calculation is to observe that

$$PT^3 = P(T^2T) = (PT^2)T \quad \text{Associative property}$$

and to use the previously calculated PT^2.

$$PT^3 = (PT^2)T = [0.6 \quad 0.4] \begin{bmatrix} 0.8 & 0.2 \\ 0.4 & 0.6 \end{bmatrix} = [0.64 \quad 0.36]$$

Thus, p(third following purchase is Brute) $= .64$. [Also, p(third following purchase is not Brute) $= .36$.] This means that three purchases after the time when the market analysis was done, 64% of the men's nutrition bars purchased will be Brute (*if current trends continue*). In other words, Brute's market share will be 64%. Brute will be doing quite well. Recall that it had a market share of only 25% at the beginning of the new advertising campaign.   ∎

Predictions with Markov Chains

1. *Create the probability matrix P. P* is a row matrix whose entries are the initial probabilities of the states.
2. *Create the transition matrix T. T* is the square matrix whose entries are the probabilities of passing from current states to first following states. The rows refer to the current states, and the columns refer to the next following states.
3. *Calculate PT^n. PT^n* is a row matrix whose entries are the probabilities of the nth following states. Be careful that you multiply in the correct order: $PT^n \neq T^nP$.

EXAMPLE 8 Sociologists have found that a strong determinant of an individual's income is the income of his or her parents.

a. Convert the data given in Figure 11.16 on family incomes in 1999 into a probability matrix.

b. Census data suggest the following (illustrated in Figure 11.17):
   - Of those individuals whose parents belong to the lower income group, 21% will become members of the middle income group, and 1% will become members of the upper income group.
   - Of those individuals whose parents belong to the middle income group, 6% will become members of the lower income group, and 4% will become members of the upper income group.
   - Of those individuals whose parents belong to the upper income group, 1% will become members of the lower income group, and 10% will become members of the middle income group.

   Convert this information into a transition matrix.

Family Income	Percent of Population
Under $15,000	12%
$15,000–$74,999	62%
$75,000 or more	26%

Figure 11.16
Income levels in the United States
Source: U.S. Bureau of the Census, Statistical Abstract of the United States, 2001.

Figure 11.17
Transitions between income groups

c. Predict the percent of U.S. families in the lower, middle, and upper income groups after one generation.

d. Predict the percent of U.S. families in the lower, middle, and upper income groups after two generations.

Solution **a.** *Creating the probability matrix P.* Recall that a probability matrix has one row, with a column for each state. Our states are "lower," "middle," and "upper." Our probability matrix is

$$P = \begin{matrix} \text{low} & \text{mid} & \text{up} \\ [0.12 & 0.62 & 0.26] \end{matrix}$$

b. *Creating the transition matrix T.* Because we have three income groups, the transition matrix T is a 3×3 matrix, with rows referring to the current state, or income group of the parents, and columns referring to the first following state, or income group of the child. We are given the following portion of T:

income group of child

$$T = \begin{bmatrix} \text{low} & \text{mid} & \text{up} \\ ? & 0.21 & 0.01 \\ 0.06 & ? & 0.04 \\ 0.01 & 0.10 & ? \end{bmatrix} \begin{matrix} \text{low} \\ \text{mid} \\ \text{up} \end{matrix} \quad \text{income group of parents}$$

Filling in the blanks is easy. If 21% of the children of lower income parents become members of the middle income group and 1% become members of the upper income group, that leaves 78% to remain members of the lower income group. Alternatively, the entries in any row must add to 1, so the missing entry is $1 - (0.21 + 0.01) = 0.78$. Similar calculations fill in the other blanks. Thus, we have

$$T = \begin{bmatrix} \text{low} & \text{mid} & \text{up} \\ 0.78 & 0.21 & 0.01 \\ 0.06 & 0.90 & 0.04 \\ 0.01 & 0.10 & 0.89 \end{bmatrix} \begin{matrix} \text{low} \\ \text{mid} \\ \text{up} \end{matrix}$$

c. *Predicting the distribution after one generation.* We are asked to calculate PT^1, the probabilities of the first following generation.

$$PT^1 = PT = \begin{bmatrix} 0.12 & 0.62 & 0.26 \end{bmatrix} \begin{bmatrix} 0.78 & 0.21 & 0.01 \\ 0.06 & 0.90 & 0.04 \\ 0.01 & 0.10 & 0.89 \end{bmatrix}$$

$$= \begin{bmatrix} 0.1334 & 0.6092 & 0.2574 \end{bmatrix}$$

$$\approx \begin{bmatrix} \underset{\text{low}}{0.13} & \underset{\text{mid}}{0.61} & \underset{\text{up}}{0.26} \end{bmatrix}$$

This means that if current trends continue, in one generation's time, the lower income group will grow from 12% of the population to 13%, the middle income group will shrink from 62% to 61%, and the upper income group will stay at 26%.

 Notice that the row adds to $0.13 + 0.61 + 0.26 = 1$, as it should.

d. *Predicting the distribution after two generations.* We are asked to calculate PT^2, the probabilities of the second following generation. The easiest way to do this calculation is to observe that $PT^2 = (PT)T$ and to use the previously calculated PT. Our answers will be more accurate if we do not round off PT before multiplying it by T.

$$PT^2 = (PT)T = \begin{bmatrix} 0.1334 & 0.6092 & 0.2574 \end{bmatrix} \cdot \begin{bmatrix} 0.78 & 0.21 & 0.01 \\ 0.06 & 0.90 & 0.04 \\ 0.01 & 0.10 & 0.89 \end{bmatrix}$$

$$= \begin{bmatrix} 0.143178 & 0.602034 & 0.254788 \end{bmatrix}$$

$$\approx \begin{bmatrix} \underset{\text{low}}{0.14} & \underset{\text{mid}}{0.60} & \underset{\text{up}}{0.25} \end{bmatrix}$$

Historical Note   Andrei Andreevich Markov, 1856–1922

© Novosti Information Agency/Sovfoto

Andrei Markov lived most of his life in St. Petersburg, Russia. During his lifetime, St. Petersburg (later renamed Leningrad) was the capital of czarist Russia, a major seaport and commercial center, and an international center of literature, theater, music, and ballet. Markov's family belonged to the middle class; his father worked for the Forestry Department and managed a private estate.

In high school, Markov showed a talent for mathematics but generally was not a good student. He studied mathematics at St. Petersburg University, where he received his bachelor's, master's, and doctorate degrees. He became a professor at St. Petersburg University in 1886. The head of the mathematics department there, P. L. Chebyshev, was a famous mathematician and statistician. Markov became a consistent follower of Chebyshev's ideas. Nominated by Chebyshev, he was elected to the prestigious St. Petersburg Academy of Sciences.

Markov's research was primarily in statistics, probability theory, calculus, and number theory. His most famous work, Markov chains, was motivated solely by theoretical concerns. In fact, he never wrote about their applications, other than in a linguistic analysis of Alexander Pushkin's *Eugene Onegin*.

During the early twentieth century, Markov participated in the liberal movement that climaxed in the Russian Revolution. When the Czar overruled the elec-

tion of author and revolutionary Maxim Gorky to the St. Petersburg Academy of Sciences, Markov wrote letters of protest to academic and state officials. In 1907, when the Czar dissolved the Duma (the Russian Parliament), Markov denounced the czarist government. In 1913, when the government celebrated the 300th anniversary of the House of Romanov (the czars' family), Markov organized a celebration of the 200th anniversary of the publishing of Jacob Bernoulli's book on probabilities, *Ars Conjectandi* (see Section 8.1). After the czar finally abdicated in 1917, Markov asked the Academy to send him to teach mathematics at a secondary school in a small country town in the center of Russia. He returned to St. Petersburg after a winter of famine. Soon after his return, his health declined rapidly, and he died of sepsis after an operation on his leg.

Brown University Library

The first page of Markov's Calculus of Probabilities.

This means that if current trends continue, in two generations' time, the lower income group will grow from 12% of the population to 14%, the middle income group will shrink from 62% to 60%, and the upper income group will shrink from 26% to 25%.

 Notice that the row adds to 1 before rounding.

$$0.143178 + 0.602034 + 0.254788 = 1$$

If it didn't, we would have to find our error.

11.1 Exercises

1. $P = \begin{bmatrix} 0.8 & 0.2 \end{bmatrix}$, and $T = \begin{bmatrix} 0.1 & 0.9 \\ 0.7 & 0.3 \end{bmatrix}$.

   a. What is the probability of moving from state 1 (as a current state) to state 2 (as a first following state)?

   b. What is state 2's current market share?

   c. Find PT.

   d. Find PT^2.

   e. Draw a tree diagram illustrating P and T.

2. $P = \begin{bmatrix} 0.75 & 0.25 \end{bmatrix}$, and $T = \begin{bmatrix} 0.6 & 0.4 \\ 0.5 & 0.5 \end{bmatrix}$.

   a. What is the probability of moving from state 1 (as a current state) to state 2 (as a first following state)?

   b. What is state 2's current market share?

   c. Find PT.

   d. Find PT^2.

   e. Draw a tree diagram illustrating P and T.

3. $P = \begin{bmatrix} 0.6 & 0.4 \end{bmatrix}$, and $T = \begin{bmatrix} 0.2 & 0.8 \\ 0 & 1 \end{bmatrix}$.

   a. What is the probability of moving from state 1 (as a current state) to state 2 (as a first following state)?

   b. What is state 2's current market share?

   c. Find PT.

   d. Find PT^2.

   e. Draw a tree diagram illustrating P and T.

4. $P = \begin{bmatrix} 0.6 & 0.4 \end{bmatrix}$, and $T = \begin{bmatrix} 0.3 & 0.7 \\ 0.4 & 0.6 \end{bmatrix}$.

   a. What is the probability of moving from state 1 (as a current state) to state 2 (as a first following state)?

   b. What is state 2's current market share?

   c. Find PT.

   d. Find PT^2.

   e. Draw a tree diagram illustrating P and T.

In Exercises 5–8, (a) rewrite the given data (and, if appropriate, their complements) in probability form, and (b) convert these probabilities into a probability matrix.

5. A Marketing analysis shows that KickKola currently commands 14% of the Cola market.

6. A marketing analysis shows that SoftNWash currently commands 26% of the fabric softener market.

7. Silver's Gym currently commands 48% of the health club market in Metropolis, Fitness Lab commands 37%, and ThinNFit commands the balance of the market.

8. Smallville has three Chinese restaurants: Asia Gardens, Chef Chao's, and Chung King Village. Currently, Asia Gardens gets 41% of the business, Chef Chao's gets 33%, and Chung King Village gets the balance.

In Exercises 9–12, (a) rewrite the given data in probability form, and (b) convert these probabilities into a transition matrix.

9. A marketing analysis for KickKola indicates that 12% of the consumers who do not currently drink KickKola will purchase KickKola the next time they buy a cola (in response to a new advertising campaign) and that 63% of the consumers who currently drink KickKola will purchase it the next time they buy a cola.

10. A marketing analysis for SoftNWash indicates that 9% of the consumers who do not currently use SoftNWash will purchase SoftNWash the next time they buy a fabric softener (in response to a free sample sent to selected consumers) and that 29% of the consumers who currently use SoftNWash will purchase it the next time they buy a fabric softener.

11. An extensive survey of gym users in Metropolis indicates that 71% of the current members of Silver's will continue their annual membership when it expires, 12% will quit and join Fitness Lab, and the rest will quit and join ThinNFit. Fitness Lab has been unable to keep its equipment in good shape, and as a result 32% of its members will defect to Silver's and 34% will leave for ThinNFit. ThinNFit's members are quite happy, and as a result, 96% plan on renewing their annual membership, with half of the rest planning on moving to Silver's and half to Fitness Lab.

12. Chung King Village recently mailed a coupon to all residents of Smallville offering two dinners for the price of one. As a result, 67% of those who normally eat at Asia Gardens plan on trying Chung King Village within the next month they eat Chinese food, and 59% of those who normally eat at Chef Chao's plan on trying Chung King Village. Also, all of Chung King Village's regular customers will return to take advantage of the special. And 30% of the people who normally eat at Asia Gardens will eat there within the next month, because of its convenient location. Chef Chao's has a new chef who isn't doing very well, and as a result, only 15% of the

people who normally eat at there are planning on returning next time.

13. Use the information in Exercises 5 and 9 to predict KickKola's market share at:
   a. the first following purchase, using probability trees
   b. the second following purchase, using probability trees
   c. the first following purchase, using matrices
   d. the second following purchase, using matrices
   e. the sixth following purchase
   f. Also, discuss the advantages and disadvantages of using trees as well of those of using matrices.

14. Use the information in Exercises 6 and 10 to predict SoftNWash's market share at:
   a. the first following purchase, using probability trees
   b. the second following purchase, using probability trees
   c. the first following purchase, using matrices
   d. the second following purchase, using matrices
   e. the seventh following purchase
   e. Also, discuss the advantages and disadvantages of using trees as well of those of using matrices.

15. Use the information in Exercises 7 and 11 to predict the market shares of Silver's Gym, Fitness Lab, and ThinNFit:
   a. in one year, using probability trees
   b. in one year, using matrices
   c. in two years, using matrices
   d. in three years
   e. in five years

Note: The answer to (b) is not given in the back of the book, but it is the same as that of (a).

16. Use the information in Exercises 8 and 12 to predict the market shares of Asia Gardens, Chef Chao's, and Chung King Village:
   a. in one month, using probability trees
   b. in one month, using matrices
   c. in two months, using matrices
   d. in three months
   e. in one year

17. A census report shows that currently 32% of the residents of Metropolis own their own home and that 68% rent. The census report also shows that 12% of the renters plan to buy a home in the next 12 months and that 3% of the homeowners plan to sell their home and rent instead. If residents follow their plans, find what percentage of Metropolis residents will be home owners, and what percent will be renters:
   a. in 1 year   **b.** in 2 years
   c. in 3 years

18. A survey shows that 23% of the shoppers in seven Midwestern states regularly buy their groceries at Safe Shop, 29% regularly shop at PayNEat, and the balance shops at any one of several smaller markets. The survey also indicates that 8% of the consumers who currently shop at Safe Shop will purchase their groceries at PayNEat the next time they shop and that 5% will switch to some other store. Also, 12% of the consumers who currently shop at PayNEat will purchase their groceries at Safe Shop the next time they shop, and 2% will switch to some other store. In addition, 13% of the consumers who currently shop at neither store will purchase their groceries at Safe Shop the next time they shop, and 10% will shop at payNEat. A typical shopper goes to a grocery store once every week. Find the percentage of the shoppers that will shop at Safe Shop and at PayNEat:
   a. in 1 week   **b.** in 2 weeks
   c. in 4 weeks

19. Sierra Cruiser currently commands 41% of the mountain bike market. A nationwide survey performed by *Get Out of My Way* magazine indicates that 31% of the bike owners who do not currently own a Sierra Cruiser will purchase a Sierra Cruiser the next time they buy a mountain bike and that 12% of the bikers who currently own a Sierra Cruiser will purchase one the next time they buy a mountain bike. If the average customer buys a new mountain bike every two years, predict Sierra Cruiser's market share in four years.

20. A marketing analysis shows that Clicker Pens currently commands 46% of the pen market. The analysis also indicates that 46% of the consumers who do not currently own a Clicker pen will purchase a Clicker the next time they buy a pen and that 37% of the consumers who currently own a Clicker will not purchase one the next time they buy a pen. If the average consumer buys a new pen every three weeks, predict Clicker's market share after six weeks.

PROJECTS

21. **a.** The Census Bureau classifies all residents of the United States as residents of a central city, a suburb, or a nonmetropolitan area. In the 1990 census, the bureau reported that the central cities held 77.8 million people, the suburbs 114.9 million people, and the nonmetropolitan areas 56.0 million people. Compute the proportion of U.S. residents for each category and present those proportions as a probability matrix. (Round off to three decimal places.)

   b. The Census Bureau (in Current Population Reports, P20–463, *Geographic Mobility, March 1990 to March 1991*) also reported the information in Figures 11.18 and 11.19 regarding migration between these three areas from 1990 to 1991. (Numbers are in thousands.)

	Moved to		
Moved from	**Central City**	**Suburb**	**Nonmetropolitan Area**
Central City	(x)	4946	736
Suburb	2482	(x)	964
Nonmetropolitan Area	741	1075	(x)

Figure 11.18
Migration data for Exercise 21
Note: Persons moving from one central city to another were not considered. This is represented by "(x)."

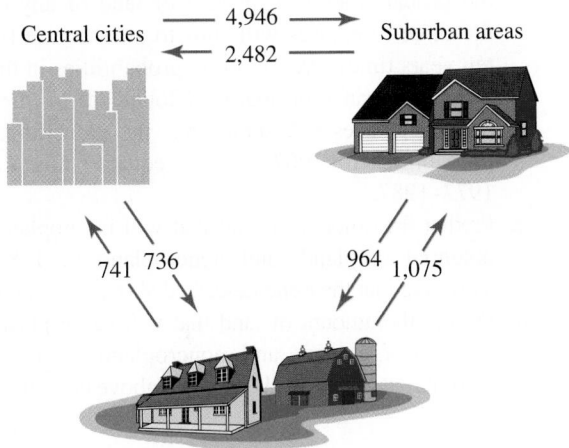

Figure 11.19
Migration data in graphic form (Numbers are in thousands.)

Use the data in Figure 11.18 and the data in part (a) to compute the probabilities that a resident of any one type of environment will move to any of the other types. Present these probabilities in the form of a transition matrix. (Round off to three decimal places.)
c. Predict the percent of U.S. residents who will reside in a central city, a suburb, and a nonmetropolitan area in 1991.
d. Predict the percent of U.S. residents who will reside in a central city, a suburb, and a nonmetropolitan area in 1992.
e. Which prediction is the stronger one; that in part (c) or that in part (d)? Why?
22. a. The Census Bureau (in *Statistical Abstract of the United States, 2000*) reported the 2000 national and regional populations given in Figure 11.20.

Region	Population (in thousands)
U.S.	281,422
Northeast	53,594
Midwest	64,393
South	100,237
West	63,198

Figure 11.20
Regional population data for Exercise 22

Compute the proportion of U.S. residents of each region, and present those proportions as a probability matrix. (Round off to three decimal places.)
b. The Census Bureau (in *Current Population Reports*, P20–465, *Geographic Mobility: March 1999 to March 2000*) also reported the information in Figures 11.21 and 11.22 regarding movement between regions from 1999 to 2000. (All numbers are in thousands.)

Region Moved from	Region Moved to			
	Northeast	**Midwest**	**South**	**West**
Northeast	(x)	83	405	127
Midwest	72	(x)	364	204
South	183	416	(x)	432
West	109	223	489	(x)

Figure 11.21
Interregional migration
Note: The Census data involved interregional movement. Thus, persons moving from one part of a region to another part of the same region were not considered. This is represented in the above chart by "(x)."

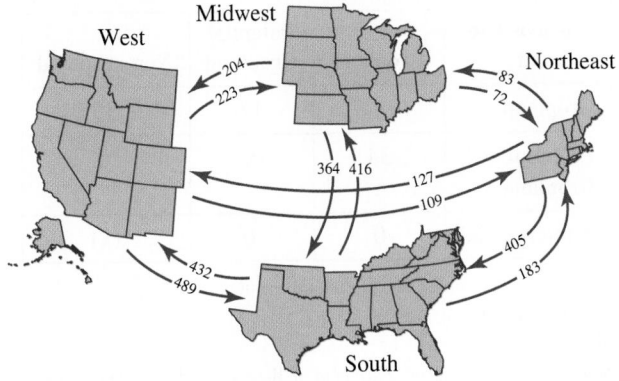

Figure 11.22
Interregional migration in graphic form

Compute the probabilities that a resident of any one region will move to any of the other regions. Present these probabilities in the form of a transition matrix. Assume that the transitional probabilities for 2000–2001 are equal to those for 1999–2000. (Round off to three decimal places.)

c. Predict the percent of U.S. residents who resided in the Northeast, the Midwest, the South, and the West in 2001.

d. Predict the percent of U.S. residents who resided in the Northeast, the Midwest, the South, and the West in 2002.

23. **Land Use.** Neil Sampson, executive vice president of the National Association of Conservation Districts, compiled data on shifting land use in the United States (*Source:* R. Neil Sampson *Farmland or Wasteland: A Time To Choose.* Emmaus, PA: Rodale Press, 1981). He stated that in 1977, there were 413 million acres of cropland, 127 million acres of land that had a high or medium potential for conversion to cropland (for example, grasslands and forests), and 856 million acres of land that had little or no potential for conversion to cropland (such as urban land or land with heavily deteriorated soil).

a. Compute the proportion of land in each category, and present those proportions as a probability matrix. (Round off to three decimal places.)

b. He also estimated that, between 1967 and 1977, the shifts in land use shown in Figures 11.23 and Figure 11.24 occurred (in millions of acres).

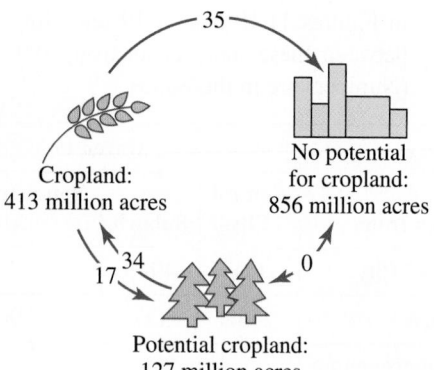

Figure 11.24
Shifting land use in graphic form

Use these data and the data in (a) to compute the probabilities that an acre of land of any of the three categories will shift to another use (in ten years time). Present these probabilities in the form of a transition matrix. (Round off to three decimal places.) Assume that the transitional probabilities for 1967–1977 are equal to those for 1977–1987.

c. Predict the amount of land that will be cropland, potential cropland, and noncropland in 1987, assuming that the trend described above continues.

d. Predict the amount of land that will be cropland, potential cropland, and noncropland in 1997, assuming that the trend described above continues.

➤ *Answer the following questions using complete sentences and your own words.*

CONCEPT QUESTIONS

24. Why does the information in Exercises 5 and 9 describe a Markov chain?

25. A coin is tossed ten times. Does this experiment and its probabilities form a Markov chain? Why or why not?

26. Are PT, PT^2, and PT^3 transition matrices or probability matrices? Why?

27. Are T^2 and T^3 transition matrices or probability matrices? Why?

28. What are the differences between a transition matrix and a probability matrix?

29. What information would a marketing analyst need to predict the future market share of a product? How would he or she obtain such information?

30. When a prediction is made using Markov chains, that prediction is based on a number of assumptions,

Previous Use of Land	New Use of Land		
	Cropland	Potential Cropland	Non-cropland
Cropland	(x)	17	35
Potential Cropland	34	(x)	0[1]
Non-cropland	0	0[1]	(x)

[1]Unknown but assumed cancel each other out.
Figure 11.23
Shifting land use

Source: R. Neil Sampson, *Farmland or Wasteland: A Time to Choose.* Emmaus, PA: Rodale Press, 1981.

one involving the trend, one involving the data summarized in the probability matrix P, and one involving the data summarized in the transition matrix T. What are these assumptions? What would make the assumption invalid?

31. Compare and contrast the tree method of Markov chains with the matrix method of Markov chains. What are the advantages of the two different methods?

32. Choose one of Exercises 13–20 and discuss why the given data describes a Markov chain.

33. What is the difference between a current state and a following state?

34. Suppose a transition matrix contains a row where one entry is a 1 and all of the other entries are zeros. What does that tell you?

HISTORY QUESTIONS

35. In creating Markov chains, was Markov motivated by theoretical concerns or by specific applications?

36. Was Markov a supporter of the czar or of the revolution?

37. How did Markov celebrate the 300th anniversary of the House of Romanov?

38. What did Markov do when the czar abdicated?

11.2 Systems of Linear Equations

Linear Equations

To graph the equation $2x + 3y = 8$, first solve the equation for y.

$$2x + 3y = 8$$
$$3y = -2x + 8$$
$$y = -\frac{2}{3}x + \frac{8}{3}$$

Comparing this equation with the slope-intercept formula $y = mx + b$, we find that this is the equation of a line with slope $m = -\frac{2}{3}$ and y-intercept $b = \frac{8}{3}$, as shown in Figure 11.25.

The graph of the equation $5x + 7y = 12$ is also a line, because we could follow the above procedure and rewrite the equation in the form $y = mx + b$. In fact, the graph of any equation of the form $ax + by = c$ is a line. (If b were 0, the above procedure wouldn't work, but the graph would still be a line.) For this reason, an equation of the form $ax + by = c$ is called a *linear equation*.

To graph an equation of the form $ax + by + cz = d$, we would need a z-axis in addition to the usual x- and y-axes; the graph of such an equation is always a plane. Despite its geometrical appearance, however, such an equation is called a linear equation owing to its algebraic similarity to $ax + by = c$. A **linear equation** is an equation that can be written in the form $ax + by = c$, the form $ax + by + cz = d$, or a similar form with more unknowns. In section 11.3, we will use linear equations to solve Markov chain problems.

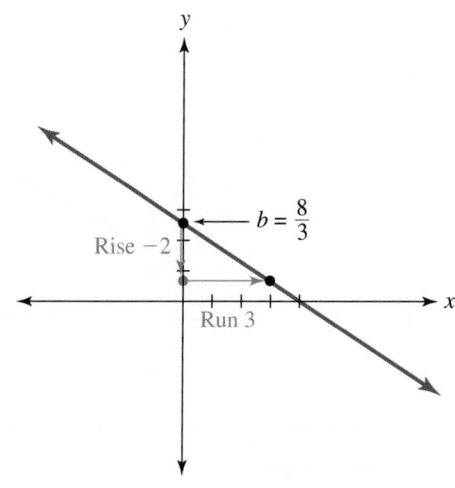

Figure 11.25
The graph of $y = -\frac{2}{3}x + \frac{8}{3}$

Systems of Equations

A **system of equations** is a set of more than one equation. **Solving a system of equations** means finding all ordered pairs (x, y) [or ordered triples (x, y, z) and so on] that will satisfy each equation in the system. Geometrically, this means finding all points that are on the graph of each equation in the system.

The system

$$x + y = 3$$
$$2x + 3y = 8$$

is a system of linear equations because it is a set of two equations and each equation is of the form $ax + by = c$. The ordered pair $(1, 2)$ is a solution to the system, because $(1, 2)$ satisfies each equation:

$$x + y = 1 + 2 = 3$$
$$2x + 3y = 2 \cdot 1 + 3 \cdot 2 = 8$$

Does this system have any other solutions? If we solve each equation for y, we get

$$y = -1x + 3$$
$$y = \frac{-2}{3}x + \frac{8}{3}$$

The first equation is the equation of a line with slope -1 and y-intercept 3; the second is the equation of a line with slope $\frac{-2}{3}$ and y-intercept $\frac{8}{3}$. Because the slopes are different, the lines are not parallel and intersect in only one point. Thus, $(1, 2)$ is the only solution to the system; it is the point at which the two lines intersect. This system is illustrated in Figure 11.26.

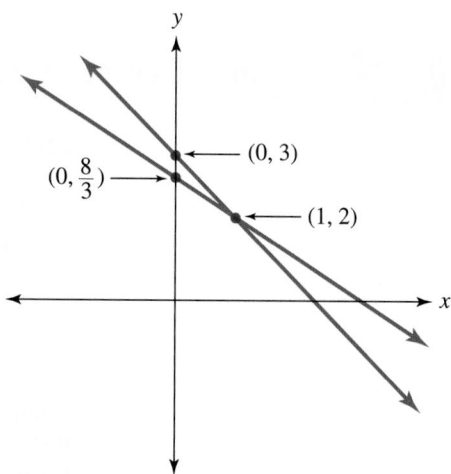

Figure 11.26
The graph of the system
$x + y = 3$
$2x + 3y = 8$

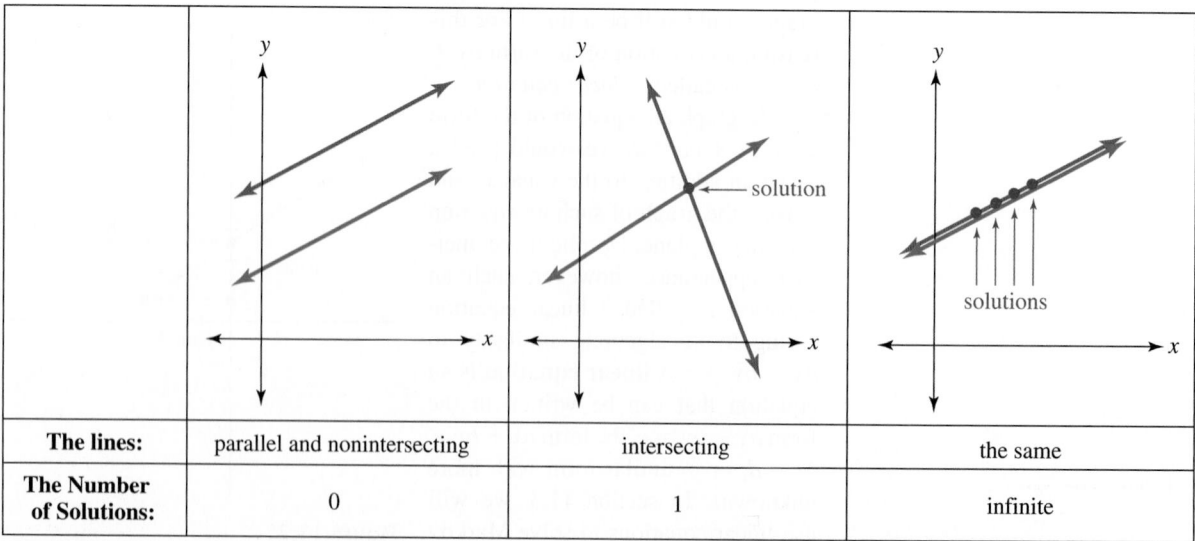

	The lines:	parallel and nonintersecting	intersecting	the same
	The Number of Solutions:	0	1	infinite

Figure 11.27
A system of two linear equations in two unknowns can have no solutions, one solution, or an infinite number of solutions.

Not all systems have only one solution. A system with two linear equations in two unknowns describes a pair of lines. The two lines can be parallel and not intersect (in which case the system has no solution), or the two lines can be the same (in which case the system has an infinite number of solutions), as shown in Figure 11.27.

A system of two linear equations in *three* unknowns (x, y, and z) describes a set of two planes. Such a system can have no solution or an infinite number of solutions, but it cannot have only one solution, as illustrated in Figure 11.28. In fact, *any system of linear equations that has fewer equations than unknowns cannot have a unique solution.*

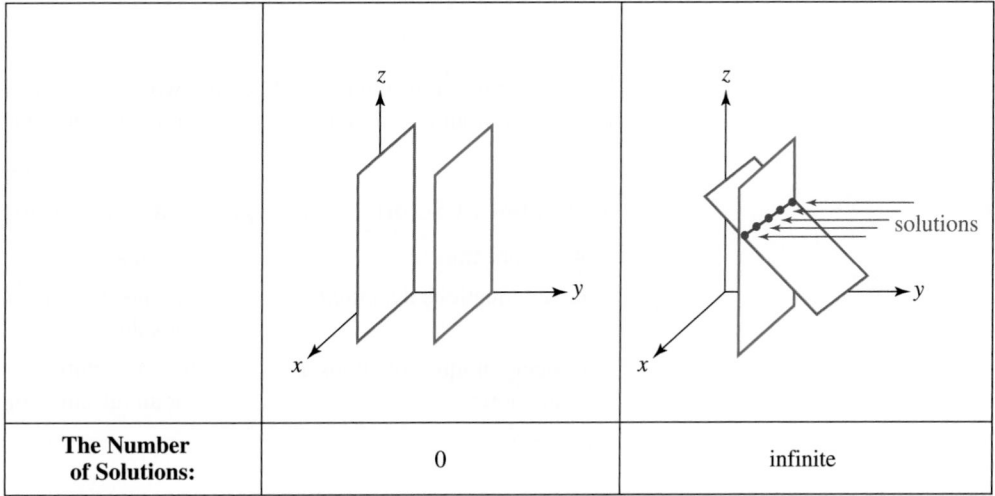

Figure 11.28
A system of two linear equations in three unknowns can have no solutions or an infinite number of solutions.

A system of *three* linear equations in three unknowns describes a set of three planes. Such a system can have no solution, only one solution, or an infinite number of solutions, as illustrated in Figure 11.29. In fact, *any system of linear equations that has as many equations as unknowns can have either no solution, one unique solution, or an infinite number of solutions.*

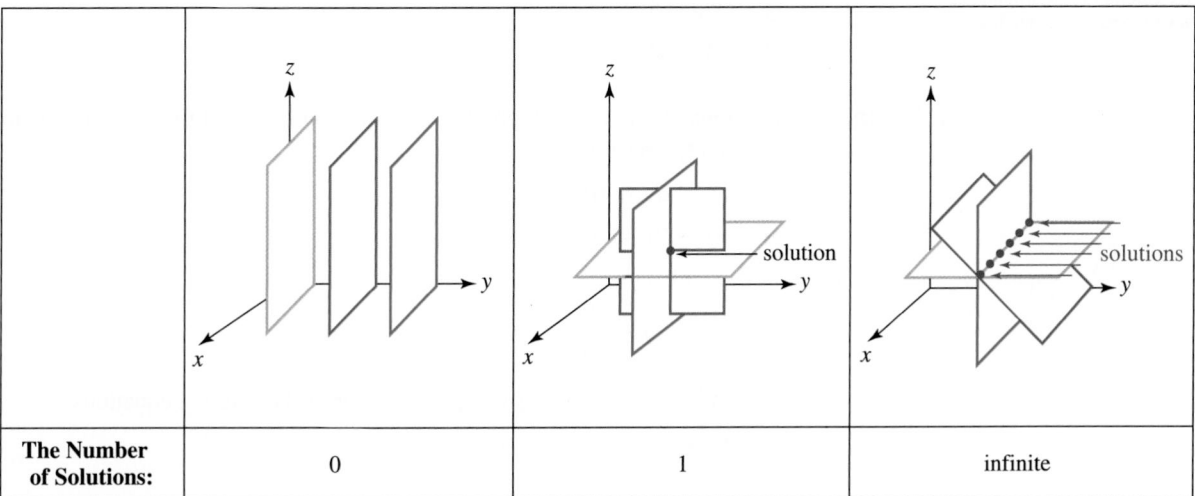

Figure 11.29
A system of three linear equations in three unknowns can have no solutions, one solution, or an infinite number of solutions.

ThomsonNOW™
for Liberal Arts Mathematics
www.thomsonedu.com/1pass.
EXAMPLE 1   How many solutions could the following system have?

$$3x - y + z = 5$$
$$2x + y - 3z = 1$$
$$20x + 10y - 30z = 10$$

Solution   This system seems to have as many equations as unknowns; however, this is not the case. The third equation is 10 times the second equation, so its presence is unnecessary. The above system is equivalent to the following system:

$$3x - y + z = 5$$
$$2x + y - 3z = 1$$

This system has three unknowns but only two *unique* equations, so it can have either no solution or an infinite number of solutions. It cannot have one solution.   ∎

Number of Solutions of a System of Linear Equations

A system that has:	will have:
fewer equations than unknowns	either no solution or an infinite number of solutions
as many unique equations as unknowns	either no solution, one solution, or an infinite number of solutions

---

Solving Systems: The Elimination Method

Our work with Markov chains will involve solving systems of equations. Many methods are used to solve systems of equations. Perhaps the easiest method for solving a small system is the elimination method (also called the addition method). With this method, you add together equations or multiples of equations in such a way that a variable is eliminated. You probably saw this method in your intermediate algebra class.

ThomsonNOW™
for Liberal Arts Mathematics
www.thomsonedu.com/1pass.
EXAMPLE 2   Use the elimination method to solve the following system:

$$3x - 2y = 5$$
$$2x + 3y = 4$$

Solution   We'll eliminate x if we multiply the first equation by 2 and the second equation by -3 and add the results.

$$
\begin{array}{ll}
6x - 4y = 10 & \text{2 times equation 1} \\
+ \ -6x - 9y = -12 & \text{-3 times equation 2} \\
\hline
-13y = -2 & \\
y = \dfrac{2}{13} &
\end{array}
$$

We can find x by substituting $y = \frac{2}{13}$ into either of the original equations.

$$3x - 2y = 5$$
$$3x - 2 \cdot \frac{2}{13} = 5$$

$$3x - \frac{4}{13} = \frac{65}{13} \qquad \text{rewriting 5 with a common denominator}$$

$$3x = \frac{69}{13}$$

$$x = \frac{23}{13}$$

Thus, the solution to the system of equations is the ordered pair $\left(\frac{23}{13}, \frac{2}{13}\right)$. ∎

We can easily check our work by substituting the solution back into the original equations.

$$3x - 2y = 3 \cdot \frac{23}{13} - 2 \cdot \frac{2}{13} = \frac{69}{13} - \frac{4}{13} = \frac{65}{13} = 5 \checkmark$$

$$2x + 3y = 2 \cdot \frac{23}{13} + 3 \cdot \frac{2}{13} = \frac{46}{13} + \frac{6}{13} = \frac{52}{13} = 4 \checkmark$$

Naturally, these calculations can be performed on a calculator.

11.2 Exercises

In Exercises 1–6, determine whether the given ordered pair or ordered triple solves the given system of equations.

1. (4, 1)
  $3x - 5y = 7$
  $2x + 2y = 10$
2. (7, −2)
  $2x - 4y = 30$
  $x + y = 5$
3. (−5, 3)
  $3x + y = 4$
  $10x - 4y = -62$
4. (−1, −2)
  $2x + 2y = -5$
  $3x - 4y = 2$
5. (4, −1, 2)
  $2x + 3y - z = 3$
  $x + y - z = 5$
  $10x - 2y = 3$
6. (0, 5, −1)
  $3x - 2y + z = -11$
  $2x + 4y + z = 19$
  $x - z = 1$

In Exercises 7–12, do the following: **a.** *Find each line's slope and y intercept.* **b.** *Use the slopes and y intercepts to determine whether the given system has no solution, one solution, or an infinite number of solutions. Do not actually solve the system.*

7. $5x + 2y = 4$
  $6x - 19y = 72$
8. $3x + 132y = 19$
  $45x + 17y = 4$
9. $4x + 3y = 12$
  $8x + 6y = 24$
10. $19x - 22y = 1$
  $190x - 220y = 10$

11. $x + y = 7$
  $3x + 2y = 8$
  $2x + 2y = 14$
12. $3x - y = 12$
  $2x + 3y = 5$
  $5x + 2y = 17$

In Exercises 13–18, determine whether the given system could have a single solution. Do not actually solve the system.

13. $3x - 2y + 5z = 1$
  $2x + y = 2$
  $5x + 7y - z = 0$
14. $8x - 4y + 2z = 10$
  $3x + y + z = 1$
  $4x - 2y + z = 5$
15. $x + y + z = 1$
  $2x + 2y + 2z = 2$
  $3x - y + 10z = 45$
16. $x + y + z = 34$
  $5x - y + 2z = 9$
  $3x + 3y + 3z = 102$
17. $x + 2y + 3z = 4$
  $5x - y = 2$
18. $9x - 21y = 476$
  $x + 3y + z = 12$

Solve the systems in Exercises 19–24 with the elimination method. Check your answers by substituting them back in. (Answers are not given at the back of the book.)

19. $2x + 3y = 5$
  $4x - 2y = 2$
20. $5x + 3y = 11$
  $2x + 7y = 16$
21. $3x - 7y = 27$
  $4x - 5y = 23$
22. $5x - 2y = -23$
  $x + 2y = 5$

23. $5x - 9y = -12$
$3x + 7y = -1$

24. $5x - 12y = 9$
$3x + 3y = 2$

▬▬▬ *Answer the following questions using complete sentences and your own words.*

CONCEPT QUESTIONS

25. By sketching lines, determine whether a system of three equations in two unknowns could have no solution, one solution, or an infinite number of solutions. What if the three equations were unique?

26. By sketching planes, determine whether a system of four equations in three unknowns could have no solution, one solution, or an infinite number of solutions. What if the four equations were unique?

27. Solving a system of two equations in two unknowns is the same as finding all points that are on the graph of both of the two equations. Why do we not solve systems by graphing them (by hand) and locating the points that are on both graphs?

28. What does the graph of a linear equation in two unknowns look like? What does the graph of a linear equation in three unknowns look like? Why is a linear equation in three unknowns called a linear equation?

29. Suppose you have a linear equation in two unknowns, and you multiply both sides of that equation by a constant (not zero). How would the graph of the new equation compare to that of the original equation? Would they be parallel lines, perpendicular lines, the same lines, or what? Why?

Solving Systems of Two Linear Equations in Two Unknowns on a Graphing Calculator

A system of two linear equations in two unknowns can be solved nicely on a graphing calculator.

> **Solving Systems of Two Linear Equations in Two Unknowns on a Graphing Calculator**
>
> - Solve each equation for y.
> - Enter the results into the graphing calculator as Y_1 and Y_2.
> - Graph the two equations on the graphing calculator.
> - Use the calculator to find the point of intersection.
>   (The details are given in Appendices C and D.)
> - Check your solution by substituting it into each of the original equations.

EXAMPLE 3   Use a graphing calculator to solve the following system:

$$3x - 2y = 5$$
$$2x + 3y = 4$$

Solution

1. *Solve each equation for y.*

$$3x - 2y = 5 \Rightarrow -2y = -3x + 5 \Rightarrow \frac{-2y}{-2} = \frac{-3x + 5}{-2} \Rightarrow y = \frac{3}{2}x - \frac{5}{2}$$

$$2x + 3y = 4 \Rightarrow 3y = -2x + 4 \Rightarrow \frac{3y}{3} = \frac{-2x + 4}{3} \Rightarrow y = -\frac{2}{3}x + \frac{4}{3}$$

2. *Enter the results into the graphing calculator as* Y_1 *and* Y_2.

Figure 11.30
A TI-83's "Y=" screen
(Other calculators are similar.)

3. *Graph the two equations on the graphing calculator.*

Figure 11.31
The graph of the system
$3x - 2y = 5$
$2x + 3y = 4$

4. *Use the calculator to find the point of intersection.*

Figure 11.32
The solution of the system
$3x - 2y = 5$
$2x + 3y = 4$

The system's solution is approximately $(1.7692308, 0.15384615)$.

5. *Check your solution by substituting it into each of the original equations.*

$$3x - 2y = 3 \cdot 1.769230 - 2 \cdot 0.15384615 = 4.9999167 \approx 5$$
$$2x + 3y = 2 \cdot 1.769230 + 3 \cdot 0.15384615 = 3.99994445 \approx 4$$

The solution checks, but not perfectly. This is because the calculator rounded off the point of intersection in step 4.

Exercises

Solve the systems in Exercises 30–35 with a graphing calculator. Check your answers by substituting them back in. (Answers are not given at the back of the book.)

30. Exercise 20
32. Exercise 22
34. Exercise 24

31. Exercise 21
33. Exercise 23
35. Exercise 19

CONCEPT QUESTIONS

36. Describe how you could use a graphing calculator to solve a system of three equations in two unknowns.

37. Why could you not use a graphing calculator to solve a system of two equations in three unknowns?

11.3 Long-Range Predictions with Markov Chains

In Section 11.1, we were given Brute's current market share (25%), and we predicted Brute's market share after one, two, and three purchases (that is, we found the probabilities of the first, second, and third following states) by computing PT, PT^2, and PT^3. In a similar way, we could predict Brute's market share after four purchases by computing PT^4. These predictions are shown in Figure 11.33 and illustrated in Figure 11.34. It appears that the trend in increasing market share created by Brute's new advertising campaign and customer satisfaction with the product will stabilize and that Brute will ultimately command close to 70% of the market.

Brute's Projected Market Share	Brute	Not Brute	Probability Matrix
Current purchase	25%	75%	P
First following purchase	50%	50%	$PT = PT^1$
Second following purchase	60%	40%	PT^2
Third following purchase	64%	36%	PT^3
Fourth following purchase	65.6%	34.4%	PT^4

Figure 11.33
Brute is gaining market share

If Brute started out with a much smaller initial market share, say, only 10% rather than 25%, a similar pattern emerges, as shown in Figure 11.35 on page 838.

$P = \begin{bmatrix} 0.1 & 0.9 \end{bmatrix}$ a new probability matrix

$T = \begin{bmatrix} 0.8 & 0.2 \\ 0.4 & 0.6 \end{bmatrix}$ the same transition matrix as before

Surprisingly, it appears that Brute's new advertising campaign and customer satisfaction with the product would have the same ultimate effect, regardless of Brute's initial market share. Under certain circumstances, PT^n will stabilize, and it will do so in a way that is unaffected by the value of P. Our goal in this section is to find the level at which the trend will stabilize. The **equilibrium matrix L,** the matrix at which the trend stabilizes, is the probability matrix $L = PT^n$ such that all following probability matrices are equal. In other words, multiplying the equilibrium matrix L by T would have no effect, and $LT = L$. In fact, we find the equilibrium matrix by solving the matrix equation $LT = L$.

Figure 11.34
As a result of a new advertising campaign and customers' satisfaction with the product, Brute is predicted to increase its market share from 25% . . .

to 65.6% after the fourth following purchase.

Brute's Projected Market Share	Brute	Not Brute	Probability Matrix
Current purchase	10%	90%	P
First following purchase	44%	56%	$PT = PT^1$
Second following purchase	57.6%	42.4%	PT^2
Third following purchase	63.04%	36.96%	PT^3
Fourth following purchase	65.22%	34.78%	PT^4

Figure 11.35
The same result, even with a different starting point

ThomsonNOW™
for Liberal Arts Mathematics
www.thomsonedu.com/1pass.

EXAMPLE 1 Make a long-range forecast for Brute's market share given the transition data from Section 11.1.

Solution We need to find the equilibrium matrix L. We find L by solving the matrix equation $LT = L$.

$$L = \begin{bmatrix} x & y \end{bmatrix} \qquad \text{L is a row matrix, since L is a probability matrix}$$

$$T = \begin{bmatrix} 0.8 & 0.2 \\ 0.4 & 0.6 \end{bmatrix} \qquad \text{from Section 11.1}$$

$$\begin{matrix} L & T & = & L \end{matrix}$$

$$\begin{bmatrix} x & y \end{bmatrix}\begin{bmatrix} 0.8 & 0.2 \\ 0.4 & 0.6 \end{bmatrix} = \begin{bmatrix} x & y \end{bmatrix}$$

$$\begin{bmatrix} 0.8x + 0.4y & 0.2x + 0.6y \end{bmatrix} = \begin{bmatrix} x & y \end{bmatrix}$$

This yields the system

$$0.8x + 0.4y = x$$
$$0.2x + 0.6y = y$$

Combining like terms gives

$$-0.2x + 0.4y = 0$$
$$0.2x - 0.4y = 0$$

These two equations are equivalent (we can multiply one equation by -1 to get the other), so we may discard one. This leaves us in a difficult spot: our single equation has an infinite number of solutions, and we are looking for a single solution. However, recall that $L = \begin{bmatrix} x & y \end{bmatrix}$ is a probability matrix, so its entries must add to 1. Thus, $x + y = 1$ is our needed second equation. Now we can solve the system

$$0.2x - 0.4y = 0$$
$$x + y = 1$$

This system can easily be solved with the elimination method or a graphing calculator.

Proceeding with the elimination method, we multiply the second equation by 0.4. The resulting system is

$$0.2x - 0.4y = 0$$
$$0.4x + 0.4y = 0.4$$

Adding gives

$$0.6x = 0.4$$
$$\frac{0.6x}{0.6} = \frac{0.4}{0.6}$$
$$x = \frac{2}{3}$$

Substituting into $x + y = 1$ gives

$$\frac{2}{3} + y = 1$$
$$y = \frac{1}{3}$$

Thus, $L = [x \quad y] = [\frac{2}{3} \quad \frac{1}{3}]$, and Brute's market share should stabilize at $\frac{2}{3}$, or approximately 67%.

 As a check, multiply L by T and see whether you get L.

$$LT = \begin{bmatrix} \dfrac{2}{3} & \dfrac{1}{3} \end{bmatrix} \begin{bmatrix} 0.8 & 0.2 \\ 0.4 & 0.6 \end{bmatrix}$$

$$= \begin{bmatrix} \dfrac{2}{3} & \dfrac{1}{3} \end{bmatrix}$$

■

Long-Range Predictions with Regular Markov Chains

1. *Create the transition matrix T. T* is a square matrix, discussed in Section 11.1, whose entries are the probabilities of passing from current states to next following states. The rows refer to the current states and the columns refer to the next following states.

2. *Create the equilibrium matrix L.* The equilibrium matrix *L* is the long-range prediction. *L* is the matrix that solves the matrix equation by *LT = L.* If there are two states, then $L = [x \quad y]$.

3. *Find and simplify the system of equations described by LT = L.*

4. *Discard any redundant equations, and include the equation x + y = 1.*

5. *Solve the resulting system.* Use the elimination method or a graphing calculator.

6. ✔ *Check your work by verifying that LT = L.*

11.3 Exercises

In Exercises 1–4, do the following.

a. *Find the equilibrium matrix L by solving LT = L for L.*

b. *Check your solution to part (b) by verifying that LT = L. (Answers are not given in the back of the book.)*

1. $T = \begin{bmatrix} 0.1 & 0.9 \\ 0.2 & 0.8 \end{bmatrix}$

2. $T = \begin{bmatrix} 0.5 & 0.5 \\ 0.6 & 0.4 \end{bmatrix}$

3. $T = \begin{bmatrix} 0.3 & 0.7 \\ 0.8 & 0.2 \end{bmatrix}$

4. $T = \begin{bmatrix} 0.4 & 0.4 \\ 0.25 & 0.75 \end{bmatrix}$

5. A marketing analysis shows that 12% of the consumers who do not currently drink KickKola will purchase KickKola the next time they buy a cola and that 63% of the consumers who currently drink KickKola will purchase it the next time they buy a cola. Make a long-range prediction of KickKola's ultimate market share, assuming that current trends continue. (See Exercise 9 in Section 11.1.)

6. A marketing analysis shows that 9% of the consumers who do not currently use SoftNWash will purchase SoftNWash the next time they buy a fabric softener and that 29% of the consumers who currently use SoftNWash will purchase it the next time they buy a fabric softener. Make a long-range prediction of SoftNWash's ultimate market share, assuming that current trends continue. (See Exercise 10 in Section 11.1.)

7. A census report shows that 32% of the residents of Metropolis own their own home and that 68% rent. The report shows that 12% of the renters plan on buying a home in the next 12 months and that 3% of the homeowners plan on selling their home and renting instead. Make a long-range prediction of the percentage of Metropolis residents that will own their own home and the percentage that will rent. Give two assumptions on which this prediction is based. (See Exercise 17 in Section 11.1.)

8. A marketing analysis indicates that 46% of the consumers who do not currently own a Clicker pen will purchase a Clicker the next time they buy a pen and that 37% of the consumers who currently own a Clicker will not purchase one the next time they buy a pen. If the average consumer buys a new pen every three weeks, make a long-range prediction of Clicker's ultimate market share, assuming that current trends continue. (See Exercise 20 in Section 11.1.)

9. A nationwide survey preformed by *Get Out of My Way* magazine indicates that 31% of the bike owners who do not currently own a Sierra Cruiser will purchase a Sierra Cruiser the next time they buy a mountain bike and that 12% of the bikers who currently own a Sierra Cruiser will purchase one the next time they buy a mountain bike. If the average customer buys a new mountain bike every two years, make a long-range prediction of Sierra Cruiser's ultimate market share, assuming that current trends continue. (See Exercise 19 in Section 11.1.)

 WEB PROJECT

10. Monopoly is "the most played board game in the world," according to Hasbro, its manufacturer. Players take turns moving around the board, according to a roll of a pair of dice. The board consists of 40 squares, 28 of which are properties such as Marvin Gardens, B&O Railroad, and Water Works. There are also three Chance squares, three Community Chest squares, Luxury Tax, Income Tax, Jail, Free Parking, Go, and Go to Jail.

This game can be analyzed by using Markov chains. There are 40 states, one for each square on the board. Read about the game's rules, and create a transition matrix for the game. Do not try to make a long-range prediction. Instead, discuss what information would be contained in such a prediction.

Some useful links for this web project are listed on the text companion web site. Go to www.thomsonedu.com/mathematics/johnson to access the web site.

The Monopoly board

11.4 Solving Larger Systems of Equations

In this section, we will discuss how to use the elimination method to solve larger systems of equations (i.e., systems of more than two equations with more than two unknowns).

In the technology portion of this section, we will discuss how to use a graphing calculator and the Gauss-Jordan method to solve larger systems of equations. The Gauss-Jordan method is a technology-friendly version of the elimination method that uses matrices rather than equations.

Solving Larger Systems with the Elimination Method

Solving a system of three equations in three unknowns with the elimination method is quite similar to solving a system of two equations in two unknowns. There are a few more steps because there are more variables to eliminate.

Solving a System of Three Equations in Three Unknowns with the Elimination Method

1. Combine two equations so that one variable is eliminated.
2. Combine two different equations so that the same variable is eliminated.
3. Steps 1 and 2 result in a system of two equations in two unknowns. Solve that system with the elimination method, as discussed in Section 11.2.
4. Substitute back into various equations to find the values of the remaining variables.
5. Check your answer by substituting it into the original equations.

ThomsonNOW™
for Liberal Arts Mathematics
www.thomsonedu.com/1pass.

EXAMPLE 1   Use the elimination method to solve the following system:

$$5x - 7y + 2z = 25$$
$$3x + 2y - 5z = -16$$
$$7x + 3y - 9z = -25$$

Solution

1. *Combine two equations so that one variable is eliminated.*
    We'll combine the first two equations, and eliminate z. To do this, multiply the first equation by 5 and the second by 2, and add the results.

    $$\begin{array}{rl} 25x - 35y + 10z = & 125 \qquad \text{5 times equation 1} \\ + \quad 6x + \;\; 4y - 10z = & -32 \qquad \text{2 times equation 2} \\ \hline 31x - 31y \qquad\quad = & 93 \end{array}$$

2. *Combine two different equations so that the same variable is eliminated.*
    We'll combine the last two equations and eliminate z. To do this, multiply the second equation by 9 and the third by -5, and add the results.

    $$\begin{array}{rl} 27x + 18y - 45z = & -144 \qquad \text{9 times equation 2} \\ + \; -35x - 15y + 45z = & 125 \qquad -5 \text{ times equation 3} \\ \hline -8x + \;\; 3y \qquad\quad = & -19 \end{array}$$

3. *Steps 1 and 2 result in a system of two equations in two unknowns. Solve that system with the elimination method, as discussed in Section 7.2.* We are left with a smaller system to solve:

    $$31x - 31y = \;\; 93 \qquad \text{the result of step 1}$$
    $$-8x + \;\; 3y = -19 \qquad \text{the result of step 2}$$

First, simplify the first equation of this new system by dividing through by 31:

$$x - y = 3 \quad \text{dividing the first equation by 31}$$
$$-8x + 3y = -19 \quad \text{the second equation}$$

We'll eliminate y if we multiply the first equation by 3 and add the result to the second equation.

$$
\begin{array}{rl}
3x - 3y = 9 & \text{3 times equation 1} \\
+ \ -8x + 3y = -19 & \text{equation 2} \\
\hline
-5x \quad\quad = -10 & \\
-5x/-5 \quad = -10/-5 & \\
x = 2 &
\end{array}
$$

4. *Substitute back into various equations to find the values of the remaining variables.* We can find y by substituting $x = 2$ into $x - y = 3$

$$2 - y = 3 \quad \text{substituting}$$
$$-y = 1 \quad \text{solving for } y$$
$$y = -1$$

We can find z by substituting $x = 2$ and $y = -1$ into $5x - 7y + 2z = 25$

$$5x - 7y + 2z = 25$$
$$5 \cdot 2 - 7 \cdot (-1) + 2z = 25 \quad \text{substituting}$$
$$17 + 2z = 25 \quad \text{solving for } z$$
$$2z = 8$$
$$z = 8/2 = 4$$

The system's solution is $(x, y, z) = (2, -1, 4)$.

5. *Check your answer by substituting it into the original equations.* It's not necessary to substitute this solution into the original system's first equation, because in step 4 we just finished forcing that equation to work.

$$3x + 2y - 5z = -16 \quad \Rightarrow \quad 3 \cdot 2 + 2 \cdot (-1) - 5 \cdot 4 = -16 \ \checkmark$$
$$7x + 3y - 9z = -25 \quad \Rightarrow \quad 7 \cdot 2 + 3 \cdot (-1) - 9 \cdot 4 = -25 \ \checkmark \quad\quad \blacksquare$$

11.4 Exercises

Solve the given system with the elimination method.

1. $5x + y - z = 17$
   $2x + 5y + 2z = 0$
   $3x + y + z = 11$

2. $9x - 2y + 4z = 29$
   $2x + 3y - 4z = 3$
   $x + y + z = 1$

3. $x + y + z = 14$
   $3x - 2y + z = 3$
   $5x + y + 2z = 29$

4. $2x + y - z = 4$
   $3x + 2y - 7z = 0$
   $5x - 3y + 2z = 20$

5. $x - y + 4z = -13$
   $2x - z = 12$
   $3x + y = 25$

6. $2x + 2y + z = 5$
   $3x - y = 5$
   $2y + 7z = -9$

7. $8x + 7y - 3z = 38$
   $4x - 3y + 2z = 11$
   $9x - 11y + 5z = 7$

8. $4x - 3y - 2z = 42$
   $5x + 5y + 4z = 78$
   $2x - 9y - 6z = 6$

9. $4x + 4y + 2z = 10$
   $3x - 2y - 2z = 18$
   $7x - y + 8z = 4$

10. $11x + 17y - 13z = -186$
    $12x + 32y + 64z = -92$
    $15x - 41y + 88z = 1132$

11. $47x + 58y + 37z = -113$
    $22x - 37y + 27z = 332$
    $47x + 15y + 52z = 145$

12. $74x - 22y - 48z = -1052$
    $49x - 75y - 47z = 506$
    $53x + 10y + 77z = 4608$

Solving Larger Systems of Linear Equations on a Graphing Calculator

We can't solve a system of equations in three unknowns by graphing the system, as we did in Section 11.2, because a graphing calculator cannot graph an equation in three unknowns.

We can, however, solve such a system using a graphing calculator and the Gauss-Jordan method. The *Gauss-Jordan method* is a technology-friendly version of the elimination method that uses matrices rather than equations.

Solving Systems on a Graphing Calculator

ThomsonNOW
for Liberal Arts Mathematics
www.thomsonedu.com/1pass.

EXAMPLE 2   Use the Gauss-Jordan method and a graphing calculator to solve the following system:

$$2x + z = y + 9$$
$$x + 2y + z = 4$$
$$4x - 3z = -4$$

Solution

1. *Write the system with all variables on the left side of the equations and with all coefficients showing.* Specifically, what needs to be done?

    - The first equation has the variable y on the right side, and this method requires that all variables be on the left side. We'll move y to the left side by subtracting y from each side.
    - In the first equation, both z and y lack coefficients. We'll insert coefficients of 1. We'll do the same thing with the second equation.
    - Finally, the third equation lacks a y term. We'll insert $0y$.
      This gives us

    $$2x - 1y + 1z = 9$$
    $$1x + 2y + 1z = 3$$
    $$4x + 0y - 3z = -4$$

2. *Rewrite the system in matrix form.* Eliminate the letters, addition symbols, and equal symbols.

    $$\begin{array}{ccc} x & y & z \\ \end{array}$$
    $$\begin{bmatrix} 2 & -1 & 1 & 9 \\ 1 & 2 & 1 & 4 \\ 4 & 0 & -3 & -4 \end{bmatrix}$$

    Notice that we've labeled the first three columns x, y, and z. This is an optional way of keeping track of the origins of the numbers.

3. *Enter the matrix into your graphing calculator,* as matrix [A] (on a TI-83/84) or as matrix A (on a TI-86). This was discussed in Section 11.0.

    Figure 11.36
    Matrix [A] on a TI-83/84
    (a TI-86 is similar)

4. *Have your calculator put the matrix in reduced row echelon form (or "rref").* To do this, make the screen read "rref([A]" on a TI-83/84, or "rref A" on a TI-86 by typing:

TI-83:	MATRX , scroll to "MATH" and select "rref(" (You will have to use the down arrow button to find it.)	MATRX 1 ENTER
TI-84:	2ND MATRIX , scroll to "MATH" and select "rref(" (You will have to use the down arrow button to find it.)	MATRX 1 ENTER
TI-86:	2ND MATRX , select "OPS" (that is, F4) and then "rref" (that is, F5)	ALPHA A ENTER
	This generates "rref (" or "rref".	*This generates "[A]" or "A".*

Once "rref([A]" ("rref A" on a TI-86) is on the screen, press ENTER , and the screen will appear as in Figure 11.37.

Figure 11.37
Matrix $[A]$ in rref (reduced row echelon form) on a TI-83/84 (a TI-86 is similar)

5. *Rewrite the matrix in system form.* This is the reverse of step 2, in which we put the system in matrix form.

The matrix is

$$\begin{array}{ccc} x & y & z \end{array}$$
$$\begin{bmatrix} 1 & 0 & 0 & 2 \\ 0 & 1 & 0 & -1 \\ 0 & 0 & 1 & 4 \end{bmatrix}$$

This translates into the system

$$1x + 0y + 0z = 2$$
$$0x + 1y + 0z = -1$$
$$0x + 0y + 1z = 4$$

The system simplifies to

$$x = 2$$
$$y = -1$$
$$z = 4$$

So the solution is $(x, y, z) = (2, -1, 4)$.

6. *Check your answer by substituting it into the system.*

$$2x - 1y + 1z = 2 \cdot (2) - 1 \cdot (-1) + 1 \cdot (4) = 9 \checkmark$$
$$1x + 2y + 1z = 1 \cdot (2) + 2 \cdot (-1) + 1 \cdot (4) = 3 \checkmark$$
$$4x + 0y - 3z = 4 \cdot (2) + 0 \cdot (-1) - 3 \cdot (4) = -4 \checkmark$$

∎

The Gauss-Jordan Method and the Calculator

If we were to solve the system

$$2x - 1y + 1z = 9$$
$$1x + 2y + 1z = 3$$
$$4x + 0y - 3z = -4$$

with the elimination method, we might start by multiplying the second equation by -2 and adding the result to the first equation.

$2x - 1y + 1z =$	9	the first equation
$+\quad -2x - 4y - 2z =$	-6	-2 times the second equation
$0x - 5y - 1z =$	3	

The Gauss-Jordan version of this is to multiply the matrix's second row by -2 and add the result to the matrix's first row.

2	-1	1	9	the first row
$+\quad -2$	-4	-2	-6	-2 times the second row
0	-5	-1	3	

If we replace the second row of the original matrix with this result, we get

The original matrix:

$$\begin{bmatrix} 2 & -1 & 1 & 9 \\ 1 & 2 & 1 & 4 \\ 4 & 0 & -3 & -4 \end{bmatrix} \Rightarrow$$

After replacing the second row:

$$\begin{bmatrix} 2 & -1 & 1 & 9 \\ 0 & -5 & -1 & 3 \\ 4 & 0 & -3 & -4 \end{bmatrix}$$

This puts a 0 in row two, column one.

When you type "rref([A]," your graphing calculator does the above calculation or one like it. It also does many more similar calculations and obtains the final matrix:

$$\begin{bmatrix} 1 & 0 & 0 & 2 \\ 0 & 1 & 0 & -1 \\ 0 & 0 & 1 & 4 \end{bmatrix}$$

Exercises

In Exercises 13–24, solve the given system with the Gauss-Jordan method and a graphing calculator.

13. The system in Exercise 1
14. The system in Exercise 2
15. The system in Exercise 3
16. The system in Exercise 4
17. The system in Exercise 5
18. The system in Exercise 6
19. The system in Exercise 7
20. The system in Exercise 8
21. The system in Exercise 9
22. The system in Exercise 10
23. The system in Exercise 11
24. The system in Exercise 12

In Exercises 25–28, solve the given system with the Gauss-Jordan method but without using a graphing calculator.

25. $2x - 3y = 22$
    $6x + 7y = 2$

26. $3x + 5y = -17$
    $x + 8y = -31$

27. $4x - 6y = -10$
    $2x + 9y = 31$

28. $3x + 5y = 7$
    $x - 7y = -15$

11.5 More on Markov Chains

In this section, we will continue to make long-range predictions with Markov chains, as we did in Section 11.3. However, the problems in this section will involve more than two unknowns. The resulting systems can be solved by using either the elimination method or the graphing calculator and the Gauss-Jordan method, as discussed in Section 11.4.

ThomsonNOW™
for Liberal Arts Mathematics
www.thomsonedu.com/1pass.

EXAMPLE 1   Sociologists have found that a strong determinant of an individual's income is the income of his or her parents. In Example 8 of Section 11.1, we used transitional data on the lower income group, the middle income group, and the upper income group to create a transition matrix T. Use T to make a long-range prediction of the levels at which the lower, middle, and upper income groups will stabilize.

Solution   We are to find the equilibrium matrix L.

1. *Create the transition matrix T. We did this in Section 11.1.*

$$T = \begin{bmatrix} 0.78 & 0.21 & 0.01 \\ 0.06 & 0.90 & 0.04 \\ 0.01 & 0.10 & 0.89 \end{bmatrix}$$

2. *Create the equilibrium matrix L. L is a row matrix, and in this problem, L has three entries: one for lower income group, one for middle income group, and one for upper income group.*

$$L = \begin{bmatrix} x & y & z \end{bmatrix}$$

3. *Find and simplify the system of equations described by LT = L.*

$$\begin{matrix} L & & T & = & L \end{matrix}$$

$$\begin{bmatrix} x & y & z \end{bmatrix} \begin{bmatrix} 0.78 & 0.21 & 0.01 \\ 0.06 & 0.90 & 0.04 \\ 0.01 & 0.10 & 0.89 \end{bmatrix} = \begin{bmatrix} x & y & z \end{bmatrix}$$

$$\begin{bmatrix} 0.78x + 0.06y + 0.01z & 0.21x + 0.90y + 0.10z \\ 0.01x + 0.04y + 0.89z \end{bmatrix} = \begin{bmatrix} x & y & z \end{bmatrix}$$

This matrix equation describes the following system:

$$0.78x + 0.06y + 0.01z = x$$
$$0.21x + 0.90y + 0.10z = y$$
$$0.01x + 0.04y + 0.89z = z$$

This system can be simplified by combining like terms.

$$-0.22x + 0.06y + 0.01z = 0$$
$$0.21x - 0.10y + 0.10z = 0$$
$$0.01x + 0.04y - 0.11z = 0$$

The system can be further simplified by multiplying by 100.

$$-22x + 6y + 1z = 0$$
$$21x - 10y + 10z = 0$$
$$1x + 4y - 11z = 0$$

4. *Discard any redundant equations, and include the equation $x + y + z = 1$.* The third equation is the negative of the sum of the first two equations, so its presence is redundant. Thus, any one of the three equations can be dropped. Also, $L = [x \quad y \quad z]$ is a probability matrix, so the sum of its entries must be 1. That is, $x + y + z = 1$.

$$-22x + 6y + 1z = 0$$
$$21x - 10y + 10z = 0$$
$$x + y + z = 1$$

5. *Solve the resulting system.*
We'll combine the first and third equations and eliminate x. To do this, multiply the third equation by 22, and add the result to the first equation.

$$\begin{array}{ll} -22x + 6y + 1z = 0 & \text{the first equation} \\ + \quad 22x + 22y + 22z = 22 & \text{22 times the third equation} \\ \hline \quad 28y + 23z = 22 & \end{array}$$

Next, we'll combine two different equations so that x is eliminated. We'll multiply the third equation by -21 and add the result to the second equation.

$$\begin{array}{ll} 21x - 10y + 10z = 0 & \text{the second equation} \\ + \quad -21x - 21y - 21z = -21 & \text{-21 times the third equation} \\ \hline \quad -31y - 11z = -21 & \end{array}$$

Finally, we'll solve the system created by the above two calculations.

$$\begin{array}{ll} 28y + 23z = 22 & \text{the result of the first calculation} \\ -31y - 11z = -21 & \text{the result of the second calculation} \end{array}$$

We'll eliminate z if we multiply the first equation by 11 and the second by 23, and we'll add the results.

$$\begin{array}{ll} 308y + 253z = 242 & \text{11 times the first equation} \\ + \quad -713y - 253z = -483 & \text{23 times the second calculation} \\ \hline \quad -405y \qquad\quad = -241 & \\ \quad -405y/-405 \; = -241/-405 & \\ \quad y = 241/405 \;\; = 0.59506\ldots \approx 0.5951 & \end{array}$$

Substitute this into one of the above equations with only y and z in it to get z.

$$\begin{array}{ll} 28y + 23z = 22 & \text{an earlier result} \\ 28 \cdot (241/405) + 23z = 22 & \text{substituting} \\ 6748/405 + 23z = 8910/405 & \text{writing with a common denominator} \\ 23z = 8910/405 - 6748/405 & \text{solving for } z \\ 23z = 2162/405 & \\ z = \dfrac{2162}{405} \cdot \dfrac{1}{23} = \dfrac{94}{405} = 0.23209\ldots \approx 0.2321 & \end{array}$$

Substitute this into one of the original equations to get x.

$$\begin{array}{ll} x + y + z = 1 & \text{an original equation} \\ x + 241/405 + 94/405 = 1 & \text{substituting} \\ x + 335/405 = 405/405 & \text{writing with a common denominator} \\ x = 70/405 = 0.17283\ldots \approx 0.1728 & \end{array}$$

Markov Chains and the Real World: The Suburbanization of America

The suburbs are a big part of American life. This wasn't always the case, though. The suburbanization of America began immediately after World War II, when returning war veterans needed housing. Cheaper building materials and government-backed mortgages allowed homebuilders to meet this demand by building new housing developments where land was cheap and plentiful—in the suburbs. In the 1950s, 20 million people moved to developments such as the three Levittowns: Levittown, New York; Levittown, Pennsylvania; and Willingboro, New Jersey.

People are still moving from the cities to the suburbs. How big is this trend? How important is it? We'll find out in Exercise 9.

In the 1970s, the Midwest hit the economic skids. Factories were abandoned, and unemployment was rampant. Manufacturers were relocating in the South. As a result, there was a great migration from the Rust Belt to the Sun Belt. We'll find out in Exercise 12 whether this is still an important trend.

The resulting solution is

$$L = [70/405 \quad 241/405 \quad 94/405] \approx [0.1728 \quad 0.5951 \quad 0.2321]$$

6. *Check your work by verifying that LT = L.*

$$
\begin{array}{ccc}
L & T & = & L
\end{array}
$$

$$[70/405 \quad 241/405 \quad 94/405] \cdot \begin{bmatrix} 0.78 & 0.21 & 0.01 \\ 0.06 & 0.90 & 0.04 \\ 0.01 & 0.10 & 0.89 \end{bmatrix} \approx [0.1728 \quad 0.5951 \quad 0.2321]$$

Our solution checks. ✓

The solution, rounded to the nearest 1%, is $\begin{bmatrix} 17\% & 60 & 23\% \end{bmatrix}$. This means that if current trends continue, the lower income group will eventually stabilize at 17% of the population, the middle income group will stabilize at 60%, and the upper income group will stabilize at 23%. (Recall that in 1999, the lower income group was 12%, the middle income group was 62%, and the upper income group was 26%, according to the U.S. Census Bureau.) It is important to remember that this prediction is based on the assumption that the current trends in the shift of people between the income groups will continue.

11.5 Exercises

In Exercises 1–4, do the following.

a. Find the equilibrium matrix L by solving LT = L for L.
b. Check your solution to part (b) by verifying that LT = L.

1. $T = \begin{bmatrix} 0.3 & 0.2 & 0.5 \\ 0.1 & 0.8 & 0.1 \\ 0.4 & 0.3 & 0.3 \end{bmatrix}$ **2.** $T = \begin{bmatrix} 0.4 & 0.3 & 0.3 \\ 0.2 & 0.7 & 0.1 \\ 0.3 & 0.3 & 0.4 \end{bmatrix}$

3. $T = \begin{bmatrix} 1/2 & 1/4 & 1/4 \\ 2/3 & 1/6 & 1/6 \\ 0 & 0 & 1 \end{bmatrix}$

4. $T = \begin{bmatrix} 1/8 & 2/8 & 5/8 \\ 3/8 & 2/8 & 3/8 \\ 1/2 & 1/8 & 3/8 \end{bmatrix}$

5. An extensive survey of gym users in Metropolis indicates that 71% of the current members of Silver's will continue their annual membership when it expires, 12% will quit and join Fitness Lab, and the rest will quit and join ThinNFit. Fitness Lab has been unable to keep its equipment in good shape, and as a result 32% of its members will defect to Silver's, and 34% will leave for ThinNFit. ThinNFit's members are quite happy, and as a result, 96% plan on renewing their annual membership, with half of the rest planning on moving to Silver's and half to Fitness Lab. Make a long-range prediction of the ultimate market shares of the three health clubs, assuming that current trends continue. (See Exercise 11 in Section 11.1.)

6. Smallville has three Chinese restaurants: Asia Gardens, Chef Chao's, and Chung King Village. Chung King Village recently mailed a coupon to all residents of Smallville offering two dinners for the price of one. As a result, 67% of those who normally eat at Asia Gardens plan on trying Chung King Village within the next month, and 59% of those who normally eat at Chef Chao's plan on trying Chung King Village. Also, all of Chung King Village's regular customers will return to take advantage of the special. And 30% of the people who normally eat at Asia Gardens will eat there within the next month because of its convenient location. Chef Chao's has a new chef who isn't doing very well, and as a result, only 15% of the people who normally eat at there are planning on returning next time. Make a long-range prediction of the ultimate market shares of the three restaurants, assuming that current trends continue. (See Exercise 12 in Section 11.1.)

7. A survey shows that 23% of the shoppers in seven Midwestern states regularly buy their groceries at Safe Shop, 29% regularly shop at PayNEat, and the balance shop at any one of several smaller markets. The survey indicates that 8% of the consumers who currently shop at Safe Shop will purchase their groceries at PayNEat the next time they shop and that 5% will switch to some other store. Also, 12% of the consumers who currently shop at PayNEat will purchase their groceries at Safe Shop the next time they shop, and 2% will switch to some other store. In addition, 13% of the consumers who currently shop at neither store will purchase their groceries at Safe Shop they next time they shop, and 10% will shop at PayNEat. Predict the percentage of Midwestern shoppers that will be regular Safe Shop customers and the percent that will be regular PayNEat customers as a result of this trend. (See Exercise 18 in Section 11.1.)

8. In October 1990, a New York Times/CBS News poll contained the information shown in Figure 11.38. For example, 78% of those registered as a Democrat reported that they would vote for a Democrat in the next election.

   a. Complete the chart.
   b. What assumption must be made in completing the chart?
   c. If the poll were interpreted as predicting a change in registered party affiliation, make a long-range prediction of the percentage of voters who will be registered as Democrats, as Republicans, and as Independents.

	Next Vote Will Be for a Democrat	**Next Vote Will Be For a Republican**	**Next Vote Will Be for an Independent**
Registered as a Democrat	78%	6%	?
Registered as a Republican	6%	75%	?
Registered as an Independent	30%	29%	?

Figure 11.38
Party membership changes
Source: New York Times, October 12 1990, page A21.

d. Do you think that the prediction in part (b) would be a good prediction? Why?

PROJECTS

9. The U.S. Census Bureau classifies all residents of the United States as residents of a central city, a suburb, or a nonmetropolitan area. In the 1990 census, the bureau reported that the central cities held 77.8 million people, the suburbs 114.9 million people, and the nonmetropolitan areas 56.0 million people. The Census Bureau (in Current Population Reports, P20–463, *Geographic Mobility, March 1990 to March 1991*) also reported the information in Figure 11.39 regarding migration between these three areas from 1990 to 1991. (Numbers are in thousands.)

	Moved to		
Moved From	**Central City**	**Suburb**	**Nonmetropolitan Area**
Central City	(x)	4946	736
Suburb	2482	(x)	964
Nonmetropolitan Area	741	1075	(x)

Note: Persons moving from one central city to another were not considered. This is represented by the "(x)" in the upper-left corner. The other "(x)'s" have similar meanings.

Figure 11.39
Migration data for Exercise 9

a. Compute the probabilities that a resident of any one type of environment will move to any of the other types. Present these probabilities in the form

of a transition matrix. (Round off to three decimal places.) (See Exercise 21 in Section 11.1.)

b. Make a long-range prediction of the ultimate percentage of U.S. residents who will reside in a central city, a suburb, and a nonmetropolitan area, assuming that the trend indicated by the above data continues.

10. Snapdragons have no color dominance. A snapdragon with two red genes will have red flowers, a snapdragon with one red gene and one white gene will have pink flowers, and a snapdragon with two white genes will have white flowers. A commercial nursery crosses many snapdragons (some with red flowers, some with pink flowers, and some with white flowers) with pink-flowered plants.

a. Use Punnett squares to create a transition matrix for this situation. (Punnett squares are covered in Section 3.2.)

b. The nursery crosses snapdragons from each generation with pink-flowered plants to produce the next generation, over and over again. Find the percentages at which the snapdragon will stabilize.

11. In May 1992, the *New York Times* reported on a University of Michigan study on intergenerational transition between income groups. That study contained the information shown in Figure 11.40. For example, 30% of those whose fathers were in the lowest income group stayed in the lowest income group.

a. Make a long-range prediction of the percent of males in the various income groups.

b. The predictions in part (a) are not 10%, 40%, 40%, and 10%. Should they be? Explain. (*Hint:* Does "the bottom 10%" mean the same income bracket for the son as it does for the father?)

	Son's Income in Bottom 10%	Son's Income from 10% to 50%	Son's Income from 50% to 90%	Son's Income in Top 10%
Father's Income in Bottom 10%	30%	52%	17%	1%
Father's Income from 10% to 50%	10%	48%	38%	4%
Father's Income from 50% to 90%	4%	38%	48%	10%
Father's Income in Top 10%	1%	17%	52%	30%

Figure 11.40
Intergenerational income group transition data for Exercise 11
Source: New York Times, May 18 1992, page D5. Data have been adjusted so that percentages add to 100%.

c. What assumptions are made in the prediction in part (a)? Consider the fact that some men have no sons, some have one son, and some have many sons. Also consider the fact that only males are involved in the study.

12. The Census Bureau (in *Statistical Abstract of the United States, 2000*) reported the 2000 national and regional populations in Figure 11.41.

Region	Population (in thousands)
United States	281,422
Northeast	53,594
Midwest	64,393
South	100,237
West	63,198

Figure 11.41
U.S. regional population
Source: U.S. Census Bureau, Statistical Abstract of the United States, 2000.

The Census Bureau also reported (*Geographical Mobility: March 1999 to March 2000*) the information in Figure 11.42 regarding the movement between the regions from 1999 to 2000 (numbers are in thousands).

Region Moved from	Region Moved to			
	Northeast	Midwest	South	West
Northeast	(x)	83	405	127
Midwest	72	(x)	364	204
South	183	416	(x)	432
West	108	223	489	(x)

Figure 11.42
Geographical mobility
Source: U.S. Census Bureau, Current Population Reports P20–538, Geographical Mobility: March 1999 to 2000.

a. Compute the probabilities that a resident of any one region will move to any of the other regions. Present these probabilities in the form of a transition matrix. Assume that the transitional probabilities for 2000–2001 are equal to those for 1999–2000. (Round off to three decimal places.) (See Exercise 22 in Section 11.1.)

b. Make a long-range prediction of the ultimate percent of U.S. residents who will reside in the Northeast, the Midwest, the South, and the West, assuming that the trend described by the above data continues.

Chapter 11   Review

TERMS

associative property of matrix multiplication	element	linear equation	solving a system of equations
column	entry	Markov chain	
column matrix	equilibrium matrix	matrix	square matrix
current states	following states	matrix multiplication	states
diagonal	identity matrix	probability matrix	system of equations
dimensions of a matrix	identity property of matrix multiplication	row	transition matrix
		row matrix	

PROCEDURES

To make **short-range predictions** with Markov chains:

1. *Create the probability matrix P, which is a row matrix whose entries are the initial probabilities of the various states.*
2. *Create the transition matrix T, which is a square matrix whose entries are the probabilities of passing from current states to first following states.*
3. *Calculate PT^n, which is a row matrix whose entries are the probabilities of the nth following states.*

To make **long-range-predictions** with Markov chains:

1. *Create the transition matrix T.*
2. *Create the equilibrium matrix L, using variables.*
3. *Find and simplify the system of equations described by LT = L.*
4. *Discard any redundant equations and include the equation $x + y = 1$ (or $x + y + z = 1$ and so on).*
5. *Solve the resulting system.*
6. *Check your work by verifying that LT = L.*

REVIEW EXERCISES

In Exercises 1–4, determine whether the given matrix is a row matrix, a column matrix, a square matrix, or neither.

1. $\begin{bmatrix} 5 & 7 \\ 3 & 9 \\ 1 & 0 \end{bmatrix}$

2. $\begin{bmatrix} 6 & 3 & 8 \\ 1 & 6 & 3 \\ 4 & 9 & 6 \end{bmatrix}$

3. $\begin{bmatrix} 8 \\ 2 \\ 4 \\ 7 \end{bmatrix}$

4. $\begin{bmatrix} 4 & 67 & 43 & 27 \end{bmatrix}$

In Exercises 5–10, find the indicated product if it exists.

5. $\begin{bmatrix} 1 & 6 \\ 8 & 4 \end{bmatrix} \cdot \begin{bmatrix} -3 & 6 \\ 0 & -5 \end{bmatrix}$

6. $\begin{bmatrix} -3 & 6 \\ 0 & -5 \end{bmatrix} \cdot \begin{bmatrix} 1 & 6 \\ 8 & 4 \end{bmatrix}$

7. $\begin{bmatrix} 6 & 3 & 5 \\ 0 & 2 & 5 \end{bmatrix} \cdot \begin{bmatrix} -3 & 6 \\ 0 & -5 \end{bmatrix}$

8. $\begin{bmatrix} -3 & 6 \\ 0 & -5 \end{bmatrix} \cdot \begin{bmatrix} 6 & 3 & 5 \\ 0 & 2 & 5 \end{bmatrix}$

9. $\begin{bmatrix} 5 & 3 & -9 \\ 1 & 0 & 5 \end{bmatrix} \cdot \begin{bmatrix} 7 & 4 & 6 & 9 \\ 3 & 5 & -7 & 2 \\ 1 & 6 & 9 & 3 \end{bmatrix}$

10. $\begin{bmatrix} 7 & 4 & 6 & 9 \\ 3 & 5 & -7 & 2 \\ 1 & 6 & 9 & 3 \end{bmatrix} \cdot \begin{bmatrix} 5 & 3 & -9 \\ 1 & 0 & 5 \end{bmatrix}$

11. Johnco Industries is planning a new promotional campaign for its Veg-O-Slicer. Use matrix multiplication and Figure 11.43 to find the cost of three newspaper ads and four television ads in New York City and the cost of a similar campaign in Washington, D.C.

	Newspaper Ad	Television Ad
New York City	$5000	$7000
Washington, D.C.	$5500	$6200

Figure 11.43
Ad data for Exercise 11

12. Is matrix multiplication commutative? Is it associative?

In Exercises 13–18, do the following.

a. Solve the system of equations.
b. Check your answer by substituting it back in.

13. $3x + 5y = -14$
$4x - 7y = 36$

14. $10x - 11y = 28$
$5x + 7y = 39$

15. $5x - 7y = -29$
$2x + 4y = 2$

16. $8x - 7y = 54$
$3x + 15y = -15$

17. $7x - 8y + 3z = -43$
$5x + 3y - z = 24$
$11x - 5y + 12z = 0$

18. $5x - 4y + 3z = 10$
$6x - 3y - 3z = 9$
$11x - 15y + 10z = 13$

19. $P = [0.63 \quad 0.37]$, and $T = \begin{bmatrix} 0.3 & 0.7 \\ 0.1 & 0.9 \end{bmatrix}$.

   a. What is the probability of moving from state 1 (as a current state) to state 2 (as a first following state)?
   b. What is state 2's current market share?
   c. Find PT.
   d. Find PT^2.
   e. Draw a tree diagram illustrating P and T.

20. $P = [0.55 \quad 0.45]$, and $T = \begin{bmatrix} 0.4 & 0.6 \\ 0.5 & 0.5 \end{bmatrix}$.

   a. What is the probability of moving from state 1 (as a current state) to state 2 (as a first following state)?
   b. What is state 2's current market share?
   c. Find PT.
   d. Find PT^2.
   e. Draw a tree diagram illustrating P and T.

In Exercises 21–26, do the following.

a. Find the equilibrium matrix L by solving $LT = L$ for L.
b. Check your answer by verifying that $LT = L$.

21. $T = \begin{bmatrix} 0.7 & 0.3 \\ 0.2 & 0.8 \end{bmatrix}$

22. $T = \begin{bmatrix} 0.8 & 0.2 \\ 0.2 & 0.8 \end{bmatrix}$

23. $T = \begin{bmatrix} 0.4 & 0.6 \\ 0.1 & 0.9 \end{bmatrix}$

24. $T = \begin{bmatrix} 0.8 & 0.1 & 0.1 \\ 0.2 & 0.3 & 0.5 \\ 0.4 & 0.3 & 0.3 \end{bmatrix}$

25. $T = \begin{bmatrix} 0.5 & 0.1 & 0.4 \\ 0.2 & 0.2 & 0.6 \\ 0.4 & 0.3 & 0.3 \end{bmatrix}$

26. $T = \begin{bmatrix} 0.4 & 0.4 & 0.2 \\ 0.2 & 0.3 & 0.5 \\ 0.1 & 0.3 & 0.6 \end{bmatrix}$

In Exercises 27–30, use the following information. Department of Motor Vehicles records indicate that 12% of the automobile owners in the state of Jefferson own a Toyonda. A recent survey of Jefferson automobile owners, commissioned by Toyonda Motors, shows that 62% of the Toyonda owners would buy a Toyonda for their next car and that 16% of automobile owners who do not own a Toyonda would buy a Toyonda for their next car. The same survey indicates that automobile owners buy a new car an average of once every three years.

27. Predict Toyonda Motors' market share after three years.
28. Predict Toyonda Motors' market share after six years.
29. Make a long-range prediction of Toyonda Motors' market share.
30. On what assumptions are these predictions based?

In Exercises 31–33, use the following information. Currently, 23% of the residences of Foxtail County are apartments, 18% are condominiums or townhouses, and the balance are single-family houses. The Foxtail County Contractors' Association has commissioned a survey that shows that 3% of the apartment residents plan on moving to a condominium or townhouse within the next two years, and 5% plan on moving to a single family house. The same survey indicates that 1% of those who currently reside in a condominium or townhouse plan on moving to an apartment, and 11% plan on moving to a single family house; 2% of the single family house dwellers plan on moving to an apartment, and 4% plan on moving to a condominium or townhouse.

31. What recommendation should the FCCA make regarding the construction of apartments, condominiums, and single-family houses in the next four years?
32. What long-term recommendation should the FCCA make?
33. What important factors does the survey ignore?

12

Linear Programming

© Tim Wimborne/Reuters/Landov

Linear programming is a method of solving problems that involve a quantity to be maximized or minimized when that quantity is subject to certain restrictions. Linear programming was invented in the 1940s by George Dantzig, as a result of an Air Force research project concerned with computing the most efficient and economical way to distribute personnel, weapons, and supplies to the various fronts during World War II. The word *programming* in the term *linear programming* means creating a plan or procedure that solves a problem; it is not a reference to computer programming.

Since its conception, linear programming has been extremely successful. The petroleum industry was the first civilian industry to use it. Now refineries use linear programming methods to blend gasoline, decide what crude oil to buy, and determine what products to produce. The Exxon Corporation uses linear programming to determine which energy-savings projects to implement. The steel industry uses it to evaluate ores and to determine what products to produce and when to build new furnaces. Airlines use it to minimize costs related to the scheduling of flights, subject to constraints such as the amount of time a pilot or a crew may fly. The federal energy authorities use linear programming to explore policy alternatives. Various businesses and government agencies use it to determine the best way to control water and air pollution, assign personnel to jobs, and achieve racial balance in schools. Supermarket chains use it to determine which warehouses should ship which products to the stores. Investment companies use it to create portfolios with the best mix of stocks and bonds.

In this chapter, we focus on how small businesses can use linear programming to determine the best way to allocate their limited resources. For example, a craftsman who manufactures coffee tables and end tables by hand would have to decide how many tables of each type he should make. He would need to consider his limited resources of time and money as well as his desire to increase profit. Linear programming analyzes the effect of such limitations on the craftsman's business and helps him make an informed decision.

In large corporations, managers must decide how to allocate their limited resources (such as raw materials, labor, and machinery) to maximize their profit and meet other objectives. You might assume that such a corporation's resources are not really limited—that a manager could just determine a production level for each of her products and obtain the resources needed to meet those production levels. However, this is frequently not the case. Sufficient resources might not be available, or they might be limited by their location or their cost. Resources are usually limited, and their allocation is a difficult decision. Linear programming analyzes the effect of such limitations and helps corporate managers make informed decisions.

Thomson NOW™ **for Liberal Arts Mathematics**
Throughout this chapter, the ThomsonNOW logo indicates an opportunity for online self-study, linking you to interactive tutorials and videos based on your level of understanding.

The method of linear programming resulted from 1940s research to determine the least expensive way to distribute personnel and equipment to the various fronts of World War II.

Corporate managers have so many different products and resources to consider that the necessary calculations must be done by computer. It is estimated that linear programming accounts for at least as much computer time as does payroll or inventory control. And some experts claim that linear programming is the most widely used form of modern mathematics.

12.0   Review of Linear Inequalities

A **linear equation** in two variables x and y is an equation that can be written in the form $ax + by = c$, where a, b, and c are constants. Such an equation is referred to as a *linear* equation because its graph is a line. We can graph most linear equations by solving for y and using the slope-intercept formula $y = mx + b$ to find the slope m and the y-intercept b of the line.

A **linear inequality** in two variables x and y is an inequality that can be written in the form $ax + by < c$ (or with $>$, $\leq$, or $\geq$ instead of $<$). In other words, a linear inequality is the result of replacing a linear equation's equals symbol with an inequality symbol.

Our goal in this chapter is to **optimize** a quantity—that is, to maximize or minimize a quantity. A craftsman who manufactures coffee tables and end tables by hand would want to know how to invest his resources of time and money to maximize his profit. His resources are restricted; he wants to work *no more than* 40 hours each week, and he has *at most* $1,000 to spend on materials each week. These restrictions are described mathematically with inequalities, because they involve the phrases "no more than" and "at most." To analyze the craftsman's restricted resources and the effect of these restrictions on his profit, we must be able to graph the inequalities that describe the restrictions.

Each restriction in a linear programming problem must be expressible as a linear inequality; this is the meaning of the word *linear* in the term *linear programming*. In this section, we discuss the graphing of linear inequalities. We can graph a linear inequality by solving the inequality for y, graphing the line described by the associated equation, and then shading the region to one side of that line.

EXAMPLE 1   Graph the linear inequality $2x + y \leq 6$.

Solution

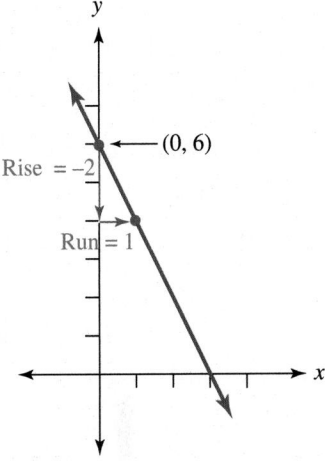

Figure 12.1
Graphing the line

1. *Solve the linear inequality for y.*

    $$2x + y \leq 6 \rightarrow y \leq -2x + 6$$

2. *Graph the line.* The equation associated with the inequality is $y = -2x + 6$. Comparing this equation with the slope-intercept formula $y = mx + b$, we find that this is the equation of a line with slope $m = -2$ and y-intercept $b = 6$. Because the slope is -2 and because slope means "rise over run," we have

    $$\frac{\text{rise}}{\text{run}} = -2 = \frac{-2}{1}$$

To graph the line, we place a point at 6 on the y-axis (since the y-intercept is 6) and then from that point rise -2 (that is, move two units down) and run 1 (that is, move one unit to the right). This takes us to a new point. We connect the points with a line. Because "=" is a part of "$\leq$," any point on the line $y = -2x + 6$ must also be a point on the graph of the inequality $y \leq -2x + 6$. We show this by using a solid line, as in Figure 12.1. (If our inequality were $y < -2x + 6$, then a point on the line would *not* be a point on the graph of the inequality. We would show this by using a dashed line.)

3. *Shade in one side of the line.* Two types of points satisfy the inequality $y \leq -2x + 6$: points that satisfy $y = -2x + 6$ and points that satisfy $y < -2x + 6$.

    Points that satisfy $y = -2x + 6$ were graphed in step 2, when we graphed the line. Points that satisfy y *is less than* $-2x + 6$ are the points *below* the line, because values of y decrease if we move down and increase if we move up, as shown in Figure 12.2.

    Thus, to graph the inequality $y \leq -2x + 6$, we make the line solid and shade in the region *below* the line, as shown in Figure 12.3. The solution of $y \leq -2x + 6$ is the set of all points on or below the line; any point on or below the line will successfully substitute into the inequality, and any point above the line will not. This region is called the **region of solutions** of the inequality.   ∎

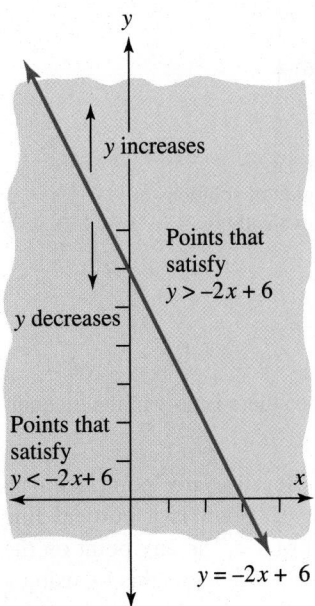

Figure 12.2
The two sides of
$y = -2x + 6$

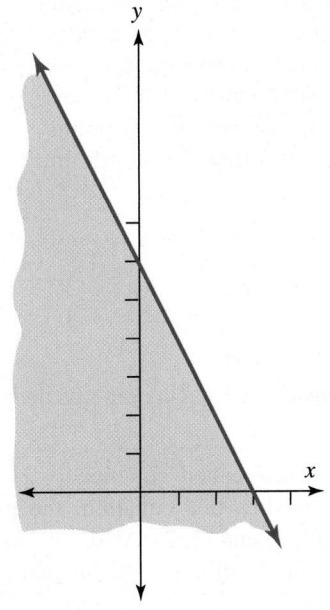

Figure 12.3
The region of solution of $y \leq -2x + 6$
(or equivalently of $2x + y \leq 6$)

EXAMPLE 2 Graph $3x - 2y < 12$.

Solution

1. *Solve the linear inequality for y.*

$$3x - 2y < 12$$
$$-2y < -3x + 12$$
$$\frac{-2y}{-2} > \frac{-3x + 12}{-2} \qquad \text{Multiplying or dividing by a negative reverses the direction of an inequality}$$
$$y > \frac{-3x}{-2} + \frac{12}{-2} \qquad \text{Distributing } -2$$
$$y > \frac{3}{2}x - 6$$

2. *Graph the line.* The associated equation is $y = \frac{3}{2}x - 6$, which is a line with slope $m = \frac{3}{2}$ and y-intercept $b = -6$.

$$m = \frac{3}{2} \rightarrow \frac{\text{rise}}{\text{run}} = \frac{3}{2}$$

To graph the line, we place a point at -6 on the y-axis and from that point rise 3 and run 2. Because "=" is *not* part of " > ," a point on the line $y = \frac{3}{2}x - 6$ is *not* a point on the graph of the inequality $y > \frac{3}{2}x - 6$. We show this by using a dashed line.

3. *Shade in one side of the line.* Because values of *y increase* if we move upward and because we want to graph where *y is greater than* $\frac{3}{2}x - 6$, we shade in the region above the dashed line. The region of solutions of the inequality is the set of all points above (but not on) the line, as shown in Figure 12.4.

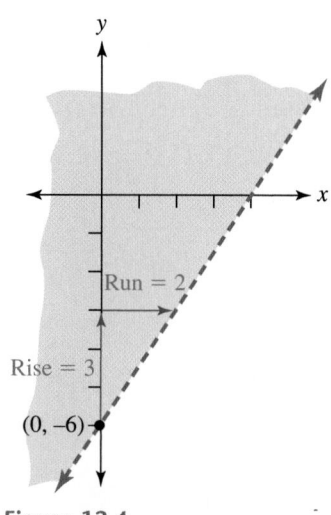

Figure 12.4
The region of solutions of $y > \frac{3}{2}x - 6$ (or equivalently of $3x - 2x < 12$) ∎

Thomson**NOW**
for Liberal Arts Mathematics
www.thomsonedu.com/1pass.

EXAMPLE 3 Graph $x \geq 3$.

Solution

1. *Solve the linear inequality for y.* This can't be done, since there is no *y* in the inequality $x \geq 3$.

2. *Graph the line.* The associated equation is $x = 3$. Because any point with an x-coordinate of 3 satisfies this equation, the graph of $x = 3$ is a vertical line through (3, 0) and (3, 1) and (3, 2). Also, "=" is part of "≥," so any point on the line $x = 3$ is a point on the graph of the inequality $x \geq 3$. We show this by using a solid line.

3. *Shade in one side of the line.* Values of *x* increase if we move to the right and decrease if we move to the left. The points on the line are the points where *x* equals 3, and the points to the right of the line are the points where *x* is greater than 3. Thus,

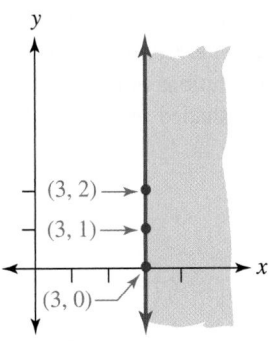

Figure 12.5
The region of solutions of $x \geq 3$

the region of solutions of the inequality is the set of all points on or to the right of the line, as shown in Figure 12.5. ■

Graphing the Region of Solutions of a Linear Inequality

1. *Solve the linear inequality for y.* This puts the inequality in slope-intercept form. If the inequality has no y, then solve the inequality for x. If you multiply or divide each side by a negative number, reverse the direction of the inequality.
2. *Graph the line.* The line is described by the equation associated with the inequality.

    If the equation is: then the line:

    in slope-intercept form $(y = mx + b)$ . . . .has slope m and y-intercept b.

    in the form $x = a$ . . . . .is a vertical line through $(a, 0)$.

    If the inequality is: then the line:

    $\leq$ or $\geq$ . . . . .is part of the region of solutions
    —use a solid line.

    $<$ or $>$ . . . . .is not part of the region of
    solutions—use a dashed line.

3. *Shade in one side of the line.*
    y increases as you move up, and decreases as you move down.
    x increases as you move to the right, and decreases as you move to the left.

Systems of Linear Inequalities

The inequalities we deal with in this chapter come from restrictions. Usually, a linear programming problem involves more than one restriction and therefore more than one inequality. A **system of linear inequalities** is a set of more than one linear inequality. The **region of solutions** of a system of linear inequalities is the set of all points that simultaneously satisfy each inequality in the system.

To graph the region of solutions of a system of linear inequalities, we graph each inequality on the same axes and shade in the intersection of their solutions.

EXAMPLE 4   Graph the following system of inequalities:

$$x + y \geq 3$$
$$-x + 2y \leq 0$$

Solution   With the aid of the chart in Figure 12.6, we can graph the two inequalities on the same axes, as shown in Figure 12.7.

Original Inequality	Slope-Intercept Form	Associated Equation	Graph of the Inequality
$x + y \geq 3$	$y \geq -x + 3$	$y = -1x + 3$	all points on or above the line with slope -1 and y-intercept 3
$-x + 2y \leq 0$	$2y \leq x \rightarrow y \leq \frac{1}{2}x$	$y = \frac{1}{2}x + 0$	all points on or below the line with slope $\frac{1}{2}$ and y-intercept 0

Figure 12.6

The region of solutions of the system is the set of all points that satisfy both the first *and* the second inequality. As we saw in Chapter 2, *intersection* is the set-theoretical word that corresponds to *and*, so the region of solutions is the intersection of the graphs of the two inequalities. This is the region shaded in Figure 12.8; it includes the solid lines bounding that region.

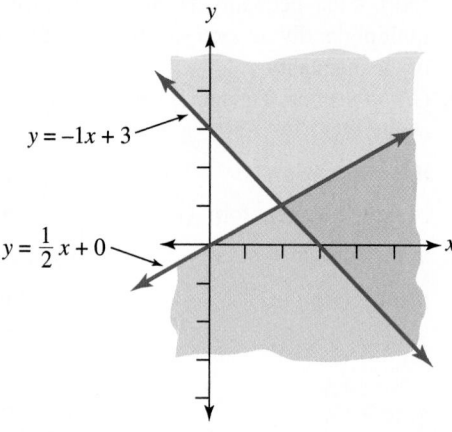

Figure 12.7
The graphs of $x + y \geq 3$ and $-x + 2y \leq 0$, on the same axes

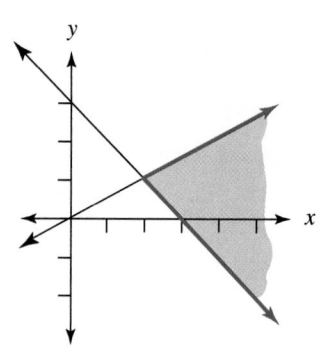

Figure 12.8
The region of solutions of the system
$$x + y \geq 3$$
$$-x + 2y \leq 0 \qquad \blacksquare$$

EXAMPLE 5   Graph the following system of inequalities:

$$2x + y \leq 8$$
$$x + 2y \leq 10$$
$$x \geq 0$$
$$y \geq 0$$

Solution   With the aid of the chart in Figure 12.9, we can graph the two inequalities on the same axes.

Original Inequality	Slope-Intercept Form	Associated Equation	Graph of the Inequality
$2x + y \leq 8$	$y \leq -2x + 8$	$y = -2x + 8$	all points on or below the line with slope -2 and y-intercept 8
$x + 2y \leq 10$	$2y \leq -x + 10 \rightarrow$ $y \leq -\frac{1}{2}x + 5$	$y = -\frac{1}{2}x + 5$	all points on or below the line with slope $-\frac{1}{2}$ and y-intercept 5
$x \geq 0$	(not applicable)	$x = 0$	all points on or to the right of the y-axis
$y \geq 0$	$y \geq 0$	$y = 0$	all points on or above the x-axis

Figure 12.9
Getting ready to graph the region of solutions

The inequalities $x \geq 0$ and $y \geq 0$ appear frequently in linear programming problems. They tell us that our graph is in the first quadrant.

The region of solutions of the system is the intersection of the graph of each individual inequality. It is shown in Figure 12.10.

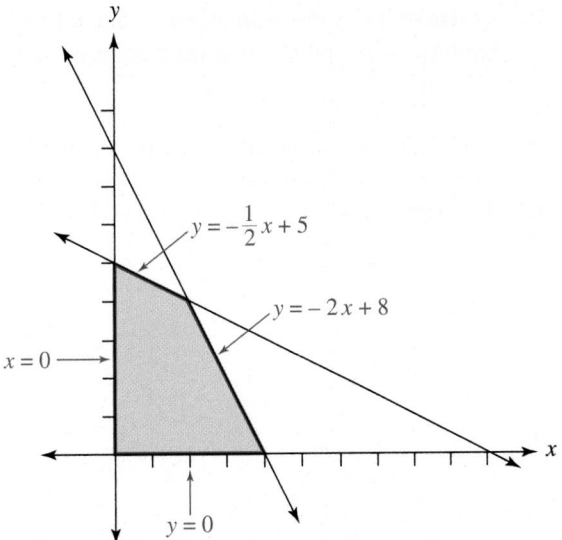

Figure 12.10
The region of solutions of the system
$$2x + y \leq 8$$
$$x + 2y \leq 10$$
$$x \geq 0$$
$$y \geq 0$$

The region of solutions in Example 4 is different from that of Example 5 in that the former is not totally enclosed. For that reason, it is called an **unbounded region.** Regions that are totally enclosed, like that in Example 5, are called **bounded regions.** See Figure 12.11. When we graph a system of inequalities as part of a linear programming problem, we must analyze unbounded regions differently than we do bounded regions.

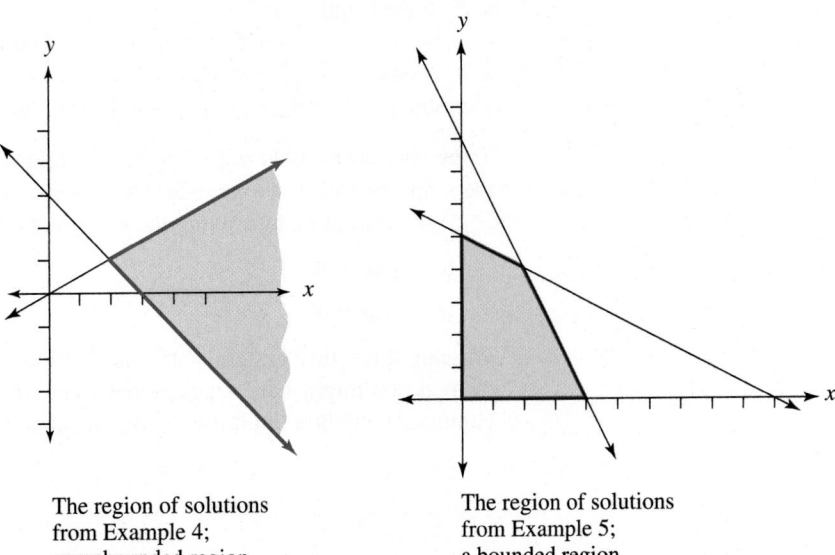

The region of solutions
from Example 4;
an unbounded region

The region of solutions
from Example 5;
a bounded region

Figure 12.11
Comparing bounded and unbounded regions.

Finding Corner Points

Our work in linear programming will involve graphing the region of solutions of a system of linear inequalities, as we did in Example 5. It will also involve finding the region's corner points. A **corner point** is a point that is at a corner of the region of solutions.

EXAMPLE 6   Find the corner points of the region of solutions in Example 5.

Solution   The region has four corner points, as shown in Figure 12.12.

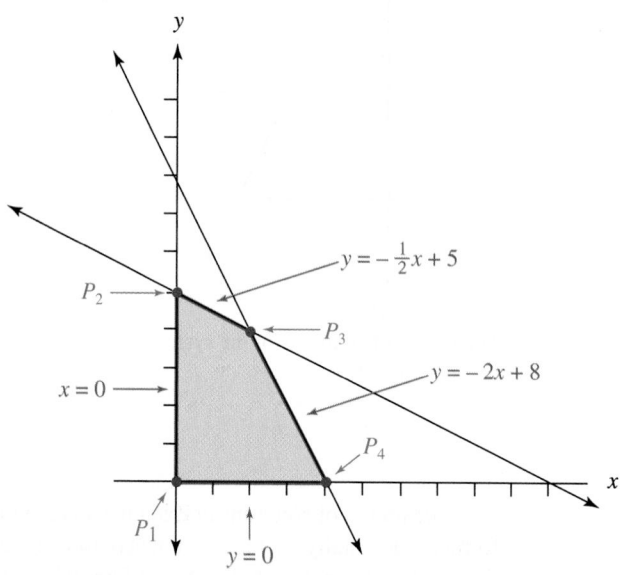

Figure 12.12
Corner points

- P_1 at the origin
- P_2 where $y = -\frac{1}{2}x + 5$ intersects $x = 0$ (the y-axis)
- P_3 where $y = -\frac{1}{2}x + 5$ intersects $y = -2x + 8$
- P_4 where $y = -2x + 8$ intersects $y = 0$ (the x-axis)

There is no work to do to find points P_1 and P_2. Point P_1 is clearly $(0, 0)$. Point P_2 is the y-intercept of $y = -\frac{1}{2}x + 5$; that y-intercept is $b = 5$, so P_2 is $(0, 5)$.

We can find P_3 by solving the system of equations

$$y = -\frac{1}{2}x + 5$$
$$y = -2x + 8$$

We can solve this system with the elimination method or with the substitution method you might have encountered in intermediate algebra. To proceed with the elimination method, multiply the second equation by -1 and add the results.

$$
\begin{array}{rl}
y = & -\dfrac{1}{2}x + 5 \\
+ \quad -y = & 2x - 8 \qquad \text{−1 times the second equation} \\
\hline
0 = & \dfrac{3}{2}x - 3
\end{array}
$$

$$\frac{3}{2}x = 3$$

$$\frac{2}{3} \cdot \frac{3}{2}x = \frac{2}{3} \cdot 3$$

$$x = 2$$

We can find y by substituting $x = 2$ into either of the original equations.

$$y = -2x + 8$$

$$y = -2 \cdot 2 + 8 = 4$$

Thus, P_3 is (2, 4).

We can find P_4 by solving the system of equations

$$y = -2x + 8$$

$$y = 0 \qquad \text{the equation of the } x\text{-axis}$$

To proceed with the elimination method, multiply the second equation by -1 and add the results.

$$y = -2x + 8$$
$$\underline{+ \;\; -y = \qquad\quad 0} \quad \text{-1 times the second equation}$$
$$0 = -2x + 8$$
$$2x = 8$$
$$x = 4$$

Since P_4 is on the x-axis, its y-coordinate must be 0. Thus, P_4 is (4, 0). The region of solutions' four corner points are

- P_1 at (0, 0)
- P_2 at (0, 5)
- P_3 at (2, 4)
- P_4 at (4, 0)   ■

12.0 Exercises

In Exercises 1–8, graph the region of solutions of the given linear inequality.

1. $3x + y < 4$
2. $8x + y > 2$
3. $4x - 3y \leq 9$
4. $5x - 2y \geq 6$
5. $x \geq 4$
6. $x \leq -3$
7. $y \leq -4$
8. $y \geq -2$

In Exercises 9–21, (a) graph the region of solutions of the given system of linear inequalities, (b) determine whether the region of solutions is bounded or unbounded, and (c) find all of the region's corner points.

9. $y > 2x + 1$
   $y \leq -x + 4$

10. $y < -2x + 6$
    $y \geq -x + 7$

11. $2x + 3y < 17$
    $3x - y \geq -2$

12. $5x - y \geq 7$
    $2x - 3y < -5$

13. $x + 2y \leq 4$
    $3x - 2y \leq -12$
    $x - y < -7$

14. $x - y + 1 \geq 0$
    $3x + 2y + 8 \geq 0$
    $3x - y < 6$

15. $2x + 5y \leq 70$
    $5x + y \leq 60$
    $x \geq 0$
    $y \geq 0$

16. $x + 20y \leq 460$
    $21x + y \leq 861$
    $x \geq 0$
    $y \geq 0$

17. $15x + 22y \leq 510$
    $35x + 12y \leq 600$
    $x + y > 10$
    $x \geq 0$
    $y \geq 0$

18. $3x + 20y \leq 2{,}200$
    $19x + 9y \leq 2{,}755$
    $2x + y \leq 120$
    $x \geq 0$
    $y \geq 0$

19. $0.50x + 1.30y \le 2.21$
$6x + y \le 9$
$0.7x + 0.6y < 3.00$
$x \ge 0$
$y \ge 0$

20. $3.70x + 0.30y \le 1.17$
$0.10x + 2.20y \le 0.47$
$x \ge 0$
$y \ge 0$

21. $x - 2y + 16 \ge 0$
$3x + y \le 30$
$x + y \le 14$
$x \ge 0$
$y \ge 0$

CONCEPT QUESTIONS

22. Make up a system of inequalities with a region of solutions that is bounded. Do not use any of the systems discussed in the text.

23. Make up a system of inequalities with a region of solutions that is unbounded. Do not use any of the systems discussed in the text.

24. Make up a system of inequalities with a region of solutions that is the triangle with corner points $(0, 0)$, $(10, 0)$, and $(0, 20)$.

25. Make up a system of inequalities with a region of solutions that has corner points $(0, 0)$, $(20, 0)$, $(10, 40)$, and $(0, 50)$.

Answer the following questions with complete sentences and your own words.

26. Why do we describe the solution of a system of linear inequalities with a graph, rather than a list of points?

27. List three industries that routinely use linear programming, and give examples of how they use it.

28. Why do large corporations have to be concerned about "limited" resources?

Graphing Linear Inequalities on a Graphing Calculator

A Texas Instruments graphing calculator can perform all of the specific tasks that are part of graphing the region of solutions of a system of linear inequalities: It can graph the lines associated with the system, shade in the appropriate sides of those lines, and locate corner points. However, the shading feature is cumbersome to use; it is much easier to do the shading by hand and to use the calculator to graph the lines associated with the inequalities and to find the corner points. If you wish to explore the use of the shading feature command, consult your calculator's operating manual.

Graphing a Linear Inequality

EXAMPLE 7 Graph $3x - 2y < 12$.

Solution

1. *Solve the linear inequality for y.*

$$3x - 2y < 12 \rightarrow -2y < -3x + 12 \rightarrow y > \frac{3}{2}x - 6 \quad \text{Multiplying or dividing by a negative reverses the direction of the inequality.}$$

2. *Graph the line,* using the procedure discussed in Appendix C. The line associated with the inequality is $y = \frac{3}{2}x - 6$. Enter 3/2*x − 6 for Y_1. Be sure to include the "*." Some calculators would interpret 3/2x − 6 as $\frac{3}{2x} - 6$. Be sure that no other equations are selected. Use the standard viewing window. The line's graph is shown in Figure 12.13.

3. *Copy the line's graph onto paper.* The inequality is a ">" inequality, so use a dashed line.

4. *By hand, shade in one side of the line.* To graph $y > \frac{3}{2}x - 6$, shade in the region above the line $y = \frac{3}{2}x - 6$. The region's graph is shown in Figure 12.14.

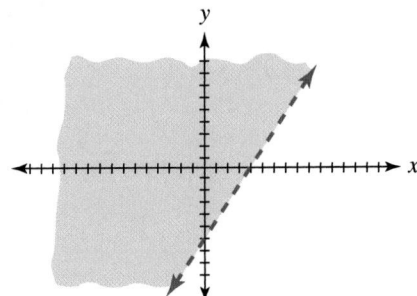

Figure 12.13
The graph of $y = \frac{3}{2}x - 6$

Figure 12.14
The graph of $y > \frac{3}{2}x - 6$ ∎

Graphing a System of Linear Inequalities

EXAMPLE 8   Graph the following system of inequalities:

$$x + y \geq 3$$
$$-x + 2y \leq 0$$

Solution

1. *Solve the linear inequalities for y.*

$$x + y \geq 3 \quad \rightarrow \quad y \geq -x + 3$$
$$-x + 2y \leq 0 \quad \rightarrow \quad y \leq \frac{1}{2}x$$

2. *Graph the lines,* using the procedure discussed in Appendix C. Enter $-x + 3$ for Y_1 and 1/2*x for Y_2. Be sure that no other equations are selected. Use the standard viewing window. See Figure 12.15.

3. *Copy the lines' graphs onto paper.* The inequalities are both "≤ or ≥" inequalities, so use solid lines.

4. *By hand, shade in the region of solutions.* The region of solutions is above the line $Y_1 = -x + 3$, because the associated inequality is $y \geq -x + 3$, and y increases as you move up. The region of solutions is also below the line $Y_2 = \frac{1}{2}x$, because the associated inequality is $y \leq \frac{1}{2}x$, and y decreases as you move down. The region of solutions is the region that is both above the line $Y_1 = -x + 3$ and below the line $Y_2 = \frac{1}{2}x$. It is shown in Figure 12.16.

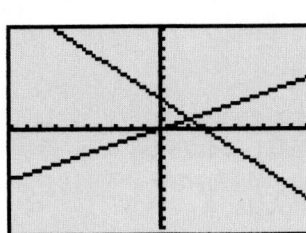

Figure 12.15
The graphs of $x + y = 3$ and
$-x + 2y = 0$

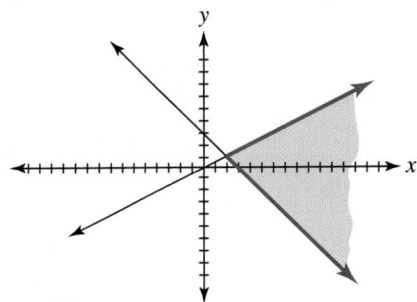

Figure 12.16
The graph of the system
$$x + y \geq 3$$
$$-x + 2y \leq 0$$ ∎

Nonstandard Viewing Windows

The previous examples were chosen so that the standard viewing window would be appropriate. In the following example, we must determine an appropriate viewing window.

EXAMPLE 9   Graph the system of inequalities:

$$2x + y \le 8$$
$$x + 2y \le 10$$
$$x \ge 0$$
$$y \ge 0$$

Solution

$$2x + y \le 8 \quad \rightarrow \quad y \le -2x + 8$$
$$x + 2y \le 10 \quad \rightarrow \quad 2y \le -x + 10 \quad \rightarrow \quad y \le -\frac{1}{2}x + 5$$

$x \ge 0$ cannot be solved for y, and $y \ge 0$ is already solved for y.

Enter $-2x + 8$ for Y_1 and $-1/2*x + 5$ for Y_2. Be sure that no other equations are selected. The last pair of inequalities tells us that the region of solutions is in the first quadrant; if we set Xmin and Ymin equal to -1, we'll leave ourselves a little extra room. By inspecting the two lines' equations, we can tell that the largest y-intercept is 8, so we'll set Ymax equal to 8. Substituting 0 for y in each equation gives x-intercepts of $(4, 0)$ and $(10, 0)$, so we'll set Xmax equal to 10. The resulting graph is shown in Figure 12.17.

The region of solutions is below the line $Y_1 = -2x + 8$, because the associated inequality is $y \le -2x + 8$, and y decreases as you move down. The region of solutions is also below the line $Y_2 = -\frac{1}{2}x + 5$, because the associated inequality is $y \le -\frac{1}{2}x + 5$. The region of solutions is the part of the first quadrant that is both on or below the line $Y_1 = -2x + 8$ and on or below the line $Y_2 = -\frac{1}{2}x + 5$. It is shown in Figure 12.18.

Figure 12.17
The graph of $2x + y = 8$ and $x + 2y = 10$

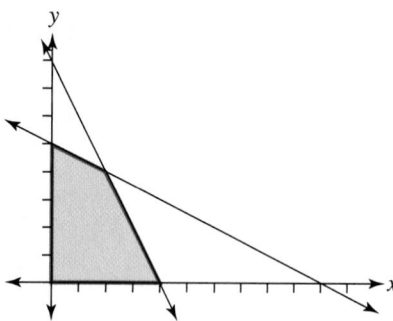

Figure 12.18
The graph of the system
$$2x + y \le 8$$
$$x + 2y \le 10$$
$$x \ge 0$$
$$y \ge 0$$

Finding Corner Points

After using a graphing calculator to graph a system's region of solutions, you can use the calculator to find the corner points as well.

EXAMPLE 10   Find the corner points for the region graphed in Example 9.

Solution This region has four corner points. One is clearly located at (0, 0). A second is the y-intercept of the line $y = -\frac{1}{2}x + 5$; by inspecting the equation, we can tell that it is (0, 5).

A third corner point is the intersection of the two lines; it can be found using the procedure discussed in Appendix D (summarized below).

📟	TI-83/84	Select "intersect" from the "CALC" menu.
	TI-86	Select "MATH" from the "GRAPH" menu, and then "ISECT" from the "GRAPH MATH" menu.
	Casio	PRESS G-SOLV and then ISCT.

Figure 12.19
Finding the intersection

This third corner point is (2, 4), as shown in Figure 12.19.

The fourth corner point is the x-intercept or root of the line $y = -2x + 8$; it can be found using the procedure discussed in Appendix C, Exercises 8 and 9 (summarized below).

📟	TI-83/84	Select "root" or "zero" from the "CALC" menu.
	TI-86	Select "MATH" from the "GRAPH" menu, and then "ROOT" from the "GRAPH MATH" menu.
	Casio	PRESS G-SOLV and then ROOT.

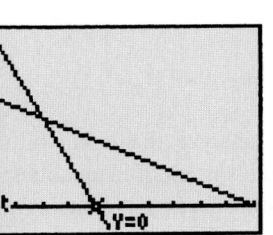

Figure 12.20

This fourth corner point is (4, 0), as shown in Figure 12.20. ∎

Exercises

In Exercises 29–36, use a graphing calculator to graph the region of solutions of the inequality given earlier in this section.

29. Exercise 1   **30.** Exercise 2
31. Exercise 3   **32.** Exercise 4
33. Exercise 5   **34.** Exercise 6
35. Exercise 7   **36.** Exercise 8

In Exercises 37–49, use a graphing calculator to graph the region of solutions of the system of

inequalities given earlier in this section, and to find the corner points.

37. Exercise 9   **38.** Exercise 10
39. Exercise 11   **40.** Exercise 12
41. Exercise 13   **42.** Exercise 14
43. Exercise 15   **44.** Exercise 16
45. Exercise 17   **46.** Exercise 18
47. Exercise 19   **48.** Exercise 20
49. Exercise 21

12.1   The Geometry of Linear Programming

As we stated in the introduction to this chapter, linear programming is a method for solving problems in which a quantity is to be maximized or minimized, when that quantity is subject to various restrictions. The following is a typical linear programming problem.

A craftsman produces two products: coffee tables and end tables. Production of one coffee table requires six hours of his labor, and the materials cost him $200. Production of one end table requires five hours of labor, and the materials cost him $100. The craftsman wants to work no more than 40 hours each week, and his financial resources allow him to pay no more than $1,000 for materials each week. If he can sell as many tables as he can make and if his profit is $240 per coffee table and $160 per end table, how many coffee tables and how many end tables should he make each week to maximize weekly profit?

Any linear programming problem has three features: *variables,* an *objective,* and *constraints.* In the problem above, the **variables** (or quantities that can vary) are the following:

- The number of coffee tables made each week
- The number of end tables made each week
- The number of hours the craftsman works each week
- The amount of money he spends on materials each week
- The weekly profit

The last three variables depend on the first two, so they are called the **dependent variables,** whereas the first two are called the **independent variables.**

The craftsman's objective is to maximize profit. The **objective function** is a function that mathematically describes the profit.

The **constraints** (or restrictions) are as follows:

- The craftsman's weekly hours ≤ 40
- The craftsman's weekly expenses ≤ $1,000

The constraints form a system of inequalities. To analyze the effect of these constraints on the craftsman's profit, we must graph the system of inequalities. The resulting graph is called the **region of possible solutions,** because it contains all the points that could *possibly* solve the craftsman's problem.

How should a craftsman allocate his time and money to maximize profit?

Creating a Model

A **model** is a mathematical description of a real-world situation. In this section, we discuss how to model a linear programming problem (that is, how to translate it into mathematical terms), how to find and graph the region of possible solutions, and how to analyze the effect of the constraints on the objective and solve the problem.

EXAMPLE 1   Model the linear programming problem from the beginning of this section and graph the region of possible solutions. The problem is summarized as follows:

A craftsman produces two products: coffee tables and end tables. Production data are given in Figure 12.21. If the craftsman wants to work no more than 40 hours each week and if his financial resources allow him to pay no more than $1,000 for materials each week, how many coffee tables and how many end tables should he make each week to maximize weekly profit?

	Labor (per table)	Cost of Materials (per table)	Profit (per table)
coffee tables	6 hours	$200	$240
end tables	5 hours	$100	$160

Figure 12.21
Data for Example 1

Solution

1. *List the independent variables.* We have already done this. If we call them x and y, the independent variables are

   x = number of coffee tables made each week

   y = number of end tables made each week

2. *List the constraints and translate them into linear inequalities.* We have already determined that the constraints (or restrictions) are as follows:

   the craftsman's weekly hours ≤ 40

   the craftsman's weekly expenses $\leq \$1,000$

   We need to translate these constraints into linear inequalities. First, let's translate the time constraint.

   hours ≤ 40

   (coffee table hours) $+$ (end table hours) ≤ 40

   $\left(\begin{array}{c}\text{6 hours per}\\\text{coffee table}\end{array}\right) \cdot \left(\begin{array}{c}\text{number of}\\\text{coffee tables}\end{array}\right) + \left(\begin{array}{c}\text{5 hours per}\\\text{end table}\end{array}\right) \cdot \left(\begin{array}{c}\text{number of}\\\text{end tables}\end{array}\right) \leq 40$

   $\quad\quad 6 \quad\quad\quad\quad\quad\quad x \quad\quad\quad + \quad\quad 5 \quad\quad\quad\quad\quad\quad y \quad\quad\quad \leq 40$

   Next, we'll translate the money constraint.

   money spent $\leq 1,000$

   (coffee table money) $+$ (end table money) $\leq 1,000$

   $\left(\begin{array}{c}\$200\text{ per}\\\text{coffee table}\end{array}\right) \cdot \left(\begin{array}{c}\text{number of}\\\text{coffee tables}\end{array}\right) + \left(\begin{array}{c}\$100\text{ per}\\\text{end table}\end{array}\right) \cdot \left(\begin{array}{c}\text{number of}\\\text{end tables}\end{array}\right) \leq 1,000$

   $\quad\quad 200 \quad\quad\quad\quad\quad\quad x \quad\quad\quad + \quad\quad 100 \quad\quad\quad\quad\quad y \quad\quad\quad \leq 1,000$

   There are two more constraints. Both x and y count things (the number of tables), so neither can be negative; therefore,

   $x \geq 0 \quad$ and $\quad y \geq 0$

   Our constraints are

   $6x + 5y \leq 40$

   $200x + 100y \leq 1,000$

   $x \geq 0 \quad$ and $\quad y \geq 0$

3. *Find the objective and translate it into a linear equation.* The objective is to maximize profit. If we let z = profit, we get

$$z = \text{(coffee table profit)} + \text{(end table profit)}$$

$$= \left(\begin{array}{c}\$240 \text{ per} \\ \text{coffee table}\end{array}\right) \cdot \left(\begin{array}{c}\text{number of} \\ \text{coffee tables}\end{array}\right) + \left(\begin{array}{c}\$160 \text{ per} \\ \text{end table}\end{array}\right) \cdot \left(\begin{array}{c}\text{number of} \\ \text{end tables}\end{array}\right)$$

$$= 240 \cdot x + 160 \cdot y$$

This equation is our objective function.

Steps 1 through 3 yield the model, or mathematical description, of the problem.

Independent Variables

x = number of coffee tables

y = number of end tables

Constraints

$6x + 5y \leq 40$ the time constraint

$200x + 100y \leq 1{,}000$ the money constraint

$x \geq 0$ and $y \geq 0$

Objective Function

$z = 240x + 160y$ z measures profit

4. *Graph the region of possible solutions.* With the aid of the chart in Figure 12.22, we can graph the two inequalities on the same axes, as shown in Figure 12.23.

Original Inequality	Slope-Intercept Form	Associated Equation	Graph of the Inequality
$6x + 5y \leq 40$	$5y \leq -6x + 40$ $\rightarrow y \leq -\frac{6}{5}x + 8$	$y = -\frac{6}{5}x + 8$	all points on or below the line with slope $-6/5$ and y-intercept 8
$200x + 100y \leq 1{,}000$	$100y \leq -200x + 1{,}000$ $\rightarrow y \leq -2x + 10$	$y = -2x + 10$	all points on or below the line with slope -2 and y-intercept 10
$x \geq 0$	(not applicable)	$x = 0$	all points on or to the right of the y-axis
$y \geq 0$	$y \geq 0$	$y = 0$	all points on or above the x-axis

Figure 12.22
Getting ready to graph the region of possible solutions

The last two constraints tell us that the region of possible solutions is in the first quadrant. The region of possible solutions is shown in Figure 12.23. ∎

Analyzing the Graph

The region of solutions in Figure 12.23 is a graph of the possible numbers of coffee tables and end tables the craftsman could make in one week without violating his constraints. That means that it consists of all the points at which the craftsman's weekly hours are no more than 40 and his weekly expenses are no more than $1000.

By inspecting the graph in Figure 12.23, we can see that (2, 1) and (3, 2) are clearly in the system's region of solutions. That means that each point should satisfy all of the constraints. Let's verify that that happens.

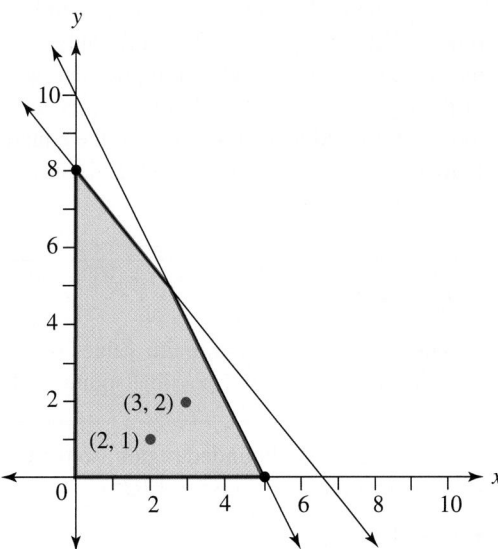

Figure 12.23
The region of possible solutions

The first constraint is that the time used, $6x + 5y$, be no more than 40 hours per week. The second constraint is that the money used, $200x + 100y$, be no more than $1000 each week. The third constraint is that x is not negative (because x counts tables), and the fourth constraint is that y is not negative. Figure 12.24 shows the calculations for our two points.

Point	(2, 1)	(3, 2)
Time Used	$6x + 5y$ $= 6 \cdot 2 + 5 \cdot 1$ $= 17 \leq 40$	$6x + 5y$ $= 6 \cdot 3 + 5 \cdot 2$ $= 28 \leq 40$
Money Spent	$200x + 100y$ $= 200 \cdot 2 + 100 \cdot 1$ $= 500 \leq 1000$	$200x + 100y$ $= 200 \cdot 3 + 100 \cdot 2$ $= 800 \leq 1000$
x not negative	$x = 2 \geq 0$	$x = 3 \geq 0$
y not negative	$y = 1 \geq 0$	$y = 2 \geq 0$

Figure 12.24
Checking the craftsman's constraints

The table in Figure 12.24 shows that if the craftsman makes two coffee tables and one end table each week, he will use 17 hours (of 40 available hours), spend $500 (of the $1,000 available), and profit $640. If he makes three coffee tables and two end tables each week, he will use 28 hours, spend $800, and profit $1,040. Each of these points represents a *possible* solution to the craftsman's problem, because each satisfies the time constraint and the money constraint. Neither represents the *actual* solution, because neither maximizes his profit; he has both money and time left over, so he should be able to increase his profit by building more tables. This is why the region in Figure 12.23 is called the region of *possible* solutions.

Common sense tells us that to maximize profit, our craftsman must use all his time and/or money and make more tables. To make more tables means to increase the value of x and/or y, which implies that we should choose points on the boundary of the region of feasible solutions. There are quite a few points on the boundary—too many to find and substitute into the equation for profit. Fortunately, the **Corner Principle** comes to our rescue. (We discuss why the Corner Principle is true later in this section.)

> ### Corner Principle
>
> The maximum and minimum values of an objective function occur at corner points of the region of possible solutions if that region is bounded.

Our region is a bounded region, so the Corner Principle applies. The region has four corner points (see Figure 12.25):

- P_1 at the origin
- P_2 where $y = -\frac{6}{5}x + 8$ intersects the y-axis
- P_3 where $y = -\frac{6}{5}x + 8$ intersects $y = -2x + 10$
- P_4 where $y = -2x + 10$ intersects the x-axis

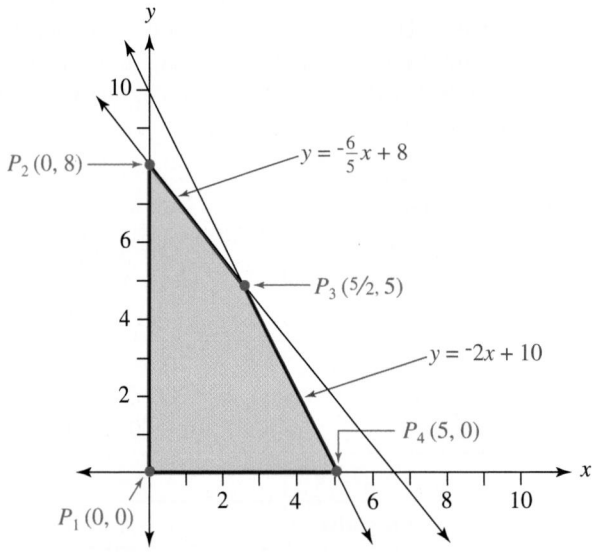

Figure 12.25
The region of possible solutions and its corner points

We can find each of these points by solving a system of equations. For example, P_3 can be found by solving the following system:

$$y = -\frac{6}{5}x + 8$$
$$y = -2x + 10$$

P_4 can be found by solving the following system:

$$y = -2x + 10$$
$$y = 0 \qquad \text{The } x\text{-axis}$$

The corner points, shown in Figure 12.25 are as follows:

P_1 (0, 0)

P_2 (0, 8)

P_3 ($\frac{5}{2}$, 5)

P_4 (5, 0)

Let's verify that the constraints are satisfied at each of these points and, more important, find the profit at each point, as shown in Figure 12.26.

Point	P_1 (0, 0)	P_2 (0, 8)	$P_3\left(\dfrac{5}{2}, 5\right)$	P_4 (5, 0)
Time Used	$6 \cdot 0 + 5 \cdot 0 = 0$	$6 \cdot 0 + 5 \cdot 8 = 40$	$6 \cdot \dfrac{5}{2} + 5 \cdot 5 = 40$	$6 \cdot 5 + 5 \cdot 0 = 30$
Money Spent	$200 \cdot 0 + 100 \cdot 0 = 0$	$200 \cdot 0 + 100 \cdot 8 = 800$	$200 \cdot \dfrac{5}{2} + 100 \cdot 5 = 1{,}000$	$200 \cdot 5 + 100 \cdot 0 = 1{,}000$
z = Profit	$240 \cdot 0 + 160 \cdot 0 = 0$	$240 \cdot 0 + 160 \cdot 8 = 1{,}280$	$240 \cdot \dfrac{5}{2} + 160 \cdot 5 = 1{,}400$	$240 \cdot 5 + 160 \cdot 0 = 1{,}200$

Figure 12.26
Verifying constraints and finding profit

At each corner point, each constraint is satisfied: the time used is at most 40 hours, and the money spent is at most \$1,000. The highest profit, \$1,400, occurs at P_3 ($\frac{5}{2}$, 5). The Corner Principle tells us that this is the point in the region of possible solutions at which the highest profit occurs. The craftsman will maximize his profit if he makes $2\frac{1}{2}$ coffee tables and 5 end tables each week (finishing that third coffee table during the next week), working his entire 40 hours per week and spending his entire \$1,000 per week on materials.

Some Graphing Tips

In solving a linear programming problem, an accurate graph of the region of possible solutions is important. A less-than-accurate graph can be the source of errors and frustration. To avoid difficulty if you are graphing without the aid of technology, follow these recommendations:

- Find the x-intercept of each line and use it (as well as the slope and y-intercept) to graph the line.
- Use graph paper and a straightedge.
- Use your graph to check your corner point computations—the computed location of a corner point should fit with the graph.

Why the Corner Principle Works

Each point in the region of possible solutions has a value of z associated with it. For example, we found that P_3 ($\frac{5}{2}$, 5) has a z-value of 1,400. Think of each point in the region as being a light bulb, with the brightness of the light given by the z-value. The point at which the craftsman's profit is maximized is the point with the brightest light.

Some points are just as bright as others. For example, all points that satisfy the equation $320 = 240x + 160y$ have a brightness of $z = 320$. If we solve this equation for y, we get

$$320 = 240x + 160y$$
$$-160y = 240x - 320$$
$$y = \frac{-3}{2}x + 2$$

This is a line with slope $\frac{-3}{2}$ and y-intercept 2. Points that satisfy the equation $640 = 240x + 160y$ have a brightness of $z = 640$. Solving this equation for y gives $y = (\frac{-3}{2})x + 4$. Points that satisfy the equation $960 = 240x + 160y$ have a brightness of 960; solving for y gives $y = (\frac{-3}{2})x + 6$. These three lines have the same slope, so they are parallel, as shown in Figure 12.27.

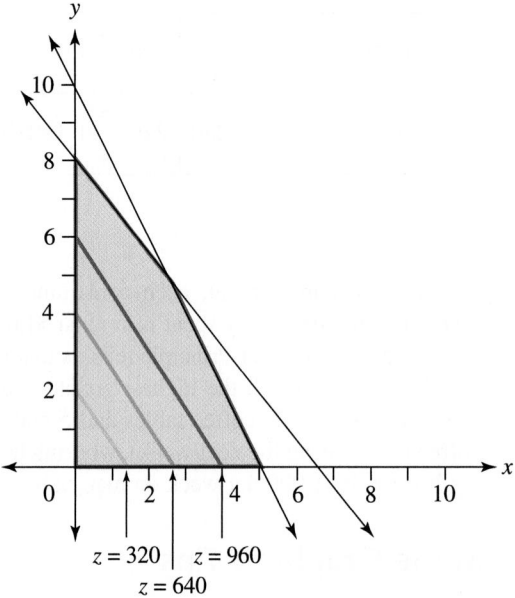

Figure 12.27
Why the corner principle works

The light bulbs that fill the region of possible solutions form parallel rows, and all bulbs in any one row are equally bright. Rows closer to the upper right corner of the region are brighter, and rows closer to the lower left corner are dimmer. The brightest bulb is at corner point P_3 $(\frac{5}{2}, 5)$, and the dimmest is at corner point P_1 $(0, 0)$.

Any linear programming problem (including our problem about the craftsman) must have constraints that are expressible as linear inequalities and an objective function that is expressible as a linear equation. This means that when a problem has two independent variables, the region of possible solutions is bounded by lines, and each value of z will correspond to a line. Thus, the graph must always look something like one of the three possibilities shown in Figure 12.28.

If the region is bounded, the maximum and minimum must be at corner points or at all points on an edge. Therefore, we can find all corner points, substitute them into the objective function, and choose the biggest or smallest. If two corners yield the same maximum z-value, we know that the maximum occurs at all points on the boundary line between those corners.

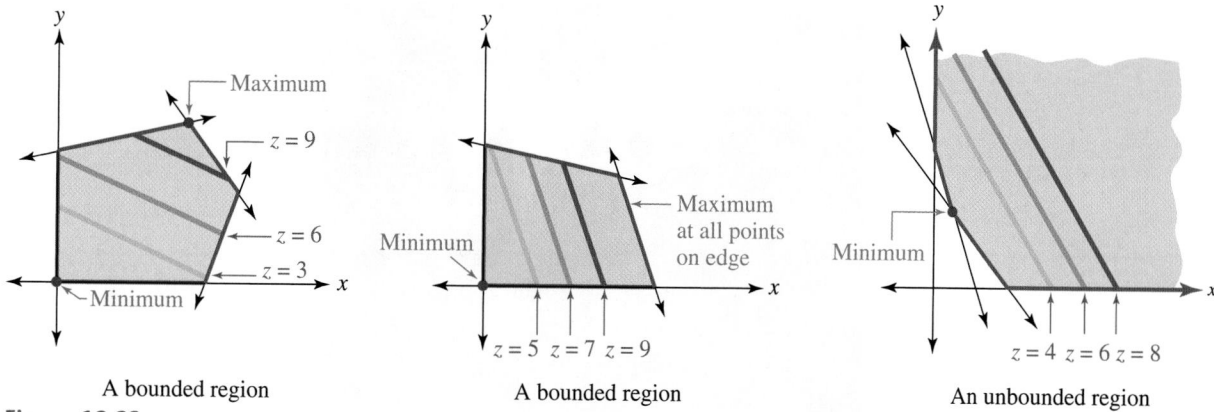

Figure 12.28
Where maximums and minimums occur

An unbounded region does not necessarily have a maximum or a minimum. Unbounded regions are explored further in the exercises.

Linear Programming Steps

1. *List the independent variables.*
2. *List the constraints and translate them into linear inequalities.*
3. *Find the objective and translate it into a linear equation.* This equation is called the objective function.
4. *Graph the region of possible solutions.* This is the region described by the constraints. Graph each line carefully, using the *x*-intercept as well as the slope and *y*-intercept, if you are graphing without the aid of technology.
5. *Find all corner points and the z-values associated with these points.*
   ✔ Check your corner point computations by verifying that a point's computed location fits with the graph, if you are graphing without the aid of technology.
6. *Find the maximum/minimum.* For a bounded region, the maximum occurs at the corner with the largest *z*-value, and the minimum occurs at the corner with the smallest *z*-value. If two corners give the same maximum (or minimum) value, then the maximum (or minimum) occurs at all points on the boundary line between those corners.

In Example 1, the craftsman maximized his profit by exhausting all of his resources. In Example 2, we will find that maximizing profit does not necessarily entail exhausting all resources.

ThomsonNOW™
for Liberal Arts Mathematics
www.thomsonedu.com/1pass.

EXAMPLE 2   Pete's Coffees sells two blends of coffee beans, Rich Blend and Hawaiian Blend. Rich Blend is one-half Colombian beans and one-half Kona beans, and Hawaiian Blend is one-quarter Colombian beans and three-quarters Kona beans. Profit on the Rich Blend is $2 per pound, while profit on the Hawaiian Blend is $3 per pound. Each day, the shop can obtain 200 pounds of Colombian beans and 60 pounds of Kona beans, and it uses that coffee only in the two blends. If the shop can sell all that it makes, how many pounds of Rich Blend and of Hawaiian Blend should Pete's Coffees prepare each day to maximize profit?

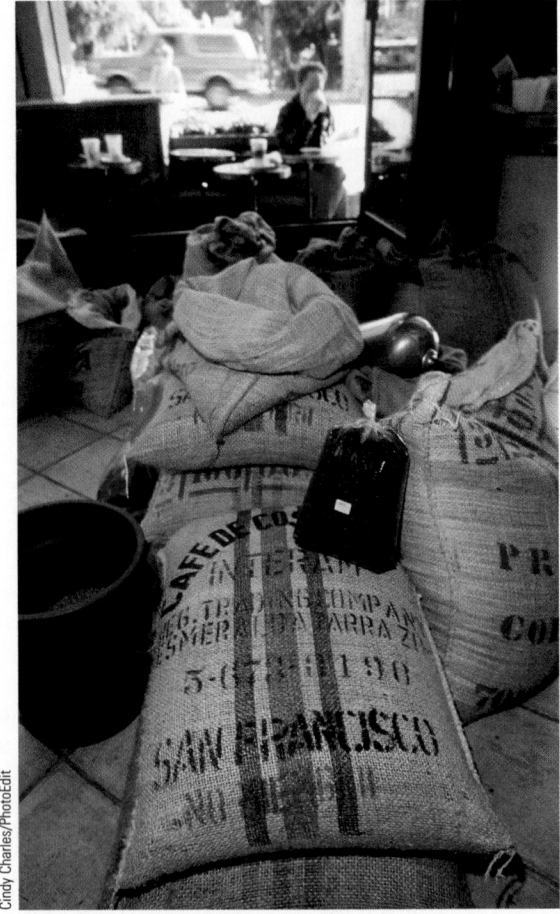

Cindy Charles/PhotoEdit

How should Pete's Coffees blend its beans to maximize its profit?

Solution

1. *List the independent variables*. The variables are as follows:

   - The amount of Rich Blend to be prepared each day
   - The amount of Hawaiian Blend to be prepared each day
   - The daily profit

   Profit depends on the amount of the two blends prepared, so profit is the dependent variable, and the amounts of Rich Blend and Hawaiian Blend are the independent variables. If we call them x and y, then

   x = pounds of Rich Blend to be prepared each day

   y = pounds of Hawaiian Blend to be prepared each day

2. *List the constraints and translate them into linear inequalities*. The constraints, or restrictions, are that there are 200 pounds of Colombian beans available each

day and only 60 pounds of Kona beans. In the blends, no more than the amount available can be used. First, let's translate the Colombian bean constraint:

Colombian beans used ≤ 200

(Colombian in Rich Blend) + (Colombian in Hawaiian Blend) ≤ 200

(one-half of Rich Blend)   + (one-fourth of Hawaiian Blend) ≤ 200

$$\frac{1}{2}x \qquad + \qquad \frac{1}{4}y \qquad \leq 200$$

Next, we'll translate the Kona bean constraint:

Kona beans used ≤ 60

(Kona in Rich Blend)   + (Kona in Hawaiian Blend) ≤ 60

(one-half of Rich Blend) + (three-quarters of Hawaiian Blend) ≤ 60

$$\frac{1}{2}x \qquad + \qquad \frac{3}{4}y \qquad \leq 60$$

Also, x and y count things (pounds of coffee), so neither can be negative.

$$x \geq 0 \qquad \text{and} \qquad y \geq 0$$

There is another implied constraint in this problem. Pete's sells its coffee in 1-pound bags, so x and y must be whole numbers. Special methods are available for handling such constraints, but these methods are beyond the scope of this class. We will ignore such constraints and accept fractional answers should they occur.

3. *Find the objective and translate it into a linear equation.* The objective is to maximize profit. If we let $z = $ profit, we get

$z = $ (Rich Blend profit) + (Hawaiian Blend profit)

$\quad = $ (\$2 per pound) (pounds of Rich Blend) + (\$3 per pound) (pounds of Hawaiian Blend)

$\quad = 2x + 3y$

Steps 1 through 3 yield the mathematical model:

Independent Variables

$x = $ pounds of Rich Blend to be prepared each day

$y = $ pounds of Hawaiian Blend to be
    prepared each day

Constraints

$$\frac{1}{2}x + \frac{1}{4}y \leq 200 \quad \text{the Colombian bean constraint}$$

$$\frac{1}{2}x + \frac{3}{4}y \leq 60 \quad \text{the Kona bean constraint}$$

$$x \geq 0 \qquad \text{and} \qquad y \geq 0$$

Objective Function

$$z = 2x + 3y$$

4. *Graph the region of possible solutions.* With the aid of the chart in Figure 12.29, we can graph the two inequalities on the same axes, as shown in Figure 12.30.

Original Inequality	Slope-Intercept Form	Associated Equation	Graph of the Inequality
$\frac{1}{2}x + \frac{1}{4}y \leq 200$	$\frac{1}{4}y \leq -\frac{1}{2}x + 200 \rightarrow$ $y \leq -2x + 800$	$y = -2x + 800$	all points on or below the line with slope -2 and y-intercept 800
$\frac{1}{2}x + \frac{3}{4}y \leq 60$	$\frac{3}{4}y \leq -\frac{1}{2}x + 60 \rightarrow$ $y \leq -\frac{2}{3}x + 80$	$y = -\frac{2}{3}x + 80$	all points on or below the line with slope $-2/3$ and y-intercept 80
$x \geq 0$	(not applicable)	$x = 0$	all points on or to the right of the y-axis
$y \geq 0$	$y \geq 0$	$y = 0$	all points on or above the x-axis

Figure 12.29
Getting ready to graph the region of possible solutions

If you are graphing the region without the aid of technology, it's easier to get an accurate graph if you find the x-intercept of each line and use it (as well as the slope and y-intercept) to graph the line. To find the x-intercept of $y = -\frac{2}{3}x + 80$, solve the system

$$y = -\frac{2}{3}x + 80$$
$$y = 0 \qquad \text{the equation of the } x\text{-axis}$$

The x-intercept is the point $(120, 0)$. The region of possible solutions is shown in Figure 12.30. It is a bounded region. Notice that $y = -2x + 800$ is not a boundary of the region of possible solutions; if we were not given the first constraint, we would have the same region of possible solutions. That constraint, which describes the limited amount of Colombian beans available to Pete's Coffees, is not really a limitation.

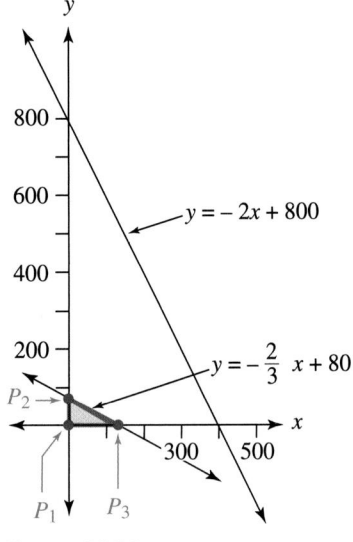

Figure 12.30
The region of possible solutions

5. *Find all corner points and the z-values associated with these points.* Clearly, $P_1 = (0, 0)$. P_2 and P_3 have already been found; they are the y- and x-intercepts of $y = -\frac{2}{3}x + 80$. P_2 is $(0, 80)$, and P_3 is $(120, 0)$.

6. *Find the maximum.* The result of substituting the corner points into the objective function $z = 2x + 3y$ is given in Figure 12.31.

The two corner points P_2 and P_3 give the same maximum value, $z = 240$. Thus, the maximum occurs at P_2 and P_3 and at all points between them on the

Point	Value of $z = 2x + 3y$
P_1 $(0, 0)$	$z = 2 \cdot 0 + 3 \cdot 0 = 0$
P_2 $(0, 80)$	$z = 2 \cdot 0 + 3 \cdot 80 = 240$
P_3 $(120, 0)$	$z = 2 \cdot 120 + 3 \cdot 0 = 240$

Figure 12.31
Finding the maximum

boundary line $y = -\frac{2}{3}x + 80$, as shown in Figure 12.32. This line includes points such as (30, 60), (60, 40), and (90, 20). (Find points like these by substituting appropriate values of x in the equation.) The meaning of these points is given in Figure 12.33. Pete's Coffees can choose to produce its Rich Blend and Hawaiian Blend in any of the amounts given in Figure 12.33 (or any other amount given by a point on the line). Because each of these choices will maximize Pete's profit at $240 per day, the choice must be made using criteria other than profit. Perhaps the Hawaiian Blend tends to sell out earlier in the day than the Rich Blend. In this case, Pete might choose to produce 30 pounds of Rich Blend and 60 pounds of Hawaiian Blend.

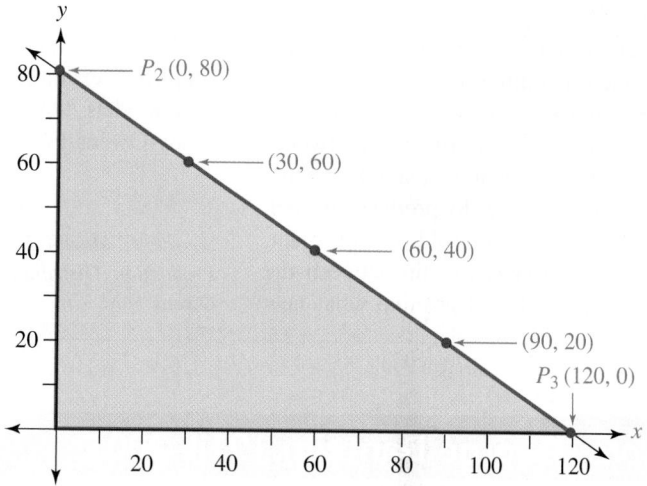

Figure 12.32
The maximum occurs at many points

Point	Interpretation
P_2 (0, 80)	Each day, prepare no Rich Blend and 80 pounds of Hawaiian Blend for a profit of $240.
P_3 (120, 0)	Each day, prepare 120 pounds of Rich Blend and no Hawaiian Blend for a profit of $240.
(30, 60)	Each day, prepare 30 pounds of Rich Blend and 60 pounds of Hawaiian Blend for a profit of $240.
(60, 40)	Each day, prepare 60 pounds of Rich Blend and 40 pounds of Hawaiian Blend for a profit of $240.
(90, 20)	Each day, prepare 90 pounds of Rich Blend and 20 pounds of Hawaiian Blend for a profit of $240.

Figure 12.33
Interpreting some typical boundary line maximums

Linear Programming in the Real World: Pilkington Libbey-Owens-Ford

We have been looking at how linear programming can be used in making product mix decisions. Our friend the craftsman had to make a product mix decision: He had to decide what mix of his products was best. And Pete's Coffees had to decide what mix of their products were best.

Pilkington Libbey-Owens-Ford, a large plate glass manufacturer, has used linear programming quite successfully in making product mix decisions. Their linear programming model deals with more than 200 products made at four different plants. The model's constraints include market demand, production capabilities, inventory storage capabilities, and the availability of railroad shipping, as well as production, inventory, and shipping costs. The model is run ten to twenty times per month and makes recommendations regarding the amount of each product to produce at each plant, the amount of inventory to hold at each plant, and the amount of each product to ship between the plants. It has also been used to determine what new products to introduce and what old products to eliminate and to determine how, when, and where to increase manufacturing capabilities. The model, which took two years to complete, has resulted in savings of more than $2 million each year.

Generally, linear programming models in business are very complex. They involve a large number of independent variables and constraints, as does the Pilkington Libbey-Owens-Ford model. The exercises in this section are realistic but simplified examples of the product mix decision models that are routinely used by businesses. Exercise 37 is a realistic but simplified example of the production-planning decision models that are routinely used by Pilkington Libby-Owens-Ford.

Source: C. Martin, D. Dent, and J. Eckhart, "Integrated Production, Distribution, and Inventory Planning at Libbey-Owens-Ford," Interfaces, 23, no. 3, 1993, pages 68–78.

Historical Note George Dantzig, 1914–2005

© Edward Souza/Stanford News Service

George Bernard Dantzig's father was both a writer and a mathematician. He had hoped that his first son would be a writer and his second a mathematician, so he named his first son after the playwright and critic George Bernard Shaw and his second after Henri Poincaré, a famous French mathematician under whom the father had studied. As it happened, both sons became mathematicians.

In 1939, George Dantzig was a graduate student at the University of California at Berkeley. One day he arrived late at a statistics class. He copied the two problems that were written on the blackboard, assuming that they were homework problems. About six weeks after turning the homework in, he was awakened early one Sunday morning by his excited professor, who wanted to send off Dantzig's work right away for publication. The two problems were not homework but famous unsolved problems in statistics. Their solutions became Dantzig's Ph.D. thesis in mathematics.

During World War II, Dantzig was hired by the Air Force to find practical ways to distribute personnel, weapons, and supplies to the various fronts. Shortly after the end of the war, he became mathematics advisor to the U.S. Air Force Comptroller at the Pentagon, where he was responsible for finding a way to mechanize this planning process. The result was linear programming. The procedure was quickly applied to a wide variety of business, economic, and environmental topics. Tjalling Koopmans of the United States and Leonid Kantorovich of the Soviet Union received the 1975 Nobel Prize in economics for their use of linear programming in developing the theory of allocation of resources. Surprisingly, Dantzig himself was not honored. George Dantzig was Professor Emeritus at Stanford University until his death in 2005.

Source: Donald J. Albers and Constance Reid, "An Interview with George B. Dantzig. The Father of Linear Programming," College Mathematics Journal, vol. 17, no. 4, September 1986.

12.1 Exercises

In Exercises 1–6, convert the information to a linear inequality. Give the meaning of each variable used.

1. A landscape architect wants his project to use no more than 100 gallons of water per day. Each shrub requires 1 gallon of water per day, and each tree requires 3 gallons of water per day.

2. A shopper wishes to spend no more than $150. Each pair of pants costs $25 and each shirt costs $21.

3. A bookstore owner wishes to generate at least $5,000 in profit this month. Each hardback book generates $4.50 in profit, and each paperback generates $1.25 in profit.

4. Dick Rudd wants to take at least 1,000 mg of vitamin C. Each tablet of Megavite has 30 mg of vitamin C, and each tablet of Healthoboy has 45 mg of vitamin C.

5. A warehouse has 1,650 cubic feet of unused storage space. Refrigerators take up 63 cubic feet each, and dishwashers take up 41 cubic feet each.

6. A coffee shop owner has 1,000 pounds of Java beans. In addition to selling pure Java beans, the shop also sells Blend Number 202, which is 32% Java beans.

In Exercises 7–16, use the method of linear programming to solve the problem.

7. A craftswoman produces two products, floor lamps and table lamps. Production of one floor lamp requires 75 minutes of her labor and materials that cost $25. Production of one table lamp requires 50 minutes of labor, and the materials cost $20. The craftswoman wishes to work no more than 40 hours each week, and her financial resources allow her to pay no more than $900 for materials each week. If she can sell as many lamps as she can make and if her profit is $39 per floor lamp and $33 per table lamp, how many floor lamps and how many table lamps should she make each week to maximize her weekly profit? What is that maximum profit?

8. Five friends, all of whom are experienced bakers, form a company that will make bread and cakes and sell them to local restaurants and specialty stores. Each loaf of bread requires 50 minutes of labor and ingredients costing $0.90 and can be sold for $1.20 profit. Each cake requires 30 minutes of labor and ingredients costing $1.50 and can be sold for $4.00 profit. The partners agree that no one will work more than 8 hours a day. Their financial resources do not allow them to spend more than $190 per day on ingredients. How many loaves of bread and how many cakes should they make each day to maximize their profit? What is that maximum profit?

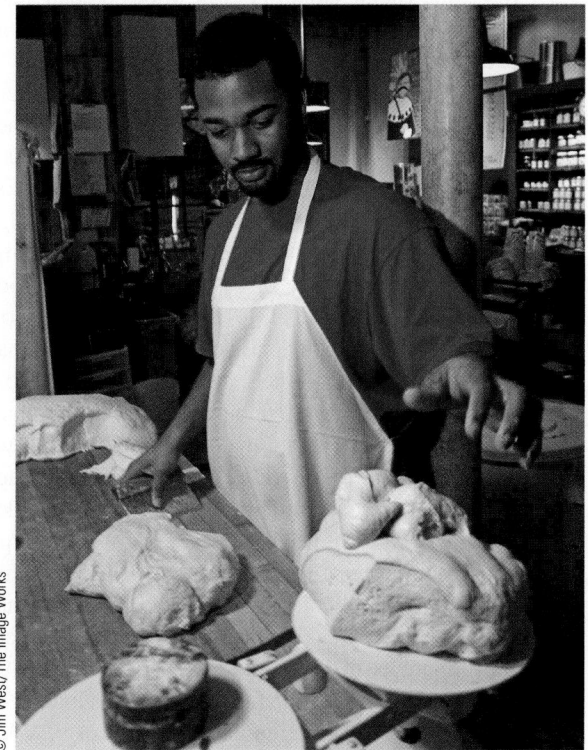

© Jim West/The Image Works

9. Pete's Coffees sells two blends of coffee beans: Morning Blend and South American Blend. Morning Blend is one-third Mexican beans and two-thirds Colombian beans, and South American Blend is two-thirds Mexican beans and one-third Colombian beans. Profit on the Morning Blend is $3 per pound, while profit on the South American Blend is $2.50 per pound. Each day, the shop can obtain 100 pounds of Mexican beans and 80 pounds of Colombian beans, and it uses that coffee only in the two blends. If the shop can sell all that it makes, how many pounds of Morning Blend and of South American Blend should Pete's Coffees prepare each day to maximize profit? What is that maximum profit?

10. Pete's Coffees sells two blends of coffee beans: Yusip Blend and Exotic Blend. Yusip Blend is one-half Costa Rican beans and one-half Ethiopian

beans, and Exotic Blend is one-quarter Costa Rican beans and three-quarters Ethiopian beans. Profit on the Yusip Blend is $3.50 per pound, while profit on the Exotic Blend is $4 per pound. Each day, the shop can obtain 200 pounds of Costa Rican beans and 330 pounds of Ethiopian beans, and it uses that coffee only in the two blends. If the shop can sell all that it makes, how many pounds of Yusip Blend and of Exotic Blend should Pete's Coffees prepare each day to maximize profit? What is that maximum profit?

11. Bake-Em-Fresh sells its bread to supermarkets. Shopgood Stores needs at least 15,000 loaves each week, and Rollie's Markets needs at least 20,000 loaves each week. Bake-Em-Fresh can ship at most 45,000 loaves to these two stores each week if it wishes to satisfy its other customers' needs. If shipping costs an average of 8¢ per loaf to Shopgood Stores and 9¢ per loaf to Rollie's Markets, how many loaves should Bake-Em-Fresh allot to Shopgood and to Rollie's each week to minimize shipping costs? What shipping costs would this entail?

12. Notel Chips manufactures computer chips. Its two main customers, HAL Computers and Peach Computers, just submitted orders that must be filled immediately. HAL needs at least 130 cases of chips, and Peach needs at least 150 cases. Owing to a limited supply of silicon, Notel cannot send more than a total of 300 cases. If shipping costs $100 per case for shipments to HAL and $90 per case for shipments to Peach, how many cases should Notel send to each customer to minimize shipping costs? What shipping costs would this entail?

13. Global Air Lines has contracted with a tour group to transport a minimum of 1,600 first-class passengers and 4,800 economy-class passengers from New York to London during a 6-month time period. Global Air has two types of airplanes, the Orville 606 and the Wilbur W-1112. The Orville 606 carries 20 first-class passengers and 80 economy-class passengers and costs $12,000 to operate. The Wilbur W-1112 carries 80 first-class passengers and 120 economy-class passengers and costs $18,000 to operate. During the time period involved, Global Air can schedule no more than 52 flights on Orville 606s and no more than 30 flights on Wilbur W-1112s. How should Global Air Lines schedule its flights to minimize its costs? What operating costs would this schedule entail?

14. Compucraft sells personal computers and printers made by Peach Computers. The computers come

in 12-cubic-foot boxes, and the printers come in 8-cubic-foot boxes. Compucraft's owner estimates that at least 30 computers can be sold each month and that the number of computers sold will be at least 50% more than the number of printers. The computers cost Compucraft $1,000 each and can be sold at a $1,000 profit, while the printers cost $300 each and can be sold for a $350 profit. Compucraft has 1,000 cubic feet of storage available for the Peach personal computers and printers and sufficient financing to spend $70,000 each month on computers and printers. How many computers and printers should Compucraft order from Peach each month to maximize profit? What is that maximum profit?

15. The Appliance Barn has 2,400 cubic feet of storage space for refrigerators. The larger refrigerators come in 60-cubic-foot packing crates, and the smaller ones come in 40-cubic-foot crates. The larger refrigerators can be sold for a $250 profit, while the smaller ones can be sold for a $150 profit.

a. If the manager is required to sell at least 50 refrigerators each month, how many large refrigerators and how many small refrigerators should he order each month to maximize profit?

b. If the manager is required to sell at least 40 refrigerators each month, how many large refrigerators and how many small refrigerators should he order each month to maximize profit?

c. Should the Appliance Barn owner require his manager to sell 40 or 50 refrigerators per month?

16. City Electronics Distributors handles two lines of televisions, the Packard and the Bell. It purchases up to $57,000 worth of television sets from the manufacturers each month and stores them in a 9,000-cubic-foot warehouse. The Packards come in 36-cubic-foot packing crates, and the Bells come in 30-cubic-foot crates. The Packards cost City Electronics $200 each and can be sold to a retailer for a $200 profit, while the Bells cost $250 each and can be sold for a $260 profit. City Electronics must stock enough sets to meet its regular customers' standing orders.

a. If City Electronics has standing orders for 250 sets in addition to orders from other retailers, how many sets should City Electronics order each month to maximize profit?

b. If City Electronics' standing orders increase to 260 sets, how many sets should City Electronics order each month to maximize profit?

17. How much of their available Mexican beans would be unused if Pete's Coffees of Exercise 9 maximizes its profit? How much of the available Colombian beans would be unused?

18. How much of their available time would be unused if the five friends of Exercise 8 maximize their profit? How much of their available money would be unused?

19. How much of her available time would be unused if the craftswoman of Exercise 7 maximizes her profit? How much of her available money would be unused?

20. How much of the available Costa Rican beans would be unused if Pete's Coffees of Exercise 10 maximizes its profit? How much of the available Ethiopian beans would be unused?

In Exercises 21–24, the region of possible solutions is not bounded: thus, there may not be both a maximum and a minimum. After graphing the region of possible solutions and finding each corner point, you can determine if both a maximum and a minimum exist by choosing two arbitrary values of z and graphing the corresponding lines, as discussed in the section "Why the Corner Principle Works."

21. The objective function $z = 2x + 3y$ is subject to the constraints

$$3x + y \geq 12$$
$$x + y \geq 6$$
$$x \geq 0 \quad \text{and} \quad y \geq 0$$

Find the following.
a. the point at which the maximum occurs (if there is such a point)
b. the maximum value
c. the point at which the minimum occurs (if there is such a point)
d. the minimum value

22. The objective function $z = 3x + 4y$ is subject to the constraints

$$2x + y \geq 10$$
$$3x + y \geq 12$$
$$x \geq 0 \quad \text{and} \quad y \geq 0$$

Find the following.
a. the point at which the maximum occurs (if there is such a point)
b. the maximum value
c. the point at which the minimum occurs (if there is such a point)
d. the minimum value

23. U.S. Motors manufactures quarter-ton, half-ton, and three-quarter-ton panel trucks. United Delivery Service has placed an order for at least 300 quarter-ton, 450 half-ton, and 450 three-quarter-ton panel trucks. U.S. Motors builds the trucks at two plants, one in Detroit and one in Los Angeles. The Detroit plant produces 30 quarter-ton trucks, 60 half-ton trucks, and 90 three-quarter-ton trucks each week, at a total cost of $540,000. The Los Angeles plant produces 60 quarter-ton trucks, 45 half-ton trucks, and 30 three-quarter-ton trucks each week, at a total cost of $360,000. How should U.S. Motors schedule its two plants so that it can fill this order at minimum cost? What is that minimum cost?

24. Eaton's Chocolates produces semisweet chocolate chips and milk chocolate chips at its plants in Bay City and Estancia. The Bay City plant produces 3,000 pounds of semisweet chips and 2,000 pounds of milk chocolate chips each day at a cost of $1,000, while the Estancia plant produces 1,000 pounds of semisweet chips and 6,000 pounds of milk chocolate chips each day at a cost of $1,500. Eaton's has an order from SafeNShop Supermarkets for at least 30,000 pounds of semisweet chips and 60,000 pounds of milk chocolate chips. How should it schedule its production so it can fill the order at minimum cost? What is that minimum cost?

▬▬▬▶ *Answer the following questions using complete sentences and your own words.*

CONCEPT QUESTIONS

25. A bicycle manufacturing firm hired a consultant who used linear programming to determine that the firm should discontinue manufacturing some of its product line. What negative ramifications might occur if this advice is followed? How could linear programming be used if the firm decided not to follow this advice?

26. Is it possible for an unbounded region to have both a maximum and a minimum? Draw a number of graphs (similar to those in Figure 12.28) to see. Justify your answer with graphs.

27. Discuss why it is that, when two corners give the same maximum (or minimum) value, the maximum (or minimum) occurs at all points on the boundary line between those corners.

28. In Example 2, Pete's Coffees mixes Colombian beans and Kona beans to make Rich Blend and Hawaiian Blend. It may be that Pete's could make slight variations in the proportions of Colombian and Kona beans in these two blends without affecting the taste of the coffee. Why would Pete's consider doing this? What effect would it have on the objective function, the constraints, the region of possible solutions, and the solution to the example?

29. In an open market, the selling price of a product is determined by the supply of the product and the demand for the product. In particular, the price falls if a larger supply of the product becomes available, and the price increases if there is suddenly a larger demand for the product. Any change in the selling price could affect the profit. However, when modeling a linear programming problem, we assume a constant profit. How could linear programming be adjusted to allow for the fact that prices change with supply and demand?

30. Make up a linear programming problem (a math problem, not a word problem) that has no maximum.

31. Make up a linear programming problem (a math problem, not a word problem) that has no minimum.

32. Make up a word problem that results in the following linear programming problem:

maximize $z = 5x + 4y$

subject to the constraints

$$10x + 20y \leq 1,000$$
$$30x + 10y \leq 2,000$$
$$x \geq 0 \quad \text{and} \quad y \geq 0$$

HISTORY QUESTIONS

33. For whom was George Dantzig named?
34. How did Dantzig choose the topic of his Ph.D. thesis?
35. What did Dantzig do during World War II?
36. What was surprising about the awarding of a Nobel Prize to two economists for their use of linear programming?

PROJECTS

37. Glassco, a large plate glass manufacturer, makes two products: half-inch-thick plate glass and three-eighths-inch-thick plate glass. It makes each of these two products at its two plants: Clear View and Panesville. For the first quarter in the next year, demand for half-inch-thick plate glass is forecast to be 5,000 sheets, and demand for three-

eighths-inch-thick plate glass is forecast to be 6,000 sheets for the first quarter in the year. Because of the difficulties inherent in storing the product, Glassco does not make a surplus of glass above and beyond the demand. The Clear View plant is capable of producing at most 7,000 sheets per quarter, and the Panesville plant can make at most 5,000 sheets per quarter. Glass made at the Clear View plant can be sold at a profit of $820 per sheet and $760 per sheet for the half-inch and three-eighths-inch products, respectively. However, the profit on glass made at the Panesville plant is smaller by $75 per sheet, owing to higher shipping costs.

a. Translate the information on demand into two constraints. Use the following variables:

v = amount of half-inch-thick glass made in the first quarter at Clear View

w = amount of three-eighths-inch-thick glass made in the first quarter at Clear View

x = amount of half-inch-thick glass made in the first quarter at Panesville

y = amount of three-eighths-inch-thick glass made in the first quarter at Panesville

b. Translate the information on plant production capabilities into two constraints.

c. Translate the information on profit into an objective function.

d. Explain why this problem cannot be solved with the geometric method of linear programming.

38. You manage the Fee-Based Technical Support Division of a large computer software firm. You have sufficient resources to provide up to 1,000 hours of technical support each week. Often, this is not enough time to meet demand. As a result, you must develop a policy to allocate technical support time. Your firm's warranty specifies that customers who have purchased software in the last year can purchase technical support for $49 per incident. Software purchased more than a year ago is not under warranty. However, your firm continues to provide technical support at the same rate after the warranty has expired, to maintain customer goodwill. Each under-warranty support incident requires an average of 30 minutes of support and each out-of-warranty support incident requires an average of one hour of support. Your firm's CEO has asked that you allot between 700 and 800 hours per week to customers whose software is still under warranty to ensure that they have sufficient access to technical

support. Your CEO has also asked that you allot between 100 and 250 hours per week to customers whose software is no longer under warranty to ensure that they still have some access to technical support.

a. Create a specific allocation policy that maximizes revenue.

b. One year after you instituted the allocation policy in part (a), demand has increased. As a result, you are considering increasing the fee to customers whose software is no longer under warranty from $49 to $149 per incident. Create a new specific allocation policy that maximizes revenue.

 WEB PROJECTS

39. Write an essay on the various theories about why George Dantzig did not receive a Nobel Prize for his invention of linear programming.

40. Research linear programming software packages. List at least five companies or software brands, and discuss their linear programming products.

Some useful links for this project are listed on the text companion web site. Go to **www.thomsonedu.com/ mathematics/johnson** to access the web site.

Chapter 12   Review

ThomsonNOW™ **for Liberal Arts Mathematics**
Preparing for an exam? Test yourself on key material by visiting ThomsonNOW at **www.thomsonedu.com/1pass.**

KEY TERMS

bounded region	linear equation	region of possible	region of solutions of a
constraint	linear inequality	solutions of a linear	system of linear
corner point	linear programming	programming problem	inequalities
Corner Principle	model	region of solutions of an	system of linear inequalities
dependent variable	objective function	inequality	unbounded region
independent variable	optimize		variable

PROCEDURES

Graphing the Region of Solutions of a Linear Inequality

1. *Solve the inequality for y.*
2. *Graph the line with a solid or dashed line.*
3. *Shade in one side of the line.*

Linear Programming Steps:

1. *List the independent variables.*
2. *List the constraints and translate them into linear inequalities.*

Linear Programming Steps: Continued

3. *Find the objective and translate it into a linear equation.*
4. *Graph the region of possible solutions, the region described by the constraints.*
5. *Find all corner points and the z-values associated with these points. Check your work.*
6. *Find the maximum/minimum.*

EXERCISES

In Exercises 1–4, graph the region of solutions of the given inequality.

1. $4x - 5y > 7$
2. $3x + 4y < 10$
3. $3x + 6y \leq 9$
4. $6x - 8y \geq 12$

In Exercises 5–10, graph the region of solutions of the given system of inequalities.

5. $8x - 4y < 10$
$3x + 5y \geq 7$

6. $x - 5y < 7$
$3x + 2y > 6$

7. $5x - y > 8$
$x \geq -3$

8. $6x + 4y \leq 7$
$y < 4$

9. $x - y \leq 7$
$5x + 3y \leq 9$
$x \geq 0$
$y \geq 0$

10. $3x - 2y \leq 12$
$5x + 3y \leq 15$
$x \geq 0$
$y \geq 0$

In Exercises 11–17, solve the given linear programming problem.

11. Minimize $z = 4x + 3y$
subject to the constraints:
$x + 2y \leq 20$
$4x + y \leq 20$
$x, y \geq 0$

12. Maximize $z = 4x + 2y$
subject to the constraints:
$7x + 3y \leq 21$
$x + y \leq 5$
$2x + y \leq 10$
$x, y \geq 0$

13. Minimize $z = 2x + 3y$
subject to the constraints:
$15x + y \geq 45$
$x + 6y \geq 40$
$8x + 7y \geq 70$
$x, y \geq 0$

14. Maximize $z = 8x + 6y$
subject to the constraints:
$x + 2y \geq 50$
$52x + 44y \geq 1144$
$x, y \geq 0$

15. The Mowson Audio Co. makes stereo speaker assemblies. They purchase speakers from a speaker manufacturing firm, and install them in their own cabinets. Mowson's model 110 speaker assembly, which sells for $200, has a tweeter and a midrange speaker. Their model 330 assembly, which sells for $350, has two tweeters, a midrange speaker, and a woofer. Mowson currently has in stock 90 tweeters, 60 midrange speakers, and 44 woofers. How many speaker assemblies should Mowson make to maximize their income? What is that maximum income?

16. The Stereo Guys store sells two lines of personal stereos: the Sunny and the Iwa. The Sunny comes in a 12-cubic-foot box and can be sold for a $220 profit, while the Iwa comes in an 8-cubic-foot box and can be sold for a $200 profit. The Stereo Guys marketing department estimates that at least 600 personal stereos can be sold each month and that, because of Sunny's quality reputation, the demand for the Sunny unit is at least twice that of the Iwa. If the Stereo Guys warehouse has 12,000 cubic feet of space available for personal stereos, how many Sunnys and Iwas should they stock each month to maximize profit? What is that maximum profit?

Answer the following questions using your own words and complete sentences.

CONCEPT AND HISTORY QUESTIONS

17. Discuss why the corner principal works.

18. What are the main elements of a linear programming model?

19. Who invented linear programming?

Using a Scientific Calculator

What Kind of Calculator Do You Have?

There are three kinds of calculators: graphing calculators, scientific calculators, and basic calculators. Graphing calculators have bigger screens, which they use to display graphs. If you have one of these, read Appendix B.

Scientific calculators and basic calculators have smaller screens; they don't graph, so they don't need the extra space on the screen. If you don't have a graphing calculator, you need to determine whether you have a scientific calculator or a basic calculator because a basic calculator is insufficient for this text. Here's how you can tell. Type

$$2 \boxed{+} 3 \boxed{\times} 4 \boxed{=}$$

If you get 14, you probably have a scientific calculator. Read this appendix to learn how to use it. If you get 20, you have a basic calculator, which is insufficient for this text.

The following discussion is meant to apply to any scientific calculator. Read it with your calculator (and its instruction booklet, if you have it) by your side. When a calculation is discussed, do that calculation on your calculator.

Some scientific calculators work like graphing calculators, except they don't graph (and thus they don't have bigger screens). If the instructions in this appendix don't always work with your calculator, try the instructions in Appendix B.

Unfortunately, scientific calculators don't all work exactly the same way. Even if you do everything correctly, your answer may have a few more or less decimal places than the one given, or may differ in the last decimal place. Occasionally, the way a calculation is performed on your calculator will differ slightly from that discussed below. If so, consult your instruction booklet, experiment a little, or consult your instructor. To paraphrase Abraham Lincoln, the following discussion applies to most scientific calculators all of the time and to all scientific calculators most of the time, but it doesn't apply to all scientific calculators all of the time.

The Equals Button

Some buttons perform operations on a pair of numbers. For example, the $\boxed{+}$ button is used to add two numbers; it only makes sense to use this button in conjunction with a pair of numbers. *When using such a button, you must finish the typing with the $\boxed{=}$ button.* That is, to add 3 and 2, type

$$3 \boxed{+} 2 \boxed{=}$$

Some buttons perform operations on a single number. For example, the $\boxed{x^2}$ button is used to square a number; it only makes sense to use this button in conjunction with a single number. *When using such a button, you do **not** type the* $\boxed{=}$ *button. That is,* to square 5, type

$$5 \ \boxed{x^2}$$

The reason for this distinction is that the calculator must be told when you are done entering information and you are ready for it to compute. If you typed

$$3 \ \boxed{+} \ 2$$

the calculator would have no way of knowing if you were ready for it to compute "3 + 2" or if you were in the middle of instructing it to compute "3 + 22." You must follow "3 + 2" with the $\boxed{=}$ button to tell the calculator that you are done entering information and are ready for it to compute. On the other hand, when you type

$$5 \ \boxed{x^2}$$

the calculator knows that you are done entering information; if you meant to square 53, you would not have pressed the $\boxed{x^2}$ button after the "5." There is no need to follow "5 $\boxed{x^2}$" with the $\boxed{=}$ button.

The Clear Button

Most calculators have two "clear" buttons (buttons that erase): one that clears just the last number you typed and one that clears everything. Frequently, the button that clears just the last number you typed is labeled $\boxed{CE}$ (for "clear entry") and the one that clears everything is labeled $\boxed{CA}$ (for "clear all"). These buttons are used when you err in typing. For example, if you typed "5 × 2" and then realized you wanted "5 × 3," you could clear the incorrect "2" with the $\boxed{CE}$ button by typing

$$5 \ \boxed{\times} \ 2 \ \boxed{CE} \ 3 \ \boxed{=}$$

This would make the display read "15," since the $\boxed{CE}$ button clears just the last entry (the "2"), and 5 × 3 = 15. However, if you typed

$$5 \ \boxed{\times} \ 2 \ \boxed{CA} \ 3 \ \boxed{=}$$

the display would read "3," since the $\boxed{CA}$ button clears everything that came before it (the "5 × 2").

Some calculators label their "clear entry" button $\boxed{CE}$ and their "clear all" button $\boxed{C}$. Some calculators label their "clear entry" button $\boxed{C}$ and their "clear all" button $\boxed{CA}$. Some calculators have a button labeled $\boxed{C}$ that functions as a "clear entry" button if you press it once and as a "clear all" button if you press it twice. Some experimentation with your clear button(s) and the above two examples will show you how yours works.

The Subtraction Symbol and the Negative Symbol

If "3 − −2" were read aloud, you could say "three subtract negative two" or "three minus minus two," and you would be understood. The expression is understandable even if the distinction between the negative symbol and the subtraction symbol is not made clear. With a calculator, however, this distinction is crucial. The subtraction button is labeled $\boxed{-}$. There is no negative button; instead, there is a $\boxed{+/-}$ button that changes the sign of whatever number is on the display. Typing

$$5 \ \boxed{+/-}$$

makes the display read "−5." Typing

$$5 \boxed{+/-} \boxed{+/-}$$

makes the display read "5," because two sign changes undo each other. Typing

$$\boxed{+/-} \ 5$$

makes the display read "5," not "−5." *You must press the sign-change button after the number itself.*

EXAMPLE 1   Calculate $3 - -2$, both (a) by hand and (b) with a calculator.

Solution   **a.** $3 - -2 = 3 + 2 = 5$

b. To do this, we must use the subtraction button, which is labeled $\boxed{-}$, and the sign-change button, which is labeled $\boxed{+/-}$. Typing

$$3 \boxed{-} 2 \boxed{+/-} \boxed{=}$$

makes the display read "5." Remember that you must press the sign-change button *after* the number itself; you press "2" followed by $\boxed{+/-}$, not $\boxed{+/-}$ followed by "2." Also, you must finish the typing with the $\boxed{=}$ button, because the $\boxed{=}$ button performs an operation on a pair of numbers.   ■

Order of Operations and Use of Parentheses

Scientific calculators are programmed so that they follow the standard order of operations. That is, they perform calculations in the following order:

1. **P**arentheses-enclosed work

2. **E**xponents

3. **M**ultiplication and
   Division, from left to right

4. **A**ddition and
   Subtraction, from left to right

This order can be remembered by remembering the word *PEMDAS,* which stands for

Parentheses/**E**xponents/**M**ultiplication/**D**ivision/**A**ddition/**S**ubtraction

Frequently, the fact that calculators are programmed to follow the order of operations means that you perform a calculation on your calculator in exactly the same way that it is written.

EXAMPLE 2   Calculate $2(3 + 4)$, both (a) by hand and (b) with a calculator.

Solution   **a.** $2(3 + 4) = 2(7)$ parentheses-enclosed work comes first
                        $= 14$

b. Type

$$2 \boxed{\times} \boxed{(} 3 \boxed{+} 4 \boxed{)} \boxed{=}$$

and the display reads "14." Notice that this is typed on a calculator exactly as it is written, with two important exceptions:

- We must press the $\boxed{\times}$ button to mean multiplication; we cannot use parentheses to mean multiplication, as is done in part (a).
- We must finish the typing with the $\boxed{=}$ button.   ■

EXAMPLE 3   Calculate $2 \cdot 3^2$, both (a) by hand and (b) with a calculator.

Solution   **a.** $2 \cdot 3^2 = 2 \cdot 9$ exponents come before multiplication
    $= 18$

b. Type

2 $\boxed{\times}$ 3 $\boxed{x^2}$ $\boxed{=}$

and the display reads "18." Notice that this is typed on a calculator exactly as it is written, except that we must finish the typing with the $\boxed{=}$ button and we must press the $\boxed{\times}$ button to mean multiplication. The $\boxed{\cdot}$ button is a decimal-point button, not a multiplication-dot button.   ∎

EXAMPLE 4   Calculate $(2 \cdot 3)^2$, both (a) by hand and (b) with a calculator.

Solution   **a.** $(2 \cdot 3)^2 = 6^2$ parentheses-enclosed work comes first
    $= 36$

b. Type

$\boxed{(}$ 2 $\boxed{\times}$ 3 $\boxed{)}$ $\boxed{x^2}$

and the display reads "36." Notice that this is typed on a calculator exactly as it is written, except that we must use the $\boxed{\times}$ button to mean multiplication. Also, we do not use the $\boxed{=}$ button, because the $\boxed{x^2}$ button performs an operation on a single number.   ∎

EXAMPLE 5   Calculate $4 \cdot 2^3$, both (a) by hand and (b) with a calculator.

Solution   **a.** $4 \cdot 2^3 = 4 \cdot 8$ exponents come before multiplication
    $= 32$

b. To do this, we must use the exponent button, which is labeled either $\boxed{y^x}$ or $\boxed{x^y}$. Type

4 $\boxed{\times}$ 2 $\boxed{y^x}$ 3 $\boxed{=}$ or   4 $\boxed{\times}$ 2 $\boxed{x^y}$ 3 $\boxed{=}$

and the display reads "32." This is typed on a calculator exactly as it is written, except that we must press the $\boxed{\times}$ button to mean multiplication and the $\boxed{=}$ button to finish the calculation.   ∎

Sometimes, you don't perform a calculation on your calculator in the same way that it is written, even though calculators are programmed to follow the order of operations.

EXAMPLE 6   Calculate $\dfrac{2}{3 \cdot 4}$ with a calculator.

Solution   **Wrong**   It is incorrect to type

2 $\boxed{\div}$ 3 $\boxed{\times}$ 4 $\boxed{=}$

According to the order of operations, multiplication and division are done *from left to right,* so the above typing is algebraically equivalent to

$= \dfrac{2}{3} \cdot 4$ first dividing then multiplying, since division is on the left and multiplication is on the right

$= \dfrac{2}{3} \cdot \dfrac{4}{1} = \dfrac{2 \cdot 4}{3}$

which is not what we want. The difficulty is that the large fraction bar in the expression $\frac{2}{3 \cdot 4}$ groups the "$3 \cdot 4$" together in the denominator; in the above typing, nothing groups the "$3 \cdot 4$" together, and only the 3 ends up in the denominator.

Right The calculator needs parentheses inserted in the following manner:

$$\frac{2}{(3 \cdot 4)}$$

Thus, it is correct to type

2 $\boxed{\div}$ $\boxed{(}$ 3 $\boxed{\times}$ 4 $\boxed{)}$ $\boxed{=}$

This makes the display read 0.166666667, the correct answer.

Also Right It is correct to type

2 $\boxed{\div}$ 3 $\boxed{\div}$ 4 $\boxed{=}$

According to the order of operations, multiplication and division are done from left to right, so the above typing is algebraically equivalent to

$$\frac{2}{3} \div 4 \qquad \text{doing the left-hand division first}$$

$$= \frac{2}{3} \cdot \frac{1}{4} \qquad \text{inverting and multiplying}$$

$$= \frac{2}{3 \cdot 4}$$

which is what we want. *When you're calculating something that involves only multiplication and division and you don't use parentheses, the $\boxed{\times}$ button places a factor in the numerator, and the $\boxed{\div}$ button places a factor in the denominator.* ∎

EXAMPLE 7 Calculate $\dfrac{2}{3/4}$ with a calculator.

Solution **Wrong** It is incorrect to type

2 $\boxed{\div}$ 3 $\boxed{\div}$ 4 $\boxed{=}$

even though that matches the way the problem is written algebraically. As discussed in Example 6, this typing is algebraically equivalent to

$$\frac{2}{3 \cdot 4}$$

which is not what we want.

Right The calculator needs parentheses inserted in the following manner:

$$\frac{2}{(3 \div 4)}$$

Thus, it is correct to type

2 $\boxed{\div}$ $\boxed{(}$ 3 $\boxed{\div}$ 4 $\boxed{)}$ $\boxed{=}$

since, according to the order of operations, parentheses-enclosed work is done first. This makes the display read 2.66666667, the correct answer. ∎

EXAMPLE 8   Calculate $\dfrac{2 + 3}{4}$ with a calculator.

Solution   **Wrong**   It is incorrect to type

2 $\boxed{+}$ 3 $\boxed{\div}$ 4 $\boxed{=}$

According to the order of operations, division is done before addition, so this typing is algebraically equivalent to

$$2 + \frac{3}{4}$$

which is not what we want. The large fraction bar in the expression $\frac{2+3}{4}$ groups the "2 + 3" together in the numerator; in the above typing, nothing groups the "2 + 3" together, and only the 3 ends up in the numerator.

Right   The calculator needs parentheses inserted in the following manner:

$$\frac{(2 + 3)}{4}$$

Thus, it is correct to type

$\boxed{(}$ 2 $\boxed{+}$ 3 $\boxed{)}$ $\boxed{\div}$ 4 $\boxed{=}$

This makes the display read 1.25, the correct answer.

Also Right   It is correct to type

2 $\boxed{+}$ 3 $\boxed{=}$ $\boxed{\div}$ 4 $\boxed{=}$

The first $\boxed{=}$ makes the calculator perform all prior calculations before continuing. This too makes the display read "1.25," the correct answer.   ∎

The Shift Button

A scientific calculator's function buttons each have two labels and two uses. For example, your calculator might have a button that is labeled $\boxed{x^2}$ on the button itself and "$\sqrt{x}$" above the button. If so, typing

4 $\boxed{x^2}$

makes the display read 16, since $4^2 = 16$. To take the square root of 4, you have to use the shift button. The shift button is so named because of its similarity to the shift key on a keyboard, which determines whether the letter typed is uppercase or lowercase. Your shift button may be labeled $\boxed{\text{SHIFT}}$, $\boxed{\text{2nd}}$, or $\boxed{\text{INV}}$. Typing

4 $\boxed{\text{SHIFT}}$ $\boxed{x^2}$ or   4 $\boxed{\text{2nd}}$ $\boxed{x^2}$ or   4 $\boxed{\text{INV}}$ $\boxed{x^2}$

makes the display read 2, since $\sqrt{4} = 2$.

Some calculators have a button that is labeled $\boxed{\sqrt{x}}$ on the button itself and "x^2" above the button. If yours is like this, then typing

4 $\boxed{\sqrt{x}}$

makes the display read 2, and typing

4 $\boxed{\text{INV}}$ $\boxed{\sqrt{x}}$

makes the display read 16. In either case, *the label on the button says what the button does without using the shift button, and the label above the button says what the button does using the shift button.*

Frequently, the two functions that share a function button are functions that "undo" each other. For example, typing

$$3 \quad \boxed{x^2}$$

makes the display read 9, since $3^2 = 9$, and typing

$$9 \quad \boxed{\text{INV}} \quad \boxed{x^2}$$

makes the display read 3, since $\sqrt{9} = 3$. This is done as a memory device; it is easier to find the various operations on the keyboard if the two operations that share a button also share a relationship.

Notice that this doesn't always work. For example, typing

$$3 \quad \boxed{+/-} \quad \boxed{x^2}$$

makes the display read 9, since $(-3)^2 = 9$, but typing

$$9 \quad \boxed{\text{INV}} \quad \boxed{x^2}$$

won't make the display read -3. Two functions that *always* undo each other are called **inverses.** The x^2 and $\sqrt{x}$ functions are not inverses, because of the above counterexample. However, there is an inverse-type relationship between the x^2 and $\sqrt{x}$ functions (they undo each other sometimes), and that is why the "shift" button is sometimes labeled $\boxed{\text{INV}}$.

Memory

The memory is a place to store a number for later use, without having to write it down. If a number is on your display, you can place it into the memory (or **store** it) by pressing the button labeled $\boxed{\text{STO}}$ or $\boxed{x \rightarrow M}$ or $\boxed{\text{M in}}$, and you can take it out of the memory (or **recall** it) by pressing the button labeled $\boxed{\text{RCL}}$ or $\boxed{\text{RM}}$ or $\boxed{\text{MR}}$. (WARNING: A button labeled $\boxed{\text{M+}}$ does *not* store; it adds the number on the display to whatever is currently stored.) In the above button labels, "M" stands for *memory* and "R" stands for *recall*.

Typing

$$5 \quad \boxed{\text{STO}} \qquad \text{or} \qquad 5 \quad \boxed{x \rightarrow M} \qquad \text{or} \qquad 5 \quad \boxed{\text{M in}}$$

makes the calculator store a 5 in its memory. If you do other calculations or just clear your display and later press

$$\boxed{\text{RCL}} \qquad \text{or} \qquad \boxed{\text{RM}} \qquad \text{or} \qquad \boxed{\text{MR}}$$

then your display will read "5."

Some calculators have more than one memory. If yours does, then pressing the button labeled $\boxed{\text{STO}}$ won't do anything; pressing $\boxed{\text{STO}}$ and then "1" will store it in memory number 1; pressing $\boxed{\text{STO}}$ and then "2" will store it in memory number 2, and so on. Pressing $\boxed{\text{RCL}}$ and then "1" will recall what has been stored in memory number 1.

EXAMPLE 9   Use the quadratic formula and your calculator's memory to solve

$$2.3x^2 + 4.9x + 1.5 = 0$$

Solution   The quadratic formula says that if $ax^2 + bx + c = 0$, then

$$x = \frac{-b \pm \sqrt{b^2 - 4ac}}{2a}$$

We have $2.3x^2 + 4.9x + 1.5 = 0$, so $a = 2.3$, $b = 4.9$, and $c = 1.5$. This gives

$$x = \frac{-4.9 \pm \sqrt{4.9^2 - 4 \cdot 2.3 \cdot 1.5}}{2 \cdot 2.3}$$

The quickest way to do this calculation is to calculate the radical, store it, and then calculate the two fractions.

Step 1   *Calculate the radical.* To do this, type

4.9 $\boxed{x^2}$ $\boxed{-}$ 4 $\boxed{\times}$ 2.3 $\boxed{\times}$ 1.5 $\boxed{=}$ $\boxed{\sqrt{x}}$ $\boxed{\text{STO}}$

This makes the display read "3.195309" and stores the number in the memory. Notice the use of the $\boxed{=}$ button; this makes the calculator finish the prior calculation before taking a square root. If the $\boxed{=}$ button were not used, the order of operations would require the calculator to take the square root of 1.5.

Step 2   *Calculate the first fraction.* To do this, type

4.9 $\boxed{+/-}$ $\boxed{+}$ $\boxed{\text{RCL}}$ $\boxed{=}$ $\boxed{\div}$ 2 $\boxed{\div}$ 2.3 $\boxed{=}$

This makes the display read "−0.3705849." Notice the use of the $\boxed{=}$ button.

Step 3   *Calculate the second fraction.* To do this, type

4.9 $\boxed{+/-}$ $\boxed{-}$ $\boxed{\text{RCL}}$ $\boxed{=}$ $\boxed{\div}$ 2 $\boxed{\div}$ 2.3 $\boxed{=}$

This makes the display read "−1.7598498."
The solutions to $2.3x^2 + 4.9x + 1.5 = 0$ are $x = -0.3705849$ and $x = -1.7598498$. These are approximate solutions in that they show only the first seven decimal places.

Step 4   *Check your solutions.* These solutions can be checked by seeing if they satisfy the equation $2.3x^2 + 4.9x + 1.5 = 0$. To check the first solution, type

2.3 $\boxed{\times}$ .3705849 $\boxed{+/-}$ $\boxed{x^2}$ $\boxed{+}$ 4.9 $\boxed{\times}$ .3705849 $\boxed{+/-}$ $\boxed{+}$ 1.5 $\boxed{=}$

and the display will read either "0" or a number very close to 0.   ∎

Scientific Notation

Typing

4000000 $\boxed{\times}$ 8000000 $\boxed{=}$

makes the display read "3.2   13" rather than "32000000000000." This is because the calculator does not have enough room on its display for "32000000000000." When the display shows "3.2   13," read it as "3.2×10^{13}," which is written in scientific notation. Literally, "3.2×10^{13}" means "multiply 3.2 by 10, thirteen times," but as a shortcut, you can interpret it as "move the decimal point in the '3.2' thirteen places to the right."

Typing

.0000005 $\boxed{\times}$ .0000007 $\boxed{=}$

makes the display read "3.5   −13" rather than "0.00000000000035," because the calculator does not have enough room on its display for "0.00000000000035." Read "3.5   −13" as "3.5×10^{-13}." Literally, this means "divide 3.5 by 10, thirteen times," but as a shortcut, you can interpret it as "move the decimal point in the '3.5' thirteen places to the left."

You can type a number in scientific notation by using the button labeled $\boxed{\text{EXP}}$ (which stands for *exponent*) or $\boxed{\text{EE}}$ (which stands for *enter exponent*). For example, typing

$$5.2 \;\boxed{\text{EXP}}\; 8 \quad \text{or} \quad 5.2 \;\boxed{\text{EE}}\; 8$$

makes the display read "5.2   8," which means "5.2×10^8," and typing

$$3 \;\boxed{\text{EXP}}\; 17 \;\boxed{+/-} \quad \text{or} \quad 3 \;\boxed{\text{EE}}\; 17 \;\boxed{+/-}$$

makes the display read "3   −17," which means "3×10^{-17}." Notice that the sign-change button is used to make the exponent negative.

Be careful that you don't confuse the $\boxed{y^x}$ button with the $\boxed{\text{EXP}}$ button. The $\boxed{\text{EXP}}$ button does *not* allow you to type in an exponent; it allows you to type in scientific notation. For example, typing

$$3 \;\boxed{\text{EXP}}\; 4$$

makes the display read "3   4," which means "3×10^4," and typing

$$3 \;\boxed{y^x}\; 4$$

makes the display read "81," since $3^4 = 81$.

Exercises

Perform the following calculations. The correct answer is given in brackets [ ]. In your homework, write down what you type on your calculator to get that answer. Answers are not given in the back of the book.

1. $-3 - -5$ [2]   **2.** $-6 - 3$ [−9]
3. $4 - -9$ [13]   **4.** $-6 - -8$ [2]
5. $-3 - (-5 - -8)$ [−6]
6. $-(-4 - 3) - (-6 - -2)$ [11]
7. $-8 \cdot -3 \cdot -2$ [−48]
8. $-9 \cdot -3 -2$ [25]
9. $(-3)(-8) - (-9)(-2)$ [6]
10. $2(3 - 5)$ [−4]   **11.** $2 \cdot 3 - 5$ [1]
12. $4 \cdot 11^2$ [484]   **13.** $(4 \cdot 11)^2$ [1,936]
14. $4 \cdot (-11)^2$ [484]   **15.** $4 \cdot (-3)^3$ [−108]

WARNING: Some calculators will not raise a negative number to a power. If yours has this characteristic, how can you use your calculator on this exercise?

16. $(4 \cdot -3)^3$ [−1,728]

17. $\dfrac{3 + 2}{7}$ [0.7142857]

18. $\dfrac{3 \cdot 2}{7}$ [0.8571429]

19. $\dfrac{3}{2 \cdot 7}$ [0.2142857]

20. $\dfrac{3 \cdot 2}{7 \cdot 5}$ [0.1714286]

21. $\dfrac{3 + 2}{7 \cdot 5}$ [0.1428571]

22. $\dfrac{3 \cdot -2}{7 + 5}$ [−0.5]   **23.** $\dfrac{3}{7/2}$ [0.8571429]

24. $\dfrac{3/7}{2}$ [0.2142857]

25. 1.8^2 [3.24]
26. $\sqrt{1.8}$ [1.3416408]
27. $47{,}000{,}000^2$ [2.209×10^{15}]
28. $\sqrt{0.0000000000027}$ [1.643168×10^{-6}]
29. $(-3.92)^7$ [−14,223.368737]
30. $(5.72 \times 10^{19})^4$ [1.070494×10^{79}]
31. $(3.76 \times 10^{-12})^{-5}$ [1.330641×10^{57}]
32. $(3.76 \times 10^{-12}) -5$ [−5]
33. Solve $4.2x^2 + 8.3x + 1.1 = 0$ for x. Check your two answers by substituting them back into the equation.
34. Solve $5.7x^2 + 12.3x - 8.1 = 0$ for x. Check your two answers by substituting them back into the equation.
35. Which of the following buttons must be used in conjunction with the $\boxed{=}$ button, and why?

$$\boxed{+}\;\boxed{-}\;\boxed{\times}\;\boxed{\div}\;\boxed{x^2}\;\boxed{\sqrt{x}}\;\boxed{y^x}\;\boxed{1/x}\;\boxed{+/-}$$

B Using a Graphing Calculator

T he following discussion was written specifically for Texas Instruments (or "TI") and Casio graphing calculators, but it frequently applies to other brands as well. Texas Instruments TI-83, TI-84, and TI-86 models and Casio CFX-9850, FX-9860, and CFX-9950 models are specifically addressed. Read this discussion with your calculator close at hand. When a calculation is discussed, do that calculation on your calculator.

To do any of the calculations discussed in this section on a Casio, start by pressing $\boxed{\text{MENU}}$. Then use the arrow buttons to select "RUN" on the main menu, and press $\boxed{\text{EXE}}$ (which stands for "execute").

The Enter Button

A TI graphing calculator will never perform a calculation until the $\boxed{\text{ENTER}}$ button is pressed; a Casio will never perform a calculation until the $\boxed{\text{EXE}}$ button is pressed. A Casio's $\boxed{\text{EXE}}$ button functions like a TI's $\boxed{\text{ENTER}}$ button; often, we will refer to either of these buttons as the $\boxed{\text{ENTER}}$ button.
To add 3 and 2, type

$\quad$ 3 $\boxed{+}$ 2 $\boxed{\text{ENTER}}$

and the display will read 5. To square 4, type

$\quad$ 4 $\boxed{x^2}$ $\boxed{\text{ENTER}}$

and the display will read 16. If the $\boxed{\text{ENTER}}$ button isn't pressed, the calculation will not be performed.

The 2nd and Alpha Buttons

Most calculator buttons have more than one label and more than one use; you select from these uses with the $\boxed{\text{2nd}}$ and $\boxed{\text{ALPHA}}$ buttons. A Casio's $\boxed{\text{SHIFT}}$ button functions like a TI's $\boxed{\text{2nd}}$ button; often, we will refer to either of these buttons as the $\boxed{\text{2nd}}$ button. For example, one button is labeled "x^2" on the button itself, "$\sqrt{\ }$" above the button, and either "I" or "K" above and to the right of the button. If it is used without the $\boxed{\text{2nd}}$ or $\boxed{\text{ALPHA}}$ buttons, it will square a number. Typing

$\quad$ 4 $\boxed{x^2}$ $\boxed{\text{ENTER}}$

makes the display read 16, since $4^2 = 16$. If it is used with the $\boxed{\text{2nd}}$ button, it will take the square root of a number. Typing

$\quad$ $\boxed{\text{2nd}}$ $\boxed{\sqrt{\ }}$ 4 $\boxed{\text{ENTER}}$

makes the display read 2, since $\sqrt{4} = 2$. If it is used with the $\boxed{\text{ALPHA}}$ button, it will display the letter I or K.

Notice that to square 4, you press the $\boxed{x^2}$ button *after* the 4, but to take the square root of 4, you press the $\boxed{\sqrt{}}$ button *before* the 4. This is because graphing calculators are designed so that the way you type something is as similar as possible to the way it is written algebraically. When you write 4^2, you write the 4 first and then the squared symbol; thus, on your graphing calculator, you press the 4 first and then the $\boxed{x^2}$ button. When you write $\sqrt{4}$, you write the square root symbol first and then the 4; thus, on your graphing calculator, you press the $\boxed{\sqrt{}}$ button first and then the 4.

Frequently, the two operations that share a button are operations that "undo" each other. For example, typing

3 $\boxed{x^2}$ $\boxed{\text{ENTER}}$

makes the display read 9, since $3^2 = 9$, and typing

$\boxed{\text{2nd}}$ $\boxed{\sqrt{}}$ 9 $\boxed{\text{ENTER}}$

makes the display read 3, since $\sqrt{9} = 3$. This is done as a memory device; it is easier to find the various operations on the keyboard if the two operations that share a button also share a relationship.

Two operations that *always* undo each other are called **inverses.** The x^2 and $\sqrt{x}$ operations are not inverses because $(-3)^2 = 9$, but $\sqrt{9} \neq -3$. However, there is an inverse-type relationship between the x^2 and $\sqrt{x}$ operations—they undo each other sometimes. When two operations share a button, they are inverses or they share an inverse-type relation.

Correcting Typing Errors

If you've made a typing error *and you haven't yet pressed* $\boxed{\text{ENTER}}$, you can correct that error with the $\boxed{\blacktriangleleft}$ button. For example, if you typed "$5 \times 2 + 7$" and then realized you wanted "$5 \times 3 + 7$," you can replace the incorrect 2 with a 3 by pressing the $\boxed{\blacktriangleleft}$ button until the 2 is flashing, and then press "3."

If you realize that you've made a typing error *after* you pressed $\boxed{\text{ENTER}}$, just press $\boxed{\text{2nd}}$ $\boxed{\text{ENTRY}}$ to reproduce the previously entered line. (With a Casio, press $\boxed{\blacktriangleleft}$.) Then correct the error with the $\boxed{\blacktriangleleft}$ button, as described above.

The $\boxed{\text{INS}}$ button allows you to insert a character. For example, if you typed "5×27" and you meant to type "5×217," press the $\boxed{\blacktriangleleft}$ button until the 7 is flashing, and then insert 1 by typing

$\boxed{\text{2nd}}$ $\boxed{\text{INS}}$ 1

The $\boxed{\text{DEL}}$ button allows you to delete a character. For example, if you typed "5×217" and you meant to type "5×27," press the $\boxed{\blacktriangleleft}$ button until the 1 is flashing, and then press $\boxed{\text{DEL}}$.

If you haven't yet pressed $\boxed{\text{ENTER}}$, the $\boxed{\text{CLEAR}}$ button erases an entire line. If you have pressed $\boxed{\text{ENTER}}$, the $\boxed{\text{CLEAR}}$ button clears everything off of the screen. With a Casio, the $\boxed{\text{AC}}$ button functions in the same way. "AC" stands "all clear."

The Subtraction Symbol and the Negative Symbol

If you read "$3 - -2$" aloud, you could say "three subtract negative two" or "three minus minus two," and you would be understood. The expression is understandable even if the distinction between the negative symbol and the subtraction symbol is not made clear. With a calculator, however, this distinction is crucial. The subtraction button is labeled "−," and the negative button is labeled "(−)."

EXAMPLE 1   Calculate $3 - -2$, both (a) by hand and (b) with a calculator.

Solution   **a.** $3 - -2 = 3 + 2 = 5$

b. Type

3 $\boxed{-}$ $\boxed{(-)}$ 2 $\boxed{\text{ENTER}}$

and the display will read 5.   ∎

If you had typed

3 $\boxed{-}$ $\boxed{-}$ 2 $\boxed{\text{ENTER}}$ or   3 $\boxed{(-)}$ $\boxed{-}$ 2 $\boxed{\text{ENTER}}$

the calculator would have responded with an error message.

The Multiplication Symbol

In algebra, we do not use "x" for multiplication. Instead, we use "x" as a variable, and we use "·" for multiplication. However, Texas Instruments graphing calculators use "×" as the label on the multiplication button, and "*" for multiplication on the display screen. (The "variable x" button is labeled either "X,T,θ,n", or "x-VAR.") This is one of the few instances in which you don't type things in the same way that you write them algebraically.

Order of Operations and Use of Parentheses

Texas Instruments and Casio graphing calculators are programmed to follow the order of operations. That is, they perform calculations in the following order:

1. **P**arentheses-enclosed work

2. **E**xponents

3. **M**ultiplication and
   Division, from left to right

4. **A**ddition and
   Subtraction, from left to right

You can remember this order by remembering the word "**PEMDAS**," which stands for

Parentheses/**E**xponents/**M**ultiplication/**D**ivision/**A**ddition/**S**ubtraction

EXAMPLE 2   Calculate $2(3 + 4)$, both (a) by hand and (b) with a calculator.

Solution   **a.** $2(3 + 4) = 2(7)$ parentheses-enclosed work comes first
                    $= 14$

b. Type

2 $\boxed{\times}$ $\boxed{(}$ 3 $\boxed{+}$ 4 $\boxed{)}$ $\boxed{\text{ENTER}}$

and the display will read 14.   ∎

In the instructions to Example 2, notice that we wrote "2(3 + 4)" rather than "2 · (3 + 4)"; in this case, it's not necessary to write the multiplication symbol. Similarly, it's not necessary to type the multiplication symbol. Example 2b could be computed by typing

$$2 \;\boxed{(}\; 3 \;\boxed{+}\; 4 \;\boxed{)}\; \boxed{\text{ENTER}}$$

EXAMPLE 3   Calculate $2 \cdot 3^3$, both (a) by hand and (b) with a calculator.

Solution   **a.** $2 \cdot 3^3 = 2 \cdot 27$ exponents come before multiplication
$$= 54$$
b. To do this, we must use the exponent button, which is labeled "∧." Type

$$2 \;\boxed{\times}\; 3 \;\boxed{\wedge}\; 3 \;\boxed{\text{ENTER}}$$

and the display will read 54.   ∎

EXAMPLE 4   Calculate $(2 \cdot 3)^3$, both (a) by hand and (b) with a calculator.

Solution   **a.** $(2 \cdot 3)^3 = 6^3$ parentheses-enclosed work comes first
$$= 216$$
b. Type

$$\boxed{(}\; 2 \;\boxed{\times}\; 3 \;\boxed{)}\; \boxed{\wedge}\; 3 \;\boxed{\text{ENTER}}$$

and the display will read 216.   ∎

In Example 3, the exponent applies only to the 3, because the order of operations dictates that exponents come before multiplication. In Example 4, the exponent applies to the (2 · 3), because the order of operations dictates that parentheses-enclosed work comes before exponents. In each example, the way you type the problem matches the way it is written algebraically, because the calculator is programmed to follow the order of operations. Sometimes, however, the way you type a problem doesn't match the way it is written algebraically.

EXAMPLE 5   Calculate $\dfrac{2}{3 \cdot 4}$ with a calculator.

Solution   **Wrong**   It is incorrect to type

$$2 \;\boxed{\div}\; 3 \;\boxed{\times}\; 4 \;\boxed{\text{ENTER}}$$

even though that matches the way the problem is written algebraically. According to the order of operations, multiplication and division are done *from left to right,* so the above typing is algebraically equivalent to

$$= \frac{2}{3} \cdot 4 \qquad \text{first dividing and then multiplying, since division is on the left and}$$
$$\text{multiplication is on the right}$$

$$= \frac{2}{3} \cdot \frac{4}{1} = \frac{2 \cdot 4}{3}$$

which is not what we want. The difficulty is that the large fraction bar in the expression $\frac{2}{3 \cdot 4}$ groups the "3 · 4" together in the denominator; in the above typing, nothing groups the "3 · 4" together, and only the 3 ends up in the denominator.

Right   The calculator needs parentheses inserted in the following manner:

$$\frac{2}{(3 \cdot 4)}$$

Thus, it is correct to type

2 ÷ [(3 × 4)] [ENTER]

This makes the display read 0.166666667, the correct answer.

Also Right   It is correct to type

2 ÷ 3 ÷ 4 [ENTER]

According to the order of operations, multiplication and division are done from left to right, so the above typing is algebraically equivalent to

$$\frac{2}{3} \div 4 \qquad \text{doing the left-hand division first}$$

$$= \frac{2}{3} \cdot \frac{1}{4} \qquad \text{inverting and multiplying}$$

$$= \frac{2}{3 \cdot 4}$$

which is what we want. *When you're calculating something that involves only multiplication and division and you don't use parentheses, the* × *button places a factor in the numerator, and the* ÷ *button places a factor in the denominator.* ■

EXAMPLE 6   Calculate $\dfrac{2}{3/4}$ with a calculator.

Solution   **Wrong**   It is incorrect to type

2 ÷ 3 ÷ 4 [ENTER]

even though that matches the way the problem is written algebraically. As was discussed in Example 5, this typing is algebraically equivalent to

$$\frac{2}{3 \cdot 4}$$

which is not what we want.

Right   The calculator needs parentheses inserted in the following manner:

$$\frac{2}{(3/4)}$$

Thus, it is correct to type

2 ÷ [(3 ÷ 4)] [ENTER]

since, according to the order of operations, parentheses-enclosed work is done first. This makes the display read 2.6666667, the correct answer. ■

EXAMPLE 7   Calculate $\dfrac{2+3}{4}$ with a calculator.

Solution   **Wrong**   It is incorrect to type

2 $\boxed{+}$ 3 $\boxed{\div}$ 4 $\boxed{\text{ENTER}}$

even though that matches the way the problem is written algebraically. According to the order of operations, division is done before addition, so this typing is algebraically equivalent to

$$2 + \dfrac{3}{4}$$

which is not what we want. The large fraction bar in the expression $\frac{2+3}{4}$ groups the "2 + 3" together in the numerator; in the above typing, nothing groups the "2 + 3" together, and only the 3 ends up in the numerator.

Right   The calculator needs parentheses inserted in the following manner:

$$\dfrac{(2+3)}{4}$$

Thus, it is correct to type

$\boxed{(}$ 2 $\boxed{+}$ 3 $\boxed{)}$ $\boxed{\div}$ 4 $\boxed{\text{ENTER}}$

This makes the display read 1.25, the correct answer.

Also Right   It is correct to type

2 $\boxed{+}$ 3 $\boxed{\text{ENTER}}$ $\boxed{\div}$ 4 $\boxed{\text{ENTER}}$

The first $\boxed{\text{ENTER}}$ makes the calculator perform all prior calculations before continuing. This too makes the display read 1.25, the correct answer.   ∎

Memory

The **memory** is a place to store a number for later use, without having to write it down. Graphing calculators have a memory for each letter of the alphabet; that is, you can store one number in memory A, a second number in memory B, and so on. Pressing

TI-83/84	5 $\boxed{\text{STO}\blacktriangleright}$ $\boxed{\text{ALPHA}}$ $\boxed{\text{A}}$ $\boxed{\text{ENTER}}$	
TI-86	5 $\boxed{\text{STO}\blacktriangleright}$ $\boxed{\text{A}}$ $\boxed{\text{ENTER}}$	
Casio	5 $\boxed{\rightarrow}$ $\boxed{\text{ALPHA}}$ $\boxed{\text{A}}$ $\boxed{\text{EXE}}$	

will store 5 in memory A. Similar keystrokes will store in memory B. Pressing $\boxed{\text{ALPHA}}$ $\boxed{\text{A}}$ will recall what has been stored in memory A.

EXAMPLE 8   Calculate $\dfrac{3 + \dfrac{5+7}{2}}{4}$ (a) by hand and (b) by first calculating $\frac{5+7}{2}$ and storing the result.

Solution   **a.** $\dfrac{3 + \dfrac{5+7}{2}}{4} = \dfrac{3 + \dfrac{12}{2}}{4} = \dfrac{3+6}{4} = \dfrac{9}{4} = 2.25$

b. First, calculate $\frac{5+7}{2}$ and store the result in memory A. The large fraction bar in this expression groups the "5 + 7" together in the numerator; in our typing, we must group the "5 + 7" together with parentheses. The calculator needs parentheses inserted in the following manner:

$$\frac{(5+7)}{2}$$

Type

What remains is to compute $\frac{3+A}{4}$. Again, the large fraction bar groups the "3 + A" together, so the calculator needs parentheses inserted in the following manner:

$$\frac{(3+A)}{4}$$

Type

　$\boxed{(}$　3　$\boxed{+}$　$\boxed{\text{ALPHA}}$　$\boxed{\text{A}}$　$\boxed{)}$　$\boxed{\div}$　4　$\boxed{\text{ENTER}}$

and the display will read 2.25.   ■

EXAMPLE 9   Calculate $\dfrac{3 + \dfrac{5+7}{2}}{4}$ using one line of instructions and without using the memory.

Solution   There are two large fractions bars, one grouping the "5 + 7" together and one grouping the "$3 + \frac{5+7}{2}$" together. In our typing, we must group each of these together with parentheses. The calculator needs parentheses inserted in the following manner:

$$\frac{\left(3 + \dfrac{(5+7)}{2}\right)}{4}$$

Type

　$\boxed{(}$　3　$\boxed{+}$　$\boxed{(}$　5　$\boxed{+}$　7　$\boxed{)}$　$\boxed{\div}$　2　$\boxed{)}$　$\boxed{\div}$　4　$\boxed{\text{ENTER}}$

and the display will read 2.25.   ■

EXAMPLE 10   **a.** Use the quadratic formula and your calculator's memory to solve $2.3x^2 + 4.9x + 1.5 = 0$.

b. Check your answers.

Solution   **a.** According to the Quadratic Formula, if $ax^2 + bx + c = 0$, then

$$x = \frac{-b \pm \sqrt{b^2 - 4ac}}{2a}$$

For our problem, $a = 2.3$, $b = 4.9$ and $c = 1.5$. This gives

$$x = \frac{-4.9 \pm \sqrt{4.9^2 - 4 \cdot 2.3 \cdot 1.5}}{2 \cdot 2.3}$$

One way to do this calculation is to calculate the radical, store it, and then calculate the two fractions.

Step 1   *Calculate the radical.* To do this, type

This makes the display read "3.195309062," and stores the number in the memory A. Notice the use of parentheses.

Step 2   *Calculate the first fraction.* The first fraction is

$$\frac{-4.9 \pm \sqrt{4.9^2 - 4 \cdot 2.3 \cdot 1.5}}{2 \cdot 2.3}$$

However, the radical has already been calculated and stored in memory A, so this is equivalent to

$$\frac{-4.9 + A}{2 \cdot 2.3}$$

Type

This makes the display read "−0.3705849866" and stores the number in memory B.

Step 3   *Calculate the second fraction.* The second fraction is

$$\frac{-4.9 - \sqrt{4.9^2 - 4 \cdot 2.3 \cdot 1.5}}{2 \cdot 2.3} = \frac{-4.9 - A}{2 \cdot 2.3}$$

Memories A and B are already in use, so we will store this in memory C. Instead of retyping the line in step 2 with the "+" changed to a "−" and the "B" to a "C," press $\boxed{2nd}$ $\boxed{ENTRY}$ ($\boxed{\leftarrow}$ with a Casio) to reproduce that line, and use the $\boxed{\blacktriangleleft}$ button to make these changes. Press $\boxed{ENTER}$ and the display will read −1.759849796, and that number will be stored in memory C. The solutions to $2.3x^2 + 4.9x + 1.5 = 0$ are $x = -0.370584966$ and $x = -1.759849796$. These are approximate solutions in that they show only the first nine decimal places.

Step 4   *Check your solutions.* These two solutions are stored in memories B and C; they can be checked by seeing if they satisfy the equation $2.3x^2 + 4.9x + 1.5 = 0$. To check the first solution, type

2.3 $\boxed{\times}$ $\boxed{ALPHA}$ $\boxed{B}$ $\boxed{x^2}$ $\boxed{+}$ 4.9 $\boxed{\times}$ $\boxed{ALPHA}$ $\boxed{B}$ $\boxed{+}$ 1.5 $\boxed{ENTER}$

and the display should read either "0" or a number very close to 0.   ∎

Scientific Notation

Typing

4000000 $\boxed{\times}$ 8000000 $\boxed{ENTER}$

makes the display read "3.2E13" rather than "32000000000000." This is because the calculator does not have enough room on its display for "32000000000000." When the display shows "3.2E13," read it as "3.2×10^{13}," which is written in scientific notation. Literally, "3.2×10^{13}" means "multiply 3.2 by 10, thirteen times," but as a shortcut, you can interpret it as "move the decimal point in the '3.2' thirteen places to the right."

Typing

.0000005 $\boxed{\times}$ .0000007 $\boxed{ENTER}$

makes the display read "3.5E-13" rather than "0.00000000000035," because the calculator does not have enough room on its display for "0.00000000000035." Read "3.5E-13" as "3.5×10^{-13}." Literally, this means "divide 3.5 by 10, thirteen times," but as a shortcut, you can interpret it as "move the decimal point in the '3.5' thirteen places to the left."

You can type a number in scientific notation by using the TI button labeled "EE" (which stands for "Enter Exponent") or the Casio button labeled "EXP" (which stands for "Exponent"). For example, typing

5.2 [EE] 8 [ENTER]

makes the display read "52000000." (If the "EE" label is above the button, you will need to use the [2nd] button.) On a Casio, type

5.2 [EXP] 8 [EXE]

Be careful that you don't confuse the [EE] or [EXP] button with the [∧] button. The [EE] or [EXP] button does *not* allow you to type in an exponent; it allows you to type in scientific notation. For example, typing

3 [EE] 4 [ENTER]

makes the display read "30000," since $3 \times 10^4 = 30,000$. Typing

3 [∧] 4 [ENTER]

makes the display read "81," since $3^4 = 81$.

Exercises

In Exercises 1–32, use your calculator to perform the given calculation. The correct answer is given in brackets []. In your homework, write down what you type to get that answer. Answers are not given in the back of the book.

1. $-3 - -5$ [2]
2. $-6 - 3$ [−9]
3. $4 - -9$ [13]
4. $-6 - -8$ [2]
5. $-3 - (-5 - -8)$ [−6]
6. $-(-4 - 3) - (-6 - -2)$ [11]
7. $-8 \cdot -3 \cdot -2$ [−48]
8. $-8 \cdot -3 - 2$ [22]
9. $(-3)(-8) - (-9)(-2)$ [6]
10. $2(3 - 5)$ [−4]
11. $2 \cdot 3 - 5$ [1]
12. $4 \cdot 11^2$ [484]
13. $(4 \cdot 11)^2$ [1,936]
14. $4 \cdot (-11)^2$ [484]
15. $4 \cdot (-3)^3$ [−108]
16. $(4 \cdot -3)^3$ [−1,728]
17. $\dfrac{3 + 2}{7}$ [0.7142857]
18. $\dfrac{3 \cdot 2}{7}$ [0.8571429]
19. $\dfrac{3}{2 \cdot 7}$ [0.2142857]
20. $\dfrac{3 \cdot 2}{7 \cdot 5}$ [0.1714286]
21. $\dfrac{3 + 5}{7 \cdot 2}$ [0.1428571]
22. $\dfrac{3 \cdot -2}{7 + 5}$ [−0.5]
23. $\dfrac{3}{7/2}$ [0.8571429]
24. $\dfrac{3/7}{2}$ [0.2142857]
25. 1.8^2 [3.24]
26. $\sqrt{1.8}$ [1.3416408]
27. $47,000,000^2$ [2.209×10^{15}]
28. $\sqrt{0.0000000000027}$ [1.643168×10^{-6}]
29. $(-3.92)^7$ [$-14,223.368737$]
30. $(5.72 \times 10^{19})^4$ [1.070494×10^{79}]
31. $(3.76 \times 10^{-12})^{-5}$ [1.330641×10^{57}]
32. $(3.76 \times 10^{-12}) - 5$ [−5]

In Exercises 33–36, perform the given calculation (a) by hand; (b) with a calculator, using the memory; and (c) with a calculator, using one line of instruction and without using memory, as shown in Examples 8 and 9. In your homework, for parts (b) and (c), write down what you type. Answers are not given in the back of the book.

33. $\dfrac{\dfrac{9-12}{5}+7}{2}$

34. $\dfrac{\dfrac{4-11}{6}+8}{7}$

35. $\dfrac{\dfrac{7+9}{5}+\dfrac{8-14}{3}}{3}$

36. $\dfrac{\dfrac{4-16}{5}+\dfrac{7-22}{2}}{5}$

In Exercises 37–38, use your calculator to solve the given equation for x. Check your two answers, as shown in Example 10. In your homework, write down what you type to get the answers, and what you type to check the answers. Answers are not given in the back of the book.

37. $4.2x^2 + 8.3x + 1.1 = 0$

38. $5.7x^2 + 12.3x - 8.1 = 0$

39. Discuss the use of parentheses in Example 10a, step 1. Why are they necessary? What would happen if they were omitted?

40. Discuss the use of parentheses in Example 10a, step 2. Why are they necessary? What would happen if they were omitted?

41. **a.** Calculate $\dfrac{\dfrac{5+7.1}{3}+\dfrac{2-7.1}{5}}{7}$ using one line of instruction and without using the memory. In your homework, write down what you type as well as the solution.

b. Use the 2nd ENTRY feature to calculate
$$\dfrac{\dfrac{5+7.2}{3}+\dfrac{2-7.2}{5}}{7}$$

c. Use the 2nd ENTRY feature to calculate
$$\dfrac{\dfrac{5+9.3}{3}+\dfrac{2-4.9}{5}}{7}$$

42. **a.** Calculate $\dfrac{3+\dfrac{5+\dfrac{6-8.3}{2}}{3}}{9}$ using one line of instruction and without using the memory. In your homework, write down what you type as well as the solution.

b. Use the 2nd ENTRY feature to calculate
$$\dfrac{3+\dfrac{5-\dfrac{6-8.3}{2}}{3}}{9}$$

c. Use the 2nd ENTRY feature to calculate
$$\dfrac{3-\dfrac{5+\dfrac{6-8.3}{2}}{3}}{9}$$

43. **a.** What is the result of typing "8.1 EE 4"?
b. What is the result of typing "8.1 EE 12"?
c. Why do the instructions in part (b) yield an answer in scientific notation, while the instructions in part (a) yield an answer that's not in scientific notation?
d. By using the MODE button, your calculator can be reset so that all answers will appear in scientific notation. Describe how this can be done.

Graphing with a Graphing Calculator

Graphing with a Texas Instruments Graphing Calculator

The Graphing Buttons

The graphing buttons on a TI graphing calculator are all at the top of the keypad, directly under the screen. The labels on these buttons vary a little from model to model, but their uses are the same. The button labels and their uses are listed in Figure A.1.

TI-83/84	Y=	WINDOW	ZOOM	TRACE	GRAPH
TI-86 labels on buttons	F1	F2	F3	F4	F5
TI-86 labels on screen in graphing mode*	y(x) =	RANGE	ZOOM	TRACE	GRAPH
use this button to tell the calculator:	what to graph	what part of the graph to draw	to zoom in or out	to give the coordinates of a highlighted point	to draw the graph

Figure A.1

Graphing a Line

To graph $y = 2x - 1$ on a TI graphing calculator, follow these steps.

1. **[For TI-86 calculators only]** *Put the calculator into graphing mode* by pressing the GRAPH button. This activates the "F" buttons and puts labels at the bottom of the screen, as shown in Figure A.2.

2. *Set the calculator up for instructions on what to graph* by pressing Y= (y(x)= or F1 on a TI-86). This produces the screen similar to that shown in Figure A.3. If

---

* **TI-86 users** Your calculator is different from the other TI models in that its graphing buttons are labeled "F1" through "F5" ("M1" through "M5" when preceded by the 2nd button). The use of these buttons varies, depending on what you're doing with the calculator. When these buttons are active, their uses are displayed at the bottom of the screen.

your screen has things written after the equals symbols, use the ▲ and ▼ buttons along with the CLEAR button to erase them.

3. *Tell the calculator what to graph* by typing "2x − 1" where the screen reads "Y₁=." To type the x symbol:

Figure A.2
The screen labels for a TI-86's "F" buttons

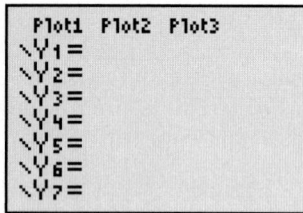

Figure A.3
A TI-84's "Y=" screen (other models' screens are similar).

After typing "2x − 1," press the ENTER button.

4. *Set the calculator up for instructions on what part of the graph to draw.*

5. *Tell the calculator what part of the graph to draw* by entering the values shown in Figure A.4. (If necessary, use the ▲ and ▼ buttons to move from line to line).
    - "*x*min" and "*x*max" refer to the left and right boundaries of the graph, respectively
    - "*y*min" and "*y*max" refer to the lower and upper boundaries of the graph, respectively
    - "*x*scl" and "*y*scl" refer to the scales on the *x*- and *y*-axes (i.e., to the location of the tick marks on the axes)

---

* **TI-86 users**   Your calculator now has a double row of labels at the bottom of the screen. The bottom row of labels refers to the current use of the F1 through F5 buttons, and the top row of labels refers to the current use of the M1 through M5 buttons (which require the use of the 2nd button, as indicated by the orange lettering). Pressing the EXIT button removes one row of labels.

Figure A.4
A TI-84's "WINDOW" screen
(other models' screens are
similar).

Figure A.5
The graph of $y = 2x - 1$.

6. *Tell the calculator to draw a graph* by pressing the GRAPH button.* This produces the screen shown in Figure A.5.

    * **TI-86 users**   Your calculator has two different GRAPH buttons:

    - One is labeled "GRAPH" on the button itself. This button puts the calculator in graphing mode, as discussed in step 1.
    - One is labeled "F5" on the button itself, and "GRAPH" on the screen, when the calculator is in graphing mode. This tells the calculator to draw a graph, as discussed in step 6.

7. *Discontinue graphing* by pressing 2nd QUIT.

Graphing with a Casio Graphing Calculator

To graph $y = 2x - 1$ on a Casio graphing calculator, follow these steps.

1. *Put the calculator into graphing mode* by pressing Menu, using the arrow buttons to highlight "GRAPH" on the main menu, and then pressing EXE.

2. *Tell the calculator what to graph* by typing "$2x - 1$" where the screen reads "Y1=" (see Figure A.6). (If something is already there, press F2, which now has the label "DEL" directly above it on the screen, and then press F1, which is labeled "YES.") To type the x symbol, press x,θ,T. After typing "$2x - 1$," press the EXE button.

Figure A.6
A Casio's "Y=" screen

3. *Tell the calculator to draw a graph* by pressing the F6 button, which now has the label "DRAW" directly above it on the screen.

4. *Set the calculator up for instructions on what part of the graph to draw* by pressing SHIFT and then the F3 button, which now has the label "V-WIN" directly above it on the screen. ("V-WIN" is short for "viewing window.")

5. *Tell the calculator what part of the graph to draw* by entering the values shown in Figure A.7. Press EXE after each entry. Then press EXIT to return to the "Y=" screen.
    - "Xmin" and "Xmax" refer to the left and right boundaries of the graph, respectively.

Figure A.7
A Casio's View Window

- "Ymin" and "Ymax" refer to the lower and upper boundaries of the graph, respectively.
- "Xscale" and "Yscale" refer to scales on the x- and y-axes, respectively (i.e., to the location of the tick marks on the axes).

6. *Tell the calculator to draw a graph* by pressing the [F6] button, which now has the label "DRAW" directly above it on the screen (see Figure A.8).

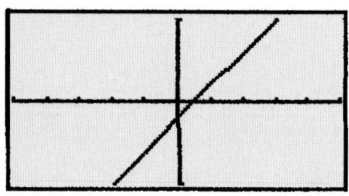

Figure A.8

7. *Discontinue graphing* by pressing [MENU].

Exercises

In the following exercises, you will explore some of your calculator's graphing capabilities. Answers are not given in the back of the book.

1. *Exploring the "Zoom Standard" command.* Use your calculator to graph $y = 2x - 1$, as discussed in this section. When that graph is on the screen, select the TI "Zoom Standard" command from the "Zoom menu" or the Casio "STD" command from the "V-WIN" menu by doing the following:

TI-83/84
- press [ZOOM]
- select option 6 "Standard" or "ZStandard" by either:
- using the down arrow to scroll down to that option and pressing [ENTER], or
- typing the number "6"

TI-86
- press [F3], which is now labeled "ZOOM"
- press [F4], which is now labeled "ZSTD"

Casio
- press [SHIFT]
- press [V-WIN] (i.e., [F3])
- press [STD] (i.e., [F3])

a. What is the result?
b. How else could you accomplish the same thing without using any zoom commands?

2. *Exploring the "RANGE," "WINDOW," or "V-WIN" screen.* Use your calculator to graph $y = 2x - 1$ as discussed in this section. When that graph is on

the screen, use the "RANGE," "WINDOW," or "V-WIN" screen described in steps 4 and 5 of this section to reset the following:
- xmin to 1
- xmax to 20
- ymin to 1
- ymax to 20

a. Why are there no axes shown?
b. Why does the graph start exactly in the lower left corner of the screen?

3. *Exploring the "RANGE," "WINDOW," or "V-WIN" screen.* Use your calculator to graph $y = 2x - 1$, as discussed in this section. When that graph is on the screen, use the "RANGE," "WINDOW," or "V-WIN" screen described in steps 4 and 5 of this section to reset the following:
- xmin to -1
- xmax to -20
- ymin to -1
- ymax to -20

Why did the TI calculator respond with an error message? What odd thing did the Casio calculator do? (Casio hint: use the [TRACE] button to investigate.)

4. *Exploring the "RANGE," "WINDOW," or "V-WIN" screen.* Use your calculator to graph $y = 2x - 1$, as discussed in this section. When that graph is on the screen, use the "RANGE" or "WINDOW" screen described in steps 4 and 5 of this section to reset the following:
- xmin to -5
- xmax to 5
- ymin to 10
- ymax to 25

a. Why was only one axis shown?
b. Why was no graph shown?

5. *Exploring the "TRACE" command.* Use your calculator to graph $y = 2x - 1$, as discussed in this

section. When that graph is on the screen, press TRACE (F4 on a TI-86, F1 on a Casio). This causes two things to happen:

- A mark appears at a point on the line. This mark can be moved with the left and right arrow buttons.
- The corresponding ordered pair is printed out at the bottom of the screen.

  a. Use the "TRACE" feature to locate the line's x-intercept, the point at which the line hits the x-axis. (You may need to approximate it.)
  b. Use algebra, rather than the graphing calculator, to find the x-intercept.
  c. Use the "TRACE" feature to locate another ordered pair on the line.
  d. Use substitution to check that the ordered pair found in part (c) is in fact a point on the line.

6. *Exploring the "ZOOM BOX" command.* Use your calculator to graph $y = 2x - 1$ as discussed in this section. When that graph is on your screen, press ZOOM (F3 on a TI-86, F2 on a Casio) and select option 1, "BOX" or "ZBOX," in the manner described in Exercise 1. This seems to have the same result as TRACE, except the mark does not have to be a point on the line. Use the four arrow buttons to move to a point of your choice (that may be on or off the line). Press ENTER. Use the arrow buttons to move to a different point, so that the resulting box encloses a part of the line. Press ENTER again. What is the result of using the "Zoom Box" command?

7. *Exploring the "ZOOM IN" command.* Use your calculator to graph $y = 2x - 1$, as discussed in this section. When that graph is on your screen, press ZOOM (F3 on a TI-86, F2 on a Casio) and select "Zoom In" or "ZIN," in the manner described in Exercise 1. (Press IN or F3 on a Casio.) This causes a mark to appear on the screen. Use the four arrow buttons to move the mark to a point of your choice, either on or near the line. Press ENTER.
  a. What is the result of the "Zoom In" command?
  b. How could you accomplish the same thing without using any zoom commands?
  c. The "Zoom Out" command is listed right next to the "Zoom In" command? What does it do?

8. *Zooming in on x-intercepts.* Use the "Zoom In" command described in Exercise 7, and the "Trace" command described in Exercise 5, to approximate the location of the x-intercept of $y = 2x - 1$ as accurately as possible. You may need to use these commands more than once.

  a. Describe the procedure you used to generate this answer.
  b. According to the calculator, what is the x-intercept?
  c. Is this answer the same as that of Exercise 5(b)? Why or why not?

9. *Calculating x-intercepts.* The TI-83, TI-84, TI-86, and Casio will calculate the x-intercept (also called a "root" or "zero") without using the "Zoom In" and "Trace" commands. First, use your calculator to graph $y = 2x - 1$ as discussed in this section. When that graph is on the screen, do the following.

On a TI-83/84:
- Press 2nd CALC and select option 2, "root" or "zero."
- The calculator responds by asking, "Lower Bound?" or "Left Bound?" Use the left and right arrow buttons to move the mark to a point slightly to the left of the x-intercept, and press ENTER.
- When the calculator asks, "Upper Bound?" or "Right Bound?", move the mark to a point slightly to the right of the x-intercept, and press ENTER.
- When the calculator asks, "Guess?", move the mark to a point close to the x-intercept, and press ENTER. The calculator will then display the location of the x-intercept.

On a TI-86:
- Press MORE until the "MATH" label appears above the F1 button, and then press that button.
- Press ROOT. This causes an ordered pair to appear on the screen and a mark to appear at the corresponding point on the line. The calculator is asking you if it is on the correct line. However, there is only one line in the exercise, so press ENTER. (If there were graphs of several different lines on the screen, you could use the up and down arrow buttons to select the correct line). The calculator will then display the location of the x-intercept.

On a Casio:
- Press G-Solv (i.e., F5).
- Press ROOT. Wait and watch.
  a. According to the calculator, what is the x-intercept of the line $y = 2x - 1$?
  b. Is this answer the same as that of Exercise 5(b)? Why or why not?

Finding Points of Intersection with a Graphing Calculator

The graphs of $y = x + 3$ and $y = -x + 9$ intersect on the standard viewing screen. To find the point of intersection with a TI or Casio graphing calculator, first enter the two equations on the "Y=" screen (one as Y_1 and one as Y_2), erase any other equations, and erase the "Y= " screen by pressing [2nd] [QUIT]. Then follow the following instructions.

On a TI-83/84
- Graph the two equations on the standard viewing screen.
- Press [2nd] [CALC] and select option 5, "intersect."
- When the calculator responds with "First curve?" and a mark on the first equation's graph, press [ENTER].
- When the calculator responds with "Second curve?" and a mark on the second equation's graph, press [ENTER].
- When the calculator responds with "Guess?", use the left and right arrows to place the mark near the point of intersection, and press [ENTER].
- Check your answer by substituting the ordered pair into each of the two equations.

On a TI-86
- Graph the two equations on the standard viewing screen.
- Press [MORE] until the "MATH" option appears, and select that option.
- Press [MORE] until the "ISECT" option appears, and select that option.
- When the calculator responds with a mark on the first equation's graph, press [ENTER].
- When the calculator responds with a mark on the second equation's graph, press [ENTER].
- Check your answer by substituting the ordered pair into each of the two equations.

On a Casio
- Graph the two equations on the standard viewing screen.
- Press [G-Solv] (i.e., [F5]).
- Press [ISCT] (i.e., [F5]). Watch and wait.
- After a pause, the calculator will display the location of the x-intercept.

Figure A.9 shows the results of computing the point of intersection of $y = x + 3$ and $y = -x + 9$.

The information on the screen indicates that the point of intersection is (3, 6). To check this, substitute 3 for x into each of the two equations; you should get 6.

$$y = x + 3 = 3 + 3 = 6. ✔$$
$$y = -x + 9 = -3 + 9 = 6. ✔$$

Figure A.9
Finding the point of intersection

Exercises

In Exercises 1–6, do the following.

a. Use the graphing calculator to find the point of intersection of the given equations.

b. Check your solutions by substituting the ordered pair into each of the two equations. Answers are not given in the back of the book.

1. $y = 3x + 2$ and $y = 5x + 5$

2. $y = 2x - 6$ and $y = 3x + 4$

3. $y = 8x - 14$ and $g(x) = 11x + 23$

HINT: You will have to change $x$min, $x$max, $y$min, and $y$max to find the point.

4. $y = -7x + 12$ and $y = -12x - 71$

HINT: You will have to change $x$min, $x$max, $y$min, and $y$max to find the point.

5. $y = x^2 - 2x + 3$ and $y = -x^2 - 3x + 12$
(Find two answers.)

6. $f(x) = 8x^2 - 3x - 7$ and $y = 2x + 4$
(Find two answers.)

E Dimensional Analysis

Most people know how to convert 6 feet into yards (it's $6 \div 3 = 2$ yards). Few people know how to convert 50 miles per hour into feet per minute. The former problem is so commonplace that people just remember how to do it; the latter problem is not so common, and people don't know how to do it. Dimensional analysis is an easy way of converting a quantity from one set of units to another; it can be applied to either of these two problems.

Dimensional analysis involves using a standard conversion (such as 1 yard = 3 feet) to create a fraction, including units (such as "feet" and "yards") in that fraction, and canceling units in the same way that variables are canceled. That is, in the fraction 2 feet/3 feet, we can cancel feet with feet and obtain 2/3, just as we can cancel x with x in the fraction $2x/3x$ and obtain 2/3.

EXAMPLE 1 Use dimensional analysis to convert 6 feet into yards.

Solution Start with the standard conversion:

1 yard = 3 feet

Create fractions by dividing each side by 3 feet:

$$\frac{1 \text{ yard}}{3 \text{ feet}} = \frac{3 \text{ feet}}{3 \text{ feet}}$$

$$\frac{1 \text{ yard}}{3 \text{ feet}} = 1$$

To convert 6 feet into yards, multiply 6 feet by the fraction 1 yard/3 feet. This is valid because that fraction is equal to 1, and multiplying something by 1 doesn't change its value.

$$6 \text{ feet} = 6 \text{ feet} \cdot 1$$
$$= 6 \text{ feet} \cdot \frac{1 \text{ yard}}{3 \text{ feet}}$$
$$= 2 \text{ yards}$$

It is crucial to include units in this work. If at the beginning of this example, we had divided by 1 yard instead of 3 feet, we would have obtained:

1 yard = 3 feet

$$\frac{\cancel{1 \text{ yard}}}{\cancel{1 \text{ yard}}} = \frac{3 \text{ feet}}{1 \text{ yard}}$$

$$1 = \frac{3 \text{ feet}}{1 \text{ yard}}$$

Multiplying 6 feet by 3 feet/1 yard would not allow us to cancel feet with feet and would not leave an answer in yards.

$$6 \text{ feet} = 6 \text{ feet} \cdot 1$$

$$= 6 \text{ feet} \cdot \frac{3 \text{ feet}}{1 \text{ yard}}$$

It's important to include units in dimensional analysis, because it's only by looking at the units that we can tell that multiplying by 1 yard/3 feet is productive and multiplying by 3 feet/1 yard isn't. ∎

EXAMPLE 2 Use dimensional analysis to convert 50 miles/hour to feet/minute.

Solution The appropriate standard conversions are

$$1 \text{ mile} = 5{,}280 \text{ feet}$$
$$1 \text{ hour} = 60 \text{ minutes}$$

This problem has two parts, one for each of these two standard conversions.

Part 1 *Use the standard conversion "1 mile = 5,280 feet" to convert miles/hour to feet/hour.* The fraction 50 miles/hour has miles in the numerator, and we are to convert it to a fraction that has feet in the numerator. To replace miles with feet, first rewrite the standard conversion as a fraction that has miles in the denominator, and then multiply by that fraction. This will allow miles to cancel.

$$1 \text{ mile} = 5{,}280 \text{ feet} \quad \text{a standard conversion}$$

$$\frac{\cancel{1 \text{ mile}}}{\cancel{1 \text{ mile}}} = \frac{5280 \text{ feet}}{1 \text{ mile}} \quad \text{placing miles in the denominator}$$

$$1 = \frac{5280 \text{ feet}}{1 \text{ mile}}$$

Now multiply 50 miles/hour by this fraction and cancel:

$$\frac{50 \text{ miles}}{\text{hour}} = \frac{50 \text{ miles}}{\text{hour}} \cdot 1$$

$$= \frac{50 \text{ \cancel{miles}}}{\text{hour}} \cdot \frac{5280 \text{ feet}}{1 \text{ \cancel{mile}}}$$

$$= \frac{50 \cdot 5280 \text{ feet}}{\text{hour}}$$

Part 2 *Use the standard conversion "1 hour = 60 minutes" to convert feet/hour to feet/minute.* To replace hours with minutes, first rewrite the standard conversion as a fraction that has hours in the numerator, and then multiply by that fraction. This will allow hours to cancel.

$$1 \text{ hour} = 60 \text{ minutes} \quad \text{a standard conversion}$$

$$\frac{1 \text{ hour}}{60 \text{ minutes}} = \frac{60 \text{ minutes}}{60 \text{ minutes}} \quad \text{placing hours in the numerator}$$

$$\frac{1 \text{ hour}}{60 \text{ minutes}} = 1$$

Continuing where we left off, multiply by this fraction and cancel:

$$\frac{50 \text{ miles}}{\text{hour}} = \frac{50 \cdot 5280 \text{ feet}}{\text{hour}} \quad \text{from part 1}$$

$$= \frac{50 \cdot 5280 \text{ feet}}{\text{hour}} \cdot 1$$

$$= \frac{50 \cdot 5280 \text{ feet}}{\text{hour}} \cdot \frac{1 \text{ hour}}{60 \text{ minutes}}$$

$$= \frac{4400 \text{ feet}}{1 \text{ minute}} \quad \text{since } 50 \cdot 5280/60 = 4400$$

Thus, 50 miles/hour is equivalent to 4,400 feet/minute.   ■

EXAMPLE 3   How many feet will a car travel in half a minute if that car's rate is 50 miles/hour?

Solution   This seems to be a standard algebra problem that uses the formula "distance = rate · time"; we're given the rate (50 miles/hour) and the time (1/2 minute), and we are to find the distance. However, the units are not consistent.

$$\text{distance} = \text{rate} \cdot \text{time}$$

$$= \frac{50 \text{ miles}}{\text{hour}} \cdot \frac{1}{2} \text{ minute}$$

None of these units cancels, and we are not left with an answer in feet. If, however, the car's rate was in feet/minute rather than miles/hour, the units would cancel, and we would be left with an answer in feet:

$$\text{distance} = \text{rate} \cdot \text{time}$$

$$= \frac{50 \text{ miles}}{\text{hour}} \cdot \frac{1}{2} \text{ minute}$$

$$= \frac{4400 \text{ feet}}{1 \text{ minute}} \cdot \frac{1}{2} \text{ minute} \quad \text{from Example 2}$$

$$= 2,200 \text{ feet}$$

The car would travel 2,200 feet in half a minute.   ■

Exercises

In Exercises 1–6, use dimensional analysis to convert the given quantity.

1. **a.** 12 feet into yards
    b. 12 yards into feet
2. **a.** 24 feet into inches
    b. 24 inches into feet
3. **a.** 10 miles into feet
    b. 10 feet into miles (Round off to the nearest ten thousandth of a mile.)
4. **a.** 2 hours into minutes
    b. 2 minutes into hours (Round off to the nearest thousandth of an hour.)
5. 2 miles into inches

    HINT: Convert first to feet, then to inches.

6. 3 hours into seconds

    HINT: Convert first to minutes, then to seconds.

In Exercises 7–12, use the following information. The metric system is based on three different units:

- the gram (1 gram = 0.0022046 pound)
- the meter (1 meter = 39.37 inches)
- the liter (1 liter = 61.025 cubic inches)

Other units are formed by adding the following prefixes to these basic units:

- kilo-, which means one thousand (for example, 1 kilometer = 1,000 meters)
- centi-, which means one hundredth (for example, 1 centimeter = 1/100 meter)
- milli-, which means one thousandth (for example, 1 millimeter = 1/1,000 meter)

7. Use dimensional analysis to convert
    a. 50 kilometers to meters
    b. 50 meters to kilometers
8. Use dimensional analysis to convert
    a. 30 milligrams to grams
    b. 30 grams to milligrams
9. Use dimensional analysis to convert
    a. 2 centiliters to liters
    b. 2 liters to centiliters
10. Use dimensional analysis to convert
    a. 1 yard to meters (Round off to the nearest hundredth.)
    b. 20 yards to centimeters (Round off to the nearest centimeter.)

11. Use dimensional analysis to convert
    a. 1 pound to grams (Round off to the nearest gram.)
    b. 8 pounds to kilograms (Round off to the nearest tenth of a kilogram.)
12. Use dimensional analysis to convert
    a. 1 cubic inch to liters (Round off to the nearest thousandth.)
    b. 386 cubic inches to liters (Round off to the nearest tenth.)
13. **a.** Use dimensional analysis to convert 60 miles per hour into feet per second.
    b. Leadfoot Larry speeds through an intersection at 60 miles per hour, and he is ticketed for speeding and reckless driving. He pleads guilty to speeding but not guilty to reckless driving, telling the judge that at that speed, he would have plenty of time to react to any cross traffic. The intersection is 80 feet wide. How long would it take Larry to cross the intersection? (Round off to the nearest tenth of a second.)
14. In the United States, a typical freeway speed limit is 65 miles per hour.
    a. Use dimensional analysis to convert this to kilometers per hour. (Round off to the nearest whole number.)
    b. At this speed, how many miles can be traveled in 10 minutes? (Round off to the nearest whole number.)
15. In Germany, a typical autobahn speed limit is 130 kilometers per hour.
    a. Use dimensional analysis to convert this into miles per hour. (Round off to the nearest whole number.)
    b. How many more miles will a car traveling at 130 kilometers per hour go in one hour than a car traveling at 65 miles per hour? (Round off to the nearest whole number.)
16. Light travels 6×10^{12} miles per year.
    a. Convert this to miles per hour. (Round off to the nearest whole number.)
    b. How far does light travel in 1 second? (Round off to the nearest whole number.)
17. In December 1999, Massachusetts' Capital Crossing Bank offered a money market account with a 5.75% interest rate. This means that Capital Crossing Bank will pay interest at a rate of 5.75% per year.
    a. Use dimensional analysis to convert this to a percent per day.

b. If you deposited $10,000 on September 1, how much interest would your account earn by October 1? (There are 30 days in September.) (Round off to the nearest cent.)

c. If you deposited $10,000 on October 1, how much interest would your account earn by November 1? (There are 31 days in October.) (Round off to the nearest cent.)

18. In December 1999, Kentucky's Republic Bank and Trust offered a money market account with a 5.36% interest rate. This means that Republic Bank will pay interest at a rate of 5.36% per year.

   a. Use dimensional analysis to convert this to a percent per day.

   b. If you deposited $10,000 on September 1, how much interest would your account earn by October 1? (There are 30 days in September.) (Round off to the nearest cent.)

   c. If you deposited $10,000 on October 1, how much interest would your account earn by November 1? (There are 31 days in October.) (Round off to the nearest cent.)

19. You cannot determine whether a person is overweight by merely determining his or her weight; if a short person and a tall person weigh the same, the short person could be overweight and the tall person could be underweight. Body mass index (BMI) is becoming a standard way of determining if a person is overweight, since it takes both weight and height into consideration. BMI is defined as (weight in kilograms)/(height in meters)^2. According to the World Health Organization, a person is overweight if his or her BMI is 25 or greater. In October 1996, Katherine Flegal, a statistician for the National Center for Health Statistics, said that according to this standard, one out of every two Americans is overweight. (*Source: San Francisco Chronicle*, 16 October 1996, page A6.)

   a. Lenny is 6 feet tall. Convert his height to meters. (Round off to the nearest hundredth.)

   b. Lenny weighs 169 pounds. Convert his weight to kilograms. (Round off to the nearest tenth.)

   c. Determine Lenny's BMI. (Round off to the nearest whole number.) Is he overweight?

   d. Fred weighs the same as Lenny, but he is 5'5" tall. Determine Fred's BMI. (Round off to the nearest whole number.) Is Fred overweight?

   e. Why does BMI use the metric system (kilograms and meters) rather than the English system (feet, inches, and pounds)?

20. According to the 1990 census, California, the most populous state, had 29,758,000 people residing in 158,706 square miles of area, while Rhode Island, the smallest state, had 1,003,000 people residing in 1,212 square miles of area. Determine which state is more crowded by computing the number of square feet per person in each state.

21. A wading pool is 4 feet wide, 6 feet long, and 11 inches deep. There are 7.48 gallons of water per cubic foot. How many gallons of water does it take to fill the pool?

22. John ate a 2,000 calorie lunch and immediately felt guilty. Jogging for one minute consumes 0.061 calorie per pound of body weight. John weighs 205 pounds. How long will he have to jog to burn off all of the calories from lunch?

Body Table for the Standard Normal Distribution

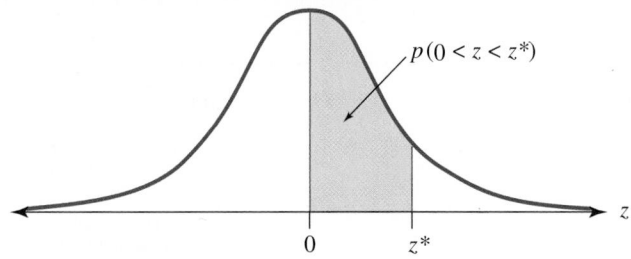

$p(0 < z < z^*)$

z*	0.00	0.01	0.02	0.03	0.04	0.05	0.06	0.07	0.08	0.09
0.0	0.0000	0.0040	0.0080	0.0120	0.0160	0.0199	0.0239	0.0279	0.0319	0.0359
0.1	0.0398	0.0438	0.0478	0.0517	0.0557	0.0596	0.0636	0.0675	0.0714	0.0753
0.2	0.0793	0.0832	0.0871	0.0910	0.0948	0.0987	0.1026	0.1064	0.1103	0.1141
0.3	0.1179	0.1217	0.1255	0.1293	0.1331	0.1368	0.1406	0.1443	0.1480	0.1517
0.4	0.1554	0.1591	0.1628	0.1664	0.1700	0.1736	0.1772	0.1808	0.1844	0.1879
0.5	0.1915	0.1950	0.1985	0.2019	0.2054	0.2088	0.2123	0.2157	0.2190	0.2224
0.6	0.2257	0.2291	0.2324	0.2357	0.2389	0.2422	0.2454	0.2486	0.2517	0.2549
0.7	0.2580	0.2611	0.2642	0.2673	0.2704	0.2734	0.2764	0.2794	0.2823	0.2852
0.8	0.2881	0.2910	0.2939	0.2967	0.2995	0.3023	0.3051	0.3078	0.3106	0.3133
0.9	0.3159	0.3186	0.3212	0.3238	0.3264	0.3289	0.3315	0.3340	0.3365	0.3389
1.0	0.3413	0.3438	0.3461	0.3485	0.3508	0.3531	0.3554	0.3577	0.3599	0.3621
1.1	0.3643	0.3665	0.3686	0.3708	0.3729	0.3749	0.3770	0.3790	0.3810	0.3830
1.2	0.3849	0.3869	0.3888	0.3907	0.3925	0.3944	0.3962	0.3980	0.3997	0.4015
1.3	0.4032	0.4049	0.4066	0.4082	0.4099	0.4115	0.4131	0.4147	0.4162	0.4177
1.4	0.4192	0.4207	0.4222	0.4236	0.4251	0.4265	0.4279	0.4292	0.4306	0.4319
1.5	0.4332	0.4345	0.4357	0.4370	0.4382	0.4394	0.4406	0.4418	0.4429	0.4441
1.6	0.4452	0.4463	0.4474	0.4484	0.4495	0.4505	0.4515	0.4525	0.4535	0.4545
1.7	0.4554	0.4564	0.4573	0.4582	0.4591	0.4599	0.4608	0.4616	0.4625	0.4633
1.8	0.4641	0.4649	0.4656	0.4664	0.4671	0.4678	0.4686	0.4692	0.4699	0.4706
1.9	0.4713	0.4719	0.4726	0.4732	0.4738	0.4744	0.4750	0.4756	0.4761	0.4767
2.0	0.4772	0.4778	0.4783	0.4788	0.4793	0.4798	0.4803	0.4808	0.4812	0.4817
2.1	0.4821	0.4826	0.4830	0.4834	0.4838	0.4842	0.4846	0.4850	0.4854	0.4857
2.2	0.4861	0.4864	0.4868	0.4871	0.4875	0.4878	0.4881	0.4884	0.4887	0.4890
2.3	0.4893	0.4896	0.4898	0.4901	0.4904	0.4906	0.4909	0.4911	0.4913	0.4916
2.4	0.4918	0.4920	0.4922	0.4925	0.4927	0.4929	0.4931	0.4932	0.4934	0.4936
2.5	0.4938	0.4940	0.4941	0.4943	0.4945	0.4946	0.4948	0.4949	0.4951	0.4952
2.6	0.4953	0.4955	0.4956	0.4957	0.4959	0.4960	0.4961	0.4962	0.4963	0.4964
2.7	0.4965	0.4966	0.4967	0.4968	0.4969	0.4970	0.4971	0.4972	0.4973	0.4974
2.8	0.4974	0.4975	0.4976	0.4977	0.4977	0.4978	0.4979	0.4979	0.4980	0.4981
2.9	0.4981	0.4982	0.4982	0.4983	0.4984	0.4984	0.4985	0.4985	0.4986	0.4986
3.0	0.4987	0.4987	0.4987	0.4988	0.4988	0.4989	0.4989	0.4989	0.4990	0.4990

Logic

1.1 Deductive versus Inductive Reasoning

1. a. Valid

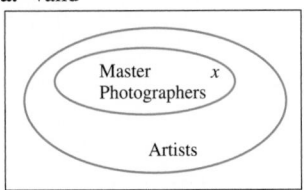

x = Ansel Adams

b. Invalid

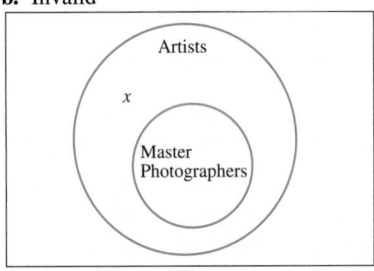

x = Ansel Adams

3. a. Invalid

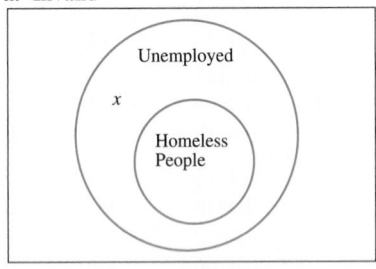

x = Bill Gates

b. Valid

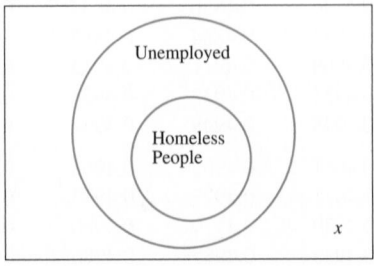

x = Bill Gates

5. Invalid

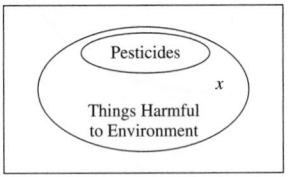

x = fertilizer

7. Valid

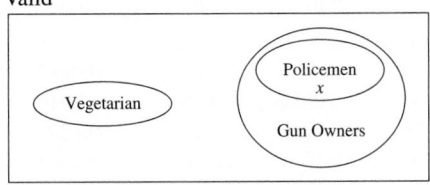

x = policeman

9. Valid

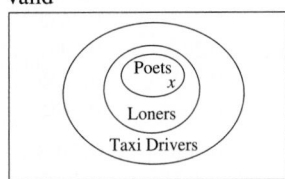

x = poet

11. Valid

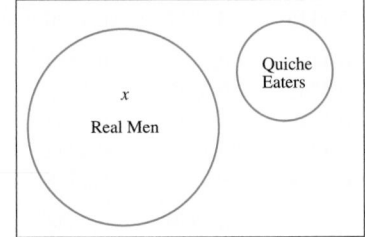

x = Clint Eastwood

13. Valid

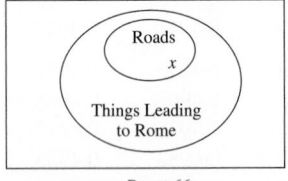

x = Route 66

15. Invalid

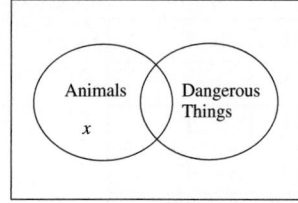

x = tiger

17. Invalid

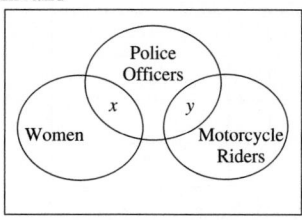

x = policewoman y = motorcycle officer

19. Valid

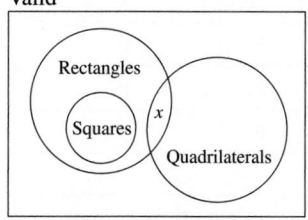

x = Quadrilateral

21. a. Inductive   **b.** Deductive

23. 23. Add 5 each time.

25. 20. Add 2 to the first term to get the second term, add 4 to the second term to get the third term, add 6 to the third term to get the fourth term, and so on.

27. 25. The first term is 1^2, the second term is 2^2, the third term is 3^2, and so on.

29. 13. Each term is a prime number.

31. 5. Think of each number in the sequence as a time (5:00, 8:00, etc.). Add three hours each time.

33. F, S. First letter of the natural numbers.

35. T, W. Each member of the sequence is the first letter of the days of the week: Friday, Saturday, . . .

37. r. Each member of the sequence is the first letter of the corresponding word.

39. f. Each member of the sequence is the first letter after the first vowel in the corresponding word.

41. f. Look at the last letter in each word. Choose the next letter in the alphabet for the sequence.

43. 13. Add 3 each time.

1. Think of the numbers as times and add 3 hours each time.

47. $x = 3 \pm \sqrt{2}$

49.

3	5	6	4	1	8	2	7	9
2	1	7	6	5	9	3	4	8
9	8	4	3	7	2	1	5	6
6	2	1	8	9	5	4	3	7
8	3	5	7	4	6	9	2	1
7	4	9	2	3	1	6	8	5
5	9	2	1	8	4	7	6	3
1	6	3	5	2	7	8	9	4
4	7	8	9	6	3	5	1	2

51.

9	5	6	3	4	1	2	8	7
8	2	3	6	5	7	1	4	9
4	7	1	9	8	2	5	3	6
6	8	7	1	3	4	9	5	2
5	9	4	2	6	8	7	1	3
3	1	2	7	9	5	4	6	8
7	6	5	8	1	9	3	2	4
2	4	8	5	7	3	6	9	1
1	3	9	4	2	6	8	7	5

53.

7	6	8	5	4	2	9	1	3
1	5	2	7	3	9	6	4	8
9	3	4	6	1	8	2	5	7
8	7	1	9	6	3	5	2	4
5	2	6	1	7	4	8	3	9
4	9	3	8	2	5	7	6	1
6	8	9	4	5	1	3	7	2
3	4	7	2	9	6	1	8	5
2	1	5	3	8	7	4	9	6

1.2 Symbolic Logic

1. Parts (a) and (b) are statements

3. (a) and (d) are negations; (b) and (c) are negations

5. a. Her dress is red.
  b. No computer is priced under \$100. (All computers are priced \$100 or more.)
  c. Some dogs are not four-legged animals.
  d. Some sleeping bags are waterproof.

7. a. $p \wedge q$ **b.** $\sim p \to \sim q$
  c. $\sim(p \vee q)$ **d.** $p \wedge \sim q$

9. a. $(p \vee q) \to r$ **b.** $p \wedge q \wedge r$
  c. $r \wedge \sim(p \vee q)$ **d.** $q \to r$

11. p: A shape is a square.
  q: A shape is a rectangle.
  $p \to q$

13. p: A number is a whole number.
  q: A number is even.
  r: A number is odd.
  $p \to (q \vee r)$

15. p: A person is born in the United States.
  q: A person is a U.S. citizen.
  $p \to q$

17. p: A car is a muscle car.
q: A car is from the 60s.
r: A car is a polluter.
$(p \wedge q) \rightarrow r$

19. p: I sleep soundly.
q: I drink coffee.
r: I eat chocolate.
$(q \vee r) \rightarrow (\sim p)$

21. p: Your check is accepted.
q: You have a driver's license.
r: You have a credit card.
$\sim(q \vee r) \rightarrow (\sim p)$

23. p: You do drink.
q: You do drive.
r: You are fined.
s: You go to jail.
$(p \wedge q) \rightarrow (r \vee s)$

25. p: You get a refund.
q: You get store credit.
r: The product is defective.
$r \rightarrow (p \vee q)$

27. a. I am an environmentalist and I recycle my aluminum cans.
b. If I am an environmentalist, then I recycle my aluminum cans.
c. If I do not recycle my aluminum cans, then I am not an environmentalist.
d. I recycle my aluminum cans or I am not an environmentalist.

29. a. If I recycle my aluminum cans or newspapers, then I am an environmentalist.
b. If I am not an environmentalist, then I do not recycle my aluminum cans or newspapers.
c. I recycle my aluminum cans and newspapers or I am not an environmentalist.
d. If I recycle my newspapers and do not recycle my aluminum cans, then I am not an environmentalist.

1.3 Truth Tables

1.

	p	q	$\sim q$	$p \vee \sim q$
1.	T	T	F	T
2.	T	F	T	T
3.	F	T	F	F
4.	F	F	T	T

3.

	p	$\sim p$	$p \vee \sim p$
1.	T	F	T
2.	F	T	T

5.

	p	q	$\sim q$	$p \rightarrow \sim q$
1.	T	T	F	F
2.	T	F	T	T
3.	F	T	F	T
4.	F	F	T	T

7.

	p	q	$\sim p$	$\sim q$	$\sim q \rightarrow \sim p$
1.	T	T	F	F	T
2.	T	F	F	T	F
3.	F	T	T	F	T
4.	F	F	T	T	T

9.

	p	q	$\sim p$	$p \vee q$	$(p \vee q) \rightarrow \sim p$
1.	T	T	F	T	F
2.	T	F	F	T	F
3.	F	T	T	T	T
4.	F	F	T	F	T

11.

	p	q	$p \vee q$	$p \wedge q$	$(p \vee q) \rightarrow (p \wedge q)$
1.	T	T	T	T	T
2.	T	F	T	F	F
3.	F	T	T	F	F
4.	F	F	F	F	T

13.

	p	q	r	$\sim q$	$\sim q \wedge r$	$p \vee (\sim q \wedge r)$
1.	T	T	T	F	F	T
2.	T	T	F	F	F	T
3.	T	F	T	T	T	T
4.	T	F	F	T	F	T
5.	F	T	T	F	F	F
6.	F	T	F	F	F	F
7.	F	F	T	T	T	T
8.	F	F	F	T	F	F

15.

	p	q	r	$q \vee r$	$\sim(q \vee r)$	$p \wedge \sim(q \vee r)$
1.	T	T	T	T	F	F
2.	T	T	F	T	F	F
3.	T	F	T	T	F	F
4.	T	F	F	F	T	T
5.	F	T	T	T	F	F
6.	F	T	F	T	F	F
7.	F	F	T	T	F	F
8.	F	F	F	F	T	F

17.

	p	q	r	~r	~r ∨ p	q ∧ p	(~r ∨ p) → (q ∧ p)
1.	T	T	T	F	T	T	T
2.	T	T	F	T	T	T	T
3.	T	F	T	F	T	F	F
4.	T	F	F	T	T	F	F
5.	F	T	T	F	F	F	T
6.	F	T	F	T	T	F	F
7.	F	F	T	F	F	F	T
8.	F	F	F	T	T	F	F

19.

	p	q	r	~r	p ∨ r	q ∧ ~r	(p ∨ r) → (q ∧ ~r)
1.	T	T	T	F	T	F	F
2.	T	T	F	T	T	T	T
3.	T	F	T	F	T	F	F
4.	T	F	F	T	T	F	F
5.	F	T	T	F	T	F	F
6.	F	T	F	T	F	T	T
7.	F	F	T	F	T	F	F
8.	F	F	F	T	F	F	T

21. *p*: It is raining.
  q: The streets are wet.
  p → q

	p	q	p → q
1.	T	T	T
2.	T	F	F
3.	F	T	T
4.	F	F	T

23. *p*: It rains.
  q: The water supply is rationed.
  ~p → q

	p	q	~p	~p → q
1.	T	T	F	T
2.	T	F	F	T
3.	F	T	T	T
4.	F	F	T	F

25. *p*: A shape is a square.
  q: A shape is a rectangle.
  p → q

p	q	p → q
T	T	T
T	F	F
F	T	T
F	F	T

27. *p*: It is a square.
  q: It is a triangle.
  p → ~q

p	q	~q	p → ~q
T	T	F	F
T	F	T	T
F	T	F	T
F	F	T	T

29. *p*: You have a drivers license.
  q: You have a credit card.
  r: Your check is approved.
  (p ∨ q) → r

p	q	r	p ∨ q	(p ∨ q) → r
T	T	T	T	T
T	T	F	T	F
T	F	T	T	T
T	F	F	T	F
F	T	T	T	T
F	T	F	T	F
F	F	T	F	T
F	F	F	F	T

31. *p*: Leaded gasoline is used.
  q: The catalytic converter is damaged.
  r: The air is polluted.
  p → (q ∧ r)

	p	q	r	q ∧ r	p → (q ∧ r)
1.	T	T	T	T	T
2.	T	T	F	F	F
3.	T	F	T	F	F
4.	T	F	F	F	F
5.	F	T	T	T	T
6.	F	T	F	F	T
7.	F	F	T	F	T
8.	F	F	F	F	T

33. *p*: I have a college degree.
   q: I have a job.
   r: I own a house.
   $p \land \sim(q \lor r)$

	p	*q*	*r*	*q* ∨ *r*	~(*q* ∨ *r*)	*p* ∧ ~(*q* ∨ *r*)
1.	T	T	T	T	F	F
2.	T	T	F	T	F	F
3.	T	F	T	T	F	F
4.	T	F	F	F	T	T
5.	F	T	T	T	F	F
6.	F	T	F	T	F	F
7.	F	F	T	T	F	F
8.	F	F	F	F	T	F

35. *p*: Proposition A passes.
   q: Proposition B passes.
   r: Jobs are lost.
   s: New taxes are imposed.
   $(p \land \sim q) \to (r \lor s)$

	p	*q*	*r*	*s*	~*q*	*p* ∧ ~*q*	*r* ∨ *s*	(*p* ∧ ~*q*) → (*r* ∨ *s*)
1.	T	T	T	T	F	F	T	T
2.	T	T	T	F	F	F	T	T
3.	T	T	F	T	F	F	T	T
4.	T	T	F	F	F	F	F	T
5.	T	F	T	T	T	T	T	T
6.	T	F	T	F	T	T	T	T
7.	T	F	F	T	T	T	T	T
8.	T	F	F	F	T	T	F	F
9.	F	T	T	T	F	F	T	T
10.	F	T	T	F	F	F	T	T
11.	F	T	F	T	F	F	T	T
12.	F	T	F	F	F	F	F	T
13.	F	F	T	T	T	F	T	T
14.	F	F	T	F	T	F	T	T
15.	F	F	F	T	T	F	T	T
16.	F	F	F	F	T	F	F	T

37. Equivalent

	p	*q*	~*q*	*p* ∨ ~*q*	*q* → *p*
1.	T	T	F	T	T
2.	T	F	T	T	T
3.	F	T	F	F	F
4.	F	F	T	T	T

39. Equivalent

41. Not equivalent

	p	*q*	*p* → *q*	*q* → *p*
1.	T	T	T	T
2.	T	F	F	T
3.	F	T	T	F
4.	F	F	T	T

43. Equivalent

	p	*q*	~*p*	~*q*	*p* → *q*	~*q* → ~*p*
1.	T	T	F	F	T	T
2.	T	F	F	T	F	F
3.	F	T	T	F	T	T
4.	F	F	T	T	T	T

45. Not equivalent

	p	*q*	~*p*	~*q*	*p* ∨ ~*q*	*q* ∧ ~*p*
1.	T	T	F	F	T	F
2.	T	F	F	T	T	F
3.	F	T	T	F	F	T
4.	F	F	T	T	T	F

49. I do not have a college degree or I am employed.
51. The television set is not broken and there is not a power outage.
53. The building contains asbestos and the original contractor is not responsible.
55. The lyrics are censored and the First Amendment has not been violated.

1.4 More on Conditionals

1. a. If she is a police officer, then she carries a gun.
   b. If she carries a gun, then she is a police officer.
   c. If she is not a police officer, then she does not carry a gun.
   d. If she does not carry a gun, then she is not a police officer.
   e. a and d; b and c

3. a. If I watch television, then I do not do my homework.
   b. If I do not do my homework, then I watch television.
   c. If I do not watch television, then I do my homework.
   d. If I do my homework, then I do not watch television.
   e. a and d; b and c

5. a. If you do not pass this mathematics course, then you do not fulfill a graduation requirement.
   b. If you fulfill a graduation requirement, then you pass this mathematics course.
   c. If you do not fulfill a graduation requirement, then you do not pass this mathematics course.

7. a. If the electricity is turned on, then the television set does work.
   b. If the television set does not work, then the electricity is turned off.
   c. If the television set does work, then the electricity is turned on.

9. a. If you eat meat, then you are not a vegetarian.
   b. If you are a vegetarian, then you do not eat meat.
   c. If you are not a vegetarian, then you do eat meat.

11. a. Premise: I take public transportation.
      Conclusion: Public transportation is convenient.
   b. If I take public transportation, then it is convenient.
   c. False statement when I take public transportation and it is not convenient.

13. a. Premise: I buy foreign products.
      Conclusion: Domestic products are not available.
   b. If I buy foreign products, then domestic products are not available.
   c. False statement when I buy foreign products and domestic products are available

15. a. Premise: You may become a U.S. senator.
      Conclusion: You are at least 30 years old and have been a citizen for nine years.
   b. If you become a U.S. senator, then you are at least 30 years old and have been a citizen for nine years.
   c. False statement when you become a U.S. senator and either you are not at least 30 years old or have not been a citizen for nine years, or both

17. If you obtain a refund, then you have a receipt, and if you have a receipt, then you will obtain a refund.

19. If $ax^2 + bx + c = 0$ has two distinct real solutions, then $b^2 - 4ac > 0$, and if $b^2 - 4ac > 0$, then $ax^2 + bx + c = 0$ has two distinct real solutions.

21. If a polygon is a triangle, then the polygon has three sides, and if a polygon has three sides, then the polygon is a triangle.

23. Equivalent

25. Equivalent

27. Equivalent

29. If I do not walk to work, then it is raining.

31. If it is not cold, then it is not snowing.

33. If you are a vegetarian, then you do not eat meat.

35. (i) and (iv); (ii) and (iii)

37. (i) and (iii); (ii) and (iv)

39. (i) and (iii); (ii) and (iv)

1.5 Analyzing Arguments

1. 1. $p \to q$
    2. p
    ∴ q

3. 1. $p \to q$
    2. $\sim q$
    ∴ $\sim p$

5. 1. $p \to q$
    2. $\sim p$
    ∴ $\sim q$

7. Valid

9. Valid

11. Invalid; the argument is invalid when you do not exercise regularly and you are healthy

13. Invalid; the argument is invalid when the Democrats have a majority and Smith is appointed and student loans are funded

15. Invalid; the argument is invalid when
(1) you do not argue with a police officer and you do get a ticket and you do break the speed limit, or
(2) you do not argue with a police officer and you do not get a ticket and you do break the speed limit

17. Rewriting the argument:

If it is a pesticide, then it is harmful to the environment.
If it is a fertilizer, then it is not a pesticide.
_____
If it is a fertilizer, then it is not harmful to the environment.

p: It is a pesticide.
q: It is harmful to the environment.
r: It is a fertilizer.

$p \to q$
$r \to \sim p$
$r \to \sim q$
$[(p \to q) \land (r \to \sim p)] \to (r \to \sim q)$

The argument is not valid if it is not a pesticide and if it is harmful to the environment and it is a fertilizer.

p	q	r	$\sim p$	$\sim q$	**1** $p \to q$	**2** $r \to \sim p$	$1 \wedge 2$	**3** $r \to \sim q$	$(1 \wedge 2) \to 3$
T	T	T	F	F	T	F	F	F	T
T	T	F	F	F	T	T	T	T	T
T	F	T	F	T	F	F	F	T	T
T	F	F	F	T	F	T	F	T	T
F	T	T	T	F	T	T	T	F	F
F	T	F	T	F	T	T	T	T	T
F	F	T	T	T	T	T	T	T	T
F	F	F	T	T	T	T	T	T	T

19. Rewriting the argument:

If you are a poet, then you are a loner.
If you are a loner, then you are a taxi driver.
_____
If you are a poet, then you are a taxi driver.

p: You are a poet.
q: You are a loner.
r: You are a taxi driver.

$p \to q$
$q \to r$
_____
$p \to r$

$\{(p \to q) \wedge (q \to r)\} \to (p \to r)$

p	q	r	**1** $p \to q$	**2** $q \to r$	$1 \wedge 2$	**3** $p \to r$	$(1 \wedge 2) \to 3$
T	T	T	T	T	T	T	T
T	T	F	T	F	F	F	T
T	F	T	F	T	F	T	T
T	F	F	F	T	F	F	T
F	T	T	T	T	T	T	T
F	T	F	T	F	F	T	T
F	F	T	T	T	T	T	T
F	F	F	T	T	T	T	T

The argument is valid.

21. Rewriting the argument:

If you are a professor, then you are not a millionaire.
If you are a millionaire, then you are literate.
_____
If you are a professor, then you are literate.

p: You are a professor.
q: You are a millionaire.
r: You are illiterate.

$p \to \sim q$
$q \to \sim r$
_____
$p \to \sim r$

$\{(p \to \sim q) \wedge (q \to \sim r)\} \to (p \to \sim r)$

p	q	r	$\sim q$	**1** $p \to \sim q$	$\sim r$	**2** $q \to \sim r$	$1 \wedge 2$	**3** $p \to \sim r$	$(1 \wedge 2) \to 3$
T	T	T	F	F	F	F	F	F	T
T	T	F	F	F	T	T	F	T	T
T	F	T	T	T	F	T	T	F	F
T	F	F	T	T	T	T	T	T	T
F	T	T	F	T	F	F	F	T	T
F	T	F	F	T	T	T	T	T	T
F	F	T	T	T	F	T	T	T	T
F	F	F	T	T	T	T	T	T	T

The argument is invalid if you are a professor who is not a millionaire and is illiterate.

23. Valid   **25.** Valid

27. Invalid; the argument is invalid when
(1) you are in a hurry and you eat at Lulu's Diner and you do not eat good food, or
(2) you are not in a hurry and you eat at Lulu's Diner and you do not eat good food.

29. Valid   **31.** Valid   **33.** Valid

35. Valid   **37.** Valid

Chapter 1 Review

1. a. Inductive   **b.** Deductive

3. 9. Think of each number in the sequence as a time (1:00, 6:00, etc.). Add five hours to get the next number in the sequence.

5. Invalid

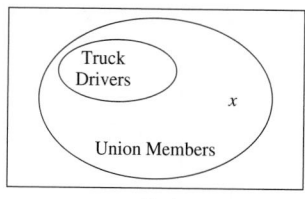

x = Rocky

7. Valid

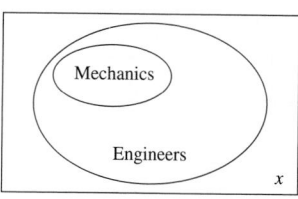

x = Casey Jones

9. Valid

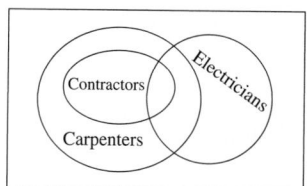

11. (a) and (b) are statements
13. a. His car is new.
  b. No building is earthquakeproof.
  c. Some children do not eat candy.
  d. Sometimes I cry in a movie theater.
15. a. $q \rightarrow p$ **b.** $\sim r \wedge \sim q \wedge p$
  c. $r \rightarrow \sim p$ **d.** $p \wedge \sim (q \vee r)$
17. a. If the movie is critically acclaimed or a box office hit, then the movie is available on videotape.
  b. If the movie is critically acclaimed and it is not a box office hit, then the movie is not available on videotape.
  c. The movie is not critically acclaimed or a box office hit and it is available on videotape.
  d. If the movie is not available on videotape, then the movie is not critically acclaimed and it is not a box office hit.

19.

	p	q	$\sim q$	$p \wedge \sim q$
1.	T	T	F	F
2.	T	F	T	T
3.	F	T	F	F
4.	F	F	T	F

21.

	p	q	$\sim q$	$p \wedge q$	$(p \wedge q) \rightarrow \sim q$
1.	T	T	F	T	F
2.	T	F	T	F	T
3.	F	T	F	F	T
4.	F	F	T	F	T

23.

	p	q	r	$\sim p$	$q \vee r$	$\sim p \rightarrow (q \vee r)$
1.	T	T	T	F	T	T
2.	T	T	F	F	T	T
3.	T	F	T	F	T	T
4.	T	F	F	F	F	T
5.	F	T	T	T	T	T
6.	F	T	F	T	T	T
7.	F	F	T	T	T	T
8.	F	F	F	T	F	F

25.

	p	q	r	$\sim r$	$p \vee r$	$q \wedge \sim r$	$(p \vee r) \rightarrow (q \wedge \sim r)$
1.	T	T	T	F	T	F	F
2.	T	T	F	T	T	T	T
3.	T	F	T	F	T	F	F
4.	T	F	F	T	T	F	F
5.	F	T	T	F	T	F	F
6.	F	T	F	T	F	T	T
7.	F	F	T	F	T	F	F
8.	F	F	F	T	F	F	T

27. Not equivalent

29. Equivalent

31. You pay the fine and you go to jail.

33. He is employed or he did apply for financial assistance.

35. The jackpot is less than $1 million.

37. a. If he is not elected, then the country is not in big trouble.
  b. If the country is in big trouble, then he is elected.
  c. If the country is not in big trouble, then he is not elected.

39. a. Premise: Unemployment goes down.
    Conclusion: The economy improves.
  b. If unemployment goes down, then the economy improves.

41. If it is a gemstone, then it is valuable.

43. Equivalent

45. Invalid

47. Invalid

49. Valid

51. Invalid

53. p: A person is a professor.
q: A person is educated.
r: An animal is a monkey.

$$p \rightarrow q$$
$$\underline{r \rightarrow \sim q}$$
$$p \rightarrow \sim r$$

$$\{[p \rightarrow q] \wedge [r \rightarrow \sim q]\} \rightarrow (p \rightarrow \sim r)$$

			1		2			3
p	q	r	$p \rightarrow q$	$\sim q$	$r \rightarrow \sim q$	$1 \wedge 2$	$\sim r$	$p \rightarrow \sim r$	$(1 \wedge 2) \rightarrow 3$
T	T	T	T	F	F	F	F	F	T
T	T	F	T	F	T	T	T	T	T
T	F	T	F	T	T	F	F	F	T
T	F	F	F	T	T	F	T	T	T
F	T	T	T	F	F	F	F	T	T
F	T	F	T	F	T	T	T	T	T
F	F	T	T	T	T	T	F	T	T
F	F	F	T	T	T	T	T	T	T

The argument is valid.

55. p: Vehicles stop.
q: The traffic light is red.
r: There is an accident.

$$q \rightarrow p$$
$$p \rightarrow \sim r$$
$$\underline{r}$$
$$\sim q$$

$$\{(q \rightarrow p) \wedge (p \rightarrow \sim r) \wedge r\} \rightarrow \sim q$$

			1		2			
p	q	r	$q \rightarrow p$	$\sim r$	$p \rightarrow \sim r$	$1 \wedge 2 \wedge r$	$\sim q$	$(1 \wedge 2 \wedge r) \rightarrow \sim q$
T	T	T	T	F	F	F	F	T
T	T	F	T	T	T	F	F	T
T	F	T	T	F	F	F	T	T
T	F	F	T	T	T	F	T	T
F	T	T	F	F	T	F	F	T
F	T	F	F	T	T	F	F	T
F	F	T	T	F	T	T	T	T
F	F	F	T	T	T	F	T	T

The argument is valid.

57.

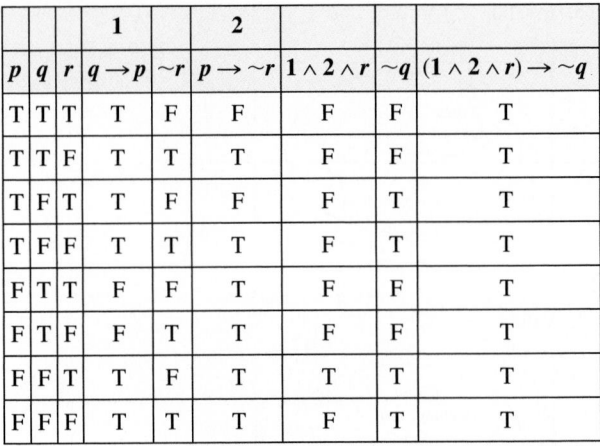

x = environmentalist

p: You own a hybrid car.
q: You are an environmentalist.

$$p \rightarrow q$$
$$\underline{q}$$
$$p$$

3	2	1		
p	q	$p \rightarrow q$	$1 \wedge 2$	$(1 \wedge 2) \rightarrow 3$
T	T	T	T	T
T	F	F	F	T
F	T	T	T	F
F	F	T	F	T

The argument is invalid.

61. If the number of statements is n, the number of rows will be 2^n.

CHAPTER 2 Sets and Counting

2.1 Sets and Set Operations

1. a. Well-defined   **b.** Not well-defined
  c. Well-defined   **d.** Not well-defined

3. Proper: $\varnothing$, {Lennon}, {McCartney}
  Improper: {Lennon, McCartney}

5. Proper: $\varnothing$, {yes}, {no}, {undecided},
  {yes, no}, {yes, undecided}, {no, undecided}
  Improper: {yes, no, undecided}

7. a. {4, 5}   **b.** {1, 2, 3, 4, 5, 6, 7, 8}
  c. {0, 6, 7, 8, 9}   **d.** {0, 1, 2, 3, 9}

9. a. { }   **b.** {0, 1, 2, 3, 4, 5, 6, 7, 8, 9}
  c. {0, 2, 4, 6, 8}   **d.** {1, 3, 5, 7, 9}

11. {Friday}

13. {Monday, Tuesday, Wednesday, Thursday}

15. {Friday, Saturday, Sunday}

17.

19.

21.

23.

25.

27. a. 21

b. 0

29. a.

b. 37.0%

31. a.

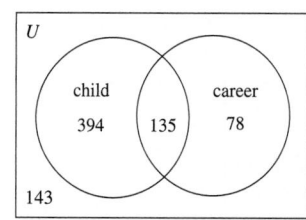

b. 18.0%

33. 42	**35.** 43	**37.** 8
39. 5	**41.** 16	**43.** 32
45. 6	**47.** 8	**49.** 0

51. a. {1, 2, 3} = A   **b.** {1, 2, 3, 4, 5, 6} = B
  c. E ⊆ F   **d.** E ⊆ F

53. a. ∅, {a}; 2
  b. ∅, {a}, {b}, {a, b}; 4
  c. ∅, {a}, {b}, {c}, {a, b}, {a, c}, {b, c}, {a, b, c}; 8
  d. ∅, {a}, {b}, {c}, {d}, {a, b}, {a, c}, {a, d}, {b, c}, {b, d}, {c, d}, {a, b, c}, {a, b, d}, {a, c, d}, {b, c, d}, {a, b, c, d}; 16
  e. Yes; $2^{n(A)}$
  f. 64

65. (d)	**67.** (c)	**69.** (e)

2.2 Applications of Venn Diagrams

1. a. 143	**b.** 16	**c.** 49	**d.** 57
3. a. 408	**b.** 1,343	**c.** 664	**d.** 149
5. a. 106	**b.** 448	**c.** 265	**d.** 159

  7. a. $x + y - z$ **b.** $x - z$
    c. $y - z$ **d.** $w - (x + y - z)$

9. a. 43.8%	**b.** 10.8%
11. a. 77%	**b.** 22%
13. a. 0%	**b.** 54.1%
15. a. 44.0%	**b.** 12.8%
17. a. 20.0%	**b.** 58.8%

19. 16 **21.** {0, 4, 5}

23. {1, 2, 3, 6, 7, 8, 9}

25.

27.

29.

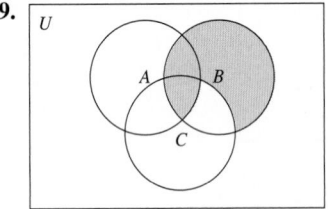

33. a. 84% **b.** 16%

35. a. $\frac{17}{20}$ = 85% **b.** $\frac{3}{20}$ = 15%

37. a. $\frac{3}{4}$ = 75% **b.** $\frac{1}{4}$ = 25%

39. B or O **41.** O

45. (d) **47.** (d) **49.** (b)

2.3 Introduction to Combinatorics

1. a. 8 **b.**

3. a. 12 **b.**

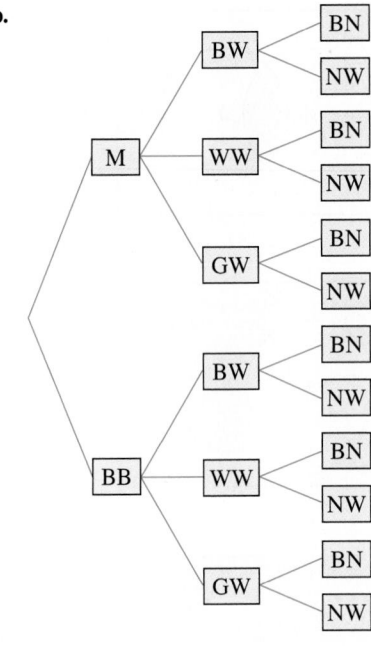

5. 24 **7.** 720

9. 2,646 **11.** 216

13. 1 billion **15.** 10^{10} or 10 billion

17. a. 128 **b.** 800

19. a. 1.17936×10^9 **b.** 6.76×10^{10} **c.** 4.2×10^{10}

21. 540,000 **23.** 24 **25.** 3,628,800

27. $2.432902008 \times 10^{18}$ **29.** 17,280

31. a. 30 **b.** 360

33. 56 **35.** 70 **37.** 3,321

39. $1.046139494 \times 10^{13}$

41. 120 **43.** 35 **45.** 1

51. (b) **53.** (d)

2.4 Permutations and Combinations

1. a. 210 **b.** 35

3. a. 120 **b.** 1

5. a. 14 **b.** 14

7. a. 970,200 **b.** 161,700

9. a. $x!$ **b.** x

11. a. $x^2 - x$ **b.** $\dfrac{x^2 - x}{2}$

13. a. 6

b. {a, b}, {a, c}, {b, a}, {b, c}, {c, a}, {c, b}

15. a. 6

b. {a, b}, {a, c}, {a, d}, {b, c}, {b, d}, {c, d}

17. a. 479,001,600 **b.** 1

19. 24

21. 91 **23.** 2,184

25. a. 420   **b.** 1,001   **c.** 406

27. 2,598,960

29. a. 4,512   **b.** 58,656

31. 123,552

33. 22,957,480

35. 376,992

37. 5/36

39. The first five rows of triangle are:

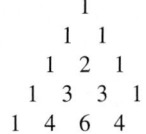

$$1$$
$$1 \quad 1$$
$$1 \quad 2 \quad 1$$
$$1 \quad 3 \quad 3 \quad 1$$
$$1 \quad 4 \quad 6 \quad 4 \quad 1$$

     a. 1
     b. 2
     c. 4
     d. 8
     e. 16
     f. Sum of entries in nth row is 2^{n-1}.
     g. $2^{6-1} = 2^5 = 32$ **h.** Yes   **i.** 2^{n-1}

41. a. Fifth
     b. $n + 1$
     c. No
     d. Yes
     e. Row $n + 1$, entry $r + 1$

43. $\dfrac{6!}{3!} = 120$ **45.** $\dfrac{8!}{3!2!} = 3360$

47. $\dfrac{7!}{2!2!2!} = 630$ **49.** $\dfrac{11!}{3!2!2!2!} = 831{,}600$

51. a. $4! = 24$ **b.** $\dfrac{4!}{2!} = 12$

53. a. $5! = 120$ **b.** $\dfrac{5!}{2!} = 60$

57. (c)   **59.** (c)   **61.** (e)

2.5 Infinite Sets

1. $n(S) = 4$, $n(C) = 4$; equivalent; state $\leftrightarrow$ capital city

3. $n(R) = 3$, $n(G) = 4$, not equivalent

5. $n(C) = 22$, $n(D) = 22$; equivalent; $3n \leftrightarrow 4n$

7. $n(G) = 250$, $n(H) = 251$; not equivalent

9. $n(A) = 62$, $n(B) = 62$; equivalent, $2n - 1 \leftrightarrow 2n + 123$

11. a. $n \leftrightarrow 2n - 1$ **b.** 918   **c.** $\frac{x+1}{2}$
     d. 1,563   **e.** $2n - 1$

13. a. $n \leftrightarrow 3n$ **b.** 312   **c.** $\frac{x}{3}$
     d. 2,808   **e.** $3n$

15. a. -344 **b.** 248   **c.** 755   **d.** $\aleph_0$

Chapter 2 Review

1. a. Well defined.   **b.** Not well defined.
     c. Not well defined.   **d.** Well defined.

3. a. $A \cup B$ {Maria, Nobuko, Leroy, Mickey, Kelly, Rachel, Deanna}
     b. $A \cap B$ {Leroy, Mickey}

5. a. 18
     b.

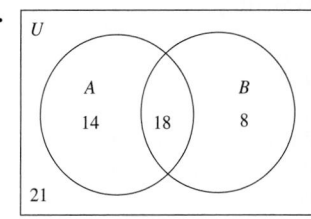

7. 29%

9. a. 85   **b.** None   **c.** 15

11. a. 12   **b.**

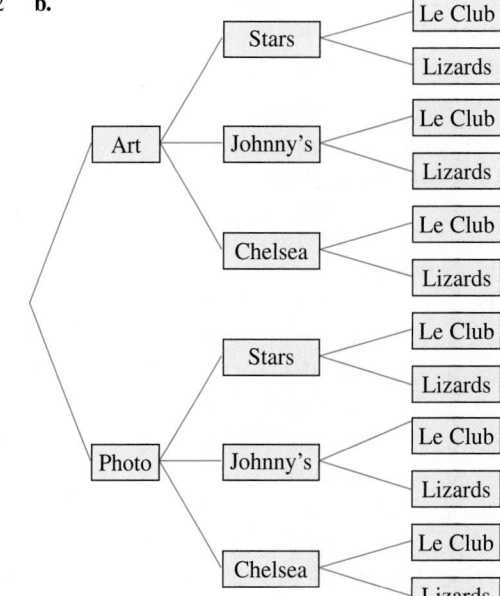

13. 1,080,000

15. a. 165   **b.** 990

17. a. 660   **b.** 1,540   **c.** 880

19. 720   **21.** 177,100,560

23. a. $7! = 5040$ **b.** $7!/2! = 2520$ **c.** $\dfrac{7!}{2!2!} = 1260$

27. a. 3   **b.** 3   **c.** 1   **d.** 1
     e. 8   **f.** $2^{n(S)} = 2^3 = 8$

29. $n(C) = 450$, $n(D) = 450$; equivalent; $2n + 1 \leftrightarrow 2n$

31. a. $n \leftrightarrow n^2$ **b.** 29   **c.** $\sqrt{x}$
     d. 20,736   **e.** n^2

35. {0} is a set with one element, while $\varnothing$ is a set with no elements.

37. If order of selection matters, we use permutations. If not, we use combinations.

CHAPTER 3 Probability

3.1 History of Probability
9. a. Win $350   **b.** Lose $10
11. a. Win $85   **b.** Win $85   **c.** Lose $5
13. a. Lose $20   **b.** Win $160   **c.** Lose $20
15. a. Lose $10   **b.** Win $50
17. a. Win $50   **b.** Lose $25
19. a. Lose $50   **b.** Win $50
21. a. Lose $45   **b.** Win $855   **c.** Win $675
23. a. Win $495   **b.** Win $525
    c. Lose $45   **d.** Lose $15
25. $3   **27.** $500
29. a. 13   **b.** $\frac{1}{4}$
31. a. 12   **b.** $\frac{3}{13}$
33. a. 4   **b.** $\frac{1}{13}$

3.2 Basic Terms of Probability
1. Selecting a jellybean
3. $\frac{12}{35}$ **5.** $\frac{18}{35}$ **7.** $\frac{5}{7}$ **9.** 0
11. $12:23$ **13.** $18:17$
15. Selecting a card
17. $\frac{1}{2}$ **b.** $1:1$ **19. a.** $\frac{1}{13}$ **b.** $1:12$
21. a. $\frac{1}{52}$ **b.** $1:51$ **23. a.** $\frac{3}{13}$ **b.** $3:10$
25. a. $\frac{10}{13}$ **b.** $10:3$ **27. a.** $\frac{3}{13}$ **b.** $3:10$
29. a. $\frac{1}{38}$ **b.** $1:37$ **31. a.** $\frac{3}{38}$ **b.** $3:35$
33. a. $\frac{5}{38}$ **b.** $5:33$ **35. a.** $\frac{6}{19}$ **b.** $6:13$
37. a. $\frac{9}{19}$ **b.** $9:10$
39. a. $\frac{143,713}{282,124} = 50.9\%$
    b. $\frac{52,294}{282,124} = 18.5\%$
41. a. $\frac{1}{25} = 4\%$ **b.** $\frac{8}{25} = 32\%$ **c.** $\frac{16}{25} = 64\%$
43. $\frac{30}{145} = \frac{6}{29}$
45. $1:4$ **47.** $\frac{3}{5}$ **49.** $a:(b-a)$
51. a. $\frac{9}{19}$ **b.** $9:10$
53. a. House odds
    b. Yankees: $\frac{2}{3}$, Boston Red Sox: $\frac{5}{7}$
    c. Boston Red Sox: $(\frac{5}{7} > \frac{2}{3})$
55. a. Intentional self-harm because it has the highest odds
    b. Intentional self-harm because it has the highest odds
57. a. $\frac{1}{17,626}$ **b.** $\frac{1}{440,952}$
59. a. $\frac{1}{17,626}$ car   **b.** $\frac{1}{440,952}$ airplane
61. a. {(b, b), (b, g), (g, b), (g, g)}   **b.** {(b, g), (g, b)}
    c. {(b, g), (g, b), (g, g)}   **d.** {(g, g)}
    e. $\frac{1}{2}$ **f.** $\frac{3}{4}$ **g.** $\frac{1}{4}$ **h.** $1:1$ **i.** $3:1$ **j.** $1:3$
63. a. {(b, b, b), (b, b, g), (b, g, b), (g, b, b), (b, g, g), (g, b, g),
    (g, g, b), (g, g, g)}
    b. {(b, g, g), (g, b, g), (g, g, b)}
    c. {(b, g, g), (g, b, g), (g, g, b), (g, g, g)}
    d. {(g, g, g)}
    e. $\frac{3}{8}$ **f.** $\frac{1}{2}$ **g.** $\frac{1}{8}$ **h.** $3:5$ **i.** $1:1$ **j.** $1:7$
65. a. $\frac{1}{4}$ **b.** $\frac{1}{2}$ **c.** $\frac{1}{4}$ **d.** Equally likely

67. Different
69. a. {(1, 1), (1, 2), (1, 3), (1, 4), (1, 5), (1, 6), (2, 1), (2, 2),
    (2, 3), (2, 4), (2, 5), (2, 6), (3, 1), (3, 2), (3, 3), (3, 4),
    (3, 5), (3, 6), (4, 1), (4, 2), (4, 3), (4, 4), (4, 5), (4, 6),
    (5, 1), (5, 2), (5, 3), (5, 4), (5, 5), (5, 6), (6, 1), (6, 2),
    (6, 3), (6, 4), (6, 5), (6, 6)}
    b. {(1, 6), (2, 5), (3, 4), (4, 3), (5, 2), (6, 1)}
    c. {(6, 5), (5, 6)}
    d. {(1, 1), (2, 2), (3, 3), (4, 4), (5, 5), (6, 6)}
    e. $\frac{1}{6}$ **f.** $\frac{1}{18}$ **g.** $\frac{1}{6}$ **h.** $1:5$ **i.** $1:17$ **j.** $1:5$
71. a. $\frac{1}{4}$ **b.** $\frac{1}{4}$ **c.** $\frac{1}{2}$
73. a. $\frac{1}{4}$ **b.** $\frac{1}{2}$ **c.** $\frac{1}{4}$
75. a. 0   **b.** $\frac{1}{2}$ **c.** 1

3.3 Basic Rules of Probability
1. Not mutually exclusive
3. Mutually exclusive
5. Not mutually exclusive
7. Mutually exclusive
9. Mutually exclusive
11. a. $\frac{1}{26}$ **b.** $\frac{7}{13}$ **c.** $\frac{25}{26}$
13. a. $\frac{1}{52}$ **b.** $\frac{4}{13}$ **c.** $\frac{51}{52}$
15. a. $\frac{2}{13}$ **b.** $\frac{5}{13}$ **c.** 0   **d.** $\frac{7}{13}$
17. a. $\frac{9}{13}$ **b.** $\frac{8}{13}$ **c.** $\frac{4}{13}$ **d.** 1
19. $\frac{12}{13}$ **21.** $\frac{10}{13}$ **23.** $\frac{11}{13}$ **25.** $\frac{9}{13}$
27. $9:5$ **29.** $2:5; 5:2$ **31.** $b:a$ **33.** $12:1$
35. $10:3$ **37.** $10:3$
39. a. $\frac{71}{175}$ **b.** $\frac{104}{175}$
41. a. $\frac{151}{700}$ **b.** $\frac{97}{140}$
43. a. $\frac{8}{25}$ **b.** $\frac{17}{25}$
45. a. $\frac{9}{25}$ **b.** $\frac{16}{25}$
47. a. $\frac{1}{6}$ **b.** $\frac{1}{9}$ **c.** $\frac{1}{18}$
49. a. $\frac{2}{9}$ **b.** $\frac{7}{18}$ **51. a.** $\frac{1}{6}$ **b.** $\frac{3}{4}$
53. a. $\frac{1}{6}$ **b.** $\frac{1}{2}$
55. a. 0.65   **b.** 0.30   **c.** 0.35
57. Relative frequency
59. a. 0.56   **b.** 0.33   **c.** 0.44
61. Relative frequencies; they were calculated from a poll.
63. a. $\frac{17}{25}$ **b.** $\frac{2}{25}$ **65. a.** $\frac{212}{1451}$ **b.** $\frac{514}{1451}$
67. a. $\frac{84}{696} = \frac{7}{58}$ **b.** $\frac{124}{696} = \frac{31}{174}$
69. a. $\frac{3}{4}$ **b.** $\frac{1}{2}$ **c.** 1
71. a. $\frac{1}{4}$ **b.** $\frac{1}{2}$ **c.** 0
73. a. 0.35   **b.** 0.25   **c.** 0.15   **d.** 0.6
75. a. 0.20   **b.** 0.95   **c.** 0.8

3.4 Combinatorics and Probability
1. 0.7063   **3.** 23
5. a. $\frac{1}{13,983,816}$ **b.** $\frac{1}{22,957,480}$
    c. $\frac{1}{18,009,460}$ **d.** About 64% more likely

7. 0.00003

9. a. 0.000003080   **b.** 0.000462062

11. a. 0.000001737   **b.** 0.000295263

13. Easiest to win: $\frac{5}{26}$, hardest to win: $\frac{6}{53}$

15. a. 0.000004   **b.** 0.000160   **c.** 0.002367
   d. 0.018303   **e.** 0.081504   **f.** 0.897662

17. a. 0.000495   **b.** 0.001981
   c. 0.000015   **d.** 0.001965

19. 0.05   **21.** 0.48

23. 0.64   **25.** 0.36

27. a. 0.09   **b.** 0.42   **c.** 0.49

29. a. 0.0000000125   **b.** 0.0000000715
   c. Winning the $\frac{6}{49}$ is 5.7 times more likely.
   d. Answers will vary.   **e.** Answers will vary.

3.5 Expected Value

1. a. −$0.053   **3. a.** −$0.053

5. a. −$0.053   **7. a.** −$0.053

9. a. −$0.053   **11.** $549

13. 1.75 books   **15.** $10.05

17. Expected value = $-\frac{10}{6}$ = −$1.67. Don't play.

19. No. Expected value of 10 games is $50.

23. Bank   **25.** 0.59 or more

27. a. 0   **b.** $\frac{1}{16}$ **c.** $\frac{3}{8}$

29. −$0.28   **31.** −$0.55

33. More than $1,200/year   **35.** Back yard

37. a. $12.32 You should buy a ticket.
   b. −$1.34 You should not buy a ticket.
   c. −$5.89 You should not buy a ticket.

39. a. −$0.47 You should not buy a ticket.
   b. −$2.74 You should not buy a ticket.
   c. −$3.49 You should not buy a ticket.

41. $7.8 million

43. a. 13,983,816   **b.** $13,983,816   **c.** 97 days

45. 6 successive losses; win $1

3.6 Conditional Probability

1. a. $p(H|Q)$ **b.** $p(H \cap Q)$

3. a. $p(S|D)$ **b.** $p(S \cap D)$ **c.** $p(D|S)$

5. a. $\frac{3}{4}$ **b.** $\frac{1}{2}$ **7. a.** $\frac{3}{8}$ **b.** $\frac{1}{8}$

9. a. 0.23   **b.** 0.53   **c.** 0.14   **d.** 0.32
   e. 0.08   **f.** 0.08

11. a. $\frac{5,954}{47,274}$ = 0.13 The probability of a person in the United States dying in a pedestrian transportation accident in 1 year is 13% of his or her chance of dying in a transportation accident in 1 year.
   b. $\frac{78}{613}$ = 0.13 The probability of a person in the United States dying in a pedestrian transportation accident in a lifetime is 13% of his or her chance of dying in a transportation accident in a lifetime.
   c. 0 No pedestrian transportation accidents are nontransportation accidents.

13. a. $\frac{65}{120,162}$ = 0.00054. In a lifetime, approximately 0.05% of those who die of a non-transportation accident die from an earthquake.

b. $\frac{4,934}{9,288,427}$ = 0.00053. In one year, approximately 0.05% of those who die of a non-transportation accident die from an earthquake.
   c. $\dfrac{(9,288,427)^{-1}}{(4934^{-1} + 5954^{-1})}$ = .00029.
   In one year, approximately 0.03% of those who die from an external cause die from an earthquake.

15. a. $\frac{1}{4}$ **b.** $\frac{4}{17}$ **c.** $\frac{1}{17}$

17. a. $\frac{1}{4}$ **b.** $\frac{13}{51}$ **c.** $\frac{13}{204}$

19. a. $p(B|A)$ **b.** $p(A')$ **c.** $p(C|A')$

21. a. $\frac{1}{6}$ **b.** $\frac{1}{3}$ **c.** 0   **d.** 1

23. a. $\frac{5}{36}$ **b.** $\frac{5}{18}$ **c.** 0   **d.** 1

25. a. $\frac{1}{12}$ **b.** $\frac{3}{10}$ **c.** 1

27. E_2 is most likely; E_3 is least likely

29. 0.46   **31.** 0.0005   **33.** 0.0020

35. 0.14   **37.** 0.20   **39.** 0.07

41. 0.00646 ≈ 0.6%   **43.** 0.01328 ≈ 1.3%

45. a. 0.2   **b.** 0.09
   c. The Democratic Party has a larger percentage of females.

47. a. 0.27   **b.** 0.08
   c. Democratic Party has a larger percentage of female governors.

49. a. p(male) = 80.9%, p(female) = 19.1%
   b. Male 24%, female 40%
   c. Male 19%, female 8%
   d. 7%, 56%   **e.** 6%, 11%
   f. Part (b) is the percentage of males that were exposed by injection drug use. Part (c) is the percentage of the total that are both male and exposed by injection drug use. Similarly for parts (d) and (e).

51. a. 19.4%   **b.** 9.2%   **c.** 27.7%   **d.** 14.6%   **e.** 23.8%
   f. Part (c) is the percentage of adult women who are obese. Part (d) is the percentage of adults who are both female and obese.

53. a. 12.6%   **b.** 6.7%   **c.** 0   **d.** 5.8%

55. $\frac{2}{3}$ **57.** 92%   **59.** 39%   **61.** 0.05

63. a. 0.86 = 86%   **b.** 0.34 = 34%
   c. 0.66 = 66%   **d.** $N'|W$

65. $A'|B$ **67.** 0.14

3.7 Independence; Trees in Genetics

1. a. Dependent   **b.** Not mutually exclusive

3. a. Dependent   **b.** Mutually exclusive

5. a. Independent   **b.** Not mutually exclusive

7. a. Dependent   **b.** Mutually exclusive

9. a. Dependent   **b.** Mutually exclusive

11. No

13. a. $\frac{1}{6}$ **b.** 0
   c. No. If you roll a 5, the probability of rolling an even number is zero.
   d. Yes. 5 is not an even number.
   e. Mutually exclusive events are always dependent.

15. a. $\frac{4}{52} = \frac{1}{13}$ **b.** $\frac{2}{26} = \frac{1}{13}$
   c. Yes. Being dealt a red card doesn't change the probability of being dealt a jack.

d. No. There are red jacks.

e. Independent events are always not mutually exclusive. 14 and 15 show that events not mutually exclusive can be dependent or independent.

17. Dependent   **19.** Dependent

21. a. Not independent   **b.** Not mutually exclusive

23. a. Not independent   **b.** Not mutually exclusive

c. Living in Bishop decreases the chances of supporting proposition 3.

25. Dependent. HAL users were more likely to quit.

27. Independent.

29. a. 0.000001   **b.** 4   **31.** 0.0101 ≈ 1%

33. a. 0.47   **b.** 0.999996   **c.** 0.53   **d.** 0.000004

e. a; c   **f.** c; d

37. a. $\frac{1}{96} \approx 1\%$ **b.** $\frac{1}{144} \approx 0.7\%$

39. 0.7225, 0.2775   **41.** $\frac{1}{3}$ **43.** $\frac{1}{2}$

45. Chestnut or shiny dark brown, each with $p = \frac{1}{2}$

47. Light brown ($\frac{1}{8}$), reddish brown ($\frac{1}{4}$), dark red ($\frac{1}{8}$), medium brown ($\frac{1}{8}$), chestnut ($\frac{1}{4}$), or auburn ($\frac{1}{8}$)

49. Light brown ($\frac{1}{16}$), reddish brown ($\frac{1}{8}$), dark red ($\frac{1}{16}$), medium brown ($\frac{1}{16}$), chestnut ($\frac{1}{8}$), auburn ($\frac{1}{16}$), dark brown ($\frac{1}{16}$), shiny dark brown ($\frac{1}{8}$), glossy dark brown ($\frac{1}{16}$), black ($\frac{1}{16}$), shiny black ($\frac{1}{8}$), glossy black ($\frac{1}{16}$)

51. a. $\frac{1}{6}$ **b.** $\frac{5}{6}$ **c.** $\frac{625}{1,296} \approx 0.48$ **d.** $\frac{671}{1,296} \approx 0.52$ **e.** \$0.04

53. \$0.04 > −\$0.017

55. a. 0.973   **b.** 0.775   **c.** 0.98

Chapter 3 Review

1. Experiment is pick one card from a deck of 52 cards. Sample space $S = $ {possible outcomes} = {jack of hearts, ace of spades, . . . }
$n(S) = 52$

3. $\frac{13}{52} = \frac{1}{4}$, 13:39 = 1:3   **5.** $\frac{16}{52} = \frac{4}{13}$, 16:36 = 4:9

7 {(h, h, h), (h, h, t), (h, t, h), (t, h, h), (h, t, t), (t, h, t), (t, t, h), (t, t, t)}

9. {(h, t, t), (t, h, t), (t, t, h), (t, t, t)}

11. $\frac{1}{2}$; 1:1   **13.** $\frac{1}{6}$ **15.** $\frac{7}{18}$ **17.** $\frac{11}{18}$

19. 0.013   **21.** 0.151   **23.** $\frac{12}{51}$ **25.** $\frac{1}{216}$

27. $\frac{2}{27}$ **29.** $\frac{1}{6}$ **31.** $\frac{1}{2}$ **33.** $\frac{1}{4}$

35. $\frac{1}{4}$ **37.** $\frac{1}{2}$ **39.** $\frac{1}{2}$ **41.** $\frac{1}{2}$

43. $\dfrac{{}_4C_2 \cdot {}_4C_3}{{}_{52}C_5} = 9.23 \times 10^{-6}$

45. a.

Winning spot	Probability
9	0.000000724
8	0.000032592
7	0.000591678
6	0.005719558
5	0.032601481
4 or less	0.961053966

b. −\$0.24

47. \$10.93   **49.** 0.40; 0.55   **51.** 0.50; 0.50

53. O'Neill; Bell   **55.** O'Neill

57. Independent; not mutually exclusive

59. Dependent; not mutually exclusive

61. 14%   **63.** 8%   **65.** $\frac{3}{7} = 43\%$

67. 0.78%   **69.** 1.817%

71. $\frac{7.8}{8.837} = 43\%$

73. No. Being made in Arkansas increases the probability of being defective.

75. Answers will vary.

CHAPTER 4 Statistics

4.1 Population, Sample, and Data

1. a.

Number of Visits to the Library	Frequency
1	9
2	8
3	2
4	5
5	6
	Total = 30

b.

Number of Visits

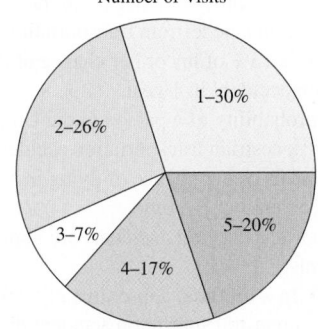

1–30%
2–26%
3–7%
4–17%
5–20%

c.

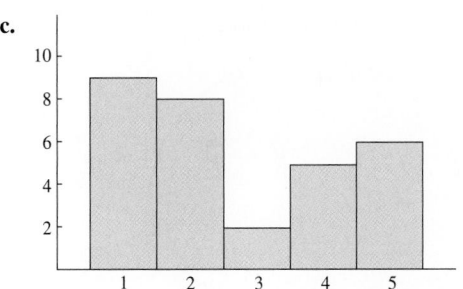

5. a.

Speed (in mph)	Tally	Frequency	Relative Frequency
$51 \leq x < 56$	II	2	0.05 = 5%
$56 \leq x < 61$	IIII	4	0.1 = 10%
$61 \leq x < 66$	||||| II	7	0.175 = 17.5%
$66 \leq x < 71$	||||| |||||	10	0.25 = 25%
$71 \leq x < 76$	||||| IIII	9	0.225 = 22.5%
$76 \leq x < 81$	||||| III	8	0.2 = 20%
		$n = 40$	1.000 = 100%

3. a.

Number of Children	Tally	Frequency	Relative Frequency	Central Angle
0	||||| III	8	0.2 = 20%	72°
1	||||| ||||| III	13	0.325 = 32.5%	117°
2	||||| IIII	9	0.225 = 22.5%	81°
3	||||| I	6	0.15 = 15%	54°
4	III	3	0.075 = 7.5%	27°
5	I	1	0.025 = 2.5%	9°
		$n = 40$	1.000 = 100%	total = 360°

b.

Speed of Cars on Interstate 40 (in mph)

b.

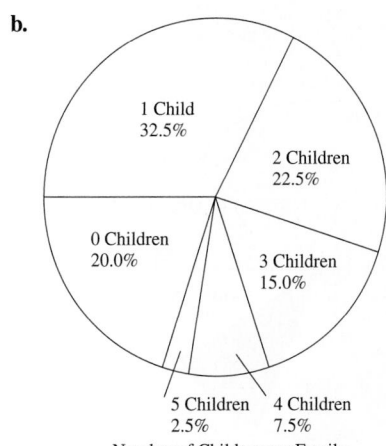

Number of Children per Family

7.

Time in minutes

c.

9.

Wages in Dollars

11.

13.

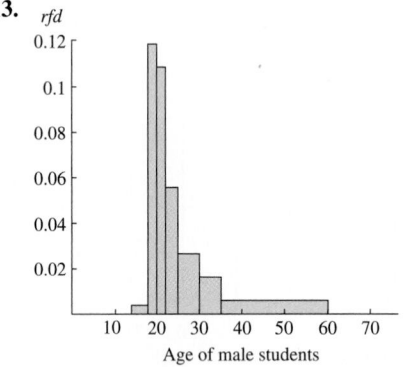

15. a. 28.5%   **b.** 32%   **c.** Not possible
d. 97.5%   **e.** 50%   **f.** 52.5%

17.

19. a.

b.

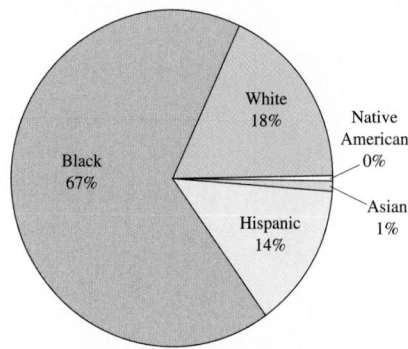
New AIDS cases - Female

c. Females have a higher percentage of blacks with new AIDS cases than males do. Males have a higher percentage of whites with new AIDS cases than females do.

d.

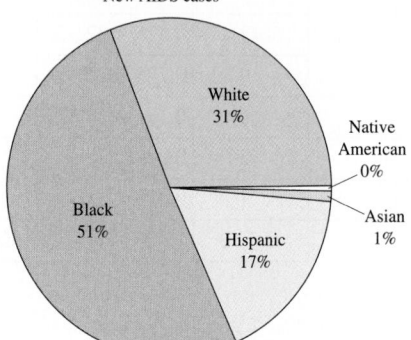
New AIDS cases

21. a.

Specialties—Male

b.

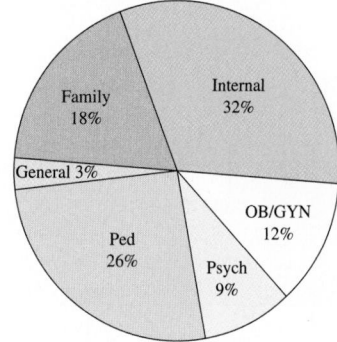
Specialties - Female

c. A higher percentage of males than females are general surgeons. A higher percentage of females than males are pediatricians.

b. Double Stuf varies more than traditional. It looks like on the average, Double Stuf is twice traditional.

d.

31. a.

29. a.

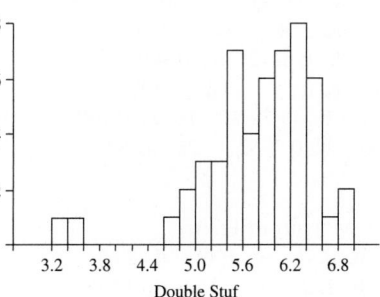

b. European and Asian are similar, but Asian varies more.

4.2 Measures of Central Tendency

1. Mean = 12.5; median = 11.5; mode = 9

3. Mean = 1.45; median = 1.5; mode = 1.7

5. a. Mean = 11; median = 10.5; mode = 9
   b. Mean = 25.5; median = 10.5; mode = 9

7. a. Mean = 7; median = 7; mode = none
   b. Mean = 107; median = 107; mode = none

9. Mean 3:14, median 3:14 mode 3:12

11. Mean = 9.85, median = 9, mode = 9 and 15

13. Mean = 37, median = 37, mode = 25, 33, 37, 45, 46

15. Mean = 7.6; median = 8; mode = 6 and 9

17. 16.028 oz

19. a. 42   **b.** 92
   c. Impossible (it would require a score of 142)

21. $51,600   **23.** 51.4 mph

25. 39 years old

27. a. $40,750   **b.** $42,000 (it does not change)

29. a. $3,523.59   **b.** 5,543.26

31. a. 35.7 years old   **b.** 36.5 years old

33. a. 23 years old   **b.** 28.5 years old

4.3 Measures of Dispersion

1. a. 16.4; 4.0   **b.** 16.4; 4.0

3. a. 0   **b.** 0

5. a. 22; 7.5   **b.** 1100; 374.2

7. a. Joey, 168; Dee Dee, 167; Joey has the higher mean.
   b. Joey, 30.9; Dee Dee, 18.2
   c. Dee Dee is more consistent because his standard deviation is lower.

9. 0.421 lb = 0:07   **11.** $s = 13.0$

13. a. 67.5; 7.1   **b.** 80%   **c.** 90%

15. a. 3.175; 1.807   **b.** 58%   **c.** 100%

17. a. 8; 1.4   **b.** 71%
   c. 94%   **d.** 100%

19. 0.285

21. 6.30

27. Answers will vary.

29. a. Double Stuf: $x = 5.76$, traditional: $\bar{x} = 2.90$
   b. Double Stuf: $s = .71$, traditional: $s = .37$

31. a. Asian: $\bar{x} = 32.9$, $s = 6.4$
   American: $\bar{x} = 31.2$, $s = 2.1$
   European: $\bar{x} = 29.9$, $s = 6.4$
   c. Asian median: 32, American median: 32, European median: 28.5

4.4 The Normal Distribution

1. Yes, it looks like a bell curve.

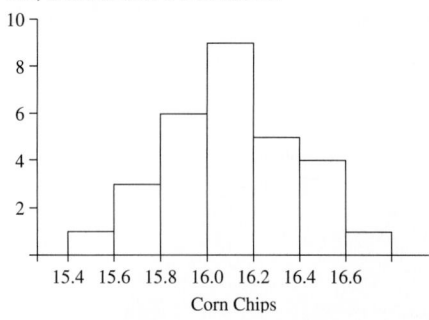

3. No, it is right-tailed.

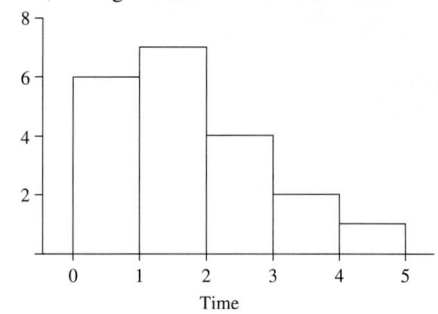

5. a. 34.13%   **b.** 34.13%   **c.** 68.26%

7. a. 49.87%   **b.** 49.87%   **c.** 99.74%

9. a. One [22.4, 27.0]; two [20.1, 29.3]; three [17.8, 31.6]
   b. 68.26%; 95.44%; 99.74%
   c.

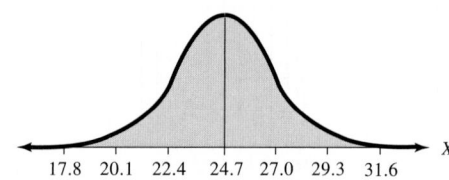

11. a. 0.4474   **b.** 0.0639   **c.** 0.5884
   d. 0.0281   **e.** 0.0960   **f.** 0.8944

13. a. 0.34   **b.** −2.08   **c.** 0.62
   d. −0.27   **e.** 1.64   **f.** −1.28

15. a. 0.42   **b.** −0.42   **c.** 2.08
   d. −1.46   **e.** 2.96   **f.** −2.21

17. a. 0.3413   **b.** 0.6272   **c.** 0.8997
   d. 0.9282   **e.** 0   **f.** 0.3300

19. a. 4.75%   **b.** 49.72%   **21. a.** 0.1908   **b.** 0.7745

23. a. 95.25%   **b.** 0.7967   **25. a.** 87   **b.** 62

27. 9.1 min

4.5 Polls and Margin of Error

1. a. 0.68   **b.** 0.31   **c.** 1.39   **d.** 2.65

3. a. 1.44   **b.** 1.28   **c.** 1.15   **d.** 2.575

5. 1.75   **7.** 1.15

9. a. 0.018 = 1.8%   **b.** 0.026 = 2.6%

11. a. 75.0% ± 3.4%   **b.** 75.0% ± 4.0%

13. a. 79.0%   **b.** 21.0%   **c.** ±1.6%
15. a. 86.0%   **b.** 70.0%
  c. ±2.8% for the men; ±2.5% for the women
17. a. 0.710 = ±71%   **b.** 0.029 = ±29.0%
  c. 0.003 = ±0.3%
19. a. ±45.0%   **b.** ±47.0%
  c. ±8.0%   **d.** ±0.2%
21. a. 36.0%   **b.** ±1.6%   **c.** ±2.3%
23. a. ±4.7%   **b.** ±3.5%   **c.** ±2.8%
25. 96.4%   **27.** 95.3%

4.6 Linear Regression
1. a. $\hat{y} = 1.0x + 3.4$ **b.** 14.4   **c.** 15.6
  d. 0.9529485   **e.** Yes
3. a. $\hat{y} = -0.087x + 8.246$ **b.** 7.811   **c.** 14.322
  d. −0.095   **e.** No, r is close to zero.
5. a. Yes

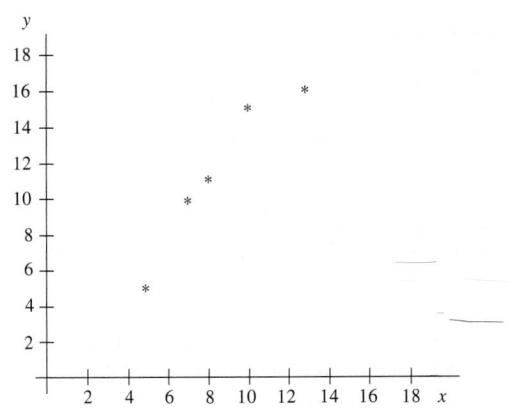

b. $\hat{y} = 1.4x - 0.3$ **c.** 12.3

d.

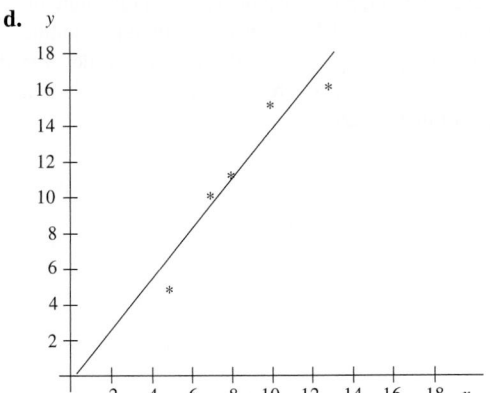

e. 0.9479459   **f.** Yes

7. a. No

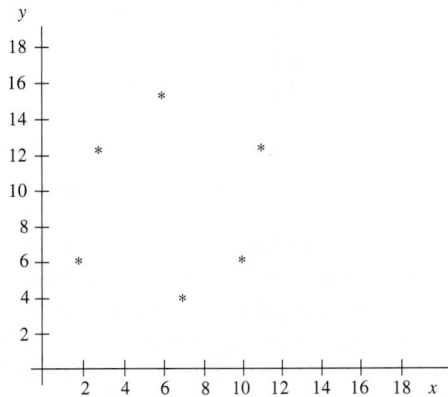

b. $\hat{y} = 0.008x + 9.1$ **c.** 9.164

d.

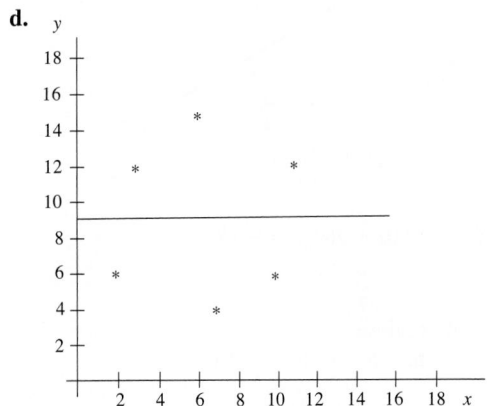

e. 0.0062782255   **f.** No

9.

Earnings	Tuition
12.49	3110
13	3229
13.47	3349
14	3501
14.53	3735
14.95	4059

a. Yes   **b.** $\hat{y} = 366.72x - 1541.52$ **c.** $3867.55
d. $15.11   **e.** $r = 0.973$ **f.** Yes, r is close to 1.

11. a. Yes

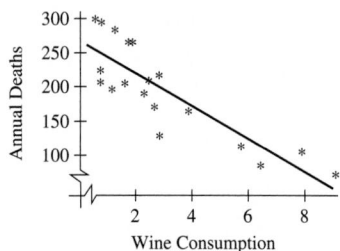

b. $\hat{y} = 1.130x - 1.5526$ **c.** 1.555 million
d. 2.699 million   **e.** 0.93344427   **f.** Yes

19. (2.5, 211), (3.9, 167), (2.9, 131), (2.4, 191), (2.9, 220),
(0.8, 297), (9.1, 71), (0.8, 211), (0.6, 300), (7.9, 107),
(1.8, 266), (1.9, 266), (0.8, 227), (6.5, 86), (1.6, 207),
(5.8, 115), (1.3, 285), (1.2, 199), (2.7, 172)

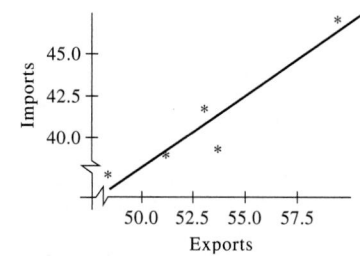

a. $\hat{y} = -24.0x + 269$; $r = -0.86$
b. Yes

Chapter 4 Review
1. a. 7.1   **b.** 7.5   **c.** 8   **d.** 2.1

3. a. 40%   **b.** 28%   **c.** 86%
   d. 14%   **e.** 50%   **f.** Cannot determine

5. 100

7. a. Timo, 98.8; Henke, 96.6; Henke has the lower mean.
   b. Timo, 6.4; Henke, 10.4
   c. Timo is more consistent because his standard deviation
   is lower.

9. a. Continuous   **b.** Neither   **c.** Discrete
   d. Neither   **e.** Discrete   **f.** Continuous

11. a. One [71, 85]; two [64, 92]; three [57, 99]
   b. 68.26%; 95.44%; 99.74%

c.

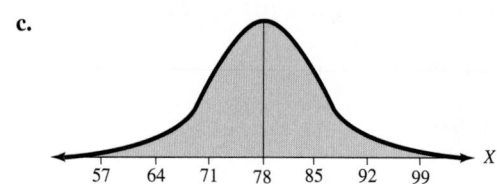

13. 401

15. a. 66.7% ± 2.4%   **b.** 66.7% ± 2.8%

17. a. ±4.1%   **b.** ±3.1%   **c.** ±2.5%

19. a. $\hat{y} = 3x + 15$ **b.** 63   **c.** 14   **d.** 1.0   **e.** Yes

21. a. Yes

Exports	Imports
48.4	37.7
51.2	39
53.7	39.4
53.1	41.9
59.5	47.3

b. $\hat{y} = 0.88x - 5.50$ **c.** 45.3 in billions of dollars
d. 59.7 in billions of dollars   **e.** $r = 0.94$
f. Yes, r is close to 1.

23. The spread or dispersion of the data.

25. Margin of error is a prediction of the largest possible error
in a sample estimate.

27. George H. Gallup founded the American Institute of
Public Opinion in 1935 to measure and report public
opinion on political and social issues. Gauss developed
the method of least squares and was a pioneer in many
areas of mathematics.

5.1 Simple Interest

1. 61   **3.** 101   **5.** $480   **7.** $25.24
9. $28.87   **11.** $4,376.48   **13.** $13,750.47
15. $1,699.25   **17.** $6,162.51   **19.** $17,042.18
21. $6,692.61   **23.** $1,058.19   **25.** $1,255.76
27. 1013 days   **29.** $233,027.39   **31.** $714.31
33. $222.86
35. a. $112,102.05   **b.** $2,102.05
37. $140.27; $2.08   **39.** $152.84; $2.73
41. a. $8,125   **b.** $146,250   **c.** $8,125   **d.** $67.71
   e. $19,500.08   **f.** $136,500   **g.** $156,000.08
43. a. $38,940   **b.** $311,520   **c.** $38,940   **d.** $356.95
   e. $95,013.60   **f.** $288,156   **g.** $383,169.60
45. a. average daily balance $989.68; finance charge $17.65;
   new balance $997.65
   b. average daily balance $988.36; finance charge $15.92;
   new balance $993.57
   c. average daily balance $983.25; finance charge $17.54;
   new balance $991.11
   d. The minimum $20 payment is reducing your debt by
   about $3 each month. It will take you roughly 28 years to
   pay off your debt. This means you will have paid the
   credit card company roughly $6720 on a $1000 debt.
   (572% of your debt was total interest)
47. a. average daily balance $979.35; finance charge $17.47;
   new balance $977.47
   b. average daily balance $958.90; finance charge $15.45;
   new balance $952.92
   c. average daily balance $932.27; finance charge $16.63;
   new balance $929.55
   d. The change in the minimum required payment policy
   has had some impact, because you're reducing your
   debt by approximately $23 per month.

5.2 Compound Interest

1. a. 0.03   **b.** 0.01   **c.** 0.000328767
   d. 0.0046153846   **e.** 0.005
3. a. 0.00775   **b.** 0.002583333   **c.** 0.000084932
   d. 0.001192307   **e.** 0.001291667
5. a. 0.02425   **b.** 0.008083333   **c.** 0.000265753
   d. 0.003730769   **e.** 0.004041667
7. a. 34 quarters   **b.** 102 months   **c.** 3,102.5 days
9. a. 120 quarters   **b.** 360 months   **c.** 10,950 days
11. $7,189.67   **13.** $9,185.46   **15.** $3,951.74
17. 8.30%   **19.** 4.34%
21. a. 10.38%   **b.** 10.47%   **c.** 10.52%
23. $583.49   **25.** $470.54
27. a. $10,083.33
   $10,167.36
   $10,252.09
   $10,337.52
   $10,423.67
   $10,510.53

b. $10,000(1 + $\frac{0.10}{12}$)^6 = $10,510.53
29. a. First: $15,900
   Second: $16,854
   Third: $17,865.24
   b. $15,000(1 + 0.06)^3$ = $17,865.24
31. $4,032,299.13
33. $9,058.33 or more
35. a. $9,664.97   **b.** $51.77
37. $19,741.51
39. a. $8,108.87   **b.** $690.65
41. First National is better: 9.55%; Citywide: 9.45%
43. a. Does not verify
   b. Does not verify
   c. The annual yield for the 5-year certificate verifies if you
   pro-rate the interest over a 360-day year, and then pay
   interest for 365 days.
45. Does not verify; true yield is 9.74%
47. a. $r = 5.12\%$ **b.** $1,025.26
   c. $25.26   **d.** 2.53%
   e. Part d is for 6 months; part a is for 1 year.
   f. You earn interest on interest; that is what compounding
   means.
49. $r = (1 + i)^n - 1$
51. 7.45%
53. a. 5.70%   **b.** 5.74%   **c.** 5.77%
   d. 5.79%   **e.** 5.78%   **f.** 5.78%
55. $i = 5.00\%$ because £1,000 · $(1 + 0.05)^{100}$ = £131,000
57. 4.57% because $4,500(1 + 0.0457)^{100}$ = $391,000
59. $8689\% = \dfrac{\$391,000}{\$4,500}$
75. a. 14.2 years   **b.** 55.80 quarters = 13.95 years
   c. 166.70 months = 13.89 years
   d. 5,060.32 days = 13.86 years
77. 1,822 days ≈ 4.99 years
   10,345 days ≈ 28.34 years
79. 2,369 days ≈ 6.49 years
   9,400 days ≈ 25.75 years

5.3 Annuities

1. $1,478.56
3. $5,422.51
5. a. $226.32   **b.** $226.32
   c. $225   **d.** $1.32
7. a. $455.46   **b.** $455.46
   c. $450   **d.** $5.46
9. a. $283,037.86   **b.** $54,600.00   **c.** $228,437.86
11. a. $251,035.03   **b.** $69,600.00   **c.** $181,435.03
13. $1,396.14   **15.** $222.40
17. $18,680.03   **19.** $51.84
21. $33.93   **23.** $33.63

25. a. $537,986.93

    b. 1 $537,986.93 $2,734.77 $650 $540,071.70
        2 $540,071.70 $2,745.36 $650 $542,167.06
        3 $542,167.06 $2,756.02 $650 $544,273.08
        4 $544,273.08 $2,766.72 $650 $546,389.80
        5 $546,389.80 $2,777.48 $650 $548,517.28

27. a. $175,384.62   **b.** $111.84

29. a. $1,484,909.47   **b.** $102,000
    c. $1,382,909.47   **d.** $979,650.96
    e. $643,631.10

31. $11,954.38   **33.** $45.82

35. a. $595.83   **b.** $5,367.01

    c. 1   2   3   4   5
    $5,367.01 $10,958.76 $16,784.67 $22,854.54 $29,178.58
      6   7   8   9   10
    $35,767.44 $42,632.21 $49,784.44 $57,236.17 $64,999.94

37. $P = \text{pymt}\,\dfrac{(1+i)^n - 1}{i(1+i)^n} = \text{pymt}\,\dfrac{1-(1+i)^{-n}}{i}$

39. $1,396.14   **41.** $222.40

51. a. 586 months = 48 years and 10 months

53. 1,171 payments = 48 years and 9.5 months

5.4 Amortized Loans

1. a. $125.62   **b.** $1,029.76

3. a. $193.91   **b.** $1,634.60

5. a. $1,303.32   **b.** $314,195.20

7. a. $378.35   **b.** $3,658.36

c. Payment Number	Principal Portion	Interest Portion	Total Payment	Balance Due
0	—	—	—	$14,502.44
1	$239.37	$138.98	$378.35	$14,263.07
2	$241.66	$136.69	$378.35	$14,021.41

9. a. $1,602.91   **b.** $407,047.60

c. Payment Number	Principal Portion	Interest Portion	Total Payment	Balance Due
0	—	—	—	$170,000.00
1	$62.28	$1,540.63	$1,602.91	$169,937.72
2	$62.85	$1,540.06	$1,602.91	$169,874.87

    d. $4,218.18 per month (assuming only home loan
    payment)

11. a. $404.72; $360.29   **b.** $4,597.24; $2,464.60
    c. Bank simple interest loan (why?)

13. a. $354.29; $1,754.44   **b.** $278.33; $2,359.84
    c. $233.04; $2,982.40

15. a. $599.55; $115,838.00   **b.** $665.30; $139,508.00
    c. $733.76; $164,153.60   **d.** $804.62; $189,663.20
    e. $877.57; $215,925.20   **f.** $952.32; $242,835.20

17. a. $877.57; $215,925.20   **b.** $404.89; $215,814.20

19. Both verify, but not exactly. Their computations
are actually more accurate than ours, since they are using
more accurate round-off rules than we do.

21. a. $19,053.71

b. Payment Number	Principal Portion	Interest Portion	Total Payment	Balance Due
0	—	—	—	$75,000.00
1	$18,569.33	$484.38	$19,053.71	$56,430.67
2	$18,689.26	$364.45	$19,053.71	$37,741.41
3	$18,809.96	$243.75	$19,053.71	$18,931.45
4	$18,931.45	$122.27	$19,053.72	$ 0.00

23. a. $23,693.67

b. Payment Number	Principal Portion	Interest Portion	Total Payment	Balance Due
0	—	—	—	$93,000.00
1	$22,986.48	$707.19	$23,693.67	$70,013.52
2	$23,161.28	$532.39	$23,693.67	$46,852.24
3	$23,337.40	$356.27	$23,693.67	$23,514.85
4	$23,514.85	$178.81	$23,693.66	$0.00

25. a. $29.91   **b.** $797.30

c.

Payment	Principal	Interest	Total Payment	Balance
0				$6,243.00
1	$767.39	$29.91	$797.30	$5,475.61
2	$771.06	$26.24	$797.30	$4,704.55
3	$774.76	$22.54	$797.30	$3,929.79
4	$778.47	$18.83	$797.30	$3,151.32
5	$782.20	$15.10	$797.30	$2,369.12
6	$785.95	$11.35	$797.30	$1,583.17
7	$789.71	$7.59	$797.30	$793.46
8	$793.46	$3.80	$797.26	$0

d. $135.36

27. a. $89.25   **b.** $1905.92

c.

Payment	Principal	Interest	Total Payment	Balance
0				$12,982.00
1	$1,816.67	$89.25	$1,905.92	$11,165.33
2	$1,829.16	$76.76	$1,905.92	$9,336.17
3	$1,841.73	$64.19	$1,905.92	$7,494.44
4	$1,854.40	$51.52	$1,905.92	$5,640.04
5	$1,867.14	$38.78	$1,905.92	$3,772.90
6	$1,879.98	$25.94	$1,905.92	$1,892.92
7	$1,892.92	$13.01	$1,905.93	$0

d. $359.45

29. $3,591.73   **31.** $160,234.64

33. a. $1,735.74   **b.** $151,437.74   **c.** $1,204.91
    d. $472,016.40   **e.** $343,404.24   **f.** Yes (why?)

35. a. $1,718.80   **b.** $172,157.40   **c.** $240,354.60
d. $573,981.98   **e.** $1,200,543.86
f. The decision is between saving more for retirement or increasing their standard of living now. They should not prepay their loan. (Why?)
37. a. $18,950   **b.** $151,600   **c.** $18,950
d. $1,501.28   **e.** $189.50
39. a. $271.61   **b.** $1,962.39   **c.** $1,501.28
53. a. $12,902.39   **b.** $768.35
c. First year "SUM(C3:C14)"
Last year "SUM (C171:C182)"
55. a. $1,529.57   **b.** $12,165.10   **c.** $724.43
57. a. $1,223.63   **b.** $13,070.29   **c.** $10,088.45
d. The amount paid is borrowed over a longer period of time, so the total monthly payment should go down. However, the interest is not evenly spread over the loan period. The lenders collect more of their interest during the first years.

5.5 Annual Percentage Rate on a Graphing Calculator
1. 14.6%
3. 11.2%
5. a. $126.50   **b.** Verifies
7. a. $228.77   **b.** Does not verify
9. Either one could be less expensive, depending on the A.P.R.
11. Really Friendly S and L will have lower payments but higher fees and/or more points.
13. $8,109.53
15. a. $819.38   **b.** $801.40
c. $8,009.75 **d.** $9,496.12
e. The RTC loan has a lower monthly payment but has higher fees.

5.6 Payout Annuities
1. a. $143,465.15   **b.** $288,000.00
3. a. $179,055.64   **b.** $390,000.00
5. a. $96.26
b. $34,653.60; she receives $253,346.40 more than she paid
7. a. $138.30   **b.** $331.39
9. a. $185,464.46   **b.** $14,000.00
c. $14,560.00   **d.** $29,495.89
11. a. $36,488.32   **b.** 576,352.60
c. $454.10   **d.** $1,079.79
13. a. $12,565.57   **b.** 10.47130674%
c. $138,977.90   **d.** $61.48
15. a. $17,502.53   **b.** 9.2721727%
c. $245,972.94   **d.** $372.99
17. $490,907.37

Chapter 5 Review
1. $8,730.15
3. $20,121.60
5. $4,067.71
7. $29,909.33
9. $23,687.72
11. $5,469.15

13. $6,943.26
15. 7.25%
17. $109,474.35
19. $275,327.05
21. a. $525.05   **b.** $6,503
23. $362,702.26
25. $3,443.70; $57.03
27. a. $10,014.44   **b.** $1,692.71
29. $255,048.53
31. $37,738.26
33. a. $102,887.14
b.

Month	Beginning	Interest Earned	Withdrawal	End
1	$102,887.14	$493.00	$1,000	$102,380.14
2	$102,380.14	$490.57	$1,000	$101,870.71
3	$101,870.71	$488.13	$1,000	$101,358.84
4	$101,358.84	$485.68	$1,000	$100,844.52
5	$100,844.52	$483.21	$1,000	$100,327.73

35. a. $1,246.53   **b.** $268,270.80
c.

Payment	Principal Portion	Interest Portion	Total payment	Balance
0				$180,480
1	$137.33	$1,109.20	$1,246.53	$180,342.67
2	$138.17	$1108.36	$1246.53	$180,204.50

37. a. $268.14   **b.** $6,085.50
c.   **d.** $1,079.50

Payment	Principal Portion	Interest Portion	Total Payment	Balance
0				$41,519.00
1	$5,817.36	$268.14	$6,085.50	$35,701.64
2	$5,854.93	$230.57	$6,085.50	$29,846.71
3	$5,892.74	$192.76	$6,085.50	$23,953.97
4	$5,930.80	$154.70	$6,085.50	$18,023.17
5	$5,969.10	$116.40	$6,085.50	$12,054.07
6	$6,007.65	$77.85	$6,085.50	$6,046.42
7	$6,046.42	$39.05	$6,085.47	$0.00

39. a. $174.50   **b.** Does not verify
41. a. $1,217.96   **b.** $7,299.98
43. a. $58,409.10   **b.** $881,764.23   **c.** $556.55

CHAPTER 6   Voting and Apportionment

6.1 Voting Systems

1. a. 2,000   **b.** Cruz   **c.** Yes
3. a. 10,351   **b.** Edelstein   **c.** No
5. a. 30   **b.** Park   **c.** 47%
   d. Beach   **e.** 53%   **f.** Beach
   g. 63   **h.** Beach   **i.** 2
7. a. 65   **b.** Coastline   **c.** 48%
   d. Coastline **e.** 63%   **f.** Coastline
   g. 150   **h.** Coastline   **i.** 2
9. a. 140   **b.** Shattuck   **c.** 40%
   d. Nirglotis **e.** 51%   **f.** Shattuck
   g. 284   **h.** Nirglotis   **i.** 2
11. a. 1,342   **b.** Jones   **c.** 44%
   d. Jones   **e.** 50.4%   **f.** Jones
   g. 3,960   **h.** Jones   **i.** 3
13. a. 31,754   **b.** Darter   **c.** 52%
   d. Darter (He has a majority, so no run off is needed.)
   e. 52%   **f.** Darter   **g.** 138,797
   h. Darter   **i.** 4
15. $6! = 720$ **17.** $_6C_2 = 15$
19. a. 75   **b.** 25
21. a. 21   **b.** 0
23. a. A has the majority of first-choice votes and therefore should win.

	A	B	C
First	7	4	2
Second	0	9	4
Third	6	0	7
Total	13	13	13

   b. B wins.

	A	B	C
First	21	12	6
Second	0	18	8
Third	6	0	7
Total	27	30	21

   c. Yes. A received 7 votes for first choice, which is a majority of the 13 votes, but A did not win.

25. a. C wins by getting 17 out of 27 first-choice votes in the modified voter preference after eliminating B.

	6	8	4	9
First	A	C	A	C
Second	C	A	C	A

   b. Eliminate B. C wins by getting 21 out of 27 first-choice votes.

	6	8	13
First	A	C	C
Second	C	A	A

   c. No. C wins the original election. All changes are in favor of C, and C wins again.

6.2 Methods of Apportionment

1. a. See table.   **b.** $d = 2000/40 = 50.00$ **c.** See table.

State	A	B	C	Total
Population (thousands)	900	700	400	2,000
Std q ($d = 50.00$)	18.00	14.00	8.00	—
Lower q	18	14	8	40
Upper q	19.00	15.00	9.00	43

3. a. See table.   **b.** $d = 39.672/15 = 2.64$ **c.** See table.

State	NY	PA	NJ	Total
Population (millions)	18.977	12.281	8.414	39.672
Std q ($d = 2.64$)	7.19	4.65	3.19	—
Lower q ($d = 2.64$)	7	4	3	14
Upper q ($d = 2.64$)	8	5	4	17

5.

State	A	B	C	Total
Population (thousands)	900	700	400	2,000
Std q ($d = 50.00$)	18.00	14.00	8.00	—
Lower q ($d = 50.00$)	18	14	8	40
Additional seats	0	0	0	0
Seats Hamilton	**18**	**14**	**8**	**40**

7.

State	NY	PA	NJ	Total
Population (millions)	18.977	12.281	8.414	39.672
Std q ($d = 2.64$)	7.19	4.65	3.19	—
Lower q ($d = 2.64$)	7	4	3	14
Additional seats	0	1	0	1
Seats Hamilton	**7**	**5**	**3**	**15**

9.

School	A	B	C	D	Total
Population	1,768	1,357	1,091	893	5,109
Std q ($d = 25.55$)	69.20	53.11	42.70	34.95	—
Lower q ($d = 25.55$)	69	53	42	34	198
Additional seats	0	0	1	1	2
Seats Hamilton	**69**	**53**	**43**	**35**	**200**

11.

Region	N	S	E	W	Total
Population	2,680	6,550	2,995	8,475	20,700
Std q ($d = 862.50$)	3.11	7.59	3.47	9.83	—
Lower q ($d = 862.50$)	3	7	3	9	22
Additional seats	0	1	0	1	2
Seats Hamilton	**3**	**8**	**3**	**10**	**24**

13.

Country	D	F	I	N	S	Total
Population (thousands)	5,413	5,215	294	4,575	8,986	24,483
Std q ($d = 1,224.15$)	4.42	4.26	0.24	3.74	7.34	—
Lower q ($d = 1,224.15$)	4	4	0	3	7	18
Additional seats	1	0	0	1	0	2
Seats Hamilton	**5**	**4**	**0**	**4**	**7**	**20**

15.

Country	C	E	G	H	N	P	Total
Population (thousands)	3,957	6,588	14,281	6,824	5,360	3,000	40,010
Std q ($d = 1,600.4$)	2.47	4.12	8.92	4.26	3.35	1.87	—
Lower q ($d = 1,600.4$)	2	4	8	4	3	1	22
Additional seats	1	0	1	0	0	1	3
Seats Hamilton	**3**	**4**	**9**	**4**	**3**	**2**	**25**

17.

State	A	B	C	Total	Comments
Population (thousands)	900	700	400	2000	
Std q ($d = 50.00$)	18.00	14.00	8.00	—	
Lower q ($d = 50.00$)	**18**	**14**	**8**	**40**	**Jefferson**
Upper q ($d = 50.00$)	19	15	9	43	Use $d_m > d$
Rounded q ($d = 50.00$)	**18**	**14**	**8**	**40**	**Webster**
Additional seats (Hamilton)	0	0	0	0	
Seats Hamilton	18	14	8	40	**Hamilton**
Modified std ($d = 52$) Upper q	17.31 **18**	13.46 **14**	7.69 **8**	— **40**	**Adams**

19.

State	NY	PA	NJ	Total	Comments
Population (millions)	18.977	12.281	8.414	39.672	
Std q ($d = 2.64$)	7.19	4.65	3.19	—	
Lower q ($d = 2.64$)	7	4	3	14	Use $d_m < d$
Upper q ($d = 2.64$)	8	5	4	17	Use $d_m > d$
Rounded q ($d = 2.64$)	7	5	3	15	Webster
Additional seats (Hamilton)	0	1	0	1	
Seats Hamilton	7	5	3	15	Hamilton
Modified std q ($d = 2.4$) Lower q	7.91 7	5.12 5	3.51 3	— 15	Jefferson
Modified std ($d = 3$) Upper q	6.33 7	4.09 5	2.80 3	15	Adams

21.

School	A	B	C	D	Total	Comments
Population	1,768	1,357	1,091	893	5,109	
Std q ($d = 25.55$)	69.20	53.11	42.70	34.95	—	
Lower q ($d = 25.55$)	69	53	42	34	198	Use $d_m < d$
Upper q ($d = 25.55$)	70	54	43	35	202	Use $d_m > d$
Rounded q ($d = 25.55$)	69	53	43	35	200	Webster
Additional seats (Hamilton)	0	0	1	1	2	
Seats Hamilton	69	53	43	35	200	Hamilton
Modified std q ($d = 25.3$) Lower q	69.88 69	53.63 53	43.12 43	35.30 35	200	Jefferson
Modified std q ($d = 25.7$) Upper q ($d = 25.7$)	68.79 69	52.8 53	42.45 43	34.75 35	200	Adams

23.

Region	N	S	E	W	Total	Comments
Population	2,680	6,550	2,995	8,475	20,700	
Std q ($d = 862.50$)	3.11	7.59	3.47	9.83	—	
Lower q ($d = 862.50$)	3	7	3	9	22	Use $d_m < d$
Upper q ($d = 862.50$)	4	8	4	10	26	Use $d_m > d$
Rounded q ($d = 862.50$)	3	8	3	10	24	Webster
Additional seats	0	1	0	1	2	
Seats Hamilton	3	8	3	10	24	Hamilton
Modified std q ($d = 800$) Lower q	3.35 3	8.19 8	3.74 3	10.59 10	24	Jefferson
Modified std q ($d = 940$) Upper q	2.85 3	6.97 7	3.19 4	9.01 10	24	Adams

25.

Country	D	F	I	N	S	Total	Comments
Population (thousands)	5,413	5,215	294	4,575	8,986	24,483	
Std q ($d = 1,224.15$)	4.42	4.26	0.24	3.74	7.34	—	
Lower q ($d = 1,224.15$)	4	4	0	3	7	18	Use $d_m < d$
Upper q ($d = 1,224.15$)	5	5	1	4	8	23	Use $d_m > d$
Additional seats	1	0	0	1	0	2	
Seats Hamilton	**5**	**4**	**0**	**4**	**7**	**20**	**Hamilton**
Modified std q ($d = 1100$) Lower q	4.92 **4**	4.74 **4**	.27 **0**	4.16 **4**	8.17 **8**	**20**	**Jefferson**
Modified std q ($d = 1425$) Upper q	3.80 **4**	3.66 **4**	.21 **1**	3.21 **4**	6.31 **7**	**20**	**Adams**
Modified std q ($d = 1200$) Rounded q	4.51 **5**	4.35 **4**	.245 **0**	3.81 **4**	7.49 **7**	**20**	**Webster**

27.

Country	C	E	G	H	N	P	Total	Comments
Population (thousands)	3,957	6,588	14,281	6,824	5,360	3,000	40,010	
Std q ($d = 1,600.4$)	2.47	4.12	8.92	4.26	3.35	1.87	—	
Lower q ($d = 1,600.4$)	2	4	8	4	3	1	22	Use $d_m < d$
Upper q ($d = 1600.4$)	3	5	9	5	4	1	28	Use $d_m > d$
Rounded q ($d = 1,600.4$)	2	4	9	4	3	2	24	Use $d_m < d$
Additional seats	1	0	1	0	0	1	3	
Seats Hamilton	**3**	**4**	**9**	**4**	**3**	**2**	**25**	**Hamilton**
Modified std q ($d = 1,400$) Lower q	2.83 **2**	4.71 **4**	10.20 **10**	4.87 **4**	3.83 **3**	2.14 **2**	**25**	**Jefferson**
Modified std q ($d = 1,786$) Upper q	2.22 **3**	3.69 **4**	7.996 **8**	3.83 **4**	3.001 **4**	1.69 **2**	**25**	**Adams**
Modified std q ($d = 1,550$) Rounded q	2.55 **3**	4.25 **4**	9.21 **9**	4.40 **4**	3.45 **3**	1.93 **2**	**25**	**Webster**

29.

State	A	B	C	Total	Comments
Population (thousands)	900	700	400	2,000	
Std q ($d = 50.00$)	18.00	14.00	8.00	—	
Lower q ($d = 50.00$)	18	14	8	40	
Upper q ($d = 50.00$)	19	15	9	41	
Geometric mean	18.49	14.49	8.49	—	
Rounded q ($d = 50.00$)	18	14	8	40	**H–H**

31.

State	NY	PA	NJ	Total	Comments
Population (millions)	18,977	12,281	8,414	39,672	
Std q ($d = 2.64$)	7.19	4.65	3.19	—	
Lower q ($d = 2.64$)	7	4	3	14	
Upper q ($d = 2.64$)	8	5	4	17	
Geometric mean	7.48	4.47	3.46	—	
Rounded q ($d = 2.64$)	**7**	**5**	**3**	**15**	**H–H**

33.

School	A	B	C	D	Total	Comments
Population	1,768	1,357	1,091	893	5,109	
Std q ($d = 25.55$)	69.20	53.11	42.70	34.95	—	
Lower q ($d = 25.55$)	69	53	42	34	198	
Upper q ($d = 25.55$)	70	54	43	35	202	
Geometric mean	69.50	53.50	42.50	34.50	—	
Rounded q ($d = 25.55$)	**69**	**53**	**43**	**35**	**200**	**H–H**

35.

Region	N	S	E	W	Total	Comments
Population	2,680	6,550	2,995	8,475	20,700	
Std q ($d = 862.50$)	3.11	7.59	3.47	9.83	—	
Lower q ($d = 862.50$)	3	7	3	9	22	
Upper q ($d = 862.50$)	4	8	4	10	26	
Geometric mean	3.46	7.48	3.46	9.49	—	
Rounded q ($d = 862.50$)	3	8	4	10	25	Use $d_m > d$
Modified std q ($d = 870$) **Rounded q**	3.08 **3**	7.53 **8**	3.44 **3**	9.74 **10**	**24**	**H–H**

37.

Country	D	F	I	N	S	Total	Comments
Population (thousands)	5,413	5,215	294	4,575	8,986	24,483	
Std q ($d = 1{,}224.15$)	4.42	4.26	0.24	3.74	7.34	—	
Lower q ($d = 1{,}224.15$)	4	4	0	3	7	18	
Upper q ($d = 1{,}224.15$)	5	5	1	4	8	23	
Geometric mean	4.47	4.47	0	3.46	7.48	—	
Rounded q ($d = 1{,}224.15$)	**4**	**4**	**1**	**4**	**7**	**20**	**H–H**

39.

Country	C	E	G	H	N	P	Total	Comments
Population (thousands)	3,957	6,588	14,281	6,824	5,360	3,000	40,010	
Std q ($d = 1,600.4$)	2.47	4.12	8.92	4.26	3.35	1.87	—	
Lower q ($d = 1,600.4$)	2	4	8	4	3	1	22	
Upper q ($d = 1,600.4$)	3	5	9	5	4	2	28	
Geometric mean	2.45	4.47	8.49	4.47	3.46	1.41	—	
Rounded q ($d = 1,600.4$)	**3**	**4**	**9**	**4**	**3**	**2**	**25**	**H–H**

41. Satellite

	Main	Satellite
Population	25,000	2,600
Instructors	465	47
HHN	2,884.30	2,996.45

43. Banach

	Agnesi	Banach	Cantor
Population	567	871	666
Teachers	21	32	25
HHN	695.86	718.41	682.39

45. The hypothetical Delaware has one more and Virginia has one fewer than the actual.
Total population = 3,615.92 in thousands
$d = \frac{3,615.92}{105} = 34.44$

State	Population (thousands)	Std q $d = 34.44$	Lower q	Additional Seats	Hamilton	Actual	Upper q	Std q_m $d_m = 36.5$	Upper q_m $d_m = 36.5$ Adams
VA	630.56	18.31	18	0	**18**	**19**	19	17.28	**18**
MA	475.327	13.8	13	1	**14**	**14**	14	13.02	**14**
PA	432.879	12.57	12	1	**13**	**13**	13	11.86	**12**
N.C	353.523	10.26	10	0	**10**	**10**	11	9.69	**10**
N.Y	331.589	9.63	9	1	**10**	**10**	10	9.08	**10**
MD	278.514	8.09	8	0	**8**	**8**	9	7.63	**8**
CT	236.841	6.88	6	1	**7**	**7**	7	6.49	**7**
S.C	206.236	5.99	5	1	**6**	**6**	6	5.65	**6**
N.J	179.57	5.21	5	0	**5**	**5**	6	4.92	**5**
N.H	141.822	4.12	4	0	**4**	**4**	5	3.89	**4**
VT	85.533	2.48	2	0	**2**	**2**	3	2.34	**3**
GA	70.835	2.06	2	0	**2**	**2**	3	1.94	**2**
KY	68.705	1.99	1	1	**2**	**2**	2	1.88	**2**
RI	68.446	1.99	1	1	**2**	**2**	2	1.88	**2**
DE	55.54	1.61	1	1	**2**	**1**	2	1.52	**2**
Total	3,615.92	—	97	8	**105**	**105**	112	—	**105**

47. Virginia has one fewer and Delaware has one more than the actual. Use $dm = 34.6$.

State	Population (thousands)	Std q ($d = 34.44$)	Rounded q	Actual	Rounded q_m $d_m = 34.6$ Webster
VA	630.56	18.31	18	19	18
MA	475.327	13.8	14	14	14
PA	432.879	12.57	13	13	13
N.C	353.523	10.26	10	10	10
N.Y	331.589	9.63	10	10	10
MD	278.514	8.09	8	8	8
CT	236.841	6.88	7	7	7
S.C	206.236	5.99	6	6	6
N.J	179.57	5.21	5	5	5
N.H	141.822	4.12	4	4	4
VT	85.533	2.48	3	2	2
GA	70.835	2.06	2	2	2
KY	68.705	1.99	2	2	2
RI	68.446	1.99	2	2	2
DE	55.54	1.61	2	1	2
Total	3615.92	—	106	105	105

6.3 Flaws of Apportionment

1. a. See table.   **b.** $d = 24.5/32 = 0.77$ **c.** See table.   **d.** See table.

State	A	B	C	Total	Comments
Population (millions)	3.5	4.2	16.8	24.5	Part (a)
Std q ($d = 0.77$)	4.55	5.45	21.82	—	Part (c)
Lower q ($d = 0.77$)	4	5	21	30	Part (c)
Upper q ($d = 0.77$)	5	6	22	33	Part (c)
Modified lower q ($d = 0.73$)	4	5	23	32	d. Jefferson

e. Yes. C has 23 seats, which is neither upper nor lower quota.

3. a. See table.   **b.** $d = 24.5/79 = 0.31$ **c.** See table.   **d.** See table.

State	A	B	C	Total	Comments
Population (millions)	3.5	4.2	16.8	24.5	Part (a)
Std q ($d = 0.31$)	11.29	13.55	54.19	—	Part (c)
Lower q ($d = 0.31$)	11	13	54	78	Part (c)
Upper q ($d = 0.31$)	12	14	55	81	Part (c)
Modified upper q ($d = 0.317$)	12	14	53	79	d. Adams

e. Yes. C has 53 seats, which is neither upper nor lower quota.

5. a. See table. **b.** $d = \frac{41.5}{200} = 0.21$ **c.** See table. **d.** See table.

State	A	B	C	D	Total	Comments
Population (millions)	1.2	3.4	17.5	19.4	41.5	Part (a)
Std q ($d = 0.21$)	5.71	16.19	83.33	92.38	—	Part (c)
Lower q ($d = 0.21$)	5	16	83	92	196	Part (c)
Upper q ($d = 0.21$)	6	17	84	93	200	Part (c)
Rounded q ($d = 0.21$)	6	16	83	92	197	
Modified rounded q ($d = 0.20715$)	6	16	84	94	200	d. Webster

e. Yes. D has 94 seats, which is neither upper nor lower quota.

7. a. See table. **b.** $d = \frac{41.5}{200} = 0.21$ **c.** See table. **d.** See table.

State	A	B	C	D	Total	Comments
Population (millions)	1.2	3.4	17.5	19.4	41.5	Part (a)
Std q ($d = 0.21$)	5.71	16.19	83.30	92.38	—	Part (c)
Lower q ($d = 0.21$)	5	16	83	92	196	Part (c)
Upper q ($d = 0.21$)	6	17	84	93	200	Part (c)
Geometric mean	5.48	16.49	83.50	92.50	—	
Rounded q ($d = 0.21$)	6	16	83	92	197	
Modified rounded q ($d_m = 0.2085$)	6	16	84	94	200	d. H–H

e. Yes D has 94 seats, which is neither upper not lower quota.

9. a. See table. **b.** $d = \frac{12,800}{120} = 106.67$. **c.** See table. **d.** See table.

State	A	B	C	Total	Comments
Population (thousands)	690	5700	6410	12800	Part (a)
Std q ($d = 106.67$)	6.47	53.44	60.09	—	
Lower q ($d = 106.67$)	6	53	60	119	
Additional seats	1	0	0	1	
Seats	7	53	60	120	c. Hamilton
New std q ($d = \frac{12,800}{121} = 105.79$)	6.52	53.88	60.59	—	
New lower q ($d = 105.79$)	6	53	60	119	
Additional seats	0	1	1	2	
Seats	6	54	61	121	d. Hamilton

e. Yes. State A has lost a seat at the expense of the two larger states even though its population didn't change.

11. a. See table. **b.** $d = \frac{5,000}{110} = 45.45$ **c.** See table.

State	A	B	C	Total	Comments
Population (thousands)	1,056	1,844	2,100	5,000	Part (a)
Std q ($d = 45.45$)	23.23	40.57	46.20	—	
Lower q ($d = 45.45$)	23	40	46	109	
Additional seats	0	1	0	1	
Seats	23	41	46	110	c. Hamilton

d. $110 \cdot \dfrac{2440}{5000} = 53.68$. Add 53 new seats.

e. New total seats = 163. New total population = 7440. New standard divisor = $\dfrac{5{,}000 + 2440}{110 + 53} = 45.64$.

State	A	B	C	D	Total
Population (thousands)	1,056	1,844	2,100	2,440	7,400
Std q ($d = 45.64$)	23.14	40.40	46.01	53.46	—
Lower q ($d = 45.64$)	23	40	46	53	162
Additional seats	0	0	0	1	1
Seats	23	40	46	54	163

f. Yes. Apportionment changes. State B is altered.

13. a. $d = \frac{6{,}000}{11} = 545.45$

	A	B	C	Total
2005	780	1,700	3,520	6,000
Std q ($d = 545.45$)	1.43	3.12	6.45	—
Lower q ($d = 545.45$)	1	3	6	10
Additional seats	0	0	1	1
2005 seats	1	3	7	11

b. $d = \frac{6{,}440}{11} = 585.45$

	A	B	C	Total
2006	820	1,880	3,740	6,440
Std q ($d = 585.45$)	1.40	3.11	6.39	—
Lower q ($d = 585.45$)	1	3	6	10
Additional seats	1	0	0	1
2006 seats	2	3	6	11

c.

	A	B	C	Total
2005	780	1,700	3,520	6,000
2006	820	1,880	3,740	6,440
Increase	$\frac{40}{780}$	$\frac{180}{1,700}$	$\frac{220}{3,520}$	
% increase	5.1	10.6	6.25	
Change in seats	+1	0	−1	

d. Yes. C lost a seat to A, but C grew at a faster rate.

Chapter 6 Review Exercises

1. a. 74   **b.** Beethoven   **c.** 34%   **d.** Vivaldi   **e.** 59%   **f.** Beethoven   **g.** 197   **h.** Tie between Beethoven and Vivaldi   **i.** 2 points each

3.

State	A	B	C	Total	Comments
Population (thousands)	1,200	2,300	2,500	6,000	
Std q ($d = 78.95$)	15.20	29.13	31.67		
Lower q ($d = 78.95$)	15	29	31	75	
Additional seats	0	0	1	1	
Seats Hamilton	**15**	**29**	**32**	**76**	**Hamilton**

5.

State	A	B	C	Total	Comments
Population (thousands)	1,200	2,300	2,500	6,000	
Std q ($d = 78.95$)	15.20	29.13	31.67		
Lower q ($d = 78.95$)	15	29	31	75	
Upper q ($d = 78.95$)	16	30	32	78	Use $d_m > d$
Modified std ($d_m = 80.1$)	14.98	28.71	31.21		
Upper q	**15**	**29**	**32**	**76**	**Adams**

7.

State	A	B	C	Total	Comments
Population (thousands)	1,200	2,300	2,500	6,000	
Std q ($d = 78.95$)	15.20	29.13	31.67		
Lower q ($d = 78.95$)	15	29	31	75	
Upper q ($d = 78.95$)	16	30	32	78	
Geometric mean ($d = 78.95$)	15.49	29.50	31.50	—	
Rounded q ($d = 78.95$)	**15**	**29**	**32**	**76**	**H–H**

9.

State	A	B	C	P	U	Total	Comments
Population (thousands)	39,145	8,724	15,824	6,191	3,399	73,283	
Std q ($d = 2{,}931.32$)	13.35	2.98	5.40	2.11	1.16	—	
Lower q ($d = 2{,}931.32$)	13	2	5	2	1	23	Use $d_m < d$
Modified std q ($d_m = 2{,}700$)	14.50	3.23	5.86	2.29	1.26	—	
Lower q	**14**	**3**	**5**	**2**	**1**	**25**	**Jefferson**

11.

State	A	B	C	P	U	Total	Comments
Population (thousands)	39,145	8,724	15,824	6,191	3,399	73,283	
Std q ($d = 2{,}931.32$)	13.35	2.98	5.40	2.11	1.16	—	
Lower q ($d = 2{,}931.32$)	13	2	5	2	1	23	
Upper q ($d = 2{,}931.32$)	14	3	6	3	2	28	
Rounded q ($d = 2{,}931.32$)	13	3	5	2	1	24	Use $d_m < d$
Modified std q ($d_m = 2{,}900$)	13.50	3.01	5.46	2.13	1.17	—	
Rounded q	**14**	**3**	**5**	**2**	**1**	**25**	**Webster**

13. Napier

	Leibniz	Maclaurin	Napier	Total
Population	987	1,242	1,763	3,992
Seats	25	31	44	100
HHN	1,498.72	1,555.00	1,569.78	

15. a. See table.   **b.** $d = \frac{20.4}{72} = 0.28$ **c.** See table.   **d.** See table.

State	A	B	C	Total	Comments
Population (millions)	1.6	3.5	15.3	20.4	Part (a)
Std q ($d = 0.28$)	5.71	12.5	54.64	—	Part (c)
Lower q ($d = 0.28$)	5	12	54	71	Part (c)
Upper q ($d = 0.28$)	6	13	55	74	Part (c)
Modifed std q ($d_m = 0.29$)	5.52	12.07	52.76	—	Use $d_m > d$
Upper q	6	13	53	72	Adams

e. Yes. C has 53 seats, which is neither an upper nor a lower quota.

17. a. See table.   **b.** $d = \frac{41.6}{201} = 0.21$ **c.** See table.   **d.** See table.

State	A	B	C	D	Total	Comments
Population (millions)	1.100	3.500	17.600	19.400	41.6	Part (a)
Std q ($d = 0.21$)	5.24	16.67	83.81	92.38	—	Part (c)
Lower q ($d = 0.21$)	5	16	83	92	196	Part (c)
Upper q ($d = 0.21$)	6	17	84	93	200	Part (c)
Geometric mean ($d = 0.21$)	5.48	16.49	83.50	92.50	—	
Rounded q ($d = 0.21$)	5	17	84	92	198	Use $d_m < d$
Modified std q ($d_m = 0.208$)	5.29	16.83	84.62	93.27	—	Part (d)
Rounded q	5	17	85	94	201	H–H

e. Yes. Neither C nor D has a number of seats that is either an upper or lower quota.

19. a. See table.   **b.** $d = \frac{5000}{110} = 45.45$ **c.** See table.

State	A	B	C	Total	Comments
Population (thousands)	1057	1942	2001	5000	
Std q ($d = 45.45$)	23.26	42.73	44.03	—	
Lower q ($d = 45.45$)	23	42	44	109	
Additional seats	0	1	0	1	
Seats Hamilton	23	43	44	110	Hamilton part (c)

d. $110\left(\frac{2,450}{5,000}\right) = 53.9$; add 53 seats.

e. New total seats = 163. New total population = 7,450.

$d = \left(\frac{5,000 + 2,450}{110 + 53}\right) = 45.71$

State	A	B	C	D	Total	Comments
Population (thousands)	1,057	1,942	2,001	2,450	7,450	
Std q ($d = 45.71$)	23.12	42.49	43.78	53.60	—	
Lower q ($d = 45.71$)	23	42	43	53	161	
Additional seats	0	0	1	1	2	
Seats Hamilton	23	42	44	54	163	Hamilton part (e)

f. Yes. The apportionment changed.

21. See answer 27 in Section 6.1
23. See answer 29 in Section 6.1
25. See answer 52 in Section 6.2
27. See answer 15 in Section 6.3

29. See answer 17 in Section 6.3
31. See answer 19 in Section 6.3
33. See answer 31 in Section 6.1

CHAPTER 7   Number Systems and Number Theory

7.1 Place Systems

1. $8 \cdot 10^2 + 9 \cdot 10 + 1$

3. $3 \cdot 10^3 + 2 \cdot 10^2 + 5 \cdot 10 + 8$

5. $3 \cdot 8^2 + 7 \cdot 8 + 2$

7. $3 \cdot 16^3 + 5 \cdot 16^2 + 9 \cdot 16 + 2$

9. $10 \cdot 16^5 + 11 \cdot 16^4 + 12 \cdot 16^3 + 13 \cdot 16^2 + 14 \cdot 16$

11. $2^6 + 2^4 + 2^3 + 1$

13. Does not exist   **15.** Does not exist

17. $4 \cdot 5^3 + 3 \cdot 5^2 + 1 \cdot 5 + 2$

19. $1 \cdot 4^2 + 2 \cdot 4 + 3$

21. 250   **23.** 13,714   **25.** 11,259,360

27. 89   **29.** Does not exist

31. 582   **33.** 27   **35.** 965   **37.** 219

39. 11,982,839   **41.** 22,424   **43.** 3,732   **45.** 59,714

47. 136   **49.** 100101010_2 **51.** 1010101100002_2

53. 12_8 **55.** 266_8 **57.** 22_{16}

59. 160_{16} **61.** 5270_8 **63.** 100035_8

65. a. $1 \cdot 2^{12}$ **b.** $1 \cdot 2^8$ **c.** $1 \cdot 2^4$

67. 101001110100010_2 **69.** 111010110000_2

71. 14_{16} **73.** BA_{16} **75.** 1,332

77. 119,665   **79.** $6, 4_{60}$ **81.** $36, 4, 5_{60}$

83. 1162_7 **85.** 221_5

87. a. 3   **b.** 5

89. a. 8   **b.** 9

91. a. 0, 1, 2, 3, 4   **b.** $5^0 \; 5^1 \; 5^2 \; 5^3$

c. $0_5 \quad 1_5 \quad 2_5 \quad 3_5 \quad 4_5$
$10_5 \quad 11_5 \quad 12_5 \quad 13_5 \quad 14_5$
$20_5 \quad 21_5 \quad 22_5 \quad 23_5 \quad 24_5$
$30_5 \quad 31_5 \quad 32_5 \quad 33_5 \quad 34_5$
$40_5 \quad 41_5 \quad 42_5 \quad 43_5 \quad 44_5$

93. a. 0, 1, 2, 3, 4, 5, 6, 7, 8, 9, A   **b.** $11^0 \; 11^1 \; 11^2 \; 11^3$

c. $0_{11} \quad 1_{11} \quad 2_{11} \quad 3_{11} \quad 4_{11} \quad 5_{11} \quad 6_{11} \quad 7_{11} \quad 8_{11} \quad 9_{11} \quad A_{11}$
$10_{11} \quad 11_{11} \quad 12_{11} \quad 13_{11} \quad 14_{11} \quad 15_{11} \quad 16_{11} \quad 17_{11} \quad 18_{11} \quad 19_{11} \quad 1A_{11}$
$20_{11} \quad 21_{11} \quad 22_{11}$

95. 16-bit color has $2^{16} = 65,536$ shades, and 24-bit color has $2^{24} = 16,777,216$ shades.

7.2 Arithmetic in Different Bases

All answers are given in the base indicated.

1. 12_8 **3.** 11_8 **5.** 10_{16} **7.** 14_{16} **9.** 10_2

11. 110_2 **25.** 4_8 **27.** 142_{16} **29.** 11_2

65.

	1_5	2_5	3_5	4_5	10_5	11_5
1_5	1_5	2_5	3_5	4_5	10_5	11_5
2_5	2_5	4_5	11_5	13_5	20_5	22_5
3_5	3_5	11_5	14_5	22_5	30_5	33_5
4_5	4_5	13_5	22_5	31_5	40_5	44_5
10_5	10_5	20_5	30_5	40_5	100_5	110_5
11_5	11_5	22_5	33_5	44_5	110_5	121_5

67.

	1_{14}	2_{14}	3_{14}	4_{14}	5_{14}	6_{14}	7_{14}	8_{14}	9_{14}	A_{14}	B_{14}	C_{14}	D_{14}	10_{14}	11_{14}
1_{14}	1_{14}	2_{14}	3_{14}	4_{14}	5_{14}	6_{14}	7_{14}	8_{14}	9_{14}	A_{14}	B_{14}	C_{14}	D_{14}	10_{14}	11_{14}
2_{14}	2_{14}	4_{14}	6_{14}	8_{14}	A_{14}	C_{14}	10_{14}	12_{14}	14_{14}	16_{14}	18_{14}	$1A_{14}$	$1C_{14}$	20_{14}	22_{14}
3_{14}	3_{14}	6_{14}	9_{14}	C_{14}	11_{14}	14_{14}	17_{14}	$1A_{14}$	$1D_{14}$	22_{14}	25_{14}	28_{14}	$2B_{14}$	30_{14}	33_{14}
4_{14}	4_{14}	8_{14}	C_{14}	12_{14}	16_{14}	$1A_{14}$	20_{14}	24_{14}	28_{14}	$2C_{14}$	32_{14}	36_{14}	$3A_{14}$	40_{14}	44_{14}
5_{14}	5_{14}	A_{14}	11_{14}	16_{14}	$1B_{14}$	22_{14}	27_{14}	$2C_{14}$	33_{14}	38_{14}	$3D_{14}$	44_{14}	49_{14}	50_{14}	55_{14}
6_{14}	6_{14}	C_{14}	14_{14}	$1A_{14}$	22_{14}	28_{14}	30_{14}	36_{14}	$3C_{14}$	44_{14}	$4A_{14}$	52_{14}	58_{14}	60_{14}	66_{14}
7_{14}	7_{14}	10_{14}	17_{14}	20_{14}	27_{14}	30_{14}	37_{14}	40_{14}	47_{14}	50_{14}	57_{14}	60_{14}	67_{14}	70_{14}	77_{14}
8_{14}	8_{14}	12_{14}	$1A_{14}$	24_{14}	$2C_{14}$	36_{14}	40_{14}	48_{14}	52_{14}	$5A_{14}$	64_{14}	$6C_{14}$	76_{14}	80_{14}	88_{14}
9_{14}	9_{14}	14_{14}	$1D_{14}$	28_{14}	33_{14}	$3C_{14}$	47_{14}	52_{14}	$5B_{14}$	66_{14}	71_{14}	$7A_{14}$	85_{14}	90_{14}	99_{14}
A_{14}	A_{14}	16_{14}	22_{14}	$2C_{14}$	38_{14}	44_{14}	50_{14}	$5A_{14}$	66_{14}	72_{14}	$7C_{14}$	88_{14}	94_{14}	$A0_{14}$	AA_{14}
B_{14}	B_{14}	18_{14}	25_{14}	32_{14}	$3D_{14}$	$4A_{14}$	57_{14}	64_{14}	71_{14}	$7C_{14}$	89_{14}	96_{14}	$A3_{14}$	$B0_{14}$	BB_{14}
C_{14}	C_{14}	$1A_{14}$	28_{14}	36_{14}	44_{14}	52_{14}	60_{14}	$6C_{14}$	$7A_{14}$	88_{14}	96_{14}	$A4_{14}$	$B2_{14}$	$C0_{14}$	CC_{14}
D_{14}	D_{14}	$1C_{14}$	$2B_{14}$	$3A_{14}$	49_{14}	58_{14}	67_{14}	76_{14}	85_{14}	94_{14}	$A3_{14}$	$B2_{14}$	$C1_{14}$	$D0_{14}$	DD_{14}
10_{14}	10_{14}	20_{14}	30_{14}	40_{14}	50_{14}	60_{14}	70_{14}	80_{14}	90_{14}	$A0_{14}$	$B0_{14}$	$C0_{14}$	$D0_{14}$	100_{14}	110_{14}
11_{14}	11_{14}	22_{14}	33_{14}	44_{14}	55_{14}	66_{14}	77_{14}	88_{14}	99_{14}	AA_{14}	BB_{14}	CC_{14}	DD_{14}	110_{14}	121_{14}

69. 133_5 Rem 11_5　　**71.** $B9_{14}$ Rem 48_{14}

7.3 Prime Numbers and Perfect Numbers

1. $2 \cdot 3 \cdot 7$　　**3.** Prime　　**5.** $2 \cdot 3^3$
7. Composite　**9.** Composite　**11.** Prime
13. 29, 31, 37, 41, 43, 47　**15.** $33 = 3(11)$; composite
17. $129 = 3(43)$; composite
19. 15 is composite because it is divisible by 3
21. 255 is composite because it is divisible by 5
23. 1, 2, 3, 6, 7, 14, 21　　**25.** 1, 2, 3, 6, 9, 18, 27
27. Abundant　　**29.** Abundant
31. Deficient　　**33.** Deficient
35. $(2^{13-1})(2^{13} - 1) = 33{,}550{,}336$
37. $(2^{13} - 1) = 8{,}191$
39. a. $(2^{127-1})(2^{127} - 1)$
　b. 1.45×10^{76}
41. 5,7　　**43.** 11,13
45. a. 110
　　11100
　　111110000
　　1111111000000
　b. The *n*th perfect number is a sequence of *m* ones, where *m* is the *n*th prime, followed by $2(n - 1)$ zeros, $n \neq 1$ (When $n = 1$, there is 1 zero.)

7.4 Fibonacci Numbers and the Golden Ratio

1. 13 total, 5 new
3. 21, 34, 55, 89, 144, 233, 377, 610, 987, 1597
5. 8　　　　　**11. a.** 832,040 and 102,334,155

13. a. $\dfrac{a + b}{a} = \dfrac{a}{b}$　　**b.** $(a + b)b = a^2$

c. $ab + b^2 = a^2$　　**d.** $\dfrac{ab}{b^2} + 1 = \dfrac{a^2}{b^2}$

e. $\dfrac{a}{b} + 1 = \left(\dfrac{a}{b}\right)^2$　　**f.** $x + 1 = x^2$

g. $x^2 - x - 1 = 0,\ x = \dfrac{1 \pm \sqrt{5}}{2},\quad x = \dfrac{1 + \sqrt{5}}{2}$

h. x is a ratio of lengths
19. 1.61803398896 is an accurate approximation to the golden ratio up to nine decimal places.

Chapter 7 Review

All answers are given in the appropriate base.
1. $5 \cdot 10^3 + 3 \cdot 10^2 + 7 \cdot 10 + 2$
3. $3 \cdot 8^2 + 2 \cdot 8 + 5$
5. Does not exist
7. $9 \cdot 16^2 + 5$
9. $10 \cdot 16^2 + 11 \cdot 16 + 12$
11. $2^7 + 2^6 + 2^4 + 2^3 + 1$
13. Does not exist　　　**15.** $39 \cdot 60^2 + 22 \cdot 60 + 54$
17. $25 \cdot 60^2 + 44 \cdot 60 + 34$
19. 332　　**21.** 53　　**23.** 48,282
25. 56　　**27.** 940　　**29.** 1002_8
31. $20EE_{16}$　　**33.** $10,54,47_{60}$　　**35.** 10111011100011_2
37. 335051_7　　**39.** 6　　**41.** 9　　**43.** 134_8
45. 1011_2　　**47.** 144_{16}　　**49.** $1, 18, 18, 17_{60}$
51. 67_8　　**53.** 11_2　　**55.** $98D_{16}$

57. $11, 49, 11_{60}$ **59.** 2024_8 **61.** 11110_2
63. 6616_{16} **65.** 3044_5
67. 23_8 Rem 3_8 **69.** 10_2 Rem 11_2
71. $5F_{16}$ Rem 3_{16} **73.** 11011_2
75. $3 \cdot 11$ **77.** Prime   **79.** 53, 59
81. 5 prime, 9 composite   **83.** 7 prime
85. 1, 2, 5, 7, 10, 14, 35

87. Abundant   **89.** Deficient
91. a. $(2^{17} - 1)(2^{17-1})$ **b.** 8.59×10^9
93. 1, 1, 2, 3, 5, 8, 13, 21, 34, 55, 89, 144, 233, 377, 610, 987, 1597, 2584, 4181, 6765
95. 5   **97.** 5 total, 2 new
99. a. 75,025 and 121,393   **b.** 196,418

CHAPTER 8   Geometry

8.1 Perimeter and Area
1. 16.1 cm^2 **3.** 12.5 in.^2
5. 21.7 ft^2 **7.** 176 m^2
9. a. 33.2 in.^2 **b.** 20.4 in.
11. a. 30 m^2 **b.** 30 m
13. a. 27.7 m^2 **b.** 24 m
15. a. 80 ft^2 **b.** 44 ft
17. a. $3,927.0 \text{ yd}^2$ **b.** 257.1 yd
19. a. 54 yd^2 **b.** 30 yd
21. a. 49.8 ft^2 **b.** 28.9 ft
23. a. $1,963.5 \text{ ft}^2$ **b.** 863.9 ft^2 **c.** $3,572.6 \text{ ft}^2$
25. a. $3,136 \text{ in.}^2$ **b.** 21.8 ft^2
27. 8.0 ft^2 **29.** 3.9 mi   **31.** 8 ft
33. a. $5,256.6 \text{ yd}^2$ **b.** 325.7 yd
35. Seven   **37.** Large
39. a. 36 million miles   **b.** $\frac{36}{93} = 0.39$ AU
   c. The distance between Mercury and the earth is 0.39 times the distance between the earth and sun.
41. a. 27,200,000 AU   **b.** 431 ly   **c.** 132 pc

8.2 Volume and Surface Area
1. a. 38.22 m^3 **b.** 72.94 m^2
3. a. 785.40 in.^3 **b.** 471.24 in.^2
5. a. 2.81 in.^3 **b.** 9.62 in.^2
7. 16.76 ft^3 **9.** 21.33 ft^3
11. 36 ft^3 **13.** 47.12 ft^3
15. a. 136.5 in.^3 **b.** 151 in.^2
17. $17,802.36 \text{ ft}^3$
19. a. Hardball, 12.31 in.^3; softball, 29.18 in.^3; volume of softball is 137% more than that of hardball
   b. Hardball, 25.78 in.^2; softball, 45.83 in.^2; surface area of softball is 78% more than that of hardball
21. 49   **23.** 147
25. No; should have paid $118.75
27. 36.82 in.^3 **29.** $175.20   **31.** $91,445.760 \text{ ft}^3$
33. 301.59 ft^3

8.3 Egyptian Geometry
1. A right triangle   **3.** Not a right triangle
5. The regular pyramid (2,025 khar > 1,554 khar)
7. 36; 18; 9; 36; 18; 9; 63; 9; 3; 63; 189; 189
9. A square with sides of 21 units   **11.** $\frac{49}{16}$
13. a. 28.4444 palms^2 **b.** 28.2743 palms^2 **c.** 0.6%
15. a. 113.7778 palms^3 **b.** 113.0973 palms^3 **c.** 0.6%

17. 11 cubits
19. The khar (1 khar = 5887 in.^3 > 1 ft^3 = $1,728 \text{ in.}^3$)
21. 1.5 setats
23. a. 2 khet   **b.** 200 cubits
25. a. 960 khar   **b.** 954.2588 khar   **c.** 0.6%

8.4 The Greeks
1. $x = 5; y = 60$ **3.** $x = 64.6; y = 2.5$
5. 36 ft   **7.** See IRM
9. 5.3 ft (rounded down so the object will fit)
11. 7.5 in.   **13.** See IRM
15. See IRM   **17.** See IRM
19. a. 3.061467459   **b.** 3.313708499
21. a. 3.121445152   **b.** 3.182597878

8.5 Right Triangle Trigonometry
1. $\theta = 60°; x = 3; y = 3\sqrt{3}$
3. $\theta = 30°; x = 14; y = 7\sqrt{3}$
5. $\theta = 45°; x = 3; y = 3\sqrt{2}$
7. $\theta = 45°; x = \frac{7\sqrt{2}}{2}, y = \frac{7\sqrt{2}}{2}$
9. $B = 53°; b = 15.9; c = 19.9$
11. $B = 35.7°; a = 7.8; c = 9.6$
13. $A = 40.1°; a = 0.59; b = 0.70$
15. $A = 80.85°; b = 249; c = 1,565.9$
17. $B = 36.875°; a = 43.52; b = 32.64$
19. $A = 85.00°; a = 11.453; c = 11.497$
21. $A = 40.7°; B = 49.3°; b = 17.4$
23. $A = 40.6°; B = 49.4°; c = 9.2$
25. $A = 74.4°; B = 15.6°; a = 0.439$
27. a. 64.0 ft   **b.** 92.3 ft
29. 150 ft   **31.** 18.4°
33. a. 176.1 ft   **b.** 26.7 ft
35. 26.2 ft   **37.** 167 ft
39. 630 ft
41. 1,454 ft
43. 24.6 ft
45. $\dfrac{1 \text{ AU}}{\tan\left(\dfrac{2}{3600}\right)^°} = 0.5$ pc
47. a. 0.9 pc = 0.9(206,162.4 AU) = 185,546 AU
   b. $x = 3600 \cdot \tan^{-1}\left(\dfrac{1}{185,546}\right) = 1.11$ seconds

8.6 Conic Sections and Analytic Geometry
1. Center $(0, 0)$; radius $= 1$

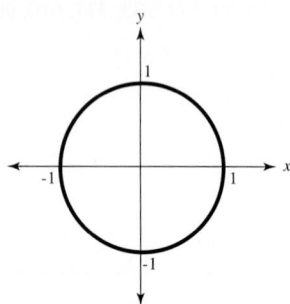

3. Center $(2, 0)$; radius $= 3$

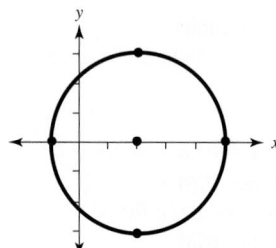

5. Center $(5, -2)$; radius $= 4$

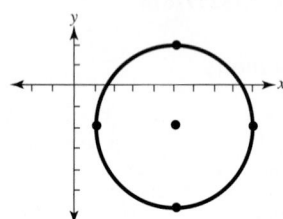

7. Center $(5, 5)$; radius $= 5$

9. Focus $(0, \frac{1}{4})$

11. Focus $(0, \frac{1}{2})$

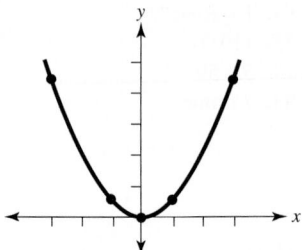

13. 2.9 ft above the bottom of the dish
15. $\frac{9}{16}$ inch above the bottom of the reflector
17. Foci $(\sqrt{5}, 0)$ and $(-\sqrt{5}, 0)$

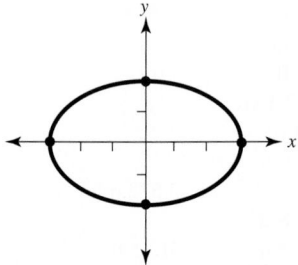

19. Foci $(0, \sqrt{21})$ and $(0, -\sqrt{21})$

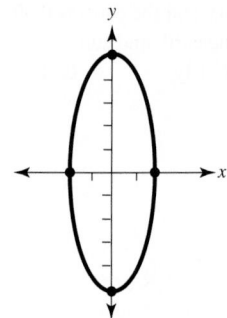

21. Foci $(0, \sqrt{7})$ and $(0, -\sqrt{7})$

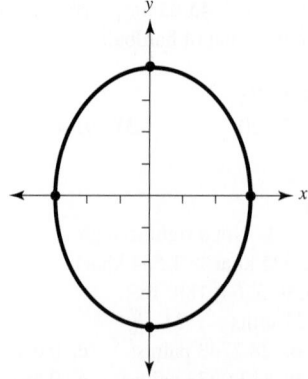

23. Foci $(1, 2 + \sqrt{3})$ and $(1, 2 - \sqrt{3})$

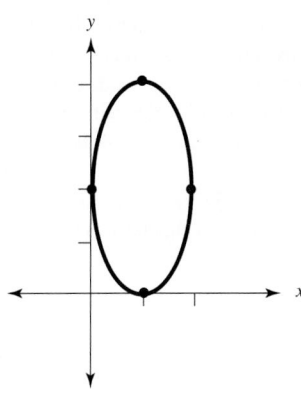

25. They should stand at the foci, which are located 9 ft from the center, in the long direction.

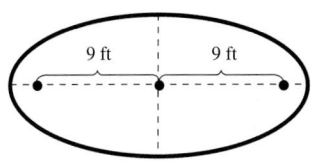

27. $\dfrac{x}{92.955^2} + \dfrac{y}{92.942^2} = 1$

29. **a.** Foci $(\sqrt{2}, 0)$ and $(-\sqrt{2}, 0)$

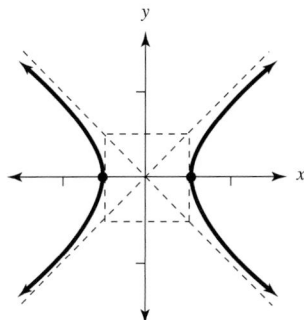

b. Foci $(0, \sqrt{2})$ and $(0, -\sqrt{2})$

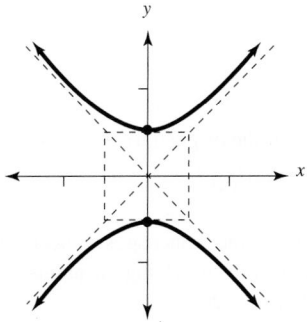

31. Foci $(\sqrt{13}, 0)$ and $(-\sqrt{13}, 0)$

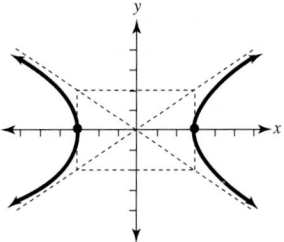

33. Foci $(0, \sqrt{29})$ and $(0, -\sqrt{29})$

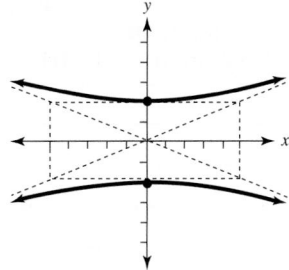

35. Foci $(3 + \sqrt{5}, 0)$ and $(3 - \sqrt{5}, 0)$

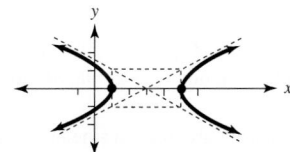

8.7 Non-Euclidean Geometry
1. One   **3.** Zero or one   **5.** Zero or one   **7.** None
9. Zero, one, two, or three   **11.** Two   **13.** Infinitely many
15. Zero or one   **17.** Zero or one

8.8 Fractal Geometry
1. b.

5. $d = 2$ **7.** $d = \frac{1}{3}$

9. $d \approx 1.6$ It's not a two-dimensional as a regular triangle is.

11. $d \approx 1.6$ It's not a two-dimensional as a regular triangle is.

13. $d = 1.5$ It's not two-dimensional.

15. $d = 3$, the same dimension that we observed before.

17. $d = 2$

b.

Step	Number of Triangles	Length of Each Side (feet)	Perimeter of One Triangle (feet)	Total Perimeter of All Triangles (feet)
1	1	1	3	3
2	3	$\frac{1}{2}$	$\frac{3}{2}$	$\frac{9}{2}$
3	9	$\frac{1}{4}$	$\frac{3}{4}$	$\frac{27}{4}$
4	27	$\frac{1}{8}$	$\frac{3}{8}$	$\frac{81}{8}$
5	81	$\frac{1}{16}$	$\frac{3}{16}$	$\frac{243}{16}$
6	243	$\frac{1}{32}$	$\frac{3}{32}$	$\frac{729}{32}$

c. 3.   **d.** $\frac{1}{2}$.   **e.** $\frac{1}{2}$.   **f.** $\frac{3}{2}$.   **g.** $3\left(\frac{3}{2}\right)^{n-1}$

h. The perimeter is infinite. The numbers get larger and larger without bound.

3. a. In step 1, there is only one square. If each side is 1 ft. in length, then the perimeter is 4 ft. In step 2, the original

3. b.

Step	Number of New Squares	Length of Each Side (feet)	Perimeter of One New Square (feet)	Total Perimeter of All New Squares (feet)	Total Perimeter of All Squares (feet)
1	1	1	4	4	4
2	1	$\frac{1}{3}$	$\frac{4}{3}$	$\frac{4}{3}$	$4+\frac{4}{3}$
3	8	$\frac{1}{9}$	$\frac{4}{9}$	$\frac{32}{9}$	$4+\frac{4}{3}+\frac{32}{9}$
4	64	$\frac{1}{27}$	$\frac{4}{27}$	$\frac{256}{27}$	$4+\frac{4}{3}+\frac{32}{9}+\frac{256}{27}$
5	512	$\frac{1}{81}$	$\frac{4}{81}$	$\frac{2048}{81}$	$4+\frac{4}{3}+\frac{32}{9}+\frac{256}{27}+\frac{2048}{81}$
6	4096	$\frac{1}{243}$	$\frac{4}{243}$	$\frac{16384}{243}$	$4+\frac{4}{3}+\frac{32}{9}+\frac{256}{27}+\frac{2048}{81}+\frac{16384}{243}$

c. 8.   **d.** $\frac{1}{3}$ **e.** $\frac{1}{3}$ **f.** $\frac{8}{3}$

g. $\frac{4}{3} \cdot \left(\frac{8}{3}\right)^{n-2}$, valid for $n > 1$.

h. The total perimeter is infinite. The numbers increase without bound.

5. a. $3 \cdot 2^{n-1}$ ft. (if the original triangle has sides of length 1 ft.)

b. ∞

8.9 The Perimeter and Area of a Fractal

1. a. In step 1, there is only one triangle. If each side is 1 ft. in length, then the perimeter is 3 ft. In step 2, the original triangle is modified by removing a center triangle. This results in three smaller triangles. Each side of the original triangle now has two triangles with sides of length $\frac{1}{2}$ ft. Thus, the perimeter of one of these smaller triangles is $\frac{1}{2} + \frac{1}{2} + \frac{1}{2} = 3 \cdot \frac{1}{2} = \frac{3}{2}$ ft. Since there are three such triangles, the total perimeter is $3 \cdot \frac{3}{2} = \frac{9}{2}$ ft.

square is divided into nine smaller squares and modified by removing the center square. The removed square contributes to the perimeter. It has sides of length $\frac{1}{3}$ ft. Thus, the perimeter of this smaller square is $4 \cdot \frac{1}{3} = \frac{4}{3}$ ft. The total perimeter is the sum of the previous perimeter and the new contribution: $4 + \frac{4}{3}$ ft.

c. $\left(\frac{2}{3}\right)^{n-1} \dfrac{\sqrt{3}}{4}$ sq. ft. (if the original triangle has sides of length 1 ft.)

d. 0

3. i. The total perimeter is infinite. It increases without bounds.

7. Inductive: noticing the patterns involving the number and length of sides and the perimeter.

Deductive: applying the general rules.

Chapter 8 Review

1. Area $= x(x - 2)$ square feet, perimeter $= 4x - 4$ feet.

3. 73.5 in.²; 41 in.   **5.** 69 yd²; 37.4 yd

7. a. 1,598,000 AU   **b.** 25.3 ly   **c.** 7.8 pc

9. 72.8 cm³; 153.9 cm²   **11.** 1,056 in.³; 664 in.²

13. Nine

15. a. 224 in.³   **b.** 200 in.²

17. a. 450.7 ft³   **b.** 268.0 ft²

19. a. 4.2140 cubits³   **b.** 4.1888 cubits³   **c.** 0.6%

21. A square with sides of 16 units

23. 6.5 ft (rounded down so the object will fit)

29. 3.105828541

31. a. Foci $(\sqrt{13}, 0)$ and $(-\sqrt{13}, 0)$

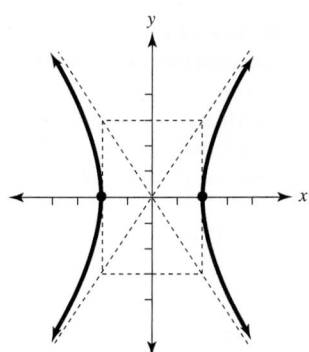

b. Foci $(0, \sqrt{13})$ and $(0, -\sqrt{13})$

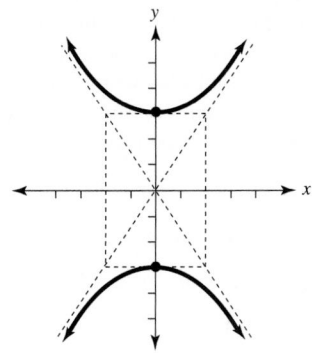

33. Foci $(0, \sqrt{5})$ and $(0, -\sqrt{5})$

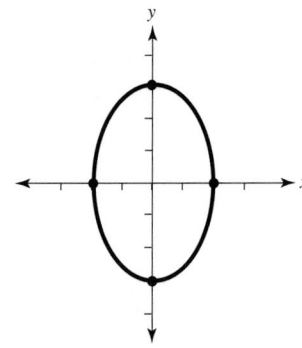

35. $B = 64.6°$; $a = 24.1$ ft; $b = 50.7$ ft

37. 213.2 ft   **39.** $\dfrac{1 \text{ AU}}{\tan\left(\dfrac{3}{3600}\right)^{\circ}} = 0.33$ pc

41. a. 0.85 pc $= 0.85(206,162.4 \text{ AU}) = 175,238$ AU

b. $x = 3600 \cdot \tan^{-1}\left(\dfrac{1}{175,238}\right) = 1.18$ Seconds

43. a. One   **b.** More than one   **c.** None

45. b. The total perimeter in steps 1–5: $4, \frac{20}{3}, \frac{100}{9}, \frac{500}{27}, \frac{2500}{81}$

c. $4 \cdot \left(\dfrac{5}{3}\right)^{n-1}$

d. The perimeter is infinite.

e. $\left(\dfrac{5}{9}\right)^{n-1}$

f. 0   **g.** $d = 1.465$

h. If we focus in on any smaller square it will look like the entire box fractal.

i. For any square that is not removed, the recursive rule is to remove the same four squares in the middle of the edges.

CHAPTER **9** **Graph Theory**

9.1 A Walk Through Konigsberg

1. Yes.There are four vertices. If you eliminate one bridge, then two of the vertices have an odd number of edges, and the rest have an even number of edges. Start at one of the odd vertices and end at the other.

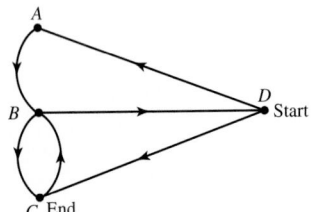

3. No
5. Yes. Start at one odd vertex and end at the other. ABDBCBADC

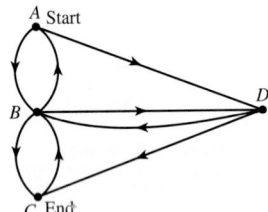

7. 4 vertices, 4 edges, 0 loops
9. 5 vertices, 5 edges (one of which is a loop), 1 loop
11. a. 7 vertices, 6 edges, 0 loops
   b. A family member is not its own child.
13.

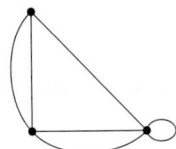

15. a. They are all graphs with 4 edges, 4 vertices, and 0 loops, and the edges in both graphs connect the same points.
   b.

17.

5. a. AB and AC   **b.** A and B
   c. A-4, B-3, C-3, D-2   **d.** Yes
7. a.   **b.**   **c.**

9. a.   **b.**

11. a.

   b.

   c.   **d.**

 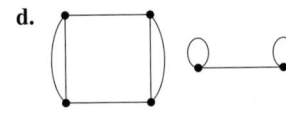

13. a. Euler trail (2 odd)   **b.** BACDA
15. a. There is an Euler trail because there are exactly two odd vertices.
   b. ACABDBD (Label the vertices AB from L to R top row and CD from L to R bottom row.)
17. a. There is an Euler trail because there are exactly two odd vertices
   b. Start at B; end at C. BADCABC
19. Answers will vary.
21. a.

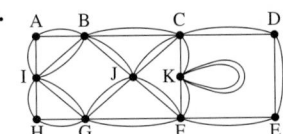

Each edge is a double edge. Start in upper left-hand corner, go clockwise one block, and detour to take Prindle, Weber, Wadsworth, Schmidt, Brooks Road, Brooks Circle, Brooks Road, and Wadsworth and Schmidt back to where you detoured. Continue along the outer border to where you started, and repeat for the other side.

   b.

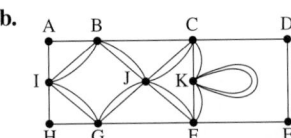

The outside border is single, and the inside edges are double. Start in upper left-hand corner and go one block clockwise. Make a detour around the inside. (See part a.) When you come back, continue along the border clockwise to your return.

17.

(figure for 17 from left column)

9.2 Graphs and Euler Trails
1. a. AB and AC   **b.** A and B
   c. A-3, B-1, C-2, D-2   **d.** Yes
3. a. AB and BD   **b.** A and B
   c. A-3, B-4, C-2, D-3   **d.** Yes

c.

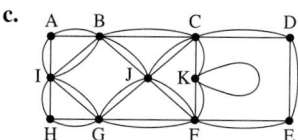

Everything is the same in part a except that when you repeat the inside circuit, you skip Cole Circle the second time.

23.

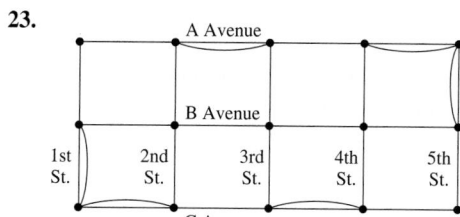

Eulerize by adding six extra edges, Avenue A from 2nd to 3rd and from 4th to 5th, Avenue C from 1st to 2nd and from 3rd to 4th, First Street from B to C, and Fourth Street from A to B.

    Start at A on 1st. Go over to 5th, down to C, over to 1st, up to B, over to 5th, back up to A, back to 4th, down to C, over to 3rd, up to A, over to 2nd, down to C, over to 1st, up to B and return back to A.

25.

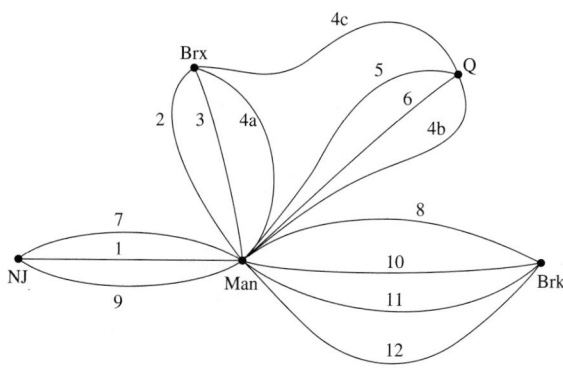

a. Manhattan has degree 13, NJ has degree 3, Brooklyn, Bronx, and Queens all have degree 4. Label the three edges of the Triborough Bridge as 4a Manhattan to Bronx, 4b Manhattan to Queens, and 4c Bronx to Queens.

b. No, there are two odds. To Eulerize, revisit any of the three bridges between NJ and Manhattan. Start in Manhattan: 7, 1, 9, 1 (or 7 or 9), 2, 3, 4a, 4c, 5, 6, 4b, 8, 10, 11, 12, ending in Manhattan.

c. There are exactly two vertices with odd degrees. Start in NJ and end in M. Here is the route NJ 1, 2, 3, 4a, 4c, 5, 6, 4b, 8, 10, 11, 12, 9, 7, M.

27. a.

b. No, there are four odd vertices. To Eulerize, revisit a ferry from Vancouver to Orcas and a ferry from Vashon to Olympic Peninsula. Now all the vertices could be considered even, and you can start and end at the same point.

c. No, there are four odd vertices. To Eulerize, revisit a ferry connecting two of the odd vertices. The other pair of odd vertices are the starting and stopping points.

29. a.

b. No, since there are only two stations with odd-degree: Pittsburg and Dublin, both with 1. Millbrae has 2 (the red and the yellow); Daly City has 6 (2 reds, 2 yellows, 1 blue and 1 green). To Eulerize, take the yellow from Pittsburg to Millbrae, then the red from Millbrae to Richmond, the orange from Richmond to Fremont, the green from Fremont to Daly City, the blue from Daly City to Dublin, and then the blue from Dublin to West Oakland and the yellow back to Pittsburg.

c. Yes, it is possible to start and end at different points since there are only two odd-degrees stations. Therefore, follow the route described in part (b) except end at Dublin. Do not proceed from Dublin back to Pittsburg.

d. They can patrol all stations efficiently. Part (b) is more useful so they can start and end their day at the same place.

9.3 Hamilton Circuits

Note: When a tie occurs, you have have to look at all the cases. This means that there may be more sequences than it appears for the repetitive nearest neighbor algorithm and cheapest edge algorithm. This might not be practical if there are too many choices. If you simply pick one choice at random, you might not get the best sequence. In some cases, answers may differ because of a tie.

1. a. 6   b. $927. A→B→D→P→A
    c. $848. A→B→P→D→A   d. $848. A→B→P→D→A
    e. 3! = 6 routes; $848 A→B→P→D→A
3. a. 4! = 24   b. $685. A→B→Po→Px→D→A
    c. Same as part (b)   d. $918. A→B→Px→Po→D→A
    e. Too many possibilities
5. 123 minutes. F→I→K→B→C→L→F
7. 108 minutes. F→K→I→C→B→L→F
9. 84 minutes. F→I→L→B→C→R→F
11. 84 minutes. F→I→L→B→C→R→F
13. a. 5 mm   b. 7 mm   c. 10 mm   d. 22 mm
15. 28 mm. (0, 0)→(2, 2)→(3, 1)→(4, 5)→(6, 4)→
    (7, 5)→(0, 0) or (0, 0)→(3, 1)→(2, 2)→(4, 5)→(6, 4)→
    (7, 5)→(0, 0)
17. 28 mm. (0, 0)→(2, 2)→(3, 1)→(4, 5)→(6, 4)→
    (7, 5)→(0, 0)
19. 28 mm. 0→5→4→3→1→2→0 (The path is abbreviated by the *x*-coordinate of the ordered pair.)
21. 26 mm. 0→1→2→3→4→5→0 (The path is abbreviated by the *x*-coordinate of the ordered pair.)
23. In some cases, answers may differ because of a tie. Cheapest route is $748. A→C→B→F→O→X→D→A, where

    A = Atlanta   B = Boston
    C = Washington   D = Denver
    F = San Francisco   O = Portland
    X = Phoenix

25. $748. A→C→B→F→O→X→D→A or
    A→D→X→O→F→B→C→A
    In some cases, answers may differ because of a tie.
27. $1,096. A→D→F→O→X→B→C→A
    In some cases, answers may differ because of a tie.
29. Exercises 23 and 25 both yielded $748.
    In some cases, answers may differ because of a tie.
31. $1,132. N→ S→C→M→H→L→N
    In some cases, answers may differ because of a tie.

9.4 Networks

1. Yes   3. No. It is not connected.
5. No. It has a circuit.   7. No. It has a circuit
9. Yes   11. No. It is not connected.
19. 65   21. 115

23. 531

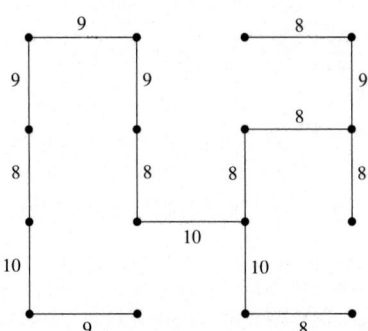

25. 2945 miles. N B C K F   27. 3442 miles. N Ph A D S Po
29. 3954 miles. S Po K C A Ph N B
31. $8 \times 7 + 9 \times 5 + 10 \times 3 = 131$

33. B because the sum of the distances is the shortest and it is given that one of them is a Steiner point.
35. a. $x + y = 100 \tan 40°$ b. $x = 100 \tan 20°$
    c. $y = 100 (\tan 40° - \tan 20°)$
37. a. 50°   b. 70°   c. 110°
39. a. 100   b. $100 \tan 25°$
    c. $100 - 100 \tan 25°$
41. a. 45°   b. 65°   c. 115°
43. Use Steiner point. $50(\sqrt{3} + \tan 40°)$
45. Length of two shortest sides. $\dfrac{560}{\sin 65°}$
47. Part c is the shortest network. $500\sqrt{3} + 600$
49. Use two Steiner points to form triangles with the shorter sides. $600\sqrt{3} + 700$
51. Part c is the shortest network. $500\sqrt{3} + 500$

9.5 Scheduling

1. Critical path: a, c, h, i, k, l, m, n, q
3. 4   7. Done drying after 2 days, 12 hours
9. 8¾ hours   11. 12¼ hours
13. Start to finish: 4½–5½ hours
15. Critical path: c, d, f, t, u   17. 3
21. Start after 1 hour, and finish before wedding starts (3 weeks and 5 hours)
23. 37 weeks
25. Start after 24 weeks, finish after 28 weeks
27. Start waiting for inspection after 35 weeks. Finish after 37 weeks.

29. Running closed ESS in lab. 1 year. End of 1999.

31. Install solar array, perform targeted lightweighting, improve reliability of motors, complete environmental control system installation, upgrade PMRF facilities, solar cell procurement, Helios prototype functional test, 1 year. End of 2001.

33. 9 years. End of 2003.

Chapter 9 Review

1. a. 3 vertices, 6 edges, 0 loops
   b. Answers vary   **c.** AB, BC   **d.** A and B
   e. A-4, B-6, C-2   **f.** Yes

3. a. 5 vertices, 6 edges, 1 loop
   b. Answers will vary.   **c.** AB, BE   **d.** A and B
   e. A-2, B-3, C-2, D-2, E-3   **f.** No

5. a. 6 vertices, 12 edges, 0 loops
   b. Answers will vary.   **c.** AB, BC   **d.** A and B
   e. A-4, B-4, C-5, D-3, E-4, F-4   **f.** Yes

7.   **9.**

11. a.   **b.**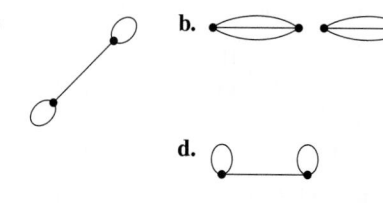

   c.   **d.**

13. a. All even vertices means that there is an Euler circuit.
   b. BABABCB

15. a. None because the graph is not connected.
   b. ABCCEBDEB

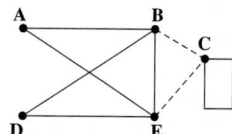

17. a. There are exactly two odd vertices. There is an Euler trail starting at C and ending at D.
   b. CAFCEBFEDCBAD

19. a. There are more than two odd vertices. There is no Euler circuit. Eulerize.

   b.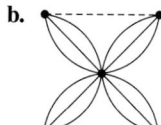

   Add two edges, one connecting the two top vertices and the other connecting the two bottom vertices.

21. a. No. There are odd vertices.

   b.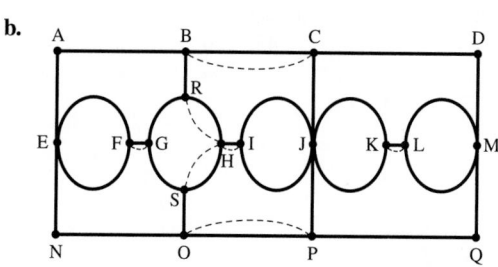

   Eulerize with seven extra edges. B-C, F-G, R-H, H-S, H-I, K-L, O-P. The route is E–F–G–R–B–C–J–P–O–S–H–I–J–K–L–M–L–K–J–I–H–R–H–S–G–F–E–A–B–C–D–M–Q–P–O–N–E

23.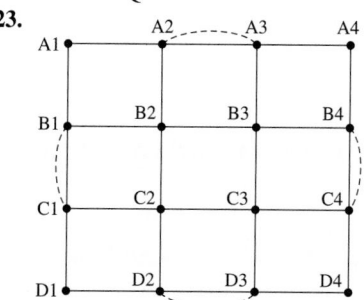

   Eulerize Avenue A from 2nd to 3rd, Avenue D from 2nd to 3rd, 4th from B to C, 1st from B to C. Start on A and 1st, go right on A one block, down to D, over to 3rd, back to A, loop to 2nd and back, over to 4th, down to D, over to 1st, up to C, over to 4th, up to B over to 1st, loop to C and back to A1.

For Problems 25–34, note that when a tie occurs, you have to look at all the cases. This means that there might be more choices than it appears for the repetitive nearest neighbor algorithm and cheapest edge algorithm. This might not be practical if there are too many choices. If you simply pick one choice at random, you might not get the best answer. In some cases, answers may differ because of a tie.

25. a. $3! = 6$
   b. $571. CMNLC. Answers may differ in case of ties.
   c. $560. CNLMC. Answers may differ in case of ties.
   d. $571. CMNLC. Answers may differ in case of ties.
   e. $560. CNLMC. Answers may differ in case of ties.

27. a. $4! = 24$
   b. $650. CMNSLC. Answers may differ in case of ties.
   c. $490. CSLNMC. Answers may differ in case of ties.
   d. $490. CSLNMC. Answers may differ in case of ties.
   e. Too many possibilities

29. $638. CMNSLHC. Answers may differ in case of ties.

31. $638. CMNSLHC. Answers may differ in case of ties.

33. 28 mm; 0, 2, 5, 6, 7, 1, 0 (The *x*-coordinate of the ordered pair is listed in the sequence.)

43. 60

45. 200

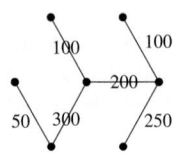

47. 1000

49. 1,866 miles. BNCKD   **51.** 2,849 miles. B N Ph C K D S

53. Use Steiner point. $125(\sqrt{3} + \tan 35°)$

55. Sum of two shortest sides. $\dfrac{300}{\cos 28°}$

57. Place two Steiner points, so the shorter sides are bases of isoceles triangles. $700\sqrt{3} + 900$

59. Critical path: a, c, h, i, j   **61.** 2

CHAPTER 10   **Exponential and Logarithmic Functions**

10.0 A Review of Exponentials and Logarithms

1. 2   **3.** –4   **5.** 25   **7.** 1   **9.** 4

11. $\frac{1}{2}$ **13.** $b^P = Q$ **15.** $b^M = N + T$

17. $b^{M+R} = N + T$ **19.** $\log_b G = F$ **21.** $\log_b G = F + 2$

23. $\log_b (E - F) = CD$ **25.** $\log_b (Z + 3) = 2 - H$

27. a. 4.05519996684   **b.** 25.1188643151

29. a. 2.14501636251   **b.** 2.34979510988

31. a. 0.301194211912   **b.** 0.063095734448

33. a. 3.93313930533   **b.** 2.87719968669

35. a. 4.51250674972   **b.** 79.2446596231

37. a. 18.0500269989   **b.** 316.978638492

39. a. 5.67969701231   **b.** 0.176065729885

41. a. 0.982078472412   **b.** 0.426511261365

43. a. -0.162518929498 **b.** -0.070581074286

45. a. 4.1   **b.** 4.1

47. a. 4.79314718056   **b.** 4.40102999566

49. a. 2.3   **b.** 2.3

51. a. 8   **b.** 8

53. a. 2.30258509299   **b.** 4.60517018599
   c. 6.90775527898

55. a. 0.434294481903   **b.** 0.868588963807
   c. 1.30288344571

57. $\log 2.9$, $\ln 2.9$, $e^{2.9}$, $10^{2.9}$

10.0 B Review of Properties of Logarithms

1. $6x$ **3.** $-0.036x$ **5.** $2x + 5$ **7.** $1 - x$

9. a. x^2 **b.** x^2

11. a. $9x^2$ **b.** $9x^2$

13. $\log x - \log 4$ **15.** $\ln 1.8 + \ln x$ **17.** $x \log 1.225$

19. $\ln 3 + 4 \ln x$ **21.** $\log 5 + 2 \log x - \log 7$

23. $\ln 20x$ **25.** $\log 3x$ **27.** $\ln 2x$

29. $\log x^3$ **31.** 0

33. a. $\ln 0.35$ **b.** $\log 0.35$

37. a. $\dfrac{1}{0.004} \ln \dfrac{17}{4}$ **b.** $\dfrac{1}{0.004} \log \dfrac{17}{4}$

41. a. $\dfrac{1}{0.0073} \ln \dfrac{16}{13}$ **b.** $\dfrac{1}{0.0073} \log \dfrac{16}{13}$

45. a. $e^{3.66}$ **b.** $10^{3.66}$

49. a. $6e^2$ **b.** 600

53. a. $\dfrac{1}{3.6} e^{3.1}$ **b.** $\dfrac{1}{3.6} 10^{3.1}$

59. 3.5; acid   **61.** 7.4; base

63. 4.9; acid   **65.** 10^{-7} moles per liter

67. a. Yes; pH = 6.5 is acceptable
   b. No; pH = 3.5 is not acceptable

69. From $10^{-8.5}$ to 10^{-7} (or 3.16×10^{-9} to 10^{-7}) moles per liter

71. a. 14.21 years, 14.21 periods
   b. 13.95 years, 55.80 periods
   c. 13.89 years, 166.70 periods
   d. 13.86 years, 5060.32 periods
   e. The doubling time decreases as the compound period decreases.

73. For 15,000: 4.99 years, 1821.68 periods
   For 100,000: 28.34 years, 10,345.07 periods

75. For 30,000: 6.49 years, 2368.12 periods
   For 100,000: 25.75 years, 9399.92 periods

77. a. $\dfrac{\log 3}{\log 2}$ **b.** 1.58

79. a. $\dfrac{\log 6}{\log 3}$ **b.** 1.63

81. a. $\dfrac{\log 8}{\log 4}$ **b.** 1.50

10.1 Exponential Growth

1. 32,473
3. July 2013
5. a. (0, 9098), (4, 9392)
   b. 4 years
   c. 294 thousand people
   d. 73.5 thousand people/year
   e. $0.008 = 0.8\%$ per year
7. a. (0, 18,323), (4, 18,710)
   b. 4 years
   c. 387 in thousands
   d. 96.75 in thousands per year
   e. $0.005 = 0.5\%$ per year
9. a. $9392 = 9098e^{4b}$
   $p = 9098e^{0.007950913484t}$
   b. $t = 9, p = 9,772,896$
   c. $t = 14, p = 10,169,239$
   d. 87.2 years from 2000
11. a. $18710 = 18323e^{4b}$
   $p = 18323e^{0.005225259747t}$
   b. $t = 7, p = 19,005,605$
   c. $t = 14, p = 19,713,639$
   d. 77.6 years from 2000
13. a. $p = 2,510e^{0.2541352t}$
   b. 14,868
   c. 2.7 days
15. a. $p = 230,000e^{.0497488314t}$
   b. October 2006
   c. $417,826
   d. $15,652 per month
17. a. $98.90 = 91.22e^{5b}$
   $y = 91.22e^{0.0161670135t}$
   b. $t = 7, y = 102.15$ quadrillion BTU's
   c. It might be an overestimate because the consumption is slowing down.
19. a. $9817 = 7398e^{5b}$
   $y = 7398e^{0.05658117666t}$
   b. $t = 9, y = $12,310$ billion
21. a. $p = 3,929,214e^{0.0300866701t}$
   b. 1810
   c. In 1810: 7,171,916. In 2000: 2,179,040,956
23. Africa: 32.9 years, North America: 55.3 years, Europe: 19.4 years
25. 49 years
27. a. Because it will not increase by the same amount each year
   b. Approximately 4.25 years
29. $18,323(1 + i)^4 = 18,710$
   $v = 18,323(1.005238935)^t$
31. $9098(1 + i)^4 = 9392$
   $v = 9098(1.007982606)^t$
33. $e^{0.007950913484} = 1.007982606 = (1 + i)$
35. 5,060.3 days = 13.86 years
37. 586 months = 48 years and 10 months

10.2 Exponential Decay

1. 3.2 g
3. 7.4 g
5. a. $Q = 50e^{-0.266595069t}$ (t in hours)   **b.** 38.3 mg
   c. 0.08 mg   **d.** −11.7 mg/hr   **e.** −23.4% per hour
   f. −2.1 mg/hr   **g.** −4.2% per hour
   h. Radioactive substances decay faster when there is more substance present. The rate of decay is proportional to the amount present.
7. a. 21.0 hr   **b.** 42.0 hr   **c.** 63.0 hr
9. 30.2 years
11. 50.3 years
13. 81,055 years
15. 99.7 sec
17. 53% of the amount expected in a living organism
19. 63.5 days   **21.** 1,441 years
23. 14,648 years   **25.** 2,949 years
27. 92.6% of the amount expected in a living organism; 78.9% of the amount expected in a living organism
29. 50% of the amount expected in a living organism
31. The claim is not justified. (62% remaining carbon-14 would indicate approximately 3,950 years.)
33. Approximately 8,300 years old
35. a.

t hours after injection (hr)	1	2	3	4	5
Q portion remaining (%)	0.891	0.794	0.708	0.631	0.562

b.

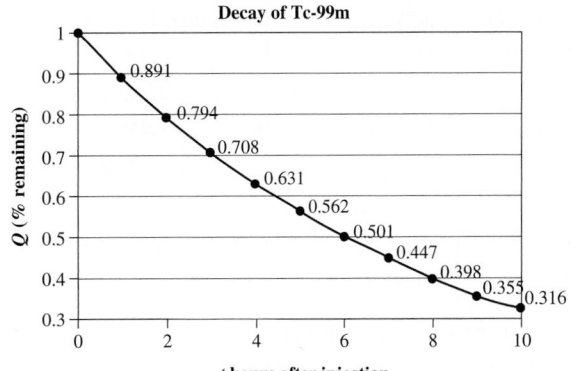

Decay of Tc-99m

37. 46.0 hours
39. a. 3.7×10^7 atomic disintegrations/sec
   b. 3.7×10^8 atomic disintegrations/sec
   c. 7.4×10^8 atomic disintegrations/sec

10.3 Logarithmic Scales

1. 7.6   **3.** 5.2
5. a. 16 times as much earth movement
   b. 55 times as much energy released

7. a. 32 times as much earth movement
  b. 150 times as much energy released
9. a. 126 times **b.** 1109 times
11. a. 158 times as much earth movement
  b. 1,549 times as much energy released
13. a. 26%   **b.** 40%
15. Energy released is magnified by a factor of 28.
17. 70 dB   **19.** 74 dB
21. 10 dB   **23.** 9.1 dB
25. four joined (five total)   **27.** one joined (two total)
29. five joined (six total)   **31.** a gain of 3 dB
33. a. The American iPod's impact is about 1000 times that of
  the European iPod.
  b. American: less than 2 minutes, European: 2 hours
35. $I_1 = 10^{-3.5}I_2$

Chapter 10 Review
1. $x = 4$ **3.** $x = 3,125$
5. $\ln(e^x) = x$ and $e^{\ln x} = x$
7. $\ln(\frac{A}{B}) = \ln A - \ln B$
9. $\ln(A^n) = n(\ln A)$
11. $\ln(AB) = \ln A + \ln B$
13. $\log(x + 2)$
15. 11.30719075   **17.** 447,213.6
19. a. $Q = 300e^{-0.130782486t}$ (t in years)   **b.** 1.5 g
  c. 263.2 g   **d.** 17.6 years
21. 3.1 on the Richter scale   **23.** 102 dB
25. Three trumpets have joined in.
27. A logarithmic function has the form $y = \log_b x$, where b is a
  positive constant. Also, $\log_b u = v$ means $b^v = u, b \neq 1$.

CHAPTER 11   Matrices and Markov Chains

11.0 Review of Matrices
1. a. 3×2 **b.** None
3. a. 2×1 **b.** Column
5. a. 1×2 **b.** Row
7. a. 3×3 **b.** Square
9. a. 3×1 **b.** Column
11. 22   **13.** 41   **15.** 3   **17.** -11 **19.** -3

21. a. $AC = \begin{bmatrix} 115 \\ 383 \\ 783 \end{bmatrix}$ **b.** CA does not exist.

23. a. AD does not exist.   **b.** DA does not exist.
25. a. CG does not exist.   **b.** GC does not exist.

27. a. HK does not exist.   **b.** $BJ = \begin{bmatrix} 2,243 \\ 1,056 \\ 52 \end{bmatrix}$

29. a. $AF = \begin{bmatrix} -10 & 50 \\ -56 & 229 \\ 0 & 153 \end{bmatrix}$ **b.** FA does not exist.

31. \$130 on sale, \$170 regular price
33. Blondie's: \$3.45; SliceMan's: \$3.70
35. 3.07; 3; 3.50
37. a. $\begin{bmatrix} -0.3 \\ 4.4 \end{bmatrix}$

  b. Change in sales from 2002 to 2003 for hotels and
  restautants

39. a. $\begin{bmatrix} 6.25 & 8.97 & 4.97 & 24.85 & 6.98 & 3.88 \\ 6.10 & 8.75 & 5.25 & 22.12 & 6.98 & 3.75 \end{bmatrix} \begin{bmatrix} 2 \\ 1 \\ 2 \\ 1 \\ 9 \\ 4 \end{bmatrix}$

  b. $\begin{bmatrix} 134.60 \\ 131.39 \end{bmatrix}$

41. a. $(53,594 \quad 64,393 \quad 100,237 \quad 63,198)$

  b. $\begin{bmatrix} 0.9885 & 0.0015 & 0.0076 & 0.0024 \\ 0.0011 & 0.9901 & 0.0057 & 0.0032 \\ 0.0018 & 0.0042 & 0.9897 & 0.0043 \\ 0.0017 & 0.0035 & 0.0077 & 0.9870 \end{bmatrix}$

  c. $(53,336 \quad 64,478 \quad 100,466 \quad 63,142)$ new population
  in 2001

43. $\begin{bmatrix} 106 \\ 68 \end{bmatrix}$ **45.** $\begin{bmatrix} 3 & -2 \\ 4 & 0 \end{bmatrix}$ **47.** Does not exist

49. $\begin{bmatrix} 19 & 7 & 34 \\ 74 & 0 & -11 \\ 13 & -2 & 44 \end{bmatrix}$

61. $\begin{bmatrix} 62 & 32 \\ -40 & 56 \end{bmatrix}$

63. a. $\begin{bmatrix} -1,178 & -2,101 & -2,378 \\ 5,970 & -3,091 & 498 \\ 5,580 & -660 & 2,340 \end{bmatrix}$

b. $\begin{bmatrix} 1,892 & -2,520 \\ 10,723 & -3,821 \end{bmatrix}$

65. a. $\begin{bmatrix} 109,348 & 23,813 & -16,663 \\ -23,840 & 1,688 & -7,720 \end{bmatrix}$

b. $\begin{bmatrix} 109,348 & 23,813 & -16,663 \\ -23,840 & 1,688 & -7,720 \end{bmatrix}$

67. a. $\begin{bmatrix} 376 & 64 & 281 \\ -932 & -1,203 & 614 \\ 952 & -808 & 293 \end{bmatrix}$

b. $\begin{bmatrix} 4,599,688 & -2,586,492 & 9,810,101 \\ -1,887,820 & -24,086,567 & 2,293,357 \\ 42,244,296 & -7,140,820 & -4,471,911 \end{bmatrix}$

11.1 Markov Chains
1. a. 0.9   **b.** 20%
    c. (0.22 0.78)   **d.** (0.568 0.432)

e.

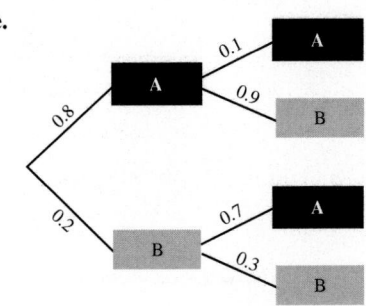

3. a. 0.8   **b.** 40%
    c. (0.12 0.88)   **d.** (0.024 0.976)

e.

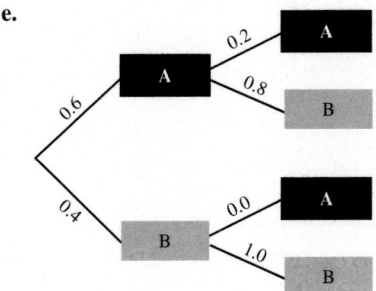

4. c. (0.34 0.66)   **d.** (0.366 0.634)
5. b. $P = [0.14 \quad 0.86]$
7. b. $P = [0.48 \quad 0.37 \quad 0.15]$

9. b. $\begin{bmatrix} 0.63 & 0.37 \\ 0.12 & 0.88 \end{bmatrix}$

11. b. $\begin{bmatrix} 0.71 & 0.12 & 0.17 \\ 0.32 & 0.34 & 0.34 \\ 0.02 & 0.02 & 0.96 \end{bmatrix}$

13. a. 19%   **b.** 22%
15. a. 46%, 19%, 35%   **c.** 39%, 13%, 48%
    d. 33%, 10%, 57%   **e.** 24%, 7%, 69%
17. a. 39%, 61%   **b.** 45%, 55%   **c.** 51%, 49%
19. 27%

23. b. $T = \begin{bmatrix} 0.874 & 0.041 & 0.085 \\ 0.268 & 0.732 & 0 \\ 0 & 0 & 1 \end{bmatrix}$

    c. 395 million acres, 110 million acres, 891 million acres
    d. 375 million acres, 97 million acres, 925 million acres
35. He was motivated purely by theoretical concerns.
37. He organized a celebration of the 200th anniversary of the publishing of Jacob Bernoulli's book on probabilities.

11.2 Systems of Linear Equations
1. Yes   **3.** No   **5.** No
7. a. $-\frac{5}{2}, 2; \frac{6}{19}, -\frac{72}{19}$ **b.** One solution
9. a. $-\frac{4}{3}, 4; -\frac{4}{3}, 4$ **b.** Infinite number of solutions
11. a. $-1, 7; -\frac{3}{2}, 4; -1, 7$ **b.** One solution
13. Yes   **15.** No   **17.** No
33. $x = -\frac{3}{2}, y = 0.5$

11.3 Long-Range Predictions with Markov Chains
1. a. $\left[\frac{2}{11} \approx 0.1818 \quad \frac{9}{11} \approx 0.8182 \right]$
3. a. $\left[\frac{8}{15} \approx 0.5333 \quad \frac{7}{15} \approx 0.4667 \right]$
5. 24%
7. 80% will own; 20% will rent; current trends continue, and residents' moving plans are realized.
9. Sierra Cruiser will eventually control 26% of the market.

11.4 Solving Larger Systems of Equations
1. $(4, -2, 1)$ **3.** $(2, 5, 7)$ **5.** $(5, 10, -2)$
7. $(3, 5, 7)$ **9.** $(4, 0, -3)$ **11.** $(5, -6, 0)$

13. $\begin{bmatrix} 1 & 0 & 0 & 4 \\ 0 & 1 & 0 & -2 \\ 0 & 0 & 1 & 1 \end{bmatrix}$ Solution is $(4, -2, 1)$.

15. $\begin{bmatrix} 1 & 0 & 0 & 2 \\ 0 & 1 & 0 & 5 \\ 0 & 0 & 1 & 7 \end{bmatrix}$ Solution is $(2, 5, 7)$.

17. $\begin{bmatrix} 1 & 0 & 0 & 5 \\ 0 & 1 & 0 & 10 \\ 0 & 0 & 1 & -2 \end{bmatrix}$ Solution is $(5, 10, -2)$.

19. $\begin{bmatrix} 1 & 0 & 0 & 3 \\ 0 & 1 & 0 & 5 \\ 0 & 0 & 1 & 7 \end{bmatrix}$ Solution is $(3, 5, 7)$.

21. $\begin{bmatrix} 1 & 0 & 0 & 4 \\ 0 & 1 & 0 & 0 \\ 0 & 0 & 1 & -3 \end{bmatrix}$ Solution is $(4, 0, -3)$.

23. $\begin{bmatrix} 1 & 0 & 0 & 5 \\ 0 & 1 & 0 & -6 \\ 0 & 0 & 1 & 0 \end{bmatrix}$ Solution is $(5, -6, 0)$.

25. $(5, -4)$ **27.** $(2, 3)$

11.5 More on Markov Chains
5. 10.8%; 4.5%; 84.6%
7. 48.6%; 37.6%; 13.8%

Chapter 11 Review
1. Neither; 3×2 **3.** Column; 4×1
5. $\begin{bmatrix} -3 & -24 \\ -24 & 28 \end{bmatrix}$ **7.** Does not exist

9. $\begin{bmatrix} 35 & -19 & -72 & 24 \\ 12 & 34 & 51 & 24 \end{bmatrix}$

11. \$43,000; \$41,300   **13.** $(2, -4)$
15. $(-3, 2)$ **17.** $(1, 7, 2)$
19. a. 0.7   **b.** 0.37
  c. $[0.226 \quad 0.774]$ **d.** $[0.1452 \quad 0.8548]$

e.

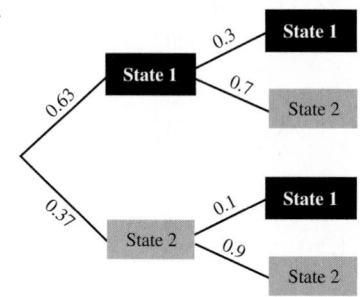

21. $[0.4 \quad 0.6]$ **23.** $[\frac{1}{7} \quad \frac{6}{7}]$ **25.** $[\frac{2}{5} \quad \frac{1}{5} \quad \frac{2}{5}]$
27. 21.5%   **29.** 29.6%
31. If current trends continue, the percentages will be 22.1%, 19.6%, and 58.3%. Thus, more condominiums and townhouses will be needed.
33. Emigration and immigration, also affordability based or economy.

CHAPTER 12 Linear Programming

12.0 Review of Linear Inequalities

1.

3.

5.

7.

9. a.

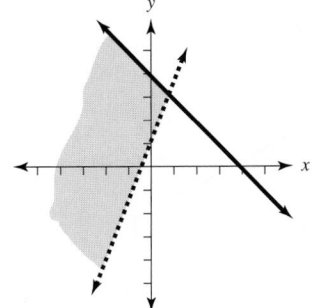

b. Unbounded **c.** (1, 3)

11. a.

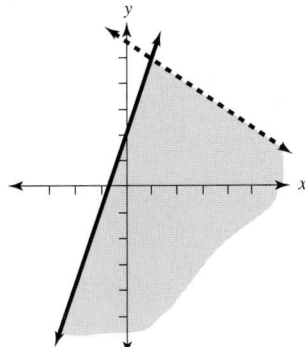

b. Unbounded **c.** (1, 5)

13. a.

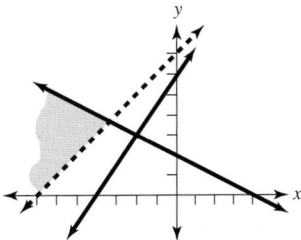

b. Unbounded **c.** $\left(-3\frac{1}{3}, 3\frac{2}{3}\right)$

15. a.

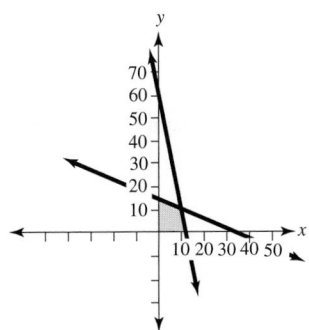

b. Bounded **c.** (0, 0), (0, 14), (10, 10), (12, 0)

17. a.

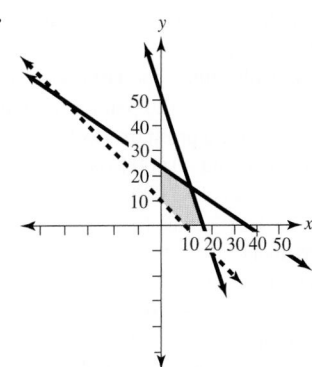

b. Bounded
c. $(0, 10), (0, 23\frac{2}{11}), (12, 15), (17\frac{1}{7}, 0), (10, 0)$

19. a.

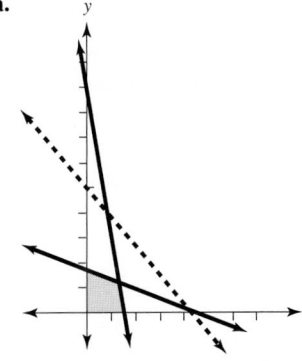

b. Bounded
c. (0, 0), (0, 1.7), (1.3, 1.2), (1.5, 0)

21. a.

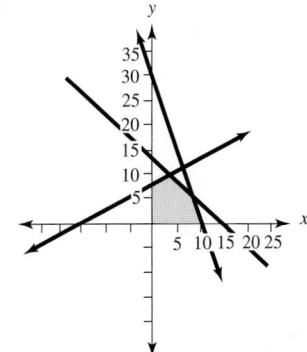

b. Bounded **c.** (0, 0), (0, 8), (4, 10), (8, 6), (10, 0)

12.1 The Geometry of Linear Programming
 1. x = number of shrubs, y = number of trees; $1x + 3y \le 100$
 3. x = number of hardbacks, y = number of paperbacks;
   $4.50x + 1.25y \ge 5,000$

5. x = number of refrigerators, y = number of dishwashers; $63x + 41y \leq 1,650$

7. No floor lamps and 45 table lamps, for a profit of $1,485

9. 60 pounds of Morning Blend and 120 pounds of South American Blend each day for a profit of $480

11. 15,000 loaves to Shopgood and 20,000 loaves to Rollie's each week for shipping costs of $3,000

13. Global has many choices, including the following:
15 Orvilles and 30 Wilburs
48 Orvilles and 8 Wilburs
(Any point on the line $y = -\frac{2}{3}x + 40$)
Each of these generates a cost of $720,000.

15. a. 20 large and 30 small, for a profit of $9,500
b. 40 large and no small, for a profit of $10,000
c. 40

17. 0 Mexican; 0 Colombian

19. 150 minutes; $0

21. a. no maximum   **b.** no maximum   **c.** (6, 0)   **d.** 12

23. 3 weeks at Detroit and 6 weeks at Los Angeles, for a minimum cost of $3,780,000

Chapter 12 Review

1.

3.

5. $(\frac{3}{2}, \frac{1}{2})$

7. $(-3, -23)$

9.

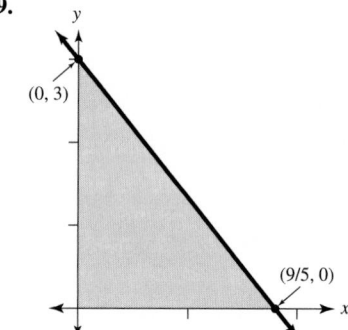

11. (0, 0) is minimum

13. $\left(\dfrac{140}{41}, \dfrac{250}{41} \right)$

15. 30 of each assembly for a maximum income of $16,500

Appendix E

1. a. 4 yd   **b.** 36 ft

3. a. 52,800 ft   **b.** 0.0019 mi

5. 126,720 in.

7. a. 50,000 m   **b.** 0.05 km

9. a. 0.02 liters   **b.** 200 cl

11. a. 454 g   **b.** 3.6 kg

13. a. 88 ft/sec   **b.** 0.9 sec

15. a. 81 mph   **b.** 16 mi

17. a. 0.0157534247% per day   **b.** $47.26
    c. $48.84

19. a. 1.83 m   **b.** 76.7 kg
    c. 23; no   **d.** 28; Yes

21. 164.56 gal

Index

All hyphenated entries (12- and 13-) pertain to sections of the book found at www.thomsonedu/mathematics/johnson.

Exponential equations and the inverse
property of logarithms, 735–737
Exponential function
calculator for natural exponential
function, 724–726
defined, 723–724
exponential functions, 723–724
logarithms (*see* Logarithms)
natural exponential function, 724–727
rational and irrational numbers
and, 724
Exponential growth, 747–763, 766
compound interest and, 756–759
delta notation, 748–750
model of, 750–756
Exponential model, 750–756
Extreme value, 250

F

Face cards, 130
Factor, 499
Factored completely, 499
Factorials, 91–93
Factorizations and primes, 499–500
Falling object, equation for, 13-62–13-71
False negative, 205
False positive, 205
Fermat, Pierre de, 126, 522, 13-25,
13-42
search for prime numbers, 503
Fibonacci numbers
defined, 510–511
plants and, 512–513
sequence and honeybees, 511
spiral, 513–514
Fibonacci sequence, 511
*Fifth Book of Euclid Proved
Algebraically, The* (Dodgson), 53
Finance, 327–409. *See* Interest,
Compound Interest and Annuities
add-on interest, 332–333
amortized loans (*see* Amortized loan
and Amortization schedule)
annual percentage rate defined, 390
annual yield, 344–347
appraisal fee for loan, 390
average daily balance, 333
charges, 390–393, 405
charges, annual percentage rate and,
390–393
credit card charges and, 333–335
credit report fee, 390
defined, 327, 390
finding the interest and present value,
343–344

future value, 328–330
interest (*see also* Interest)
loans (*see also* Loans)
national debt, 330–331
points on loan, 390
prepaid charges and estimation of,
393–394
present value, 332
simple interest, 328–339
term or number of days, 334
First simplex matrix, 12-4
Five geometric postulates, 601
Flaws of apportionment, 460–472.
See Apportionment and Voting
Systems
Alabama paradox, 462–464
Balinski-Young impossibility theorem,
469–470
new states paradox, 464–466
population paradox, 466–469
quota rule, 460–462
Flaws of voting systems, 422–425
Fleury's algorithm to find Euler trails and
Euler circuits and, 657–659, 715
Florida, map of, 701
Fluxions, 13-43
Foci, 594
Focus, 593
Following states, 816
Formulae of Plane Trigonometry
(Dodgson), 53
Formulas
add-on, 332–333
annual payout annuity with COLA,
399–405
annual yield, 345, 405
annuity due, 405
Binet's, 521
cardinal number, for the union of sets,
67–69
circle, area of, 531
circle, circumference of, 530–531
combination, 101
compound interest, 341, 405
cone, volume of, 544
dimension, 623
earthquake energy, 788
falling object, 13-52
fractal dimension, 623
future value, 329
Heron's formula for the area of a
triangle, 526–529
magnitude comparison of two
earthquakes, 787
margin or error, 298

ordinary annuity, present value of,
359, 405
parallelogram, area of, 525
payout annuity, 399
permutation, 97
point-slope, 13-39
present value of annuity, 363
present value, 332
pyramid, volume of, 544
quadratic, 3
rectangle, area of, 525
sample variance, alternative for, 263
simple interest, 328, 405
simple interst amortized loan, 367, 405
sphere, surface area of, 542
sphere, volume of, 542
surface area, 540
trapezoid, area of, 525
triangle, area of, 525
truncated pyramid, 552–554
unpaid loan balance, 374, 405
volume of a sphere, 542
volume, 538–540
Fractal dimension, applications of, 625
*Fractal Geometry of Nature,
The* (Mandelbrot), 611, 621
Fractal geometry, 522, 611–632
applications of, 620, 625, 637–638
area of Koch snowflake, 625–637
computer graphics in movies, 621
dimension of, 622–625
Koch snowflake, 617–619, 623–625
logarithms and, 745
Mandelbrot, Benoit, 522, 611,
626–629
Menger sponge, 619–620
perimeter of Koch snowflake,
623–641
recursive processes, 614
self-similarity, 613–614
Sierpinski Gasket, 612–618 (*see also*
Sierpinski gasket)
Fractions on a graphing calculator,
163–164
Franklin, Benjamin, 347
Frequency distribution, 220–223
construction of, 223
Frequency, relative, 220
Function, 723, 13-10–13-11
area, 13-74–13-77
objective, 868
Functional notation, 13-10
Fundamental principle of counting,
88–91, 104–105
Future value, 328–330